# Sustainable Swine Nutrition

# Sustainable Swine Nutrition

## Second Edition

*Edited by*
Lee I. Chiba
Department of Animal Sciences
Auburn University
Auburn, Alabama, USA

# WILEY Blackwell

*Library of Congress Cataloging-in-Publication Data*
Names: Chiba, Lee, editor.
Title: Sustainable swine nutrition / edited by Lee I. Chiba, Professor
   Department of Animal Sciences Auburn University Auburn, Alabama, USA.
Description: Second edition. | Hoboken, NJ, USA : Wiley-Blackwell, 2023. |
   Includes bibliographical references and index.
Identifiers: LCCN 2022007979 (print) | LCCN 2022007980 (ebook) | ISBN
   9781119583899 (cloth) | ISBN 9781119583981 (adobe pdf) | ISBN
   9781119583936 (epub)
Subjects: LCSH: Swine—Nutrition. | Swine—Feeding and feeds.
Classification: LCC SF396.5 .S87 2022 (print) | LCC SF396.5 (ebook) | DDC
   636.4/0852—dc23/eng/20220223
LC record available at https://lccn.loc.gov/2022007979
LC ebook record available at https://lccn.loc.gov/2022007980

# Contents

# Contributors

**Olayiwola Adeola, Ph.D.**

Department of Animal Sciences
Purdue University
West Lafayette, Indiana, United States

**Jason K. Apple, Ph.D.**

Department of Animal Science and Veterinary Technology
Texas A&M University-Kingsville
Kingsville, Texas, United States

**Knud Erik Bach Knudsen, Ph.D.**

Department of Animal Science
Faculty of Technical Sciences
Aarhus University
Tjele, Denmark

**Eduardo Beltranena, Ph.D.**

Department of Agricultural, Food and Nutritional Science
University of Alberta
Edmonton, Alberta, Canada

**Werner G. Bergen, Ph.D.**

Department of Animal Sciences
Auburn University
Auburn, Alabama, United States

**Roger G. Campbell, Ph.D.**

RG Campbell Advisory PTY Ltd.
Adelaide, South Australia, Australia

**Lee I. Chiba, Ph.D.**

Department of Animal Sciences
Auburn University
Auburn, Alabama, United States

**Daniel A. Columbus, Ph.D.**

Prairie Swine Centre
Saskatoon, Saskatchewan, Canada &
Department of Animal and Poultry Science
University of Saskatchewan
Saskatoon, Saskatchewan, Canada

**Kara M. Dunmire, Ph.D.**            Feed Innovation Deployment
                                      Purina Animal Nutrition, LLC.
                                      Gray Summit, Missouri, United States

**Sandra A. Edwards, Ph.D.**          School of Natural and Environmental Sciences
                                      Newcastle University
                                      Newcastle upon Tyne, United Kingdom

**Charmaine D. Espinosa, Ph.D.**      Department of Animal Sciences
                                      University of Illinois at Urbana-Champaign
                                      Urbana, Illinois, United States

**Ming Z. Fan, Ph.D.**                Department of Animal Biosciences
                                      Ontario Agricultural College
                                      University of Guelph
                                      Guelph, Ontario, Canada

**Robert D. Goodband, Ph.D.**         Animal Sciences and Industry
                                      Kansas State University
                                      Manhattan, Kansas, United States

**Jeffrey A. Hansen, Ph.D.**          NutriQuest
                                      Burgaw, North Carolina, United States

**Gretchen M. Hill, Ph.D.**           Department of Animal Science
                                      Michigan State University
                                      East Lansing, Michigan, United States

**Mark S. Honeyman, Ph.D.**           College of Agriculture and Life Sciences
                                      Iowa State University
                                      Ames, Iowa, United States

**Alemu Regassa, Ph.D.**              Department of Animal Science
                                      University of Manitoba
                                      Winnipeg, Manitoba, Canada

**Rajesh Jha, Ph.D.**                 Human Nutrition, Food and Animal Sciences
                                      University of Hawaii
                                      Honolulu, Hawaii, United States

**Lee J. Johnston, Ph.D.**            West Central Research and Outreach Center
                                      University of Minnesota
                                      Morris, Minnesota, United States

**Henry Jørgensen, Ph.D.**  
Department of Animal Science  
Faculty of Technical Sciences  
Aarhus University  
Tjele, Denmark

**Brian J. Kerr, Ph.D.**  
National Laboratory for Agriculture and the Environment  
ARS, USDA  
Ames, Iowa, United States

**Sung Woo Kim, Ph.D.**  
Department of Animal Science  
North Carolina State University  
Raleigh, North Carolina, United States

**Etienne Labussière, Ph.D.**  
Physiology, Environment, and Genetics for the Animal and Livestock Systems  
INRAE  
Rennes, France

**Helle Nygaard Lærke, Ph.D.**  
Department of Animal Science  
Faculty of Technical Sciences  
Aarhus University  
Tjele, Denmark

**Peter J. Lammers, Ph.D.**  
School of Agriculture  
University of Wisconsin-Platteville  
Platteville, Wisconsin, United States

**J. Paola Lancheros, M.S.**  
Department of Animal Sciences  
University of Illinois at Urbana-Champaign  
Urbana, Illinois, United States

**Charlotte Lauridsen, Ph.D.**  
Department of Animal Science  
Aarhus University  
Tjele, Denmark

**Su A. Lee, Ph.D.**  
Department of Animal Sciences  
University of Illinois at Urbana-Champaign  
Urbana, Illinois, United States

**Rodrigo Manjarin, Ph.D.**  
Animal Science Department  
California Polytechnic State University  
San Luis Obispo, California, United States

**J. Jacques Matte, Ph.D.**  
Agriculture and Agri-Food Canada  
Sherbrooke, Quebec, Canada

**Mariana B. Menegat, DVM, Ph.D.**
Holden Farms
Northfield, Minnesota, United States

**Maximiliano Müller, Ph.D.**
Centre for Nutrition and Food Sciences
Queensland Alliance for Agriculture and Food Innovation
The University of Queensland
Brisbane, Queensland, Australia

**Marta Navarro, Ph.D.**
Centre for Nutrition and Food Sciences
Queensland Alliance for Agriculture and Food Innovation
The University of Queensland
Brisbane, Queensland, Australia

**Jean Noblet, Ph.D.**
Physiology, Environment, and Genetics for the Animal and Livestock Systems
INRAE
Rennes, France (Retired)

**Charles M. Nyachoti, Ph.D.**
Department of Animal Science
University of Manitoba
Winnipeg, Manitoba, Canada

**Maryane S. Oliveira, Ph.D.**
Department of Animal Sciences
University of Illinois at Urbana-Champaign
Urbana, Illinois, United States

**Cormac J. O'Shea, Ph.D.**
Division of Animal Sciences
School of Biosciences
University of Nottingham, Sutton Bonington Campus
Loughborough, United Kingdom

**Monique D. Pairis-Garcia, DVM, Ph.D., DACAW**
College of Veterinary Medicine
North Carolina State University
Raleigh, North Carolina, United States

**Chan Sol Park, Ph.D.**
Department of Animal Sciences
Purdue University
West Lafayette, Indiana, United States

**Rachel M. Park, M.S.**
College of Veterinary Medicine
North Carolina State University
Raleigh, North Carolina, United States

**John F. Patience, Ph.D.**
Department of Animal Science
Iowa State University
Ames, Iowa, United States

**Chad B. Paulk, Ph.D.**
Department of Grain Science and Industry
Kansas State University
Manhattan, Kansas, United States

**David Renaudeau, Ph.D.**
Physiology, Environment, and Genetics for the Animal and
Livestock Systems
INRAE
Rennes, France

**Lucas A. Rodrigues, M.S., DVM**
Department of Animal and Poultry Science
University of Saskatchewan
Saskatoon, Saskatchewan, Canada

**Eugeni Roura, Ph.D.**
Centre for Nutrition and Food Sciences
Queensland Alliance for Agriculture and Food Innovation
The University of Queensland
Brisbane, Queensland, Australia

**Michael Ryoo, B.S.**
Centre for Nutrition and Food Sciences
Queensland Alliance for Agriculture and Food Innovation
The University of Queensland
Brisbane, Queensland, Australia

**Charles R. Stark, Ph.D.**
Department of Grain Science and Industry
Kansas State University
Manhattan, Kansas, United States

**Hans H. Stein, Ph.D.**
Department of Animal Sciences
University of Illinois at Urbana-Champaign
Urbana, Illinois, United States

**Steven L. Trabue, Ph.D.**
National Laboratory for Agriculture and the Environment
ARS, USDA
Ames, Iowa, United States

**Nathalie L. Trottier, Ph.D.**
Department of Animal Science
Cornell University
Ithaca, New York, United States

**Andrew G. Van Kessel, Ph.D.**
Department of Animal and Poultry Science
University of Saskatchewan
Saskatoon, Saskatchewan, Canada

**Jaap van Milgen, Ph.D.**
Physiology, Environment, and Genetics for the Animal and
Livestock Systems
INRAE
Rennes, France

**Li Fang Wang, Ph.D.**
Department of Agricultural, Food and Nutritional Science
University of Alberta
Edmonton, Alberta, Canada

**Weijun Wang, Ph.D.**
Canadian Food Inspection Agency – Ontario Operation
Guelph, Ontario, Canada

**Michael O. Wellington, Ph.D.**
Department of Animal and Poultry Science
University of Saskatchewan
Saskatoon, Saskatchewan, Canada

**Hayden E. Williams, Ph.D.**
Eichelberger Farms
Wayland, Iowa, United States

**Tofuko A. Woyengo, Ph.D.**
Department of Animal Science
Aarhus University
Tjele, Denmark

**Zeyu Yang, M.S.**
Department of Animal Biosciences
Ontario Agricultural College
University of Guelph
Guelph, Ontario, Canada

**Xindi Yin, Ph.D.**
Department of Animal Biosciences
Ontario Agricultural College
University of Guelph
Guelph, Ontario, Canada

**Sai Zhang, Ph.D.**
Institute of Animal Science
Guangdong Academy of Agricultural Sciences
Guangzhou, Guangdong, China

**Ruurd T. Zijlstra, Ph.D.**
Department of Agricultural, Food and Nutritional Science
University of Alberta
Edmonton, Alberta, Canada

# Preface

The field of swine nutrition is dynamic and changing rapidly. New information is continuously generated and added to the body of expanding fundamental knowledge on swine nutrition. All the information would be, obviously, imperative for successful and sustainable commercial swine production. To utilize the information effectively, all those recent developments or current advances in swine nutrition must be put into a proper context simply because of the diversity of such information. There are many books that cover various aspects of swine nutrition, but, unfortunately, not many books are specifically designed to address pertinent issues necessary for "successful and sustainable swine production." Developing optimum, environmentally friendly feeding strategies for such a purpose involves consideration of a multitude of factors. The first edition of "*Sustainable Swine Nutrition*" successfully presented comprehensive reviews on those pertinent factors and issues.

As a comprehensive book on swine nutrition, it is, obviously, important to include the latest reviews on the basic aspects of nutrition, i.e. water, protein or amino acids, lipids, carbohydrates, energy, vitamins, minerals, and nutritional immunology. In addition, it is essential to dedicate the remaining portion of the book to the discussion of specific, pertinent issues that can be beneficial to the ultimate goal of the present book. That is, to provide a comprehensive review on the pertinent issues and subjects necessary for developing the optimum feeding strategies for successful and sustainable swine production. In the second edition of the book, additional chapters on the digestive physiology, primary energy and protein sources, antinutritional factors, feed processing, housing, behavior, welfare, and organic swine production have been included. Furthermore, several previous chapters have been combined or divided to address some pertinent issues more thoroughly. The latest up-to-date fundamental information, additional chapters, and modified chapters enhance the value of the book and contribute greatly to "*nutrition for successful and sustainable swine production.*"

This second edition of the book would not have been possible without the generous support of my colleagues, and I would like to thank our contributors for the first and second editions and new collaborators for their willingness to participate in this endeavor. I sincerely appreciate their time and dedicated effort on this book project.

**Lee I. Chiba**

# Editor

**Lee I. Chiba** is a professor of animal science in the Department of Animal Sciences at Auburn University, Auburn, Alabama, United States. He received his B.S. degree in animal science and M.S. and Ph.D. degrees in nonruminant nutrition from the University of Nebraska, Lincoln, Nebraska, United States. Dr. Chiba is responsible for teaching undergraduate courses in animal nutrition and swine production and graduate courses in nonruminant nutrition and vitamin and mineral metabolism. His research interests are in the areas of dietary manipulations to improve the growth performance of weaning pigs, leanness and efficiency of growing pigs, organoleptic characteristics of pork, and nutritional management to improve reproductive performance of sows. Dr. Chiba served as a member of the Editorial Board for three terms, as an associate editor for two terms, and as a division editor of the *Journal of Animal Science* for two terms. He is currently serving as a section editor of the *Livestock Science*.

# Part I
# Fundamental Nutrition

# 1 Digestive Physiology and Nutrition of Swine

Eugeni Roura, Maximiliano Müller, Roger G. Campbell,
Michael Ryoo, and Marta Navarro

## Introduction

The domestic pig (*Sus scrofa*) belongs to the order of Artiodactyla suborder Suiformes (Simpson 1945). Many of the physio-anatomical features of the species *S. scrofa* can be related to dietary adaptations typical of a plant-dominated omnivore, characterized by a high fermentative capacity in the proximal hindgut.

In brief, the digestion of nutrients starts in the mouth and continues in the stomach and small intestine with the help of both mechanic and enzymatic mechanisms. Most dietary nutrients are digested into single moieties and absorbed before the gut content reaches the end of the ileum and enters the large intestine where fermentation occurs. Undigested compounds such as complex carbohydrates, reaching the large intestine, are fermented and utilized by the microbiota releasing mainly short-chain fatty acids (SCFA). This chapter reviews the main anatomical and physiological features relevant to the digestion and absorption of nutrients in the pig. A comprehensive understanding of the role of gut microbiota in the digestive physiology of pigs has not been included in this chapter.

## Anatomy of the Porcine Digestive System

### The Oral Cavity

The mouth is the necessary point of entrance for all solid and liquid nutrients essential to maintain metabolic homeostasis. The anatomical and physiological features of the oral cavity respond to three main functions relevant to eating: (i) food selection and apprehension, (ii) physical breakdown and grinding of ingredients including predigestion, and (iii) lubrication and swallowing. These functions have been related to three main organs: the tongue, the dental, and the salivary system.

### The Tongue
The pig tongue consists of a muscle appendix protruding from the base of the oral cavity (lower mandibula) characterized for a high mobility essential to move oral contents and a dorsal epithelial layer containing mechanical and sensory papillae. The tongue plays a fundamental role in food acquisition (particularly the front mobile tip), food selection (through taste), and oral manipulation as part of the mastication and the act of swallowing (Lærke and Hedemann 2012). The mechanical

*Sustainable Swine Nutrition*, Second Edition. Edited by Lee I. Chiba.
© 2023 John Wiley & Sons Ltd. Published 2023 by John Wiley & Sons Ltd.

papillae (known as filiform) facilitate a central role in moving the food bolus around the oral cavity for optimal processing (e.g., grinding, crushing, and lubrication) prior to and including swallowing.

In addition, the tongue plays a fundamental role in sensing dietary nutrients through the taste system. The basic organelle is the taste bud, with bundles of 100–150 taste sensory cells embedded in defined epithelial structures named as taste papillae. There are three types of taste bud-containing taste papillae: fungiform, foliate, and circumvallate. Pigs have many fungiform papillae covering the upper lingual surface. The foliate are located distal at each side of the tongue and is in contact with the oropharynx. The circumvallate papillae are two mushroom-shaped macroscopic structures, located and also in contact with the oropharynx but as part of the dorsal epithelia in pigs (Roura et al. 2016). The circumvallate papillae contain the highest amount of taste buds. A detailed explanation of the sense of taste and nutrient sensing mechanisms in the pig can be found in the section on "Nutrient Digestion."

*The Dental System*

The apprehension and oral processing of solid food occurs mainly due to the combined function of the lips, the tongue, and the dental system. The pig is a dental heterodont consisting of four types of specialized teeth: incisors (I), canines (C), premolars (Pm), and molars (M). The array of teeth is arranged in four identical and symmetric quarters two in each half of the upper and lower jaws. The number of tooth types differs in the various orders of mammals and is closely related to their feeding habits. Incisors and canines are primarily involved in food acquisition (stripping and rooting), while premolars and molars are involved in oral processing of foods (crushing, grinding, and shearing). Specific to the pig, the dentition consists of 44 permanent teeth with a basic dental formula in each quarter of 3-1-4-3 indicating the number of I, C, Pm, and M, respectively. The lower tusks are kept sharp by friction against the upper ones making them formidable weapons as well. Numerous tubercles make the occlusal surface of the molars very irregular and ideal for grinding and crushing food (Figure 1.1) (Sova et al. 2018).

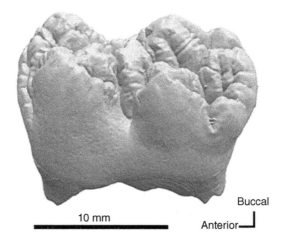

10 mm

Buccal

Anterior

**Figure 1.1** Pig molars have a complex fragmentary pattern with high crushing and grinding power. The enamel surface shows rounded, convex cusps with narrow intervening fissures. Porcine molars are difficult to distinguish from human molars. Source: Sova et al. (2018), Figure 03 [p. 04]/Frontiers Media S.A/CC BY 4.0.

*The Salivary System*

Adult pigs produce about 15 L of saliva daily. Saliva is a viscid, slightly acidic, and watery fluid secreted by the salivary glands present in the oral cavity that facilitates the mastication, digestion, lubrication, and deglutition of foods. The enzymatic digestion occurring before and after swallowing thanks to enzymatic activities present in saliva is further presented in the section "Nutrient Digestion."

The salivary glands are exocrine secretory acini with an intralobular system of intercalated and striated ducts. Pigs have three paired major salivary glands (parotid, submandibular, and sublingual) as well as hundreds of minor salivary glands embedded as part of the oral epithelia (Figure 1.2). Salivary glands produce three types of secretions referred to as serous, mucous, or seromucous (mixed). The main proteins present in serous secretions are alpha-amylase and the pig-specific enzyme ptyalin, both involved in the breaking down of starch into maltose and glucose. In mucous secretions, the main protein secreted is mucin acting mainly as a lubricant.

### Pharynx, Esophagus, Stomach, and Omentum

When leaving the oral cavity, foods travel through the pharynx and esophagus before reaching the stomach. The esophageal submucosa is rich in bicarbonate and mucin-producing glands that lubricate the walls and the food protecting the epithelium, while also buffering contents with extreme pH.

The stomach is a food conditioning chamber preceding the main site of digestion and absorption of nutrients, the small intestine (duodenum). The size of the stomach represents 0.6% of the body weight (BW) in an adult pig reaching a volume of up to 6–8 L. The relative size decreases with the

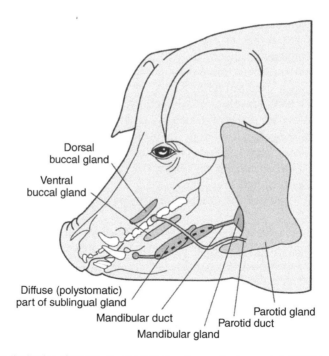

**Figure 1.2**    Schema of the distribution of the salivary glands in pigs. Source: Ducharme et al. (2017) / with permission of Elsevier.

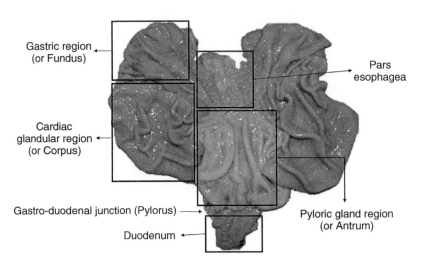

**Figure 1.3**   The stomach of a pig highlighting the four epithelial/glandular regions (*pars esophagea*, cardiac, gastric, and pyloric regions) and the gastro-duodenal junction (pylorus sphincter) gate to the duodenum.

age of the animal to the extent that in young pigs the relative size is around twice as much as in the adult (see the section on "Developing GIT in Piglets"). Based on the structure–function of the gastric mucosa, the porcine stomach can be divided into four main regions (Figure 1.3): the stratified squamous region named as *pars esophagea*, the cardiac glandular region (or *corpus*), the proper gastric region (or *fundus*), and the pyloric gland region (or *antrum*).

The most cranial region (*pars esophagea*) is a nonglandular extension of the esophagus located at the entrance of the stomach from the esophagus. In pigs, this region has a high risk in developing ulcers due to fine feed particle size or because of dealing with stressing events. The cardiac region contains mainly mucus-producing cells that protect the gastric mucosa from the entering bolus and the low pH gastric environment. Compared to other mammalian and avian species, the pig stomach has a very large cardiac region that is associated with having one of the lowest pH of around 1.5–2.5 (Smith 1965; Stevens 1977) (Figure 1.4). The fundus is also referred to as the proper gastric region because of the high density of HCl and mucus-producing glands (fundus glands). Fundus gland secretions are high in protease activity and contain pepsinogen (see "Nutrient Digestion" in this chapter below for additional descriptions). The stomach fundus is also rich in sensory cells identified as part of the endocrine/paracrine system described later in the section on "Nutrient Sensing and Gut-brain Axis." The pyloric region includes mucus; chief (zymogenic) cells, which oversee the production of pepsinogen and gastric lipase; and enteroendocrine cells (EEC).

The main function of the stomach is to precondition foods for further digestion in the small intestine, using mechanical and chemical processing. Mechanical digestion occurs thanks to the peristaltic movements that mix and soften the digesta. The apical entrance of the stomach is regulated by the cardiac valve that prevents the accidental regurgitation of highly acidic stomach contents. These movements increase in force as they reach the pylorus. The pylorus acts as a filter allowing the passage of only small particles into the duodenum. The digested food then slowly pushes into the duodenum through a process known as gastric emptying, regulated by neural and hormonal mechanisms involving ghrelin, cholecystokinin (CCK), glucagon-like peptide 1 (GLP-1), or peptide tyrosine-tyrosine (PYY) (see the section on "Nutrient Sensing and the Gut-brain Axis"). The chemical

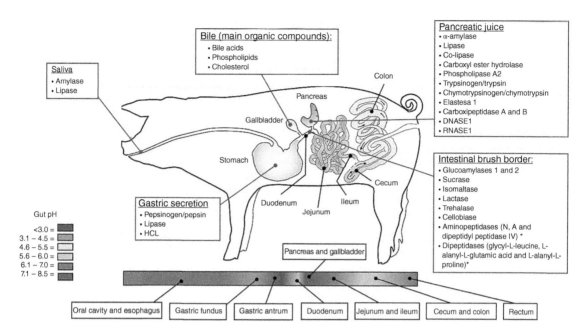

**Figure 1.4** The porcine gastrointestinal tract: main enzymatic activities secreted by tissue and pH variations along the tract using color coding.

processing primarily consists of the secretion of hydrochloric acid that lowers the pH below 5. The acidic pH allows for the activation of pepsinogen to pepsin. In addition, gastric lipase is also secreted in the stomach. A more detailed accounting of the chemical digestion of foods in the stomach is presented in the next section.

The capacity of the stomach to absorb nutrients has often been underestimated. Absorption of simple molecules occurs including water and water-soluble vitamins, short- and medium-chain fatty acids, hormones ($\alpha$-ketoglutarate, estrogens), and some AAs among others (Zebrowska et al. 1982; Ruoff and Dziuk 1994; Buddington et al. 2004).

The omentum is a membrane structure that covers the stomach and part of the liver in pigs. It helps anchoring the gastrointestinal tract to the body wall (along the spine) keeping the stomach in place. In addition, the omentum is one of the main fat body stores creating a network of adipocytes sitting around the blood vessels.

### Small intestine

The small intestine is a tubular organ approximately 20 m long in the adult pig, extending from the stomach to the large intestine and located in the right abdomen. It is divided into three segments: the duodenum, jejunum, and ileum representing around 5, 90, and 5% of the total length, respectively (Adeola and King 2006). The small intestine consists of four membrane layers: in a distal to medial direction (from outside to inside the intestinal lumen), first is the serosa, then the muscularis, followed by the submucosa and the mucosa (Figure 1.5). The serosa or outermost layer has a squamous epithelium forming the mesentery that contains large blood vessels, connective tissue, and nerves. The muscular layer includes two sublayers, the outer consisting of longitudinal muscle

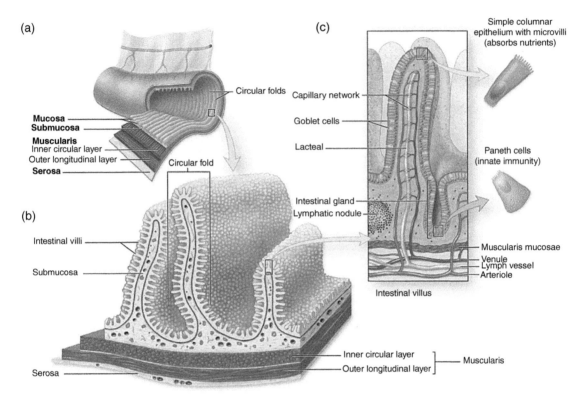

**Figure 1.5**   Anatomy of the small intestinal mucosa. The gut has five concentric layers (a). The most inner layer is the mucosa that consists of circular folds (b) and a covering of villi (c). Each villus contains connective tissue with microvasculature and lymphatics called lacteals and is covered with a simple columnar epithelium composed of enterocytes and goblet cells. The microvilli present in the apical cell membrane of enterocytes dramatically increase the absorptive area. Crypts are short tubular glands between villi with stem cells and Paneth cells. Source: Mescher (2009) / with permission of McGraw Hill LLC.

fibers and the inner consisting of circular muscles. Collectively, the synchronic contraction and relaxation of the muscles in the two sublayers define the rhythmic waves of intestinal motility also referred to as peristalsis that determines food passage rates (see the section on "Gut motility, transit time and the ileal brake"). The submucosa is made up of connective tissue and contains blood and lymphatic vessels and nerves. The mucosa, which is the most inner layer, is divided into three smaller layers: muscularis mucosa, lamina propria, and epithelium. Like the muscular and submucosa layers, the muscularis mucosa possess both inner longitudinal and outer encircling muscles, whereas the lamina propria contains blood vessels, gut-associated lymphatic tissue (GALT), and neurons. The epithelium consists of folds and villi that increase the contact between the chyme and the epithelial cells and crypts with high secretory capacity (Figure 1.5). The average height of the villi is constant throughout the duodenum and jejunum while slightly declining in the ileum (Table 1.1). In addition, the apical membrane of enterocytes consists of microvilli enabling higher digestive capacity (through brush border enzymes) and subsequent absorption of nutrients from the lumen. In addition to the enterocytes, the intestine has four specialized epithelial cell types: goblet, tuft, Paneth, and EEC.

**Table 1.1** Anatomy and physiology of different cell and tissue types in porcine small intestine, colon, and rectum.

| Item | Duodenum | Jejunum | Ileum | Colon | Rectum |
|---|---|---|---|---|---|
| Goblet cells[a] | Low density | Medium density | Medium-high density | High density increasing caudally | Very high density |
| Tuft cells[b] | 0.5% of total cells | 0.5% of total cells | 0.5% of total cells | 0.5% of total cells | — |
| Paneth cells[c] | — | Lieberkühn crypts | — | — | — |
| Enteroendocrine cells[d] | High in I cells; low in EC and L cells | Medium-high in EC and L cells in distal section | High in EC and L cells | High in EC and L cells | Very high in EC and L cells |
| GALT[e] | Low density | Medium density | Very high density | Lymphoid follicles only | Lymphoid follicles only |
| Villus height (μm)[f,h] | 420±15.0 | 480±53.8 | 372±43.0 | — | — |
| Crypt depth (μm)[g,h] | 367±49.7 | 429±59.9 | 316±32.9 | — | — |
| Surface area (m²)[i] | 0.090 | 60.000 | 60.000 | 0.250 | 0.015 |

GALT = gut-associated lymphoid tissue consisting of lymphoid follicles and PPs.
References: [a] Gonzalez et al. 2013; Vertzoni et al. 2019; Zhang and Wu 2020
[b] Banerjee et al. 2018
[c] Myer 1982; Kong et al. 2018
[d] Gunawardene et al. 2011; Ripken and Hendriks 2015
[e] Jung et al. 2010; Ruth and Field 2013; Urmila et al. 2019
[f] Lin et al. 2020
[g] Metzler-Zebeli et al. 2017
[h] Tian et al. 2020
[i] Kararli 1995.

*Duodenum*

The adult porcine duodenum extends from the pylorus of the stomach until the duodenal-jejunal junction located 70–95 cm down the tract divided in four segments: superior, descending, inferior, and ascending. After the chemical digestion occurring in the oral cavity and by stomach, the duodenum is the third organ involved in enzymatic/chemical digestion of foods. While accounting only for 5% of the length of the small intestine, the duodenum has a crucial role in receiving the main digestive juices essential for food digestion: the bile (i.e., bile acids) and pancreatic juices (i.e., pancreatic enzymes). Pancreatic enzymes and bile acids enter the duodenum through the common hepatopancreatic duct forming the major duodenal papilla located 2–5 cm from the stomach pylorus. Pigs have a secondary pancreatic duct (accessory) that opens at a minor duodenal papilla located 8–10 cm distally from the major duodenal papilla. The hepatopancreatic sphincter (Sphincter of Oddi) controls the flow of digestive juices into the duodenum. The optimization of the enzymatic digestion requires of a pH of around 6.4 (Figure 1.4). At the start of the duodenum, the pH is 6.1 and steadily rises to 6.4 thanks to the alkaline secretions released from Brunner's glands present in the duodenal submucosa facilitate the neutralization of the acidic digested contents received from the stomach via the gastric pylorus. Unique to the duodenal epithelium is the presence of CCK-secreting I cells (Table 1.1). The duodenum also plays a minor role in absorbing a small part of free nutrients.

*Jejunum*

The main site of digestion and absorption of nutrient is the jejunum. The length of a mature pig jejunum is between 14 and 19 m (this is 90% of the small intestine) that begins at the duodenojejunal junction transitioning into the ileum without any specific anatomical feature. This is the longest segment in the middle of the small intestine that is red in color (indicating very high vascularization) and consists of a relatively thick intestinal wall. At the beginning of the jejunum, the pH is 6.4 that rises to 6.6 before transitioning into the ileum back at 6.4 (Figure 1.4). Deep crypts, known as Lieberkühn, and highly developed villous are characteristics of the jejunal mucosa. The crypts contain a high density of Paneth cells with antimicrobial functions (Table 1.1). In addition, the many parallel circular folds in the mucosa (valves of Kerckring) further increase the surface area for nutrient absorption. Brunner's glands are also widely distributed up to the proximal part of the jejunum to ensure that the acidity has been neutralized by the alkaline secretions before comprehensive absorption occurs. The porcine jejunal submucosa is densely populated with Peyer's patches (PPs), lymphoid follicles rich B and T cells forming part of the so-called gut-associated lymphoid tissue (GALT).

*Ileum*

The ileum is the most distal tubular part of the small intestine. In an adult pig, the length of the ileum is between 0.70 and 1 m and has a pink color indicative of less vascularization than the jejunum. The ileocecal valve is a sphincter that regulates the movement of digested contents from the small to the large intestine that begins at the cecum. The primary role of this sphincter is to prevent the reflux of microbe-rich enteric fluid from the large intestine to the small intestine. At the entry point of the ileum, the pH was 6.4 increasing to 6.7 in the middle segments to fall back to 6.4 at the terminal end near the ileocecal junction (Figure 1.4). Compared to the duodenum and jejunum, digestion is minimal in the ileum, while nutrient absorption is maximized thanks to a relatively thin intestinal wall that reduces diffusion distance. Partially digested foods that have not been fully broken down to simple nutrients, particularly lipids at the proximal part of the small intestine have the capacity to induce the ileal brake. In pigs, the ileal submucosa also contains PPs with a unique composition consisting of 90% of the immune cells being B lymphocytes.

*Epithelial Cells*

Intestinal epithelia consist mainly of enterocytes and four highly specialized types of cells: goblet, tuft, Paneth, and EEC (Table 1.1). Enterocytes are nutrient-absorptive cells that line the surface of the finger-like projections referred as intestinal villi. These structures collectively amplify the intestinal surface area by 30-fold. Enterocytes differentiate from stem cells at the bottom of the crypts and mature as they migrate toward the tip of the villi. As the enterocytes move above the basal third of the villi, the digestive functions are activated such as border brush enzymes. When the enterocytes reach the upper to mid-level of the villi, the absorptive role begins to develop. When the mature cells reach the top of the villi, their active life terminates and are sloughed off into the lumen. Goblet cells are mucus-secreting cells that are distributed in the epithelial layer surrounded by enterocytes. There is a higher frequency of goblet cells found in the distal (ileum) compared with the proximal (duodenum/jejunum) small intestine. Tuft cells are chemosensory cells in the epithelial lining of small and large intestines. Their main function is to detect physiological disturbances associated with parasitic infections in the intestinal lumen and signal the immune system. In the intestine, tuft cells are the sole source of secreted interleukin 25. During helminth infection, IL-25 production initiates a feed-forward mechanism involving the activation of group 2 innate lymphoid cells (ILC2) that promote type 2 inflammation with the generation of IL-5, IL-9, and IL-13. During a helminth infection, tuft cell populations can become elevated to ten times their original frequency. Paneth cells are epithelial cells located at the bottom of the crypts of Lieberkühn. They produce peptides such as defensins and lysozymes and proteins with antimicrobial activity that mediate the host–microbe interactions, including innate immune protection from enteric pathogens. EEC have the primary role of secreting gut peptides that facilitate the dialogue with the gut-brain axis of the central nervous system in response to specific nutrients present in the lumen. A description of the types and roles of EEC can be found in the section on "Enteroendocrine System."

### Large intestine

The large intestine consists of a highly lobulated tube that facilitates the development of microbiota and the fermentation of undigested material leaving the small intestine. It consists of the cecum, the colon, and the rectum. Following dietary adaptations where fibrous components are likely to increase with age, the development of the large intestine, particularly the colon, continues long after the small intestine has already been fully developed, reaching a total length of around 7.5 m long in mature individuals. Thus, the colon becomes the longest proportion of the large intestine. Consequently, pigs like humans have been defined as mainly colonic fermenters. The undigested chyme from the small intestine is released into the ileo-cecal junction passing into the ceca or directly into the colon. A series of pouches derived from longitudinal muscular bands from the cecum and proximal part of the spiral colon are referred to as haustra. These pouches compartmentalize the passage of luminal contents while facilitating the proliferation of the microbiota. Villi are absent in the mucosa of the large intestine a differential feature compared to the small intestinal mucosa. In contrast, the large intestine mucosa consists of columnar epithelial cells with microvilli that form straight tubular crypts. The epithelium of the large intestine consists of enterocytes as the main cell type, and goblet, tuft and EEC (Table 1.1). Goblet cells play an important role in the large intestine by secreting a sulfated carbohydrate-protein complex, which lubricate columnar epithelial cells that absorb water from the chyme.

### Cecum

The cecum is a highly lobulated cylindrical blind sac with three teniae running along its length, located at the proximal end of the colon, and lying transversely on the left side of the abdomen. The apex of the cecum points caudoventrally. The cecum's pH is kept at 6.1, which is a noticeable decline from the 6.4 at the terminal end of the ileum (Figure 1.4). There are no villi throughout the cecum epithelium. Segmental contractions induce absorptive and microbial activity, which is followed by mass movements that transfer the chyme to the colon. In pigs, soluble fiber is fermented in the cecum, while insoluble fiber is mainly fermented in the colon. A thick mucous membrane makes up the internal wall of the cecum through which water and salts are absorbed. While the mucosa has mucous glands, these are absent in the submucosa. The lamina muscularis particularly in the distal section of the cecum has large lymphatic nodules (PPs).

### Colon

The porcine colon begins at the ascending colon, which is an elongated, coiled, and cone-shaped structure. It is attached to the dorsal abdominal wall, where the apex is oriented ventrally, that ascends cranially (anterior direction) until encountering the liver where it turns at a right angle to move horizontally. This bend is referred to as the hepatic flexure marking the start of the transverse colon. Over a short distance along the colon, there are three teniae coli of smooth muscle (mesocoloic, free, and omental coli) that run longitudinally along the surface of the large intestine, which contract to form rows of sacs known as the haustra. The pH rises from 6.1 to 6.4 across the proximal segment of the colon (Figure 1.4). The transverse colon then extends horizontally until it turns at a right angle at the spleen to descend caudally (posterior direction). The splenic flexure is the term for this bend between the transverse colon and the descending colon. The pH across the middle section of the colon remains at 6.5 from start to end. The descending colon moves caudally toward the pelvis until it begins to turn medially and transition into the sigmoid colon. The sigmoid colon is an "S" shaped part of the colon in the lower left quadrant of the left abdomen that is attached to the dorsal pelvic wall of a mesentery called the sigmoid mesocolon that provides mobility to the sigmoid colon. The pH across the distal section of the colon slightly rises and remains at 6.6 (Figure 1.4). Eventually, in the spiral colon region, there are two teniae and two haustra that are no longer identifiable in the centrifugal section. No villi are found in the colon, but there are microvilli that line the simple columnar epithelium of the mucosa that assist in the absorption of water that leads to the thickening of the stool. Anaerobic fermentation of indigestible components such as the insoluble fiber cellulose by the intestinal microflora, which inhabit the crypts of Lieberkühn, is another major role played by the colon. Large lymphatic nodules are also found in the submucosa of the colon. The temporary storage and transport of feces across the large intestine is enhanced by a layer of mucus secreted by an abundance of goblet cells distributed throughout the colon that provides lubrication. The teniae coli of the smooth muscle in the sigmoid colon broaden to form a complete layer within the rectum at the rectosigmoid junction in the large intestine that marks the end of the sigmoid colon and the start of the rectum.

*Rectum, Anal Canal, and Anus*
The rectum follows the posterior end of the large intestine and is the last straight part of the digestive system connecting with the anus in the anal canal. It is surrounded by bands of muscles required to move the waste product of digestion (feces) out of the body. Goblet cells are abundant throughout the rectal mucosa to lubricate the feces as they pass into the anus. The median pH of the porcine rectum is 6.3. The ampulla is the last segment of the rectum that relaxes to accumulate and transiently store feces. The recto-anal junction functions thanks to the longitudinal layer of the tunica muscularis and the lamina muscularis in the rectum. As feces accumulate in the ampulla, pressure sensitive receptor cells detect increasing distension that initiates the defecation reflex. The combination of both voluntary and involuntary process involves the rectum to forcefully contract and subsequently relax the internal anal sphincter regulating defecation. The anal sinuses are furrows in the anal canal that contain glands in charge of secreting mucus to lubricate the feces as it passes through to the external environment. The median pH of the porcine anal canal, like the rectum, is 6.3. The anus is referred to the exterior opening surrounded by the anal sphincter.

### The Pancreas

The pancreas is both an endocrine and an exocrine organ, associated with the digestive system located in the inner side of the duodenum loop (part of the proximal small intestine). In the pig, the pancreas is composed of two glandular lobes and a pancreatic duct that drains into the duodenum 20–25 cm from the pylorus and closely distal to the bile duct from the gallbladder. Acinar cells, present in the exocrine pancreas, oversee the production and release of enzymes and zymogens into the duodenum when stimulated. The pancreatic enzymes are essential to digest proteins, carbohydrates, lipids, and nucleic acids. The endocrine pancreas produces hormones that regulate the utilization of the absorbed digested products and it is structured in islets named after Professor Paul Langerhans (islets of Langerhans). The islets consist of α-cells producing the peptide glucagon, β-cells producing insulin and amylin, δ-cells producing somatostatin, and F-cells producing pancreatic polypeptide. More detail on the endocrine and exocrine roles of pancreatic secretions can be found in sections "Nutrient Digestion and Nutrient Sensing and the Gut-brain Axis."

### The Liver

The liver is a complex organ with several essential functions in the maintenance of metabolic home-ostasis including detoxification of hazardous metabolites, synthesis of a wide range of proteins, and other compounds necessary for digestion and growth. The liver is an accessory but crucial organ to the digestive system that produces bile, an alkaline fluid containing cholesterol and cholesterol-derived bile acids required for the breakdown of fat. The liver in pigs is divided into six lobes by deep interlobular fissures: the left lateral and medial, the right lateral and medial, the quadrate, and the caudate lobes. The gallbladder (3–5 cm, 25 mL), a small pouch that sits just under the liver, stores the bile produced before it is released into the small intestine to complete digestion. Hepatocytes are the main functional liver cells that produce the bile juice (0.6–1.1 mL kg$^{-1}$ BW/h) (Swindle 2015). The bile juice is moved through the intrahepatic bile ducts to the extrahepatic biliary system consisting of a cystic' duct, connected to the gallbladder, and a bile duct, that drains in the duodenum, approximately 2–5 cm distal to the pylorus. Different from the pancreas, bile juices are produced constantly and stored in the gallbladder until nutrient stimulation. The presence of nutrients in the proximal small intestine, particularly lipids, triggers the sphincter of Oddi to relax, and the gallbladder contracts allowing the release of bile juices into the duodenum.

### The Gut-Associated Lymphoid Tissue (GALT)

The GALT is the specialized intestinal immune system that protects the gut mucosa avoiding microbial invasion. The distribution of GALT through small and large intestine has been outlined in Table 1.1. The GALT includes PPs, the mesenteric lymph nodes (MLNs), and isolated lymphoid follicles. PPs are transmucosal clusters with dendritic cells in the dome and interfollicular areas as antigen-presenting cells. PPs are present in the small intestine (Jejunum -JPP- and Ileum -IPP-) and the large intestine, with isolated lymphoid follicles (ILF; Liebler-Tenorio and Pabst 2006). In pigs, pancreatic polypeptides in the jejunum are richer in T cells than B cells (65% vs. 35%, respectively), while in the ileal submucosa, pancreatic polypeptides consist of 90% of B cells. This suggests that the jejunum has a stronger role in cell-mediated immunity compared to the ileum that is dependent on humoral (antibody mediated) immunity. Microfold cells are specialized epithelial cells, which facilitate the transport of antigens from pathogenic microorganisms to the B lymphocytes, that elicit the immune response. In addition, individual lymphoid nodules have been observed in the submucosa and lamina propria of the gastric cardia region in pigs.

### Gut Motility, Transit Time, and the Ileal Brake

The transit time of the food bolus and chyme through the gastrointestinal tract determines the efficiency of the digestion and absorption of nutrients ultimately affecting appetite and feeding behavior. Gut motility is an essential feature to understand the passage rate of gastrointestinal contents and is mainly a result of (i) tissue-specific sequences of contraction–relaxation patterns of the smooth muscle layers, (ii) the presence of two valves (epiglottis and cardia) and four sphincters (pyloric, ileo-cecal, and internal and external anal) with a "stop-go" flow function in critical points of the gastrointestinal tract determine bolus and chyme passage rates (Figure 1.5).

*Smooth Muscle Layers*
Smooth muscle layers are present in two locations throughout the gastrointestinal tract, in the mucosa outside the lamina propria (muscularis mucosae) and surrounding the submucosa (external muscular layers). The muscularis mucosae is a continuous thin layer separating mucosa from submucosa, consisting of crossed fibers of smooth muscle that facilitate the expel and mixing of glandular crypt secretions into the lumen and a gentle agitation of the epithelial surface and gut contents. The external muscular layers trigger the sequential waves that will further mix and move the bolus or chyme down the gastrointestinal tube, also referred to as gut peristalsis. There are two main types of external smooth muscle layers, the middle circular just outside but in contact with the submucosa and the outermost longitudinal. Circular muscles result in slow waves that promote the mixing of particles and digestive secretions while preventing the reflux of the chyme. Longitudinal muscles stimulate the aboral movement of the chyme through combining contraction–relaxation sequences that result in shortening–lengthening of the tube. The result of the two external muscle layers working together is the mixing and propelling of gut contents.

*Enteric Myogenesis and Enteric Nervous System (ENS)*
The regulation of gut motility is a combination of myogenic and neural impulses generated by the enteric nervous system (ENS), the second largest accumulation of nerve cells in the body. Enteric myogenesis refers to the capacity of the myocyte itself to initiate a contraction and is responsible for initiating intestinal motor patterns. Smooth muscle fibers are electrically connected by gap junctions. In addition, the gastrointestinal tract contains interstitial pacemaker cells (cells of Cajal) between muscle layers that generate electrical rhythmicity in the smooth muscle cells allowing for coordinated generation of slow waves. The ENS, in turn, is based on an autonomic neural system (thus capable of operating independently from the central nervous system) with thousands of small, interconnected ganglia that form the myenteric and submucosal plexus coordinating muscle activity and mucosal secretion and absorption of nutrients, respectively. In addition, the vagal innervation of the gut has the capacity to modulate the autonomic nervous system to help regulate digestive functions and to synchronize the digestive system with other vital systems (e.g., blood flow). In the pig, the vagal innervation is dominant in the motility of esophagus and stomach, while the autonomic system is dominant in small and large intestines. Furthermore, gut peptides also play a relevant role capable of modifying the electrical and contractile pattern within the gut. For example, motilin and ghrelin promote the gastrointestinal motility, while CCK and GLP-1 inhibit gastric emptying and proximal small intestine motility as part of the ileal brake (see additional details below).

Tissue-specific functions related to gut motility are briefly explained next.

*Deglutition and Esophagus*
Deglutition (or swallowing) is the passing of foods from the oral cavity (oral phase) to the esophagus through the pharynx (pharyngeal and esophageal phases). The oral phase is based on skeletal muscle (mainly tongue), while the pharyngeal and esophageal phases are regulated by the autonomic nervous system following activation with mechanoreceptors. The act of deglutition requires synchronization with the closure of the trachea thanks to the epiglottis preventing the accidental passage of foods to the upper respiratory system. In addition, swallowing is associated with the primary peristaltic contractions

resulting from the circular muscles of the esophagus that lead the bolus to a rapid transit into the stomach. Secondary peristaltic contractions are triggered by the distension caused by material left in the esophagus. The primary contractions are mediated by central mechanisms through the vagal nerve, while the secondary contractions are locally mediated by mechanoreceptors in the esophageal lining.

*Stomach*
The bolus enters the stomach through the cardiac valve that functions as a barrier to prevent reflux of the potentially damaging highly acidic gastric content. The primary motility is also governed through the ENS. Characteristic of the stomach is the inner oblique muscle layer that coordinates the mechanical digestion promoting the grinding and mixing of large feed particles with gastric juices to release a more liquified state, also known as chyme. A second main function of the contraction of smooth muscle in the stomach is the forced passage of small quantities of chyme through the pyloric sphincter into the duodenum, a process also known as gastric emptying. Distal gastric contractions and gastric emptying through the pyloric valve are triggered by the volume of feed as well as by the distension of the stomach through the stimulation of mechanoreceptors. Large grain particles modulate gastric emptying by delaying the passage rate into the duodenum. The middle circular muscular layer forms part of the pyloric sphincter, while the longitudinal layer participates in coordinating a radially symmetrical contraction and relaxation of muscles that stimulates the aboral movement of food into the pyloric region and duodenum. In addition, CCK and gastrin contribute to the modulation of gastric motility by stimulating the ENS (vagal nerve).

*Small intestine and the Ileal Brake*
The arrival of the chyme to the duodenum triggers chemical and mechanical stimulus that mediate the onset of peristaltic and secretory reflexes. The chemical signals are sensed through EEC in the intestinal mucosa that release the neurotransmitter serotonin that in turn stimulates the vagal nerve (see the section on "Nutrient sensing and the gut-brain axis" for more information on nutrient sensing). The mechanical sensing activates the ENS. The motility consists of slow waves that allow for the mixing of the chyme with intestinal, pancreatic, and biliary secretions as well as longitudinal or peristaltic waves responsible for the aboral passage. Peristaltic waves do not propagate along the entire intestine, but instead help propel the chyme only a few centimeters into the duodenum. In pigs, the number of duodenal waves is positively correlated with the flow rate of the chyme (Ruckebusch and Bueno 1976). The physical distension and the presence of undigested nutrients into the duodenum, jejunum, and ileum trigger the so-called brakes consisting of negative feedback mechanism increasing the retention time in proximal parts of the gut (Roura and Navarro 2018). The ileal brake has the strongest impact slowing down the passage rate of the chyme by inhibiting intestinal motility, delaying gastric emptying, digestive secretions, and hunger, altogether resulting in feed intake suppression.

*Ileo-cecal Junction, Large intestine, Rectum, and Anal Canal*
The ileum generates strong contractions to propel batches of digesta into the cecum at 7–12 minutes intervals. At the ileo-cecal junction, there is a sphincter consisting of a circular smooth muscle layer with the main function of preventing reflux of cecal contents into the ileum. The motility pattern in the colon is characterized by segmentation contractions within the haustra that mix luminal contents and peristaltic contractions that facilitate the movement of digesta through haustra and into the rectum. As the fecal matter reaches the distal colon, contractions speed up, but their force declines leading to a slower passage to facilitate water absorption. In addition, the motility of the large intestine includes reversed peristalsis contractions that propel the digesta back toward the ileum resulting

in the accumulation of fecal matter in the proximal colon. This allows for additional capacity to absorb water and electrolytes and extends the time for microbial fermentation. The ejection of fecal matter through the anal sphincters (defecation) involves giant contractions (or mass movements) involving most of the large intestine and providing the major force required. In pigs, the mass movements occur at intervals of around seven minutes in coordination with the ileal strong contractions (Hipper and Ehrlein 2001). Passed the rectum, fecal content accumulates in the anal canal where the internal anal sphincter is formed by an involuntary circular layer of the tunica muscularis in the upper two-thirds of the anal canal, whereas the external anal sphincter is a voluntary muscle composed of skeletal muscles that surrounds the lower two-thirds of the anal canal. Mechanoreceptors in the anal canal will result in the relaxation of the internal sphincter leaving the defecation as a voluntary act to relax the external sphincter.

*Transit Time*

The time required for food to travel through the gastrointestinal tract varies with the age and size of the animal and the physicochemical characteristics of the diet. In adult-size pigs, the mean retention time (MRT) has been estimated in 1, 4, and 38 hours for stomach, small intestine, and large intestine, respectively (Figure 1.6). Thus, the passing time in cecum and colon represents 70–85% of the total. However, depending on the composition of the bolus/chyme, the contents may stay up to 24 hours in stomach, 15–20 hours in small intestine, and up to 50–55 hours in the large intestine. Importantly, the amount of chyme present in the small intestine at any given time is relatively small indicating the relevance of a high enzyme-to-substrate ratio for an effective digestion. Dietary factors that increase transit time in the stomach include large food particles (>2 cm) or food with high water holding capacity. Diets rich in soluble fiber can also dramatically increase the passage rate of the chyme through the small and large intestines, attributed to their water holding capacity (Jørgensen et al. 1997). The size of the large intestine increases with age (and weight), which, in turn, results in longer MRT in the whole gastrointestinal tract. MRT increased from 32 to 44 hours and above 80 hours for a 28, 85 kg pig and mature sows, respectively (Holzgraefe et al. 1985; Potkins et al. 1991). In contrast, increasing insoluble fiber content in feeds does not affect gastric transit time while potentially decreasing retention time in small and large intestines in pigs. As previously mentioned, the activation of the ileal brake following the presence of undigested nutrients into the small intestine increases the retention time in proximal parts of the gut through negative feedback mechanisms. As to the effect of animal age and size, older pigs have a longer retention time compared to young pigs associated with a longer intestinal tract and a better digestion and absorption of nutrients (Ratanpaul et al. 2019). In addition, sows have a much longer MRT than growing and finishing pigs potentially lasting more than 80 hours.

## Nutrient Digestion

Nutrient digestion is defined as the process by which food is dissected into simple chemical compounds (nutrients) that can be absorbed along the gastrointestinal tract. In pigs, enzymatic digestion takes place mainly in the small intestine and fermentation in the large intestine. The composition of the diet determines the type of digestion, absorption capacity, and secretion of gut peptides involved in the regulation of energy homeostasis. The main digestive enzymes, their sources, and roles are summarized in Table 1.2 and illustrated in Figure 1.4 together with variations of pH along the gastrointestinal tract.

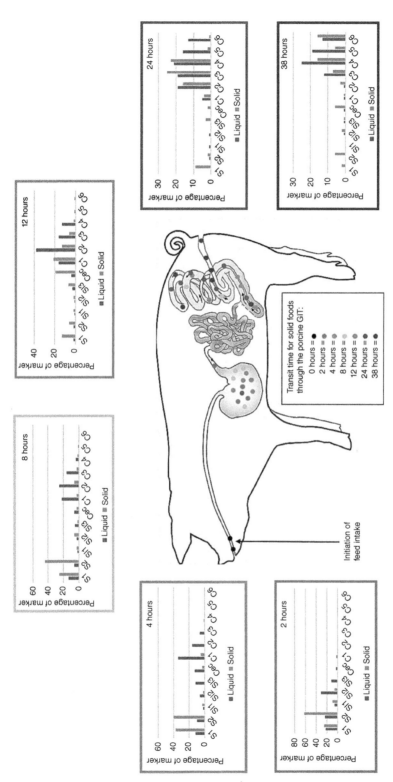

**Figure 1.6** Transit time of liquid and solid particles through the gastrointestinal tract in young pigs. Dietary components may stay up to 24 hours in the stomach depending on particle size among other aspects. The amount of chyme in the small intestine is relatively small at any given time. The transit time in cecum and colon represents 70–85% of the total. Source: Adapted from Stevens and Hume (1995). Abbreviations: S1 = stomach cranial, S2 = stomach caudal, SI1 = duodenum, SI2 = jejunum, SI3 = ilium, Cec = cecum, C1 = colon cranial, C2 = colon medium 1, C3 = colon medium 2, C4 = colon medium 3, C5 = colon distal, C6 = rectum.

**Table 1.2** Main digestive enzymes with main location in the gastrointestinal tract of the pig, substrate and end products, and hydrolysis site.

| Enzyme | Location/secretory glands or organs | Nutrient target and product | Hydrolysis site |
|---|---|---|---|
| *Carbohydrases* | | | |
| α-amylase | Parotid gland and pancreas | Polysaccharides → oligosaccharides (mainly di- and tri-saccharides) | α-1,4 internal links |
| Glucoamylases (1 and 2) | Brush border of enterocytes | Oligosaccharides → glucose | α-1,4 internal links |
| Sucrase | Brush border of enterocytes | Sucrose → glucose and fructose | α-1,2 internal links |
| Isomaltase | Brush border of enterocytes | Dextrins and maltose → glucose | α-1,4 and α-1,6 internal links |
| Lactase | Brush border of enterocytes | Lactose → glucose and galactose | α-1,4 internal links |
| Trehalase | Brush border of enterocytes | Trehalose → glucose | α-1,4 internal links |
| Cellobiose | Brush border of enterocytes | Cellobiose → glucose | β-1,4 internal links |
| *Lipases* | | | |
| Lipase | Von Ebner's glands, stomach, and pancreas | Triglycerides → FFA and glycerol | Ester bonds (sn-1 and sn-3) |
| Co-lipase | Pancreas | Triglycerides → FFA and glycerol | Cofactor for pancreatic lipase |
| Carboxylic/cholesterol-ester hydrolase | Pancreas | Triglycerides, phospholipids, and cholesteryl esters → Non-esterified cholesterol, FFA, and glycerol | Ester bonds (sn-2) |
| Phospholipase A2 | Pancreas | Phospholipids → Lysophospholipids and FFA | Ester bonds (sn-2) |
| *Proteases* | | | |
| Pepsinogen/Pepsin | Stomach | Proteins → polypeptides | Peptide bonds-aromatic AA |
| Trypsinogen/Trypsin | Pancreas | Proteins and polypeptides → Free AA and oligopeptides | Peptide bonds-basic AA |
| Chymotrypsin/-ogen | Pancreas | Proteins and polypeptides → Free AA and oligopeptides | Peptide bonds-aromatic AA |
| Elastase 1 | Pancreas | Proteins and polypeptides → Free AA and oligopeptides | Peptide bonds-hydrophobic AA |
| Carboxypeptidases A/B | Pancreas | Polypeptides → Free AA and small oligopeptides (tri- and di-peptides) | Peptide bonds, C-terminalaromatic, basic or BC AA |
| Aminopeptidases* | Cytoplasm and brush border of enterocytes | Polypeptides → Free AA and small oligopeptides (tri- and di-peptides) | Peptide bonds, N-terminal |
| Dipeptidases** | Cytoplasm and brush border of enterocytes | Dipeptides → Free AA | Di-peptide bonds |
| *Nucleases* | | | |
| DNASE1 | Pancreas | DNA → nucleic acids | 5′-phosphooligonucleotide |
| RNASE1 | Pancreas | RNA → nucleic acids | 3′-Cy and Ur nucleosides |

Abbreviations: AA = amino acids; BCAA = branched chain amino acids; FFA = free fatty acids; C-terminal = carboxyl terminal; N-terminal = amino terminal; RNA = ribonucleic acid; DNA = deoxyribonucleic acid; sn-1, -2,-3 = stereospecific numbering system; Cy = cytidine nucleoside; and Ur = uridine nucleosides.

* The main aminopeptidases studied in pig are N, A, and dipeptidyl peptidase IV.

** The main dipeptidases studied in pigs are glycyl-L-leucine, L-alanyl-L-glutamic acid, and L-alanyl-L-proline.

*Protein and Amino Acid Digestion*

Protein digestion in pigs starts in the stomach, through the action of pepsin and hydrochloric acid (HCl). HCl is required to lower the stomach pH below 5, to transform pepsinogen into its active form, pepsin. Pepsin, in turn, increases the susceptibility of the ingested protein to the activity of pancreatic proteases by altering their quaternary and tertiary structure and exposing their AA residues.

Once the digesta reaches the proximal small intestine, pancreatic and intestine enzymes break down proteins and polypeptides further (Table 1.2). Among pancreatic enzymes, trypsin, chymotrypsin, and elastase are responsible for the initial breakdown of protein and polypeptides into small peptides and free AA within the small intestine. Nevertheless, these enzymes are secreted in their inactive forms. Pancreatic trypsinogen is activated within the duodenum through the action of brush border enzymes (enteropeptidase) into trypsin, which activates other zymogens including other trypsinogen proenzymes (trypsinogen, chymo-trypsinogen, procarboxypeptidase, and pepsinogen). Smaller peptides are then digested by carboxypeptidases secreted by the pancreas as well as by aminopeptidases present in the brush border of the small intestinal enterocytes. Thus, the final product of lumen digestion is free AA, di- and tripeptides that can be easily absorbed and further processed by cytosolic enzymes within the enterocytes (Krehbiel and Matthews 2003).

The rate-limiting step in the digestion of protein seems to be the rate at which the protein source is dissolved and hydrolyzed. For example, skim milk powder has a higher and more rapid digestion than soybean meal which in turn has a higher and more rapid digestibility than corn gluten meal (Asche et al. 1989). Digestion of protein from diets based on barley and soybean meal fed to growing pigs occurs rapidly based on the net portal appearance (NPA) of amino acids (AAs), and this can be affected by the inclusion of materials such as betaine and conjugated linoleic acid (CLA) in the diet. The NPA of essential, nonessential, and total AAs for pigs following a single feed is shown in Figure 1.7.

*Nonfiber Carbohydrates Digestion*

Although carbohydrate breakdown starts in the mouth, the digestion that occurs in the oral cavity by salivary $\alpha$-amylase produced by the sublingual glands is negligible compared to the digestion processes that occur in the intestinal lumen, partly due to the rapid inactivation of $\alpha$-amylase by the low pH in the stomach.

Within the stomach, carbohydrate digestion is limited for both sugars and starches. The gastric digestion of carbohydrates is mainly due to fermentation and leads to the production of SCFA and lactic acid, which can only be absorbed in limited amounts by the gastric mucosa.

In the small intestine, most of the starch consumed is degraded by pancreatic $\alpha$-amylase into maltose, isomaltose, maltotriose, alpha limit-dextrins, and some free glucose. These new di-, tri-, and oligosaccharides are then further processed by brush border enzymes (oligosaccharidases) as enterocytes membrane transporters cannot absorb molecules bigger than monosaccharides (Table 1.2). Among oligosaccharidases, glycoamylases can remove 1,4 linked glucose residues, whereas $\alpha$-dextrinases (isomaltases) and sucrases can cleave the nonreducing terminal $\alpha$-1,6 and $\alpha$-1,2 links, respectively. The products of carbohydrate enzymatic digestion are monosaccharides mainly glucose, fructose, and galactose. Glucose is then actively transported inside enterocytes by the sodium-potassium ATPase system, which pumps the intracellular $Na^+$ across the basolateral membrane (Drochner 1993). The $Na^+$/glucose cotransporter (SGLT1) is the major route for the

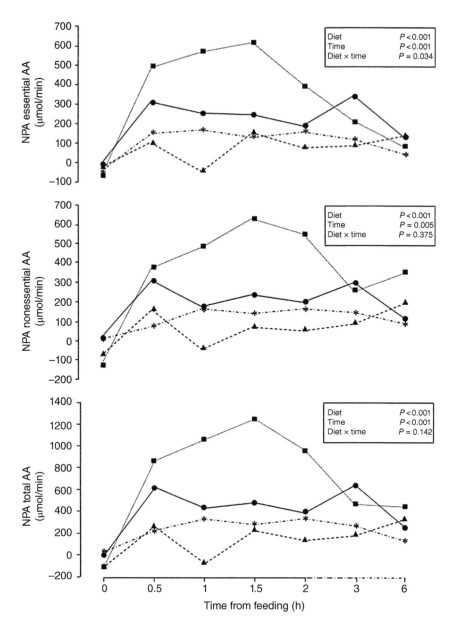

**Figure 1.7** From top to bottom: Net portal appearance (NPA) of essential, nonessential, and total amino acids (AA) of Iberian growing pigs fed control (●), betaine (■), CLA (▲), or betaine + CLA diets (*). Values are means for four pigs. Source: Lachica et al. (2021) / with permission of Elsevier.

transport of dietary sugars from the lumen of the intestine into enterocytes. Regulation of this protein is essential for the provision of glucose to the body and avoidance of intestinal malabsorption. It has been demonstrated that dietary sugars and artificial sweeteners increase SGLT1 expression and the capacity of the gut to absorb monosaccharides. Fructose is transported by the $Na^+$-independent fructose transporter, GLUT5, transporting fructose from the lumen of the intestine into enterocytes down its concentration gradient. These monosaccharides, when accumulated in the enterocytes, exit the cell across the basolateral membrane into the systemic circulation by another Na$^+$-independent monosaccharide transporter, GLUT2, a bidirectional transporter that can move glucose out of or into the cell depending on its concentration gradient (Shirazi-Beechey 1995).

Starch digestibility is affected by molecular features such as the degree of porosity of the granular surface, particle size, and degree of gelatinization such as after grain processing (i.e., pelleting or extrusion), all the above ultimately affecting the accessibility of digestive enzymes to the glucose polymers. A more detail explanation on the effects of grain processing and starch structure on digestibility have been explained in the section on "Effects of grain processing and starch structure on nutrient digestibility."

### Fiber Digestion

Degradation of dietary fiber in the stomach and small intestine is limited compared to that of the cecum and large intestine. Dietary fiber will pass the ileo-cecal sphincter mainly intact, becoming substrate for the microbiome in the large intestine. Bacteria in the cecum and colon use the soluble fiber through fermentation producing SCFA including propionate, acetate, and butyrate, as well as other products such as lactic acid, methane, carbon dioxide, hydrogen, water, and heat. The final content of SCFA depends on the degradability of the substrate that reaches the large intestine. While oligosaccharides are easily fermented by the microflora in the colon, other elements such as cellulose show a much slower breakdown, resulting in only a moderate production of SCFA in comparison (Drochner 1993).

As illustrated by Pekas (1991), the size and weight of the digestive system increase with age and BW, with a longer hindgut in 270 kg BW pigs (7.5 m) than in 70 kg BW pigs (5.4 m) or in 30 kg BW pigs (4.3 m). Le Goff et al. (2002) reported neutral detergent fiber digestibility was 8–17% higher for pigs weighing 77 kg than those weighing 34 kg. Digestibility of NDF did not differ between pigs of different weight when a low fiber control diet was fed. In contrast, the difference in NDF digestibility for the same higher fiber diets when fed to sows was 18.5 and 9.4% higher than that measured in pigs weighing 77 kg when fed diets containing maize or wheat bran, respectively. There was no difference in NDF digestibility between the heavy pigs and sows when either were fed the lower NDF control diet or a diet containing sugar beet pulp as the fiber source.

In an earlier study, Noblet and Shi (1994) measured the digestibility of maize bran, wheat bran, and sugar beet pulp in grower pigs, finisher pigs, and in sows. The results for organic matter and neutral detergent fiber digestibility are shown in Figure 1.8. They demonstrate how the animal and the source of fiber affect energy and nutrient digestibility.

### Fat Digestion

Like carbohydrate digestion, lipid breakdown (triglycerides) starts, to a small extent, in the oral cavity with the activity of lingual lipase. Digestion is then continued in the stomach by gastric lipase, where a part of the emulsification of dietary fats and fat-soluble vitamins occurs (the physical breakdown of fat globules into smaller size droplets).

Lipids then enter the duodenum mixing with bile salts and pancreatic juices that contain lipase, colipase, phospholipase A2, and carboxyl ester hydrolase (Table 1.2). While bile salts promote the solubilization of lipids, facilitating their digestion, pancreatic lipolytic enzymes (lipase, phospholipase

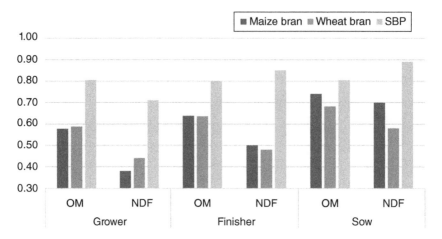

**Figure 1.8**   Effects of animal type (grower, finisher, and sow) and fiber source on the digestibility coefficient for pigs of organic matter (OM) and neutral detergent fiber (NDF). SBP: sugar beet pulp. Source: Based on Noblet and Shi (1994).

A2, and carboxyl ester hydrolase) help in the breakdown of triglycerides into more polar and smaller molecules such as free fatty acids (FFA), free cholesterol, and lysophospholipids. Colipase prevents the inhibition of lipase by bile salts within the duodenum. FFA (mono- and diacylglycerols) can then be easily absorbed by enterocytes and used for the biosynthesis of new triglycerides within the cell. Absorption of FFA is nearly complete at the distal ileum and has been attributed to both passive diffusion as well as to protein-dependent models (fatty acid transporters). Within enterocytes, triglycerides can then, together with cholesterol, be packaged into phospholipid vesicles (chylomicrons) enabling their transport/movement through the lymphatic and circulatory system. Any fatty acid that is not absorbed in the small intestine becomes substrate for microbial hydrogenation in the cecum and colon. Saturated fatty acids have lower ileal digestibility than unsaturated fatty acids.

### Nutrient During Digestion

Dietary fiber has the biggest effect on the digestibility of other nutrients. This interaction has been associated to fiber's physicochemical properties such as solubility (thus affecting chyme's viscosity) and consists of partially interfering with the digestion of mainly proteins and fats. It has been estimated that each 1% of NDF in the diet reduces by 0.5% the ileal apparent digestibility of dietary protein and the coefficient of energy digestibility by 1% in pigs. In contrast, the effects on insoluble fiber on ileal nutrient digestibility are less relevant (Dégen et al. 2007). A mechanism of action contributing to how soluble fiber effects fat digestibility in pigs has been described by Gunness et al. (2016a, b) showed that soluble fiber (e.g., as β-glucans and arabinoxylan) inhibit bile acid availability.

The impact of dietary fiber affects the digestibility of the overall diet. This is shown in Figure 1.9 for OM in pigs fed a low fiber and a high fiber diet over four periods when weight increased from 35.8 to 74.3 kg. Both diets were based on corn, barley, wheat, and soybean meal. The low fiber diet contained 10% NDF and 3.4% ADF. The high fiber diet contained 20% NDF and 9.5% ADF. The high fiber diet contained 15% wheat bran, 10% soybean hulls, and 5% sugar beet pulp. Both diets had the same crude protein level and digestible lysine to NE ratio. The results show that OM digestibility increased with period/weight and was lower for the high fiber diet. The digestibility of most other nutrients also increased with period/weight. The digestibility of energy and N was also

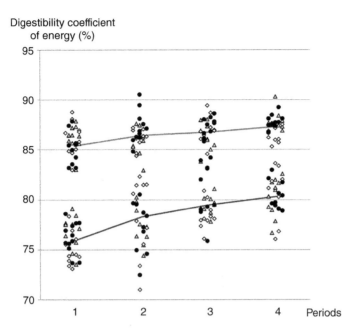

**Figure 1.9** Evolution of digestibility coefficient of energy (in %) across four periods for a low-fiber (bold line on the top) or high-fiber (bold line on the bottom) diets in Duroc (▲), Large-White (◇), and Piertrain (●), pigs. Source: Le Sciellour et al. (2018) / PLOS ONE / CC BY 4.0.

reduced in pigs fed the high fiber diet. The digestibility of fiber components was higher for pigs fed the high fiber diet. The negative effect of NDF on OM and protein digestibility is quite dramatic and results in reduced growth performance (Le Sciellour et al. 2018).

Divalent minerals can also reduce the digestibility and bioavailability of liposoluble gut contents. Increasing dietary Ca reduces linearly the digestibility of fat by forming soaps that are excreted in feces. In particular, an increase from 0 to 6 g kg$^{-1}$ of dietary Ca led to a fourfold increase ($P \leq 0.05$) in excreted palmitic and stearic acids in diets containing tallow or palm olein as major fat sources. More than 80% of these excreted fatty acids were present as soaps in feces. For the tallow-based diets, stearic acid digestibility decreased from 91 to 66% ($P \leq 0.01$), while palmitic acid digestibility dropped from 96 to 83% ($P \leq 0.01$). In contrast, dietary Ca excess did not influence fatty acid (or soap) excretion when the main oil sources where olive oil or soybean oil.

### *Effects of Grain Processing and Starch Structure on Nutrient Digestibility*

The main limiting step in the digestion of grains is the access of digestive enzymes to their target substrates such as starch or proteins, which depends on the physical structure, including the intactness of cell walls. Thus, the digestibility of grains (the main energy source in pig diets) and other ingredients can be affected by particle size and how they are processed, and this is not always considered when allocating energy values to ingredients in diet formulation. Reducing particle size disrupts plant cell walls improving the digestion of starch due to the increased surface area of interaction with digestive enzymes (termed "outside-in" hydrolysis). The removal of the large particle size fraction in feeds significantly improves feed utilization in pigs (Al-Rabadi et al. 2017). In addition, the structure and porosity of the starch granules also determines enzyme access and the degree of resistance to digestion. Starch granules with smooth continuous surfaces can be surprisingly resistant to enzyme hydrolysis, whereas other starches have granules with surface pores that allow for the entry of enzymes into the granules' interior (termed "inside-out" hydrolysis). Figure 1.10 illustrates the hydrolysis behavior of potato and maize starch granules. Maize starch has surface pores allowing the "inside-out" rapid digestion independent of granule size, whereas potato starch has no surface pores allowing only the slower "outside-in" hydrolysis, the latter being highly dependent on granule size.

The apparent total tract digestibility (ATTD) of gross energy (GE) improves in most grains when particle size is reduced to less than 400 microns with the improvement being more evident in grower pigs than weaner pigs and for corn compared to wheat. In contrast, the effect on protein digestibility is generally small and inconsistent. The effect of reducing diet particle size on performance is reflected mainly in feed efficiency, which is consistent with the process improving grain and diet

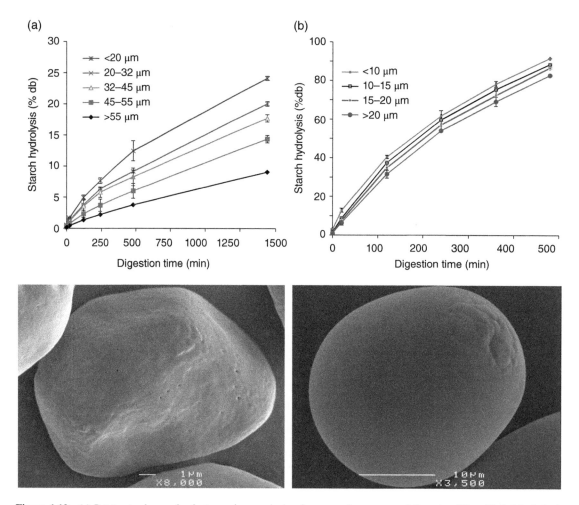

**Figure 1.10**   (a) Potato starch granule electron micrograph showing no surface pores and the rate of "outside-in" hydrolysis highly influenced by particle size; (b) maize starch granule electron micrograph showing surface pores allowing "inside-out" hydrolysis and rate of hydrolysis independent of particle size. Source: Dhital et al. (2010), Figure 02 & 03[p. 483 & 484]/with permission from Elsevier.

energy availability (Lancheros et al. 2020). The optimal particle size depends on the type of cereal that, in turn, reflects the molecular structure and the accessibility or diffusion rate of digestive enzymes in pigs as previously mentioned when comparing maize to potato starches. In ground oats, a reduction in the geometric mean of the particles to c. 600 micron improved ATTD of NDF, hemi-cellulose, and GE, while a further reduction to c. 570 micron reduced the ATTD of Ca and P but not NDF, hemicellulose, and GE. The effects of particle size on the DE, ME, and NE of the ground oats are shown in Figure 1.11. The values determined for the coarse ground oats were the closest to those derived from current prediction equations, but the DE, ME, and NE values for the medium ground oats were 4.6, 5.4, and 12.1% higher than those for the coarse ground oats. Finer grinding did not further improve the energy values of the ground oats.

Pelleting pig diets alters the structure of starch also making it more accessible to digestive enzymes and improving energy digestibility particularly in weaner and grower pigs where the ATTD of GE was improved 10.2 and 5.8%, respectively, compared with pigs offered the same diets in the mash form. The apparent discrepancy between energy digestibility and feed efficiency may be associated with the effects of pelleting on protein digestibility and on reducing feed wastage (Lancheros et al. 2020).

Extrusion consists of continuous cooking under pressure, moisture, and elevated temperature (Vatansever et al. 2020), which involves the use of shear force, and may be applied to feed ingredi-ents to increase absorption and utilization of nutrients (Li et al. 2011). Extrusion improves the digestibility of energy and improves the digestibility of AAs in corn-soybean diets and in peas. In contrast to the effect observed in pelleting, the effects of extruding diets on growth performance were more evident in grower than weaner pigs, and the average improvement in feed efficiency for grower pigs was 11.1% (Lancheros et al. 2020).

Overall, reducing particle size of pig feed improves the digestibility of starch in the absence of surface pores such as in potato starch granules but not in highly porous starches such as maize that allow the "inside-out" hydrolysis to occur. Small particle sizes enable quite marked increases in NDF and energy availability. The responses differ with the ingredients involved, the age/weight of the pig, and what equipment is used to reduce particle size. Pelleting pig feeds have a more consistent effect on starch digestion and a larger effect than reducing particle size on pig feed efficiency. The latter is probably associated with the effect of pelleting diets on reducing feed wastage.

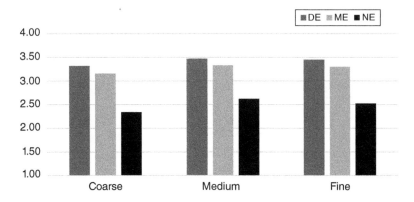

**Figure 1.11** Effects of oats ground coarse, medium, or fine on the measured digestible energy (DE), metabolizable energy (ME), and net energy value (NE) expressed as Mcal/kg. The geometric mean particle size of the course, medium, and fine ground oats was 765, 619, and 569 microns, respectively.

## *Nutrient Sensing and the Gut-Brain Axis*

Nutritional chemosensing is a scientific discipline that studies the perception of nutrients in biological systems linking molecular sensing mechanisms to genomic, metabolic, physiologic, and behavioral responses (Roura et al. 2016). Nutrient sensing in mammals starts in the nasal and oral cavities, as both taste and smell provide critical information to the central neuronal system (CNS) regarding the nutritional value of foods. The sense of smell is essential to the perception of food flavors. The retronasal smell relevant to foods is a response to volatile compounds released in the oral cavity reaching the olfactory epithelium (OE) through the nasopharynx. Pigs have an extremely sensitive olfactory system with one of the largest OE known in mammalian species. The OE, located in the upper wall of the nasal cavity, is composed of three main cell types: olfactory sensory neurons (OSN), supporting cells, and basal or progenitor cells. Food odorants stimulate olfactory receptors present on the plasma membrane of the cilia projected by the dendrite of OSN within the OE (Figure 1.12). OSN transduce the chemical stimulus into an electrical stimulus through membrane depolarization, reaching the olfactory bulb in the first instance and transferring the stimulus to mitral cells that will reach the primary olfactory cortex in the CNS (Roura and Tedo 2009).

The sense of taste is the result of perceiving nonvolatile compounds present in food in the oral cavity. Taste has been classified into five categories based on human sensing and believed to be conserved in pigs (Table 1.3): sweet (evoked by simple carbohydrates such as sugars), umami (associated mainly to AAs and peptides), salty (triggered by mineral salts), sour (sensing acids), and bitter (related to potentially toxic or antinutritional compounds). The taste system in the pig is based on c. 19 900 taste buds located in the oral cavity clustered in groups of 100–150 specialized epithelial cells projecting microvilli into the mucus layer of the tongue and grouped in macroscopic structures known as taste papillae (Figure 1.12). In pigs, there are three types of taste papillae: fungiform (scattered uniformly through the upper surface of the rear half of the pig tongue), circumvallate or vallate (two mushroom-shaped epithelial structures carved in the rear epithelia of the tongue in contact with the oropharynx and surrounded by deep crypts with the presence of von Ebner's

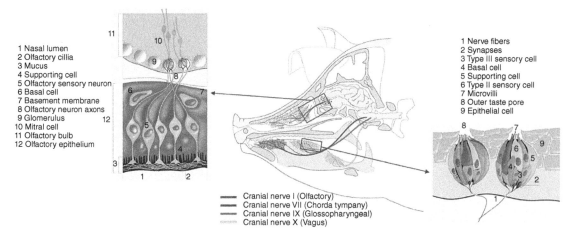

1 Nasal lumen
2 Olfactory cillia
3 Mucus
4 Supporting cell
5 Olfactory sensory neuron
6 Basal cell
7 Basement membrane
8 Olfactory neuron axons
9 Glomerulus
10 Mitral cell
11 Olfactory bulb
12 Olfactory epithelium

1 Nerve fibers
2 Synapses
3 Type III sensory cell
4 Basal cell
5 Supporting cell
6 Type II sensory cell
7 Microvilli
8 Outer taste pore
9 Epithelial cell

Cranial nerve I (Olfactory)
Cranial nerve VII (Chorda tympany)
Cranial nerve IX (Glossopharyngeal)
Cranial nerve X (Vagus)

**Figure 1.12** Drawing showing the four cranial nerves responsible for smell (I) or taste (VII, IX, and X) sensing. The olfactory epithelium and olfactory bulb showing how sensory neurons reach the lumen of the nasal cavity through the mucosa by setting a net of olfactory cilia. The axons of the neurons reach the glomerulus in the olfactory bulb where mitral cells will take the signal to the primary olfactory cortex. Lingual epithelium showing the main taste types and other functional structures in taste buds. Source: Adapted from Roura and Tedo (2009); Original artwork by Joaquim Roura.

**Table 1.3** Main taste-types and taste-active compounds known in pigs.

| Taste-type | Chemical groups | Main ligands known | Taste-receptors |
|---|---|---|---|
| Umami | Free AA | L-AA: Lys, Glu, Asp, Ala, Gln, and Asn | T1R1/T1R3 |
| Unknown taste | Peptides and free AA | Peptone, L-Hyp, L-Ser, L-Thr, aromatic, and basic AA | CASR, GPRC6A, GPR92, mGluR1 and 4 |
| Sweet | Sugars, alcohols, free AA, and other | Monosaccharides (e.g., glucose, fructose, galactose, and mannose); oligosaccharides (e.g., maltose, lactose, melibiose, sucrose, and trehalose); polyols (e.g., xylitol and sorbitol); D-AA (e.g., Ala, Asn, Gln, Phe, Ser, Thr, Trp) and Gly; high-intensity sweeteners (e.g., sucralose, rebaudioside A, acesulfame-K, saccharine, dulcin) | T1R2/T1R3, SLGT1/SLGT3, GLUT4 |
| Non-sweet | High-intensity sweeteners | Aspartame, cyclamate, NHDHC, perillartine, monellin, and thaumatin | None |
| Starchy "taste" | Glucose polymers | Malto-oligosaccharides | Unknown |
| Fatty "taste" | Triglycerides and FFA | Long-, medium-, and short-chain fatty acids (saturated and unsaturated) | FFAR1, 2, 3 and 4; GPR84 |
| Bitter | Diverse | Caffeine, quinine, denatonium benzoate, L-Trp | 16 T2Rs |
| Sour | Acids | Citric, ascorbic, tartaric, formic, lactic, and phosphoric acids | Otop1 |
| Salty | Salts | NaCl, LiCl, KCl | ENAC$\alpha$, $\beta$, $\gamma$ |

Abbreviations: AA = amino acids; FFA = free fatty acids; NHDHC = neo-hesperedin dihydrochalcone.

salivary glands), and foliate (bilateral striated cuts in the back of the tongue). Some taste buds have also been found in the epiglottis and the soft palate epithelium. Taste buds are composed of four types of cells: taste cells I, II, III, and a basal/progenitor cell type. Taste cells I can sense sour compounds, whereas cell types II oversee umami, sweet, and bitter sensing. Type III cells perceive salty taste together with playing a fundamental role as presynaptic cells that connect taste sensory cells to sensory neurons of cranial nerves VII, IX, or X (Figure 1.12).

The lingual nerves connect directly with the primary gustatory cortex (i.e., insular cortex). Subsequently, multimodal sensory signals (e.g., smell, taste, and somatosensing) convey in secondary brain centers such as the orbitofrontal cortex involved in hedonic responses. In pigs, quinine (bitter) and sucrose (sweet) both promote the activation of the gustatory cortex while differentially activating other regions. Sucrose activates the putamen, which is involved in the reward circuits of perception. In contrast, quinine strongly activates the amygdala, a brain region involved in associative conditioning such as taste aversion. The sweet and bitter brain activation patterns found in pigs are concordant to results from human studies indicating the similarity between the two species in terms of taste perception (Coquery et al. 2018).

The activation of taste cells occurs when nutrients interact with the specific taste receptors (TR) located in the apical transmembrane (Figure 1.12 and Table 1.3). Taste perception of sweet and umami tastes in the tongue is mediated by two dimers of the T1R family of G-protein-coupled receptors (GPCR). The T1R1/T1R3 dimer perceives umami taste responding to L-Ala, L-Asn, L-Asp, L-Glu, L-Gln, Pro, and Thr (Roura and Fu 2017). The T1R2/T1R3 dimer elicits sweet taste responding to mono and oligosaccharides, polyols, some D-AA, and some sweet compounds known to be of high intensity (HIS) in humans such as sucralose or rebaudioside A, among others. In contrast, other HIS such as aspartame and cyclamate have failed to elicit sweet responses in pigs. In addition, other AAs associated with taste-type responses in pigs include L-Hyp, L-Ser, L-Thr, aromatic and basic AA, and peptones. Based on other mammalian data (mostly mice), the receptors responsible for the oral sensing of other AA may be G-protein-coupled receptors such as CaSR, GPRC6A, T1R1/T1R3, mGluR1, mGluR4, GPR92, as well as AA transporters/systems including the Pept1 receptor. Similarly, based on rodent data, some monosaccharides, such as glucose, galactose, and fructose, seemed to be sensed by a pathway independent to the T1R2/T1R3 involving the sodium glucose-linked transporter SGLT-1, SGLT-2, GLUT2, or GLUT5.

Other taste types and their ligands described in pigs include bitter, salty, sour, fatty, and starchy tastes (Table 1.3). The perception of bitter compounds is performed by a second family of TR known as T2R. The T2R family in pigs consists of 16 functional receptors (da Silva et al. 2014). Salty taste is evoked mainly by ionic sodium when in contact with the epithelium sodium channel receptors EnaC-$\alpha$, -$\beta$, -$\gamma$, whereas sour taste has been linked to the abundance of hydrogen ions perceived through the ion channel Otop 1. FFA sensing has been linked with a variety of GPCR mainly FFAR1, 2, 3, 4 and GPR84 (Table 1.3).

The molecular mechanisms involved in nutrient/taste sensing in the mouth are also present throughout the intestinal tract. These so-called chemosensors are strategically positioned as part of the gut mucosa. They are continuously monitoring the presence and concentration of the nutrients, digestion products, and microbial metabolites in the gut playing a pivotal role in the regulation of gastrointestinal digestive secretions, passage rates and absorption, as well as the hunger/satiety cycle. Nutrient sensors are expressed by EEC and mediate the secretion of gut peptides that orchestrate feeding behavior.

### Enteroendocrine System (EES)

The EES is a network of EEC present in the gastrointestinal mucosa that responds to foods by releasing gut peptides that coordinate digestion and absorption of nutrients involving biliary and pancreatic secretion and the modulation of gut motility. In addition, the EES senses toxic/harmful substances to initiate protective responses such as the delay of gastric emptying, the induction of vomiting and diarrhea, and the increase of intestinal secretions. The sensing of nutrients by EEC occurs through the same transmembrane receptors described as part of the taste system in the oral cavity (see Table 1.3). EEC represent approximately 1% of the intestinal epithelia accounting for the largest endocrine organ in the body. EEC are classified based on the expression and secretion of gut peptides. Gut peptides interact with neural circuits including the hypothalamus (the main organ controlling food intake) orchestrating the hunger-satiety cycle a pathway known as the gut-brain axis. The brain receives input via gut peptides secreted by EEC in response to ingested nutrients. Gut peptides are believed to stimulate brain structures (such as the hypothalamus) directly via blood or indirectly through efferent vagal nerve signals. A full classification can be found in the review by Fothergill and Furness (2018). Gut peptides are often coexpressed in the same cell, such as PYY with GLP-1 or gastric inhibitory peptide (GIP) in duodenal L-cells in pigs. The main gut peptides studied to date in pigs are ghrelin, CCK, the two incretins GLP-1 and GIP, and PYY (Table 1.4).

### Ghrelin

Ghrelin is an appetite stimulatory (or orexigenic) peptide produced and secreted mainly by enteroendocrine A/X-like cells located in the fundus and corpus of the stomach and in decreasing amounts from duodenum to colon (Peeters 2005). Ghrelin has a positive impact on appetite increasing eating time and growth by activating the growth hormone secretagogue receptor (GHS-R) in pigs (Reynolds et al. 2010). In addition, ghrelin has been linked to GIT motility and pancreatic

**Table 1.4** Main gut hormones, associated enteroendocrine cell (EEC) types, and physiological impact in pig.

| Digestive system hormones | EEC-type | Nutrients known to stimulate EEC hormone release (in pigs) | Gut tissue | Feeding and gastrointestinal effects |
|---|---|---|---|---|
| Ghrelin | X/A | Peptones, L-AA (L-Trp and L-Phe), SCFA, LCFA, LPA, bitter compounds | Stomach | ↑ Feed intake, ↑ gastric acid secretion and emptying, ↑ pancreatic enzyme secretion |
| Serotonin | EC | Casein, LC FA | Stomach, small and large intestine | ↓ Feed intake[b], ↑ intestinal motility[b] |
| Somatostatin | D | Unknown | Stomach, small intestine | ↓ Gastric acid secretion and emptying |
| Gastrin | G | Peptones, L-AA (Trp and Phe), LPA | Stomach, duodenum | ↑ Gastric emptying and acid secretion |
| Secretin | S | LCFA | Proximal small intestine | ↑ Biliary and pancreatic bicarbonate secretion |
| Cholecystokinin | I | Mono/disaccharides, polyols, di/tripeptides, L-AA (Trp, Phe, Arg, Leu, Ile, and Lys), LCFA, LPA, bitter compounds | Proximal small intestine | ↓ Feed intake, ↓ gastric emptying, ↑ gall bladder contraction, ↑ pancreatic enzyme secretion |
| GIP | K | Mono/disaccharides, polyols, peptones, L-AA (Trp and Phe), LCFA | Proximal small intestine | ↓ Feed intake[b], ↑ insulin, ↓ gastric acid secretion and emptying[b], ↓ intestinal motility[b] |
| Motilin | M | Unknown | Proximal small intestine | ↑ Gastric emptying and pepsin secretion[b], ↑ intestinal motility[b] |
| Insulin | β-cells[a] | Mono and disaccharides, polyols, L-AA, LCFA, bitter compounds | Pancreas | ↓ Feed intake |
| Neurotensin | N | Unknown | Distal small intestine | ↓ Feed intake[b], ↑ Pancreatic enzyme secretion |
| GLP-1 | L | Mono/disaccharides, polyols, peptones, di/tripeptides, L-AA (Trp and Phe), SCFA, LCFA, LPA, BA, bitter compounds | Distal small intestine, colon | ↓ Feed intake, ↓ gastric emptying and acid secretion, ↓ intestinal motility, ↓ pancreatic enzyme secretion, ↑ insulin |
| GLP-2 | L | LCFA, BA | Distal small intestine, colon | ↑ Pancreatic enzyme secretion, ↑ intestinal enzyme activity, ↑ epithelial proliferation, ↑ barrier function, ↑ blood flow[b] |
| PYY | L | Mono and disaccharides, polyols, peptones, L-AA (Trp and Phe), SCFA, BA | Distal small intestine, colon | ↓ Feed intake, ↓ gastric emptying and acid secretion, ↓ intestinal motility, ↓ pancreatic enzyme secretion |
| Oxyntomodulin | L | Unknown | Distal small intestine, colon | ↓ Feed intake[b], ↓ gastric emptying[b] |

Abbreviations: AA = amino acids; BA = bile acids; SCFA = short-chain fatty acids; GIP = gastric inhibitory polypeptide; GLP-1,2 = glucagon-like peptide 1 or 2; LCFA = long-chain fatty acids; LPA = lysophosphatidic acid; PYY = peptide tyrosine-tyrosine.

[a] β-cells have been described in the pancreas and are not part of the enteroendocrine system.

[b] Indicates effects observed in laboratory rodents not confirmed in pigs.

enzyme secretion, cardiovascular functions, and cell proliferation and reproduction, among other activities in the pig (Dong et al. 2009). Zinc oxide and tryptophan are strong secretagogues of ghrelin in pigs (Zhang et al. 2007; Yin et al. 2009).

*Cholecystokinin (CCK)*
The CCK is a gut peptide strongly associated with meal termination/satiation produced by enteroendocrine I-cells located mainly in the proximal segments of the small intestine (duodenum and proximal jejunum) in pigs. Recent studies have shown that CCK function is related to at least two isoforms CCK-8 and CCK-58 associated with short- or long-term activities, respectively. CCK modulates appetite through the activation of the CCK-1R located mainly in vagal afferents nerves or by acting directly on the CNS in pigs (Ripken et al. 2015). CCK secretion is stimulated by the intraluminal presence of products from the hydrolysis of carbohydrates, proteins, and lipids. Free AA including Arg, aromatic AA (Phe and Trp), and branched-chain AA (Leu and Ile) have been described to significantly stimulate CCK secretion through interaction with GPCR, such as CaSR and T1R1/T1R3 in ECC. In addition, CCK has been associated with improved nutrient absorption via stimulating GI secretion and motility, gallbladder contraction, pancreatic enzyme secretion, and/ or delayed gastric emptying (Müller et al. 2021).

*Glucagon-like Peptide 1 (GLP-1) and Gastric Inhibitory Peptide (GIP)*
The GLP-1 and GIP are often referred to as incretin hormones because of the role as stimulants of insulin secretion. As such, they are both considered anorexigenic hormones (thus stimulating meal termination or satiation) (Renner et al. 2018). GLP-1 arises from the processing of proglucagon and is secreted by L-cells in a biphasic manner following meal ingestion. First, GLP-1 is released and linked with duodenal neurohumoral mechanisms, whereas the second peak of secretion seems to be mediated by the activation of L-cells in the ileum and colon in response to the intraluminal presence of nutrients such as glucose and sucrose, lipids, and proteins in pigs (Steinert et al. 2013; Ripken et al. 2014). The GLP-1 receptors (GLP-1R) are expressed in many organs including pancreas, brainstem, hypothalamus, and vagal afferents in the gut. In addition, the incretin hormones have been related to the activation of the ileal brake, which may contribute to nutrient digestion and absorption as well as satiety by delaying gastric emptying and reducing proximal intestinal motility. GLP-1 and GIP are often coexpressed with PYY in enteroendocrine L-cells located in the porcine distal gut.

*Peptide Tyrosine-Tyrosine (PYY)*
The PYY is an anorexigenic gut peptide that activates the Y2 receptors located in vagal afferents and the CNS. Glucose, lipids, and proteins/AA have been shown to trigger PYY release in the proximal small intestine (duodenum/jejunum) where 40–50% of the EEC have been identified as PYY-producing cells in pigs (Cho et al. 2015; Wewer Albrechtsen et al. 2016). In addition, PYY also inhibits gastric acid and pepsin secretion, pancreatic exocrine secretion, gastric emptying, and intestinal motility. Thus, like GLP-1, PYY participates in the ileal brake system, reducing the speed at which nutrient transport occurs in the GIT (Sheikh et al. 1989).

### Developing GIT in Piglets

Pigs are born with a markedly immature digestive system. This affects feeding behavior and feed preferences early in life particularly under commercial conditions where piglets are weaned very young and abruptly compared to the natural process in the wild. In brief, the immature digestive system implies impaired digestive and absorptive capacities compared to a grown pig, which have profound implications in terms of nutrient supply and feeding and feed formulation practices.

The innate dentition of piglets consists of 8 deciduous teeth that increase to 32 early in life, reaching the full set of 44 permanent teeth by 18 months of age. Other differences relevant to anatomical metrics, small intestine development (villi height and crypt depths), and digestive enzyme activities between weaner and finisher pigs are highlighted in Table 1.5. The size of the stomach seems to increase linearly with BW. Similarly, as the pig's BW increases, an increment in small intestine weight, length, and mucosal weight occurs proportionally. The relative length of the small intestine segments changes with age. At birth they are 7–10%, 78–84%, and 9–12% for duodenum, jejunum, and ileum, respectively, while in older pigs, the corresponding proportions are 4–5%, 88–91%, and 4–5% (Adeola and King 2006). Relevant to the nutrient digestion and absorption capacity, immature pigs have shorter villi and decreased crypt depths than older pigs. In addition, the most dramatic change relates to digestive enzyme secretion. Compared to adult pigs, the relatively low enzymatic activities seen in piglets are compatible with an immature digestive system that explain the constraint in digesting feeds in early life. The shortage in enzymatic activity is particularly dramatic for the protease trypsin (Table 1.5).

**Table 1.5** Main anatomical and physiological differences relevant to the gastrointestinal tract between weaner (approximately 5 kg) and finisher (approximately 100 kg) pigs.

| | Length (m) | | Weight (g) | |
|---|---|---|---|---|
| Organ size: | Weaner[a] | Finisher[a] | Weaner[b] | Finisher[c] |
| Stomach | — | — | 20–35 | 600–680 |
| Small intestine | 7.65–9.10 | 16.5–19.73 | 180.5–230 | 1530–2000 |
| Large intestine | 1.2–1.5 | 4.13–5.0 | 53.12 | 1280[c] |

| Villus height and crypt depths (µm): | Duodenum | | Jejunum | | Ileum | |
|---|---|---|---|---|---|---|
| | Weaner[d] | Finisher | Weaner[d] | Finisher[g] | Weaner[d] | Finisher[g] |
| Villus height | 328 ± 16.2 | 420 ± 15.0[e] | 332 ± 13.4 | 480 ± 53.8 | 299 ± 22.7 | 372 ± 43.0 |
| Crypt depth | 207 ± 18.6 | 367 ± 49.7[f] | 208 ± 17.0 | 429 ± 59.9 | 202 ± 18.4 | 316 ± 32.9 |

| Enzyme activities (U g-1 digesta): | Weaner[h] | Finisher[i] |
|---|---|---|
| Lipase | 10 | 175.5 |
| Trypsin | 0.25 | 957.7 |
| Amylase | 25 | 2,807 |

Abbreviations: GI = gastrointestinal.
References: [a] McCance 1974
[b] Marion et al. 2002
[c] Lærke and Hedemann 2012
[d] Yu et al. 2010
[e] Lin et al. 2020
[f] Metzler-Zebeli et al. 2017
[g] Tian et al. 2020
[h] Hedemann and Jensen 2004
[i] Pluschke et al. 2018.
The measures of enzyme activities were obtained from the proximal small intestine (caudal to the hepatopancreatic duct).

## Digestive Secretions in the Young Pig

Some of the most relevant differences affecting pig nutrition practices relate to the low production of digestive juices, particularly enzymes in young pigs. The weaning process is particularly disrupting causing a decline in gastric protease and pancreatic trypsin, chymotrypsin, amylase, and lipase, all fell the week after weaning with amylase declining by 72% and chymotrypsin by 59% (Jensen et al. 1997).

Gastric protease chymosin, a milk clotting enzyme with low proteolytic activity, is one of the few enzymes highly expressed in the stomach during the first weeks following birth. Chymosin may initially help gastric development by stimulating gastric distention through milk clotting. With the increase in pepsinogen and HCl secretion during the weeks following weaning, pepsin replaces chymosin, becoming the dominant protease in the stomach of growing pigs (Rezaei et al. 2013). However, parietal cells in young pigs are not able to secret high amounts of HCl. The result is an inefficient conversion of pepsinogen into the active form pepsin (which requires a pH below 4.0). The low production of gastric pepsin together with the low secretion of pancreatic enzymes such as trypsin, chymotrypsin, and carboxypeptidases A and B is the cause of the low protein digestibility observed in piglets and is often associated with early weaning stress and diarrhea (Hedemann and Jensen 2004).

During lactation, the lactic acid produced from the lactose-rich maternal milk helps maintain a low stomach pH in piglets (Suiryanrayna and Ramana 2015). Thus, under commercial early-weaning practices, keeping a source of lactose seems to alleviate digestive problems in the early postweaning phase. Lactase levels are rather high during the suckling period but rapidly decrease around weaning partly related to an ontogenic decline of the enzyme activity in the intestinal brush border (Heo et al. 2013). Digestion of main complex carbohydrates (i.e., starch) is also impaired by the limited secretion of pancreatic α-amylase in young pigs (Pierzynowski et al. 1993).

In addition to the limited protein and carbohydrate digestive capabilities, lipase, colipase, and carboxyl ester hydrolase activity is also low in young pigs. During the suckling period, carboxyl ester hydrolase is the most prevalent fat hydrolase secreted by the pig. However, in the post weaned pig, colipase and lipase are the main enzymes. The low activity of pancreatic fat hydrolases described at weaning has been attributed to the abrupt changes in the dietary fat content (sow milk vs. weaner feed) and relative low feed intake (Lindemann et al. 1986). In addition, age and BW seem to also be major contributors in the changes of pancreatic lipase, colipase, and carboxyl ester hydrolase activity in young pigs as concentrations of these enzymes increase in the weeks following weaning. Gastric lipase activity increases at weaning and then remains relatively constant significantly contributing to the hydrolysis of triacylglycerols (Jensen et al. 1997).

A high level ($173\,g\,d^{-1}$) compared to a low level ($45\,g\,d^{-1}$) of DM intake in the five days immediately after weaning at 14 days of age resulted in greater villus length and higher maltase and glucoamylase activities, while lactase or sucrase activities were not affected. However, despite having a more developed GIT and higher enzyme activities, the digestion of DM, CP, OM, and CHO was significantly higher for pigs fed at the lower level. The digestibility values ranged from 87.1% for CP to 94.7% for CHO illustrating the digestive capacity of piglets even in the period immediately after weaning can be high and that enzyme activities were able to cope with the substrates consumed even by pigs on the high DM intake treatment (Kelly et al. 1991).

## Postweaning Feeding Practices

The exposure to a complete feed (changing from the sow's milk to a solid diet) contributes to an increase in HCl and digestive enzyme secretions, indicating the adaptive capacity of the gastrointestinal tract in young pigs (Marion et al. 2003). The decline in most digestive enzyme activities in the period immediately after weaning is well established as is the fact that piglet feed intake during this period is often very low. These factors have resulted in the use of specialized diets during the first 10–12 days after weaning. The main aspects of a postweaning diet for optimal nutrition can be summarized as (partially adapted from Pluske et al. 2018):

- Use of highly palatable feed ingredients to promote early feed intake.
- If allowed, these diets may benefit from high levels of ZnO that increase feed intake and reduce the incidence and severity of diarrhea.
- Low crude protein content diets using highly digestible protein sources (e.g., skim milk powder, fish meal and whey powder, and plasma protein) balanced with synthetic AAs.
- Increase sulfur amino acid to lysine ratio to 65%, the Trp to Lys ratio to 22%, and the Thr to Lys ratio to 70%.
- Low inclusion of soluble nonstarch polysaccharide (NSP) and add a source of insoluble NSP.
- Use processed starch sources to improve the digestibility.
- Use enzyme supplementation including carbohydrases (xylanase, β-glucanase), where appropriate, and consider phytase super-dosing.
- Low levels of PUFA (particularly keeping a low n-6 to n-3 ratio) and inclusion of medium-chain triglycerides/fatty acids.
- Reduce Ca levels by 10–20% and supplement with organic acids (to reduce buffering capacity in the stomach).
- Consider other additives that will have impacts on mitigating inflammation and immune function such as antioxidants and plant-derived extracts/essential oils.

## References

Adeola, O., and D. E. King. 2006. Developmental changes in morphometry of the small intestine and jejunal sucrase activity during the first nine weeks of postnatal growth in pigs. J. Anim. Sci. 84:112–118.

Al-Rabadi, G. J., B. J. Hosking, P. J. Torley, B. A. Williams, W. L. Bryden, S. G. Nielsen, J. L. Black, and M. J. Gidley. 2017. Regrinding large particles from milled grains improves growth performance of pigs. Anim. Feed Sci. Technol. 233:53–63.

Asche, G. L., A. J. Lewis, and E. R. Peo Jr. 1989. Protein digestion in weanling pigs: effect of dietary protein source. J. Nutr. 119:1093–1099.

Banerjee, A., E. T. McKinley, J. von Moltke, R. J. Coffey, and K. S. Lau. 2018. Interpreting heterogeneity in intestinal tuft cell structure and function. J. Clin. Investig. 128:1711–1719.

Buddington, R. K., A. Pajor, K. K. Buddington, and S. Pierzynowski. 2004. Absorption of α-ketoglutarate by the gastrointestinal tract of pigs. Comp. Biochem. Physiol. Part A Mol. Integr. Physiol. 138:215–220.

Cho, H. J., S. Kosari, B. Hunne, B. Callaghan, L. R. Rivera, D. M. Bravo, and J. B. Furness. 2015. Differences in hormone localisation patterns of K and L type enteroendocrine cells in the mouse and pig small intestine and colon. Cell Tissue Res. 359:693–698.

Coquery, N., P. Meurice, R. Janvier, E. Bobillier, S. Quellec, M. Fu, E. Roura, H. Saint-Jalmes, and D. Val-Laillet. 2018. fMRI-based brain responses to quinine and sucrose gustatory stimulation for nutrition research in the minipig model: a proof-of-concept study. Front. Behav. Neurosci. 12:151.

da Silva, E. C., N. de Jager, W. Burgos-Paz, A. Reverter, M. Perez-Enciso, and E. Roura. 2014. Characterization of the porcine nutrient and taste receptor gene repertoire in domestic and wild populations across the globe. BMC Genom. 15:1057.

Dégen, L., V. Halas, and L. Babinszky. 2007. Effect of dietary fibre on protein and fat digestibility and its consequences on diet formulation for growing and fattening pigs: a review. Acta Agric. Scand. A Anim. Sci. 57:1–9.

Dhital, S., A. K. Shrestha, and M. J. Gidley. 2010. Relationship between granule size and in vitro digestibility of maize and potato starches. Carbohydr. Polym. 82:480–488.

Dong, X. Y., J. Xu, S. Q. Tang, H. Y. Li, Q. Y. Jiang, and X. T. Zou. 2009. Ghrelin and its biological effects on pigs. Peptides 30:1203–1211.

Drochner, W. 1993. Digestion of carbohydrates in the pig. Arch. Tierernahr. 43:95–116.

Ducharme, N. G., A. Desrochers, S. L. Fubini, A. P. Pease, L. A. Mizer, W. Walker, A. M. Trent, J. P. Roy, M. Rousseau, R. M. Radcliffe, & A. Steiner. 2017. Chapter 14: Surgery of the bovine digestive system. In: S. L Fubini, and N. G. Ducharme, editor, Farm animal surgery. 2nd ed. Elsevier, St Louis, MO. p. 223–343.

Fothergill, L. J., and J. B. Furness. 2018. Diversity of enteroendocrine cells investigated at cellular and subcellular levels: The need for a new classification scheme. Histochem. Cell Biol. 150:693–702.

Gonzalez, L. M., I. Williamson, J. A. Piedrahita, A. T. Blikslager, and S. T. Magness. 2013. Cell lineage identification and stem cell culture in a porcine model for the study of intestinal epithelial regeneration. PloS ONE 8:e66465.

Gunawardene, A. R., B. M. Corfe, and C. A. Staton. 2011. Classification and functions of enteroendocrine cells of the lower gastrointestinal tract. Int. J. Exp. Pathol. 92:219–231.

Gunness, P., B. M. Flanagan, J. P. Mata, E. P. Gilbert, and M. J. Gidley. 2016a. Molecular interactions of a model bile salt and porcine bile with (1,3:1,4)-β-glucans and arabinoxylans probed by 13C NMR and SAXS. Food Chem. 197:676–685.

Gunness, P., B. A. Williams, W. J. J. Gerrits, A. R. Bird, O. Kravchuk, and M. J. Gidley. 2016b. Circulating triglycerides and bile acids are reduced by a soluble wheat arabinoxylan via modulation of bile concentration and lipid digestion rates in a pig model. Mol. Nutr. Food Res. 60:642–651.

Hedemann, M. S., and B. B. Jensen. 2004. Variations in enzyme activity in stomach and pancreatic tissue and digesta in piglets around weaning. Arch. Anim. Nutr. 58:47–59.

Heo, J. M., F. O. Opapeju, J. R. Pluske, J. C. Kim, D. J. Hampson and C. M. Nyachoti. 2013. Gastrointestinal health and function in weaned pigs: A review of feeding strategies to control post-weaning diarrhoea without using in-feed antimicrobial compounds. J. Anim. Physiol. Anim. Nutr. (Berl). 97:207–237.

Hipper, K., and H. J. Ehrlein. 2001. Motility of the large intestine and flow of digesta in pigs. Res. Vet. Sci. 71:93–100.

Holzgraefe, D., G. Fahey and A. H. Jensen. 1985. Influence of dietary alfalfa: orchardgrass hay and lasalocid on in vitro estimates of dry matter digestibility and volatile fatty acid concentrations of cecal contents and rate of digesta passage in sows. J. Anim. Sci. 60:1235–1246.

Jensen, M. S., S. K. Jensen, and K. Jakobsen. 1997. Development of digestive enzymes in pigs with emphasis on lipolytic activity in the stomach and pancreas. J. Anim. Sci. 75:437–445.

Jørgensen, H., T. Larsen, X. Q. Zhao, and B. O. Eggum. 1997. The energy value of short-chain fatty acids infused into the caecum of pigs. Br. J. Nutr. 77:745–756.

Jung, C., J. P. Hugot and F. Barreau. 2010. Peyer's patches: The immune sensors of the intestine. Int. J. Inflam. 2010:823710.

Kararli, T. T. 1995. Comparison of the gastrointestinal anatomy, physiology, and biochemistry of humans and commonly used laboratory animals. Biopharm. Drug Dispos. 16:351–380.

Kelly, D., J. A. Smyth, and K. J. McCracken. 1991. Digestive development of the early-weaned pig. 2. Effect of level of food intake on digestive enzyme activity during the immediate post-weaning period. Br. J. Nutr. 65:181–188.

Kong, S., Y. H. Zhang, and W. Zhang. 2018. Regulation of intestinal epithelial cells properties and functions by amino acids. Biomed. Res. Int. 2018:1–10.

Krehbiel, C. R., and J. C. Matthews. 2003. Chapter 3: Absorption of amino acids and peptides. In J. P. F. D'Mello, editor, Amino acids in animal nutrition. CABI Publishing, Wallingford, UK. p. 41–70.

Lachica, M., M. L. Rojas-Cano, L. Lara, A. Haro and I. Fernández-Fígares. 2021. Net portal appearance of proteinogenic amino acids in Iberian pigs fed betaine and conjugated linoleic acid supplemented diets. Anim. Feed. Sci. Tech. 273:1–9.

Lærke, H. N., and M. S. Hedemann. 2012. The digestive system of the pig. In K. E. B. Knudsen, N. J. Kjeldsen, H. D. Poulsen, and B. B. Jensen, editors, Nutritional physiology of pigs. Videncenter for Svineproduktion, Foulum, Denmark. p. 1–27.

Lancheros, J. P., C. D. Espinosa, and H. H. Stein. 2020. Effects of particle size reduction, pelleting, and extrusion on the nutritional value of ingredients and diets fed to pigs: A review. Anim. Feed Sci. Tech. 268:1–9.

Le Goff, G., J. Milgen, and J. Noblet. 2002. Influence of dietary fibre on digestive utilization and rate of passage in growing pigs, finishing pigs and adult sows. Anim. Sci. 74:503–515.

Le Sciellour, M., E. Labussière, O. Zemb and D. Renaudeau. 2018. Effect of dietary fiber content on nutrient digestibility and fecal microbiota composition in growing-finishing pigs. PloS ONE 13:e0206159.

Li, M., P. Liu, W. Zou, L. Yu, F. Xie, H. Pu, H. Liu, and L. Chen. 2011. Extrusion processing and characterization of edible starch films with different amylose contents. J. Food Eng. 106:95–101.

Lin, B., J. Yan, Z. Zhong, and X. Zheng. 2020. A study on the preparation of microbial and nonstarch polysaccharide enzyme synergistic fermented maize cob feed and its feeding efficiency in finishing pigs. Biomed Res. Int. 2020:8839148.

Liebler-Tenorio, E., and R. Pabst. 2006. MALT structure and function in farm animals. Vet Res. 37:257–280.

Lindemann, M., S. Cornelius, S. E. Kandelgy, R. Moser, and J. Pettigrew. 1986. Effect of age, weaning and diet on digestive enzyme levels in the piglet. Journal of animal science. J. Anim. Sci. 62:1298–1307.

Marion, J., M. Biernat, F. Thomas, G. Savary, Y. Le Breton, R. Zabielski, I. Le Huërou-Luron, and J. Le Dividich. 2002. Small intestine growth and morphometry in piglets weaned at 7 days of age. Effects of level of energy intake. Reprod. Nutr. Dev. 42:339–354.

Marion, J., V. Romé, G. Savary, F. Thomas, J. Le Dividich, and I. Le Huërou-Luron. 2003. Weaning and feed intake alter pancreatic enzyme activities and corresponding mRNA levels in 7-d-old piglets. J. Nutr. 133:362–368.

McCance, R. A. 1974 The effect of age on the weights and lengths of pigs' intestines. J Anat. 117:475–479.

Mescher, A. L. 2009. Junqueira's basic histology. Text and atlas, 12th ed. McGraw-Hill Medical Publishing.

Metzler-Zebeli, B. U., P. G. Lawlor, E. Magowan, U. M. McCormack, T. Curião, M. Hollmann, R. Ertl,J. R. Aschenbach, and Q. Zebeli. 2017 Finishing pigs that are divergent in feed efficiency show small differences in intestinal functionality and structure. PLoS ONE 12:e0174917.

Müller, M., M. Ryoo, and E. Roura. 2021. Gut sensing of dietary amino acids, peptides and proteins, and feed-intake regulation in pigs. Anim. Prod. Sci. doi:https://doi.org/10.1071/AN21185

Myer, M. S. 1982. The presence of Paneth cells confirmed in the pig. Onderstepoort. J. Vet. Res. 49:131–132.

Noblet, J., and X. S. Shi. 1994. Effect of body weight on digestive utilization of energy and nutrients of ingredients and diets in pigs. Livest. Prod. Sci. 37:323–338.

Peeters, T. L. 2005. Ghrelin: A new player in the control of gastrointestinal functions. Gut 54:1638–1649.

Pekas, J. C. 1991. Digestion and absorption capacity and their development. In: E. R. Miller, D. E. Ullrey, and A. J. Lewis, editors, Swine nutrition. Butterworth-Heinemann, Boston, MA. p. 37–73.

Pierzynowski, S. G., B. R. Weström, C. Erlanson-Albertsson, B. Ahre'n, J. Svendsen, and B. W. Karlsson. 1993. Induction of exocrine pancreas maturation at weaning in young developing pigs. J. Pediatr. Gastroenterol. Nutr. 16:287–293.

Pluschke, A. M., B. A. Williams, D. Zhang, and M. J. Gidley 2018. Dietary pectin and mango pulp effects on small intestinal enzyme activity levels and macronutrient digestion in grower pigs. Food Funct. 9:991–999.

Pluske, J. R., D. L. Turpin, and J. C. Kim. 2018. Gastrointestinal tract (gut) health in the young pig. Anim. Nutr. 4:187–196.

Potkins, Z., T. Lawrence, and J. Thomlinson. 1991. Effects of structural and nonstructural polysaccharides in the diet of the growing pig on gastric emptying rate and rate of passage of digesta to the terminal ileum and through the total gastrointestinal tract. Br. J. Nutr. 65:391–413.

Ratanpaul, V., B. A. Williams, J. L. Black, and M. J. Gidley. 2019. Review: Effects of fibre, grain starch digestion rate and the ileal brake on voluntary feed intake in pigs. Animal 13:2745–2754.

Renner, S., A. Blutke, B. Dobenecker, G. Dhom, T. D. Müller, B. Finan, C. Clemmensen, M. Bernau, I. Novak, B. Rathkolb, S. Senf, S. Zöls, M. Roth, A. Götz, S. M. Hofmann, M. Hrab de Angelis, R. Wanke, E. Kienzle, A. M. Scholz, R. DiMarchi, M. Ritzmann, M. H. Tschöp, and E. Wolf. 2018. Metabolic syndrome and extensive adipose tissue inflammation in morbidly obese Göttingen minipigs. Mol. Metab. 16:180–190.

Reynolds, C. B., A. N. Elias, and C. S. Whisnant. 2010. Effects of feeding pattern on ghrelin and insulin secretion in pigs. Domest. Anim. Endocrinol. 39:90–96.

Rezaei, R., W. Wang, Z. Wu, Z. Dai, J. Wang, and G. Wu. 2013. Biochemical and physiological bases for utilization of dietary amino acids by young Pigs. J. Anim. Sci. Biotechnol. 4:1–12.

Ripken, D., N. van der Wielen, H. Wortelboer, J. Meijerink, R. Witkamp, and H. Hendriks. 2014. Stevia glycoside rebaudioside A induces GLP-1 and PYY release in a porcine ex vivo intestinal model. J. Agric. Food Chem. 62:8365–8370.

Ripken, D., N. van der Wielen, J. van der Meulen, T. Schuurman, R. F. Witkamp, H. F. Hendriks, and S. J. Koopmans. 2015. Cholecystokinin regulates satiation independently of the abdominal vagal nerve in a pig model of total subdiaphragmatic vagotomy. Physiol. Behav. 139:167–176.

Ripken, D., and H. Hendriks. 2015. Porcine ex vivo intestinal segment model. In: K. Verhoeckx, P. Cotter, I. López-Expósito, C. Kleiveland, T. Lea, A. Mackie, T. Requena, D. Swiatecka, and H. Wichers, editors, The impact of food bioactives on health: in vitro and ex vivo models. Springer, Cham, Switzerland. p. 255–262.

Roura, E., and G. Tedo. 2009. Feed appetence in pigs: An oronasal sensing perspective. In: D. Torrallardona, and E. Roura, editors, Voluntary feed intake in pigs. Wageningen Acad. Publ., Wageningen, The Netherlands. p. 105–140.

Roura, E., S. J. Koopmans, J. P. Lallès, I. Le Huerou-Luron, N. de Jager, T. Schuurman, and D. Val-Laillet. 2016. Critical review evaluating the pig as a model for human nutritional physiology. Nutr. Res. Rev. 29:1–31.

Roura, E., and M. Fu. 2017. Taste, nutrient sensing and feed intake in pigs (130 years of research: Then, now and future). Anim. Feed Sci. Tech. 233:3–12.

Roura, E., and M. Navarro. 2018. Physiological and metabolic control of diet selection. Anim. Prod. Sci. 58:613–626.

Ruckebusch, Y., and L. Bueno. 1976. The effect of feeding on the motility of the stomach and small intestine in the pig. Br. J. Nutr. 35:397–405.

Ruoff, W. L., and P. J. Dziuk. 1994. Absorption and metabolism of estrogens from the stomach and duodenum of pigs. Domest. Anim. Endocrinol. 11:197–208.

Ruth, M. R., and C. J. Field. 2013. The immune modifying effects of amino acids on gut-associated lymphoid tissue. J. Animal. Sci. Biotechnol. 4:27.

Sheikh, S. P., J. J. Holst, C. Orskov, R. Ekman, and T. W. Schwartz. 1989. Release of PYY from pig intestinal mucosa; luminal and neural regulation. Regul. Pept. 26:253–266.

Shirazi-Beechey, S. 1995. Molecular biology of intestinal glucose transport. Nutr. Res. Rev. 8:27–41.

Simpson, G. G. 1945. The principles of classification and a classification of mammals. Bull. Am. Mus. Nat. Hist. 85:1–350.

Smith, H. W. 1965. The development of the flora of the alimentary tract in young animals. J. Pathol. Bacteriol. 90:495–513.

Sova, S. S., L. Tjäderhane, P. A. Heikkilä and J. Jernvall. 2018. A microCT study of three-dimensional patterns of biomineralization in pig molars. Front. Physiol. 9:71.

Steinert, R. E., C. Feinle-Bisset, N. Geary, and C. Beglinger. 2013. Digestive physiology of the pig symposium: secretion of gastrointestinal hormones and eating control. J. Anim Sci. 91:1963–1973.

Stevens, C. E. 1977. Comparative physiology of the digestive system. In: M. J. Swenson, editor, Duke's physiology of domestic animals. 9th ed., Cornell University Press, Ithaca, New York, NY. p. 216–232.

Stevens, C. E., and I. D. Hume. 1995. Digesta transit and retention. In: C. E. Stevens, and I. D. Hume, editors, Comparative physiology of the vertebrate digestive system. 2nd ed. Cambridge University Press, New York, NY. p. 118–150.

Suiryanrayna, M. V. A. N., and J. V. Ramana. 2015. A review of the effects of dietary organic acids fed to swine. J. Anim. Sci Biotechnol. 6:45.

Swindle, M. M. 2015. Liver and biliary system. In: M. M. Swindle, and A. C. Smith, editors, Swine in the laboratory: Surgery, anesthesia, imaging, and experimental techniques. 3rd ed. CRC Press, Boca Raton, FL. p. 134–154.

Tian, Z., Y. Cui, H. Lu, and X. Ma. 2020 Effects of long-term feeding diets supplemented with Lactobacillus reuteri 1 on growth performance, digestive and absorptive function of the small intestine in pigs. J. Funct. Foods. 71:104010.

Urmila, T. S., P. J. Ramayya, M. S. Lakshmi, and A. V. N. S Kumar. 2019. Histomorphological studies on gut associated lymphoid tissue of pig (Sus scrofa). Pharma Innovation. 8:97–101.

Vatansever, S., M. Tulbek, and M. Riaz. 2020. Low- and high-moisture extrusion of pulse proteins as plant-based meat ingredients: A review. Cereal Food World. 65:4.

Vertzoni, M., P. Augustijns, M. Grimm, M. Koziolek, G. Lemmens, N. Parrott, C. Pentafragka, C. Reppas, J. Rubbens, J. Van Den Abeele, W. Weitschies and C. G. Wilson. 2019. Impact of regional differences along the gastrointestinal tract of healthy adults on oral drug absorption: An UNGAP review. Eur. J. Pharm. Sci. 134:153–175.

Wewer Albrechtsen, N. J., R. E. Kuhre, S. Toräng, and J. J. Holst. 2016. The intestinal distribution pattern of appetite- and glucose regulatory peptides in mice, rats and pigs. BMC Res. Notes. 9:60.

Yin, J., X. Li, D. Li, T. Yue, Q. Fang, J. Ni, X. Zhou, and G. Wu. 2009. Dietary supplementation with zinc oxide stimulates ghrelin secretion from the stomach of young pigs. J. Nutr. Biochem. 20:783–790.

Yu, J., P. Yin, F. Liu, G. Cheng, K. Guo, A. Lu, X. Zhu, W. Luan and J. Xu. 2010. Effect of heat stress on the porcine small intestine: A morphological and gene expression study. Comp. Biochem. Physiol. A Mol. Integr. Physiol. 156:119–128.

Zebrowska, T., O. Simon, R. Muchmeyer, H. Bergnerm, and H. Zebrowska. 1982. Investigations on the amino acid secretion and absorption in the stomach of growing pigs. Arch. Tierernahr. 32:703–710.

Zhang, H., J. Yin, D. Li, X. Zhou, and X. Li. 2007. Tryptophan enhances ghrelin expression and secretion associated with increased food intake and weight gain in weanling pigs. Domest. Anim. Endocrinol. 33: 47–61.

Zhang, M., and C. Wu. 2020. The relationship between intestinal goblet cells and the immune response. Biosci. Rep. 40:BSR20201471.

# 2    Water in Swine Nutrition

Charles M. Nyachoti, Alemu Regassa, and John F. Patience

## Introduction

Water is a crucial nutrient for the sustenance of life. In pigs and other mammals, a 10% loss of body water results in death (Maynard et al. 1979; Patience 2012a). Pigs require water for various important body functions such as body temperature regulation, nutrient digestion and absorption, and elimination of waste products of digestion and metabolism. Water is a major component of secretions made by the pig such as milk and saliva.

However, surprisingly little research has been done to better understand how the use of water can be optimized. The lack of research on water and its use in swine production originates from the fact that until recently, water supply in most parts of the world has been plentiful and inexpensive and is easily taken for granted. Because of this reason, water has been described as the forgotten nutrient (Thalin and Brumm 1991). These days, however, access to good quality water is becoming increasingly limited thus presenting major challenges to the growth and development of pork production. Additionally, poor quality water negatively affects pig performance and may encourage excessive water usage. The latter will create manure handling and disposal problems due to increased slurry volume (McLeese et al. 1992).

Therefore, to optimize performance, pigs should be supplied with adequate amounts of good quality water. Any factors that influence the supply or the quality of drinking water will certainly impact on pig performance. Moreover, understanding of important factors that influence water consumption by swine and devising strategies to enhance water consumption is also important. Pigs obtain water from three sources: through direct drinking of water, feedstuffs (10–12% of air dry feed), and metabolic break down of fat, protein, and carbohydrates. Pigs lose body water through a number of avenues including respiration, evaporation, urination, and feces. Although the amount lost through each of these avenues varies considerably, urination is the major route for water loss (NRC 2012).

This review discusses factors affecting water quality and the possible management strategies that can be put in place to deal with such concerns. Water content of the body and water balance; water requirement by different classes of pigs; and factors affecting water consumption also will be discussed.

*Sustainable Swine Nutrition*, Second Edition. Edited by Lee I. Chiba.
© 2023 John Wiley & Sons Ltd. Published 2023 by John Wiley & Sons Ltd.

## Water Content of the Body

The water molecule is by far the most abundant in the pig's body, representing some 99% of the total (Shields et al. 1983). By weight, water ranges from about 82.5% at birth to 53% of the body at market weight; the difference explained largely by declining lean and increasing lipid in the carcass (Shields et al. 1983). Water in the body is distributed among three pools: the intracellular space, representing about 69% of the total, the interstitium, representing about 22% of the total, and the remainder in the vascular system (Mroz et al. 1995). Maintaining proper water balance for the total body, as well as within cells and tissues, is a critical requirement of life in all terrestrial species. This is intimately related to electrolyte balance within and among cells and organs, another essential homeostatic process (Patience et al. 1989).

Regulation of drinking in the pig is not well understood. While hypovolemia and hypertonicity appear to be involved, other signals related to food consumption must also exist (Mroz et al. 1995). Furthermore, behavioral stimulation is well known in the pig, leading to luxury consumption of water during periods of boredom, hunger, and other stressors (Fraser et al. 1991).

Water is absorbed from, and secreted into, all sections of the intestinal tract, except the stomach. Absorption occurs by both active and passive processes (Argenzio 1984). As the chime passes progressively through the small and large intestine, the osmotic gradient increases, allowing for removal of most water by the terminal colon. The osmotic balance can be disturbed, for example, by the presence of large quantities of osmotically active ions in the intestine. This is the cause of the diarrhea, which occurs when sulfate-rich water is offered to pigs (Fraser et al. 1991).

## Water Balance

### Water Intake

While drinking represents the most important way for the pig to obtain water, it is by no means the only source. Feed contains free water, which is obligatorily ingested during meals. Oxidation of amino acids, carbohydrates, and lipids also contributes a substantial portion of the pig's daily needs. However, understanding drinking behavior has proven to be a very complex topic because there are so many factors that influence the pig;s need and demand for water (Fraser et al. 1991). These factors include the need to satisfy physiological, biochemical, and nutrition requirements, which themselves are influenced by environment, health, diet, and the quality of the drinking water. But the pig also will use water to satisfy a variety of behavioral needs, if water is freely available to it.

Schiavon and Emmans (2000) have proposed a simplified model to predict water intake of the growing pig (Table 2.1). Their model indicates that water intake will be increased by the water needed to support digestive processes, the quantity of water lost via the feces and urine, and the water retained during growth. In turn, water intake will be reduced by water obtained from the feed, water produced by oxidative processes, and water released during protein and lipid synthesis. However, the authors concluded that additional experimentation is required to refine estimates, for example, of the quantity of water required to excrete excess nitrogen and electrolytes from the body, the partitioning of mineral excretion between urine and feces, and the water required for osmotic regulation, among other topics.

The largest source of daily water intake for the pig is derived from drinking. Indeed, many publications indicate that the only management required in the supply of water is to ensure it is readily available and of good quality. It is widely viewed that under such conditions, the pig will correctly regulate its own water supply according to its need. However, as Fraser et al. (1991) point out, this

**Table 2.1** Estimated water balance of a 45 kg growing pig.

| Intake | mL day$^{-1}$ | Excretion | mL day$^{-1}$ |
|---|---|---|---|
| Drinking water[a] | 5552 | Feces[d] | 672 |
| Water from metabolism[b] | 788 | Urine[d] | 2839 |
| Water in feed[b] | 252 | Digestion[e] | 185 |
| Tissue synthesis[c] | 74 | Other[f] | 2335 |
| | | Total water excreted | 6,031 |
| | | Retained with bodyweight gain[g] | 635 |
| Total water supply | 6666 | Total water excreted or retained | 6666 |

[a] The pig weighs 45 kg, consumes 2.1 kg feed per day, gains 0.98 kg d$^{-1}$, and drinks 5.55 kg water d$^{-1}$ (Shaw et al. 2006). Although not measured, it was assumed that protein accretion rate was 160 g d$^{-1}$, ash accretion was 35 g d$^{-1}$, and lipid accretion was 150 g d$^{-1}$ (Oresanya et al. 2008).

[b] The diet contains 12% moisture, 5% ether extract (85% digested and efficiency of digestible lipid accretion as body lipid is 90%), 18% crude protein [of which 80% is digested and 80% is actual protein (20% is non-protein nitrogen) and 35% of digestible protein is retained and the rest is catabolized]. This results in 9 g lipid, 157 g protein, and 1260 g carbohydrate being oxidized per day, generating 1190, 450, and 560 mL water kg$^{-1}$, respectively (NRC 1981).

[c] From Schiavon and Emmans (2000).

[d] Assumes diet digestibility of 82% and fecal moisture of 64%.

[e] From Schiavon and Emmans (2000).

[f] Water lost that is not accounted for by the model, the majority of which will be evaporation.

[g] Tissue accretion rates: 150 g lipid, 35 g ash, and 160 g protein for a total of 345 g d$^{-1}$; total body weight gain was 980 g d$^{-1}$, resulting in 635 g water d$^{-1}$.

is definitely not the case, as pigs will exhibit considerable drive to consume additional water beyond that required for physiological need (Vermeer et al. 2009). However, the main factors affecting drinking water intake are body weight, feed intake, and temperature (Mroz et al. 1995).

It is critically important to the body that water balance remains under tight control because dehydration and overhydration are both fatal. The hypothalamic region of the brain is considered to be the center for the control of thirst and drinking behavior (Koeppen and Stanton 2001). Osmoreceptors located in the hyptothalamus detect changes in the osmolality of extracellular fluids, and a rise in plasma osmolality of only 10 mosm kg$^{-1}$ is sufficient to induce the sensation of thirst, resulting in drinking (Anderson and Houpt 1990). Hypovolemia also serves as a signal for thirst, such that a 6–7% fall in blood volume will also induce thirst (Anderson and Houpt 1990). However, based on drinking patterns, other signals must be involved. Mroz et al. (1995) have suggested mucosal blood flow, vascular stretch or distention, and dryness of the mouth as possibilities.

The literature contains many estimates of the drinking water intake of pigs under ad libitum conditions. These estimates sometimes refer to water "disappearance" as opposed to water intake because no allowance was made for waste. Wasted drinking water has financial implications, especially as it relates to manure volumes and annual slurry hauling costs. Consequently, the selection of drinker design and location is generally given considerable weight to minimize wastage (Brumm 2010).

*Feed Water*
The pig obtains a certain amount of water from the feed. The actual amount consumed with feed would be a function of the quantity of feed eaten and of the percent moisture in that feed. Quantitatively, this is not a large portion of the pig's daily intake, representing something less than 5% of the total.

*Metabolic Water*
The oxidation of 1 g of lipid, protein, or carbohydrate, on average, releases 1.10, 0.44, and 0.60 g of water. Of course, the exact quantity will be a function of the structure of the specific fatty acid, carbohydrate, and amino acid (Patience et al. 1989).

*Water Released by Tissue Synthesis*
Water is released by the synthesis of body constituents. Thus, 1 g of protein retained in the body releases 0.16 g water, while 1 g of lipid releases 0.07 g water (Schiavon and Emmans 2000).

## Water Excretion

*Renal Excretion*
The quantity of water eliminated from the body via the urine will be a function of the solutes present in the urine and the ability of the kidney to concentrate the urine, which has been estimated at 1 mosm $L^{-1}$ in the pig (Brooks and Carpenter 1990). The solutes of greatest importance in this regard will be nitrogen, primarily but not exclusively as urea, calcium, phosphorus, sodium, chloride, magnesium, and potassium. These fixed cations and anions will be accompanied by metabolizable anions and cations, respectively (Patience et al. 1989).

The permeability of the renal tubules is under the influence of the antidiuretic hormone (ADH), which is released from the pituitary gland. The ADH is released when receptors in the atria of the heart detect a decrease in blood volume. In response to ADH, the kidney reabsorbs more water, thus returning blood volume to a desirable level (Berdanier 1995). In addition to ADH, the renninangiotensin system plays a role in maintaining fluid volume, by stimulating ADH and aldosterone release, enhancement of sodium and chloride resorption, and vasoconstriction. Aldosterone is secreted by the adrenal glands and serves to conserve sodium and chloride reserves (Berdanier 1995).

*Fecal Excretion*
Water lost with the feces can be estimated in a number of ways. The simplest, but least precise, is to assume a typical moisture content of feces (Table 2.1). More sophisticated approaches look at the individual constituents of the feces and determine the quantity of moisture associated with each. Unfortunately, there are insufficient data available to undertake this approach with any reasonable degree of precision (Schiavon and Emmans 2000).

## Water Consumption by Different Classes of Pigs

Water consumption by different classes of pigs has been discussed in details elsewhere (NRC 1998; Thacker 2001) and therefore is only briefly covered in the current review. In general, the actual amount of water required by pigs is not known because of the difficulties involved in quantifying requirements. However, the amount of water required by pigs is mainly determined by the amount required to maintain the body water pool, which tends to remain constant at any stage of growth (Thacker 2001) and are commonly based on water to feed ratios (Shaw et al. 2006). For example, Shaw et al. (2006) indicated 2 : 1 to 3 : 1 water to feed ratio is normal for nursery and grow-finish pigs, respectively, and declining as pigs grow, whereas Brumm (2006) reported water to feed ratios with nipple drinkers to be 3.35 : 1 for nursery pigs and 2.5 : 1 for growing pigs. Moreover, Jongbloed et al. (1997) have demonstrated that water to feed ratio of 2 : 1 for pregnant sows kept

**Table 2.2** Estimated water intake levels of different classes of pigs.

| Class of pigs | Water consumption (kg day$^{-1}$) |
| --- | --- |
| Gestating sows | 11.5–20 |
| Lactating sows | 12–40 |
| Suckling pigs | 0–0.2 |
| Weaned pigs | 0.5–1.5 |
| Growing-finishing pigs (fed ad libitum) | 5–7.5 |
| Growing-finishing pigs (restricted feed) | 6–9 |

Source: Adapted from NRC (2012).

at an ambient temperature of 18–20°C has no detrimental effect on health and nutrient digestibility, but reduces urine production by 3.6 liter (L) per day, as compared to ad libitum water consumption. Estimates of water consumed by different classes of pigs are shown in Table 2.2.

### Gestating Sows

The water intake of gestating sows increases proportionally to dry matter intake (Friend 1971). Similarly, a positive correlation between feed intake and water consumption has been observed in gestating sows until day 60 of pregnancy (Kruse et al. 2010). The water requirements for gestating sows range from 11 to 23 L of water per day (Brumm 2010) and gilts in advanced stage of pregnancy consume about 20 L of water daily (Bauer 1982). In addition to satisfy their physiological needs, water consumption by gestating sows is also influenced by behavioral characteristics. Therefore, we need to clearly differentiate between water consumption and requirement in gestation sows. Because gestating sows are limit fed, they consume additional water so as to feel satiated. If individually housed, gestating sows may experience some degree of boredom, which they often try to offset by excessive drinking. In gestation, restricted-fed sows consume most of the water between meals, and there is no relation between water and feed intake. Despite these problems, gestating sows should be provided with water for ad libitum intake as this may play an important role in fulfilling their welfare requirements (Mroz et al. 1995).

### Lactating Sows

Lactating sows need large amount of water to replace 8–16 kg of milk produced daily and to remove excess metabolic end products. Daily water consumption by lactating sows provided with free access to drinking water varies widely among individual sows. Consequently, it has been recommended that lactating sows should be allowed between 15 and 20 L day$^{-1}$ of drinking water, depending on size and milk production levels (Fraser et al. 1991). Latest findings have shown that water requirements for lactating sows range from 19 to 38 L day$^{-1}$ (Kruse et al. 2011).

Adequate water intake is required for optimal milk production, which in turn impacts on litter performance (Fraser and Phillips 1989). In lactating sows, restricted water intake coupled with summer conditions can dramatically decrease sow feed and nutrient intake. Reduced energy intake in turn increases the interval between weaning and estrus in primiparous (Johnston et al. 1989) and second-litter sows (Reese et al. 1982). Low protein intake has also been demonstrated to decrease the percentage of sows returning to estrus within seven days after weaning (Brendemuhl et al. 1987).

Provision of chilled water during the times of thermal stress helps to increase sow feed intake and water consumption, average daily gain (ADG), and weaning weights of litters (Jeon et al. 2006). In addition to milk production, lactating sow water intake is influenced by dietary factors such as salt content. For instance, a study in lactating sows, Seynaeve et al. (1996) found that feeding a diet containing 0.4% salt over a four-week lactation period led to a significantly higher water consumption than those fed a low salt (0.1%) containing diet (13.9 vs. 12.4 L).

However, production parameters like milk composition, sow body weight loss during lactation, and piglet performance were not affected by water intake levels. This observation suggests that any impact on litter performance as result of inadequate water intake by lactating sows is likely due to reduced milk yield as opposed to altered nutritional quality of milk. This is because early piglet weight gain and survival were slightly correlated with the sow's water intake during the same period (Fraser and Phillips 1989). In a study that evaluated the effect of extra water and nesting material on growth and reproductive performance and plasma protein concentration (Phengvilaysouk et al. 2018), it was reported that provision of water ad libitum together with nesting material significantly improves the reproductive and growth performances in sows and piglets, respectively. Interestingly, ad libitum water intake and nesting material reduced piglet mortality and plasma protein concentration compared with sows that received nesting material only and control sows showing the importance of water in immune development. However, from practical pork production stand point, reduced water consumption by sows is important as it relates to urine production and the associated environmental challenges.

### Suckling Pigs

It may appear that suckling pigs do not drink water. A common assumption is suckling pigs satisfy their water needs because they drink milk that contains more than 80% water. However, water consumption by suckling pigs is critical for optimal performance, and, therefore, it is important to ensure that they have free access to a good quality drinking water (Thacker 2001). Water consumption at this stage is closely related to milk consumption, effective environmental conditions in the creep area, and creep feed intake. High water intake may encourage creep feed consumption, but may negatively impact on milk intake levels (Mroz et al. 1995; Thacker 2001). The results of a study by Fraser et al. (1988) indicated that on average suckling piglets can drink around 46 g of water per day within four days after birth with the largest daily consumption of 200 g. They can consume 36 mL of water per day on the first days after birth and gradually increase consumption to around 403 mL daily on day 28 (Nagai et al. 1994).

In general, suckling pigs should be provided with water for ad libitum intake in the farrowing crate to encourage creep feed consumption. This is particularly critical if milk intake is low, in which case water intake may help prevent dehydration and increase survival rate of piglets (Fraser et al. 1988). Early water consumption by newly born piglets depends on the type of water dispenser in use. Piglets discover exposed water surfaces such as bowls or cups earlier (24 hours) than nipple drinkers (72 hours) to increase water consumption (Phillips et al. 1990; Phillips and Fraser 1991). Newborn litters consume more water from a larger bowl, sufficient to accommodate three piglets at once. This may allow piglets to learn to drink by imitating their litter-mates (Phillips and Fraser 1991).

### Weaned Pigs

During the first few days following weaning, water intake is reduced as piglets learn to seek and drink water. This is undesirable as it might compromise the process of digestion and absorption thus leading to increased incidences of diarrhea (Stockill 1990). It has been suggested that because feed intake is low soon after weaning, piglets tend to increase water intake so as to achieve the feeling of being satisfied (McLeese et al. 1992). However, Torrey et al. (2008) reported that early weaned pigs do not obtain the sense of satiety through water consumption. Eating and drinking times in newly weaned pigs are positively correlated (Dybkjaer et al. 2006). The relationship between feed intake and water consumption in weanling pigs has been mathematically described (Brooks et al. 1984) as water intake (L/d) = 0.149 + (3.053 × kg daily dry feed intake). Clean, fresh, and safe water must be available at all times to achieve higher feed intake. Weaned pigs should be encouraged to drink water because this is an important factor determining feed intake levels (Brooks et al. 1984).

One of the factors that affect the survival of weaned piglet is long distance transportation from the farrowing site to the finishing site. In addition to feed, water should be provided to weaned pigs if they are to be transported for over 24 hours (Garcia et al. 2016). For instance, pigs transported without water lost markedly more weight than those transported with water, and the neutrophil to lymphocyte ratio also was noticeably higher in male pigs transported without water after 32 hours of transport period (Garcia et al. 2016). The higher neutrophil to lymphocyte ratio in pigs transported for longer period of time may indicate intestinal inflammation due to stressful situations associated with weaning and transportation.

### Growing-Finishing Pigs

It is advisable to give free access to water located near to feed dispensers for growing-finishing pigs. Water intake in growing-finishing pigs is essential for lean muscle growth as lean meat is 72% water (Kober 1993). The minimum water requirement for pigs between 20 and 90 kg body weight is approximately 2 kg for each kilogram of feed consumed. The amount of water consumed per day by a growing-finishing pig depends largely on the feeding program. If pigs are allowed ad libitum feed intake, their water consumption will be around 2.5 kg kg$^{-1}$ of feed, whereas pigs with restricted feed intake may consume up to 3.7 kg of water/kg of feed (NRC 1998). This variation in water intake is most likely due to the feeling of satiation of the pigs as stated previously for gestating sows. Water should therefore be available ad libitum for pigs given ad libitum access to feed. The pigs receiving restricted feed intake should also have access to water ad libitum as their welfare may be impaired with restricted water intake. Availability of water within a feeder decreases the time spent on eating, but increases average daily feed intake (ADFI) and ADG in growing/finishing pigs (Gonyou and Lou 2000).

### Boars

Although there are no enough data on the water requirement of boars, free access to good quality water is advisable. Elevated ambient temperature (Wettemann and Bazer 1985) and nutrition (Close and Roberts 1991; Wilson 2004) are among the most common factors affecting spermatogenesis in boars. Thermal stress during hot summer reduces feed intake and may indirectly inhibit spermatogenesis. Therefore, sufficient supply of chilled and good quality water may increase feed intake and reduce the problem when ambient temperature is high. A temperature-related water intake study indicated that growing boars consume more water (15 L day$^{-1}$) at 25°C compared with 10 L day$^{-1}$ at 15°C (Straub et al. 1976).

**Factors Affecting Water Consumption in Pigs**

Pigs need to drink water regularly as their bodies lose water constantly via urine, respiration, feces, and skin. Additionally, water losses can occur in sows during gestation (amniotic fluids), at farrowing (blood) and during lactation (milk). In growing pigs and fetuses, water supplies are also needed for tissue growth and body fluids. Water intake varies over time and between individuals (Renaudeau et al. 2013; Rousselière et al. 2016). It is extremely difficult to define a precise water requirement for the pig. Water cannot be studied like other nutrients because many complex and highly dynamic processes take place within the body. For example, drinking water deficit potentially dehydrates the body with consequent renal compensation with no visible impact on performance, at least in the short term (Fraser et al. 1991). This can be achieved by a mechanism known as production of hypertonic urine. However, this adaptation is limited by the ability of the kidneys to concentrate urine (Andreoli 2000). Therefore, in addition to simple recording of growth performance, indicators of the hydration status of the body must be involved in growth studies that aimed to evaluate water requirements.

Unfortunately, such measurements bring their own unique challenges. We need to be careful to distinguish between water requirements and water consumption in pigs (Fraser et al. 1993), as most watering systems for pigs can result in significant amounts of water wastage owing to behavioral patterns at watering points and drinker type. This water loss is not a biological requirement for production but a behavioral response to confinement that can be overcome by good housing design. It can lead to significant overestimations of water consumption if it is not taken into account. The best estimates of pig water requirements are obtained by measuring water turnover rates, using labeled water. Yang et al. (1981) determined the water requirements of pigs under confined and dry feeding conditions as being approximately 120 mL kg$^{-1}$ for growing pigs (30–40 kg) and 80 mL kg$^{-1}$ of body weight for nonlactating adult pigs (157 kg).

Water consumption by pigs is influenced by several factors. These factors often contribute to the main water concerns encountered on swine farms. Key among these factors are body weight, the quality of the water provided, diet composition, physiological status of the pig, environmental conditions, social factors, and equipment design and placement (Nyachoti and Kiarie 2010). A thorough understanding of these factors is important to provide the amount of water that meets the requirements of the animals and control water wastage. Table 2.3 provides summary of the factors that increase or decrease water consumption by pigs.

Like all nutrients, as the pig grows, its daily requirement for water increases. Unfortunately, there are insufficient data in the literature to develop a credible relationship between body weight and water requirement. Schiavon and Emmans (2000) reported that the coefficient of correlation ($R^2$) between body weight and water intake was only 0.45; this was measured under highly controlled conditions and one would reasonably assume that under commercial conditions, the relationship would be even less powerful.

**Table 2.3**  Factors affecting water consumption by pig.

| Increase water consumption | Decrease water consumption |
| --- | --- |
| Hunger | Cold stress |
| Boredom | Warm water temperature |
| Heat stress | High mineral levels in water |
| Dietary minerals in the water | |
| Pelted feed | |
| Cold water temperature | |
| Acidification | |

*Waterer Type and Adjustment*

Nipple drinkers are the most common watering devices, particularly in North America. However, there is an increasing use of wet-dry feeders, swinging drinkers, and dish drinkers with the primary motivation to conserve water (Patience 2012b). Swinging nipple drinkers are also becoming increasingly common in the North American pig industry. Both are designed to provide ad libitum water to pigs, and there is no evidence that growth or reproductive performance is influenced by waterer type. Boe and Kjelvik (2011) concluded that water supply from nipple drinkers is a better alternative for weaned piglets because of higher water intake, better water quality, and improved ADG. Yet, waterer adjustment to the proper height is important as pigs grow to encourage water consumption and reduce spillage. Cups and nipples mounted at a 90° angle should be adjusted at shoulder height, whereas nipples mounted at a 45° angle should be set at 2–3 inches over shoulder height. It is also desirable to mount the drinkers so that pigs can drink "straight on" rather than at an angle (Figure 2.1). Mounting drinkers at corners may help to accomplish this and reduce water wastage (Gonyou 1996).

There is a wide variation in the amount of water wastage due to waterer type and management. Nipples waste more water than any other watering devices. Depending on pig size, nipple height, and water-flow rates, nipples can waste 15–42% of the water provided (Li et al. 2005). Whereas Fraser and Phillips (1989) reported 23–80% waste in sows, and Brooks (1994) reported 60% wastage in growing-finishing pigs in a poorly managed nipple drinkers. The challenge with nipple drinkers is although they provide a continuous supply of fresh water; pigs can easily activate the nipple for recreation or by unintentional leaning on the nipple resulting in water spillage. The water spill goes directly into the manure pit and increases slurry volume. Therefore, nipple height and its water flow rate should be properly adjusted to minimize water wastage. With cup drinkers, water spill flows into a bowl placed beneath the water delivery devices and is available for pigs to drink. However, cup drinkers may also accumulate feed, urine, and feces in the bowl and may impact the well-being of piglets if not routinely monitored. Use of automatic stainless bowl drinkers may help to reduce water wastage (Bekaert and Daelemans 1970) if carefully managed to avoid excessive fouling, which occurs when improperly positioned in the pen. Bowl drinkers have also been shown to encourage water intake in nursing piglets. Wet-dry feeders, which allow the pig to eat feed either in

**Figure 2.1**  Straight positioned nipple drinkers helping pigs to drink with ease (https://www.nationalhogfarmer.com/animal-health/5-commandments-starting-wean-pigs; Accessed 13 January 2020).

dry form or to mix it with water prior to consumption, also reduce wastage and tend to increase feed intake (Gonyou and Lou 2000). Other types of waterers representing a very small portion of the market include troughs and straw drinkers, the latter requiring pigs to suck water from the water line.

### *Water Flow Rate and Pressure*

In addition to drinker type, water flow rate and pressure in the pipeline can impact water consumption and wastage. For younger pigs, nipple flow rate is important because they are hesitant drinkers. If water is flowing too fast, the pigs may be discouraged to drink. On the other hand, for older pigs, if water is flowing too slowly, pigs on the higher end of the social hierarchy become more aggressive and stay on the drinker; hence, the pigs at the lower end of the social hierarchy may have less chance to drink. However, Phillips et al. (1990) reported that water consumption was not affected because of changes in water flow rate and the number of nipples. We also need to make sure that water is flowing equally in all nipples in the pig house. There is a chance that the first nipple has a greater water supply than the last one. This may hinder the delivery of the desired amount of water to each pig and could also result in incorrect dosages of medicine delivered via water. Therefore, in order to achieve the desired water flow rate from the first to last nipple of the supply system, a special water pressure regulator can be mounted on the nipple. While waterers should be checked daily for maintenance, a monthly or quarterly water flow check may improve pig performance and water utilization. Low water flow rate of drinker increases the time spent at the drinker and discourages optimal water consumption by the pigs. On the other hand, high water flow rate increases water wastage. Therefore, selection of the optimal water flow rates will help to encourage adequate intake without excessive wastage of water (Table 2.4).

Because feed intake is positively correlated with sow milk production, one can argue that reduced nipple water flow rate would indirectly affect the performance of suckling piglets. However, reports from literature are inconsistent. No measurable effects of water flow rate on body weight gain, feed intake, or feed conversion efficiency was observed in a study where nursery pigs consumed water from nipple waterers with flow rates of 100, 350, 600, 850, and 1100 mL min$^{-1}$ (Nienaber and Hahn 1984). Elsewhere, nipples with 70 mL min$^{-1}$ flow rate reduced sows feed intake and body weight gain compared with nipples with 700 mL min$^{-1}$ flow rate during the first 21 days of lactation in summer. However, litter size and litter weight were not affected by the difference in nipple flow rates (Leibbrandt et al. 2001). This variation could be attributed to the ambient temperature at which

**Table 2.4**  Recommended water flow rates for different classes of pigs.

| Class | Flow rate | | Time to fill a 473 mL bottle (s) |
|---|---|---|---|
| | mL min$^{-1}$ | mL min$^{-1}$ | |
| Nursery | 500 | 237–473 | 60–120 |
| Grower-finisher | 750 | 473–946 | 30–60 |
| Gestating sows | 1,000 | 947 | 30 |
| Lactating sows | 1,500 | NA | NA |

Source: Brumm (2006, 2010) and Gonyou (1996).
Min = minute; s = second; NA = data not available.

these studies were conducted as high temperature coupled with water restriction reduces sows feed intake and piglet performance (Leibbrandt et al. 2001).

Water pressure also influences the activation of water delivery devices by the pigs and the amount of water wastage. A general recommendation is that water pressure in supply lines be limited to 20 psi. This facilitates the activation of water device by the pig while controlling water wastage from drinking devices.

The location and color of dispenser appear to affect water consumption by the newly born piglets. For instance, data from a study that examined the effect of drinker location and color within the farrowing crate on drinking behavior and water intake of newborn pigs during the first two days of their birth have indicated that water consumption was higher for back left and red and blue colored dispenser but lower for front left and green colored dispenser lives (Deligeorgis et al. 2006).

### *Waterer to Pig Ratio*

Waterer to pig ratio of 1 : 25 is typical for nursery and grow-finishers in the industry. The general recommendation is to have a single waterer for every 10–15 pigs, typically in a 1 : 12 waterer to pig ratio. A study with nursery pigs indicated that smaller pig to waterer ratios give better chances for pigs to drink and gain more body weight, and bowl drinker is preferred to nipple drinker (Ciara and Anna 2007). Providing access to water at the recommended level may improve growth performance of nursery (Sadler et al. 2008) and grow-finish pigs (Vier et al. 2018).

### **Feed Form and Feeder Type**

Water consumption is greater for pigs fed meal diets compared to pellet diets, resulting in a similar water to feed ratio when accounting for differences in feed efficiency between the feed forms (Laitat et al. 1999). A liquid feeding system is a practice of mixing dry feed with water to deliver a minimum amount of water that the animal has to consume. It is prepared by suspending dry feed in water using dilution rate, expressed as the amount of water (kg or l) per kg of dry feed. Dilution rate may vary from farm to farm, production types, and season. For example, in France, it ranges from 2.5 to 4.8 L kg$^{-1}$ for pregnant sows and 2.4–3.2 L kg$^{-1}$ for fattening pigs (Roy et al. 2007; Royer et al. 2007; Massabie et al. 2014). The water-feed mixture should be homogenous, and its distribution usually is computer controlled. The feed is mixed with water in a tank, and the mixture is transported by means of a pump via a set of pipes that extend to reach trough in each pen. The use of wet-dry feeders reduces water wastage compared to dry feeders and waterers (Brumm et al. 2000). This is particularly important for lactating sows in that it provides the sows the choice when and how much to eat and improve feed intake and litter growth performance (Peng et al. 2007).

Liquid feeding system is important also for newly weaned pigs as it creates smooth transition from milk to feed, allows the pigs to consume 35% more water, and improves feed intake than dry feeding system (Russell et al. 1996). However, in group-housed pigs, there is no assurance that all pigs satisfy their water requirements through the liquid diet alone. Therefore, additional water should be provided from watering devices that can lead to water spillage particularly when the design of the watering device is faulty (Torrey et al. 2008). Moreover, depending on cleanliness of the trough, the quality of water provided, the ambient temperature, and the duration of time the water remains in the trough, this feeding system can be associated with increased risk of bacterial contamination due to residual water in the trough.

## *Diet Composition*

Water consumption increases when dietary composition increases the need to eliminate metabolites or surplus nutrients. Dietary protein and mineral concentration and daily protein and mineral intake have a relatively large effect on water intake and excretion (Mroz et al. 1995; NRC 1998; Shaw et al. 2006). When pigs are fed diets with a protein concentration that exceeds their requirements for maintenance and growth or production purposes, the excess protein is broken down and excreted as urea (NRC 1998). This process exerts an additional need for water to excrete excess nitrogen, which explains why pigs consuming high protein diets have high water intake levels (NRC 1998; Figure 2.2). For instance, Shaw et al. (2006) reported increased water consumption in pigs fed diets with excess salt, proteins, or minerals. Reduction in water intake and ADG was reported in growing pigs fed a diet containing less than 0.2% salt as compared to those fed 0.27 and 0.48% salt (Hagsten and Perry 1976). Increased nitrogen excretion, which explains increased renal water loss, was observed when barrows fed a diet formulated based on apparent ileal digestible protein relative to pigs fed a diet formulated based on standardized ileal digestible amino acid (Lee et al. 2017).

Similarly, increasing water intake has been observed in a study where pigs were fed a diet containing increasing levels of dietary D-xylose (Huntley and Patience 2018). On the other hand, significantly higher nitrogen excretion was observed in barrows fed high crude protein diet compared with those fed low crude protein diet (Wu et al. 2018), indicating excess nitrogen in the diet leading to higher water loss in the form of urine. Moreover, the acidification of drinking water has been implicated to increase water consumption in piglets. For instance, the addition of DL-methionine hydroxy analog-free acid (Kaewtapee et al. 2010) and an organic acid blend consisting of formic acid, acetic acid, and ammonium formate (Mesonero et al. 2016) into drinking water increased water intake in nursery pigs. Mroz et al. (1995) have suggested that many studies relating water intake and dietary protein content were confounded by concurrent changes in dietary mineral levels.

It is also well known that increasing the salt concentration in the diet will result in elevated water consumption (Seynaeve et al. 1996). Interestingly, pigs also consume greater quantities of water when the water itself is high in minerals (Maenz et al. 1994). Therefore, modification of dietary composition may be part of the strategies to reduce excessive water utilization by the pig industry.

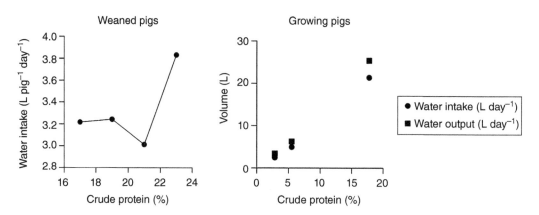

**Figure 2.2**    The effect of dietary crude protein on water intake in weaned pigs (Nyachoti et al. 2005) and water usage in growing pigs (Shaw et al. 2003).

As occurs in many species, hunger increases water consumption. For example, Yang et al. (1981) reported that providing limit-fed pigs with increasing amounts of feed reduced the observed water to feed ratio from 5.1 : 1 to 3.3 : 1. It is widely accepted that pigs consume luxury amounts of water because of hunger-induced or stress-induced polydypsias or even for play, such that simply measuring water disappearance may introduce errors into the estimation of water requirements (Fraser et al. 1991; Vermeer et al. 2009).

### Environmental Conditions

Aside from body weight, dietary composition, stocking density, water flow rate, health status, and stress level of the animal, the amount of water consumed by a pig can be affected by ambient temperature, humidity, and ventilation. Knowledge of the daily water consumption patterns of pigs can serve as an indicator of unfavorable environmental conditions and a predictor of the onset of health challenges (Brumm 2006). Unlike other livestock species, pigs do not have functional sweat glands that help them to efficiently remove body heat. Hence, respiration is the major route that pigs use to cool themselves. Accordingly, the water requirements for pigs increase in hot environments due to the greater amount of water vapor that is released from the lungs. For example, an increase in ambient temperature from 10 to 25 °C has been observed to increase water consumption from 2.2 to 4.2 L day$^{-1}$ in finisher pigs (Agriculture and Food 2019). Similarly, Schiavon and Emmans (2000) suggested that for every 1 °C increase in the air temperature, water intake increased by 0.12 L d$^{-1}$. Vandenheede and Nicks (1991) reported that water intake increased from 2.2 to 4.2 L d$^{-1}$ in finishing pigs when the temperature increased from 10 to 25 °C, a difference that supports the relationship established by Schiavon and Emmans (2000). Mount et al. (1971) reported that raising the temperature from 12–15 to 30–35 °C increased water consumption by 57% in 33.5 kg pigs, while Straub et al. (1976) reported a 63% increase in 90 kg pigs.

It is thus very important that adequate supply of drinking water be available. High humidity in itself does not have a negative effect on swine performance. However, combined with high temperatures, it enhances the negative effects of the high temperatures. Since the pig must rely heavily on evaporative heat loss in order to stay cool when it is hot, humidity level is very important. The higher the humidity level in the air, the less effective is the process of evaporative cooling (i.e., less moisture can evaporate into humid air than dry air). It has been estimated that at 30 °C, an increase of 18% in humidity is equivalent to an increase in air temperature of 1 °C (Huynh et al. 2005). In a hot-humid environment, water as a cooling agent for pig becomes very important through the application of water sprinkler, fogging, and dripping systems. Whereas numerous studies have shown these systems to be beneficial in maintaining pig performance in hot summers and their use should not override the provision of adequate floor spaces, insulation, and ventilation as well as diet reformulation strategies (low fiber and high energy) as these might save on the amount of water used for cooling purposes and thus manure volumes in the long run. For example, studies at the University of Kentucky with finishing pigs (weighing 50–110 kg and with a space allowance or 10 square feet/pig) and conducted in two summers months of 1996 and 1997 (temperature range: 22–34 °C) demonstrated the importance of cooling (either by fan or sprinkler) on performance (Cromwell 1999). Whereas combining fans and sprinkler resulted in better performance, data from the fan only suggest that having well-insulated buildings with good air movement might be as good as or better than using sprinklers.

**Factors Affecting Water Quality**

Numerous factors including the concentration of dissolved minerals and bacterial contamination can affect the quality of drinking water for swine. Poor water quality may inhibit the pig's desire to drink and hence impact water intake and subsequently growth performance. Dealing with a water quality problem can be a major challenge and often water treatment does not correspond to improved animal performance. The levels of two main components namely chemical elements and composition and level of bacteria characterize the quality of water. Chemical elements dissolved in water have a significant influence on the quality of water for swine depending on their concentration. The effects of the physical aspects of water quality such as color, turbidity, and odor on pork production may be minimal. However, it may have health and or productivity implications if color, odor, or turbidity is high.

*Chemical Factors*

Chemical factors affecting the quality of water include pH, hardness, total dissolved solids (TDS), nitrates and nitrites, sulfates, iron, and lead (Kober 1993). Each of these factors is discussed briefly later.

*pH*

Water pH is the measure of its acidity or alkalinity. Water with pH ranging from 6.5 to 8.5 is considered acceptable to pigs (NRC 2012). A low pH (less than 6.5) is undesirable as it may corrode and dissolve metals from the plumbing and cause precipitation of medication delivered via the water system (NRC 1994). This might enhance the proliferation and survivability of the target pathogens and the precipitation of medication that could impact on pork quality by causing deposits of drug residues in meat (Russell 1985). High water pH (greater than 8.5) on the other hand gives the water a slippery feeling and leaves scaly deposits (Kober 1993) that reduce water flow. Basic water (pH above 7) is considered a risk factor for *E. coli* caused diarrhea, and water pH should be reduced via acidification in a situation where *E. coli* infection is persistent. Another practical problem with high water pH is that it reduces the efficacy of chlorine used for water treatment. However, wide deviations in water pH are uncommon.

*Hardness*

Water hardness is a measure of the sum of divalent cations in the water, which for the most part in practical circumstances means calcium and magnesium salts. Other divalent ions in the water are quantitatively lower and therefore irrelevant to the determination of hardness. Water hardness is expressed as calcium carbonate ($CaCO_3$) equivalents (Kober 1993), although one needs to know the actual amount of calcium and magnesium to be able to predict how animal performance might be impacted. Water is considered soft if the concentration of calcium (Ca) and magnesium (Mg) is below 60 ppm and hard if above 120 ppm (NRC 2012). Water with low levels of Ca and Mg (i.e., <60 ppm) has no effect on pig performance, but excess Ca intake might interfere with phosphorus (P) utilization. Thus, dietary phosphorous levels may require adjustment when using water with high $CaCO_3$

(>300 ppm) levels (Kober 1993). In pork production, the most significant practical problem associated with the use of hard water relates to the deposition of scales in the watering system, which causes serious water delivery problems and requires considerable labor resources to manage.

### Total Dissolved Solids

Total dissolved solids (TDS, also known as the salinity of water) is the amount of soluble salts (commonly magnesium, calcium and sodium bicarbonate, chloride, or sulfate forms) in the water (Kober 1993). Pigs consuming water with high levels of TDS have transitory diarrhea, but health, growth, and reproductive performances are not usually affected. The recommended maximum TDS level in water is 3000 ppm (NRC 2012; Table 2.5). There is contradictory evidence as to the level of TDS at which pig performance is impaired. For example, growth rates, feed intake, and feed efficiency were better for weanling pigs fed water with 217 ppm compared with water with 4390 ppm TDS (McLeese et al. 1992). However, TDS at between 1000–6000 ppm have, generally, not been shown to consistently affect pig performance (Anderson and Stothers 1978). Furthermore, the performance of growing-finishing pigs was not influenced by TDS levels up to 7000 ppm in the form of sulfates (Anderson et al. 1994). As experimental period in this experiment was only four days long, the long-term effects of high TDS levels on pig performance cannot be substantiated. A detailed study of water quality on a commercial farm (Patience et al. 2004) has indicated that sulfate levels of 1650 mg L$^{-1}$ did not impair either the rate or the efficiency of growth in weanling pigs, although diarrhea was clearly present (Figure 2.3). However, in another trial, pigs that consumed

**Table 2.5** Evaluation of water quality for pigs based on total dissolved solids.

| Total dissolved solids (ppm) | Rating | Comment |
| --- | --- | --- |
| <1000 | Safe | No risk for pigs |
| 1000–2999 | Satisfactory | Mild diarrhea in pigs not adapted to it |
| 3000–4999 | Satisfactory | May cause temporary refusal of water |
| 5000–6999 | Reasonable | Higher levels for breeding stock should be avoided |
| >7000 | Unfit | Risky for breeding stock and pigs exposed to heat stress |

Source: NRC (2012).

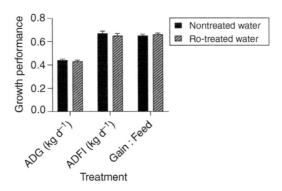

**Figure 2.3** The effect of reverse osmosis (RO) treatment of drinking water high in total dissolved solids and sulfates on ADG (SEM = 0.011 kg d$^{-1}$), average daily feed intake (ADFI; SEM = 0.019 kg d$^{-1}$), and G:F (SEM = 0.013) evaluated under commercial farm conditions. Source: Based on Patience et al. (2004).

water that contained 3000 mg L⁻¹ sodium sulfate were reported to have reduced ADG and gain to feed ratio (G:F) compared with pigs that consumed water without sodium sulfate (Flohr et al. 2014).

### Nitrates and Nitrites

Contamination of groundwater with nitrates and nitrites mainly occurs through leaching from the soil or through surface water runoff that has been exposed to materials containing high nitrogen levels such as animal wastes, nitrogen fertilizers, decaying organic matter, silage juices, and others. A high level of either nitrites or nitrates may indicate bacterial contamination (Kober 1993). Nitrites are ten times more toxic than nitrates in monogastric animals (Emerick 1974). The Canadian Task Force on Water Quality (1987) recommend a limit for nitrite-N and nitrate-N at 10 and 100 ppm, respectively, in water for livestock. Symptoms of nitrate poisoning include high respiration rate, increased incidence of diarrhea, reduced feed intake, poor growth, increased abortions among sows, and reduced vitamin A utilization (Grizzle et al. 1996; Meek 1996; Thacker 2001). Water nitrates levels less than 1500 (~333 ppm nitrate-N) ppm might not affect pig performance (Kober 1993) although a higher respiration rate was observed with the inclusion of 300-ppm nitrate in water fed to weaned pigs (Anderson and Stothers 1978).

### Sulfates

Of all mineral contaminants, sulfates are the cause of most water quality problems with respect to pig production in North America (NRC 2012). It is very expensive to remove sulfates from drinking water, including both the initial capital cost for equipment and ongoing operating costs, making such systems prohibitive to install on most farms. There is considerable inconsistency in the literature regarding the effect of water sulfate content on pig performance. According to Environment Canada, a water sulfate level of less than 1000 ppm may not affect pig performance (Table 2.6). However, Patience et al. (2004) concluded that weanling pigs can tolerate drinking water containing high concentrations of sulfates (1650 mg L⁻¹), and poor performance or diarrhea in nursery pigs should not be attributed to water quality until other possible contributing factors are ruled out. For example, growth and ADG were not affected when piglets consumed water that contained 1260 or 75 mg sulfates L⁻¹ (Tremblay et al. 1989). In another study, sulfate level up to 3300 ppm caused a laxative effect and increased the water intake in pigs. Higher levels

**Table 2.6** Effect of water sulfate on the performance of weanling pigs.

| Item | Control (0 ppm) | Sulfate (2402 ppm) |
|---|---|---|
| Avg. daily gain, kg d⁻¹ | 0.40 | 0.33 |
| Avg. daily feed intake, kg d⁻¹ | 0.79 | 0.71 |
| Feed/gain | 1.89 | 2.02 |
| Avg. daily water intake, L d⁻¹ | 1.15 | 1.35 |
| Scour days (wk 1 only) | 3.5 | 6.0 |

Source: Modified from Anderson and Stothers (1978).
Avg. = average; d = day; wk = week.
Sum of sodium and magnesium sulfate.

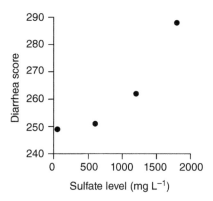

**Figure 2.4** Effect of high water sulfate levels on incidences of diarrhea in piglets. Source: Based on Veenhuizen et al. (1992).

cause diarrhea (Veenhuizen et al. 1992; Figure 2.4) and make water unsuitable for pigs (Anderson and Stothers 1978; AAFRD 1993; NRC 2012).

Nonetheless, pigs may tolerate higher sulfate levels than poultry as drinking water with 3000 ppm had no effect on pigs weaned at 28 days of age (Paterson et al. 1979). Other data indicate that water with less than 1500 ppm sulfate may affect the performance of early-weaned pigs (Kober 1993; NRC 1998). Water with 3500 ppm of sulfate is not good for sows and generally water with more than 4500 ppm should not be used for any livestock (Kober 1993). Sulfate in the drinking water will inevitably lead to diarrhea in most classes of swine. Younger pigs are most susceptible, but even sows may sustain a transient diarrhea if rapidly switched to water high in sulfates. For young piglets, levels as low as 750 ppm can be problematic, while older animals can tolerate even higher levels. Because sulfate causes diarrhea of an osmotic nature, care must be taken to not assume that performance is impaired, just because pigs have loose stools, because in many experiments, growth was unaffected. On the other hand, diarrhea even of osmotic origin can increase the pig's susceptibility to secondary causes of gastrointestinal disturbance that might have a much more profound impact on performance and health. Recommended maximum sulfate level in water is 1000 ppm (NRC 2012).

### Bacterial Contamination

Bacterial contamination is currently viewed as a serious problem relative to the quality of water for both human and livestock use. The main types of bacteria that have been associated with water quality problems include *Cryptosporidium*, *enterotoxigenic Escherichia coli*, *Salmonella*, and *Leptospira* (Fraser et al. 1993; Meek 1996; NRC 1998; Thacker 2001). Other microorganisms such as Protozoa as well as eggs and cysts of intestinal worms might affect the water quality for pigs (Fraser et al. 1993). The degree of water pollution by bacteria is traditionally estimated by measuring the level of coliforms, which represents a group of generally pathogenic bacteria, as an indicator (Meek 1996). A count of 5000 total coliform/100 mL is normally used as a guideline for maximum levels in water for pig production (Meek 1996; NRC 2012). However, it must be emphasized that the actual level that can impact water quality differs depending on the virulence of specific bacteria present. Some pathogens may be harmful below this level, whereas other benign species may be tolerated at higher levels. The effect of waterborne bacteria on pig performance is poorly investigated.

When dealing with bacteria in the drinking water, it is important to consider the cleanliness of the water lines. Therefore, regular maintenance and cleaning of drinking water supply systems are very important. Biofilm in pipes can be a considerable source of microorganisms in the water, but chlorine-based disinfectants have little to no effect on biofilms. It is important to flush out all buildup regularly. For example, washing of the water lines with alkaline detergent such as AH-tec (Desintec, Germany) followed by a high-pressure micro-impulse system and careful washing with clean water removes all organic deposits and slime from the internal surfaces of water pipes and prevents the development of undesirable microorganisms (Vorozhbitov and Trukhman 2012).

## Solving Water Quality Problem

To ensure a supply of good quality water, regular monitoring of the water itself and supply system is critical. While waterers should be checked daily for maintenance (National Pork Board 2018), a monthly or quarterly water flow check may improve pig performance and water utilization. Water quality should be monitored at least once a year and should always include a measurement of bacterial (coliform) contamination (Kober 1993). Ideally, water test results should be used to adjust feed formulas to compensate for any excessive amounts of minerals supplied by the water (Flipot and Ouellet 1988). However, the challenge here is that the bio-availability of many of the water minerals is unknown although one will expect that if a mineral is soluble, then it is also available (Christensen 2001).

Considering the wide variability in the levels at which pig performance is impaired by water contaminants, water analysis results alone may not be sufficient to justify changes in a production system. Therefore, any steps to institute water quality corrective measures should be guided by the impact the water has on animal performance (Veenhuizen 1993). This is particularly important considering that water treatment methods can be costly.

### Water Treatment Techniques

The swine industry spends considerable resources on water treatment, particularly in the nursery. Although in some cases pigs may be able to acclimatize to water of poor quality (Veenhuizen 1993), there are situations in which the quality of the water is so poor to be utilized economically for pork production. Therefore, either an alternative source of water has to be found or the water has to be treated before use (Meek 1996). A major challenge with respect to water treatment is to identify a suitable treatment system that is not only effective but also affordable. Various water treatment methods (e.g., chlorination, coagulation, filtration, and pH adjustment) are available to the livestock industry, but their impact on animal performance is largely unknown. The following is a brief description of these methods.

### Chlorination

Chlorination is now an accepted practice of water disinfection within the pork industry (particularly for nursery pigs). However, the impact of different levels of chlorination on water intake and pig performance is not well understood. The amount of chlorine required to effectively disinfect water varies depending on the quality of the water. In general, more chlorine will be required to effectively disinfect water with high levels of contaminants (e.g., nitrite, iron, organic matter, etc.) or high pH because these are known to reduce the efficacy of chlorine. The effectiveness of chlorine is also dependent on the length of the contact time and the type of bacterial contaminant; protozoa and enteroviuses are more resistant to chlorination than bacteria (Patience et al. 1995). For practical pork production, there is no merit in increasing the free chlorine level in water, but it is important to know how pigs will respond in the event that high chlorine is inadvertently added to the water. For instance, Tofant et al. (2010) indicated that over-chlorination of drinking water (2.11 mg L$^{-1}$) resulted in significant reproductive failure explained by increased number of spontaneous abortion, return to oestrus number and percentage of stillbirths, and reduced farrowing and total number of piglets in gilts and sows (Table 2.7) indicating the need for regular monitoring of free chlorine level in drinking water.

### Acidification

Water from many sources is slightly basic (i.e., alkaline) with a pH value between 7 and 8. As the pH increases, the level of carbonates compared to bicarbonates also increases. This causes physiological and digestive upsets, especially in younger pigs. This is attributed to the fact that newly weaned pigs produce insufficient quantity of gastric acid (Kil et al. 2011) that is required to lower gastric pH and activate pepsin. Results from studies that examined the effect of drinking water acidification on the performance of pigs are inconsistent. McLeese et al. (1992) noted that although there is anecdotal evidence supporting water acidification in nursery pig production, such benefits have not been demonstrated in controlled research studies. In support of this concept, in a study conducted in a commercial facility, acidification of surface water did not affect the performance of nursery pig (Nyachoti et al. 2005).

However, acidification of the water supply has been reported to increase piglet performance (Walsh et al. 2007), by decreasing the pH of the water down to neutral pH. Similarly, acidification of drinking water with organic acid blend consisting of formic acid, acetic acid, and ammonium formate at the rate of 2 L/1000 L of drinking water significantly improved final body weight and average daily weight gain in newly weaned pigs (Mesonero et al. 2016). Daily weight gain was also improved in nursery pigs that consumed water acidified with a mixture of organic acids containing lactic, formic, propionic, and acetic acid (De Busser et al. 2011). Adding liquid DL-methionine hydroxy analog free acid into drinking water at 0.10% also significantly increased ADG and ADFI in nursery pigs (Kaewtapee et al. 2010). Such discrepancies could be attributed to the type of the acid product and its concentration that was used to acidify the water and the pH of the control water. Therefore, an in depth investigation is required as it is difficult to draw any conclusion based on these few data.

**Table 2.7** The effect of over chlorination on reproductive performance of gilts and sows.

| Parameters | Not over chlorinated | | Over chlorinated | | |
| | $n$ | Mean ± SD | $n$ | Mean ± SD | $P$-value |
| --- | --- | --- | --- | --- | --- |
| Number of spontaneous abortions | 15 | 1.87 ± 1.41 | 1 | 91.00 ± 0.00 | <0.001 |
| Return to estrus number | 11 | 65.9 ± 23.1 | 5 | 203.2 ± 65.1 | <0.001 |
| Number of farrowing | 12 | 204 ± 33 | 4 | 81.3 ± 33.5 | <0.001 |
| Total number of piglets | 12 | 2056 ± 387 | 4 | 777.3 ± 339.7 | <0.001 |
| Stillborn piglet (%) | 12 | 6.77 ± 1.60 | 4 | 12.57 ± 3.39 | <0.001 |

Source: Adapted from Tofant et al. (2010); $n$ = number of months.

### Coagulation and Other Methods

Water with a high turbidity can be unsuitable for use in pork production, as the suspended particles reduce effective disinfection with chlorine by preventing accessibility of the microorganisms. Coagulation, which involves lumping together the fine particles into large particles, which are in turn settled and then removed from the water, offers a viable means for treating water with high turbidity.

Slow sand filtration (SSF) also is a quite effective water treatment technology. It is efficient in removing coliforms including *Giardia, Cryptosporidium, Salmonella, Escherichia coli, fecal streptococci, bacteriophage*, and *MS2* virus from water/wastewater (Verma et al. 2017). In addition to pathogenic microorganisms, SSF can effectively remove suspended solid particles and toxic metals (Verma et al. 2017). The drawback of this treatment method is the slow flow rate, which may not produce sufficient amount of treated water for big swine farms. However, it can be the best option in remote areas where fecal contamination and suspension of solid particles are the biggest challenges. Construction of big reservoir for SSF treated water could alleviate the quantity issue. However, results from the assessment of the effect of coagulation, chlorine levels, and acidification of surface water did not improve piglet performance (Nyachoti et al. 2005; Table 2.8). But coagulating the surface water reduced the number of most contaminants measured (Data not shown). The absence of the effect of these treatment methods in that study could be attributed to the fact that the level of contaminants in the untreated surface water may not have been high enough to affect piglet performance such that any reduction in the levels of these compounds could not lead to any measurable benefits. Second, pigs seem to handle water of variable quality reasonably well, depending on the presence of other factors as evidenced by various studies (Anderson and Stothers 1978; NRC 1998; Patience et al. 2004). A two-step postweaning waterline cleaning protocol that used mechanical draining of alkaline or enzymatic detergents followed by acidification of water have been shown to improve water quality and reduce total microflora concentration in the treated water (Brilland et al. 2017).

**Table 2.8**  Effect of water type and treatment on short-time nursery pig performance.

|                          | Surface water (control) | Well water | Acidified surface water |
|--------------------------|:-----------------------:|:----------:|:-----------------------:|
| Trial 1, d 0 to 46       |                         |            |                         |
| Number of observations   | 24                      | 12         | 12                      |
| ADFI (g d$^{-1}$)        | 553                     | 573        | 539                     |
| ADG (g d$^{-1}$)         | 398                     | 406        | 401                     |
| FC                       | 1.36                    | 1.43       | 1.43                    |
| Trial 2, d 0 to 42       |                         |            |                         |
| Number of observations   | 24                      | 12         | 12                      |
| ADFI (g d$^{-1}$)        | 518                     | 524        | 504                     |
| ADG (g d$^{-1}$)         | 394                     | 384        | 394                     |
| FC                       | 1.33                    | 1.39       | 1.30                    |

d = day; ADFI = average daily feed intake; ADG = average daily gain; FC = feed conversion.
This is a short-time study and may not reflect the long-time effect.Source: Nyachoti et al. (2005).

## Softening

Simply installing a water softener will reduce the hardness of the water. However, the softening of the water results in a higher level of sodium (i.e., Na is exchanged with Ca and Mg), which may again produce an intake of salt that is too high for optimal performance (Roush and Mylet 1986). Other available techniques for water treatment include ion exchange or reverse osmosis treatment systems and ultraviolet radiation. However, many of these techniques are quite costly and therefore of limited application in commercial pork production. Furthermore, reducing the concentration of mineral contaminants in water may not always lead to improved animal performance. For example, in a study by Patience et al. (1997), reducing TDS and sulfate levels from 3086 and 1634 ppm to 93 and 15 ppm, respectively, by reverse osmosis had no beneficial effect on piglet growth performance. This observation and that of Nyachoti et al. (2005) indicate that water treatment may have no effect on performance unless water quality is an issue. However, reverse osmosis is the only method available to reduce sulfate content in water, and thus reduce or eliminate associated diarrhea, especially in newly weaned pigs. Some producers choose to utilize reverse osmosis in nurseries as a preemptive approach to maintaining piglet health; however, the value of such installations remains speculative. It might be considered an insurance policy against health problems that can be exacerbated by osmotic diarrhea; however, because of the cost of reverse osmosis systems, in terms of both installation and operation, it tends to be very expensive insurance.

## Practical Management Approaches to Conserve Water for Pork Production

### Adjusting Diets for Minerals in the Water

This is something related to adjusting diets to account for minerals obtained from water or where possible blending water from different sources (Veenhuizen et al. 1992; Patience et al. 1995). For example, when using water with high sodium levels, reducing dietary salt content may be considered. However, care should be taken to ensure adequate chloride levels as poor quality water often contains large quantities of sodium with relatively small amounts of chloride. Also, as mineral content of water varies widely, and the availability of minerals in water is not well established yet, caution should be taken when formulating diets to compensate for poor quality water. The specific mineral profile of a water supply and how it varies over time must be known before appropriate adjustment in ration formulation can be made.

### Animal Management

Factors such as boredom, season of the year, and the stage of the breeding cycle can influence the water needs of the pig and therefore potential wastage. For instance, group-housing gestating sows rather than confining them in individual crates may reduce boredom, which has been associated with excessive intake of water. High stocking densities in hot weather affect the environmental temperature felt by the pigs and influence their water needs for cooling purposes. Therefore, we need to be careful with stocking density when housing pigs in groups as it may contribute to heat stresses. If stocking densities are too high, the temperature inside the barn may rise as more metabolic heat is added to the barn, and this in turn leads to increased water consumption by the pigs to chill themselves and dissipate heat from the body.

### *Wastewater Treatment and Modification of Floor Type*

Reclaiming of wastewater for reuse may be plausible provided that it is combined with advanced treatment technology. It may be one of the ways forward to alleviate the forthcoming challenge facing livestock production in line with the availability of good quality water. However, due to the perception and health concerns associated with direct or indirect reuse of recycled water for human consumption and the involved high cost, this technology is not widely practiced. On top of this, if all of the wastewater is reused, the concentrations of dissolved ions or TDS will increase in the recycle loop over time. If allowed to accumulate to intolerable levels, the health and performance of the animals can be adversely affected. Although inclusion of a significant portion of the drinking water as recycled water from animal waste did not reduce growth performance and other responses (Bull et al. 2005), its effect on animal health and performance must be further evaluated before providing any recommendation. However, the use of recycled water for cleaning swine buildings and equipment would help to reduce the amount of potable water that could otherwise be used for cleaning.

Modification of floor from solid to partially and totally slatted type during cool season has been shown to reduce water wastage without affecting the ADFI and feed efficiency in grower-finisher pigs (Su et al. 2018). Combination of microbial fuel cells with flocculation also has been implicated to reduce the concentrations of pollutants in swine waste water (Ding et al. 2017).

### Summary and Conclusions

Water is universally recognized as an important nutrient. However, there have been surprisingly few studies conducted on the water requirements of swine. Pigs require sufficient supply of good quality water for body temperature regulation, nutrient digestion and absorption, and elimination of waste products of digestion and metabolism. Lack of good quality water negatively affects the health of pigs and hence reduces their growth performance.

The quality of drinking water for pigs varies widely depending on the concentration and type of contaminants in the water. Although pigs can tolerate some water quality problems, there is no doubt that poor quality water can negatively impact their performance. For most water contaminants, there is not a single level at which performance is impacted. This is partly due to the fact that effects on animal performance at any one level vary depending on the presence of other contaminants. Therefore, whenever assessing water quality problems caused by a specific factor, consideration for other factors present is also critical. Before implementing costly water treatment procedures, it is prudent to first ensure that indeed water quality is impacting on the performance. Finally, there is a need for further research to better characterize the individual and collective effects of factors that determine the water quality, the benefits of various water treatment methods, and surface water quality and its effects on performance.

High concentration of minerals in the water is one of the factors affecting water quality. Therefore, determination of the availability and amounts of minerals in the water may help to account for mineral supply from water and adjust feed formulas without compromising animal performance.

It is important to differentiate between requirement and consumption in determining water requirement. Different classes of pig require different quantity of water and several factors, including dietary composition, physiological status, and environmental temperature affect the water requirements of pigs. Therefore, these factors should be critically considered in determining water requirement, which is generally overestimated, because of the failure to consider wastage. However, owing to the difficulty to make these measurements, water consumption is used to estimate water requirements. In general, water requirements are commonly based on water to feed ratios.

Drinker type, water flow rate, and pressure in the pipeline can impact water consumption and wastage. Therefore, the most efficient and water saving device must be selected and placed so that it is easily accessible to pigs. The number of drinkers in the pig house should be carefully determined and the flow rate of the drinkers regularly monitored so that water is available at all times. Furthermore, surveillance of pigs for signs of dehydration such as thirst, lack of appetite, constipation, nervousness, and lack of muscular control can help to inspect and correct faulty drinkers and the general water supply system.

Although 75% of our planet is covered by water body, the vast majority of this water is not potable because of high alkalinity. Additionally, the quality of the remaining small percentage of fresh water of the world is deteriorating because of chemical and microbiological contaminations. On the other hand, human population is alarmingly increasing and the demand for animal protein will increase as well. This in turn will put tremendous pressure on livestock industries to produce more animal products. Therefore, future research should focus on the economical use of the available fresh water and generating technologies to conserve and reuse it for both livestock production and human use.

## References

AAFRD (Alberta Agriculture, Food and Rural Development). 1993. Agri-facts, water analysis interpretation, Agdex 400/716-2. Edmonton, AB, Canada.

Agriculture and Food. 2019. Water: the forgotten nutrient for pigs. Department of Primary Industries and Regional Development. https://www.agric.wa.gov.au/water/water-forgotten-nutrient-pigs (Accessed 22 October 2019.)

Anderson, C. R., and T. R. Houpt. 1990. Hypertonic and hypovolemic stimulation of thirst in pigs. Am. J. Physiol. 258:R149–R154.

Anderson, D. M., and S. C. Stothers. 1978. Effects of saline water high in sulfates, chlorides and nitrates on the performance of your weanling pigs. J. Anim. Sci. 47:900–907.

Anderson, J. S., D. M. Anderson, and J. M. Murphy. 1994. The effect of water quality on nutrient availability for grower/finisher pigs. Can. J. Anim. Sci. 74:141–148.

Andreoli, T. E. 2000. Water: normal balance, hyponatremia, and hypernatremia. Ren. Fail. 22:711–735.

Argenzio, R. A. 1984. Intestinal transport of electrolytes and water. In: M. J. Swenson, editor, Duke's physiology of domestic animals. Cornell University Press, Ithaca, NY. p. 311–326.

Bauer, W. 1982. Water consumption by non-pregnant, pregnant, and lactating gilts. Archives of Vet. Sci. Med. 36:823–827.

Bekaert, H., and J. Daelemans. 1970. Dedrinkwatervoorziening bij zogende biggen. Landbouwtijdschrift. 6–7:925–939.

Berdanier, C. D. 1995. Advanced nutrition: macronutrients. CRC Press, Inc, Boca Raton, FL.

Boe, K. E., and O. Kjelvik. 2011. Water nipples or water bowls for weaned piglets: effect on water intake, performance, and plasma osmolality. Acta Agriculturae Scandinavica. Section A, Anim. Sci. 61:86–91.

Brendemuhl, J. H., A. J. Lewis, and E. R. Peo, Jr. 1987. Effect of protein and energy intake by primiparous sows during lactation on sow and litter performance and sow serum thyroxine and urea concentrations. J. Anim. Sci. 64:1060–1069.

Brilland, S., P. Gambade, C. Belloc, and M. Leblanc-Maridor 2017. Assessment of the efficiency of waterline cleaning protocols in post-weaning rooms. J. de la Recherche Porcine en France 49:221–222.

Brooks, P. H. 1994. Water: Forgotten nutrient and novel delivery system. In: P. Lyons, and K. A. Jaques, editors, Biotechnology in the feed industry. Nottingham Press, Leicestershire, UK. p. 211–234.

Brooks, P. H., S. J. Russell, and J. L. Carpenter. 1984. Water intake of weaned piglets from three to seven weeks old. The Vet. Rec. 115:513–515.

Brumm, M. 2006. Patterns of drinking water use in pork production facilities. Nebraska Swine Reports. 221. https://digitalcommons.unl.edu/coopext_swine/221 (Accessed 14 January 2020.)

Brumm, M. C. 2010. Water recommendations and systems for swine. In: National swine nutrition guide. U.S. Pork Center of Excellence, Des Moines, IA.

Brumm, M. C., J. M. Dahlquist, and J. M. Heemstra. 2000. Impact of feeders and drinker devices on pig performance, water use and manure volume. Swine Health Prod. 8:51–57.

Bull, L. S., C. M. Williams, J. M. Rice, S. Liehr, and D. G. Rashash. 2005. Water recycling and use of recovered nutrients in animal production systems. Animal and Poultry Waste Management Center Annual Report. North Carolina State Univ., Raleigh, NC. p. 41–42.

Canadian Task Force on Water Quality. 1987. Canadian water quality guidelines. In land Waters Directorate, Ottawa, ON, Canada.

Christensen, D. A. 2001. Water quality effect on diet formulation. Proc. 62nd Minnesota Nutr. Conf. Minnesota Corn Growers Assoc. Tech. Symp., September 11–12, Bloomington, MN.

Ciara, J., and K. J. Anna. 2007. Drinking behavior in nursery aged pigs. MS Thesis. Iowa State Univ., Ames, IA, US.

Close, W. H., and F. G. Roberts. 1991. Nutrition of the working boar. Rec. Adv. Anim. Nutr. 2:21–44.

Cromwell, G. 1999. Water for swine: quantity and quality important. The Farmers Pride, KPPA News 11:11.

De Busser, E. V., J. Dewulf, L. DeZutter, F. Haesebrouck, J. Callens, T. Meyns, W. Maes, and D. Maes. 2011. Effect of administration of organic acids in drinking water on faecal shedding of E. coli, performance parameters and health in nursery pigs. Vet. J. 188:184–188.

Deligeorgis, S. G., K. Karalis, and G. Kanzouros. 2006. The influence of drinker location and colour on drinking behaviour and water intake of newborn pigs under hot environments. Appl. Anim. Behav. Sci. 96:233–244.

Ding, W. J., S. Cheng, L. Yu, and H. Huang. 2017. Effective swine wastewater treatment by combining microbial fuel cells with flocculation. Chemosphere 182:567–573.

Dybkjaer, L., A. P. Jacobsen, F. A. Togersen, and H. D. Poulsen. 2006. Eating and drinking activity of newly weaned piglets: Effects of individual characteristics, social mixing, and addition of extra zinc to the feed. J. Anim. Sci. 84:702–711.

Emerick, R. J. 1974. Consequences of high nitrate levels in feed and water supplies. Fed. Proc. 33:1183–2287.

Flipot, P. M., and G. Ouellet. 1988. Mineral and nitrate content of swine drinking water in four Quebec regions. Can. J. Anim. Sci. 68:997–1000.

Flohr, J. R., M. D. Tokach, S. S. Dritz, J. M. DeRouchey, R. D. Goodband, and J. L. Nelssen. 2014. The effects of sodium sulfate in the water of nursery pigs and the efficacy of non-nutritive feed additives to mitigate those effects. J. Anim. Sci. 92:3624–3635.

Fraser, D., and P. A. Phillips. 1989. Lethargy and low water intake by sows during early lactation: A cause of low piglet weight gains and survival? Appl. Anim. Behav. Sci. 24:13–22.

Fraser, D., J. F. Patience, and J. M. McLeese. 1991. Water for piglets and sows: quantity, quality and quandaries. In: W. Haresign, and D. J. A. Cole, editors, Recent advances in animal nutrition. Butterworths, Sevenoaks, UK. p. 137–160.

Fraser, D., J. F. Patience, P. A. Phillips, and J. M. McLeese. 1993. Water for piglets and lactating sows: Quantity, quality, and quandaries. In: D. J. A. Cole, W. Haresign, and P. C. Garn Worthy, editors, Recent developments in pig nutrition. Nottingham University Press, UK. p. 201–224.

Fraser, D., W. B. Peters Weem, P. A. Philips, and B. K. Thompson. 1988. Use of water by piglets in the first days after birth. Can. J. Anim. Sci. 68:603–610.

Friend, D. W. 1971. Self-selection of feeds and water by swine during pregnancy and lactation. J. Anim. Sci. 32:658–666.

Garcia, A., M. Sutherland, G. Pirner, G. Picinin, M. May, B. Backus, J. McGlone. 2016. Impact of providing feed and/or water on performance, physiology, and behavior of weaned pigs during a 32-h transport. Animal 6:31.

Gonyou, H. 1996. Water use and drinker management: a review. Annu. Res. Rep. Prairie Swine Centre Inc., Saskatoon, SK, Canada. p. 74–80.

Gonyou, H. W., and Z. Lou. 2000. Effects of eating space and availability of water in feeders on productivity and eating behavior of grower/finisher pigs. J. Anim. Sci. 78:865–870.

Grizzle, J., T. Armbrust, M. Bryan, and A. Saxton. 1996. Water quality I: The effect of water nitrate and pH on broiler growth performance. Appl. Poult. Res. 5:330–336.

Hagsten, I. B. and T. W. Perry. 1976. Evaluation of dietary salt levels for swine. I. Effect on gain, water consumption and efficiency of feed conversion. J. Anim. Sci. 42:1187–1190.

Huntley, N. F. and J. F. Patience. 2018. The effect of xylose on water and energy balance in pigs. J. Anim. Sci. 96:162–163.

Huynh, T.T.T., A. J. A. Aarnink, M. W. A. Verstegen, W. J. J. Gerrits, M. J. W. Heetkamp, B. Kemp and T. T. Canh. 2005. Effects of increasing temperatures on physiological changes in pigs at different relative humidity. J. Anim. Sci. 83:1385–1396.

Jeon, J. H., S C.Yeon, Y. H. Choi, W. Min, S. Kim, P. J. Kim, and H. H. Chang. 2006. Effects of chilled drinking water on the performance of lactating sows and their litters during high ambient temperatures under farm conditions. Livest. Sci. 105:86–93.

Johnston, L. J., R. L. Fogwell, W. C. Weldon, N. K. Ames, D. E. Ullrey, and E. R. Miller. 1989. Relationship between body fat and post weaning interval to estrus in primiparous sows. J. Anim. Sci. 67:943–950.

Jongbloed, A. W., N. P. Lenis, and Z. Mroz. 1997. Impact of nutrition on reduction of environmental pollution by pigs: An overview of recent research. Vet. Quart. 19:30–134.

Kaewtapee. C., N. Krutthai, K. Poosuwan, T. Poeikhampha, S. Koonawootrittriron, and C. Bunchasak. 2010. Effects of adding liquid DL-methionine hydroxy analogue-free acid to drinking water on growth performance and small intestinal morphology of nursery pigs. J. Anim. Physiol. Anim. Nutr. 94:395–404.

Kil, D. Y., W. B. Kwon, and B. G. Kim. 2011. Dietary acidifiers in weanling pig diets: A review. Rev. Colomb. Cienc. Pecu. 24:231–247.

Kober, J. A. 1993. Water: The most limiting nutrient. Agri-Practice 14:39–42.

Koeppen, B. M., and B. A. Stanton. 2001. Renal physiology. 3rd ed. Mosby Inc., St. Louis, MO.

Kruse, S., E. Stamer, I. Traulsen, and J. Krieter. 2010. Relationship between feed, water intake, and body weight in gestating sows. Livest. Sci. 137:37–41.

Kruse, S., I. Traulsen, J. Krieter. 2011. Analysis of water, feed intake and performance of lactating sows. Livest. Sci. 135:177–183.

Laitat, M., M. Vandenheede, A. Desiron, B. Canart, and B. Nicks. 1999. Comparison of performance, water intake and feeding behaviour of weaned pigs given either pellets or meal. J. Anim. Sci. 69:491–499.

Lee, S. A., H. Jo, C. Kong, and B. K. Kim. 2017. Use of digestible rather than total amino acid in diet formulation increases nitrogen retention and reduces nitrogen excretion from pigs. Livest. Sci. 197:8–11.

Leibbrandt, V. D., L. J. Johnston, G. C. Shurson, J. D. Crenshaw, G. W. R. Libal, and D. Arthur. 2001. Effect of nipple drinker water flow rate and season on performance of lactating swine. J. Anim. Sci. 79:2770–2775.

Li, Y. Z., L. Chénard, S. P. Lemay, and H. Gonyou. 2005. Water intake and wastage at nipple drinkers by growing-finishing pigs. J. Anim. Sci. 83:1413–1422.

Maenz, D. D., J. F. Patience, and M. S. Wolynetz. 1994. The influence of the mineral level in drinking water and the thermal environment on the performance and intestinal fluid flux of newly-weaned pigs. J. Anim. Sci. 72:300–308.

Massabie, P., H. Roy, A. L. Boulestreau-Boulay, and A. Dubois. 2014. La consommation d'eau en élevage de porcs. Des leviers pour réduire la consommation d'eau en élevage de porcs. French Pork and Pig Institute (IFIP). https://www.ifip.asso.fr/sites/default/files/pdf-documentations/abreuvement-elevages-porc-ifip.pdf. (Accessed 14 January 2020.)

Maynard, L. A., H. F. Loosli, H. F. Hinz, and R. G. Warner. 1979. Animal nutrition, 7th ed. McGraw Hill Inc., New York, NY.

McLeese, J. M., M. L. Tremblay, J. F. Patience, and G. I. Christison. 1992. Water intake patterns in the weanling pig: Effect of water quality, antibiotics and probiotics. Anim. Prod. 54:135–142.

Meek, A. J. 1996. Water quality concerns for swine. Small Farm Today 13:51.

Mesonero, J. A., Y. van der Horst, J. Carr, and D. Maes. 2016. Implementing drinking water feed additive strategies in post-weaning piglets, antibiotic reduction and performance impacts: case study. Porcine Health Manag. 2:25.

Mount, L. E., C. W. Holmes, W. H. Close, S. R. Morrison, and I. B. Start. 1971. A note on the consumption of water by the growing pig at several environmental temperatures and feeding levels. Anim. Prod. 13:561–563.

Mroz, Z., A. W. Jongbloed, N. P. Lenis, and K. Vreman. 1995. Water in pig nutrition: Physiology, allowances and environmental implications. Nutr. Res. Rev. 8:137–164.

Nagai, M., K. Hachimura, and K. J. Takahashi. 1994. Water consumption in suckling pigs. J. Vet. Med. Sci. 56:181–183.

National Pork Board. 2018. Swine care handbook. https://library.pork.org/?mediaId=B75B3A6A-75B3-441B-9A316C342353D356. (Accessed 14 January 2020.)

Nienaber, J. A., and G. L. Hahn. 1984. Effects of water flow restriction and environmental factors on performance of nursery-age pigs. J. Anim. Sci. 59:1423–1429.

NRC. 1994. Nutrient requirements of swine. 9th rev. ed. Natl. Acad. Press, Washington, DC.

NRC. 1998. Nutrient requirements of swine. 10th rev. ed. Natl. Acad. Press, Washington, DC.

NRC. 2012. Nutrient requirements of swine. 11th rev. ed. Natl. Acad. Press, Washington, DC.

Nyachoti, C. M., J. F. Patience, and I. R Seddon. 2005. Effect of water source (ground versus surface) and treatment on nursery pig performance. Can. J. Anim. Sci. 85:405–44.

Nyachoti, M., and E. Kiarie. 2010. Water in swine production: A review of its significance and conservation strategies. Manitoba Swine Seminar, Winnipeg, MB, Canada.

Paterson, D. W., R. C. Wahlstrom, G. W. Libal, and O. E. Olson. 1979. Effects of sulfate in water on swine reproduction and young pig performance. J. Anim. Sci. 49:664–667.

Patience, J. F. 2012a. Water in swine nutrition. In: L. I. Chiba, editor, Sustainable swine nutrition. John Wiley and Sons, Inc., Hoboken, NJ. p. 3–22.

Patience, J. F. 2012b. The importance of water in pork production. Anim. Front. 2:28–35.

Patience, J. F., J. McLeese, and M. L. Tremblay. 1989. Water quality: Implications for pork production. Proc. 10th West. Nutr. Conf., Saskatoon, Saskatchewan, Canada.

Patience, J. F., A. D. Beaulieu, and D. A. Gillis. 2004. The impact of ground water high in sulfates on the growth performance, nutrient utilization, and tissue mineral levels of pigs housed under commercial conditions. J. Swine Health Prod. 12:228–236.

Patience, J. F., N. Possberg, and D. Gills. 1997. Water quality and weanling pig performance. Annual Research Report. Prairie Swine Centre Inc., Saskatoon, SK, Canada.

Patience, J. F., P. A. Thacker, and C. F. M. de Lange. 1995. Swine nutrition guide. 2nd ed. Prairie Swine Centre Inc., Saskatoon, SK, Canada.

Peng, J. J., S. A. Somes, and D. W. Rozeboom. 2007. Effect of system of feeding and watering on performance of lactating sows. J. Anim. Sci. 85:853–860.

Phengvilaysouk, A., J. E. Lindberg, V. Sisongkham, P. Phengsavanh, A. Jansson. 2018. Effects of provision of water and nesting material on reproductive performance of native Moo Lath pigs in Lao PDR. Trop. Anim. Health. Prod. 50:1139–1145.

Phillips, P. A., and D. Fraser. 1991. Discovery of selected water dispensers by newborn pigs. Can. J. Anim. Sci. 71:233–236.

Phillips, P. A., D. Fraser, B. K. Thompson. 1990. The influence of water nipple flow rate and position, and room temperature on sow water intake and spillage. Appl. Eng. Agric. 6:75–78.

Reese, D. E., B. D. Moser, E. R. Peo, Jr, A. J. Lewis, D. R. Zimmerman, J. E. Kinder, and W. W. Stroup. 1982. Influence of energy intake during lactation on the interval from weaning to first estrus in sows. J. Anim. Sci. 55:590–598.

Renaudeau, D., G. Frances, S. Dubois, H. Gilbert, and J. Noblet. 2013. Effect of thermal heat stress on energy utilization in two lines of pigs divergently selected for residual feed intake. Anim. Sci. 91:1162–1175.

Roush, W. B. and M. Mylet. 1986. Effect of water softening, watering devices, and dietary salt level on the performance of caged single Comb While leghorn laying hens. Poult. Sci. 65:1866–1871.

Rousselière, Y, A. Hemonic, and M. Marcon. 2016. Suivi individuel du comportement d'abreuvement du porcelet sevré. J. Rech. Porc. 48:355–356.

Roy, H., C. Calvar, B. Landrain, and E. Royer. 2007. Le point sur l'utilisation et les possibilités du matériel de distribution de l'aliment en soupe en élevage de porcs: matériels techniques et informatiques, problèmes rencontrés, améliorations possibles. Report from Chambres d'agriculture de Bretagne. http://www.bretagne.synagri.com.

Royer, E., V. Ernandorena, and F. Escribano. 2007. Effects of the water-feed ratio and of a rheological sepiolite on some physical parameters of liquid feed and performances of pigs. Proc. 58th Annu. Meet. Eur. Assoc. Animal Prod. (EAAP), Dublin, Ireland. p. 26–29.

Russell, I. D. 1985. Some fundamentals of water medications. Poult. Digest. 44:422–423.

Russell, P. J., T. M. Geary, P. H. Brooks, and A. Campbell. 1996. Performance, water use and effluent output of weaner pigs fed ad libitum with either dry pellets or liquid feed and the role of microbial activity in the liquid feed. J. Sci. Food Agric. 72:8–16.

Sadler, L. J., J. R. Garvey, T. J. Uhlenkamp, C. J. Jackson, K. J. Stalder, A. K. Johnson, L. A. Karriker, R. A. Edler, J. T. Holck, and P. R. DuBois. 2008. Drinker to nursery pig ratio: effects on drinking behavior and performance. Animal Industry Report. AS 654, ASL R2335. Iowa State University. doi:https://doi.org/10.31274/ans_air-180814-704.

Schiavon, S., and G. C. Emmans. 2000. A model to predict water intake of a pig growing in a known environment on a known diet. Br. J. Nutr. 84:873–883.

Seynaeve, M., R. Wilde, G. Janssens, and B. De Smet. 1996. The influence of dietary salt level on water consumption, farrowing, and reproductive performance of lactating sows. J. Anim. Sci. 74:1047–1055.

Shields, R. G., Jr., D. C. Mahan, and P. L. Graham. 1983. Changes in swine body composition from birth to 145 kg. J. Anim. Sci. 57:43–54.

Stockill, P. 1990. Water: Why it should not be the neglected nutrient for pigs. Feed Int. 11:10–18.

Straub, G., J. H. Weniger, E. S. Tawfik, and D. Steinhauf. 1976. The effect of high environmental temperatures on fattening performance and growth of boars. Livest. Prod. Sci. 3:65–74.

Su, T., Y. Weng, C. Chung, T. Hsiao, and M. Cheng. 2018. Effects of floor types on growth performance of grower-finisher pig and pig house's, wastewater quantity and quality during cool season. Taiwan. Livest. Res. 51:75–83.

Thacker, P. A. 2001. Water in swine nutrition. In: A. J. Lewis and L. L. Southern, editors, Swine nutrition. 2nd ed. CRC Press, New York, NY. p. 381–398.

Thalin, P. A. and M. C. Brumm. 1991. Water: The forgotten nutrient. In: E. R. Miller, D. E. Ullrey, and A. J. Austin, editors, Swine nutrition. Butterworths-Heinemann, Stoneham, MA. p. 315–324.

Tofant, A., M. Ostovic, S. Wolf, A.Ekert Kabalin, Z. Pavicic, and J. Grizelj. 2010. Association between over-chlorinated drinking water and adverse reproductive outcomes in gilts and sows. Vet. Med. 55:394–398.

Torrey, S., E. L. Tamminga, and T. M. Widowski. 2008. Effect of drinker type on water intake in newly weaned piglets. J. Anim. Sci. 86:1439–1445.

Tremblay M. L., G. I. Christison, and J. F. Patience. 1989. The effect of sulfate in water on weanling pig performance. Can. J. Anim Sci. 69:1118. (Abstr.).

Vandenheede, M., and B. Nicks. 1991. L'approvisionnement en eau des porcs: un element a ne pas negliger. J. Ann. Med. Vet. 135:123–128.

Veenhuizen, M. F. 1993. Association between water sulfate and diarrhea in swine on Ohio farms. J. Am. Vet. Med. Assoc. 202:1255–1260.

Veenhuizen, M. F., G. C. Shurson, and E. M. Kohler. 1992. Effect of concentration and source of sulfate on nursery pig performance and health. J. Am. Vet. Med. Assoc. 201:1203–1208.

Verma, S., A. Daverey, and A. Sharma. 2017. Slow sand filtration for water and wastewater treatment: - A review. Environ. Technol. Rev. 6:47–58.

Vermeer, H. M., N. Kujiken, and H. A. M. Spoolder. 2009. Motivation for additional water use of growing-finishing pigs. Livest. Sci. 124:112–118.

Vier, C. M., S. S. Dritz, M. D. Tokach, M. A. Gonçalves, F. Gomez, D. Hamilton, J. C. Woodworth, R. D. Goodband, and J. M. DeRouchey. 2018. Determining the effects of cup waterer on growth performance of growing and finishing pigs. Kansas Agricultural Experiment Station Research Reports. Kansas State Univ., Manhattan, KS.

Vorozhbitov, O. and S. Trukhman. 2012. It is beneficial to provide pigs with high quality drinking water. Svinovodstvo. 6:29–30.

Walsh, M. C., D. M. Sholly, R. B. Hinson, K. L. Saddoris, A. L. Sutton, J. S. Radcliffe, R. Odgaard, J. Murphy, and B. T. Richert. 2007. Effects of water and diet acidification with and without antibiotics on weanling pig growth and microbial shedding. J. Anim. Sci. 85:1799–1808.

Wettemann, R. P., and F. W. Bazer. 1985. Influence of environmental temperature on prolificacy in pigs. J. Reprod. Fertil (Suppl.) 33:199–208.

Wilson, M. E. 2004. Boar nutrition for optimum sperm production. Proc. Banff Pork Seminar. 15:295–306.

Wu, L., X. Zhang, Z. Tang, Y. Li, T. Li, and Q. Xu. 2018. Low-protein diets decrease porcine nitrogen excretion but with restrictive effects on amino acid utilization. J. Agric. Food. Chem. 66:8262–8271.

Yang, T. S., B. Howard, and W.V. McFarlane 1981. Effects of food on drinking behaviour of growing pigs. Appl. Anim. Ethol. 7:259–270.

# 3  Energy and Energy Metabolism in Swine

Jean Noblet, Etienne Labussière, David Renaudeau, and Jaap van Milgen

## Introduction

The cost of feed represents an important part of the total cost in swine production (>60%). Within that cost, energy is the most expensive component. This economic importance and the effects of energy on animal performance have led to the development of different systems to express the energy value of feeds and the energy requirements of animals. In addition, the competition for landscape use and feed ingredients among different animal-production sectors and the use of these ingredients for biogas and biofuel production and human nutrition can occur alongside efficient production systems with low environmental impact. However, this requires the definition of energy values of feeds and energy requirements of animals to provide effective facilitation for improved sustainability.

The objectives of this chapter are(i) to describe the different steps of energy utilization in swine with a description of available energy systems for evaluating the feeds, (ii) to quantify the different components of energy requirements in swine production and the response of growing pigs and reproductive sows to energy intake, and (iii) to consider some aspects of energy intake and its regulation by feed characteristics, animal characteristics, and environmental factors. The international unit for expressing energy is the joule (J), and this unit will be used in this chapter, even though some nutritionists feel more comfortable in expressing energy as calorie (1 cal = 4.184 J).

## Energy Utilization in Swine

### Methodological Aspects

Not all gross energy (GE) that is consumed will be retained by the animal; there will be losses via feces, urine, gas, and heat. Based on these losses in the process of energy utilization, different energy values and energy systems have been defined: digestible energy (DE) is the difference between GE intake and energy losses in feces; metabolizable energy (ME) is the difference between DE intake and energy losses in urine and gases from digestive fermentation; and net energy (NE) is the difference between ME intake and heat increment (HI) Noblet et al. 2022.

*Gross Energy*

The heat of combustion, or GE, is the most basic form in which energy can be expressed and is a property of the feed itself. The GE content of a feedstuff can be measured in a bomb calorimeter. To measure GE, a small quantity of feed is completely oxidized and the heat release is measured. The GE content of raw materials varies greatly and ranges from about 15 kJ $g^{-1}$ DM for sugarcane molasses to 39 kJ $g^{-1}$ DM for oils and fats (Sauvant et al. 2004). The difference in GE content between feeds is due to differences in chemical composition and chemical bonds. Of all organic components, carbohydrates (i.e., starch, sugars, and dietary fiber (DF)) have a relatively low GE content, whereas fat has a very high GE content. In the absence of a bomb calorimeter, the GE values may be estimated from the chemical composition by prediction equations. The INRA-AFZ Tables (Sauvant et al. 2004) proposed the following equation:

$$GE = 17.3 + 0.0617\,CP + 0.2193\,EE + 0.0387\,CF - 0.1867\,Ash \qquad \text{(Eq. 3.1)}$$

where GE is in MJ/kg DM and CP, EE, CF, and ash are the crude protein, ether extract (fat), crude fiber, and ash fractions in the diet as percentages of DM. Alternatively, GE (kJ) can be predicted directly by an equation that includes all energy-providing nutrients (g). The following equation was obtained from data from Noblet et al. (2004):

$$GE = 23.0\,CP + 38.9\,EE + 17.4\,starch + 16.5\,sugars + 18.8\,NDF + 17.7\,residue$$

$$\text{(Eq. 3.2)}$$

where residue is the difference between OM and the other identified fractions in the equation. As can be seen from this equation, the energy values are lowest for carbohydrates, intermediate for proteins, and highest for lipids. Although Eq. 3.2 is an empirical equation, it reflects the energy value of individual nutrients very well. For example, the difference in energy values between starch and sugars is mainly related to the degree of polymerization of carbohydrates. Glucose has an energy value of 15.7 kJ $g^{-1}$ (180 g $mole^{-1}$). A long-chain polymer of glucose will have the same energy value per glucose unit, but will weigh less due to the release of water during the polymerization (180 − 18 g $mol^{-1}$). The theoretical energy value of a long-chain glucose polymer would, thus, be 15.7*180/(180 − 18) = 17.4 kJ $g^{-1}$. Some variation in energy values can exist depending on the amino acid composition of protein and, to a lesser extent, the fatty acid composition of lipids. For amino acids, the GE values range from 14 kJ $g^{-1}$ for aspartate to 31.6 kJ $g^{-1}$ for leucine, isoleucine, and phenylalanine (van Milgen 2002).

*Digestible Energy*

The DE content of a feed corresponds to its GE content minus energy losses after digestion in the digestive tract and is obtained as GE minus the energy lost in the feces. Even though they are related to digestion, the energy of gas and heat originating from hindgut fermentation is not considered in the calculation of DE. The ratio between DE and GE corresponds to the digestibility coefficient (DCe) of energy. The DE content is usually measured in pigs kept in digestibility cages; the quantity of feces is either obtained from total collection over a minimum of four days or estimated by using indigestible markers in the feed. For complete feeds or ingredients that can be fed alone (e.g., cereals), a direct measurement of DE content is possible. However, many ingredients can only be included in limited amounts in a feed, either to ensure toleration by the pig or to ensure practical levels of inclusion. In these instances, either the difference method or the regression method is used. With the difference method, the DE contents of two diets are measured. A control diet is used providing the majority of the ingredients. A second diet is prepared based on the control diet and includes the ingredient to be evaluated, using a single level of inclusion. It is assumed that the difference in the measured DE contents between both diets is due only to the test ingredient. It is also assumed that the minerals and vitamins (MV) fraction of the diet do not provide energy, even though the DCe depends on the ash content in the diet (as discussed further on). Therefore, it is important to have a constant MV fraction in the control and experimental diets. The DCe of the test ingredient, then, is calculated as follows:

$$DCe, \% = 100\left[\left(DEexp - DEcrtl \times \%crtl / (1 - MV0)\right)\right] / \left[\left(GEexp - GEcrtl \times \%crtl / (1 - MV0)\right)\right]$$

(Eq. 3.3)

where GEexp and DEexp are the GE and DE of the experimental diet (MJ/kg DM), GEcrtl and DEcrtl are GE and DE of the control diet (MJ/kg DM), MV0 is the percentage of MV in the control diet (DM as % of DM), and %crtl is the percentage of the control diet (i.e., control diet minus its MV content, or MV0) in the experimental diet. The DE value of the test ingredient is then calculated as its GE as measured in the laboratory multiplied by DCe estimated according to Eq. 3.3. In such trials, the same control diet can be used for several experimental diets containing different ingredients to be tested or the same ingredient at different inclusion levels. Finally, as for feeds when no calorimeter is available, the GE content of feces can be calculated from the proximate composition of the feces. The following equation has been proposed by Noblet and Jaguelin (unpublished data):

$$GE\,feces = 18.73 - 0.192\,Ash + 0.223\,EE + 0.065\,CP$$

(Eq. 3.4)

with GE as MJ/kg of DM and chemical composition as a percentage of DM.

*Metabolizable Energy*
The ME content of a feed is equivalent to the difference between the DE content and energy losses in urine and gases (mainly methane in pigs). The energy content of urine can be measured in pigs kept in metabolism crates. However, these experiments are laborious and too time consuming to be used on a routine basis. Equations for predicting urinary energy (MJ per kg feed DM) have been proposed for growing pigs and adult sows (Le Goff and Noblet 2001; Noblet et al. 2004):

$$Urinary\ energy = 0.19 + 0.031\ Nuri$$

(Eq. 3.5)

$$Urinary\ energy = 0.22 + 0.031\ Nuri$$

(Eq. 3.6)

where Nuri is the N content in the urine, expressed as g of N per kg DM feed intake. The excretion of N in the urine depends on the difference between digestible N and retained N or, in other words, the quantity of protein in the feed and the capacity of the pig to retain energy as protein. The urinary energy can therefore vary according to the physiological stage of the pig and diet characteristics. For practical purposes and to apply a single ME value to a feed or raw material, it is suggested to calculate standardized urinary energy losses and standardized ME values for a urinary N loss calculated as a constant proportion of digestible N or of total N. This also indicates that most ME values of ingredients published in original papers should be considered with caution, especially with high protein ingredients for which the N excretion is in excess of what would occur with balanced and low protein diets with subsequent underestimated ME contents. For example, the ME value of a supplemented free amino acid would be close to its DE (or GE) value if the amino acid is deposited as protein. However, if the amino acid is catabolized, the resulting urea will be excreted in the urine and the ME value of the amino acid will be (much) lower than its DE value.

The measurement of methane production necessitates the pig to be housed in a respiration chamber. In addition, the energy loss as methane is small in piglets and growing pigs and can therefore be neglected in most situations. However, in adult pigs where hindgut fermentation is more important (as discussed further on), methane production is four to five times greater than in growing pigs and thus deserves consideration in ME evaluation.

*Net Energy*
Net energy is defined as the ME content minus HI associated with feed utilization (i.e., the energy cost of ingestion, digestion, and metabolic utilization of ME) and the energy cost corresponding to a "normal" level of physical activity (Figure 3.1). The NE-to-ME ratio (or k) corresponds to the efficiency of ME utilization for NE; it also corresponds to $1 - (\Delta HI/\Delta ME)$. However, the HI-to-ME ratio of a given feed depends on the ME intake and also on several physiological factors. For instance, the HI is lower for ME supplied below the maintenance-energy requirement than for ME supplied above maintenance energy requirement (Noblet et al. 1990; 1994a,b; Birkett and de Lange 2001). The HI is also lower when ME is used for fat deposition compared with protein deposition (Noblet et al. 1999). As the proportion of fat deposition typically increases more rapidly

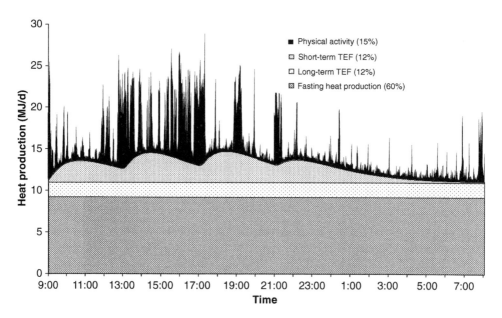

**Figure 3.1**  Components of heat production in a growing pig (60 kg) offered 2.4 MJ ME/kg BW$^{0.60}$/d) in four meals at 0900, 1300, 1700, and 2100 hours (TEF = thermic effect of feeding; adapted from van Milgen and Noblet 2000).

than the protein deposition with increasing ME intake, the HI/ME should, at least theoretically, be lower at higher levels of ME intake. Therefore, to maintain the concept of a single NE value for a given feed or raw material, it is necessary to determine this value under standardized conditions: at protein and amino acid supplies meeting the requirement, at a constant composition of the gain, and(or) at a given physiological stage.

For growing pigs, NE intake is usually calculated as the sum of retained energy (RE) at a given production level and the fasting heat production at zero activity (FHP; Noblet et al. 1994a). This NE value and the corresponding k value then correspond to a combined utilization of energy for meeting requirements for maintenance and growth. The RE is either measured by the comparative slaughter technique or, more frequently, calculated as the difference between ME intake and HP estimated by calorimetry. The FHP is either measured directly in fasting animals or obtained from literature data. It can also be calculated by extrapolating HP measured at different feeding levels to zero ME intake (Figure 3.2; FHPr). However, even though it has been widely used in the past, the latter method has important limitations. First, it consists of extrapolating HP measured at feed intake levels typically ranging between 60 and 100% of ad libitum to HP at zero feed intake, with subsequent inaccuracies in the slope and intercept. Second and more importantly, the measured FHP is not constant and is affected by the feeding level prior to fasting, especially in growing animals (Koong et al. 1982; de Lange et al. 2006; Labussière et al. 2011). Apparently, the animal adapts its basal energy expenditure to the level of feed intake and(or) growth intensity. These authors observed that FHPr was markedly lower than the measured FHP with subsequent lower values for NE and k and a higher HI (Figure 3.2). They also observed that HI, calculated as HP minus the measured FHP and expressed per unit of ME, is constant for different feeding levels. Furthermore, the degree of adaptation of FHP and HP to feeding level also depends on animal characteristics such as the genotype (Renaudeau et al. 2006; Barea et al. 2010). Overall, these observations question the use of FHPr as an estimate of FHP for calculating NE values. The measurement of FHP according to indirect calorimetry methods immediately after a fed period is highly preferable (Noblet et al. 2010). If it is not possible to obtain these measurements, literature values of FHP can be used as an alternative (as indicated

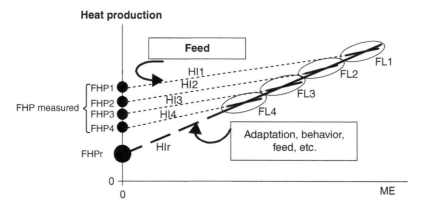

**Figure 3.2** Schematic representation of the effect of feeding level (FLi) on heat production and fasting heat production (FHP) in nonruminant animals. Each FHPi corresponds to the FHP measured on animals receiving the FLi during the immediately preceding period. The FHPr (r for regression) is obtained from the regression between HP and ME. The slope is the "regression" heat increment (HIr), and the slope between each FHPi and HPi corresponds to the measured heat increment (HIi). (Source: Adapted from Koong et al. 1982; de Lange et al. 2006; Labussière et al. 2011.)

further on). The HP also depends on climatic factors with an increased HP and reduced RE if the animals are kept below thermoneutrality and reduced HP and RE above the upper limit of the thermoneutral zone. It is, therefore, recommended to keep the animals above thermoneutrality during fasting to avoid bias in estimating NE and k.

From a practical point of view, and to avoid bias in the calculation of NE for different feeds, it is necessary to carry out energy balance measurements in similar animals (i.e., same sex, same breed, and in the same body- weight range), keep these animals within their thermoneutral zone, minimize variation in behavior, and feed the animals at about the same feed intake level with balanced diets so that the animals can express their growth potential. Under these circumstances, an erroneous estimate of FHP will affect the absolute NE value, but not the ranking between feeds. This also means that NE should not be measured in animals fed only an ingredient for which the chemical characteristics are very different from those of a complete balanced diet.

While measurements of DE and, to a lesser extent, ME are relatively easy and can be undertaken on a large number of feeds at a reasonable cost, the actual measurement of NE is more complex and expensive. The best alternative is then to use reliable NE prediction equations established from measurements carried out under similar and standardized conditions with balanced complete diets. Those based on DE or ME content and information on chemical characteristics (Noblet et al. 1994a; as discussed further on) can be used directly from measured chemical composition and digestibility data provided by direct measurements, in vitro or near infra-red methods, or by feeding tables.

Heat production can be measured directly through direct calorimetry, estimated from gas exchanges through indirect calorimetry, or calculated as the difference between ME intake and energy gain obtained by the comparative slaughter technique. The latter technique can easily be used in small animals such as poultry, but is much more difficult to perform and less accurate in large animals. As such, the most commonly used method for pigs is indirect calorimetry, which consists of measuring oxygen consumption, and carbon dioxide and methane production in respiration chambers. These measurements, combined with the urinary energy production, are used to calculate HP (Brouwer 1965). This method allows measurements over a short period of time (i.e., three to five complete days following a one-day adaptation to the chamber) with possibilities of combination of successive measurements under different feeding, housing, and physiological conditions on the same animal. Modeling methods can also be implemented on indirect calorimetry measurements to partition the total daily HP between different components, which can be used in the further interpretation of energy balance data (van Milgen et al. 1997; Figure 3.1) or correction for difference in level of physical activity.

In conclusion, the NE value of a feed and the corresponding k value should be evaluated according to standardized and adequate methods. The values are dependent on assumptions and calculations methods (FHP), conditions of measurement (e.g., climate, activity), and the composition of the energy gain. This means that data on NE and k available in the literature for pigs should be interpreted with caution and may not be directly comparable across studies. The same comment can be applied to ME values, which depend of the importance of protein catabolism and, to a lesser extent, the inclusion of gas energy losses.

### Digestive Utilization of Energy

#### Effect of Diet Composition
For most pig diets, DCe varies between 70 and 90%, but the variation is larger for feed ingredients (10–100%; Sauvant et al. 2004; Lyu et al. 2018; Navarro et al. 2019). Most of the variation of DCe

is related to the presence of DF, which is less digestible than other nutrients (<50% vs. 80–90% for fat or protein and 100% for starch and sugars; Tables 3.1 and 3.2), and reduces the apparent fecal digestibility of other dietary nutrients such as crude protein and fat (Noblet and Perez 1993; Le Goff and Noblet 2001). Consequently, DCe of a feed is linearly and negatively related to its DF content (Tables 3.1 and 3.3). The coefficients relating DCe to DF (Table 3.1; Eq. 3.13 in Table 3.3) are such that NDF, or total DF, essentially dilutes the diet, at least in growing pigs. Although DF is partly digested by the growing pig (Tables 3.1 and 3.2), it provides very little DE to the animal.

The digestive utilization of DF varies with its botanical origin (Tables 3.1 and 3.2), with subsequent variable effects on energy digestibility. Therefore, the DCe prediction equations presented in Table 3.2 represent average equations for mixed feeds that should not be applied to raw materials where specific relationships are to be used (Noblet et al. 2003). Eq. 3.14 (Table 3.3) indicates that minerals present in the diet have a negative effect on DCe. This effect is partly related to minerals associated with DF in some ingredients but also to a direct effect (perhaps the abrasion of gut tissues) of minerals provided by calcium carbonate or phosphates (−0.5% of DCe per 1% additional ash; INRA, unpublished data).

**Table 3.1**  Effect of fiber origin on its digestibility coefficient in growing pigs[a].

| Item | Fiber source | | | |
|---|---|---|---|---|
|  | SBP | SBH | WB | WS |
| Digestibility, % |  |  |  |  |
| NDF | 60.1 | 67.9 | 40.4 | 15.0 |
| ADF | 54.0 | 62.2 | 19.0 | 11.2 |
| NSP | 69.5 | 79.1 | 45.8 | 16.3 |
| Change in DCe[b] | −0.80 | −0.83 | −1.25 | −1.77 |

[a] Adapted from Chabeauti et al. (1991).
[b] Decrease in digestibility coefficient of energy (DCe, %) per 1% increase in NSP.
Starch in a basal diet was partly replaced by the fiber source (SBP = sugar beet pulp, SBH = soybean hulls, WB = wheat bran, WS = wheat straw) in the experimental diets, NDF = neutral detergent fiber, and ADF = acid detergent fiber. DCe = digestibility coefficient of energy, and NSP = nonstarch polysaccharides.

**Table 3.2**  Digestibility coefficient of fiber fractions and energy in high fiber ingredients in growing pigs (G) and adult sows (S)[a,b].

| Digestibility, % | WB | | CB | | SBP | |
|---|---|---|---|---|---|---|
|  | G | S | G | S | G | S |
| NSP | 46 | 54 | 38 | 82 | 89 | 92 |
| NCP | 54 | 61 | 38 | 82 | 89 | 92 |
| Cellulose | 25 | 32 | 38 | 82 | 87 | 91 |
| Dietary fiber | 38 | 46 | 32 | 74 | 82 | 86 |
| Energy | 55 | 62 | 53 | 77 | 70 | 76 |

[a] Adapted from Noblet and Bach-Knudsen (1997).
[b] WB = wheat bran, CB = corn bran, SBP = sugar beet pulp, NSP = nonstarch polysaccharides, NCP = Noncellulose polysaccharide, and dietary fiber = NSP+lignin.

**Table 3.3** Effect of diet composition (% DM) on energy digestibility coefficient (DCe, %), ME:DE coefficient (%), and efficiency of utilization of ME for NE of mixed diets for growth ($k_g$, %) or maintenance ($k_m$, %)[a].

| Equation | | Source[b] |
|---|---|---|
| 3.7 | DCe = 98.3 − 0.90 NDF | 1 |
| 3.8 | DCe = 102.6 − 1.06 Ash − 0.79 NDF | 1 |
| 3.9 | DCe = 96.7 − 0.64 NDF | 1 |
| 3.10 | ME/DE = 100.3 − 0.21 CP | 1 |
| 3.11 | $k_g$ = 74.7 + 0.36 EE + 0.09 ST − 0.23 CP − 0.26 ADF | 2 |
| 3.12 | $k_m$ = 67.2 + 0.66 × EE + 0.16 × ST | 3 |

[a] CF = crude fiber, CP = crude protein, NDF = neutral detergent fiber, EE = ether extract, ST = starch, and ADF = acid detergent fiber.
[b] Sources: 1 = Le Goff and Noblet (2001; $n$ = 77 diets; Equations 3.7 and 3.8 in 60 kg growing pigs and Equation 3.9 in adult sows), 2 = Noblet et al. (1994b; $n$ = 61 diets; 45 kg pigs), and 3 = Noblet et al. (1993c; $n$ = 14 diets; adult sows fed maintenance).

### Effect of Technology

Digestibility of energy can be modified by technological treatments (Rojas and Stein 2017). Pelleting, for instance, increases the energy digestibility of feeds by about 1% (Skiba et al. 2002; Le Gall et al. 2009). However, for some feeds, the improvement can be more important and depends on the chemical and physical (particle size) characteristics of feeds. In the examples given in Table 3.4, the improvement in energy digestibility is mainly due to an improved digestibility of fat provided by corn, full-fat rapeseed, or linseed. Consequently, the energy values of these ingredients depend greatly on the technological treatment. In the specific situation of a high-oil corn (7.5% oil), pelleting increased the DE content by approximately 0.45 MJ per kg (Noblet and Champion 2003).

**Table 3.4** Effect of pelleting on digestibility coefficient (%) of fat and energy in growing pigs.

| Item | Mash | Pellet |
|---|---|---|
| Corn-soybean meal diets[a] | | |
| Fat | 61.0 | 77.0 |
| Energy | 88.4 | 90.3 |
| Wheat-soybean meal-full fat rapeseed diets[b] | | |
| Fat | 27.0 | 84.0 |
| Energy | 73.1 | 87.4 |
| Wheat-corn-barley-soybean meal diets | | |
| Energy[c] | 75.8 | 77.3 |
| Energy | | |
| Corn | 87.0 | 90.0 |
| Full-fat rapeseed | 35.0 | 83.0 |
| Linseed (extruded)[d] | 51.0 | 84.0 |

[a] Mean of three diets containing 81% corn and 15.5% soybean meal (Noblet and Champion 2003).
[b] One diet containing 60% wheat, 15% soybean meal, and 20% full fat rapeseed; rapeseed was coarsely ground (Skiba et al. 2002).
[c] Mean of four diets containing variable amounts of fiber-rich ingredients (wheat bran, and sugar beet pulp; Le Gall et al. 2009).
[d] From Noblet et al. (2008).

For coarsely ground full-fat rapeseed, the DE values were 10.0 and 23.5 MJ DE/kg DM as mash and after pelleting, respectively (Skiba et al. 2002). Unfortunately, there is insufficient information in the literature to quantify the changes of DCe by pelleting, extrusion, acidification, particle size (Casas et al. 2017), or enzyme addition (Cozannet et al. 2012; Agyekum et al. 2016) on most ingredients used in pig feeds. In addition, information on the impact of technology, especially the addition of enzymes (carbohydrase, phytases, proteases), on the changes in the site of digestion (i.e., small intestine or hindgut) would be needed. It should also be noted that some effects of technology might be negative. For example, overheating during the drying procedures of wet products such as distillers dried grains with solubles (DDGS) can result in a Maillard reaction, thereby reducing the digestibility of organic matter and amino acids (lysine) (Cozannet et al. 2010).

*Effect of Body Weight (BW) or Physiological Stage*

Energy digestibility is affected by factors other than those related to the diet itself. In growing pigs, DCe may vary among genotypes and, more importantly, increases with increasing body weight (BW) (Noblet 2007; Noblet et al. 2013; Le Sciellour et al. 2018; Table 3.5). The largest combined effect of BW and feeding level is observed when adult sows fed slightly above maintenance level are compared with growing pigs offered feed close to ad libitum (Fernandez et al. 1986; Noblet and Shi 1993; Le Goff and Noblet 2001; Table 3.6). The difference is most important for diets or ingredients with a high DF content (Eqs. 3.13 and 3.15 in Table 3.3; Table 3.5). The negative effect of DF on DCe becomes smaller for heavier pigs or adult sows and, unlike young growing pigs, DF will have a positive contribution to the energy supply in heavier pigs. From a large data set of

**Table 3.5** Effect of diet composition and BW on energy digestibility in pigs (%)[a,b,c].

| BW, kg | Control diet | + Corn starch | + Dietary fiber | + Rapeseed oil |
|---|---|---|---|---|
| 44 | 85.3 | 90.6 | 71.6 | 86.0 |
| 103 | 87.2 | 91.6 | 75.6 | 88.7 |
| 148 | 87.2 | 92.2 | 78.0 | 88.9 |

[a] Adapted from Noblet and Shi (1994).

[b] BW = body weight.

[c] The control diet contained cereals and soybean meal. The other diets were the control diet + 30% corn starch or 8% rapeseed oil or 30% of a mixture of fibrous ingredients (¼ wheat bran, ¼ soybean hulls, ¼ sugar beet pulp, and ¼ wheat straw).

**Table 3.6** Effects of body weight and physiological stage on energy digestibility in pigs[1].

| Item | Trial 1[2] | | Trial 2[3] | |
|---|---|---|---|---|
| | G-pig | Dry sow | G-pig | Lac. sow |
| BW, kg | 60 | 227 | 62 | 246 |
| Feed intake, g DM/day | 2044 | 2119 | 2062 | 4850 |
| Energy digestibility, % | 77.2[a] | 80.5[b] | 79.9[a] | 84.9[b] |

[1] G-pig = growing pigs, Lac. sow = lactating sow, and BW = body weight.

[2] Mean of three diets based on corn, wheat, barley, peas, soybean meal and variable proportions of wheat bran, sunflower meal, corn gluten feed, and animal fat (INRA data).

[3] Mean of three diets based on corn, wheat, barley, peas, soybean meal and variable proportions of wheat bran, sunflower meal, corn gluten feed, and animal fat (Etienne et al. 1997).

[a,b] Within a trial, means without a common superscript differ ($P < 0.05$).

measurements (77 diets), Le Goff and Noblet (2001) calculated that 1 g of NDF provided 3.4 and 6.8 kJ in 60-kg growing pigs and mature sows, respectively. Using the same data, it was also shown that the difference in DE values between adult sows and growing pigs is proportional to the amount of indigestible organic matter measured in the growing pig (4.2 kJ $g^{-1}$ on average; Noblet et al. 2003; Noblet and van Milgen 2004; Figure 3.3; EvaPig 2008).

The improvement in energy digestibility with increasing BW is due to the greater digestibility of the DF fraction (Table 3.1), which is related to a greater digestive capacity of the hindgut in heavier pigs and, more importantly, a reduced rate of passage in the digestive tract (80 hours in adult sows versus 35 hours in growing pigs; Le Goff et al. 2002). The depressive effect of DF on protein and fat digestibility (i.e., endogenous losses) is also smaller in adult than in growing pigs, which also contributes to the reduced effect of DF on DCe in adult pigs (Le Goff and Noblet 2001). In lactating sows with a high feed intake capacity (6–10 kg day$^{-1}$), energy digestibility is also higher than in the growing pig (Table 3.7), and the results indicate that the difference does not seem to depend on the

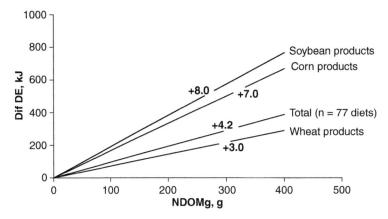

**Figure 3.3**   Relationship between DE value in adult sows and DE value in growing pigs (dif DE) and nondigestible organic matter in growing pigs (NDOM) for a set of 77 diets and some ingredients. (Source: Adapted from Noblet et al. 2003.)

**Table 3.7**   Digestible energy value of some ingredients for growing pigs and adult sows (as-fed)[a].

| | DE, MJ/kg | | |
| Ingredient | Growing pig | Adult pig | a[b] |
| --- | --- | --- | --- |
| Wheat | 13.85 | 14.10 | 3.0 |
| Barley | 12.85 | 13.18 | 2.5 |
| Corn | 14.18 | 14.77 | 7.0 |
| Pea | 13.89 | 14.39 | 6.0 |
| Soybean meal | 14.73 | 15.61 | 8.0 |
| Rapeseed meal | 11.55 | 12.43 | 3.5 |
| Sunflower meal | 8.95 | 10.25 | 3.5 |
| Wheat bran | 9.33 | 10.29 | 3.0 |
| Corn-gluten feed | 10.80 | 12.59 | 7.0 |
| Soybean hulls | 8.37 | 11.46 | 8.0 |

[a] Adapted from Sauvant et al. (2004).
[b] Kilojoule difference in DE between adult sows and growing pigs per g of indigestible organic matter in the growing pig (also see Figure 3.3).

physiological status or the feeding level of the adult sow. This means that values obtained in adult dry sows at pregnancy feeding levels can be used for both pregnant and lactating sows and that these values are higher than those obtained in growing pigs (Table 3.7).

The DCe or the DE differences between sows and growing pigs, for a given level of DF, also depend on the origin of DF or on the physicochemical properties of DF. This is illustrated in Table 3.1 and in Figure 3.3 where the effects of DF from wheat bran, corn bran, and sugar beet pulp are compared. Noblet and Le Goff (2001) presented detailed information on the effect of botanical origin of DF on DCe in both growing pigs and adult sows. These results indicate that growing pigs have a limited ability to digest DF with small differences between fiber sources, while adult sows digest DF more efficiently but the improvement depends on the chemical characteristics of DF (e.g., level of lignin). The examples presented in Table 3.7 also illustrate the effect of botanical origin with smaller differences between physiological stages for Gramineae (e.g., wheat, barley, wheat bran), Brassicaceae (e.g., rapeseed), or Compositae (e.g., sunflower) and more pronounced differences for Leguminosae (e.g., pea, soybean, lupin), especially for the hull fraction of these grains. The consequence is that the DE difference between adult sows and growing pigs is proportional to indigestible organic matter in growing pigs, but with specific coefficients for each (botanical) family of ingredients (Noblet et al. 2003; EvaPig 2008; Figure 3.3).

Little information concerning the comparative digestibility in piglets and growing pigs is available. Considering that piglets are usually fed highly digestible, low-fiber diets, from a practical point of view, piglets can be considered as growing pigs concerning the digestive utilization of energy. For growing pigs, especially when they are raised up to heavy BW (i.e., late finishing pigs), we should theoretically use energy values adapted to each stage of growth. However, the extent of the improvement is limited and, for practical reasons, it is recommended to use the same values for growing pigs and piglets, irrespective of their BW. This means that, in practice, two different DE values can be given to feeds: one for piglets and growing pigs and one for adult sows (Table 3.7; Sauvant et al. 2004). This proposal is especially justified for fibrous ingredients. A consequence of the effect of BW on DCe is that digestibility trials should be carried out at approximately 60 kg BW to be most representative for the overall weaning-growing-finishing period.

### Utilization of DE for ME

The ME content of a feed equals the DE content minus the material energy losses in the urine and combustible gases resulting from fermentation (i.e., methane and hydrogen). The energy losses in the urine (mainly as urea) are mainly due to the deamination of amino acids given in excess of what can be deposited. At a given stage of production, urinary nitrogen excretion depends mainly on the protein content of the diet. Consequently, the ME:DE ratio is linearly related to the dietary protein content (Eq. 3.10 in Table 3.3). In most situations, the ME:DE of complete feeds is approximately 0.96. However, this mean value cannot be applied to single feed ingredients and Eq. 3.10 cannot be applied beyond the range of typical CP contents of pig diets (10–20%). The most appropriate solution is then to estimate urinary energy (kJ/kg DM intake) from urinary nitrogen (g/kg DM intake) according to Eqs. 3.1 and 3.2 and for a retention coefficient of N equal to 50% of digestible N (or 40% of total N) which is representative of most practical situations in pigs fed common diets. This approach has been applied in the INRA&AFZ feeding tables (Sauvant et al. 2004). This also suggests that the ME values available from literature studies should be considered with caution, and the ME:DE of complete feeds or of ingredients when calculated according to the difference method being directly dependent on the dietary CP level of the complete diet.

In growing pigs, the average energy loss in methane is about 0.4% of DE intake and with little variability (Noblet et al. 1994a). In sows fed at maintenance level, methane production represents a much greater proportion of DE intake (1.5%; Noblet and Shi 1993) and may reach up to 3% of DE intake in sows fed very high fiber diets (Ramonet et al. 2000). More generally, methane production increases with BW and DF level in the diet (Noblet and Shi 1993; Jorgensen et al. 2001). A compilation of literature data by Le Goff et al. (2002) and unpublished data from our laboratory (Noblet et al. 2004) suggests that methane energy is equivalent to 0.67 and 1.33 kJ per g of fermented DF in growing pigs and adult sows, respectively.

### *Metabolic Utilization of Energy*

#### *Effect of Physiological Stage*
The utilization of ME is associated with HI or heat loss originating from the energy cost of ingestion, digestion, and some physical activity, in addition to the energy loss associated with metabolic transformations (e.g., the synthesis of lipid from glucose). The efficiency of ME utilization (1 − HI, or k) is either directly measured or, in most situations, obtained according to regression methods. First, k depends on the final utilization of energy with a higher value for fat energy deposition ($k_f$; approximately 80%) than for protein deposition ($k_p$; approximately 60%). Also, the efficiency of using ME below maintenance differs from the utilization above maintenance energy requirements (Noblet et al. 1990; 1994b; 1999; Table 3.8). In the case of using ME for maintenance functions, the situation is rather complex. During fasting, animals mobilize body reserves (i.e., glycogen, lipid, and protein) to supply energy for maintenance functions and include the cost of energy mobilization. When the ME supply meets the maintenance requirements (MEm), all nutrients are provided by the diet, and the maintenance energy requirement then includes the cost of intake, digestion, and absorption of these nutrients. This means that the $k_m$ (the slope of the line connecting FHP to MEm; Figure 3.2) is a relative efficiency value (i.e., the efficiency of using dietary energy for maintenance relative to the efficiency of using energy from body reserves) rather than an absolute value. The consequence of this is that $k_m$ values can exceed unity.

**Table 3.8**  Efficiencies of utilization of ME in swine (%)[a].

| Stage | Production | ME component | Efficiency | Source[b] |
|---|---|---|---|---|
| Adult | Maintenance | ME | 77 | 1 |
| Growth/pregnancy | Protein gain | ME | 60 | 2,4 |
|  | Fat gain | ME | 80 | 2,4 |
|  | BW gain | ME | 74 | 3 |
| Pregnancy | Uterus gain | ME | 50 | 4 |
| Lactation | Milk | ME | 72 | 5 |
| Growth | BW gain + maintenance | Fat | 90 | 3 |
|  |  | Starch | 82 | 3 |
|  |  | Protein | 58 | 3 |
|  |  | Dietary fiber | 58 | 3 |

[a] BW = body weight.
[b] 1 = Noblet et al. (1993c), 2 = Noblet et al. (1999), 3 = Noblet et al. (1994a), 4 = Noblet and Etienne (1987b), and 5 = Noblet and Etienne (1987a).

## *Effect of Diet Composition*

In a standardized situation in terms of animal characteristics and feeding level (see the "Methodological Aspects" section), k varies with diet composition (Eq. 3.11 in Table 3.3). It increases with the addition of dietary fat and starch and decreases with the addition of fiber and protein. The variation in k in growing pigs is due to differences in efficiencies of ME utilization between nutrients with the highest values for fat (approximately 90%) and starch (approximately 82%) and the lowest (approximately 60%) for DF and crude protein (Schiemann et al. 1972; Just et al. 1983; Noblet et al. 1994a; van Milgen et al. 2001). The average k value obtained in 61 diets was 74%, and the k value of a standard cereals and soybean meal diet was 75% (Noblet et al. 1994a). Similar average values were obtained by Li et al. (2018) for a large set of diets measured at China Agricultural University in Beijing. These differences in efficiencies between nutrients also mean that HI per unit of energy associated with metabolic utilization of energy is higher for crude protein and DF than for starch or ether extract (Table 3.9). Measurements conducted in pigs, which differ in BW and in composition of BW gain, indicate that the efficiency of utilization of ME is little affected by the composition of BW gain under most practical conditions with a similar ranking between nutrients (Noblet et al. 1993b). Furthermore, the measurements conducted in adult sows fed at maintenance energy level indicate that the ranking of k values of nutrients is similar to that observed in growing pigs with absolute values slightly greater (Eq. 3.12 in Table 3.3; Noblet et al. 1990; 1993b). An interesting aspect of energy efficiency is illustrated in the results of van Milgen et al. (2001) and concerns the HP associated to the utilization of dietary protein for protein deposition or for lipid deposition. The data show that the heat increment associated with both pathways is similar and efficiencies are equivalent. From a practical point of view, this means that the NE value of dietary CP is constant, irrespective of its final utilization.

Overall, these results indicate that an increase in dietary crude protein content results in an increased HP (Noblet et al. 1987; Le Bellego et al. 2001; Le Bellego et al. 2002a), while the inclusion of fat contributes to a reduction of HP (Noblet et al. 2001b). Diets with low crude protein and/or high fat contents can then be considered as low heat increment diets. The effect of DF on HP remains unclear. In some trials and in agreement with the low efficiency of ME from DF, HP is increased when DF is increased (Noblet et al. 1989; Ramonet et al. 2000; Sølund Olesen et al. 2001; Rijnen et al. 2003; Lyu et al. 2018) whereas in other trials, HP remains constant or even decreases (Rijnen et al. 2001; Le Goff et al. 2002). From a biochemical perspective, HP should increase and most results are consistent with this. However, addition of DF may change the behavior of animals

**Table 3.9** Energy value of starch, crude protein, and fat according to energy systems[a].

| Item | Starch | Crude protein[c] | Crude fat[c] |
| --- | --- | --- | --- |
| Energy values[b] kJ/g | | | |
| DE | 17.5 (100) | 20.6 (118) | 35.3 (202) |
| ME | 17.5 (100) | 18.0 (103) | 35.3 (202) |
| NE | 14.4 (100) | 10.2 (71) | 31.5 (219) |
| Heat production, kJ/g | 3.1 | 7.8 | 3.8 |

[a] Adapted from Noblet et al. (1994a; $n = 61$ diets.)

[b] Parentheses = % of starch.

[c] Crude protein and crude fat are assumed to be 90% digestible, whereas starch is assumed to be 100% digestible.

(i.e., reduced physical activity) or the overall metabolism, thereby decreasing HP (Schrama et al. 1998). Furthermore, the effects of DF probably also depend on the nature of DF.

### Interaction with Climatic Factors

At ambient temperatures above the lower critical temperature (LCT), pigs dissipate their heat to maintain their body temperature with adjustments of feed intake (see Energy Requirements for Thermoregulation). Below the LCT, pigs reduce the dissipation of heat to their environment by mainly behavioral changes but have to increase their HP to maintain homeothermy. In that situation, HI is not a "loss" of energy and participates to covering the energy requirement to maintain homeothermy (Quiniou et al. 2001). A higher HI is observed with high protein or high DF feeds. High protein diets are seldom used because of their high cost and their negative environmental impact. Therefore, only high DF feeds can be a potential solution for meeting thermoregulation energy requirements. Data in Table 3.10 confirm this possibility with a special interest in gestating sows that are able to degrade high DF feeds (see "Digestive Utilization of Energy"), are fed below their voluntary feed intake capacity, and frequently housed at temperatures below their LCT (<20 °C).

### Energy Evaluation Systems

### Digestible and Metabolizable Energy

Apart from direct measurement in pigs, the DE value of raw materials can be obtained from feeding tables (Sauvant et al. 2004; NRC 2012; Rostagno 2017) or from the literature, which has been particularly abundant over the last 10–15 years (e.g., China Agricultural University, Beijing; University of Illinois, Urbana; University of Alberta, Edmonton) and allows a regular update of tabulated values or a production of nutritional values for new ingredients. However, the utilization of table values should be restricted to ingredients having chemical characteristics similar to those actually used. The effect of variation in chemical composition can be taken into account by using prediction equations of DCe or DE content of families of ingredients [Noblet et al. 2003; EvaPig 2008]. As illustrated in the previous section, DCe is affected by BW of the animals. It is therefore appropriate to use DE values adapted to each situation. From a practical point of view, it is suggested to use two DE values, one for "60 kg" pigs, which can be applied to piglets and growing-finishing pigs, and one for adult pigs applicable to both pregnant and lactating sows. Values given in most available feeding tables are typically obtained in the 30- to 60 kg pig and are therefore not applicable to adult pigs. A methodology based on the fact that the difference in DE between adult and growing pigs is proportional to the amount of indigestible OM in the growing pig has been

**Table 3.10**  Effect of dietary fiber level and ambient temperature on utilization of energy in pregnant sows[a,b,c].

| Item | Control diet | + Straw | + Alfalfa |
|------|------|------|------|
| ME intake, MJ/d | 29.6 (100) | 32.0 (108) | 34.0 (115) |
| Heat production, MJ/d | | | |
| At 21.5 °C | 26.2 (100) | 27.1 (103) | 26.9 (103) |
| At 10.5 °C | 34.9 (100) | 34.6 (99) | 34.5 (99) |

[a] Adapted from Noblet et al. (1989).

[b] Parentheses = % of the control diet.

[c] All sows received the same amount of the control diet; sows on the straw or alfalfa treatments received a daily supplement (600 g) of straw or alfalfa. Mean body weight of sows was 205 kg.

proposed (Noblet et al. 2003; EvaPig 2008) for estimating DE values in adult pigs from DE values in growing pigs (Figure 3.3). It has been implemented in the tables proposed by Sauvant et al. (2004), and the EvaPig software and some examples are given in Table 3.7.

The DE content of compound feeds can be obtained by adding the DE contributions of ingredients and assuming that there is no interaction, an assumption that seems to hold in most instances (Noblet and Shi 1994). When the actual composition of the feed is unknown, it is possible to use prediction equations based on chemical criteria (Noblet and Perez 1993; Le Goff and Noblet 2001):

$$\text{DE, MJ / kg DM} = 17.69 + 0.146 \text{ EE} + 0.071 \text{ CP} - 0.132 \text{ NDF} - 0.341 \text{ Ash} \quad \text{(Eq. 3.13)}$$

where chemical criteria are expressed as % of DM and EE and CP are ether extract and crude protein, respectively. But such an equation obtained and applicable for complete feeds cannot be used for ingredients.

Other possibilities with near infrared or in vitro methods (Boisen and Fernandez 1997; Noblet and Jaguelin-Peyraud 2007) have been proposed for estimating the DE value of feeds. Some tables provide estimates of the content of digestible nutrients obtained as the product of nutrient content and (a constant) digestibility coefficient. When this information is available, the DE content can be estimated according to the following equation:

$$\text{DE, MJ / kg DM} = 0.232 \text{ DCP} + 0.383 \text{ DEE} + 0.174 \text{ Starch} + 0.162 \text{ Sugars} + 0.178 \text{ DRes}$$

$$\text{(Eq. 3.14)}$$

where DCP and DEE are the digestible crude protein and digestible crude fat contents, respectively, and DRes is the digestible residue calculated as the difference between digestible OM and the sum of other nutrients considered in the equation (% of DM; Le Goff and Noblet 2001). The DE can also be estimated directly from the average contribution of all crude nutrients in OM of feed according to the following equation (Le Goff and Noblet 2001):

$$\text{DE} = 0.225 \text{ CP} + 0.317 \text{ EE} + 0.172 \text{ Starch} + 0.032 \text{ NDF} + 0.163 \text{ Residue} \quad \text{(Eq. 3.15)}$$

with "Residue" as the difference between OM content and the other nutrients considered in the equation (% of DM). In all equations or predictions, DF has an important impact on the accuracy of the prediction. Eqs. 3.14 and 3.15 can be applied to raw materials and compound feeds but with inaccuracies due to their inability to consider the nature of DF and, to a smaller extent, the composition of fat.

The approaches for predicting ME value of pig feeds are similar to those described for DE. However, because direct ME measurements are not carried out routinely and ME values depend on protein catabolism, it is suggested to calculate ME values from DE values and standardized urinary energy losses (Eq. 3.5 for growing pigs).

*Net Energy*

All published NE systems for pigs combine the utilization of ME for maintenance and for growth (Just et al. 1983; Noblet et al. 1994a) or for fattening (Schiemann et al. 1972) by assuming similar efficiencies for maintenance and energy retention. The system used in the Netherlands (CVB, 2018) has been adapted from recalculations on the data of Noblet et al. (1994a) and literature data. The NE systems proposed by NRC (2012) and Rostagno (2017) are based on the equations published origi-

**Table 3.11** Equations for prediction of DE, ME, and NE in feeds for growing pigs (NEg; 61 diets; MJ/kg dry matter and % of DM)[a].

| Number | Equation[b] |
|--------|-------------|
| 3.16 | DE = 0.232×DCP+0.387×DEE+0.174 ST+0.168 SU+0.167 DRes |
| 3.17 | ME = 0.204 DCP+0.393 DEE+0.174 ST+0.165 SU+0.154×DRes |
| 3.18 | NEg = 0.121 DCP+0.350 DEE+0.143 ST+0.119 SU+0.086 DRes |
| 3.19 | NEg = 0.703 DE − 0.041 CP+0.066 EE − 0.041 CF+0.020 ST |
| 3.20 | NEg = 0.700 DE − 0.038 CP+0.067 EE − 0.037 ADF+0.020 ST |
| 3.21 | NEg = 0.730×ME − 0.028×CP+0.055×EE − 0.041×CF+0.015×ST |

[a] Adapted from Noblet et al. (1994a).
[b] CP = crude protein, EE = ether extract, ST = starch, SU = sugars, DCP = digestible CP, DEE = digestible EE, and DRes = digestible residue (i.e., difference between digestible organic matter and other digestible nutrients considered in the equation).

nally by Noblet et al. (1994a). Emmans (1994) proposed a generic model based on corrections applied to the ME content. Similarly, Boisen and Verstegen (1998) suggested new concepts for estimating the NE value of pig feeds (so-called physiological energy) based on the combination of in vitro digestion methods for estimating digestible nutrients and biochemical coefficients for evaluating the ATP production potential from the nutrients.

The system proposed by Noblet et al. (1994a) is based on a large set of measurements (61 diets) and regression analysis; FHP was 750 kJ kg$^{-1}$ BW$^{0.60}$/d in the calculation of NE, which was calculated as FHP+retained energy. The prediction equations are listed in Table 3.11. These equations have been validated in further calorimetry trials conducted in our laboratory (Le Bellego et al. 2001; van Milgen et al. 2001; Noblet 2007) and also in the numerous studies conducted at China Agricultural University in Beijing (Li et al. 2018). The equations are based on information available in conventional feeding tables, and they are applicable to single ingredients and compound feeds and at any stage of pig production. It is important to point out that different DE values or digestible nutrient contents should be used in growing-finishing pigs and adult sows with two subsequent NE values. Reliable information on digestibility of energy or of nutrients is, therefore, necessary for prediction of the NE content of feed for pigs. In fact, this information is the most limiting factor for predicting energy values of pig feeds.

*Comparison of Energy Systems*
The efficiency of ME utilization for NE differs greatly between nutrients (Table 3.9; Eqs. 3.16–3.18 in Table 3.11). It is, therefore, logical that the hierarchy among feeds obtained in DE or ME systems can be different from that obtained in the NE system. Because NE represents the best compromise between the energy value (a property of the feed) and energy requirement (a property of the animal), the energy value of protein or fibrous feeds will be overestimated when expressed on a DE (or ME) basis. On the other hand, fat or starch sources are underestimated in a DE (or ME) system (Table 3.12).

The quality of a nutritional evaluation system is given by its ability to predict the response of the animals in terms of units of feed per unit of performance and independently of the diet composition (or specific effects of nutrients). With regard to energy evaluation systems, data presented in Table 3.13 illustrate the relationship between the energy system and energy cost of BW gain and confirm that NE as calculated according to Noblet et al. (1994a) is a better predictor of performance than DE or ME. In other words, the NE value is the most satisfactory estimate of the energy value of feeds (Noblet and van Milgen 2004).

**Table 3.12** Relative DE, ME, and NE values of ingredients for growing pigs[a,b].

| Item | DE | ME | NE | NE/ME, % |
|---|---|---|---|---|
| Animal fat | 243 | 252 | 300 | 90 |
| Corn | 103 | 105 | 112 | 80 |
| Wheat | 101 | 102 | 106 | 78 |
| *Reference diet* | *100* | *100* | *100* | *75* |
| Pea | 101 | 100 | 98 | 73 |
| Soybean (full-fat) | 116 | 113 | 108 | 72 |
| Wheat bran | 68 | 67 | 63 | 71 |
| Soybean meal | 107 | 102 | 82 | 60 |

[a] Adapted from Sauvant et al. (2004).

[b] Within each system, values are expressed as percentages of the energy value of a diet containing wheat, soybean meal, fat, wheat bran, peas, and minerals and vitamins.

**Table 3.13** Performance of growing-finishing pigs according to energy system and diet characteristics[a,b].

| Item | DE | ME | NE |
|---|---|---|---|
| Added fat, % (Trial 1) | | | |
|     0 | 100 | 100 | 100 |
|     2 | 100 | 100 | 100 |
|     4 | 99 | 99 | 100 |
|     6 | 98 | 98 | 100 |
| Crude protein (30–100 kg; Trial 2) | | | |
|     Normal | 100 | 100 | 100 |
|     Low | 96 | 97 | 100 |
| Crude protein (90–120 kg; Trial 3) | | | |
|     Normal | 100 | 100 | 100 |
|     Low | 97 | 98 | 100 |

[a] Adapted from Noblet (2006) and unpublished data.

[b] Energy requirements [or energy cost of body weight (BW) gain] for similar daily BW gain and composition of BW gain; values are expressed relative to the energy requirement (or energy cost of BW gain) in the control treatment (considered as 100).

## Conclusion

The energy value of a feed depends primarily on its chemical characteristics. At the DE level, it is mainly determined by its DF content, which acts more or less as a diluent and by fat, which is very energy dense. At the ME level, the change is essentially related to the dietary CP content, which affects the urinary N and energy losses. Finally, at the NE stage, most of the difference originates from CP. The impact of nutrient composition on energy values is illustrated in the coefficients of equations presented in Table 3.14, with the highest contribution to NE of fat and the lowest for CP. The energy value also depends on technological treatments and enzymes addition to the feed. Finally, energy value depends on the type of animal the feed is offered where the DF fraction makes a greater contribution to the energy value in adult pigs than in growing pigs.

**Table 3.14**  Actual contribution of dietary nutrients to energy supply in growing pigs (kJ/g)[a,b].

| Number | | CP | Fat | Starch | Sugars | Residue |
|---|---|---|---|---|---|---|
| 3.22 | GE | 22.6 | 38.8 | 17.5 | 16.7 | 18.6 |
| 3.23 | DE | 22.5 | 31.8 | 18.3 | 16.1 | 0.5 |
| 3.24 | ME | 19.7 | 32.2 | 18.2 | 15.9 | 0.5 |
| 3.25 | NE | 11.8 | 28.9 | 14.8 | 11.5 | −0.9 |

[a] From recalculations of data of Noblet et al. (1994a).
[b] Measurements were collected from 61 diets fed to 45-kg pigs, and coefficients are obtained from multiple linear regression equations (without intercept). Residue corresponds to the difference between organic matter and the sum of CP, fat, starch, and sugars.

## Energy Requirements

### *Introduction*

Energy requirements are expressed on different bases. In pigs offered feed ad libitum, they consist mainly in fixing the diet energy density according to regulation of feed intake (appetite), the growth potential of the pig, climatic factors, or economical considerations. In that situation, it is difficult to precisely define an energy requirement because the animal will attempt to regulate its feed intake to meet its energy requirements. In restrictively fed growing pigs or in reproductive sows, it is necessary to define a feeding level according to anticipated performance (e.g., BW gain, carcass composition) or estimated requirements. These recommendations represent average values, which are unable to account for the effects of genotype, production level, climatic environment, behavior of the animal, or animal genetic variability. In more sophisticated analytical approaches (factorial approach or modeling), the components of energy requirements (e.g., maintenance, physical activity, thermoregulation, inflammatory and immune response, growth, and milk production) are determined. This section will deal mainly with the last approach.

Most trials and recommendations from literature were conducted using DE or ME as a basis for the expression of requirements. These recommendations have been obtained mostly with conventional feeds: cereals/soybean meal-based diets with an efficiency of ME utilization for NE in growing pigs of about 75% (or 72% for DE). Consequently, the NE requirements (as diet energy density, daily energy requirements, components of energy requirements, etc.) can be obtained by multiplying the DE or ME requirements by 0.72 or 0.75, respectively. Our calorimetry studies have shown that absolute values of efficiencies of ME for NE differed slightly according to BW or genotype in growing pigs (Noblet et al. 1994c) or were higher in adult sows fed at maintenance level than in growing pigs (Noblet et al. 1993b). The difference did not depend on diet characteristics and the magnitude of the difference in the different situations was identical for all nutrients (Noblet 2006). This means that NE requirements can be calculated similarly for any stage of pig production, including pregnant or lactating sows or growing pigs with different growth potentials. Because the most reliable and accurate NE equations have been obtained in growing pigs, it is proposed to use these NE equations at all stages of pig production; requirements are then expressed according to a "growing pig" NE value (Noblet 2006). However, these growing pig NE values should differ according to BW or physiological stage; for simplification, only two

NE values should be used, one for growing pigs including piglets and one for adult sows, either pregnant or lactating (Sauvant et al. 2004).

### Maintenance Energy Requirements

The energy requirement for maintenance (MEm, expressed as ME) is assumed to be proportional to metabolic BW ($BW^b$). The most appropriate b value is 0.60 in growing pigs (Noblet et al. 1994a, 1999); this exponent is preferred over the commonly used 0.75 exponent, which has been developed for interspecies comparisons or estimations for adult animals within one species. The value to be considered for growing pigs raised indoor and at an environmental temperature within the thermoneutral zone is about $1.00\,MJ\ ME/kg\ BW^{0.60}/d$ (Noblet et al. 1999). Using an average efficiency value of ME for NE of 75%, this corresponds to an NE requirement for maintenance of $0.750\,MJ\ NE/kg\ BW^{0.60}/d$, which is the value used in the calculation of NE of feeds for growing pigs (Noblet et al. 1994a) and also the average value of FHP measured in several trials in our research group with castrated pigs (Le Bellego et al. 2001; van Milgen et al. 2001; Noblet et al. 1994b; Le Goff et al. 2002; de Lange et al. 2006; Lovatto et al. 2006; Barea et al. 2010), and in other research groups (Koong et al. 1982; Tess et al. 1984; Li et al. 2018).

The energy requirements for maintenance, when expressed per kg of metabolic BW ($BW^{0.60}$), are almost constant over the growth period with small differences between breeds or sexes. Differences are substantial only for extreme breeds with lower values for slow growing and(or) fat pigs such as the Meishan breed and higher values for fast growing and lean-type pigs (Noblet et al. 1999), entire males (Labussière et al. 2013) or pigs treated with somatotropin (Noblet et al. 1992) or between growing pigs selected for low and high residual feed intake (RFI; Barea et al. 2010). In addition, FHP and MEm are reduced at low feeding levels (de Lange et al. 2006; Labussière et al. 2011) and in hot conditions (Renaudeau et al. 2013). Therefore, for most pigs, MEm can be considered as constant under standardized practical conditions (i.e., conventional housing, thermoneutral, and feeding level close to ad libitum). This maintenance energy requirement includes a standard level of physical activity, which has an average energy cost of $0.200\,MJ\ ME/kg\ BW^{0.60}/d$. Approximately half of this cost is due to standing (approximately four hours per day), while the other half is due to movements during lying (van Milgen and Noblet 2003). The value of $1\,MJ\ ME/kg\ BW^{0.60}/d$ for maintenance energy requirements has been obtained in respiration chambers and therefore with a reduced level of physical activity and at thermoneutrality. A slightly greater value of MEm of $1.05\,MJ\ ME/kg\ BW^{0.60}/d$ can be used to account for the greater activity in normal housing production systems (Table 3.15). Additionally, a $0.1\,MJ\ ME/kg\ BW^{0.60}/d$ can be further added to account for the activation of inflammatory system in case of pigs exhibiting subclinical signs of inflammation (Campos et al. 2014). These figures may not be applicable in suckling piglets (Noblet and Etienne 1987a) or in early-weaned piglets (Noblet and Le Dividich 1982); these specific stages should be revisited according to more appropriate methodologies and to the recent data on thermoregulation and physical activity at these early stages of growth.

In reproductive sows, the MEm requirement is proportional to metabolic BW, using the classical value for the exponent of 0.75. The values measured for pregnant and lactating sows at thermoneutral and at "standard" activity levels are given in Table 3.15. The MEm value in lactating sows is higher than in pregnant sows, probably due to a greater production level. Inversely, MEm in pregnant sows is rather variable in connection with animal variability in levels of physical activity that interact with housing facilities (Noblet et al. 1990; see section on *Energy Cost of Physical Activity*).

**Table 3.15**  Energy requirements in swine[a].

| Stage | Energy requirement, MJ | Source[b] |
|---|---|---|
| Growth | $MEm = 1.05 \times kg\ BW^{0.60}$ | 1 |
| | Energy gain $= 23.0 \times$ protein gain, kg $+ 39.9 \times$ fat gain, kg | 2 |
| | Energy content lean tissue gain $= 8.5$ to $10.5\,MJ\,kg^{-1}$ | 2 |
| | Energy content adipose tissue gain $= 31$ to $33\,MJ\,kg^{-1}$ | 2 |
| | ME thermoregulation, see Figure 3.5 | 3 |
| Pregnancy | $MEm = 0.440 \times kg\ BW^{0.75}$ | 4 |
| | Energy maternal gain $= 9.7 \times BW$ gain, kg $+ 54 \times P2$ gain, mm | 5 |
| | Energy uterus gain $= 4.8 \times$ fetus BW gain, kg | 6 |
| | ME per 100 min standing $= 0.035 \times BW^{0.75}$ | 7 |
| | ME thermoregulation/°C $= 0.010 - 0.020 \times BW^{0.75}$ | 8 |
| Lactation | $MEm = 0.460 \times BW^{0.75}$ | 9 |
| | Energy in milk $= 20.6 \times$ litter BW gain, kg $- 0.376 \times$ litter size | 10 |

[a] See Table 8 for efficiencies.

[b] 1 = Noblet et al. (1999), 2 = Noblet et al. (1999) and Karege (1991), 3 = Quiniou et al. (2001), 4 = Noblet and Etienne (1987b), 5 = Dourmad et al. (1998), 6 = Noblet et al. (1985b), 7 = Noblet et al. (1993a), 8 = Noblet et al. (1989) (below 20 °C), 9 = Noblet and Etienne (1987a), and 10 = Noblet and Etienne (1989).

## *Energy Requirements for Growth*

From a nutritional point of view, growth corresponds to the deposition of protein, fat, minerals, and water with a subsequent ME requirement for protein and lipid deposition. The ME requirements for body protein or lipid deposition can be estimated from the quantities of deposited protein or lipid and the efficiencies of utilization of ME for energy deposited as protein and fat ($k_p$ and $k_f$, respectively). For a conventional cereals/soybean meal diet, Noblet et al. (1999) proposed 60 and 80% for $k_p$ and $k_f$, respectively. The energy content of body proteins and lipids is approximately 23.8 and 39.5 kJ g$^{-1}$, respectively. The calculated ME requirements are then 40 and 50 kJ ME per g of protein and fat, respectively.

From technical and economical points of view, the growth of tissues (lean, adipose, viscera, etc.) is the most important with efforts to increase the importance of lean tissues in the carcass and concomitantly to reduce the importance of adipose tissues. Measurements of chemical composition of lean and adipose tissues weight gain, and the associated feed energy costs calculated according to the previous discussion, indicate that the feed cost of adipose tissue gain is about 3.5 times the cost of lean tissue gain (Table 3.16). The consequence of these major differences in energy content and energy requirement for tissues gain in pigs is that the ME requirement for BW gain depends directly on the lean to adipose ratio in BW gain or lipid content since the protein content of BW gain is relatively constant (16–17%) in most practical situations of pig production.

The chemical and tissue composition of BW gain in growing pigs depends on several factors that will not be reviewed in detail here. In brief, the energy content of BW gain is lower in lean-type pigs than in obese-type pigs, lower in males than in females or in barrows, lower in lighter than in heavier growing-finishing pigs, and lower in energy-restricted than in ad libitum fed pigs (Campbell and Taverner 1988; Bikker et al. 1996; Noblet et al. 1994a; Quiniou et al. 1999a; Table 3.16). Overall, the energy content of empty BW gain over the 20–100 kg BW phase ranges from 10 MJ kg$^{-1}$ in very lean animals to 20–22 MJ kg$^{-1}$ in obese types of pigs (Noblet et al. 1994c).

**Table 3.16** Chemical composition of tissues and body weight gain in growing pigs and consequences on energy requirements for tissues gain[a].

| Composition | Entire males | | | Castrated males | | |
| --- | --- | --- | --- | --- | --- | --- |
| | Lean[b] | Adipose | eBW | Lean[b] | Adipose | eBW |
| Water, % | 69.9 | 18.7 | 58.5 | 65.6 | 14.9 | 51.0 |
| Ash, % | 1.0 | 0.2 | 3.1 | 1.0 | 0.2 | 3.0 |
| Protein, % | 17.9 | 5.4 | 16.7 | 18.2 | 4.1 | 16.0 |
| Fat, % | 10.2 | 75.4 | 21.1 | 15.3 | 81.8 | 30.4 |
| Energy, kJ g$^{-1}$ | 8.5 | 31.3 | 12.3 | 10.4 | 33.3 | 15.6 |
| ME requirement[3], kJ g$^{-1}$ | 12.2 | 39.8 | 17.2 | 14.9 | 42.6 | 21.6 |

[a] Adapted from Noblet et al. (1994c) and J. Noblet (unpublished data); over the 20–95 kg body weight (BW) period and based on the comparative slaughter technique.
[b] Lean = including intermuscular fat; eBW = empty BW.
[c] Calculated as 40 and 50 kJ ME/g of protein and fat, respectively.

Overall, this emphasizes the efforts for reducing the body fatness of growing pigs by either genetic selection or by nutritional manipulations.

### Energy Requirements for Reproduction

The energy requirements of pregnant sows have been reviewed by Noblet et al. (1989, 1997) and Dourmad et al. (2008). The energy requirements during pregnancy correspond to the sum of requirements for maintenance, uterine growth, and reconstitution of body reserves. Under specific conditions, additional requirements related to physical activity, or exposure to low temperatures, have to be taken into account. The basis for estimating energy requirements of pregnant sows is given in Table 3.15. The LCT of pregnant sows is 20–22 °C in single-housed females and is lower when straw-bedding or group-housing is used. In most practical situations, about two-thirds of the energy needed to meet the requirements of pregnant sows corresponds to the sow's maintenance requirement. The specific pregnancy requirement (i.e., uterine tissues gain) represents a negligible proportion of energy gain. However, in practice, it is higher if we consider the additional requirements for maintenance of the sow related to her additional metabolic BW because of uterine growth and the requirement for development of the mammary gland. The requirement for maternal tissues depends on the objective to realize a certain BW gain and its composition. This objective can be quite variable in multiparous sows according to their body condition at weaning.

Overall, energy requirements of pregnant sows can be variable according to their BW, the housing conditions, and their body condition at mating (Dourmad et al. 2008; NRC 2012; INRAPorc software 2009). Consequently, feeding pregnant sows of a herd the same quantities of feed may result in large variations of performance, especially body condition at farrowing. Indeed, requirements for maintenance and uterine growth and possible requirements for thermoregulation and physical activity are priority. The energy deposited in maternal tissues will depend directly on the difference between feed allowance and these priority requirements. Energy deposition can even be lower because physical activity may be increased in sows with poor body condition or kept at low ambient temperatures (Noblet et al. 1997). Therefore, changes in behavior or physical activity can affect markedly the energy balance in pregnant sows. It is generally accepted that uterus growth follows an exponential curve (Noblet et al. 1985b) which demonstrates low energy requirements for uterine tissues during the first two-thirds of pregnancy and more during the last third. According to the increase in BW during pregnancy, the requirements for maintenance will increase progressively. The consequence is that if daily feed supply is kept constant during pregnancy, the daily deposition of energy in maternal tissues will decrease progressively with the advancement of pregnancy and may even become negative during the last two to three weeks before farrowing (Dourmad et al. 1998; Young et al. 2004). This increased energy expenditure over the last third of pregnancy in sows indicates that an increase in energy supply during this period may be considered. The increased feed allowance at the end of pregnancy may also prepare the digestive tract to the rapid increase in feed intake after farrowing. These observations are illustrated in Table 3.17 and have been implemented in NRC (2012) or in the INRAPorc model and software tool (Dourmad et al. 2008).

In lactating sows, the most important factor of variation is clearly the level of milk production of the sow. Milk production is difficult to measure in sows, and it is usually estimated from litter growth (Noblet and Etienne 1989; Hansen et al. 2012; Table 3.15). Milk production depends on the genetic potential of the sow, litter size, and stage or duration of lactation (Etienne et al. 1998; Noblet et al. 1998). The energy requirements during lactation correspond to the sum of maintenance requirements and energy requirements for milk production; the efficiency of ME for milk energy averages 72% (Table 3.15). The variation in energy requirements with litter gain is illustrated in Figure 3.4. At very high production levels (>3000 g day$^{-1}$ average litter BW gain), the feed requirement is greater than 8 kg day$^{-1}$. The calculation method also indicates that the additional ME requirement because of an additional litter BW gain is proportional to the litter weight gain difference. It averages 26 MJ ME per kg litter gain or the equivalent of about 2 kg of a conventional feed per kg of additional kg litter BW gain. This approach is an easy and convenient technique to estimate the feed energy requirement in lactating sows, which is the equivalent to the sum of the requirement for maintenance (1.9–2.2 kg of feed per day for 200–250 kg sows) and the requirement for milk production (2 kg of feed per kg litter BW gain). Under most practical situations, lactating sows are unable to consume enough feed to meet their energy requirements, and they lose BW during lactation. The energy deficit and the subsequent BW loss are generally more important in primiparous sows. It is, therefore, critical to use all available techniques to maximize energy intake in

**Table 3.17** Effect of stage of pregnancy on energy utilization and activity in sows[a].

| | Stage of gestation, wk | | |
|---|---|---|---|
| Item | 5–6 | 9–10 | 14–15 |
| Body weight, kg | 182 | 207 | 224 |
| Energy balance, MJ d$^{-1}$ | | | |
| ME intake | 28.6 | 28.4 | 28.8 |
| Heat production | 22.6 | 23.1 | 26.4 |
| Retained energy | | | |
| Total | 6.0 | 5.2 | 2.4 |
| In uterus | 0.4 | 1.3 | 2.6 |
| In maternal tissues | 5.7 | 4.2 | 0.1 |
| As protein | 2.5 | 2.1 | 2.7 |
| As fat | 3.5 | 3.2 | −0.1 |
| Duration of standing, min | 288 | 263 | 247 |
| Activity heat production, MJ/d | 5.7 | 6.2 | 6.9 |

[a] Adapted from Young et al. (2004); $n = 12$ sows.

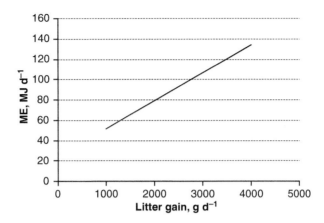

**Figure 3.4** Effect of litter body weight gain on energy requirements (MJ ME/d) of lactating sows. Calculations for a sow weighing 220 kg nursing 6 (1000 g d$^{-1}$) to 13 (>3000 g d$^{-1}$) piglets for 21 days.

lactating sows. This is beyond the scope of this chapter, but it is clear that ad libitum feeding is highly recommended. Energy intake can be increased by using energy-dense diets by reducing the DF content and increasing the fat content. Although sows that are fed these diets usually consume less feed, they will consume more energy. Nevertheless, a large fraction of the additional energy supply is excreted as fat in the milk with little direct benefit for the sow (Noblet et al. 1998).

### *Energy Requirements for Physical Activity*

As mentioned previously, energy losses associated with physical activity are difficult to estimate directly. From a methodological point of view, these losses represent an uncontrolled source of variation of HP and may lead to inaccurate estimates of energy requirements, especially under conditions that affect the behavior of the animals. Physical activity in swine represents a considerable proportion of energy expenditure, despite the low duration of standing in pigs and the reduced activity and locomotion when animals are kept indoors. This is due to a four to five times greater energy expenditure per "unit" of physical activity in swine than in most other domestic species (Noblet et al. 1990). Results obtained in our group are summarized in Table 3.18. Even though a minimal level of physical activity is inevitable and is included in the estimate of ME$_m$, specific energy requirements for physical activity should be considered, for example, in pregnant sows with stereotypic activities or in pigs kept outdoors. The most critical stage of pig production where physical activity is high and variable is the pregnancy period. Our studies indicate that the HP is increased by about 0.30 kJ kg$^{-1}$ BW$^{0.75}$ per additional minute standing (Noblet et al. 1990; Ramonet et al. 2000; Le Goff et al. 2002; Young et al. 2004). For instance, in the study presented in Table 3.17, the duration of standing ranged from 50 to 500 minutes per day among animals. This difference corresponds to a difference in feed requirement of approximately 700 g per day. In general, activity represents a high (20% of ME intake) and variable (10–40% of ME intake) proportion of the energy expenditure in pregnant sows. This variability is the major source of variability in body condition of pregnant sows at farrowing. In growing pigs that are usually offered feed close to ad libitum intake, the activity of HP is less variable and represents a lower fraction of ME intake (8–10%; Table 3.18).

**Table 3.18**  Heat production related to physical activity in swine.

| Stage: | | Piglet | Growing pig | | | | Pregnant sow |
|---|---|---|---|---|---|---|---|
| Item | Housing:<br>Feeding: | Group<br>Ad libitum | Group<br>Ad libitum | Group<br>Ad libitum | Single<br>Controlled | | Single<br>Restricted |
| Ambient temperature, °C | | 23 | 19–22 | 12 | 24 | | 24 |
| Body weight, kg | | 27 | 62 | 61 | 62 | | 260 |
| ME intake, MJ d⁻¹ | | 21.7 | 31.1 | 33.5 | 29.2 | | 35.6 |
| Heat production, MJ d⁻¹ | | 11.2 | 17.9 | 19.7 | 16.9 | | 29.5 |
| Activity heat production | | | | | | | |
|   MJ/d | | 2.0 | 2.3 | 3.3 | 2.5 | | 6.7 |
|   % heat production | | 17.9 | 12.8 | 16.9 | 14.7 | | 22.6 |
|   % ME intake | | 9.2 | 7.4 | 10.0 | 8.5 | | 18.7 |
| Source[a] | | 1 | 2 | 2 | 3 | | 4 |

[a] 1 = Collin et al. (2001), 2 = Quiniou et al. (2001), 3 = Le Bellego et al. (2001), and 4 = Ramonet et al. (2000).

## *Energy Requirements for Thermoregulation*

When pigs are kept below their LCT, HP is increased to maintain body temperature. The concepts of thermoregulation and values of LCT in swine production have been reviewed by Mount and Monteith (1979) and Noblet et al. (2001a). However, the increased HP at temperatures below LCT is usually compensated for by higher feed intake, so that BW gain is usually maintained at low temperatures when pigs are offered feed ad libitum (Quiniou et al. 2001; see section "Regulation of Energy Intake in Pigs"). The LCT is particularly high in newborn pigs (32–34 °C) and during the first days after weaning (26–28 °C) with a relative high susceptibility to cold stress during these periods (Noblet and Le Dividich 1981, 1982). During other periods of swine production, LCT is lower (20–24 °C) and requirements for thermoregulation depend on housing conditions (e.g., indoor vs. outdoor, floor type, and group size) and feeding management. Additional feed requirements to maintain performance are illustrated in Figure 3.5 for growing-finishing pigs. Pregnant sows are frequently exposed to temperatures below their LCT because of their relatively high LCT (>22 °C)

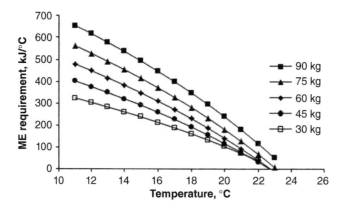

**Figure 3.5**  Energy requirement for thermoregulation in growing pigs according to their body weight (30–90 kg). (Source: Adapted from Quiniou et al. 2001.)

and simplified housing conditions. The HP can be increased by $10–20\,kJ\,kg^{-1}\,BW^{0.75}$ per °C decrease of ambient temperature below LCT (Geuyen et al. 1984). The higher rate is observed in individually housed sows and/or in poorly insulated thin sows (Noblet et al. 1997). In a 200-kg sow, the increase of HP due to cold stress should be compensated for by approximately 70 g of feed per 1 °C. The LCT of lactating sows is low (<15 °C) due to the high levels of feed intake and production, which are associated with a high rate of HP. In addition, heating is provided to the suckled piglets to improve comfort and survivability, so that cold is rarely a problem for lactating sows. More generally, the problem of cold stress has diminished in many countries due to improved insulation and quality of the buildings. On the other hand, heat stress has become increasingly important in tropical or subtropical areas of the world, or during summer periods in temperate countries (Mayorga et al. 2019). The heat increment is not necessarily a loss of energy and can contribute to meeting the energy requirement for thermoregulation during the cold periods (Table 3.10; Quiniou et al. 2001). Therefore, from a practical point of view, high heat-increment diets (i.e., high fiber diets) are, therefore, energetically more efficient under cold condition than under thermoneutrality of hot conditions (Noblet et al. 1985a, 1989, 2001a).

**Response to Energy Intake**

Growth of pigs depends on factors related to the animal itself (e.g., BW, sex, and genotype), the supply of nutrients, and the climatic environment. Their response is also characterized by the partitioning of energy gain between protein and fat during the growing phase (Campbell and Taverner 1988; Quiniou et al. 1999a; van Milgen et al. 2008) or during pregnancy (Dourmad et al. 1996, 2008). The response during lactation is rather specific with a priority given to milk production at the expense of body reserves, and lactating sows are able to maintain milk production under conditions of a negative energy balance as long as their body reserves are not depleted too much (Noblet et al. 1998).

In the classical factorial view on energy utilization, energy will first be used to meet the maintenance energy requirements, second for protein gain, and last for fat gain. However, this view of a succession of priorities is not necessarily appropriate because there is a relation between protein gain and lipid gain. When energy intake is restricted, both protein gain and lipid gain can be affected simultaneously and there seems to be an "obligatory" lipid gain. Most data demonstrate that the response of protein gain to energy intake can be described by a linear-plateau or curvilinear-plateau relationship. The response of lipid gain to energy intake is practically close to linear and the theoretical increase in lipid gain when the maximum protein deposition is reached is difficult to detect (Figure 3.6). As previously mentioned, the deposition rates of protein and lipid are associated with deposition rates of lean and adipose tissues and BW gain (Table 3.19). The important aspects of these relationships are (i) that the increase of gain (g per MJ ME) is much more pronounced for lipid or adipose tissue gain than for protein or lean tissue gains, so that adiposity of the body will increase with energy level (Table 3.20); (ii) the slope for protein gain in the linear response phase is higher in leaner pigs or in boars (versus barrows) or in younger pigs (versus older pigs); and (iii) with increasing energy intake, protein deposition will reach a plateau (PDmax) and energy given in excess of that required to reach the plateau will be used for fat gain. These aspects have been reviewed by Black et al. (1986), Quiniou et al. (1999a), van Milgen and Noblet (2003), van Milgen et al. (2008) and are partly illustrated in Table 3.19. Similar results are given in Table 3.19 for pregnant sows, for which the protein response is lower than in finishing pigs. The protein response (slope and PDmax) to energy intake is also affected by the ambient temperature. At high ambient

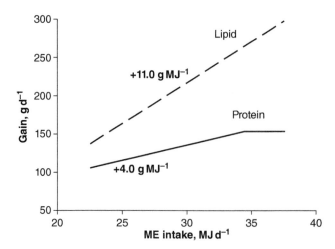

**Figure 3.6** Response of protein and lipid gains to ME intake in 45- to 100-kg growing barrows (combined results of data obtained with two types of barrows). (Source: Adapted from Quiniou et al. 1996a.)

**Table 3.19** Response of swine to energy intake.

| Item | Growing pig | | Pregnant Sow |
|---|---|---|---|
| | Male | Barrow | |
| Body weight range, kg | 45–100 | 45–100 | 205[a] |
| Variation, g MJ⁻¹ of additional ME | | | |
| Protein | 6.1 | 4.7 | 2.3 |
| Fat | 13.2 | 13.2 | NA[b] |
| Lean | 21.0 | 16.5 | 10.0 |
| Adipose | 9.7 | 9.7 | 12.0 |
| Body weight gain | 36.0 | 28.5 | 24.0 |
| Source[c] | 1 | 1 | 2 |

[a] Body weight at mating.
[b] NA = not available.
[c] 1 = Quiniou et al. (1996a,b, Large White × Piétrain crossbreds), and 2 = Dourmad et al. (1996).

**Table 3.20** Effect of energy supply on growth performance and body composition of growing pigs (45–100 kg BW)[a,b].

| Item | Energy supply, MJ ME/d | | | | |
|---|---|---|---|---|---|
| | 22.6 | 26.7 | 29.4 | 32.2 | 37.6 |
| BW gain, g d⁻¹ | 622 | 738 | 820 | 931 | 1013 |
| Feed cost, MJ ME/kg BW gain | 36.4 | 36.2 | 35.8 | 34.6 | 37.1 |
| Body protein content,[c] % | 17.2 | 16.6 | 16.5 | 16.3 | 16.0 |
| Body lean content[c] | 56.1 | 54.2 | 53.6 | 53.7 | 52.6 |
| Body lipid content[c] | 18.6 | 21.0 | 22.0 | 22.4 | 22.8 |
| Adipose tissues content[c] | 12.0 | 14.2 | 15.1 | 15.2 | 15.7 |

[a] Adapted from Quiniou et al. (1996a, b).
[b] BW = body weight.
[c] As a percentage of the empty BW (i.e., BW – gut fill) at slaughter, and empty BW is equivalent to 95% of live BW.

temperatures, the slope for protein gain is reduced with a commensurate slight increase in the slope for fat gain (Le Bellego et al. 2002b). Therefore, despite the lower feed intake at high ambient temperatures, which should favor a leaner carcass, the actual adiposity of the carcass is similar in ad libitum fed pigs raised at thermoneutrality or in hot conditions.

## Feed Efficiency in Growing Pigs

From a technical point of view, an important criterion for evaluating the efficiency of pig production is the feed efficiency calculated as the quantity of feed or energy per kg of BW gain (or F:G). This F:G criterion can also be presented as

$$F:G = \left( ME\,intake\,/\left[ME\right]\right) / \left( Energy\,gain\,/\left[E\right]_{ADG}\right) \; or$$
$$F:G = \left( ME\,intake\,/\left[ME\right]\right) / \left( \left(ME\,intake - MEm\right) \times k_g \,/\left[E\right]_{ADG}\right) \; or \qquad (Eq.\ 3.21)$$
$$F:G = \left(1\,/\left[ME\right]\right) \times \left(1\,/\,k_g\right) \times \left[E\right]_{ADG} \times \left(FL\,/\left[FL - 1\right]\right)$$

where [ME] is the ME concentration of the feed, $k_g$ is the efficiency of ME for energy gain (see Table 3.8), $[E]_{ADG}$ is the energy concentration of BW gain and FL is feeding level as a multiple of MEm. This formula indicates that F:G is reduced when [ME] of the feed is increased and also when FL is increased with a lower relative contribution of energy intake used for maintenance. However, as indicated above, $[E]_{ADG}$ is increased at higher energy intakes with a subsequent increase of F:G. Therefore, these two effects of FL on $[E]_{ADG}$ and on FL/[FL − 1] are opposite and F:G remains relatively constant over a rather large range of FL (Table 3.20). However, at high feed intake, especially those above the ME intake required to attain PDmax (Figure 3.6), $[E]_{ADG}$ increases rapidly and the effect of FL/[FL − 1] becomes smaller, so that F:G increases (Table 3.20). On the other hand, at very low feed intake, the effect of FL/[FL − 1] is important with a subsequent increased F:G value. Practically, this means that the ME intake required to minimize the F:G ratio is usually below ad libitum feed intake, especially in pigs with a lower potential for protein gain and/or a high appetite. As such, a slight energy restriction, especially during the finishing phase, may be recommended. This also means that, at a given FL value, the F:G is as low as the $[E]_{ADG}$ is low in connection with a reduced fat content in BW gain. In conclusion, the best solution for improving F:G is to reduce the adiposity of the carcass by minimizing the fat-to-protein ratio in BW gain.

Depending on the ME content of the feed and the BW range, the F:G in 25–100 kg pigs ranges between 2.5 and 3.0 in most practical situations with lower values for boars than for barrows and intermediate values for females (Table 3.21). This also means that the BW gain achieved by 1 kg of feed ranges between 350 and 400 g, which is equivalent to 25–30 g BW gain per MJ ME of feed

**Table 3.21** Comparative growth performance of boars, barrows, and gilts[a,b]

| Item | Boars | Gilts | Barrows |
| --- | --- | --- | --- |
| Feed intake, kg d$^{-1}$ | 2.41 | 2.45 | 2.70 |
| BW gain, g d$^{-1}$ | 1069 | 988 | 1032 |
| Feed cost, kg/kg BW gain | 2.26 | 2.48 | 2.62 |

[a] Adapted from Quiniou et al. (2010); 63–152 days of age.
[b] BW = body weight.

**Table 3.22**  Effect of castration on the efficiency of pig growth (40–100 kg BW)[a,b].

| Item | Boars | Barrows[c] | |
|---|---|---|---|
| ME intake, MJ d$^{-1}$ | 35.1 | 37.1 | (106) |
| BW gain, g d$^{-1}$ | 1096 | 1014 | (92) |
| Protein gain, g d$^{-1}$ | 150 | 144 | (96) |
| Lipid gain, g d$^{-1}$ | 232 | 255 | (110) |
| BW gain, g MJ$^{-1}$ ME | 31.2 | 27.3 | (87) |
| Lean gain, g MJ$^{-1}$ ME | 16.9 | 15.0 | (89) |
| Body energy gain, MJ MJ$^{-1}$ ME | 0.39 | 0.40 | (101) |
| Lean energy gain, MJ MJ$^{-1}$ ME | 0.14 | 0.13 | (91) |

[a] Adapted from Quiniou et al. (1995) and J. Noblet (unpublished data).
[b] ME = metabolizable energy; BW = body weight.
[c] Parentheses = % of boars.

(Table 3.22). However, pigs are raised mainly for producing lean meat, and it is important to maximize the quantity of lean gain or the energy gain in lean tissues per unit of feed. Indicative values are given in Table 3.22 for boars and barrows (15–17 g lean gain per MJ ME). Calculations given in Table 3.22 also indicate that about 40 and 14% of ME intake are retained in the BW gain or in the lean-tissue gain in growing pigs, respectively.

In most parts of the world, male pigs are (surgically) castrated to avoid boar taint problems in meat products. However, boars are raised in some areas (e.g., Australia and UK), and there is a tendency for not castrating the males for both welfare and economical reasons. Boars can also be raised up to a few weeks before slaughter with a late immunocastration (Dunshea et al. 2001; Dunshea et al. 2013). The comparison of gilts, barrows, and boars, in terms of energy utilization, indicates that feed intake depends on gender (boar = gilt < barrow) with the lowest feed cost in boars (Table 3.21). In fact, castration reduces the potential of the pig for protein gain with a subsequent higher fat gain that is accentuated by the relative hyperphagy of barrows. The BW gain, lean gain, or energy gain in lean tissues per unit of feed energy intake is higher for boars. However, the body energy gain per unit of energy intake is better in barrows due a greater fraction of energy gain that is retained as fat (Table 3.22). This example of the castration of males indicates that an improvement in feed efficiency (boars versus barrows) does not necessarily correspond to an improvement in overall energy efficiency. The same conclusion will hold for the impact of genetic improvement for leaner carcasses and/or faster growth.

### Regulation of Energy Intake in Pigs

Under ad libitum conditions, it is important to evaluate the ability of the pig to consume enough feed or energy to meet the requirements or the objectives in terms of rate of growth, protein gain, and fat gain. It is not the purpose of this chapter to consider all aspects of feed intake regulation in swine. In this section, we only want to describe briefly some general aspects of energy intake in pigs in connection with major animal factors such as BW, physiological stage, or gender and major environmental factors such as feed energy concentration (Quiniou and Noblet 2012) and ambient temperature (Renaudeau et al. 2011).

Feeding patterns have been described in piglets, growing pigs, and lactating sows kept under conventional environmental conditions (Table 3.23). In brief, these studies indicate that the number

**Table 3.23**  Feeding behavior in swine.

| Item | Stage:<br>Breed:<br>Housing: | Piglet<br>Crossbred<br>Group | Growing pig<br>Meishan<br>Single | Growing pig<br>Piétrain<br>Single | Growing pig<br>Crossbred<br>Group | Growing pig<br>Crossbred<br>Group | Lactating sow<br>Crossbred<br>Single |
|---|---|---|---|---|---|---|---|
| Body weight range, kg | | 20–30 | 20–60 | 20–60 | 30–90 | 30–90 | 270 |
| Temperature, °C | | 23 | 24 | 24 | 19–22 | 29 | 22 |
| Feed intake, g d⁻¹ | | 1502 | 1659 | 1622 | 2395 | 1820 | 6600 |
| Number of meals/d | | 14.4 | 14.4 | 7.3 | 11.2 | 10.1 | 7.4 |
| Meal size, g | | 114 | 125 | 250 | 248 | 205 | 972 |
| Diurnal feed intake, % | | 67 | 61 | 64 | 65 | 62 | 80 |
| Source[a] | | 1 | 2 | 2 | 3 | 3 | 4 |

[a] 1 = Collin et al. (2001), 2 = Quiniou et al. (1999b), 3 = Quiniou et al. (2000a), and 4 = Quiniou et al. (2000b).

of meals per day decreases when BW increases and confirm that pigs, at any stage of production, are predominantly diurnal (with less than one-third during the night). This diurnal behavior is even more pronounced in heavier pigs or in lactating sows with two main consumption peaks, one in the morning and one in the late afternoon (Figure 3.7). However, this diurnal behavior consumption can interact with climatic environment such that under hot temperatures during the day and cooler temperatures at night, there may be an increased proportion of the feed during the nocturnal period. That is particularly noticeable in lactating sows (Quiniou et al. 2000a; Renaudeau et al. 2003). Pregnant sows are usually fed restrictively and consume their feed immediately after the distribution, unless it is a high-fiber feed in large volume, distributed once a day.

In growing pigs, the voluntary feed intake increases curvilinearly with BW (NRC 2012; Figure 3.8), but the rate of increase is affected by growth potential (e.g., genotype, sex) of the pig (Quiniou et al. 1999b). The rate of increase is also highly dependent on ambient temperature with

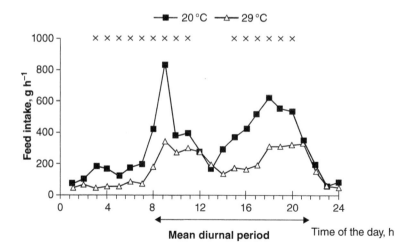

**Figure 3.7**  Effect of temperature on kinetics of daily feed intake in large white lactating sows. (Source: Adapted from Renaudeau et al. 2002.)

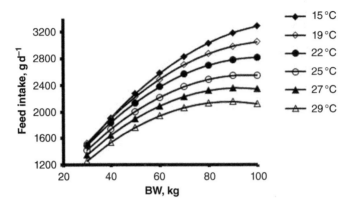

**Figure 3.8** Effects of body weight (BW) and ambient temperature on voluntary feed intake in growing pigs. (Source: Adapted from Quiniou et al. 2000a.)

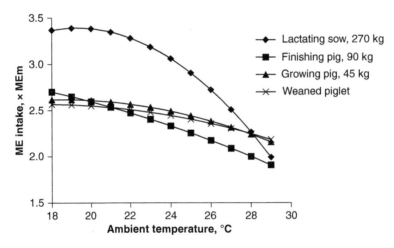

**Figure 3.9** Effect of ambient temperature on feed intake in piglets (Collin et al. 2001), growing pigs (Quiniou et al. 2000a), and lactating sows (Quiniou and Noblet 1999). Feed intake is expressed as a multiple of the ME requirement for maintenance.

smaller rates of increase at high ambient temperatures, which means that heavier pigs are more sensitive to heat stress than lighter pigs (Nienaber et al. 1996; Renaudeau et al. 2011; Figure 3.8). In lactating sows, voluntary feed intake is dependent on body size or parity number with lower intakes in primiparous sows (O'Grady et al. 1985; Dourmad, 1991; Dourmad et al. 1994; Neil et al. 1996). As growing pigs, lactating sows are particularly susceptible to heat stress by reducing markedly their voluntary feed intake at high ambient temperatures (Schoenherr et al. 1989). They are even more affected by ambient temperature changes according to their particularly high voluntary feed intake at thermoneutrality (Figure 3.9). In addition, the reduction in voluntary feed intake per °C change is as high as ambient temperature is high with a reduction averaging 200 g/°C between 20 and 25 °C and up to 500 g/°C between 25 and 30 °C in lactating sows (Quiniou and Noblet 1999); corresponding values would be 10 and 30 g/°C in 25 kg piglets (Collin et al. 2001) and 40 and 70 g/°C in 60-kg growing pigs (Quiniou et al. 2000). These negative effects of high temperatures on voluntary feed intake in sows can even be worse under high relative humidities in tropical areas (Renaudeau et al. 2003). More generally, the exposure to high ambient temperatures is associated with a reduced ability of the growing pig or the lactating sow to dissipate their HP and a potential risk of hyperthermy. As a result, the best adaptation consists of reducing energy intake and the inevitable heat associated with the ingestion and the metabolic utilization of feed energy. The magnitude of these effects is the most pronounced for lactating sows.

Diet energy density can be modified by including either fiber-rich ingredients that reduce the energy concentration or fat-rich ingredients that increase the energy concentration (Table 3.12). Because one limiting factor of pig growth in practical conditions is energy intake, a lot of attention has been focused on the relationship between feed intake, growth, feed efficiency, body composition, and energy concentration of the feed. For reviewing this aspect, some literature data have been compiled; in each study, at least four energy densities were compared and protein-energy ratios were as constant as possible. In most studies, pigs were kept individually and(or) under favorable climatic conditions. An increase in energy concentration is usually associated with a reduction in voluntary feed intake, but the reduction is less important than the energy density, so that energy intake is almost systematically increased (Figure 3.10). However, in most studies, there is a plateau DE intake at the highest energy densities or the increase becomes negligible when the lowest energy concentration in the study is quite high. In agreement with this effect of energy concentration on energy intake, BW gain is increased at higher energy concentrations with a plateau for BW gain at the highest energy densities. Such an effect of diet energy density on energy intake has also been demonstrated in lactating sows but with a limited interest for the sow because most additional ingested energy as fat is exported in the milk (Noblet et al. 1998). Finally, this beneficial effect of high dietary energy densities on energy intake can be utilized in growing pigs or lactating sows exposed to heat stress. High-energy diets can attenuate the effect of high ambient temperature on pig performance (Le Bellego et al. 2002a; Renaudeau et al. 2002).

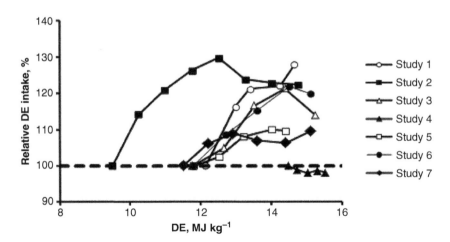

**Figure 3.10** Effect of diet energy concentration on voluntary energy intake in growing pigs (as a percentage of the DE intake at the lowest DE content in each study). (Source: Adapted from: Campbell and Taverner 1986; Chadd 1990; Stein et al. 1996; Smith et al. 1999; Quiniou and Noblet 2012.)

## Conclusion

Energy systems are based on the concept that an energy value can be attributed to a feed so that it can be compared with a requirement that is expressed on the same scale. In this chapter, we confirm that the situation is far more complex with different energy values attributed to the same feed according to the type of pig receiving the diet and the implementation of a feed technology. Different energy systems have been proposed and based on the steps of energy utilization. The NE system is probably as far as we can go nowadays, while maintaining the concepts of "value" and "require-ment" in feed formulation. In reality, there are interactions between energy supply, the environment, and the animal. The only way to deal with the complexity of these interactions is through modeling (Whittemore and Fawcett 1976; Black et al. 1986; Birkett and de Lange 2001; van Milgen et al. 2008). Although considerable progress has been made in this area, nutritional models of swine nutrition vary widely in scope. To be implemented on a large scale in the field, nutritional models should provide a compromise between "scientific truth" (or scientific perception) and robustness of the system. In that respect, it is probably too early to bury the classical concepts of energy nutrition outlined in this chapter.

## References

Agyekum, A. K., A. Regassa, E. Kiarie, and C. M. Nyachoti. 2016. Nutrient digestibility, digesta volatile fatty acids, and intestinal bacterial profile in growing pigs fed a distillers dried grains with solubles containing diet supplemented with a multi-enzyme cocktail. Anim. Feed Sci. Technol. 212:70–80.

Barea, R., S. Dubois, H. Gilbert, P. Sellier, J. van Milgen, and J. Noblet. 2010. Energy utilization in pigs selected for high and low residual feed intake. J. Anim. Sci. 88:2062–2072.

Bikker, P., M. W. A. Verstegen, and R. G. Campbell, 1996. Performance and body composition of finishing gilts (45–85 kg) as affected by energy intake and nutrition in earlier life. II. Protein and lipid accretion in body components. J. Anim. Sci. 74:817–826.

Birkett, S., and K. de Lange, 2001. Limitations of conventional models and a conceptual framework for a nutrient flow representative of energy utilization by animals. Br. J. Nutr. 86:647–659.

Black, J. L., R. G. Campbell, I. H. Williams, K. J. James, and G. T. Davies, 1986. Simulation of energy and amino acid utilization in the pig. Res. Dev. Agric. 3:121–145.

Boisen, S., and J. A. Fernandez. 1997. Prediction of the total tract digestibility of energy in feedstuffs and pig diets by in vitro analyses. Anim. Feed Sci. Technol. 68:277–286.

Boisen, S, and M. W. A. Verstegen. 1998. Evaluation of feedstuffs and pig diets. Energy or nutrient-based evaluation systems? II. Proposal for a new nutrient-based evaluation system. Acta Agric. Scandinavica, Sec. A Anim. Sci. 40:86–94.

Brouwer, E. 1965. Report of sub-committee on constants and factors. In: K. L. Blaxter, editor, Proc. 3rd symp. Energy metabolism. Academic Press, London, UK. p. 441.

Campbell, R. G., and M. R. Taverner. 1986. The effects of dietary fiber, source of fat and dietary energy concentration on the voluntary food intake and performance of growing pigs. Anim. Prod. 43:327–333.

Campbell, R. G., and M. R. Taverner. 1988. Genotype and sex effects on the relationship between energy intake and protein deposition in growing pigs. J. Anim. Sci. 66:676–686.

Campos, P. H. R. F., E. Labussière, J. Hernandez-Garcia, S. Dubois, D. Renaudeau, and J. Noblet, 2014. Effects of ambient temperature on energy and nitrogen utilization in lipopolysaccharide-challenged growing pigs. J. Anim. Sci. 92: 4909–4920.

Casas, G. A., C. Huang, and H. H. Stein. 2017. Nutritional value of soy protein concentrate ground to different particle sizes and fed to pigs. J. Anim. Sci. 95:827–836.

Chabeauti, E., J. Noblet, and B. Carré, 1991. Digestion of plant cell walls from four different sources in growing pigs. Anim. Feed Sci. Technol. 32:207–213.

Chadd, S. A. 1990. Voluntary feed intake of hybrid pigs. PhD Diss. University of Nottingham, Nottingham, UK.

Collin, A., J. van Milgen, and J. Le Dividich, 2001. Modelling the effect of high, constant temperature on feed intake in young growing pigs. Anim. Sci. 72:519–527.

Cozannet, P., Y. Primot, C. Gady, J. P. Métayer, M. Lessire, F. Skiba, and J. Noblet. 2010. Energy value of wheat distillers grains with solubles for growing pigs and adult sows. J. Anim. Sci. 88:2382–2392.

Cozannet, P., A. Preynat, and J. Noblet. 2012. Digestible energy values of feed ingredients with or without addition of enzymes complex in growing pigs. J. Anim. Sci. 90:209–211.

CVB. 2018. CVB Feed table 2018: Chemical composition and nutritional values of feedstuffs (http://www.cvbdiervoeding.nl/).

de Lange, K., J. van Milgen, J. Noblet, S. Dubois, and S. Birkett, 2006. Previous feeding level influences plateau heat production following a 24 h fast in growing pigs. Br. J. Nutr. 95:1082–1087.

Dourmad, J. Y. 1991. Effect of feeding level in the gilt during pregnancy on voluntary feed intake during lactation and changes in body composition during gestation and lactation. Livest. Prod. Sci. 27:309–319.

Dourmad, J. Y., M. Etienne, A. Prunier, and J. Noblet. 1994. The effect of energy and protein intake of sows on their longevity. Livest. Prod. Sci. 40:87–97.

Dourmad, J. Y., M. Etienne, and J. Noblet, 1996. Reconstitution of body reserves in multiparous sows during pregnancy: Effect of energy intake during pregnancy and mobilization during the previous lactation. J. Anim. Sci. 74:2211–2219.

Dourmad, J. Y., J. Noblet, M. C. Père, and M. Etienne, 1998. Mating, pregnancy and pre-natal growth. In: I. Kyriazakis, editor, A quantitative biology of the pig. CAB International, Wallingford, UK. p. 129–153.

Dourmad, J. Y., M. Étienne, A. Valancogne, S. Dubois, J. van Milgen, and J. Noblet. 2008. INRAPorc: A model and decision support tool for the nutrition of sows. Anim. Feed Sci. Technol. 143:372–386.

Dunshea, F. R., C. Colantoni, K. Howard, I. McCauley, P. Jackson, K. A. Long, S. Lopaticki, E. A. Nugent, J. A. Simons, J. Walker, and D. P. Hennessy. 2001. Vaccination of boars with a GnRH vaccine (Improvac) eliminates boar taint and increases growth performance. J. Anim. Sci. 79:2524–2535.

Dunshea, F. R., J. R. D. Allison, M. Bertram, D. D. Boler, L. Brossard, R. Campbell, J. P. Crane, D. P. Hennessy, L. Huber, C. de Lange, N. Ferguson, P. Matzat, F. McKeith, P. J. U. Moraes, B. Mullan, J. Noblet, N. Quiniou, and M. Tokach. 2013. The effect of immunization against GnRF on nutrient requirements of male pigs: A review. Animal 7:1769–1778.

Emmans, G. 1994. Effective energy: A concept of energy utilization applied across species. Br. J. Nutr. 71:801–821.

Etienne, M., J. Noblet, J. Y. Dourmad, and J. Castaing. 1997. Digestive utilization of feeds in lactating sows: comparison with growing pigs. In: J. P. Laplace, C. Février, and A. Barbeau, editors, Digestive physiology in pigs. INRA, Paris, France. p. 583–586.

Etienne, M., J. Y. Dourmad, and J. Noblet. 1998. The influence of some sow and piglet characteristics and of environmental conditions on milk production. In: M. W. A. Verstegen, P. J. Moughan, and J. W. Schrama, editor, The lactating sow. Wageningen Pers, Wageningen, The Netherlands. p. 285–299.

EvaPig. 2008. Pig feed evaluation made easy. INRAE, METEX NØØVISTAGO, and AFZ. https://en.evapig.com.

Fernandez, J. A., H. Jorgensen, and A. Just. 1986. Comparative digestibility experiments with growing pigs and adult sows. Anim. Prod. 43:127–132.

Geuyen, T. P. A., J. M. F. Verhagen, and M. W. A. Verstegen. 1984. Effect of housing and temperature on metabolic rate of pregnant sows. Anim. Prod. 38:477–485.

Hansen, A. V., A. B. Strathe, E. Kebreab, J. France, and P. K. Theil. 2012. Predicting milk yield and composition in lactating sows: A Bayesian approach. J. Anim. Sci. 90:2285–2298.

INRAPorc. 2009. Software. INRA, Saint-Gilles, France. https://inraporc.inra.fr/inraporc/.

Jorgensen, H., K. E. Bach Knudsen, and P. K. Theil. 2001. Effect of dietary fibre on energy metabolism of growing pigs and pregnant sows. In: A. Chwalibog, and K. Jakobsen, editos. Energy metabolism in animals. EAAP Publ. No. 103. Wageningen Pers, Wageningen, The Netherlands. p. 105–108.

Just, A., H. Jorgensen, and J. A. Fernandez. 1983. Maintenance requirement and the net energy value of different diets for growth in pigs. Livest. Prod. Sci. 10:487–506.

Karege, C. 1991. Influence de l'âge et du sexe sur l'utilisation de l'énergie et la composition corporelle du porc en croissance. PhD. Diss. Université de Montpellier, France.

Koong, L. J., J. A. Nienaber, J. C. Pekas, and J. T. Yen, 1982. Effects of plane of nutrition on organ size and fasting heat production in pigs. J. Nutr. 112:1638–1642.

Labussière, E., J. van Milgen, C. F. M. de Lange, and J. Noblet. 2011. Maintenance energy requirements of growing pigs and calves are influenced by feeding level. J. Nutr. 141:1855–1861.

Labussière, E., S. Dubois, J. van Milgen, and J. Noblet. 2013. Partitioning of heat production in growing pigs as a tool to improve the determination of efficiency of energy utilization. Front. Physiol. 4:146.

Le Bellego, L., J. van Milgen, and J. Noblet. 2001. Energy utilization of low protein diets in growing pigs. J. Anim. Sci. 79:1259–1271.

Le Bellego, L., J. van Milgen, and J. Noblet. 2002a. Effect of high temperature and low protein diets on performance of growing-finishing pigs. J. Anim. Sci. 80:691–701.

Le Bellego, L., J. van Milgen, and J. Noblet. 2002b. Effects of high temperature on protein and lipid deposition and energy utilization in growing pigs. Anim. Sci. 75:85–96.

Le Gall, M., M. Warpechowski, Y. Jaguelin-Peyraud, and J. Noblet. 2009. Influence of dietary fibre level and pelleting on digestibility of energy and nutrients in growing pigs and adult sows. Animal 3:352–359.

Le Goff, G., and J. Noblet. 2001. Comparative digestibility of dietary energy and nutrients in growing pigs and adult sows. J. Anim. Sci. 79:2418–2427.

Le Goff, G., L. Le Groumellec, J. van Milgen, and J. Noblet. 2002. Digestive and metabolic utilization of dietary energy in adult sows: influence of level and origin of dietary fibre. Br. J. Nutr. 87:325–335.

Le Sciellour, M., E. Labussière, O. Zemb, and D. Renaudeau. 2018. Effect of dietary fiber content on nutrient digestibility and fecal microbiota composition in growing-finishing pigs. PLoS ONE 13:e0206159.

Li, Z. C., H. Liu, Y. K. Li, Z. Q. Lv, L. Liu, C. H. Lai, J. J. Wang, F. L. Wang, D. F. Li, and S. Zhang. 2018. Methodologies on estimating the energy requirements for maintenance and determining the net energy contents of feed ingredients in swine: a review of recent work. J. Anim. Sci. Biotech. 9:39.

Lovatto, P., D. Sauvant, J. Noblet, S. Dubois, and J. van Milgen. 2006. Effects of feed restriction and subsequent re-feeding on energy utilization in growing pigs. J. Anim. Sci. 84:3329–3336.

Lyu, Z. Q., C. F. Huang, Y. K. Li, P. L. Li, H. Liu, Y. F. Chen, D. F. Li, and C. H. Lai. 2018. Adaptation duration for net energy determination of high fiber diets in growing pigs. Anim. Feed Sci. Technol. 241:15–26.

Mayorga, E. J., D. Renaudeau, B. C. Ramirez, J. W. Ross, and L. H. Baumgard. 2019. Heat stress adaptations in pigs. Anima. Front. 9:54–61.

Mount, L. E., and J. L. Monteith. 1979. The concept of thermoneutrality. In: J. L. Monteith, and L. E. Mount, editors, Heat loss from animals and man. Butterworths, London, UK. p. 425–439.

Navarro, D. M. D. L., E. M. A. M. Bruininx, L. de Jong, and H. H. Stein. 2019. Effects of inclusion rate of high fiber dietary ingredients on apparent ileal, hindgut, and total tract digestibility of dry matter and nutrients in ingredients fed to growing pigs. Anim. Feed Sci. Technol. 248:1–9.

Neil, M., B. Ogle, and K. Anner. 1996. A two diet system and ad libitum lactation feeding of the sow. 1. Anim. Sci. 62:337–347.

Nienaber, J. A., G. L. Hahn, T. P. McDonald, and R. L. Korthals. 1996. Feeding patterns and swine performance in hot environments. Trans. ASAE 39:195–202.

Noblet, J., and J. Le Dividich. 1981. Energy metabolism of the newborn pig during the first 24 hrs of life. Biol. Neonate 40:175–182.

Noblet, J., and J. Le Dividich. 1982. Effect of environmental temperature and feeding level on energy balance traits in early weaned piglets. Livest. Prod. Sci. 9:619–632.

Noblet, J., J. Le Dividich, and T. Bikawa, 1985a. Interaction between energy level in the diet and environmental temperature on the utilization of energy in growing pigs. J. Anim. Sci. 61:452–459.

Noblet, J., W. H. Close, R. P. Heavens, and D. Brown. 1985b. Studies on the energy metabolism of the pregnant sow. Uterus and mammary tissue development. Br. J. Nutr. 53:251–265.

Noblet, J., and M. Etienne. 1987a. Metabolic utilization of energy and maintenance requirements in lactating sows. J. Anim. Sci. 64:774–781.

Noblet, J., and M. Etienne. 1987b. Metabolic utilization of energy and maintenance requirements in pregnant sows. Livest. Prod. Sci. 16:243–257.

Noblet, J., Y. Henry, and S. Dubois. 1987. Effect of protein and lysine levels in the diet on body gain composition and energy utilization in growing pigs. J. Anim. Sci. 65:717–726.

Noblet, J., and M. Etienne. 1989. Estimation of sow milk nutrient output. J. Anim. Sci. 67:3352–3359.

Noblet, J., J. Y. Dourmad, J. Le Dividich, and S. Dubois. 1989. Effect of ambient temperature and addition of straw or alfafa in the diet on energy metabolism in pregnant sows. Livest. Prod. Sci. 21:309–324.

Noblet, J., J. Y. Dourmad, and M. Etienne. 1990. Energy utilization in pregnant and lactating sows: modelling of energy requirements. J. Anim. Sci. 68:562–572.

Noblet, J., P. Herpin, and S. Dubois. 1992. Effect of recombinant porcine somatotropin on energy and protein utilization in growing pigs: interaction with capacity for lean tissue growth. J. Anim. Sci. 70:2471–2484.

Noblet, J., and J. M. Perez. 1993. Prediction of digestibility of nutrients and energy values of pig diets from chemical analysis. J. Anim. Sci. 71:3389–3398.

Noblet, J., and X. S. Shi. 1993. Comparative digestibility of energy and nutrients in growing pigs fed ad libitum and adult sows fed at maintenance. Livest. Prod. Sci. 34:137–152.

Noblet, J., X. S. Shi, and S. Dubois. 1993a. Energy cost of standing activity in sows. Livest. Prod. Sci. 34:127–136.

Noblet, J., X. S. Shi, and S. Dubois. 1993b. Metabolic utilization of dietary energy and nutrients for maintenance energy requirements in pigs: Basis for a net energy system. Br. J. Nutr. 70:407–419.

Noblet, J., and X. S. Shi. 1994. Effect of body weight on digestive utilization of energy and nutrients of ingredients and diets in pigs. Livest. Prod. Sci. 37:323–338.

Noblet, J., H. Fortune, X. S. Shi, and S. Dubois. 1994a. Prediction of net energy value of feeds for growing pigs. J. Anim. Sci. 72:344–354.

Noblet, J., X. S. Shi, and S. Dubois. 1994b. Effect of body weight on net energy value of feeds for growing pigs. J. Anim. Sci. 72:648–657.

Noblet, J., C. Karege, and S. Dubois. 1994c. Prise en compte de la variabilité de la composition corporelle pour la prévision du besoin énergétique et de l'efficacité alimentaire chez le porc en croissance. Journées Rech. Porcine en France 26:267–276.

Noblet, J., and K. E. Bach-Knudsen, 1997. Comparative digestibility of wheat, maize and sugar beet pulp non-starch polysaccharides in adult sows and growing pigs. In: J. P. Laplace, C. Février, and A. Barbeau, editors, Digestive physiology in pigs. INRA, Paris, France. p. 571–574.

Noblet, J., J. Y. Dourmad, M. Etienne, and J. Le Dividich, 1997. Energy metabolism in pregnant sows and newborn pigs. J. Anim. Sci. 75:2708–2714.

Noblet, J., M. Etienne, and J. Y. Dourmad, 1998. Energetic efficiency of milk production. In: M. W. A. Verstegen, P. J. Moughan, and J. W. Schrama, editors, The lactating sow. Wageningen Pers, Wageningen, The Netherlands. p. 113–130.

Noblet, J., C. Karege, S. Dubois, and J. van Milgen, 1999. Metabolic utilization of energy and maintenance requirements in growing pigs: Effect of sex and genotype. J. Anim. Sci. 77:1208–1216.

Noblet, J., and G. Le Goff, 2001. Effect of dietary fibre on the energy value of feeds for pigs. Anim. Feed Sci. Technol. 90:35–52.

Noblet, J., J. Le Dividich, and J. van Milgen, 2001a. Thermal environment and swine nutrition. In: A. J. Lewis, and L. L. Southern, editors, Swine nutrition. 2nd ed. CRC Press, Boca raton, FL. p. 519–544.

Noblet, J., L. Le Bellego, J. van Milgen, and S. Dubois, 2001b. Effects of reduced dietary protein level and fat addition on heat production and nitrogen and energy balance in growing pigs. Anim. Res. 50:227–238.

Noblet, J., and M. Champion. 2003. Effect of pelleting and body weight on digestibility of energy and fat of two corns in pigs. J. Anim. Sci. 81 (Suppl. 1):140.

Noblet, J., V. Bontems, and G. Tran. 2003. Estimation de la valeur énergétique des aliments pour le porc. INRA Prod. Anim. 16:197–210.

Noblet, J., and J. van Milgen. 2004. Energy value of pig feeds: Effect of pig body weight and energy evaluation system. J. Anim. Sci. 82(E. Suppl):E229–E238.

Noblet J., B. Sève, and C. Jondreville. 2004. Nutritional values for pigs. In: D. Sauvant, J. M. Perez, and G. Tran, editors, Tables of composition and nutritional value of feed materials: Pigs, poultry, cattle, sheep, goats, rabbits, horses, fish. Wageningen Academic Publishers, Wageningen and INRA, Versailles, The Netherlands. p. 25–35.

Noblet, J. 2006. Recent advances in energy evaluation of feeds for pigs. In: P. C. Garnsworthy, and J. Wiseman, editors, Recent advances in animal nutrition. Nottingham Univ. Press, Nottingham, UK. p. 1–26.

Noblet, J. 2007. Recent developments in energy and amino acid nutrition of pigs. In: J. A. Taylor-Pickard, and P. Spring, editors. Gaining the edge in pork and poultry production: Enhancing efficiency, quality and safety. Wageningen Academic Publishers, Wageningen, The Netherlands. p. 21–48.

Noblet, J., and Y. Jaguelin-Peyraud. 2007. Prediction of digestibility of organic matter and energy in the growing pig from an in vitro method. Anim. Feed Sci. Technol. 134:211–222.

Noblet, J., Y. Jaguelin-Peyraud, B. Quemeneur, and G. Chesneau. 2008. Valeur énergétique de la graine de lin chez le porc: Impact de la technologie de cuisson-extrusion. Journées Rech. Porcine en France 40:203–208.

Noblet, J., J. Van Milgen, and S. Dubois. 2010. Utilisation of metabolisable energy of feeds in pigs and poultry: interest of net energy systems? Proc. Aust. Poult. Sci. Symp. 21:26–35.

Noblet, J., H. Gilbert, Y. Jaguelin-Peyraud, and T. Lebrun. 2013. Evidence of genetic variability for digestive efficiency in the growing pig fed a fibrous diet. Animal 7:1259–1264.

Noblet, J., S.B. Wu, and M. Choct. 2022. Methodologies for energy evaluation of pig and poultry feeds: a review. Anim. Nutr. 8:185–203.

NRC. 2012. Nutrient requirements of swine. 11th rev. ed. Natl. Acad. Press, Washington, DC.

O'Grady, J. F., P. B. Lynch, and P. A. Kearney. 1985. Voluntary feed intake by lactating sows. Livest. Prod. Sci. 12:355–365.

Quiniou, N., J. Noblet, J. van Milgen, and J. Y. Dourmad. 1995. Effect of energy intake on performance, nutrient and tissue gain and protein and energy utilisation in growing boars. Anim. Sci. 61:133–143.

Quiniou, N., J. Y. Dourmad, and J. Noblet. 1996a. Effect of energy intake on the performance of different types of pig from 45 to 100 kg body weight. 1. Protein and lipid deposition. Anim. Sci. 63:277–288.

Quiniou, N., J. Y. Dourmad, and J. Noblet. 1996b. Effect of energy intake on the performance of different types of pig from 45 to 100 kg body weight. 2. Tissue gain. Anim. Sci. 63:289–296.

Quiniou, N., and J. Noblet. 1999. Influence of high ambient temperatures on performance of multiparous lactating sows. J. Anim. Sci. 77:2124–2134.

Quiniou, N., S. Dubois, Y. Le Cozler, J. F. Bernier, and J. Noblet. 1999a. Effect of growth potential (body weight and breed/castration combination) on the feeding behaviour of individually-kept growing pigs. Livest. Prod. Sci. 61:13–22.

Quiniou, N., J. Noblet, J. Y. Dourmad, and J. van Milgen. 1999a. Influence of energy supply on growth characteristics in pigs and consequences for growth modelling. Livest. Prod. Sci. 60:317–328.

Quiniou, N., J. Noblet, and S. Dubois. 2000a. Voluntary feed intake and feeding behaviour of group-housed growing pigs are affected by ambient temperature and body weight. Livest. Prod. Sci. 63:245–253.

Quiniou, N., D. Renaudeau, S. Dubois, and J. Noblet. 2000b. Influence of high ambient temperatures on food intake and feeding behaviour of multiparous lactating sows. Anim. Sci. 70:471–479.

Quiniou, N., J. Noblet, J. van Milgen, and S. Dubois, 2001. Modelling heat production and energy balance in group-housed growing pigs exposed to cold or hot ambient temperatures. Br. J. Nutr. 85:97–106.

Quiniou, N., V. Courboulay, Y. Salaün, and P. Chevillon. 2010. Conséquences de la non castration des porcs mâles sur les performances de croissance et le comportement: comparaison avec les mâles castrés et les femelles. Journées Rech. Porcine en France 42:113–118.

Quiniou, N., and J. Noblet. 2012. Effect of the dietary net energy content on feed intake and performance of growing-finishing pigs housed individually. J. Anim. Sci. 90:4362–4372.

Ramonet, Y., J. van Milgen, J. Y. Dourmad, S. Dubois, M. C. Meunier-Salaün, and J. Noblet. 2000. The effect of dietary fibre on energy utilisation and partitioning of heat production over pregnancy in sows. Br. J. Nutr. 84:85–94.

Renaudeau, D., N. Quiniou, S. Dubois, and J. Noblet. 2002. Effects of high ambient temperature and dietary protein level on feeding behaviour of multiparous lactating sows. Anim. Res. 51:227–243.

Renaudeau, D., J.-L. Weisbecker, and J. Noblet. 2003. Effects of season and dietary fibre on feeding behaviour of lactating sows in a tropical climate. Anim. Sci. 77:429–437.

Renaudeau, D., B. Bocage, and J. Noblet. 2006. Influence of energy intake on protein and lipid deposition in Creole and Large White growing pigs. Anim. Sci. 82:937–945.

Renaudeau, D., J. L. Gourdine, and N. R. St-Pierre. 2011. A meta-analysis of the effect of high ambient temperature on growing-finishing pigs. J. Anim. Sci. 89:2220–2230.

Renaudeau, D., G. Frances, S. Dubois, H. Gilbert, and J. Noblet. 2013. Effect of thermal heat stress on energy utilization in two lines of pigs divergently selected for residual feed intake. J. Anim. Sci. 91:1162–1175.

Rijnen, M. M. J.A., M. W. A. Verstegen, M. J. W. Heetkamp, J. Haaksma, and J. W. Schrama. 2001. Effect of dietary fermentable carbohydrates on energy metabolism in group-housed sows. J. Anim. Sci. 79:148–154.

Rijnen, M. M. J. A., M. W. A. Verstegen, M. J. W. Heetkamp, and J. W. Schrama. 2003. Effects of two different dietary fermentable carbohydrates on activity and heat production in group-housed growing pigs. J. Anim. Sci. 81:1210–1219

Rojas, O. J., and H. H. Stein. 2017. Processing of ingredients and diets and effects on nutritional value for pigs. J. Anim. Sci. Biotech. 8:48.

Rostagno, H. S. 2017. Brazilian tables for poultry and swine: composition of feedstuffs and nutritional requirements. 4th ed. Dept. of Anim. Sci., Fed. Univ. of Viçosa, Viçosa, MG, Brazil.

Sauvant, D, J. M. Perez, and G. Tran. 2004. Tables of composition and nutritional value of feed materials: Pigs, poultry, cattle, sheep, goats, rabbits, horses, fish. Wageningen Academic Publishers, Wageningen, The Netherlands.

Schiemann R, K. Nehring, L. Hoffmann, W. Jentsch, and A. Chudy. 1972. Energetische Futterbevertung und Energienormen. VEB Deutscher Landwirtschatsverlag, Berlin, Germany.

Schoenherr, W. D., T. S. Stahly, and G. L. Cromwell. 1989. The effects of dietary fat or fiber addition on yield and composition of milk from sows housed in warm or hot environment. J. Anim. Sci. 67:482–495.

Schrama, J. W., M. W. Bosch, M. W. A. Verstegen, A. H. P. M. Vorselaars, J. Haaksma, and M. J. W. Heetkamp. 1998. The energetic value of nonstarch polysaccharides in relation to physical activity in group-housed growing pigs. J. Anim. Sci. 76:3016–3023.

Skiba F., J. Noblet, P. Callu, J. Evrard, and J. P. Melcion. 2002. Influence du type de broyage et de la granulation sur la valeur énergétique de la graine de colza chez le porc en croissance. Journées Rech. Porcine en France 34:67–74.

Smith, J. W., M. D. Tokach, P. R. O'Quinn, J. L. Nelssen, and R. D. Goodband. 1999. Effects of dietary energy density and lysine:calorie ratio on growth performance and carcass characteristics of growing-finishing pigs. J. Anim. Sci. 77:3007–3015.

Sølund Olesen, C., H. Jørgensen, and V. Danielsen. 2001. Effect of dietary fibre on digestibility and energy metabolism in pregnant sows. Acta Agric. Scand. Sec. A - Anim. Sci. 51:200–207.

Stein, H. H., J. D. Hahn, and R. A. Easter. 1996. Effects of decreasing dietary energy concentration in finishing pigs on carcass composition. J. Anim. Sci. 74(Suppl. 1):65.

Tess, M. W., G. E. Dickerson, J. A. Nienaber, J. T. Yen, and C. L. Ferrell. 1984. Energy costs of protein and fat deposition in pigs fed ad libitum. J. Anim. Sci. 58:111–122.

van Milgen, J., J. Noblet, S. Dubois, and J. F. Bernier, 1997. Dynamic aspects of oxygen consumption and carbon dioxide production swine. Br. J. Nutr. 78:397–410.

van Milgen, J., and J. Noblet, 2000. Modelling energy expenditure in pigs. In: J. P. McNamara, J. France, and D.E. Beever, editors, Modelling nutrient utilization in farm animals. CAB International, Oxon, UK. p. 103–114.

van Milgen, J., J. Noblet, and S. Dubois, 2001. Energetic efficiency of starch, protein, and lipid utilization in growing pigs. J. Nutr. 131:1309–1318.

van Milgen, J. 2002. Modeling biochemical aspects of energy metabolism in mammals. J. Nutr. 132:3195–3202.

Van Milgen J., and J. Noblet. 2003. Partitioning of energy intake to heat, protein and fat in growing pigs. J. Anim. Sci. 81(E. Suppl. 2):E86–E93.

Van Milgen, J., A. Valancogne, S. Dubois, J. Y. Dourmad, B. Sève,and J. Noblet. 2008. InraPorc: A model and decision support tool for the nutrition of growing pigs. Anim. Feed Sci. Technol. 143:387–405.

Whittemore, C. T., and R. H. Fawcett, 1976. Theoretical aspects of a flexible model to simulate protein and lipid growth in pigs. Anim. Prod. 22:87–96.

Young, M., M. D. Tokach, J. Noblet, F. X. Aherne, S. S. Dritz, R. D. Goodband, J. Nelssen, J. van Milgen, and J. C. Woodworth. 2004. Influence of carnichrome on energy balance of gestating sows over pregnancy. J. Anim. Sci. 82:2013–2022.

# 4    Lipids and Lipid Utilization in Swine

Werner G. Bergen

## Introduction

Overall lipid metabolism processes in swine have been studied and identified. As in all other animals, lipid metabolism in swine can be summarized in Figure 4.1. Fatty acids can arise from dietary consumption and from de novo synthesis in the adipose tissue in pigs. Fatty acids are rich energy sources and are metabolized in body tissues such as muscle to fulfil energy needs. Fatty acids are central players in lipid metabolism, but free fatty acids are not stored in tissue but rather in the esterified state as triacylglycerol. Similarly, cholesterol is also not stored in the free form in tissues but as cholesterol esters. Both absorbed fatty acids and fatty acids synthesized from glucose de novo can be used for energy expenditures; however, a positive energy balance will incur the accretion of fat in the adipose tissues.

In the previous chapter, "Lipids and Lipid Utilization in Swine," in the first edition, an excellent review on lipids and lipid metabolism in neonatal and adult swine was presented (Lin et al. 2013). In this edition, a different approach will be used and built on the previous chapter. This chapter features information on the genomics of fatty acid synthesis that may be exploited to enhance sustainability, the putative role of genetics in fatty acid composition in pork lipids, the role of bioactive compounds on lipid metabolism regulation, nutritional genomics (Bergen 2018), dietary modifications to influence carcass composition, and regulation of feed efficiency.

## Lipid Transport Between and into Tissues

Endogenous fatty acids, after adipose tissue triacyl glycerol (TAG) lipolysis, can be transported in the circulation bound to blood proteins as nonesterified fatty acids (NEFA) to tissues for oxidation purposes. Exogenous or dietary fatty acids after absorption are transported as blood lipoproteins (chylomicrons) to the various tissues. NEFA can enter cells directly with a transporter, while TAG-bound exogenous fatty acids in chylomicrons must be released by lipoprotein lipase before tissue uptake. Fat digestion, de novo fatty acid synthesis, lipolysis, lipoprotein transport, and NEFA transport are all critical events in lipid metabolism.

During the intestinal absorption process, fatty acids (as TAG) are bound to chylomicrons and channeled through the lymph to the circulating blood (Figure 4.2). In pigs, de novo fatty acid synthesis from glucose occurs principally in the adipose tissue (Bergen and Mersmann 2005), and if not oxidized, they are stored as triacyl-glycerol. Stored fats in adipose tissue release fatty acids for utilization after lipolysis and transported in the blood as non-NEFA, while bound on blood proteins.

*Sustainable Swine Nutrition*, Second Edition. Edited by Lee I. Chiba.
© 2023 John Wiley & Sons Ltd. Published 2023 by John Wiley & Sons Ltd.

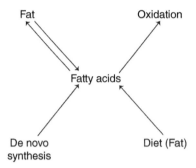

**Figure 4.1**  Fatty acids are the central currency of porcine lipid metabolism. Exogenous (consumed) fatty acids and endogenous fatty acids (de novo synthesis, adipose storage) can either be oxidized for energy purposes or stored as triglycerides mostly in the adipose tissue.

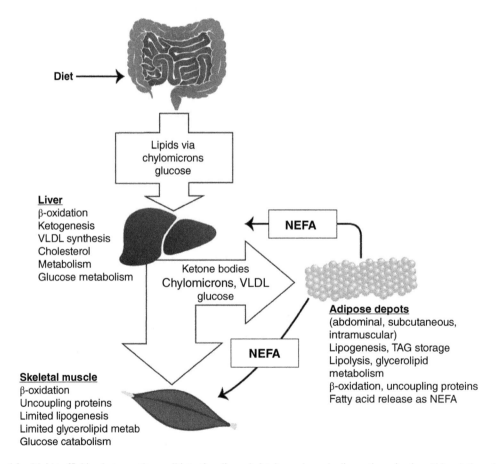

**Figure 4.2**  Lipid trafficking between the small intestine, liver, skeletal muscle, and adipose tissue in pigs. Abbreviations: NEFA, nonesterified fatty acids; VLDL, very low-density lipoprotein; and TAG, triacyl glycerol.

Fatty acids for tissue utilization from chylomicrons are released by lipoprotein lipases, and the resulting NEFA is taken up mostly by skeletal muscle, adipose tissue, and liver. Once in the liver, any excess NEFA is esterified to TAG and released by the liver as very high-density lipoproteins.

Similarly, the uptake mechanisms for fatty acids into across cell membranes and mitochondria maybe amenable to genetic regulation for improvement (Figure 4.3; Schwenk et al. 2010). Blood protein-bound very-long chain fatty acids, once released by blood proteins, can be transported directly across the bilayer membrane by fatty acid transport protein 1 (FATP1). Other NEFAs are released from blood proteins and lipoproteins and concentrated near the membrane by fatty acid-binding protein (FABP). In combination with the integral protein fatty acid transporter (CD36), NEFA then transverse the membrane and are reattached to FABP after reaching the cytoplasm.

In this regard, researchers are using genome-wide association studies in farm animals (e.g., Huang et al. 2019a) to discover genes that respond to physiological and pathological effects of lipids during ketosis in dairy cows. Similar research may lead to a better understanding of such physiological processes may lead to enhanced efficiency and sustainability of swine production. Enhancing the digestibility of dietary components has been a long favored approach in nutrition research to improve overall efficiency in animal nutrition.

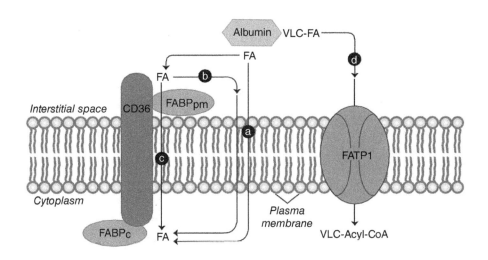

**Figure 4.3** Fatty acid transport/absorption into cells. Abbreviations: CD36, integral fatty acid transporter; FA, fatty acids; FABPpm, fatty acid-binding protein-plasma membrane; FABPc, fatty acid-binding protein-cytoplasm; VLC-FA, very long-chain fatty acids, FATP1, fatty acid transport protein, and VLC-Acyl-CoA, Very long-chain acyl-coenzyme A. (Source: Adapted from Schwenk et al. 2010.)

**Carcass Lipids in Pigs**

For time immemorial, pork had been considered a high-fat food with sausage and bacon being the highest. Carcass lipid content of traditional swine breeds was high with thick back fat measures. As the war on fat in animal agriculture emerged in the 1970s, great effort was expanded to reduce fat content of pork. Initially, this could be accomplished pharmacologically with growth hormone (somatotropin) injections or feeding of ractopamine (beta- adrenergic agonists). Concurrently to work on pharmacological agents, others proceeded to lower fat by genetic selection and cross breeding. After 20 plus years what emerged were growthy, lean pigs as Yorkshires, Pietrain crossbreds and large white composites. Over the years, as pigs continually became leaner, pharmacological approaches became less and less applicable and somatotropin was never approved for pigs in the US and elsewhere (Hausman et al. 2018). As total carcass fat declined, intramuscular fat (marbling) also declined. To many consumers a lowered marbling has negative effects on organoleptic properties of pork (Cisneros et al. 1996; Gerbens et al. 2001). Consequently, the idea to have more fat in pork has re-emerged, despite all warnings by the medical community and human nutritionist for overconsumption of animal/saturated fats.

Proliferation and differentiation of intramuscular adipocytes appear to be similar to adipocytes in other fat depots. The putative regulatory processes that specifically cause deposition of intramuscular fat have not been identified. What we do know is that marbling is always associated with finish (fattening). Extensive studies have been conducted to uncover the reason(s) for the inability to have a desirable level of marbling in lean or not highly finished pigs. Much was learned about adipocyte determination, proliferation, and finally differentiation, and the role of numerous transcription factors but regulation of intramuscular fat has not been discovered yet. Parenthetically, today's beef is also much leaner and marbling content is an issue as well.

Utilizing a variety of diet formulations, carcass composition may be changed in growing animals. For example, feeding of low-energy diets will reduce carcass fat or feeding low protein diets will result in lowered lean gains. Fatty acid profiles in pork adipose will reflect either endogenous fatty acid synthesis (saturated and monounsaturated fatty acids) or dietary/exogenous fatty acids. This was made aptly clear when in the past swine were fed cull peanuts [high in polyunsaturated fatty acids (PUFA)] and developed "soft pork."

When including dietary lipids in pigs feeding programs, usually an increase in bodyweight gain is achieved, but often intramuscular fat is not enhanced (Adhikari et.al. 2017; Huang et al. 2019b). Another approach to modify carcass composition has been the application of compensatory gain principles to enhance the efficiency of nutrient utilization. Here, feeding restricted protein or restricted indispensable amino acid diets resulted in lowered lean gain, but after realimentation from the restriction period resulted in compensatory growth in lean; if the restriction was too excessive or lasted too long, compensatory growth and improved utilization of nutrients are not achieved. The principle of compensatory growth in pigs subjected to dietary amino acid restrictions during the early growth phase has been applied to alter growth performance and carcass composition of pigs (Chiba 1995; Chiba et al. 1999; Fabian et al. 2002, 2003, 2004; Kamalakar et al. 2009; Adhikari et al. 2017). In their work with amino acid-restricted diets, they showed a slight trend for an compensatory effect at the 80% restriction with a positive effect on nutrient utilization (Adhikari et al. 2017). Reports by Cisneros et al. (1996) and Palma-Granados et al. (2019) suggested that an amino acid restriction resulted in an increase in intramuscular fat in pork. Because feeding lipids have been reported to depress de novo lipogenesis in animals (Jump 2004), it has been hypothesized that adding lipids to the diet will also discourage depot fat deposition, while stimulating intramuscular fat development (e.g., Adhikari et al. 2017; Huang et al. 2019b). In an initial study, to amino acid-restricted diets, they added 5% lipids (3% flaxseed oil and 2% poultry fat). Results showed that dietary lipids improved bodyweight, gain: feed ratio, and increased pork off flavor. There was no effect on intramuscular fat content. Differential expression of lipogenic genes measured in muscle and adipose tissue between all dietary treatments are not different (Adhikari et al., unpublished data). In a follow-up study, Huang et al. (2019b) fed complete finisher diets to Yorshire pigs containing either 0, 2, 4, or 6% lipids and either 11 or 230 IU of vitamin E/kg of diet. All lipid-containing diets contained 1% flaxseed oil. Poultry fat was added to lipid-containing diets to achieve the 2, 4, and 6% dietary treatment concentrations. Dietary lipid treatments had no effect on subjective marbling scores of loin muscle and tissue fat content (Huang et al. 2019b). Differential expression of FAS, SCD, PPARγ, ME1, SREBP-1c, PPARα, leptin, adiponectin, DGAT1, and DGAT2 (all lipid metabolisms and/or lipogenic genes) was assessed in subcutaneous adipose and longissimus muscle using quantitative real-time PCR. There were no differences in differential gene expression irrespective of diet fed (Huang et al. 2020; Tables 4.1 and 4.2). To what degree amino acid restrictions or supplementation of lipids and T-3 fatty acid sources (with vitamin E as an antioxidant) in swine finishing diets can affect pork composition and fatty acid profiles is not clear from the studies in the South Eastern United States (e.g., Adhikari et al. 2017; Huang et al. 2019b), while several European reports on lysine/protein-restricted feeding have resulted in increased intramuscular fat content in pork.

### Utilization of Dietary Omega (T)-3 PUFA Tin Pigs

The essential fatty acids in swine diets are 18 : 2 T-6 and 18 : 3 T-6 PUFA, 18 : 2 T-6 is plentiful in such diets, but 18 : 3 T-6 is usually present in quantities below bodily needs. This is true as well in humans' diets. The most reliable source of T-3 fatty acids is fish, but average fish intake across the world's population cannot supply enough 18 : 3 T-3 for optimal nutrition. Thus, research has been ongoing to increase 18 : 3 T-3 PUFA content in dairy foods (Nguyen et al. 2019), pork (Huang et al. 2019b), and poultry (de Tonnac et al. 2018). In this regard, ruminants have a special issue with feeding of lipids and PUFA. The rumen microbiota produces conjugated linoleic acids from PUFA or totally saturates PUFA by biohydrogenation (Harvatine et al. 2009). This topic will not be discussed further in this review.

One approach to increasing T-3 PUFA in pork products has been the feeding of certain plant oils. Huang et al. (2020) showed a modest increase in alpha linolenic acid (18 : 3 T-3) in total extracted fat in muscle samples. Similar results have been reported by other workers. The 18 : 3 T-3 in itself does not fullfil the T-3 needs in animals, but the linolenic acid has to be processed (elongated and desaturated) into the T-3 highly unsaturated fatty acids (HUFA), eicosapentaenoic acid (EPA), and docosahexanoic (DHA). The T-6 fatty acid linoleic acid (18 : 2 T-6) is typically much more abundant in swine feeds than linolenic acid. Processing of 18 carbon T-3 and T-6 fatty acids utilizes the same elongases and desaturases (Figure 4.4; Jump 2004; Leonard et al. 2004).

Hence, linoleic acid out-competes linolenic acid for the available elongases and desaturases. (Leonard et al. 2004; Lee et al. 2019; de Tonnac et al. 2018). In addition, in some organisms, the elongase activity would be the rate limiting for elongation of 18 carbon fatty acids. The role of elongases and desaturase in humans is very similar to the above concepts. Thus, even when a reasonable amount of linolenic acid is consumed, most likely little will be processed to EPA and DHA. As an alternative, pigs may be fed EPA and DHA directly from micro-algae. To that end, a number

**Table 4.1** Effects of dietary lipid and vitamin E (VE) supplementation on mRNA abundance in the adipose tissue at the end of the finisher-2 phase[a,b,c].

| Item lipid, %: VE, IU/kg: | 0 | | 2 | | 4 | | 6 | | SEM[d] | P-value Lipid, Ln | Lipid, Qd | VE | Lipid×VE | Flax, 0 vs. 1%[e] |
|---|---|---|---|---|---|---|---|---|---|---|---|---|---|---|
| | 11 | 220 | 11 | 220 | 11 | 220 | 11 | 220 | | | | | | |
| FAS | 1.08 | 0.58 | 0.71 | 0.64 | 1.04 | 0.44 | 1.05 | 0.71 | 0.21 | 0.142 | 0.792 | 0.482 | 0.514 | 0.223 |
| SCD | 1.00 | 1.01 | 1.68 | 1.21 | 0.84 | 1.04 | 0.66 | 0.95 | 0.34 | 0.208 | 0.570 | 0.113 | 0.324 | 0.529 |
| PPAR-γ | 1.00 | 0.30 | 0.72 | 0.57 | 0.28 | 0.56 | 0.55 | 0.68 | 0.22 | 1.000 | 0.637 | 0.422 | 0.572 | 0.741 |
| MEI | 1.05 | 1.50 | 0.88 | 0.78 | 0.99 | 0.71 | 1.08 | 1.41 | 0.34 | 0.962 | 0.946 | 0.116 | 0.009 | 0.899 |
| SREBP1c | 1.13 | 0.90 | 1.69 | 0.96 | 0.91 | 1.13 | 1.18 | 1.48 | 0.42 | 0.036 | 0.684 | 0.351 | 0.430 | 0.070 |
| PPAR-α | 1.00 | 0.84 | 0.61 | 1.25 | 0.86 | 0.96 | 1.06 | 1.54 | 0.30 | 0.209 | 0.492 | 0.057 | 0.134 | 0.505 |
| Leptin | 1.13 | 0.98 | 0.61 | 1.00 | 1.06 | 1.13 | 1.65 | 1.00 | 0.36 | 0.066 | 0.718 | 0.216 | 0.544 | 0.326 |
| Adiponectin | 1.06 | 0.45 | 1.49 | 0.84 | 0.73 | 1.01 | 0.80 | 1.18 | 0.32 | 0.379 | 0.999 | 0.510 | 0.217 | 0.487 |
| DGAT1 | 1.00 | 0.50 | 2.21 | 1.33 | 0.61 | 1.29 | 0.78 | 1.32 | 0.41 | 0.148 | 0.966 | 0.147 | 0.735 | 0.378 |
| DGAT2 | 1.00 | 0.77 | 1.10 | 1.13 | 1.16 | 0.81 | 0.80 | 1.49 | 0.37 | 0.178 | 0.425 | 0.100 | 0.565 | 0.163 |

[a] Reported by Huang et al. (2019c).

[b] Lipid = 0% flaxseed oil and poultry fat for the diet with 0% lipids and 1% flaxseed oil+1, 3, or 5% poultry fat for the diets supplemented with 2, 4, or 6% lipids, respectively; Ln = linear, Qd = quadratic, flax = flaxseed oil; FAS = fatty acid synthase; SCD = stearoyl CoA desaturate; PPAR-γ = peroxisome proliferator-activated receptor γ; MEI = malic enzyme 1; SREBP1c = sterol regulatory element-binding protein-1c; PPAR-α = peroxisome proliferator-activated receptor α; DGAT1 = diacylglycerol acyltransferase 1, and DGAT2 = diacylglycerol O-acyltransferase 2.

[c] Data were expressed relative to pigs fed the diet containing 0% lipids and 11 IU VE/kg. Least square means based on 12 pigs per treatments because of the uniqueness of each assay.

[d] SEM = pooled standard error of the mean.

[e] Flax, 0 vs. 1% = 0% flaxseed oil and poultry fat vs. 1% flaxseed oil+1, 3, or 5% poultry fat for the diets supplemented with 2, 4, or 6% lipids, respectively.

**Table 4.2** Effects of dietary lipid and vitamin E (VE) supplementation on mRNA abundance in the muscle tissue at the end of the finisher-2 phase[a,b,c].

| Item | lipid, %: 0 | | 2 | | 4 | | 6 | | SEM[d] | P-value Lipid, Ln | Lipid, Qd | VE | Lipid x VE | Flax, 0 vs.1%[e] |
|---|---|---|---|---|---|---|---|---|---|---|---|---|---|---|
| VE, IU/kg: | 11 | 220 | 11 | 220 | 11 | 220 | 11 | 220 | | | | | | |
| *FAS* | 1.00 | 1.07 | 0.77 | 0.60 | 0.70 | 0.55 | 0.33 | 0.52 | 0.38 | 0.906 | 0.194 | 0.060 | 0.112 | 0.614 |
| *SCD* | 1.00 | 1.39 | 1.31 | 0.85 | 1.86 | 1.70 | 1.34 | 0.98 | 0.46 | 0.704 | 0.016 | 0.240 | 0.179 | 0.127 |
| *PPAR-γ* | 1.00 | 1.34 | 0.87 | 0.90 | 1.48 | 0.97 | 1.73 | 0.27 | 0.40 | 0.458 | 0.018 | 0.976 | 0.396 | 0.055 |
| *MEI* | 0.97 | 1.35 | 2.33 | 0.97 | 0.78 | 0.52 | 0.84 | 2.19 | 0.68 | 0.117 | 0.473 | 0.852 | 0.582 | 0.234 |
| *SREBP1c* | 1.00 | 1.84 | 2.22 | 1.27 | 0.86 | 0.97 | 1.61 | 2.44 | 0.41 | 0.420 | 0.925 | 0.852 | 0.577 | 0.450 |
| *PPAR-α* | 1.00 | 1.26 | 0.72 | 0.82 | 0.55 | 0.51 | 0.87 | 0.80 | 0.40 | 0.153 | 0.924 | 0.379 | 0.813 | 0.176 |

[a] Reported by Huang et al. (2019c).

[b] Lipid = 0% flaxseed oil and poultry fat for the diet with 0% lipids and 1% flaxseed oil + 1, 3, or 5% poultry fat for the diets supplemented with 2, 4, or 6% lipids, respectively; Ln = linear, Qd = quadratic; flax = flaxseed oil; *FAS* = fatty acid synthase; *SCD* = stearoyl CoA desaturate; *PPAR-γ* = peroxisome proliferator-activated receptor γ; *ME1* = malic enzyme 1; *SREBP1c* = sterol regulatory element-binding protein-1; and *PPAR-α* = peroxisome proliferator-activated receptor α.

[c] Data were expressed relative to pigs fed the diet containing 0% lipids and 11 IU VE/kg. The results of primer test indicated that mRNA abundance for other genes was low; thus, the qPCR quantification for those genes was not conducted. Least square means based on 12 pigs per treatments because of the uniqueness of each assay.

[d] SEM = pooled standard error of the mean.

[e] Flax, 0 vs. 1% = 0% flaxseed oil and poultry fat vs. 1% flaxseed oil + 1, 3, or 5% poultry fat for the diets supplemented with 2, 4, or 6% lipids, respectively.

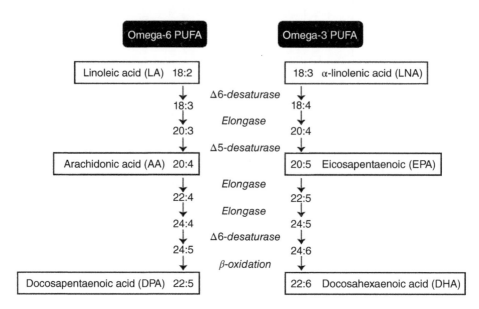

**Figure 4.4**   Omega (T)-6 and T-3 polyunsaturated fatty acids elongation and desaturatioin pathways. Elongase and desaturase enzymes are shared by T-6 and T-3 fatty acids. Abbreviation: PUFA, polyunsaturated fatty acids. (Adapted from Jump 2004; Leonard et al. 2004.)

of feed supplement manufacturers have developed supplement of micro-algae rich in DHA (Moran et al. 2018; Lee et al. 2019). Including PUFA/HUFA often requires additions of antioxidants (vitamin E) to the diet.

### Potential of Genetic Manipulations to Enhance T-3 PUFA Content in Pork Products

Mammals do not possess a Delta 12 or Delta 15 desaturase; genes for these enzymes would have to be inserted with biotechnical approaches (transgenics) into the genome of pigs. While many transgenic animals have been produced by such methods, the subjects of almost all genetic engineering were nonfood species. Based on some previous experiences, many consumers would likely not feel such transgenic foods are safe to eat or healthy to consume. One alternative to this conundrum would be to feed plant origin 18 : 3 T-3 and select for pigs with higher activities of elongases and delta 5/6 desaturases to stimulate synthesis of 20, 22, and 24 carbon T-3 fatty acids (Figure 4.4). These type of studies would require an efficient, robust, sensitive, and easily to apply laboratory approach. Thus, routinely, RNA sequencing would not be valuable in the short term, this procedure could evaluate longer time progress to assess whether increase in elongases and desaturases mRNA had occurred. Palmquist et al. (2004) have used ratios of fatty acids to estimate activity of desaturases from fatty acid analyses data. After tissue fat extraction, fatty acid analysis by gas chromatography could, for example, note increases in 20- to 24-carbon T-3 HUFA in depot fats or liver in selection experiments. From a practical perspective, as we have little knowledge on the genetics of elongases and desaturases in pigs, how would you start such work and with what type of pigs? Some would advance the argument that when a workable assessment procedure had been established, using classical selection approaches would take a rather long time. This argument was also advanced as animal breeders started to select for leaner pigs over 50 years ago; in retrospect, however, those selection efforts were very effective.

**Insuring Desirable Marbling in Pork Products Without Large Increases in Adipose Depot Fat**

Extent of intramuscular fat deposition (marbling) is related to the level of "finish" or fat in food animal carcasses (Hausman et al. 2018). Intramuscular fat (IMF) is very minimal in young pigs, and development of visible intramuscular adipose depots do not become evident until large depot fat stores (subcutaneous and visceral) start to fill in food animals (Hausman et al. 2018). Any process that enhances the carcass growth and protein deposition (lean gain) will lower the extent of marbling. This was clearly noted in swine with the use of beta agonists (enhance lean gain; lower fat gain) and contemporary pigs that were selected for many generations to deposit less carcass fat (Bergen and Merkel 1991).

Over the years, pharmacological or genetic selection approaches have resulted in a major decrease in carcass fat content of pigs. As already noted above, often marbling is not even evident in lean, grown pigs. Our contemporary approach to producing pork includes a finishing period after the growth phase. Using for example, large white crossbreds (Landrace) or Yorshires, this finishing period will not result in the traditional amount of carcass fat cover and is typically accompanied by less marbling. An obvious, but technically difficult to explore, hypothesis arising is, e.g., reasonably low fat in the carcass is desirable, can IMF fat deposition be stimulated independently during the finishing period by utilizing molecular regulation?

Fat cells arise from the mesochyme, followed by a determination into preadipocytes and finally differentiation into adipocytes (Hausman et al. 2009). This process has been well studied with cells in culture, and numerous transcription factors have been identified that are involved in the molecular regulation of adipogenesis. These regulatory processes apply to both depot fat adipocytes and IMF adipocytes (Hausman et al. 2009). Thus, there is a possibility that the proliferation of adipocytes in fat depots maybe attenuated, while the proliferation, differentiation process of cells in IMF may be enhanced via molecular regulation. The time needed to this type of discovery research to routine application in pork production would be expected to fairly long and expensive (and maybe against mother nature). This time frame should be compared to genetic selection of enhanced marbling in low carcass fat pigs.

Spiegelman and coworkers (Rosen et al. 2000) summarized work on adipose cell differentiation and presented the following regulatory pathway as follows:

Hormonal cues > stimulation of the CCAAT/enhancer-binding proteins (C/EBPα,β,delta) > SREBP-1c (ADD1) > PPARgamma. PPAR gamma promotes adipogenesis including fatty acid synthesis, triacylglycerol production, and fatty acid-binding protein 4 (FABP4; for fatty acid trafficking).

Sul and coworkers (Wang and Sul 2009) expanded this regulatory differentiation pathway by showing that the transcription factor Pref1 (aka DLK1) induces transcription factor Sox9. This transcription factor promotes proliferation of early adipose precursors cells and preadipocytes. As the signal for Pref-1 expression wanes during development, Sox9 expression is attenuated, and the preadipocytes are differentiated to adipocytes. Furthermore, Sox9 ablation promotes adipogenesis (Gulyaeva et al. 2018) in animals. From these studies, it may be concluded that inhibiting Sox9 expression only in IMF may increase marbling without changing fat storage in depot adipose tissues.

A much more cost-effective alternative for enhancing IMF may be the development of biomarkers. Much of the IMF is derived from circulating fatty acids rather than from de novo synthesis.

Thus, dietary fatty acids are directly deposited in IMF. Since pigs primarily consume corn, linoleic deposition in IMF adipocytes may be elevated and serve as a biomarker for genetic selection of marbling. Typically, the heritability of IMF is in the 0.30–0.40 range (Gol et al. 2019), but this increased genetic effect needs to be achieved without changing the deposition of backfat. Gol et al. (2019) utilized 18 : 2 T-6 as a source for IMF fat deposition and biomarker. They found an $h^2$ for 18 : 2 deposition in IMF at greater than 0.40, while uniquely in this study IMF 18 : 2 T-6 deposition was negatively correlated to back fat deposition rates.

## Regulation of Feed Efficiency

Because feed is the number one cost item in swine production, efforts to improve feed conversion into salable pork products have continued to be a major goal of swine nutrition research. A most common measure of feed efficiency is the feed intake to gain ratio, but this value is influenced by feed intake, diet composition, and animal size. Thus, more satisfactory determinations of animal efficiency were developed. From growth and intake data, ratio-metrics such as energy conversion ratio (ECR; energy intake/average daily gain, [ADG]), Kleiber ratio (KR; ADG/metabolic body-weight) and residual metrics such as residual energy intake (REI) and residual feed intake (RFI) were developed. The RFI system has been studied extensively in pigs and cattle, and the difference between the actual feed intake and the predicted feed intake from a regression analysis of ADG and feed intake (residuals) have been measured. A negative RFI implies a higher efficiency of feed usage. Calderon Diaz et al. (2017) measured seven feed efficiency metrics (three ratio metrics and four residual metrics) in pigs and reported a strong correlation between ECR and residual metrics. This means that improvements in any measure/trait of feed efficiency will also signal improvements in other efficiency metrics. While to date it has been difficult to pinpoint exactly which functional biological processes influence traits associated with feed efficiency, many efficiency metrics were related to cellular development, negative regulation of apoptosis, protein synthesis and degradation, signal transduction, hormonal and immune responses, nutrient absorption, and various aspects of energy metabolism (Horodyska et al. 2018; Messad et al. 2019; Wang et al. 2019). In addition, Horodyska et al. (2019) observed a negative influence of feed efficiency on quality of pork muscle with lower IMF, lower 45 minutes postmortem muscle pH values, and higher Warner–Bratzler shear values. Sensory evaluation did not confirm such differences in tenderness. Finally, sensory analysis revealed that meat from high-efficiency pigs was more salty and had a tendency toward an increased barny/earthy/animal stable flavor (Horodyska et al. 2019). Future efforts to improve feed efficiency of pigs should include an assessment of pork quality.

## Additional Potential Innovations

The genome of pigs has been sequenced, and the data have been mostly applied to classical selection programs. The sequenced genome is an extremely fertile ground to gather new knowledge in mechanisms of protein synthesis, degradation, fat metabolism, immunology and nutrition, and major systemic diseases. Research approaches such as genome-wide association studies can be exploited to this end. Marker genes to track polygenic traits in swine genomic biology to enhance efficiency of pork production may also be a fruitful endeavor. Alternatives to antibiotics such as biologically active molecules from herbs or spices are starting to be explored, particularly in Southeast Asia. While some of such bioactives may indeed be a replacement for antibiotics in the future, feeding such compounds may be detrimental to the organoleptic characteristics of pork.

Progress toward new innovations to enhance sustainability of pork production will depend on basic-bench level research. Innovations in feed ingredients processing and digestibilities have already been discovered and applied to pork production for many years.

## References

Adhikari, C. K, L. I. Chiba, S. D. Brotzge, M. S. Viera, C. Huang, W. G. Bergen, C. L. Bratcher, S. P. Rodning, and E. G. Welles. 2017. Early dietary amino acid restrictions and flaxseed oil supplementations on the leanness of pigs and quality of pork: Growth performance, serum metabolites, carcass characteristics, and physical and sensory characteristics of pork. Livest. Sci. 198:182–190.

Bergen, W. G. 2018. Nutrigenomics in food producing animals. In: F. Toldrá, and L. M. L. Nollet, editors, Advanced technologies for meat processing. 2nd ed. CRC Press, Boca Raton, FL. p. 355–369.

Bergen, W. G., and H. J. Mersmann. 2005. Comparative aspects of lipid metabolism: Impact on contemporary research and use of animal models. J. Nutr. 135:2499–2502.

Bergen, W. G., and R. A. Merkel. 1991. Body composition of animals treated with partitioning agents: implications for human health. FASEB J. 14:2951–2957.

Calderon Diaz, J. A., D. P. Berry, N. Rebeiz, B. U. Metzler-Zebbelli, E. Magowan, G. E. Gardner, and F. G. Lawlor. 2017. Feed effieciency metrics in growing pigs. J. Anim. Sci. 95:3037–3046.

Chiba, L. I. 1995. Effects of nutritional history on the subsequent and overall growth performance and carcass traits of pigs. Livest. Prod. Sci. 41:151–161.

Chiba, L. I., H. W. Ivey, K. A. Cummins, and B. E. Gamble. 1999. Growth performance and carcass traits of pigs subjected to marginal dietary restrictions during the grower phase. J. Anim. Sci. 77:1769–1776.

Cisneros, F., M. Ellis, D. H. Baker, R. A. Easter and F. K. Keith. 1996. The influence of short term feeding of amino acid-deficient diets and high dietary leucine levels on the intramuscular fat content of pig muscle. J. Anim. Sci. 63:517–522.

de Tonnac, A., M. Guillevic, and J. Mourot. 2018. Fatty acid composition of several muscles and adipose tissues in pigs fed PUFA rich diets. Meat Sci. 140:1–8.

Fabian, J., L. I. Chiba, L. T. Frobish, W. H. McElhenny, D. L. Kuhlers, and K. Nadarajah. 2004. Compensatory growth and nitrogen balance in grower-finisher pigs. J. Anim. Sci. 82:2579–2587.

Fabian, J., L. I. Chiba, D. L. Kuhlers, L. T. Frobish, K. Nadarajah, and W. H. McElhenny. 2003. Growth performance, dry matter and nitrogen digestibilities, serum profile, and carcass and meat quality of pigs with distinct genotypes. J. Anim. Sci. 81:1142–1149.

Fabian, J., L. I. Chiba, D. L. Kuhlers, L. T. Frobish, K. Nadarajah, C. R. Kerth, W. H. McElhenney, and A. J. Lewis. 2002. Degree of amino acid restrictions during the grower phase and compensatory growth in pigs selected for lean growth efficiency. J. Anim. Sci. 80:2610–2618.

Gerbens, F., F. J. Verburg, H. T. B. Van Moerkerk, B. Engel, W. Buist, J. H. Veerkamp and M. F. W. te Pas. 2001. Association of heart and adipocyte fatty acid binding protein gene expression with intramuscular far content in pigs. J. Anim. Sci. 79:347–354.

Gol, S., R. Gonzales-Prendes, L. Bosch, J. Reixach, R. N. Pena, and J. Estany. 2019. Linoleic acid metabolic pathways allow for an efficient increase of intramuscular fat content in bpigs. J. Anim. Sci. Biotech. 10:33.

Gulyaeva, O., H. Nguyen, A. Sambeat, K. Heydari, and H.-K. Sul. 2018. Sox9-Meis 1 inactivation is required for adipogenesis, advancing Pref-1 to PDGFRα cells. Cell Rep. 25:1002–1017.

Harvatine K.J., Y. R. Boisclair, and D. E. Bauman. 2009. Recent advances in the regulation of milk fat synthesis. Animal 3:40–54.

Hausman, G. J., M. V. Dodson, K. Ajuwon, M. Azain, K. M. Barnes, L. L. Guan, Z. Jiang, S. P. Poulos, R. D. Sainz, S. Smith, M. Spurlock, J. Novakofski, M. E. Fernyhough, and W. G. Bergen. 2009. Board-invited review: the biology and regulation of preadipocytes and adipocytes in meat animals. J. Anim. Sci. 87:1218–1246.

Hausman, G. J., W. G. Bergen, T. D. Etherton, and S. B. Smith. 2018. The history of adipocyte and adipose tissue research in meat animals. J. Anim. Sci. 96:473–486.

Horodyska, J., R. M. Hamill, H. Reyer, N. Trakooljul, P. G. Lawlor, U. M. McCormack, and K. Wimmers. 2019. RNA-seq of liver from pigs divergent in feed efficiency highlights shifts in macronutrient metabolism, hepatic growth and immune response. Front. Genet. 10:117.

Horodyska, J. K. H. Wimmers N. Reyer A. M. Trakooljul P. G. L. Muller, and R. M. Harrill. 2018. RNA-seq of muscle from pigs divergent in feed efficiency and product quality identifies differences in immune response, growth, and macronutrient and connective tissue metabolism. BMC Genomics 19:791.

Huang, H., J. Cao, Q. Hanif, Y. Wang, Y. Yu, S. Zhang, and Y. Zhang. 2019a. Genome-wide association study identifies energy metabolism genes for resistance to ketosis in Chinese Holstein cattle. Anim. Genet. 50:376–380.

Huang, C., L. I. Chiba, W. E. Magee, Y. Wang, D. A. Griffing, L. M. Torres, S. P. Rodning, C. L. Bratcher, W. G. Bergen, and E. A. Spangler. 2019b. Effect of flaxseed oil, animal fat, and vitamin E supplementation on growth performance, serum metabolites, and carcass characteristics of finisher pigs, and physical characteristics of pork. Livest. Sci. 220:143–151.

Huang, C., L. I. Chiba, W. E. Magee, Y. Wang, S. P. Rodning, C. L. Bratcher, W. G. Bergen, and E. A. Spangler. 2020. Effect of flaxseed oil, poultry fat, and vitamin E supplementation on physical and organoleptic characteristics and fatty acid profile of pork, and expression of genes associated with lipid metabolism. Livest. Sci. 231:103849.

Jump, D. B. 2004. Fatty acid regulation of gene transcription. Crit. Rev. Clin. Lab. Sci. 41:41–78.

Kamalakar, R. B., L. I. Chiba, K. C. Divakala, S. P. Rodning, E. G. Welles, W. G. Bergen, C. R. Kerth, D. L. Kuhlers, and N. Nadarajah. 2009. Effect of the degree and duration of early dietary amino acid restrictions on subsequent and overall pig performance and physical and sensory characteristics of pork. J. Anim. Sci. 87:3596–3606.

Lee, S. A., N. Whenham, and M. R. Bedford. 2019. Review on docosahexaenoic acid in poultry and swine nutrition: Consequences of enriched animal products on performance and health characteristics. Anim. Nutr. 5:11–21.

Leonard, A. E., S. L. Pereira, H. E. Sprecher, and Y.-S. Huang. 2004. Elongation of long-chain fatty acids. Prog. Lipid Res. 43:36–54.

Lin, X., M. Azain, and J. Odle. 2013 Lipids and lipid utilization in swine. In: L. I. Chiba, editor, Sustainable swine nutrition. John Wiley & Sons, Inc., Hoboken, NJ. p. 59–79.

Messad, F., I. Louveau, B. Koffi, H. Gilbert, and F. Gondret. 2019. Investigation of muscle transcriptomes using gradient boosting machine learning identifies predictors of feed efficiency in growing pigs. BMC Genomics 20:659.

Moran, C. A., M. Morlacchini, J. D. Keegan, and G. Fusconi. 2018. Dietary supplementation of finishing pigs with the docosahexaenoic acid-rich microalgae, Aurantiochytrium limacinum: Effects on performance, carcass characteristics and tissue fatty acid profile. Asian-Australs. J. Anim. Sci. 31:712–720.

Nguyen, Q. V., B. S. Malau-Aduli, J. Cavalieri, P. D. Nichols, and A. E. O. Malau-Aduli. 2019. Enhancing omega 3 long-chain polyunsaturated fatty acid content of dairy-derived foods for human consumption. Nutrients 11:743–766.

Palma-Granados, P. I. R. Seiquer C. O. Benitez, and R. Nierto. 2019. Effects of lysine deficiency on carcass composition and activity and gene expression of lipogenic enzymes imuscles and backfat adipose tissue of fatty and lean pigs. Animal 7:1–13.

Palmquist, D. L., N. St-Pierre, and K. McClure. 2004. Tissue fatty acid profiles can be used to quantify endogenous rumenic acid synthesis in lambs. J. Nutr. 134:2407–2414.

Rosen, E. D., C. J. Walkey, P. Puigserver, and B. M. Spiegelmam. 2000. Transcriptional regulation of adipogenesis. Genes Dev. 14:1293–1307.

Schwenk, R. W., G. P. Holloway, J. J. F. P. Luiken, A. Bonnen, and J. F. C. Glatz. 2010. Fatty acid transport across the cell membrane: Regulation by fatty acid transporters. Prostaglandins Leucotriens Essent. Fatty Acids 82:149–154.

Wang, N., S. Li, J. Wu, R. Ding, J. Quan, E. Zheng, J. Yang, and Z. Wu. 2019. A transcriptome analysis identifies biological pathways and candidate genes for feed effiency in DLY pigs. Genes 10:725–739.

Wang, Y., and H. S. Sul. 2009. Pref-1regulates mesenchymal cell commitment and differentiation through SoX9. Cell Metab. 9:287–302.

# 5 Amino Acids and Amino Acid Utilization in Swine

Sai Zhang, Rodrigo Manjarin, and Nathalie L. Trottier

## Introduction

Extensive research has been conducted to improve the efficiency of amino acids (AA) utilization in an effort to maximize growth and lactation performances, improve environmental sustainability, and minimize costs of production. Nutritionists have emphasized that prediction of AA requirements should be based on their individual ratio relative to lysine. Such efforts have led to the development of diets aiming to contain an optimal protein content and fixed AA ratios. Dietary AA ratios, however, are dynamic because physiological processes differentially affect individual AA utilization and deposition in tissues.

More recent understanding of these processes highlights the limitations behind the traditional approach of matching dietary AA profiles to the AA composition of the animal's tissue. For instance, efficiency utilization coefficients are largely unknown across both AA and tissue pools, leading to inaccuracies in factorial estimates of dietary AA requirement and dietary profile when based on tissue AA composition alone. Dietary AA are utilized by the intestinal epithelia and microflora, giving rise to a different AA profile following first-pass metabolism. Recent discovery of novel AA transporter systems and the regulation of their molecular entities have provided insight behind AA interactions at the cell membrane interface and the resulting impact on cellular utilization. Ultimately, these processes dictate the individual AA efficiency values.

The goal of this chapter is to discuss processes regulating and impacting AA utilization during various stages of the animal's life cycle. Determination of AA efficiency values during lactation and implementation of thereof to predict requirements are also discussed.

## Cellular Amino Acid Transport: Passport to Amino Acid Utilization

The intracellular availability of dietary AA is controlled by a coordinated activity of AA carrier proteins located in the cellular membrane and responsible for channeling AA across the cell membranes (Palacin et al. 1998; Shennan and Peaker 2000; Broër 2008). Regulation of AA transport is complex because many transporters not only handle multiple AA but also cotransport them in and out of the cells (Shennan and Peaker 2000). A review of mechanisms of AA transport across eukariotic cells is beyond the scope of this book chapter. Therefore, the discussion focusses on current knowledge of systems and transporters of relevance to the pig and specific tissues, with extrapolation from knowledge gained from other animal species.

*Sustainable Swine Nutrition*, Second Edition. Edited by Lee I. Chiba.
© 2023 John Wiley & Sons Ltd. Published 2023 by John Wiley & Sons Ltd.

Movement of AA across epithelial cells occurs via either through the transcellular or paracellular route. Slowly transported hydrophilic compounds are absorbed through the tight junctions via the paracellular route (Urakami et al. 2003). Recognizing that the barrier properties of tight junctions vary widely among tissues in both magnitude, typically quantified as electrical resistance and charge selectivity, most epithelia are "tight" thus favoring AA transport via the transcellular pathway (Colegio et al. 2002). The transcellular route of AA transport is composed of a noncarrier-mediated, freely diffusible component, and carrier-mediated component is made up of a large number of carriers (or transporter proteins) that are either energy-dependent (active transport) or independent (passive transport). The passive-mediated transport processes include a simple diffusion and a facilitated diffusion component. The rate of AA movement across the membrane lipid bilayer via the simple diffusion component is dependent on the solubility of a specific AA in the hydrophobic core of the lipid bilayer membrane, which is also referred to as the permeability coefficient, and by the concentration gradient of the AA across the membrane. Small, uncharged AA may pass freely, while charged AA, such as lysine, arginine, histidine, ornithine, aspartate, and glutamate, are transferred through channels or pores. The relative importance of AA transport via the simple diffusion process likely varies across tissues and organs, depending on the transport capacity of the carrier-mediated processes in that same tissue or organ.

The other component of the passive transport is the facilitated diffusion process, characterized by a carrier or a transport protein, which mediates the uptake of an AA. The active transport process is also characterized by a carrier or transport process, but, unlike the facilitated diffusion process, it is energy-dependent. Therefore, AA carriers or transporters of the facilitated diffusion and active transport processes have been classified into systems based on their transport mechanism (energy and nonenergy dependent) and substrate specificity (Hyde et al. 2003; Broër et al. 2004; Hundal and Taylor 2009). Energy-dependent transporters of the active transport processes function as secondary active transporters, promoting the concentrative uptake and intracellular accumulation of specific AA by coupling AA transport to the inward movement of $Na^+$. Thus, $Na^+$-dependent AA carriers are symport systems as they move two solutes (i.e., AA and $Na^+$) in the same direction. Nonenergy-dependent AA transporters are $Na^+$-independent and are either antiport or uniport systems.

The antiport system is composed of tertiary active exchangers moving two AA in opposite directions. It facilitates the uptake of extracellular AA in exchange for cytoplasmic AA accumulated via secondary active transporters (Hundal and Taylor 2009). For instance, in Figure 5.1, the $b^{0,+}$ carrier facilitates the uptake of extracellular lysine in exchange for intracellular leucine, which was inwardly accumulated via the $B^{0,+}$, a $Na^+$-dependent symport system. Similar to the active transport processes, the facilitated diffusion processes of AA are stereospecific, but unlike the active transport processes, require no metabolic energy. Therefore, AA transport is passively mediated by uniport transporters' down concentration or electrochemical gradient and can operate bidirectionally.

The ability of an AA to enter into organ cells depends both on the affinity of a specific AA to a transporter domain, as defined by $K_m$, and the number of functional transporters on the cell wall, as defined by $V_{max}$ (Souba and Pacitti 1992). Amino acid transport systems are characterized by their transport products (Souba and Pacitti 1992). Several systems have ubiquitous expression, while some systems are limited to specific tissues (Souba and Pacitti 1992).

### Cationic Amino Acid Transport

Cationic AA participating in protein synthesis include arginine, histidine, and lysine. These AA are also referred as basic AA. They carry a net positive charge at physiological pH as their amino and

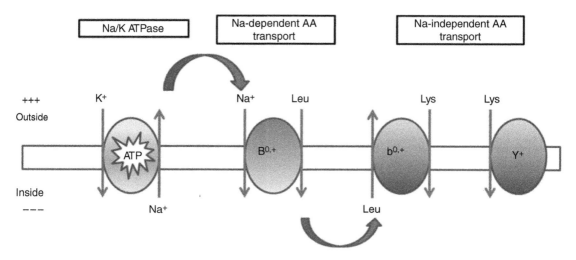

**Figure 5.1** Primary (Na/K ATPase), secondary (Na$^+$-dependent AA transporter), and tertiary (Na$^+$-independent AA transporter) active transport mechanisms in the cellular membrane. Secondary active transporters (e.g., System B$^{0,+}$) generate net movement of amino acids from the extracellular (outside of cell) to the intracellular (inside of cell) pool, whereas tertiary active transport (Systems b$^{0,+}$ and Y$^+$ allow for redistribution of individual amino acids without affecting total pool sizes. +++ and ——— indicate that the extracellular compartment carries a net positive charge and that the intracellular compartment carries a net negative charge, respectively. (Source: Adapted from Hundal and Taylor 2009.)

carboxylic groups and side chain are protonated. Their isoelectric point ranges from 7.59 (histidine) to 10.76 (arginine), which are pH values higher than the physiological pH of 7.4. Cationic AA transporter (CAT) proteins are defined as transporters exhibiting affinities and translocation rates for cationic AA. There are different cationic AA transport systems, and the CAT proteins specifically are members of the y$^+$ system of cationic transporters. The CAT transporters are typically pH independent, and transport activity is stimulated by membrane hyperpolarization. These transporters seem to be stimulated by the presence of AA on the trans side of the membrane (Devés and Boyd 1998; Closs 2002). The Na$^+$-independent system y$^+$ is ubiquitously expressed and specifically transport cationic AA.

### Neutral Amino Acid Transport

Neutral AA are those for which the isoelectric point ranges from 5.07 (cysteine) to 6.30 (proline). These AA carry a neutral charge at physiological pH: their amino group is protonated and thus has a net positive charge and the carboxyl group is deprotonated and thus has a net negative charge. Neutral AA transporter proteins exclusively transport neutral AA. Three families or systems of neutral AA transporters, notably ASC, A, and L, have been studied extensively so far. The ASCT2 protein (encoded by SLC1A5 gene), member of family ASC, is closely associated with glutamine cellular uptake and is highly expressed in proliferative and glutamine-demanding cells such as inflammatory cells, stem cells, and even human cancer cells (Fuchs and Bode 2005; Scalise et al. 2018). Family A has three isoforms, i.e., SNAT1, SNAT2, and SNAT4, with SNAT2 (SLC38A2) being the most widely expressed and extensively regulated by glucocorticoids, growth factors, and insulin (Hyde et al. 2002; Hoffmann et al. 2018).

### Cationic and Neutral Amino Acid Shared Transport

Lysine can be transported via other carrier proteins that do not have a unique affinity for cationic AA as that of the CAT proteins. Lysine can be transported in the absence or the presence of $Na^+$ (Vilella et al. 1990; Wilson and Webb 1990). The $Na^+$-dependent transporter utilizes $Na^+$ ions to increase transporter affinity (Vilella et al. 1990; Souba and Pacitti 1992; Soriano-Garcia et al. 1999). The $Na^+$-dependent lysine uptake occurs via system $B^{0,+}$ (Souba and Pacitti 1992). The $ATB^{0,+}$ transporter of system $B^{0,+}$ carries cationic and neutral hydrophobic (dipolar) AA through a $Na^+$- and $Cl^-$-dependent mechanism (Broër 2008), with greater affinity for neutral than for cationic AA (Sloan and Mager 1999). The $ATB^{0,+}$ demonstrates upregulation in vitro in Xenopus oocytes cells in response to AA starvation of cells and downregulation in response to AA supplementation (Taylor et al. 1996).

Sodium-independent systems sharing the transport of lysine and neutral AA include $b^{0,+}$ and $y^+L$. These two systems are widely expressed in epithelial cells. System $b^{0,+}$ is characterized by the molecular entity heteromeric cationic transporter $rBAT/b^{0,+}AT$. The $rBAT/b^{0,+}AT$ is a broad specificity transporter composed of a catalytic light subunit $b^{0,+}AT1$ (light chain) and the covalently associated type II glycoprotein heavy subunit rBAT (heavy chain) linked by a disulfide bridge (Dave et al. 2004; Broër 2008). Under physiological conditions, $rBAT/b^{0,+}AT$ acts as a tertiary active transporter, inducing the absorption of lysine and arginine coupled to the efflux of neutral AA (Bauch et al. 2003). Lysine/neutral AA counter-transport by $rBAT/b^{0,+}AT$ is ensured by intracellular availability of neutral AA. Cellular accumulation of neutral AA can be realized by the concentrative neutral AA transporter $B^0AT1$ ($Na^+$ cotransport), the molecular entity of System $B^0$. Transporter $B^0AT1$ is a $Na^+$-dependent AA transporter that actively transports the large branched neutral AA leucine and valine, but not anionic nor cationic AA. It has been localized on the apical membrane of kidney and intestinal epithelial cells (Broër et al. 2004), where it contributes to reabsorption of neutral AA from the lumen into the cells.

The system $y^+L$ is characterized by the molecular entities $y^+LAT1/4F2hc$ and $y^+LAT2/4F2hc$. Both $y^+LAT1/4F2hc$ and $y^+LAT2/4F2hc$ are composed of a catalytic light chain ($y^+LAT$) and a heavy subunit (4F2hc) linked by a disulfide bond (Torrents et al. 1998). The $y^+LAT1/4F2hc$ is expressed mainly in kidney and intestine epithelial cells, whereas $y^+LAT2/4F2hc$ has a wider tissue distribution, including brain, heart, testis, kidney, small intestine, and parotids (Broër et al. 2000a). Both transporters mediate the efflux of cationic AA in exchange for extracellular neutral AA (Broër 2008).

In the kidney and intestinal cells, $rBAT/b^{0,+}AT$ and $y^+LAT1/4F2hc$ act as a functional unit for the reabsorption of cationic AA in exchange for the secretion of neutral AA into the lumen (Chillaron et al. 1996; Bauch et al. 2003; Sperandeo et al. 2008).

### Anionic Amino Acid Transport

Anionic AA are aspartic acid and glutamic acid and are also referred to as acidic AA. They carry a net negative charge at physiological pH with their isoelectric point falling below 7.4, with a value of 2.77 for aspartic acid and 3.22 for glutamic acid. Thus, both their amino and carboxylic groups are dissociated at physiological pH consequently carrying a net negative charge. The anionic AA transporters, also referred as the excitatory AA transporters (EAAT), are $Na^+$-dependent and responsible for cellular transport of glutamate and aspartate. Glutamate can serve as an excitatory neurotransmitter and thus EAAT is widely expressed in dendrites and axon terminal of neurons (Roberts et al. 2014). The porcine mammary gland was reported to express EAAT (Shu et al. 2012; Chen et al. 2018), with expression of EAAT3 greater during lactation compared to late pregnancy (Chen et al. 2018), and continued upregulation through d 17 of lactation (Shu et al. 2012), suggesting that EAAT3 may play a role in milk protein synthesis.

### Intestinal Amino Acid Utilization

#### *Mechanisms of Intestinal Amino Acid Absorption and Transport*

Utilization of dietary protein involves a series of steps, including protein digestion in the stomach and small intestine, as well as absorption of peptides and free AA by the small intestine. The first step in AA utilization following protein digestion occurs at the luminal-apical membrane interface (or the epithelial brush-border) where tri- and dipeptides may be either absorbed intact or hydrolyzed into their constituents dipeptide and single AA, prior to trans apical transport (Figure 5.2). The intracellular fates of AA following brush border membrane transport include (i) metabolism, (ii) *in situ* protein synthesis (e.g., peptidases, apoproteins, mucins, etc.), (iii) efflux back to the intestinal lumen in exchange for luminal AA influx, and (iv) final efflux to the portal blood via the basolateral membrane. In addition, intracellular AA availability and utilization is dependent on AA influx from the mesenteric arterial supply across the basolateral membrane. As discussed earlier, the movement of AA across epithelial cells occurs via either the transcellular or paracellular route. However, in the "leaky" epithelia of the small intestine, the paracellular pathway is a major component of nutrient transport (Colegio et al. 2002), and thus, presumably, AA transport. The paracellular route, albeit less well characterized and understood, may constitute a component of dietary AA utilization because of its contribution to intact absorption of small molecular weight-proteins. For the purpose of this chapter, the discussion focuses on transcellular mechanisms of peptide and AA transport.

#### *Peptides*
Earlier studies indicated intraluminal peptidase activity was insufficient to account for the appearance of more than a small proportion of the released free AA (Adibi 1971; Silk et al. 1976). Later, a close to equal distribution of tripeptidase or dipeptidase activity in the human intestine was found between the brush-border and soluble fractions leading to the conclusion that two distinct groups of mucosal peptidases exist, one located within the cytoplasmic compartment and the other at the brush-border membrane of the cell (Silk et al. 1985). Thus, a substantial proportion of tripeptides and dipeptides is hydrolyzed at the brush border before the uptake of their constituent dipeptides and AA by a peptide and a free AA transport mechanism, respectively.

Physiological and molecular studies have shown that the intestinal oligopeptide transporter, designated PepT-1, is the exclusive oligopeptide transporter of the brush-border membrane of the intestinal mucosa (Adibi 2003) and is not shared by free AA (Adibi 2003). It has been proposed that the transport of peptides with a high affinity for brush-border peptidases are predominantly hydrolyzed by surface brush-border membrane enzymes and absorbed as free AA, whereas those with a low affinity for the surface brush border membrane enzymes are absorbed predominantly intact and hydrolyzed by cytoplasmic peptidases (Silk et al. 1985; Figure 5.2).

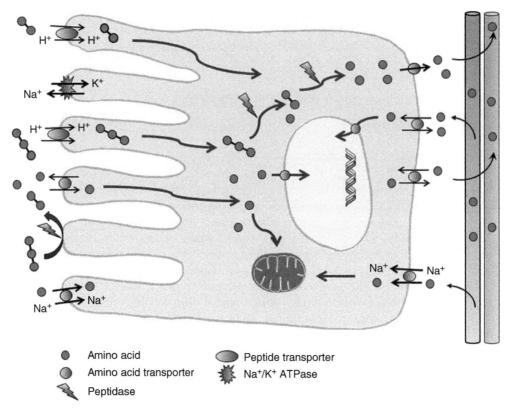

**Figure 5.2**  Mechanisms of peptide and amino acid transmembrane transport and absorption by the intestinal columnar epithelial cell. See text for description.

In pigs, of all tissues (i.e., semitendinosus muscle, longissimus dorsi muscle, kidney, liver, stomach, cecum, colon, and the small intestine) that were characterized using northern blot, the presence of peptide transporter PepT-1 message was found only in the small intestine, with highest abundance in the jejunum, followed by the duodenum and the ileum. Because the majority of AA appearing in the portal blood are in their free form, basolateral AA transport following apical uptake and intracellular metabolism is likely mediated via single AA transport mechanisms. Hence, PepT-1 protein localization would be expected at the apical membrane. Furthermore, PepT-1 is pH dependent and, thus, is stimulated by $H^+$ ions at the luminal-brush border membrane interface. Transport across the basolateral membrane of enterocytes has not been characterized as well as that of the apical side because of the difficulties in isolating the basolateral membrane into vesicles for uptake studies. Nonetheless, peptide transport across the basolateral membrane of epithelial columnar cells is likely, but the quantitative and nutritional importance of peptide-derived intestinal absorption is unknown. Recent studies also showed that PepT-1 expression in pig intestine is regulated by dietary AA level (Zhang et al. 2013; Boudry et al. 2014). Zhang et al. (2013) indicated downregulation of PepT-1 in piglets fed low protein+branched chain amino acids (BCAA) (17.9% CP) compared to a low protein diet without BCAA (17.1% CP). Boudry et al. (2014) found that compared to normal protein formula (50 g CP l⁻¹), feeding higher protein formula (77 g CP l⁻¹) enhanced colonic PepT-1 activity in low birth weight pig neonates.

*Single Amino Acids*

The mechanisms and anatomical sites of intestinal AA uptake have been studied in many animal models such as murine, equine, ovine, bovine, chicken, and porcine.

### *Apical Transport*

All of apical transporters are ion-dependent and capable of concentrative transport, except for system $b^{0,+}$. System $b^{0,+}$ has received much attention because a defect in the human kidney system $b^{0,+}$ results in inherited hyperaminoaciduria cystinuria (Feliubadaló et al. 1999). The heteromeric transporter rBAT/$b^{0,+}$AT is the major transporter for cationic AA and cystine in kidney and intestine. The ontogenetic and anatomical expression of transporter rBAT/$b^{0,+}$AT have been characterized in swine intestinal tissue (Xiao et al. 2004; Feng et al. 2008; Wang et al. 2009). In the Tibetan pig, $b^{0,+}$AT is expressed at high level in small intestine and kidney, with weaker expression in the heart, brain, lung, and dorsal muscle tissue (Wang et al. 2009). The presence of $b^{0,+}$AT in the domestic pig (*sus scrofa*) was also confirmed in the duodenum, jejunum, and ileum, as well as in the colon (Xiao et al. 2004; Feng et al. 2008). As shown in Figure 5.3, system $b^{0,+}$ AA transport activity is $Na^+$-independent, mediating apical uptake of the basic AA arginine, histidine, and lysine ($AA^+$) and of cysteine (CSSC), in exchange for intracellular neutral AA, such as serine and threonine (represented as $AA^0$ in Figure 5.3; Broër 2008). The segmental distribution and developmental expression of the $b^{0,+}$AT protein (*SLC7A9* gene) along the intestine and in the kidney and muscle of the Tibetan suckling piglet at d 7 and 21 of age are shown in Figure 5.4.

The pattern of expression changed with lower expression in the jejunum and high expression in the ileum at d 21 compared to d 7. Similar pattern of expression was reported by Xiao et al. (2004), with higher abundance of $b^{0,+}$AT mRNA in the ileum and jejunum than in the duodenum, and leveling off in duodenum and jejunum by d 35. Feng et al. (2008) reported the mRNA abundance of $b^{0,+}$AT increased linearly from d 1 to 150 in ileum. System $b^{0,+}$ also facilitates the movement of lysine into epithelial cells of the chicken jejunum (Soriano-Garcia et al. 1999; Angelo et al. 2002). In addition to $b^{0,+}$AT, the presence of an inducible transporter for lysine in the presence of $Na^+$ has

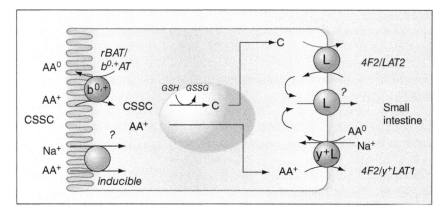

**Figure 5.3** Apical and basolateral transport of cationic ($AA^+$) and neutral ($AA^o$) amino acids in the intestinal cell. See text for description. (Source: Reproduced from Broër 2008 / with permission of American Physiological Society.)

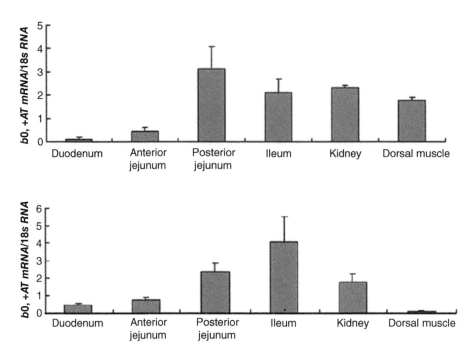

**Figure 5.4**    Relative mRNA abundance of b⁰,⁺AT in different tissues from Tibetan suckling piglet at day 7 (top) and d 21 (bottom). Abundance was determined by Northern blot and bands quantified with densitometry. (Source: Reproduced from Wang et al. 2009 / with permission of Elsevier.)

been suggested (Broër 2008). Along the bovine small intestine, lysine is transported via both $Na^+$ and $Na^+$-independent processes, with increasing capacity for uptake in the distal region such as the ileum but higher affinity in the jejunal region (Wilson and Webb 1990). The bulk of apical transport of neutral AA occurs against intracellular concentration gradient and, thus, is $Na^+$-dependent. Two major systems have been proposed, including $B^0$ and ASC, based on the respective identification of their molecular entities, namely $B^0AT$ and ASCT1 proteins.

*Basolateral Transport*
Transporter proteins of the $Na^+$-independent system $y^+$ are believed to play a role in lysine uptake across intestinal tissue in chicken (Angelo et al. 2002), bovine (Wilson and Webb 1990), and pigs (Wang et al. 2012), although the protein may be present and functional only on the basolateral membrane (Figure 5.5). Cationic AA efflux is mediated via CAT-1, a uniport transporter, in counter-exchange for neutral AA by y⁺LAT1 and y⁺LAT2. Cationic AA transporter CAT-1 mRNA was found in the horse's small intestinal mucosa (Woodward et al. 2010). Given its intracellular binding preference for arginine as demonstrated in oocytes and cultured brain cells (as discussed by Broër et al. 2000a) and the high blood concentration of glutamine, it has been proposed that the predominant function of y⁺LAT2 on the small intestinal basolateral membrane is to transport arginine from the intracellular enterocyte space to the abluminal side into portal blood in exchange for glutamine, thus explaining in part the high extraction rate of glutamine by the small intestinal epithelia.

The bulk of neutral AA are thought to be transported down-concentration gradient via ion-independent AA exchangers, including the ASC-1, LAT-2, y⁺LAT-1, and y⁺LAT-2 tranporter proteins (Krehbiel and Matthews 2003; Figure 5.5). The LAT-2 is one of the molecular entities of system L,

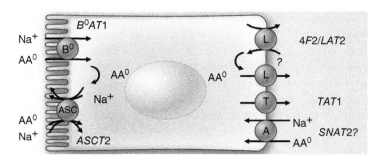

**Figure 5.5** Apical and basolateral transport of neutral amino acids (AA°) in the intestinal cell. See text for description. (Source: Reproduced from Broër 2008 / with permission of American Physiological Society)

and the mRNA of its encoding gene (SLC7A8) has been reported in the basolateral membrane of mouse and human intestinal epithelial cells (Rossier et al. 1999) and whole intestinal mucosa of the horse (Woodward et al. 2010). Aromatic AA seem to be selectively transported down concentration gradient via the Na+-independent uniporter TAT1. Finally, a symporter capable of concentrative transport and function to transport neutral AA from mesenteric blood supply into the cell appears via the Na+-dependent ATA2 and SNAT2 transport proteins. These systems of lysine uptake are likely utilized throughout the small intestine as evidenced by the presence of competitive inhibition among dietary AA in bovine (Wilson and Webb 1990) and eel cells (Vilella et al. 1990).

The affinity and capacity of Na+-independent and Na+-dependent transport systems and their transporters vary in different segments of the small intestinal tract (Wilson and Webb 1990), and thus presumably compensates for possible inefficiencies in lysine uptake resulting from competitive inhibition. For example, both Na+-dependent and Na+-independent jejunal systems of lysine transport showed greater affinity but lower transport capacity for lysine, while the same ileal systems have lower affinity but higher capacity for lysine transport (Wilson and Webb 1990). Whether these systems are under a coordinated regulation in response to dietary AA balance to ensure maximal lysine uptake is unknown.

*Luminal versus Arterial Amino Acid Utilization*
The first-pass metabolism has been recognized as substantial metabolism of AA (Stoll et al. 1998) in the gut, intestine, and liver. The intestinal enterocytes have high demand of nutrients from both the arterial blood across its basolateral membrane and the intestinal lumen across its brush-border membrane (Fang et al. 2010). It is without any doubt that the intestine is mainly supplied on the luminal side during feeding. However, AA are also available of the adluminal side of the basolateral membrane for counter exchange and efflux mediation of AA into the venous capillary system, particularly during interprandial periods. The mechanisms that determine the relative contribution of luminal and arterial AA directed toward meeting intestinal AA requirements are unknown. A clear picture of the partition between dietary and arterial AA utilization by the intestine is complicated by the fact that the relative contributions of these two sources vary continuously during the postprandial and interprandial periods, differing among the individual AA, and are affected by nutritional conditions (Bos et al. 2003). Stoll et al. (2000) reported that the luminal availability and mucosal uptake of enteral leucine were substantial in the proximal jejunum, but progressively diminished along the length of the small intestine. On the other hand, based on the notion that there is considerable protein accretion in the distal intestine despite reduced luminal AA availability, Stoll

et al. (1998) proposed that the distal intestine derives a larger proportion of its AA needed for protein synthesis from the circulation than from the diet. This explains in part the reduced fractional rate of protein turnover in the distal compared to the proximal intestine (Stoll et al. 2000). However, higher rate of protein turnover in the proximal region may be related to higher activities and that proximal region also relies on arterial supply rather than luminal supply.

Unlike free fatty acids, AA have small molecular weight and, thus, have a low reflection coefficient and can diffuse readily from the extracellular space into the venous capillaries via capillary pores. Thus, portal AA appearance represents the net absorption of AA.

*Large Intestine Amino Acid Utilization*

There is no doubt as to the role of the small intestine in AA utilization and absorption (Metges 2000), but there is enough evidence to suggest that the pig's large intestine has the ability to absorb AA. Labeled lysine was measured in the venous blood after introduction of $^{15}$N-labeled bacteria distal to the jejunum (Niiyama et al. 1979), indicating a possible role of microbially synthesized lysine absorption by the large intestine. In the pig's cecum, in situ absorption of asparagine, serine, threonine, tyrosine, arginine, histidine, lysine, and aspartic acid has been demonstrated (Olszewski and Buraczewski 1978). In the pig proximal colon, which lies immediately distal to the cecum, hydrophobic neutral AA were more readily absorbed than hydrophilic neutral or basic AA from any transport system (Sepúlveda and Smith 1979), characteristic of system $B^{0,+}$. Presence of $B^{0,+}$ system in the colon was also reported in the mouse based on absorption of radiolabeled glycine following direct administration to the luminal surface of the mouse colon in vitro (Ugawa et al. 2001). Expression of SLC6A14 mRNA was, however, evident in the fourth segment of the small intestine, cecum, and colon (Hatanaka et al. 2001). Abundance of mRNA was greater in the colon and cecum than the distal small intestine. In contrast, SLC15A1 mRNA (PepT-1) was detectable in all four segments of the small intestine, but not in the cecum and colon. These data show that the expression of SLC6A14 mRNA encoding $ATB^{0,+}$ is restricted to the distal region of the mouse intestinal tract. Quantitative estimate, however, points to approximately 10% of the microbially synthesized AA being absorbed from the large intestine, indicating that the large intestine of the pig, at least those fed conventional grain-based diets, is of little nutritional significance (Torrallardona et al. 2003a,b).

**Intestinal Epithelial Amino Acid Metabolism and Ontogeny of Utilization: From Neonatal to Early Postweaning Life**

*Glutamine and Glutamate*

Sow milk contains high concentrations of free and peptide-bound glutamine and glutamate, which are crucial for the growth, development, and function of the piglet small intestine (Wang et al. 2008; Kim and Wu 2009). Both glutamate and glutamine are entirely metabolized in the nursing pig small intestine (Reeds et al. 1996) as energy substrates for the enterocytes and for the synthesis of specific AA (proline, arginine, ornithine, and citruline), DNA, and proteins (Bertolo and Burrin 2008). Following transport across the brush-border membrane, the enzyme glutaminase (GSE) catalyzes the deamination of glutamine into glutamate, and the enzymes glutamate-oxalacetate amino transferase (GOT), glutamate-pyruvate transaminase (GPT), and glutamate deshydrogenase (GDH) catalyze the transamination of glutamate into $\alpha$-ketoglutarate. The resulting $\alpha$-ketoglutarate, then, enters the tricarboxylic acid cycle, where it is completely metabolized into $CO_2$ and ATP (Burrin and Stoll 2009). Glutamine and glutamate are also used for the synthesis of citruline, an arginine intermediate. The enzyme pyrroline-5-carboxylate synthase (P5CS) catalyzes the first step in the reaction, converting glutamate to $\Delta^1$-L-pyrroline-5-carboxylate (P5C). Then, $\Delta^1$-L-pyrroline-5-carboxylate is either exported to the cytosol, where it is converted to proline by the enzyme pyrroline carboxylate reductase (P5CR), or remaining in the mitochondria and available for transamination to ornithine by the enzyme ornithine $\delta$-aminotransferase (OAT). During most part of the nursing period, both pathways are absent in the piglet enterocyte because P5CS activity is low (Wu 1997; Davis and Wu 1998). Finally, glutamate can be used for the synthesis of N-acetylglutamate (NAG) by the enzyme N-acetylglutamate synthase (NAGS). The NAG is an allosteric activator of the enzyme ornithine transcarbomoylase (OTC), and thus plays a regulatory role in arginine and proline metabolism (Figure 5.6).

*Arginine*

Arginine is an essential AA for maximal growth (Southern and Baker 1983). Porcine milk provides less than 40% of the arginine requirement estimated for the one-week-old pig (Wu et al. 2004). Thus, endogenous synthesis of arginine plays a crucial role in maintaining arginine homeostasis in nursing piglets (Flynn and Wu 1996). Arginine is synthesized from dietary proline in the small intestine of nursing piglets and from endogenous citrulline in the kidney of postweaning pigs (Wu and Morris 1998; Bertolo and Burrin 2008). This age-related variation is due to developmental changes in the activity and expression of enzymes involved in arginine metabolism in the pig enterocytes, such as arginase (ARG I and II) and P5CR (Wu et al. 1996;

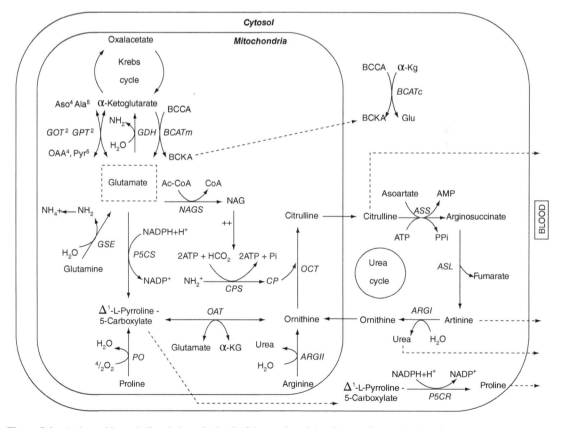

**Figure 5.6** Amino acid metabolism in intestinal cell of the nursing piglet. See text for description of pathways and symbols.

Wu 1997; Wu et al. 2004). Dietary proline is first converted into $\Delta^1$-L-pyrroline-5-carboxylate (P5C) by the enzyme proline oxidase (PO). Then, ornithine aminotransferase (OAT) catalyzes the transamination of P5C with glutamate to form ornithine, which is subsequently converted to citrulline by carbamoyltransferase (OCT). Citrulline passes from the mitochondria to the cytosol, where it is converted to arginosuccinate by the enzyme argininosuccinate synthetase (ASS). The argininosuccinate is then cleaved by the enzyme argininossucinase (ASL) to form free arginine and fumarate, the former released into the portal vein and the later entering the mitochondria to join the pool of citric acid cycle intermediates. Finally, portal arginine bypasses the liver and kidney and is available for whole body metabolism (Wu and Morris 1998; Bertolo and Burrin 2008).

In contrast to nursing piglets, enterocytes of postweaning pigs have very low activity of argininosuccinate synthetase (ASS) and argininossucinase (ASL), which catalyze the conversion of citruline into argininine, and very high activity of arginase (ARG I and II), that degrades the arginine (Wu 1997). Therefore, postweaning pigs cannot synthesize arginine directly in the small intestine. Instead, dietary arginine, glutamine, glutamate, and proline are converted to citruline in the enterocytes, and then released into the portal blood (Wu 1997; Bertolo and Burrin 2008). Subsequently, citrulline is converted to arginine by the kidney cells, which show abundant activities of argininosuccinate synthetase (ASS) and argininossucinase (ASL; Brosnan 2003). The net synthesis of citrulline by the small intestine provides an effective mechanism to bypass catabolism of dietary arginine by hepatic arginase, that otherwise would catabolize arginine completely (Bertolo and Burrin 2008). The conversion of dietary arginine to citrulline is catalyzed by both cytosolic and mitochondrial arginases (ARGI and II). In contrast, the synthesis of citrulline from glutamine involves the mitochondrial enzymes glutaminase (GSE), which converts glutamine to glutamate, and P5CS, which converts glutamate to $\Delta^1$-L-pyrroline-5-carboxylate (P5C). Finally, postweaning pig enterocytes can also synthesize proline from dietary arginine through the enzymes (OAT) and $\Delta^1$-L-pyrroline-5-carboxylate reductase (cytosol). For instance, inhibition of OAT was shown to decrease proline synthesis from arginine by 80–85% in pig enterocytes (Wu et al. 1996).

*Other Indispensable Amino Acids*
Enterocytes are an important site for substantial degradation of BCAA but not other essential amino acid (EAA) in the gut of nursing piglets (Chen et al. 2009). In nursing piglets, 40% of leucine, 30% of isoleucine, and 40% of valine are extracted by the portal drained viscera during first-pass metabolism (Stoll et al. 1998). Once absorbed by the enterocytes, BCAA are transaminated to the corresponding $\alpha$-keto acids (BCKA) by the enzyme branched-chain amino transferase (BCAT), located in both the cytosol and the mitochondria. Following transamination, mitochondrial branched-chain $\alpha$-keto acid dehydrogenase complex (BCKD) catalyzes the oxidative decarboxylation of all three BCKA producing the acyl-CoA derivatives. However, the activity of BCKAD is very low in piglet enterocytes, and consequently most of the BCKA are released into the extracellular space (Chen et al. 2009). It is, therefore, unlikely that the BCAA are quantitatively important energy substrates for the piglet small intestine. It has been proposed that the mucosal catabolism of BCAA may function to provide nitrogen for the synthesis of both alanine and glutamate (Figure 5.6) and to generate BCKA. The role of BCKA is the enterocytes is unknown, but it was also proposed that BCKA decreases proteolysis in enterocytes as reported in the chick's skeletal muscle (Nakashima et al. 2007).

Stoll et al. (1998) reported high catabolism of histidine, lysine, methionine, phenylalanine, threonine, and tryptophan by enterocytes of 0- to 21-day old piglets. Others (Chen et al. 2007, 2009) have shown a lack of substantial oxidation of these indispensable AA in enterocytes of pigs because of the absence of the key enzymes responsible for their degradation, including threonine dehydrogenase, threonine dehydratase, saccharopine dehydrogenase, and phenylalanine hydroxylase. Consequently, it is possible that the metabolism of histidine, lysine, methionine, phenylalanine, threonine, and tryptophan by the small intestine may result from the action of luminal microbes in the intestinal mucosa and thus accounting in part for their disappearance from the intestinal lumen. Chen et al. (2009) suggested that microbial modification and utilization of these indispensable AA may offer a mechanism by which dietary supplementation with antibiotics enhances protein deposition in skeletal muscle of young pigs (Bergen and Wu 2009). Limited degradation of methionine and phenylalanine may be catalyzed by the BCAT and possibly glutamine transaminases L and K (Wu and Thompson 1989a), however, only when BCAA and other AA are absent as shown in the rat and chicken (Wu and Thompson 1989b; Wu et al. 1991). For instance, 2 mM addition of leucine, isoleucine, and valine to incubation medium completely inhibited the transamination of methionine and phenylalanine in enterocytes of both preweaning (Chen et al. 2009) and postweaning pigs (Chen et al. 2007).

## Mammary Gland Amino Acid Utilization

### Mammary Amino Acid Transport: Mechanisms and Regulation

Amino acid transfer processes across the porcine mammary tissue have been studied both in vivo (Trottier 1997; Guan et al. 2002; Nielsen et al. 2002; Guan et al. 2004) and in vitro (Hurley et al. 2000; Jackson et al. 2000; Chen et al. 2018) (Table 5.1). Milk demand by growing piglets increases as lactation advances (Hartmann et al. 1997; Zhang et al. 2018), and it appears that the net uptake of AA by the porcine mammary gland increases with the progression of lactation via increases in arteriovenous differences, indicating some regulation of AA uptake at the level of transport per se (Nielsen

**Table 5.1** Major amino acid transporters identified in porcine mammary tissue[a].

| Systems | Genes | Proteins | Substrates | Energy sources |
|---------|-------|----------|------------|----------------|
| Cationic amino acid transporters | | | | |
| y$^+$ | SLC7A1 | CAT-1 | Lys, Arg, His, Orn | Na$^+$ independent |
| y$^+$ | SLC7A2 | CAT-2b | Lys, Arg, His, Orn | Na$^+$ independent |
| Anionic amino acid transporters | | | | |
| | SLC1A1 | EAAT3 | Glu, Asp | Na$^+$ dependent |
| Neutral amino acid transporters | | | | |
| ASC | SLC1A4 | ASCT1 | Ala, Ser, Cys | Na$^+$ dependent |
| | SLC1A5 | ASCT2 | Ala, Ser, Cys, Thr, Gln | Na$^+$ dependent |
| A | SLC38A2 | SNAT2 | Gly, Pro, Ala, Ser, Cys, Gln, Asn, Ser, Met | Na$^+$ dependent |
| L | SLC7A8 | LAT2 | Leu, Ile, Val, Trp | Na$^+$ independent |
| Cationic and neutral amino acid shared transporters | | | | |
| y$^+$L | SLC7A7/SLC3A2 | y$^+$LAT1/4F2hc | Cationic: Lys, Arg, His | Na$^+$ independent |
| | SLC7A6/SLC3A2 | y$^+$LAT2/4F2hc | Neutral: Gln, Leu, Met, Ala, Cys | Na$^+$ independent |
| b$^{0,+}$ | SLC7A9/SLC3A1 | b$^{0,+}$AT1/rBAT | Cationic: Lys, Arg, Orn | Na$^+$ independent |
| | | | Neutral: Cys, Leu | Na$_+$ independent |
| B$^{0,+}$ | SLC6A14 | ATB$^{0,+}$ | Cationic: Lys, Arg, His | Na$^+$ dependent |
| | | | Neutral: Leu, Ile, Val, Met, Ala, Ser, Thr | Na$^+$ dependent |

[a] Adapted from Manjarín et al. (2014) and Zhang et al. (2018).

et al. 2002). The role of blood flow in contributing to the change in net mammary AA transport with progression of lactation is less well understood. Transcript abundance of genes SLC1A4 and SLC6A14, which respectively encode ASCT-1 and ATB$^{0,+}$ transporters, increased over 2- and 1.3-folds, respectively, from d 4 to 18 of lactation (Pérez Laspiur et al. 2004). Chen et al. (2018) also reported the upregulation of SLC7A6/SLC3A2 and SLC1A1 during peak lactation. Lysine is the first limiting AA for milk protein synthesis, particularly when diets are based on corn and soybean meal as the main protein sources (Richert et al. 1997). The limiting availability of lysine for lactation was first biologically demonstrated by its highest mammary extraction rate among of all of the essential AA throughout lactation (Trottier et al. 1997). As a consequence, up or downregulation of genes encoding for lysine transporters may regulate lysine utilization via modulation of lysine uptake by mammary cells and, thus, may impact total dietary nitrogen utilization during lactation. A number of studies have shed light on the nature of the lysine transport system in porcine mammary tissue. Studies using porcine mammary explants have shown lysine to be transported via a Na$^+$-independent system that differs from the classical y$^+$ system (Shennan et al. 1994; Hurley et al. 2000). This notion is based on two observations: (i) lysine uptake in porcine mammary tissue occurs via a Na$^+$-independent transport mechanism with a K$_m$ of approximately 1.4 mM (Hurley et al. 2000), which is 3–10-fold greater than the reported K$_m$ for y$^+$ systems in any other tissues (Devés and Boyd 1998) and (ii) lysine uptake in porcine mammary tissue is not specific for lysine because it can be inhibited by 50% in the presence of L-leucine, L-alanine, and L-methionine at supra-physiological concentrations (Calvert and Shennan 1996; Hurley et al. 2000). In the mouse mammary gland, arginine was shown to be transported via two systems, one specific for cationic AA (i.e., the classical y$^+$ system, as discussed earlier in this chapter) and the other capable of interacting with both cationic and neutral AA (Sharma and Kansal 2000). In fact, the K$_m$ for arginine uptake in mouse mammary tissue via the y$^+$ system was reported to be 0.76 mM (Sharma and Kansal 2000), which is approximately 10-fold greater than the reported K$_m$ for arginine uptake in other tissues (Devés and Boyd 1998). It may be argued that the K$_m$ for lysine uptake via the y$^+$ system in porcine mammary tissue may also be much greater than that found in other tissues. The fact that lysine uptake is only partly inhibited by supra-physiological concentrations of neutral AA and strongly inhibited by physiological concentrations of arginine (Hurley et al. 2000; Shennan and Boyd 2014) indicates that system y$^+$ is of physiological and nutritional importance in lactating sow mammary gland. Pérez Laspiur et al. (2004) reported expression of genes SLC7A1 and SLC7A2, respectively, encoding for system y$^+$ AA transporters CAT-1 and CAT-2b in porcine mammary cells, with CAT-2b responding to AA availability in an adaptive regulation pattern (Pérez Laspiur et al. 2009).

Numerous studies support the notion that an interaction for transport exist between cationic and the BCAA. Uptake of valine from lactating sow mammary tissue was strongly inhibited by leucine and lysine present at physiological concentrations (Hurley et al. 2000; Jackson et al. 2000). Uptake of lysine from lactating rat mammary tissue was also strongly inhibited by leucine present at physiological concentration (Shennan et al. 1994; Calvert and Shennan 1996). Interactions between cationic and neutral AA in the mammary gland may have important nutritional implications. For instance, purified AA are commonly used to supplement lactation diets. Over supplementation with purified lysine in sow diets has led to deficiencies in valine (Richert et al. 1996, 1997). Conversely, over supplementation of purified valine decreased in vivo lysine mammary transmembrane transport in lactating sows (Guan et al. 2002). Knowledge of interaction between cationic AA and BCAA is necessary to understand potential interactions among other AA as more purified AA are commercially available for sows fed low protein diets. In lactating sows, arteriovenous differences of AA across the mammary glands increased as concentration of dietary protein increased to meet requirement, but decreased except for that of leucine and isoleucine when protein concentration was fed in excess of requirement (Guan et al. 2004). Leucine and isoleucine arteriovenous concentrations increase in response to increase in dietary intake of these AA. Thus, when feeding high protein diets, mammary glands appears to adaptively respond by decreasing transport of cationic and other neutral AA, but not by decreasing leucine or isoleucine transport. The mechanisms behind potential interactions between the cationic and BCAA in the lactating sow are unknown, although as discussed earlier in this chapter, AA share common transporters. Pérez Laspiur et al. (2004) reported the existence of gene SLC6A14 for transporter ATB$^{0,+}$ (System B$^{0,+}$) in porcine mammary tissue. However, SLC6A14 mRNA abundance remained unaffected by changes in dietary protein intake and stage of lactation (Pérez Laspiur et al. 2009), indicating that the nature of an inhibition or decrease in lysine uptake via ATB$^{0,+}$ in the presence of high levels of neutral AA would be competitive rather than noncompetitive.

The earlier described heteromeric cationic transporters rBAT/b$^{0,+}$AT (system b$^{0,+}$), y$^{+}$LAT1/4F2hc, and y$^{+}$LAT2/4F2hc (system y$^{+}$L) may explain the nature of lysine, arginine, and large neutral BCAA interaction at the mammary apical interface. These transporter proteins are expressed in pig mammary tissue during lactation (Manjarín et al. 2012). In conditions of protein excess, y$^{+}$LAT1/4F2hc or y$^{+}$LAT2/4F2hc or both may be upregulated, increasing the net uptake of neutral AA and decreasing the uptake of lysine by mammary cells (Figure 5.7). This hypothesis is based on the earlier discussion whereby excess of protein in diet selectively increased the uptake of leucine and isoleucine by the pig mammary gland, whereas the transport of lysine decreased (Guan et al. 2004). Moreover, an excess of leucine (Calvert and Shennan 1996) and valine (Richert et al. 1997) selectively inhibited the uptake of lysine in sow mammary tissue. In conditions of AA deprivation such as low protein intake or high milk demand, rBAT/b$^{0,+}$AT may be upregulated ensuring sufficient amounts of lysine uptake by mammary cells to support milk protein synthesis. To support lysine/neutral AA counter-transport, high intracellular concentration of neutral AA would be ensured by concentrative (Na$^{+}$ cotransport) neutral AA transport activities via ATB$^{0,+}$ or B$^{0}$AT1 of system B$^{0}$ as described earlier in this chapter. Again, B$^{0}$AT1 is a Na$^{+}$-dependent AA transporter that actively transports large BCAA, such as leucine and valine, but not anionic or cationic AA.

Finally, Pérez Laspiur et al. (2004) and Chen et al. (2018) reported the existence of gene SLC1A4 which encodes ASCT1 transporter protein. Transporter ASCT1 is a Na$^{+}$-dependent AA transporter for small neutral AA, such as alanine, serine, and cysteine (Scalise et al. 2018). However, ASCT1 cannot contribute to the net transport of neutral AA across the apical membrane because of an obligatory exchange of substrate AA against each other (Broër et al. 2000b). Nevertheless, Pérez Laspiur et al. (2009) found an increase of SLC1A4 (ASCT1) mRNA abundance throughout lactation, indicating that it may play a role in the regulation of cellular AA utilization by the porcine mammary cells.

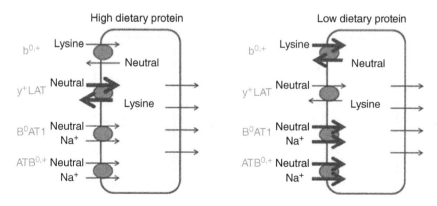

**Figure 5.7** Hypothetical model for regulation of specific lysine transporters in conditions of dietary protein excess (left) and in conditions of protein deprivation or high milk demand (right).

## Amino Acid Utilization During Growth

### *Insulin Signaling Pathway Is Shared by Amino Acids*

The neonatal period is characterized by rapid growth rate, due to high levels of protein synthesis in all tissues of the animal in response to feeding (Suryawan et al. 2006). Gain in protein mass is more pronounced in skeletal muscle, where it is controlled by the postprandial rise in insulin and AA. The ability of skeletal muscle to respond to both anabolic stimuli contributes to a more efficient AA utilization and, therefore, higher rate of protein synthesis than other organs (Davis et al. 2002). The molecular mechanisms, by which insulin and AA control protein synthesis, are just beginning to be understood. Numerous studies have shown that both hormonal and nutrient stimuli share the mammalian target of rapamycin (mTOR) pathway to induce gene expression and protein synthesis in the neonate (Kimball and Jefferson 2006; Avruch et al. 2008; Wang and Proud 2009; Wu 2009).

Insulin activates mTOR via the PI-3-kinase-Akt pathway (Figure 5.8). Binding of insulin to its receptor activates it, which in turn stimulates the enzyme phosphatidylinositol 3-kinase (PI3K). The PI3K, then, catalyzes the conversion of membrane-bound $PIP_2$ (phosphatidylinositol

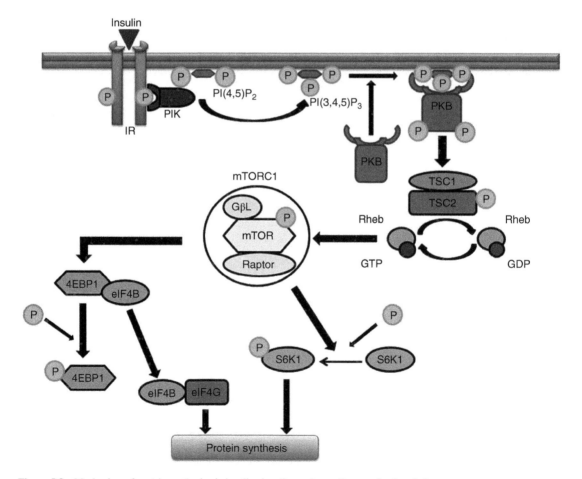

**Figure 5.8**    Mechanism of protein synthesis via insulin signaling pathway. See text for description.

(4,5)-bisphosphate) to PIP$_3$ (phosphatidylinositol (3,4,5)-triphosphate). The PIP3 recruits the protein kinase B (PKB), and this leads to the activation of the enzyme. The PKB activates mTOR directly by phosphorylation of the protein and indirectly by inhibiting the actions of the TSC1/TSC2 (Tuberous Sclerosis Complex). The TSC1 and TSC2 form a functional complex that, in its active form, inactivate Rheb (Ras Homolog Enriched in Brain), an mTOR activator. Activation of mTOR results in the formation of mTOR Complex-1, containing mTOR, raptor (regulatory associated protein of mTOR), and G-BetaL (G-protein beta-subunit-like protein). The mTOR Complex-1 mediates the phosphorylation of the eIF4EBP1 (Eukaryotic Translation Initiation Factor-4E-Binding Protein-1) and the ribosomal protein S6K1 (S6 Kinase). Unphosphorylated 4EBP1 binds to eIF4E (Eukaryotic Translation Initiation Factor-4E) and inhibits the initiation of protein synthesis. Phosphorylation of 4EBP1 by mTOR reduces its affinity for eIF4E, and the 2 proteins dissociate. The EIF4E is then able to associate with other components of eIF4F, forming an active complex and beginning protein translation. On the other hand, the activation of S6K1 leads to phosphorylation of the 40S ribosomal S6 protein, facilitating the recruitment of the 40S ribosomal subunit into actively translating polysomes and, therefore, increasing protein synthesis.

The molecular mechanisms, by which AA or leucine modulates the activation of mTOR pathway in vivo, are unclear. AA stimulate mTOR Complex-1 independent of PKB and TSC1/2, as indicated by the fact that fed AA levels have no effect on PKB or TSC2 phosphorylation (Suryawan et al. 2008). AA may induce protein synthesis by increasing S6K1 and 4EBP1 activation (Suryawan et al. 2008), or by regulating the interaction of raptor with mTOR (Hara et al. 2002; Corradetti and Gua 2006), resulting in the activation of mTORC1.

### *Developmental Regulation of Protein Synthesis in the Growing Pig*

Rate of protein synthesis in skeletal muscle is the greatest among all tissues in the neonate and decline more rapidly than in the rest of the body during the early postnatal period (Davis et al. 2002). Such developmental decline in muscle protein synthesis is more accentuated in fast twitch, glycolytic muscles, and is accompanied by a decrease in ribosome number in the cells (Davis et al. 2002) and a decrease of the efficiency with which ribosomes translate mRNA in response to the postprandial rise in insulin and AA (Suryawan et al. 2006). Additionally, activation of many of the AA and insulin-signaling components that are involved in the regulation of the protein synthesis in skeletal muscle is developmentally regulated. Activation of positive regulators of protein synthesis, mTOR, S6K1, and 4EBP1, decreased with age in muscle, whereas activation of negative regulators of protein synthesis such as TSC2 was higher in younger pigs (Suryawan et al. 2006). Finally, raptor abundance and the association of raptor to mTOR were greater in 7 than in 26-day-old pigs, indicating a decrease not only in protein activity but also in expression of genes in the muscle protein synthesis machinery associated with the age of the piglet (Suryawan et al. 2006). In a nutshell, muscle AA utilization is impacted by the aging process via the tissue's ability to respond to hormonal (e.g., insulin) and nutritional (e.g., leucine) stimuli after meals on the level of gene expression and protein activity (O'Connor et al. 2003; Escobar et al. 2005; Suryawan and Davis 2014).

### *Utilization and Requirement of Amino Acids Affected by Pig Health Status*

The antibiotics are being phased out as feed additives in many countries due to increasing concerns of resistant bacteria in the food chain (Zeng et al. 2015). The restriction of antibiotics use poses swine herd at lower health status and prone to health challenge. The impact of lower health status on AA requirements has not been clearly demonstrated, and results have been inconsistent. While it is traditionally believed that lower health status increases AA requirement, a number of studies suggest that decreasing health status barely affects requirements, and a few suggest lowered requirements.

The immune system stimulation (ISS) generally results in reduction in skeletal muscle protein synthesis and increases in muscle protein breakdown, leading to lower whole-body protein deposition and compromised growth in animals (Johnson 1997; Klasing 2007; Lang et al. 2007). Reduction of muscle protein synthesis may due to lack of mTOR response to leucine in muscle during ISS (Frost and Lang 2011). On the other hand, the increase in muscle protein breakdown may be due to higher demand of free AA for immunologically related protein synthesis, such as acute phase proteins (APP), immunoglobulins, and glutathione (McGilvray et al. 2019b). The increased demand for immunologically related protein synthesis has been suggested to potentially impact the AA requirement both qualitatively (AA profile) and quantitatively (Rakhshandeh and de Lange 2011; McGilvray et al. 2019a).

*Nitrogen*

Glutamine serves as either fuel source or, more likely, major nitrogen donor for the immune system, and consequently organisms have higher glutamine demand during ISS (Reeds and Jahoor 2001). Glutamine is a dispensable AA (NRC 2012), and higher requirement for glutamine may increase the demand for nitrogen. McGilvray et al. (2019a) reported alteration of dietary nitrogen utilization, AA flux and pool size when growing pigs are exposed to ISS following administration of LPS (30 and $36 \mu g \, kg^{-1}$ BW). In that study, ISS exposure reduced the apparent ileal digestibility of dietary nitrogen by 57–76% and tended to decrease the efficiency of nitrogen utilization by 53–59% (McGilvray et al. 2019a). However, Huntley et al. (2018) found ISS did not impact nutrient digestibility and N balance of weaned pigs, but increased $ME_m$ (23%), resulting in reduced lipid deposition (−30%) and average daily gain (−18%). The increased glutamine demand for immune activation may not be high enough to affect the global N requirement.

*Lysine and Threonine*

Lysine is traditionally recognized as the first limiting AA for pigs when fed corn-soybean-based diet (NRC 2012) and may be one of the limiting factor for protein synthesis, such as immunoglobulin synthesis during ISS. Williams et al. (1997) and McGilvray et al. (2019a) indicate ISS lowered lysine requirement, pointing to less body protein deposition (Waterlow 2006). Kampman-van De Hoek et al. (2015) reported that lysine flux was unaffected with exposure to ISS in pigs.

Threonine is essential for hepatic synthesis of APP and immunoglobulins during ISS (Reeds and Jahoor (2001) and for intestinal synthesis of mucus (Montagne et al. 2003), which is composed of threonine-rich glycoproteins (Munasinghe et al. 2017; Pluske et al. 2018). Intestinal mucus secretion is greater during ISS (Rakhshandeh et al. 2013). Rémond et al. (2009) reported a sevenfold increase in threonine flux across the portal drained viscera in pigs and suggested threonine requirements are likely to be augmented based on these observations. Stuart et al. (2015) reported a 1.5-fold increase in plasma threonine flux in pigs challenged with porcine reproductive and respiratory syndrome virus. Potential interaction between dietary fiber and ISS on threonine requirement was suggested based on the fact that high dietary fiber increases threonine requirement via increased endogenous threonine loss (Wellington et al. 2018). While ISS may increase threonine requirement, fiber and ISS effects on threonine requirement are not additive (Wellington et al. 2018).

*Branched-Chain Amino Acids*

Although inflammation enhances muscle protein degradation, a significant portion of the inflammation-induced muscle waste is due to a failure of muscle to maintain protein synthesis (Klasing 2007; Lang et al. 2007; Frost and Lang 2011). The BCAA are in close relationship to muscle protein synthesis and breakdown. McGilvray et al. (2019a) reported reduced metabolic demand for isoleucine and no change in metabolic demand for leucine during ISS in growing pigs. In addition, leucine supplementation above requirement in growing pigs exposed to ISS did not affect protein turnover, thus there were no impact on protein synthesis nor degradation (Rudar et al. 2017). Possible mechanisms include impaired transport of leucine into muscle and reduced responsiveness of the mTOR pathway (Frost and Lang 2011)

*Aromatic Amino Acids*

The utilization of aromatic AA (phenylalanine, tyrosine, and tryptophan) is expected to increase during the acute phase response of ISS because of the high turnover of aromatic AA-enriched APP during ISS (Reeds et al. 1994). It has been reported that tryptophan utilization increases during ISS (Melchior et al. 2004; Rakhshandeh and de Lange 2011; de Ridder et al. 2012), based on the notion that tryptophan catabolism in greater during inflammatory and immune responses. McGilvray et al. (2019a) found that ISS had no effect on tryptophan utilization and utilization declined, which was probably attributed to the reduced protein synthesis (McGilvray et al. (2019a). Furthermore, lower utilization of phenylalanine and ISS may also lead to lower flux of tyrosine, since proinflammatory cytokines induced from ISS can downregulate the phenylalanine-hydroxylase, which is essential to the conversion from phenylalanine to tyrosine (Capuron et al. 2011). Taken together, apparent utilization of aromatic AA during ISS is affected by (i) high turnover of aromatic AA-rich APP leading to increased aromatic AA utilization and (ii) reduced phenylalanine-hydroxylase activity and protein synthesis, resulting in decreased aromatic AA utilization. In fact, McGilvray et al. (2019a) recently suggested the utilization of phenylalanine for APP turnover during ISS may not be significant enough to affect phenylalanine flux, considering the reduced phenylalanine utilization associated with decelerated protein synthesis, and thus phenylalanine requirement likely remains unaffected.

*Sulfur Containing AA*

The sulfur-containing AA (SAA) are essential for the synthesis of immune system metabolites (e.g., APP, glutathione, taurine, etc.), and the demand of SAA is considered to be higher during ISS (Rakhshandeh and de Lange 2010; Litvak et al. 2013). The ISS induces production of free radicals with strong antimicrobial properties (Colditz 2002). Animal cells are also prone to these free radicals, and thus have developed protective strategies to eliminate excessive free radicals, such as production of SAA-derived glutathione (Brosnan and Brosnan 2009). Increases in glutathione turnover is suggested to have an impact on dietary cysteine requirements and may contribute to muscle protein mobilization during ISS (Rakhshandeh and de Lange 2011). Rakhshandeh and de Lange (2011) also suggested approximately 20% increase in SAA demand during the ISS and the need of methionine:cystine increased from 0.57 to 0.62, pointing to greater demand for methionine than cystine during immune response in pigs. However, Kampman-van De Hoek et al. (2015) and McGilvray et al. (2019a) reported that ISS did not affect SAA flux, suggesting that increased immunological demand for methionine may not be significant enough to affect methionine flux, utilization, and therefore dietary requirements.

While AA requirements may increase during ISS in response to augmented production of immunologically related proteins, the increase in muscle protein breakdown and decrease in muscle accretion may offset most of the changes in requirements (Goodband et al. 2014). Humans are a good model to help understand the biological rationale. For instance, Reeds and Jahoor (2001) pointed out that additional AA required for new protein synthesis (~10 mg kg$^{-1}$ day$^{-1}$) during immune response is small compared to whole body protein turnover (5000 mg kg$^{-1}$day$^{-1}$) and nitrogen loss (150 mg kg$^{-1}$ day$^{-1}$). Increased demand of AA for immune response can be mostly satisfied by increased proteolysis during ISS (Goodband et al. 2014). Additional studies are needed to determine the impact of ISS on tryptophan and SAA dietary needs, and the contribution of proteolysis to increase availability of these AA depends on the magnitude of the immune response (Pluske et al. 2018).

**Amino Acid Partitioning During Gestation**

Limited information is available on the AA utilization of the pregnant sow. Nonetheless, several studies have characterized temporal changes in protein and AA accretion in fetal tissue during gestation. Such information provides potential valuable tool for designing nutritional strategies aimed at maximizing the efficiency of dietary AA utilization by the pregnant sow. AA during pregnancy are partitioned to fulfill maternal needs, which include maintenance and mammary tissue growth, and extra-mammary tissue accretion, and to fulfill needs associated with the products of conception, which includes fetal and placental growth.

*Fetal Growth*

Knowledge of fetal AA metabolism is fundamental to our understanding of amino utilization during pregnancy. Current knowledge, however, has not culminated in the development and adoption of new protein and AA feeding strategies of the pregnant sow because many of the processes are not quantified. Earlier studies reported that pregnant sows can be fed a diet containing as low as 0.5% protein early in gestation before negatively affecting the birth weight of her progeny (Atinmo et al. 1976). Wu et al. (1998) demonstrated that pregnant sows fed negligible protein (0.5%) maintained plasma AA concentrations likely though increased maternal body protein degradation and decreased AA oxidation rate to spare AA for fetal utilization. In that same study, amniotic fluid and fetal plasma AA concentrations decreased, indicating a reduction in placental AA transfer, potentially via downregulation of placental apical AA transporter genes as reported in protein malnourished pregnant rats (Malandro et al. 1996). Protein deficiency until day 60 of gestation decreased amniotic fluid concentrations of arginine and ornithine by as much as 37 and 48%, respectively, with a corresponding increase in fetal plasma ammonia concentration, possibly indicating an important role for arginine and ornithine on in utero ammonia detoxification.

Fetal growth significantly accelerates during the second half of pregnancy (Figure 5.9; Wu et al. 1999; McPherson et al. 2004; Ji et al. 2005). Prior to d 70 of gestation, daily protein accretion per fetus is 0.25 g and increases to 4.63 g day$^{-1}$ after d 70 of gestation. Between d 40 and 110 of gestation, protein accretion accounts for as much as 64–51%, while fat accounts for only 16–12.4%. Intestinal growth increases noticeably, accounting for 2.5–6.2% of body weight for d 40 to 110 of gestation, respectively, and represents the largest increase in organ size (McPherson et al. 2004). Fetal AA composition changes with gestation, with a substancial increase in glycine and hydroxyproline and a notable increase in proline and arginine (Table 5.2). All other AA decrease as a proportion of total AA. Glutamine and glutamate are not presented in Table 5.2. The sum of these two AA represent 13.5% of fetal AA composition and do not change with gestational age.

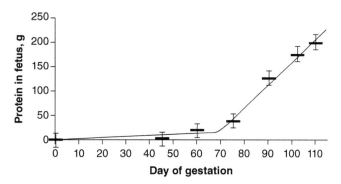

**Figure 5.9**  Relationship between day of gestation and fetal protein content (g). Breakpoint occurred at 68.5 d of gestation. The relationship before d 68.5 is described as $y = 0.249 \times (x - 68.5) + 17.078$ and after d 68.5 is described as $y = 4.629 \times (x - 68.5) + 17.078$. (Source: Reproduced from Ji et al. 2005 / with permission of Oxford University Press.)

**Table 5.2**  Amino acid composition of the fetal pig (g amino acid/100 g total amino acids)[a].

| Amino acid | Gestational age, d | | | | |
|---|---|---|---|---|---|
| | 40 | 60 | 90 | 110 | 114 |
| Alanine | 6.06 | 6.13 | 6.45 | 6.83 | 6.76 |
| Arginine | 6.30 | 6.60 | 6.93 | 6.80 | 6.78 |
| Glycine | 6.16 | 7.34 | 9.88 | 10.7 | 11.30 |
| Histidine | 2.57 | 2.14 | 2.15 | 2.15 | 2.17 |
| Hydroxyproline | 0.84 | 1.85 | 3.15 | 3.50 | 3.64 |
| Isoleucine | 3.64 | 3.48 | 3.15 | 3.04 | 3.03 |
| Leucine | 8.42 | 7.81 | 7.29 | 7.13 | 7.07 |
| Lysine | 8.60 | 6.97 | 6.26 | 6.04 | 6.04 |
| Methionine | 2.32 | 2.14 | 2.03 | 1.99 | 1.95 |
| Phenylalanine | 4.64 | 4.14 | 3.81 | 3.65 | 3.60 |
| Proline | 6.15 | 8.29 | 8.20 | 8.21 | 8.19 |
| Threonine | 4.21 | 3.87 | 3.66 | 3.47 | 3.39 |
| Tryptophan | 1.20 | 1.24 | 1.24 | 1.20 | 1.16 |
| Valine | 5.68 | 4.91 | 4.74 | 4.51 | 4.41 |

[a] Adapted from Wu et al. (1999). The number of animals per gestational age is 6 for d 40, 60, and 90, 5 for d 110, and 4 for d 114.

Based on the hydroxyproline figure, it is estimated that collagen represents 7% of the total fetal protein content on d 40 of gestation and up to 29% between d 110 and 114 of gestation. Uterine uptake of arginine and proline plus hydroxyproline meets requirements for fetal growth during late gestation only marginally. Proline extraction by the uterine vein is second to highest of all dispensable AA, while arginine extraction is nearly half of that of proline on d 110 of pregnancy (Figure 5.9).

On the other hand, although citrulline and ornithine exhibit relatively low uterine extraction rates (not shown in Figure 5.10), their corresponding net uterine uptake was estimated to be 55- and 15-fold greater, respectively, than their accretion in fetal tissue, indicative of fetal citrulline and ornithine utilization for arginine synthesis (Figure 5.11). Thus, citrulline and ornithine seem to spare arginine. Of the indispensable AA, lysine and methionine demonstrate the highest extraction

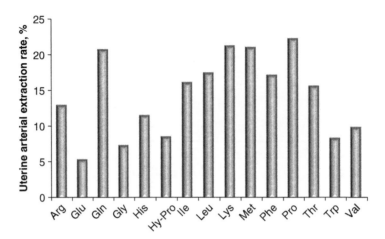

**Figure 5.10** Uterine arterial amino acid extraction. Values are expressed as percentage calculated from Wu et al. (1999) uterine arteriovenous difference concentration × 100/arterial concentration obtained on days 110–114 of gestation.

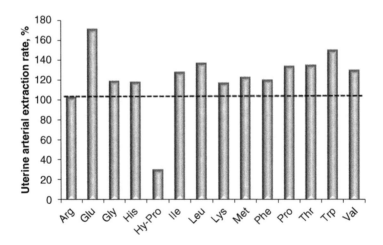

**Figure 5.11** Net uterine amino acid uptake relative to fetal amino acid accretion on days 110–114 of gestation. Values obtained from Wu et al. (1999). Dashed-line indicates 100% of uterine uptake corresponding to fetal accretion. Citrulline and ornithine are 5537 and 1502%, respectively, and, thus, not presented in the figure.

rates, followed by leucine and phenylalanine. In terms of uterine arteriovenous differences, leucine demonstrates the largest, followed by lysine, arginine, and threonine (Wu et al. 1999). Therefore, leucine, lysine, threonine, methionine, and phenylalanine seem to be crucial indispensable AA used in fetal development. It is unknown whether or not the extraction rate and pattern change with advancement of gestation. Such change would reflect the relative utilization of AA during pregnancy for fetal growth. Uterine uptake of arginine, proline, and hydroxyproline is marginal to meet fetal arginine requirement, indicating that arginine is synthesized by the fetal pig. Of all indispensable AA, lysine demonstrates the closest to 100% utilization of uterine uptake to fetal accretion, followed by phenylalanine and methionine (Figure 5.11).

### Mammary Gland Growth

The mammary gland of swine develops extensively during gestation for both the primiparous and multiparous sows and as well as during lactation for the primiparous sow. Understanding mammary gland development during gestation is critical to maximize development of the milk producing mammary epithelial cells that define the mammary parenchyma because litter growth rate in lactation is directly correlated with mammary gland size (Nielsen et al. 2001). During gestation, mammary growth is characterized by a marked increase in parenchymal tissue mass and mammary tissue DNA concentration during the last third of the pregnancy period (Kensinger et al. 1982; Sørensen et al. 2002), which coincide with the characteristic increase in fetal protein accretion rate. Although one could propose that these two events are under some shared coordinated regulation, such as mammogenic hormones (Ji et al. 2006), mammary tissue development may in part be regulated by fetal signaling (Kensinger et al. 1986). The mammary gland contains substantial amount of fat with ether extract concentration linearly decreasing from 94 to 58% from d 45 to 112 of pregnancy. In contrast, protein concentration increases from 5 to 39%, with the abdominal glands containing more proteins and less fat compared to those of the thoracic and inguinal regions (Ji et al. 2006). Individual mammary gland daily protein accretion rate during the slow developmental phase of pregnancy (d 1 to 75) averages 0.08 g and during the rapid developmental phase (d 75 to 112) averages 1.05 g. Assuming a total of 12 fetuses and 14 mammary glands, these measurements translate into dietary AA partitioning between fetal and mammary pools to meet the need of tissue protein accretion, with 27 and 73% utilized in mammary and fetal protein synthesis, respectively, during the slow developmental phase of gestation. During the rapid developmental phase, a shift occurs in favor of fetal protein accretion, with a partitioning of 21% to mammary and 79% to fetal pools. Although relatively minor, such shift may translate into higher channeling of dietary AA toward the mammary tissue pool in the early up to two thirds relative to last third of gestation. This observation sheds some lights into the importance of minimizing mammary fat pad accretion and maximizing AA utilization for mammary protein accretion, in particular during the slow developmental phase of gestation. Such maximization may be dependent on the appropriate balance of AA. For instance, the transition between slow and rapid developmental phase coincides with changes in histological structures, with abundant network of collagenous tissue defining the fat pad adipocyte lobules in early to mid-gestation to elongation of lactiferous ducts and development of the epithelial structures in latter period (Hovey et al. 1999). Thus, mammary protein composition may change with stromal cell proteins containing more hydroxyproline compared to epithelial cell proteins. Amino acid composition and profile of the mammary gland between early and late pregnancy are not known, but such knowledge would provide additional tool for the factorial estimate of dietary AA profile for pregnancy. In addition, the factors that govern AA uptake by the mammary gland during gestation have not been researched. Given that mammary development markedly accelerates during the last third of pregnancy, it is likely that mammary AA utilization and possibly the efficiency therefore increases.

Finally, energy intake during pregnancy can negatively impact mammary parenchymal tissue protein accretion during the rapid developmental phase between d 75 and 105, and, thus, likely AA utilization. Excessive energy intake in gilts decreased DNA accretion, parenchymal RNA, and total mammary parenchymal protein. Although greater dietary protein intake does not increase total mammary parenchymal protein, it seems to reduce mammary extraparenchymal stromal weight. Thus, increasing dietary protein between d 75 and 105 of gestation does not benefit mammary development, but high dietary energy is detrimental to the development of mammary secretory tissue (Weldon et al. 1991).

**Amino Acid Partitioning During Lactation**

Lactating sow must be provided with an adequate amount and proportion of AA to maximize dietary protein utilization. Approximately one-third of the total circulating AA is extracted by the mammary gland during lactation (Guan et al. 2004) and directly used for milk protein synthesis and mammary tissue accretion and metabolism. Imbalances created by excesses or deficiencies of dietary AA reduce the efficiency of dietary nitrogen utilization by the animal, limiting milk protein synthesis (Pérez Laspiur et al. 2009) and increasing the release of nitrogen products into the environment (Otto et al. 2003). For years, the focus of lactating sow AA nutrition has been to maximize piglet growth. Amino acid nutrition of the sow should factor in AA needs, function, and metabolism of the mammary gland. Amino acid utilization from arterial supply into the mammary epithelial cells for milk protein synthesis and secretion is not 100% efficient. Thus, knowledge of mammary efficiency of AA utilization is needed to better define AA requirement for lactation.

*Definition of Utilization Efficiency Value for Amino Acids*

Knowledge of accurate efficiency values for individual AA is needed for future model prediction of dietary AA requirements and feed formulation. It is also important to note that the efficiency is only quantifiable at least so far for those AA that are strictly essential, or in other words, those that cannot be synthesized by the animal.

The apparent efficiency value can be estimated as follows:

$$\text{Apparent AA utilization efficiency} = \frac{\text{Milk AA output} \left(\text{g d}^{-1}\right)}{\text{Dietary SID AA intake} \left(\text{g d}^{-1}\right)} \qquad \text{(Eq. 5.1)}$$

The apparent efficiency includes AA contribution from body protein mobilization. Thus, the "milk AA output" in the numerator does not only originate from the diet. "Dietary AA intake" in the denominator is not specific for milk alone since AA are partitioned to both milk and maternal needs (Zhang and Trottier 2019). The NRC (2012) estimated a "true" efficiency by correcting the numerator and denominator to be specific for "milk AA output from diet" and "dietary SID AA intake for milk," respectively, as follows (Zhang and Trottier 2019):

$$\text{True AA utilization efficiency} = \frac{\text{Milk AA output from diet} \left(\text{g d}^{-1}\right)}{\text{Dietary SID AA intake for milk} \left(\text{g d}^{-1}\right)} \qquad \text{(Eq. 5.2)}$$

In the true utilization efficiency calculation, the numerator specifies "from diet" to indicate that AA contribution from body protein losses, if any, is corrected for, and the denominator specifies "for milk" to indicate that SID AA required for maintenance is corrected for (Zhang and Trottier 2019). Milk yield is estimated based on piglet ADG (NRC 2012). Thus,

True AA utilization efficiency

$$= \frac{\text{AA output in milk}\left(\text{g / d}\right) - \text{AA mobilized from body protein}\left(\text{g d}^{-1}\right)}{\text{SID AA intake}\left(\text{g / d}\right) - \text{AA for maintenance}\left(\text{g d}^{-1}\right)} \quad \text{(Eq. 5.3)}$$

Biological efficiency values (BEV) of AA for lactation have been shown to vary depending on the dietary CP concentration and level of AA intake, AA balance (Huber et al. 2015, 2016; Zhang et al. 2019b), and environmental temperature (Zhang et al. 2019a). Maximal biological efficiency values (MBEV) are assumed when used to predict AA requirement. In order to generate MBEV, dietary AA must be marginally limiting to push utilization at its maximum level. Feeding reduced CP diet with better AA balance improves BEV and is a useful tool to experimentally generate MBEV. Huber et al. (2015, 2016) and Zhang et al. (2019b) compared the BEV of AA in nonreduced CP and reduced CP diets. Across dietary CP concentrations and CAA inclusion rates, Huber et al. (2016) showed that AA efficiencies generally increase and quite considerably for some AA (arginine, histidine, isoleucine, and leucine) with improvement in dietary AA balance. Zhang et al. (2019b) showed that the MBEV of N and EAA increased in reduced CP diet except that of lysine, methionine, threonine, valine showed no significant improvement. It is proposed based on these data that lysine, methionine, threonine, and valine are the top four limiting AA in corn and soybean meal-based lactation diets. In the study by Zhang et al. (2019b), MBEV of lysine was similar to that of NRC (0.67) in nonreduced CP diets. Although marginally limiting in lysine, nonreduced CP diets contain excessive nitrogen and other EAA, which may affect lysine efficiency based on the molecular activity of the shared transporter systems and proteins, as discussed earlier in this chapter. When Zhang et al. (2019b) tested whether the presence of other EAA in excess affected lysine efficiency for lactation, no apparent interactions between lysine and other EAA on lysine utilization were observed.

Among the EAA, pertinent efficiency values for arginine and histidine remain debatable because of the *de novo* synthesis of arginine and the extensive recycling of 3-methyl-histidine between muscle protein and blood pools with possible milk secretion of histidine arising from mammary metabolism (Trottier et al. 1997).

### The Scope of Amino Acid Utilization Efficiency

The apparent utilization efficiency of AA, by definition (Eq. 5.1), considers individual sow as a "black box" including maintenance cost and mobilization. However, lactating sows have relatively high maintenance cost due to greater BW (NRC 2012) and generally mobilize body protein and lipid to align with high lactation demand (Theil 2015; Strathe et al. 2017). According to Figure 5.12, the inefficiency occurs mostly in the process where (i) AA are transported from blood into the mammary gland and (ii) AA are metabolized by the mammary gland.

The mechanism by which AA efficiency increases with optimization of AA balance is not only due to the simple fact that less AA are available. There is a consistent increase in piglet litter gain and milk AA output associated with optimization of AA balance (Manjarín et al. 2012; Huber et al. 2015; Chamberlin 2017; Zhang et al. 2019b). There are likely interactions among AA at the mammary basolateral membrane interface that affect their efficiency of transport across the mammary cells and ultimately their utilization by the mammary gland (Guan et al. 2002; Guan et al. 2004; Manjarín et al. 2012). On the other hand, the relatively low utilization efficiency (0.56)

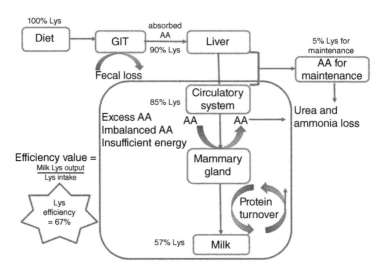

**Figure 5.12**   Steps of amino acid inefficiency towards lactation.

of valine (Guan et al. 2002; Zhang et al. 2019b) is likely associated with valine mammary metabolism for mammary protein remodeling in situ (Trottier 1995) or synthesis of glutamate AA family in mammary gland (Li et al. 2009). So far, it is difficult to tease out the contribution of inefficiency derived from mammary AA transporter system and mammary AA metabolism and turnover.

### *Hormonal Regulation of Amino Acid Utilization*

Thus far, AA transport processes have been compared between rodent mammary tissue and pregnant and lactating animals (Verma and Kansal 1993; Sharma and Kansal 1999, 2000) and between suckled and nonsuckled glands (Shennan et al. 1994; Trottier et al. 1997). In mammary tissue obtained from pregnant, compared to that of lactating mice, $V_{max}$ for AA transport systems was less (Verma and Kansal 1993), indicating that lactation induces change in capacity transport, hence an increase in expression of genes encoding for AA transport proteins. It is well recognized that milk yield is limited by the rate of milk removal (Mepham 1983) and is stimulated by the suckling action and activity of the nursing piglets (Auldist et al. 1995; King et al. 1997). Transport systems $y^+$ (Sharma and Kansal 2000) and L (Sharma and Kansal 1999) remained suppressed in nonlactating mammary tissue and were upregulated by lactogenic hormones, including insulin and prolactin, at the onset of lactation (Sharma and Kansal 1999, 2000), indicating that insulin and prolactin play a role in mediating cationic AA transport in mammary cells. Prolactin responds to milk demand by partitioning nutrients away from adipose tissue in favor of the mammary gland where it affects the synthesis of milk proteins, including $\beta$-casein, whey acidic protein, and $\alpha$-lactalbumin (Ben-Jonathan et al. 2006). In lactating sows, prolactin is the key lactogenic hormone and is essential not only for the initiation of lactation but, unlike for the dairy cow, is also indispensable for maintenance of lactation (Farmer 2001).

In lactating rodents, milk accumulation for a few hours decreased A-V differences of AA across the mammary gland (Viña et al. 1981). Similarly, Shennan et al. (1994), using rodent mammary explants, showed that the uptake of amino-isobutyric acid (AIB), an AA analog with substrate specificity for the L transporter system, decreased in glands not suckled for 24 hours compared to suckled glands. In the same study, Shennan et al. (1994) made the observation that AIB uptake by the rat mammary gland via system L was upregulated by prolactin. Theil et al. (2005) showed that prolactin receptor gene expression decreased in porcine glands that remained unsuckled for 72 hours. Earlier, Shennan et al. (1994) had suggested that milk accumulation may decrease AA uptake by lowering the number of prolactin receptors in the basolateral aspect of mammary epithelial cells. Based on those reports, it is suggested that prolactin binding may mediate the uptake and utilization of AA in lactating mammary tissue.

### Mammary Gland Growth

During lactation, there is considerable mammary tissue growth as demonstrated by a twofold increase in total mammary DNA from parturition to d 21 of lactation in gilts (Kim et al. 1999a). Kim et al. (1999a) also indicated that the mammary gland of gilts undergoes both hypertrophy and hyperplasia during lactation. In contrast, albeit reported in only one study, no increase in DNA concentration in mammary tissue was found from parturition to d 21 of lactation in multiparous sows, indicating that is little net mammary gland growth in multiparous sows (Manjarín et al. 2012). In that same study, total RNA concentration linearly increased from parturition until d 21 of lactation, indicating increasing cellular protein synthetic activity (Manjarín et al. 2012). AA entering the mammary gland are utilized for the synthesis of both milk and constitutive proteins. A quantitative estimate of the latter is not available, but it is reported that the sum of indispensable AA uptake by the lactating mammary gland exceeds the sum in secreted milk (Trottier et al. 1997; Trottier and Guan 2000; Guan et al. 2004). Approximately 26% of the indispensable AA are retained by the lactating mammary gland. Kim et al. (1999b) reported that for every suckling piglet, there is a net 0.7 g of indispensable AA accreted per day, which accounts for approximately 14% of the retained indispensable AA. Consequently, there is a substantial use of indispensable AA in mammary metabolic pathways, including oxidation and synthesis of dispensable AA and other compounds. For example, of a total of 188.5 g of essential AA taken up by the sow mammary gland daily, 49 g is retained, accounting for about 25% of the total uptake (Trottier et al. 1997). Indispensable AA retained by the mammary cells may be used to synthesize dispensable AA to provide energy for lactose and fatty acid synthesis, or to support structural protein synthesis and mammary gland remodeling (Spires et al. 1975). The indispensable AA arginine, leucine, isoleucine, and valine were reported to have the largest accumulation in mammary gland, whereas the output of dispensable AA such as proline, aspartate, and asparagine in sow's milk exceed uptake (Trottier et al. 1997).

### Amino Acid Metabolism in Porcine Mammary Gland During Lactation

A substantial proportion of apparently retained AA by the mammary gland during lactation is used in metabolic pathways as illustrated in Figure 5.13. Whether these pathways represent obligate losses or are regulated as part of the global milk synthetic process is not known. Nonetheless, mapping the entire AA metabolic processes may allow in the future to target genes of interests and increase funneling of AA utilization into products of nutritional values for the nursing piglet.

#### Arginine

Several studies have shown that arginine is catabolized in lactating porcine mammary tissue to form proline, ornithine, and urea via the arginase pathway, and small amounts of polyamines and nitric

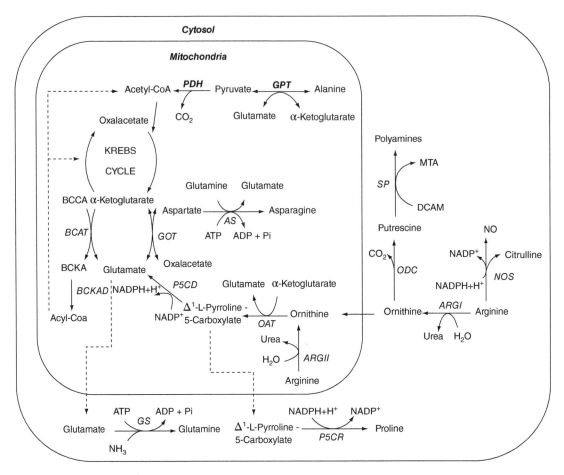

**Figure 5.13**  Amino acid metabolism in the mammary epithelial cell of the sow during lactation. (See text for description of pathways and symbols).

oxide (NO) via the arginase and NO synthase pathways (O'Quinn et al. 2002). There are two different arginases in the lactating porcine mammary tissue: arginase I (a cytosolic enzyme) and arginase II (a mitochondrial enzyme). Both enzymes cleave arginine to yield urea and ornithine. The ornithine produced in the cytosol can be either utilized for polyamine synthesis by ornithine decarboxylase (ODC) and spermidine synthase (SP; Wu and Morris 1998), or can be transported into the mitochondria and converted to $\Delta^1$-L-pyrroline-5-carboxylate by the enzyme OAT. Then, $\Delta^1$-L-pyrroline-5-carboxylate is either converted to glutamate by the enzyme $\Delta^1$-L-pyrroline-5-carboxylate dehydrogenase (P5CD) or exported to the cytosol and converted to proline by the enzyme $\Delta^1$-L-pyrroline-5-carboxylate reductase (P5CR). According to this, the activity of P5CR was 56-fold greater than that of P5CD in lactating porcine mammary tissue, thus favoring the conversion of arginine-derived P5C into proline rather than into glutamate or glutamine. Moreover, porcine mammary gland lacks the enzyme $\Delta^1$-L-P5CS, and, therefore, proline cannot be synthesized from glutamine or glutamate by this tissue (O'Quinn et al. 2002).

Nitric oxide synthesis is quantitatively a minor pathway for arginine degradation in lactating mammary gland (O'Quinn et al. 2002). Nitric oxide is produced from arginine and molecular oxygen in a reaction catalyzed by the enzyme, NO synthase. Once it is synthesized, it rapidly diffuses into the tissue, regulating blood flow and the uptake of plasma nutrients by the mammary cells (Meininger and Wu 2002; Kim and Wu 2009).

Due to extensive degradation of arginine by arginase I and II in mammary epithelial cells (Rezaei et al. 2016), milk arginine alone is not sufficient to support neonate's growth (Wu 2013). The piglet small intestine, however, is equipped with the ability to convert proline to arginine, as discussed earlier, to meet arginine requirement. In fact, proline is the most abundant AA found in milk.

### Branched Chain Amino Acids (BCAA)

Uptake of BCAA leucine, valine, and isoleucine by porcine mammary gland ($76\,g\,d^{-1}$ on d 13–20 of lactation) is much greater than their secretion in milk protein ($46\,g\,d^{-1}$; Trottier et al. 1997). Thus, the lactating porcine mammary gland may catabolize approximately 30 g of BCAA per day. Several studies indicate that BCAA catabolism in mammary cells resembles catabolism of BCAA in other organs, involving 2 initial enzymatic steps (Li et al. 2009). The first step is the transamination of leucine, isoleucine, and valine by the enzyme branched-chain amino transferase (BCAT). There are two mammalian BCAT isozymes: a mitochondrial (BCATm) expressed ubiquitously and a cytosolic (BCATc). Although several studies have shown that BCATc is found almost exclusively in nervous tissue (Sweatt et al. 2004; Hutson et al. 2005), Li et al. (2009) reported the presence of both mitochondrial and cytosolic isoforms of BCAT in mammary tissue. Therefore, transamination of BCAA in porcine mammary gland may occur in the mitochondria and the cytoplasm of mammary cells. In this transamination reaction, the $\alpha$-amino group of leucine, isoleucine, and valine is transferred to $\alpha$-ketoglutarate to form glutamate, leaving behind the corresponding BCKA ($\alpha$-ketoisocaproate, $\alpha$-keto-ß-methylvalerate, and $\alpha$-ketoisovalerate, respectively). Then, the branched-chain BCKD complex catalyzes oxidative decarboxylation of all three BCKA, producing the acyl-CoA derivative. The branched-chain BCKD is a multienzyme complex located on the inner surface of the mitochondrial membrane (Harper et al. 1984). Therefore, if transamination of BCAA occurs in the cytoplasm by the cytosolic isoform of BCAT, the BCKA produced may need to be transported to the mytochondria to complete oxidation. The next step in the oxidation of BCAA is oxidation of the acyl-CoA, catalyzed by two different dehydrogenases. After this step, the individual BCAA catabolic pathways diverge, producing acetyl-CoA (leucine and isoleucine) and succinyl-CoA (valine and isoleucine), which are finally incorporated into the Krebs cycle (Nelson and Cox 2008).

### Glutamate/Glutamine and Aspartate/Asparagine

Glutamate/glutamine and aspartate/asparagine may have considerable nutritional importance, as they are the most abundant free and protein-bound AA in sow milk at peak of lactation (Wu and Knabe 1994). Lack of glutaminase activity in lactating mammary glands partially explains the high concentration of glutamine in milk (Li et al. 2009). Additionally, glutamate and glutamine have the highest extraction rate by the mammary gland during lactation (Trottier et al. 1997).

Extraction of aspartate/asparagine is lower than their output in milk, indicating their synthesis by the mammary cells (Trottier et al. 1997). Li et al. (2009) showed that most milk aspartate is derived from transamination of glutamate, a reaction catalyzed by the enzyme glutamate oxalacetate transaminase (GOT). Alternatively, glutamate can be converted into glutamine by the cytosolic enzyme glutamine synthetase (GS). Although the activity of GOT was shown to be higher than the activity of GS, glutamine synthesis was higher than aspartate synthesis in porcine mammary tissue (Li et al. 2009). Finally, glutamate can be transminated with pyruvate by the enzyme GPT to form alanine and $\alpha$-ketoglutarate. Glutamate synthesis predominates over alanine synthesis, indicating that the transamination reaction moves toward the formation of glutamate in the lactating sow (Li et al. 2009).

## Conclusion

Extensive research has been performed to improve utilization efficiency of AA to maximize growth and lactation performance, improve environmental sustainability, and minimize production cost. Accurate estimation of utilization efficiency to predict requirements is key to achieve these goals.

This chapter emphasized discussing mechanisms (i.e., transporter and regulation) of AA utilization in different production phases (i.e., growth, gestation, and lactation). Utilization of AA in response to immune challenge was also covered. Transport systems, interactions, regulations, and the resulting impact on cellular utilization in the intestine and mammary gland, the key organs in nutrient uptake in growing pigs and neonates were reviewed. Continued research progress in this area is critical in order to improve AA nutrition of swine in a sustainable way.

## References

Adibi, S. A. 1971. Intestinal transport of dipeptides in man: Relative importance of hydrolysis and intact absorption. J. Clin. Invest. 50:2266–2275.

Adibi, S. A. 2003. Regulation of expression of the intestinal oligopeptide transporter (Pept-1) in health and disease. Am. J. Physiol. 285:G779–G788.

Angelo, S., A. M. Rojas, H. Ramirez, and R. Deves. 2002. Epithelial cells isolated from chicken jejunum: An experimental model for the study of the functional properties of amino acid transport system b$^{0,+}$. Comp. Biochem. Physiol. A 132:637–644.

Atinmo, T., C. Baldijão, W. G. Pond, and R. H. Barnes. 1976. Decreased dietary protein or energy intake and plasma growth hormone levels of the pregnant pig, its fetuses and developing progeny. J. Nutr. 106:940–946.

Auldist, D. E., D. Carlson, L. Morrish, C. Wakeford, and R. H. King. 1995. The effect of increased suckling frequency on mammary development and milk yield of sows. In: D. P. Hennessy, and P. D. Cranwell, editor, Manipulating pig production V. Proc. Australasian Pig Sci. Assoc., Werribee, Australia. p. 137.

Avruch, J., X. Long, S. Ortiz-Vega, J. Rapley, A. Papageorgiou, and N. Dai. 2008. Amino acid regulation of TOR complex 1. Am. J. Physiol. Endocrinol. Metab. 296:592–602.

Bauch, C., N. Forster, D. Loffing-Cueni, V. Summa, and F. Verrey. 2003. Functional cooperation of epithelial heteromeric amino acid transporters expressed in madin-darby canine kidney cells. Biochem. J. 278:1316–1322.

Ben-Jonathan, N., E. R. Hugo, T. D. Brandebourg, and C. R. LaPensee. 2006. Focus on prolactin as a metabolic hormone. Trends Endocrinol. Metab. 17:110–116.

Bergen, W. G., and G. Wu. 2009. Intestinal nitrogen recycling and utilization in health and disease. J. Nutr. 139:821–825.

Bertolo, R. F., and D. G. Burrin. 2008. Comparative aspects of tissue glutamine and proline metabolism. J. Nutr. 138:2032–2039.

Bos, C., B. Stoll, H. Fouillet, C. Gaudichon, X. Guan, M. A. Grusak, P. J. Reeds, D. Tomé, and D. G. Burrin. 2003. Intestinal lysine metabolism is driven by the enteral availability of dietary lysine in piglets fed a bolus meal. Am. J. Physiol. Endocrinol. Metab. 285:E1246–E1257.

Boudry, G., V. Rome, C. Perrier, A. Jamin, G. Savary and I. L. Huerou-Luron. 2014. A high-protein formula increases colonic peptide transporter 1 activity during neonatal life in low-birth-weight piglets and disturbs barrier function later in life. Br. J. Nutr. 112:1073–1080.

Broër, A., C. A. Wagner, F. Lang, and S. Broër. 2000a. The heterodimeric amino acid transporter 4F2hc/y$^+$LAT2 mediates arginine efflux in exchange with glutamine. Biochem. J. 349:787–795.

Broër, A., C. A. Wagner, F. Lang, and S. Broër. 2000b. Neutral amino acid transporter ASCT2 displays substrate-induced Na$^+$ exchange and a substrate-gated anion conductance. Biochem. J. 346:705–710.

Broër, A., K. Klingel, S. Kowalczuk, J. E. J. Rasko, J. Cavanaugh, and S. Broër. 2004. Molecular cloning of mouse amino acid transport system B$^0$, a neutral amino acid transporter related to Hartnup disorder. Biochem. J. 279:24467–24476.

Broër, S. 2008. Amino acid transport across mammalian intestinal and renal epithelia. Physiol. Rev. 88:249–286.

Brosnan, J. T., and M. E. Brosnan. 2009. Glutathione and the sulphur containing amino acids. In: R. Masella, and G. Mazza, editors, Glutathione and sulphur amino acids in human health and disease. John Wiley & Sons, Inc., Hoboken, NJ. p. 3–18.

Brosnan, J. T. 2003. Interorgan amino acid transport and its regulation. J. Nutr. 133:S2068–S2072.

Burrin, D. G., and B. Stoll. 2009. Metabolic fate and function of dietary glutamate in the gut. Am. J. Clin. Nutr. 90(Suppl):850S–856S.

Calvert, D. T., and D. B. Shennan. 1996. Evidence for an interaction between cationic and neutral amino acids at the blood-facing aspect of lactating rat mammary epithelium. J. Dairy Res. 63:25–33.

Capuron, L., S. Schroecksnadel, C. Féart, A. Aubert, D. Higueret, P. Barberger-Gateau, S. Layé, and D. Fuchs. 2011. Chronic low-grade inflammation in elderly persons is associated with altered tryptophan and tyrosine metabolism: role in neuropsychiatric symptoms. Biol. Psychiatry 70:175–182.

Chamberlin, D. P. 2017. Impacts of reducing dietary crude protein with crystalline amino acid supplementation on lactating sow performance, nitrogen utilization and heat production. MS. Thesis. Michigan State Univ., East Lansing, MI, US.

Chen, L. X., Y. L. Yin, W. S. Jobgen, S. C. Jobgen, D. A. Knabe, W. X. Hu, and G. Wu. 2007. In vitro oxidation of essential amino acids by intestinal mucosal cells of growing pigs. Livest. Prod. Sci. 109:19–23.

Chen, L. X., P. Li, J. Wang, X. L. Li, H. Gao, Y. Yin, Y. Hou, and G. Wu. 2009. Catabolism of nutritionally essential amino acids in developing porcine enterocytes. Amino Acids 37:143–152.

Chen, F., S. Zhang, Z. Deng, Q. Zhou, L. Cheng, S. W. Kim, J. Chen, and W. Guan. 2018. Regulation of amino acid transporters in the mammary gland from late pregnancy to peak lactation in the sow. J. Anim. Sci. Biotechnol. 9:35.

Chillaron, J., R. Estevez, C. Mora, C. A. Wagner, H. Suessbrich, F. Lang, J. L. Gelpi, X. Testar, A. E. Busch, A. Zorzano, and M. Palacin. 1996. Obligatory amino acid exchange via systems b$^{o,+}$-like and y$^+$L-like. A tertiary active transport mechanism for renal reabsorption of cystine and dibasic amino acids. Biochem. J. 271:17761–17770.

Closs, E. I. 2002. Expression, regulation and function of carrier proteins for cationic amino acids. Curr. Opin. Nephrol. Hypertens. 11:99–107.

Colegio, O. R., C. M. Van Itallie, H. J. McCrea, C. Rahner, and J. M. Anderson. 2002. Claudins create charge-selective channels in the paracellular pathway between epithelial cells. Am. J. Physiol. Cell Physiol. 283:C142–C147.

Colditz, I. G. 2002. Effects of the immune system on metabolism: Implications for production and disease resistance in livestock. Livest. Prod. Sci. 75:257–268.

Corradetti, M. N., and K. L. Gua. 2006. Upstream of the mammalian target of rapamycin: do all roads pass through mTOR? Oncogene 25:6347–6360.

Dave, M. H., N. Schulz, M. Zecevic, C. A. Wagner, and F. Verrey. 2004. Expression of heteromeric amino acid transporters along the murine intestine. Am. J. Physiol. 258:597–610.

Davis, P. K., and G. Wu. 1998. Compartmentation and kinetics of urea cycle enzymes in porcine enterocytes. Comp. Biochem. Physiol. B Biochem. Mol. Biol. 119:527–537.

Davis, T. A., M. L. Fiorotto, D. G. Burrin, P. J. Reeds, H. V. Nguyen, P. R. Beckett, R. C. Vann, and P. M. O'Connor. 2002. Stimulation of protein synthesis by both insulin and amino acids is unique to skeletal muscle in neonatal pigs. Am. J. Physiol. Endocrinol. Metab. 282:E880–E890.

de Ridder, K., C. L. Levesque, J. K. Htoo, and C. F. M. de Lange. 2012. Immune system stimulation reduces the efficiency of tryptophan utilization for body protein deposition in growing pigs. J. Anim. Sci. 90:3485–3491.

Devés, R., and C. A. R. Boyd. 1998. Transporters for cationic amino acids in animal cells: Discovery, structure, and function. Physiol. Rev. 78:487–545.

Escobar, J., J. W. Frank, A. Suryawan, H. V. Nguyen, S. R. Kimball, L. S. Jefferson, and T. A. Davis. 2005. Physiological rise in plasma leucine stimulates muscle protein synthesis in neonatal pigs by enhancing translation initiation factor activation. Am. J. Physiol. Endocrinol. Metab. 288:E914–E921.

Fang, Z., F. Huang, J. Luo, H. Wei, L. Ma, S. Jiang, and J. Peng. 2010. Effects of DL-2-hydroxy-4-methylthiobutyrate on the first-pass intestinal metabolism of dietary methionine and its extra-intestinal availability. Br. J. Nutr. 103:643–651.

Farmer, C. 2001. The role of prolactin for mammogenesis and galactopoiesis in swine. Livest. Prod. Sci. 70:105–113.

Feliubadaló, L., M. Font, J. Purroy, F. Rousaud, X. Estivill, V. Nunes, E. Golomb, M. Centola, I. Aksentijevich, Y. Kreiss, B. Goldman, M. Pras, D. L. Kastner, E. Pras, P. Gasparini, L. Bisceglia, E. Beccia, M. Gallucci, L. de Sanctis, A. Ponzone, G. F. Rizzoni, L. Zelante, M. T. Bassi, A. L. George Jr, M. Manzoni, A. De Grandi, M. Riboni, J. K. Endsley, A. Ballabio, G. Borsani, N. Reig, E. Fernández, R. Estévez, M. Pineda, D. Torrents, M. Camps, J. Lloberas, A. Zorzano, and M. Palacín. 1999. Non-type I cystinuria caused by mutations in SLC7A9, encoding a subunit (b$^{o,+}$AT) of rBAT. Nat. Genet. 23:52–57.

Feng, D. Y., X. Y. Zhou, J. J. Zuo, C. M. Zhang, Y. L. Yin, X. Q. Wang, and T. Wang. 2008. Segmental distribution and expression of two heterodimeric amino acid transporter mRNAs in the intestine of pigs during different ages. J. Sci. Food Agric. 88:1012–1018.

Flynn, N. E., and Wu, G. 1996. An important role for endogenous synthesis of arginine in maintaining arginine homeostasis in neonatal pigs. Am. J. Physiol. Regul. Integr. Comp. Physiol. 271:R1149–R1155.

Frost, R.A., and C. H. Lang. 2011. mTor signaling in skeletal muscle during sepsis and inflammation: Where does it all go wrong? Physiology 26:83–96.

Fuchs, B. C., and B. P. Bode 2005. Amino acid transporters ASCT2 and LAT1 in cancer: partners in crime? Semin. Cancer Biol. 15:254–266.

Goodband, B., M. Tokach, S. Dritz, J. DeRouchey, and J. Woodworth. 2014. Practical starter pig amino acid requirements in relation to immunity, gut health and growth performance. J. Anim. Sci. Biotechnol. 5:12.

Guan, X., B. J. Bequette, G. Calder, P. K. Ku, K. N. Ames, and N. L. Trottier. 2002. Amino acid availability affects amino acid transport and protein metabolism in the porcine mammary gland. J. Nutr. 132:1224–1234.

Guan, X., J. E. Pettigrew, P. K. Ku, K. N. Ames, B. J. Bequette, and N. L. Trottier. 2004. Dietary protein concentration affects plasma arteriovenous difference of amino acids across the porcine mammary gland. J. Anim. Sci. 82:2953–2963.

Hara, K., Y. Maruki, X. Long, K. Yoshino, N. Oshiro, S. Hidayat, C. Tokunaga, J. Avruch, and K. Yonezawa. 2002. Raptor, a binding partner of target of rapamycin (TOR), mediates TOR action. Cell 110:177–189.

Harper, A. E., R. H. Miller, and K. P. Block. 1984. Branched-chain amino acid metabolism. Annu. Rev. Nutr. 4:409–454.

Hartmann, P. E., N. A. Smith, M. J. Thompson, C. M. Wakeford, and P. G. Arthur. 1997. The lactation cycle in the sow: physiological and management contradictions. Livest. Prod. Sci. 50:75–87.

Hatanaka, T., T. Nakanishi, W. Huang, F. H. Leibach, P. D. Prasad, V. Ganapathy, and M. E. Ganapathy. 2001. Na$^+$- and Cl$^-$-coupled active transport of nitric oxide synthase inhibitors via amino acid transport system B$^{0,+}$. J. Clin. Invest. 107:1035–1043.

Hoffmann, T. M., E. Cwiklinski, D. S. Shah, C. Stretton, R. Hyde, P. M.Taylor, and H. S. Hundal. 2018. Effects of sodium and amino acid substrate availability upon the expression and stability of the SNAT2 (SLC38A2) amino acid transporter. Front. Pharmacol. 9:63.

Hovey, R. C., T. B. McFadden, and R. M. Akers. 1999. Regulation of mammary gland growth and morphogenesis by the mammary fat pad: A species comparison. J. Mammary Gland Biol. Neoplasia. 4:53–68.

Huber, L., C. F. M. de Lange, U. Krogh, D. Chamberlin, and N. L. Trottier. 2015. Impact of feeding reduced crude protein diets to lactating sows on nitrogen utilization. J. Anim. Sci. 93:5254–5264.

Huber, L., C. F. de Lange, C. W. Ernst, U. Krogh, and N. L. Trottier. 2016. Impact of improving dietary amino acid balance for lactating sows on efficiency of dietary amino acid utilization and transcript abundance of genes encoding lysine transporters in mammary tissue. J. Anim. Sci. 94:4654–4665.

Hundal, H. S., and P. M. Taylor. 2009. Amino acid transceptors: gate keepers of nutrient exchange and regulators of nutrient signaling. Am. J. Physiol. Endocrinol. Metab. 296:E603–E613.

Huntley, N. F., C. M. Nyachoti, and J. F. Patience. 2018. Lipopolysaccharide immune stimulation but not $\beta$-mannanase supplementation affects maintenance energy requirements in young weaned pigs. J. Anim. Sci. Biotechnol. 9:47.

Hurley, W. L., H. Wang, J. M. Bryson, and D. B. Shennan. 2000. Lysine uptake by mammary gland tissue of the lactating sow. J. Anim. Sci. 78:391–395.

Hutson, S. M., A. J. Sweatt, and K. F. Lanoue. 2005. Branched-chain amino acid metabolism: Implications for establishing safe intakes. J. Nutr. 135:1557S–1564S.

Hyde, R., K. Peyrollier, and H. S. Hundal. 2002. Insulin promotes the cell surface recruitment of the SAT2/ATA2 System A amino acid transporter from an endosomal compartment in skeletal muscle cells. J. Biol. Chem. 277: 13628–13634.

Hyde, R., P. M. Taylor, and H. S. Hundal. 2003. Amino acid transporters: roles in amino acid sensing and signalling in animal cells. Biochem. J. 373:1–18.

Jackson, S. C., J. M. Bryson, H. Wang, and W. L. Hurley. 2000. Cellular uptake of valine by lactating porcine mammary tissue. J. Anim. Sci. 78:2927–2932.

Ji, F., G. Wu, J. R. Blanton, Jr., and S. W. Kim. 2005. Changes in weight and composition in various tissues of pregnant gilts and their nutritional implications. J. Anim. Sci. 83:366–375.

Ji, F., W. L. Hurley, and S. W. Kim. 2006. Characterization of mammary gland development in pregnant gilts. J. Anim. Sci. 84:579–587.

Johnson, R. W. 1997. Inhibition of growth by pro-inflammatory cytokines: An integrated view. J. Anim. Sci. 75:1244–1255.

Kampman-van De Hoek, E., P. Sakkas, W. J. J. Gerrits, J. J. G. C. van den Borne, C. M. C. van der Peet-Schwering, and A. J. M. Jansman. 2015. Induced lung inflammation and dietary protein supply affect nitrogen retention and amino acid metabolism in growing pigs. Br. J. Nutr. 113:414–425.

Kensinger, R. S., R. J. Collier, F. W. Bazer, C. A. Ducsay, and H. N. Becker. 1982. Nucleic acid, metabolic and histological changes in gilt mammary tissue during pregnancy and lactogenesis. J. Anim. Sci. 54:1297–1308.

Kensinger, R. S., R. J. Collier, F. W. Bazer, and R. R. Kraeling. 1986. Effect of number of conceptuses on maternal hormone concentrations in the pig. J. Anim. Sci. 62:1666–1674.

Kim, S. W., W. L. Hurley, I. K. Han, and R. A. Easter. 1999a. Changes in tissue composition associated with mammary gland growth during lactation in sows. J. Anim. Sci. 77:2510–2516.

Kim, S. W., W. L. Hurley, I. K. Han, H. H. Stein, and R. A. Easter. 1999b. Effect of nutrient intake on mammary gland growth in lactating sows. J. Anim. Sci. 77:3304–3315.

Kim, S. W., and G. Wu. 2009. Regulatory role for amino acids in mammary gland growth and milk synthesis. Amino Acids 37:89–95.

Kimball, S. R., and L. S. Jefferson. 2006. New functions for amino acids: effects on gene transcription and translation. Am. J. Clin. Nutr. 83:500S–507S.

King, R. H., B. P. Mullan, F. R. Dunshea, and H. Dove. 1997. The influence of piglet body weight on milk production of sows. Livest. Prod. Sci. 47:169–174.

Klasing, K. C. 2007. Nutrition and the immune system. British Poult. Sci. 48:525–537.

Krehbiel, C. R., and J. C. Matthews. 2003. Absorption of amino acids and peptides. In: J. P. F. D'Mello, editor, Amino acids in animal nutrition. 2nd ed. CAB International, Waalingford, UK. p. 60.

Lang, C. H., R. A. Frost, and T. C. Vary. 2007. Regulation of muscle protein synthesis during sepsis and inflammation. American J. Physiol. Endocrinol. Metabolism 293:E453–E459.

Li, P., D. A. Knabe, S. W. Kim, C. J. Lynch, S. M. Hutson, and G. Wu. 2009. Lactating porcine mammary tissue catabolizes branched-chain amino acids for glutamine and aspartate synthesis. J. Nutr. 139:1502–1509.

Litvak, N., J. K. Htoo, and C. F. M. de Lange. 2013. Restricting sulfur amino acid intake in growing pigs challenged with lipopolysaccharides decreases plasma protein and albumin synthesis. Can. J. Anim. Sci. 93:505515.

Malandro, M. S., M. J. Beveridge, M. S. Kilberg, and D. A. Novak. 1996. Effect of low-protein diet-induced intrauterine growth retardation on rat placental amino acid transport. Am. J. Physiol. Cell Physiol. 271:C295–C303.

Manjarín, R., V. Zamora, G. Wu, J. P. Steibel, R. N. Kirkwood, N. P. Taylor, E. Wils-Plotz, K. Trifilo, and N. L. Trottier. 2012. Effect of amino acids supply in reduced crude protein diets on performance efficiency of mammary uptake, and transporter gene expression in lactating sows. J. Anim. Sci. 90:3088–3100.

Manjarín, R., B. J. Bequette, G. Wu, and N. L. Trottier. 2014. Linking our understanding of mammary gland metabolism to amino acid nutrition. Amino Acids 46:2447–2462.

McGilvray, W. D., D. Klein, H. Wooten, J. A. Dawson, D. Hewitt, A. R. Rakhshandeh, C. F.M. de Lange, and A. Rakhshandeh. 2019a. Immune system stimulation induced by Escherichia coli lipopolysaccharide alters plasma free amino acid flux and dietary nitrogen utilization in growing pigs. J. Anim. Sci. 97:315–326.

McGilvray, W. D., H. Wooten, A. R. Rakhshandeh, A. Petry, and A. Rakhshandeh. 2019b. Immune system stimulation increases dietary threonine requirements for protein deposition in growing pigs. J. Anim. Sci. 97:735–744.

McPherson, R. L., F. Ji, G. Wu, J. R. Blanton Jr, and S. W. Kim. 2004. Growth and compositional changes of fetal tissues in pigs. J. Anim. Sci. 82:2534–2540.

Meininger, C. J., and G. Wu. 2002. Regulation of endothelial cell proliferation by nitric oxide. Methods Enzymol. 352:280–295.

Melchior, D., N. Meziere, B. Seve, and N. Floc'h. 2004. Does an inflammatory response decrease tryptophan availability in pigs? Journées de la Recherche 36:165–171.

Mepham, T. B. 1983. Physiological aspects of lactation. In: T. B. Mepham, editor, Biochemistry of lactation. Elsevier Science Publishers, Amsterdam, The Netherlands. p. 4–28.

Metges, C. C. 2000. Contribution of microbial amino acids to amino acid homeostasis of the host. J. Nutr. 130:1857S–1864S.

Montagne, L., J. R. Pluske, D. J. Hampson. (2003). A review of interactions between dietary fibre and the intestinal mucosa, and their consequences on digestive health in young non-ruminant animals. Anim. Feed Sci. Technol. 108:95–117.

Munasinghe, L. L., J. L. Robinson, S. V. Harding, J. A. Brunton, and R. F. Bertolo. 2017. Protein synthesis in mucin-producing tissues is conserved when dietary threonine is limiting in piglets. J. Nutr. 147:202–210.

Nakashima, K., Y. Yakabe, A. Ishida, M. Yamazaki, and H. Abe. 2007. Suppression of myofibrillar proteolysis in chick skeletal muscles by $\alpha$-ketoisocaproate. Amino Acids. 33:499–503.

Nelson, D. L., and M. M. Cox. 2008. Lehninger principles of biochemistry. 5th ed. W. H. Freeman and Co., New York, NY.

Nielsen, O. L., A. R. Pedersen, and M. T. Sørensen. 2001. Relationships between piglet growth rate and mammary gland size of the sow. Livest. Prod. Sci. 67:273–279.

Nielsen, T. T., N. L. Trottier, H. H. Stein, C. Bellaver, and R. A. Easter. 2002. The effect of litter size and day of lactation on amino acid uptake by the porcine mammary gland. J. Anim. Sci. 80:2402–2411.

Niiyama, M., E. Deguchi, K. Kagota, and S. Namioka. 1979. Appearance of [15]N labeled intestinal microbial amino acids in the venous blood of the pig colon. Am. J. Vet. Res. 40:716–718.

NRC. 2012. Nutrient requirements of swine. 11th rev. ed. Natl. Acad. Press, Washington, DC.

O'Connor, P. M. J., J. A. Bush, A. Suryawan, H. V. Nguyen, and T. A. Davis. 2003. Insulin and amino acids independently stimulate skeletal muscle protein synthesis in neonatal pigs. Am. J. Physiol. Endocrinol. Metab. 284:E110–E119.

Olszewski, A., and S. Buraczewski. 1978. Absorption of amino acids in isolated pig caecum in situ. Effect of concentration of enzymatic casein hydrolysate on absorption of amino acids. Acta Physiol. Pol. 29:67–77.

O'Quinn, P. R., D. A. Knabe, and G. Wu. 2002. Arginine catabolism in lactating porcine mammary tissue. J. Anim. Sci. 80:467–474.

Otto, E. R., M. Yokoyama, P. K. Ku, N. K. Ames, and N. L. Trottier. 2003. Nitrogen balance and ileal amino acid digestibility in growing pigs fed diets reduced in protein concentration. J. Anim. Sci. 81:1743–1753.

Palacin, M., R. Estevez, J. Bertran, and A. Zorzano. 1998. Molecular biology of mammalian amino acid transporters. Physiol. Rev. 78:969–1054.

Pérez Laspiur, J., J. L. Burton, P. S. D. Weber, R. N. Kirkwood, N. L. Trottier. 2004. Short communication: Amino acid transporters in porcine mammary gland during lactation. J. Dairy Sci. 87:3235–3237.

Pérez Laspiur, J., J. L. Burton, P. S. D. Weber, J. Moore, R. N. Kirkwood, and N. L. Trottier. 2009. Dietary protein intake and stage of lactation differentially modulate amino acid transporter mRNA abundance in porcine mammary tissue. J. Nutr. 139:1677–1684.

Pluske, J. R., J. C. Kim, and J. L. Black. 2018. Manipulating the immune system for pigs to optimise performance. Anim. Prod. Sci. 58:666–680.

Rakhshandeh, A., and C. F. M. de Lange. 2010. Immune system stimulation increases reduced glutathione synthesis rate in growing pigs. In: G. Matteo Crovetto, editor, Energy and protein metabolism and nutrition. EAAP Publication No. 127. Wageningen Academic Publishers, Wageningen, The Netherlands. p. 501–502.

Rakhshandeh, A., and C. F. M. de Lange. 2011. Immune system stimulation in the pig: effect on performance and implications for amino acid nutrition. In: R. J. Van Barnevled, editor, Manipulating pig production XIII. Australas. Pig Sci. Assoc. Incorp., Werribee, VIC, Austrilia. p. 31–46.

Rakhshandeh, A., T. E. Weber, J. C. M. Dekkers, C. K. Tuggle, B. J. Kerr, and N. K. Gabler. 2013. Impact of systemic immune system stimulation on intestinal integrity and function in pigs. FASEB J. 27(Suppl. 1):867.2.

Reeds, P. J., C. R. Fjeld, and F. Jahoor. 1994. Do the differences in anion acid composition of acute phase and muscle pro- teins have a bearing on nitrogen loss in traumatic states? J. Nutr. 124:906–910.

Reeds, P. J., L. J. Wykes, J. E. Henry, M. E. Frazer, D. G. Burrin, and F. Jahoor. 1996. Enteral glutamate is almost completely metabolized in first pass by the gastrointestinal tract of infant pigs. Am. J. Physiol. 270:E413–E418.

Reeds, P. J., and F. Jahoor. 2001. The amino acid requirements of disease. Clin. Nutr. 20:15–22.

Rémond, D.,C. Buffière, J. P. Godin, P. P. Mirand, C. Obled, I. Papet, D. Dardevet, G. Williamson, D. Breuillé, and M. Faure.2009. Intestinal inflammation increases gastrointestinal threonine uptake and mucin synthesis in enterally fed minipigs. J. Nutr. 139:720–726.

Rezaei, R., Z. Wu, Y. Hou, F. W. Bazer, and G. Wu. 2016. Amino acids and mammary gland development: Nutritional implications for milk production and neonatal growth. J. Anim. Sci. Biotechnol. 7:20.

Richert, B. T., M. D. Tokash, R. D. Goodband, J. L. Nelssen, J. E. Pettigrew, R. W. Walker, and L. J. Johnston. 1996. Valine requirement of the high-producing sow. J. Anim. Sci. 74:1307–1313.

Richert, B. T., M. D. Tokach, R. D. Goodband, J. L. Nelssen, R. G. Campbell, and S. Kershaw. 1997. The effect of dietary lysine and valine fed during lactation on sow and litter performance. J. Anim. Sci. 75:1853–1860.

Roberts, R. C., J. K. Roche, and R. E. McCullumsmith. 2014. Localization of excitatory amino acid transporters EAAT1 and EAAT2 in human postmortem cortex: a light and electron microscopic study. Neuroscience 277:522–540.

Rossier, G., C. Meier, C. Bauch, V. Summa, B. Sordat, F. Verrey, and L. C. Kuhn. 1999. LAT2, a new basolateral 4F2hc/ CD98-associated amino acid transporter of kidney and intestine. Biochem. J. 274:34948–34954.

Rudar, M., C. L. Zhu, and C. F. de Lange. 2017. Dietary leu- cine supplementation decreases whole-body protein turn- over before, but not during, immune system stimulation in pigs. J. Nutr. 147:45–51.

Scalise, M., L. Pochini, L. Console, M. A. Losso, and C. Indiveri. 2018. The human SLC1A5 (ASCT2) amino acid transporter: From function to structure and role in cell biology. Front. Cell Dev. Biol. 6:96.

Sepúlveda, F. V., and M. W. Smith. 1979. Different mechanisms for neutral amino acid uptake by new-born pig colon. J. Physiol. 286:479–490.

Sharma, R., and V. K. Kansal. 1999. Characteristics of transport systems for L-alanine in mouse mammary gland and their regula- tion by lactogenic hormones: Evidence for two broad spectrum systems. J. Dairy Res. 66:385–398.

Sharma, R., and V. K. Kansal. 2000. Heterogeneity of cationic amino acid transport systems in mouse mammary gland and their regulation by lactogenic hormones. J. Dairy Res. 67:21–30.

Shennan, D. B., S. A. McNeillie, E. A. Jamison, and D. T. Calvert. 1994. Lysine transport in lactating rat mammary tissue: Evidence for an interaction between cationic and neutral amino acids. Acta Physiol. Scand. 151:461–466.

Shennan, D. B., and M. Peaker. 2000. Transport of milk constituents by the mammary gland. Physiol. Rev. 80:925–951.

Shennan, D. B., and C. A. R. Boyd. 2014. The functional and molecular entities underlying amino acid and peptide transport by the mammary gland under different physiological and pathological conditions. J. Mammary Gland Biol. Neiplasia 19:19–33.

Shu, D. P., B. L. Chen, J. Hong, P. P. Liu, D. X. Hou, X. Huang, F. T. Zhang, J. L.Wei, and W. T.Guan. 2012. Global transcriptional profiling in porcine mammary glands from late pregnancy to peak lactation. Omics 16:123–137.

Silk, D. B., A. Nicholson, and Y. S. Kim. 1976. Hydrolysis of peptides within lumen of small intestine. Am. J. Physiol. 231:1322–1329.

Silk, D. B. A., G. K. Grimble, and R. G. Rees. 1985. Protein digestion and amino acid and peptide absorption. Proc. of the Nutr. Soc. 44:63–72.

Sloan, J. L., and S. Mager. 1999. Cloning and functional expression of a human Na$^+$ and Cl$^-$-dependent neutral and cationic amino acid transporter B$^{0,+}$. J. Biol. Chem. 274:23740–23745.

Sørensen, M. T., K. Sejrsen, and S. Purup. 2002. Mammary gland development in gilts. Livest. Prod. Sci. 75:143–148.

Soriano-Garcia, J. F., M. Torras-LlortMoreto, and R. Ferrer. 1999. Regulation of L-methionine and L-lysine uptake in chicken jejunal brush border by dietary methionine. Am. J. Physiol. 2777:R1654–R1661.

Souba, W. W., and A. J. Pacitti. 1992. How amino acids get into cells: mechanisms, models, menus and mediators. J. Parenter. Enteral. Nutr. 16:569–578.

Southern, L. L., and D. H. Baker. 1983. Arginine requirement of the young pig. J. Anim. Sci. 57:402–412.

Sperandeo, M. P., G. Andria, and G. Sebastio. 2008. Lysinuric protein intolerance: update and extended mutation analysis of the SLC7A7 gene. H. Mutation. 29:14–21.

Spires, H. R., J. H. Clark, R. G. Derrig, and C. L. Davis. 1975. Milk production and nitrogen utilization in response to post-ruminal infusion of sodium caseinate in lactating cows. J. Nutr. 105:1111–1121.

Stoll, B., J. Henry, P. J. Reeds, H. Yu, F. Jahoor, and D. G. Burrin. 1998. Catabolism dominates the first-pass intestinal metabolism of dietary essential amino acids in milk protein-fed piglets. J. Nutr. 128:606–614.

Stoll, B., X. Chang, M. Z. Fan, P. J. Reeds, and D. G. Burrin. 2000. Enteral nutrient intake determines the rate of intestinal protein synthesis and accretion in neonatal pigs. Am. J. Physiol. 279:G288–G294.

Strathe, A. V., T. S. Bruun, and C. F. Hansen. 2017. Sows with high milk production had both a high feed intake and high body mobilization. Animal 11:1913–1921.

Stuart, W., T. E. Burkey, N. K. Gabler, K. Schwartz, C. F. M. de Lange, D. Klein, J. A. Dawson, and A. Rakhshandeh. 2015. Infection with porcine reproductive and respiratory syndrome virus (PRRSV) affects body protein deposition and alters amino acid metabolism in growing pigs. J. Anim. Sci. 93:855.

Suryawan, A., J. Escobar, J. W. Frank, H. V. Nguyen, and T. A. Davis. 2006. Developmental regulation of the activation of signaling components leading to translation initiation in skeletal muscle of neonatal pigs. Am. J. Physiol. Endocrinol. Metab. 291:E849–E859.

Suryawan, A., A. S. Jeyapalan, R. A. Orellana, F. A. Wilson, H. V.Nguyen, and T. A. Davis. 2008. Leucine stimulates protein synthesis in skeletal muscle of neonatal pigs by enhancing mTORC1 activation. Am. J. Physiol. Endocrinol. Metab. 295:E868–E875.

Suryawan, A., and T. A.Davis. 2014. Regulation of protein degradation pathways by amino acids and insulin in skeletal muscle of neonatal pigs. J. Anim. Sci. Biotechnol. 5:8.

Sweatt, A., M. Wood, A. Suryawan, R. Wallin, M. C. Willingham, and S. M. Hutson. 2004. Branched-chain amino acid catabolism: Unique segregation of pathway enzymes in organ systems and peripheral nerves. Am. J. Physiol. 286:E64–E76.

Taylor, P. M., S. Kaur, B. Mackenzie, and G. J. Peter. 1996. Amino-acid dependent modulation of amino acid transport in Xenopus laevis oocytes. J. Exp. Biol. 199:923–931.

Theil, P. K., R. Labouriau, K. Sejrsen, B. Thomsen, and M. T. Sorensen. 2005. Expression of genes involved in regulation of cell turnover during milk stasis and lactation rescue in sow mammary glands. J. Anim. Sci. 83:2349–2356.

Theil, P. K. 2015. Transition feeding of sows. In: C. Farmer, editor, The gestating and lactating sow. Wageningen Academic Publishers, Wageningen, The Netherlands. p. 147–172.

Torrallardona, D., C. I. Harris, and M. F. Fuller. 2003a. Pigs' gastrointestinal microflora provide them with essential amino acids. J. Nutr. 133:1127–1131.

Torrallardona, D., C. I. Harris, and M. F. Fuller. 2003b. Lysine synthesized by the gastrointestinal microflora of pigs is absorbed, mostly in the small intestine. Am. J. Physiol. Endocrinol. Metab. 284:E1177–E1180.

Torrents, D., R. Estevez, M. Pineda, E. Fernandez, J. Lloberas, Y. B. Shi, A. Zorzano, and M. Palacin. 1998. Identification and characterization of a membrane protein (y+L amino acid transporter-1) that associates with 4F2hc to encode the amino acid transport activity y+L: A candidate gene for lysinuric protein intolerance. Biochem. J. 273:32437–32445.

Trottier, N. L. 1995. Protein metabolism in the lactating sow. PhD Diss. Univ. Illinois, Urbana–Champaign.

Trottier, N. L. 1997. Nutritional control of amino acid supply to the mammary gland during lactation in the pig. Proc. Nutr. Soc. 56:581–591.

Trottier, N. L., C. F. Shipley, and R. A. Easter. 1997. Plasma amino acid uptake by the mammary gland of the lactating sow. J. Anim. Sci. 75:1266–1278.

Trottier, N. L., and X. Guan. 2000. Research paradigms behind amino acid requirements of the lactating sow: Theory and future application. J. Anim. Sci. 78(Suppl. 3):48–58.

Ugawa, S., Y. Sunouchi, T. Ueda, E. Takahashi, Y. Saishin, and S. Shimada. 2001. Characterization of a mouse colonic system B0,+ amino acid transporter related to amino acid absorption in colon. Am. J. Physiol. Gastrointest. Liver Physiol. 281:G365–G370.

Urakami, M., R. Ano, Y. Kimura, M. Shima, R. Matsuno, T. Ueno, and M. Akamatsu. 2003. Relationship between structure and permeability of tryptophan derivatives across human intestinal epithelial (Caco-2) cells. Z. Naturforsch. 58c:135–142.

Verma, N., and V. K. Kansal. 1993. Characterization of the routes of methionine transport in mouse mammary glands. Indian J. Med. Res. [B] 98:297–304.

Vilella, S., G. A. Ahearn, G. Cassano, M. Maffia, and C. Storelli. 1990. Lysine transport by brush-border membrane vesicles of eel intestine: Interaction with neutral amino acids. Am. J. Physiol. 259:R1181–R1188.

Viña, J. R., I. R. Puertes, and J. Viña. 1981. Effect of premature weaning on amino acid uptake by the mammary gland of lactating rats. Biochem. J. 200:705–709.

Wang, W., W. Gu, X. Tang, M. Geng, M. Fan, T. Li, W. Chu, C. Shi, R. Huang, H. Zhang, and Y. Yin. 2009. Molecular cloning, tissue distribution and ontogenetic expression of the amino acid transporter b$^{0,+}$ cDNA in the small intestine of Tibetan suckling piglets. Comp. Biochem. Physiol. B154:157–164.

Wang, J., L. X. Chen, P. Li, X. L. Li, H. J. Zhou, F. L. Wang, D. F. Li, Y. L. Yin, and G. Wu. 2008. Gene expression is altered in piglet small intestine by weaning and dietary glutamine supplementation. J. Nutr. 138:1025–1032.

Wang, X., and C. G. Proud. 2009. Nutrient control of TORC1, a cell-cycle regulator. Trends Cell. Biol. 19:260–267.

Wang, X., P. Zeng, Y. Feng, C. Zhang, J. Yang, G. Shu, and Q. Jiang. 2012.Effects of dietary lysine levels on apparent nutrient digestibility and cationic amino acid transporter mRNA abundance in the small intestine of finishing pigs, *Sus scrofa*. Anim. Sci. J. 83:148–155.

Waterlow, J. C. 2006. Models and their analysis. In: Protein turnover. CAB International, Oxfordshire, UK. p. 7–19.

Weldon, W. C., A. J. Thulin, O. A. MacDougald, L. J. Johnston, E. R. Miller, and H. A. Tucker. 1991. Effects of increased dietary energy and protein during late gestation on mammary development in gilts. J. Anim. Sci. 69:194–200.

Wellington, M. O., J. K. Htoo, A. G. Van Kessel, and D. A. Columbus. 2018. Impact of dietary fiber and immune system stimulation on threonine requirement for protein deposition in growing pigs. 96:5222–5232.

Wilson, J. W., and K. E. Webb Jr. 1990. Lysine and methionine transport by bovine jejunal and ileal brush border membrane vesicles. J. Anim. Sci. 68:504–514.

Williams, N. H., T. S. Stahly, and D. R. Zimmerman. 1997. Effect of chronic immune system activation on body nitrogen retention, partial efficiency of lysine utilization, and lysine needs of pigs. J. Anim. Sci. 75:2472–2480.

Woodward, A. D., S. J. Holcombe, J. P. Steibel, W. B. Staniar, C. Colvin, and N. L. Trottier. 2010. Cationic and neutral amino acid transporter transcript abundances are differentially expressed in the equine intestinal tract. J. Anim. Sci. 88:1028–1033.

Wu, G., and J. R. Thompson. 1989a. Is methionine transaminated in skeletal muscle? Biochem. J. 257:281–284.

Wu, G., and J. R. Thompson. 1989b. Methionine transamination and glutamine transaminases in skeletal muscle. Biochem. J. 262:690–691.

Wu, G., J. R. Thompson, and V. E. Baracos. 1991. Glutamine metabolism in skeletal muscle from the broiler chick (Gallus domesticus) and the laboratory rat (Rattus norvegicus). Biochem. J. 274:769–774.

Wu G., and D. A. Knabe. 1994. Free and protein-bound amino acids in sow's colostrum and milk. J. Nutr. 124:415–424

Wu, G., D. A. Knabe, N. E. Flynn, W. Yan, and S. P. Flynn. 1996. Arginine degradation in developing porcine enterocytes. Am. J. Physiol. Gastrointest. Liver Physiol. 271:G913–G919.

Wu, G. 1997. Synthesis of citrulline and arginine from proline in enterocytes of postnatal pigs. Am. J. Physiol. Gastrointest. Liver Physiol. 272:G1382–G1390.

Wu, G., and S. M. Morris, Jr. 1998. Arginine metabolism: Nitric oxide and beyond. Biochem. J. 336:1–17.

Wu, G., W. G. Pond, T. Ott, and F. W. Bazer. 1998. Maternal dietary protein deficiency decreases amino acid concentrations in fetal plasma and allantoic fluid of pigs. J. Nutr. 128:894–902.

Wu, G., T. L. Ott, D. A. Knabe, and F. W. Bazer. 1999. Amino acid composition of the fetal pig. J. Nutr. 129:1031–1038.

Wu, G., D. A. Knabe, and S. W. Kim. 2004. Arginine nutrition in neonatal pigs. J. Nutr. 134:S2783–S2790.

Wu, G. 2009. Amino acids: Metabolism, functions, and nutrition. Amino Acids 37:1–17.

Wu, G. 2013. Amino acids: Biochemistry and nutrition. CRC Press, Boca Raton, FL.

Xiao, X. J., E. A. Wong, and K. E. Webb. 2004. Developmental regulation of fructose and amino acid transporter gene expression in the small intestine of pigs. FASEB J. 18:269.

Zeng, Z., S. Zhang, H. Wang, and X. Piao. 2015. Essential oil and aromatic plants as feed additives in non-ruminant nutrition: A review. J. Anim. Sci. Biotechnol. 6:7.

Zhang, S., S. Qiao, M. Ren, X. Zeng, X. Ma, Z. Wu, P. Thacker, and G. Wu. 2013. Supplementation with branched-chain amino acids to a low-protein diet regulates intestinal expression of amino acid and peptide transporters in weanling pigs. Amino Acids 45:1191–1205.

Zhang, S., F. Chen, Y. Zhang, Y. Lv, J. Heng, T. Min, L. Li, and W. Guan. 2018. Recent progress of porcine milk components and mammary gland function. J. Anim. Sci. Biotechnol. 9:77.

Zhang, S., J. S. Johnson, and N. L. Trottier. 2019a. Effects of dietary near ideal amino acid profile on amino acid utilization efficiency for milk production in sows under thermoneutral or heat stress conditions. J. Anim. Sci. 97(E-Suppl. 3):27.

Zhang, S., M. Qiao, and N. L. Trottier. 2019b. Feeding a reduced protein diet with a near ideal amino acid profile improves amino acid efficiency and nitrogen utilization for milk production in sows. J. Anim. Sci. 97:3882–3897.

Zhang, S., and N. L. Trottier. 2019. Review: Dietary protein reduction improves the energetic and amino acid efficiency in lactating sows. Anim. Prod. Sci. 59:1980–1990.

# 6    Carbohydrates and Carbohydrate Utilization in Swine

Knud Erik Bach Knudsen, Helle Nygaard Lærke, and Henry Jørgensen

## Introduction

Carbohydrates are naturally occurring compounds that consist of carbon, hydrogen, and oxygen in the ratio of $C_n$:$H_{2n}$:$O_n$. Carbohydrates are the single most abundant feed energy in diets for pigs, comprising 60–70% of total energy intake. Dietary carbohydrate are classified according to their degree of polymerization into sugars, oligosaccharides, and polysaccharides, the latter consisting of starches with different degrees of resistant to complete digestion and of nonstarch polysaccharides (NSP) (Cummings and Stephen 2007; Englyst et al. 2007; Bach Knudsen and Lærke 2018). It is now clear that the dietary carbohydrates have different fates and physiological properties in the intestinal tract and the body (Bach Knudsen and Jørgensen 2001). Starch and sugars are digested (hydrolyzed by the pigs' own enzymes) to monosaccharides in the small intestine and absorbed and metabolized. Nondigestible oligosaccharides resistant starch (RS) and NSP are on the other hand fermented to varying degree by the microbiota primarily in the large intestine (cecum and colon). Here they are converted into short-chain fatty acids (SCFA) and lactic acids (LA), which also are absorbed and metabolized in colonic epithelium, hepatic, adipose, and muscle cells (Bergman 1990). Although all products deriving from the assimilation of carbohydrates can be metabolized in the cells to provide energy or stored in the body as glycogen or lipids for later use, the amount of energy provided as ATP per mole of hexose depends on whether the absorption product is a hexose or a SCFA (Table 6.1). Therefore, the composition of the carbohydrate fraction has a considerable impact on the energy value of the feed due to differences both in digestibility and utilization (Just 1982; Noblet and Perez 1993; Jha and Berrocoso 2015).

   The composition of dietary carbohydrates ingested throughout the pigs lifespan varies widely (Bach Knudsen and Jørgensen 2001). The carbohydrates in sows' milk are mainly lactose, while diets provided for growing pigs and adult sows have a more complex composition in terms of chemical structure and organization. On top of the inherent development in intestinal morphological structure, digestive enzyme activity, and microbial hydrolytic activity, there is an influence of the dietary changes occurring over the pigs lifespan (Kidder and Manners 1980; Efird et al. 1982; van Beers-Schreurs et al. 1998; Janczyk et al. 2007). The main purpose of this chapter is to provide an overview of the carbohydrates present in the diet for pigs and how the carbohydrates are digested, absorbed, and utilized.

**Table 6.1**  Classes of feed carbohydrates and their likely degradation end products available for absorption in the intestinal tract of pigs[a].

| Class | DP | Example | Endogenous enzymes | Absorbed molecules | ATP per mole molecule |
|---|---|---|---|---|---|
| Monosaccharides | 1 | Glucose | | Glucose | 38 |
| | 1 | Fructose | | Fructose | 38 |
| Disaccharides | 2 | Sucrose | + | Glucose + fructose | 38 |
| | 2 | Lactose | + | Glucose + galactose | 38 |
| Oligosaccharides | 3 | Raffinose | − | SCFA | 24 |
| | 4 | Stachyose | − | SCFA | 24 |
| | 3–9 | Fructooligosaccharides | − | SCFA | 24 |
| Polysaccharides | ≥10 | Starches | + | Glucose | 38 |
| | | Resistant starch | + | SCFA | 24 |
| | ≥10 | Non-starch polysaccharides | − | SCFA | 24 |

[a] DP, degree of polymerization; SCFA, short-chain fatty acids.

## Chemistry of Dietary Carbohydrates and Lignin

The dietary carbohydrates make up a diverse group of substances with a range of chemical, physico-chemical, and physiological properties. The primarily chemical classification is by composition of sugar monomers, the type of linkage (α or β), and molecular size (degree of polymerization, DP) (Cummings and Stephen 2007; Englyst et al. 2007; Bach Knudsen and Lærke 2018). This approach divides the carbohydrates into three main groups: sugars (DP, 1-2), oligosaccharides (DP, 3-9), and polysaccharides (DP, ≥10) (Table 6.1).

### Sugars

Sugars (DP, 1–2) are water-soluble components composing of mono- and disaccharides (Figure 6.1). The three principal monosaccharides are glucose, fructose, and galactose, which are the building blocks not only for sugars but also for di-, oligo-, and polysaccharides (Table 6.1, Figure 6.1) Sucrose is the most abundant sugar in plant products (Bach Knudsen 1997), whereas lactose consti-tutes the major part of the carbohydrates in milk. A disaccharide like maltose can also occasionally be found in feeds if starchy grains have germinated, but mostly monosaccharides and maltose are present in low concentrations.

### Oligosaccharides

Oligosaccharides (DP, 3–10) are water-soluble compounds composed of three to nine monomers linked together by either α or β bonds (Figure 6.1). The most commonly present oligosaccharides in feedstuffs are α-galactosides (raffinose, stachyose, and verbascose) (Bach Knudsen and Li 1991; Carlsson et al. 1992) and fructooligosaccharides (Pollock and Cairns 1991; van Loo et al. 1995). Oligosaccharides are present in roots, tubers, and seeds and coproducts of many legumes, mallow, composite, and mus-tard species (Bach Knudsen and Li 1991; Pollock and Cairns 1991; Carlsson et al. 1992; van Loo et al. 1995) but can also be incorporated as ingredients (van Loo et al. 1995; Flickinger et al. 2003). Other types of oligosaccharides such as xylooligosaccharides and trans-galactooligosaccharides are occasionally used as ingredients in diets for pigs (Grizard and Barthomeuf 1999), and maltooligosac-charides can be present in case of sprouted grains.

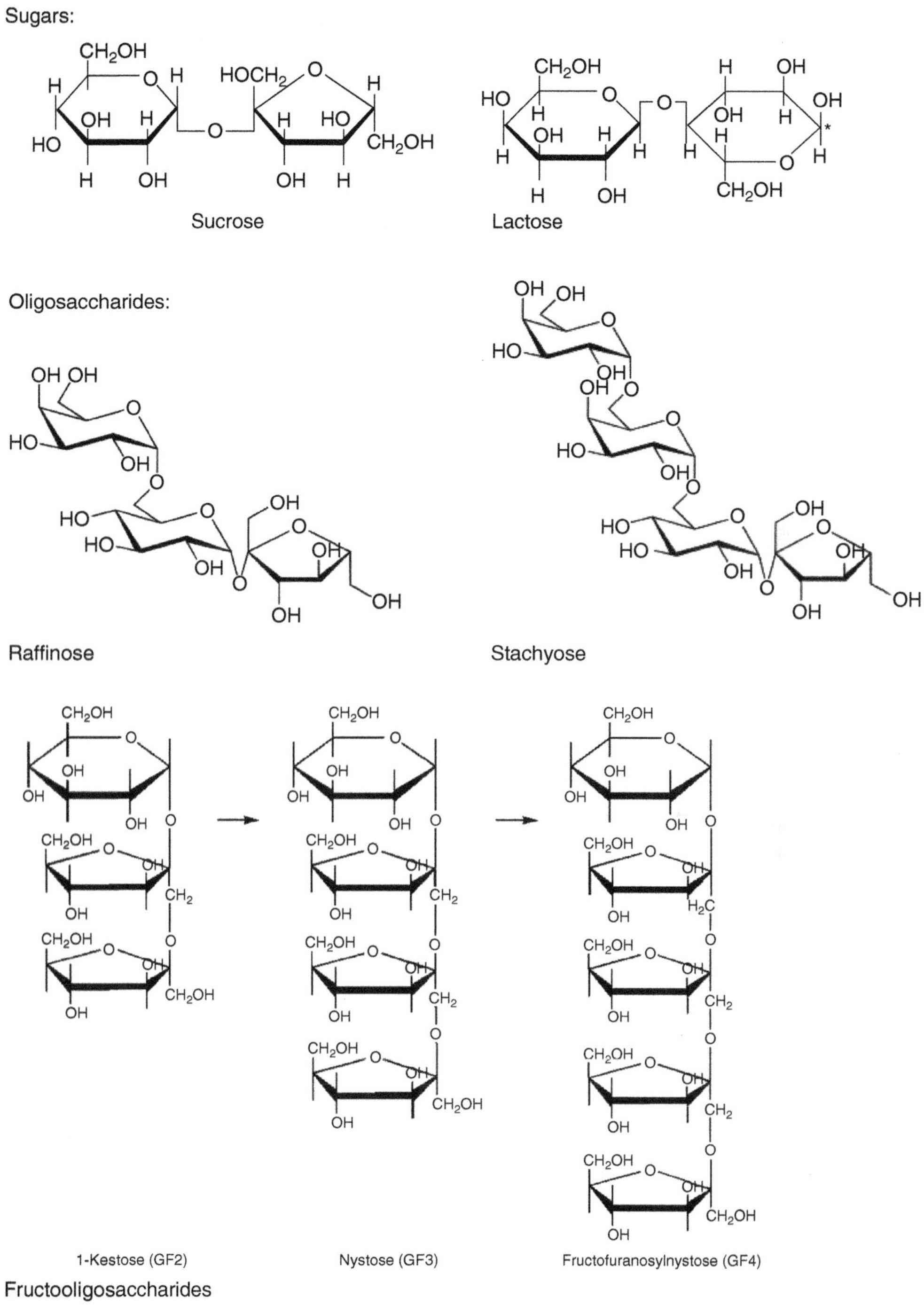

Sugars:

Sucrose

Lactose

Oligosaccharides:

Raffinose

Stachyose

1-Kestose (GF2)  Nystose (GF3)  Fructofuranosylnystose (GF4)

Fructooligosaccharides

**Figure 6.1**  Example of common di- and oligosaccharides.

## *Starch*

Native starch is water-insoluble semi-crystalline storage compounds synthesized as roughly spherical granules in many plant tissues (Gallant et al. 1992; Jenkins and Donald 1995; Martens et al. 2018)
(Figures 6.2 and 6.3) of which, cereals, peas, and beans are the most important sources in the nutrition of pigs (Theander et al. 1989; Bach Knudsen 1997). Pure starch consists predominantly of
α- glucan (approximately 99% of dry matter) in the form of amylose and amylopectin. Amylose is
roughly a linear α(1-4)-molecule (approximately 99%) with a molecular weight of approximately
$1 \times 10^5$ to $1 \times 10^6$, while amylopectin is a much larger molecule (molecular weight, approximately
$1 \times 10^7$ to $1 \times 10^9$) that is heavily branched and consist of approximately 95% of α(1-4)- and approximately 5% of α(1-6)-linkages (Biliaderis 1991). The two α-glucans are present in various proportions
in the starch granules, and the starches are defined as waxy when the ratio of amylose to amylopectin

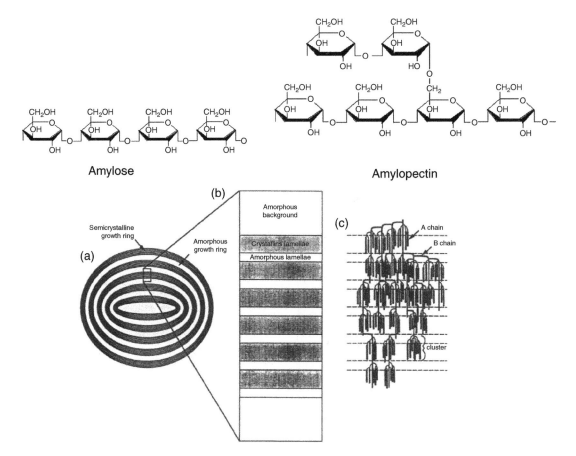

**Figure 6.2**   The two major polymers of starch (amylose and amylopectin) making up the starch granules and their organization
within the starch granule. (a) A single granule, comprising concentric rings of alternating amorphous and semi-crystalline
composition. (b) Expanded view of the internal structure. The semi-crystalline growth ring contains stacks of amorphous and
crystalline lamellae. (c) The currently accepted cluster structure for amylopectin with semi-crystalline growth ring. A-chain
sections of amylopectin form double helices, which are regularly packed into crystalline lamellae. B-chains of amylopectin provide
intercluster connections. Branching points for both A and B chains are predominantly located within the amorphous lamellae.
Source: From Jenkins et al. 1994 / with permission of John Wiley & Sons.

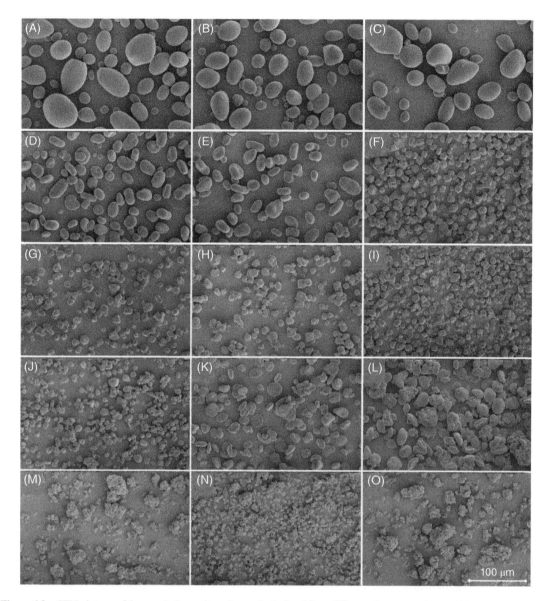

**Figure 6.3** SEM pictures of the morphology of starch samples isolated from different plants (magnitude 1000 times). A = regular potato, B = waxy potato, C = HMT potato, D = pea A, E = pea B, F = waxy corn, G = corn A, H = corn B, I = high amylose corn A, J = high amylose corn B, K = barley, L = wheat, M = waxy rice, N = rice A, O = rice B. Source: From Martens et al. 2018 / Springer Nature / CC BY 4.0.

is low (approximately < 15%), normal when amylose represents approximately 16–35% and high-amylose (or amylo-) when amylose exceeds approximately 36% (Ring et al. 1988). By X-ray diffraction studies, the starch can be divided into types A, B, and C (Biliaderis 1991). Starch with A-type crystallinity is dominant in cereals and has in general an open structure, while B-type crystallinity is more compact and found in tubers as, e.g., potato. Type C starch is a combination of A and B starch

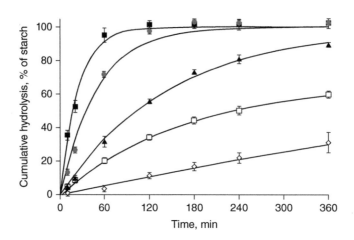

**Figure 6.4** *In vitro* digestion kinetics for purified cereal starch (wheat B, maize B E, and high amylose maize B G), legume starch (pea B H), and tuber starch (potato, A). Symbols indicate the average of triplicate measured values, and line represents the first-order kinetic model fitted to these data. Source: From Martens et al. 2018 / Springer Nature / CC BY 4.0.

and is present in legumes (Würsch et al. 1986; Dhital et al. 2016). It is important here to note also that starch is often also separated into large A granule and small B granules (Zhang et al. 2016).

All unmodified starch, if solubilized, can potentially be hydrolyzed by α-amylase to a variety of glucose, maltose, maltotriose, and α-limited dextrins but at a variable rate as indicated in Figure 6.4 where the in vitro hydrolysis of cereal, legume, and tuber starches is shown (Martens et al. 2018). A number of physical and chemical properties, however, may interfere with the rate and extent of starch digestion in the small intestine and make starch resistant to digestion (known as resistant starch; RS) (Sun et al. 2006; Martens et al. 2019). A division of RS into four different categories was proposed by Englyst et al. (1992): $RS_1$ is physically inaccessible starch entrapped within intact cell wall structures as found in peas and fava beans; $RS_2$ is raw starch granules that are resistant because of the starch structure as found in raw potato starch; $RS_3$ is retrograded amylose that is formed by irreversible recrystallization during cooling of gelatinized starch, and $RS_4$ is chemically modified starch. The latter is normally not used in animal feed.

### *Nonstarch Polysaccharides*

The NSP consist of a series of soluble and insoluble polysaccharides primarily present in plant cell walls but some may also be present intracellular as storage NSP (Selvendran 1984; Carpita and Gibeaut 1993; McDougall et al. 1996; Vincken et al. 2003; Loix et al. 2017) (Figure 6.5). The cell wall consists of middle lamella and primary and secondary cell walls. The middle lamella is the outermost layer of the cell walls and is a pectin-rich region between primary walls of adjacent cells and cement adjacent cells together (O'Neill and York 2003; Cosgrove and Jarvis 2012). The primary cell walls are flexible and highly hydrated. They are mainly composed of polysaccharides, structural glycoproteins, phenolic esters, ionically and covalently bonded minerals, and enzymes (Cosgrove 1997; Loix et al. 2017). The composition and quantity of matrix polysaccharides vary significantly according to plant species and cell type (Carpita 2000). There are two distinct types of primary walls:

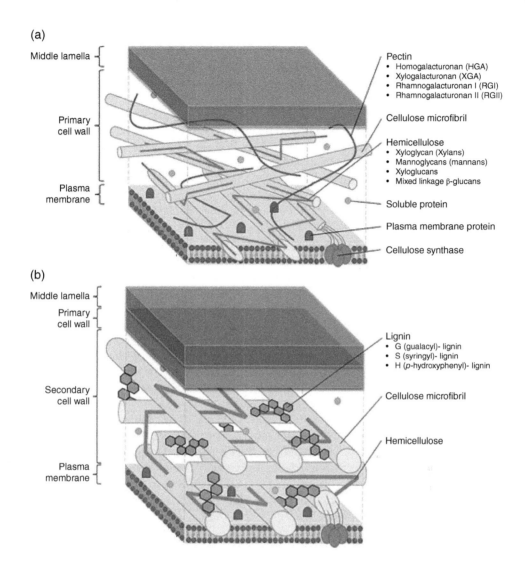

**Figure 6.5**   Structure and composition of the primary and secondary cell wall of plants. (a) The primary cell wall is located outside of the plasma membrane and consists of cellulose microfibrils, which are constructed by cellulose synthase complexes, hemicellulose, lignin, and soluble proteins. Hemicellulose binds to the surface of the cellulose microfibrils and can be divided into four groups; xyloglycan (xylans), mannoglycans (mannans), xyloglucans, and mixed linkage β-glucans. Pectins form a hydrated gel between the cellulose-hemicellulose network and consists of four pectin domains: homogalacturonan (HGA), xylogalacturonan (XGA), rhamnogalacturonan I (RGI), and rhamnogalacturonan II (RGII). (b) The secondary cell wall is constructed between the primary cell wall and the plasma membrane. Between the typically more arranged cellulose microfibrils, lignin molecules are impregnated, thereby replacing pectin molecules. Lignin is a complex phenolic polymer consisting of three monolignol subunits: G (guaiacyl)-, S (syringyl)-, and H (p-hydroxyphenyl)-lignin. Source: From Loix et al. 2017 / Frontiers Media S.A. / CC BY 4.0.

Type I cell walls are found in dicotyledons and nongraminaceous monocotyledons, whereas Type II cell walls are found in the Graminae family and some other monocotyledons (Carpita and Gibeaut 1993). Secondary cell walls, located between plasma membrane and primary cell walls, are formed in plant tissues that have ceased growing and are rich in cellulose thick, rigid, and lignified (Albersheim et al. 1994).

The building blocks of NSP are the pentoses arabinose and xylose, the hexoses glucose, galactose, and mannose, the 6-deoxyhexoses rhamnose and fucose, and the uronic acids glucuronic and galacturonic acids (or their 4-O-methyl esters). The monosaccharides making up the polysaccharides can exist in two ring (pyranose and furanose) forms. They can be linked through glycosidic bonds at any one of their three, four, or five available hydroxyl groups and in two (α or β) orientations. As a result, NSP can adopt a huge number of three-dimensional shapes and, thereby, offer a vast range of functional surfaces. The NSP can also be linked to lignin and suberin, which creates hydrophobic surfaces. In addition, charged groups on monomers, i.e., the acid group of uronic acids, can affect the ionic properties and be esterified to different degrees.

The most important NSP present in plant and plant coproducts used in feeds for animals are cellulose, mixed linkage $(1 \rightarrow 3; 1 \rightarrow 4)$-β-glucan (β-glucan), arabinoxylan (AX), xyloglucans, and pectic substances (homogalacturonan, rhamnogalacturonan type I and II, xylogalacturonan, and arabinogalactans type I and II) (Figure 6.6). Cellulose is a linear homopolymer of D-glucopyranosyl residues linked via consecutive β-$(1 \rightarrow 4)$ linkages (Figure 6.6). Because the hydrogen of the –OH groups at C-3 is in close proximity to the ring oxygen of the adjacent residue, the hydrogen bonds between O-3 and O-5 stabilizes the glucan chains intermolecularly (Selvendran 1984). β-Glucan is a linear homopolymer of D-glucopyranosyl residues linked mostly via two to three consecutive β-$(1 \rightarrow 4)$ linkages that are separated by a single β-$(1 \rightarrow 3)$ linkage (Izydorczyk and Dexter 2008; Wood 2010); trisaccharide (DP 3) and tetrasaccharide units (DP 4) typically account for 90–95% of total oligosaccharides after hydrolysis of β-glucan with lichenase (Izydorczyk and Dexter 2008). AX represent a family of polysaccharides formed from a linear backbone of $(1 \rightarrow 4)$-β-D-xylopyranosyl units substituted to varying degrees with mainly α-arabinofuranosyl residues at the O-2 position, the O-3 position or both, which results in four structural elements in the molecular structure of AX: monosubstituted xylose (X) at O-2 or O-3, disubstituted X at O-2,3 and unsubstituted X (Voragen et al. 1992; Izydorczyk and Dexter 2008). In cereals, the relative amount and the sequence of distribution of the structural elements in AX vary with the source and type of tissue (Saulnier et al. 2007; Izydorczyk and Dexter 2008). Xyloglucan consists of a backbone of β-$(1 \rightarrow 4)$-linked D-glucose units, which is heavily branched with xylose and β-galactose residues attached to some of the xylose units. Pectins represent a heterogeneous group of cell wall polysaccharides with the main structural elements: homogalacturonan, rhamnogalacturonan types I and II, xylogalacturonan, and arabinogalactans types I and II (Visser and Voragen 1996; Vincken et al. 2003; Caffall and Mohnen 2009).

### Lignin

Lignin is not a carbohydrate, but will be mentioned here as it is tightly associated to cell wall polysaccharides. Furthermore, many of the earlier and still commonly used analytical methods for fiber determination include lignin as a part of the fiber fraction. Therefore, it is difficult to discuss the physiochemical properties and degradation of carbohydrates in the gastrointestinal tract without inclusion of lignin in the description. Lignin is formed by the polymerization of coniferyl, p-coumaryl, and sinapyl alcohols (Liyama et al. 1994; Davin et al. 2008). These phenylpropane units are linked in an irregular three-dimensional pattern by ether and carbon–carbon bonds in

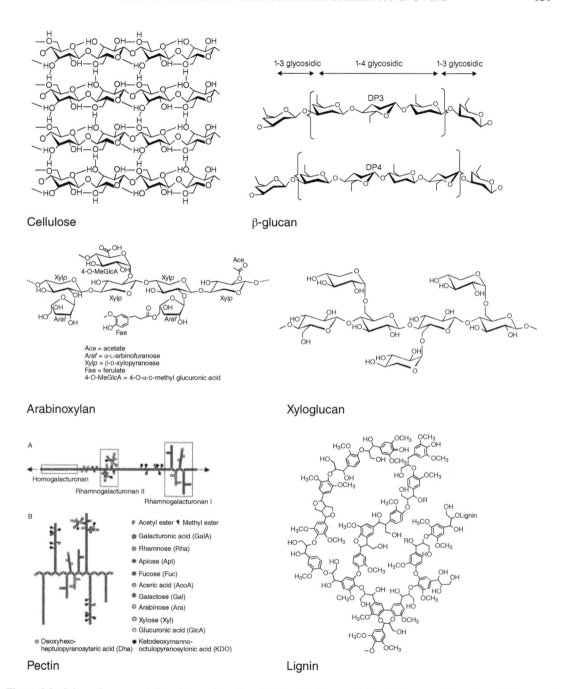

**Figure 6.6** Schematic representation of the major cell wall polysaccharides – cellulose, β-glucan, arabinoxylan, xyloglucan and pectin – and the non-carbohydrate polyphenolic ether lignin.

which any of the carbons may be part of the aromatic ring. The main function of lignin is to cement and anchor the cellulose microfibrils and other matrix polysaccharides, and in this way stiffen the walls making it very rigid and with the purpose for the plant to protect itself. This structure, however, makes it difficult to degrade by the microorganisms in the large intestine (Figures 6.5 and 6.6).

### Physicochemical Properties of Fiber

The physicochemical properties, hydration properties, and viscosity of fiber are linked to the type of polymers that makes up the cell wall and their intermolecular association (McDougall et al. 1996). The hydration properties of fiber are characterized by the swelling capacity, solubility, water-holding capacity, and water binding capacity (WBC). The latter two have been used interchangeably in the literature because both reflect the ability of a fiber source to immobilize water within its matrix. Due to the many multi-OH groups present in polysaccharides, they have a strong affinity to water, but also strong interactions among polysaccharide molecules via hydrogen bonding. Therefore, the balance between molecule–molecule interaction and molecule–water interaction is the key to understand the polysaccharide solubility and most nonstarch polysaccharides is in amorphous state (Guo et al. 2017). The first part of the solubilization process of polymers is swelling, in which incoming water spreads the macromolecules until they are fully extended and dispersed (e.g., the cell wall in Figure 6.5 expands in the three-dimensional space) (Thibault et al. 1992). For polysaccharides that adopt regular, ordered structures (e.g., cellulose or linear arabinoxylan), solubilization is not possible because the linear structure increases the strength of the noncovalent bonds, which stabilize the ordered conformation. Under these conditions, only swelling can occur (Thibault et al. 1992). The majority of polysaccharides result in viscous solutions if dissolved in water (Morris 1992). The viscosity is dependent on the primary structure, molecular weight of the polymer, and concentration. Large molecules increase the viscosity of diluted solutions and their ability to do so depends primarily on the volume they occupy. Although a range of polysaccharides by analytical definitions are soluble, their in vivo solubility may be restricted in the feed matrix, thus limiting their viscosity elevating properties (Wood 1990; Johansen et al. 1997).

### Classification and Terminology of Carbohydrates Based on Physiology

A challenge in classification of dietary carbohydrates by their chemical composition is to relate it to nutritional impact and particularly how to the various chemical divisions relate to the site of digestion (Table 6.1) (Cummings and Stephen 2007; Bach Knudsen and Lærke 2018). Another classification of the carbohydrate fraction is therefore a division into carbohydrates that can potentially be digested by endogenous enzymes (digestible carbohydrates, DC) and carbohydrates that potentially can be fermented by the microbiota (nondigestible carbohydrates, NDC) (Cummings and Stephen 2007; Bach Knudsen and Lærke 2018). Digestible carbohydrates include sugars and starch, whereas NDC include NSP, RS, and nondigestible oligosaccharides (NDO). Nondigestible carbohydrates plus lignin are named dietary fibers as defined by the European Commission as: *"carbohydrate polymers with ten and more monomeric units which are neither digested nor absorbed in the human small intestine."* The European Commission definition is based on the Codex Alimentarius (Cummings et al. 2009; Phillips and Cui 2011) but extended to include NDOs with DP 3-9. Although lignin is not mentioned directly in the definition, it is stated in the text that lignin is included in the DF definition (Cummings et al. 2009; Phillips and Cui 2011). The term "Fiber" is used for total NSP and lignin and represents the part of the feed that is traditionally considered fibers of the feed (Theander et al. 1989; Chesson 1991).

## Measurements of Dietary Carbohydrates and Lignin

The conventional system, now in use for more than 150 years, is the so-called proximate system of analysis according to Weende (Henneberg and Stohmann 1859). It consists of the analysis of dry matter, ash, fat, crude protein ($N \times 6.25$), and crude fiber, where the latter is determined as the residue after reflux of fat-extracted residue with 1.25% sulfuric acid and 1.25% sodium hydroxide and corrected for ash in the insoluble residue. A nitrogen-free extract (NFE) is then calculated as the dry matter not accounted for by the sum of ash, protein ($N \times 6.25$), fat, and crude fiber (Figure 6.7). The principle in the Van Soest method for determination of fiber is the quantification of nonsolubilized neutral detergent fiber (NDF), which includes almost entirely insoluble fiber components, $RS_3$ and cell wall-bound protein, after removal of the cell content (sugars, starch, protein but also pectins and other soluble fiber components) by a neutral detergent solution (Van Soest 1963a,b; Van Soest 1982; van Soest et al. 1991). Further treatment of the NDF fraction with 1M $H_2SO_4$ solution solubilize the hemicellulose, and the remaining acid detergent fiber (ADF) fraction includes mainly lignocellulose plus insoluble minerals. Further treatment of the ADF fraction with 12M $H_2SO_4$ solubilizes the cellulose and yields a crude lignin fraction (acid detergent lignin, ADL).

With modern analytical techniques based on specific enzymes, colorimetric, and chromatographic assays (Figure 6.8), the carbohydrate fraction can be divided according to the chemical and nutritional classifications in Table 6.1 and Figures 6.1, 6.2 and 6.6. Commonly used methods include enzymatic or chromatographic methods to determine sugars and oligosaccharides, enzymatic methods to determine starch and RS, and gravimetric or enzymatic-chemical-gravimetric methods to determine dietary fiber as a whole or separated into soluble and insoluble polysaccharides and lignin; for an overview see (Bach Knudsen and Lærke 2018). Because the different analytical methods used for the determination of fiber vary widely in terms of analytical principles, the values reported in the literature will vary too (Bach Knudsen and Lærke 2018). For all classes of feedstuffs, DF gives the highest values because it includes all DF components including NDO, fructans, and RS.

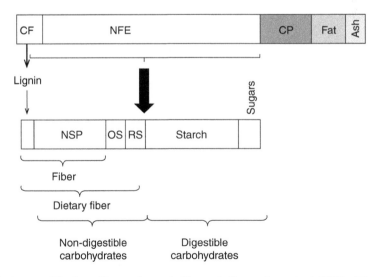

**Figure 6.7** Carbohydrates and lignin making up the crude fiber and nitrogen free extract (NFE) of the proximate (Weende) analysis. CP, crude protein; NFE, nitrogen free extract; NSP, non-starch polysaccharides; OS, oligosaccharides; RS, resistant starch.

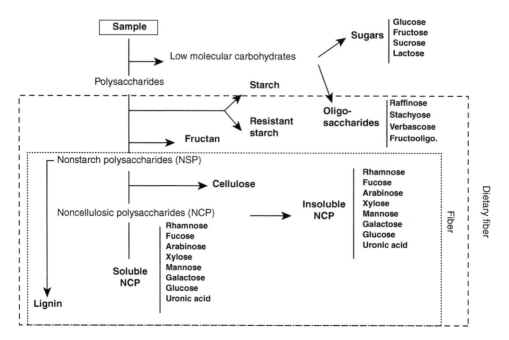

**Figure 6.8** The principles in the characterization of carbohydrates and lignin by specific enzymatic-colorimetric and enzymatic chemical-gravimetric methods. The fiber is represented by non-starch polysaccharides and lignin and dietary fiber by non-starch polysaccharides, lignin, resistant starch, fructans and oligosaccharides. NCP = non-cellulosic polysaccharides.

NDF values are consistently lower than the total fiber values but relatively close to the insoluble fiber components (sum of lignin, cellulose, and I-NCP). The ADF values are close to the sum of cellulose and lignin, whereas the values for crude fiber are consistently the lowest for all groups of feedstuffs (Bach Knudsen and Lærke 2018).

### Carbohydrates and Lignin in Feedstuffs

The carbohydrates in the feeds are not present as pure chemical entities but as a mix of sugars, oligosaccharides, and polysaccharides, the latter mostly linked to other biopolymers such as proteins and lignin (Figure 6.9). The modern pig industry relies on relatively few feedstuffs, mostly from cereals (corn, wheat, barley, oats, rye, and rice), cereal coproducts (different milling fractions, residues from biofuel and alcohol industries, etc.), cereal substitutes (e.g., tapioca and maniocca), legumes (e.g., peas, beans, and lupins), protein concentrates (e.g., meal or cakes of soybean, rape, sunflower, and cotton), and coproducts from the sugar and starch industries.

Whole grain cereals and cereal coproducts are used worldwide as energy and protein sources for pigs. Cereals have a high content of digestible carbohydrates primarily in the form of starch and an intermediate level DF in the form of insoluble cellulose and lignin and insoluble and soluble AX and β-glucan. Cereal brans, hulls, and coproducts from the bioethanol and alcohol industry have a lower concentration of starch and concomitant higher concentration of DF than the corresponding whole grain cereals. In these feeds, the DF is primarily in the form of insoluble components – cellulose, insoluble AX, and lignin. Protein crops and feedstuffs are included in diets for pigs because of their high protein and amino acid content. However, the carbohydrate fraction is also an

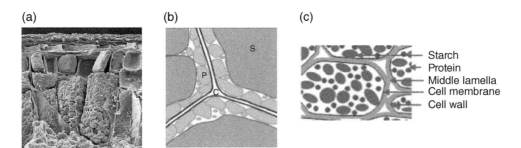

**Figure 6.9** (a) Scanning electron microscope photo of the outer layers of the wheat kernel. Source: Eye of Science/Science Photo Library; (b) Model illustrating location of starch granules in plant cells. S, P and C refer to starch granules, protein bodies and cell wall matrices, respectively; Source: Dhital et al. 2017 / with permission of Taylor & Francis Ltd. (c) Schematic diagram showing starch granules and protein embedded in the cell wall structure of legume cotyledon. Source: Adapted from Dhital et al. 2016.

important energy source even though the composition is somewhat different to that of cereals. While all cereals contain high levels of starch, this polysaccharide is only present in measurable quantities in peas and fava beans. In contrast, the protein concentrates contain relatively high levels of sugars and NDO in the form of α-galactosides. The content of the main NSP – arabinan (peas and rapeseed), arabinogalactans (soybeans and rapeseed), and galactans (lupins) present either free or linked to rhamnogalacturonans – present in protein-rich crops and feedstuffs vary widely. This is because the protein crops and feedstuffs derive from different botanical families – leguminosae (soybean meal, peas, fava beans, and lupins), cruciferae (rapeseed), compositiae (sunflower), etc. – and due to the diverse composition of the individual tissues making up the seed (for more details see (Bach Knudsen and Lærke 2018). Coproducts from the vegetable food and agro industries are mainly residues from fruit and vegetables deriving primarily from the industrial production of sugar, starch, and pectin. These coproducts represent a very heterogeneous group as they come from different plant families and botanical origins (tubers, roots, fruits, culms, shells, and hulls). A factor contributing to the diversity in composition and functionality is the exposure of the plant materials during processing to a wide variety of different physical and chemical treatments for the extraction of the economically important components. The coproducts will consequently have different matrices depending on the botanical origin but all with plant cell walls in the form of NSP and lignin constituting most of the DM.

The diverse composition of the carbohydrates in the feedstuffs (Table 6.2) makes it possible to produce compound feeds with very different composition. For instance, diets very low in dietary fiber can be produced from rice as the primary source of carbohydrates (Hopwood et al. 2004), whereas, in contrast, coproducts from the vegetable food and agricultural industries can be used to produce feeds very high in dietary fiber (Serena et al. 2008b).

### Processing of Feedstuffs and Common Feeds

Starches in feedstuffs are always present in grains and legumes in association with proteins, many of which are relatively hydrophobic, and the protein-starch network is surrounded by cell walls (Figure 6.9). During the digestion processes, starch, therefore, tends to be maintained in the interior of the ingested particles protected from water. Starches from tubers and legumes are particularly well protected from the polar environment of luminal fluids (Würsch et al. 1986; Dhital et al. 2016),

**Table 6.2** Typical carbohydrate and lignin contents (g kg$^{-1}$ DM) of feedstuffs[a].

| Feedstuff | Digestible CHO | | | Nondigestible CHO | | | | | | Dietary | | |
|---|---|---|---|---|---|---|---|---|---|---|---|---|
| | Sugars | Starch[b] | Total | OS | Fructans[c] | RS | S-NCP[d] | I-NCP[e] | Cellulose[e] | KL[e] | Fiber[f] | fiber[g] |
| Rice | 2 | 837 | 839 | 2 | <1 | 3 | 9 | 1 | 3 | 8 | 22 | 27 |
| Corn | 17 | 680 | 697 | 3 | 6 | 10 | 9 | 66 | 22 | 11 | 108 | 127 |
| Wheat | 13 | 647 | 660 | 6 | 15 | 4 | 25 | 74 | 20 | 19 | 138 | 163 |
| Rye | 25 | 611 | 636 | 7 | 31 | 2 | 42 | 94 | 16 | 21 | 174 | 214 |
| Barley | 16 | 585 | 601 | 6 | 4 | 2 | 56 | 88 | 43 | 35 | 221 | 233 |
| Oats | 13 | 466 | 479 | 5 | 3 | 2 | 40 | 110 | 82 | 66 | 298 | 308 |
| Wheat middling | 24 | 572 | 596 | 12 | 23 | 3 | 71 | 101 | 19 | 11 | 201 | 239 |
| Wheat bran | 37 | 220 | 257 | 16 | 20 | 2 | 29 | 273 | 72 | 75 | 449 | 487 |
| Corn feed meal | 35 | 560 | 595 | 5 | 2 | 6 | 10 | 114 | 33 | 18 | 174 | 187 |
| Corn bran | 26 | 374 | 400 | 26 | 4 | 2 | 32 | 240 | 83 | 25 | 379 | 411 |
| Barley hull meal | 21 | 172 | 193 | 12 | 7 | 2 | 20 | 267 | 192 | 115 | 594 | 615 |
| DDGS - corn | ND | 35 | — | ND | ND | ND | 25 | 183 | 68 | 47 | 323 | — |
| DDGS - wheat | ND | 92 | — | ND | ND | ND | 55 | 135 | 61 | 86 | 337 | — |
| Soybean meal | 77 | 27 | 104 | 60 | — | — | 63 | 92 | 62 | 16 | 233 | 293 |
| Peas - white | 39 | 432 | 471 | 49 | — | 22 | 52 | 76 | 53 | 12 | 192 | 263 |
| Fava beans | 32 | 375 | 407 | 54 | — | 32 | 50 | 59 | 81 | 20 | 210 | 296 |
| Rape seed cake | 72 | 15 | 87 | 16 | — | — | 43 | 103 | 59 | 90 | 295 | 311 |
| Cotton seed cake | 12 | 18 | 30 | 54 | — | — | 61 | 103 | 92 | 83 | 340 | 394 |
| Pea hull | 15 | 84 | 99 | ND | 5 | 4 | 121 | 148 | 452 | 9 | 721 | 730 |
| Potato pulp | <1 | 122 | 122 | ND | ND | 127 | 280 | 95 | 202 | 35 | 612 | 739 |
| Sugar beet pulp | 38 | 5 | 43 | ND | 0 | ND | 290 | 27 | 203 | 37 | 737 | 737 |
| Alfalfa | 21 | 68 | 89 | 2 | 6 | ND | 77 | 113 | 139 | 128 | 457 | 465 |
| Chicory roots | 156 | ND | 156 | ND | 470 | ND | 76 | 24 | 48 | 11 | 158 | 628 |

[a] CHO, carbohydrates; OS, oligosaccharides; RS, resistant starch; S-NCP, soluble noncellulosic polysaccharides; I-NCP, insoluble noncellulosic polysaccharides; KL, Klason lignin; DDGS, distiller's dried grains with solubles; ND, not determined.
[b] Digestible starch is calculated as total starch-RS.
[c] Fructans are a mix of oligosaccharides (DP 3-9) and polysaccharides (DP > 10).
[d] S-NCP is synonymous with soluble fiber.
[e] Insoluble fiber = I-NCP + cellulose + KL.
[f] Fiber = S-NCP + I-NCP + cellulose + KL.
[g] Dietary fiber = Fiber + OS + fructans + RS.

but even in cereals, the starch may not be accessible to α-amylase unless they have been physically altered (Dhital et al. 2017). The principal processes facilitating starch availability for water penetration and consequent α-amylase digestion are physical processing (grinding, cracking, and roller milling) and heating (pelleting, expanding, and extrusion cooking) of the grains and legumes (Chae and Han 1998). Grinding is a physical process that reduces the size of the particles and increases the surface area inducing more contact points between digestive enzymes and the substrate. For instance, a cubic, measuring 1 cm on each side, will have a surface area of 6 cm$^2$. If the cubic is then divided into pieces each measuring 0.1 cm, then the surface area of the divided cubic will increase to 60 cm$^2$ or a 10-fold increase. Another form of physical treatment is hydrothermal treatment, which alters the physical form of the starch from a crystalline to a gel structure (Figure 6.10). This enlarges the surface and promotes efficient entry into the polar solution for interaction with α-amylases (Biliaderis 1991). Chilling after cooking re-alters the polysaccharide's physical state (retrograde), which may reduce the digestibility (Figure 6.7). Starches with a high proportion of amylose (high amylose) are usually more susceptible to reduction in digestibility

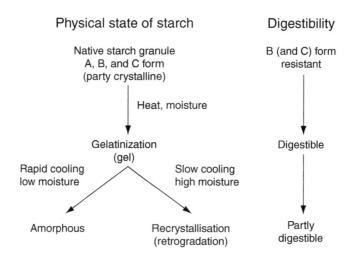

**Figure 6.10**   Physical state of starch in relation to its digestibility. Source: From Cummings (1997); based on data from Biliaderis (1991) and Cummings and Englyst (1995).

after heat processing than the waxy types (Brown et al. 2001) because retrogradation of amylase is an irreversible process leading to a compact structure (similar to cellulose).

Co-products from the vegetable food and agro industries represent yet another form of processing of the feedstuffs prior to its use as feed for animals. For instance, co-products from the industrial production of oil, biofuel, sugar, starch, beer, and pectin have all undergone physical and chemical processing in one or another way (Serena and Bach Knudsen 2007; Pedersen et al. 2014). These types of feedstuffs will in general, represent a very heterogeneous group of plant residues, coming from different plant families and botanical origin (cereals, tubers, roots, fruits, seeds, culms, shells, and hulls). During the processing steps, they are exposed to a wide variety of different physical and chemical treatments for the extraction of the economically important components. The residue will consequently have a different matrix with a high content of protein (e.g., soybean meal, rape seed meals, etc.) and plant cell walls as NSP and lignin.

### Digestion of Carbohydrates in the Small Intestine

The transport processes in the intestinal enterocyte cannot accommodate anything larger than monosaccharides. Consequently, the majority of carbohydrates present in the feed need to be degraded to low-molecular-weight compounds prior to absorption. For disaccharides and starches, this involves the α-amylase secreted in salivary and pancreatic juice and various oligosaccharidases, sucrase, and lactase present on the intestinal brush boarder (Kidder and Manners 1980; Gray 1992). A contributing carbohydrate hydrolytic effect, however, comes from the microflora permanently colonizing these sites of the gastrointestinal tract. Jensen and Jørgensen (1994) reported a gradual increase in total anaerobic bacteria from $10^7$ to $10^9$ viable counts in stomach to $10^9$ viable counts in distal small intestine. Substantial levels of LA and SCFA have also been reported in digesta collected from the stomach and the more distal parts of the small intestine (Argenzio and Southworth 1974; Bach Knudsen et al. 1991).

**Table 6.3** Typical values for the digestibility of sugars and oligosaccharides in the gastrointestinal tract of pigs.

| Source | Sugars | | | Oligosaccharides | | | | |
| --- | --- | --- | --- | --- | --- | --- | --- | --- |
| | Glucose | Fructose | Sucrose | Raffinose | Stachyose | Verbascose | Fructans | Total |
| Glucose | 98.3 | — | — | — | — | — | — | — |
| Fructose | — | 86.6 | — | — | — | — | — | — |
| Sucrose | — | — | 98.3 | — | — | — | — | — |
| Peas | | | | | | | | |
| Dried | — | — | 96 | 68 | 74 | 58 | — | 66 |
| Toasted | — | — | 94 | 65 | 53 | 23 | — | 42 |
| Soybean meal | — | — | 87 | 38 | 49 | — | — | 49 |
| Lupins | | | | | | | | |
| Blue | — | — | 93 | 84 | 93 | 77 | — | — |
| Yellow | — | — | 92 | 64 | 88 | 74 | — | — |
| Inulin | | | | | | | 40[a] | 40[a] |
| Chicory roots[b] | — | — | — | — | — | — | 33[c] | 33[c] |

[a] Average valued of three diets varying 53–208 g kg⁻¹ DM
[b] The fructans in chicory roots are a mix of oligosaccharides and polysaccharides.
[c] Average values from three diets varying 79–156 g kg⁻¹ DM.
Sources: Ly 1992; Gdala et al. 1994; Canibe and Bach Knudsen 1997b; Hedemann and Bach Knudsen 2009.

## Sugars

Sucrose is readily degraded and absorbed from the gut lumen of pigs, and there is no limitation in the digestibility even at very high levels (Table 6.3) (Ly 1992, 1996). Glucose is directly available for absorption in the small intestine and 100% absorbable, while the absorption of fructose is slower and in some cases incomplete. The lactase activity is high in unweaned piglets, and, therefore, lactose, the disaccharide deriving from milk, is efficiently digested and absorbed. However, the activity is markedly reduced after weaning to a level that is not sufficient to degrade high loads of lactose in the diet. In a study with catheterized pigs, it was shown the absorption coefficient was reduced to almost half when the intake level was doubled (Rérat et al. 1984b; Oksbjerg et al. 1988).

## Oligosaccharides

Although pigs lack the enzymes capable of cleaving the bonds in most oligosaccharides, studies with raffinose from soyabean, lupins and peas, fructans (a mix of oligosaccharides and inulin), fructooligosaccharides, and transgalactoologosaccharides show highly variable and relatively high digestibility coefficients in the small intestine (Gdala et al. 1994; Gdala and Buraczewska 1997; Gdala et al. 1997; Canibe and Bach Knudsen 1997b; Houdijk et al. 1999; Houdijk et al. 2002) (Table 6.3) due to microbial activity. The digestibility coefficients also appear to be dose related as indicated by the much greater digestibility coefficient at a low (Houdijk et al. 1999) compared to a higher inclusion level (Hedemann and Bach Knudsen 2009).

## *Starch*

The only carbohydrase secreted in the endogenous digestive juice is salivary and pancreatic α- amylase, which digests α-1,4-glucosidic linkages in starches (Kidder and Manners 1980; Moran 1985; Gray 1992). Quantitatively, salivary α-amylase contributes little to the degradation of starch, since it is acid labile and rapidly degraded in the stomach where the pH typically will be in the range of 2–4 (Argenzio and Southworth 1974; Duke 1986). The vast majority of starch is therefore degraded by pancreatic α-amylase in the intestinal lumen. Bicarbonate secreted in the pancreatic juice raises the pH to a level of 5–6 in duodenum, and the pH further increases along the length of the small intestine to reach neutrality at the ileum (Argenzio and Southworth 1974; Duke 1986). The high endogenous secretion also dilutes the feed bolus to a level of approximately 10% dry matter, which facilitates the penetration of the polar solution the feed particles, and thereby ensuring an efficient cleavage of starch. The α-amylase is not able to break the α-1,6 linkages present in amylopectin, and the capacity to break α-1,4 links adjacent to the branching point is sterically hindered. Consequently, end-products of α-amylase digestion are maltose, maltotriose, and α-limit dextrins. These oligosaccharides are then degraded further to glucose by oligosaccharidases present as large glycoprotein components on the surface of the intestinal brush-border membrane (Lentze 1995). Glycoamylase (amyloglucosidase, maltase-glucoamylase) is capable of cleaving single α-1,4-linked glucose residues sequentially from the nonreducing end of α-limited dextrin, but is blocked when an α-1,4-linked glucose is located at the terminal end of the saccharide. The α-dextrinase is the only carbohydrase capable of cleaving the nonreducing terminal α-1,6 link after it has been uncovered by action of the other enzymes (Moran 1985; Gray 1992). The shorter α-1,4 linked oligosaccharides, such as maltose and maltotriose, are then cleaved to glucose by maltase, which is a highly efficient α-1,4 glucosidase. The final glucose product is transported by the specific glucose carrier SLGT1, an integrated brush-boarder glycoprotein expressed only in the small intestine and with a high affinity for the monosaccharide. The actual driving force for uphill transport of glucose into the enterocyte is provided by the sodium-potassium ATPase, which pumps the intracellular $Na^+$ across the apical membrane (Lentze 1995). Glucose probably diffuses from the basolateral (serosal) surface via GLUT2 to the capillaries of the villous core and further to the portal vein.

The digestibility of starch at the end of the small intestine is influenced by the crystalinity (A, B, C), the presence of cell walls, and chemical composition of the starch (amylose, amylopectin) (Table 6.4). In finely ground cereals, the finely ground particles enable easy access to starch for hydrolysis by α-amylases, and most studies show that the bulk of the starch has been digested and absorbed by the time the digesta reach the end of the small intestine (Martens et al. 2019). A factor, however, that may reduce the digestibility of starch in the small intestine is the particle size of the feed, as coarse particles encapsulate intracellular nutrients and, thereby, withhold the content from digestion by endogenous enzymes. For instance, the digestibility of starch in coarse barley was approximately four absolute units lower than in finely ground barley (Table 6.4). Hydrothermal treatments such as pelleting and extrusion cooking alter the native structure of starch and makes the starch from cereals even more digestible as demonstrated in studies with breads and extrusion cooking that all show starch digestibility above 98% (Glitsø et al. 1998; Bach Knudsen and Canibe 2000; Bach Knudsen et al. 2005; Sun et al. 2006; Stein and Bolkhe 2007; Le Gall et al. 2009b; Martens et al. 2019). Encapsulation of starch in cell walls (Würsch et al. 1986) is another factor that influences the digestibility of starch and partly responsible for the lower digestibility of legume starches (82–89%) than cereal starches (95–99%) (Table 6.4). Hydrothermal treatment in the form of toasting does not seem to be sufficiently efficient in increasing the digestibility of pea starch (Canibe and Bach Knudsen 1997b), whereas extrusion cooking increases the digestibility to a level intermediate between raw peas and cereals (Sun

**Table 6.4** Intake and digestibility of starch and nonstarch polysaccharides in the small intestine of pigs fed various types of raw and processed starches and fibers[a].

| Item | Starch | | | | NSP | |
| --- | --- | --- | --- | --- | --- | --- |
| | type | Form | Intake, g d⁻¹ | Digestibility, % | intake, g d⁻¹ | Digestibility, % |
| Cereal mix | A | R | 556 | 96.0 | 182 | 20.0 |
| +Wheat bran | A | R | 310 | 95.7 | 351 | 11.0 |
| +Sugar beet pulp | A | R | 193 | 95.3 | 633 | 37.0 |
| Wheat flour | A | R | 978 | 99.4 | 45 | 30.0 |
| +Wheat bran | A | R | 1003 | 98.7 | 86 | 10.0 |
| +Oat bran | A | R | 1086 | 98.6 | 77 | 36.0 |
| Oat groats | A | R | 885 | 97.0 | 123 | 21.0 |
| Oat flour | A | R | 878 | 98.6 | 81 | 25.0 |
| Oat bran | A | R | 814 | 98.9 | 202 | 15.0 |
| Wheat-fine, 2.9% >2 mm | A | R | 874 | 96.3 | 212 | 8.0 |
| Wheat-coarse, 12.0% >2 mm | A | R | 832 | 96.3 | 215 | 6.0 |
| Barley-fine, 0.7% >2 mm | A | R | 819 | 96.5 | 286 | 18.0 |
| Barley-coarse, 23.3% >2 mm | A | R | 832 | 92.2 | 303 | −7.0 |
| Rye flour bread | A | G | 838 | 98.9 | 116 | 19.0 |
| Whole grain rye bread | A | G | 777 | 98.0 | 202 | 7.0 |
| High fiber wheat bread | A | G | 841 | 98.8 | 350 | 21.8 |
| High fiber rye bread | A | G | 721 | 98.3 | 321 | 22.7 |
| Peas-dried | C | R | 457 | 88.9 | 225 | 40.0 |
| Peas-toasted | C | G | 443 | 85.7 | 197 | 24.0 |
| Faba bean 1 | C | R | — | 81.5 | — | — |
| Faba bean 2 | C | R | — | 86.4 | — | — |
| Potato | B | R | 712 | 39.8 | — | — |
| Potato | B | G | 751 | 98.3 | — | — |

[a] NSP, nonstarch polysaccharides; A, type A starch; B, type B starch; C, type C starch; R, raw; G, gelatinized.
Sources: Graham et al. 1986; Bach Knudsen and Hansen 1991; Bach Knudsen et al. 1993; Canibe and Bach Knudsen 1997b; Gdala and Buraczewska 1997; Glitsø et al. 1998; Bach Knudsen et al. 2005; Sun et al. 2006.

et al. 2006). The crystalline nature of raw potato starch (type B) makes it the least digestible starch for pigs with values in the range of 30–50% (van der Meulen et al. 1997a; Sun et al. 2006) but can also be partly responsible for the lower digestibility of legume starches (type C). Raw potato starch, however, is only occasionally used at low levels in practical swine production. The amylose:amylopectin ratio is a another factor of importance, where starches with higher amylose content generally have a lower ileal digestibility (Regmi et al. 2011a; Nielsen et al. 2015; Martens et al. 2019).

### Nonstarch Polysaccharides

Although no NSPases are secreted to or present on the intestinal brush-border membrane to cleave the bondings in NSP, it has since long been known that there is a significant loss (20–25%) and modification of NSP during passage of the small intestine. It is also known that the different NSPs are degraded to a variable extent with a higher digestibility of the linear and relatively soluble β- glucan than of cellulose and soluble and insoluble arabinoxylan (Bach Knudsen and Hansen 1991; Bach Knudsen et al. 1993; Glitsø et al. 1998; Bach Knudsen et al. 2005; Le Gall et al. 2009b). Pectin polysaccharides have shown high degradation in some studies (Canibe and Bach Knudsen 1997b) and low in others (Jørgensen et al. 1996).

*Physical Effects*

The fiber fraction may potentially interact with the digestion processes in the foregut as it represents the entity of feed not degraded by the endogenous enzymes (Bach Knudsen and Jørgensen 2001). Therefore, its concentration and the flow at ileum increases with increasing dietary fiber concentration (Table 6.5). As the digesta move along stomach and small intestine, a proportion of NSP is solubilized, which will enhance the viscosity of the liquid phase (Figure 6.11). However, the ability of the different fiber sources to increase luminal viscosity depends on the chemical and structural composition, molecular weight, and resistance to degradation. Thus, although β-glucan has a higher molecular weight than arabinoxylan (Izydorczyk and Biliaderis 1995; Wood 2010), the heavy depolymerization of β-glucan (Johansen et al. 1997; Kasprzak et al. 2012) make it relatively less influential on luminal viscosity than the more resistant arabinoxylan (Bach Knudsen et al. 2005; Lærke et al. 2008; Le Gall et al. 2009a,b; Kasprzak et al. 2012). Nevertheless, a compilation of results with 78 diets that all have been studied in ileal cannulated pigs (Bach Knudsen et al., unpublished) showed that neither content of soluble nor insoluble fiber had any major impact on the digestibility of starch, except in some very special situations. Moreover, the ileal digestibility of starch in sows is at the same level as in growing pigs and not influenced by either soluble or insoluble fiber even when provided at very high levels (429–455 g/kg DM) (Serena et al. 2008b).

## Fermentation of Carbohydrates in the Large Intestine

The large intestine can be considered an anaerobic fermentation chamber with a low oxygen concentration, low flow rate, and high moisture content, all being conditions that favor bacterial growth (Figures 6.12 and 6.13). In numbers, the microbial ecosystem may reach $10^{11}$–$10^{12}$ viable counts per

**Table 6.5** Digesta flow, marker index, and concentration of carbohydrates in diet and ileal digesta[a].

| Item | Digesta flow, g d$^{-1}$ | Marker index | Dig CHO | | NDC | |
| | | | sugars | Starch | fructans | NSP |
|---|---|---|---|---|---|---|
| Growing pigs | | | | | | |
| Low dietary fiber | | | | | | |
| Diet | | 100 | 6 | 517 | — | 56 |
| Ileum | 2126 | 652 | 7 | 17 | — | 366 |
| Medium dietary fiber | | | | | | |
| Diet | | 100 | 7 | 454 | — | 97 |
| Ileum | 2584 | 472 | 8 | 12 | — | 372 |
| High dietary fiber | | | | | | |
| Diet | | 100 | 29 | 492 | 14 | 211 |
| Ileum | 3785 | 345 | 8 | 28 | 20 | 514 |
| Adult sows | | | | | | |
| Low dietary fiber | | | | | | |
| Diet | | 100 | 21 | 501 | 9 | 140 |
| Ileum | 5560 | 347 | 10 | 59 | 3 | 267 |
| High dietary fiber | | | | | | |
| Diet | | 100 | 23 | 210 | 6 | 363 |
| Ileum | 9816 | 187 | 3 | 33 | 1 | 507 |

[a] Dig, digestible; CHO, carbohydrates; NDC, nondigestible carbohydrates; NSP, nonstarch polysaccharides. Sources: Bach Knudsen and Canibe 2000; Bach Knudsen et al. 2005; Serena et al. 2008b.

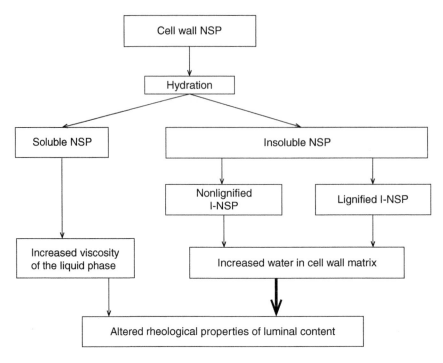

**Figure 6.11**   Impact of soluble and insoluble non-starch polysaccharides on the rheological properties of digesta.

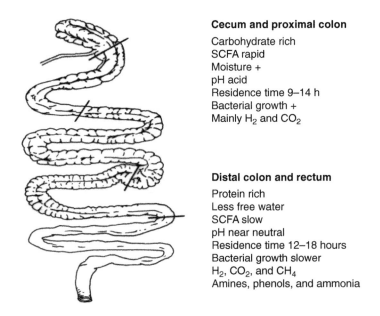

**Cecum and proximal colon**

Carbohydrate rich
SCFA rapid
Moisture +
pH acid
Residence time 9–14 h
Bacterial growth +
Mainly $H_2$ and $CO_2$

**Distal colon and rectum**

Protein rich
Less free water
SCFA slow
pH near neutral
Residence time 12–18 hours
Bacterial growth slower
$H_2$, $CO_2$, and $CH_4$
Amines, phenols, and ammonia

**Figure 6.12**   Fermentation and fermentation parameters in different segments of the large intestine.

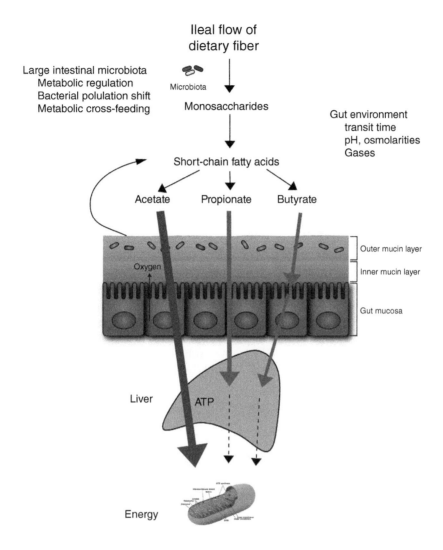

**Figure 6.13** Fermentation of dietary fiber residues into short-chain fatty acids, acetate, propionate and butyrate, in the large intestine. A part of butyrate will be used as energy for the epithelial cells lining the large intestine but the major part absorbed to the portal vein and, like propionate, cleared in the liver. Acetate, propionate and butyrate will all contribute energy to body cells.

gram fresh material and with each species occupying a particular niche and with numerous inter-relationships between them (Louis et al. 2007; Flint et al. 2008). The microbiota of pigs mainly consists of the Firmicutes and Bacteroidetes phyla (Leser et al. 2002; Kim et al. 2015) with the main bacterial groups comprising the following bacteria: *Streptococcus* spp., *Lactobacillus* spp., *Eubacterium* spp., *Fusobacterium* spp., Bacteroides spp., *Peptostreptococcus* spp., *Bifidobacterium* spp., *Selenomonas* spp., *Clostridium* spp., *Butyrivibrio* spp., *Escherichia* spp., *Prevotella*, and *Ruminococcus* spp.(Leser et al. 2002; Kim et al. 2015). The products of fermentation in the large intestine are SCFA, which are absorbed into the portal vein, used as substrate for intestinal cell growth and renewal, or excreted in feces (Bergman 1990), gases that are excreted through flatus and expiration (Jensen and Jørgensen 1994) and microbial biomass. The most important substrate for the bacteria in the large intestine is the dietary residues that escape digestion in the small intestine. Consequently, there is a nutrient gradient in distal direction of the large intestine (Figure 6.12) with the greatest concentration of carbohydrates in cecum and the proximal colon and the lowest in distal colon and rectum. The high nutrient concentration in cecum and proximal colon leads to high microbial growth and SCFA production and consequently high SCFA concentration and low pH (Bach Knudsen et al. 1993; Jensen and Jørgensen 1994; Glitsø et al. 1998; Nielsen et al. 2015). As the digesta move aborally and most of the readily available carbohydrates are broken down, the bacterial growth is reduced resulting in reduced SCFA production, lower SCFA concentration, and pH near neutrality (Bach Knudsen et al. 1991; Bach Knudsen et al. 1993; Nielsen et al. 2015). In some occasions, the profile of SCFA may also change, usually in the way that acetate increase at the expense of propionate from cecum/proximal colon to the distal colon (Bach Knudsen et al. 1991).

### Sugars and Oligosaccharides

All sugars except lactose and fructose at high doses will be digested almost completely in the small intestine (Ly 1992, 1996; Bach Knudsen and Canibe 2000). The amount that passes to the large intestine is consequently very low and in any case rapidly degraded in cecum and proximal colon. For oligosaccharides, the amounts reaching the large intestine may in some cases be substantial but also for this group of carbohydrates, the degradation is rapid in the large intestine. High doses of rapidly fermentable carbohydrates provided as unabsorbed sugars, starch, and inulin, however, may cause a rapid lactic acid production and consequently a substantial drop in pH (Bach Knudsen et al. 2003; Petkevicius et al. 2003).

### Starch

Residual starches ($RS_1$) that reach the large intestine from a finely ground diet is rapidly degraded in cecum and proximal colon (Bach Knudsen et al. 1993). The same is presumably the case with $RS_2$ from peas, whereas the degradation in the large intestine is somewhat slower when a sorghum-acorn-based diet is fed (Morales et al. 2002). Also diets where digestible wheat starch is substituted by $RS_1$ from raw potato and high-amylose maize starch (Nielsen et al. 2015) or a diet with $RS_3$ from tapioca (Jonathan et al. 2013) result in slow microbial degradation. For the latter two experiments, RS also reduced the degradation of NSP (Jonathan et al. 2013; Nielsen et al. 2015). Incomplete total tract digestion of starch can sometime be observed when intact kernels are present in the diets (Bach Knudsen, personal communication).

### Nonstarch Polysaccharides

When NSP arrive in the large intestine, the polysaccharides have already been modified and partly degraded by the microbiota in the foregut (Johansen et al. 1997; Kasprzak et al. 2012; Ivarsson et al. 2014), which makes the substrate more prone for degradation. Additionally, the higher swelling capacity of soluble as compared to insoluble fiber (Zhou et al. 2018) will make the former NSPs more prone for microbial degradation. Consequently, β-glucan, soluble arabinoxylan, and pectins (Bach Knudsen et al. 1993; Canibe and Bach Knudsen 1997a; Glitsø et al. 1998; Le Gall et al. 2009b; de Vries et al. 2014; Nielsen et al. 2015) are all degraded in cecum and proximal colon while the more insoluble NSP, e.g., cellulose and insoluble arabinoxylan, are degraded more slowly and at more distal locations of the colon (Bach Knudsen et al. 1993; Gdala et al. 1997; Canibe and Bach Knudsen 1997a; Glitsø et al. 1998; Le Gall et al. 2009b; de Vries et al. 2014; Nielsen et al. 2015). This is illustrated by the results in Figure 6.14 showing the progression in the degradation of NSP residues from ileum to feces of piglets that were fed a diet consisting of cereals and soybean meal as sole plant materials (Gdala et al. 1997). Based on the composition of the plant ingredients, galactose and uronic acids can be regarded as markers for pectin polysaccharides deriving predominantly from soybean meal, xylose as marker for arabinoxylan from cereals, and glucose as marker of β-glucan and cellulose (Bach Knudsen 1997, 2014). The impact of the chemical composition of the dietary fiber fraction is further illustrated in Figure 6.15 and Table 6.6 that summarizes studies where various plant materials have been fed. The total tract degradation (fermentation) of cellulose and arabinoxylan is much greater in nonlignified materials (wheat flour, rolled oats, rye flour, oat bran, and sugar beet pulp) than in lignified materials (pericarp/testa from rye and wheat, and wheat bran) (Graham et al. 1986; Bach Knudsen and Hansen 1991; Bach Knudsen et al. 1993; Longland et al. 1993; Glitsø et al. 1998). Moreover, because of the close association of polysaccharides and lignin in secondary lignified tissues (Figure 6.5), the whole polysaccharide lignin complex becomes very insoluble, and the main cell wall polysaccharides are virtually degraded to the same extent as illustrated by the rather similar degradation of rye pericarp/testa as compared to other rye fractions (Glitsø et al. 1998). This is in contrast to nonlignified materials where cellulose is less degraded than to noncellulosic (hemicellulose) polysaccharides (Bach Knudsen and Hansen 1991; Glitsø et al. 1998). The inclusion level can also reduce the degradation of NSP as illustrated for crystalline cellulose from solka-floc in Table 6.6, whereas the degradation of the NSP in sugar beet pulp is not influenced by the inclusion level (Longland et al. 1993). Moreover, Stanogias and Pearce (1985)

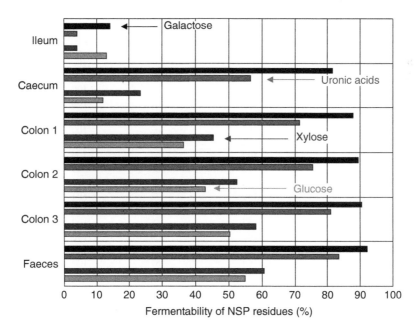

**Figure 6.14** Fermentation of non-starch polysaccharide residues from ileum to feces of piglets fed a cereal-soybean meal as sole plant materials. Source: Data from Gdala et al. (1997). Colon 1, 2, and 3 defines the proximal, mid, and, distal third of the colon.

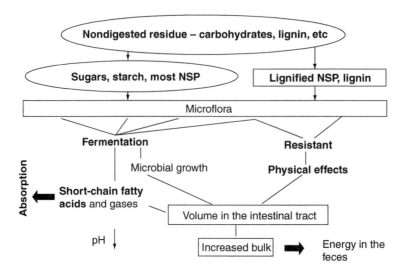

**Figure 6.15** Schematic illustration of carbohydrate degradation in the large intestine and influence on colonic and fecal weight, bulk and energy. NSP, non-starch polysaccharides.

**Table 6.6** Total tract degradation (fermentation) of total nonstarch polysaccharides, β-glucan, cellulose, and arabinoxylan measured in experiments with plant materials of different fiber content[a].

| Plant source | Fiber, g kg⁻¹ DM | Digestibility, % | | | |
|---|---|---|---|---|---|
| | | Total NSP | β-glucan | Cellulose | AX |
| Cereal mix-soyabean meal | 210 | 58 | 100 | 45 | 44 |
| Barley-soyabean meal | 148 | 74 | 100 | 56 | 66 |
| Wheat flour | 35 | 83 | 100 | 60 | 85 |
| +"Wheat aleurone" | 55 | 67 | 100 | 47 | 68 |
| +"Wheat pericarp/testa" | 62 | 50 | 100 | 24 | 50 |
| +Wheat bran | 62 | 62 | 100 | 44 | 62 |
| Rolled oats | 93 | 90 | 100 | 78 | 82 |
| +Oat bran | 109 | 92 | 100 | 83 | 84 |
| Whole grain rye | 156 | 67 | ND | 28 | 65 |
| Rye flour | 94 | 87 | ND | 84 | 83 |
| Rye aleurone | 180 | 73 | ND | 35 | 73 |
| Rye pericarp/testa | 177 | 14 | ND | 10 | −1 |
| Wheat-fine, 2.9% >2 mm | 154 | 68 | 100 | — | 71 |
| Wheat-coarse, 12.0% >2 mm | 148 | 64 | 100 | — | 68 |
| Barley-fine, 0.7% >2 mm | 185 | 61 | 100 | — | 62 |
| Barley-coarse, 23.3% >2 mm | 148 | 57 | 100 | — | 53 |
| Semi-purified + solka-floc – low | 128 | 50 | ND | 51 | ND |
| Semi-purified + solka-floc – high | 229 | 12 | ND | 9 | ND |
| Semi-purified + sugar beet pulp – low | 123 | 97 | ND | 93 | ND |
| Semi-purified + sugar beet pulp – high | 211 | 96 | ND | 89 | ND |

[a] NSP = nonstarch polysaccharides; AX = arabinoxylan; ND = not determined.

Sources: Graham et al. 1986; Bach Knudsen and Hansen 1991; Bach Knudsen et al. 1993; Longland et al. 1993; Canibe and Bach Knudsen 1997b; Glitsø et al. 1998; Bach Knudsen et al. 2005.

failed to demonstrate any consistent relationship between the fiber inclusion level and the degradation of fiber from natural sources. Long-term feeding of high-fiber diets may also have an effect on the degradation of NSP. Longland et al. (1993) concluded that pigs fed high-fiber diets may adapt to the diets in terms of N and energy balance after one week, but three to five weeks may be necessary before adapting to the degradation of resistant NSP residues.

Several factors favor a more extensive degradation of fibrous components in sows or adult animals compared to piglets and growing pigs. Adult animals usually have a lower feed intake per unit of body weight, a slower digesta transit, a greater intestinal volume, and a higher cellulolytic activity (Varel and Pond 1985; Glitsø et al. 1998; Serena et al. 2008a). Past and recent studies have generally also shown a greater degradation of fiber and a greater content of metabolizable energy in sows than in growing pigs fed the same diets (Fernández et al. 1986; Shi and Noblet 1993; Jørgensen et al. 2007). The greatest differences have been observed for some cereal coproducts and most roughage (Fernández et al. 1986; Jørgensen et al. 2007). Moreover, fiber polysaccharides with a complex composition, i.e., arabinoxylan from corn bran are far more degraded in sows than in growing pigs (Noblet and Bach Knudsen 1997).

### Physical Effects

An increased intake of dietary fiber will inevitably influence the bowel not only due to SCFA production, lowering of pH, and stimulation of microbial growth but also because of the mechanical action of the nondegradable residues that influence the water-holding properties (Figure 6.16). The consequence is an increased bulk in colon and feces, stimulation of bowel movements, leading to a reduction in the transit time (Glitsø et al. 1998; Wilfart et al. 2007; Serena et al. 2008a). The effect of the various fiber sources on fecal weight and energy excretion, however, is tightly correlated to the type of polymers entering the large intestine. Soluble fibers like arabinoxylan, β-glucan, and pectins are extensively degraded in the large intestine (Table 6.6), providing only modest influence on the fecal dry and wet weight. This is in contrast to insoluble fiber components such as cellulose, insoluble noncellulosic polysaccharides (hemicelluloses), and lignin that are very resistant to microbial degradation (Glitsø et al. 1998; Serena et al. 2008b; Le Gall et al. 2009b) and with a high impact

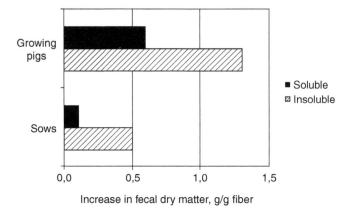

**Figure 6.16** Effect of soluble and insoluble fiber on the fecal bulking in growing pigs and sows. Source: Adapted from Bach Knudsen and Hansen (1991) and Serena et al.(2008b).

on fecal dry and wet weight. From Figure 6.16, it is also clear that sows have a greater capacity to degrade fibers in the gut than growing pigs.

## Quantitative Digestion and Fermentation of Nutrients in the Small and Large Intestines

The importance of the dietary fiber concentration on the quantitative digestion of nutrients at ileum and over the total tract is illustrated in Table 6.7. The bulk of sugars (close to 100%), starch (97%), protein (75%), and fat (72%) disappear during the passage of the small intestine, whereas NSP and lignin for most parts are recovered in ileal effluent. The dietary concentration of fiber is therefore the most important contributor to the amount of organic matter that passes from the small to the large intestine. Approximately half of the OM that enters the large intestine is fermented as it passes along the large intestine. However, there is a substantial difference between the nutrients; no fat disappears, whereas 37% of crude protein, 59% of the NSP, 71% of nonidentified residues, and 90% of starch disappear. The table also shows that the amount of organic matter degraded in the large intestine increases in response to the fiber concentration, i.e., the degradation is 170 g OM/d with a fiber level of 150 g/kg DM and 286 g OM/d with a fiber concentration of 200 g/kg dry matter. For sows fed diets with 429–455 g/kg DM, the degradation of OM can reach levels of 355–503 g/d (Serena et al. 2008b).

## Absorption of Products Deriving from Carbohydrate Assimilation

Several studies have shown that the supply of energy from carbohydrate deriving nutrients to the body takes place in two phases: a phase with rapid and high influxes of nutrients (absorptive phase) lasting four to five hours after a meal and a phase with low influx of nutrients (post absorptive phase) lasting until the next feeding (Giusi-Peerier et al. 1989; Rérat 1996; Bach Knudsen et al. 2000; Bach Knudsen et al. 2005) (Figure 6.17). In the absorptive phase, reducing sugars are dominating

**Table 6.7**  Intake and recovery of nutrients (g per day) at ileum and in feces and the effects of fiber on the recovery of nutrients at ileum and in feces[a].

| Item | Intake, g d⁻¹ | Recovery ileum, g d⁻¹ | Effect of fiber Intercept | Slope | R² | Recovery feces, g d⁻¹ | Effect of fiber Intercept | Slope | R² |
|---|---|---|---|---|---|---|---|---|---|
| Dry matter | 2000 | 536 | 113 | 3.1 | 0.75 | 273 | −25 | 2.2 | 0.79 |
| Organic matter | 1903 | 475 | 88 | 2.8 | 0.78 | 231 | −38 | 2.0 | 0.80 |
| Protein (Nx6.25) | 351 | 88 | 39 | 0.4 | 0.29 | 56 | 10 | 0.34 | 0.65 |
| Fat | 130 | 36 | 25 | 0.1 | 0.06 | 35 | 21 | 0.1 | 0.15 |
| Carbohydrates: | | | | | | | | | |
| Sugars | 99 | ND[c] | — | — | — | ND[c] | — | — | — |
| Starch | 984 | 31 | 13 | 0.11 | 0.08 | 3 | −1 | <0.1 | 0.15 |
| NSP | 244 | 191 | 5 | 1.3 | 0.76 | 79 | −49 | 0.9 | 0.69 |
| Lignin[b] | 36 | 36[c] | −2 | 0.3 | 0.54 | 36[c] | −2 | 0.27 | 0.34 |
| Residue | 59 | 100 | 6 | 0.7 | 0.31 | 29 | −16 | 0.3 | 0.21 |

[a] The data in this table were compiled from 21 published and one unpublished work representing 78 diets. The intake was calculated based on 2000 g of dry matter and converted to macronutrients from the reported chemical compositions. The recovery at ileum and in feces was calculated based on the digestibility coefficients reported in the papers (K. E. Bach Knudsen et al., unpublished data). ND, not determined; NSP, nonstarch polysaccharides.

[b] It is assumed that lignin is not broken down during passage of the gut.

[c] Sugar residues in ileum and feces will be part of the residue fraction.

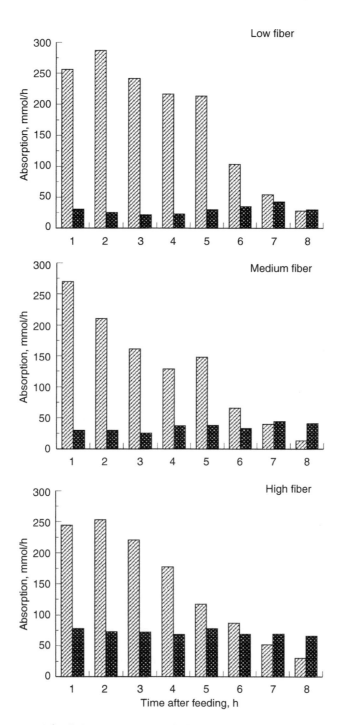

**Figure 6.17** Absorption (mmol h$^{-1}$) of glucose and short-chain fatty acids after consumption of a diet low in dietary fiber, medium in dietary fiber and high in dietary fiber. The pigs were fed three times daily at eight hours interval. Source: Adapted from Bach Knudsen et al. (2000) and (2005).

products deriving from carbohydrate assimilation with levels that are four to eight times higher than of SCFA. In the postabsorptive phase, however, SCFA become increasingly important, and the amount of SCFA may equal the amount of reducing sugars in the last hours before the next feeding.

The dietary composition of carbohydrates influences the rate and type of products deriving from carbohydrate assimilation. Thus, glucose and sucrose are absorbed more rapidly than starch and, particularly, lactose (Rérat et al. 1984a, b). It also seems that maltose is absorbed to the portal vein more rapidly than starch (Rérat et al. 1993). The type of starch has also been found to influence the rate of glucose absorption (van der Meulen et al. 1997a; Regmi et al. 2011b; Ingerslev et al. 2014), while the fiber level only seems to have an influence on the rate of glucose absorption when added to the diet as a fiber isolate or concentrate. Viscous guar-gum and oat β-glucan added to a semi-synthetic diet or an arabinoxylan concentrate reduce the postprandial appearance of glucose in the portal vein (Ellis et al. 1995; Hooda et al. 2010; Christensen et al. 2013). In contrast, neither insoluble (wheat bran) nor soluble fiber (sugar-beet fiber, oat bran and rye) sources or soluble fiber in a feed matrix seem to modify nutrient absorption (Michel and Rerat 1998; Bach Knudsen et al. 2000; Bach Knudsen et al. 2005; Theil et al. 2011; Ingerslev et al. 2014).

Lactic acid appears in the portal vein in the early phase of absorption and is better synchronized with the absorption profile of starch than with SCFA (Bach Knudsen et al. 2000; Serena et al. 2007). In some studies, the estimated absorption of lactic acid is close to the absorption of SCFA, but it cannot be excluded that the absorption of LA in experiments with catheterized pigs is overestimated as a substantial proportion of LA can derive from glucose oxidation in the gut (Vaugelade et al. 1994).

The absorption of SCFA occurs at a slower rate than glucose (Figure 6.17). Diurnal variations in portal concentrations of SCFA have only been reported when high levels of readily fermentable carbohydrates (lactose or sugar alcohols) were fed (Giusi-Peerier et al. 1989; Rérat et al. 1993) or after prolonged fasting (Rérat et al. 1987; Bach Knudsen et al. 2005). A considerable increase in absolute and relative contribution of SCFA to the animals' energy supply is observed when the level of carbohydrates fermented in the large intestine is increased either by fibers, RS, poorly absorbed sugars, or sugar alcohols (Giusi-Peerier et al. 1989; Rérat et al. 1993; Bach Knudsen et al. 2000; Bach Knudsen et al. 2005; Ingerslev et al. 2014). For instance, the portal flux of SCFA increases ($r = 0.90$) and that of glucose decreases ($r = -0.70$) in response to more carbohydrates being fermented in the large intestine (Table 6.8). This also influences the proportion of energy absorbed as glucose or as SCFA; only approximately 4% of absorbed energy derives from SCFA with a low NDC maize starch diets compared to 44% when a high NDC potato diet was fed. An even greater energy contribution from SCFA was observed in sows that were fed a high-fiber diet with 429 g/kg DM fiber and where 52% of the energy derived from SCFA compared with 12% in a low-fiber diet containing 177 g/kg DM fiber (Serena et al. 2009).

The type of carbohydrates that reach the large intestine can also modify the molar composition of SCFA in the portal vein. Thus, replacing wheat bran by sugar beet fiber increased the proportion of acetate and decreased propionate (Michel and Rerat 1998), RS increased butyrate at the expense of acetate (van der Meulen et al. 1997a; Regmi et al. 2011b; Ingerslev et al. 2014), and oat bran and rye increased the proportion of butyrate at the expense of acetate in portal blood (Bach Knudsen et al. 2000; Bach Knudsen et al. 2005; Ingerslev et al. 2014).

**Utilization of Absorption Products from Carbohydrate Assimilation**

The site of absorption of carbohydrate-derived products influences the utilization of energy. From the fermentation equation for the conversion of carbohydrates into SCFA, it can be estimated that approximately 25% of the energy is lost in $H_2$ and $CH_4$. While the loss as $H_2$ is relatively low, the loss as $CH_4$

**Table 6.8**  Effect of meal size and intake of digestible starch and nondigestible carbohydrates on portal concentrations and fluxes of glucose and short-chain fatty acids and the proportion of energy absorbed as glucose and short-chain fatty acids[a].

| Diet | Intake, g | | | | Glucose | | SCFA | | Absorbed energy, % | |
| | Meal size | Dig starch | NDC | | | | | | | |
| | | | RS | NSP | mmol l$^{-1}$ | mmol h$^{-1}$ | μmol l$^{-1}$ | mmol h$^{-1}$ | Glu | SCFA |
|---|---|---|---|---|---|---|---|---|---|---|
| LF wheat bread | 1300 | 746 | 4 | 77 | 8.10 | 175 | 775 | 30 | 93.0 | 7.0 |
| HF wheat bran | 1300 | 663 | 3 | 140 | 7.69 | 127 | 854 | 30.8 | 90.5 | 9.5 |
| HF oat bran | 1300 | 605 | 3 | 140 | 7.66 | 132 | 908 | 37.1 | 89.1 | 10.9 |
| HF rye bread | 1250 | 676 | 13 | 254 | 6.60 | 157 | 1140 | 76.9 | 82.4 | 17.6 |
| HF wheat bread | 1250 | 610 | 7 | 275 | 6.43 | 117 | 1001 | 66.5 | 80.2 | 19.8 |
| Maize starch | 860 | 536 | 9 | 39 | 8.85 | 146 | 459 | 13.9 | 96.0 | 4.0 |
| Pea starch | 860 | 535 | 15 | 36 | 6.90 | 105 | 454 | 17.8 | 93.1 | 6.9 |
| Maize starch | 1250 | 762 | 20 | 66 | 8.14 | 185 | 480 | 19.1 | 95.7 | 4.3 |
| Maize:potato (1 : 1) starch | 1250 | 609 | 189 | 66 | 6.94 | 109 | 1240 | 60.3 | 80.6 | 19.4 |
| Potato starch | 1250 | 361 | 458 | 66 | 5.97 | 49 | 1620 | 88.9 | 55.9 | 44.1 |

[a] Dig, digestible; NDC, nondigestible carbohydrate; SCFA, short-chain fatty acids; RS, resistant starch; NSP, nonstarch polysaccharides; Glu, glucose; LF, low fiber; HF, high fiber.
Sources: van der Meulen et al. 1997a, b; Bach Knudsen et al. 2000; Bach Knudsen et al. 2005.

is substantial and much greater in sows (1.5–3.5% of digested energy) than in growing pigs (0.5–1.2% of digested energy) (Jørgensen 2007). The theoretical production of adenosine triphosphate (ATP) in the intermediate metabolism per mole of released hexose (glucose) from the small intestine or as SCFA from the large intestine are 38 and 24, respectively (Table 6.1, Figure 6.18) which is the reason for the lower utilization of metabolizable energy absorbed from the large intestine relative to the small intestine (Table 6.9). The difference is a consequence of loss of energy in fermentation gases and a less efficient utilization of SCFA in the intermediary metabolism. In adult sows, however, the utilization is somewhat greater than in growing pigs, probably because the energy primarily is used for maintenance rather than for growth. By infusion of SCFA into the large intestine of growing pigs, a partial utilization of the energy in SCFA was found to be 82% (Jørgensen et al. 1997) compared to a relative energy value of energy derived from fermentation in the large intestine of 73% relative to energy being enzymatically digested in the small intestine (Jørgensen et al. 1996). Other experiments have shown lower efficiency compared to that found by Jørgensen et al. (1997) when the SCFA either were given orally as fermented feed (Jentsch et al. 1968) or infused into the cecum (Gädeken et al. 1989; Müller et al. 1991; Roth et al. 1991) of sows or growing pigs (Table 6.10).

The effect of a change in carbohydrate composition from digestible to nondigestible carbohydrates on energy value are twofold. First, the digestibility of energy and consequently metabolizable energy will decrease as not all potential energy present in fiber polysaccharides will be degraded. Second, the utilization and thereby the net energy as proportion of metabolizable energy will be reduced because a larger fraction of absorbed energy will derive from SCFA, which is utilized less efficiently than glucose.

**Implication**

The intestinal tract of growing pigs and adult sows has a high hydrolytic activity from endogenous and microbial enzymes to hydrolyze simple and complex carbohydrates present in feedstuffs. This ensures that the bulk of sugars and starch are digested and absorbed in the small intestine.

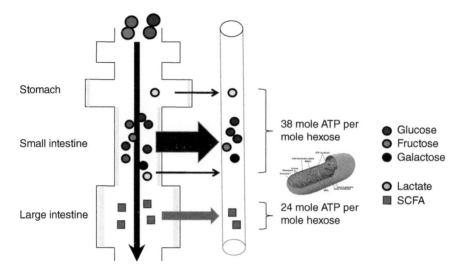

**Figure 6.18**  Theoretical production of adenosine triphosphate (ATP) in the intermediate metabolism per mole of released hexose as glucose from the small intestine or as short-chain fatty acids from the large intestine.

**Table 6.9**  Utilization of energy fermented in the large intestine[a].

| Diet component | Body weight range, kg | Energy fermented large intestine, % of DE | Efficiency (RE/ME) in relation to small intestine, % | Author |
|---|---|---|---|---|
| Potato starch, cellulose | 60–90 | 18–33 | 51 | Just et al. (1983) |
| Potato, sugar beet, grass, Lucerne meal | 90–180 | 9–40 | 66 | Hoffmann et al. (1990) |
| Maize starch, cellulose, soya hull | 30–105 | 13–27 | 63 | Bakker et al. (1994) |
| Beet pulp, corn distillers grain, sunflower meal, etc | 38–47 | 3–27 | 82 | Noblet et al. (1994) |
| Pea fiber, pectin | 40–125 | 7–29 | 73 | Jørgensen et al. (1996) |
| Barley straw, barley hull, wheat bran, potato fiber, soya fiber | 50–70 | 4–29 | 76 | Jørgensen (unpublished) |
| Wheat bran, sugar beet pulp, seed residue, brewers spent grain, pea hull, potato pulp, pectin residue, sugar beet pulp, sugar beet pulp | 46–125  160–243 | 8–35  5–40 | 69  90 | Jørgensen (unpublished)  Jørgensen (unpublished) |

[a] RE, retained energy; ME, metabolizable energy.

However, high loads of fructose and lactose may lead to malabsorption of these sugars. Although no endogenous enzymes are secreted or present in the small intestine to cleave the bonds in most oligosaccharides and NSP, around 40% of oligosaccharides and 20% of NSP disappear during passage of the stomach and small intestine. The main site for degradation of fiber polysaccharide is the large intestine where the dietary fibers are converted into SCFA, gases, and microbial growth. The utilization of the energy absorbed as SCFA from the large intestine is, however, lower than

**Table 6.10**  Utilization of infused short-chain fatty acids into the large intestine.

| Diet type | Body weight range, kg | Infusate | Utilization, % of ME | Author |
|---|---|---|---|---|
| Barley/Fishmeal | 140–180 | Ethanol<br>Lactic acid<br>Acetic acid | Ethanol: 72<br>Lactic acid: 75<br>Acetic acid: 60 | Jentsch et al. (1968) |
| Barley/Soybean meal | 160–200 | Acetic acid<br>Propionic acid | Acetic acid: 79<br>Propionic acid: 75 | Roth et al. (1988) |
| Grain/Oat meal by-product/Wheat bran, Soybean meal/Fish meal | 55–120 | Acetic acid, propionic acid, and butyric acid | Acetic acid: 65<br>Propionic acid: 71<br>Butyric acid: 67 | Gädeken et al. (1989) |
| Barley/Soybean meal | 179±17 | Mixture of: Acetic acid+propionic acid | Mixture: 70 | Müller et al. (1991) |
| Barley/Wheat starch/ Fish meal/Casein | 60–120 | Mixture of: Acetic acid+propionic acid+butyric acid | Mixture: 82 | Jørgensen et al. (1997) |

that of energy absorbed as glucose from the small intestine primarily due to losses of energy as fermentative gases and a lower utilization of SCFA in the intermediate metabolism. Careful choices of feedstuffs and processing conditions are ways by which the energy value of diets for pigs can be improved.

# References

Albersheim, P., J. An, G. Freshour, M. S. Fuller, R. Guillen, K. S. Ham, M. G. Hahn, J. Huang, M. O'Neill, A. Whitcombe, and M. V. Williams 1994. Structure and function studies of plant cell wall polysaccharides. Biochem. Soc. Trans. 22:374–378.

Argenzio, R., and M. Southworth. 1974. Site of organic acid production and absorption in gastrointestinal tract of pigs. Am. J. Physiol. 228:454–460.

Bach Knudsen, K. E. 1997. Carbohydrate and lignin contents of plant materials used in animal feeding. Anim. Feed Sci. Technol. 67:319–338.

Bach Knudsen, K. E. 2014. Fiber and non-starch polysaccharide content and variation in common crops used in broiler diets. Poult. Sci. 93:2380–2393.

Bach Knudsen, K. E., and N. Canibe. 2000. Breakdown of plant carbohydrates in the digestive tract of pigs fed on wheat or oat based rolls. J. Sci. Food Agric. 80:1253–1261.

Bach Knudsen, K. E., N. Canibe, and H. Jørgensen. 2000. Quantification of the absorption of nutrients deriving from carbohydrate assimilation: Model experiment with catheterised pigs fed on wheat and oat based rolls. Br. J. Nutr. 84:449–458.

Bach Knudsen, K. E., and I. Hansen. 1991. Gastrointestinal implications in pigs of wheat and oat fractions 1. Digestibility and bulking properties of polysaccharides and other major constituents. Br. J. Nutr. 65:217–232.

Bach Knudsen, K. E., B. B. Jensen, J. O. Andersen, and I. Hansen. 1991. Gastrointestinal implications in pigs of wheat and oat fractions 2. Microbial activity in the gastrointestinal tract. Br. J. Nutr. 65:233–248.

Bach Knudsen, K. E., B. B. Jensen, and I. Hansen. 1993. Digestion of polysaccharides and other major components in the small and large intestine of pigs fed diets consisting of oat fractions rich in ß-D-glucan. Br. J. Nutr. 70:537–556.

Bach Knudsen, K. E., and H. Jørgensen. 2001. Intestinal degradation of dietary carbohydrates - from birth to maturity. In: J. E. Lindberg and B. Ogle, editors, Digestive physiology in pigs. CABI Publishing, Wallingford, UK. p. 109–120.

Bach Knudsen, K. E., and H. N. Lærke. 2018. Carbohydrates and lignin in the feed – from sugars to complex composed fibres. In: P. Moughan and W. Hendriks, editors, Feed evaluation science. Wageningen Academic Publisher, Wageningen, The Netherlands. p. 109–140.

Bach Knudsen, K. E., and B. W. Li. 1991. Determination of oligosaccharides in protein-rich feedstuffs by gas-liquid chromatography and high-performance liquid chromatography. J. Agric. Food Chem. 39:689–694.

Bach Knudsen, K. E., S. Petkevicius, H. Jørgensen, and K. D. Murrell. 2003. A high load of rapidly fermentable carbohydrates reduces worm burden in infected pigs. In: J. E. Paterson, editor, Manipulating pig production. Australas. Pig Sci. Assoc., Werribee, Vic., Australia. p. 169.

Bach Knudsen, K. E., A. Serena, A. B. K. Kjær, H. Jørgensen, and R. Engberg. 2005. Rye bread enhances the production and plasma concentration of butyrate but not the plasma concentrations of glucose and insulin in pigs. J. Nutr. 135:1696–1704.

Bakker, G. C. M., R. A. Dekker, R. Jongbloed, and A. W. Jongbloed. 1994. The effect of starch, fat and non-starch polysaccharides on net energy and on the proportion of digestible organic matter of energy disappeared in the hindgut. In: J. F. Aguilera, editor, Energy metabolism of farm animals. CSIB Publishing, Madrid, Spain. p. 163–166.

Bergman, E. N. 1990. Energy contributions of volatile fatty acids from the gastrointestinal tract in various species. Phys. Rev. 70:567–590.

Biliaderis, C. G. 1991. The structure and interactions of starch with food constituents. Can. J. Physiol. 69:60–78.

Brown, I. L., K. J. McNaught, D. Andrews, and T. Morita. 2001. Resistant starch: Plant breeding, applications development and commercial use. In: B. V. McCleary and L. Prosky, editors, Advanced dietary fibre technology. Blackwell Science Ltd, Oxford, UK. p. 401–412.

Caffall, K. H., and D. Mohnen. 2009. The structure, function, and biosynthesis of plant cell wall pectic polysaccharides. Carb. Res. 344:1879–1900.

Canibe, N., and K. E. Bach Knudsen. 1997a. Apparent digestibility of non-starch polysaccharides and short-chain fatty acids production in the large intestine of pigs fed dried or toasted peas. Acta Agric. Scand. 47:106–116.

Canibe, N., and K. E. Bach Knudsen. 1997b. Digestibility of dried and toasted peas in pigs. 1. Ileal and faecal digestibility of carbohydrates. Anim. Feed Sci. Technol. 64:293–310.

Carlsson, N. G., H. Karlsson, and A. S. Sandberg. 1992. Determination of oligosaccharides in foods, diets, and intestinal contents by high-temperature gas chromatography and gas chromatography/mass spectrometry. J. Agric. Food Chem. 40:2404–2412.

Carpita, N. 2000. The cell wall. In: N. Carpita, editor, Biochemistry and molecular biology of plants. Am. Soc. Plant Biol. p. 52–109.

Carpita, N. C., and D. M. Gibeaut. 1993. Structural models of primary cell walls in flowering plants: Consistency of molecular structure with the physical properties of the walls during growth. Plant J. 3:1–30.

Chae, B. J., and I. K. Han. 1998. Processing effects of feeds in swine - review. Asian-Australas. J. Anim. Sci 11:597–607.

Chesson, A. 1991. Mechanistic models of forage cell wall degradation. In: H. G. Jung, D. R. Buxton, R. D. Hatfield, and J. Ralph, editors, International symposium on cell wall structure and digestibility. Am Soc Agron, Madison, WI. p. 347–376.

Christensen, K. L., M. S. Hedemann, H. N. Lærke, H. Jørgensen, S. J. Mutt, K. H. Herzig, and K. E. Bach Knudsen. 2013. Concentrated arabinoxylan but not concentrated beta-glucan in wheat bread has similar effects on postprandial insulin as whole-grain rye in porto-arterial catheterized pigs. J. Agric. Food Chem. 61:7760–7768.

Cosgrove, D. J. 1997. Assembly and enlargement of the primary cell wall in plants. Annu. Rev. Cell. Dev. Biol. 13:171–201.

Cosgrove, D. J., and M. C. Jarvis. 2012. Comparative structure and biomechanics of plant primary and secondary cell walls. Front. Plant Sci. 3:204.

Cummings, J. H. 1997. The large intestine in nutrition and disease. Institute Danone, Bruxelles, Belgium.

Cummings, J. H., and H. N. Englyst. 1995. Gastrointestinal effects of food carbohydrate. Am. J. Clin. Nutr. 61:938S–945S.

Cummings, J. H., J. I. Mann, C. Nishida, and H. H. Vorster. 2009. Dietary fibre: an agreed definition. Lancet 373(9661):365–366.

Cummings, J. H., and A. M. Stephen. 2007. Carbohydrate terminology and classification. Eur. J. Clin. Nutr. 61(Suppl. 1):S5–S18.

Davin, L. B., M. Jourdes, A. M. Patten, K. W. Kim, D. G. Vassao, and N. G. Lewis. 2008. Dissection of lignin macromolecular configuration and assembly: comparison to related biochemical processes in allyl/propenyl phenol and lignan biosynthesis. Nat. Prod. Rep. 25:1015–1090.

de Vries, S., A. M. Pustjens, C. van Rooijen, M. A. Kabel, W. H. Hendriks, and W. J. Gerrits. 2014. Effects of acid extrusion on the degradability of maize distillers dried grain with solubles in pigs. J. Anim. Sci. 92:5496–5506.

Dhital, S., R. R. Bhattarai, J. Gorham, and M. J. Gidley. 2016. Intactness of cell wall structure controls the in vitro digestion of starch in legumes. Food Funct. 7:1367–1379.

Dhital, S., F. J. Warren, P. J. Butterworth, P. R. Ellis, and M. J. Gidley. 2017. Mechanisms of starch digestion by alpha-amylase-structural basis for kinetic properties. Crit. Rev. Food Sci. Nutr. 57:875–892.

Duke, G. E. 1986. Alimentary canal: Secretion and digestion, special digestion functions and absorption. In: P. D. Sturkie, editor, Avian physiology. Springer-Verlag, New York, NY. p. 289–302.

Efird, R. C., W. D. Armstrong, and D. L. Herman. 1982. The development of digestive capacity in young pigs: Effects of age and weaning system. J. Anim. Sci. 55:1380–1387.

Ellis, P. R., F. G. Roberts, A. G. Low, and L. M. Morgan. 1995. The effect of high-molecular-weight guar gum on net apparent glucose absorption and net apparent insulin and gastric inhibitory polypeptide production in the growing pig: relationship to rheological changes in jejunal digesta. Br. J. Nutr. 74:539–556.

Englyst, H. N., S. M. Kingman, and J. H. Cummings. 1992. Classification and measurement of nutritionally important starch fractions. Eur. J. Clin. Nutr. 46:S33–S50.

Englyst, K. N., S. Liu, and H. N. Englyst. 2007. Nutritional characterization and measurement of dietary carbohydrates. Eur. J. Clin. Nutr. 61(Suppl. 1):S19–S39.

Fernández, J. A., H. Jørgensen, and A. Just. 1986. Comparative digestibility experiments with growing pigs and adult sows. Anim. Prod. 43:127–132.

Flickinger, E. A., J. Van Loo, and G. C. Fahey, Jr. 2003. Nutritional responses to the presence of inulin and oligofructose in the diets of domesticated animals: A review. Crit. Rev. Food Sci. Nutr. 43:19–60.

Flint, H. J., E. A. Bayer, M. T. Rincon, R. Lamed, and B. A. White. 2008. Polysaccharide utilization by gut bacteria: potential for new insights from genomic analysis. Nat. Rev. Microbiol. 6:121–131.

Gädeken, D., G. Breves, and H. J. Oslage. 1989. Efficiency of energy utilization of intracaecally infused volatile fatty acids in pigs. Proc. 11th Symp. Energy Metabolism of Farm Animals. Eur. Fed. Anim. Sci., Rome, Italy. p. 115–118.

Gallant, D. J., B. Bouchet, A. Buléon, and S. Pérez. 1992. Physical characteristics of starch granules and susceptibility to enzymatic degradation. Eur. J. Clin. Nutr. 46:S3–S16.

Gdala, J., and L. Buraczewska. 1997. Ileal digestibility of pea and faba bean carbohydrates in growing pigs. J. Anim. Feed Sci. 6:235–245.

Gdala, J., L. Buraczewska, A. M. J. Jansman, J. Wasilewko, and P. Leeuwen. 1994. Ileal digestibility of amino acids and carbohydrates in lupins for young pigs. In: W.-B. Souffrant and H. Hagemesiters, editors, VIth International symposium on digestive physiology in pigs. Eur. Fed. Anim. Sci., Rome, Italy. p. 93–96.

Gdala, J., H. N. Johansen, K. E. Bach Knudsen, I. Knap, P. Wagner, and O. B. Jørgensen. 1997. The digestibility of carbohydrates, protein and fat in the small and large intestine of piglets fed non-supplemented and enzyme supplemented diets. Anim. Feed Sci. Technol. 65:15–33.

Giusi-Peerier, A., M. Fiszlewicz, and A. Rérat. 1989. Influence of diet composition on intestinal volatile fatty acid and nutrient absorption in unanesthetized pigs. J. Anim. Sci. 67:386–402.

Glitsø, L. V., G. Brunsgaard, S. Højsgaard, B. Sandström, and K. E. Bach Knudsen. 1998. Intestinal degradation in pigs of rye dietary fibre with different structural characteristics. Br. J. Nutr. 80:457–468.

Graham, H., K. Hesselman, and P. Åman. 1986. The influence of wheat bran and sugar-beet pulp on the digestibility of dietary components in a cereal-based pig diet. J. Nutr. 116:242–251.

Gray, G. M. 1992. Starch digestion and absorption in nonruminants. J. Nutr. 122:172–177.

Grizard, D., and C. Barthomeuf. 1999. Non-digestible oligosaccharides used as prebiotic agents: Mode of production and beneficial effects on animal and human health. Reprod. Nutr. Dev. 39:563–588.

Guo, M. Q., X. Hu, C. Wang, and L. Ai. 2017. Polysaccharides: Structure and solubility, solubility of polysaccharides. In: Z. Xu, editor, Solubility of polysaccharides. IntechOpen, London, UK.

Hedemann, M. S., and K. E. Bach Knudsen. 2009. Dried chicory root has minor effects on the digestibility of nutrients and the composition of the microflora at the terminal ileum and in faeces of growing pigs. Livest. Sci. 134:53–55.

Henneberg, W., and F. Stohmann. 1859. Über das Erhaltungsfutter volljährigen Rindviehs. J. Landwirtsch. 3:485–551.

Hoffmann, L., W. Jentsch, and R. Schiemann. 1990. Energiumsatzmessungen am adulten Schwein bei Verfütterung von Rationen mit Kartoffelstärke, Kartoffeln, Rüben, Pessschitzeln und Grobfuttermitteln als Zulagen zu einer Grundrationen. 1. Energiumsatz und Energiverwertung. Arch. Anim. Nutr. 40:191–207.

Hooda, S., J. J. Matte, T. Vasanthan, and R. T. Zijlstra. 2010. Dietary oat beta-glucan reduces peak net glucose flux and insulin production and modulates plasma incretin in portal-vein catheterized grower pigs. J. Nutr. 140:1564–1569.

Hopwood, D. E., D. W. Pethick, J. R. Pluske, and D. J. Hampson. 2004. Addition of pearl barley to a rice-based diet for newly weaned piglets increases the viscosity of the intestinal contents, reduces starch digestibility and exacerbates post-weaning colibacillosis. Br. J. Nutr. 92:419–427.

Houdijk, J. G., R. Hartemink, M. W. Verstegen, and M. W. Bosch. 2002. Effects of dietary non-digestible oligosaccharides on microbial characteristics of ileal chyme and faeces in weaner pigs. Arch. Tierernahr. 56:297–307.

Houdijk, J. G. M., M. W. Bosch, S. Tamminga, M. W. A. Verstegen, E. B. Berenpas, and H. Knoop. 1999. Apparent ileal and total-tract nutrient digestion by pigs as affected by dietary nondigestible oligosaccharides. J. Anim. Sci. 77:148–158.

Ingerslev, A. K., P. K. Theil, M. S. Hedemann, H. N. Lærke, and K. E. Bach Knudsen. 2014. Resistant starch and arabinoxylan augment SCFA absorption, but affect postprandial glucose and insulin responses differently. Br. J. Nutr. 111:1564–1576.

Ivarsson, E., S. Roos, H. Y. Liu, and J. E. Lindberg. 2014. Fermentable non-starch polysaccharides increases the abundance of Bacteroides-Prevotella-Porphyromonas in ileal microbial community of growing pigs. Animal 8:1777–1787.

Izydorczyk, M. S., and C. G. Biliaderis. 1995. Cereal arabinoxylans: advances in structure and physicochemical properties. Carbohydr. Polym. 28:33–48.

Izydorczyk, M. S., and J. E. Dexter. 2008. Barley β-glucans and arabinoxylans: Molecular structure, physicochemical properties, and uses in food products–a review. Food Res. Int. 41:850–868.

Janczyk, P., R. Pieper, H. Smidt, and W. B. Souffrant. 2007. Changes in the diversity of pig ileal lactobacilli around weaning determined by means of 16S rRNA gene amplification and denaturing gradient gel electrophoresis. FEMS Microbiol. Ecol. 61:132–140.

Jenkins, P., and A. Donald. 1995. The influence of amylose on starch granule structure. Int. J. Biol. Macromol. 17:315–321.

Jenkins, P. J., Comerson, R. E., Donald, A. M., Bras, W., Derbyshire, G. E., Mant, G. R., and Ryan, A. J. 1994. In situ simultaneous small and wide angle X-ray scattering: A new technique to study starch gelatinization. J. Polym. Sci. Polym. Phys. Educ. 32: 1579–1583.

Jensen, B. B., and H. Jørgensen. 1994. Effect of dietary fiber on microbial activity and microbial gas production in various regions of the gastrointestinal tract of pigs. Appl. Environ. Microbiol. 60:1897–1904.

Jentsch, W., R. Schiemann, and L. Hoffmann. 1968. Modellversuche mit Schweinen zur Bestimmung der energetischen Verwertung von Alkohol, Essig- und Milchsäure. Arch. Tierernahr. 18:352–357.

Jha, R., and J. D. Berrocoso. 2015. Review: Dietary fiber utilization and its effects on physiological functions and gut health of swine. Animal 9:1441–1452.

Johansen, H. N., K. E. Bach Knudsen, P. J. Wood, and R. G. Fulcher. 1997. Physico-chemical properties and the digestibility of polysaccharides from oats in the gastrointestinal tract of pigs. J. Sci. Food Agric. 73:81–92.

Jonathan, M. C., D. Haenen, C. Souza da Silva, G. Bosch, H. A. Schols, and H. Gruppen. 2013. Influence of a diet rich in resistant starch on the degradation of non-starch polysaccharides in the large intestine of pigs. Carbohydr. Polym. 93:232–239.

Jørgensen, H. 2007. Methane emission by growing pigs and adult sows as influenced by fermentation. Livest. Sci. 109:216–219.

Jørgensen, H., T. Larsen, X.-Q. Zhao, and B. O. Eggum. 1997. The energy value of short-chain fatty acids infused into the carcum of pigs. Br. J. Nutr. 77:745–756.

Jørgensen, H., A. Serena, M. S. Hedemann, and K. E. Bach Knudsen. 2007. The fermentative capacity of growing pigs and adult sows fed diets with contrasting type and level of dietary fibre. Livest. Sci. 109:111–114.

Jørgensen, H., X.-Q. Zhao, and B. O. Eggum. 1996. The influence of dietary fibre and environmental temperature on the development of the gastrointestinal tract, digestibility, degree of fermentation in the hind-gut and energy metabolism in pigs. Br. J. Nutr. 75:365–378.

Just, A. 1982. The influence of crude fibre from cereals on the net energy value of diets for growth in pigs. Livest. Prod. Sci. 9:569–580.

Just, A., J. A. Fernández, and H. Jørgensen. 1983. The net energy value of diets for growing pigs in relation to the fermentative processes in the digestive tract and the site of absorption of nutrients. Livest. Prod. Sci. 10:171–186.

Kasprzak, M. M., H. N. Laerke, and K. E. Knudsen. 2012. Changes in molecular characteristics of cereal carbohydrates after processing and digestion. Int. J. Mol. Sci. 13:16833–16852.

Kidder, D. E., and M. J. Manners. 1980. The level and distribution of carbohydrases in the small intestine mucosa of pigs from three weeks of age to maturity. Br. J. Nutr. 43:141–153.

Kim, J., S. G. Nguyen, R. B. Guevarra, I. Lee, and T. Unno. 2015. Analysis of swine fecal microbiota at various growth stages. Arch. Microbiol. 197:753–759.

Lærke, H. N., C. Pedersen, M. A. Mortensen, P. K. Theil, T. Larsen, and K. E. Bach Knudsen. 2008. Rye bread reduces plasma cholesterol levels in hypercholesterolaemic pigs when compared to wheat at similar dietary fibre level. J.Sci. Food Agric. 88:1385–1393.

Le Gall, M., K. Eybye, and K. E. Bach Knudsen. 2009a. Molecular weight changes of arabinoxylans incurred by the digestion processes in the upper gastrointestinal tract of pigs. Livest. Sci. 134:72–75.

Le Gall, M., A. Serena, H. Jorgensen, P. K. Theil, and K. E. Bach Knudsen. 2009b. The role of whole-wheat grain and wheat and rye ingredients on the digestion and fermentation processes in the gut - a model experiment with pigs. Br. J. Nutr. 102:1590–1600.

Lentze, M. J. 1995. Molecular and cellular aspects of hydrolysis and absorption. Am. J. Clin. Nutr. 61(4 Suppl):946S–951S.

Leser, T. D., J. Z. Amenuvor, T. K. Jensen, R. H. Lindecrona, M. Boye, and K. Moller. 2002. Culture-independent analysis of gut bacteria: The pig gastrointestinal tract microbiota revisited. Appl. Environ. Microbiol. 68:673–690.

Liyama, K., T. B.-T. Lam, and B. A. Stone. 1994. Covalent cross-links in the cell wall. Plant Physiol. 104:315–320.

Loix, C., M. Huybrechts, J. Vangronsveld, M. Gielen, E. Keunen, and A. Cuypers. 2017. Reciprocal interactions between cadmium-induced cell wall responses and oxidative stress in plants. Front. Plant Sci. 8:1867.

Longland, A. C., A. G. Low, D. B. Quelch, and S. P. Bray. 1993. Adaptation to the digestion of non-starch polysaccharide in growing pigs fed on cereal or semi-purified basal diets. Br. J. Nutr. 70:557–566.

Louis, P., K. P. Scott, S. H. Duncan, and H. J. Flint. 2007. Understanding the effects of diet on bacterial metabolism in the large intestine. J. Appl. Microbiol. 102:1197–1208.

Ly, J. 1992. Studies of the digestibility of pigs fed dietary sucrose, fructose or glucose. Arch. Anim. Nutr. 42:1–9.

Ly, J. 1996. The pattern of digestion and metabolism in high sugar feeds for pigs. Cuban J. Agric. Sci. 30:117–129.

Martens, B. M. J., T. Flecher, S. de Vries, H. A. Schols, E. Bruininx, and W. J. J. Gerrits. 2019. Starch digestion kinetics and mechanisms of hydrolysing enzymes in growing pigs fed processed and native cereal-based diets. Br. J. Nutr. 121:1124–1136.

Martens, B. M. J., W. J. J. Gerrits, E. Bruininx, and H. A. Schols. 2018. Amylopectin structure and crystallinity explains variation in digestion kinetics of starches across botanic sources in an in vitro pig model. J. Anim. Sci. Biotechnol. 9:91.

McDougall, G. J., I. M. Morrison, D. Stewart, and J. R. Hillman. 1996. Plant cell walls as dietary fibre: Range, structure, processing and function. J. Sci. Food Agric. 70:133–150.

Michel, P., and A. Rerat. 1998. Effect of adding sugar beet fibre and wheat bran to a starch diet on the absorption kinetics of glucose, aminonitrogen and volatil fatty acids in the pig. Reprod. Nutr. Dev. 38:49–68.

Morales, J., J. F. Perez, S. M. Martin-Orue, M. Fondevila, and J. Gasa. 2002. Large bowel fermentation of maize or sorghum-acorn diets fed as a different source of carbohydrates to Landrace and Iberian pigs. Br. J. Nutr. 88:489–498.

Moran, E. T., Jr. 1985. Digestion and absorption of carbohydrates in fowl and events through perinatal development. J. Nutr. 115:665–674.

Morris, E. R. 1992. Physico-chemical properties of food polysaccharides. In: T. F. Schweizer and C. A. Edwards, editors, Dietary fibre - a component of food: Nutritional function in health and disease. Springer-Verlag, London, UK. p. 41–55.

Müller, H. L., M. Kirchgessner, and F. X. Roth. 1991. Energetic efficiency of a mixture of acetic and propionic acid in sows. J. Anim. Phys. Anim. Nutr. 65:140–145.

Nielsen, T. S., P. K. Theil, S. Purup, N. P. Nørskov, and K. E. Bach Knudsen. 2015. Effects of resistant starch and arabinoxylan on parameters related to large intestinal and metabolic health in pigs fed fat-rich diets. J. Agric. Food Chem. 63:10418–10430.

Noblet, J., and K. E. Bach Knudsen. 1997. Comparative digestibility of wheat, maize and sugar beet pulp non-starch polysaccharides in adult sows and growing pigs. In: J. P. Laplace, C. Fevrier, and A. Barbeau, editors, Digestive physiology in pigs. INRA, Saint Malo, France. p. 571–574.

Noblet, J., H. Fortune, X. S. Shi, and S. Dubois. 1994. Prediction of net energy value of feeds for growing pigs. J. Anim. Sci. 71:3389–3398.

Noblet, J., and J. M. Perez. 1993. Prediction of digestibility of nutrients and energy values of pig diets from chemical analysis. J. Anim. Sci. 71:3389–3398.

Oksbjerg, N., H. Jørgensen, H.P. Mortensen, J. A. Fernández, and A. Madsen 1988. The feeding value of hydrolysed permeate Lactose in growing and finishing pigs. Acta Agric. Scand. 38:253–260.

O'Neill, M. A., and W. S. York. 2003. The composition and structure of plant primary cell walls. In: J. K. C. Rose, editor, The plant cell wall. Blackwell Publisher, CRC Press, Ithaca, New York, NY. p. 1–54.

Pedersen, M. B., S. Dalsgaard, K. Bach Knudsen, S. Yu, and H. N. Lærke. 2014. Compositional profile and variation of distillers dried grains with solubles from various origins with focus on non-starch polysaccharides. Anim. Fedd Sci. Technol. 197:130–141.

Petkevicius, S., K. E. Bach Knudsen, K. D. Murrell, and H. Wachmann. 2003. The effect of inulin and sugar beet fibre on oesophagostomum dentatum infection in pigs. Parasitology 127(Pt 1):61–68.

Phillips, G. O., and S. W. Cui. 2011. An introduction: evolution and finalisation of the regulatory definition of dietary fibre. Food Hydrocoll. 25:139–143.

Pollock, C., and A. Cairns. 1991. Fructan metabolism in grasses and cereals. Ann. Rev. Plant Biol. 42:77–101.

Regmi, P. R., B. U. Metzler-Zebeli, M. G. Ganzle, T. A. van Kempen, and R. T. Zijlstra. 2011a. Starch with high amylose content and low in vitro digestibility increases intestinal nutrient flow and microbial fermentation and selectively promotes bifidobacteria in pigs. J. Nutr. 141:1273–1280.

Regmi, P. R., T. A. van Kempen, J. J. Matte, and R. T. Zijlstra. 2011b. Starch with high amylose and low in vitro digestibility increases short-chain fatty acid absorption, reduces peak insulin secretion, and modulates incretin secretion in pigs. J. Nutr. 141:398–405.

Rérat, A. 1996. Influence of the nature of carbohydrate intake on the absorption chronology of reducing sugars and volatile fatty acids in the pigs. Reprod. Nutr. Dev. 1996:3–19.

Rérat, A., M. Fiszlewicz, A. Giusi, and P. Vaugelade. 1987. Influence of meal frequency on postprandial variations in the digestive tract of conscious pigs. J. Anim. Sci. 64:448–456.

Rérat, A., A. Giusi-Périer, and P. Vaissade. 1993. Absorption balances and kinetics of nutrients and bacterial metabolites in conscious pigs after intake of maltose- or maltitol-rich diets. J. Anim. Sci. 71:2473–2488.

Rérat, A. A., P. Vaissade, and P. Vaugelade. 1984a. Absorption kinetics of some carbohydrates in conscious pigs. 1. Qualitative aspects. Br. J. Nutr. 51:505 515.

Rérat, A. A., P. Vaissade, and P. Vaugelade. 1984b. Absorption kinetics of some carbohydrates in conscious pigs. 2. Quantitative aspects. Br. J. Nutr. 51:517–529.

Ring, S. G., J. M. Gee, M. Whittam, P. Orford, and I. T. Johnson. 1988. Resistant starch: Its chemical form in foodstuffs and effect on digestibility *in vitro*. Food Chem. 28:97–109.

Roth, F. X., M. Kirchgessner, and H. L. Müller. 1988. Energetische Verwertung von intracaecal infundierten Essig- und Propionsäure bei Sauen. J. Anim. Physiol. Anim. Nutr. 59:211–217.

Roth, F. X., M. Kirchgessner, and H. L. Müller. 1991. Energetische Verwertung von intracaecal infundierten Essig- und Propionsäure bei Sauen. J. Anim. Physiol. Anim. Nutr. 59:211–217.

Saulnier, L., F. Guillon, P.-E. Sado, and X. Rouau. 2007. Plant cell wall polysaccharides in storage organs: Xylans (food application), comprehensive glycoscience, Vol. 2. Elsevier, Amsterdam, The Netherlands. p. 653–689.

Selvendran, R. R. 1984. The plant cell wall as a source of dietary fibre: Chemistry and structure. Am. J. Clin. Nutr. 39:320–337.

Serena, A., and K. E. Bach Knudsen. 2007. Chemical and physicochemical characterisation of co-products from the vegetable food and agro industries. Anim. Feed. Sci. Technol. 139:109–124.

Serena, A., M. S. Hedemann, and K. E. Bach Knudsen. 2008a. Influence of dietary fiber on luminal environment and morphology in the small and large intestine of sows. J. Anim. Sci. 86:2217–2227.

Serena, A., H. Jorgensen, and K. E. Bach Knudsen. 2008b. Digestion of carbohydrates and utilization of energy in sows fed diets with contrasting levels and physicochemical properties of dietary fiber. J. Anim. Sci. 86:2208–2216.

Serena, A., H. Jorgensen, and K. E. Bach Knudsen. 2009. Absorption of carbohydrate-derived nutrients in sows as influenced by types and contents of dietary fiber. J. Anim. Sci. 87:136–147.

Serena, A., H. Jørgensen, and K. E. Bach Knudsen. 2007. The absorption of lactic acid is more synchronized with the absorption of glucose than with the absorption of short-chain fatty acids — A study with sows fed diets varying in dietary fibre. Livest. Sci. 109:118–121.

Shi, X. S., and J. Noblet. 1993. Digestible and metabolizable energy values of ten feed ingredients in growing pigs fed ad libitum and sows fed at maintenance level; comparative contribution of the hindgut. Anim. Feed Sci. Technol. 42:223–236.

Stanogias, G., and G. R. Pearce. 1985. The digestion of fibre by pigs 1. The effects of amount and type of fibre on apparent digestibility, nitrogen balance and rate of passage. Br. J. Nutr. 53:513–530.

Stein, H. H., and R. A. Bolkhe. 2007. The effects of thermal treatment of field peas (*Pisum sativum* L.) on nutrient and energy digestibility by growing pigs. J. Anim. Sci. 85:1424–1431.

Sun, T., H. N. Lærke, H. Jørgensen, and K. E. Bach Knudsen. 2006. The effect of heat processing of different starch sources on the *in vitro* and *in vivo* digestibility in growing pigs. Anim. Feed Sci. Technol. 131:66–85.

Theander, O., E. Westerlund, P. Åman, and H. Graham. 1989. Plant cell walls and monogastric diets. Anim. Feed Sci. Technol. 23:205–225.

Theil, P. K., H. Jorgensen, A. Serena, J. Hendrickson, and K. E. Bach Knudsen. 2011. Products deriving from microbial fermentation are linked to insulinaemic response in pigs fed breads prepared from whole-wheat grain and wheat and rye ingredients. Br. J. Nutr. 105:373–383.

Thibault, J.-F., M. Lahaye, and F. Guillon. 1992. Physico-chemical properties of food plant cell walls. In: T. F. Schweizer and C. A. Edwards, editors, Dietary fibre - a component of food: Nutritional function in health and disease. Springer-Verlag, London, UK. p. 21–39.

van Beers-Schreurs, H. M. G., M. J. A. Nabuurs, L. Vellenga, H. J. Kalsbeek-van der Valk, T. Wensing, and H. J. Breukink. 1998. Weaning and the weanling diet influence the villous height and crypt depth in the small intestine of pigs and alter the concentrations of short-chain fatty acids in the large intestine and blood. J. Nutr. 128:947–953.

van der Meulen, J., G. C. M. Bakker, J. G. M. Bakker, H. D. Visser, A. W. Jongbloed, and H. Everts. 1997a. Effect of resistant starch on net portal-drained viscera flux of glucose, volatile fatty acids, urea, and ammonia in growing pigs. J. Anim. Sci. 75:2697–2704.

van der Meulen, J., J. G. M. Bakker, B. Smits, and H. D. Visser. 1997b. Effect of source of starch on net portal flux of glucose, lactate, volatile fatty acids and amino acids in the pig. Br. J. Nutr. 78:533–544.

van Loo, J., P. Coussement, L. de Leenheer, H. Hoebregs, and G. Smits. 1995. On the presence of inulin and oligofructose as natural ingredients in the western diet. Crit. Rev. Food Sci. Nutr. 35:525–552.

Van Soest, P. J. 1963a. Use of detergents in the analysis of fibrous feeds. I. Preparation of fiber residue of low nitrogen content. J. AOAC 46(5):825–829.

Van Soest, P. J. 1963b. Use of detergents in the analysis of fibrous feeds. II. A rapid method for the determination of fiber and lignin. J. AOAC 46(5):829–835.

Van Soest, P. J. 1982. Analytical systems for evaluation of feeds. Nutritional Ecology of the Ruminant. Chapter 6. O. & B. Brooks Inc. p. 75–94.

van Soest, P. J., J. B. Robertson, and B. A. Lewis. 1991. Methods for dietary fiber, neutral detergent fiber, and nonstarch polysaccharides in relation to animal nutrition. J. Dairy Sci. 74:3583–3597.

Varel, V. H., and W. G. Pond. 1985. Enumeration and activity of cellulolytic bacteria from gestating swine fed various levels of dietary fibre. Appl. Environ. Microbiol. 49:858–862.

Vaugelade, P., L. Posho, B. Darcy-Vrillon, F. Bernard, M.-T. Morel, and P.-H. Duée. 1994. Intestinal oxygen uptake and glucose metabolism during nutrient absorption in the pig. Proc. Soc. Exp. Biol. Med. 207:309–316.

Vincken, J.-P., H. A. Schols, R. Oomen, M. C. McCann, P. Ulvskov, A. G. J. Voragen, and R. G. F. Visser. 2003. If homogalacturonan were a side chain of Rhamnogalacturonan I. Implications for cell wall architecture. Plant Physiol. 132:1781–1789.

Visser, J., and A. G. J. Voragen. 1996. Pectins and pectinases. In: J. Visser and A. G. J. Voragen, editors, Progress in biotechnology 14. Elsevier, Amsterdam, The Netherlands. p. 990.

Voragen, A. G. J., H. Gruppen, M. A. Verbruggen, and R. J. Viëtor. 1992. Characterization of cereal arabinoxylans. In: J. Visser, G. Beldman, M. A. K.-V. Someren, and A. G. J. Voragen, editors, Xylans and xylanases. Elsevier, Amsterdam, The Netherlands. p. 51–67.

Wilfart, A., L. Montagne, H. Simmins, J. Noblet, and J. Milgen. 2007. Digesta transit in different segments of the gastrointestinal tract of pigs as affected by insoluble fibre supplied by wheat bran. Br. J. Nutr. 98:54–62.

Wood, P. J. 1990. Physicochemical properties and physiological effects. In: I. Furda and C. J. Brine, editors, New developments in dietary fibre. Plenum Press, New York, NY. p. 119–127.

Wood, P. J. 2010. Review: Oat and rye β-Glucan: Properties and function. Cereal Chem. J. 87:315–330.

Würsch, P., S. Del Vedovo, and B. Koellreutter. 1986. Cell structure and starch nature as key determinants of the digestion rate of starch in legume. Am. J. Clin. Nutr. 43:23–29.

Zhang, Y., Q. Guo, N. Feng, J.-R. Wang, S.-J. Wang, and Z.-H. He. 2016. Characterization of A- and B-type starch granules in Chinese wheat cultivars. J. Integr. Agric. 15:2203–2214.

Zhou, P., P. K. Theil, D. Wu, and K. E. B. Knudsen. 2018. in vitro digestion methods to characterize the physicochemical properties of diets varying in dietary fibre source and content. Anim. Feed Sci. Technol. 235:87–96.

# 7    Vitamins and Vitamin Utilization in Pigs

J. Jacques Matte and Charlotte Lauridsen

## Introduction

The history of the discovery of vitamins was concentrated mainly in the first half of the twentieth century. It extends from the first use of the term vitamin (short for "vital amine") in 1915, to describe the factor extracted from the cuticle of rice that cures beriberi (thiamine), to the most recent discovery of the vitamin $B_{12}$, in 1948. During this time, the approach used in both human and animal nutrition research focused on symptoms of vitamin deficiency and hence the associated diseases. The consequences of this approach are still present in vitamin research nowadays. In the field of animal nutrition, in particular, estimation of vitamin requirements is still often based on the levels required to prevent the appearance of deficiency symptoms rather than those required to optimize production performance. In addition, current recommended vitamin intakes, particularly for B-complex vitamins, are the result of research carried out mostly in the 1950s and 1960s. At that time, it was assumed that a significant portion of animals' vitamin intake occurred through coprophagia (ARC 1981; McDowell 2000) since several B-complex vitamins are synthesized in the digestive tract by bacterial microflora and excreted in the feces. Today, however, it is known that, under current livestock housing conditions, the coprophagia behavior is negligible and cannot provide a reliable supply of B-complex vitamins (Greer and Lewis 1978; de Passillé et al. 1989; Bilodeau et al. 1989). In addition, the widespread use of slatted floors can only further marginalize the importance of coprophagia as a source of B-complex vitamins in swine.

For several vitamins, the requirements for swine have not been clearly established (ARC 1981; INRA 1984; NRC 1998; McDowell 2000; BSAS 2003; NRC 2012) mostly because of the lack of up-to-date information that could be used to support reliable recommendations. The high level of productivity characterizing today's intensive farm management calls for intensive anabolism to maintain growth, gestation, and lactation at an optimal level. Nutrient allowances may need to be adjusted to meet the increased level of production, but much of the limited information available on this topic is outdated. Not only was most of this information published before the 1980s, literature reports since 1998 have focused mainly on specific vitamins mostly vitamin E and biotin. (Table 7.1). Since the first edition of the present book (i.e., 2012), vitamin D has been in focus in pig nutrition.

Thirteen substances are considered as vitamins and are generally divided in two main groups, fat- and water-soluble vitamins (Table 7.2). Some of the historic, metabolic, and dietary characteristics related to vitamin nutrition in pigs are presented in Table 7.2. Vitamins A, D, E, and K,

**Table 7.1** Approximate number of citations related to vitamins in pigs reported by ARC (1981), NRC (1998) and since 1998.

| Vitamins | ARC (1981) | NRC (1998) | NRC (2012) additions vs. NRC (1998)[a] |
|---|---|---|---|
| *Fat-soluble* | | | |
| Vitamin A[b] | 107 | 24 | 2 |
| Vitamin D | 32 | 14 | 2 |
| Vitamin E | 90 | 43 | 15 |
| Vitamin K | 5 | 6 | — |
| *Water-soluble* | | | |
| Thiamine | 13 | 11 | 1 |
| Biotin | 22 | 35 | — |
| Riboflavin | 20 | 15 | — |
| Pantothenic acid | 23 | 19 | 1 |
| Pyridoxine | 23 | 14 | 4 |
| Niacin | 34 | 22 | 1 |
| Choline | 22 | 20 | — |
| Folic acid | 12 | 12 | 2 |
| Vitamin $B_{12}$ | 86 | 15 | 4 |
| Vitamin C | 42 | 13 | 4 |

[a] Citations (not included) were reported on cocktail of all vitamins (n = 4), of fat-soluble vitamins (n = 1) or of water-soluble vitamins (n = 4).
[b] including β-carotene.

so-called fat-soluble, can be stored in reserves when the dietary provision is in excess of the animal requirements. Vitamin C and B-complex vitamins are water soluble and are not stored in significant amount except for vitamin $B_{12}$. Therefore, a regular (daily) provision corresponding to the animal requirements is necessary for an optimal metabolic activity of those micronutrients. These last two characteristics are particularly important during critical physiological stages where metabolic needs may change rapidly in particular, gestation and weaning. In such cases, the animals can count (for fat-soluble) or not (most of the B-vitamins) on tissue reserves. The withdrawal of vitamin supplementation during the finishing period which has been proposed to reduce feed costs relies partly on the importance of these tissue reserves. Although growth performance is generally not altered by this practice (McGlone 2000; Shaw et al. 2002), tissue reserves are depleted enough to affect concentrations of vitamins (in particular, water-soluble) in meat (Shaw et al. 2002) and, in turn, the nutritional value of pork (Leonhardt et al. 1996; Lombardi-Boccia et al. 2005). In such situations, the natural content of some water-soluble vitamin in feed ingredients is critical, but the accurate information is very limited and outdated especially in view of the variety of ingredients used nowadays in swine diets. A rare report was recently released on that matter (Chen et al. 2019). They mentioned that although vitamin concentrations vary greatly from one feed ingredient to another, the contribution to a balanced diet may be far from negligible in some cases especially if high levels of these ingredients are used.

The present chapter aims to integrate recent scientific advances in vitamin nutrition of pigs in an attempt to complement the basic information already available in significant publications on that subject, such as ARC (1981), NRC (2012), McDowell (2000), Lewis and Southern (2001), and BSAS (2003).

**Table 7.2** Historical, dietary, and metabolic characteristics of vitamins.

| Vitamin[a] | History discovery/ isolation/synthesis[a] | Vitamers (biologically active forms)[a] | Monthly loss (%)[b] | Recommended provision NRC (2012)[c] | | | | | | Current provision (mg/kg)[c,d] | | | | | |
|---|---|---|---|---|---|---|---|---|---|---|---|---|---|---|---|
| | | | | W[e] | G[e] | F[e] | P[e] | L[e] | B[e] | W | G | F | P | L | B |
| A | 1909/1931/1947 | Retinol, retinoic acid, retinal, carotenoids | 3.5/9 | 2.2 | 1.3 | 1.3 | 4.0 | 2.0 | 4.0 | 12.1 | 5.9 | 5.2 | 9.8 | 10.0 | 10.6 |
| E[f] | 1922/1936/1938 | Tocopherol, tocotrienols | 3.0/4.5 | 16 | 11 | 11 | 44 | 44 | 44 | 71 | 25 | 22 | 66 | 67 | 72 |
| D | 1918/1932/1959 | Ergocalciferol (vitamin D$_2$) and cholecalciferol (vitamin D$_3$) | 1.6/1.1 | 220 | 150 | 150 | 800 | 800 | 200 | 1910 | 985 | 875 | 1531 | 1557 | 1608 |
| K | 1929/1936/1939 | Menadione, phylloquinone, menaquinone | 6.0/10 | 0.5 | 0.5 | 0.5 | 0.5 | 0.5 | 0.5 | 4.8 | 2.4 | 2.2 | 3.5 | 3.5 | 3.5 |
| C[g] | 1912/1928/1933 | Ascorbic acid | 1–2 | — | — | — | — | — | — | — | — | — | 250 | 250 | 250 |
| Thiamin | 1890/1910/1936 | Thiamin | 2.6/7.9 | 1.0 | 1.0 | 1.0 | 1.0 | 1.0 | 1.0 | 2.9 | — | — | 2.1 | 2.1 | 2.1 |
| Riboflavin | 1920/1933/1935 | FMN (flavin mononucleotide) and FAD (flavin adenine dinucleotide) | 3.3/2.7 | 5.0 | 5.0 | 2.0 | 3.8 | 3.8 | 3.8 | 9.7 | 4.8 | 4.2 | 7.5 | 7.7 | 7.7 |
| Niacin | 1935/1935/1949 | NAD (nicotinamide-adenine dinucleotide) et le NADP (nicotinamide-adenosine dinucleotide phosphate) | 3.5/3.2 | 30 | 30 | 30 | 10 | 10 | 10 | 51 | 25 | 22 | 41 | 41 | 41 |
| Pantothenic acid | 1931/1938/1940 | Pantothenic acid | 0/0 | 10 | 8.0 | 7 | 12 | 12 | 12 | 36 | 17 | 15 | 27 | 27 | 28 |
| Pyridoxine | 1934/1938/1939 | Pyridoxine, pyridoxal and pyridoxamine | 5.9/8.6 | 7.0 | 1.0 | 1.0 | 1.0 | 1.0 | 1.0 | 4.0 | — | — | 4.0 | 4.0 | 3.7 |
| Biotin | 1931/1935/1943 | Biotin | 4.4/2.9 | 50 | 50 | 50 | 200 | 200 | 200 | 370 | 70 | 70 | 260 | 290 | 310 |
| Choline | 1932 | Acetylcholine and phosphatidylcholine | 2.1/4.9 | 500 | 300 | 300 | 1250 | 1000 | 1250 | 224 | — | — | 645 | 479 | 638 |
| Folic acid | 1941/1941/1946 | Méthyltétrahydrofolates et tétrahydrofolates | 2.1/5.6 | 0.3 | 0.3 | 0.3 | 1.3 | 1.3 | 1.3 | 1.8 | — | — | 1.7 | 1.7 | 1.8 |
| Vitamin B$_{12}$ | 1926/1948/1972 | Methylcobalamin and adenosylcobalamin | 2.0/5.4 | 15 | 10 | 5 | 15 | 15 | 15 | 46 | 23 | 20 | 34 | 35 | 39 |

[a] Adapted from Le Grusse and Watier (1993) and Charlton and Ewing (2007).

[b] Adapted from Shurson et al. 2011, the first value is for vitamin premix only and the second for combined vitamin and trace mineral premixes. The value for vitamin C is reported by Charlton and Ewing (2007).

[c] Values are all in mg/kg of diet (90% of dry matter) except for vitamin A (kIU), vitamin D(IU), biotin (µg/kg), and vitamin B$_{12}$ (µg/kg).

[d] Adapted from Flohr et al. (2016).

[e] Body weight (kg) intervals for weaner (W), grower (G), and finisher (F) are 7–11, 23–55, and 55–100. P is for pregnancy, L for lactation, and B for boars.

[f] Current provisions of vitamin E may be higher for lactating sows (150 mg kg$^{-1}$), or higher for growing-finishing pigs (meat quality stabilizer).

[g] Vitamin C may be provided in stressful conditions where synthesis is limited.

**Vitamins and Reproduction in Pigs**

*Male Reproduction*

Traditionally, the research on male fertility has dealt with genetic, behavior, and housing aspects. Very little research has been done on boar nutrition, and this limited interest is probably linked to animal husbandry practices (natural service) in previous decades. Indeed, in such situation, nutrition of the boar has little effect on sows' reproductive performance or fertilization rates because a boar ejaculate contains 5–20 times more spermatozoa than required to fertilize all the ova (n = 15–20). However, artificial insemination (AI), which is worldwide used nowadays, 80% or more of services in many countries (Khalifa et al. 2014), has changed completely the importance of husbandry conditions for boars. This includes adequate and specific nutrition to maximize AI doses per ejaculate. In this context, the effect of micronutrients on reproductive performance in male pigs is poorly documented (Audet et al. 2004).

Testicular degeneration has long been recognized as one of the manifestations of a vitamin E deficiency in the male. However, there are not that many studies available regarding the requirement of vitamin E for male fertility in swine. Semen is rich in unsaturated fatty acids, which increase the susceptibility to peroxidation and may cause structural damage to sperm and subsequently alter sperm motility. Brzezinska-Slebodzinska et al. (1995) suggested that dietary vitamin E may serve as an antioxidant in boar semen, whereas addition of tocopherol to the ejaculate did not protect the sperm from peroxide damage (Jones and Mann 1977). When 1000 IU vitamin E/day was fed to boars for a seven-week period the lipid peroxidation of the collected semen was reduced (Brzezinska-Slebodzinska et al. 1995). Probably, α-tocopherol, which is the biologically most important form of vitamin E, needs to be incorporated in between the fatty acid methyl esters of the phospholipids in the cellular membranes in order to exert efficient antioxidant protection, which is possible through dietary means (see also Marin-Guzman et al. 1997, 2000).

Preliminary findings with a limited number of animals suggested that dietary vitamin concentrations as recommended by NRC (1998) were not sufficient to maximize semen production in situation of intensive semen collection (Audet et al. 2004). Dietary supplements of vitamins (3–10×NRC recommendation) did not influence hormonal status but affected blood and seminal pools of vitamins (Audet et al. 2009a). Except for vitamin $B_6$, most of the vitamins were transferred, though with different efficiencies, from blood to seminal plasma (Audet et al. 2009b). However, no marked effect of dietary vitamin supplements was observed on semen production and semen quality either in situation of long-term intense semen collection or in AI commercial conditions (Audet et al. 2009b).

Under stressful conditions such as excessive heat, a cocktail of antioxidative micronutrients such as zinc, selenium, and vitamins C and E was beneficial for some aspects of sperm quality such as motility (Horký et al. 2016). In terms of vitamin D, Lin et al. (2017) showed that semen production, sperm quality, hormone synthesis, and gene expression related vitamin D metabolism were optimized at 2000 IU/kg of diet in a titration study with three levels of supplementation at 200, 2000, and 4000 IU/kg. It should be noted that the current vitamin D recommendation is 200 IU/kg for boars (Table 7.2).

*Female Reproduction*

Prolificacy in female pig lines has increased substantially during the last 10–20 years after several decades of stagnation. Intensive genetic selection programs based on litter size have been set up with occidental breeds in order to develop the so-called "hyperprolific" lines of pigs. In such case, a considerable increase in ovulation rate (four to five) has been achieved but, at the expense of embryo survival at the beginning of gestation. Nevertheless, the litter size is increased by one to two piglets per litter (Driancourt et al. 1998; Foxcroft 2012). Although it has been known for many years that an increase in ovulation rate in pigs is detrimental for embryo survival, the precise mechanisms involved are not well understood. The poor quality of oocytes produced from these supplementary mature follicles along with its consequence on either fertilization of ova and/or the ability of the embryo cells for development and differentiation has been raised as a main related issue (Driancourt et al. 1998).

Hyperprolificacy has also induced heterogeneity of piglet weight within the litter (Tribout et al. 2003; Le Cozler et al. 2004). A lack of uniformity in piglet weights has carry-over effects after birth in terms of growth performance and quality of carcass and meat (Gondret et al. 2005). Several postnatal strategies (adoption, split-weaning) have been proposed to decrease within-litter variation of piglet weights but with limited success (Matte et al. 1991). A prenatal strategy appears essential, but a reliable improvement of this production trait requires a better understanding of the related "in utero" physiological causes of the phenomenon. Several fat- and water-soluble vitamins have been identified and related to those "in utero" physiological events, especially in early gestation (Mahan and Vallet 1997; Matte et al. 2006). During this period, embryo losses may represent between 15 and 40% of the initial number of fertilized ova and the optimal embryo development relies on the so called « feto-maternal dialogue » (Cox 1997). In such situation, uterine secretions, also called "uterine milk" (Solymosi and Horn 1994), provide both the right amount and balance of nutrients, hormones, and growth factors. Viable embryos also contribute to the composition of this medium by producing hormones and cytokines which, in turn, influence the composition of uterine secretions. Later in gestation and lactation, fetuses and piglets are entirely dependent for their vitamin requirements on the transfer (in utero, colostrum, and milk) from the dam. In fact, in term of duration, the combined gestation and lactation periods (approximately 135–140 days) are as long as the rest of the postweaning life of a slaughter pig (approximately 130 days). This emphasizes the importance of an adequate maternal transfer of these micronutrients. Using specific data from the relevant literature and from a trial with sows during late gestation and early lactation (Matte and Audet 2020), an attempt was made to estimate the efficiency of prenatal (in utero) transfer from dams to piglets using prefarrowing plasma concentrations of vitamers in dams and the corresponding pre-colostral values in piglets. Similarly, the efficiency of the postnatal (colostrum-milk) transfer was estimated from plasma values of pre- and postcolostral vitamers in piglets (Table 7.3). It appears that the efficiency of in utero-placental transfer of vitamins from dams to fetuses and to piglets varies from a vitamin to another, while the importance of the early postnatal transfer through colostrum is crucial for the early status of most vitamins in the newborn animal (Table 7.3).

### Fat-Soluble Vitamins and Vitamin C

*Vitamin A*

Vitamin A is essential for reproduction, although the role for this biological function is relatively unknown. It is required for maturation of ovarian follicles, for the proper functioning of corpora lutea and epithelial cells of the oviducts, the uterine environment, and the cervix, and for embryonic development. It stimulates estrogen synthesis in tertiary follicles and progesterone synthesis in corpora lutea (Nunetz et al. 1995). Lack of retinol leads not only to a decline in general health, but more specifically to a decrease in ovarian size and to testicular atophy (Palludan 1963). The vitamin A reserves of the sow make it difficult to establish requirements (NRC 2012). Braude et al. (1941)

**Table 7.3**  Estimation of efficiency of "in utero"-placental and colostral milk transfers of different vitamins from dams to newborn piglets.

| Vitamin[a] | Pre-colostral (piglets)/ pre-farrowing (dam)[c] | Relative importance of "in utero" transfer | Postcolostral/ pre-colostral (piglets)[d] | Relative importance of colostral-milk transfer |
|---|---|---|---|---|
| A[a] | 0.26 | - - - - | 2.7 | + + + |
| E[a] | 0.44 | - - | 4.7 | + + + + + |
| D[a] | 0.09 | - - - - - - - - - - | 2.8 | + + + |
| C[a] | 2.9 | + + + | 4.1 | + + + + |
| B$_9$[a] | 0.54 | - - | 3.1 | + + + |
| B$_{12}$[a] | 4.4 | + + + + | 2.0 | + + |
| B$_2$[b] | 0.68 | - - | 3.0 | + + + |
| B$_3$[b] | 1.0 | + | 4.0 | + + + + |
| B$_6$[b] | 5.1 | + + + + | 0.34 | - - - |
| B$_8$[b] | 7.9 | + + + + + + + + | 0.23 | - - - - |

[a] Adapted from Hakansson et al. (2001) for vitamin A; from Hakansson et al. (2001), Loudenslager et al. (1986), and Pinelli-Saavedra and Scaife (2005) for vitamin E; Amundson et al. (2017) for vitamin D; Pinelli-Saavedra and Scaife (2005) and Yen and Pond (1983) for vitamin C; Barkow et al. (2001) and Matte and Girard (1989) for vitamin B$_9$; and Simard et al. (2007) for vitamin B$_{12}$.

[b] Adapted from Matte and Audet (2020).

[c] Ratio between plasma vitamer concentrations of pre-colostral values in piglets and the prefarrowing values in dams.

[d] Ratio between plasma vitamer concentrations of pre-colostral and postcolostral values in piglets.

reported that mature sows fed diets without supplemental vitamin A completed three pregnancies normally; only in the fourth pregnancy did signs of vitamin A deficiency appeared.

Reports from Eastern Europe in the mid 1960s indicated that injected retinol would increase litter size (ARC 1981). Likewise, subsequent studies (Coffey and Britt 1993) showed that dietary supplementation of vitamin A and/or β-carotene before mating and during early pregnancy can have beneficial effects on litter size at birth and at weaning. However, in the study by Pusateri et al. (1999), a single injection ($1 \times 10^6$ IU vitamin A) at any time from weaning to farrowing did not influence total litter size, litter weight, pig weight, number of runts, or number of mummies.

A study from Lindemann et al. (2008) showed that the injection of high doses (intramuscular injection with 250 000 IU vitamin A or intramuscular injection with 500 000 IU of vitamin A at weaning and breeding) to young sows (parity 1 and 2) increased linearly the subsequent number of pigs born and weaned per litter, whereas for sows of parity 3 to 6, litter sizes were not affected by the vitamin A treatments. The high-dose treatments were compared against a diet containing 11 000 IU vitamin A/kg diet, and the study concluded that vitamin A requirement for maximal performance may vary with age (Lindemann et al. 2008).

Vitamin A may influence both ovarian steroidogenesis and the uterine environment by affecting ovarian progesterone production (Chew 1993). The pig uterus secretes a large amount of several proteins in response to progesterone (Roberts and Bazer 1988). These uterine proteins are very important to the nutriture of the conceptus (Buhi et al. 1979). This is especially true in the pig, because the porcine throphoblast does not invade the uterine epithelium, it rather remains in superficial attachment to the uterine surface. The presence of vitamin A-carrier proteins, the so-called retinol-binding protein (RBP), that can transport vitamin A from the maternal uterine endometrium to the conceptus has been demonstrated in uterine secretions from pigs in the luteal phase of the estrous cycle, and in the pig conceptus (Chew 1993). Adams et al. (1981) reported that total vitamin A in uterine secretions increased in progesterone treated pigs, thus indicating an increased local transport of retinoids by RCP to the developing conceptus. Antipatis et al. (2008) demonstrated that reduced vitamin A during conception and early pregnancy, but not during later pregnancy, was associated with increased within-litter uniformity in birth weight, and the mechanism was proposed to be ascribed to alteration in progesterone production. Moderate reductions in maternal vitamin A at either stage of pregnancy did not affect pregnancy rate, litter size, progesterone secretion, and allometric relationships between fetal or neonatal organ and total body size (Antipatis et al. 2008).

Research on the possible specific role of β-carotene (and other carotenoids) has been previously hampered by the assumption that its sole function in animals is to provide vitamin A. Consequently, knowledge of the role of β-carotene (and other carotenoids) on reproduction as well as on immune responses is relatively scarce compared with the knowledge existing on vitamin A. Cross-breed gilts injected weekly with 228 mg of β-carotene from mating trough weaning had lower embryonic mortality, larger litter size, and heavier litter weight at birth and at weaning than did unsupplemented gilts (Brief and Chew 1985). This is in general agreement with the study of Coffey and Britt (1989) who reported a linear increase in litter size at birth in multiparous sows injected once at weaning with increasing doses of β-carotene (0, 50, 100, or 200 mg of β-carotene). It was unclear whether this was due to increased ovulation rate or to decreased embryonic mortality (Chew 1993). Even though some studies show improved reproductive performance with supplemental β-carotene, it remains unclear whether β-carotene plays a direct role in regulating certain reproductive processes or whether it merely serves as a source of vitamin A (Chew 1993). Studies on the possible role of β-carotene on male reproduction are lacking.

*Vitamin D*

Vitamin D recommendation for sows according to NRC (1998) during gestation and lactation was not based on scientific reports, and, in general, very little evidence was available regarding vitamin D in relation to reproduction in swine. However, more scientific studies have been conducted since that time and the recommendation of vitamin D for gestating and lactating swine was increased to 800 IU in NRC (2012) (Table 7.2), and some of the scientific studies that formed the basis for this increase in vitamin D recommendation are presented later. Besides the classical actions, the $1,25(OH)_2D_3$ has well-known immunomodulatory and antiproliferative properties (Hewison et al. 2001). In contrast, the impact of $1,25(OH)_2D_3$ on reproduction is poorly understood despite evidence relating vitamin D with reproductive function in both males and females. For instance, female fertility seems to be markedly reduced in vitamin D-deficient murine models (Vigiano et al. 2003). The influence of vitamin D on the reproductive capacity has been linked to calcium-independent mechanisms (Kwiecinski et al. 1989). In this way, it has been found that $1\alpha$-OHase is expressed in placenta from normal pregnancies (Zehnder et al. 2002). In view of this, and the fact that reproduction in females is markedly diminished in states of vitamin D deficiency, it has been postulated that local synthesis of $1,25(OH)_2D_3$ may play a role in implantation and/or placentation (Halloran and Deluca 1983; Hickie et al. 1983; Kwiecinski et al. 1989), either through the established immunomodulatory effects of $1,25(OH)_2D_3$ or via the regulation of specific target genes associated with implantation (Zehnder et al. 2001). The study by Jang et al. (2018) seems to confirm that calcitriol may play an important role in the establishment and maintenance of the pregnancy by regulating endometrial function in pigs.

A study reported by Lauridsen et al. (2010) was undertaken to obtain information on the dose–response pattern of two vitamin D sources, vitamin $D_3$, and $25(OH)D_3$ with respect to early reproduction of reproducing female sows. In experiment 1, 160 gilts were randomly assigned from the first estrus until d 28 of gestation to dietary treatments containing 4 concentrations of one of the two different vitamin D sources (200, 800, 1400, and 2000 IU/kg from cholecalciferol or corresponding of 5, 20, 35, and 50 μg kg$^{-1}$ feed from 25(OH)D$_3$. In a concurrent experiment, the same eight dietary treatments were provided to 160 multiparous sows from the first day of mating until weaning. Reproductive performance of sows was not influenced by dietary vitamin D treatments, except for a lower number of still born piglets with the high doses of vitamin D (1400 and 2000 IU vitamin D giving 1.17 and 1.13 still born piglets per litter, respectively) compared with the low doses of vitamin D (200 and 800 IU vitamin D giving 1.98 and 1.99 still born piglets per litter, respectively). More recent studies have been conducted on the effect of dietary 25(OH)D$_3$ on sow milk quality and health parameters of sows. In the study by Zhou et al. (2017), sows were, immediately after mating, randomly allotted to one of two diets supplemented with 50 μg 25(OH)D$_3$/kg or basal diets without 25(OH)D$_3$ (containing 50 μg kg$^{-1}$ vitamin D3). Interestingly, the 25(OH)D$_3$-fed sows had one more piglet at farrowing and 1.17 more piglets at weaning than sows fed basal diets. Furthermore, beneficial effects were obtained on the studied milk quality and bone health parameters of the offspring. A more recent study (Zhang et al. 2019) comparing 2000 IU cholecalciferol with 50 μg 25(OH)D$_3$/ kg feed for sows from day 107 of gestation until weaning on day 21 of lactation concluded that the addition of 25(OH)D$_3$ to the diets improved total litter weight gain, calcium digestibility, milk fat content, and milk IgG concentration and milk 25(OH)D$_3$ content of the sows. From the above-mentioned studies, it may be concluded that addition of high levels of vitamin D and addition of the 25(OH)D$_3$ exposed beneficial effect for reproducing and lactating sows, which also confirmed the rationale of increasing the vitamin D recommendation by NRC (2012) as also explained by Lauridsen (2014). However, the effect regarding transfer of vitamin D to the offspring seems less efficient at least it was concluded by Lauridsen et al. (2010) that, irrespective to the dietary dose and form of vitamin D for the sows, very little vitamin D was transferred to the progeny, i.e., 25(OH)D$_3$ was not detectable in most of the suckling piglets, and in those piglets where 25(OH)D$_3$ was detectable, the concentration was below 5 nmol l$^{-1}$ plasma (Lauridsen et al. 2010). In the study by Zhou et al. (2017), clear effects of the additional 25(OH)D$_3$ treatment of sows were obtained with regard to the 25-(OH)D$_3$ concentrations in sow colostrum and milk and in sow and piglet plasma. In the latter study, vitamin D concentration was analyzed by ELISA, whereas in the study by Lauridsen et al. (2010), HPLC was used. Likewise in a more recent study (Flohr et al. 2014) using RIA for analyzing concentrations of 25-(OH)D$_3$ in body fluids, it was shown that an increase in maternal vitamin D (from 2000 to 9600 mg vitamin D3/kg feed) only increased the serum 25-(OH)D$_3$ in piglets from 1.4 to 5.7 ng ml$^{-1}$, while the concentration of 25-(OH)D$_3$ in serum of piglets suckling sows provided 800 mg vitamin D3/kg was below detection limit.

Among domestic farm animals tested, baby pigs are born with the lowest plasma concentration of 25(OH)D$_3$, which increases their susceptibility to vitamin D deficiency (Horst and Littledike 1982). In fact, an increased risk of hyperkyposis was observed in off-spring when maternal diets deficient in vitamin D was used in a porcine model (Halanski et al. 2018). The fact that swine differs from other husbandry may be partly due to the production systems; for instance, the concentration of plasma vitamin D and most of its metabolites in pigs exposed to sunlight was, respectively, 2.2–20.3 times the concentration of pigs kept in confinement (Engstrom and Littledike 1986). In a study by Goff et al. (1984), five sows received 5 million IU cholecalciferol intramuscularly 20 days prepartum, and the effect of the sow treatment was investigated on sow and piglet plasma concentrations of vitamin D$_3$ and its metabolites analysed by RIA. Though a limited number of sows in this experiment, a high degree of correlation was found with regard to the concentrations of 25(OH)D$_3$, 24,25-(OH)$_2$D$_3$, and 25,26-(OH)$_2$D$_3$ in plasma of both sows and piglet at birth. However, piglet neonatal plasma concentration of 1,25(OH)$_2$D$_3$ was low and did not correlate with maternal plasma. It was concluded that the treatment provided an effective means of supplementing pigs with D$_3$ via sow's milk (Goff et al. 1984). Irrespective of analytical technique used for determination of vitamin D concentration, the studies somehow come to that same conclusion that the more 25(OH)D$_3$ transfer via milk in sows fed 25(OH)D$_3$ was not enough to significantly increase serum 25(OH)D$_3$ levels of the off-spring. Hence, in order to increase vitamin D status of piglets other strategies than maternal transfer via sow milk may be needed.

*Vitamin E*

Severe vitamin E deficiency in reproducing animal species results in fetal death and resorption (Nielsen et al. 1979). Most swine studies have demonstrated an increased litter size at birth when vitamin E was supplemented in cereal grain-based diets (Adamstone et al. 1949). Mahan (1994) investigated increasing levels of dl-α-tocopheryl acetate (22, 44, or 66 IU/kg diet) over a five-parity period on sow reproductive performance and on α-tocopherol concentration in serum, colostrum, and milk of the sows. With increasing dietary vitamin E, there was an increased number of piglet born and decreased incidence of mastitis, metritis, and agalactia (Mahan 1994). In addition, the concentration of α-tocopherol in colostrum and milk increased with increasing dietary levels of vitamin E (Mahan 1994). The pig is born almost deficient in vitamin E because as stated earlier (see also Table 7.3) there is a low placental transfer of α-tocopherol to the developing fetus (Mahan 1991). However, within a short period (four days) after birth, suckling pigs are capable of increasing their plasma α-tocopherol status with a factor 83 (Lauridsen et al. 2002b). Increasing levels of dietary vitamin E to lactating sows are reflected in the pig plasma and tissue status at weaning (Mahan 1991; Lauridsen and Jensen 2005).

Studies have investigated the effect of different vitamin E sources (RRR- and all-rac-α-tocopheryl acetate) on sow reproductive performance and α-tocopherol status of dam and the progeny. There was no effect of the vitamin E source on various sow reproductive measurements, litter size, or the incidences of mastitis-metritis-agalactia or fluid discharges from the vagina (Mahan et al. 2000). Feeding the natural source (RRR-α-tocopheryl acetate) compared with all-rac-α-tocopheryl acetate resulted in higher α-tocopherol concentrations in serum, colostrum and milk, as well as in serum and liver of 21-day old nursing pigs (Mahan et al. 2000). Quantitative measurements of the comparative efficacy of the two vitamin E forms are difficult because the newly absorbed α-tocopherol replaces the circulating α-tocopherol (Traber et al. 1998), thereby preventing quantitative estimates for the total dose incorporated into plasma or milk. The use of stable isotope-labeled α-tocopherol is required and has major advantages over nonlabeled compounds for determination of the relative activities of natural and synthetic vitamin E forms can be ingested simultaneously for intraindividual comparisons. By feeding sows capsules with labeled vitamin E forms (d$_3$-RRR-α-) and d$_6$-all-rac-α-tocopheryl acetate to pregnant sows it was found (Lauridsen et al. 2002a,b) that swine discriminate between RRR- and all-rac-α-tocopherol with a preference for RRR-α-tocopherol; thus, the official bioequivalence factor of 1.36:1 for RRR- to all-rac-α-tocopherol is underestimated. Sow plasma and milk d$_3$-α- to d$_6$-α-tocopherol concentrations were 2:1, leading to a 2:1 ratio in suckling piglet plasma and tissues (Lauridsen et al. 2002a,b). In a subsequent sow-study, a HPLC method was used to separate the eight steroisomers of α-tocopherol into five peaks (peak 1: all four 2S-forms, peak 2: 2RSS-α-tocopherol, peak 3: 2RRS-α-tocopherol, peak 4: 2RRR-α-tocopherol (=natural α-tocopherol), and peak 5: 2RSR-α-tocopherol. The concentration of α-tocopherol and its stereoisomer distribution was studied in sow milk and plasma of the sows and their progeny. Sows were provided increasing levels (70, 150, or 250 IU of all-rac-α-tocopheryl acetate kg, as fed basis) one week prior to parturition and during the lactation of 28 days, and milk and plasma samples were obtained on day 2, 16, and 28 of the sows and at day 2, 16, and 28 of the piglets (for dctails see Lauridsen and Jensen 2005). Table 7.4 shows the distribution of the stereoisomer forms in milk of sows and plasma of piglets. As seen the RRR-stereoisomer form was the most predominant form α-tocopherol, whereas the 2S-forms were

**Table 7.4**   Total concentration and stereoisomer forms of α-tocopherol in sow milk (mg/kg, d 2 of lactation) and piglet plasma (mg/l, day 4 of age). Sows were fed increasing levels (70, 150, and 250 IU/kg) of all-rac-α-tocopheryl acetate from day 108 of gestation until weaning (at day 28 after farrowing).

| Item | Total α-tocopherol[a] | 2S | RSS | RSS | RSR | RRR |
|---|---|---|---|---|---|---|
| 70 IU |  |  |  |  |  |  |
|   Sow milk | 6.39 | 0.25 | 1.08 | 1.50 | 1.26 | 2.28 |
|   Piglet plasma | 7.24 | 0.13 | 1.24 | 1.63 | 1.33 | 2.21 |
| 150 IU |  |  |  |  |  |  |
|   Sow milk | 10.2 | 0.69 | 1.73 | 2.37 | 2.07 | 3.40 |
|   Piglet plasma | 6.53 | 0.27 | 1.42 | 1.97 | 1.56 | 2.79 |
| 250 IU |  |  |  |  |  |  |
|   Sow milk | 16.5 | 1.41 | 3.22 | 3.72 | 3.35 | 4.79 |
|   Piglet plasma | 7.27 | 0.23 | 1.25 | 1.81 | 1.38 | 2.38 |

[a] Adapted from Lauridsen and Jensen (2005).

only present in limited proportions, irrespective that the proportion of the RRR-form and the 2S-forms in the dietary all-rac-α-tocopheryl acetate is 12.5 and 50%, respectively. As expected from previous studies, increasing dietary level of all-rac-α-tocopheryl acetate increased the concentration of total α-tocopherol in the sow milk but this effect was not so clear for piglet plasma. However, when samples were lipid standardized, the influence of the dietary treatment of sows on α-tocopherol concentration was highly significant (Lauridsen and Jensen 2005). Overall, the data in Table 7.4 indicate that the sow discriminates between natural and synthetic vitamin E with a preference for the RRR-α-tocopherol, which is reflected in the sow milk and suckling piglets.

*Vitamin C*
Currently, the conditions in which supplemental vitamin C may be beneficial for reproduction are not well defined, and therefore no vitamin C requirement estimate is given for pigs (ARC 1981, NRC 2012). However, the importance of adequate tissue levels of ascorbic acid for fetal development is clear from studies on sows with a genetically determined defect of synthesis. When Wegger and Palludan (1994) supplemented feed with 50 mg ascorbic acid/kg body weight, maturation of oocytes, fertilization, and embryonic and fetal development occurred normally. Plasma concentrations of ascorbate were $0.58 \pm 0.04$ and $1.27 \pm 0.09$ mg dl$^{-1}$ in supplemented sows and fetuses, respectively, whereas the corresponding values for unsupplemented animals were $17 \pm 8$ μg dl$^{-1}$ and $18 \pm 6$ μg dl$^{-1}$, respectively. After stopping supplementation with ascorbic acid for 24–38 days at various stages of pregnancy, edema, and subcutaneous and subperiosteal hemorrhages developed in the fetuses. Calcification of the skeleton was greatly reduced. Thus, ascorbic acid is required for correct functioning of the ovaries, specifically for maturation of tertiary follicles and maintenance of the function of corpora lutea. With regard to male reproduction, ascorbic acid is also required for proper development, maturation, and maintenance of function of sperm and for the synthesis of testosterone in the interstitial cells. The effect of supplemental vitamin C (1 g day$^{-1}$) for boar reproduction is controversial (Audet et al. 2004; Lechowski et al. 2018).

Like vitamin E, ascorbic acid plays a role in protection against oxidative processes, which is important for both protection of sperm, and the breakdown of highly reactive oxygen molecules and radicals in granulosa and luteal cells and also immune function. In a study reported by by Pinelli-Saavedra et al. (2008), the effects of dietary supplementation with vitamin E and vitamin C from beginning of pregnancy until weaning at 21 days of age was studied in relation to α-tocopherol deposition in colostrum, milk piglet plasma, and tissues, and immune responses. When ascorbic acid was fed alone (10 g vitamin C/day), increased lymphocyte response to ConA and PHA was observed, whereas no effect was seen on piglet lymphocyte activity. Vitamin E supplementation (500 mg kg$^{-1}$ feed) affected the content of $\alpha$-tocopherol, and the combined vitamin E and C supplementation (500 mg vitamin E and 10 g vitamin C/kg feed) increased the concentration of IgA and IgG in piglet plasma. In another sow study (Pinelli-Saavedra and Scaife 2005), no effect of vitamin C supplementation (up to 10 g day$^{-1}$) was seen on sow reproductive performance and piglet growth performance. The efficacy of the placental transfer decreased as maternal serum concentrations of vitamin C increased. The main supply of vitamin C to the new born piglet was via the mammary gland rather than the placenta. A significant positive correlation between piglet plasma ascorbic acid and maternal milk ascorbic acid concentration has been shown (Hidiroglou and Batra 1995).

### Water-Soluble Vitamins

#### Thiamin and Pantothenic Acid

To the best of our knowledge, nothing has been reported in the literature on thiamine and pantothenic acid in relation to reproduction of pig species for several decades. After reporting only one experiment on thiamine in reproducing swine (Ensminger et al. 1947), ARC (1981) stated: "... There appears, however, to be no good evidence from work with other animals to suggest that the requirement for thiamin for reproduction is likely to be substantially different from that for optimum growth." For pantothenic acid, the absence of information is also apparent as most recent references are from around fifty years ago (Davey and Stevenson 1963; Teague et al. 1971).

#### Biotin

Several studies have shown beneficial effects of supplementation of this vitamin on reproductive performance, while others have not found any effect (see reviews by Brooks 1986 and Kornegay 1986). In general, experiments have either involved a limited number of animals which have not allowed significant differences to be clearly discerned or have been carried out on animals with a low reproductive capacity. In a trial using over 300 litters on several parities, Lewis et al. (1991) reported that a dietary supplement of 330 µg kg$^{-1}$ did not influence litter size at birth but increased the number of piglets weaned at 21 days of age. In other large-scale trials (with more than 500 sows and gilts), Greer et al. (1991) reported a clear reduction of foot lesions after a 12 months' supplementation of biotin at 500 µg kg$^{-1}$ feed but no tangible effect on reproductive performance. The lack of effect on reproduction was also observed by Watkins et al. (1991) in another study using 223 litters from 90 sows over a period covering 5 parities (three years) with fortification of 440 µg biotin/kg feed.

To the best of our knowledge, there has been only one report published after the above-mentioned initial studies on biotin in reproducing sows (Garcia-Castillo et al. 2006). No effect was observed on reproductive performance of gilts after massive supplements of 10 and 28 (mg kg$^{-1}$) of biotin vs a control treatment of 0.07 mg kg$^{-1}$ from the prepubertal period (70 kg of body weight) until the end of the first lactation. Nevertheless, this new information is unlikely to be of major impact or biological significance considering the limited number of animals, only 7 gilts per treatment.

#### Riboflavin

Before 1988, researchers had not suspected a role for riboflavin on swine reproduction. In the early 1980s, it was shown that the sow's uterus secreted large quantities of riboflavin approximately one week after mating (Moffatt et al. 1980). A dietary supplement of 100 mg day$^{-1}$, given from day 4 to day 10 of gestation, increased litter size at farrowing, probably due to the higher survival rate of embryos (Bazer and Zavy 1988). However, the results of this experiment have never been reproduced in later studies (Luce et al. 1990; Tilton et al. 1991; Wiseman et al. 1991; Pettigrew et al. 1996). The supplement used by Bazer and Zavy (1988) appears to have been effective in animals with a low reproductive capacity (<10 live-born piglets per litter). Results obtained in a more recent study on the subject using 60 mg riboflavin/day during the entire gestation (Rivera-Alejandro and Jiménez-Cabán (2014) confirm the initial study of Bazer and Zavy (1988), where the number of live piglets at parturition was increased from 8.1 to 10.5 in unsupplemented and supplemented sows (daily bolus of 100 mg from days 4 to 10 of gestation), respectively. The factors explaining these divergent results need to be better understood before making more definitive recommendations on riboflavin requirements in reproducing sows. Along with the traditional reproductive performance criteria, metabolic criteria have been used in order to determine optimal levels of dietary riboflavin. Such criteria include blood levels of total riboflavin (FAD, FMN, and riboflavin) and glutathione reductase activity in erythrocytes (often called EGRAC). This technique, developed in human nutrition, has also been used commonly in swine nutrition. The result is expressed in the form of a coefficient, with values of 1.0–1.2 representing an adequate intake of riboflavin; 1.2–1.3, marginal intake; and >1.3, inadequate intake (Le Grusse and Watier 1993). Table 7.5 shows the effects of dietary riboflavin levels on the evolution of EGRAC activity according to the stage of gestation in two separate experiments (Frank et al. 1984; Pettigrew et al. 1996). The experiments can be compared since one common level of dietary riboflavin supplement was used and the response, in terms of EGRAC values, was similar. A daily intake of 10 mg of riboflavin appears to be sufficient in stabilizing and minimizing EGRAC values. However, even though EGRAC levels appear to be a reliable criterion for identifying riboflavin deficiency, some doubts have been raised upon its validity as a sensitive index of riboflavin requirements for both human and pigs. In fact, EGRAC values are not correlated with total riboflavin concentrations in the blood or liver of pigs (Giguère et al. 2002), while, in humans, alternative indicators are required (Hoey et al. 2009).

*Pyridoxine and Niacin*

The effect of pyridoxine on litter size is not well documented, and most studies were carried out before the 1960s. NRC (1998) and ARC (1981) have not made formal recommendations, only suggestions, for requirements. Two studies from the early 1980s suggest that dietary concentrations of

**Table 7.5**  Riboflavin status estimated from glutathione reductase activity in erythrocytes (EGRAC) according to dietary supplement of vitamin B2 and stage of gestation in sows.

| Dietary supplement of riboflavin (mg/j) | EGRAC (3 weeks) | EGRAC (7 weeks) | EGRAC (14 weeks) |
|---|---|---|---|
| 1.5[a] | 1.45 | 1.91 | 2.82 |
| 5.5[a] | 1.37 | 1.42 | 1.64 |
| 9.5[a] | 1.16 | 1.21 | 1.20 |
| 10[b] | 1.23 | 1.22 | 1.17 |
| 60[b] | 1.20 | 1.22 | 1.18 |
| 110[b] | 1.20 | 1.21 | 1.19 |
| 160[b] | 1.20 | 1.23 | 1.18 |

[a] Adapted from Frank et al. (1984).
[b] Pettigrew et al. (1996).

2.0–3.0 mg kg$^{-1}$ would be required to meet tissue needs and to increase reproductive performance (Easter et al. 1983; Russell et al. 1985). Knights et al. (1998) reported that dietary levels of 16 (vs. 2.6 mg kg$^{-1}$) after weaning increased pyridoxine status and the apparent nitrogen retention of sows during gestation, decreased weaning to oestrus interval without any other effects on reproductive performance. In reports studying the interaction between selenium (Se) and pyridoxine in early gestation, Bueno-Dalto et al. (2015a,b and 2016) showed the importance of the supply of this vitamin for the endogenous antioxidant system glutathione peroxidase (GPx). This metabolism is particularly important in early gestation because ovarian metabolism may generate an excess of reactive oxygen species (ROS) during the peri-estrus period, which may impair ovulatory functions and early embryo development. However, later in gestation, placentation raises embryo oxygen tension and may induce a higher expression of ROS markers and eventually embryo losses. Pyridoxine is crucial for the regulation of the flow of one-carbon units leading to the transsulfuration pathway and activation of the GPx system (Bueno Dalto and Matte 2017). However, although this metabolism is fully functional in adults, here the sow, it is inactive in conceptuses and embryos even after placentation. In this line, while beneficial effects for the GPx system and ovulatory process were reported shortly after estrus, with provision of supplementary pyridoxine (10 mg kg$^{-1}$), it was not the case for embryo development and integrity later in gestation at 30 days of gestation, i.e., after placentation (Bueno Dalto and Matte 2017). A better understanding of the role of vitamin B$_6$ in the foeto-maternal connection in pig species is required for reliable assessments of the most adequate dietary levels of this vitamin.

For niacin, Ivers et al. (1993) concluded that a diet devoid of exogenous niacin and containing low concentration (0.12%) of tryptophan (precursor of niacin) was sufficient to meet niacin requirements during both gestation and lactation.

In a descriptive study, Mosnier et al. (2009) showed that niacin and pyridoxal status decreased shortly after farrowing in multiparous sows in spite of a high feed intake during the first week of lactation (6.4 kg d$^{-1}$) and dietary supplies of niacin, tryptophan, and vitamin B$_6$ at 45 mg kg$^{-1}$ 0.22% and 3 mg kg$^{-1}$, respectively. It was suggested that niacin and pyridoxine might have been transiently suboptimal in early lactation. The importance of the partition of tryptophan for niacin synthesis remains to be better understood during that period in order to make reliable estimate of niacin required for lactating sows.

*Choline*

Maternal supplementation of methyl donors including choline (400 mg kg$^{-1}$), betaine (3 g kg$^{-1}$), folic acid (15 mg kg$^{-1}$), and of vitamin B$_{12}$ (150 µg kg$^{-1}$) provided throughout gestation enhanced birthweight and postnatal [up to 110 kg of body weight (BW)] growth rate of offspring, this, in parallel with an greater expression of the IGF-1 genes and altered DNA methylation of IGF-1 gene promotor (Jin et al. 2018).

For the period of lactation, two reports (Donovan et al. 1997; Mudd et al. 2016) presented the partition of the different forms of choline in sow colostrum and milk from 12 hours pre-partum to 28 days postpartum. Most of the choline content (75–85%) in the colostrum and milk is bound to phospholipids that is likely to be highly available for the piglet. Although inadequate (517 mg kg$^{-1}$) vs adequate dietary choline (1591 mg kg$^{-1}$) did not elicit a global impact on milk nutrient composition, certain milk nutrients such as betaine, select fatty acids, and free amino acids were sensitive to maternal choline intake in sows (Mudd et al. 2016). In fact, the dietary provision of choline from sow milk will be partitioned by piglet metabolism between oxidation to betaine (remethylation) and direct utilization of the preformed vitamer (membrane integrity and neurotransmitter functions), this last role sparing endogenous synthesis of choline and reducing methylation needs. This partition of milk dietary choline may contribute to the regulation of the development of hyperhomocysteinemia observed in piglets (Ballance and House 2005; Simard et al. 2007; Côté-Robitaille et al. 2015).

*Folic Acid*

Folic acid has attracted, during the last decades, a great interest in human nutrition mainly because of its protective effects against cardiovascular disease and congenital malformations. The metabolic roles played by folic acid are closely linked to those of vitamin $B_{12}$. The vitamin is critical for the transfer of single-carbon units, which is fundamental for the synthesis of purine and pyrimidine bases and for the remethylation pathway (methionine cycle). Therefore, folic acid plays a crucial role in protein deposition and tissue synthesis. In the remethylation pathway of homocysteine to methionine, folic acid is a precursor providing a methyl group and vitamin $B_{12}$ is the enzymatic cofactor (Le Grusse and Watier 1993). Thus, one of the beneficial roles of folic acid is related to the control of homocysteine levels in the organism. This substance, an intermediary metabolite from the normal metabolism of methionine, is a powerful oxidant which, at high levels, is harmful to blood vessel integrity and normal embryo development (Pietrzik and Brönstrup 1997); the organism must maintain a level of homocysteine as low as possible. A deficiency in folic acid and/or vitamin $B_{12}$ results in local or systemic homocysteine accumulation (Bässler, 1997), which would be a key factor increasing the risks of cardiovascular disease, abortion and congenital malformations (Pietrzik and Brönstrup 1997).

In pig species, it is now well acknowledged that providing folic acid supplements to gestating sows increases prolificacy by roughly 10% (Matte et al. 1984; Kovčin et al. 1988; Lindemann and Kornegay 1989; Thaler et al. 1989; Friendship and Wilson 1991; Lindemann 1993; van Wettere et al. 2013). Supplements given during lactation were also efficient in increasing postnatal growth of piglets (Matte et al. 1992; Wang et al. 2011) possibly through a higher transfer of the vitamin through colostrum and milk (Barkow et al. 2001). Such a response was also obtained by Jin et al. (2018), who supplemented folic acid with choline. The effect of folic acid on prolificacy is probably due to the decrease in embryo mortality during the first month of gestation (Tremblay et al. 1989). Folic acid likely acts at two levels: directly on embryo development (DNA, proteins, and estrogen secretion) (Matte et al. 1996; Guay et al. 2002a) and indirectly by stimulating the uterine secretion of growth promoters such as the cytokine transforming growth factor $\beta_2$ (TGF$\beta_2$) and eicosanoids beneficial to the acceptance of embryos by the uterus such as prostaglandin $E_2$ (PGE$_2$) (Matte et al. 1996; Giguère et al. 2000; Guay et al. 2004a; Guay et al. 2004b). However, it is also likely that the embryonic and uterine responses to folic acid also depend on animals' parity (Matte et al. 2006). In fact, responses to dietary folic acid, both in terms of prolificacy (Lindemann and Kornegay 1989) and uterine secretions of PGE$_2$ (Duquette et al. 1997) and TGF$\beta_2$ (Guay et al. 2004b), are more marked in multiparous sows than in nulliparous sows (Table 7.6). Several hypotheses have been raised to explain this parity-based difference in response. It is now believed that the mitigated effects of folic acid in gilts is probably linked to the limiting metabolic availability of another vitamin mentioned earlier, $B_{12}$ (see the next section for more specific details on vitamin $B_{12}$), which would interfere with homocysteine homeostasis. According to Matte et al. (2006), it appears that at least part of the parity effect of folic acid on sow physiology and reproductive performance (Table 7.5) is related to homocysteine homeostasis. In terms of physiology, it is supported by the importance of this intermediary amino acid in (i) altering trophoblast integrity (DiSimone et al. 2004; Kim et al. 2009) and (ii) favoring extracellular release of arachidonic acid (C20:4n − 6) and then inducing an intracellular depletion of this precursor of PGE$_2$ and accumulations of thromboxane B2 (TXB2) and reactive oxygen species (ROS), leading to an unbalance in the redox state of cells (Signorello et al. 2002). In terms of performance, the report of van Wettere et al. (2013) using folic

**Table 7.6** Prolificity at parturition and embryo mortality and composition of allantoic fluid at 30 days of gestation according to parity and dietary supplements of folic acid during gestation[a].

| Item | Sow:<br>Supplement of $B_9$: | Nulliparous | | Multiparous | |
|---|---|---|---|---|---|
| | | − | + | − | + |
| Piglets born alive per litter | | 9.1 | 9.3 | 11.5 | 13.5 |
| Embryonic mortality %) | | 14.4 | 12.8 | 39.2 | 32.6 |
| Total allantoic $PGE_2$ (ng) | | 1574 | 1750 | 1890 | 2318 |
| Total allantoic $TGF\beta_2$ (ng) | | 81.1 | 66.0 | 66.7 | 138.1 |

[a] From Matte et al. (2006).

acid (5 vs. 20 mg kg$^{-1}$) and vitamin $B_{12}$ (20 vs 150 µg kg$^{-1}$) supplementations with an extensive population of sows (n = 1079) with parity ranking and homocysteine data is clearly supportive of the concept of a parity-dependent response to folic acid and vitamin $B_{12}$ supplementations related to homocysteine homeostasis.

In experiments studying the effects of folic acid fortifications (30 vs. 1.3 mg kg$^{-1}$) throughout gestation on early life (28 days of age), metabolism of intrauterine growth retarded (IUGR) suckling piglets, reported that, in high- vs low-supplemented litters, some aspects of hepatic one-carbon metabolism were improved in IUGR piglets (Liu et al. 2011), whereas their altered antioxidant capacity and lipid metabolism along with their gene-related expression were normalized to the level of normal birth weight piglets (Liu et al. 2012, 2014). Liu et al. (2014) associated their treatment responses to the regulation of DNA methylation of promoter by folic acid supplementation during pregnancy. Nevertheless, in terms of BW at birth or at weaning, there was no dietary treatment effect either on IUGR or normal birth weight piglets. The fetal programming and lactocrine consequences of those responses on metabolism of IUGR pigs need to be further investigated.

In terms of optimal dietary concentrations of folic acid, a whole body balance approach with 5 supplement levels ranging from 0 to 20 mg kg$^{-1}$ was used by Matte and Girard (1999) who showed that the metabolic use of folic acid in pregnant sows would be optimal at approximately 10 mg kg$^{-1}$ in sows with litter sizes of 12–13 live-born piglets.

*Vitamin $B_{12}$*

The role of vitamin $B_{12}$ in the reproductive process is poorly known. Vitamin $B_{12}$ requirements are studied in a different way than other water-soluble vitamins since the withdrawal of the vitamin supplement or of other vitamin $B_{12}$ source (animal-origin ingredients only) in the feed must be done over a prolonged period (at least a year) to induce symptoms of deficiency such as abortion shortly before farrowing (Ensminger et al. 1951; Cunha et al. 1944; Frederick and Brisson 1961). As mentioned earlier, recent studies on folic acid have identified a possible influence of homocysteine on the response of the reproductive function to folic acid, but they have also raised concerns about vitamin $B_{12}$, which has been neglected in swine nutrition during the last 30–40 years. In fact, the last NRC publication (1998) reported that there have been few experiments on vitamin $B_{12}$ in reproducing sows since the 1940s (Anderson and Hogan 1950; Frederick and Brisson 1961; Teague and Grifo 1966). Guay et al. (2002a) showed that fasting concentrations of plasma $B_{12}$ in nulliparous sows were two to three times lower than in multiparous sows. This low $B_{12}$ status in nulliparous sows is probably linked to the fact that, as for other micronutrients, $B_{12}$ needs for reproduction are in competition with requirements for growth and maintenance. Once maturity is reached, more of the $B_{12}$ supply becomes available for the reproductive function. Furthermore, $B_{12}$ requirements for the reproductive function appear to be particularly large in early pregnancy considering the massive transfer of vitamin $B_{12}$ to the uterus at that time (Guay et al. 2002b). Indeed, the total vitamin $B_{12}$ concentration in uterine horns at 15 days of gestation represents between 180 and 300% of the total concentration in plasma, with the total volume of plasma being estimated at 4% of body weight (Matte and Girard 1996). The physiological mechanism underlying this transfer is not well understood in pigs although one of the known $B_{12}$ carriers, transcobalamin I, has been identified in porcine endometrial tissues (Pearson et al. 1998). Moreover, Guay et al. (2002b) showed that a dietary supplement of 160 ppb similar to that used by Frederick and Brisson (1961) could be an efficient way to increase $B_{12}$ status as measured by the variation of plasma $B_{12}$ in early gestation (up to day 15). Nevertheless, an optimal level of dietary $B_{12}$ needs to be estimated considering the limited information that was used to establish the present estimate of requirements at 15 ppb (NRC 1998). In this respect, the statement made by ARC (1981) is quite evocative: "The proposed figure of $15\,\mu g \cdot kg^{-1}$ of dietary DM, must be considered very tentative and may need to be increased tenfold or more according to the results (. . .)." More recently, attempts have been made to estimate the optimal level of dietary $B_{12}$ for nulliparous sows (Simard et al. 2007). The dietary concentrations of cyanocobalamin that maximized plasma vitamin $B_{12}$ and minimized plasma homocysteine of sows during gestation were estimated at 164 and $93\,\mu g \cdot kg^{-1}$, respectively. The biological significance of such concentrations of cyanocobalamin needs to be validated with performance criteria using large numbers of animals during several parities. The supplementation of vitamin $B_{12}$ during gestation also affected the transfer of this vitamin to the piglets, which occurred both in utero and through colostrum intake (see Table 7.3), and it prevented accumulation of homocysteine in piglets during lactation (Simard et al. 2007).

## Vitamins and Growth in Pigs

As mentioned previously, the duration of the postweaning period in modern pig production corresponds roughly to about half the life existence of a pig (from fertilization of ova to slaughter). A logical first step for a reliable estimation of the optimal provision of vitamins for the early postweaning period could be to establish, on a standard basis (per day and per kilogram of BW), the provision of vitamins brought by colostrum and milk during lactation as compared to the postweaning dry feed. Concentrations of some vitamins (as fed basis) in colostrum, milk, and dry feed are presented in Table 7.7 along with an estimation of the daily provision of those vitamins to piglets per kilogram of BW. Vitamins are more concentrated by 1.4 (vitamin $B_{12}$) to 4 (vitamin E) times in colostrum than in milk. This cannot be explained entirely by differences in total solid contents that are approximately 20–25% lower in milk than in colostrum (Pond and Houpt 1978; Yang et al. 2009). For the daily provision of those vitamins to piglets per kilogram of BW, the importance of colostrum intake is critical bringing between 3.6 and 18.3 times more vitamins $B_{12}$ and E, respectively, for the newborn piglets than milk, later in lactation. As mentioned earlier, the transfer of vitamin from dams to piglets via the colostral route is a crucial because the lack of efficiency for prenatal transfer of some vitamins (Matte and Audet 2020). Therefore, one can argue that the colostral provision would correspond to the maximum amount required by a pig during his ex-utero life.

For the postweaning feed, the dietary vitamin level recommended by NRC (2012) corresponded, in terms of daily provision per kilogram BW, to values generally lower (except for D, $B_9$ and $B_3$) than for colostrum but generally higher (except for A, D, $B_2$, and $B_8$) than for milk (Table 7.7). The average dietary vitamin levels currently used in the USA swine industry (Flohr et al. 2016) provided more vitamins (per day per kilogram of BW) than both milk and colostrum (except for A, E, $B_2$, and

**Table 7.7** Vitamin concentrations in colostrum, milk, and postweaning dry feed and corresponding daily provisions to piglets per kilogram body weight (BW).

| Vitamin | Colostrum[a] Amount (/l) | Colostrum[a] Provision to piglets (/d/kg BW)[c] | Milk[a] Amount (/l) | Milk[a] Provision to piglets (/d/kg BW)[c] | Postweaning feed as recommended by NRC (2012) Amount (/kg) | Postweaning feed as recommended by NRC (2012) Provision to piglets (/d/kg BW)[c] | Postweaning feed (currently used according to Flohr et al. (2016) Amount (/kg) | Postweaning feed (currently used according to Flohr et al. (2016) Provision to piglets (/d/kg BW)[c] |
|---|---|---|---|---|---|---|---|---|
| A (mg) | 2.5[a] | 0.77 | 0.7 | 0.09 | 0.67 | 0.03 | 3.6 | 0.18 |
| E (mg) | 13.2[a] | 4.1 | 2.5 | 0.31 | 16 | 0.8 | 71.3 | 3.6 |
| D (µg) | 0.53[a] | 0.16 | 2.7 | 0.33 | 5.5 | | 47.8 | 2.4 |
| C (mg) | 24.7[a] | 7.6 | 11.1 | 1.4 | N/A[d] | N/A[d] | N/A[d] | N/A[d] |
| $B_9$ (µg) | 44.6[a] | 13.7 | 13.4 | 1.6 | 300 | 15.0 | 1800 | 90.0 |
| $B_{12}$ (µg) | 6.1[a] | 1.9 | 6.1 | 0.75 | 17.5 | 0.88 | 46 | 2.3 |
| $B_2$ (mg) | 4.4[b] | 1.4 | 2.2[e] | 0.27 | 3.5 | 0.18 | 9.7 | 0.49 |
| $B_3$ (mg) | 0.56[b] | 0.17 | 7.0[e] | 0.86 | 30.0 | 1.5 | 51.3 | 2.6 |
| $B_6$[b] (mg) | 4.4[b] | 1.4 | 1.7[e] | 0.21 | 7.0 | 0.35 | 4.0 | 0.2 |
| $B_8$[b] (µg) | 23.5[b] | 7.21 | 46[e] | 5.7 | 50.0 | 2.5 | 370 | 18.5 |

[a] Adapted from Hakansson et al. (2001) for vitamin A; from Håkansson et al. (2001), Loudenslager et al. (1986), Pinelli-Saavedra and Scaife (2005), and Pinelli-Saavedra et al. (2008) for vitamin E; Amundson et al. (2017) for vitamin D; Pinelli-Saavedra and Scaife (2005) and Yen and Pond (1983) for vitamin C; Barkow et al. (2001) and Matte and Girard (1989) for vitamin $B_9$; and Simard et al. (2007) and Audet et al. (2015) for vitamin $B_{12}$.
[b] Adapted from Matte and Audet (2020).
[c] Based on a colostrum daily intake of 0.43 l and 1.4 kg BW at 0–1 d of age, a milk daily intake of 0.8 l and 6.5 kg BW at 14–21 d of age, and a feed daily intake of 0.5 kg and 10 kg BW at 21–35 d of age.
[d] N/A = data not available. However, a dietary supplement at 1 g kg⁻¹ would correspond to 50 mg d⁻¹ kg⁻¹ BW.
[e] Adapted from Davis et al. (1951) for vitamins $B_2$ and $B_3$; Côté-Robitaille et al. (2015) for vitamin $B_6$, and from Bryant et al. (1985) for vitamin $B_8$.

$B_6$ which are still below the maximal colostral level). In addition, some countries such as the Danish pig production officially recommend higher levels of some specific vitamins, for example, the recommendation of vitamins E and D for weaners is 140 and 800 IU/kg feed, respectively (SEGES, Svineproduction 2019).

Some reports have been published on growth performance and various criteria responses to dietary provision of "multi-vitamins." The dietary vitamin provision (riboflavin, niacin, pantothenic acid vitamin $B_{12}$, and folic acid) needed to optimize some aspects of growth performance were higher (>470% of NRC 1998) in high lean than in moderate lean strain of pigs between 9 and 28 kg of BW (Stahly et al. 2007). More recently, it has been reported that provision of these 5 B-vitamins (>270% of NRC 1998) for crossbred conventional pigs is beneficial for several aspects of growth performance during the three growth phases corresponding to 10–20 kg, 20–50 kg, and 50–105 kg of BW (Cho et al. 2017). These last authors mentioned that responses are not consistent (see also Zhang et al. 2013) and may depend upon several factors related to basic vitamin levels in ingredients, basic vitamin status of animals, and conditions of husbandry. In fact, other studies reported that during the growing-finishing period (33–110 kg of BW), supplements of vitamins (riboflavin, pyridoxine, pantothenic acid, and biotin) at 400 and 800% of GfE (1987) or NRC (1998) did not influence performance but enriched pork meat by 37, 58, and 129% for vitamin $B_6$, pantothenic acid, and biotin, respectively (Böhmer and Roth-Maier 2007).

### *Fat-Soluble Vitamins and Vitamin C*

Although the risks of vitamin deficiency are practically nonexistent today, it should be noted that the postweaning period may be critical especially for fat-soluble vitamins due to limited fat absorption capability. The apparent digestibility of fat by suckling piglets is high (96%; Cranwell and Moughan 1989), but at weaning it decreases to 65–80% (Cera et al. 1988). Furthermore, weaning age has declined over time in order to increase sow productivity, and the European ban on antibiotic growth promoters are factors of importance for the development of postweaning stress in pigs. Besides, during the postweaning phase, there is a greater relative growth response for weanling pigs, particularly for muscle tissue, thus requiring a higher need for vitamins. It should be noted that the antioxidant activity of especially vitamins E and C calls for special attention in relation to the interaction with other antioxidants (selenium), potential prooxidants (iron and copper and unsaturated fatty acids), both in relation to the dietary supplementation of weaners and growers; however, these issues will be addressed elsewhere.

### *Vitamin A*

Swine is able to store vitamin A in the liver, which makes the vitamin available during periods of low intake. Requirements for vitamin A depend on the criteria evaluated; weight gain is less sensitive than cerebrospinal fluid pressure, liver storage, or plasma levels. For pigs during the first eight weeks of life, 75–605 µg retinyl acetate/kg of diet is required depending on response criteria used. In growing-finishing pigs, the requirement varies from 35 to 130 µg retinyl acetate/kg when daily gain was used as criterion, and from 344 to 930 µg retinyl acetate/kg, when liver storage and cerebrospinal fluid pressure was used as the criterion (NRC 2012). However, vitamin A fortification in commercial swine starter diets is generally added in excess of NRC (1998, 2012) standards. In weaners, no differences between three levels of vitamin A (2200; 13 200, or 26 400 IU/kg) were obtained on performance with regard to a 35 days postweaning period (Ching et al. 2002). In addition, no effect on the performance of growers was obtained with high doses of vitamin A (10 000, 20 000, or 40 000 IU/vitamin A/kg feed) in comparison with the control treatment at 5000 IU/kg feed (Hoppe et al. 1992). It should be noted that high dietary levels of vitamin A could have detrimental effect on the young pig's vitamin E status during the postweaning period and could be detrimental to the antioxidant status (Ching et al. 2002). Symptoms corresponding to hypervitaminosis A (sudden lameness) appeared in pigs three to four days after introduction of a feed with a vitamin A-content ten times higher than the claim (195 000 IU/kg) (Reiner et al. 2004).

As mentioned before, there is a poor absorption of intact β-carotene in the pig. However, β-carotene plays an important role in immunoregulation and may therefore call for special attention with regard to weaners, whose immune system may be depressed. It has been shown that concentrations of plasma immunoglobulins (IgG) were higher among piglets born from gilts injected with β-carotene (Brief and Chew 1985) although concentrations of immunoglobulins in colostrum were not changed. Increased mitogen proliferation has been reported in pigs supplemented with β-carotene (Hoskinson et al. 1992). In pigs (50–55 kg) injected once with 0, 20, or 40 mg of β-carotene, the vitamer was found in all subcellular fractions of lymphocytes with highest concentration in nuclei, intermediate in mitochondria and microsomes, and lowest in cytosol (Chew et al. 1991a). The treatment did not alter concentrations of retinol or α-tocopherol in plasma (Chew et al. 1991b). Supplementation of β-carotene did not appear to influence the serum IgG concentration in sows and piglets (Kostoglou et al. 2000). The role of β-carotene supplementation on the immune function of pigs remains to be elucidated.

*Vitamin D*

The fact that the piglet is born with the lowest plasma concentration of 25-OHD$_3$ may predispose piglets to neonatal rickets, a condition of vitamin D deficiency associated with retarded skeletal growth and myopathy. Vitamin D deficiency reduces retention of calcium, phosphorus, and magnesium (Miller et al. 1965). Thus, in mature swine, a mild deficiency reduced bone mineral content (osteomalacia), whereas pigs may exhibit signs of calcium and magnesium deficiency, including tetany, after a severe vitamin D deficiency. Bethke et al. (1946) and NRC (2012) suggested a minimum requirement of 150–200 IU/kg diet for pigs during the weaning and growing pigs (Table 7.2), but the official British (BSAS 2003) recommendations are generally higher varying from 800 (for pigs until 60 kg live weight) and 600 IU (from 60 to 90 kg live weight). Danish swine production recommendations (Norms for Nutrients. 2019. 10th ed. National Committee for Pigs, Copenhagen, Denmark) vary from 800 (for piglets, approximately 6–9 kg) to 500 (for piglets, approximately 9–30 kg) and 400 IU (for pigs from 25 to 100 kg live weight). Some guidelines have incorporated the added value of increased bioavailability of vitamin D when using 25-hydroxyvitamin D3 instead of vitamin D3 for enhancing the vitamin D status of the pig (Lauridsen 2014), and the potency of various doses of the 25-hydroxyvitamin D-3 with regard to biological responses such as growth performance, immune function and antioxidative capacity in piglets post weaning (Konowalchuk et al. 2013; Yang et al. 2019). Since the active form of vitamin D, 1,25(OH)$_2$D$_3$, is dependent on the presence of 25(OH)D3, and the fact that pigs are in general low in vitamin D status at birth and weaning, other than nutritional strategies have been investigated to increase the concentration of serum and plasma 25(OH)D3 of piglets, and to enrich pork with vitamin D. In brief, these methods have been injection (Matte et al. 2017) or administration of vitamin D3 to the drinking water (Jang et al. 2018) and use of UVB light (Matte et al. 2017; Kolp et al. 2017; Barnkob et al. 2019) or sunlight (Larson-Meyer et al. 2017) to weaned, growing, or finishing pigs.

The efficacy of feeding high amounts of vitamin D$_3$ to finishing pigs to increase muscle calcium and subsequently to improve meat tenderness has also been investigated. Wiegand et al. (2002) fed 250 000 or 500 000 of vitamin D$_3$ per day to finishing pigs three days prior to slaughter and although the high dietary supplementation resulted in elevated plasma calcium concentrations, no improvement was observed on pork tenderness. However, supranutritional levels of 40 000 or 80 000 IU of supplemental vitamin D$_3$ per kilogram of diet (as fed basis) for at least 44 days improved pork color and increased pH, but muscle calcium concentrations were not affected (Wilborn et al. 2004). Likewise, Lahucky et al. (2007) investigated the effect of 500 000 IU vitamin D$_3$/day for five days before slaughter for growing-finishing pigs and observed a higher plasma calcium concentration and some effects on the redness of muscle. In addition to the effect of feeding supplemental vitamin D on pork, the impact on the nutritional quality of the meat has been studied. A study by Jakobsen et al. (2007) identified a dose–response effect between the total concentration of vitamin D (i.e., the sum of vitamin D$_3$ and its metabolite, 25-hydroxyvitamin D$_3$) in the meat and in liver with the total dietary concentration in the feed. The combination of supplementing finisher pigs with both phytase and 25-OH-D$_3$ was recently addressed (Duffy et al. 2018),; however, no benefit was obtained on pig performance, bone parameters, or pork quality.

Besides toxicity resulting from excess vitamin A, vitamin D is the vitamin most likely to be toxic for both humans and livestock. Excessive intake of vitamin D produces a variety of effects, all associated with abnormal elevation in blood calcium. The estimated upper dietary level of vitamin D$_3$ in swine has been estimated at 33 000 and 2200 IU/kg diet depending on the exposure time, less and above 60 days, respectively (McDowell 2000).

*Vitamin E*

The early postweaning period is a critical state for vitamin E as a nutrient for growth and health. As tocopherols (vitamin E) are absorbed in the small intestine as free alcohols alone or in combination with emulsified fat products, the commercial available all-rac-α-tocopheryl acetate must be hydrolyzed before absorption, a process that may be limited in weaned pigs (Chung et al. 1992; Hedemann and Jensen 1999, Lauridsen et al. 2001). Although piglets nursing sows fed adequate or increased dietary vitamin E had higher reserves upon weaning (Mahan 1991, 1994; Lauridsen and Jensen 2005), the tissue (heart and muscle) tocopherol declines rapidly during the postweaning period (Lauridsen and Jensen 2005). However, the concentration of α-tocopherol in liver did not change in piglets from 35 to 49 days of age. Previous studies have indicated that the porcine liver has a very high short-term storage capacity for α-tocopherol (Jensen et al. 1990; Lauridsen et al. 2002b). Thus, the decrease observed in other tissues (heart and muscle) may indicate that a dietary level of 70 IU/kg of all-rac-α-tocopheryl acetate in the weaner diet or the bioavailability of the all-rac-α-tocopheryl acetate source was insufficient to induce exportation of α-tocopherol from the liver to other tissues (Lauridsen and Jensen 2005). In a subsequent experiment with weaners fed increasing levels of different dietary all-rac-α-tocopheryl acetate (85, 150, and 300 mg all-rac-α-tocopheryl acetate/kg diet) at a 5% dietary fat level of either animal fat, sunflower oil, or fish oil, it was observed that the concentration of α-tocopherol in serum decreased during the first week after weaning and that dietary vitamin E supplementation had no influence on plasma concentration before day 42 of age (Lauridsen 2010). Thereafter, serum α-tocopherol concentration reflected the dietary vitamin E level and stayed between 1.5 and 2.0 mg l$^{-1}$ when piglets were fed 150 and 300 mg all-rac-α-tocopheryl acetate/kg diet (Lauridsen 2010). A plasma or serum concentration of 1.5 mg to 2.0 mg l$^{-1}$ may be used as a guideline for establishing a satisfactory vitamin E status of the pig (Wilburn et al. 2008). In addition, a long-term high-dose enriched diet (250 mg kg$^{-1}$) was recommended for improving oxidative status of piglets in the postweaning period (Rey et al. 2017).

Several studies have been performed with the overall purpose of improving pork quality through dietary vitamin E supplementation (of review see Jensen et al. 1998). Lipid oxidation is one of the primary processes of quality deterioration in meat and meat products. The changes in quality are manifested by adverse changes in flavor, color, texture, and nutritive value and by possible production of toxic products. Dietary supplementation of vitamin E above requirement levels has been found to be effective in reducing lipid oxidation in meat and meat products. The amount of vitamin E accumulated depends on muscle characteristics, supplementation level, and duration of supplementation (Jensen et al. 1998). It should be noted that vitamin E toxicity has not yet been demonstrated in swine, and dietary levels as high as 550 mg kg$^{-1}$ (Bonette et al. 1990) and 700 mg kg$^{-1}$ (Jensen et al. 1997) have been fed to growing pigs without toxic effects. Actually, deposition of α-tocopherol does not reach saturation levels in muscle tissue of pigs fed 700 mg α-tocopheryl acetate (Jensen et al. 1997). The major benefit attained is protection against oxidative changes, thereby improving storage stability. Other meat quality parameters are affected such as meat color and drip loss, though, the efficiency of supplemental vitamin E to control color deterioration in pork meat varies considerably (Jensen et al. 1998). Relationship between endogenous vitamin E and drip loss reduction has been shown in pre-frozen (Asghar et al. 1991; Monahan et al. 1994) and raw (Cheah et al. 1995) pork, although the mechanism behind needs to be fully explained (Jensen et al. 1998). Subsequently, it has been shown that supranutritional vitamin E supplementation in a strategic finishing diet formulated to reduce muscle glycogen levels amplified effects on glycogen levels and tended to reduce the water binding capacity of meat as compared to a control diet (Rosenvold et al. 2002). It has in addition been shown that drip losses from porcine meat were reduced by elevating the dietary inclusion levels of vitamins and trace minerals 150–200% of the recommended levels (Apple 2007). Other antioxidants and antioxidant blends with vitamin E have been investigated for their potential to attenuate oxidative stress reactions leading to impaired carcass and pork quality (see for instance Lu et al. 2014).

*Vitamin C*

Vitamin C is not routinely added to pig feed because pigs are capable of synthesizing vitamin C within one week of age (Braude et al. 1950). However, during stressful situations such as weaning, the presence of l-gulono-γ-lactone oxidase (GLO), a critical enzyme for the biosynthesis for vitamin C, might be low (Ching et al. 2001). In addition, upon weaning a decline in plasma ascorbic acid (Yen and Pond 1981, Mahan and Saif 1983) has been observed, which may have been an indication of inadequate synthesis of the vitamin or stress associated with weaning. Mahan (1994) observed improved performance in weanling pigs fed a stable source of vitamin C during the first two weeks after weaning. De Rodas et al. (1998) evaluated the efficacy of a stable source of vitamin C for improving performance and iron status in early weaned pigs (14 days of age). It was concluded that 75 ppm L-ascorbyl-2-poly-phosphate was adequate to meet the dietary vitamin C of early-weaned pigs. In a study reported by Fernandez-Duenas et al. (2008), supplementation of vitamin C at 150 mg kg$^{-1}$ and/or β-carotene at 350 mg kg$^{-1}$ had no effect on animal performance or antioxidant status.

The interaction between vitamin E and C is particularly relevant due to the synergism between the two vitamins, a sparing action of vitamin C on vitamin E (as reviewed by Burton et al. 1990). This is illustrated by the fact that the highest response of α-tocopherol in immune cells (Figure 7.1), liver, and muscle tissue in weanling piglets after dietary vitamin E was observed when vitamin C was included in the weanling feed at 500 mg of vitamin C per kilogram of feed (Lauridsen and Jensen 2005). In ddition, the vitamin C supplementation caused an increase in the relative contribution of the RRR-α-tocopherol in the immune cells of the piglets on the expense of the RRS-tocopherol and consistently increased the concentration of IgM throughout the weaner period compared with piglets with no added dietary vitamin C. In the study by Zhao et al. (2002), plasma levels of IgG in weanling piglets showed a linear increase with increasing levels of vitamin C

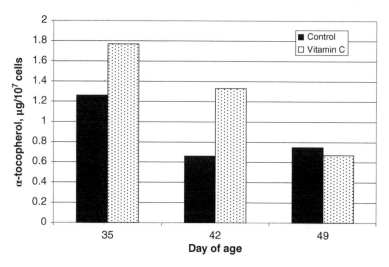

**Figure 7.1** Concentration of α-tocopherol in immune cells of weanling piglets fed control (without vitamin C supplementation or with 500 mg vitamin C/kg feed. Source: Adapted from Lauridsen and Jensen (2005).

(up to 300 mg vitamin C per kilogram feed). Whether or not supplemental vitamin C plays a role in stimulating immune function in pigs requires further studies.

The interaction of vitamins C and E is also important in relation to meat quality. When 300 mg vitamin C/kg or 200 mg dl-α-tocopheryl acetate, or a combination of both vitamins, was fed to barrows from 25 to 106 kg live weight, some synergistic actions between the two vitamins were observed in terms of enhanced vitamin E concentration in all investigated tissues (except ham) on the dietary treatment with vitamin C alone (Eichenberger et al. 2004). However, the oxidative stability measured as TBARS was not influenced by dietary vitamin C supplementation, whereas it was improved with the dietary vitamin E supplementation. In another study (Pion et al. 2004), on-farm vitamin C supplementation to drinking water at 0, 1000, and 2000 mg l⁻¹ for 48 hours before slaughter did not influence pork quality; however, the importance of the timing of the supplementation relative to slaughter needs to be investigated.

Ascorbic acid is also essential for the hydroxylation of proline and lysine, which are integral constituents of collagen. Collagen is essential for the growth of cartilage and bone. Vitamin C enhances the formation of both bone matrix and tooth dentin. The role of vitamin C in the prevention or alleviation of osterochondrosis in swine has been investigated since osteochondrosis could be related to insufficient collagen cross-linking because of reduced hydroxylation of lysine (Nakano et al. 1983; NRC 1998). However, dietary supplementation with vitamin C was ineffective in preventing osteochondrosis. In a study with 47 day-old-pigs receiving 1500 or 1000 mg ascorbic acid/kg diet for four months, no influence on bone formation marker (except osteocalcin) and various plasma and urine indices of bone metabolism (Pointillart et al. 1997). Based on this long-term study, the authors concluded that high intakes of ascorbic acid had no positive influence on bone metabolism or bone characteristics in pigs. Furthermore, in a study with a high prevalence of foreleg lesions, no association of low levels of vitamin C in pigs and incidence of foreleg defects could be found (Armocida et al. 2001).

*Vitamin K*

The pig's (weanling and growing) requirement for vitamin K is estimated to be 0.5 mg kg⁻¹ diet. Dietary vitamin K may be provided as the natural forms; $K_1$ (phylloquinone) and $K_2$ (menaquinone-4), but these forms are costly. Therefore, the synthetic forms, $K_3$ (menadione) or its derivative, menadione sodium bisulfate (MSB), are used as vitamin K sources, and additionally Marchetti et al. (2000) also concluded that the synthetic form, menadione nicotinamide bisulfite (MNB) is a good source of vitamin K for the pig. Vitamin K is required for proper blood coagulation. It is necessary for the conversion of four so-called clotting factors (II, VII, IX, X) in blood from their precursor forms into their active forms. Hence, in deficiency, clotting factors remain inactive resulting in prolonged bleeding time. Diagnosis is based on the estimation of prothrombin time. Liver stores of vitamin K can be depleted very rapidly during even very short periods of vitamin K-deficient diet consumption (Kindberg and Suttie 1989). According to NRC (1998), vitamin K levels up to 1000 times the requirement are tolerated by the animals. Vitamin K deficiency increases prothrombin and clotting times and may result in internal hemorrhages and death (NRC 1998). Some factors, which interfere with blood clotting, may increase the pig's requirement for vitamin K. These are excess calcium (Hall et al. 1991) and mycotoxin or mold contaminated feed ingredients (Hoppe 1988; NRC 1998). It is not known whether a high level of dietary calcium reduced the synthesis of vitamin K by intestinal microbes, reduces the absorption of vitamin K from the gut, or destroys the activity of vitamin K. Further antagonistic effects on vitamin K in pigs may, besides mouldy corn, be exerted by antibiotics and imbalance in fat-soluble vitamins. There appears to be a tissue-specific interaction between vitamins E and K, when vitamin E is supplemented to rat diets (Tovar et al. 2006), since phylloquinone concentrations were lower in the vitamin E supplemented compared to the vitamin E restricted group. In addition, the four natural occurring tocopherols and β-carotene tended to cause relatively strong and weak, respectively, haemorrhage effects with regard to prothrombin and partial troboplastin time indicies (Takahashi 1995).

## Water-Soluble Vitamins

### Thiamin and Biotin

The number of reports specifically related to thiamine for pigs is very limited since both ARC (1981) and NRC (1998, 2012). Woodworth et al. (2000) showed that dietary supplementation of thiamin up to $5.5\,mg\,kg^{-1}$ (2.5–5 times the ARC (1981) and NRC (1998) levels, respectively) had no effect on growth performance of weanling pigs (5–25 kg of body weight) as compared to the unsupplemented corn-soybean meal-dried whey-based diet. A case report was released recently (Hough et al. 2015) for postweaned pigs in relation to a neurological defect identified as polioencephalomalacia which occurred in several herds in USA in 2012 and responded to administration of vitamins preparations only when thiamin was included. According to the authors, this apparent transient thiamin deficiency was probably related to a feed manufacturing problems. Although this case report illustrates the importance of an adequate levels of this vitamin for young modern pigs, the level of thiamin in an basal diet (in particular organic) containing a variety of cereals and home-grown (Germany) ingredients is sufficient to cover thiamin requirements of all classes of pigs (Witten and Aulrich 2018, 2019).

For biotin, the most known effect of supplemental biotin in swine diet is related to its role in maintenance of hoof integrity and in increasing hooves' resistance to lesions. According to the recommendations of NRC (1998) and INRA (1984), biotin requirements range from 50 to $100\,\mu g\,kg^{-1}$. Although NRC (1998) stated that "A considerable portion of the pig's biotin requirement is presumed to come from bacterial synthesis in the gut," it is likely to be unavailable for the pig since the main site of intestinal absorption is the small intestine (Mosenthin et al. 1990). Therefore, for biotin, the dietary provision is particularly critical; this is all the more important that, in contrast to other vitamins such as folic acid and vitamin $B_{12}$, the importance of the enterohepatic cycle for biotin homeostasis is negligible, biotin biliary excretion representing less than 2.2% of the dietary biotin intake (Zempleni et al. 1997).

Kopinski and Leibholz (1989), in an elegantly designed study on requirements in growing-finishing pigs, suggested that the highest level of the range $50–100\,\mu g\,kg^{-1}$ was required to effectively prevent hoof lesions. Also, levels of $100\,\mu g\,kg^{-1}$ optimized biotin concentrations in various organs (Figure 7.2). Such responses could be a useful criterion for determining biotin requirements in swine.

In another study, Partridge and McDonald (1990) demonstrated a trend toward the improvement of food conversion efficiency in pigs receiving a supplement of $500\,\mu g\,kg^{-1}$ between 15 and 88 kg of body weight. According to these authors, the effect may be caused by the more efficient metabolic use of polyunsaturated fatty acids in animals receiving the biotin supplements. The process of fatty acid elongation requires the participation of two-carbon molecules such as acetate and malonate, with malonate being 20–30 times more effective in long-chain fatty acid elongation than acetate (Roland and Edwards 1971). Malonate synthesis requires the presence of biotin (Watkins 1989; Watkins and Kratzer 1987). Because dietary supplements of biotin may modify the fatty acid profile in pigs (Martelli et al. 2005), it seems likely that the type of fats in the carcass could be, at least partly, related to the animal's biotin status.

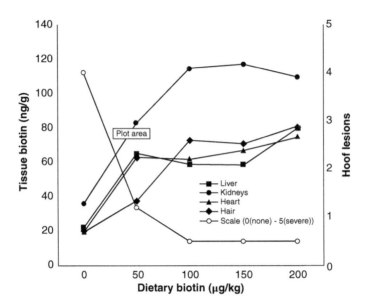

**Figure 7.2**   Tissue concentrations of biotin in different organs and severity of hoof lesions according to the level of dietary biotin. Source: Adapted from Kopinski and Leibholz (1989).

*Vitamin $B_6$, Riboflavin, and Niacin*

In piglets, pyridoxine concentrations in plasma are very low at weaning (Matte et al. 1997, 2001), probably due to the fact that sow's milk is a poor source of vitamin $B_6$, around $0.40 \mu g \ ml^{-1}$ according to Benedikt et al. (1996). This amount would represent roughly half the daily requirement needed for growth in piglets (Coburn 1994). After weaning, the deficit is exacerbated by the fact that the *in vivo* interconversion and oxidation of amino acids increases substantially since the protein content in feed is greater and less adequately balanced (in terms of amino acids) as compared to sow's milk. Pyridoxal-5-P (one of the active metabolites of pyridoxine) is an essential enzyme cofactor in these protein metabolism reactions (Le Grusse and Watier 1993). The metabolic use of pyridoxal-5-P is dependent upon the growth rate of the young piglet (Matte et al. 1997, 2001).

Several aspects of tryptophan metabolism are dependent upon pyridoxine (vitamin $B_6$): the tryptophan oxidation pathway releases alanine (a glucogenic amino acid), acetyl-CoA (total oxidation), or nicotinamide nucleotides (Le Grusse and Watier 1993). Following an inadequate provision of $B_6$, intermediate metabolites of tryptophan oxidation (such as kynurenine (Kyn) and xanthurenic acid) may chronically accumulate in circulation and impair tryptophan uptake into the brain by competing for the carriage of this amino acid through the blood-brain barrier and hence reduced serotonine synthesis (cited by Bender 1987). The synthesis of this amine that is also derived from the $B_6$-dependent catabolism (decarboxylation) of tryptophan can be stimulated, in rats, by dietary $B_6$ supplements (Hartvig et al. 1995), but the effects are more marked after concomitant supplements of $B_6$ and tryptophan (Lee et al. 1988). In $B_6$-supplemented piglets, it has been shown that plasma disappearance of tryptophan increased along with a concomitant transient production of Kyn after a gastric load of tryptophan (Matte et al. 2001) suggesting that the oxidation pathway of tryptophan can be stimulated by $B_6$. However, no information is available on dietary $B_6$ and serotonin in pigs.

In early-weaned piglets, impacts of interactions between vitamin $B_6$, tryptophan, and other nutrients (riboflavin) have been investigated in relation to glucose homeostasis. (Matte et al. 1997, 2001, 2005, 2016). In such case, either the decarboxylation of tryptophan to serotonin, a $B_6$-dependent enzymatic reaction, or the role of pyridoxine in tryptophan oxidation toward the synthesis of niacin might be involved. Indeed, as mentioned before, a $B_6$ deficiency could block the sequence of reactions to an intermediate metabolite, xanthurenic acid, which is believed, besides its above-mentioned detrimental metabolic effects, to also reduce considerably insulin efficiency (Kotake et al. 1968). Beyond growth performance, the interaction between pyridoxine and tryptophan was also investigated in relation to development of immune competence in postweaned piglets (Matte et al. 2011). As far as the immune status is concerned, separate effects of treatments were observed on in vivo antibody response (tryptophan supplementation) and on immune cell populations and on in vitro responses to a mitogen agent suggesting that each nutrient exerts a role through independent mechanisms, at least for the immunological indicators chosen in that experiment. In spite of late (eight to nine weeks of age) and short effects of supplementary dietary levels of Trp (2.2 vs. 2.6 g kg$^{-1}$) and $B_6$ (2.8 vs. 6.0 mg kg$^{-1}$) on feed conversion ratio, it was concluded that vitamin $B_6$ is not a major factor for the partition of this amino acid between protein deposition (growth responses) and catabolism and therefore for tryptophan requirements of postweaning piglets.

Pyridoxine is also particularly important as an enzymatic cofactor in the transsulphuration pathway for the disposal of homocysteine to cysteine (Le Grusse and Watier 1993). Zhang et al. (2009) showed that $B_6$ deficiency in weaned piglets induced rapidly (within three weeks) depressed growth performance and pyridoxal-P status along with a severe homocysteinemia. Using the same growth and metabolic criteria in a repletion experiment with different levels of dietary supplements of pyridoxine, they reached a conclusion similar to Matte et al. (2005) and Woodworth et al. (2000) that there was no advantage, in terms of performance, to exceed a total $B_6$ intake of 8 mg kg$^{-1}$ for weanling piglets. This value is five times greater than the NRC recommendation (1998; 2012) and corresponds to average supplementation levels used by the industry in North America (BASF 2001; Flohr et al. 2016)). Based on criteria such as growth performance and/or nitrogen retention, it had been thought for many years that dietary levels of 2–5 ppm of $B_6$ were required for postweaning pigs (5–30 kg in weight) (Adams et al. 1967; Kösters and Kirchgessner 1976; ARC 1981; Bretzinger 1991; NRC 1998).

As mentioned earlier, vitamin $B_6$ metabolism is closely related to other B-vitamins, in particular, riboflavin ($B_2$). The biologically active metabolites of $B_2$, FMN, and FAD are involved in the conversion of $B_6$ in its biologically active vitamers, pyridoxal phosphate, and its excretory form, 4-pyridoxic acid (Le Grusse and Watier 1993). Riboflavin is an enzymatic cofactor involved with a niacin component of niacin (NADPH) in the glutathione reductase that regenerates oxidized glutathione to reduced glutathione. The latter becomes available for the Se-dependent glutathione peroxidase (GSH-Px), critical for tissue peroxidation in several species including pigs. Decreased hepatic GSH-Px activity and Se retention were reported in young pigs on a $B_2$-deficient diet (Brady et al. 1979). A supplementation of 2 mg per day (vs. none) of riboflavin increased GSH-Px in several organs and increased Se retention particularly when the Se source was selenite (Parsons et al. 1985).

The link between tryptophan and niacin, as mentioned earlier, is well known (Le Grusse and Watier 1993) for several species including pigs (Markant et al. 1993). The conversion of tryptophan to niacin is dependent upon several dietary and hormonal factors (Fukuwatari and Shibata 2007) including species (60, 45, 37–50, and 170 mg of tryptophan equivalent to 1 mg of niacin in human, chicks, pigs, and ducks) (Firth and Johnson 1956; Fukuwatari and Shibata 2007; Chen et al. 1996; Matte et al. 2016) but is independent of dietary niacin content (Fukuwatari and Shibata 2007).

For pigs, the genotype has been evoked as a factor of variation for the partition of tryptophan between catabolism (toward synthesis of niacin) and protein deposition (growth responses) (Le Floc'h et al. 2017) but it remains to be better understood. In early weaned piglets, it has been shown that supplements of 0.05% of tryptophan increased by over 45% the concentration of plasma nicotinamide, a biologically active metabolite of niacin (Matte et al. 2011). In terms of growth performance criteria during the postweaning, growing and finishing period, results are inconsistent. Ivers et al. (2012) estimated that a supplementation of 14 mg kg$^{-1}$ of niacin is likely sufficient to maximize N utilization and growth performance between 14 and 39 kg of BW, whereas in a previous report, they (Ivers et al. 1993) did not observe any treatment effect on any of growth performance criteria within a range of supplementation between 0 and 81 mg kg$^{-1}$ In growing-finishing pigs, Real et al. (2002) concluded that 13–55 mg kg$^{-1}$ added dietary niacin improved average daily gain, gain:feed ratio, and meat quality (drip loss, pH and color). This last report was the basis for the recommended dietary requirement for niacin at 30 mg kg$^{-1}$ in NRC (2012) for all postweaning, growing, and finishing periods (5–135 kg of BW).

For postweaned piglets (3–10 weeks of age), using growth performance criteria, it was recently suggested (Matte et al. 2016) that a basal dietary level of niacin corresponding approximately to the NRC (2012) recommendation, was suboptimal. Moreover, high levels of dietary niacin (79–87 mg kg$^{-1}$) may attenuate Trp oxidation toward niacin metabolites.

*Pantothenic Acid*

Pantothenic acid has received some attention for the recent decades. In early weaned piglets, linear response on ADG and ADFI was observed within the range of 0–120 mg kg$^{-1}$ (Grinstead et al. 1998). As a matter of comparison for earlier in life, the pantothenic provision to suckling piglets during lactation corresponds to 47 mg kg$^{-1}$ DM of milk. In growing-finishing pigs, at dietary concentrations higher than that required to maximize body weight gain (up to 120 mg kg$^{-1}$), it would modify body composition through a partition of energy from fat accretion toward protein deposition (Stahly and Lutz 2001; Autrey et al. 2002; Santoro et al. 2006). However, it seems that this effect of pantothenic acid on fat and protein partitioning of meat is controversial because several studies did not report any response with dietary levels of supplementation between 30 and 90 mg kg$^{-1}$ (Radcliffe et al. 2003; Yang et al. 2004; Saddoris et al. 2005; Groesbeck et al. 2007). Nevertheless, supplemental dietary pantothenic acid at levels of 110 mg kg$^{-1}$ from 95 to 165 kg of BW may have valuable applications in niche markets such as heavy pig production by improving carcass quality (leanness) without affecting fatty acid composition (Minelli et al. 2013).

*Choline and Betaine*

Choline is considered as a B-vitamin although its typical dietary concentration is much higher than other micronutrients (Emmert et al. 1998). Moreover, it does not correspond to the classical definition of vitamins because it can be metabolically synthesized by the pig through methylation brought by the conversion of S-adenosyl-methionine (SAM) to S-adenosyl-homocysteine (SAH); three molecules of SAM (3 methyl groups) are required for synthesis of phosphatidylcholine (Stead et al. 2006). This vitamer is involved mainly as a structural components being the predominant phospholipids (>50%) in most mammalian membranes (Zeisel 2006). Another choline vitamer, acetylchloline, is well known for its neurotransmitter functions. Phosphatidylcholine (also called lecithin) can become a methyl donor via its oxidation in betaine. The importance of betaine for nutrition and physiology of livestock have been reviewed (Eklund et al. 2005). Briefly, betaine is the trimethyl derivative of the amino acid glycine that originates in the body from either choline oxidation or from nutritional sources. It is recognized as a powerful osmoprotective agent and as a methyl donor in transmethylation reactions (Lipiński et al. 2012). This last metabolic pathway, which regulates homocysteine in circulation, is called betaine-homocysteine methyl transferase (BHMT) pathway (Finkelstein 1990). However, it seems that the BHMT pathway is poorly efficient in depressing systemic homocysteine in growing-finishing pigs (Skomial et al. 2004) probably because of its limited tissue distribution, only in liver and kidneys in pigs (Delgado-Reyes et al. 2001) and also because of the relative insensitivity of hepatic and renal BHMT enzymatic activities to dietary intakes of sulfur amino acids, choline, or betaine. Recent advances in the functions of choline and betaine as well as the interactions between these two methyl donors and methionine have been reviewed for several species including pigs (Simon 1999; Cronje 2018). Some studies has been reported since NRC (1998) using plant lecithin as an alternative to dietary choline chloride which is often considered as harmful for the oxidative stability of some other nutrients in the diet. In fact, in contrast to human where lecithin is absorbed more slowly than choline chloride, there was no difference in diurnal variations between the two sources of choline in pigs (Jakob et al. 1998). In terms of growth performance and meat quality, there was no significant impact of supplements of choline chloride (0 vs. 0.2%) or lecithin (0 vs. 1.0, 1.5, 2.0, or 2.5%) (Kuhn et al. 1998).

*Folic Acid and Vitamin B$_{12}$*

In starter, grower, and finisher swine, the parenteral or oral administration of folic acid has been found to have positive effects on growth performance in some cases (Lindemann and Kornegay 1986; Matte et al. 1990; Matte et al. 1993; Yu et al. 2010) or has had no effect at all (Gannon and Leibholz 1990; Letendre et al. 1991). The impact of interactions between folic acid, vitamin B$_{12}$, and methionine on growth performance and meat quality were evaluated in growing-finishing pigs (Giguère et al. 2008). Although supplements had no impact on growth performance, meat quality, and meat oxidative stability, the dietary supplement of folic acid enriched meat in folates by 22% and decreased its content in homocysteine, a response that was also observed by Yu et al. (2010); The enrichment of pork with folates deserves to be better investigated with regard to its importance (mentioned earlier) for human nutrition and health (Giguère et al. 2008). Links between folic acid and vitamin B$_{12}$ metabolism and health status of growing-finishing pigs (16–24 weeks of age) have been suggested by Grutzner et al. (2015). They showed that concentrations of serum folates and vitamin B$_{12}$ were lower and their related metabolites (homocysteine and methyl-malonic acid) were higher in animals that were acutely, chronically, or subclinically infected by *Lawsonia intracellularis*. The consequences of this link between folic acid and B$_{12}$ nutrition to health response have been recently addressed in an experiment where two populations of piglets (birth to eight weeks of age) with high or low homocysteine levels in blood plasma were generated by altering the dietary provision of folates (1 vs. 10 mg kg$^{-1}$) and B$_{12}$ (20 vs. 200 µg kg$^{-1}$) to sows during gestation and lactation and by direct intramuscular (i.m.) injections of B$_{12}$ to piglets (Audet et al. 2015). Detrimental effects on indicators of cell-mediated immunity suggested a weakened immune competence. Unexpectedly, however, plasma homocysteine was positively correlated with growth rate and feed intake of piglets after weaning. Therefore, the only logical interpretation to such results would be that young "high performing" piglets that also generate high levels of plasma homocysteine are immunologically more fragile (Audet et al. 2015) and speculatively, less resistant to disease challenges. This concept might be important as an explanation to the opposite relation often reported in concrete husbandry situations between superior performance and disease resistance.

Estimates of vitamin B$_{12}$ requirements in growing-finishing pigs are based on studies carried out before 1966. Requirements are known to increase with the animal's potential for protein deposition and protein concentrations in feed. Cyanocobalamin supplements may alleviate for some imbalances in the intake of amino acids such as methionine and lysine (cited in ARC 1981). In piglets weaned at 26 days of age, an intramuscular injection of vitamin $B_{12}$ corresponding to a dietary intake 100 times greater than that recommended by the NRC (1998) at $20 \mu g$ $kg^{-1}$, resulted in an increase in weight gain and in feed intake of approximately 14 and 12% during the four weeks after weaning (Wilson et al. 1991). More recently, using performance criteria and metabolic criteria (mainly homocysteine), House and Fletcher (2003) showed that a supplement of $35 \mu g$ of crystalline cyanocobalamin/kg of feed without natural cobalamin was required for 5–10 kg weaner pigs. In a study mentioned above with growing-finishing pigs (Giguère et al. 2008), a vitamin $B_{12}$ supplement over $25 \mu g$ $kg^{-1}$ had no effect on growth performance, meat quality criteria, or metabolic criteria linked to methionine remethylation, particularly homocysteine status. However, the highest supplement in vitamin $B_{12}$ (150 ppb) increased the $B_{12}$ content by 55% in pork. These results were confirmed by bioavailability measurements of dietary vitamin $B_{12}$ and metabolic fate of homocysteine in growing pigs that were surgically prepared to assess portal appearance of these metabolites after a meal (Matte et al. 2010, 2012).

## Conclusion

Vitamins are nutrients essential to growth, maintenance, and health of pigs, just as amino acids and fatty acids are. Each vitamin plays well-defined metabolic roles, the importance of which varies depending on the animal's physiological status (starter, grower, or breeder).

The high level of productivity characterizing today's intensive farm management calls for intensive anabolism to maintain growth, gestation, and lactation at an optimal level. Nutrient allowances may need to be adjusted to meet the increased level of production but much of the limited information available on this topic is outdated. The levels used and/or recommended by private and public agencies vary widely. The lack of information on vitamins is responsible for this empirical or anecdotal approach that persists even today. Although the risks of a vitamin deficiency are practically nonexistent today, determining optimal levels for the productivity of swine production operations is a research challenge for the future. Besides the cost of feed vs reproductive and growth performance as the classical economical criteria to optimize the vitamin intake, other relevant issues should be also taken into account. For example, the vitamin provision to gestating and lactating sows and its efficient transfer to fetuses and piglets are particularly critical because pigs are entirely dependent for half of their existence (ova fertilization to slaughter) on the transfer of vitamins from their mother. Vitamin provision may have impact on health of pigs when taking into account that all vitamins have direct or indirect influence on the immune response. Further, vitamin supplementation above recommended level may influence human nutrition when considering effects on meat quality both in terms of stability (e.g. vitamin E) and meat enrichment (e.g., some B-vitamins and vitamin D), the latter being a potential marketing tool for the promotion of pork commodities.

## References

Adams, C. R., C. E. Richardson, and T. J. Cunha. 1967. Supplemental biotin and vitamin $B_6$ for swine. J. Anim. Sci. 26:903.

Adams, K. L., F. W. Bazer, and R. M. Roberts. 1981. Progesterone-induced secretion of a retinol-binding protein in the pig uterus. J. Reprod. Fertil. 62:39–47.

Adamstone, F. B., J. L. Krider, and M. F. James. 1949. Response of swine to vitamin E-deficient rations. Ann. N.Y. Acad. Sci. 52:260–268.

ARC. 1981. The nutrient requirements of pigs. Agricultural Research Council. Commonwealth Agricultural Bureaux, Slough, UK.

Amundson, L.A., Hernandez, L.L. and T. D. Crenshaw. 2017. Serum and tissue 25-OH vitamin D3 concentrations do not predict bone abnormalities and molecular markers of vitamin D metabolism in the hypovitaminosis D kyphotic pig model. Brit. J. Nutr. 118:30–40.

Anderson, G. C., and A. C. Hogan. 1950. Adequacy of synthetic diets for reproduction of swine. Proc. Soc. Exp. Biol. Med. 75:288–290.

Antipatis, C., A. M. Finch, and C. J. Ashworth. 2008. Effect of controlled alterations in maternal dietary retinol on foetal and neonatal retinol status on pregnancy outcome in pigs. Livest. Sci. 118:247–254.

Apple, J. K. 2007. Effects of nutritional modifications on the water-holding capacity of fresh pork: A review. J. Anim. Breed. Gen. 124:43–58.

Armocida, A., P. Beskow, P. Amcoff, A. Kallner, and S. Ekman. 2001. Vitamin C plasma concentrations and leg weakness in the forelegs of growing pigs. J. Vet. Med. Ser. A-Phys Path. Clin. Med. 48:165–178.

Asghar, A., J. I. Gray, A. M. Booren, E. A. Gomaa, M. M. Abouzied, E. R. Miller, and J. Buckley. 1991. Effects of supranutritional dietary vitamin E levels on subcellular deposition of $\alpha$-tocopherol in the muscle and on pork quality. J. Agric. Sci. 57:31–41.

Audet, I., J. -P. Laforest, G. P. Martineau, and J. J. Matte. 2004. Effect of vitamin supplements on some aspects of performance, vitamin status, and semen quality in boars. J. Anim Sci. 82:626–633.

Audet, I., N. Bérubé; J. L. Bailey, J. -P. Laforest, H. Quesnel, and J. J. Matte. 2009a. Effects of dietary vitamin supplementation and semen collection frequency on hormonal profile during ejaculation in the boar. Theriogenology 71:334–341.

Audet, I., N. Bérubé; J. L. Bailey, J. -P. Laforest, and J. J. Matte. 2009b. Effects of dietary vitamin supplementation and semen collection frequency on reproductive performance and semen quality in boars. J. Anim. Sci. 87:1960–1970.

Audet, I., C. L. Girard, M. Lessard, L. Lo Verso, F. Beaudoin, and J. J. Matte. 2015. Homocysteine metabolism, growth performance, and immune responses in suckling and weanling piglets. J. Anim. Sci. 93:147–157.

Autrey, B. A., T. S. Stahly, and T. R. Lutz. 2002. Efficacy of pantothenic acid as a modifier of body composition in pigs. J. Anim. Sci. 80 (Suppl. 1):163. (Abstr.).

Ballance, D. M., and J. D. House. 2005. Development of the enzymes of homocysteine metabolism from birth through weaning in the pig. J. Anim. Sci. 83 (Suppl. 1):160. (Abstr.).

Barkow, B., J. J. Matte, H. Böhme, and G. Flachowsky. 2001. Influence of folic acid supplements on the carry-over of folates from the sow to the piglet. Br. J. Nutr. 85:179–184.

Barnkob, L. L., P. M. Petersen, J. P. Nielsen, and J. Jakobsen. 2019. Vitamin D enhanced pork from pigs exposed to artificial UVB light in indoor facilities. Eur. Food Res. Tech. 245:411–418.

Bässler, K. H. 1997. Enzymatic effects of folic acid and vitamin $B_{12}$. Internat. J. Vit. Nutr. Res. 67:385–388.

BASF. 2001. Fortification en vitamines des aliments porcs et volailles au Canada. In: M. Duval, editor, Séminaire technique. St-Pie de Bagot, Québec, Canada. p. 17.

Bazer, F. W., and M. T. Zavy. 1988. Supplemental riboflavin and reproductive performance of gilts. J. Anim. Sci. 66(Suppl. 1):324. (Abstr.)

Bender, D. A. 1987. Oestrogens and vitamin B6 - actions and interactions. World Rev. Nutr. Diet. 51:140–188.

Benedikt, J., D. A. Roth-Maier, and M. Kirchgessner. 1996. Influence of dietary vitamin $B_6$ supply during gravidity and lactation on total vitamin $B_6$ concentration (Pyridoxine, Pyridoxal and Pyridoxamine) in blood and milk. Int. J. Vit. Nutr. Res. 66:146–150.

Bethke, R. M., M. Burroughs, O. H. M. Wilder, B. H. Edgington, and W. L. Robison. 1946. The comparative efficacy of vitamin D from irradiated yeast and cod liver oil for growing pigs, with observations on their vitamin D requirements. Ohio Agric. Exp. Stn. Bulletin 667. Ohio Agric. Exp. Stn., Wooster, OH. p. 1–29.

Bilodeau, R., J. J. Matte, A. -M. B. de Passillé, C. L. Girard, and G. J. Brisson. 1989. Effects of floor type on serum folates, serum vitamin $B_{12}$, plasma biotin and on growth performances of pigs. Can. J. Anim. Sci. 69:779–788.

Brady, P. S., L. J. Brady, M. J. Parsons, D. E.Ullrey, and E. R. Miller. 1979. Effects of riboflavin deficiency on growth and glutathione peroxidase system enzymes in the baby pig. J. Nutr. 109:1615–1622.

Böhmer, B. M., and D. A. Roth-Maier. 2007. Effects of high-level dietary B-vitamins on performance, body composition and tissue vitamin contents of growing finishing pigs. J Anim. Physiol. Anim. Nutr. 91:6–10.

Bonette, E. D., E. T. Kornegay, M. D. Lindemann, and C. Hammerberg. 1990. Humoral and cell-mediated immune response and performance of weaned pigs fed four supplemental vitamin E levels and housed at two nursery temperatures. J. Anim. Sci. 58:1337–1345.

Braude, R., A. S. Foot, K. M. Henry, S. K. Kon, S. Y. Thompson, and T. H. Mead. 1941. Vitamin A studies with rats and pigs. Biochem. J. 35:693–707.

Braude, R., S. K. Kon, and J. W. G. Porter. 1950. Studies in the vitamin Cmetabolism of the pig. Br. J. Nutr. 4:186–197.

Bretzinger, J. 1991. Pyridoxine supply of early weaned piglets. PhD Diss. Vet. Med., Ludwig-Maximillians Univ. of Munich, Munich, Germany.

Brief, S., and B. P. Chew. 1985. Effects of vitamin A and $\beta$-carotene on reproductive performance in gilts. J. Anim. Sci. 60:998–1004.

Brooks, P. H. 1986. The role of biotin in intensive systems of pig production. Proc. 6th Int. Conf. on Production Disease in Farm Animals, Belfast, Northern Ireland, UK.

Bryant, K. L., E. T. Kornegay, J. W. Knight, K. E. Webb, Jr., and D. R. Notter. 1985. Supplemental biotin for swine. II. Influence of supplementation to corn- and wheat-based diets on reproductive performance and various biochemical criteria of sows during four parities. J. Anim. Sci., 60:145–153.

Brzezinska-Slebodzinska, E., A. B. Slebodzinski, B. Pietras, and G. Wieczorek. 1995. Antioxidant effect of vitamin E and glutathione on lipid peroxidation in boar semen plasma. Biol. Trace Elem. Res. 47:69–74.

Bueno Dalto, D., and J. J. Matte. 2017. Pyridoxine (vitamin B6) and the glutathione peroxidase system: a link between one-carbon metabolism and antioxidation. Nutrients 9:189.

Bueno-Dalto, D., I. Audet, J. Lapointe, and J. J. Matte. 2016. The importance of pyridoxine for the impact of the dietary selenium sources on redox balance, embryo development, and reproductive performance in gilts. J. Trace Elem. Med. Biol. 34:79–89.

Bueno-Dalto, D., S. Tsoi, I. Audet, M. K. Dyck, G. R. Foxcroft, and J. J. Matte. 2015a. Gene expression of porcine blastocysts from gilts fed organic or inorganic selenium and pyridoxine. Reproduction 149:31–42.

Bueno-Dalto, D., M. Roy, I. Audet, M.-F. Palin, F. Guay, J. Lapointe, and J. J. Matte. 2015b. Interaction between vitamin B6 and source of selenium on the response of the selenium-dependent glutathione peroxidase system to oxidative stress induced by oestrus in pubertal pig. J. Trace Elem. Med. Biol. 32 21–29.

Buhi, W., F. W. Bazer, C. Ducsay, P. W. Chun, and R. M. Roberts. 1979. Iron content, molecular weight and possible function of the progesterone-induced purple glycoprotein. Fed. Proc. 38:733–733.

Burton, G. W., U. Wronska, L. Stone, D. O. Foster, and K. U. Ingold. 1990. Biokinetics of dietary RRR-α-tocopherol in the male guinea pig at three dietary levels of vitamin C and two levels of vitamin E. Evidence that vitamin C does not "spare" vitamin E in vivo. Lipids 25:199–209.

BSAS. 2003. Nutrient requirement standards for pigs. BSAS, Penicuik, Midlothian, UK.

Cera, K. R., D. C. Mahan, and G. A. Reinhart. 1988. Weekley difgestibilities of diets supplemented with corn oil, lard, or tallow by weanling swine, J. Anim. Sci., 66:1430–1437.

Charlton S. J., and W. N. Ewing. 2007. The vitamins directory. Context. Products Ltd., Leicestershire, UK.

Cheah, K. S., A. M. Cheah, and D. T. Krausgrill. 1995. Effect of dietary supplementation of vitamin E on pig meat quality. Meat Sci. 39:255–264.

Chen, B. J., T. F. Shen, and R. E. Austic. 1996. Efficiency of tryptophan-niacin conversion in chickens and ducks. Nutr. Res. 16:91–104.

Chen, Y. F., C. F. Huang, L. Liu, C. H. Lai, and F. L. Wang. 2019. Concentration of vitamins in the 13 feed ingredients commonly used in pig diets. Anim. Feed Sci. Technol. 247:1–8.

Chew, B. P., T. S. Wong, J. J. Michal, F. R. Standaert, and L. R. Heriman. 1991a. Kinetic characteristics of β-carotene uptake after an injection of β-carotene in pigs. J. Anim. Sci. 69:4883–4891.

Chew, B. P., T. S. Wong, J. J. Michal, F. R. Standaert, and L. R. Heriman. 1991b. Subcellular distribution of beta-carotene, retionol, and alpha-tocopherol in porcine lymphocytes after a single injection of beta-carotene. J. Anim. Sci. 69:4892–4897.

Chew, B. P. 1993. Effects of supplemental β-carotene and vitamin A on reproduction in swine. J. Anim. Sci. 71:247–252.

Ching, S., Mahan, D.C. and K. Dabrowski. 2001. Liver L-gulonolactone oxidase activity and tissue ascorbic acid concentrations in nursing pigs and the effect of various weaning ages. J Nutr. 131:2002–2006.

Ching, S., D. C. Mahan, T. G. Wisemann, and N. D. Fastinger. 2002. Evaluating the antioxidant status of weanling pigs fed dietary vitamin A and E. J. Anim. Sci. 80:2396–2401.

Cho, J. H., N. Lua, and M. D. Lindemann. 2017. Effects of vitamin supplementation on growth performance and carcass characteristics in pigs. Livest. Sci. 204:25–32.

Chung, Y. K., D. C. Mahan, and A. J. Lepine. 1992. Efficacy of dietary D-α-tocopherol and Dl-α-tocopheryl acetate in for weanling pigs. J. Anim. Sci. 70:2485–2492.

Coburn, S. P. 1994. A critical review of minimal vitamin B-6 requirements for growth in various species with a proposed method of calculation. Vitam. Horm. 48: 259–300.

Coffey, M. T., and J. H. Britt. 1989. Effect of β-carotene injection on reproductive performance of sows. J. Anim. Sci. 67(Suppl. 1):251. (Abstr.).

Coffey, M. T., and J. H. Britt. 1993. Enhancement of sow reproductive performance by β-carotene or vitamin A. J. Anim. Sci. 71:1198–1202.

Côté-Robitaille, M.-É., C. L. Girard, F. Guay, and J. J. Matte. 2015. Oral supplementations of betaine, choline, creatine and vitamin B6 and their influence on the development of homocysteinaemia in neonatal piglets. J. Nutr. Sci. 4:e31.

Cox, N. M. 1997. Control of follicular development and ovulation rate in pigs. J. Reprod. Fertil. Suppl. 52:31–46.

Cranwell, P. D., and P. J. Moughan. 1989. Biological limitations imposed by the digestive system to the growth performance of weaned pigs. In: J. L. Barnett, and D. P. Hennessy, editors, Manipulating pig production. Vol. II. Australian Pig Sci. Assoc., Victoria, Australia. p. 149.

Cronje, P. B. 2018. Essential role of methyl donors in animal productivity. Anim. Prod. Sci. 58:655–665.

Cunha, T. J., O. B. Ross, P. H. Phillips, and G. Bohstedt. 1944. Further observations on the dietary insufficiency of a corn-soybean ration for reproduction of swine. J. Anim. Sci. 3:415–421.

Davey, R. J., and J. W. Stevenson. 1963. Pantothenic acid requirement of swine for reproduction. J. Anim. Sci. 22:9–13.

Davis, V. E., A. A.HeidebrechtOklahoma Agricultural Experiment Station, Stillwater Search for other works by this author on: A. A. Heidebrecht, R. MacVicar, O. B. Ross, and C. K. Whitehair. 1951. Oxford academic the composition of swine milk: II. Thiamine, riboflavin, niacin and pantothenic acid. J. Nutr. 44:17–27.

De Passille, A. M. B., Girard, C. L., Matte, J. J. and R. R. Bilodeau. 1989. A Study on the occurrence of coprophagy behavior and its relationship to B-vitamin status in growing-finishing pigs. Can J. Anim. Sci. 69:299–306.

De Rodas, B. Z., C. V. Maxwell, M. E. Davis, S. Mandali, E. Broekman, and B. J. Stoeker. 1998. L-ascorbyl-2-polyphosphate as a vitamin C source for segrated and conventionally weaned pigs. J. Anim. Sci. 76:1636–1643.

Delgado-Reyes, C. V., M. A. Wallig, and T. A. Garrow. 2001. Immunohistochemical detection of betaine-homocysteine S-methyltransferase in human, pig, and rat liver and kidney. Archiv. Biochem. Biophys. 393:184–186.

DiSimone, N., P. Riccardi, N. Maggiano, A. Piacentani, M. D'Asta, A. Capelli, and A. Caruso, 2004. Effect of folic acid on homocysteine-induced trophoblast apoptosis. Molecul. Hum. Reprod. 10:665–669.

Donovan, S. M., M.-H. Mar, and S. H. Zeisel. 1997. Choline and choline ester concentrations in porcine milk throughout lactation Nutr. Biochem. 8:603–607.

Driancourt, M. A., F. Martinat-Botté, and M. Terqui. 1998. Contrôle du taux d'ovulation chez la truie:l'apport des modèles hyper-prolifiques. INRA Prod. Anim. 11:221–226.

Duffy, S. K., A. K. Kelly, G. Rajauria, L. C. Clarke, V. Gath, F. J. Monahan, and J. V. O'Doherty. 2018. The effect of 25-hydroxyvitamin D-3 and phytase inclusion on pig performance, bone parameters and pork quality in finisher pigs. J. Phys. Anim. Nutr. 102:1296–1305.

Duquette, J., J. J. Matte, C. Farmer, C. L. Girard, and J. P. Laforest. 1997. Pre- and post-mating dietary supplements of folic acid and uterine secretory activity in gilts. Can. J. Anim. Sci. 77:415–420.

Easter, R. A., P. A. Anderson, E. J. Michel, and J. R. Corley. 1983. Response of gestating gilts and starter, grower and finisher swine to biotin, pyridoxine, folacin and thiamine additions to corn-soybean meal diets. Nutr. Rep. Int. 28:945–950.

Eichenberger, B., H. P. Pfirter, C. Wenk, and S. Gebert. 2004. Influence of dietary vitamin E and C supplementation on vitamin E and C content and thiobarbituric acid reactive substances (TBARS) in different tissues of growing pigs. Arch. Anim. Nutr. 58:195–208.

Eklund, M., E. Bauer, J. Wanatu, and R. Mosenthin. 2005. Potential nutritional and physiological functions of betaine in livestock. Nutr. Res. Rev. 18:31–48.

Emmert, J. L., D. M. Webel, R. R. Biehl, M. A. Griffiths, L. S. Garrow, T. A. Garrow, and D. H. Baker. 1998. Hepatic and renal betaine-homocysteine methyltransferase activity in pigs as affected by dietary intakes of sulfur amino acids, choline, and betaine. J. Anim. Sci. 76:606–610.

Engstrom, G. W., and T. Littledike. 1986. Vitamin D metabolism in the pig. In: M. E. Tumbleson, editor, Swine in biomedical research. Plenum, New York, NY. p. 1091–1112.

Ensminger, M. E., J. P. Bowland, and T. J. Cunha. 1947. Observations on the thiamine, riboflavin, and choline needs of sows for reproduction. J. Anim. Sci. 6:409–423.

Ensminger, M. E., R. W. Colby, and T. J. Cunha. 1951. Effect of certain B-complex vitamins on gestation and lactation in swine. Washington Agric. Exp. Sta., Sta. Circ. 134:1–35.

Fernandez-Duenas, D. M., G. Mariscal, E. Ramirez, and J. A. Cuaron. 2008. Vitamin C and β-carotene in diets for pigs at weaning. Anim. Feed. Sci. Tech. 146:313–326.

Finkelstein, J. D. 1990. Methionine metabolism in mammals. J. Nutr. Biochem. 1:228–237.

Firth, J., and C. Johnson. 1956. Quantitative relationships of tryptophan and nicotinic acid in the baby pig. J. Nutr. 59:223–234.

Flohr, J. R., M. D. Tokach, S. S. Dritz, J. M. DeRouchey, R. D. Goodband, J. L. Nelssen, and J. R. Bergstrom. 2014. An evaluation of the effects of added vitamin D3 in maternal diets on sow and pig performance. J. Anim. Sci. 92:594–603.

Flohr, J. R., J. M. DeRouchey, J. C. Woodworth, M. D. Tokach, R. D. Goodband, and S. S. Dritz. 2016. A survey of current feeding regimens for vitamins and trace minerals in the US swine industry. J. Swine Health Prod. 24:290–303.

Foxcroft, G. R. 2012. Reproduction in farm animals in an era of rapid genetic change: Will genetic change outpace our knowledge of physiology? Reprod. Domest. Anim. 47:313–319.

Frank, G.R., R. A. Easter, and J. M. Bahr. 1984. Riboflavin requirement of gestating swine. J. Anim. Sci. 59:1567–1572.

Frederick, G. L., and G. J. Brisson. 1961. Some observations on the relationship between vitamin B$_{12}$ and reproduction in swine. Can. J. Anim. Sci. 41:212–219.

Friendship, R. M., and M. R. Wilson. 1991. Effects of intramuscular injections of folic acid in sows on subsequent litter size. Can. Vet. J. 32:565–566.

Fukuwatari, T., and K. Shibata. 2007. Effect of nicotinamide administration on the tryptophan-nicotinamide pathway in humans. Intern. J. Vit. Nutr. Res. 77:255–262.

Garcia-Castillo, R. F., J. L. Jasso-Pitol, R. Morones-Reza, J. R. Kawas-Garza, and J. Salinas-Chavira. 2006. Adicion de altos niveles de biotina en dietas para cerdas puberes y gestantes. Agronomia Mesoamericana. 17:1–5.

Gannon, N. J., and J. Leibholz. 1990. Manipulating pig production II. Proc. of the Biennal Conf. of the Australasian Pig Sci. Assoc., November 27–29, 1989. Sydney, Australia. p. 136.

Giguère, A., C. L. Girard, and J. J. Matte. 2002. Erythrocyte glutathione reductase activity and riboflavin nutritional status in early-weaned piglets. Int. J. Vit. Nutr. Res. 72:383–387.

Giguère, A., C. L. Girard, and J. J. Matte. 2008. Methionine, folic acid and vitamin b-12 in growing-finishing pigs: Impact on growth performance and meat quality. Arch. Anim. Nutr. 62:193–206.

Giguère, A., C. L. Girard, R. Lambert, J. P. Laforest, and J. J. Matte. 2000. Reproductive performance and uterine prostaglandin secretion in gilts conditioned with dead semen and receiving dietary supplements of folic acid. Can. J. Anim. Sci. 80:467–472.

Gesellschaft für Ernärungsphysiologie (GfE). 1987. Energie-und Nährstoffbedarf landwirtschaftlicher Nutztiere, Schweine. DLG-Verlag, Frankfurt/Main, Germany.

Goff, J. P., R. L. Horst, and E. T. Littledike. 1984. Effect of sow vitamin D status at parturition on the vitamin D status of neonatal piglets. J. Nutr. 114:163–169.

Gondret, F., Lefaucheur, L., Louveau, I., Lebret, B., Pichodo, X. and Y. Le Cozler. 2005. Influence of piglet birth weight on post-natal growth performance, tissue lipogenic capacity, and muscle histological traits at market weight. Livest. Prod. Sci. 93:137–146.

Greer, E. B., and C. E. Lewis. 1978. Mineral and vitamin supplementation of diets for growing pigs. I. Wheat-based diets and the effect of preventing cross-coprophagy on the response to supplementation. Aust. J. Exp. Agric. Anim. Husb. 18:688–697.

Greer, E. B., J. M. Leibholz, D. I. Pickering, R. E. Macoun, and W. L. Bryden. 1991. Effect of supplementary biotin on the reproductive performance, body condition and foot health of sows on 3 farms. Aust. J. Agric. Res. 42:1013–1021.

Grinstead, G. S., R. D. Goodband, J. L. Nelssen, M. D. Tokach, S. S. Dritz, and R. Stott. 1998. Effects of increasing pantothenic acid on growth performance of segregated early-weaned pigs. Kansas State Swine Day Rep. Kansas State Univ. Agric. Exp. Stn. Coop. Ext. Serv., Manhattan, KS. p. 87–89.

Groesbeck, C. N., R. D. Goodband, M. D. Tokach, S. S. Dritz, J. L. Nelssen, and J. M. Derouchey. 2007. Effects of pantothenic acid on growth performance and carcass characteristics of growing-finishing pigs fed diets with or without ractopamine hydrochloride. J. Anim. Sci. 85:2492–2497.

Grutzner, N., Gebhart, C. J.,Lawhorn, B. D., Suchodolski, J. S., and J. M. Steiner, 2015. Serum folate, cobalamin, homocysteine and methylmalonic acid concentrations in pigs with acute, chronic or subclinical Lawsonia intracellularis infection. Vet. J. 203:320–325.

Guay, F., J. J. Matte, C. L. Girard, M.-F. Palin, A. Giguère, and J.-P. Laforest. 2002a. Effect of folic acid and glycine supplementation on embryo development and folate metabolism during early pregnancy in pigs. J. Anim. Sci. 80:2134–2143.

Guay, F., J. J. Matte, C. L. Girard, M.-F. Palin, A. Giguère, and J.-P. Laforest. 2002b. Effect of folic acid and vitamin $B_{12}$ supplements on folate and homocysteine metabolism in pigs during early pregnancy in pigs. Br. J. Nutr. 88:253–263.

Guay, F., J. J. Matte, C. L. Girard, M.-F. Palin, A. Giguère, and J.-P. Laforest. 2004a. Effects of folic acid supplement on uterine prostaglandin metabolism and interleukin-2 expression on day 15 of gestation in white breed and crossbred Meishan sows. Can J. Anim. Sci. 84:63–72.

Guay, F., J. J. Matte, C. L. Girard, M. F. Palin, A. Giguère, and J.-P. Laforest. 2004b. Effect of folic acid plus glycine supplement on uterine prostaglandin and endometrial granulocyte-macrophage colony-stimulating factor expression during early pregnancy in pigs. Theriogenology 61:485–498.

Håkansson, J., J. Hakkarainen, and N. Lundeheim. 2001. Variation in vitamin E, glutathione peroxidase and retinol concentrations in blood plasma of primiparous sows and their piglets, and in vitamin E, selenium and retinol contents in sows' milk. Acta Agric. Scand. A. Anim. Sci. 51:224–234.

Halanski, M.A., Hildahl, B., Amundson, L.A., Leiferman, E., Gendron-Fitzpatrick, A., Chaudhary, R., Hartwig-Stokes, H.M., McCabe, R., Lenhart, R., Chin, M., Birstler, J. and T.D. Crenshaw. 2018. Maternal diets deficient in vitamin D increase the risk of kyphosis in offspring: A novel Kyphotic Porcine Model. J Bone Joint Surg. Am. 7:406–415.

Hall, D. D., G. L. Cromwell, and T. S. Stahly. 1991. Effects of dietary calcium, phosphorus, calcium:phosphorus ratio and vitamin K on performance, bone strength and blood clotting status of pigs. J. Anim. Sci. 69:646–655.

Halloran, B. P., and H. F. DeLuca. 1983. Effect of vitamin D deficiency on fertility and reproductive capacity in female in the rat. J Nutr. 110:1573–1580.

Hartvig, P., K. G. Lindner, P. Bjurling, B. Langstrom, and J. Tedroff. 1995. Pyridoxine effect on synthesis rate of serotonin in the monkey brain measured with positron emission tomography. J. Neural. Transm. [Gen. Sect.] 102:91–97.

Hedemann, M. S., and S. K. Jensen. 1999. Vitamin E status in newly weaned piglets is correlated to the activity of carboxyl ester hydrolase. Proc. Aust. Pig Sci. Assoc. Adelaide, Australia. p. 181. (Abstr.).

Hewison, M., M. A. Gacad, J. Lemire, and J. S. Adams. 2001. Vitamin D as a cytokine and a hematopoietic factor. Rev. Endocrinol. Metab. Disorders 2:217–227.

Hickie, J. P., D. M. Lavigne, and W.D. Woodward. 1983. Reduced fecundity of vitamin D deficient rats. Comp. Biochim. Physiol. A. 74:923–925.

Hidiroglou, M., and T. R. Batra. 1995. Concentration of vitamin C in milk of sows and in plasma of piglets. Can. J. Anim. Sci. 75:275–277.

Hoey, L., H. Mcnulty, and J. J. Strain. 2009. Studies of biomarker responses to intervention with riboflavin: A systematic review. Am. J.Clin. Nutr. 89:1960S–1980S.

Hoppe, P. P. 1988. Lack of vitamin K can kill weaners. Pig Int. June:16–18.

Hoppe, P. P., F. J. Schoner, and M. Frigg. 1992. Effects of dietary retinol on hepatic retinol storage and on plasma and tissue α-tocopherol in pigs. Internat. J. Vit. Nutr. Res. 62:121–129.

Horký, P., L. Zeman, J. Skladanka, P. Nevrkla, and P. Slama. 2016. Effect of selenium, zinc, vitamin C and E on boar ejaculate quality at heat stress. Acta Univ. Agric. Silvic. Mend. Brun. 64:1167–1172.

Horst, R. L., and E. T. Littledike. 1982. Comparison of plasma concentrations of vitamin D and its metzabolites in young and aged animals. Comp. Biochem. Physiol. 73B:485–489.

Hoskinson, C. D., B. P. Chew, and T. S. Wong. 1992. Effects of injectable beta-carotene and vitamin A on mitogen induced lymphocyte proliferation in the pig in vivo. Biol. Neon. 62:325–336.

Hough, S.D., S.H. Jennings and G.W. Almond. 2015. Thiamine-responsive neurological disorder of swine. J. Swine Health Prod. 23: 143–151.

House, J. D., and C. M. T. Fletcher. 2003. Response of early weaned piglets to graded levels of dietary cobalamin. Can. J. Anim. Sci. 83:247–255.

INRA. 1984. Alimentation des animaux monogastriques: Porcs, lapins, volailles. Institut National de la Recherche Agronomique, Paris, France.

Ivers D. J., S. L. Rodhouse, M. R. Ellersieck, and T. L. Veum. 1993. Effect of supplemental niacin on sow reproduction and sow and litter performance. J. Anim. Sci. 71:651–655.

Ivers, D. J., and T. L. Veum. 2012. Effect of graded levels of niacin supplementation of a semipurified diet on energy and nitrogen balance, growth performance, diarrhea occurrence, and niacin metabolite excretion by growing swine. J Anim Sci. 90:282–288.

Jakob, S., R. Mosenthin, G. Huesgen, J. Kinkeldei, and K.-J. Poweleit. 1998. Diurnal pattern of choline concentrations in serum of pigs as influenced by dietary choline or lecithin intake. Z. Ernähr. 37:353–357.

Jakobsen, J., H. Maribo, A. Bysted, H. M. Sommer, and O. Hels. 2007. 25-Hydroxyvitamin $D_3$ affects vitamin D status similar to vitamin $D_3$ in pigs – but the meat produced has a lower content of vitamin D. Br. J. Nutr. 98:908–913.

Jang, Y. D., J. Y. Ma, N. Lu, J. Lim, H. J. Moneque, R. L. Stuart, and M. D. Lindemann. 2018. Administration of vitamin D-3 by injection or drinking water alters 25-hydroxycholecalciferol concentrations of nursery pigs. Asian-Australas. J. Anim. Sci. 32:278–286.

Jensen, M., A. Lindholm, and R. V. J. Hakkareinen. 1990. The vitamin E distribution in serum, liver, adipose and muscle tissues in the pig during depletion and repletion. Acta Vet. Scand. 31:129–136.

Jensen, C., J. Guidera, I. M. Skovgaard, H. Staun, L. H. Skibsted, S. K. Jensen, A. J. Møller, J. Bucklery, and G. Bertelsen. 1997. Effects of dietary alpha-tocopheryl acetate supplementation on alpha-tocopherol deposition in porcine m psoas major and m longissimus dorsi and on drip loss, colour stability and oxidative stability of pork meat. Meat Sci. 55:491–500.

Jensen, C., C. Lauridsen, and G. Bertelsen. 1998. Dietary vitamin E: Quality and storage stability of pork and poultry. Trends in Food Sci. Technol. 9:62–72.

Jin, C., Y. Zhuo, J. Wang, Z. Yang, X. Yuedong, L. Daolin, L. Hong, Z. Pan, F. Zhengfeng, C. Lianqiang, X. Shengyu, F. Bin, L. Jian, J. Xuemei, L. Yan, and W. De. 2018. Methyl donors dietary supplementation to gestating sows diet improves the growth rate of offspring and is associating with changes in expression and DNA methylation of insulin- like growth factor-1 gene. J. Anim. Physiol. Anim. Nutr. 102:1340–1350.

Jones, R. and T. Mann. Toxicity of exogenous fatty acid peroxides towards spermatozoa. 1977 J. Reprod Fertil. 50:255–260.

Khalifa, T., C. Rekkas, F. Samartzi, A. Lymberopoulos, K. Kousenidis, and T. Dovenski. 2014. Highlights on artificial insemination (AI) technology in the pigs. Mac. Vet. Rev. 37:5–34.

Kim, E. S., J. S. Seo, J. H. Eum, J. E. Lim, D. H. Kim, T. K. Youn, D. R. Lee. 2009. The effect of folic acid on in vitro maturation and subsequent embryonic development of porcine immature oocytes. Mol. Reprod. Dev. 76:120–121.

Kindberg, C. G., and J. W. Suttie. 1989. Effect of various intakes of phylloquinone on signs of vitamin K deficiency and serum liver phylloquinone concentrations in the rat. J. Nutr. 119:175–180.

Knights, T. E. N., R. R. Grandhi, and S. K. Baidoo. 1998. Interactive effects of selection for lower backfat and dietary pyridoxine levels on reproduction, and nutrient metabolism during the gestation period in Yorkshire and Hampshire sows. Can. J. Anim. Sci. 78:167–173

Kwiecinski, G. C., G. I. Petri, and H. F. DeLuca. 1989. 1,25-dihydroxyvitamin D3 restores fertility of vitamin D-deficient female rats. Am. J. Physiol., 256:E483–E487.

Kolp, E., M. R. Wilkens, W. Pendl, B. Eichenberger, and A. Liesegang. 2017. Vitamin D metabolism in growing pigs: influence of UVB irradiation and dietary vitamin D supply on calcium homeostasis, its regulation and bone metabolism, J. Anim. Phys. Anim. Nutr. 101:79–94.

Konowalchuk, J. D., A. M. Rieger, M. D. Kiemele, D. C. Ayres, and D. R. Barreda. 2013. Modulation of weanling pig cellular immunity in response to diet supplementation with 25- hydroxyvitamin D3. Vet. Immu. Immunopathol. 155:5766.

Kopinski, J. S., and J. Leibholz. 1989. Biotin studies in pigs. II. The biotin requirement of the growing pig. Br. J. Nutr. 62:761–772.

Kornegay, E. T. 1986. Biotin in swine production: A review. Livest. Prod. Sci. 14:65–89.

Kostoglou, P., S. C. Kyriakis, A. Papasteriadis, N. Roumpies, C. Alexopoulos, and K. Saoulidis. 2000. Effect of β-carotene on health status and performance of sows and their litters. J. Anim. Physiol. A. Anim. Nutr. 83:150–157.

Kösters, W. W., and M. Kirchgessner. 1976. Change in feed intake of early-weaned piglets in response to different vitamin $B_6$ supply. Z. TierphysiologieTiernährung und Futtermittelkde 37:247–254.

Kotake, Y., T. Sotokawa, A. Hisatake, M. Abeand, and Y. Ikeda. 1968. Studies on xanthurenic acid insulin complex. II. Physiological activities. J. Biochem. (Tokyo) 63:578–581.

Kovčin, S., S. Živkovivć, M. Beuković, and M. Lalić. 1988. Effect of folic acid on reproduction of sow. Proceedings No. 17–18. Institute of Animal Husbandry, Novi Sad, Servia.

Kuhn, M., B. Frohmann, A. Petersen, K. Rubesam, and C. Jatsch. 1998. Utilization of crude soybean lecithin as a native choline source in feed rations of fattening pigs. Fett/Lipid 100:78–84.

Lahucky, R., I. Bahelka, U. Kuechenmeister, K. Vasickova, K. Nuernberg, K. Ender, and G. Nuernberg. 2007. Effects of dietary supplementation of vitamin D3 and E on quality characteristics of pigs and longissimus muscle antioxidative capacity. Meat Sci. 77:264–268.

Larson-Meyer, D. E., B. C. Ingold, S. R. Fenstseifer, K. J. Austin, P. J. Wechsler, B. W. Hollis, A. J. Makowskim, and B. M. Alexander 2017. Sun exposure in pigs increases the vitamin D nutritional quality of pork. PLoS One 12:e0187877.

Lauridsen, C., M. S. Hedemann, and S. K. Jensen. 2001. Hydrolysis of tocopherol and retinyl esters by porcine carboxyl ester hydrolase is affected by their carboxylate moiety and bile acids. J. Nutr. Biochem. 12:219–224.

Lauridsen, C., H. Engel, A. M. Craig, and M. G. Traber. 2002a. Relative bioactivity of dietary RRR- and all-rac-α-tocopheryl acetates in swine assessed with deuterium-labeled vitamin E. J. Anim. Sci. 80:702–707.

Lauridsen, C., H. Engel, S. K. Jensen, A. M. Craig, and M. G. Traber. 2002b. Lactating sows and suckling piglets preferentially incorporate RRR- over all-rac-α-tocophol into milk, plasma and tissues. J. Nutr. 132:1258–1264.

Lauridsen, C., and S. K. Jensen. 2005. Influence of supplementation of all-rac-α-tocopheryl acetate preweaning and vitamin C postweaning on α-tocopherol and immune responses of piglets. J. Anim. Sci. 83:1274–1286.

Lauridsen, C., T. Larsen, U. Halekoh, and S. K. Jensen. 2010. Reproductive performance and bone status markers of gilts and lactating sows supplemented with two different forms of vitamin D. J. Anim. Sci., 88:202–213.

Lauridsen, C. 2010. Evaluation of the effect of increasing dietary vitamin E in combination with different fat sources on performance, humoral immune responses and antioxidant status of weaned pigs. Anim. Feed. Sci. Tech. 158:85–94.

Lauridsen, C. 2014. Triennial growth symposium: Establishment of the 2012 vitamin D requirements in swine with focus on dietary forms and levels of vitamin D. J. Anim. Sci. 92:910–916.

Le Cozler, Y., X. Pichodo, H. Roy, C. Guyomarc'h, H. Pellois, N. Quiniou, I. Louveau, B. Lebret, L. Lefaucheur and F. Gondret. 2004. Influence du poids individuel et de la taille de la portée à la naissance sur la survie du porcelet, ses performances de croissance et d'abattage et la qualité de la viande. Journées Rech. Porcine en France 36:443–450.

Lee, N. S., G. Muhs, G. C. Wagner, R. D. Reynolds, and H. Fisher. 1988. Dietary pyridoxine interaction with tryptophan or histidine on brain serotonin and histamine metabolism. Pharmacol. Biochem. Behav. 29:559–564.

Le Grusse, J., and B. Watier. 1993. Les vitamines. Données biochimiques, nutritionnelles et cliniques. Centre d'Études et d'Information sur les Vitamines. Produits Roche, Neuilly sur Seine, France.

Lechowski, J., A. Kasprzyk, and B. Trawińska. 2018. Variability of semen in boars treated with vitamin C in food ration. Med. Weter. 74:48–53.

Le Floc'h, N., A. Simongiovanni, E. Corrent and J. J. Matte. 2017. Comparison of growth and plasma tryptophan related metabolites in crossbred Piétrain and Duroc pigs. J. Anim. Sci. 95:1606–1613.

Letendre, M., C. L. Girard, J. J. Matte and J.-F. Bernier. 1991. Effects of intramuscular injections of folic acid on folates status and growth performance of weanling pigs. Can. J. Anim. Sci. 71:1223–1231.

Leonhardt, M., S. Gerbert, and C. Wenk. 1996. Stability of α-tocopherol, thiamine, riboflavin and retinol in pork muscle and liver during heating as affected by dietary supplementation. J. Food Sci. 61:1048–1052.

Lewis, A. J., and L. L. Southern. 2001. Swine nutrition. CRC Press, Boca Raton, FL.

Lewis, A. J., G. L. Cromwell, and J. E. Pettigrew. 1991. Effects of supplemental biotin during gestation and lactation on reproductive performance of sows: a cooperative study. J. Anim. Sci. 69:207–214.

Lindemann, M. D. 1993. Supplemental folic acid: a requirement for optimizing swine reproduction. J. Anim. Sci. 71:239–246.

Lindemann, M. D., and E. T. Kornegay. 1986. Effect of folic acid additions to weanling pig diets. In: E. T. Kornegay, J. Gerken, J. Knight, and D. Notter, editors, VA. Tech. Livest. Res. Rep. p. 20–22.

Lin, Y., G. Lv, H. Dong, D. Wu, Z. Tao, S. Xu, L. Che, Z. Fang, S. Bai, B. Feng, J Li, and X. Xu. 2017. Effects of the different levels of dietary vitamin D on boar performance and semen quality. Livest. Sci. 203:63–68.

Lindemann, M. D., and E. T. Kornegay. 1989. Folic acid supplementation to diets of gestating and lactating swine over multiple parities. J. Anim. Sci. 67:459–464.

Lindemann, M. D., J. H. Brendemuhl, L. I. Chiba, C. S. Darroch, C. R. Dove, M. J. Estienne, and A. F. Harper. 2008. A regional evaluation of injections of high levels of vitamin A on reproductive performance of sows. J. Anim. Sci. 86:333–338.

Lipiński, K., E. Szramko, H. Jeroch, and P. Matusevičius. 2012. Effects of betaine on energy utilization in growing pigs – a review. Ann. Anim. Sci.12:291–300.

Liu, J., D. Chen, B. Yu, and X. Mao. 2011. Effect of maternal folic acid supplementation on hepatic one-carbon unit associated gene expressions in newborn piglets Mol. Biol. Rep. 38:3849–3856.

Liu, J., B. Yu, X. Mao, J. He, J. Yu, P. Zheng, Z. Huang, and D. Chen. 2012. Effects of intrauterine growth retardation and maternal folic acid supplementation on hepatic mitochondrial function and gene expression in piglets. Arch. Anim. Nutr. 66:357–371.

Liu, J., B. Yu, X. Mao, Z. Huang, P. Zheng, J. Yu, J. He, and D. Chen. 2014. Maternal folic acid supplementation and intrauterine growth retardation on epigenetic modification of hepatic gene expression and lipid metabolism in piglets. J. Anim. Plant Sci. 24:63–70.

Lombardi-Boccia, G., S. Lanzi, and A. Aguzzi. 2005. Aspects of meat quality: trace elements and B vitamins in raw and cooked meats. J. Food Comp. Anal. 18:39–46.

Loudenslager, M. J., P. K. Ku, P. A. Whetter, D. E. Ullrey, C. K. Whitehair, H. D. Stowe, and E. R. Miller. 1986. Importance of diet of dam and colostrum to the biological antioxidant status and parenteral iron tolerance of the pig. J. Anim. Sci. 63:1905–1914.

Luce, W. G., R. D. Geisert, M. T. Zavy, A. C. Clutter, F. W. Bazer, C. W. Maxwell, M. D. Wiltmann, R. M. Blair, M. Fairchild, and J. Wiford. 1990. Effect of riboflavin supplementation on reproductive performance of bred sows. Animal Science Report. Oklahoma Agric. Exp. Stn., Stillwater, OK, US.

Lu, T., A. F. Harper, J. T. Dibner, J. M. Scheffler, B. A. Carl, M. J. Estienne, J. Zhao, and R. A. Dalloul. 2014. Supplementing antioxidants to pigs fed diets high in oxidants: II. Effects on carcass characteristics, meat quality, and fatty acid profile. J. Anim. Sci. 92:5464–5475.

Mahan, D. C., and J. L. Vallet. 1997. Vitamin and mineral transfer during fetal development and the early postnatal period in pigs. J. Anim. Sci. 75:2731–2738.

Mahan, D. C. 1991. Assessment of the influence of dietary vitamin E on sows and offspring in three parities: reproductive performance, tissue tocopherol, and effects on progeny. J. Anim. Sci. 69:2904–2917.

Mahan, D. C. 1994. Effects of dietary vitamin E on sow reproductive performance over a five-parity period. J. Anim. Sci. 72:2870–2879.

Mahan, D. C., Y. Y. Kim, and R. L. Stuart. 2000. Effect of vitamin E sources (RRR- or all-rac-α-tocopheryl acetate) and levels on sow reproductive performance, serum, tissue, and milk α-tocopherol contents over a five-perity period, and the effects on the progeny. J. Anim. Sci. 78:110–119.

Mahan, D. C., and L. J. Saif. 1983. Efficacy of vitamin C-supplementation for weanling swine. J. Anim. Sci. 56:631–639.

Markant, A., M. Kuhn, O. P. Walz, and J. Pallauf. 1993. The intermediate relationship of nicotinamid and tryptophan in piglets. J. Anim. Physiol. Anim. Nutr. 70:225–235.

Marin-Guzman, J., D. C. Mahan, Y. K. Chung, J. L. Pate, and W. F. Pope. 1997. Effects of dietary selenium and vitamin E on boar performance and tissue responses, semen quality and subsequent fertilization rates in mature gilts. J. Anim. Sci. 75:2994–3003.

Marin-Guzman, J., D. C. Mahan, and J. L. Pate. 2000. Effect of dietary selenium and vitamin E on spermatogenic development in boars. J. Anim. Sci. 78:1537–1543.

Marchetti, M., M. Tassinari, and S. Marchetti. 2000. Menadione nicotinamide bisulphite as a source of vitamin K and niacin activities for the growing pig. Anim. Sci., 71:111–117.

Martelli, G., L. Sardi, P. Parishini, A. Badiani, P. Parazza and A. Mordenti. 2005. The effects of a dietary supplement of biotin on Italian heavy pigs' (160 kg) growth slaughtering parameters, meat quality and the sensory properties of cured hams. Livest. Prod. Sci. 93:117–124

Matte, J. J., and I. Audet. 2020. Maternal perinatal transfer of vitamins and trace elements to piglets. Animal 14:31–38.

Matte J. J., and C. L. Girard. 1989. Effects of intramuscular injections of folic acid during lactation on folates in serum and milk and performance of sows and piglets. J. Anim. Sci. 67:426–431.

Matte, J. J., and C. L. Girard. 1996. Changes of serum and blood volumes during gestation and lactation in multiparous sows. Can. J. Anim. Sci. 76:263–266.

Matte, J. J. and C. L. Girard. 1999. An estimation of the requirement for folic acid in gestating sows: the metabolic utilization of folates as a criterion of measurement. J. Anim. Sci. 77:159–165.

Matte, J. J., W. H. Close, and C. Pomar. 1991. The effect of interrupted suckling and split-weaning on reproductive performance of sows: a review. Livest. Prod. Sci. 30:195–212.

Matte, J. J., A. Giguère, and C. Girard. 2005. Some aspects of the pyridoxine (vitamin $B_6$) requirement in weanling piglets. Br. J. Nutr. 93:723–730.

Matte, J. J., C. L. Girard, and G. J. Brisson. 1984. Folic acid and reproductive performances of sows. J. Anim. Sci. 59:1020–1025.

Matte, J. J., C. L. Girard, and G. J. Brisson. 1992. The role of folic acid in the nutrition of gestating and lactating primiparous sows. Livest. Prod. Sci. 32:131–148.

Matte, J. J., C. L. Girard, and B. Sève. 2001. Effects of long term parenteral administration of vitamin $B_6$ on $B_6$ status and some aspects of the glucose and protein metabolism of early-weaned piglets. Br. J. Nutr. 85:11–21.

Matte, J. J., C. L. Girard, and G. F. Tremblay. 1993. Effect of long-term addition of folic acid on folate status, growth performance, puberty attainment, and reproductive capacity of gilts. J. Anim. Sci. 71:151–157.

Matte, J. J., F. Guay, and C. L. Girard. 2006. Folic acid and vitamin $B_{12}$ in reproducing sows: New concepts. Can J. Anim. Sci. 86:197–205.

Matte, J. J., F. Guay, and C. L. Girard. 2012. The contribution of portal drained viscera to circadian homocysteinemia in pigs. J. Anim. Sci. 90(Suppl. 4):68–70.

Matte, J. J., A. A. Ponter, and B. Sève. 1997. Effects of chronic parenteral pyridoxine and acute enteric tryptophan on pyridoxine status, glycemia and insulinemia stimulated by enteric glucose in weanling piglets. Can. J. Anim. Sci. 77:663–668.

Matte, J. J., C. Farmer, C. L. Girard and J.-P. Laforest. 1996. Dietary folic acid, uterine function and early embryonic development in sows. Can. J. Anim. Sci. 76:427–433.

Matte, J. J., C. L. Girard, R. Bilodeau and S. Robert. 1990. Effects of intramuscular injections of folic acid on serum folates, haematological status and growth performance of growing-finishing pigs. Reprod. Nutr. Develop. 30:103–114.

Matte, J. J., E. Corrent, A. Simongiovanni, and N. Le Floc'h. 2016. Tryptophan metabolism, growth responses and post-prandial insulin metabolism in weaned piglets according to the dietary provision of niacin (vitamin B3) and tryptophan. J. Anim. Sci. 94:1961–1971.

Matte, J. J., F. Guay, N. Le Floc'h, and C. L. Girard. 2010. Bioavailability of dietary cyanocobalamin (vitamin $B_{12}$) in growing pigs. J. Anim. Sci. 88:3936–3944.

Matte, J. J., N. LeFloc'h, Y. Primot, and M. Lessard. 2011. Interaction between dietary tryptophan and pyridoxine on tryptophan metabolism, immune responses and growth performance in post-weaning pigs. Anim. Feed Sci. Technol. 170:256–264.

Matte, J. J., I. Audet, B. Ouattara, N. Bissonnette, G. Talbot, J. Lapointe, G. Guay, L Lo Verso, and M. Lessard. 2017. Effects of sources and routes of administration of copper and vitamins A and D on postnatal status of these micronutrients in suckling piglets [Effets des sources et voies d'administration du cuivre et des vitamines A et D sur le statut postnatal de ces micronutriments chez les porcelets sous la mère]. Journées Rech. Porcine en France. 49:69–74.

Miller, E. R., D. A. Schmidt, J. A. Hoefer, and R. W. Luecke. 1965. Comparisons of casein and soy protein upon mineral balance and vitamin $D_2$ requirement of the baby pig. J. Nutr. 85:347–353.

Minelli, G., P. Macchioni, M. C. Ielo, P. Santoro, and D. P. Lo Fiego. 2013. Effects of dietary level of pantothenic acid and sex on carcass, meat quality traits and fatty acid composition of thigh subcutaneous adipose tissue in Italian heavy pigs. Ital. J. Anim. Sci. 12:329–335.

Monahan, F. J., A. Asghar, J. I. Gray, D. J. Buckley, and P. A. Morrissey. 1994. Effect of oxidized dietart-lipid and vitamin E on the colour stability of pork chops. Meat Sci. 37:205–215.

Mosenthin, R., W. C. Sauer, L. Völker, and M. Frigg. 1990. Synthesis and absorption of biotin in the large intestine of pigs. Livest. Prod. Sci. 25:95–103.

McDowell, L. R. 2000. Vitamins in animal nutrition. Academic Press Inc. San Diego, CA.

McGlone, J. J. 2000. Deletion of supplemental minerals and vitamins during the late finishing period does not affect pig weight gain and feed intake. J. Anim. Sci. 78:2797–2800.

Moffatt, R. J., F. A. Murray, A. P. Grifo, Jr., L. W. Haynes, J. E. Kinder, and G. R. Wilson. 1980. Identification of riboflavin in porcine uterine secretions. Biol. Reprod. 23:331–335.

Mosnier, E., J. J. Matte, M. Etienne, P. Ramaekers, B. Sève, and N. Le Floc'h. 2009. Tryptophan metabolism and related B vitamins in the multiparous sow fed *ad libitum* after farrowing. Arch. Anim. Nutr. 63:467–478.

Mudd, A. T., L. S. Alexander, S. K. Johnson, C. M. Getty, O. V. Malysheva, M. A. Caudill, and R. N. Dilger. 2016. Perinatal dietary choline deficiency in sows influences concentrations of choline metabolites, fatty acids, and amino acids in milk throughout lactation. J. Nutr. 146:2216–2223.

Nakano, T., F. X. Aherne, and J. R. Thompson. 1983. Effect of dietary supplementation of vitamin C on pig performance and the incidence of osteochondrosis in elbow and stiftle joints in young growing swine. Can. J. Anim. Sci. 63:421–428.

NRC. 1998. Nutrient requirements of swine. 10th rev. ed. Natl. Acad. Press, Washington, DC.

NRC. 2012. Nutrient requirements of swine. 11th rev. ed. Natl. Acad. Press, Washington, DC.

Nielsen, H. E., V. Danielsen, M. G. Simesen, G. Gissel-Nielsen, W. Hjarde, T. Leth, and A. Basse. 1979. Selenium and vitamin E deficiency in pigs. I. Influence on growth and reproduction. Acta Vet. Scand. 20:276–288.

Nunetz, S. B., J. A. Medin, H. Keller, K. Wang, K. Ozato, W. Wahli, and J. Segars. 1995. Retinoid X receptor b and peroxisome proliferator activated receptor activate an estrogen responsive element. Rec. Prog. Horm. Res. 50:465–469.

Palludan, B. 1963. Vitamin A deficiency and its effect on the sexual organs of the boar. Acta Vet. Scand. 4:166–155.

Parsons, M. J., P. K. Ku, D. E. Ullrey, H. D. Stowe, P. A. Whetter, and E. R. Miller. 1985. Effects of riboflavin supplementation and selenium source on selenium metabolism in the young pig. J. Anim. Sci. 60:451–461.

Partridge, I. G., and M. S. McDonald. 1990. A note on the response of growing pigs to supplemental biotin. Anim. Prod. 50:195–197.

Pearson, P. L., H. G. Klemcke, R. K. Christenson, and J. L. Vallet. 1998. Uterine environment and breed effects on erythropoiesis and liver protein secretion in late embryonic and early fetal swine. Biol. Reprod. 58:911–918.

Pettigrew, J. E., S. M. El-Kandelgy, L. J. Johnston, and G. C. Shurson. 1996. Riboflavin nutrition of sows. J. Anim. Sci. 74:2226–2230.

Pietrzik, K., and A. Brönstrup. 1997. Folate in preventive medicine: A new role in cardiovascular disease, neural tube defects and cancer. Ann. Nutr. Metabol. 41:331–343.

Pinelli-Saavedra, A., and J. R. Scaife. 2005. Pre- and postnatal transfer of vitamins E and C to piglets in sows supplemented with vitamin E and vitamin C. Livest. Prod. Sci. 97:231–240.

Pinelli-Saavedra, A., A. M. C. de la Barca, J. Hernandez, R. Valenzuela, and J. R. Scaife. 2008. Effect of supplementing sows' feed with alpha-tocopherol acetate and vitamin C on transfer of alpha-tocopherol to piglet tissues, colostrum, and milk: Aspects of immune status of piglets. Res. Vet. Sci. 85:92–100.

Pion, S.J., E. van Heugten, M. T. See, D. K. Larik, and S. Pardue. 2004. Effects of vitamin C supplementation on plasma ascorbic acid and oxalate concentrations and meat quality in swine. J. Anim. Sci. 84:2004–2012.

Pointillart, A., T. Denis, C. Colin, and H. Lacroix. 1997. Vitamin C supplementation does not modify bone mineral content or mineral absorption in growing pigs. J. Nutr. 127:1514–1518.

Pond, W. G., and K. A. Houpt. 1978. The biology of the pig. Cornell Univ. Press, Ithaca, NY.

Pusateri, A. E., M. A. Diekman, and W. L. Singleton. 1999. Failure of vitamin A to increase litter size in sows receiving injections at various stages of gestation. J. Anim. Sci. 77:1532–1535.

Radcliffe, J. S., B. T. Richert, L. Peddireddi, and S. A. Trapp. 2003. Effects of supplemental pantothenic acid during all or part of the growing-finishing period on growth performance and carcass composition. J. Anim. Sci. 81(Suppl. 1):255. (Abstr.).

Real, D. E., J. L. Nelssen, J. A. Unruh, M. D. Tokach, R. D. Goodband, S. S. Dritz, J. M. DeRouchey, and E. Alonso. 2002. Effects of increasing dietary niacin on growth performance and meat quality in finishing pigs reared in two different environments. J. Anim. Sci. 80:3203–3210.

Reiner, G., B. Hertrampf, and K. Kohler. 2004. Vitamin A-intoxification in the pig. Tierarztliche Praxis Ausgabe Grosstiere Nuttztiere 32:218–224.

Rey, A. I., C. J. Lopez-Bote, and G. Litta. 2017. Effects of dietary vitamin E (dl-alpha-tocopheryl acetate) and vitamin C combination on piglets oxidative status and immune response at weaning. J. Anim. Feed Sci. 26:226–235.

Rivera-Alejandro, N., and E. Jiménez-Cabán. 2014. Reproductive performance of gestating gilts supplemented with riboflavin. J. Agric. Univ. P.R. 98:119–128.

Roberts, R. M., and F. W. Bazer. 1988. The functions of uterine secretions. J. Reprod. Fertil. 82:875–892.

Roland, D. A., Sr., and H. M. Edwards, Jr. 1971. Effect of essential fatty acid deficiency and type of dietary fat supplementation on biotin-deficient chicks. J. Nutr. 101:811–818.

Rosenvold, K., H. N. Lærke, S. K. Jensen, A. Karlsson, K. Lundström, and H. J. Andersen. 2002. Manipulation of critical quality indicators and attributes in pork trough vitamin E supplementation level, muscle glycogen, reducing finishing feeding and preslaughter stress. Meat Sci. 62:485–496.

Russell, L. E., R. A. Easter, and P. J. Bechtel. 1985. Evaluation of the erythrocyte aspartate aminotransferase activity coefficient as an indicator of the vitamin B-6 status of postpubertal gilts. J. Nutr. 115:1117–1123.

Saddoris, K. L., L. Peddireddi, S. A. Trapp, B. T. Richert, B. Harmon, and J. S. Radcliffe. 2005. The effects of supplemental pantothenic acid in grow-finish pig diets on growth performance and carcass composition. Prof. Anim. Sci. 21:443–448.

Santoro, P., P. Macchioni, L. Franchi, F. Tassone, M. C. Ielo, and D. P. Lo Fiego. 2006. Effect of dietary pantothenic acid supplementation on meat and carcass traits in the heavy pig. Vet. Res. Comm. 30(Suppl. 1):383–385.

Shaw D.T., Rozeboom D.W., Hill G.M., Booren A.M., and J. E. Link. 2002. Impact of vitamin and mineral supplement withdrawal and wheat middling inclusion on finishing pig growth performance, fecal mineral concentration, carcass characteristics, and the nutrient content and oxidative stability of pork. J. Anim. Sci. 80:2920–2930.

Shurson, G. C., T. M. Salzerb, D. D. Koehlerc and M. H. Whitneyd. 2011. Effect of metal specific amino acid complexes and inorganic trace minerals on vitamin stability in premixes. Anim. Feed Sci. Technol. 163:200–206.

Signorello, M.G., Pascale, R. and G. Leoncini. 2002. ROS accumulation induced by homocysteine in human platelets. Ann. N. Y. Acad. Sci. 973, 546–549.

Simard, F., F. Guay, C. L. Girard, A. Giguere, J.-P. Laforest, and J. J. Matte. 2007. Effects of concentrations of cyanocobalamin in the gestation diet on some criteria of vitamin B-12 metabolism in first-parity sows. J. Anim. Sci. 85:3294–3302.

Simon, J. 1999. Choline, betaine and methionine interactions in chickens, pigs and fish (Including crustaceans). Worlds Poult. Sci. J. 55:353–374.

Skomial, J., M. Gagucki, and E. Sawosz. 2004. Urea and homocysteine in the blood serum of pigs fed diets supplemented with betaine and an enhanced level of B group vitamins. J. Anim. Feed Sci. 13(Suppl. 2):53–56.

Solymosi, F., and P. Horn. 1994. Protein content and amino acid composition of the uterine milk in swine and cattle. Acta Veterinaria Hungarica 42:487–494.

Stahly, T. S., and T. R. Lutz. 2001. Role of pantothenic acid as a modifier of body composition in pigs. J. Anim. Sci. 79(Suppl. 1):68. (Abstr.).

Stahly, T. S., N. H. Williams, T. R. Lutz, R. C. Ewan, and S. G. Swenson. 2007. Dietary B vitamin needs of strains of pigs with high and moderate lean growth. J. Anim. Sci. 85:188–195.

Stead, L. M., J. T. Brosnan, M. E. Brosnan, D. E. Vance, and R. L. Jacobs. 2006. Is it time to reevaluate methyl balance in humans? Am. J. Clin. Nutr. 83:5–10.

Takahashi, O. 1995. Hemorrhagic toxicity of a large dose of alpha-tocopherol, beta-tocopherol, gamma-tocopherol, and delta-tocopherol, ubiquinone, beta-carotene, retinol actate and l-ascorbic acid in the rat. Food Chem. Tox. 33:121–128.

Teague, H. S., and A. P. Grifo. 1966. Vitamin $B_{12}$ supplementation of sow rations. J. Anim. Sci. 25:895.

Teague, H. S., A. P. Grifo, Jr., and W. M. Palmers. 1971. Pantothenic acid deficiency in the sow. J. Anim. Sci. 33(Suppl. 1):239. (Abstr.).

Thaler, R. C., J. R. Nelssen, R. D. Goodband, and G. L. Allee. 1989. Effect of dietary folic acid supplementation on sow performance through two parities. J. Anim. Sci. 67:3360–3369.

Tilton, S. L., R. O. Bates, and R. J. Moffatt. 1991. Effect of riboflavin supplementation during gestation on reproductive performance of sows. J. Anim. Sci. 69 (Suppl. 1):482. (Abstr.).

Tovar, A., C. K. Ameho, J. B. Blumberg, J. W. Peterson, D. Smith, and S. L. Booth. 2006. Extrahepatic tissue concentrations of vitamin K are lower in rats fed a high vitamin E diet. Nutr. Metab. 3:29.

Traber, M. G., A. Elsner, and R. Brigelium-Flohe. 1998. Synthetic as compared with natural vitamin E is preferentially excreted as α-CEHC in human urine: studies using deuterated α-tocopheryl acetates. Am. J. Clin. Nutr. 60:397–402.

Tremblay, G. F., J. J. Matte, J. J. Dufour, and G. J. Brisson. 1989. Survival rate and development of foetuses during the first 30 days of gestation after folic acid addition to a swine diet. J. Anim. Sci. 67:724–732.

Tribout, T., J. C. Caritez, J. Gogue, J. Gruand, Y. Billon, M. Bouffaud, H.Lagant, J. LeDividich, F. Thomas, H. Quesnel, R. Guéblez, and J. P. Bidanel. 2003. Estimation, par utilisation de semence congelée, du progrès génétique réalisé en France entre 1977 et 1998 dans la race porcine Large White: résultats pour quelques caractères de reproduction femelle. Journées Rech. Porcines en France. 35:285–292.

van Wettere, W. H. E. J., R. J. Smits, and P. E. Hughes. 2013. Methyl donor supplementation of gestating sow diets improves pregnancy outcomes and litter size. Anim. Prod. Sci. 53:1–7.

Vigiano, P., S. Mangioni, F. Pompei, and I. Chiodo. 2003. Maternal-conceptus cross talk – a review. Placenta 24:S56–S61.

Wang, S. P., Y. L. Yin, Y. Qian, L. L. Li, F. N. Li, B. E. Tan, X. S. Tang, and R. L. Huang. 2011. Effects of folic acid on the performance of suckling piglets and sows during lactation. J. Sci. Food Agric. 91:2371–2377.

Watkins, B. A. 1989. Influence of biotin deficiency and dietary trans-fatty acids on tissue lipids in chickens. Br. J. Nutr. 61:99–111.

Watkins, B. A., and F. H. Kratzer. 1987. Dietary biotin effects on polyunsaturated fatty acids in chick tissue lipids and prostaglandin $E_2$ levels in freeze-clamped hearts. Poult. Sci. 66:1818–1828.

Watkins, K. L., L. L. Southern, and J. E. Miller. 1991. Effect of dietary biotin supplementation on sow reproductive performance and soundness and pig growth and mortality. J. Anim. Sci. 69:201–206.

Wegger, I., and B. Palludan. 1994. Vitamin C deficiency causes hematological and skeletal abnormalities during fetal development. J. Nutr. 124:241–248.

Wiegand, B. R., J. C. Sparks, D. C. Beitz, F. C. Parrish, Jr., R. L. Horst, A. H. Trenckle, and R. C. Ewan. 2002. Short-term feeding of vitamin D3 improves color but does not chnge tenderness of pork-loin chops. J. Anim. Sci. 80:2116–2121.

Wilborn, B. S., C. R. Kerth, W. F. Owsley, W. R. Jones, and L. T. Frobish. 2004. Improving pork quality by feeding supranutritional concentrations of vitamin $D_3$. J. Anim. Sci. 82:218–224.

Wilburn, E. E., D. C. Mahan, D. A. Hill, T. E. Shipp, and H. Yang. 2008. An evaluation of natural (RRR-α-tocopheryl acetate) and synthetic (all-rac-α-tocopheryl acetate) vitamin E fortification in the diet or drinking water of weanling pigs. J. Anim. Sci. 86:584–591.

Wilson, M. E., J. E. Pettigrew, and R. D. Walker. 1991. Provision of additional vitamin B$_{12}$ improved growth rate of weanling pigs. J. Anim. Sci. 69(Suppl. 1):359. (Abstr.).

Wiseman, S. L., J. R. Wenninghoff, R. D. Sauer, and D. M. Danielson. 1991. The effect of supplementary riboflavin fed during the breeding and implantation period on reproductive performance of gilts. J. Anim. Sci. 69(Suppl. 1):359. (Abstr.).

Witten, S., and K. Aulrich. 2018. Effect of variety and environment on the amount of thiamine and riboflavin in cereals and grain legumes. Anim. Feed Sci. Technol. 238:39–46.

Witten, S., and K. Aulrich. 2019. Exemplary calculations of native thiamine (vitamin B1) and riboflavin (vitamin B2) contents in common cereal-based diets for monogastric animals. Org. Agric. 9:155–164.

Woodworth, J. C., R. D. Goodband, J. L. Nelssen, M. D. Tokach, and R. E. Musser. 2000. Added dietary pyridoxine, but not thiamin, improves weanling pig growth performance. J. Anim. Sci. 78:88–93.

Yang, H., J. Lopez, T. Radle, M. Cecava, D. Holzgraefe, and J. Less. 2004. Effects of adding pantothenic acid into reduced protein diets on performance and carcass traits of growing-finishing pigs. J. Anim. Sci. 82(Suppl. 2):39. (Abstr.).

Yang, Y. X., S. Heo, Z. Jin, J. H. Yun, J. Y. Choi, S. Y. Yoon, M. S. Park, B. K. Yang, and B. J. Chae. 2009. Effects of lysine intake during late gestation and lactation on blood metabolites, hormones, milk composition and reproductive performance in primiparous and multiparous sows. Anim. Reprod. Sci. 112:199–214.

Yang, J. W., G. Tian, D. W. Chen, P. Zheng, J. Yu, X. B. Mao, J. He, Y. H. Luo, J. Q. Luo, Z. Q. Huang, and B. Yu. 2019. Effects of 25-hydroxyvitamin D-3 supplementation on growth performance, immune function, and antioxidative capacity in weaned piglets. Arch. Anim. Nutr. 73:44–51.

Yen, J. T., and W. G. Pond. 1981. Effect of dietary vitamin C addition on performance, plasma vitamin C and hematic iron status in weanling pigs. J. Anim. Sci. 5:1292–1296.

Yu, B., Yang, G., Liu, J., and D. Chen. 2010. Effects of folic acid supplementation on growth performance and hepatic folate metabolism-related gene expressions in weaned piglets. Front. Agric. China. 4:494–500.

Yen, J. T., and W. G. Pond. 1983. Response of swine to periparturient vitamin supplementation. J. Anim. Sci. 56:621–624.

Zeisel, S. H. 2006. Choline: Critical role during fetal development and dietary requirements in adults. Ann. Rev. Nutr. 26:229–250.

Zempleni, J., G. M. Green, A. W. Spannagel, and D. M. Mock. 1997. Biliary excretion of biotin and biotin metabolites is quantitatively minor in rats and pigs. J. Nutr. 127:1496–1500.

Zehnder, D., R. Bland, M. C. Williams, R. W. McNinch, A. J. Howie, P. M. Stewart, and M. Hewison. 2001. Extra-renal expression of 25-hydroxyvitamin D$_3$-1α-hydrolase. J. Clin. Endocrinol. Metab. 86:888–894.

Zehnder, D., N. K. Evans, M. D. Kilby, N. J. Bulmer, B. A. Innes, P. M. Stewart, and M. Hewison. 2002. The ontogeny of 25-dihydroxyvitamin D3, 1α-hydroxylase expression in human placenta and deciduas. Am. J. Pathol. 161:105–114.

Zhao, J. M., D. F. Li, X. S. Piao, W. J. Yang, and F. L. Wang. 2002. Effects of vitamin C supplementation on performance, iron status and immune function of weaned pigs. Arch. Anim. Nutr. 56:33–40.

Zhang, Z., E. Kebreab, M. Jing, J. C. Rodriguez-Lecompte, R. Kuehn, M. Flintoft, and J. D. House. 2009. Impairments in pyridoxine-dependent sulphur amino acid metabolism are highly sensitive to the degree of vitamin B6 deficiency and repletion in the pig. Animal 3:826–837.

Zhang, L., Li, M., Shang, Q., Hu, J., Long, S. and X. Piao. 2019. Effects of maternal 25-hydroxycalciferol on nutrient digestibility, milk composition and fatty-acid profile of lactating sows and gut bacteria metabolites in the hindgut of sucking piglets. Anim. Nutr. 4:271–286.

Zhang, Z. F., J. Li, J. C. Park, and I. H. Kim. 2013. Effect of vitamin levels and different stocking densities on performance, nutrient digestibility, and blood characteristics of growing pigs. Asian-Australas. J. Anim. Sci. 26: 241–246.

Zhou, H., Y. Chen, Y. Zhuo, G. Lv, Y. Lin, B. Feng, Z. Fang, L. Che, S. Xu, and D. Wu. 2017. Effects of 25-hydroxycholecalciferol supplementation in maternal diets on milk quality and serum bone status markers of sows and bone quality of piglets. Anim. Sci. J. 88:476–483.

# 8    Minerals and Mineral Utilization in Swine

Gretchen M. Hill

## Introduction

Minerals in swine diets continue to be a challenge in swine production. Research revealing requirements for specific genotypes, age, and physiological state has not been determined for specific minerals, yet it has been known for over 10 years that genotypes differ in body composition and differing growth patterns (Hinson et al. 2009; Alexander et al. 2010; Pomar et al. 2014). The determination of a mineral in a feed ingredient or diet does not guarantee that the presence of this nutrient is useful to the animal to perform a function, hence, bioavailable. Some ultratrace elements do not have a confirmed role as a physiological function, hence may not really be a required nutrient that should be added to diets (O'Dell and Sunde 1997. As noted previously (Hill 2013), F, Cr, V, Si, Ni, As, Li, Pb, and B have not been proven as a catalyst or a regulator confirming its physiological function. Additionally, Zn and Cu are used by the swine industry as pharmacological agents that enhance growth at concentrations exceeding requirements for these nutrients.

Macro minerals are those required by the animal in large amounts (g or kg/d) and include Ca, P, Mg, and S as well as the electrolytes Cl, K, and Na. Trace or microelements (μg/d) include Zn, Cu, Mn, Se, Mo, I, and Co. Although Fe is required in diets in an intermediate amount between macro and microelements, it is usually considered as a trace element.

As noted by Mahan and Vallet (1997), ". . . There are periods during pregnancy when sows may have a temporally high requirement for certain vitamins and minerals." Using markers to study bone metabolism, van Riet et al. (2016) reported that primiparous and multiparous sows differed, but not during the reproductive cycle. However, Crenshaw et al. (2013) found that the percent of bone ash increased with parity. Genetic differences and this research help to remind us why it is not possible to quantify definitely the nutrient requirement of sows and probably their offspring.

## Sulfur

This element is a part of certain vitamins and amino acids and is found in skin, nail, and hair tissue where there is a high S-amino acid content. Animals that require thiamin, biotin, cysteine, cysteine, taurine, and methionine need S. Plants and micro-organisms utilize S to make methionine, an essential amino acid for swine, the pig's microorganisms cannot synthesize in adequate amounts to meet their needs.

In recent years, the role of S in its interactions with Se, Cu, and Fe is probably of greatest concern in swine diets because of its presence in dried distillers grains with solubles (DDGS) where sulfuric

*Sustainable Swine Nutrition*, Second Edition. Edited by Lee I. Chiba.
© 2023 John Wiley & Sons Ltd. Published 2023 by John Wiley & Sons Ltd.

acid is used in processing. As noted previously (Hill 2013), determination of S is difficult because it is an anion, and consistent results are only available from reputable laboratories who perform S analysis routinely.

## Calcium

This important macroelement is associated with its essentiality in the skeleton as hydroxyapatite, whitlockite, carbonate, or phosphate, but the other 1% of the body's Ca is also essential in the blood and cells. The amount of Ca must be carefully controlled because of its relationship with P, Mg, etc. Hence, measured Ca in swine diets does not guarantee its bioavailability to serve a function in the body, and its absorption is carefully controlled resulting in plasma Ca homeostasis, meeting the needs of the body. This regulation in plasma Ca concentration prevents hypercalcemia or hypocalcemia and is the role of calcitonin and parathyroid hormone (PTH) to prevent tetany, paresthesia, muscle weakness, anorexia, etc.

The role of Ca bone and tissue health was previously discussed (Hill 2013).

The research of Rortvedt and Crenshaw (2012) to characterize the cause of kyphosis (spine curvature that is outward and abnormal) illustrates the interaction of Ca, P, and vitamin D and reminds us why absolute requirements are not possible without considering other nutrients.

### Calcium Absorption and Transport

Active transport is utilized in the duodenum when Ca intake is low, but when Ca intake is high, passive paracellular transport in the jejunum and ileum is utilized. The hormone derived from vitamin D, calcitriol, controls active transport under the control of PTH. Calbindin is involved in the transfer of Ca in absorption. The $Ca^{2+}$-$Mg^{2+}$ ATPase is utilized in the duodenum with the exchange of Mg for Ca. Because Ca absorption is limited, excess Ca is excreted in the feces.

Calcitriol activity in the brush border is also involved in P absorption via alkaline phosphatase and other carriers. Numerous other roles of calcitriol are known in bone metabolism and other tissues for cell differentiation, proliferation, and growth. Fiber and phytate (myoinositol hexaphosphate) decrease the absorption of Ca and other minerals.

As noted previously (Hill 2013), the amount of Ca in plasma is tightly regulated with about half of the Ca in blood being free and the remainder bound to protein (albumin or pre-albumin) and sulfate, phosphate, or citrate. Hence, plasma Ca is not a good indicator of Ca status.

### Digestibility and Metabolism

The mobilization of Ca from bone and utilization from dietary sources are important for bone health, milk production, and longevity. Crenshaw's laboratory (Darriet et al. 2017) determined that diets with acidogenic mineral supplements increased urinary and fecal Ca excretion and decreased Ca retention. Greater amount of Ca (0.95% with 0.63% total P) has not been shown to improve sow lactation or pig performance, which is similar to the results in growing pigs reported by Stein's laboratory (Merriman et al. 2017). The retention of Ca and P changes during gestation with increased digestibility in the later stages (Lee et al. 2019). This means of compensation by the sow may be its means of "managing" different housing systems and increased productivity as genetics change (Tan et al. 2016). The observed homeostatic regulations of Ca and P in neonates via transcriptional changes of regulatory genes (Gagliardi et al. 2017) may be a means utilized by sows during gestation and lactation.

## *Dietary Needs*

Because of the cost of P sources and the importance of bone health, research has continued relative to Ca:P and use of phytase. Using Ca:total P of 1 : 1 in a low P diet with phytase (500 units Natuphos/kg), one laboratory (Liu et al. 2014) found increased apparent absorption of P from the small intestine and P absorption from the cecum with Ca unaffected. The true total tract digestibility (TTTD) of limestone and dicalcium phosphate or their combination were similar with 1 : 1 ratio and increasing concentrations of Ca at 4.0, 5.0, and 6.0 g/kg in young pigs (Zhang and Adeola 2017). With semi-purified diets, TTTD of calcitic limestone was reported to be 76.8% and monosodium phosphate was 88.8% (She et al. 2015). Using breaking strength and ash (Ross et al. 1984), calcitic limestone (2 sources), oyster shell flour, gypsum, marble dust, and aragonite had bioavailabilities of 93–102%, but dolomitic limestone was less bioavailable (51 and 78%). With Ca:P from 0.6:1 to 1.3:1 did not affect average daily gain (ADG) in growing pigs (Fan and Archbold 2012). Standardized total tract digestible (STTD) Ca requirement of 11–25-kg pigs to maximize bone ash was 0.48%. However, greater STTD of Ca was found to decrease ADG and G:F (González-Vega et al. 2016). Researchers have also investigated the cost/benefit on Ca and P needs by altering depletion and repletion feeding sequences (Gonzalo et al. 2017).

## *Phytase*

Research on the use of phytase supplements in nonruminant diets has been investigated for some time (Hill 2013). There are new products (Torrallardona et al. 2012; Torrallardona and Ader 2016; Arredondo et al. 2019) and techniques of utilizing phytase such as "super dosing," as well as studies to determine the effect of dietary ingredients on intestinal organisms production of phytase (Heyer et al. 2016). Adding phytase to diets increases apparent total tract digestibility (ATTD) and STTD of Ca and the ATTD of P regardless of Ca source (González-Vega et al. 2015). Additionally, regression equations for the use of predicting STTD of P are not always accurate for all DDGS products (Almeida and Stein 2012).

Recently, Moran et al. (2019) reported that exogenous inositol improved the efficiency of gain initially after weaning similar to super-dosing phytase. They also reported that super-dosing phytase (2500 FTU/kg) initially increased serum Zn (21d not 42d), Cu, and decreased Fe.

## Phosphorus

The body's need for P, its relationship with Ca and Mg, high costs in diet formulation, and high fecal excretion make P the mineral element most likely to be of concern in swine production. Like Ca, a majority of P is found in the skeleton, but it is essential in intermediary metabolism, as a nucleic acid and cell membrane component, in oxygen delivery, and a component of acid–base balance cannot be overlooked.

### *Digestion, Absorption, and Transport*

As previously noted (Hill 2013), inorganic P is absorbed in the inorganic form throughout the small intestine. Although the exact mechanisms are not known, it is believed that most is absorbed primarily in the duodenum and jejunum. Without phytase, phytic acid negatively affects absorption and makes the bound P unavailable. Alkaline phosphatase and phospholipase hydrolyze organic P so carrier-mediated active transport or diffusion can occur. Calcitriol stimulates absorption. Organic trace minerals have been shown to increase digestibility of P in corn–soybean diets in a research setting. Known interactions with Mg, Al, and Ca may reduce P absorption and may bind to P in plasma. However, most P is transported as phospholipids or bound to other proteins.

Phosphorus is the second most abundant mineral in the body, and most of it is found in the skeleton associated with Ca and laid down on collagen in the ossification process during the bone formation. Calcitriol and PTH as with Ca influence P metabolism. The P not in bone is found in soft tissue and extracellular fluids where it is part of nucleic acids, energy metabolism (high-energy phosphate bonds), acid–base balance, oxygen delivery, and enzymatic activities. With heat stress and differing P concentrations, Weller et al. (2013) used gene expression profile of nine genes encoding electron transport proteins in muscle and reported that P and temperature were important in regulation of oxidative phosphorylation. Immunological castrates are similar to boars (entire males) in P utilization (Elsbernd et al. 2015). The digestibility and absorption of P from monocalcium phosphate are not altered by the amount of P in the diet (Stein et al. 2008), and it is more bioavailable than dicalcium phosphate (Petersen et al. 2011).

### *Excretion*

In high grain diets, most of the dietary P is bound in the phytate molecule, which is not available without phytase. Hence, P is excreted in the feces. However, the absorbed P is excreted in the urine in the inorganic form and is involved in P balance. It has been known for some time that pharmacological Zn reduces digestibility of Ca and P, but the addition of microbial phytase increases ATTD and STTD if Ca and ATTD of P (Blavi et al. 2017).

Bikker et al. 2016 using semi-purified low-P diets reported that increasing feeding level increased endogenous phosphorus losses from the digestive tract and fecal excretion in grow-finish pigs. The phosphorus excretion per day when expressed on a BW basis was less for grow-finish pigs than sows (4.6 vs. 5.6 mg/kg BW). Bikker et al. (2017); however, noted that using endogenous P loss determined in grow-finish pigs may underestimate P needs in sows.

Phytase above 4000 FTU/kg has been shown to supply adequate Ca and P when fed deficient diets to nursery pigs, resulting in performance similar to pigs fed adequate Ca and P. This practice would need to be evaluated further before use in perspective breeding animals. When phytase is added to sow diets that are deficient in Ca and P, weight loss is similar to that of sows fed adequate Ca and P (Wealleans et al. 2015). Phytase has been shown to increase urinary P and retained P when adequate P diets are fed (Olsen et al. 2019).

Ingredient source may affect efficacy of phytase as noted by Bournazel et al. (2018) who reported that dehulled rapeseed meal may improve the release of P in stomach by microbial phytase. The ATTD and STTD of P increased in four differing canola meal diets as phytase in the diet increased (She et al. 2017). Rojas et al. (2013) reported that phytase added to corn, hominy feed, bakery meal, and corn germ meal would improve P digestibility. Research indicates that using total or apparent P digestibility may overfeed P in brown rice diets (Yang et al. 2007). If Ca is reduced in the diet by changing the Ca:P from 1.9 to 1.3, the efficiency of phytase is not improved (Létourneau-Montminy et al. 2010). A prebiotic, inulin, has been shown to not interact with phytase relative to mineral status or growth performance (Jolliff and Mahan 2012). Zinc balance was improved with supplemental Zn and phytase when 0 or 100 ppm Zn was fed, but as expected due to the Cu X Zn interaction, Cu balance was reduced by Zn supplementation (Adeola et al. 1995). Hinson et al. (2009) found that a diet formulated with reduced crude protein and additional amino acids, low phytic acid corn and phytase maintained growth performance, and carcass characteristics compared to a typical corn-soy diet. In sows, phytase supplementation increased apparent total tract digestibility of P but not crude protein or Ca during lactation in low P diet (Nasir et al. 2012).

## Electrolytes

The anions and cations are how the body fluids maintain pH balance and buffering capacity necessary for cellular function. The important cations are sodium, potassium, calcium, and magnesium; the anions are chloride, bicarbonate, and proteins/amino acids, organic acids, phosphate, and sulfate. A brief description can be found in Hill (2013).

### *Sodium*

Salt is 40% Na and has a high absorption rate. The kidney's glomerulus filters Na and as the major cation in extracellular fluid is reabsorbed by the proximal tubule, loop of Henle, distal convoluted tubule, and collecting ducts. In the blood, Na is unbound and its concentration is tightly regulated.

### *Chloride*

Usually, the concentration of Cl in extracellular fluid is the opposite of Na and hence passively follows Na in absorption from the small intestine. It can be reabsorbed actively by parts of the kidney.

### *Potassium*

Because K is the essential and primary cation in intracellular fluid, it is carefully balanced with excretion primarily by the kidneys equaling absorption. The hormone, aldosterone, is involved in K and Na and hence fluid balance.

### *Salt*

Although Na, K, and Cl are known to be required electrolytes, little work has been completed to define the role of salt (NaCl). Earlier work was reviewed by Hill (2013). Research indicates that a range of 166–250 mEq/kg is appropriate for weanling pigs (Lei et al. 2017). When dietary electrolytes were studied as mEq in sows (DeRouchey et al. 2003) weaning weight, milk composition, and rebreeding were not affected. Changing dietary electrolyte balance has been shown to not affect growth performance, meat quality, carcass yield, or esophageal ulcer scores in finishing pigs, and the electrolyte blood components returned to prior concentrations after a 10-day withdrawal period (Edwards et al. 2010).

## Iron

### Iron Regulation

Work in other mammalian species indicates that Fe homeostatic regulation develops around weaning (Lönnerdal 2017). Hence, the development of small intestinal Fe transporters (divalent metal-ion transporter 1 and ferroportin) occurs overtime. The relative expressions of both duodenal divalent transporter 1(DMT1) and solute carrier family 39 member 14 (ZIP14) were increased in weaned pigs fed 20 ppm Fe compared to those fed 520 ppm Fe (Hansen et al. 2009). The excretion of Fe is not controlled, but the liver, which also controls the release to other organs, regulates absorption in the proximal small intestine. Hepcidin, a peptide secreted by the liver, responds to Fe parameters (Darshan and Anderson 2007). Rincker et al. (2005) reported the role of Fe-regulated proteins (IRP) in storage and transport proteins in neonatal and nursery pigs. Additional information is available in Hill (2013).

### Iron Metabolism in the Pig

Pigs are born with about 30% of the total Fe in the body as humans yet their rate of growth is much greater, and colostrum and milk are very poor sources. Hence, Fe must be supplied in an available form. Giving an Fe injection at two to three days of age results in almost three times the total body Fe of an uninjected pig. In today's swine production, usually 200 mg of Fe is injected as Fe dextran within three days of birth. Jolliff and Mahan (2011) reported that additional 100 mg of injected Fe prior to weaning increased hematological status but with no increase in growth. Gut closure occurs early in life and prevents the absorption of large molecules. As the parity of the sow increases, the hepatic Fe concentrations decrease in the neonate (Hill et al. 1983). Spears laboratory (Hansen et al. 2009) also found that weaned pigs fed adequate (120 ppm) or high dietary Fe (520 ppm) had greater relative hepatic hepcidin expression than those fed low Fe.

Wan et al. (2017) reported that feeding 80 mg Fe as ferrous N-carbamyglycinate chelate during late gestation increased the liter weight of live pigs and Fe saturation, but reduced total iron-binding capacity and hepatic and renal Fe in neonates compared to supplementing the sows with Fe sulfate (Wan et al. 2018).

Lactoferrin, an Fe binding protein, has antibacterial and antiviral activity as a nonimmune natural defense (Valenti and Antonini 2005). Two peptide fragments of lactoferrin that are released during digestion were shown to increase growth and decrease the concentration of *E. coli* in the ileum, cecum, and colon (Tang et al. 2009). Wang et al. (2006) reported that dietary lactoferrin increased villus height and decreased crypt depth while improving growth and relative abundance of mRNA of PR-39 and protegrin-1 genes in nursery pigs. Bacterial growth and hence susceptibility to infection will occur with excessive oral or injectable Fe in the neonate. Details of Fe needs and utilization are reviewed in the latest NRC (2012).

*Dietary Fe*

The Fe concentration in feedstuffs and Fe sources is not an indication of its usefulness to the pig. For example, 80 ppm provided by dietary ingredients will not support adequate growth in growing pigs (Rincker et al. 2004). In general, the Fe in animal protein sources is more available for absorption if the animal needs metabolic Fe. Rincker et al. (2005) and Kerr et al. (2008) recently reported Fe concentrations in feedstuffs. Phytase (2500 FTU/kg) added to a low Fe diet (50 ppm) increased growth performance compared to 300 ppm Fe in a wheat-based soy diet (Laird et al. 2018).

## Zinc

The essentially of Zn in mammals is not questioned due to its role in many structures and enzymes in the body. As noted by McMahon and Cousins (1998), "it is small, hydrophilic, and a highly charged species which cannot cross biological membranes by passive diffusion." Hence, transporters with unique time, temperature, and pH requirements are involved in the absorption, utilization, and release of Zn and continue to be studied.

### *Transporters and Metallothionein*

Two families of proteins are involved in Zn transport: (i) ZnT (solute-inked carrier 30 or SLC30A) and (ii) Zip (Zrt-and Irt-like proteins (slc39A). The role of ZnT transporters is to lower Zn in cells and vesicles, while Zip proteins transport Zn from extracellular fluid or vesicles into cytoplasm (Cousins et al. 2006). There are 10 members in the ZnT family and 14 in the SLC39 family.

The first recognized transporter, ZNT-1, is found in many tissues of the body and is involved in transporting Zn out of the duodenal and jejunal intestinal cells where it is highly expressed. Hence, as expected when 2500 ppm Zn was fed to nursery pigs, mRNA of ZnT1 was increased (Martin et al. 2013). Because ZnT-1 is not decreased during low Zn intakes, it is likely that it is not a control point in Zn status regulation.

Until recently, ZnT-2 was only known to import Zn into vesicles and was found in secretory tissues such as the pancreas, prostrate, placenta, ovary, and mammary. Lee et al. (2015) reported that ZnT-2 is expressed in mammary epithelial cells. Without ZnT-2, there were defects in the mammary glands that resulted in reduced milk production that had less protein, fat, and lactose content.

The next Zn transporter found in mice was ZnT-3 that is expressed in the brain and testis in synapses and axons of specific neurons. The ZnT-4 is involved in Zn transport from the mammary gland into milk as well as a high level of expression in the brain. Zip 5 responds to Zn intake in the intestine. Zip 10 has a role in the liver, and there is an upregulation of Zip4 as well as Zip2 during Zn deficiency while Zip3 is decreased when Zn is supplemented (Cousins et al. 2006). Many of the Zn transporter genes are regulated by hormones and cytokines and as a result there is often an increase in hepatic Zn and hypozincemia during sepsis and inflammation.

The DCT1 transporter is high in the crypts and lower in the villi of enterocytes. Its expression increases during Fe deficiency, and this multi-element transporter may explain some of the interactions between Zn, Fe, Cd, Mn, and Cu. Metallothionein (MT) is high in the intestine, pancreas, kidney, and liver. The 20 cysteines that are a part of MT are responsible for the metal binding that is characteristic of this protein. Its role during cellular stress is as a redox unit.

## *Immunity*

Using regression models and k-means cluster analysis, Brugger and Windisch (2019) were able to confirm statistically that tissues essential for acute survival (heart, skeletal muscle, and immune tissues) were spared during Zn deficiency. The addition of Zn (37, 60, and 120 ppm) had no effect on an antibody response to bovine serum albumin or serum IL-1 or IL-2 concentrations (Guo-jun et al. 2009). Heat stress negatively affects the intestinal barrier, and Sanz-Fernandez et al. (2014) reported that supplementing Zn as an amino acid complex can reduce the damage to the intestinal integrity. In contrast, excessive intake of Zn was found to reduce lymphocyte stimulation as well as chemotaxis and phagocytosis of bacteria by polymorphonuclear leukocytes (Chandra 1984). Using mice, Philcox et al. (2000) were able to show with mice that MT reduces intestinal Zn loss during an acute endotoxin inflammation, but not during starvation or dietary Zn restriction. Hence, the importance of MT relative to Zn in immunity maybe important when pigs are weaned, not eating, have depressed antioxidant status, and impaired intestinal barrier and mitochondrial function (Cao et al. 2018).

## *Pharmacological Zn*

Since the reports of pharmacological Zn (2000–3000 ppm) fed as Zn oxide improved growth performance (Hill 2013), Carlson et al. (1999) were to first to report improved changes in gut morphology. Even with the negative impact of transmissible gastroenteritis on normal gut architecture, pharmacological Zn improved gut health and performance (Stanger et al. 1998). Even though the major site of Zn absorption is not the ileum, Grilli et al. (2015) reported that is part of the gut had improved structure and integrity and tumor necrosis factor-*a* protein concentration was reduced and suggested that a microencapsulated Zn oxide might produce similar responses.

Several studies have reported the impact on the microbiota of pharmacological Zn (Hill 2013). Recently using experimental animals, Zackular et al. (2016) reported that mice infected with *Clostridium difficile* had reduced disease activity and increased toxin activity. When pharmacological Zn was fed to nursery pigs, Kreuzer-Redmer et al. (2018) reported that after one week, the level of activated T-helper cells was increased as well as higher transcript amounts of interferon γ and T-box 21 compared to pigs fed 57 or 164 ppm Zn. However, after two weeks of dietary intervention, they reported "higher relative cell counts of CD4+CD25 regulatory T-helper cells and higher expression of forkhead box P3 (FOXP3) transcripts."

Long-term feeding of pharmacological Zn has been known to alter Cu status and interact with other nutrients as well (Hill et al. 1983). Recently, Walk et al. (2015) reported that Ca, P, Na, K, and Cu digestibility was reduced and that source of P and phytase can alter the effect. Feed preference studies (Reynolds et al. 2010) showed that pigs did not prefer diets supplemented with Zn oxide at pharmacological concentrations compared to basal diets (100 ppm).

## *Reproduction*

The first report of the relationship of Zn deficiency and defective reproduction resulting from the practice of geophagia and consuming a wheat diet led Prasad and associates (Prasad et al. 1961) to conclude that Zn deficiency might explain hypogonadism. More recently, Hunt et al. (1992) noted that serum testosterone concentrations, seminal volume, and total seminal Zn loss per ejaculate is sensitive in Zn depletion even if short term. In rat studies, it appears that prepubertal Zn deficiency damage in male reproduction can be repaired with Zn supplementation, but postpubertal degenerative changes due to Zn deficiency may not be reversible (Mason et al. 1982). More recently, Croxford et al. (2011), using mice as the experimental animal, found that Zip5 imports Zn into sertoli cells and spermatocytes assisted by Zip8 and Zip10. Zip 14 was found in undifferentiated spermatogonia and Leydig cells. Sutovsky et al. (2019) have recently reviewed improvement of sperm capacitation management relative to Zn.

Hostetler et al. (2003) noted that the level and concentration of Zn, Cu, and Mn are much greater in the conceptus than reproductive tissue indicating the important role of transporters and adequate trace mineral status for reproduction. Interactions of trace elements due to deficiencies and dietary excesses on reproductive success or lack thereof are well documented in experimental animals (Reinstein et al. 1984; Cottin et al. 2018). Zinc transporters in pregnancy and lactation have been determined in rodents (Liuzzi et al. 2003) as has the importance of MT (Andrews and Geiser 1999). Since pro- and antioxidative environment is important for health and providing an appropriate environment for the embryo (Guérin et al. 2001), the impact of deficiency and excess has been studied. Tian and Diaz (2013) reported that at the end of oocyte development, available Zn is essential. The importance of Zn to improve preimplantation embryonic development has been studied in pigs by Jeon et al. (2014). In humans, it has been found that Zn transporters, MT, and metal regulatory transcription factors are in oocytes but not in cumulus cells (Ménézo et al. 2011).

Following the findings of Hill et al. (1983) and Vallet et al. (2014), Johnston et al. (2019) recently reported research carried out in a commercial setting. Adding 365 ppm Zn as sulfate from day 75 to farrowing reduced the number of low birth weight pigs, and mortality was reduced from a 15 to 12.2% when 595 ppm Zn was added as sulfate to a diet containing 125 ppm Zn.

## Copper

Seldom in swine production in agriculture do we see overt Cu deficiency, but health and growth may be impaired. The Cu requirement for the growing pig is estimated to be 5–10 ppm, but no research has been carried out to determine the requirement for today's genetics and of all ages and in all physiological states. Neonates are born with a higher amount of Cu on a weight basis than they will have in later life, and Cu, Zn, and Fe concentrations in colostrum are higher than in mature milk, which supports the nursing pig.

When the dietary Cu concentration was increased beyond the 1998 NRC, the Cu content of the colostrum, liver, and empty sow body increased (Peters et al. 2013) and further increased if additional Ca and P were added to their diets. The opposite was true for Fe. The Cu concentration in mature milk decreased quadratically, and a linear decline in Fe concentration was observed as parity advanced (Peters et al. 2013).

### *Absorption and Transport*

As noted previously (Hill 2013), Cu and Fe must be bound so no "free" Cu or Fe is available to cause oxidative stress. The "free" Cu is released from the dietary sources of Cu by stomach acid and protease enzymes. The primary copper transporter, CTR1, can then be involved in cellular uptake of $Cu^{+1}$ by enterocytes in the upper small intestine and ultimately other cells. Divalent metal transporter 1 (DMT1), which the major transporter of Fe and likely Mn, may also be involved in limited Cu transport. Copper is exported from the upper small intestine into the blood by ATP7a, a Cu

ATPase (Fry et al. 2012). From the liver, the primary Cu storage organ, Cu is incorporated into ceruloplasmin (Cp) for transport to other tissues. There is more than one form of Cp and its form differs with age, and because it is influenced by estrogen, its concentration may vary with age and sex (Hill 2013). Although Cp has a role in the antioxidant system, so does Cu/Zn SOD, which may be the best indicator of Cu status.

### Grower-Finisher Pigs

Mahan's laboratory (Gowanlock et al. 2015) reported that when Cu, Zn, Fe, and Mn were added to basal corn-soy diets, the hepatic and duodenal MT increased, but jejunum MT did not change. Plasma minerals were not affected. Hepatic Zn increased as dietary Zn concentrations increased, but tissue concentrations of Cu, Mn, and Fe were not changed with dietary additions. This study indicates that when minerals are fed according to NRC during the nursery stage, the innate Cu, Fe, and Mn may be adequate for market animals. Similar results were found by Shaw et al. (2006). Hernández et al. (2008) reported that decreasing Cu from 156 ppm to 27 ppm Cu and Zn from 170 ppm to 56 ppm from various sources, performance, and mineral status were not altered.

Ruiz-Ascacibar et al. (2019) studied the mineral composition and deposition in the empty body of pigs from birth to 140 kg. They reported that the relative deposition rate decreased with increasing empty body weight except for Zn. This may explain the observation about the need for added Zn in finisher diets of Mahan's laboratory.

### Pharmacological Cu

As noted previously (Hill 2013), it was observed in England that pigs that licked Cu pipes grew faster, but the mysteries about the need pharmacologically continues. Like Kornegay et al. (1989), most believe that concentrations of 150–250 ppm of Cu that increase growth is not an immunological response, but more likely microbiota or gut physiology changes (Shurson et al. 1990; Zhao 2007). While the use of 250 ppm Cu as sulfate in grow/finish diets was the initial Cu pharmacological agent, many other forms of Cu successfully stimulate growth. However, it should be noted that like pharmacological Zn, it does not always result in a uniform response.

Research of Shelton et al. (2011) revealed that if 3000 ppm Zn as oxide was fed initially from weaning for 14 days followed by 125 ppm Cu as sulfate for 28 days growth performance equaled or exceeded if both stimulates were fed. Spear's laboratory (Fry et al. 2012) reported the feeding 225 ppm Cu as tribasic Cu chloride may cause less oxidative stress than when Cu in the sulfate form was fed to weanling pigs. Their work showed that at the transcription level, several Cu transporters and chaperones were increased.

### Manganese

Previously, we noted that the need for Mn was greater for reproduction than for growth although this mineral is found throughout the body without a main storage organ (Hill 2013). Since Mn is found in plants, typically it is found in livestock diets, and questions about its value as an organic dietary form are unanswered. It is known that biliary excretion hence fecal excretion is the route used by the body to control Mn status in pigs. As noted earlier, it is imported and exported into and out of cells by using Fe transporters such as DMT1 and ferroportin. Hence, its interaction with Fe is important and an example is the increase in duodenal Mn concentrations when pigs are fed a low Fe diet, and hepatic Mn is lower when pigs are fed a high Fe diet (Hansen et al. 2009). The mitochondrial form of SOD requires Mn to function in the antioxidant system and may be used as a status indicator.

Of interest in swine production, Damo et al. (2013) reported that Mn is necessary for maximal growth inhibition of several pathogenic bacteria by S100A8/S100A9 heterodimer calprotectin as part of nutritional immunity.

Excess dietary Mn (150 ppm) reduces Se concentration when the Mn concentration is increased in the *longissimus thoracis et lumborum* tissue, decreased ADG, increased catalase activity, and caused increased lipid oxidation while tenderness was increased (Schwarz et al. 2017).

## Selenium

Initially, Se was known to be toxic, but in 1972, it was found to be a component of glutathione peroxidase. It has both metallic and nonmetallic properties and hence is unique in biology. Analysis of tissue Se concentrations is difficult and should be accomplished by a quality laboratory. In humans, the tissues with the highest concentration of Se are kidneys, liver, spleen, pancreas, and testes.

As previously noted (Hill 2013), FDA controls the amount of Se that can be added to livestock diets due to the narrow range between deficiency and toxicity. Although most species are believed to require 0.1 ppm, pigs are thought to need 0.1–0.3 ppm. Thus, 0.3 ppm is the maximum that can be added to swine diets in the United States. Glutathione peroxidase 1 (GPX1) is usually used in determining Se status of animals. However, there are many forms of GPX (1, 2, 3, and 4), and it is found in many tissues. The three iodothyronine 5′ deiodiase enzymes are involved in the removal of 5 or 5′I from thyroid hormones. There are other important Se-containing proteins. It should be noted that Se is not managed the same in all species and tissues and hence status indicators are not the same in all species.

When Se was not fed to nonpregnant females, they were able to maintain serum Se, but it decreased in pregnant gilts (Piatkowski et al. 1979). Ultimately, this early work provided evidence that 0.1 ppm Se and 22 IU/kg vitamin E were necessary to maintain tissue Se.

Mahan's laboratory has shown in numerous studies and with pigs of various ages and physiological states that organic Se is more useful to the pigs than inorganic sources. Recently, they reported (Peters et al. 2013) that organic Se increased Se in colostrum and mature milk more than inorganic sources.

## Chromium

Although there are many sources of Cr available to swine producers for use in diets, Cr does not meet the criteria that were established for essentially as previously noted (Hill 2013). Additionally, Cr is very difficult to analyze; hence, published values in the literature are often inaccurate. Although some research in rats indicates that Cr III improves "glucose tolerance factor," which is thought to facilitate the uptake in insulin-sensitive cells, this has not been validated in swine. It has been suggested that the result of some Cr studies is due to insulin and not to other physiological roles of Cr in swine. Chromium may increase muscle tissues, decrease fatty tissues, decrease stress, and improve reproduction. However, those effects have not been validated or the possible mechanisms identified. Recently, tolerance of heat stress was studied to determine if Cr propionate could assist in production (Mayorga et al. 2019). However, feeding 200 ppb Cr did not improve ADG or circulating glucose, insulin, NEFA, cholesterol, triglycerides, or lipolysaccharide-binding protein.

# References

Adeola, O., B. V. Lawrence, A. L. Sutton, and T. R. Cline. 1995. Phytase-induced changes in mineral utilization in zinc-supplemented diets for pigs. J. Anim. Sci. 73:3384–3391.

Alexander, L. S., A. Qu, S. A. Cutler, A. Mahajan, and M. F. Rothschild. 2010. A calcitonin receptor (CALCR) single nucleotide polymorphism is associated with growth performance and bone integrity in response to dietary phosphorus deficiency. J. Anim. Sci. 88:1009–1016.

Almeida, F. N., and H. H. Stein. 2012. Effects of graded levels of microbial phytase on the standardized total tract digestibility of phosphorus in corn and corn co-products fed to pigs. J. Anim. Sci. 90:1262–1269.

Andrews, G. K., and J. Geiser. 1999. Expression of the mouse metallothionein-I and –I genes provide a reproductive advantage during maternal dietary zinc deficiency. J. Nutr. 129:1643–1648.

Arredondo, M. A., G. A. Casas, and H. H. Stein. 2019. Increasing levels of microbial phytase increases the digestibility of energy and minerals in diets fed to pigs. Anim. Feed Sci. Technol. 248:27–36.

Bikker, P., H. van Laar, V. Sips, C. Walvoort, and W. J. J. Gerrits. 2016. Basal endogenous phosphorus losses in pigs are affected by both body weight and feeding level. J. Anim. Sci. 94:294–297.

Bikker, P., C. M. C. van der Peet-Schwering, W. J. J. Gerrits, V. Sips, and C. Walvoort. 2017. Endogenous phosphorus losses in growing-finishing pigs and gestating sows. J. Anim. Sci. 95:1637–1643.

Blavi, L., D. Sola-Oriol, J. F. Perez, H. H. Stein. 2017. Effects of zinc oxide and microbial phytase on digestibility of calcium and phosphorus in maize-based diets fed to growing pigs. J. Anim. Sci. 95:847–854.

Bournazel, M., M. Lessiere, M.J. Duclos, M. Magnin, N. Même, C. Peyronnet, E. Recoule, A. Quinsac, E. Labussière, A. Narcy. 2018. Effects of rapeseed meal fiber content on phosphorus and calcium digestibility in growing pigs fed diets without or with microbial phytase. Animal 12:34–42.

Brugger, D., and W. M. Windisch. 2019. Adaption of body zinc pools in weaned piglets challenged with subclinical zinc deficiency. Br. J. Nutr. 121:849–858.

Cao, S. T., C. C. Wang, H. Wu, Q. H. Zhang, L. F. Jiao, and C. H. Hu. 2018. Weanng disrupts intestinal antioxidant status, impairs, intestinal barrier and mitochondrial function, and triggers mitophagy in piglets. J. Anim. Sci. 96:1073–1083.

Carlson M.S., G. M. Hill, J. E. Link. 1999. Early and traditionally weaned nursery pigs benefit from phase-feeding pharmacological concentrations of zinc oxide: effect on metallothionein and mineral concentrations. J. Anim. Sci. 77:1199–207.

Chandra, R. K. 1984. Excessive intake of zinc impairs immune response. JAMA 252:1443–1446.

Cottin, S. C., G. Roussel, L. Gambling, H. E. Hayes, V. J. Currier, and H. J. McArdle. 2018. The effect of maternal iron deficiency on zinc and copper levels on genes of zinc and copper metabolism during pregnancy in the rat. Br. J. Nutr. 121:121–129.

Cousins, R. J., J. P. Liuzzi, and L. A. Lichten. 2006. Mammalian zinc transport, trafficking, and signals. J. Biol. Chem. 281:24085–24089.

Crenshaw, T. D., D. K. Schneider, C. S. Carlson, J. B. Parker, J. P. Sonderman, T. L. Ward, and M. E. Wilson. 2013. Tissue mineral concentrations and osteochondrosis leisions in prolific sows across parities 0 through 7. J. Anim. Sci. 91:1255–1269.

Croxford, T. P., N. H. McCormick, and S. L. Kelleher. 2011. Moderate zinc deficiency reduces testicular Zip 6 and Zip 10 abundance and impairs spermatogenesis in mice. J. Nutr. 141:359–365.

Damo, S. M., T. E. Kehl-Fie, N. Sugitani, M. E. Holt, S. Rathi, W. J. Murphy, Y. Zhang, C. Betz, L. Hench, G. Fritz, E. P. Skaar, and W. J. Chazin. 2013. Molecular basis for manganese sequestration by calprotectin and roles in the innate immune response to invading bacterial pathogens. Proc. Natl. Acad. Sci. USA 110:3841–3846.

Darriet, C., D. E. Axe, and T. D. Crenshaw. 2017. Acidogenic mineral additions increased Ca mobilization in prepartum sows. J. Anim. Sci. 95:212–225.

Darshan, D., and G. J. Anderson. 2007. Liver-gut axis in the regulation of iron homeostasis. World J. Gastroenterol 13:4737–4745.

DeRouchey, J. M., J. D. Hancock, R. H. Hines, K. R. Cummings, D. J. Lee, C. A. Malone, D. W. Dean, J. S. Park, and H. Cao. 2003. Effects of dietary electrolyte balance on the chemistry of blood and urine in lactating sows and sow litter performance. J. Anim. Sci. 81:3067–3074.

Edwards, L. N., T. E. Engle, M. A. Paradis, J. A. Correa, and D. A. Anderson. 2010. Persistence of blood changes associated with alteration of the dietary electrolyte balance in commercial pigs after feed withdrawal, transportation, and lairage, and the effects on performance and carcass quality. J. Anim. Sci. 88:4068–4077.

Elsbernd, A. J., K. J. Stalder, L. A. Karriker, and J. F. Patience. 2015. Comparison among gilts, physical castrates, entire males, and immunological castrates in terms of growth performance, nitrogen and phosphorus retention, and carcass fat iodine value. J. Anim. Sci. 93:5702–5710.

Fan, M. Z., and T. Archbold. 2012. Effects of dietary true digestible calcium to phosphorus ratio on growth performance and efficiency of calcium and phosphorus use in growing pigs fed corn and soybean meal-based diets. J. Anim. Sci. 90 (Suppl. 4): 254–256.

Fry, R. S., M. S. Ashwell, K. E. Lloyd, A. T. O'Nan, W. L. Flowers, K. R. Stewart, and J. W. Spears. 2012. Amount and source of dietary copper affects small intestine morphology, duodenal lipid peroxidation, hepatic oxidative stress, and mRNA expression of hepatic copper regulatory proteins in weanling pigs. J. Anim. Sci. 90:3112–3119.

Gagliardi, R., W. Zhang, R. L. Murray, L. Zhao, K. Kroscher, R. P. Rhoads, and C. H. Stahl. 2017. Transcriptional regulation of genes involved in calcium and phosphate metabolism in neonatal pigs fed with different levels of dietary calcium and phosphate. J. Anim. Sci. 95 (Suppl. 4):61.

González-Vega, J. C., C. L. Walk, and H. H. Stein. 2015. Effects of microbial phytase on apparent and standardized total tract digestibility of calcium in calcium supplements fed to growing pigs. J. Anim. Sci. 93:2255–2264.

González-Vega, J. C., C. L. Walk, M. R. Murphy, and H. H. Stein. 2016. Requirement for digestible calcium by 25 to 50 kg pigs at different dietary concentrations of phosphorus as indicated by growth performance, bone ash concentration, and calcium and phosphorus balances. J. Anim. Sci. 94:5272–5285.

Gonzalo, E., M. P. Létourneau-Montminy, A. Narcy, J. F. Bernier, and C. Pomar. 2017. Consequences of dietary calcium and phosphorus depletion and repletion feeding sequences on growth performance and body composition of growing pigs. Animal 11:1–9.

Gowanlock, D. W., D. C. Mahan, J. S. Jolliff, and G. M. Hill. 2015. Evaluating the influence of National Research Council levels of copper, iron, manganese, and zinc using organic (Bioplex) minerals on resulting tissue mineral concentrations, metallothionein, and liver antioxidant enzymes in grower-finisher swine diets. J. Anim. Sci. 93:1149–1156.

Grilli, E., B. Tugnoli, F. Viotari, C. Domeneghini, M. Morlacchini, A. Piva, and A. Prandini. 2015. Select low doses of microencapsulated zinc oxide improve performance and modulate the ileum architecture, inflammatory cytokines, and tight junctions expression of weaned pigs. Animal 9:1760–1768.

Guérin, P., S. El Mouatassim, and Y. Ménézo. 2001. Oxidative stress and protection against reactive oxygen species in the pre-implantation embryo and its surroundings. Hum Reprod Update 7:175–189.

Guo-jun, S., C Dai-wen, Z. Ke-ying, and Y. Bing. 2009. Effects of dietary Zn level and an inflammatory challenge on performance and immune response of weanling pigs. Asian-Australas. J. Anim. Sci. 22:1303–1310.

Hansen, S. L., N. Trakooljul, H. Liu, A. J. Moeser, and J. W. Spears. 2009. Iron transporters are differentially regulated by dietary iron, and modifications are association with changes in manganese metabolism in young pigs. J. Nutr. 139:1474–1479.

Heyer, C. M. E., S. Schmucker, E. Weiss, M. Eklund, and T. Aumiller, E. Graeter, T. Hofmann, M. Rodehutscord, L. E. Hoelzle, J. Seifert, V. Stefanski, and R. Mosenthin. 2016. Effect of supplemented mineral phosphorus and fermentable substrates on gut microbiota composition and metabolites, phytate hydrolysis, and health status of growing pigs. J. Anim. Sci. 94 (Suppl. 5):844.

Hernández, A., J. R. Pluske, D. N. D'Souza, and B.P. Mullan. 2008. Select levels of copper and zinc in diets for growing and finishing pigs can be reduced without detrimental effects on production and mineral status. Animal 12:1763–1771.

Hill, G. M., E. R. Miller, and H. D. Stowe. 1983. Effect of dietary zinc levels on health and productivity of gilts and sows through two parities. J. Anim. Sci. 57:114–122.

Hill, G. M. 2013. Minerals and mineral utilization in swine. In: L. I. Chiba, editor, Sustainable swine nutrition. Wiley-Blackwell, John Wiley & Sons, Inc., Hoboken, NJ.

Hinson, R. B., A. P. Schinkel, J. S. Radcliffe, G. L. Allee, A. L. Sutton, and B. T. Richert. 2009. Effect of feeding reduced crude protein and phosphorus diets on weaning-finishing pig growth performance, carcass characteristics, and bone characteristics. J. Anim. Sci. 87:1502–1517.

Hostetler, C., R. Kincaid, and M. A. Mirando. 2003. The role of essential trace elements in embryonic and fetal development in livestock. Vet. J. 166:125–139.

Hunt, C. D., P. E. Johnson, J. Herbel, and L. K. Mullen. 1992. Effects of dietary zinc depletion on seminal volume and sinc loss, serum testosterone concentrations, and sperm morphology in young men. Am. J. Clin. Nutr. 56:148–157.

Jeon, Y., J. D. Yoon, L. Cai, S. U. Hwang, E. Kim, Z. Zheng, E. Lee, D.Y. Kim, and S. H. Hyun. 2014. Supplementation of zinc on oocyte in vitro maturation improves preimplatation embryonic development in pigs. Theriogenol. 82:866–874.

Johnston, L., J. Holen. J. C. Jang, P. Urriola, J. Shurson, and M. Schwartz. 2019. Zinc in late-gestation diets for sows improves piglet survival. Natl. Hog Farmer. June 13, 2019.

Jolliff, J. S., and D. C. Mahan. 2011. Effect of injected and dietary iron in young pigs on blood hematology and postnatal pig growth performance. J. Anim. 89:4068–4080.

Jolliff, J. S., and D. C. Mahan. 2012. Effect of dietary inulin and phytase on mineral digestibility and tissue retention in weanling and growing swine. J. Anim. Sci. 90:3012–3022.

Kerr, B. J., C. J. Ziemer, T. E. Weber, S. L. Trabue, B. L. Bearson, G. C. Shurson, and M. H. Whitney. 2008. Comparative sulfur analysis using thermal combustion or inductively coupled plasma methodology and mineral composition of common livestock feedstuffs. J. Anim. Sci. 86:2377–2384.

Kornegay, E. T., P. H. G. van Heugten, M. D. Lindemann, and D. J. Blodgett. 1989. Effects of biotin and high copper levels on performance and immune response of weanling pigs. J. Anim. Sci. 67:1471–1477.

Kreuzer-Redmer, S., D. Aremds, J. N. Schulte, D. Karweina, P. Korkuc, N. Wöltje, D. Hesse, R. Pieper, V. Gerdts, J. Zentek, F. Meurens, and G. A. Brockmann. 2018. High dosage of zinc modulates T-cells in a time-dependent manner within porcine gut-associated lymphatic tissue. Br. J. Nutr. 120:1349–1358.

Laird, S., I. Kühn, and H. M. Miller. 2018. Super-dosing phytase improves the growth performance of weaner pigs fed a low iron diet. Anim. Feed. Sci. Technol. 242:150–160.

Lee, S., S. R. Hennigar, S. Alam, K. Nishida, and S. L. Kelleher. 2015. Essential role of zinc transporter 2 (ZnT2)-mediated zinc transport in mammary gland development and function during lactation. J. Biol. Chem. 290:13064–13078.

Lee, Su A, L. V. Lagos, C. L. Walk, and H. H. Stein. 2019. Basal endogenous loss, standardized total tract digestibility of calcium in calcium carbonate, and retention of calcium in gestating sows change during gestation, but microbial phytase reduces basal endogenous loss of calcium. J. Anim. Sci. 97:1712–1721.

Lei, X. J., J. Y. Chung, J. H. Park, and I. H. Kim. 2017. Evaluation of different dietary electrolyte balance in weanling pigs diets. Anim. Feed Sci. Technol. 226:98–102.

Létourneau-Montminy, M. P., A. Narcy, M. Magnin, D. Sauvant, J. F. Bernier, C. Pomar, and C. Jondreville. 2010. Effect of reduced dietary calcium concentration and phytase supplementation on calcium and phosphorus utilization in weanling pigs with modified mineral status. J. Anim. Sci. 88:1706–1717.

Liu, Y., L. Ma, J. M. Zhao, M. Vazquez-Añón, and H. H. Stein. 2014. Digestibility and retention of zinc, copper, manganese, iron, calcium, and phosphorus in pigs fed diets containing inorganic and organic minerals. J. Anim. Sci. 92:3407–3415.

Liuzzi, J. P., J. A. Bobo, L. Cui, R. J. McMahon, and R. J. Cousins. 2003. Zinc transporters 1, 2 and 4 are differentially expressed and localized in rats during pregnancy and lactation. J. Nutr. 133:342–351.

Lönnerdal, B. 2017. Development of iron homeostasis in infants and young children. Am. J. Clin. Nutr. 106:1575S–1580S.

Mahan, D. C., and J. L. Vallet. 1997. Vitamin and mineral transfer during fetal development and the early postnatal period in pigs. J. Anim. Sci. 75:2731–2738.

Martin, L., R. Pieper, N. Schunter, W. Vahjen, and J. Zentek. 2013. Performance, organ zinc concentration, jejunal brush border membrane enzyme activities and mRNA expression in piglets fed with different levels of dietary zinc. Arch. Anim. Nutr. 67:248–261.

Mason, K. E., W. A. Burns, and J. C. Smith, Jr. 1982. Testicular damage associated with zinc deficiency in pre- and post-pubertal rats: Response to zinc repletion. J. Nutr. 112:1019–1028.

Mayorga, E. J., S. K. Kvidera, J. T. Seibert, E. A. Horst, and M. Abuajamieh, M. Al-Qaisi, S. Lei, J. W. Ross, C. D. Johnson, B. Kremer, L. Ochoa, R. P. Rhoads, and L. H. Baumgard. 2019. Effects of dietary chromium propionate on growth performance, metabolism, and immune biomarkers in heat-stressed finishing pigs. J. Anim. Sci. 97:1185–1197.

McMahon, R. J., and R. J. Cousins. 1998. Mammalian zinc transporter. J. Nutr. 128:667–670.

Ménézo, Y., L. Pluntz, J. Chouteau, T. Gurgan, A. Demirol, A. Dalleac, and M. Benkhalifa. 2011. Zinc concentrations in serum a follicular fluid during ovarian stimulation and expression of Zn2+ transporters in human oocytes and cumulus cells. Reprod. Biomed. 22:647–652.

Merriman, L. A., C. L. Walk, M. R. Murphy, C. M. Parsons, and H. H. Stein. 2017. Inclusion of excess dietary calcium in diets for 100- to 130-kg growing pigs reduces feed intake and daily gain if dietary phosphorus is at or below the requirement. J. Anim. Sci. 95:5439–5446.

Moran, K., P. Wilcock, A. Elsbernd, C. Zier-Rush, R. D. Boyd, and E. van Heugten. 2019. Effects of super-dosing phytase and inositol on growth performance and blood metabolites of weaned pig housed under commercial conditions. J. Anim. Sci. 97:3007–3015.

Nasir, Z., J. Broz, and R. T. Zijlstra. 2012. The effects of supplementations of novel bacterial 6-phytse on mineral digestibility and plasma mineral in lactating sows. J. Anim. Sci. 90:116–118.

NRC. 2012. *Nutrient requirements of swine. 11th rev. ed..* National Academy of Science, Washington, DC.

O'Dell, B. L. and R. A. Sunde. 1997. Handbook of nutritionally essential mineral elements. Marcel Dekker, Inc., New York, NY.

Olsen, K. M., S. A. Gould, C. L. Walk, N. V. L. Serão, S. L. Hansen, J. F. Patience. 2019. Evaluating phosphorus release by phytase in diets fed to growing pigs that are not deficient in phosphorus. J. Anim. Sci. 97:327–337.

Petersen, G. I., C. Pedersen, M. D. Lindemann, and H.H. Stein. 2011. Relative bioavailability of phosphorus in inorganic phosphorus sources fed to growing pigs. J. Anim. Sci. 89:460–466.

Peters, J. C., D. C. Mahan, T. G. Wiseman, and N. D. Fastinger. 2013. Effect of dietary organic and inorganic micromineral source and level on sow body, liver colostrum, mature milk, and progeny mineral compositions over six parities. J. Anim. Sci. 88:626–637.

Philcox, J. C., M. Sturkenboom, P. Coyle, and A. M. Rofe. 2000. Metallothionein in mice reduced intestinal zinc loss during acute endotoxin inflammation, but not during starvation or dietary zinc restriction. J. Nutr. 130:1901–1909.

Piatkowski, T. L., D. C. Mahan, A. H. Cantor, A. L. Moxon, J. H. Cline, and A. P Goifo, Jr. 1979. Selenium and vitamin E in semi-purified diets for gravid and nongravid gilts. J. Anim. Sci. 48:1357–1365.

Pomar, C., J. Pomar, J. Rivest, L. Cloutier, M.-P. Letourneau-Montminy, I. Andretta, and L. Hauschild. 2014. Estimating real-time individual amino acid requirements in growing-finishing pigs: Towards a new definition of nutrient requirements in growing-finishing pigs? In: N. K. Sakomura, R. M. Gous, I. Kyriazakis, and L. Hauschild, editors, Nutritional modeling in pigs and poultry. CABI, Wallingford, Oxon, UK.

Prasad, A. S., J. A. Haldssted, and M. Nadimi, 1961. Syndrome of iron deficiency anemia, hepatosplenomegaly, hypogonadism, dwarfism, and geophagia. Am. J. Med. 31:532–546.

Reinstein, N. H., B. Lönnerdal, C. L. Keen, and L. S. Hurley. 1984. Zinc-copper interactions in the pregnant rat: Fetal outcome and maternal and fetal zinc, copper, and iron. J. Nutr. 114:1266–1279.

Reynolds, F. H., J. M. Forbes, and H. H. Miller. 2010. Does the newly weaned piglet select a zinc oxide supplemented feed, when given the choice? Animal 4:1359–1367.

Rincker, M.J., G. M. Hill, J. E. Link and J. Rowntree. 2004. Effects of dietary iron on supplementation on growth performance, hematological status, and whole body mineral concentrations of nursery pigs. J. Anim. Sci. 82:3189–3197.

Rincker, M. J., S. L. Clarke, R. S. Eisenstein, J. E. Link, and G. M. Hill. 2005. Effects of Fe supplementation on binding activity of iron regulatory proteins and the subsequent impact on growth performance and indices of hematological and mineral status of young pigs. J. Anim. Sci. 83:2137–2145.

Rojas, O. J., Y. Liu, and H. H. Stein. 2013. Phosphorus digestibility and concentration of digestible and metabolizable energy in corn, corn coproducts, and bakery meal fed to growing pigs. J. Anim. Sci. 91:5326–5335.

Rortvedt, L. A., and T. D. Crenshaw. 2012. Expression of kyphosis in pigs is induced by a reduction of supplemental vitamin D in maternal diets and vitamin D, Ca, and P concentrations in nursery diets. 2012. J. Anim. Sci. 90:4905–4915.

Ross, R. D., G. L. Cromwell, and T. S. Stahly. 1984. Effects of source and particle size on the biological availability of calcium in calcium supplements for growing pigs. J. Anim. Sci. 59:125–134.

Ruiz-Ascacibar, I., P. Stoll, G. Bee, and P. Shlegel. 2019. Dynamics of the mineral composition and deposition rates in the empty body of entire males, castrates, and female pigs. Animal 13:950–958.

Sanz-Fernandez, M. V., S. C. Pearce, N. K. Gabler, J. F. Patience, M. E. Wilson, M. T. Socha, J.L. Torrison, R. P. Rhoads, and L.H. Baumgard. 2014. Effects of supplemental zinc amino acid complex o gut integrity in hear-stressed growing pigs. Animals 8:43–50.

Schwarz, C., K. M. Ebner, F. Furtner, S. Duller, W. Wetscherek, W. Wernert, W. Kandler, and K. Schelde. 2017. Influence of high inorganic selenium and manganese diets for fattening pigs on oxidative stability and pork quality products. Animal 11:345–353.

Shaw, D. T., D. W. Rozeboom, G. M. Hill, M. W. Orth, D. S. Rosenstein, and J. E. Link. 2006. Impact of supplement withdrawal and wheat middling inclusion on bone metabolism, bone strength, and the incidence of bone fractures occurring at slaughter in pigs. J. Anim. Sci. 84:1138–1146.

She, Y., Y. B. Su, L. Liu, C. F. Huang, J. T. Li, P. Li, D.F. Li, and X. S. Piao. 2015. Effects of microbial phytase on coefficient of standardized total tract digestibility of phosphorus in growing pigs fed corn and corn co-products, wheat and wheat co-products and oilseed meals. Anim. Feed Sci. Technol. 208:132–144.

She, Y., Y. Liu, and H. H. Stein. 2017. Effects of graded levels of microbial phytase on apparent total tract digestibility of calcium and phosphorus and standardized total tract digestibility of phosphorus in four sources of canola meal and in soybean meal fed to growing pigs. J. Anim. Sci. 95:2061–2070.

Shelton, N. W., M. D. Tokach, J. L. Nelssen, R. D. Goodband, S. S. Dritz, J. M. DeRouhey, and G. M. Hill. 2011. Effects of coper sulfate, tri-basic copper chloride, and zinc oxide on weanling pig performance. J. Anim. Sci. 89:2440–2451.

Shurson, G. C., P. K. Ku, G. L. Waxler, M. T. Yokoyama, and E. R. Miller. 1990. Physiological relationships between microbiological status and dietary copper levels in the pig. J. Anim. Sci. 68:1061–1071.

Stanger, B. R., G. M. Hill, J. E. Link, J. R. Turk, M. S. Carlson, and D. W. Rozeboom. 1998. Effect of high Zn diets on TGE-challenged early-weaned pigs. J. Anim. Sci. 76 (Suppl. 2):53 (Abstr.).

Stein, H. H., C. T. Kadzere, S. W. Kim, and P. S. Miller. 2008. Influence of dietary phosphorus concentration on the digestibility of phosphorus in monocalcium phosphate by growing pigs. J. Anim. Sci. 86:1861–1867.

Sutovsky, P., K. Kerns, M. Zigo, and D. Zuidema. 2019. Boar semen improvement through sperm capacitation management, with emphasis on zinc ion homeostasis. Theriogenol. 137:50–55.

Tan, F. P. Y., S. A. Kontulainen, and A. D. Beaulieu. 2016. Effects of dietary calcium and phosphorus on reproductive performance and markers of bone turnover in stall- or group-housed sows. J. Anim. Sci. 94:4205–4216.

Tang, Z., Y. Yin, Y. Zhang, R. Huang, Z. Sun, T. Li, W. Chu, X. Kong, L. Li, M. Geng, and Q. Tu. 2009. Effects of dietary supplementation with an expressed fusion peptide bovine lactoferricin-lactoferrampin on performance, immune function and intestinal mucosal morphology in piglets weaned at age 21 d. Br. J. Nutr. 101:998–1005.

Tian, X., and F. J. Diaz. 2013. Acute dietary zinc deficiency before conception compromises oocyte epigenetic programming and disrupts embryonic development. Dev. Biol. 376:51–61.

Torrallardona, D., and O. Ader. 2016. Effects of a novel 6-phytase (EC 3.1. 3.26) on performance, phosphorus and calcium digestibility, and bone mineralization in weaned piglets. Anim. Sci. J. 94:194–197.

Torrallardona, D., N. Andrés-Elias, S. López-Soria, I. Badiola, and M. Cerdà-Cuéllar, 2012. Effect of feeding different cereal-based diets on the performance and gut health of weaned piglets with or without previous access to creep feed during lactation. J. Anim. Sci. 90:31–33.

Valenti, P., and G. Antonini. 2005. Lactoferrin. Cell. Molec. Life Sci. 62:1576.

Vallet, J. L. L. A. Rempe, J. R. Miles and S. K. Webel. 2014. Effect of essential fatty acid and zinc supplementation during pregnancy on birth intervals, neonatal piglet brain myelination, stillbirth, and pre-weaning mortality. J. Anim. Sci. 92:2422–2432.

Van Riet, M. M. J., S. Millet, A. Liesegang, E. Nalon, B. Ampe, F. A. M. Tuyttens, D. Maes, and G. P. J. Janssens. 2016. Impact of parity on bone metabolism throughout the reproductive cycle in sows. Animal 10:1714–1721.

Walk, C. L., P. Wilcock, and E. Magowan. 2015. Select evaluation of the effects of pharmacological zinc oxide and phosphorus source on weaned piglet growth performance, plasma minerals, and mineral digestibility. Animal 9:1145–1152.

Wan, D., Y M Zhang, X. Wu, X. Lin, X. G. Shu, X. H. Zhou, H. T. Du, W. G. Xing, H. N. Liu, L. Li, Y. Li, and Y. L. Yin. 2018. Maternal dietary supplementation with ferrous N-carbamylglycinate chelate affects sow reproductive performance and iron status of neonatal piglets. Animal 12:1372–1379.

Wang, Y., T. Shan, Z. Xu, J. Liu, and J. Feng. 2006. Effect of lactoferrin on the growth performance, intestinal morphology, and expression of PR-39 and protegrin-1 genes in weaned piglets. J. Anim. Sci. 84:2636–2641.

Wan, D., Y. M. Zhang, X. Wu, X. Lin, X. G. Shu, X. H. Zhou, H. T. Du, W. G. Xing, H. N. Liu, L. Liu, L. Li, Y. Li, and Y. L. Yin. 2017. Maternal dietary supplementation with ferrous N-carbamylglycinate chelate affects sow reproductive performance and iron status of neonatal piglets. Animal 12:1372–1379.

Wealleans, A. L., R. M. Bold, Y. Dersjant-Li, and A. Awati. 2015. The addition of a *Buttiauxella* sp. phytase to lactating sow diets deficient in phosphorus and calcium reduces weight loss and improves nutrient digestibility. J. Anim. Sci. 93:5283–5290.

Weller, M. M. D. C. A., L. Alebrante, P. H. R. F. Campos, A. Saraiva, B. A. N. Silva, J. L. Donzele, R.F.M. Oliveira, F. F. Silva, E. Gasparino, P. S. Lopes, and S. E. F. Guimarães. 2013. Effect of heat stress and feeding phosphorus levels on pig electron transport chain gene expression. Animal 7:1985–1993.

Yang, H., A. N. D. Li, Y Lin, J. Li, Z. R. Wang, G. Wu, R. L. Huang, X. F. Kong, C. B. Yang, P. Kang, J. Deng, S. X. Wang, B. E. Tan, Q. Hu, F. F. Xing, X. Wu, Q. H. He, K. Yao, Z. J. Liu, Z. R. Tang, F. G. Yin, Z. Y. Deng, M. Y. Xie, and M. Z. Fan. 2007. True phosphorus digestibility and endogenous phosphorus outputs associated with brown rice for weanling pigs measured by simple regression analysis technique. Animal 1:213–220.

Zackular, J. P., J. L. Moore, A. T. Jordan, L. J. Juttukonda, M. J. Noto, M. R. Nicholson, J. D. Crews, M. W. Semler, Y. Zhang, L. B. Ware, M. K. Washington, W. J. Chazin, R. M. Caprioli, and E. P. Skaar. 2016. Dietary zinc alters the microbiota and decreases resistance to *Clostrium difficile* infection. Nat. Med. 22:1330–1334.

Zhang, F., and O. Adeola. 2017. True is more additive than apparent total tract digestibility of calcium in limestone and dicalcium phosphate for twenty-kilogram pigs fed semipurified diets. J. Anim. Sci. 95:5466–5473.

Zhao, J., A. F. Harper, M. J. Estienne, K. E. Webb, A. P. Mcelroy, and D. M. Denbow. 2007. Growth performance and intestinal morphology responses in early weaned pigs to supplementation of antibiotic-free diets with an organic copper complex and spray-dried plasma protein in sanitary and nonsanitary environments. J. Anim. Sci. 85:1302–1310.

# 9    Nutrition and Immunology in Swine

Lucas A. Rodrigues, Michael O. Wellington, Andrew G. Van Kessel, and Daniel A. Columbus

## Introduction

Despite extensive efforts in swine nutrition research towards the nutritional characterization of dietary ingredients (e.g., nutrient content and digestibility) and the determination of nutrient requirements for maximum output (e.g., lean gain, milk production), there is a remarkable gap between growth potential and realized performance in commercial herds. While several factors contribute to this variability in performance (Patience et al. 2015), immunological status likely plays a significant role (Laanen et al. 2013). In today's highly intensive swine production systems, pigs are continuously exposed to pathogens and immune stimulatory antigens that negatively impact productivity. Pigs exposed to immune challenge show reduced appetite and growth and less efficient use of nutrients compared to healthy animals (van Heugten et al. 1996; Rakhshandeh and de Lange 2012). This decrease in performance can have a substantial impact on profitability of producers. Largely due to a concerted effort to reduce antibiotic usage in livestock production, there is increased interest in furthering our understanding of the role that nutrition plays in supporting the immune response (Pluske et al. 2018). This information will be critical to development of programs that support production and robustness of pigs under a variety of stressful conditions. Among the nutrients whose utilization is impacted by immune status, amino acids are likely to have critical roles in supporting both aspects of the immune response and maintaining growth performance in animals exposed to a pathogen challenge and other enteric challenges associated with ingredient selection (Wu 2013; Le Floc'h et al. 2018). As such, this chapter will discuss the impact of immune status on animal production and the interaction between diet composition and animal health with a primary focus on the functional roles of amino acids during times of immune challenge.

## An Overview of Immune Response

There have been many excellent reviews on the immune system (Bourne 1973; Salmon 1987; Rothkötter et al. 2002; Bailey 2009; Rakhshandeh and de Lange 2012), and readers are directed to these for a more comprehensive discussion. Briefly, the immune system is composed of both innate (nonspecific, rapid) and adaptive (specific, slow) responses (Sompayrac 2019) that combine to provide remarkable protection against a vast microbial world employing a plethora of infection strategies. The innate system is composed of chemical (e.g., defensins) and physical (e.g., mucin, epithelium) barriers as well as a number of cell types (e.g., granulocytes, natural killer cells, macrophages, and dendritic cells) capable of nonspecifically preventing pathogens and other harmful compounds from

*Sustainable Swine Nutrition*, Second Edition. Edited by Lee I. Chiba.
© 2023 John Wiley & Sons Ltd. Published 2023 by John Wiley & Sons Ltd.

gaining access to the systemic environment. Identification of microorganisms that penetrate the innate immune system occurs through recognition of specific chemical structures (e.g., lipopolysaccharide [LPS], flagellin) known as microbe-associated molecular patterns (MAMPs). These chemical structures are common in bacterial, viral, and fungal particles but are absent in mammalian cells, allowing for a rapid identification of potentially pathogenic organisms (Medzhitov 2007). Identification of MAMPs is mediated by several families of pattern recognition receptors (PRR) present on epithelia and cells of the innate system such that activation of these receptors by their microbial-origin ligands indicates possible microbial penetration of defenses and danger resulting in stimulation of inflammatory cytokines and/or viral defense mediators. These cytokines and immune mediators activate both systemic and local inflammatory responses that include fever, inappetence, hepatic synthesis of acute-phase proteins (e.g., haptoglobin, C-reactive protein, and albumin), leukocyte proliferation, increased blood flow, and infiltration of immune cells and plasma proteins to the local site of PRR activation (Medzhitov 2007). In this way, the innate system serves as the first line of defence against external stimuli at tissue and cell levels and triggers the subsequent immune response.

The majority of infections occur through mucosal surfaces (e.g., gastrointestinal tract, respiratory tract, and reproductive tract) as opposed to the skin. Along these surfaces the innate immune system employs a variety of unique innate mechanisms to prevent the microbial penetration of what is often a single layer of epithelial cells (Cesta 2006). Such mechanisms include the secretion of mucin, antimicrobial peptides, and immunoglobulins (i.e., secretory immunoglobulin A), which prevent exposure of the epithelial barrier to pathogenic organisms (Mantis et al. 2011; Robinson et al. 2015). Furthermore, the epithelium is lined with specialized mucosa-associated lymphoid tissue (MALT) including cells of the innate and acquired immune system. These tissues play a critical role in defence against pathogens and are markedly influenced by nutritional management, as further discussed later.

While the innate immune system provides a rapid, nonspecific first line of defense to pathogens, the adaptive immune system mediates specific immune responses. Immune cells, such as macrophages and dendritic cells, that are key mediators of the innate immune system also serve as antigen-presenting cells (APCs) that ultimately activate T and B cell population-generating antigen (pathogen)-specific cytotoxic T cells and immunoglobulins (Goerdt and Orfanos 1999). Critically, some activated B and T cells become memory cells, remaining within tissues after resolution of the pathogen challenge and producing a more rapid and higher magnitude response on subsequent exposure (O'Leary et al. 2006).

## Gut Health, Microbiome, and Immunity

The gastrointestinal tract (GIT) performs the crucial role of nutrient absorption while acting as a barrier to luminal threats, such as toxins and microorganisms from entering the body. Immune challenge results in significant physiological alterations of the gut, including changes in gut motility, digestive enzyme secretion, absorptive capacity, mucin production, and gut barrier permeability (Faure et al. 2007; Turner 2009). As such, gut health, as defined by the combined ability of the gut to perform digestive, absorptive, and barrier functions, has often become synonymous with overall health of the animal. Overall, maintenance of gut and animal health is achieved through the combined effects of a number of mechanisms as outlined in Figure 9.1 and contributing to the creation of bacterial, physical, chemical, and immunological barriers (Hooper 2009).

The selective permeability of the GIT is achieved by a single layer of epithelial cells that separate the lumen from the underlying tissues of the body (Wells et al. 2010, 2011). The functions of physical barrier and regulation of nutrient transport are facilitated by the presence of transmembrane

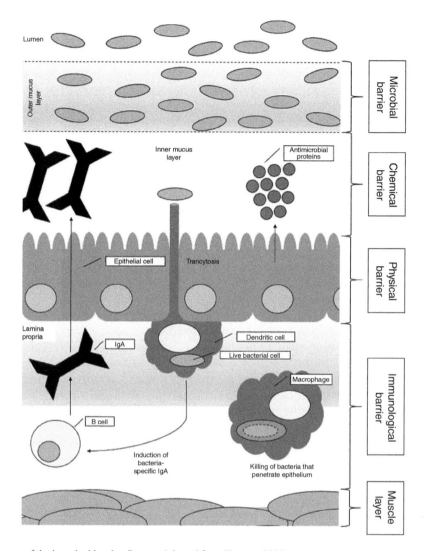

**Figure 9.1** Layers of the intestinal barrier. Source: Adapted from Hooper (2009).

protein networks as well as the intestinal mucus layer. Desmosomes, adherens junctions, and tight junctions (i.e., occludin, claudins, junctional adhesion molecules) act to physically bind epithelial cells together, regulating both transcellular and paracellular transport of nutrients from the intestinal lumen (Groschwitz and Hogan 2009) and preventing invasion of the host by intestinal bacteria, viruses, and parasites. The maintenance of gut barrier function is aided by the production of secreted mucin by specialized epithelial cells called goblet cells (Li et al. 2007; Dharmani et al. 2009) as well as transmembrane mucins located in the apical enterocyte cell wall. The resulting mucus layer, via gel formation between transmembrane mucins and secreted mucin, may serve as an anchor for the mucin layer, acting as the first component of gut barrier function. The mucus layer consists of an inner, dense layer adhered to the epithelium which is resistant to bacterial penetration and serves to retain high concentrations of secreted defense proteins (e.g., antimicrobial peptides, secretory IgA) and an outer, loosely adhered layer (Hooper 2009; Broom and Kogut 2018). Together, the mucus layers act to prevent contact of pathogens and other luminal threats with the underlying epithelium through maintenance of a physical barrier and through continuous removal of pathogens via peristalsis (Johansson et al. 2011; Faderl et al. 2015). Studies have demonstrated the stimulatory effects of immune challenge on mucin secretion in the gut (Faure et al. 2007; Dharmani et al. 2009; Wellington et al. 2020a), supporting its importance in host defence.

The gut microbiome has been shown to play a significant role in maintaining gut homeostasis, gut health, and improved growth performance in pigs (Lallès et al. 2007; Mann et al. 2014; Guevarra et al. 2018). The microbiome represents the combined genome of the entire gastrointestinal microbial community, representing hugely diverse metabolic activities and contributing to an equally diverse gut metabolome that alters host responses through a variety of receptor-mediated (e.g., PRR, short-chain fatty acid receptors) and metabolic (e.g., butyrate metabolism) pathways (Bäckhed et al. 2012). In pigs, the microbiome is established at birth during the interaction of the neonate with the environment immediately after birth (Mach et al. 2015), although there remains a controversial suggestion that colonization is initiated in utero (Perez-Muñoz et al. 2017). Significant changes occur with age, mainly influenced by diet and environment (Isaacson and Kim 2012; te Pas et al. 2020) and, to a lesser extent, by genetics (Pajarillo et al. 2014; Roubos-van den Hil et al. 2017). Although the sow may not have a primary influence on neonatal microbial succession (Mach et al. 2015), there is evidence of sow microbiome influence on the neonate (Cheng et al. 2018). The composition of gut microbial communities and the metabolites generated through fermentation of dietary and endogenously secreted substrates are markedly influenced by diet composition. As detailed below, the resulting changes in gut luminal composition have significant implications for barrier function and for nutritional requirements. Finally, since recent work has identified an association between the gut microbiome and the microbiome and health of other mucosal surfaces (Surendran Nair et al. 2019; Enaud et al. 2020), the argument that nutritional modification of the gut microbiome could also influence the health of distant mucosal surfaces could be reasoned.

**Impact of Immune System Activation on Performance**

While the immune system is important for elimination of invading pathogens and maintaining the health of the animal, mounting an immune response has consequences for the animal both in the short- (e.g., production of immune cells) and long-term (e.g., recovery of damaged tissues) (Schokker et al. 2015). Animals exposed to immune challenge, without exhibiting any clinical signs of disease, show reduced appetite and growth and less efficient use of nutrients compared to healthy animals (Le Floc'h et al. 2004). Previous studies have estimated a reduction in lean growth of 20–35% and feed efficiency of 10–20% in growing pigs at subclinical levels of disease (Williams et al. 1997a,b; Le Floc'h et al. 2009). It was recently reported that pigs reared in suboptimal housing conditions had 19, 23, and 20% poorer daily gain, feed intake, and nitrogen utilization efficiency, respectively, when compared to animals kept in clean pens (Jayaraman et al. 2015a; Jayaraman et al. 2017). The observed reduction in growth is explained, in part, by modifications to nutrient utilization and metabolism during pathogen exposure which redirect nutrients from diet and body reserves to support immune function (Reeds et al. 1994). These alterations are brought about by the release of pro-inflammatory cytokines by stimulated immune cells and have a direct effect on liver, brain, muscle and fat tissue, resulting in reduced feed intake, fever, growth depression, increased proteolysis/decreased protein synthesis in muscle, negative nutrient balance, and lethargy (Dionissopoulos et al. 2006). In pigs, where receptors for specific cytokines were blocked, researchers observed greater than 30% increase in whole body protein deposition and growth rate (Dionissopoulos et al. 2006), demonstrating the key role cytokine release plays in the growth response.

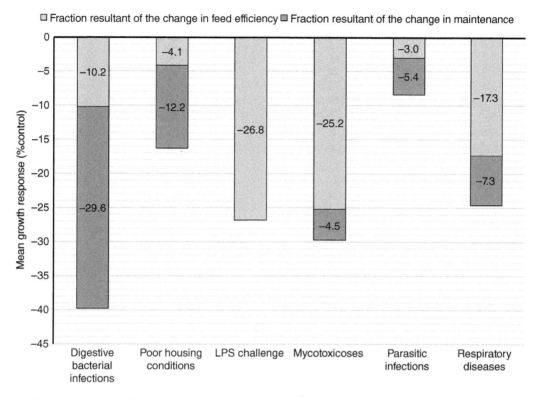

□ Fraction resultant of the change in feed efficiency ▣ Fraction resultant of the change in maintenance

**Figure 9.2** The influence of different health challenge models on the proportion of reduction in growth rate as a result of change in maintenance requirement (i.e., nutrient utilization) or feed efficiency (i.e., reduced feed intake). Source: Pastorelli et al. (2012) / with permission of Elsevier.

In general, the reduction in growth performance observed with exposure to immune challenge is related to a decrease in feed intake and/or a reduction in nutrient utilization efficiency (Figure 9.2; Pastorelli et al. 2012). When the infection involves the gut, there are additional losses associated with reduced barrier function, increased mucin production, and decreased digestive function in the gut, further impairing nutrient availability (Kim et al. 2012a). An understanding of the extent to which these factors contribute to the reduction in growth performance is critical to development of nutritional strategies aimed at preventing and/or mitigating the effects of immune challenge. In a meta-analysis of studies examining various immune-stimulating conditions (i.e., digestive bacterial infections, poor housing conditions, mycotoxicosis, parasitic infections, respiratory diseases, LPS challenge), Pastorelli et al. (2012) quantified the extent to which reduced performance was due to a depression in feed intake vs. increased maintenance nutrient requirements. Both the magnitude of the response in growth performance and the contribution of the two factors to the response were dependent on the immune challenge, with parasitic infections and poor housing conditions showing the lowest reduction in growth performance and digestive bacterial infections, mycotoxicosis, respiratory diseases, and LPS challenge having the greatest impact. Decreased feed intake had the greatest contribution to reduced performance in mycotoxicosis, respiratory diseases, and LPS challenge with alterations in maintenance nutrient requirements having the greatest contribution to digestive bacterial infections and poor housing conditions. Interestingly, Pastorelli et al. (2012) reported no contribution of increased maintenance requirements to reduced growth performance with LPS challenge. However, there have been multiple studies demonstrating changes in requirements for growth of various amino acids during LPS challenge (Rakhshandeh et al. 2010; de Ridder et al. 2012; Litvak et al. 2013b; Wellington et al. 2018) suggesting that maintenance requirements/nutrient utilization efficiency is affected by LPS. Overall, these findings reveal that the effects of immune challenge as assessed by experimental models can differently affect growth performance, nutrient utilization, and feed intake patterns based on the model used; therefore, caution should be used when interpreting results across various conditions. Moreover, under commercial conditions, multiple strategies may be needed to maintain pig performance due to the potential for exposure to various sources of immune stimulation.

Given the potentially significant effects on animal productivity as a result of exposure to pathogens and other stressors, Pluske et al. (2018) suggested several strategies that should be considered to improve the health and productivity of commercially raised pigs. These include (i) decreasing the presence of immune-stimulating agents (e.g., pathogens), (ii) selecting (genetics) for enhanced immunity, (iii) regulating the immune response for specific conditions, and (iv) reducing the negative effects of the immune response while maintaining disease control. Sustainable pig production will employ multiple strategies while reducing reliance on strategies with potential negative impacts on human health (e.g., antibiotics) or environment (e.g., zinc oxide) (Marquardt and Li 2018). While there are several nutritional strategies available (e.g., diet acidification, plant-based compounds, probiotics, etc.), only the general effects of dietary macronutrients (i.e., protein, fiber) and functional amino acid supplementation will be discussed here. Readers are directed to Chapter 18 for information on various feed additives.

### The Role of Protein and Fiber on Gut Environment

Both dietary protein and dietary fiber content can have a significant impact on animal health, mainly through effects on gastrointestinal health (Jha and Berrocoso 2016). These factors are becoming increasingly important in swine production as coproducts and other alternative feedstuffs commonly used in swine diets to reduce feed costs tend to have a high fiber content and vary in protein content and amino acid bioavailability (Pieper et al. 2012a; Zijlstra and Beltranena 2013), largely due to damage associated with heat processing (Jha et al. 2011).

While adequate dietary protein is required to meet amino acid requirements for maintenance and growth functions, high dietary protein intake can negatively impact animal health (Windey et al. 2012; Diether and Willing 2019). High protein diets result in an increase in the amount of undigested protein flow into the hindgut, and the fermentation of this undigested protein by colonic microbiota may have detrimental effects on gut health (Jha and Berrocoso 2016; Zhao et al. 2020). Metabolites of protein fermentation, including branched-chain fatty acids, ammonia, biogenic amines, hydrogen sulfide, and phenolic and indolic compounds (Pieper et al. 2012b; Yao et al. 2016; Gilbert et al. 2018), have been associated with toxic and proinflammatory effects on the gut epithelium (Windey et al. 2012; Diether and Willing 2019). These include compromised colonic epithelial cell structure and metabolic functions, thinning of the mucus barrier, and increased colonic permeability (Gaskins 2000; Hughes et al. 2008; Yao et al. 2016). Diets high in protein are a predisposing factor in the development of postweaning diarrhea due to *E. coli* (Prohászka and Baron 1980; de Lange et al. 2010). Multiple studies have demonstrated that feeding a low protein diet that is fortified with adequate amounts of essential amino acids will reduce the amount of substrate available for the proliferation of pathogenic bacteria and minimizing proteolytic fermentation and the production of associated toxic metabolites (Nyachoti et al. 2006; Opapeju et al. 2008). Of note, the negative effects, including impaired growth performance and increased intestinal permeability, are generally associated with feeding plant, non-digestible, presumably fermentable protein sources that contain higher content of antinutritional factors (e.g., trypsin inhibitors, lectins, and P34 protein) (Zeamer et al. 2021).

In general, greater inclusion of dietary fiber results in reduced growth performance in swine (Jha et al. 2019) mainly through impacts on nutrient availability, although many of these negative effects can be overcome through adjustments in feed formulation (e.g., formulation on dietary net energy basis). Dietary fiber content can also have substantial impacts on animal health primarily due to effects on the gut epithelial integrity, availability of fermentable substrate for gut microbes, formation of fermentation end products, and stimulation of proliferation of beneficial bacteria (de Lange et al. 2010; Bach Knudsen et al. 2012). Unlike with protein, fermentation of fiber is generally considered to have beneficial effects on gut health in the pig, with many of these effects being attributable to the inclusion of fermentable (i.e., soluble) dietary fiber (Jha and Berrocoso 2015). The main end products of fiber fermentation, short-chain fatty acids (such as acetate, propionate, and butyrate), have been shown to stimulate intestinal epithelial proliferation and barrier function (Bach Knudsen et al. 2012), inhibit undesirable bacterial growth through reduction of luminal pH (May et al. 1994), and impact mucosal immunity (Kanauchi et al. 2001). Butyrate, in particular, has been shown to have a positive impact on gut mucosal health (Ohira et al. 2017; Feng et al. 2018). The provision of fermentable dietary fibers has also been shown to result in a more stable, diverse microbiota, and the inclusion of fermentable fiber has also been suggested to preferentially promote the proliferation of nonpathogenic and potentially beneficial bacteria, such as *Lactobacilli* and *Enterococci* (Mikkelsen et al. 2003; Konstantinov et al. 2004; Bikker et al. 2006), while reducing harmful bacteria, such as some *Clostridia* spp. and *E. coli* (Jeaurond et al. 2008; Liu et al. 2008a). However, the actual impact of inclusion of fiber in diets on animal health, and, particularly, pathogen susceptibility, has not been consistent (Pluske et al. 2003; de Lange et al. 2010; Bach Knudsen et al. 2012), likely due to differences in the physicochemical properties and fermentability of different fiber sources (Bach Knudsen et al. 2012; Taciak et al. 2017). The increase in intestinal mucous secretion and enterocyte sloughing with high-fiber diets affect gut integrity (Mariscal-Landín et al. 1995; Schmidt-Wittig et al. 1996; Hedemann et al. 2006) and may result in increased pathogen susceptibility. Pluske et al. (2003) noted increased incidence of clinical swine dysentery in growing pigs and diarrhea in weanling pigs fed diets high in fermentable fiber (i.e., nonstarch polysaccharides) due to increased water secretion in the gut. Moreover, nursery pig diets supplemented with beta-glucan resulted not only in increased growth performance but also increased the susceptibility to *Streptococcus suis* infection (Dritz et al. 1995).

While the inclusion of fermentable fiber has generally positive effects on the gut environment, fiber can also act as an antinutritional factor through effects on nutrient digestibility and increased endogenous protein secretions (e.g., mucin) which contribute to an increase in supply of undigested protein to the colon (Yao et al. 2016) and reduce amino acid availability for growth. Likewise, fermentation of dietary protein also results in the formation of beneficial metabolites such as short-chain fatty acids (Wong et al. 2006). In general, however, the inclusion of dietary fermentable fiber and/or the reduction in dietary protein will result in an increase in beneficial metabolites while reducing negative metabolites (Bikker et al. 2006; Nyachoti et al. 2006; Htoo et al. 2007). For example, Wellington et al. (2020b) showed, in growing pigs, an attenuation of the negative effects of high dietary protein content when high dietary fiber was offered concomitantly, mainly through alteration to the gastrointestinal production of short- and branched-chain fatty acids. This is likely a result of gastrointestinal microbes favoring the energetic fermentation of fiber over protein sources as well as a greater incorporation of nitrogen-containing substrates such as amino acids and ammonia into the microbial biomass (Jeaurond et al. 2008; Pieper et al. 2012b; Yao et al. 2016).

## Amino Acids: Feeding the Pig's Immune System

Stimulation of the immune system alters protein and amino acid metabolism and utilization, with amino acids redirected from growth toward supporting the immune response (Reeds et al. 1994). This is exacerbated by the reduced feed intake that generally accompanies infection and subsequent immune response, further reducing lean tissue gain and limiting the exogenous supply of amino acids to meet changing requirements (Quiniou et al. 1996; Reeds and Jahoor 2001; Le Naou et al. 2012). Thus, the increase in amino acid requirement to support the immune response is largely met through the mobilization of endogenous sources of amino acids (i.e., muscle protein) (Reeds et al. 1994). The amino acid profile of muscle protein differs significantly from that of proteins involved in the immune response, resulting in an amino acid imbalance and a disproportionate use of some amino acids during immune challenge that leads to an obligatory increase in whole-body amino acid catabolism and reduction in body protein growth (Figure 9.3 Reeds et al. 1994; Klasing 2007). Goodband et al. (2014) and Rakhshandeh and de Lange (2011) suggested that mobilization of endogenous sources (i.e., body protein catabolism) is sufficient to provide the additional amino acids required for the immune response for the majority of amino acids, however, due to the distinct amino acid profile in immune system components vs. muscle protein, considerable protein stores need to be mobilized in order to meet immune system requirements. For example, Rakhshandeh and de Lange (2011) estimated that 6 g of muscle protein would need to be catabolized in order to meet the cysteine requirements for just 1 g of acute-phase protein (APP) production. Moreover, this limits the efficiency with which mobilized amino acids can be reutilized for muscle protein synthesis (Reeds and Jahoor 2001). Overall, adjustment of diet formulations, through

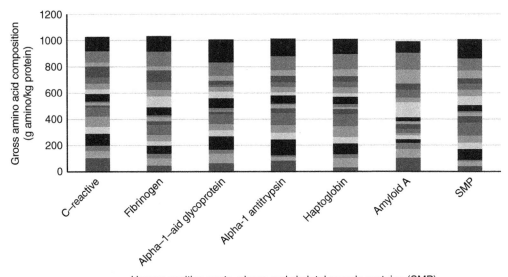

**Figure 9.3**  Amino acid profile of human acute-phase proteins and skeletal muscle protein (SMP). Source: Adapted from Reeds et al. (1994).

provision of supplemental amino acids and/or altered amino acid profile, may be needed in order to support the immune response and gut health as well as maintain pig growth performance (i.e., lean gain) or, at the very least, limit the negative impact of immune stimulation on body protein deposition. This will be even more significant in the post-antibiotic era, as there is evidence suggesting that an altered essential amino acid profile in low protein, antibiotic-free diets improves growth performance, intestinal development, nutrient utilization, and gut health of weaned pigs (Spring et al. 2020; Zhou et al. 2020).

Of the amino acids, threonine (Thr), aromatic amino acids (AAA), and sulfur amino acids (SAA) are of particular importance as precursors for the synthesis of many acute-phase proteins, immunoglobulins, and other critical components of the immune response (Reeds et al. 1994; Reeds and Jahoor 2001; Klasing 2007). Previous studies have demonstrated an increased requirement for growth for methionine (Met) and cysteine (Cys) (Litvak et al. 2013b; Rakhshandeh et al. 2014), threonine (Thr) (Jayaraman et al. 2015a; Mathai et al. 2016), tryptophan (Trp) (Le Floc'h et al. 2009; de Ridder et al. 2012), and arginine (Arg) (Li et al. 2012; Zhu et al. 2013) in response to immune stimulation. Likewise, studies have shown that pigs housed in low sanitary conditions (van der Meer et al. 2016) or inoculated with *Salmonella typhimurium* (Rodrigues et al. 2021) had greater weight gain and feed efficiency when fed diets supplemented with Met, Thr, and Trp above requirements for growth, suggesting that they may benefit from the supplementation of a combination of amino acids. In addition, several amino acids generally considered as nonessential (e.g., glutamine (Gln), Arg, Cys) may become essential during immune challenge, necessitating their inclusion in diet formulations (Rezaei et al. 2013). Moreover, traditional determination of amino acid essentiality has been based on outcome variables related to growth (i.e., average daily gain, feed efficiency) with little consideration given to the nonproteinogenic functions of amino acids. This has led to a necessary shift in terminology in order to account for the "functional" aspects of amino acids with respect to nutrition and health (Wu 2013). Indeed, while growth performance may not necessarily be influenced by supplementation of functional amino acids, immune status, gut health, and overall pig robustness may be improved (Klasing 1988; Defa et al. 1999; Wang et al. 2006). The remainder of this chapter is dedicated to discussing key functional amino acids of importance to pig health (Table 9.1).

### Sulfur-containing Amino Acids (SAA)

Both Met and Cys have been shown to play critical roles in the development of the immune system and in maintaining the immune response, and the metabolism of these amino acids is closely linked (Shoveller et al. 2004; Lewis 2009). As reviewed by Ruan et al. (2017), Met is involved in the immunity development at organ, nonspecific, humoral, and cellular levels. Methionine utilization increases during the immune response due to its key role as a methyl donor, which is important for both cell proliferation (including immune cells) and in the production of polyamines, choline, and carnitine (Lu 2000; Grimble 2008) as well as acute-phase proteins (Grimble 2008; Métayer et al. 2008). Methionine also contributes to the maintenance of gut barrier function through the upregulation of tight junction proteins (Chen et al. 2014) and can act as an antioxidant through the methionine sulfoxide reductase system for the repair of damaged proteins (Oien and Moskovitz 2007). A key role of Met and Cys during immune stimulation is the production of glutathione and taurine, major cellular antioxidants required for theprotection of proteins that can be damaged during pathogenesis (Oien and Moskovitz 2007; Wang et al. 2009; Wu et al. 2013). Glutathione and taurine (a product of SAA metabolism) have also been shown to increase cell

**Table 9.1** Summary of effects of amino acids influencing intestinal health, overall immunity, or growth in healthy and immune challenged pigs[a].

| Amino acid | Effect[b] | Status | References |
|---|---|---|---|
| SAA | ↑ Intestinal epithelial growth | Healthy | Bauchart-Thevret et al. (2009) |
| | ↑ Efficiency of AA utilization | LPS | Kim et al. (2012b) |
| | ↑ Plasma protein and albumin synthesis | LPS | Litvak et al. (2013a) |
| | ↑ Protein deposition | LPS | Litvak et al. (2013b) |
| | ↑ Integrity and barrier function of the small-intestinal mucosa | Healthy | Chen et al. (2014) |
| | ↑ Protein deposition | LPS | Rakhshandeh et al. (2014) |
| | ↑ Villus development and ↓ oxidative stress | Healthy | Shen et al. (2014) |
| | ↑ Feed efficiency | LSC | van der Meer et al. (2016) |
| | ↑ Proinflammatory IFN-γ response and protein utilization | ETEC | Capozzalo et al. (2017a) |
| | ↑ Daily weight gain | ETEC | Capozzalo et al. (2017b) |
| | ↓ Plasma urea nitrogen and ↑ villus height | LSC | Kahindi et al. (2017) |
| | ↓ Ear biting | LSC | van der Meer et al. (2017) |
| Trp | ↓ Plasma haptoglobin levels, IDO activity, and lung weight | CFA | Le Floc'h et al. (2008) |
| | ↑ Plasma levels of Trp | LSC | Le Floc'h et al. (2009) |
| | ↑ Feed intake | ETEC | Trevisi et al. (2009) |
| | ↓ Histopathological inflammation | DDS | Kim et al. (2010) |
| | ↑ Substrate for protein deposition | LPS | de Ridder et al. (2012) |
| | ↑ Daily weight gain | LSC | Capozzalo et al. (2013) |
| | ↓ Oxidative stress of the liver | Diquat | Mao et al. (2014b) |
| | ↑ Feed efficiency and plasma levels of Trp | ETEC | Capozzalo et al. (2015) |
| | ↓ Intestinal inflammatory and immune response | ETEC | Jayaraman et al. (2015b) |
| | ↑ Feed efficiency | LSC | van der Meer et al. (2016) |
| | ↑ Average daily gain | LSC | Jayaraman et al. (2017) |
| | ↓ Ear biting | LSC | van der Meer et al. (2017) |
| Thr | ↑ Serum IgG concentration | Healthy | Wang et al. (2006) |
| | ↑ Production of mucus | Healthy | Law et al. (2007) |
| | ↑ Protein synthesis in muscle and intestine and mucins | Healthy | Wang et al. (2007) |
| | ↑ Gut barrier function | Healthy | Wang et al. (2010) |
| | ↑ Downregulation on the expression of TLR in tissues | PRLV | Mao et al. (2014a) |
| | ↓ Intestinal damage and immune responses | ETEC | Ren et al. (2014) |
| | ↑ Feed efficiency | LSC | Jayaraman et al. (2015a) |
| | ↓ Fever response | ETEC | Trevisi et al. (2015) |
| | ↑ Feed efficiency | LSC | van der Meer et al. (2016) |
| | ↑ Protein synthesis in gastrointestinal tissues | Healthy | Munasinghe et al. (2017) |
| | ↓ Ear biting | LSC | van der Meer et al. (2017) |
| | ↑ Protein deposition | LPS | Wellington et al. (2018) |
| | ↑ Protein deposition | LPS | McGilvray et al. (2019) |
| | ↑ Growth performance | ST | Wellington et al. (2019) |
| | ↑ Mucin production | LPS/ST | Wellington et al. (2020) |
| Arg | ↑ Intestinal protein synthesis and ↓ intestinal permeability | PRV | Corl et al. (2008) |
| | ↓ Expression of intestinal proinflammatory cytokines | LPS | Liu et al. (2008b) |
| | ↑ mTOR signal activation in skeletal muscle | Healthy | Yao et al. (2008) |
| | ↑ Microvascular development in the small intestine | Healthy | Zhan et al. (2008) |
| | ↑ Cellular and humoral immunity | Healthy | Tan et al. (2009) |
| | ↑ Protein synthesis for mTOR and TLR4 signaling pathways | LPS | Tan et al. (2010) |
| | ↑ Intestinal gene expression, growth, and integrity | Healthy | Wu et al. (2010) |
| | ↓ Stress-induced metabolites | Weaning | He et al. (2011) |
| | ↓ Activation of the TLR4-Myd88 signaling pathway | SC | Chen et al. (2012) |
| | ↓ Release of liver proinflammatory cytokines and free radicals | LPS | Li et al. (2012) |
| | ↑ Antioxidant capacity and ↓ expression of inflammatory cytokines | Diquat | Zheng et al. (2013) |

*(Continued)*

**Table 9.1** (Continued)

| Amino acid | Effect[b] | Status | References |
|---|---|---|---|
| | ↑ Intestinal mucosal immune barrier function and integrity | ETEC | Zhu et al. (2013) |
| | ↑ mRNA levels for key intestinal transporters | DON | Wu et al. (2015) |
| | ↑ Jejunal morphology and AA levels in the serum and intestine | DON | Wu et al. (2015) |
| Gln/Glu | ↓ Villous atrophy and intestinal morphology impairment | ETEC | Yi et al. (2005) - Gln |
| | ↑ Hsp70 expression | Healthy | Zhong et al. (2011) - Gln |
| | ↑ Plasma IgG concentration | LPS | Hsu et al. (2012) - Gln |
| | ↑ Feed efficiency | Healthy | Cabrera et al. (2013) - Gln+Glu |
| | ↓ Abnormalities of intestinal structure | MYC | Duan et al. (2014) - Glu |
| | ↑ Intestinal cell growth and membrane integrity | Diquat | Jiao et al. (2015) - Glu |
| | ↓ Intestinal permeability and ↑ villus height | Healthy | Wang et al. (2015b) - Gln |
| | ↑ Antioxidative capacity and energy metabolism | LPS | Wang et al. (2015c) - Glu |
| | ↑ Antioxidant status and expression of AA transporters | Diquat | Yin et al. (2015) - Glu |
| | ↓ Growth suppression, oxidative stress, and ↑ serum AA pool | HP | Duan et al. (2016) - Glu |
| | ↑ Immunity and intestinal antioxidative capacity | Healthy | Lv et al. (2018) - Gln |
| Ast/Asn | ↓ Expression of hepatic proinflammatory cytokines | LPS | Leng et al. (2014) - Asp |
| | ↑ Intestinal integrity and energy status | LPS | Pi et al. (2014) - Asp |
| | ↑ Villus height and villus height:crypt depth ratio | LPS | Wang et al. (2015a) - Asn |
| | ↓ Gut proinflammatory cytokine and enterocyte apoptosis | LPS | Chen et al. (2016) - Asn |
| | ↓ Expression of muscle proinflammatory cytokines | LPS | Wang et al. (2017) - Ast |
| | ↑ Intestinal claudin-1 protein expression and ↓ villous atrophy | LPS | Zhu et al. (2017) - Asn |
| BCAA | ↑ Performance, intestinal development, and expression of AA transporters | Healthy | Zhang et al. (2013) - BCAA |
| | ↑ Mucin production and goblet cell numbers in the jejunal mucosa | PRV | Mao et al. (2015) - Leu |
| | ↑ Growth performance and intestinal immunity | Healthy | Ren et al. (2015) - BCAA |
| | ↑ Intestinal development and whole-body growth | Healthy | Sun et al. (2015) - Leu |
| | ↑ mTOR complex 1-dependent translation | LPS | Hernandez-García et al. (2016) - Leu |
| | ↑ Intestinal β-defensin expression | Healthy | Ren et al. (2016) - BCAA |
| | ↑ Efficiency of protein deposition | Healthy | Rudar et al. (2017) - Leu |
| | ↑ Intestinal morphology and cell proliferation | Healthy | Duan et al. (2018) - BCAA |
| | ↑ Repartitioning of AA from visceral to peripheral protein deposition | LPS | Rudar et al. (2019) - Leu |

[a] AA, amino acid; Ast/Asn, aspartate/asparagine; Arg, arginine; BCAA, branched-chain amino acids; CFA, complete Freund's adjuvant; DDS, dextran sodium sulfate-induced colitis; DON, deoxynivalenol; ETEC, enterotoxigenic *E. coli*; Gln/Glu, glutamine/glutamate; HP, hydrogen peroxide; Hsp70 = heat shock protein 70; IDO, indoleamine 2,3-dioxygenase; IgG, immunoglobulin G; IFN-γ, interferon gamma; Leu, leucine; LPS, lipopolysaccharide; LSC, low sanitary condition; mTOR, mammalian/mechanistic target of rapamycin; Myd88, myeloid differentiation primary response protein MyD88; MYC, mycotoxin; PRLV, pseudorabies live vaccine; PRV, porcine rotavirus; SAA, sulfur-containing amino acids; Thr, threonine; TLR, Toll-like receptor; Trp, tryptophan; SC, *Salmonella* enterica serovar *Choleraesuis*; ST, *Salmonella* enterica serovar *Typhimurium*.
[b] ↓: decreased, impaired, or suppressed; ↑: increased, improved, or supported.

proliferation (Redmond et al. 1998; Wu et al. 2004), which may be critical in maintaining gut barrier as well as acute-phase protein production.

The importance of Met and Cys, which is produced via trans-sulfuration of Met, during immune stimulation has been demonstrated by a number of previous studies. Malmezat et al. (2000) reported increased contribution of trans-sulfuration to both Met and Cys flux in plasma of rats after infection with enterotoxigenic *E. coli*. Additionally, an increased whole-body flux of Cys was observed in mini pigs after experimental ileitis (Rémond et al. 2011) and in starter pigs after immune system stimulation by LPS (Rakhshandeh et al. 2020), possibly related to an increased pool of glutathione in the liver (Budziński et al. 2011). Supplementation of SAA has also been shown to increase the intracellular glutathione pool and maintain gut redox status and integrity in weaned pigs (Bauchart-Thevret et al. 2009; Chen et al. 2014; Shen et al. 2014). Albumin production has also been shown to be reduced in immune-stimulated pigs with a restricted dietary SAA intake (Litvak et al. 2013a).

Early work by Rakhshandeh et al. (2010) demonstrated that during immune stimulation, SAA are preferentially retained in nonprotein pools, such as glutathione, as indicated by a reduction in nitrogen:sulfur-balance. Rakhshandeh et al. (2014) further demonstrated that while the efficiency of utilization of total SAA for whole-body protein deposition was not influenced by immune stimulation, estimates of maintenance requirements for total SAA were increased. An increased requirement for Met+Cys has been demonstrated in pigs (as Met+Cys:Lys ratio) in pigs following LPS injection (Kim et al. 2012b), infected with *E.coli* (Capozzalo et al. 2017a,b) or *Salmonella typhimurium* (Rodrigues et al. 2021), and under commercial (unsanitary) conditions (van der Meer et al. 2016; Kahindi et al. 2017; van der Meer et al. 2017). The increased requirement for total SAA may be mostly due to an increased in Met requirement, as demonstrated by Litvak et al. (2013b) who showed an increased requirement for Met while maintaining the dietary Met to total SAA ratio (i.e., Met:Met+Cys). This is further supported by an increase in rates of transmethylation (Yu et al. 1993) and trans-sulfuration (Malmezat et al. 2000) of Met during immune challenge. Indeed, Met is highly unstable and rapidly oxidizes to Cys, releasing free radicals in the process (Grimble 2008). Therefore, intracellular Cys is maintained at low levels (Anderson and Meister 1987) limiting free Cys availability for the production of glutathione and APP. As a result, the provision of SAA during times of immune challenge is most likely better met through provision of Met.

### *Aromatic Amino Acids*

As demonstrated by Reeds et al. (1994), many APP are rich in aromatic amino acids (i.e., phenylalanine, tyrosine, and tryptophan) indicating the potential for an increased need for these amino acids during disease challenge, although to date only Trp has been studied extensively. For example, plasma levels of Trp have been shown to decrease during inflammation (Le Floc'h et al. 2012) indicating increased utilization or catabolism of Trp and potential need for dietary supplementation to support its use. A major function of Trp during the immune response is as a precursor for kynurenine production, via upregulation of indoleamine 2,3-dioxygenase (IDO) during immune challenge (Popov and Schultze 2008). Increased IDO activity and Trp catabolism are well-established outcomes of immune system activation resulting in decreased plasma Trp and increased kynurenine concentrations (Le Floc'h et al. 2018). Kynurenine is further utilized to produce a number of metabolites including quinolinic acid, niacin, and nicotinamide (Le Floc'h et al. 2012). Tryptophan is also a precursor for melatonin, which, along with two end-products of the kynurenine pathway, 3-hydroxyanthranilic acid and 3-hydroxykynurenine, have been shown to have antioxidant properties (Christen et al. 1990; Goda et al. 1999; Le Floc'h and Seve 2007). Indeed, Trp supplementation above requirements has been shown to reduce oxidative stress in piglets (Mao et al. 2014b). Tryptophan also appears to have an immunomodulatory function, acting to reduce the severity of the immune response. For example, Trp has been shown to result in reduced IDO activity (Le Floc'h et al. 2008) and reduced overall inflammation (Kim et al. 2010). Supplemental Trp may also act to reduce the severity of immune stimulation through improvements in feed intake (Capozzalo et al. 2015).

Studies have demonstrated a beneficial effect of Trp supplementation in pigs and an increased requirement for Trp during immune stimulation. de Ridder et al. (2012) estimated a 7% increase in the Trp requirement during immune stimulation via administration of LPS. This is in line with Jayaraman et al. (2017) who estimated an increase in optimal Trp:lysine (Lys) ratio of 4 percentage points in piglets kept under low vs. high sanitary conditions and further demonstrated that provision of an increased dietary Trp:Lys ratio reduced the intestinal inflammatory and immune response in *E. coli* challenged pigs (Jayaraman et al. 2015b). Le Floc'h et al. (2009) observed improved growth performance with greater dietary Trp in weaned pigs housed in low sanitary condition. In pigs challenged with *E. coli*, Trevisi et al. (2009) and Capozzalo et al. (2013, 2015) demonstrated improved feed intake and growth performance with additional dietary Trp.

### Threonine

Threonine is a critical amino acid during times of immune stimulation, mainly as a precursor for production of mucins in the gut but also as a key constituent of many APP and immune cells (Faure et al. 2007; Munasinghe et al. 2017). Wang et al. (2006) and Defa et al. (1999) demonstrated that greater dietary Thr was required to maximize immune status (i.e., plasma IgG concentration) than was required to optimize growth. Adequate levels of Thr have been shown to be crucial for mucin synthesis and maintenance of gut function in piglets (Law et al. 2007) and maintenance of integrity of nonspecific defenses of the gut wall. It was estimated that Thr represents approximately 16% of total amino acids in mucins (Lien et al. 1997) and that Thr uptake is critical for adequate mucin production, particularly in young pigs (Law et al. 2007). Wang et al. (2010) recommended an increase in dietary Thr to 0.89% in post-weaned pigs based on improved intestinal structure and health. Likewise, Ren et al. (2014) reported a positive effect of greater dietary Thr on intestinal morphology and immune status of weaned piglets challenged with *E. coli*.

In addition to its importance in maintaining gut integrity, Thr is an important precursor for the synthesis of many acute-phase proteins involved in the immune response, such as C-reactive protein, fibrinogen, and haptoglobin (Reeds et al. 1994). Threonine is also the most abundant essential amino acid in immunoglobulin protein (Tenenhouse and Deutsch 1966; Li et al. 2007), and higher dietary Thr is required to optimize serum IgG in both healthy (Cuaron et al. 1984; Hsu et al. 2001) and immune-challenged pigs (Defa et al. 1999; Wang et al. 2006). There was a positive relationship between Thr intake and serum IgG levels in weaned piglets after injection with bovine serum albumin or swine fever-attenuated vaccine (Defa et al. 1999). Moreover, after challenging pigs with *E. coli*, supplementation of Thr improved jejunal levels of IgG and interleukin-1β (Ren et al. 2014) as well as increased secretion of IgM in pigs (Trevisi et al. 2015). Piglets fed increasing levels of Thr showed reduced elevation of serum interferon-γ and higher expression of different Toll-Like Receptors following pseudorabies vaccination (Mao et al. 2014a).

Immune stimulation by LPS injection has been shown to increase the Thr requirement for protein deposition (Wellington et al. 2018). This increase has been demonstrated to be the result of an increase in the maintenance Thr requirement and not a decrease in the efficiency of utilization of Thr for whole-body protein synthesis (Wang et al. 2007; McGilvray et al. 2019). Dietary supplementation of Thr above the requirements for growth has been shown to increase growth performance in pigs exposed to poor sanitary conditions (Jayaraman et al. 2015a) and when challenged with *Salmonella* (Wellington et al. 2019) or *E. coli* (Trevisi et al. 2015). Likewise, supplemental Thr has been shown to enhance gut health (Koo et al. 2020a), nitrogen utilization, and body protein deposition (Koo et al. 2020b) in pigs fed simple diets.

## *Arginine*

Arginine serves as a precursor for several compounds including polyamine synthesis, hormones (e.g., IGF, insulin, and growth hormone), and creatine (Wu and Morris 1998; Suchner et al. 2002) and is a major component of intestinal alkaline phosphatase (Le Floc'h et al. 2018). A major metabolite of Arg metabolism, nitrous oxide, acts as a vasodilator, signaling molecule, and mediator of the immune response (Le Floc'h et al. 2018).

Dietary supplementation of Arg increases expression or production of a number of immune-related proteins (e.g., Il-8, TNF-α, immunoglobulins) (Li et al. 2007; Tan et al. 2009). The positive effects of Arg appear to be mainly as a result of its effects on gut function and health. Arginine increases protein synthesis in the small intestine and muscle via the mTOR signaling pathway (Fumarola et al. 2005; Yao et al. 2008), and supplementation with Arg has been shown to be effective at activating intestinal protein synthesis in pigs exposed to an enteric pathogen (Corl et al. 2008) and protecting against enterocyte death in LPS-challenged pigs (Tan et al. 2010). Arginine has also been shown to reduce the intestinal inflammatory response in pigs challenged with LPS (Liu et al. 2008b), improves microvascular development and tissue architecture in the small intestine, and mitigates the negative impact as well as suppress the negative effects of harmful metabolites, such as oxidative stress, on intestinal function (Zhan et al. 2008; Wu et al. 2010; He et al. 2011; Zheng et al. 2013). Indeed, Arg improves intestinal barrier function and integrity and overall immunity in healthy weaned pigs (Tan et al. 2009) as well as in pigs following either *E. coli* LPS (Zhu et al. 2013) or *Salmonella* (Chen et al. 2012) challenge and was able to mitigate the deoxynivalenol-related reduction in expression of intestinal nutrient transporters (i.e., glucose and amino acids) (Wu et al. 2015). Additionally, Arg supplementation was reported to exert protective effects on the liver of weaned piglets after LPS injection, possibly through the inhibition on the TLR4 signaling pathway (Li et al. 2012).

## *Glutamine and Glutamate*

Glutamine (Gln), the most abundant free amino acid in the body, and glutamate (Glu) have multiple functions important in maintaining animal growth and health. Both Gln and Glu (via Gln) are key precursors for production of the major cellular antioxidant glutathione (Reeds et al. 1997; Humbert et al. 2007). They have also been shown to regulate gene expression, particularly those related to production of tight junction proteins (Jiao et al. 2015; Wang et al. 2015b), heat shock proteins (Zhong et al. 2011), and antioxidants (Yin et al. 2015). Glutamine also serves as a precursor for purine and pyrimidine biosynthesis (Wu 1998), as an energy substrate for enterocytes (Wu et al. 2011), and as nitrogen donor (Le Floc'h et al. 2004). Overall, both Glu and Gln play key roles in the maintenance of protein synthesis and cell proliferation and in the maintenance of gut integrity and support of the immune response.

Although considered nonessential amino acids, several studies have demonstrated positive effects of dietary supplementation with Gln and Glu, although few studies have been conducted to demonstrate these effects in specific disease models. Weaned piglets supplemented with Gln for 25 days showed lower inflammatory response in the gut and improved immunity and growth potential at weaning (Hsu et al. 2012). Improved weight gain and feed efficiency in Gln-supplemented weaned piglets may be linked to enhanced intestinal antioxidative capacity, morphology, and health (Yi et al. 2005; Cabrera et al. 2013; Lv et al. 2018). Both Gln and Glu have been shown to reduce villus atrophy (Wu et al. 1996; Rezaei et al. 2013). Glutamate supplementation was able to alleviate antioxidative stress and damage to intestinal architecture caused by mycotoxins in growing pigs (Duan et al. 2014). Pigs challenged with *E. coli* had improved immune function, as indicated by lymphocyte function, when fed a diet supplemented with Gln (Yoo et al. 1997). Likewise, dietary supplementation with Glu or α-ketoglutarate (a Glu precursor) improved antioxidant balance in weaned piglets that either injected intraperitoneally with hydrogen peroxide (Duan et al. 2016) or LPS (Wang et al. 2015c).

### *Aspartate and Asparagine*

Aspartate has been shown to be involved in the regulation of hepatic proinflammatory cytokine expression as well as in TLR4 and NOD signaling and also enhanced liver integrity in weanling pigs, at short- and long-term in response to LPS challenge (Leng et al. 2014). Similarly, TLR4, NODs/NF-κB, and p38 signaling pathways were inhibited, while intestinal architecture was maintained when LPS-challenged piglets were fed diets supplemented with Asp (Wang et al. 2017). Aspartate's key role in intestinal health is due, in part, to improved energy status via increased availability of ATP and ADP with a concomitant decrease in the AMP/ATP ratio and improvement in the activities of tricarboxylic acid cycle enzymes (Pi et al. 2014) or diamine oxidases (Wang et al. 2017). Asparagine supplementation decreased the expression of TLR4, NODs, and p38 and suppressed enterocyte apoptosis in LPS-challenged weanling pigs (Chen et al. 2016). In the same way, Asn supplementation decreased intestinal damage caused by LPS injection and regulated mast cell activation in piglets (Zhu et al. 2017). Tissue-specific beneficial effects of feeding Asn were further confirmed by the modulation of the AMPK signaling pathway and improved energy status in LPS-challenged piglets (Wang et al. 2015a).

### *Branched-Chain Amino Acids (BCAA)*

The branched-chain amino acids (BCAA) may be beneficial during times of immune challenge. Leucine (Leu) in particular may play a role in supporting growth and gut health during immune challenge due to its in major protein synthetic and protein degradation pathways (e.g., mTOR) (Columbus et al. 2015; Lynch and Adams 2014). A primary result of disease challenge is an increase in protein turnover, mainly due to a decrease in protein synthesis in muscle tissue and increase in viscera (e.g., liver, spleen) (Orellana et al. 2002) largely through a decrease in translation initiation (Orellana et al. 2011) and increase in protein degradation signaling, which is reversed with Leu supplementation (Hernandez-García et al. 2016). Leucine may also support a redistribution of amino acids toward muscle protein synthesis, as demonstrated by Rudar et al. (2019) who observed an increase in protein synthesis in muscle tissue and decrease protein synthesis in viscera (e.g., liver) in LPS-challenged pigs provided supplemental Leu. The overall benefit of Leu supplementation is unclear; however, as Rudar et al. (2017) demonstrated that feeding high Leu levels to nursery pigs was only able to improve whole-body protein turnover before and not during an LPS challenge.

In addition to their role in muscle growth, BCAA are required for optimal expression of intestinal amino acid and peptide transporters, including rBAT and PepT-1, further enhancing intestinal development and amino acid absorption and availability (Zhang et al. 2013). Optimal dietary BCAA levels also support gut health. Duan et al. (2018) demonstrated improved intestinal villus height and villus height:crypt depth as well as increased goblet cell number with increasing dietary Val and Ile ratio to Leu. Likewise, Sun et al. (2015) supplemented Leu to suckling piglets and reported increased duodenum villus height, greater ileum and duodenum villus height:crypt depth ratio, and enhanced expression of Leu transporters in the jejunum. Supplementation with Leu increased mucin production and goblet cell numbers in the jejunal mucosa of weaned pigs challenged by porcine rotavirus (Mao et al. 2015). Ren et al. (2015) reported the maintenance of growth performance and intestinal immunity in weaned piglets when fed a low protein diet supplemented with BCAA. Comparatively, an in vitro study revealed positive effects of BCAA on intestinal defensin production (Ren et al. 2016).

## Summary

Traditional swine nutrition has focused on meeting the demands for maximum growth; however, it is now clear that nutrition will play a key role in promoting animal health. A greater understanding of the interaction between nutrition and health and the development of nutritional strategies to maintain pig health is particularly critical with legislative and societal pressure to reduce antibiotic usage in livestock agriculture. While several nutrients have key roles in supporting gut health and immune status, this review has focused on the functional roles of amino acids in maintaining pig health. The immune response is supported via a repartitioning of amino acids away from growth toward the needs of the immune system. There also exists ample evidence of nonproteinogenic functions of amino acids, such as regulation of the immune response and support of gut health. The future of swine nutrition will necessarily involve a reclassification of amino acids based on functionality as well as a determination of the nutrient requirements of disease challenge.

## References

Anderson, M. E., and A. Meister. 1987. Intracellular delivery of cysteine. Enzymol. 143:313–325. doi: https://doi.org/10.1016/0076-6879(87)43059-0.

Bach Knudsen, K. E., M. S. Hedemann, and H. N. Lærke. 2012. The role of carbohydrates in intestinal health of pigs. Anim. Feed Sci. Technol. 173:41–53. doi: https://doi.org/10.1016/j.anifeedsci.2011.12.020.

Bäckhed, F., C. M. Fraser, Y. Ringel, M. E. Sanders, R. B. Sartor, P. M. Sherman, J. Versalovic, V. Young, and B. B. Finlay. 2012. Defining a healthy human gut microbiome: Current concepts, future directions, and clinical applications. Cell Host Microbe 12:611–622. doi: https://doi.org/10.1016/j.chom.2012.10.012.

Bailey, M. 2009. The mucosal immune system: Recent developments and future directions in the pig. Dev. Comp. Immunol. 33:375–383. doi: https://doi.org/10.1016/j.dci.2008.07.003.

Bauchart-Thevret, C., B. Stoll, S. Chacko, and D. G. Burrin. 2009. Sulfur amino acid deficiency upregulates intestinal methionine cycle activity and suppresses epithelial growth in neonatal pigs. Am. J. Physiol.-Endoc. M. 296:E1239–E1250. doi: https://doi.org/10.1152/ajpendo.91021.2008.

Bikker, P., A. Dirkzwager, J. Fledderus, P. Trevisi, I. le Huërou-Luron, J. P. Lallès, and A. Awati. 2006. The effect of dietary protein and fermentable carbohydrates levels on growth performance and intestinal characteristics in newly weaned piglets. J. Anim. Sci. 84: 3337–3345. doi: https://doi.org/10.2527/jas.2006-076.

Bourne, F. J. 1973. The immunoglobulin system of the suckling pig. P. Nutr. Soc. 32:205–215. doi: https://doi.org/10.1079/PNS19730041.

Broom, L. J., and M. H. Kogut. 2018. The role of the gut microbiome in shaping the immune system of chickens. Vet. Immunol. Immunop. 204:44–51. doi: https://doi.org/10.1016/j.vetimm.2018.10.002.

Budziński, G., A. Suszka-Świtek, A. Caban, G. Oczkowicz, M. Heitzman, W. Wystrychowski, B. Dolińska, F. Ryszka, and L. Cierpka. 2011. Evaluation of cysteine effect on redox potential of porcine liver preserved by simple hypothermia. Transplant. Proc. 43:2897–2899. doi: https://doi.org/10.1016/j.transproceed.2011.08.065.

Cabrera, R. A., J. L. Usry, C. Arrellano, E. T. Nogueira, M. Kutschenko, A. J. Moeser, and J. Odle. 2013. Effects of creep feeding and supplemental glutamine or glutamine plus glutamate (Aminogut) on pre- and post-weaning growth performance and intestinal health of piglets. J. Anim. Sci. Biotechnol. 4:29. doi: https://doi.org/10.1186/2049-1891-4-29.

Capozzalo, M. M., J. C. Kim, J. K. Htoo, C. F. M. de Lange, B. P. Mullan, C. F. Hansen, J. W. Resink, and J. R. Pluske. 2017a. Pigs experimentally infected with an enterotoxigenic strain of *Escherichia coli* have improved feed efficiency and indicators of inflammation with dietary supplementation of tryptophan and methionine in the immediate post-weaning period. Anim. Prod. Sci. 57:935–947. doi: https://doi.org/10.1071/AN15289.

Capozzalo, M. M., J. C. Kim, J. K. Htoo, C. F. M. de Lange, B. P. Mullan, C. F. Hansen, J.-W. Resink, P. A. Stumbles, D. J. Hampson, and J. R. Pluske. 2015. Effect of increasing the dietary tryptophan to lysine ratio on plasma levels of tryptophan, kynurenine and urea and on production traits in weaner pigs experimentally infected with an enterotoxigenic strain of *Escherichia coli*. Arch. Anim. Nutr. 69:17–29. doi: https://doi.org/10.1080/1745039X.2014.995972.

Capozzalo, M. M., J. C. Kim, J. K. Htoo, P. A. Stumbles, D. J. Hampson, N. Ferguson, and J. R. Pluske. 2013. Pigs kept under commercial conditions respond to a higher dietary tryptophan: lysine ratio immediately after weaning. Manipulating Pig Production XIV. Proceedings of the 14th Australasian Pig Science Association (APSA) Biennial Conference, 24–27 November, Melbourne, Australia. p. 91.

Capozzalo, M. M., J. W. Resink, J. K. Htoo, J. C. Kim, F. M. de Lange, B. P. Mullan, C. F. Hansen, and J. R. Pluske. 2017b. Determination of the optimum standardised ileal digestible sulphur amino acids to lysine ratio in weaned pigs challenged with enterotoxigenic *Escherichia coli*. Anim. Feed Sci. Technol. 227:118–130. doi: https://doi.org/10.1016/j.anifeedsci.2017.03.004.

Cesta, M. F. 2006. Normal structure, function, and histology of mucosa-associated lymphoid tissue. Toxicol. Pathol. 34:599–608. doi: https://doi.org/10.1080/01926230600865531.

Chen, S., Y. Liu, X. Wang, H. Wang, S. Li, H. Shi, H. Zhu, J. Zhang, D. Pi, C.-A. A. Hu, X. Lin, and J. Odle. 2016. Asparagine improves intestinal integrity, inhibits TLR4 and NOD signaling, and differently regulates p38 and ERK1/2 signaling in weanling piglets after LPS challenge. Innate Immun. 22:577–587. doi: https://doi.org/10.1177/1753425916664124.

Chen, Y., D. Chen, G. Tian, J. He, X. Mao, Q. Mao, and B. Yu. 2012. Dietary arginine supplementation alleviates immune challenge induced by *Salmonella* enterica serovar *Choleraesuis* bacterin potentially through the Toll-like receptor 4-myeloid differentiation factor 88 signalling pathway in weaned piglets. Brit. J. Nutr. 108:1069–1076. doi: https://doi.org/10.1017/S0007114511006350.

Chen, Y., D. Li, Z. Dai, X. Piao, Z. Wu, B. Wang, Y. Zhu, and Z. Zeng. 2014. L-Methionine supplementation maintains the integrity and barrier function of the small-intestinal mucosa in post-weaning piglets. Amino Acids. 46:1131–1142. doi: https://doi.org/10.1007/s00726-014-1675-5.

Cheng, C., H. Wei, C. Xu, X. Xie, S. Jiang, and J. Peng. 2018. Maternal soluble fiber diet during pregnancy changes the intestinal microbiota, improves growth performance, and reduces intestinal permeability in piglets. Appl. Environ. Microbiol. 84. doi: 10.1128/AEM.01047-18.

Christen, S., E. Peterhans, and R. Stocker. 1990. Antioxidant activities of some tryptophan metabolites: Possible implication for inflammatory diseases. Proc. Natl. Acad. Sci. USA. 87:2506–2510.

Columbus, D. A., M. L. Fiorotto, and T. A. Davis. 2015. Leucine is a major regulator of muscle protein synthesis in neonates. Amino Acids 47(2):259–270. doi: https://doi.org/10.1007/s00726-014-1866-0.

Corl, B. A., J. Odle, X. Niu, A. J. Moeser, L. A. Gatlin, O. T. Phillips, A. T. Blikslager, and J. M. Rhoads. 2008. Arginine activates intestinal p70s6k and protein synthesis in piglet Rotavirus enteritis. J. Nutr. 138:24–29. doi: https://doi.org/10.1093/jn/138.1.24.

Cuaron, J. A., R. P. Chapple, and R. A. Easter. 1984. Effect of lysine and threonine supplementation of sorghum gestation diets on nitrogen balance and plasma constituents in first-litter gilts. J. Anim. Sci. 58:631–637. doi: https://doi.org/10.2527/jas1984.583631x.

de Lange, C. F. M., J. Pluske, J. Gong, and C. M. Nyachoti. 2010. Strategic use of feed ingredients and feed additives to stimulate gut health and development in young pigs. Livest. Sci. 134:124–134. doi: https://doi.org/10.1016/j.livsci.2010.06.117.

de Ridder, K., C. L. Levesque, J. K. Htoo, and C. F. M. de Lange. 2012. Immune system stimulation reduces the efficiency of tryptophan utilization for body protein deposition in growing pigs. J. Anim. Sci. 90:3485–3491. doi: https://doi.org/10.2527/jas.2011-4830.

Defa, L., X. Changting, Q. Shiyan, Z. Jinhui, E. W. Johnson, and P. A. Thacker. 1999. Effects of dietary threonine on performance, plasma parameters and immune function of growing pigs. Anim. Feed Sci. Technol. 78:179–188. doi: https://doi.org/10.1016/S0377-8401(99)00005-X.

Dharmani, P., V. Srivastava, V. Kissoon-Singh, and K. Chadee. 2009. Role of intestinal mucins in innate host defense mechanisms against pathogens. J. Innate Immun. 1:123–135. doi: https://doi.org/10.1159/000163037.

Diether, N. E., and B. P. Willing. 2019. Microbial fermentation of dietary protein: an important factor in diet–microbe–host interaction. Microorganisms 7:19. doi: https://doi.org/10.3390/microorganisms7010019.

Dionissopoulos, L., C. E. Dewey, H. Namkung, and C. F. M. D. Lange. 2006. Interleukin-1ra increases growth performance and body protein accretion and decreases the cytokine response in a model of subclinical disease in growing pigs. Anim. Sci. 82:509–515. doi: https://doi.org/10.1079/ASC200654.

Dritz, S. S., J. Shi, T. L. Kielian, R. D. Goodband, J. L. Nelssen, M. D. Tokach, M. M. Chengappa, J. E. Smith, and F. Blecha. 1995. Influence of dietary β-glucan on growth performance, nonspecific immunity, and resistance to *Streptococcus suis* infection in weanling pigs. J. Anim. Sci. 73:3341–3350. doi: https://doi.org/10.2527/1995.73113341x.

Duan, J., J. Yin, W. Ren, T. Liu, Z. Cui, X. Huang, L. Wu, S. W. Kim, G. Liu, X. Wu, G. Wu, T. Li, and Y. Yin. 2016. Dietary supplementation with l-glutamate and l-aspartate alleviates oxidative stress in weaned piglets challenged with hydrogen peroxide. Amino Acids 48:53–64. doi: https://doi.org/10.1007/s00726-015-2065-3.

Duan, J., J. Yin, M. Wu, P. Liao, D. Deng, G. Liu, Q. Wen, Y. Wang, W. Qiu, Y. Liu, X. Wu, W. Ren, B. Tan, M. Chen, H. Xiao, L. Wu, T. Li, C. M. Nyachoti, O. Adeola, and Y. Yin. 2014. Dietary glutamate supplementation ameliorates mycotoxin-induced abnormalities in the intestinal structure and expression of amino acid transporters in young pigs. PLOS One 9:e112357. doi: https://doi.org/10.1371/journal.pone.0112357.

Duan, Y., B. Tan, J. Li, P. Liao, B. Huang, F. Li, H. Xiao, Y. Liu, and Y. Yin. 2018. Optimal branched-chain amino acid ratio improves cell proliferation and protein metabolism of porcine enterocytes in in vivo and in vitro. Nutrition 54:173–181. doi: https://doi.org/10.1016/j.nut.2018.03.057.

Enaud, R., R. Prevel, E. Ciarlo, F. Beaufils, G. Wieërs, B. Guery, and L. Delhaes. 2020. The gut-lung axis in health and respiratory diseases: A place for inter-organ and inter-kingdom cross talks. Front. Cell. Infect. Microbiol. 10. doi: https://doi.org/10.3389/fcimb.2020.00009.

Faderl, M., M. Noti, N. Corazza, and C. Mueller. 2015. Keeping bugs in check: The mucus layer as a critical component in maintaining intestinal homeostasis. IUBMB Life 67:275–285. doi: https://doi.org/10.1002/iub.1374.

Faure, M., F. Choné, C. Mettraux, J.-P. Godin, F. Béchereau, J. Vuichoud, I. Papet, D. Breuillé, and C. Obled. 2007. Threonine utilization for synthesis of acute phase proteins, intestinal proteins, and mucins is increased during sepsis in rats. J. Nutr. 137:1802–1807. doi: https://doi.org/10.1093/jn/137.7.1802.

Feng, W., Y. Wu, G. Chen, S. Fu, B. Li, B. Huang, D. Wang, W. Wang, and J. Liu. 2018. Sodium butyrate attenuates diarrhea in weaned piglets and promotes tight junction protein expression in colon in a gpr109a-dependent manner. Cell. Physiol. Biochem. 47:1617–1629. doi: https://doi.org/10.1159/000490981.

Fumarola, C., S. L. Monica, and G. G. Guidotti. 2005. Amino acid signaling through the mammalian target of rapamycin (mTOR) pathway: Role of glutamine and of cell shrinkage. J. Cell. Physiol. 204:155–165. doi: https://doi.org/10.1002/jcp.20272.

Gaskins, H. R. 2000. Intestinal bacteria and their influence on swine growth. In: Swine nutrition, 2nd ed. CRC Press, Boca Raton, FL. p. 585–608. doi: https://doi.org/10.1201/9781420041842.

Gilbert, M. S., N. Ijssennagger, A. K. Kies, and S. W. C. van Mil. 2018. Protein fermentation in the gut; implications for intestinal dysfunction in humans, pigs, and poultry. Am. J. Physiol.-Gastr. L. 315:G159–G170. doi: https://doi.org/10.1152/ajpgi.00319.2017.

Goda, K., Y. Hamane, R. Kishimoto, and Y. Ogishi. 1999. Radical scavenging properties of tryptophan metabolites. In: G. Huether, W. Kochen, T. J. Simat, and H. Steinhart, editors. Tryptophan, serotonin, and melatonin: Basic aspects and applications. Springer US, Boston, MA. p. 397–402.

Goerdt, S., and C. E. Orfanos. 1999. Other functions, other genes: Alternative activation of antigen-presenting cells. Immunity 10:137–142. doi: https://doi.org/10.1016/S1074-7613(00)80014-X.

Goodband, B., M. Tokach, S. Dritz, J. DeRouchey, and J. Woodworth. 2014. Practical starter pig amino acid requirements in relation to immunity, gut health and growth performance. J. Anim. Sci. Biotechnol. 5:12. doi: https://doi.org/10.1186/2049-1891-5-12.

Grimble, R. 2008. Sulfur amino acids, glutathione, and immune function. In: Glutathione and sulfur amino acids. In: M. Roberta, M. Guiseppe, and N. J. Hoboken, editors, Human Health and Disease. John Wiley & Sons, Ltd., Hoboken, NJ. p. 273–288.

Groschwitz, K. R., and S. P. Hogan. 2009. Intestinal barrier function: Molecular regulation and disease pathogenesis. J. Allergy Clin. Immun. 124:3–20. doi: https://doi.org/10.1016/j.jaci.2009.05.038.

Guevarra, R. B., S. H. Hong, J. H. Cho, B.-R. Kim, J. Shin, J. H. Lee, B. N. Kang, Y. H. Kim, S. Wattanaphansak, R. E. Isaacson, M. Song, and H. B. Kim. 2018. The dynamics of the piglet gut microbiome during the weaning transition in association with health and nutrition. J. Anim. Sci. Biotechnol. 9:54. doi: https://doi.org/10.1186/s40104-018-0269-6.

He, Q., H. Tang, P. Ren, X. Kong, G. Wu, Y. Yin, and Y. Wang. 2011. Dietary supplementation with l-arginine partially counteracts serum metabolome induced by weaning stress in piglets. J. Proteome Res. 10:5214–5221. doi: https://doi.org/10.1021/pr200688u.

Hedemann, M. S., M. Eskildsen, H. N. Lærke, C. Pedersen, J. E. Lindberg, P. Laurinen, and K. E. B. Knudsen. 2006. Intestinal morphology and enzymatic activity in newly weaned pigs fed contrasting fiber concentrations and fiber properties. J. Anim. Sci. 84:1375–1386. doi: https://doi.org/10.2527/2006.8461375x.

Hernandez-García, A. D., D. A. Columbus, R. Manjarín, H. V. Nguyen, A. Suryawan, R. A. Orellana, and T. A. Davis. 2016. Leucine supplementation stimulates protein synthesis and reduces degradation signal activation in muscle of newborn pigs during acute endotoxemia. Am. J. Physiol.-Endoc. M. 311:E791–E801. doi: https://doi.org/10.1152/ajpendo.00217.2016.

Hooper, L. V. 2009. Do symbiotic bacteria subvert host immunity? Nat. Rev. Microbiol. 7:367–374. doi: https://doi.org/10.1038/nrmicro2114.

Hsu, C. B., S. P. Cheng, J. C. Hsu, and H. T. Yen. 2001. Effect of threonine addition to a low protein diet on IgG levels in body fluid of first-litter sows and their piglets. Asian-Austral. J. Anim. Sci. 14:1157–1163. doi: 2001.14.8.1157.

Hsu, C. B., J. W. Lee, H. J. Huang, C. H. Wang, T. T. Lee, H. T. Yen, and B. Yu. 2012. Effects of supplemental glutamine on growth performance, plasma parameters and LPS-induced immune response of weaned barrows after castration. Asian-Austral. J. Anim. Sci. 25:674–681. doi: https://doi.org/10.5713/ajas.2011.11359.

Htoo, J. K., B. A. Araiza, W. C. Sauer, M. Rademacher, Y. Zhang, M. Cervantes, and R. T. Zijlstra. 2007. Effect of dietary protein content on ileal amino acid digestibility, growth performance, and formation of microbial metabolites in ileal and cecal digesta of early-weaned pigs. J. Anim. Sci. 85:3303–3312. doi: https://doi.org/10.2527/jas.2007-0105.

Hughes, R., M. J. Kurth, V. McGilligan, H. McGlynn, and I. Rowland. 2008. Effect of colonic bacterial metabolites on CaCo-2 cell paracellular permeability in vitro. Nutr. Cancer 60:259–266. doi: https://doi.org/10.1080/01635580701649644.

Humbert, B., P. Nguyen, L. Martin, H. Dumon, G. Vallette, P. Maugère, and D. Darmaun. 2007. Effect of glutamine on glutathione kinetics in vivo in dogs. J. Nutr. Biochem. 18:10–16. doi: https://doi.org/10.1016/j.jnutbio.2006.02.002.

Isaacson, R., and H. B. Kim. 2012. The intestinal microbiome of the pig. Anim. Health Res. Rev. 13:100–109. doi: https://doi.org/10.1017/S1466252312000084.

Jayaraman, B., J. K. Htoo, and C. M. Nyachoti. 2017. Effects of different dietary tryptophan:lysine ratios and sanitary conditions on growth performance, plasma urea nitrogen, serum haptoglobin and ileal histomorphology of weaned pigs. Anim. Sci. J. 88:763–771. doi: https://doi.org/10.1111/asj.12695.

Jayaraman, B., J. Htoo, and C. M. Nyachoti. 2015a. Effects of dietary threonine: Lysine ratioes and sanitary conditions on performance, plasma urea nitrogen, plasma-free threonine and lysine of weaned pigs. Anim. Nutr. 1:283–288. doi: https://doi.org/10.1016/j.aninu.2015.09.003.

Jayaraman B., J. Htoo, and C. M. Nyachoti. 2015b. Effects of increasing standardized ileal digestible tryptophan:Lysine ratio on performance and ileal expression of cytokine mRNA in weaned pigs challenged with E. coli K88. J. Anim. Sci. 98(Suppl. S3):201–202.

Jeaurond, E. A., M. Rademacher, J. R. Pluske, C. H. Zhu, and C. F. M. de Lange. 2008. Impact of feeding fermentable proteins and carbohydrates on growth performance, gut health and gastrointestinal function of newly weaned pigs. Can. J. Anim. Sci. 88:271–281. doi: https://doi.org/10.4141/CJAS07062.

Jha, R., and J. D. Berrocoso. 2015. Review: Dietary fiber utilization and its effects on physiological functions and gut health of swine. Animal 9:1441–1452. doi: https://doi.org/10.1017/S1751731115000919.

Jha, R., and J. F. D. Berrocoso. 2016. Dietary fiber and protein fermentation in the intestine of swine and their interactive effects on gut health and on the environment: A review. Anim. Feed Sci. Technol. 212:18–26. doi: https://doi.org/10.1016/j.anifeedsci.2015.12.002.

Jha, R., J. Bindelle, A. Van Kessel, and P. Leterme. 2011. In vitro fibre fermentation of feed ingredients with varying fermentable carbohydrate and protein levels and protein synthesis by colonic bacteria isolated from pigs. Anim. Feed Sci. Technol. 165:191–200. doi: https://doi.org/10.1016/j.anifeedsci.2010.10.002.

Jha, R., J. M. Fouhse, U. P. Tiwari, L. Li, and B. P. Willing. 2019. Dietary fiber and intestinal health of monogastric animals. Front. Vet. Sci. 6. doi: https://doi.org/10.3389/fvets.2019.00048.

Jiao, N., Z. Wu, Y. Ji, B. Wang, Z. Dai, and G. Wu. 2015. L-glutamate enhances barrier and antioxidative functions in intestinal porcine epithelial cells. J. Nutr. 145:2258–2264. doi: https://doi.org/10.3945/jn.115.217661.

Johansson, M. E. V., D. Ambort, T. Pelaseyed, A. Schütte, J. K. Gustafsson, A. Ermund, D. B. Subramani, J. M. Holmén-Larsson, K. A. Thomsson, J. H. Bergström, S. van der Post, A. M. Rodriguez-Piñeiro, H. Sjövall, M. Bäckström, and G. C. Hansson. 2011. Composition and functional role of the mucus layers in the intestine. Cell. Mol. Life Sci. 68:3635. doi: https://doi.org/10.1007/s00018-011-0822-3.

Kahindi, R., A. Regassa, J. Htoo, and M. Nyachoti. 2017. Optimal sulfur amino acid to lysine ratio for post weaning piglets reared under clean or unclean sanitary conditions. Anim. Nutr. 3:380–385. doi: https://doi.org/10.1016/j.aninu.2017.08.004.

Kanauchi, O., T. Iwanaga, A. Andoh, Y. Araki, T. Nakamura, K. Mitsuyama, A. Suzuki, T. Hibi, and T. Bamba. 2001. Dietary fiber fraction of germinated barley foodstuff attenuated mucosal damage and diarrhea, and accelerated the repair of the colonic mucosa in an experimental colitis. J. Gastroen. Hepatol. 16:160–168. doi: https://doi.org/10.1046/j.1440-1746.2001.02427.x.

Kim, C. J., J. A. Kovacs-Nolan, C. Yang, T. Archbold, M. Z. Fan, and Y. Mine. 2010. l-Tryptophan exhibits therapeutic function in a porcine model of dextran sodium sulfate (DSS)-induced colitis. J. Nutr. Biochem. 21:468–475. doi: https://doi.org/10.1016/j.jnutbio.2009.01.019.

Kim, J. C., C. F. Hansen, B. P. Mullan, and J. R. Pluske. 2012a. Nutrition and pathology of weaner pigs: Nutritional strategies to support barrier function in the gastrointestinal tract. Anim. Feed Sci. Technol. 173:3–16. doi: https://doi.org/10.1016/j.anifeedsci.2011.12.022.

Kim, J. C., B. P. Mullan, B. Frey, H. G. Payne, and J. R. Pluske. 2012b. Whole body protein deposition and plasma amino acid profiles in growing and/or finishing pigs fed increasing levels of sulfur amino acids with and without *Escherichia coli* lipopolysaccharide challenge. J. Anim. Sci. 90:362–365. doi: https://doi.org/10.2527/jas.53821.

Klasing, K. C. 1988. Nutritional aspects of leukocytic cytokines. J. Nutr. 118:1436–1446. doi: https://doi.org/10.1093/jn/118.12.1436.

Klasing, K. C. 2007. Nutrition and the immune system. Br. Poult. Sci. 48:525–537. doi: https://doi.org/10.1080/00071660701671336.

Konstantinov, S. R., A. Awati, H. Smidt, B. A. Williams, A. D. L. Akkermans, and W. M. de Vos. 2004. Specific response of a novel and abundant *Lactobacillus* amylovorus-like phylotype to dietary prebiotics in the guts of weaning piglets. Appl. Environ. Microbiol. 70:3821–3830. doi: https://doi.org/10.1128/AEM.70.7.3821-3830.2004.

Koo, B., J. Choi, C. Yang, and C. M. Nyachoti. 2020a. Diet complexity and l-threonine supplementation: Effects on growth performance, immune response, intestinal barrier function, and microbial metabolites in nursery pigs. J. Anim. Sci. 98. doi: https://doi.org/10.1093/jas/skaa125.

Koo, B., J. Lee, and C. M. Nyachoti. 2020b. Diet complexity and l-threonine supplementation: effects on nutrient digestibility, nitrogen and energy balance, and body composition in nursery pigs. J. Anim. Sci. 98. doi: https://doi.org/10.1093/jas/skaa124.

Laanen, M., D. Persoons, S. Ribbens, E. de Jong, B. Callens, M. Strubbe, D. Maes, and J. Dewulf. 2013. Relationship between biosecurity and production/antimicrobial treatment characteristics in pig herds. Vet. J. 198:508–512. doi: https://doi.org/10.1016/j.tvjl.2013.08.029.

Lallès, J.-P., P. Bosi, H. Smidt, and C. R. Stokes. 2007. Nutritional management of gut health in pigs around weaning. P. Nutr. Soc. 66:260–268. doi: https://doi.org/10.1017/S0029665107005484.

Law, G. K., R. F. Bertolo, A. Adjiri-Awere, P. B. Pencharz, and R. O. Ball. 2007. Adequate oral threonine is critical for mucin production and gut function in neonatal piglets. Am. J. Physiol.-Gastr. L. 292:G1293–1301. doi: https://doi.org/10.1152/ajpgi.00221.2006.

Le Floc'h, N., L. LeBellego, J. J. Matte, D. Melchior, and B. Sève. 2009. The effect of sanitary status degradation and dietary tryptophan content on growth rate and tryptophan metabolism in weaning pigs. J. Anim. Sci. 87:1686–1694. doi: https://doi.org/10.2527/jas.2008-1348.

Le Floc'h, N. L., F. Gondret, J. J. Matte, and H. Quesnel. 2012. Towards amino acid recommendations for specific physiological and patho-physiological states in pigs. P. Nutr. Soc. 71:425–432. doi: https://doi.org/10.1017/S0029665112000560.

Le Floc'h, N., D. Melchior, and C. Obled. 2004. Modifications of protein and amino acid metabolism during inflammation and immune system activation. Livest. Prod. Sci. 87:37–45. doi: https://doi.org/10.1016/j.livprodsci.2003.09.005.

Le Floc'h, N., D. Melchior, and B. Sève. 2008. Dietary tryptophan helps to preserve tryptophan homeostasis in pigs suffering from lung inflammation. J. Anim. Sci. 86:3473–3479. doi: https://doi.org/10.2527/jas.2008-0999.

Le Floc'h, N., and B. Seve. 2007. Biological roles of tryptophan and its metabolism: Potential implications for pig feeding. Livest. Sci. 112:23–32. doi: https://doi.org/10.1016/j.livsci.2007.07.002.

Le Floc'h, N., A. Wessels, E. Corrent, G. Wu, and P. Bosi. 2018. The relevance of functional amino acids to support the health of growing pigs. Anim. Feed Sci. Technol. 245:104–116. doi: https://doi.org/10.1016/j.anifeedsci.2018.09.007.

Le Naou, T., N. Le Floc'h, I. Louveau, H. Gilbert, and F. Gondret. 2012. Metabolic changes and tissue responses to selection on residual feed intake in growing pigs. J. Anim. Sci. 90:4771–4780. doi: https://doi.org/10.2527/jas.2012-5226.

Leng, W., Y. Liu, H. Shi, S. Li, H. Zhu, D. Pi, Y. Hou, and J. Gong. 2014. Aspartate alleviates liver injury and regulates mRNA expressions of TLR4 and NOD signaling-related genes in weaned pigs after lipopolysaccharide challenge. J. Nutr. Biochem. 25:592–599. doi:https://doi.org/10.1016/j.jnutbio.2014.01.010.

Lewis, A. J. 2009. Methionine-cystine relationships in pig nutrition. In: A. J. P. F. D'Mello, editor, Amino acids in animal nutrition. CABI Publishing, Wallingford, UK. p. 143–155. doi: https://doi.org/10.1079/9780851996547.0000.

Li, P., Y.-L. Yin, D. Li, S. W. Kim, and G. Wu. 2007. Amino acids and immune function. Brit. J. Nutr. 98:237–252. doi: https://doi.org/10.1017/S000711450769936X.

Li, Q., Y. Liu, Z. Che, H. Zhu, G. Meng, Y. Hou, B. Ding, Y. Yin, and F. Chen. 2012. Dietary L-arginine supplementation alleviates liver injury caused by *Escherichia coli* LPS in weaned pigs. Innate Immun. 18:804–814. doi: https://doi.org/10.1177/1753425912441955.

Lien, K. A., W. C. Sauer, R. Mosenthin, W. B. Souffrant, and M. E. R. Dugan. 1997. Evaluation of the 15N-isotope dilution technique for determining the recovery of endogenous protein in ileal digestion of pigs: Effect of dilution in the precursor pool for endogenous nitrogen secretion. J. Anim. Sci. 75:148–158. doi: https://doi.org/10.2527/1997.751148x.

Litvak, N., J. K. Htoo, and C. F. M. de Lange. 2013a. Restricting sulfur amino acid intake in growing pigs challenged with lipopolysaccharides decreases plasma protein and albumin synthesis. Can. J. Anim. Sci. 93:505–515. doi: https://doi.org/10.4141/cjas2013-014.

Litvak, N., A. Rakhshandeh, J. K. Htoo, and C. F. M. de Lange. 2013b. Immune system stimulation increases the optimal dietary methionine to methionine plus cysteine ratio in growing pigs. J. Anim. Sci. 91:4188–4196. doi: https://doi.org/10.2527/jas.2012-6160.

Liu, P., X. S. Piao, S. W. Kim, L. Wang, Y. B. Shen, H. S. Lee, and S. Y. Li. 2008a. Effects of chito-oligosaccharide supplementation on the growth performance, nutrient digestibility, intestinal morphology, and fecal shedding of *Escherichia coli* and *Lactobacillus* in weaning pigs. J. Anim. Sci. 86:2609–2618. doi: https://doi.org/10.2527/jas.2007-0668.

Liu, Y., J. Huang, Y. Hou, H. Zhu, S. Zhao, B. Ding, Y. Yin, G. Yi, J. Shi, and W. Fan. 2008b. Dietary arginine supplementation alleviates intestinal mucosal disruption induced by *Escherichia coli* lipopolysaccharide in weaned pigs. Br. J. Nutr. 100:552–560. doi: https://doi.org/10.1017/S0007114508911612.

Lu, S. C. 2000. S-Adenosylmethionine. Int. J. Biochem. Cell B. 32:391–395. doi: https://doi.org/10.1016/S1357-2725(99)00139-9.

Lv, D., X. Xiong, H. Yang, M. Wang, Y. He, Y. Liu, and Y. Yin. 2018. Effect of dietary soy oil, glucose, and glutamine on growth performance, amino acid profile, blood profile, immunity, and antioxidant capacity in weaned piglets. Sci. China Life Sci. 61:1233–1242. doi: https://doi.org/10.1007/s11427-018-9301-y.

Lynch, C. J., and S. H. Adams. 2014. Branched-chain amino acids in metabolic signalling and insulin resistance. Nat. Rev. Endocrinol. 10:723–736. doi: https://doi.org/10.1038/nrendo.2014.171.

Mach, N., M. Berri, J. Estellé, F. Levenez, G. Lemonnier, C. Denis, J.-J. Leplat, C. Chevaleyre, Y. Billon, J. Doré, C. Rogel-Gaillard, and P. Lepage. 2015. Early-life establishment of the swine gut microbiome and impact on host phenotypes. Env. Microbiol. Rep. 7:554–569. doi: https://doi.org/10.1111/1758-2229.12285.

Malmezat, T., D. Breuillé, C. Pouyet, C. Buffière, P. Denis, P. P. Mirand, and C. Obled. 2000. Methionine transsulfuration is increased during sepsis in rats. Am. J. Physiol.-Endoc. M. 279:E1391–E1397. doi: https://doi.org/10.1152/ajpendo.2000.279.6.E1391.

Mann, E., S. Schmitz-Esser, Q. Zebeli, M. Wagner, M. Ritzmann, and B. U. Metzler-Zebeli. 2014. Mucosa-associated bacterial microbiome of the gastrointestinal tract of weaned pigs and dynamics linked to dietary calcium-phosphorus. PLOS One 9. doi: https://doi.org/10.1371/journal.pone.0086950.

Mantis, N. J., N. Rol, and B. Corthésy. 2011. Secretory IgA's complex roles in immunity and mucosal homeostasis in the gut. Mucosal Immunol. 4:603–611. doi: https://doi.org/10.1038/mi.2011.41.

Mao, X., X. Lai, B. Yu, J. He, J. Yu, P. Zheng, G. Tian, K. Zhang, and D. Chen. 2014a. Effects of dietary threonine supplementation on immune challenge induced by swine *Pseudorabies* live vaccine in weaned pigs. Arch. Anim. Nutr. 68:1–15. doi: https://doi.org/10.1080/1745039X.2013.869988.

Mao, X., M. Liu, J. Tang, H. Chen, D. Chen, B. Yu, J. He, J. Yu, and P. Zheng. 2015. Dietary leucine supplementation improves the mucin production in the jejunal mucosa of the weaned pigs challenged by porcine *Rotavirus*. PLOS One 10:e0137380. doi: https://doi.org/10.1371/journal.pone.0137380.

Mao, X., M. Lv, B. Yu, J. He, P. Zheng, J. Yu, Q. Wang, and D. Chen. 2014b. The effect of dietary tryptophan levels on oxidative stress of liver induced by diquat in weaned piglets. J. Anim. Sci. Biotechnol. 5:49. doi: https://doi.org/10.1186/2049-1891-5-49.

Mariscal-Landín, G., B. Sève, Y. Colléaux, and Y. Lebreton. 1995. Endogenous amino nitrogen collected from pigs with end-to-end ileorectal anastomosis is affected by the method of estimation and altered by dietary fiber. J. Nutr. 125:136–146. doi: https://doi.org/10.1093/jn/125.1.136.

Marquardt, R. R., and S. Li. 2018. Antimicrobial resistance in livestock: Advances and alternatives to antibiotics. Anim. Front. 8:30–37. doi: https://doi.org/10.1093/af/vfy001.

Mathai, J. K., J. K. Htoo, J. E. Thomson, K. J. Touchette, and H. H. Stein. 2016. Effects of dietary fiber on the ideal standardized ileal digestible threonine:lysine ratio for twenty-five to fifty kilogram growing gilts. J. Anim. Sci. 94:4217–4230. doi: https://doi.org/10.2527/jas.2016-0680.

May, T., R. I. Mackie, G. C. Fahey, J. C. Cremin, and K. A. Garleb. 1994. Effect of fiber source on short-chain fatty acid production and on the growth and toxin production by *Clostridium difficile*. Scand. J. Gastroenterol. 29:916–922. doi: https://doi.org/10.3109/00365529409094863.

McGilvray, W. D., H. Wooten, A. R. Rakhshandeh, A. Petry, and A. Rakhshandeh. 2019. Immune system stimulation increases dietary threonine requirements for protein deposition in growing pigs. J. Anim. Sci. 97:735–744. doi: https://doi.org/10.1093/jas/sky468.

Medzhitov, R. 2007. Recognition of microorganisms and activation of the immune response. Nature 449:819–826. doi: https://doi.org/10.1038/nature06246.

Métayer, S., I. Seiliez, A. Collin, S. Duchêne, Y. Mercier, P.-A. Geraert, and S. Tesseraud. 2008. Mechanisms through which sulfur amino acids control protein metabolism and oxidative status. J. Nutr. Biochem. 19:207–215. doi: https://doi.org/10.1016/j.jnutbio.2007.05.006.

Mikkelsen, L. L., M. Jakobsen, and B. B. Jensen. 2003. Effects of dietary oligosaccharides on microbial diversity and fructo-oligosaccharide degrading bacteria in faeces of piglets post-weaning. Anim. Feed Sci. Technol. 109:133–150. doi: https://doi.org/10.1016/S0377-8401(03)00172-X.

Munasinghe, L. L., J. L. Robinson, S. V. Harding, J. A. Brunton, and R. F. Bertolo. 2017. Protein synthesis in mucin-producing tissues is conserved when dietary threonine is limiting in piglets. J. Nutr. 147:202–210. doi: https://doi.org/10.3945/jn.116.236786.

Nyachoti, C. M., F. O. Omogbenigun, M. Rademacher, and G. Blank. 2006. Performance responses and indicators of gastrointestinal health in early-weaned pigs fed low-protein amino acid-supplemented diets. J. Anim. Sci. 84:125–134. doi: https://doi.org/10.2527/2006.841125x.

Ohira, H., W. Tsutsui, and Y. Fujioka. 2017. Are short chain fatty acids in gut microbiota defensive players for inflammation and atherosclerosis? J. Atheroscler. Thromb. 24:660–672. doi: https://doi.org/10.5551/jat.RV17006.

Oien, D. B., and J. Moskovitz. 2007. Substrates of the methionine sulfoxide reductase system and their physiological relevance. Curr. Top. Dev. Biol. 80:93–133. doi: https://doi.org/10.1016/S0070-2153(07)80003-2.

O'Leary, J. G., M. Goodarzi, D. L. Drayton, and U. H. von Andrian. 2006. T cell– and B cell–independent adaptive immunity mediated by natural killer cells. Nat. Immunol. 7:507–516. doi: https://doi.org/10.1038/ni1332.

Opapeju, F. O., M. Rademacher, G. Blank, and C. M. Nyachoti. 2008. Effect of low-protein amino acid-supplemented diets on the growth performance, gut morphology, organ weights and digesta characteristics of weaned pigs. Animal 2:1457–1464. doi: https://doi.org/10.1017/S175173110800270X.

Orellana, R. A., P. M. O'Connor, H. V. Nguyen, J. A. Bush, A. Suryawan, M. C. Thivierge, M. L. Fiorotto, and T. A. Davis. 2002. Endotoxemia reduces skeletal muscle protein synthesis in neonates. Am. J. Physiol.-Endoc. M. 283:E909–E916. doi: https://doi.org/10.1017/S0007114512004321.

Orellana, R. A., F. A. Wilson, M. C. Gazzaneo, A. Suryawan, T. A. Davis, and H. V. Nguyen. 2011. Sepsis and development impede muscle protein synthesis in neonatal pigs by different ribosomal mechanisms. Pediatr. Res. 69:473–478. doi: https://doi.org/10.1203/PDR.0b013e3182176da1.

Pajarillo, E. A. B., J. P. Chae, M. P. Balolong, H. B. Kim, K.-S. Seo, and D.-K. Kang. 2014. Pyrosequencing-based analysis of fecal microbial communities in three purebred pig lines. J. Microbiol. 52:646–651. doi: https://doi.org/10.1007/s12275-014-4270-2.

Pastorelli, H., J. van Milgen, P. Lovatto, and L. Montagne. 2012. Meta-analysis of feed intake and growth responses of growing pigs after a sanitary challenge. Animal 6:952–961. doi: https://doi.org/10.1017/S175173111100228X.

Patience, J. F., M. C. Rossoni-Serão, and N. A. Gutiérrez. 2015. A review of feed efficiency in swine: Biology and application. J. Anim. Sci. Biotechnol. 6:33. doi: https://doi.org/10.1186/s40104-015-0031-2.

Perez-Muñoz, M. E., M.-C. Arrieta, A. E. Ramer-Tait, and J. Walter. 2017. A critical assessment of the "sterile womb" and "in utero colonization" hypotheses: Implications for research on the pioneer infant microbiome. Microbiome. 5:48. doi: https://doi.org/10.1186/s40168-017-0268-4.

Pi, D., Y. Liu, H. Shi, S. Li, J. Odle, X. Lin, H. Zhu, F. Chen, Y. Hou, and W. Leng. 2014. Dietary supplementation of aspartate enhances intestinal integrity and energy status in weanling piglets after lipopolysaccharide challenge. J. Nutr. Biochem. 25:456–462. doi: https://doi.org/10.1016/j.jnutbio.2013.12.006.

Pieper, R., J. Bindelle, G. Malik, J. Marshall, B. G. Rossnagel, P. Leterme, and A. G. V. Kessel. 2012a. Influence of different carbohydrate composition in barley varieties on *Salmonella Typhimurium* var. Copenhagen colonisation in a "Trojan" challenge model in pigs. Arch. Anim. Nutr. 66:163–179. doi: https://doi.org/10.1080/1745039X.2012.676814.

Pieper, R., S. Kröger, J. F. Richter, J. Wang, L. Martin, J. Bindelle, J. K. Htoo, D. von Smolinski, W. Vahjen, J. Zentek, and A. G. Van Kessel. 2012b. Fermentable fiber ameliorates fermentable protein-induced changes in microbial ecology, but not the mucosal response, in the colon of piglets. J. Nutr. 142:661–667. doi: https://doi.org/10.3945/jn.111.156190.

Pluske, J. R., B. Black, D. W. Pethick, B. P. Mullan, and D. J. Hampson. 2003. Effects of different sources and levels of dietary fibre in diets on performance, digesta characteristics and antibiotic treatment of pigs after weaning. Anim. Feed Sci. Technol. 107:129–142. doi: https://doi.org/10.1016/S0377-8401(03)00072-5.

Pluske, J. R., J. C. Kim, and J. L. Black. 2018. Manipulating the immune system for pigs to optimise performance. Anim. Prod. Sci. 58:666–680. doi: https://doi.org/10.1071/AN17598.

Popov, A., and J. L. Schultze. 2008. IDO-expressing regulatory dendritic cells in cancer and chronic infection. J. Mol. Med. 86:145–160. doi: https://doi.org/10.1007/s00109-007-0262-6.

Prohászka, L., and F. Baron. 1980. The predisposing role of high dietary protein supplies in enteropathogenic E. coli infections of weaned pigs. Zbl. Vet. Med. B. 27:222–232. doi: https://doi.org/10.1111/j.1439-0450.1980.tb01908.x.

Quiniou, N., J.-Y. Dourmad, and J. Noblet. 1996. Effect of energy intake on the performance of different types of pig from 45 to 100 kg body weight. 1. Protein and lipid deposition. Anim. Sci. 63:277–288. doi: https://doi.org/10.1017/S1357729800014831.

Rakhshandeh, A., and C. F. M. de Lange. 2011. Immune system stimulation in the pig: Effect on performance and implications for amino acid nutrition. In: R. J. van Barneveld, editor, Manipulating pig production XIII. Aust. Pig Sci. Assoc., Werribee, Australia. p. 31–46.

Rakhshandeh, A., and C. F. de Lange. 2012. Evaluation of chronic immune system stimulation models in growing pigs. Animal. 6:305–310.

Rakhshandeh, A., C.F.M. de Lange, J.K. Htoo, and A.R. Rakhshandeh. 2020. Immune system stimulation increases the irreversible loss of cystein to taurine, but not sulfate, in starter pigs. J. Anim. Sci. 98:skaa001. doi:10.1093/jas/skaa001.

Rakhshandeh, A., J. K. Htoo, N. Karrow, S. P. Miller, and C. F. M. de Lange. 2014. Impact of immune system stimulation on the ileal nutrient digestibility and utilisation of methionine plus cysteine intake for whole-body protein deposition in growing pigs. Br. J. Nutr. 111:101–110. doi: https://doi.org/10.1017/S0007114513001955.

Rakhshandeh, A., J. K. Htoo, and C. F. M. de Lange. 2010. Immune system stimulation of growing pigs does not alter apparent ileal amino acid digestibility but reduces the ratio between whole body nitrogen and sulfur retention. Livest. Sci. 134:21–23. doi: https://doi.org/10.1016/j.livsci.2010.06.085.

Redmond, H. P., P. P. Stapleton, P. Neary, and D. Bouchier-Hayes. 1998. Immunonutrition: The role of taurine. Nutrition 14:599–604. doi: https://doi.org/10.1016/S0899-9007(98)00097-5.

Reeds, P. J., D. G. Burrin, B. Stoll, F. Jahoor, L. Wykes, J. Henry, and M. E. Frazer. 1997. Enteral glutamate is the preferential source for mucosal glutathione synthesis in fed piglets. Am. J. Physiol.-Endoc. M. 273:E408–E415. doi: https://doi.org/10.1152/ajpendo.1997.273.2.E408.

Reeds, P. J., C. R. Fjeld, and F. Jahoor. 1994. Do the differences between the amino acid compositions of acute-phase and muscle proteins have a bearing on nitrogen loss in traumatic states? J. Nutr. 124:906–910. doi: https://doi.org/10.1093/jn/124.6.906.

Reeds, P. J., and F. Jahoor. 2001. The amino acid requirements of disease. Clin. Nutr. 20:15–22. doi: https://doi.org/10.1054/clnu.2001.0402.

Rémond, D., C. Buffière, C. Pouyet, I. Papet, D. Dardevet, I. Savary-Auzeloux, G. Williamson, M. Faure, and D. Breuillé. 2011. Cysteine fluxes across the portal-drained viscera of enterally fed minipigs: Effect of an acute intestinal inflammation. Amino Acids 40:543–552. doi: https://doi.org/10.1007/s00726-010-0672-6.

Ren, M., X. T. Liu, X. Wang, G. J. Zhang, S. Y. Qiao, and X. F. Zeng. 2014. Increased levels of standardized ileal digestible threonine attenuate intestinal damage and immune responses in *Escherichia coli* K88+ challenged weaned piglets. Anim. Feed Sci. Technol. 195:67–75. doi: https://doi.org/10.1016/j.anifeedsci.2014.05.013.

Ren, M., S. H. Zhang, X. F. Zeng, H. Liu, and S. Y. Qiao. 2015. Branched-chain amino acids are beneficial to maintain growth performance and intestinal immune-related function in weaned piglets fed protein restricted diet. Asian-Australas. J. Anim. Sci. 28:1742–1750. doi: https://doi.org/10.5713/ajas.14.0131.

Ren, M., S. Zhang, X. Liu, S. Li, X. Mao, X. Zeng, and S. Qiao. 2016. Different lipopolysaccharide branched-chain amino acids modulate porcine intestinal endogenous β-defensin expression through the sirt1/erk/90rsk pathway. J. Agric. Food Chem. 64:3371–3379. doi: https://doi.org/10.1021/acs.jafc.6b00968.

Rezaei, R., W. Wang, Z. Wu, Z. Dai, J. Wang, and G. Wu. 2013. Biochemical and physiological bases for utilization of dietary amino acids by young Pigs. J. Anim. Sci. Biotechnol. 4:7. doi: https://doi.org/10.1186/2049-1891-4-7.

Robinson, K., Z. Deng, Y. Hou, and G. Zhang. 2015. Regulation of the intestinal barrier function by host defense peptides. Front. Vet. Sci. 2. doi: https://doi.org/10.3389/fvets.2015.00057.

Rodrigues, L. A., M. O. Wellington, J. C. González-Vega, J. K. Htoo, A. G. Van Kessel, and D. A. Columbus. 2021. Functional amino acid supplementation, regardless of dietary protein content, improves growth performance and immune status of weaned pigs challenged with Salmonella Typhimurium. J. Anim. Sci. 99:skaa365. doi: https://doi.org/10.1093/jas/skaa365.

Rothkötter, H. J., E. Sowa, and R. Pabst. 2002. The pig as a model of developmental immunology. Hum. Exp. Toxicol. 21:533–536. doi: https://doi.org/10.1191/0960327102ht293oa.

Roubos-van den Hil, P. J., R. Litjens, A.-K. Oudshoorn, J. W. Resink, and C. H. M. Smits. 2017. New perspectives to the enterotoxigenic E. coli F4 porcine infection model: Susceptibility genotypes in relation to performance, diarrhoea and bacterial shedding. Vet. Microbiol. 202:58–63. doi: https://doi.org/10.1016/j.vetmic.2016.09.008.

Ruan, T., L. Li, X. Peng, and B. Wu. 2017. Effects of methionine on the immune function in animals. Health. 09:857. doi: https://doi.org/10.4236/health.2017.95061.

Rudar, M., L.-A. Huber, C. L. Zhu, and C. F. M. de Lange. 2019. Effects of dietary leucine supplementation and immune system stimulation on plasma AA concentrations and tissue protein synthesis in starter pigs. J. Anim. Sci. 97:829–838. doi: https://doi.org/10.1093/jas/sky449.

Rudar, M., C. L. Zhu, and C. F. de Lange. 2017. Dietary leucine supplementation decreases whole-body protein turnover before, but not during, immune system stimulation in pigs. J. Nutr. 147:45–51. doi: https://doi.org/10.3945/jn.116.236893.

Salmon, H. 1987. The intestinal and mammary immune system in pigs. Vet. Immunol. Immunop. 17:367–388. doi: https://doi.org/10.1016/0165-2427(87)90155-3.

Schmidt-Wittig, U., M.-L. Enss, M. Coenen, K. Gärtner, and H. J. Hedrich. 1996. Response of rat colonic mucosa to a high fiber diet. Ann. Nutr. Metab. 40:343–350. doi: https://doi.org/10.1159/000177936.

Schokker, D., J. Zhang, S. A. Vastenhouw, H. G. H. J. Heilig, H. Smidt, J. M. J. Rebel, and M. A. Smits. 2015. Long-lasting effects of early-life antibiotic treatment and routine animal handling on gut microbiota composition and immune system in pigs. PLOS One 10:e0116523. doi: https://doi.org/10.1371/journal.pone.0116523.

Shen, Y. B., A. C. Weaver, and S. W. Kim. 2014. Effect of feed grade L-methionine on growth performance and gut health in nursery pigs compared with conventional DL-methionine. J. Anim. Sci. 92:5530–5539. doi: https://doi.org/10.2527/jas.2014-7830.

Shoveller, A. K., J. D. House, J. A. Brunton, P. B. Pencharz, and R. O. Ball. 2004. The balance of dietary sulfur amino acids and the route of feeding affect plasma homocysteine concentrations in neonatal piglets. J. Nutr. 134:609–612. doi: https://doi.org/10.1093/jn/134.3.609.

Sompayrac, L. M. 2019. How the immune system works. John Wiley & Sons, Hoboken, NJ.

Spring, S., H. Premathilake, C. Bradway, C. Shili, U. DeSilva, S. Carter, and A. Pezeshki. 2020. Effect of very low-protein diets supplemented with branched-chain amino acids on energy balance, plasma metabolomics and fecal microbiome of pigs. Sci. Rep. 10:15859. doi: https://doi.org/10.1038/s41598-020-72816-8.

Suchner, U., D. K. Heyland, and K. Peter. 2002. Immune-modulatory actions of arginine in the critically ill. Br. J. Nutr. 87:S121–S132. doi: https://doi.org/10.1079/BJN2001465.

Sun, Y., Z. Wu, W. Li, C. Zhang, K. Sun, Y. Ji, B. Wang, N. Jiao, B. He, W. Wang, Z. Dhai, and G. Wu. 2015. Dietary l-leucine supplementation enhances intestinal development in suckling piglets. Amino Acids 47:1517–1525. doi: https://doi.org/10.1007/s00726-015-1985-2.

Surendran Nair, M., T. Eucker, B. Martinson, A. Neubauer, J. Victoria, B. Nicholson, and M. Pieters. 2019. Influence of pig gut microbiota on *Mycoplasma hyopneumoniae* susceptibility. Vet. Res. 50:86. doi: https://doi.org/10.1186/s13567-019-0701-8.

Taciak, M., M. Barszcz, E. Święch, A. Tuśnio, and I. Bachanek. 2017. Interactive effects of protein and carbohydrates on production of microbial metabolites in the large intestine of growing pigs. Arch. Anim. Nutr. 71:192–209. doi: https://doi.org/10.1080/1745039X.2017.1291202.

Tan, B., X. G. Li, X. Kong, R. Huang, Z. Ruan, K. Yao, Z. Deng, M. Xie, I. Shinzato, Y. Yin, and G. Wu. 2009. Dietary l-arginine supplementation enhances the immune status in early-weaned piglets. Amino Acids 37:323–331. doi: https://doi.org/10.1007/s00726-008-0155-1.

Tan, B., Y. Yin, X. Kong, P. Li, X. Li, H. Gao, X. Li, R. Huang, and G. Wu. 2010. L-Arginine stimulates proliferation and prevents endotoxin-induced death of intestinal cells. Amino Acids 38:1227–1235. doi: https://doi.org/10.1007/s00726-009-0334-8.

te Pas, M. F. W., A. J. M. Jansman, L. Kruijt, Y. van der Meer, J. J. M. Vervoort, and D. Schokker. 2020. Sanitary conditions affect the colonic microbiome and the colonic and systemic metabolome of female pigs. Front. Vet. Sci. 7. doi: https://doi.org/10.3389/fvets.2020.585730.

Tenenhouse, H. S., and H. F. Deutsch. 1966. Some physical-chemical properties of chicken γ-globulins and their pepsin and papain digestion products. Immunochemistry 3:11–20. doi: https://doi.org/10.1016/0019-2791(66)90277-1.

Trevisi, P., E. Corrent, M. Mazzoni, S. Messori, D. Priori, Y. Gherpelli, A. Simongiovanni, and P. Bosi. 2015. Effect of added dietary threonine on growth performance, health, immunity and gastrointestinal function of weaning pigs with differing genetic susceptibility to *Escherichia coli* infection and challenged with E. coli K88ac. J. Anim. Physiol. Anim. Nutr. 99:511–520. doi: https://doi.org/10.1111/jpn.12216.

Trevisi, P., D. Melchior, M. Mazzoni, L. Casini, S. De Filippi, L. Minieri, G. Lalatta-Costerbosa, and P. Bosi. 2009. A tryptophan-enriched diet improves feed intake and growth performance of susceptible weanling pigs orally challenged with *Escherichia coli* K88. J. Anim. Sci. 87:148–156. doi: https://doi.org/10.2527/jas.2007-0732.

Turner, J. R. 2009. Intestinal mucosal barrier function in health and disease. Nat. Rev. Immunol. 9:799–809. doi: https://doi.org/10.1038/nri2653.

van der Meer, Y., W. J. J. Gerrits, A. J. M. Jansman, B. Kemp, and J. E. Bolhuis. 2017. A link between damaging behaviour in pigs, sanitary conditions, and dietary protein and amino acid supply. PLOS One 12:e0174688. doi: https://doi.org/10.1371/journal.pone.0174688.

van der Meer, Y., A. Lammers, A. J. M. Jansman, M. M. J. A. Rijnen, W. H. Hendriks, and W. J. J. Gerrits. 2016. Performance of pigs kept under different sanitary conditions affected by protein intake and amino acid supplementation. J. Anim. Sci. 94:4704–4719. doi: https://doi.org/10.2527/jas.2016-0787.

van Heugten, E., M. T. Coffey, and J. W. Spears. 1996. Effects of immune challenge, dietary energy density, and source of energy on performance and immunity in weanling pigs. J. Anim. Sci. 74:2431–2440. doi: https://doi.org/10.2527/1996.74102431x.

Wang, H., Y. Liu, H. Shi, X. Wang, H. Zhu, D. Pi, W. Leng, and S. Li. 2017. Aspartate attenuates intestinal injury and inhibits TLR4 and NODs/NF-κB and p38 signaling in weaned pigs after LPS challenge. Eur. J. Nutr. 56:1433–1443. doi: https://doi.org/10.1007/s00394-016-1189-x.

Wang, H., C. Zhang, G. Wu, Y. Sun, B. Wang, B. He, Z. Dai, and Z. Wu. 2015b. Glutamine enhances tight junction protein expression and modulates corticotropin-releasing factor signaling in the jejunum of weanling piglets. J. Nutr. 145:25–31. doi: https://doi.org/10.3945/jn.114.202515.

Wang, L., Y. Hou, D. Yi, Y. Li, B. Ding, H. Zhu, J. Liu, H. Xiao, and G. Wu. 2015c. Dietary supplementation with glutamate precursor α-ketoglutarate attenuates lipopolysaccharide-induced liver injury in young pigs. Amino Acids 47:1309–1318. doi: https://doi.org/10.1007/s00726-015-1966-5.

Wang, W. W., S. Y. Qiao, and D. F. Li. 2009. Amino acids and gut function. Amino Acids 37:105–110. doi: https://doi.org/10.1007/s00726-008-0152-4.

Wang, W., X. Zeng, X. Mao, G. Wu, and S. Qiao. 2010. Optimal dietary true ileal digestible threonine for supporting the mucosal barrier in small intestine of weanling pigs. J. Nutr. 140:981–986. doi: https://doi.org/10.3945/jn.109.118497.

Wang, X., Y. Liu, S. Li, D. Pi, H. Zhu, Y. Hou, H. Shi, and W. Leng. 2015a. Asparagine attenuates intestinal injury, improves energy status and inhibits AMP-activated protein kinase signalling pathways in weaned piglets challenged with *Escherichia coli* lipopolysaccharide. Br. J. Nutr. 114:553–565. doi: https://doi.org/10.1017/S0007114515001877.

Wang, X., S. Y. Qiao, M. Liu, and Y. X. Ma. 2006. Effects of graded levels of true ileal digestible threonine on performance, serum parameters and immune function of 10–25 kg pigs. Anim. Feed Sci. Technol. 129:264–278. doi: https://doi.org/10.1016/j.anifeedsci.2006.01.003.

Wang, X., S. Qiao, Y. Yin, L. Yue, Z. Wang, and G. Wu. 2007. A Deficiency or excess of dietary threonine reduces protein synthesis in jejunum and skeletal muscle of young pigs. J. Nutr. 137:1442–1446. doi: https://doi.org/10.1093/jn/137.6.1442.

Wellington, M. O., A. K. Agyekum, K. Hamonic, J. K. Htoo, A. G. Van Kessel, and D. A. Columbus. 2019. Effect of supplemental threonine above requirement on growth performance of *Salmonella typhimurium* challenged pigs fed high-fiber diets. J. Anim. Sci. 97:3636–3647. doi: https://doi.org/10.1093/jas/skz225.

Wellington, M. O., K. Hamonic, J. E. C. Krone, J. K. Htoo, A. G. Van Kessel, and D. A. Columbus. 2020a. Effect of dietary fiber and threonine content on intestinal barrier function in pigs challenged with either systemic E. coli lipopolysaccharide or enteric *Salmonella typhimurium*. J. Anim. Sci. Biotechnol. 11:38. doi: https://doi.org/10.1186/s40104-020-00444-3.

Wellington, M. O., R. B. Thiessen, A. G. Van Kessel, and D. A. Columbus. 2020b. Intestinal health and threonine requirement of growing pigs fed diets containing high dietary fibre and fermentable protein. Animals 10:2055. doi: https://doi.org/10.3390/ani10112055.

Wellington, M. O., J. K. Htoo, A. G. Van Kessel, and D. A. Columbus. 2018. Impact of dietary fiber and immune system stimulation on threonine requirement for protein deposition in growing pigs. J. Anim. Sci. 96:5222–5232. doi: https://doi.org/10.1093/jas/sky381.

Wells, J. M., L. M. P. Loonen, and J. M. Karczewski. 2010. The role of innate signaling in the homeostasis of tolerance and immunity in the intestine. Int. J. Med. Microbiol. 300:41–48. doi: https://doi.org/10.1016/j.ijmm.2009.08.008.

Wells, J. M., O. Rossi, M. Meijerink, and P. van Baarlen. 2011. Epithelial crosstalk at the microbiota–mucosal interface. Proc. Natl. Acad. Sci. 108:4607–4614. doi: https://doi.org/10.1073/pnas.1000092107.

Williams, N. H., T. S. Stahly, and D. R. Zimmerman. 1997a. Effect of level of chronic immune system activation on the growth and dietary lysine needs of pigs fed from 6 to 112 kg. J. Anim. Sci. 75:2481–2496. doi: https://doi.org/10.2527/1997.7592481x.

Williams, N. H., T. S. Stahly, and D. R. Zimmerman. 1997b. Effect of chronic immune system activation on body nitrogen retention, partial efficiency of lysine utilization, and lysine needs of pigs. J. Anim. Sci. 75:2472–2480. doi: https://doi.org/10.2527/1997.7592472x.

Windey, K., V. D. Preter, and K. Verbeke. 2012. Relevance of protein fermentation to gut health. Mol. Nutr. Food Res. 56:184–196. doi: https://doi.org/10.1002/mnfr.201100542.

Wong, J. M. W., R. de Souza, C. W. C. Kendall, A. Emam, and D. J. A. Jenkins. 2006. Colonic health: Fermentation and short chain fatty acids. J. Clin. Gastroenterol. 40:235.

Wu, G. 1998. Intestinal mucosal amino acid catabolism. J. Nutr. 128:1249–1252. doi: https://doi.org/10.1093/jn/128.8.1249.

Wu, G. 2013. Functional amino acids in nutrition and health. Amino Acids 45:407–411. doi: https://doi.org/10.1007/s00726-013-1500-6.

Wu, G., F. W. Bazer, G. A. Johnson, D. A. Knabe, R. C. Burghardt, T. E. Spencer, X. L. Li, and J. J. Wang. 2011. Triennial growth symposium: Important roles for L-glutamine in swine nutrition and production. J. Anim. Sci. 89:2017–2030. doi: https://doi.org/10.2527/jas.2010-3614.

Wu, G., Y.-Z. Fang, S. Yang, J. R. Lupton, and N. D. Turner. 2004. Glutathione metabolism and its implications for health. J. Nutr. 134:489–492. doi: https://doi.org/10.1093/jn/134.3.489.

Wu, G., S. A. Meier, and D. A. Knabe. 1996. Dietary glutamine supplementation prevents jejunal atrophy in weaned pigs. J. Nutr. 126:2578–2584. doi: https://doi.org/10.1093/jn/126.10.2578.

Wu, G., and S. M. Morris. 1998. Arginine metabolism: Nitric oxide and beyond. Biochem. J. 336:1–17. doi: https://doi.org/10.1042/bj3360001.

Wu, G., Z. Wu, Z. Dai, Y. Yang, W. Wang, C. Liu, B. Wang, J. Wang, and Y. Yin. 2013. Dietary requirements of "nutritionally non-essential amino acids" by animals and humans. Amino Acids. 44:1107–1113. doi: https://doi.org/10.1007/s00726-012-1444-2.

Wu, L., P. Liao, L. He, Z. Feng, W. Ren, J. Yin, J. Duan, T. Li, and Y. Yin. 2015. Dietary l-arginine supplementation protects weanling pigs from deoxynivalenol-induced toxicity. Toxins 7:1341–1354. doi: https://doi.org/10.3390/toxins7041341.

Wu, X., Z. Ruan, Y. Gao, Y. Yin, X. Zhou, L. Wang, M. Geng, Y. Hou, and G. Wu. 2010. Dietary supplementation with l-arginine or N-carbamylglutamate enhances intestinal growth and heat shock protein-70 expression in weanling pigs fed a corn- and soybean meal-based diet. Amino Acids 39:831–839. doi: https://doi.org/10.1007/s00726-010-0538-y.

Yao, C. K., J. G. Muir, and P. R. Gibson. 2016. Review article: Insights into colonic protein fermentation, its modulation and potential health implications. Aliment. Pharm. Ther. 43:181–196. doi: https://doi.org/10.1111/apt.13456.

Yao, K., Y.-L. Yin, W. Chu, Z. Liu, D. Deng, T. Li, R. Huang, J. Zhang, B. Tan, W. Wang, and G. Wu. 2008. Dietary arginine supplementation increases mTOR signaling activity in skeletal muscle of neonatal pigs. J. Nutr. 138:867–872. doi: https://doi.org/10.1093/jn/138.5.867.

Yi, G. F., J. A. Carroll, G. L. Allee, A. M. Gaines, D. C. Kendall, J. L. Usry, Y. Toride, and S. Izuru. 2005. Effect of glutamine and spray-dried plasma on growth performance, small intestinal morphology, and immune responses of *Escherichia coli* K88+-challenged weaned pigs. J. Anim. Sci. 83:634–643. doi: https://doi.org/10.2527/2005.833634x.

Yin, J., M. Liu, W. Ren, J. Duan, G. Yang, Y. Zhao, R. Fang, L. Chen, T. Li, and Y. Yin. 2015. Effects of dietary supplementation with glutamate and aspartate on diquat-induced oxidative stress in piglets. PLOS One 10:e0122893. doi: https://doi.org/10.1371/journal.pone.0122893.

Yoo, S. S., C. J. Field, and M. I. McBurney. 1997. Glutamine supplementation maintains intramuscular glutamine concentrations and normalizes lymphocyte function in infected early weaned pigs. J. Nutr. 127:2253–2259. doi: https://doi.org/10.1093/jn/127.11.2253.

Yu, Y. M., J. F. Burke, and V. R. Young. 1993. A kinetic study of L-2H3-methyl-1-13C-methionine in patients with severe burn injury. J. Trauma. 35:1–7. doi: https://doi.org/10.1097/00005373-199307000-00001.

Zeamer, K. M., R. S. Samuel, B. S. Pierre, R. C. Thaler, T. A. Woyengo, T. Hymowitz, and C. L. Levesque. 2021. Effects of a low allergenic soybean variety on gut permeability, microbiota composition, ileal digestibility of amino acids, and growth performance in pigs. Livest. Sci. 243:104369. doi: https://doi.org/10.1016/j.livsci.2020.104369.

Zhan, Z., D. Ou, X. Piao, S. W. Kim, Y. Liu, and J. Wang. 2008. Dietary arginine supplementation affects microvascular development in the small intestine of early-weaned pigs. J. Nutr. 138:1304–1309. doi: https://doi.org/10.1093/jn/138.7.1304.

Zhang, S., S. Qiao, M. Ren, X. Zeng, X. Ma, Z. Wu, P. Thacker, and G. Wu. 2013. Supplementation with branched-chain amino acids to a low-protein diet regulates intestinal expression of amino acid and peptide transporters in weanling pigs. Amino Acids 45:1191–1205. doi: https://doi.org/10.1007/s00726-013-1577-y.

Zhao, Y., G. Tian, D. Chen, P. Zheng, J. Yu, J. He, X. Mao, Z. Huang, Y. Luo, J. Luo, and B. Yu. 2020. Dietary protein levels and amino acid supplementation patterns alter the composition and functions of colonic microbiota in pigs. Anim. Nutr. 6:143–151. doi: https://doi.org/10.1016/j.aninu.2020.02.005.

Zheng, P., B. Yu, J. He, G. Tian, Y. Luo, X. Mao, K. Zhang, L. Che, and D. Chen. 2013. Protective effects of dietary arginine supplementation against oxidative stress in weaned piglets. Br. J. Nutr. 109:2253–2260. doi: https://doi.org/10.1017/S0007114512004321.

Zhong, X., X. H. Zhang, X. M. Li, Y. M. Zhou, W. Li, X. X. Huang, L. L. Zhang, and T. Wang. 2011. Intestinal growth and morphology is associated with the increase in heat shock protein 70 expression in weaning piglets through supplementation with glutamine. J. Anim. Sci. 89:3634–3642. doi: https://doi.org/10.2527/jas.2010-3751.

Zhou, J., Y. Wang, X. Zeng, T. Zhang, P. Li, B. Yao, L. Wang, S. Qiao, and X. Zeng. 2020. Effect of antibiotic-free, low-protein diets with specific amino acid compositions on growth and intestinal flora in weaned pigs. Food Funct. 11:493–507. doi: https://doi.org/10.1039/C9FO02724F.

Zhu, H., Y. Liu, X. Xie, J. Huang, and Y. Hou. 2013. Effect of l-arginine on intestinal mucosal immune barrier function in weaned pigs after *Escherichia coli* LPS challenge. Innate Immun. 19:242–252. doi: https://doi.org/10.1177/1753425912456223.

Zhu, H., D. Pi, W. Leng, X. Wang, C.-A. A. Hu, Y. Hou, J. Xiong, C. Wang, Q. Qin, and Y. Liu. 2017. Asparagine preserves intestinal barrier function from LPS-induced injury and regulates CRF/CRFR signaling pathway. Innate Immun. 23:546–556. doi: https://doi.org/10.1177/1753425917721631.

Zijlstra, R. T., and E. Beltranena. 2013. Swine convert co-products from food and biofuel industries into animal protein for food. Anim. Front. 3:48–53. doi: https://doi.org/10.2527/af.2013-0014.

# Part II
# Nutrition for Successful and Sustainable Swine Production

# 10    Diet Formulation and Feeding Programs

Sung Woo Kim and Jeffrey A. Hansen

## Diet Formulation

### Purpose of Formulation

#### Nutritional Plane

Feed formulation for swine can serve many purposes, but the first purpose served should be that of the animal needs. As an initial consideration, the stage of growth, age of an animal, or weight is typically used to describe nutrient requirements in most reference publications (ARC 1981; NRC 2012). Alternatively, in the simplest terms, one can assess requirements or needs based on the nutrients required for maintenance and the nutrients required for some productive/nonmaintenance function. For many nutrients, one might include a need for endogenous losses that categorically do not fit in the other two areas because it is a predictable outcome, such as endogenous losses of amino acids associated with digestion. In all cases, one must identify and establish some baseline requirements for inclusion in the formulation software in order to determine how to meet those requirements.

One can further break down the maintenance and productive functions into more specific entities, such as those for lean growth, fat growth, fetal growth, milk production, and activity. These compartmental methods have coefficient estimates published in various publications such as the NRC (2012) or ARC (1981). From a formulation standpoint, the use of such models offers the potential to improve formulation accuracy from a directional standpoint, but not likely from an absolute standpoint. For example, energy consumption is often described in many species with the following equation:

$$\text{Energy Intake} = \text{Energy Maintenance} + \text{Energy Production Function}$$

For weaned pigs, the formulator must consider how to transition the piglets from nursing their mother to ultimately a dry feed-based diet made up primarily of grains. The newly weaned piglet is expected to rapidly adapt to a new environment, new feed and water source, new social order, separation from its mother, etc. This transition is significant due to known food allergies (soy protein sensitivity; Li et al. 1991; Kim et al. 2010; Taliercio et al. 2014), lack of enzymes for digestion of the new diet, emotional stress of weaning, potential disease stresses, and the like. The formulator has many different techniques available to help the animal cope from specialty ingredients to setting

*Sustainable Swine Nutrition*, Second Edition. Edited by Lee I. Chiba.
© 2023 John Wiley & Sons Ltd. Published 2023 by John Wiley & Sons Ltd.

nutrient minimums and maximums, each of these constraints contributes significantly to the final feed recipe.

### Purchasing Support

*Provide General Pricing Targets*
Most feed formulators apply the practice of least-cost feed formulation, where multiple ingredients are allowed to enter into a formula but not all are needed to create a feasible solution to the problem. This method is of primary interest to those persons associated with commercial livestock production or feed production, where a true value proposition is created as an outcome of the formulation exercise. The role of feed formulation in commercial enterprises is multidimensional, including the purpose of meeting the animals' needs, but also the purpose of supporting the purchasing agent. Purchasing support is poorly understood by the student and academic areas primarily due to overemphasis of the nutritional plane/requirement component. In commercial enterprises, the formulator is often required to establish purchase price points, relative valuations, quantification of savings amounts, and many other financial aspects. Many purchasing agents need the ability to make decisions independent of the formulation software, and therefore need simple tools or relationships to make a purchasing decision, again the formulator is more often than not the skilled worker providing these data.

*Establish Alternative Price Points (Alternative Ingredient Strategies)*
In principle feed formulation does not vary dramatically between a commercial feed manufacturer and a livestock feeder, but in practice these operations vary significantly. More often than not the livestock feeder has fewer rations and fewer ingredients to solve for in their feed mill. In reality, both formulators tend to have two or three primary ingredients that provide the bulk of the energy (corn or wheat) and protein/amino acids (soybean meal). These ingredients are typically readily available in most locations substantially all of the time and in large quantities. Alternatives to the base ingredients are often dictated by the time of year (harvest season), proximity of the feed mill to other food/feed manufacturing facilities, and origination locations. For example, a feeder located next to a commercial flour mill most likely has wheat middlings readily available at an attractive price, whereas some located in very rural areas may have no access to the same product. Likewise, barley is a good alternative to corn/wheat in locations where it is grown, but may only be available at harvest time in limited quantities. The formulator will need to help the purchasing agent determine how best to value the reasonable alternative ingredients.

*Develop Purchasing Strategies*
Beyond providing the pricing or purchasing tools for the purchasing agent, the feed formulator can dramatically influence purchasing decisions by how they implement the nutrient matrix or the formulation technique and program. Each feed ingredient has variance associated with it due to processing, agronomics, and the like. It should be obvious that variance in the nutrient levels can lead to performance levels less than expected, particularly on those nutrients that are limiting in the formula. The formulator has several options for addressing the risk of not meeting the key nutrient requirements in the formula due to nutrient variation in the ingredients, this is often referred to as stochastic formulation. Stochastic formulation incorporates knowledge of ingredient variance coupled with a probability of achieving the target nutrient levels. From a practical standpoint the

formulator can account for the variance either in the individual ingredient nutrient specs or at the time of formulation (true stochastic formulation). Roush et al. (2007) argues in favor of real-time stochastic formulation versus ingredient modification, this argument is moot if the true probabilities are compared. Indeed, applying nutrient variance to individual ingredients offers the potential to more accurately value ingredients prior to purchase versus a method that only intends to account for the variance, thereby truly improving the purchasing decision.

### Other Objectives

Outside of meeting the animals' needs and productive function, solving for costs of production/ purchasing support, the formulator may be faced with incorporating other objectives into the formulation process. Incorporating constraints into the formulation process for these objectives often allows the formulator to discover the cost of managing these elements. Examples of other objectives may include

- Environmental regulations, which may require the formulator to impose restrictions on nutrients such as nitrogen or phosphorus, maximum mycotoxin or other toxins in the ingredients, and specific functional characteristics such as pellet quality, flow characteristics, bulk density, moisture content, or the like.

## Ingredient Matrix Development

### Initial Matrix

Commercial feed formulation software may be offered to the formulator with some base matrix from NRC (1998), Feedstuffs magazine, or other origins. The formulator is encouraged to establish the initial matrix using information from their local geography and on those ingredients more likely to present themselves as opportunities. Ideally, the formulator would have some amount of actual results to establish nutrient means and variances for each ingredient. Each user is encouraged to recognize that processed ingredients tend to have fairly predictable results, but these results are often inherently linked to the inbound product stream and the processing methodology. The reader is cautioned against assuming these two components are the same or similar enough among vendors to create a single ingredient for the formula, many like ingredients should be valued separately in the formulation.

Many nutrients are less than 100% available to animal and availability may differ by ingredients. Thus, the use of available nutrient profiles is recommended for those nutrients of economic importance and that enough data exist for to establish reliable availabilities among many ingredients. The formulator must reconcile that available nutrients represent an incremental improvement over total and is almost always directionally correct. In brief, the value proposition is increased with no significant increase in risk. There is a point of diminishing returns, such is the case of energy availability. While it is easily understood that the ideal energy system for swine is a Net Energy System, it should also be easily understood that the availability and quality of information available to establish Net Energy values are limited and does not offer substantial improvement over a "modified." Metabolizable Energy method. Whether this is true at the moment is not the point, the point is that while a Net Energy System represents a directional improvement, the lack of information does not allow us to take advantage of the system or increases our risk to the point of not offering improvement.

### *Formulation Methodology*

#### *Least-Cost Methods*

#### *Least-Cost per Unit of Feed*
Most feed formulators and software vendors focus on solving feed formulas to the least cost per unit of feed. Typically, the formulator supplies a set of constraints consistent with the nutrient requirements for key nutrients such as amino acids, major minerals, and energy. The software then solves a series of simultaneous equations to an optimum cost solution for the given ingredients and nutrients. Care must be given to allow for a feasible solution, the fewest possible constraints typically allow for a range of feasible outcomes at the lowest possible cost, adding constraints typically increases costs.

#### *Least-Cost per Unit of Gain*
Least-cost per kilogram of gain typically represents some combination of formulation methodologies. One will not likely find commercially available software to provide this functionality, but certain companies provide customized programs that match the formulator's objective. More likely than not, the goal will be to tie some production function outcome to a specific/variable nutrient to predict a cost function. This technology typically provides and incremental improvement in formulation that is directional correct, and it should not be used as a predictor of future performance.

#### *Key Software Concerns*
#### *Nutrient Factoring and Ratioing.*
Ratioing nutrients in the formula specification is a key consideration. For example, setting the "optimum" ratio of amino acids allows the ratios to be maintained. In many systems, most major nutrients are ratioed to energy density of the formula, this tends to work well because most species modify consumption based on energy density. Typically, ratioing to energy represents a significant directional improvement in formulation.

Factoring is not necessarily the same as ratioing. Factoring will often set a nutrient at one specific level of the factored nutrient, whereas ratioing tends to maintain the ratio regardless of nutrient level. For example, if we set a lysine:energy factor of 1.5 per 1500 kcal, we can be assured lysine is 1.5 at 1500, but if energy is 1600, this does not mean lysine will be 1.6, it just depends on how you get there. It is important to understand how factoring is used versus ratios, these are not typically handled the same in software.

#### *Production Minimums*
It often occurs that the mechanical systems of a mill cause the formulator to implement a production minimum. For example, a particular mill may not be able to accurately weigh less than 1.82 kg/ton of an ingredient on its minor scale. Thus, you could set a production minimum of 1.82 kg, forcing the formula to require either 0 or 1.82 + kilogram, of the ingredient.

#### *Rounding*
One key issue is how software engineers approach ingredient rounding when solving formulations. This aspect is particularly relevant for highly potent ingredients like phytase, where a small amount of rounding greatly contributes to a nutrient level. In some software rounding is applied after a feasible solution is found, which can be costly. Ideally, the software solution applies rounding as part of the feasible solution thereby maximizing value.

**Feeding Program**

*Principles of Feeding Program*

This section will discuss general aspects of feeding programs for pigs at different ages or physiological stages. Detailed nutrient requirements for pigs are discussed in other chapters (Chapter 22 for sows, Chapter 23 for nursery pigs, Chapter 23 for finishing pigs). In swine production, the main goals of feeding program: are (i) to provide nutrients meeting the requirements for optimal productive performance and health, (ii) to maximize economic benefits, and also (iii) to minimize nutrient excretion by improving efficiency of nutrient utilization. A sound feeding program would utilize phase feeding by stage of production or physiological status to achieve these three goals.

*Phase Feeding by Different Ages or Physiological Status*

*Sow*

*Lactation*
Feeding management of lactating sows primarily targets to improve milk production and to minimize maternal tissue loss, which is also related to reproductive performance in the subsequent parity. However, a major hurdle to achieve these targets would be insufficient voluntary feed intake. Thus, the feeding program for lactating sows should consider both improving nutrient intake and enhancing efficiency of nutrient utilization. It is well demonstrated that gestational body condition affects voluntary feed intake during lactation (Williams 1998; Kim and Easter 2003). Obese sows due to overfeeding during gestation are shown to eat less during lactation than normal sows. It is suggested that reduced voluntary feed intake is related to increase insulin resistance and altered blood levels of insulin, leptin, and probably grhelin (Weldon et al. 1994; Père and Etienne 2007).

In order to prevent the occurrence of obesity at farrowing, feed restriction is commonly practiced during gestation. Traditionally, in the United States and many other countries, sows are housed individually in gestation stalls and fed individually in order to control energy intake. Societal requests of the removal of gestation stalls, however, have influenced the traditional practice of using gestation stalls to utilize more group sow housing in pens. Increasing number of hog farmers has voluntarily removed gestation stalls (Kaufmann 2007). Group housing of gestating sows is often adequate to have 5–20 sows in small pens. These pens often have feeders with partial dividers that allow individual feeding and thus to control energy intake. However, it would be important to group sows by body condition to lessen behavioral stress (Zhao et al. 2013). Alternative group feeding uses a large pen with electronic feeders allowing individual sows to enter and eat a given amount of feeds (Spoolder et al. 1997; Bates et al. 2003). However, the initial investment and maintenance cost of electronic feeders should be considered.

Feeding sows during lactation should consider helping to increase nutrient intake. Popular feeding program could be a step increase in feed allowance during the first three to five days after farrowing and then frequent provisions of fresh feed (two to four times per day) to encourage appetite during the remainder of lactation period. Feeding lactating sows during hot and humid climate can be challenging to maintain appetite due to heat stress (Black et al. 1993). Sows with heat stress have increased systemic oxidative stress and reduced voluntary feed intake (Zhao and Kim 2020). Altering time of feed provision to early morning and late evening may encourage sows eat more in hot and humid climate. Another consideration for feeding lactating sows is to provide nutrient dense feeds. When feed intake is not sufficient, increasing nutrient concentration would help to prevent a severe catabolic status during lactation. Increased use of supplemental fat by reducing the amount of starch and reduction of crude protein, whereas providing sufficient essential amino acids is sound dietary strategy for lactating sows under heat stress condition (Rosero et al. 2012; Zhang et al. 2020).

When determining nutrient requirements for lactating sows, these components need to be considered: (i) nutrients for milk production, (ii) nutrients for mammary tissue synthesis, and (iii) nutrients needed for maternal growth and maintenance (Kim et al. 2009). Nutrient requirements of lactating sows can be affected by nutrient needs from mammary glands to produce milk and to build mammary parenchymal tissues (Kim et al. 1999a,b). A sow will mobilize her body tissue to provide nutrients for mammary glands when dietary intake is not sufficient (Kim et al. 2001).

When nutrient intake is not sufficient during lactation, maternal body tissues would be mobilized to offset the difference between nutrient output through mammary glands and nutrient input from dietary intake. When nutrient intake changes, the amount of nutrients contributed from maternal tissue mobilization will be adjusted to balance the nutrient needs for milk production, which related to litter sizes. Amino acids contributed from maternal tissue mobilization have different profile from those used for milk production, and thus ideal dietary amino acid profile would be altered as the contribution of amino acids from maternal tissue mobilization changes (Kim et al. 2001; Table 10.1). Young, high prolific lean-type sows often have insufficient voluntary feed intake with higher nutrient needs for maternal gain, whereas old sows often have sufficient voluntary feed intake with limited nutrient needs for maternal use. Considering different needs of nutrients in quantity and quality between young and old sows, parity feeding can be applied in feeding lactating sows (Kim et al. 2013). Diets for young sows with insufficient feed intake and maternal growth can contain higher nutrient concentrations than diets for old sows with sufficient feed intake without significant needs for maternal growth.

*Gestation*
Highly prolific sows need to support the growth of as many as 14–20 fetuses. Fetal growth occurs mostly after day 70 of gestation (McPherson et al. 2004; Ji et al. 2005). Accretion of protein in

**Table 10.1**  Ideal amino acid patterns and the order of limiting amino acids for lactating sows (Kim et al. 2009 / with permission of Oxford University Press.).

| Item | Estimated 21-day weight loss (kg)[a]: | 75–80 | 33–45 | 12–15 | 6–8 | 0 | 7–0 |
|---|---|---|---|---|---|---|---|
| | Level of tissue mobilization (%)[b]: | 50 | 40 | 20 | 5 | 0 | NRC (1998)[c] |
| Ideal amino acid pattern (% of Lys) | | | | | | | |
| Lys | | 100 | 100 | 100 | 100 | 100 | 100 |
| Thr | | 75 | 69 | 63 | 60 | 59 | 62 |
| Val | | 78 | 78 | 78 | 77 | 77 | 85 |
| Leu | | 128 | 123 | 118 | 115 | 115 | 114 |
| Ile | | 60 | 59 | 59 | 59 | 59 | 56 |
| Arg | | 22 | 38 | 59 | 69 | 72 | 56 |
| Order of limiting amino acids[d] | | | | | | | |
| First | | Thr | Lys | Lys | Lys | Lys | Lys |
| Second | | Lys | Thr | Thr | Val | Val | Val |
| Third | | Val | Val | Val | Thr | Thr | Thr |

[a] These values refer to weight loss of sows during lactation estimated based on amount of protein loss and tissue composition measured by Kim et al. (2001).

[b] These values refer to the percentage of AA in milk output that derive from tissue protein catabolism in the sow.

[c] The NRC (1998) estimates do not consider tissue protein mobilization.

[d] This assumes that a typical corn-soybean meal diet (0.90%) is fed during lactation.

**Table 10.2** Amino acid needs for maternal gain and maintenance (g/day; Kim et al. 2009). Average BW of sows was 160 kg at breeding, 195 kg at day 70 of gestation, and 220 kg at day 114 of gestation (Ji et al. 2005). Values for Trp and Met were adapted from finishing pigs (Mahan and Shields 1998).

| Amino acid | Day 0–70 | | | Day 70 to farrowing | | |
|---|---|---|---|---|---|---|
| | Sum | Maintenance | Gain | Sum | Maintenance | Gain |
| Lys | 6.41 | 1.64 | 4.77 | 8.06 | 1.78 | 6.28 |
| Thr | 5.19 | 2.48 | 2.71 | 6.78 | 2.69 | 4.09 |
| Trp | 0.93 | 0.43 | 0.50 | 1.17 | 0.46 | 0.71 |
| Met | 1.60 | 0.46 | 1.14 | 2.02 | 0.50 | 1.52 |
| Val | 4.12 | 1.10 | 3.02 | 4.66 | 1.19 | 3.47 |
| Leu | 5.58 | 1.15 | 4.43 | 6.23 | 1.25 | 4.98 |
| Ile | 3.80 | 1.23 | 2.57 | 4.68 | 1.34 | 3.34 |
| Arg | 5.77 | 1.23 | 4.54 | 7.96 | 1.34 | 6.62 |

**Table 10.3** Lysine-based ideal amino acid ratio for sows with various fetal numbers (Data from Kim 2010).

| Number of fetus | Day of gestation | Lys | Thr | Trp | Met | Val | Leu | Ile | Arg |
|---|---|---|---|---|---|---|---|---|---|
| 6 | 0–70 | 1.00 | 0.80 | 0.15 | 0.25 | 0.65 | 0.88 | 0.59 | 0.90 |
| | 70–114 | 1.00 | 0.73 | 0.15 | 0.26 | 0.65 | 0.92 | 0.56 | 0.95 |
| 8 | 0–70 | 1.00 | 0.80 | 0.15 | 0.25 | 0.65 | 0.88 | 0.59 | 0.90 |
| | 70–114 | 1.00 | 0.72 | 0.16 | 0.27 | 0.66 | 0.93 | 0.56 | 0.96 |
| 10 | 0–70 | 1.00 | 0.80 | 0.15 | 0.25 | 0.65 | 0.88 | 0.59 | 0.90 |
| | 70–114 | 1.00 | 0.72 | 0.16 | 0.27 | 0.66 | 0.94 | 0.56 | 0.97 |
| 12 | 0–70 | 1.00 | 0.79 | 0.15 | 0.25 | 0.65 | 0.88 | 0.59 | 0.90 |
| | 70–114 | 1.00 | 0.71 | 0.16 | 0.27 | 0.66 | 0.95 | 0.56 | 0.97 |
| 14 | 0–70 | 1.00 | 0.79 | 0.15 | 0.25 | 0.65 | 0.88 | 0.59 | 0.90 |
| | 70–114 | 1.00 | 0.71 | 0.16 | 0.27 | 0.66 | 0.96 | 0.55 | 0.98 |
| 16 | 0–70 | 1.00 | 0.79 | 0.15 | 0.25 | 0.65 | 0.88 | 0.59 | 0.90 |
| | 70–114 | 1.00 | 0.70 | 0.16 | 0.27 | 0.67 | 0.97 | 0.55 | 0.98 |
| 18 | 0–70 | 1.00 | 0.79 | 0.15 | 0.25 | 0.65 | 0.89 | 0.59 | 0.90 |
| | 70–114 | 1.00 | 0.70 | 0.16 | 0.28 | 0.67 | 0.97 | 0.55 | 0.99 |

fetal tissues increases at least 19 fold after day 70 of gestation, whereas accretion of fat is reasonably constant during gestation (Kim et al. 2009). Sows also need to support the growth of mammary glands which occurs mostly after day 70 of gestation (Ji et al. 2006). Accretion of protein in mammary tissues increases at least 24 fold after day 70 of gestation, whereas accretion of fat is fairly constant during gestation (Kim et al. 2009; Table 10.2). It has been shown that ideal dietary amino acid profiles for gestating sows differ between early and late gestation due to the changes in rates of protein accretion among different types of tissues (Kim et al. 2009; Table 10.3). Considering significant increase in protein needs and altered ideal amino acid profiles for sows during late gestation, phase feeding can be applied in feeding gestating sows by providing low protein diet during early gestation and high protein diet toward late gestation (Kim et al. 2013). Phase feeding with varied protein concentrations can allow energy restriction without compromising increased protein needs and altered ideal amino acid profiles for sows in late gestation. However, increase of feed allowance (so called "bump feeding") may have limited benefits due to increase in energy supply negatively influencing lactation feed intake (Weldon et al. 1994).

*Nursing Pigs*

In typical US swine production, sows nurse piglets for 14–28 days after parturition. During this short nursing period, piglets gain about 150–250 g per day, and sow milk is the only source of nutrients to support this rapid growth if creep feed is not provided. Pigs are born with minimal nutrient storage and thus sufficient milk intake is essential for optimal growth of nursing piglets. Colostrum and mature milk from a sow provide nutrients that are highly available to piglets (Lin et al. 2009; Mavromichalis et al. 2001). However, sows with poor body condition or/and poor feed intake may have problems with mammary glands providing quality nutrients to nursing piglets (Kim et al. 1999a,b). Moreover, research shows that milk yield is a limiting factor to support the growth of nursing piglets (Aherne 1980; Zijlstra et al. 1996). Creep feed can be a practical way of supporting growth of nursing pigs if milk production from sows is not sufficient.

*Nursery Pigs*

In swine production, pigs are typically weaned at day 14–28 of age. Newly weaned pigs undergo a high stress period due to sudden changes in diet types from milk to solid-type feed, separation from sows, and changes in the environment (Maxwell and Carter 2000; Zheng et al. 2021). Pigs often experience a postweaning growth slump due to weaning stresses. Sudden changes in diet types often cause diarrhea because newly weaned pigs have insufficient secretion of endogenous enzymes and gastric HCl to digest feed ingredients with complex structures (Pluske et al. 1997; Lalles et al. 2004) as well as antinutritional compounds causing allergenic immune response in the small intestine (Taliercio and Kim 2013). It is, therefore, important to use ingredients that are highly digestible and lacking antinutritional compounds to newly weaned young pigs and then gradually changing to conventional ingredients such as corn and soybean meal. Dairy coproducts such as whey permeate, whey powder, lactose, whey protein concentrates are commonly used in the diets for newly weaned young pigs because of similar structural property of nutrients in dairy co-products to sow milk (Mahan et al. 2004; Cromwell et al. 2008; Jang et al. 2021). Feed ingredients derived from animal products including blood plasma, blood meal, fish meal, meat meal, meat and bone meals, poultry meal, etc. are excellent protein supplements for newly weaned young pigs (de Rodas et al. 1995; Kim and Easter 2001; Adedokun and Adeola 2005) because of high digestibility compared to plant proteins with complex structures and potential antinutritional compounds. However, some plant proteins, after processing to hydrolyze nutrients and antinutritional compounds, can be fed to newly weaned young pigs (Kim et al. 2003 and 2010; Goebel and Stein 2011; Oliveira and Stein 2016).

In order to help nutrient digestion and maintenance of intestinal health, feed additives are often used in nursery diets. Acidifiers and feed enzymes can help to enhance nutrient digestibility and utilization. Prebiotics, direct-fed microbials, postbiotics, and phytobiotics can help to maintain intestinal health of newly weaned pigs. Extensive research has been done to investigate effective types and dose levels of feed additives to achieve the goal. There are numerous papers reviewing the efficacy of feed additives (Liao and Nyachoti 2017; Baker et al. 2021; Zheng et al. 2021). Selective use of feed additives can help the growth and health of newly weaned pigs.

Phase feeding program can be well applied in feeding nursery pigs. Typically, three or four phases are used in feeding nursery pigs from wean to 9 or 10 weeks of age when pigs reach 22–25 kg body weight. First phase of nursery diet can include dairy coproducts and animal proteins gradually changing to corn and soybean meal toward the last phase of nursery diet. Phase feeding program should be adjusted based on weaning age or body weight at weaning.

*Finishing Pigs*

In commercial pig production, nursery pigs are moved to finisher unit (or grower-finisher unit) when they reach around 22–25 kg, which is typically 9–10 weeks of age. In this stage, pigs will eat well and thus feed intake and growth are not typical problems. But, instead, increasing fat gain and reduced feed efficiency (expressed as gain/feed) are major challenges to the producers.

A split-sex feeding regime is one of practice used in swine production because barrows and gilts have different growth rate and lean gain potentials. Barrows eat more energy and grow faster than gilts with the same age (NRC 2012). However, barrows have faster fat gain than gilts. If both barrows and gilts are raised in a same pen consuming a diet with same nutrient compositions, barrows will grower faster than gilts, whereas gilts will be leaner than barrows. Therefore, it can be beneficial to house barrows and gilts separately to provide different diets. Barrows can be fed a lower protein diet compared with a diet for gilts. Thus, a split-sex feeding could benefit the producers by enhancing lean growth of barrows, weight gain of gilts, and uniformity of the herds. However, mixed-sex feeding is often practiced in pig production as this allows the increased use of pen space by housing more pigs during an earlier phase. As pigs grow, barrows can be marketed early providing space for gilts until marketing.

As pigs grow, their feed intake increases and fat gain accelerates. In order to encourage lean gain, feed should be designed to contain ideally balanced amino acids (=ideal protein) needed for protein gain. Protein synthesis would stop if any essential amino acids are limited (=limiting amino acids) and thus balancing dietary amino acids for protein synthesis is important to enhance lean gain. Ideal protein has been investigated and characterized (Wang and Fuller 1989; Chung and Baker 1991). Use of the ideal protein concept in feed formulation will also benefit pig production by reducing nitrogen excretion into the environment and reducing feed costs. Ideal protein for growing pigs is summarized in Table 10.4 (NRC 2012). Typical supplemental amino acids used in swine feeds are L-Lys, L-Thr, L/DL-Met, and L-Trp, whereas L-Val and L-Ile are also considered occasionally.

In a typical pig production facility in the United States, pigs are fed dry rations. Pigs have access to dry feeds in feeders, whereas water is supplied from a waterer separated from the feeder. In the meantime, liquid feeding systems or wet feeding systems have been shown to improve feed intake. Liquid feeding usually provides feed at 20–30% dry matter after mixing with water to create slurry in a pan or provide feed using liquid ingredients typically from food dairy process-

**Table 10.4**  Lysine-based ideal amino acid ratio for growing pigs.

| Amino acid | Body weight (kg) | | | | | | |
|---|---|---|---|---|---|---|---|
| | 5–7 | 7–11 | 11–25 | 25–50 | 50–75 | 75–100 | 100–135 |
| Lys | 1.00 | 1.00 | 1.00 | 1.00 | 1.00 | 1.00 | 1.00 |
| Thr | 0.59 | 0.59 | 0.59 | 0.60 | 0.61 | 0.63 | 0.66 |
| Trp | 0.17 | 0.16 | 0.16 | 0.17 | 0.18 | 0.18 | 0.18 |
| Met | 0.29 | 0.29 | 0.29 | 0.29 | 0.28 | 0.29 | 0.30 |
| Val | 0.63 | 0.64 | 0.63 | 0.65 | 0.65 | 0.66 | 0.67 |
| Leu | 1.00 | 1.00 | 1.00 | 1.01 | 1.00 | 1.01 | 1.02 |
| Ile | 0.51 | 0.51 | 0.51 | 0.52 | 0.53 | 0.53 | 0.54 |
| Phe | 0.59 | 0.59 | 0.59 | 0.60 | 0.60 | 0.60 | 0.61 |
| His | 0.35 | 0.34 | 0.34 | 0.35 | 0.46 | 0.34 | 0.34 |
| Arg | 0.45 | 0.45 | 0.46 | 0.46 | 0.46 | 0.45 | 0.46 |

Source: Adapted from NRC (2012).

ing and liquid fermentation. Pigs with liquid feeding have shown improved weight gain mainly due to increased feed intake (Gonyou and Lou 2000; de Lange et al. 2006). Usually liquid feeding does not improve feed efficiency due to increased feed intake. As pigs in liquid feeding have greater feed intake, pigs could gain more fat compared with pigs in dry feeding. When liquid feeding is used in a farm, feeder management is important to prevent spoilage and mold problems from wastage in the feeder.

Antimicrobial growth promotors (AGP) and ractopamine have been shown to enhance lean gain of pigs. Therefore, until recently, AGP and ractopamine have been used in swine production. However, recent ban or voluntary removal of AGP and ractopamine has adversely influenced days to market and lean growth of finishing pigs.

Particle size of feedstuffs affects the pig's ability to digest nutrients in the feed (Wondra et al. 1995; Acosta et al. 2020). It is typical to grind corn to 600–800 μm when a diet is fed as a mash or meal form. Digestibility of nutrients in corn can be improved if corn is ground finely, many producers are grinding to 300–400 μm or finer. However, when corn is ground finely, it is suggested to pellet the feed in order to prevent bridging in a feeder and a feed bin.

## Summary

Feed formulation for swine should consider the stage of growth, age of an animal, or body weight, which affects nutrient requirements. Nutrient requirements can be assessed based on the nutrients needed for maintenance and for production (or gain). Most feed formulators apply the practice of least-cost feed formulation, where multiple ingredients are allowed to enter into a formula. The role of feed formulation in commercial enterprises is multidimensional, including the purpose of meeting the pig's need, but also the purpose of supporting the purchasing agent. Most feed formulators and software vendors focus on solving feed formulas to the least-cost per unit of feed. Ratioing nutrients in the formula specification is a key consideration. Most major nutrients are ratioed to energy density of the formula.

Main goals of feeding program are to provide nutrients meeting the requirements for optimal productive performance but at the same time to maximize economic benefits and minimize nutrient excretion. Basic principles of a sound feeding program would include phase feeding of pigs by stage of growth or physiological status. For sows, the formulator should consider maximizing nutrient intake during lactation to support intensive milk production. However, nutrient intake, especially energy, should be controlled during gestation to prevent obesity at farrowing. For weaned pigs, the formulator must consider how to transition the piglets from nursing their mother to ultimately a dry feed-based diet made up primarily of grains and legume seed meals. For finishing pigs, formulation to improve feed efficiency should be a key target.

## References

Acosta, J. A., A. L. Petry, S. A. Gould, C. K. Jones, C. R. Stark, A. Fahrenholz, and J. F. Patience. 2020. Effects of grinding method and particle size of wheat grain on energy and nutrient digestibility in growing and finishing pigs. Transl. Anim. Sci. 16:txaa062. doi: https://doi.org/10.1093/tas/txaa062.

Adedokun, S. A., and O. Adeola. 2005. Metabolizable energy value of meat and bone meal for pigs. J. Anim. Sci. 83:2519–2526. doi: https://doi.org/10.2527/2005.83112519x.

Aherne, F. X. 1980. Management and nutrition of the newly weaned pig. Univ. of Illinois Pork Industry Conf. Urbana, IL. p. 55.

ARC. 1981. The nutrient requirements of pigs. Commonwealth Agricultural Bureaux, Farnham Royal, UK.

Baker, J. T., M. E. Duarte, D. M. Holanda, and S. W. Kim. 2021. Friend or foe? Impacts of dietary xylans, xylooligosaccharides, and xylanases on intestinal health and growth performance of monogastric animals. Animals 11:609. doi: https://doi.org/10.3390/ani11030609.

Bates, R. O., D. B. Edwards, and R. L. Korthals. 2003. Sow performance when housed either in groups with electronic sow feeders or stalls. Livest. Prod. Sci. 79:29–35. doi: https://doi.org/10.1016/S0301-6226(02)00119-7.

Black, J. L., B. P. Mullan, M. L. Lorschy, and L. R. Giles, 1993. Lactation in the sow during heat stress. Livest. Prod. Sci. 35:153–170. doi: https://doi.org/10.1016/0301-6226(93)90188-N.

Chung, T. K., and D. H. Baker. 1991. A chemically defined diet for maximal growth of pigs. J. Nutr. 121:979–982. doi: https://doi.org/10.1093/jn/121.7.979.

Cromwell, G. L., G. L. Allee, and D. C. Mahan. 2008. Assessment of lactose level in the mid- to late-nursery phase on performance of weanling pigs. J. Anim. Sci. 86:127–133. doi: https://doi.org/10.2527/jas.2006-831.

de Lange, C. F. M., C. H. Zhu, S. Niven, D. Columbus, and D. Woods. 2006. Swine liquid feeding: Nutritional considerations. Proc. 27th Western Nutr. Conf. Dept. of Anim. Sci., Univ. of Manitoba, Winnipeg, MB, Canada. p. 37–50.

de Rodas, B. Z., K. S. Sohn, C. V. Maxwell, and L. J. Spicer. 1995. Plasma protein for pigs weaned at 19 to 24 days of age: effect on performance and plasma insulin-like growth factor I, growth hormone, insulin, and glucose concentrations. J. Anim. Sci. 73:3657–3665. doi: https://doi.org/10.2527/1995.73123657x.

Goebel, K. P., and H. H. Stein. 2011. Phosphorus digestibility and energy concentration of enzyme-treated and conventional soybean meal fed to weanling pigs. J. Anim. Sci. 89:764–772. doi: https://doi.org/10.2527/jas.2010-3253.

Gonyou, H. W., and Z. Lou. 2000. Effects of eating space and availability of water in feeders on productivity and eating behavior of grower-finisher pigs. J. Anim. Sci. 78:865–870. doi: https://doi.org/10.2527/2000.784865X.

Jang, K. B., J. M. Purvis, and S. W. Kim. 2021. Dose–response and functional role of whey permeate as a source of lactose and milk oligosaccharides on intestinal health and growth of nursery pigs, J. Anim. Sci. 99:skab008. doi: https://doi.org/10.1093/jas/skab008.

Ji, F., G. Wu, J. R. Blanton, Jr., and S. W. Kim. 2005. Changes in weight and composition in various tissues of pregnant gilts and their nutritional implications. J. Anim. Sci. 83:366–375. doi: https://doi.org/10.2527/2005.832366x.

Ji, F., W. L. Hurley, and S. W. Kim. 2006. Characterization of mammary gland development in pregnant gilts. J. Anim. Sci. 84:579–587. doi: https://doi.org/10.2527/2006.843579x.

Kaufmann, M. 2007. Largest pork processor to phase out crates. The Washington Post, 26 January 26, 2007. https://www.washingtonpost.com/wp-dyn/content/article/2007/01/25/AR2007012501785.html.

Kim, S. W. 2010. Recent advances in sow nutrition. Revista Brasileira de Zootecnia 39:303–310. doi: https://doi.org/10.1590/S1516-35982010001300033.

Kim, S. W., and R. A. Easter. 2001. Nutritional value of fish meals in the diet for young pigs. J. Anim. Sci. 79:1829–1839. doi: https://doi.org/10.2527/2001.7971829x.

Kim, S. W., and R. A. Easter. 2003. Amino acid utilization for reproduction in sows. In: J. P. F. D'Mello, editor, Amino acids in animal nutrition. CABI Publishing, Wallingford, Oxfordshire, UK. p. 203–222.

Kim, S. W., W. L. Hurley, I. K. Han, and R. A. Easter. 1999. Changes in tissue composition associated with mammary gland growth during lactation in the sow. J. Anim. Sci. 77:2510–2516. doi: https://doi.org/10.2527/1999.7792510x.

Kim, S. W., D. H. Baker, and R. A. Easter. 2001. Dynamic ideal protein and limiting amino acids for lactating sows: Impact of amino acid mobilization. J. Anim. Sci. 79:2356–2366. doi: https://doi.org/10.2527/2001.7992356x.

Kim, S. W., D. L. Knabe, K. J. Hong, and R. A. Easter. 2003. Use of carbohydrases in corn-soybean meal-based nursery diets. J. Anim. Sci. 81:2496–2504. doi: https://doi.org/10.2527/2003.81102496x.

Kim, S. W., W. L. Hurley, G. Wu, and F. Ji. 2009. Ideal amino acid balance for sows during gestation and lactation. J. Anim. Sci. 87:E123–E132. doi: https://doi.org/10.2527/jas.2008-1452.

Kim, S. W., E. van Heugten, F. Ji, C. H. Lee, and R. D. Mateo. 2010. Fermented soybean meal as a vegetable protein source for nursery pigs: I. Effects on growth performance of nursery pigs. J. Anim. Sci. 88:214–224. doi: https://doi.org/10.2527/jas.2009-1993.

Kim, S. W., A. C. Weaver, Y. B. Shen, and Y. Zhao. 2013. Improving efficiency of sow productivity: Nutrition and health. J. Anim. Sci. Biotechnol. 4:26. doi: https://doi.org/10.1186/2049-1891-4-26.

Lalles, J., G. Boudry, C. Favier, N. LeFloc, I. Luron, L. Montagne, I. P. Oswald, S. Pié, C. Piel, and B. Sève. 2004. Gut function and dysfunction in young pigs: physiology. Anim. Res. 53:301–316. doi: https://doi.org/10.1051/animres:2004018.

Li, D. F., J. L. Nelssen, P. G. Reddy, F. Blecha, R. D. Klemm, D. W. Giesting, J. D. Hancock, G. L. Allee, and R. D. Goodband. 1991. Measuring suitability of soybean products for early-weaned pigs with immunological criteria. J. Anim. Sci. 69:3299–3307. doi: https://doi.org/10.2527/1991.6983299x.

Liao, S. F., and M. Nyachoti. 2017. Using probiotics to improve swine gut health and nutrient utilization. Anim. Nutr. 3:331–343. doi: https://doi.org/10.1016/j.aninu.2017.06.007.

Lin, C., D. C. Mahan, G. Wu, and S. W. Kim. 2009. Protein digestibility of colostrums by neonatal pigs. Livest. Sci. 121:182–186. doi: https://doi.org/10.1016/j.livsci.2008.06.006.

Mahan, D. C., N. D. Fastinger, and J. C. Peters. 2004. Effects of diet complexity and dietary lactose levels during three starter phases on postweaning pig performance. J. Anim. Sci. 82:2790–2797. doi: https://doi.org/10.2527/2004.8292790x.

Mahan, D. C., R. G. Shields. 1998. Macro- and micromineral composition of pigs from birth to 145 kilgrams of body weight. J. Anim. Sci. 76:506–512.

Mavromichalis, I., T. M. Parr, V. M. Gabert, and D. H. Baker. 2001. True ileal digestibility of amino acids in sow's milk for 17-day-old pigs. J. Anim. Sci. 79: 707–713. doi: https://doi.org/10.2527/2001.793707x.

Maxwell, C. V., and S. D. Carter. 2000. Feeding the weaned pig. In: A. J. Lewis, and L. L. Southern, editors, Swine nutrition 2nd ed. CRC Press, Boca Raton, FL. p. 691–716.

McPherson, R. L., F. Ji, G. Wu, and S. W. Kim. 2004. Fetal growth and compositional changes of fetal tissues in the pigs. J. Anim. Sci. 82:2534–2540. doi: https://doi.org/10.2527/2004.8292534x.

NRC. 1998. Nutrient requirements of swine. 10th rev ed.. Natl. Acad. Press, Washington, DC.

NRC. 2012. Nutrient requirements of swine. 11th rev. ed. Natl. Acad. Press, Washington, DC.

Oliveira, M. S., and H. H. Stein. 2016. Digestibility of energy, amino acids, and phosphorus in a novel source of soy protein concentrate and in soybean meal fed to growing pigs. J. Anim. Sci. 94:3343–3352. doi: https://doi.org/10.2527/jas.2016-0505.

Père M.-C., and M. Etienne. 2007. Insulin sensitivity during pregnancy, lactation, and postweaning in primiparous gilts. J. Anim. Sci. 85:101–110. doi: https://doi.org/10.2527/jas.2006-130.

Pluske, J. R., D. J. Hampson, and I. H. Williams. 1997. Factors influencing the structure and function of the small intestine in the weaned pigs: A review. Livest. Prod. Sci. 51:215–236. doi: https://doi.org/10.1016/S0301-6226(97)00057-2.

Rosero, D. S., E. van Heugten, J. Odle, R. Cabrera, C. Arellano, and R. D. Boyd. 2012. Sow and litter response to supplemental dietary fat in lactation diets during high ambient temperatures. J. Anim. Sci. 90:550–559. doi: https://doi.org/10.2527/jas.2011-4049.

Roush, W. B., J. Purswell, and S. L. Branton. 2007. An adjustable nutrient margin of safety comparison using linear and stochastic programming in an excel spreadsheet. J. Appl. Poult. Res. 16:514–520. doi: https://doi.org/10.3382/japr.2007-00033.

Spoolder, H., J. Burbidge, S. Edwards, A. Lawrence, and P. Simmins. 1997. Effects of food level on performance and behaviour of sows in a dynamic group-housing system with electronic feeding. Anim. Sci. 65:473–482. doi: https://doi.org/10.1017/S1357729800008675.

Taliercio, E., and S. W. Kim. 2013. Epitopes from two soybean glycinin subunits are antigenic in pigs. J. Sci. Food Agric. 93:2927–2932. doi: https://doi.org/10.1002/jsfa.6113.

Taliercio, E., T. M. Loveless, M. J. Turano, and S. W. Kim. 2014. Identification of epitopes of the β subunit of soybean β-conglycinin that are antigenic in pigs, dogs, rabbits and fish. J. Sci. Food Agric. 94:2289–2294. doi: https://doi.org/10.1002/jsfa.6556.

Wang, T. C., and M. F. Fuller. 1989. The optimum dietary amino acid pattern for growing pigs. 1. Experiments by amino acid deletion. Br. J. Nutr. 62:77–89. doi: https://doi.org/10.1079/bjn19890009.

Weldon, W. C., A. J. Lewis, G. F. Louis, J. L. Kovar, and P. S. Miller. 1994. Postpartum hypophagia in primiparous sows: II. Effects of feeding level during gestation and exogenous insulin on lactation feed intake, glucose tolerance, and epinephrine-stimulated release of nonesterified fatty acids and glucose. J. Anim. Sci. 72:395–403. doi: https://doi.org/10.2527/1994.722395x.

Wondra, K. J., J. D. Hancock, K. C. Behnke, R. H. Hines, and C. R. Stark. 1995. Effects of particle size and pelleting on growth performance, nutrient digestibility, and stomach morphology in finishing pigs. J Anim. Sci. 73:757–763. doi: https://doi.org/10.2527/1995.733757x.

Williams, I. H. 1998. Nutritional effects during lactation and during the interval from weaning to estrus. In: M. W. A. Verstegen, and P. S. Moughan, editors, The lactating sow. Wageningen University Press, Wageningen, The Netherllads. p. 159–182.

Zijlstra, R. T., K. Y. Whang, R. A. Easter, and J. Odle. 1996. Effect of feeding a milk replacer to early-weaned pigs on growth, body composition, and small intestinal morphology, compared with suckled littermates. J. Anim. Sci. 74:2948–2959. doi: https://doi.org/10.2527/1996.74122948x.

Zhang, S., J. S. Johnson, and N. L. Trottier. 2020. Effect of dietary near ideal amino acid profile on heat production of lactating sows exposed to thermal neutral and heat stress conditions. J. Anim. Sci. Biotechnol. 11:75. doi: https://doi.org/10.1186/s40104-020-00483-w.

Zhao, Y., and S. W. Kim. 2020. Oxidative stress status and reproductive performance of sows during gestation and lactation under different thermal environments. Asian- Australas. J. Anim. Sci. 33:722–731. doi: https://doi.org/10.5713/ajas.19.0334.

Zhao, Y., B. Flowers, A. Saraiva, K.-J. Yeum, and S. W. Kim. 2013. Effect of social ranks and gestation housing systems on oxidative stress status, reproductive performance, and immune status of sows. J. Anim. Sci. 91:5848–6388. doi: https://doi.org/10.2527/jas.2013-6388.

Zheng, L., M. E. Duarte, A. Sevarolli Loftus, and S. W. Kim. 2021. Intestinal health of pigs upon weaning: challenges and nutritional intervention. Front. Vet. Sci. 8:628258. doi: https://doi.org/10.3389/fvets.2021.628258.

# 11 Cereal Grains, Cereal By-products, and Other Energy Sources in Swine Diets

Rajesh Jha and Tofuko A. Woyengo

## Introduction

Pigs require energy to maintain normal body processes, growth, reproduction and lactation processes, and physical activity. Feeds supplying energy are major components of all swine diets, of which carbohydrates (sugar, starch, and fiber), lipids, and protein/amino acids components provide energy. Carbohydrates are the most abundant energy source in swine diets. Although fat and oils contribute on average 2.25 times more gross energy than carbohydrates, these make a smaller overall contribution to the total energy of swine diets as are included in diets in lower quantities. Protein usually contributes between 15 and 20% of the total dietary energy, primarily when diets are formulated to have excess protein than the requirement for protein synthesis.

Energy is the most expensive component in swine diets, accounting for about 50% of the total farrow-to-finish production cost and more than 30% of the total cost of raising a pig to market. Thus, it is very important to develop a feeding program that provides energy to swine diets cost-effectively. Traditionally, carbohydrates from cereal grains have been the most abundant energy source in swine diets. Among the cereals, corn and wheat have been primarily used as energy sources in different parts of the world. But, the demand and supply chain of conventional feedstuffs is stretched because of the competition between food, feed, and fuel, as well as the decrease in cultivable land to meet the need of the increased human population (Avalos 2014). Therefore, there is a need for cost-effective alternative energy feedstuffs for swine diets.

For sustainable swine production, alternative feedstuffs play a key role for three main reasons. First, for economic sustainability, alternative feedstuffs, and coproducts in particular, have become an important option to control rapidly increasing feed costs (Zijlstra and Beltranena 2009). In part for biofuel production, the novel industrial demand for feed grains has elevated the long-term price forecasts for feed grains to another price plateau. Alternative feedstuffs are a short-term solution for commercial swine production to control feed costs, with proper risk management strategies, including modern feed quality evaluation, as key components. Second, for agronomic sustainability, alternative crops with unique agronomic features might be important (Miller et al. 2002). For example, the drought tolerance of sorghum and triticale may support a switch from traditional feed grains that are less drought-tolerant such as wheat and corn. Triticale requires 14% less crop inputs than wheat (Davis-Knight and Weightman 2008). An important criterion for sustainable success is that the newly included crops have a market as alternative feedstuff if quality targets for the primary market cannot be met. Finally, for societal and environmental sustainability, the use of coproducts as feedstuffs for swine addresses the argument that pigs compete with humans for food (Nonhebel 2004).

*Sustainable Swine Nutrition*, Second Edition. Edited by Lee I. Chiba.
© 2023 John Wiley & Sons Ltd. Published 2023 by John Wiley & Sons Ltd.

The conversion of inedible residues from the food, biofuel, and bioprocessing industries into high-quality animal protein food mitigates the impact of these industries on the environment. For example, behind every food product in a supermarket, there is at least one useful coproduct that is overlooked, even though certain global regions already use coproducts from the food industry effectively to produce pork that is less reliant on feed grains. Pigs, as an omnivorous species, are ideally suited to consume a wide variety of feedstuffs and, thus, can be an integral part of sustainable livestock production systems.

The use of alternative feedstuffs in the swine industry is not new. In traditional swine production, pigs were housed in small numbers, and high growth rates were less important. Pigs were fed feedstuffs that currently are regarded as alternatives, such as leftover human food products (Pond and Lei 2001). Such traditional production is still common practice in global small-scale swine production, particularly in Asia (Chen 2009). The development of modern swine-production systems that demand high growth rates, safe, and consistent pork products resulted from a reliance on a supply of affordable feed grains and a few protein sources to produce pork competitively (Pond and Lei 2001). Currently in North America, the inclusion of alternative sources of dietary energy in the commercial swine industry is considered advantageous solely during periods of price increases for common feed grains. Dietary inclusion of coproducts provides opportunities for diversifying the feedstuff matrix by using local feedstuffs, reducing feed costs, and producing value-added pork (Jha et al. 2013). In the last decade, only one alternative feedstuff, corn dried distiller's grains with solubles (**DDGS**), has reached commodity status within the North American swine industry (Patience et al. 2007). Across the world, few regions have a solid logistical system in place for the commercial swine industry to rely on coproducts as main feedstuffs in swine diets. However, some European countries with a small land base, such as The Netherlands, have historically been heavily dependent on a large array of alternative feedstuffs (FEFAC 2005). Finally, alternative feedstuffs are considered for the organic pork production (Partanen et al. 2006), perhaps to either avoid the use of corn and soybeans that have been genetically modified for herbicide resistance and other traits or the required use of homegrown organic feedstuffs.

This chapter describes three categories of dietary energy sources: (i) developments in traditional crops, (ii) alternative crops, and (iii) coproducts resulting from the biofuel and food industry and crop fractionation. Details of alternative dietary energy sources in the diets of pigs have been summarized previously (Thacker and Kirkwood 1990; Myer and Brendemuhl 2001; Sauber and Owens 2001). So, the focus of the present chapter will be on new developments within the last decade, with emphasis on the coproducts available for use in the diets of pigs.

### Energy Evaluation Systems

Swine diets are formulated based primarily on using cereals as the main source of energy. The competition for food, feed, and fuel of these conventional feedstuffs has forced animal nutritionists and the feed industry to formulate animal diets using alternative feedstuffs including coproducts from cereals. Feed quality information is essential as feed accounts for 65–70% of the total cost involved in pig production (Woyengo et al. 2014). The basis of any commercial farming system is to formulate a diet combining ingredients with the least cost in order to give a better return on investment. However, the nutritional content varies widely within and between crops depending upon the type of feedstuffs (Jha et al. 2011a, 2011b), within and between fibrous and starchy feedstuffs (Tiwari and Jha 2016, 2017), and harvesting condition and processing method (Hernot et al. 2008). In addition, a large range in nutrient content exists within and between these feedstuffs because of variable agronomic conditions, genetic variation, and processing techniques in these feedstuffs and

**Table 11.1**   Relative energy value of the selected feedstuffs used in swine diet.

| Feedstuff | DE | ME | NE | NE:ME |
|---|---|---|---|---|
| *Reference diet* | *100* | *100* | *100* | *75* |
| Animal fat | 243 | 252 | 300 | 90 |
| Corn | 103 | 105 | 112 | 80 |
| Wheat | 101 | 102 | 106 | 78 |
| Barley | 94 | 94 | 96 | 77 |
| Pea | 101 | 100 | 98 | 73 |
| Soybean (full-fat) | 116 | 113 | 108 | 72 |
| Wheat bran | 68 | 67 | 63 | 71 |
| Dried distiller's grains | 82 | 80 | 71 | 67 |
| Soybean meal | 107 | 102 | 82 | 60 |
| Canola meal | 84 | 81 | 64 | 60 |

Source: Adapted from Sauvant et al. (2004).

coproducts. The wide variation among the nutritional values of these feedstuffs creates a necessity for developing routine feed evaluation techniques to detect these variations in order to formulate balanced diets and achieve optimal animal performance.

Indisputably, the choice of an energy evaluation system will alter the relative values placed on feeds (Noblet et al. 1993). For example, the relative energy value of the selected feedstuffs used in the swine diet (Sauvant et al. 2004) is presented in Table 11.1. For energy evaluation, the digestible energy (DE) and metabolizable energy (ME) systems overestimate the energy contribution to support maintenance and growth (Black 1995), while the net energy (NE) system offers a more accurate ranking of feedstuffs (Whittemore 1997). The feed industry in The Netherlands has been relying on the NE system since 1970 (CVB 1993), partly to manage the risk of a wide ingredient matrix (Zijlstra and Payne 2007). The difference in approach to energy evaluation among scientists and countries is reflected in the selected approach in research deliverables.

Traditionally, feedstuffs are subjected to different protocols of laboratory analysis for nutrient profiling from representative samples to be analyzed followed by digestibility and energy utilization determination using animal studies to determine their inclusion level, effects and adverse effects (if any) on the performance of animals. The new feedstuff is then incorporated in the commercial feeding program if it is found to be comparable to the conventional feedstuffs. Also, several published table values and prediction equations have been used to determine the nutrient profile before incorporating it in feeding programs traditionally.

Table values or reference values of the nutrient profile of feed ingredients are available from different sources like NRC (National Research Council, United States), INRA (French National Institute for Agricultural Research), CVB (Centraal Veevoeder Bureau, the Netherlands), FEDNA (Fundación Española para el Desarrollo de la Nutrición Animal, Spain), and NARO (National Agriculture and Food Research Organization, Japan). It is the easiest method of getting the nutrient value of feedstuffs, but the table values vary widely within and between the tables as the source of data differs. It also varies among batches and countries due to several factors, including agronomical conditions and genetic variations. This discrepancy among tables makes it inferior in accuracy and of limited value for getting quality outcomes. As an example, table values and analyzed values of the same feed ingredients were found to vary widely, as presented in Table 11.2. Lower accuracy is the limiting factor in the use of table values. Due to this variation, a potential lower safety margin in using table values for feed formulation can affect the performance of pigs.

**Table 11.2** Comparison of nutrient profile of wheat by-products between analyzed values (Jha et al. 2012) and Table values (NRC 2012)[a].

| | Analyzed values | | | | | | Table values | | |
|---|---|---|---|---|---|---|---|---|---|
| | Shorts | | Millrun | | Middlings | Bran | Shorts | Middlings | Bran |
| Item | A | B | A | B | | | | | |
| Dry matter | 90.1 | 89.9 | 89.1 | 90.4 | 88.6 | 91.0 | 87.9 | 89.1 | 87.3 |
| Ash | 7.3 | 5.5 | 6.5 | 6.2 | 5.3 | 6.7 | NA | 2.1 | 4.2 |
| ether extract | 2.9 | 3.4 | 4.8 | 4.1 | 4.1 | 3.0 | 4.6 | 3.2 | 4.7 |
| Crude protein | 27.8 | 24.9 | 18.7 | 19.0 | 22.1 | 15.9 | 16.7 | 15.8 | 15.1 |
| Crude fiber | 7.9 | 5.2 | 8.3 | 9.9 | 7.1 | 12 | NA | 5.2 | 7.8 |

[a] NA = not available.

Several predictions equations have been developed to predict the energy values of feedstuffs in swine and are found to predict with very high accuracy ($R^2 = 0.97$) (Noblet and Perez 1993; Noblet et al. 1994). All the prediction equations are based on the basic nutrient values of feedstuffs. Obtaining an accurate prediction of the NE content of alternative feedstuffs is considered important (Smits and Sijtsma 2007) to assure equivalent growth performance following the introduction of alternative feedstuffs or coproducts.

### Animal Studies

The in vivo method is so far the best and most robust model to determine the digestibility of feedstuffs in animals. However, the accuracy of the digestibility of the in vivo method depends upon the marker used and the method by which the amount of marker is quantified. Also, there has been wide variation reported by different researchers while determining the nutritional value of feedstuffs using animal studies. Moreover, logistical limitations, skilled expertise required, and longer time and higher costs involved to conduct animal studies are the limiting factors in using animal studies in the routine feed evaluation program.

### In vitro Studies

For defining the energy value of individual samples within a feedstuff, getting an accurate measurement or prediction of the ATTD of energy is most important (Zijlstra and Payne 2007). The main reason is that for defining the DE, ME, or NE content, the ATTD is the most variable component and the GE content (especially of cereal grains) is the most consistent component. A novel aspect in describing the energy value of feedstuffs for swine is a description of the kinetics of digestion or fermentation of carbohydrates, the main energy-contributing macronutrient. For starch, the objective is to define the kinetics of starch digestion, whereas for fiber the kinetics of fiber fermentation is regarded to be important (Zijlstra et al. 2010). The in vitro studies provide the opportunities to overcome the practical limitations of the commonly used approaches (table values, prediction equations, and animal studies) used to determine energy value of feedstuffs.

*In vitro Enzymatic Digestion Study*

In vitro enzymatic digestion method simulates the activity taking place in the whole gastrointestinal tract (stomach, small intestine, and large intestine) of animals to determine the digestibility of nutrients (Boisen and Fernández 1997), while the in vitro fermentation method simulates the microbial fermentation taking place in the large intestine, in addition to the enzymatic digestion in the stomach and small intestine (Jha et al. 2011b, 2011c). In vitro digestion techniques can be used to screen large set of samples in a short period of time and are noninvasive to animals and relatively very cheaper than in vivo methods.

It is imperative that any simulation should be reproducible and should correlate well with in vivo parameters for diverse feedstuffs. A close linear relationship ($R^2 = 0.94$) between the in vitro enzyme digestibility of organic matter and in vivo total tract digestibility of energy was found for 90 samples of 31 different feedstuffs (Boisen and Fernández 1997). A similar finding was reported by Regmi et al. (2009) for single samples of 8 feedstuffs ($R^2 = 0.97$). Within feedstuff, variability was predicted well for cereal grains and but poorly for canola meal and corn DDGS (Regmi et al. 2009). However, a medium to poor correlation and prediction accuracy was found when other coproducts [canola meal, corn DDGS, soybean meal (SBM), and wheat millrun] were evaluated (combined average $R^2 = 0.69$), with the highest for wheat millrun ($R^2 = 0.79$) and lowest for corn DDGS ($R^2 = 0.29$) (Wang 2014). Thus, further refinement in the in vitro digestibility study, as initiated by some workers (Regmi et al. 2008, 2009; Wang 2014), might be helpful to get better predictions using this model.

*In vitro Microbial Fermentation Study*

Some energy feedstuffs (like barley and oats) and coproducts are typically rich in fiber content, which needs to be considered while using in the swine feeding program. The fiber in feedstuffs negatively affects the nutrients digestibility (Jha et al. 2010), but it may also play an important role in improving the gut health of pigs (Jha and Berrocoso 2015) and the environment (Jha and Berrocoso 2016). Some fibers in concentrated form, such as barley and oat β-glucan, have a prebiotic effect in pigs (Pieper et al. 2008). The fibers and some of the starches that are not digested by endogenous enzymes of pigs become available for microbial fermentation, primarily in the large intestine, and produce metabolites like volatile fatty acids (**VFA**; Jha et al. 2019; Tiwari et al. 2019), which can contribute nutritional value to animals, directly by providing energy. The VFA can be used to predict the extent of energy digestion in the large intestine (McBurney and Sauer 1993). The energy produced from VFA may contribute up to 15.0% of the ME requirements of growing pigs (Dierick et al. 1989) and even up to 30.0% of the ME requirements of gestating sows (Varel and Yen 1997). Use of purified enzymes for this purpose is promising (Regmi et al. 2008; Regmi et al. 2009); however, their validation with in vivo study is important. In vitro fermentation models using pig fecal inoculum (Jha et al. 2011b,c; Jha et al. 2012) have been used to determine both fermentation metabolites and intestinal microbiota. Moreover, this method allows the evaluation of the role of exogenous enzymes in enhancing nutrient utilization (Jha et al. 2015). In vitro fermentation method is validated with high correlation in pigs for nonstarch polysaccharide (NSP) degradation (r = 0.96) (Anguita et al. 2006) and organic matter (r = 0.77) (Christensen et al. 1999). The fermentation characteristics of different conventional feedstuffs and coproducts were studied both in vivo and in vitro (Jha et al. 2010 2011b, c; Jha and Leterme 2012) and had similar fermentation characteristics and metabolites produced. Thus, in vitro fermentation models can serve as an option for rapid technique while considering the evaluation of energy and functional properties of any feedstuff. However, the in vitro model does not consider simultaneous production-absorption of the metabolites as happens in the large intestine in vivo. Thus, it may overestimate the energy contribution from VFA in animals.

## *Near-Infrared Spectrophotometry*

The near-infrared spectrophotometry (NIRS) has been adopted widely as a rapid feed evaluation technique and is becoming increasingly popular. The use of NIRS technology to determine basic nutrients such as moisture, protein, fat, and fiber of major feed ingredients and finished feeds have been around for many years (Valdes and Leeson 1992). With the advancement in technology, NIRS is being used for a range of measurements that are based on using laboratory methods to provide reference values for establishing calibration. For nutritionists that rely on digestible as opposed to total nutrient values for feed formulation, the additional cost of obtaining a large sample set with determined digestibility values has been cost-prohibitive to establish NIRS calibrations, but some research calibrations have been established (Zijlstra et al. 2011). Also, industries have been working to develop a rapid NIRS technology-based tool to estimate the bioavailability of nutrients from different feedstuffs. The idea is to assist the industry to adopt and use standards based on the functional utility of the grain purchased rather than relying on the current grain physical grades to determine value in use. Every feed ingredient has its spectral properties, which can be utilized to determine the nutritional value. Robustness is the ability of calibration to accurately predict DE in samples from a wide range of sources. Thus, it is important to have a good calibration database to have better prediction equations. Agronomic conditions, cleanliness of sample, and genetic origin will vary among feed samples and will affect the ability of NIRS to predict swine DE content accurately (Zijlstra et al. 2010). Also, NIRS penetrates deeper into the sample because of its wavelength; however, the depth of penetration depends upon the particle size and particle density (DeThomas and Brimmer 2002). Thus, not only the feed ingredient or complete feed type but also their other physical properties need to be considered. Moreover, the substitution of in vivo determinations of DE content with a reliable in vitro method would be the key for cost-effective NIRS application.

## Dietary Energy Sources

### *Cereals*

Traditionally, the driver for a competitive swine industry has been extensive, low-cost grain production. The grain standard will differ locally because of agronomic conditions. Withingrains, alternative cultivars are being developed mostly not only to enhance yield but also to enhance the density of the digestible nutrients.

### *Corn*
Corn is globally the cereal standard grain and is the basis for commercial swine production in the United States and Latin America. The nutritional value of hybrid yellow, dent corn is well defined (e.g., Sauber and Owens 2001) and has been updated over the years. It has high starch and low fiber content resulting in most nutrients being easily digested by pigs. The apparent total tract digestibility (**ATTD**) of dry matter and starch in corn is close to 90 and 95%, respectively (Rojas and Stein 2015). Corn can be included in the diets of all categories of pigs as the sole energy source cereal, except in the initial two to three weeks of the postweaned pigs. However, some fiber sources should be added to the corn-based diet of gestating sows to avoid constipation.

Within the last decade, breeding efforts improved, apart from yield, the nutritional or agronomic characteristics of alternative corn cultivars with unique traits such as short season (Opapeju et al. 2006), low phytate (Spencer et al. 2000; Veum et al. 2001; Hill et al. 2009), phytase-containing (Nyannor et al. 2007), herbicide-resistance (Hyun et al. 2004), rootworm resistance (Hyun et al. 2005), or enhanced oil (energy) and amino acid (AA) density (Pedersen et al. 2007). Although breeding programs have placed emphasis on yield increases for small grains, yield increases in corn have been much larger than for small grains like barley, oat, wheat, and sorghum (Alston et al. 2009) in recent decades. The yield of corn in areas with sufficient heat and water is much greater than barley or wheat and is the main reason that the production of small grains in the United States has been largely replaced with corn.

*Wheat*

Wheat is the major cereal grain used as an energy source in the swine diets in western Canada, northern Europe, and Australia, primarily because of their local production and price competitiveness. The nutritional value of wheat has been well defined (Sauber and Owens 2001; Jha et al. 2011a) and found to have comparable DE content to corn. The range in chemical characteristics of wheat, especially CP, starch, and NSP, causes variation in chemical composition and DE content of wheat for pigs and perhaps feed processing quality (Zijlstra et al. 1999; Jha et al. 2011a). The impact of variability in wheat quality on subsequent growth performance in young pigs can be reduced by enzyme application (Cadogan et al. 2003) and processing (Jha et al. 2011a). Thus, like corn, wheat can be efficiently utilized by pigs of all ages. However, consideration must be given to nutrient composition, method of processing, and quality and price of wheat for use in pig diets.

*Barley*

Second to wheat, barley is a cereal grain used as a dietary source of energy for swine in western Canada, northern Europe, and Australia as it is locally grown and widely available at a competitive price. Similar to wheat, the perception exists that barley has higher variability in DE content (Fairbairn et al. 1999; Jha et al. 2010) than corn. Similar to corn, unique traits have been developed, such as low phytate in barley (Veum et al. 2002; Htoo et al. 2007a,b) and starch profile in barley (Bird et al. 2004), but these advances have been achieved to a lesser extent compared with corn.

The inclusion of barley grain in young pig's diets is limited due to its high fiber content that is associated with lower nutrient digestibility and NE value (Lynch et al. 2007, Jha et al. 2010). Whether young pigs fed barley grain instead of wheat in diets formulated to equal NE value can maintain growth performance requires investigation.

*Oat*

Traditionally, oat was not used in the regular swine feeding program due to its high fiber content and relatively lower nutritional value. However, it has been tested as a potential energy source in swine diets for long. Although the gross energy (**GE**) value of oat is higher than that for corn (4.125 vs. 3.959 Mcal/kg), NE value is much lower (1.401 vs. 2.329 Mcal/kg) in oat than corn (De Goey and Ewan 1975). This is primarily due to the high fiber content in oat, which reduces nutrient utilization (Jha et al. 2010). The nutritional value of oat has been defined; the average energy value of the oat is given as 80% of the energy value of corn while using in swine diets. Despite lower energy value, oat can be used effectively in swine diets as oats are highly palatable and are higher in protein and lysine content than corn. However, the inclusion of oat in young pigs' diets should be limited due to their high fiber content (Jha et al. 2010). Oat can constitute up to 25% of the starter pig diet and up to 40% in the grower-finisher diet. Processing like grinding, rolling, and pelleting enhances the nutrient utilization of oat and should be considered. Oat is getting more attraction in recent years due to its competitive price and higher availability of improved varieties with reduced fiber content (Jha et al. 2011b). Also, oat fibers (especially β-glucans) are found to improve the intestinal health of pig by modulating gut microbiota (Pieper et al. 2008; Tiwari et al. 2019).

*Sorghum*

Sorghum (*Sorghum bicolor*), also known as milo, is grown in a number of countries around the world and is a major energy feedstuff in swine diets in Mexico due to the high amount of local production. It has higher starch content than corn and wheat but lower fat content than corn (Jaworski et al. 2015). The digestibility of starch in sorghum is lower than that of corn and wheat (Cervantes-Pahm et al. 2014b). The GE, DE, and ME content of sorghum is similar to corn and wheat (Cervantes-Pahm et al. 2014a; Bolarinwa and Adeola 2016), and the feeding value of sorghum compared to corn is 98–99% (Tokach et al. 2012). Thus, it can be used as the sole cereal grain in the diets of pigs without any adverse effect on the growth performance of weanling or grower-finisher pigs.

*Rye*

Rye (*Secale cereal*) is a grass grown extensively as a grain, a forage crop, and a cover crop in some parts of Europe. It is one of the most important cereal crops in central Europe; it is closely related to wheat; however, it has several economic advantages (Allen 2002). It has about 90% of the energy content of corn for growing-finishing pigs. Previously, rye was not commonly used in swine feeding programs as older cultivars were susceptibility to ergot infection and had high fiber content, especially viscous NSP components. However, recently developed new types of hybrid rye are characterized not only by increased yield potential, resistance to fungus and pests, and low production costs but also the lower content of antinutritive substances in these varieties (Schwarz et al. 2014). Thus, rye has a competitive advantage compared to corn and wheat to be grown in organic farming as there is a reduced requirement of fertilizer and pesticide. Also, rye has gained increased interest as an energy feedstuff for pigs because of its lower price, winter hardiness, and ability to grow on sandy soils with low fertility (Bushuk 2001). However, rye has a high concentration of NSP, which has been correlated with antinutritive properties, especially in young pigs (Jürgens et al. 2012). So, the use of rye should be limited in your pigs, while it can be a good alternative energy source for use in grower-finisher pig diets and becomes increasingly attractive as the age of the pig increases. Hybrid rye may be included in diets fed to weanling, growing, and finishing pigs by 10, 25, and 50%, respectively, without causing reductions in growth performance or carcass quality (Schwarz et al. 2014).

*Triticale*

Water and nitrogen are key drivers for successful grain production. In semi-arid areas such as parts of western Canada and Australia, water supply and drought are recurring issues. Enhanced drought tolerance and N-use efficiency in corn and small grains might be an approach to increase grain yield. However, the cultivation of feed grains with a higher yield and lower crop input requirements (Davis-Knight and Weightman 2008) in areas with marginal growing conditions should also be part of a solution package to maximize pork produced per hectare. Crops such as triticale, a hybrid of wheat and rye (Radecki and Miller 1990), may improve feed grain yield in marginal growing conditions (McLeod et al. 2001) and requires 14% less crop inputs compared to wheat (Davis-Knight and Weightman 2008). Traditionally, the growth performance of pigs fed triticale has been assumed lower because studies conducted in the distant past indicated that triticale might reduce the growth performance of young pigs relative to corn (Hale and Utley 1985). Ergot tolerance has also been enhanced (Salmon 2004). However, modern triticale cultivars are low in trypsin inhibitors, and the palatability of pigs fed triticale is, thus, less of a concern (Radecki and Miller 1990). Indeed, weaned pigs fed diets containing either 60% of wheat or 60% of modern varieties of triticale achieved an identical growth performance (Beltranena et al. 2008).

**Table 11.3**   Nutrient profile and in vitro digestibility of selected tubers (% DM basis; adapted from Tiwari and Jha 2016)[a].

| Feedstuff | DM | Ash | CP | EE | ADF | NDF | Starch | GE[b] | DE[b] | ME[b] | NE[b] | IVDMD | IVDGE |
|---|---|---|---|---|---|---|---|---|---|---|---|---|---|
| Purple sweet potato | 42.98 | 1.95 | 4.79 | 2.77 | 5.68 | 7.95 | 47.02 | 4.134 | 4.135 | 4.120 | 2.882 | 86.8 | 87.5 |
| Okinawa sweet potato | 40.83 | 2.79 | 5.30 | 2.04 | 8.14 | 9.67 | 51.67 | 4.154 | 4.116 | 4.100 | 2.869 | 81.6 | 82.3 |
| Taro | 37.43 | 2.39 | 8.84 | 1.87 | 10.38 | 11.46 | 38.43 | 4.333 | 4.117 | 4.099 | 2.860 | 70.3 | 64.9 |
| Cassava | 41.87 | 4.13 | 3.73 | 1.05 | 6.53 | 11.30 | 60.85 | 4.193 | 4.095 | 4.080 | 2.863 | 82.1 | 83.1 |

[a] DM = dry matter; CP = crude protein; EE = ether extract; ADF = acid detergent fiber; NDF = neutral detergent fiber; GE = gross energy; DE = digestible energy; ME = metabolizable energy; NE = net energy; IVDMD = in vitro digestibility of DM, IVDGE = in vitro digestibility of GE.
[b] Mcal/kg.

### Roots and Tubers

Roots and tubers are rich in starch and can serve as a good source of energy in pig diets. Among the roots and tubers, the largest at the global level is potatoes. Also, sweet potatoes, cassava, and taro are available for animal feeding in different parts of the world, especially in developing countries and can be a cost-effective source of energy in the diets of pigs. The GE contents and energy digestibility of tubers (sweet potato, cassava, and taro) range from 4.1 to 4.3 Mcal/kg and 74 to 87%, respectively (Table 11.3; Tiwari and Jha 2016). However, some of the tubers like cassava and taro contain antinutritional factors, which need to be considered while using those tubers in the diets of pigs.

### Coproducts

#### Corn DDGS

The DDGS is a coproduct from dry-grind cereal grain ethanol industry. Several cereal grains, including corn, wheat, sorghum and triticale, are used for the production of ethanol, leading to the availability of DDGS for livestock feeding. Of these cereals, corn is the most widely used cereal grain used for the production of ethanol by the dry milling process, and hence corn DDGS is the most widely available form of DDGS for livestock feeding. Thus, the focus of this chapter will be on the nutritive value of corn DDGS.

Production of ethanol from corn involves saccharification of corn grain by starch degrading enzymes, fermentation of the hydrolyzed starch by yeast to produce ethanol, and distillation to obtain ethanol. The remaining material, which is known as whole stillage, is processed into DDGS. Most ethanol plants extract some oil from the liquid phase (solubles) of the whole stillage during the production of DDGS from the whole stillage, leading to the production of low-fat DDGS. Some ethanol plants also separate fiber from corn before fermentation (front-end fraction) or after fermentation (tail-end fraction), leading to the production of DDGS with relatively high protein (high-protein DDGS). Thus, DDGS has generally lower content of starch and greater content of CP, fiber, fat, and P than parent corn grain. The nutrient compositions of regular DDGS, low-fat DDGS, and high-protein DDGS are presented in Table 11.4. The CP, fiber, and P content of regular and low-fat DDGS is approximately three times greater than that for corn grain. The fat content in low-fat DDGS is close to that of corn grain, whereas the protein content in the high-protein DDGS is around

**Table 11.4**    Mean nutrient composition, energy values for selected coproduct feedstuffs for pigs[a].

| Feedstuff[b] | Content, % | | | | | | | | | Energy, kcal/kg | |
|---|---|---|---|---|---|---|---|---|---|---|---|
| | DM | CP | EE | Starch | NDF | ADF | TDF | IDF | SDF | DE | NE |
| Regular DDGS | 88.88 | 28.59 | 8.43 | 4.11 | 29.51 | 12.67 | 36.86 | 35.49 | 1.37 | 3483 | 2206 |
| HP corn DDGS | 89.20 | 38.01 | 8.43 | 1.87 | 30.64 | 14.64 | 34.20 | 31.80 | 2.40 | 4277 | 2994 |
| Wheat middlings | 88.48 | 15.96 | 4.23 | 19.38 | 36.27 | 10.93 | 36.35 | 33.69 | 2.69 | 2840 | 1911 |
| Wheat bran | 90.10 | 17.24 | 3.90 | 14.08 | 44.76 | 13.77 | 46.15 | 42.80 | 3.35 | 2478 | 1024 |
| Full-fat rice bran | 94.90 | 14.92 | 19.40 | 26.16 | 15.93 | 8.37 | 20.08 | 19.21 | 0.87 | 3984 | 3142 |
| Defatted rice bran | 89.74 | 17.20 | 3.18 | 28.09 | 22.75 | 11.92 | 28.92 | 26.43 | 2.50 | 2849 | 2115 |
| Sugar beet pulp | 91.23 | 8.96 | 1.31 | 3.37 | 43.79 | 21.35 | 54.60 | 40.83 | 13.62 | | 470 |

[a] DM = dry matter; CP = crude protein; EE = ether extract; NDF = neutral detergent fiber, ADF = acid detergent fiber; TDF = total dietary fiber; IDF = insoluble dietary fiber; SDF = soluble dietary fiber; DE = digestible energy; NE = net energy; DDGS = dried distiller's grains with solubles; and HP = high protein.
[b] Source: Regular DDGS = Avelar et al. (2010), Wu et al. (2016), Acosta et al. (2017), Coble et al. (2017), Jaworski and Stein (2017), Rho et al. (2017), Espinosa and Stein (2018), and Abelilla and Stein (2019); HP corn DDGS = Rho et al. (2017), and Espinosa and Stein (2018); Wheat middlings = Berrocoso et al., 2015; Garcia et al., 2015; Wu et al. 2016; Acosta et al. 2017; Casas and Stein, 2017; Casas et al. (2018), Jaworski and Stein (2017), Navarro et al. (2018), Abelilla and Stein (2019), Navarro et al. (2019); Wheat bran = Jaworski et al. (2015), and Jaworski et al. (2016); Full-fat rice bran = Casas and Stein (2016a,b), and Casas et al. (2019); Defattd rice bran = Casas and Stein (2016a,b), Huang et al. (2018), and Casas et al. (2019); Sugar beet pulp = Serena and Bach Knudsen (2007), Murray et al. (2008), Eklund et al. (2014), Berrocoso et al. (2015), Wang et al. (2016), Navarro et al. (2019), and Nguyen et al. (2019).

four times greater than that in corn grain. Corn has a high leucine content, and hence corn DDGS has a very high leucine content (NRC 2012).

More than 90% of total dietary fiber in corn DDGS is insoluble (Table 11.4). The major NSP in corn are arabinoxylans (Bach Knudsen 2004). Thus, the major NSP in corn DDGS are also arabinoxylans. Arabinoxylans constituted approximately 40% of total NSP in corn DDGS (Pedersen et al. 2014). The DGGS also contains some yeast-derived fiber components because of the presence of yeast in DDGS. Yeast fiber is rich in mannanoligosaccharides and β-glucans (Böttger and Südekum 2018; Shurson 2018). However, the actual contribution of yeast to total fiber in DDGS is variable and has not been well established (Böttger and Südekum 2018). Starch present in DDGS is the starch that escapes digestion by starch degrading enzymes and fermentation by yeast during ethanol production. Thus, starch present in DDGS can be considered to contain more resistant starch as a proportion of total starch than corn. The high proportion of fat in DDGS is unsaturated because linoleic acid (unsaturated fatty acids) constitutes a high proportion (>40%) of total fat in corn (NRC 2012).

As shown in Table 11.4, DDGS has a lower NE value than corn grains because of its lower DE value and starch content and greater CP and fiber content. The DE and starch values of a feedstuff are positively correlated with NE value of the feedstuff, whereas CP and fiber values of a feedstuff are negatively correlated with NE value of the feedstuff (Noblet et al. 1994). The lower DE value for DDGS than for corn grain is due to the lower digestibility of GE in DDGS than in corn because DDGS has a greater GE value than corn grain (NRC 2012). For example, the ATTD of GE for DDGS was 65% (Jaworski and Stein 2017), whereas that for corn was 84% (Opapeju et al. 2007). The lower ATTD of GE for DDGS than for corn is due to the greater fiber content in the former than in the latter. The DDGS can increase the weight of visceral organs in pigs, likely due to the high fiber content in the DDGS. For example, Agyekum et al. (2012) reported lower empty visceral organ weight of growing pigs fed corn-SBM-based diet with 0% DDGS than of growing pigs fed corn-SBM-based diet with 30% DDGS; however, the empty visceral organ weight for the diet with

30% DDGS was reduced to that of a diet with 0% DDGS by supplementation with fiber-degrading enzymes. The DDGS can also modulate the immune system in pigs. For instance, dietary inclusion of corn DDGS promoted differentiation of goblet cells that secrete mucin (Saqui-Salces et al. 2017), whereas supplementation of corn DDGS-based diets with fiber degrading enzymes improved intestinal barrier function in weaned pigs (Li et al. 2018). An increase in visceral organ weight results in increased dietary energy requirement for maintenance at the expense of skeletal tissue deposition due to increased energy expenditure in the visceral organs. Also, immune response results in the increased dietary requirement of energy and nutrients such as AA for maintenance at the expense of skeletal tissue deposition.

The effects of dietary inclusion of DDGS on growth performance and carcass traits of pigs have been reported. Woyengo et al. (2014) after reviewing results from several studies on the effects of dietary inclusion of DDGS on growth performance and carcass traits of pigs concluded that the level of DDGS could be increased up to 30% in diets for finishing pigs without compromising growth performance. However, the growth performance of nursery pigs is reduced by the dietary inclusion of DDGS (Avelar et al. 2010). Also, the dressing percentage of finishing pigs was reduced by dietary inclusion of DDGS at 30%. Recently, increasing dietary inclusion levels of high protein DDGS from 0 to 30% linearly reduced the average daily gain of nursery pigs by 19% (Yang et al. 2019). With regard to grow-finish pigs, inclusion of 30% of corn DDGS in corn-based diets (formulated to similar NE value and SID AA content) for pigs from 29 to 120 kg body weight reduced final body weight by 1.8 kg and carcass yield by 1.9% but did not affect backfat depth (Wu et al. 2016). Also, inclusion of 30% of corn DDGS in corn-based diets (formulated to similar NE value and SID AA content) for pigs from 106 to 125 kg body weight reduced final body weight by 1.0 kg and carcass yield by 0.4% but did not affect backfat depth (Coble et al. 2017). The reduced growth performance and carcass yield of pigs can partly be attributed to increased expenditure of energy and nutrients in the gastrointestinal tract and the high content of leucine in DDGS. Excess dietary leucine has an antagonist effect on isoleucine and valine, and hence the very high content of leucine in corn DDGS can result in reduced bioavailability of isoleucine and valine in pigs (Yang et al. 2019), leading to reduced growth performance and carcass yield. However, the nutrient utilization of DDGS can be enhanced by the use of some exogenous enzymes, thereby minimizing the negative effects of DDGS inclusion in swine diets (Tiwari et al. 2018).

In summary, the inclusion of large amount of DDGS in diets for pigs results in reduced growth performance, likely due to its higher fiber content that can increase the dietary requirement of energy and AA for maintenance, and its very high leucine content that can antagonize isoleucine and valine metabolism.

*Wheat Milling Coproducts*
Wheat bran and wheat millruns are coproducts from the wheat milling industry to obtain wheat flour mainly for the human food industry. Wheat consists of endosperm that is rich in starch; the germ that is rich in protein, fat, minerals, and vitamins; aleurone layer that are rich in fiber and protein; and seed coat that is rich in fiber (Tervila-Wilo et al. 1996; Bach Knudsen 2004). The endosperm is located at the center of the grain, and it constitutes a high proportion of the seed (Tervila-Wilo et al. 1996; Bach Knudsen 2004). The germ is located at the inner and lower parts of the grain (Bach Knudsen 2004), whereas the aleurone layer and seed coat are the outer parts of the grain (Tervila-Wilo et al. 1996; Bach Knudsen 2004). During flour milling, the starchy endosperm is separated from the other parts of grains. The isolated starch endosperm is then processed into the flour, whereas the other remaining parts of the wheat grain are processed into various coproducts that are used mainly in the livestock feed industry. The most important wheat milling coproducts

used to formulate swine diets include wheat bran and wheat middlings (Jha et al. 2012). Wheat bran is composed of the seed coat, aleurone layer, germ, and small amounts of starchy endosperm, whereas wheat middlings is composed of fine particles of bran, germ, and starchy flour (Rosenfelder et al. 2013). Thus, both wheat bran and wheat middlings have a greater content of CP, fiber, minerals, and fat and lower content of starch than wheat (Table 11.4). However, wheat bran has lower starch content and higher fiber content than wheat middlings due to the presence of some flour in the wheat middlings. Also, wheat bran has higher total dietary fiber content than corn DDGS; wheat middlings and corn DDGS have similar total dietary fiber content.

A high proportion (>90%) of total dietary fiber in wheat milling coproducts is insoluble (Table 11.4). However, wheat coproducts have a slightly higher soluble fiber content than corn DDGS (Table 11.4). Arabinoxylans constitute a high proportion of total NSP in wheat grain (Bach Knudsen 2004), and hence they are the major NSP present in wheat coproducts. Arabinoxylans are composed of xylan backbones with branches that are composed of arabinose units (Bach Knudsen 2004). The ratio of arabinose to xylose in arabinoxylans of wheat and its coproducts is lower than that of corn and its coproducts. For instance, the ratio of arabinose to xylose in arabinoxylans of wheat was 0.62, whereas that for corn was 0.74. (Bach Knudsen 2004). The digestibility of arabinoxylans is partly dependent on the ratio of arabinose to xylose in arabinoxilans because the arabinose cross-link xylans with lignin, leading to increased resistance of arabinoxylans to enzymatic degradation (Huisman et al. 2000; Appeldoorn et al. 2010). Thus, the digestibility of NSP in wheat and its coproducts is greater than that of corn and its coproducts. For instance, wheat middlings had greater apparent ileal digestibility (AID; 46.6 vs. 1.5%) and ATTD (72.4 vs. 57.2%) of NSP than regular corn DDGS (Jaworski and Stein 2017). Also, the AID of GE for wheat middlings was greater than that for regular corn DDGS (42.4 vs. 29.2%); however, the two feedstuffs did not differ in ATTD of GE (66.3 vs. 64.8%; Jaworski and Stein 2017). The NE value for wheat middlings is greater than that for wheat bran (Table 11.4) likely because of the high fiber and lower starch content in the latter than in the former. The NE value for wheat middlings is lower than that for regular corn DDGS (Table 11.1), likely due to the lower fat content and hence lower GE content for wheat middlings than for regular corn DDGS because wheat middlings has a lower content of CP and greater content of starch than regular corn DDGS (Table 11.4), and both wheat middlings and regular corn DDGS are similar in ATTD of GE. However, the NE value of wheat middlings that is estimated based on its DE value and macronutrient composition might be lower than its actual NE value because of higher the ileal digestibility of NSP and hence GE for the wheat middlings. The energy that is digested in the small intestine is utilized more efficiently than the energy that is digested in the hindgut.

The effects of dietary inclusion of wheat coproducts on growth performance and carcass traits of pigs have been reported. The increasing inclusion level of wheat middlings from 0 to 20% in wheat-based diets formulated to similar NE and SID AA content did not affect the growth performance of nursery pigs (Garcia et al. 2015). However, inclusion of 30% wheat middlings in wheat-based diets (formulated to similar ME value and SID AA content) for pigs from 85 to 125 kg body weight reduced final body weight by 5.2 kg and hot carcass weight by 5.6 kg, increased full visceral weight by 0.8 kg; but did not affect dressing percentage (Stewart et al. 2013). Thus, there is a need to establish if higher dietary levels (>20%) of wheat middlings can affect the growth performance of nursery pigs. Also, there is a need to identify the optimal dietary level of wheat coproducts in diets for grow-finish pigs with regard to growth performance and carcass traits.

*Rice Bran*

Rice bran is a coproduct of the rice milling industry. The structure of whole rice grain (after the removal of hulls) is generally similar to that of wheat grain. During rice milling, the seed coat, aleurone layer, and germ are isolated from the whole rice grain to obtain brown rice. Sometimes the brown outer layer of the brown rice is removed by a process known as rice polishing, which gives rise to white rice that is rich in starch and rice polish. Rice bran is the major coproduct of rice milling, and it consists of seed coat, aleurone layer, germ, and small amounts of starchy endosperm. Thus, rice bran, like wheat barn, has a relatively high content of CP, fiber, and fat (Table 11.1). Rice bran can be defatted or not. Full-fat rice bran has a fat content that is greater than that in wheat coproducts and corn-DDGS (Table 11.4). Defatted rice bran has lower fat content than wheat coproducts or corn DDGS. Arabinoxylans are the major NSP in rice bran (Casas et al. 2019). However, the arabinose to xylose ratio in rice bran arabinoxylans (0.98; Casas et al. 2019) is greater than that in wheat coproducts or corn DDGS, implying that rice bran NSP are more resistant to digestion by fiber degrading enzymes. However, the ATTD of GE for full-fat rice bran (75.6%, Casas and Stein 2016b) and defatted rice bran (75.6–83.2%) (Casas and Stein 2016b; Huang et al. 2018) was greater than the values for wheat middlings and regular corn DDGS. It will be interesting to establish if the greater ATTD of GE for rice bran is due to greater GE digestibility in the small intestine, or hindgut, or both.

The NE value for full-fat rice bran and defatted rice barn for pigs have not been reported. However, based on the CP, starch, EE, and ADF values (Table 11.1) and the fore-mentioned ATTD of GE values of the rice brans, the NE value for full-fat rice bran is greater than that for wheat bran, wheat middlings, and corn DDGS (Table 11.1), whereas that for defatted rice bran is similar to that of wheat middlings and corn DDGS, but greater than that for wheat bran. An increase in dietary inclusion of full fat or defatted rice bran in diets from 0 to 30% through partial replacement of corn and SBM resulted in a quadratic decrease in the average daily gain of for nursery pigs such that an increase in the dietary level of full fat or defatted rice bran from 0 to 10% resulted in an increase in the average daily gain, whereas a further increase in the dietary level of these feedstuffs to 30% resulted in a decrease in the average daily gain (Casas and Stein 2016a). In their (Casas and Stein 2016a) study, the average daily gain of the pigs fed diet with 20% rice bran did not differ from that of pigs fed a diet with 0% rice bran. Thus, up to 20% of defatted or full-fat rice bran could be included in diets for nursery pigs to partially replace corn and SBM without compromising performance. Information is lacking on the effect of dietary inclusion of rice bran on growth performance and carcass traits of grow-finish pigs.

*Sugar Beet Pulp*
Sugar beet pulp is a coproduct from sugar beet processing plants to obtain sugar. It has a higher content of fiber and lower content of starch, amino acids, and fat than corn DDGS, wheat bran, wheat middlings, and rice bran (Table 11.4). One of the major NSP in sugar beet pulp is pectin (Serena and Bach Knudsen 2007). Pectin is more soluble than most NSP (Serena and Bach Knudsen 2007; Slama et al. 2018). Hence, NSP in sugar beet pulp is more soluble than those in the aforementioned cereal coproducts (Table 11.4). The AID of NSP values (11.5 and 22.9%) for sugar beet pulp were lower than the values for wheat middlings (38.8 and 50.1%; Navarro et al. 2019). However, the ATTD of NSP for (88.7 and 91.4%) sugar beet pulp values were greater than the values for wheat middlings (56.9 and 62.5%; Navarro et al. 2019) likely because of the higher content of soluble NSP (that is highly fermentable) in the sugar beet pulp. The NE value for sugar beet pulp is lower than that of rice bran, wheat bran, wheat middlings, and corn DDGS (Table 11.4) because of the lower content of fat and starch and higher content of fiber in sugar beet pulp than in the other afore-mentioned feedstuffs. Inclusion of sugar beet pulp in diets for nursery pigs at 6% did not affect the growth performance of the pigs; however, the inclusion of sugar beet pulp in diets for nursery pigs at 12, 18, or 24% reduced the growth performance of the pigs (Wang et al. 2016).

*Fat*

The major fats that are included in swine diets include soybean oil, tallow, and choice white grease. All these fats have approximately three times higher content of NE than cereal grains. Thus, they are included in diets for swine to increase dietary energy density and hence feed efficiency. Also, heat increment for fat is lower than that of other energy sources in swine diets, such as starch, NSP, and AA, and hence fat can be included in diets for pigs housed in environments with high temperatures to alleviate the heat stress. Soybean oil has a relatively high content of unsaturated fatty acids, whereas tallow and choice white grease have a relatively high content of saturated fatty acids (NRC 2012). Digestibility of fat increases with an increase in the proportion of unsaturated fatty acids in the fat (Su et al. 2015). Also, unsaturated fatty acids have been shown to have an extra-caloric effect because they can increase amino acid digestibility probably by slowing down the flow of digesta in the gastrointestinal tract (Cervantes-Pahm and Stein 2008; Woyengo et al. 2016). Thus, the addition of soybean oil in diets for pigs can result in increased nutrient digestibility. Indeed, the inclusion of soybean oil in SBM-based diet for growing pigs at 7.55% increased the standardized ileal digestibility of indispensable AAs by a mean of 2.9% (Cervantes-Pahm and Stein 2008). Also, Li and Sauer (1994) included canola oil in diets for growing pigs at 3.2, 6.2, 9.2, and 12.2% and observed a linear increase in AID of AAs with an increase in the dietary level of canola oil. However, the NE value for soybean oil was lower than that for choice white grease (4679 vs. 5900 kcal/kg), and this lower NE for soybean oil was attributed to the high heat increment for unsaturated fatty acids than for saturated fatty acids (Kil et al. 2011). Thus, the NE value of soybean oil can be lower than that for choice white grease or tallow.

Inclusion of tallow or soybean oil in diets for grow-finish pigs (28–114 kg body weight) at 5% did not affect growth performance; however, soybean oil increased the iodine value of longissimus muscle from 58.7 to 65.1 g/100 g (Apple et al. 2009). Inclusion of tallow or soybean oil in diets for grow-finish pigs (46–130 kg body weight) at 4% increased growth performance. In their (Stephenson et al. 2016) study, however, soybean oil increased the jowl iodine value from 67.8–75.9 g/100 g, which was alleviated by the withdrawal of soybean oil from diet or replacement of soybean oil with tallow 42 days prior to slaughter. Dietary inclusion of soybean oil or choice white grease in diets for grow-finish pigs (44–125 kg body weight) at 5% increased feed efficiency; also, dietary soybean oil increased average daily gain (Benz et al. 2011). However, the jowl fat iodine value was increased from 67.1 to 71.5 g/100 g by the dietary choice white grease and to 82.0 g/100 g by the dietary soybean oil (Benz et al. 2011). The jowl fat iodine value for pigs fed a diet with soybean oil was reduced from 82.0 to 73.3 g/100 g by removal of soybean oil from the diet 58 days prior to slaughter (Benz et al. 2011). Since the acceptable upper limit of jowl fat iodine value is 73 g/100 g (Benz et al. 2011), tallow or choice white grease can be included in diets for grow-finish pigs up to around 5% to improve growth performance without compromising carcass iodine value. However, soybean oil should be removed from diets of grow-finish pigs or be replaced with fat rich in saturated fatty acids (such as tallow and choice white grease) several days before slaughter weight to avoid the negative effects of unsaturated fatty acids on carcass quality. The actual number of days by which soybean oil should be removed from the diets or be replaced by fat rich in saturated fatty acids needs to be established.

## Minor Feedstuffs

*Crude Glycerol*

Crude glycerol, a sugar alcohol, is a coproduct of oil biodiesel industry. Ingested glycerol can be absorbed in the small intestine and be used to generate energy via glycolysis and tricarboxylic acid cycle or be used for the synthesis of glucose (Tao et al. 1983). Glycerol, like any other carbohydrate, can be fermented in the gastrointestinal tract by microorganisms to generate VFA that serve as energy for animals (Avila-Stagno et al. 2014). Thus, glycerol serves as a source of energy for pigs. Crude glycerol (87% glycerol) contained 3344 kcal/kg of DE and 3207 kcal/kg of ME (Lammers et al. 2008), which are close to the DE and ME values that were reported by NRC (2012) for corn (3451 and 3395 kcal/kg, respectively) and wheat 3313 and 3215 kcal/kg, respectively). Partial replacement of wheat with glycerol at 0, 4, or 8% in diets for nursery pigs (9–22 kg) resulted in a linear increase in average daily gain by 7.8% (Zijlstra et al. 2009). Also, the inclusion of glycerol at 0, 3, or 6% in diets for nursery pigs through a reduction in corn content and a slight increase in soybean content of the diets resulted in a linear increase in average daily gain of nursery pigs (11–27 kg) by 7.4% (Groesbeck et al. 2008). Thus, it appears that crude glycerol is a more efficient source of energy than corn or wheat and can be included up to 8% in diets for nursery pigs to improve the growth performance.

*Starch Fractions*
Starch is the major source of energy in practical swine diets, and it is included as a component of the feedstuffs. Starch concentrates may be produced by dry or wet fractionation. Starch concentrates that were produced from field pea and zero tannin faba bean by air classification had NE values (2610 and 2680 kcal/kg, respectively; Gunawardena et al. 2010) that were close to that of corn grain (2672; NRC 2012). Pulses such as field pea and zero tannins have antinutritional factors that limit nutrient utilization in pigs (Woyengo et al. 2017). Hence, field pea and zero tannin faba bean starch concentrates that have negligible amounts of antinutritional factors can be a good source of energy in diets for pigs, especially for nursery pigs that have poorly developed digestive system. Also, cooked rice was more digestible than cooked corn (Parera et al. 2010), implying that cooked rice can be a good source of energy in nursery pig diets. In addition to these highly digestible starch products, resistant starch (which is starch that escapes the enzymatic digestion in the small intestine but is highly fermented in the hindgut) can be included in diets for nursery pigs to improve gut health. Some of the sources of resistant starch that improved gut health when they were included in diets for nursery pigs include raw potato starch (Bhandari et al. 2009), high amylose starch (Fouhse et al. 2015), and enzymatically modified waxy cornstarch (Newman et al. 2016). However, it should be noted that resistant starch is not a good source of energy because VFA, which are the end product of carbohydrate digestion in the hindgut, are less efficient sources of energy than simple sugars (monosaccharides) that are the end products of carbohydrate digestion in the small intestine. Also, resistant starch can reduce nutrient digestibility in the small intestine (De Schrijver et al. 1999) and reduce recalcitrant fiber fermentation in the hindgut (de Vries et al. 2016). Thus, there is a need to identify the optimal level of inclusion of resistant starch in diets for nursery pigs.

*Fiber Fractions*
Various cereal grains, pulses, and oilseeds are dehulled to reduce their fiber content before their inclusion in swine diets. Thus, hulls from these fore-mentioned feedstuffs are available for livestock feeding. Inclusion of hulls in diets for sows that have restricted access to feed caused satiety and reduced stereotypic behavior (Holt et al. 2006). However, hulls are bulky and have a high content of insoluble fiber that is poorly fermented by pigs and reduce the ileal digestibility of other nutrients. For instance, the ATTD of GE (46.9%) for soybean hulls by growing pigs was lower than that for corn DDGS (64.8%) and wheat middlings (66.3%; Jaworski and Stein 2017). Inclusion of oat hulls in corn-SBM-based diets for growing pigs at 10% reduced AID of fat (Ndou et al. 2019). Thus, the inclusion of hulls in diets for weaned and grower-finisher pigs should be limited. In addition to hulls, fiber fractions, including

β-glucans and arabinoxylans that are extracted from cereal grains, can be included in pig diets to improve gut health. For example, the inclusion of arabinoxylan rich fraction from wheat in diets for growing pigs increased fermentation in the hindgut and lowered colonocyte DNA damage (Belobrajdic et al. 2012).

# References

Abelilla, J. J., and H. H. Stein. 2019. Degradation of dietary fiber in the stomach, small intestine, and large intestine of growing pigs fed corn- or wheat-based diets without or with microbial xylanase. J. Anim. Sci. 97:338–352.

Acosta, J. A., R. D. Boyd, and J. F. Patience. 2017. Digestion and nitrogen balance using swine diets containing increasing proportions of coproduct ingredients and formulated using the net energy system. J. Anim. Sci. 95:1243–1252.

Agyekum, A. K., B. A. Slominski, and C. M. Nyachoti. 2012. Organ weight, intestinal morphology, and fasting whole-body oxygen consumption in growing pigs fed diets containing distillers dried grains with solubles alone or in combination with a multienzyme supplement. J. Anim. Sci. 90:3032–3040.

Allen, T. 2002. The world supply of fall (winter) rye. In: D. B. Fowler, editor, In winter cereal production. Univ. of Saskatchewan, Saskatoon, SK, Canada. http://www.usask.ca/agriculture/plantsci/winter_cereals/winter_rye/production1.

Alston, J. M., J. M. Beddow, and P. G. Pardey. 2009. Agricultural research, productivity, and food prices in the long run. Science 325:1209–1210.

Anguita, M., N. Canibe, J. F. Perez, and B. B. Jensen. 2006. Influence of the amount of dietary fiber on the available energy from hindgut fermentation in growing pigs: use of cannulated pigs and in vitro fermentation. J. Anim. Sci. 84:2766–2778.

Appeldoorn, M. M., M. A. Kabel, D. Van Eylen, H. Gruppen, and H. A. Schols. 2010. Characterization of oligomeric xylan structures from corn fiber resistant to pretreatment and simultaneous saccharification and fermentation. J. Agric. Food Chem. 58:11294–1130.

Apple, J. K., C. V. Maxwell, D. L. Galloway, S. Hutchison, and C. R. Hamilton. 2009. Interactive effects of dietary fat source and slaughter weight in growing-finishing swine: I. Growth performance and longissimus muscle fatty acid composition. J. Anim. Sci. 87:1407–1422.

Avalos, F., 2014. Do oil prices drive food prices? The tale of a structural break. J. Int. Money Financ. 42, 253–271.

Avelar, E., R. Jha, E. Beltranena, M. Cervantes, A. Morales, and R. T. Zijlstra, 2010. The effect of feeding wheat distillers dried grain with solubles on growth performance and nutrient digestibility in weaned pigs. Anim. Feed Sci. Technol. 160:73–77.

Avila-Stagno, J., A. V. Chaves, G. O. Ribeiro Jr., E. M. Ungerfeld, and T. A. McAllister. 2014. Inclusion of glycerol in forage diets increases methane production in a rumen simulation technique system. Br. J. Nutr. 111:829–835.

Bach Knudsen, K. E. 2004. Fiber and nonstarch polysaccharide content and variation in common crops used in broiler diets. Poult. Sci. 93: 2380–2393.

Belobrajdic, D. P., A. R. Bird, M. A. Conlon, B. A. Williams, S. Kang, C. S. McSweeney, D. Zhang, W. L. Bryden, M. J. Gidley, and D. L. Topping. 2012. An arabinoxylan-rich fraction from wheat enhances caecal fermentation and protects colonocyte DNA against diet-induced damage in pigs. Br. J. Nutr. 107:1274–1282.

Beltranena, E., D. F. Salmon, L. A. Goonewardene, and R. T. Zijlstra. 2008. Triticale as a replacement for wheat in diets for weaned pigs. Can. J. Anim. Sci. 88:631–635.

Benz, J. M., M. D. Tokach, S. S. Dritz, J. L. Nelssen, J. M. DeRouchey, R. C. Sulabo, and R. D. Goodband. 2011. Effects of choice white grease and soybean oil on growth performance, carcass characteristics, and carcass fat quality of growing-finishing pigs. J. Anim. Sci. 89:404–413.

Berrocoso, J. D., D. Menoyo, P. Guzmán, B. Saldaña, L. Cámara, and G. G. Mateos. 2015. Effects of fiber inclusion on growth performance and nutrient digestibility of piglets reared under optimal or poor hygienic conditions. J. Anim. Sci. 93:3919–3931.

Bhandari, S. K., C. M. Nyachoti, and D. O. Krause. 2009. Raw potato starch in weaned pig diets and its influence on postweaning scours and the molecular microbial ecology of the digestive tract. J. Anim. Sci. 87:984–993.

Bird, A. R., M. Jackson, R. A. King, D. A. Davies, S. Usher, and D. L. Topping. 2004. A novel high-amylose barley cultivar (*Hordeum vulgare* var. Himalaya 292) lowers plasma cholesterol and alters indices of large-bowel fermentation in pigs. Br. J. Nutr. 92:607–615.

Black, J. L. 1995. Modelling energy metabolism in the pig- critical evaluation of a simple reference model. In: P. J. Moughan, M. W. A. Verstegen, and M. Visser-Reyneveld, editrs, Modelling Growth in the Pig. Wageningen Press, Wageningen, The Netherlands. p. 87–102.

Boisen, S., and J. A. Fernández. 1997. Prediction of the total tract digestibility of energy in feedstuffs and pig diets by in vitro analyses. Anim. Feed Sci. Technol. 68:277–286.

Bolarinwa, O. A., and O. Adeola. 2016. Regression and direct methods do not give different estimates of digestible and metabolizable energy values of barley, sorghum, and wheat for pigs. J. Anim. Sci. 94:610–618.

Böttger, C., and K.-H. Südekum. 2018. Review: Protein value of distillers dried grains with solubles (DDGS) in animal nutrition as affected by the ethanol production process. Anim. Feed Sci. Technol. 244:11–17.

Bushuk, W. 2001. Rye production and uses worldwide. Cereal Food World 46:70–73.

Cadogan, D. J., M. Choct, and R. G. Campbell. 2003. Effects of storage time and exogenous xylanase supplementation of new season wheats on the performance of young male pigs. Can. J. Anim. Sci. 83:105–112.

Casas, G. A., and H. H. Stein. 2016a. Effects of full fat or defatted rice bran on growth performance and blood characteristics of weanling pigs. J. Anim. Sci. 94:4179–4187.

Casas, G. A., and H. H. Stein. 2016b. Effects of microbial xylanase on digestibility of dry matter, organic matter, neutral detergent fiber, and energy and the concentrations of digestible and metabolizable energy in rice coproducts fed to weanling pigs. J. Anim. Sci. 94:1933–193.

Casas, G. A., and H. H. Stein. 2017. The ileal digestibility of most amino acids is greater in red dog than in wheat middlings when fed to growing pigs. J. Anim. Sci. 95:2718–2725.

Casas, G. A., D. A. Rodriguez, and H. H. Stein. 2018. Nutrient composition and digestibility of energy and nutrients in wheat middlings and red dog fed to growing pigs. J. Anim. Sci. 96:215–224.

Casas, G. A., N. H. Lærke, K. E. Bach Knudsen, and H. H. Stein. 2019. Arabinoxylan is the main polysaccharide in fiber from rice coproducts, and increased concentration of fiber decreases in vitro digestibility of dry matter. Anim. Feed Sci. Technol. 247:255–261.

Cervantes-Pahm, S. K., and H. H. Stein. 2008. Effect of dietary soybean oil and soybean protein concentration on the concentration of digestible amino acids in soybean products fed to growing pigs. J. Anim. Sci. 86:1841–1849.

Cervantes-Pahm, S. K., Y. Liu, and H. H. Stein. 2014a. Comparative digestibility of energy and nutrients and fermentability of dietary fiber in eight cereal grains fed to pigs. J. Sci. Food. Agric. 94:841–849.

Cervantes-Pahm, S. K., Y. Liu, and H. H. Stein. 2014b. Digestible indispensable amino acid score (DIAAS) and digestible amino acids in eight cereal grains. Br. J. Nutr. 111:1663–1672.

Chen, D. 2009. Nutrition and feed strategies for sustainable swine production in China. Front. Agric. China 3:471–477.

Christensen, D. N., K. E. Bach Knudsen, J. Wolstrup, and B. B. Jensen. 1999. Integration of ileum cannulated pigs and in vitro fermentation to quantify the effect of diet composition on the amount of short-chain fatty acids available from fermentation in the large intestine. J. Sci. Food Agric. 79:755–762.

Coble, K. F., J. M. DeRouchey, M. D. Tokach, S. S. Dritz, R. D. Goodband, and J. C. Woodworth. 2017. Effects of distillers dried grains with solubles and added fat fed immediately before slaughter on growth performance and carcass characteristics of finishing pigs. J. Anim. Sci. 95:270–278.

CVB. 1993. Net energy of feedstuffs for swine. CVB Report No. 7. CVB, Lelystad, The Netherlands.

Davis-Knight, H. R., and R. M. Weightman. 2008. The potential of triticale as a low input cereal for bioethanol production. Project Report 434. Home-Grown Cereals Authority, Kenilworth, UK.

De Goey, L. W. and R. C. Ewan. 1975. Effect of level of intake and diet dilution on energy metabolism in the young pig. J. Anim. Sci. 40(6): 1045–1051.

DeThomas, F. A., and P. J. Brimmer. 2002. Monochromators for near-infrared spectroscopy. In: J. M. Chalmers and P. R. Griffiths, editors, Handbook of vibrational theory. Vol. 1. John Wiley & Sons Ltd., Chichester, UK. p. 383–392.

De Schrijver, R., K. Vanhoof, and J. Van de Ginste. 1999. Nutrient utilization in rats and pigs fed enzyme resistant starch. Nutr. Res. 19:1349–1361.

de Vries, S., W. J. J. Gerrits, M. A. Kabel, T. Vasanthan, and R. T. Zijlstra. 2016. β-Glucans and resistant starch alter the fermentation of recalcitrant fibers in growing pigs. PLOS One 11:e0167624.

Dierick N. A., I. J, Vervaeke, D. I. Demeyer, and J. A. Decuypere. 1989. Approach to the energetic importance of fibre digestion in pigs. I. Importance of fermentation in the overall energy supply. Anim. Feed Sci. Technol. 23:141–167.

Eklund, M., M. Rademacher, W. C. Sauer, R. Blank, and R. Mosenthin. 2014. Standardized ileal digestibility of amino acids in alfalfa meal, sugar beet pulp, and wheat bran compared to wheat and protein ingredients for growing pigs. J. Anim. Sci.92:1037–1043.

Espinosa, C. D., and H. H. Stein. 2018. High-protein distillers dried grains with solubles produced using a novel front-end– back-end fractionation technology has greater nutritional value than conventional distillers dried grains with solubles when fed to growing pigs. J. Anim. Sci. 96:1869–1876.

Fairbairn, S. L., J. F. Patience, H. L. Classen, and R. T. Zijlstra. 1999. The energy content of barley fed to growing pigs: characterizing the nature of its variability and developing prediction equations for its estimation. J. Anim. Sci. 77:1502–1512.

FEFAC. 2005. Feed and food statistical yearbook 2005. Euro. Feed Manufac. Fed., Brussels, Belgium.

Fouhse, J. M., M. G. Ganzle, P. R. Regmi, T. A. van Kempen, and R. T. Zijlstra. 2015. High amylose starch with low in vitro digestibility stimulates hindgut fermentation and has a bifidogenic effect in weaned pigs. J. Nutr. 145:2464–2470.

Garcia, H., L. F. Wang, J. L. Landero, E. Beltranena, M. Cervantes, A. Morales, and R. T. Zijlstra. 2015. Effect of feeding wheat millrun on diet nutrient digestibility and growth performance in starter pigs. Anim. Feed Sci. Technol. 207:283–288.

Groesbeck, C. N., L. J. McKinney, J. M. DeRouchey, M. D. Tokach, R. D. Goodband, S. S. Dritz, J. L. Nelssen, A. W. Duttlinger, A. C. Fahrenholz, and K. C. Behnke. 2008. Effect of crude glycerol on pellet mill production and nursery pig growth performance. J. Anim. Sci. 86:2228–2236.

Gunawardena, C. K., R. T. Zijlstra, and E. Beltranena. 2010. Characterization of the nutritional value of air-classified protein and starch fractions of field pea and zero-tannin faba bean in grower pigs. J. Anim. Sci. 88:660–670.

Hale, O. M., and P. R. Utley. 1985. Value of Beagle 82 triticale as a substitute for corn and soybean meal in the diet of pigs. J. Anim. Sci. 60:1272–1279.

Hernot, D. C., T. W. Boileau, L. L Bauer, K. S. Swanson, and G. C. Fahey. 2008. In vitro digestion characteristics of unprocessed and processed whole grains and their components. J. Agric. Food Chem. 56:10721–10726.

Hill, B. E., A. L. Sutton, and B. T. Richert. 2009. Effects of low-phytic acid corn, low-phytic acid soybean meal, and phytase on nutrient digestibility and excretion in growing pigs. J. Anim. Sci. 87:1518–1527.

Holt, J. P., L. J. Johnston, S. K. Baidoo, and G. C. Shurson. 2006. Effects of a high-fiber diet and frequent feeding on behavior, reproductive performance, and nutrient digestibility in gestating sows. J. Anim. Sci. 84:946–955.

Htoo, J. K., W. C. Sauer, Y. Zhang, M. Cervantes, S. F. Liao, B. A. Araiza, A. Morales, and N. Torrentera. 2007a. The effect of feeding low-phytate barley-soybean meal diets differing in protein content to growing pigs on the excretion of phosphorus and nitrogen. J. Anim. Sci. 85:700–705.

Htoo, J. K., W. C. Sauer, J. L. Yanez, M. Cervantes, Y. Zhang, J. H. Helm, and R. T. Zijlstra. 2007b. Effect of low-phytate barley or phytase supplementation to a barley-soybean meal diet on phosphorus retention and excretion by grower pigs. J. Anim. Sci. 85:2941–2948.

Huang, B., C. Huang, Z. Lyu, Y. Chen, P. Li, L. Liu, and C. Lai. 2018. Available energy and amino acid digestibility of defatted rice bran fed to growing pigs. J. Anim. Sci. 96:3138–3150.

Huisman, M. M. H., H. A. Schols, and A.G. J. Voragen, 2000. Glucuronoarabinoxylans from maize kernel cell walls are more complex than those from sorghum kernel cell walls. Carbohydr. Polym. 43:269–279.

Hyun, Y., G. E. Bressner, M. Ellis, A. J. Lewis, R. Fischer, E. P. Stanisiewski, and G. F. Hartnell. 2004. Performance of growing–finishing pigs fed diets containing Roundup Ready corn (event nk603), a nontransgenic genetically similar corn, or conventional corn lines. J. Anim. Sci. 82:571–580.

Hyun, Y., G. E. Bressner, R. L. Fischer, P. S. Miller, M. Ellis, B. A. Peterson, E. P. Stanisiewski, and G. F. Hartnell.2005. Performance of growing-finishing pigs fed diets containing YieldGard Rootworm corn (MON 863), a nontransgenic genetically similar corn, or conventional corn hybrids. J. Anim. Sci. 83:1581–1590.

Jaworski, N. W., and H. H. Stein. 2017. Disappearance of nutrients and energy in the stomach and small intestine, cecum, and colon of pigs fed corn-soybean meal diets containing distillers dried grains with solubles, wheat middlings, or soybean hulls. J. Anim. Sci. 95:727–739.

Jaworski, N. W., D. W. Liu, D. F. Li, and H. H. Stein. 2016. Wheat bran reduces concentrations of digestible, metabolizable, and net energy in diets fed to pigs, but energy values in wheat bran determined by the difference procedure are not different from values estimated from a linear regression procedure. J. Anim. Sci. 94:3012–3021.

Jaworski, N.W., H. N. Lærke, K. E. Bach Knudsen, and H. H. Stein. 2015. Carbohydrate composition and in vitro digestibility of dry matter and non-starch polysaccharides in corn, sorghum, and wheat and co-products from these grains. J. Anim. Sci. 93:1103–1113.

Jha, R., A. Owusu-Asiedu, P. H. Simmins, A. Pharazyn, and R. T. Zijlstra. 2012. Degradation and fermentation characteristics of wheat coproducts from flour milling in the pig intestine studied in vitro. J. Anim. Sci. 90:173–175.

Jha, R., B. Rossnagel, R. Pieper, A. Van Kessel, and P. Leterme. 2010. Barley and oat cultivars with diverse carbohydrate composition alter ileal and total tract nutrient digestibility and fermentation metabolites in weaned piglets. Animal 4:724–731.

Jha, R., D. N. Overend, P. H. Simmins, D. Hickling, and R. T. Zijlstra. 2011a. Chemical characteristics, feed processing quality, voluntary feed intake, growth performance, and energy digestibility among wheat classes fed in pelleted diets fed to weaned pigs. Anim. Feed Sci. Technol. 170:78–90.

Jha, R., J. Bindelle, B. Rossnagel, A. Van Kessel, and P. Leterme. 2011b. In vitro evaluation of the fermentation characteristics of the carbohydrate fractions of hulless barley and other cereals in the gastrointestinal tract of pigs. Anim. Feed Sci. Technol. 163:185–193.

Jha, R., J. Bindelle, A. Van Kessel, and P. Leterme. 2011c. In vitro fibre fermentation of feed ingredients with varying fermentable carbohydrate and protein levels and protein synthesis by colonic bacteria isolated from pigs. Anim. Feed Sci. Technol. 165:191–200.

Jha, R., and J. D. Berrocoso. 2015. Review: Dietary fiber utilization and its effects on physiological functions and gut health of swine. Animal 9:1441–1452.

Jha, R., and J. F. D. Berrocoso. 2016. Dietary fiber and protein fermentation in the intestine of swine and their interactive effects on gut health and on the environment: A review. Anim. Feed Sci. Technol. 212:18–26.

Jha, R., J. K. Htoo, M. G. Young, E. Beltranena, and R. T. Zijlstra. 2013. Effect of increasing co-product inclusion and reducing dietary protein on growth performance, carcass characteristics, and jowl fatty acid profile of grower-finisher pigs. J. Anim. Sci. 91:2178–2191.

Jha, R., T. A. Woyengo, J. Li, M. R. Bedford, T. Vasanthan, and R. T. Zijlstra. 2015. Enzymes enhance degradation of the fiber-starch-protein matrix of distillers dried grains with solubles as revealed by a porcine in vitro fermentation model and microscopy. J. Anim. Sci. 93:1039–1051.

Jha, R., and P. Leterme. 2012. Feed ingredients differing in fermentable fibre and indigestible protein content affect fermentation metabolites and faecal nitrogen excretion in growing pigs. Animal. 6:603–611.

Jha, R., J. M. Fouhse, U. P. Tiwari, L. Li, B. P. Willing. 2019. Dietary fiber and intestinal health of monogastric animals. Front. Vet. Sci. 6:48.

Jürgens, H., G. Jansen, C. B. Wegener. 2012. Characterisation of several rye cultivars with respect to arabinoxylans and extract viscosity. J. Agri. Sci. 4:1–12.

Kil, D. Y., F. Ji, L. L. Stewart, R. B. Hinson, A. D. Beaulieu, G. L. Allee, J. F. Patience, J. E. Pettigrew, and H. H. Stein. 2011. Net energy of soybean oil and choice white grease in diets fed to growing and finishing pigs. J. Anim. Sci. 89:448–459.

Lammers, P. J., B. J. Kerr, T. E. Weber, W. A. Dozier III, M. T. Kidd, K. Bregendahl, and M. S. Honeyman. 2008. Digestible and metabolizable energy of crude glycerol for growing pigs. J. Anim. Sci. 86:602–608.

Li, Q., N. K. Gabler, C. L. Loving, S. A. Gould, and J. F. Patience. 2018. A dietary carbohydrase blend improved intestinal barrier function and growth rate in weaned pigs fed higher fiber diets. J. Anim. Sci. 96:5233–5243.

Li, S. and W. C. Sauer. 1994. The effects of dietary fat content on amino acid digestibility in young pigs. J. Anim. Sci. 72:1737–1743.

Lynch, M. B., I. Sweeney, I. J. Callan, I.V. O'Doherty. 2007. Effects of increasing the intake of dietary beta-glucans by exchanging wheat for barley on nutrient digestibility, nitrogen excretion, intestinal microflora, volatile fatty acid concentration and manure ammonia emissions in finishing pigs. Animal 1:812–819.

McBurney, M. I., and W. C. Sauer. 1993. Fiber and large bowel energy absorption: validation of the integrated ileostomy-fermentation model using pigs. J. Nutr. 123:721–727.

McLeod, J. G., W. H. Pfeiffer, R. M. DePauw, and J. M. Clarke. 2001. Registration of "AC Ultima" spring triticale. Crop Sci. 41:924–925.

Miller, P. R., B. G. McConkey, G. W. Clayton, S. A. Brandt, J. A. Staricka, A. M. Johnston, G. P. Lafond, et al. 2002. Pulse crop adaptation in the Northern Great Plains. Agron. J. 94:261–272.

Murray, J. M.D., A. Longland, P. M. Hastie, M. Moore-Colyer, and C. Dunnett. 2008. The nutritive value of sugar beet pulp-substituted lucerne for equids. Anim. Feed Sci. Technol. 140:110–124.

Myer, R. O., and J. H. Brendemuhl. 2001. Miscellaneous feedstuffs. In: A. J. Lewis, and L. L. Southern, editors, Swine nutrition. 2nd ed. CRC Press, Boca Raton, FL. p. 839–864.

Navarro, D. M. D. L., E. M. A. M. Bruininx, L. de Jong, and H. H. Stein. 2019. Effects of inclusion rate of high fiber dietary ingredients on apparent ileal, hindgut, and total tract digestibility of dry matter and nutrients in ingredients fed to growing pigs. Anim. Feed Sci. Technol. 248:1–9.

Navarro, D. M. D. L., E. M. A. M. Bruininx, L. de Jong, and H. H. Stein. 2018. The contribution of digestible and metabolizable energy from high-fiber dietary ingredients is not affected by inclusion rate in mixed diets fed to growing pigs. J. Anim. Sci. 96:1860–1868.

Ndou, S. P., E. Kiarie, N. Ames, and C. M. Nyachoti. 2019. Flaxseed meal and oat hulls supplementation: impact on dietary fiber digestibility, and flows of fatty acids and bile acids in growing pigs. J. Anim. Sci. 97:291–301.

Newman, M. A., Q. Zebeli, K. Velde, D. Grull, T. Molnar, W. Kandler, and B. U. Metzler Zebeli. 2016. Enzymatically modified starch favorably modulated intestinal transit time and hindgut fermentation in growing pigs. PLOS One 11:e0167784. doi: https://doi.org/10.1371/journal.pone.0167784.

Nguyen, N., M. Jacobs, J. Li, C. Huang, D. Li, D. M. D. L. Navarro, H. H. Stein, and N. W. Jaworski. 2019. Technical note: concentrations of soluble, insoluble, and total dietary fiber in feed ingredients determined using Method AOAC 991.43 are not different from values determined using Method AOAC 2011.43 with the AnkomTDF Dietary Fiber Analyzer. J. Anim. Sci. 97:3972–3983.

Noblet, J., H. Fortune, C. Dupire, and S. Dubois. 1993. Digestible, metabolisable and net energy value of 13 feedstuffs for growing pigs: Effect of energy system. Anim. Feed Sci. Technol. 42:131–149.

Noblet, J., and J. M. Perez. 1993. Prediction of digestibility of nutrients and energy values of pig diets from chemical analysis. J. Anim. Sci. 71:3389–3398.

Noblet, J., H. Fortune, X. S. Shi, and S. Dubois. 1994. Prediction of net energy value of feeds for growing pigs. J. Anim. Sci. 72:344–354.

Nonhebel, S. 2004. On resource use in food production systems: The value of livestock as rest-stream upgrading system. Ecol. Econ. 48:221–230.

NRC. 2012. Nutrient requirements of swine. 11th rev. ed. Natl. Acad. Press, Washington, DC.

Nyannor, E. K. D., P. Williams, M. R. Bedford, and O. Adeola. 2007. Corn expressing an Escherichia coli-derived phytase gene: A proof-of-concept nutritional study in pigs. J. Anim. Sci. 85:1946–1952.

Opapeju, F. O., C. M. Nyachoti, J. D. House, H. Weiler, and H. D. Sapirstein. 2006. Growth performance and carcass characteristics of pigs fed short-season corn hybrids. J. Anim. Sci. 84:2779–2786.

Opapeju, F. O., C. M. Nyachoti, and J. D. House. 2007. Digestible energy, protein and amino acid content in selected short season corn cultivars fed to growing pigs. Can. J. Anim. Sci. 87:221–226.

Parera, N, R. P. Lazaro, M. P. Serrano, D. G. Valencia, G. G. Mateos. 2010. Influence of the inclusion of cooked cereals and pea starch in diets based on soy or pea protein concentrate on nutrient digestibility and performance of young pigs. J. Anim. Sci. 88:671–679.

Partanen, K., H. Siljander-Rasi, and T. Alaviuhkola. 2006. Feeding weaned piglets and growing-finishing pigs with diets based on mainly home-grown organic feedstuffs. Agric. Food Sci. 15:89–105.

Patience, J. F., P. Leterme, A. D. Beaulieu, and R. T. Zijlstra. 2007. Utilization in swine diets of distillers dried grains with solubles derived from corn or wheat used in ethanol production. In: J. Doppenberg, and P. van der Aar, editors, Biofuels: Implications for the feed industry. Wageningen Academic Press, Wageningen, The Netherlands. p. 89–102.

Pedersen, C., M. G. Boersma, and H. H. Stein. 2007. Energy and nutrient digestibility in NutriDense corn and other cereal grains fed to growing pigs. J. Anim. Sci. 85:2473–2483.

Pedersen, M. B., S. Dalsgaard, K. E. Bach Knudsen, S. Yu, and H. N. Lærke. 2014. Compositional profile and variation of distillers dried grains with solubles from various origins with focus on non-starch polysaccharides. Anim. Feed Sci. Technol. 197:130–141.

Pieper, R., R. Jha, B. Rossnagel, A. Van Kessel, W. B. Souffrant, and P. Leterme. 2008. Effect of barley and oat cultivars with different carbohydrate compositions on the intestinal bacterial communities in weaned piglets. FEMS Microbiol. Ecol. 66:556–566.

Pond, W. G., and X. G. Lei. 2001. Of pigs and people. In: A. J. Lewis, and L. L. Southern, editors, Swine nutrition. 2nd ed. CRC Press, Boca Raton, FL. p. 3–18.

Radecki, S. V. and E. R. Miller. 1990. Triticale. In: P. A. Thacker and R. N. Kirkwood, editors, Nontraditional feed sources for use in swine production. Butterworths, Stoneham, MA. p. 493–499.

Regmi, P. R., W. C. Sauer, and R. T. Zijlstra. 2008. Prediction of in vivo apparent total tract energy digestibility of barley in grower pigs using an in vitro digestibility technique. J. Anim. Sci. 86:2619–2626.

Regmi, P. R., N. S. Ferguson, and R. T. Zijlstra. 2009. In vitro digestibility techniques to predict apparent total tract energy digestibility of wheat in grower pigs. J. Anim. Sci. 87:3620–3629.

Rho, Y., C. Zhu, E. Kiarie, and C. F. M de Lange. 2017. Standardized ileal digestible amino acids and digestible energy contents in high-protein distiller's dried grains with solubles fed to growing pigs. J. Anim. Sci. 95:3591–3597.

Rojas, O. J., and H. H. Stein. 2015. Effects of reducing the particle size of corn grain on the concentration of digestible and metabolizable energy and on the digestibility of energy and nutrients in corn grain fed to growing pigs. Livest. Sci. 181:187–193.

Rosenfelder, R., M. Eklund, and R. Mosenthin. 2013. Nutritive value of wheat and wheat by-products in pig nutrition: A review. Anim. Feed Sci. Technol. 185:107–125.

Saqui-Salces, M., Z. Huang, M. Ferrandis Vila, J. Li, J. A. Mielke, P. E. Urriola, and G. C. Shurson. 2017. Modulation of intestinal cell differentiation in growing pigs is dependent on the fiber source in the diet. J. Anim. Sci. 95:1179–1190.

Salmon, D. F. 2004. Production of triticale on the Canadian Prairies. In: M. Mergoum and H. G´omez MacPherson, editors. Triticale Improvement and Production. Plant Production and Protection Paper 179. FAO, Rome, Italy. p 154.

Sauber, T. E., and F. N. Owens. 2001. Cereal grains and by-products for swine. In: A. J. Lewis, and L. L. Southern, editors, Swine nutrition. 2nd ed. CRC Press, Boca Raton, FL. p. 85–802.

Sauvant, D., J. M. Perez, and G. Tran. 2004. Tables of composition and nutritional value of feed materials: Pig, poultry, sheep, goats, rabbits, horses, and fish. Wageningen Academic Publishers, Wageningen, the Netherlands and INRA, Paris, France.

Schwarz, T., W. Kuleta, A. Turek, R. Tuz, J. Nowicki, B. Rudzki, and P. M. Bartlewski. 2014. Assessing the efficiency of using a modern hybrid rye cultivar forpig fattening with emphasis on production costs and carcass quality. Anim. Prod. Sci. 55:467–473.

Serena, A., and K. E. Bach Knudsen. 2007. Chemical and physicochemical characterisation of co-products from the vegetable food and agro industries. Anim. Feed Sci. Technol. 139:109–124.

Shurson, G. C. 2018. Yeast and yeast derivatives in feed additives and ingredients: sources, characteristics, animal responses, and quantification methods. Anim. Feed Sci. Technol. 235:60–76.

Slama, J., K. Schedle, G. K. Wurzer, and M. Gierus. 2018. Physicochemical properties to support fibre characterization in monogastric animal nutrition. J. Sci. Food Agric. 99:3895–3902.

Smits, C., and R. Sijtsma. 2007. A decision tree for co-product utilization. In: R. O. Ball, and R. T. Zijlstra, editors, Advances in pork production, proc. banff pork seminar. Vol. 18, Univ. of Alberta, Edmonton, Alberta, Canada. p. 213–221.

Spencer, J. D., G. L. Allee, and T. E. Sauber. 2000. Growing–finishing performance and carcass characteristics of pigs fed normal and genetically modified low-phytate corn. J. Anim. Sci. 78:1529–1536.

Stephenson, E. W., M. A. Vaughn, D. D. Burnett, C. B. Paulk, M. D. Tokach, S. S. Dritz, J. M. DeRouchey, R. D. Goodband, J. C. Woodworth, and J. M. Gonzalez. 2016. Influence of dietary fat source and feeding duration on finishing pig growth performance, carcass composition, and fat quality. J. Anim. Sci. 94:2851–2866.

Stewart, L. L., D. Y. Kil, F. Ji, R. B. Hinson, A. D. Beaulieu, G. L. Allee, J. F. Patience, J. E. Pettigrew, and H. H. Stein. 2013. Effects of dietary soybean hulls and wheat middlings on body composition, nutrient and energy retention, and the net energy of diets and ingredients fed to growing and finishing pigs. J. Anim. Sci. 91:2756–2765.

Su, Y., Y. She, Q. Huang, C. Shi, Z. Li, C. Huang, X. Piao, and D. Li. 2015. The Effect of inclusion level of soybean oil and palm oil on their digestible and metabolizable energy content determined with the difference and regression method when fed to growing pigs. Asian Australas. J. Anim. Sci. 28:1751–1759.

Tao, R. C., R. E. Kelley, N. N. Yoshimura, and F. Benjamin. 1983. Glycerol: Its metabolism and use as an intravenous energy source. J. Parenter. Enteral Nutr. 7:479–488.

Tervila-Wilo, A., T. Parkkonen, A. Morgan, M. Hopeakoski-Nurminen, K. Poutanen, P. Heikkinen, and K. Autio. 1996. In vitro digestion of wheat microstructure with xylanase and cellulase from Trichoderma reesei. J. Cereal Sci. 24:215–225.

Thacker, P. A., and R. N. Kirkwood. 1990. Nontraditional feed sources for use in swine production. Butterworth, Stoneham, MA.

Tiwari, U. P., A. K. Singh, and R. Jha. 2019. Fermentation characteristics of resistant starch, arabinoxylan, and β-glucan and their effects on the gut microbial ecology of pigs: A review. Anim. Nutr. 5:217–226.

Tiwari, U. P., H. Chen, S. W. Kim, and R. Jha. 2018. Supplemental effect of xylanase and mannanase on nutrient digestibility and gut health of nursery pigs studied using both in vivo and in vitro model. Anim. Feed Sci. Technol. 245:77–90.

Tiwari, U. P., and R. Jha. 2016. Nutrient profile and digestibility of tubers and agro-industrial coproducts determined using an in vitro model of swine. Anim. Nutr. 2:357–360.

Tiwari, U. P., and R. Jha. 2017. Nutrients, amino acid, fatty acid and non-starch polysaccharide profile and in vitro digestibility of macadamia nut cake in swine. Anim. Sci. J. 88: 1093–1099.

Tokach, M., R. Goodband, and J. DeRouchy. 2012. Sorghum in swine production. Feeding Guide. United Sorghum Checkoff Program. http://texassorghum.org/wp-content/uploads/2011/09/Swine-Feeding-Guide.pdf (accessed 2 December 2019).

Valdes, E. V., and S. Leeson. 1992. Near infrared reflectance analysis as a method to measure metabolizable energy in complete poultry feeds. Poul. Sci. 71:1179–1187.

Varel, V. H. and J. T. Yen. 1997. Microbial perspective on fiber utilization by swine. J. Anim. Sci. 75:2715–2722.

Veum, T. L., D. R. Ledoux, D. W. Bollinger, V. Raboy, and A. Cook. 2002. Low-phytic acid barley improves calcium and phosphorus utilization and growth performance in growing pigs. J. Anim. Sci. 80:2663–2670.

Veum, T. L., D. R. Ledoux, V. Raboy, and D. S. Ertl. 2001. Low-phytic acid corn improves nutrient utilization for growing pigs. J. Anim. Sci. 79:2873–2880.

Wang, L. 2014. Mid-infrared spectroscopy estimates nutrient digestibility in pigs to improve in vitro digestion models. PhD Diss. Univ. of Alberta, Edmonton, AB, Canada.

Wang, L. F., E. Beltranena, and R. T. Zijlstra. 2016. Diet nutrient digestibility and growth performance of weaned pigs fed sugar beet pulp. Anim. Feed Sci. Technol. 211:145–152.

Whittemore, C. T. 1997. An analysis of methods for the utilisation of net energy concepts to improve the accuracy of feed evaluation in diets for pigs. Anim. Feed Sci. Technol. 68:89–99.

Woyengo, T. A., E. Beltranena, and R. T. Zijlstra. 2014. Controlling feed cost by including alternative ingredients into pig diets: A review. J. Anim. Sci. 92:1293–1305.

Woyengo, T. A., E. Beltranena, and R. T. Zijlstra. 2017. Effect of anti-nutritional factors in oilseed co-products on feed intake of pigs and poultry: Invited review. Anim. Feed Sci. Technol. 233:76–86.

Woyengo, T. A., R. Patterson, and C. L. Levesque. 2016. Nutritive value of extruded or multi-enzyme supplemented cold-pressed soybean cake for pigs. J. Anim. Sci. 94:5230–5238.

Wu, F., L. J. Johnston, P. E. Urriola, A. M. Hilbrands, and G. C. Shurson. 2016. Effects of feeding diets containing distillers' dried grains with solubles and wheat middlings with equal predicted dietary net energy on growth performance and carcass composition of growing–finishing pigs. J. Anim. Sci. 94:144–154.

Yang, Z., P. E. Urriola, A. M. Hilbrands, L. J. Johnston, and G. C. Shurson. 2019. Growth performance of nursery pigs fed diets containing increasing levels of a novel high-protein corn distillers dried grains with solubles. Transl. Anim. Sci. 3:350–358.

Zijlstra, R. T., C. F. M. de Lange, and J. F. Patience. 1999. Nutritional value of wheat for growing pigs: Chemical composition and digestible energy content. Can. J. Anim. Sci. 79:187–194.

Zijlstra, R. T., and E. Beltranena. 2009. Regaining competitiveness: alternative feedstuffs for swine. In: R. O. Ball, editor, Advances in pork production. Proc. banff pork seminar. Vol. 20, University of Alberta, Edmonton, AB, Canada. p. 237–243.

Zijlstra, R. T., K. Menjivar, E. Lawrence, and E. Beltranena. 2009. The effect of feeding crude glycerol on growth performance and nutrient digestibility in weaned pigs. Can. J. Anim. Sci. 89:85–89.

Zijlstra, R. T., and R. L. Payne. 2007. Net energy system for pigs In: J. E. Patterson and J. A. Barker, editors, Manipulating pig production XI. Australasian Pig Science Association, Werribee, Vic, Australia. p. 80–90.

Zijlstra, R. T., M. Swift, L. Wang, P. Regmi, J. H. Helm, and R. Jha. 2010. Rapid methods for prediction of energy values of feedstuffs for pigs. In; Proceedings of Western Nutrition Conference (Sept 21-23, 2010), Saskatoon, SK, Canada. p. 235–242.

Zijlstra, R. T., M. L. Swift, L. F. Wang, T. A. Scott, and M. J. Edney. 2011. Near infrared reflectance spectroscopy accurately predicts the digestible energy content of barley for pigs. Can. J. Anim. Sci. 91:301–304.

# 12 Major Protein Supplements in Swine Diets

Lee I. Chiba

## Introduction

Feed costs account for the largest economic input in most swine enterprises, and improving the efficiency of feed utilization is the utmost importance for successful and sustainable pig production. Wasteful usage of protein supplements is likely to increase the cost of production in any swine enterprise because protein sources, along with energy sources, account for a major portion of total feed costs. In addition, feeding excess protein to animals can have an adverse impact on the environment. The management of wastes and odors has become a major issue facing the pig industry, and large amounts of N excreted in animal wastes can lead to contamination of water and odorous emissions. Furthermore, the competition between humans and animals, particularly nonruminant species, for quality sources of protein is likely to increase continuously because of the ever-increasing world population. Therefore, efficient feed utilization not only improves profitability of pig enterprises and has a positive impact on the environment but it also helps ensure continuous availability of quality sources of nutrients for future pig production.

The objective of this chapter is to review briefly the major protein supplements used in pig production. In addition to some traditional protein supplements, some information on potential protein sources or proteinaceous feed additives has been included. The relative feeding values and suggested maximum incorporation rates of some protein supplements are presented in Table 12.1. For the complete information on the composition of feed ingredients, readers are referred to a recent NRC (2012) publication, Jurgens et al. (2012), various publications by Feedipedia (2012-2020), and others. Excellent reviews on protein supplements have been presented over the years [e.g., Aherne and Kennelly (1985), Thacker and Kirkwood (1990), Seerley (1991), Church and Kellems (1998), Chiba (2001), and Chiba (2010a,b)], and the readers are referred to those publications for further information.

## Diet Formulation

The main objective of diet formulation and feeding strategy in commercial pig production is to maximize profits, which does not necessarily imply maximal animal performance. To maximize economic efficiency, supplying indispensable nutrients as close as possible to meeting but not exceeding the requirements of the pig is advantageous. In addition, it will have a positive impact on today's environmentally conscious society by reducing the excretion of unutilized nutrients. Such optimum feeding strategies involve consideration of a multitude of factors, but two

*Sustainable Swine Nutrition*, Second Edition. Edited by Lee I. Chiba.
© 2023 John Wiley & Sons Ltd. Published 2023 by John Wiley & Sons Ltd.

**Table 12.1**  Relative feeding values and recommended maximum inclusion rates of selected protein sources[a]

| Protein sources | Relative feeding value[b,c] | Lys (g/100 g)[d] | Maximum recommended inclusion rate (% of diet)[b,e] | | | |
|---|---|---|---|---|---|---|
| | | | Starter | Grower-finisher | Gestation | Lactation |
| Alfalfa meal, dehy | — | 4.55 | 0 | 10 | 25 | 0 |
| Blood meal, spray-dried | 220–230 | 9.70 | 3 | 6 | 5 | 5 |
| Canola meal | 70–85 | 5.52 | 0 | 15 | 15 | 15 |
| Corn DDGS | 45–55 | 2.44–3.29 | 5 | 20 | 40 | 10 |
| Corn gluten feed | 40–55 | 3.62 | 5 | 25 | 40 | 10 |
| Corn gluten meal | 55–70 | 1.60 | 5 | 5 | 5 | — |
| Cottonseed meal | — | 3.82 | 0 | 10 | 20 | 0 |
| Fababeans | 65–75 | 6.08 | 10 | 20 | 10 | — |
| Fish meal, menhaden | 160–170 | 7.21[f] | 20 | 6 | 6 | 6 |
| Flaxseed | — | 4.03 | 5 | 5 | 5 | — |
| Meat and bone meal | 105–115 | 5.17 | 7.5 | 7.5 | 10 | 5 |
| Meat meal | 120–140 | 5.67 | 0 | 5 | 10 | 5 |
| Peas | 65–75 | 7.15[f] | — | 20–35 | 40 | — |
| Plasma protein, spray-dried | 205–215 | 8.86 | 10 | T | T | T |
| Skim milk, dried | 100–115 | 6.58 | 30 | T | T | T |
| Soy protein concentrate | 135–145 | 6.27 | 20 | T | T | T |
| Soy protein isolate | — | 6.12 | 10 | T | T | T |
| Soybean meal | 100 | 6.28 | 15 | 25 | 25 | 20 |
| Soybean meal, dehulled | 105–110 | 6.20 | 15 | 25 | 25 | 20 |
| Soybeans, full-fat, heated | 85–100 | 5.94 | 0 | 20 | 10 | 10 |
| Sunflower meal | 50–65 | 3.66 | 0 | 20 | 10 | 0 |
| Whey, dried | 55–65 | 7.62 | 30–40 | 15 | 5 | 5 |

[a] CP = crude protein, dehy = dehydrated, and DDGS = dried distillers grains with solubles.

[b] Based on: Reese et al. (1995), Hill et al. (1998), Reese et al. (2000), Reese et al. 2010, and Simpson (2012).

[c] Soybean meal (44% CP) = 100%. Values apply when ingredients are fed at no more than maximum recommended percent of complete diet. A range is provided to compensate for quality variation.

[d] Based on values reported by NRC (2012).

[e] A symbol (T) indicates no nutritional limitation in a diet balanced for indispensable amino acids, minerals, and vitamins, but the economical consideration or some other factors may preclude the use of an ingredient for a particular class of swine.

[f] Based on the combined data or the average.

concepts that may contribute greatly to the formulation of efficient and environmentally friendly diets are the ideal protein concept and formulation of diets based on available amino acids (AA).

The historical account of ideal protein and how the fundamental concepts of AA nutrition can be integrated into practical modeling approaches for the nutrition of growing pigs and sows have been described by van Milgen and Dourmad (2015). The body uses mixtures of AA collectively for protein synthesis, thus the balance of AA is obviously important for optimum utilization of protein. Any deviations in the bodily functions from a desirable pattern of AA may lead to a reduction in pig performance or some aberrations, depending on the degree of departure. The efficiency of AA utilization can be increased by, for instance, using the high-quality CP source with a desirable AA balance or formulating the diet to achieve the ideal protein. In the commercial pig production, using high-quality protein sources to satisfy the AA needs may not always be possible. Thus, using a mixture of various protein sources with complementary AA compositions, or supplementing diets with synthetic AA to simulate the ideal AA pattern is, perhaps, more practical means to incorporate this concept into the diet formulation.

All the nutrients are not available to pigs, thus expressing the requirements and formulating diets based on the available nutrients would be more effective in precisely satisfying the pig's nutritional needs. However, it is questionable whether there is sufficient information on the nutritive value of individual feed ingredients to achieve such an objective. Unfortunately, there seems to be no agreement on how to address the availability issue in practice (Batterham et al. 1990; Parsons 1996; Mosenthin et al. 2000) and the question remains whether using available nutrient values will improve the precision of diet formulation to meet the needs of the industry. Nevertheless, formulating diets based on available AA should be the improvement over formulation on a total AA basis. Obviously, to formulate pig diets based on available nutrients, further progress must be made in describing a true nutritional value of feed ingredients.

## Plant Protein Supplements

### Oilseed Meals in General

The major protein sources used for food animal production are oilseed meals. The production of oilseeds has increased from approximately 524 million metric tons (mmt) in 2015–2016 to approximately 581 mmt (preliminary data) in 2019–2020 [Foreign Agricultural Service (FAS), United States Department of Agriculture (USDA)]. The total oilseed production was only approximately 287 mmt in 1997–1998 (Chiba 2001). Many oilseeds are grown for their seeds to extract oils for human consumption and other purposes, whereas some oilseeds are grown for, e.g., fiber production. The soybean is clearly the prominent oilseed produced in the world (Figure 12.1), and soybean meal (SBM) accounted for 70.7% of the world production of protein meals in 2019–2020 (Table 12.2). Major producers of soybeans and SBM are presented in Tables 12.3 and 12.4, respectively.

Although moderate heating is generally necessary for inactivating antinutritional factors present in oilseed meals, overheating of meals can greatly reduce the amount of digestible or available Lys and other AA (Church and Kellems 1998). Fortunately, the potential problems are well recognized by the oilseed processors, which are reflected in today's production of high-quality

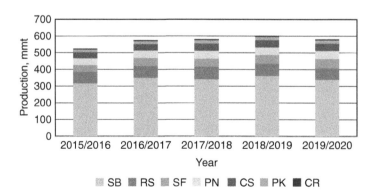

**Figure 12.1** Annual production of major oilseeds. SB = soybean, RS = rapeseed, SF = sunflower, CS = cottonseed, PN = peanut, PK = palm kernel, and CR = copra. Source: Foreign Agricultural Service, United States Department of Agriculture (USDA), Washington, DC, US; mmt = million metric ton.

**Table 12.2**   World production of major protein meals (million metric tons).

| Item | Year | | | | |
|---|---|---|---|---|---|
| | 2015/2016 | 2016/2017 | 2017/2018 | 2018/2019 | 2019/2020 |
| Soybean meal | 215.97 | 225.93 | 232.35 | 236.38 | 244.49 |
| Rapeseed meal | 38.61 | 38.8 | 39.53 | 39.47 | 39.41 |
| Sunflower meal | 16.51 | 19.34 | 19.89 | 21.13 | 22.15 |
| Cottonseed meal | 13.10 | 13.44 | 15.73 | 15.72 | 15.69 |
| Palm kernel meal | 8.18 | 8.91 | 9.77 | 10.11 | 9.96 |
| Peanut meal | 6.64 | 7.09 | 7.32 | 6.93 | 7.69 |
| Fish meal | 4.51 | 4.87 | 4.98 | 4.70 | 4.62 |
| Copra meal | 1.77 | 1.81 | 1.95 | 1.96 | 1.90 |
| *Total* | *305.27* | *320.19* | *331.52* | *336.41* | *345.91* |

Source: Foreign Agricultural Service, United States Department of Agriculture (USDA), Washington, DC, US.

**Table 12.3**   Soybean production by main producers (thousand metric tons).

| Item | Year | | | | |
|---|---|---|---|---|---|
| | 2015/2016 | 2016/2017 | 2017/2018 | 2018/2019 | 2019/2020 |
| Brazil | 96 500 | 114 600 | 122 000 | 117 000 | 128 500 |
| United States | 106 869 | 116 931 | 120 065 | 123 664 | 96 667 |
| Argentina | 58 800 | 55 000 | 37 800 | 56 000 | 48 800 |
| China | 12 360 | 13 644 | 15 200 | 15 900 | 18 100 |
| Paraguay | 9217 | 10 336 | 10 300 | 9000 | 10 100 |
| India | 6929 | 10 992 | 8350 | 11 500 | 9300 |
| Canada | 6456 | 6597 | 7717 | 7300 | 6145 |
| Others | 19 434 | 22 482 | 20 112 | 21 711 | 21 807 |
| *Total* | *316 565* | *350 582* | *341 544* | *362 075* | *339 419* |

Source: Foreign Agricultural Service, United States Department of Agriculture (USDA), Washington, DC, US.

**Table 12.4**   Soybean meal production by main producers (thousand metric tons).

| Item | Year | | | | |
|---|---|---|---|---|---|
| | 2015/2016 | 2016/2017 | 2017/2018 | 2018/2019 | 2019/2020 |
| China | 64 548 | 69 696 | 71 280 | 68 112 | 72 468 |
| United States | 40 525 | 40 630 | 44 657 | 44 583 | 46 358 |
| Argentina | 33 211 | 33 280 | 27 930 | 31 800 | 29 870 |
| Brazil | 30 750 | 31 280 | 34 500 | 33 100 | 35 650 |
| European Union | 11 811 | 11 376 | 11 811 | 12 877 | 12 324 |
| India | 4400 | 7200 | 6160 | 7600 | 6720 |
| Mexico | 3480 | 3635 | 4152 | 4350 | 4750 |
| Others | 27 247 | 28 837 | 31 861 | 33 958 | 36 353 |
| *Total* | *215 972* | *225 934* | *232 351* | *236 380* | *244 493* |

Source: Foreign Agricultural Service, United States Department of Agriculture (USDA), Washington, DC, US.

meals. Most oilseed meals are high in crude protein (CP) content, but except SBM (Aherne and Kennelly 1985), they are generally low in Lys. The extent of dehulling affects the CP and fiber contents, whereas the method of oil extraction affects the ether extract content (Aherne and Kennelly 1985), thus, the energy content of the meal. Oilseed meals are generally low in Ca, but high in P. The biological availability of minerals in plant sources such as oilseeds is generally low, and this is especially true for P.

## Alfalfa Meal

Alfalfa (*Medicago species*) is one of the most popular forage crops grown throughout the world (Thacker 1990a), and it is an excellent source of many nutrients. The CP content of alfalfa ranges from 12 to 23% (Seerley 1991; Feedipedia 2012-2020), and it has a reasonable amount of Lys and a good AA balance (NRC 2012; Feedipedia 2012-2020). Sun cured alfalfa is high in Ca and a reasonably good source of other minerals, except P (Thacker 1990a). Alfalfa meal is a good source of vitamins A, D, E, and K and some B vitamins such as riboflavin, pantothenic acid, biotin, and niacin (Thacker 1990a; Seerley 1991). The crude fiber content of alfalfa can range from 21 to 30% (Seerley 1991; Feedipedia 2012-2020), which can reduce the digestibility of energy and protein (Thacker 1990a). On the average, energy digestibility of dehydrated alfalfa is 45.2% in growing pigs, whereas its N digestibility is 39.6% (Feedipedia 2012-2020). In addition to high in fiber, alfalfa contains potential toxins such as saponins and tannins (Thacker 1990a), and it may also contain a trypsin inhibitor and a photosensitizing agent. Those components can, therefore, depress feed intake, inhibit digestive enzymes, interfere with cellular metabolism, or have other adverse effects. Despite its shortcomings, however, there is still considerable interest in using alfalfa or alfalfa meal in pig diets.

It is commonly recommended that alfalfa meal should not be used in diets for weanling pigs (Thacker 1990a), but the dietary inclusion of alfalfa may affect cecal microbiota composition and short-chain fatty acids (FA), decrease diarrhea incidence, and improve weanling pig performance (Adams et al. 2019). The rate and efficiency of weight gain in grower-finisher pigs decreased progressively as the content of alfalfa meal increased from 0 to 60% (Powley et al. 1981). The corn-SBM diet containing 10% alfalfa meal did not affect AA digestion, but increasing alfalfa meal from 0 to 20% linearly reduced apparent ileal digestibility and standardized ileal digestibility (SID) of most AA (Chen et al. 2015). Although Chen et al. (2014) indicated that alfalfa meal should be limited to less than 5% of the diet, it has been assumed that grower-finisher pigs can utilize 2.5–10% alfalfa meal (Seerley 1991). The use of alfalfa in gestating sows has a long history, and most research has indicated the improved reproductive performance (e.g., Pollmann et al. 1981), which may be related to the increased ketogenic substances in sows. More than 60% alfalfa in the gestation diet may have, however, adverse effects, and except during the preparturition and early lactation phase to prevent or alleviate constipation, alfalfa meal should not be included in the lactation diet (Thacker 1990a).

Because of its high fiber content and palatability problems, alfalfa should not be used in weanling pig diets, but it may reduce diarrhea incidence. Grower-finisher pig diets may contain approximately 5–10% alfalfa meal, and greater inclusion rate may reduce growth performance. Although alfalfa may have its greatest potential for use in gestating sows, it should be limited to 25% of the diet. Although alfalfa can be fed to sows during preparturition and early lactation phase to prevent or alleviate constipation, it is not recommended for lactating sows.

### Canola Seed (Rapeseed) and Meal

World annual rapeseed production was 11.4 mmt in 1980–1981 (Aherne and Kennelly 1985), and it increased to 34.3 mmt in 1997–1998, 68.7 mmt in 2015–2016, and 69.0 mmt in 2019–2020 (FAS, USDA). The leading countries in rapeseed production are Canada, China, and India. The world production of rapeseed meal ranged from 38.6 to 39.4 mmt during 2015–2016 to 2019–2020 (Table 12.2). In the mid-1970s, Canada's rapeseed production shifted to cultivars that contain low erucic acid and glucosinolate in the oil and meal, respectively, and those cultivars were trade-named canola (Aherne and Bell 1990). The oil must contain less than 2% erucic acid, whereas the meal must contain less than 30 μmol glucosinolates/g (Thacker 1990b). Canola seeds contain only about 15% glucosinolates compared with the old rapeseed. Generally, canola seed contains about 40% ether extract (Aherne and Bell 1990), and the CP content of canola meal can range from 36 to 39% (Canola Council of Canada 2019). Although Lys content of canola meal is lower than SBM, it has relatively comparable AA profile (NRC 2012). The SID of Lys in canola meal is approximately 77% (Canola Council of Canada 2019). The DE and ME contents in solvent extracted meal are 2605–3180 and 2303–2925 kcal/kg for DE and ME, respectively, and canola meal is relatively a good source of essential minerals, choline, and B complex vitamins (NRC 2012; Canola Council of Canada 2019).

Tannins and sinapine are two other compounds found in canola seed or meal, but their effects on pigs are not really clear (Thacker 1990b). The max tolerance of glucosinolate seem to be approximately 2 μmol/g in growing pig diet (Schöne et al. 1997a,b). It has been reported earlier that canola meal can have negative effects on growth performance of weanling pigs (Aherne and Bell 1990) because of, perhaps, the low palatability and reduced feed intake. The inclusion rates of up to 5% canola meal in weanling pig diets and 10% in grower-finisher and sow diets have been recommended (Thacker 1990b); however, Wang et al. (2017) indicated that 20% canola meal can be included in weanling pig diets. For grower-finisher pigs, canola meal can be included up to 25% of the diet (Little et al. 2015), and the high-protein and conventional canola meal may fully replace SBM as a protein supplement without impairing the pig performance or carcass quality (Little et al. 2015). Reproductive performance of sows or their litter performance was not affected by including 25% canola meal during gestation and 30–35% canola meal during lactation (Liu et al. 2018).

Some recent research indicated that weanling pig and grower-finisher diets may contain 20 and 25% canola meal, respectively. Some researchers indicated that canola meal can supply all the supplemental protein need for finisher pigs and replace SBM completely in grower-finisher pig diets. Approximately 25 and 30–35% canola meal can be included in the gestation and lactation diet, respectively, without loss of performance of sows or their litter.

### Copra (Coconut) Meal

Coconut (*Cocos nucifera*) is widely distributed in many tropical areas of the world. The primary product is copra or its dry kernel, which is used for coconut oil production (Thorne et al. 1990). The production of copra was 4.9 mmt in 1992–1993, 5.5 mmt in 1997–1998, 5.3 mmt in 2015–2016, and 5.8 mmt in 2019–2020 (FAS, USDA). The leading oil-producing countries are Indonesia, Philippines, and India (Thorne et al. 1990). Copra meal or cake is the main by-product of the oil production, and the world copra meal production ranged from 1.8 to 1.9 mmt during 2015–2016 to 2019–2020, (Table 12.2). The terms copra cake and copra meal refer to the mechanically extracted and solvent extracted products, respectively, but they are often used interchangeably (Feedipedia 2012-2020).

The CP content of copra meal typically ranges from 20 to 26%, and it is deficient in Lys (Church and Kellems 1998) with Lys being only 1.91–2.1% of CP (Feedipedia 2012-2020; Stein et al. 2015). The average N digestibility in copra mea was 67.9% (Feedipedia 2012-2020), whereas the SID of Lys was 72.8% in growing pigs (Stein et al. 2015). Although the gross energy content of copra meal is greater, its digestible (DE) and metabolizable energy (ME) content are lower than corn because of the crude fiber content (NRC 2012; Feedipedia 2012-2020). Coconut meal can be subjected to mold growth such as *Aspergillus* spp. Although there have been no reports of aflatoxicosis in pigs fed copra mea, in the past, aflatoxin contaminations in pig diets have been reported in the field (Thorne et al. 1990).

Considering its deficiency in Lys, low available AA, and high fiber content, copra meal may not be a suitable protein source for pig diets. However, copra meal may represent the largest quantity of locally available source of protein in many tropical areas (Siebra et al. 2008; Stein et al. 2015). Increasing dietary copra meal from 0 to 15% reduced growth performance of weanling pigs (Jaworski et al. 2014). For grower-finisher pigs, growth performance, digestibility, and carcass quality decreased progressively as dietary copra meal increased from 0 to 50% (Thorne et al. 1990). O'Doherty and McKeon (2000) concluded that grower-finisher pig diets can contain 20% copra meal, but dressing percentage may be reduced. Supplementation of diets with synthetic AA or high-quality protein sources may increase the use of copra meal (Feedipedia 2012-2020; Stein et al. 2015), but adding Lys or other AA or both (Thorne et al. 1990) was not really effective in alleviating the reduced pig performance. With β-mannanase supplementation, 12% or even 25–30% copra meal can be included in the grower-finisher pig diet (Diarra 2016; Jang et al. 2020). Stein et al. (2015) concluded that copra meal should be less than 15% in diets fed to weanling pigs and less than 25% in diets for grower-finishing pigs.

It has been suggested in the past that copra meal should be limited to 10–20% of pig diets for optimum performance. Coconut meal is not a good protein source for weanling pigs, but up to 25–30% of copra meal can be included in the grower-finisher pig diets by adding β-mannanase. Supplementation with synthetic indispensable AA seems to be a plausible way to utilize copra meal in pig diets, but such an approach has not been fully explored.

### Cottonseed Meal

Cotton (*Gossypium* spp.) has been grown for several thousand years as a source of textile fiber, and the cotton plant also yields cottonseed (Tanksley 1990). The world cottonseed production was 34.7 mmt in 1997–1998, 35.8 mmt in 2015–2016, and 44.4 mmt in 2019–2020 (FAS, USDA), and the major producers are India, China, United States, and Brazil (ERS, USDA). Typical cottonseed processing can yield 50% cottonseed meal, and the world production of cottonseed meal increased from 13.1 mmt in 2015–2016 to 15.7 mmt in 2019–2020 (Table 12.2). The CP content of cottonseed meal varies from 35 to 51% (Tanksley 1990; Ma et al. 2018). The Lys content ranges from 1.51 to 2.44% and its SID ranges from 55.0 to 76.0% (Ma et al. 2018). The crude fiber content of cottonseed meal is greater than SBM (2012; Ma et al. 2018), and its energy value is negatively related to the fiber content (Ma et al. 2018). Cottonseed meal compares favorably with SBM in most B-vitamins, but it is a poorer source vitamin D, β-carotene, and minerals (Seerley 1991). Gossypol, a phenolic compound produced by the glands in the cottonseed plant (≥100 ppm), can cause toxicity in animals (Tanksley 1990; Church and Kellems 1998). Cottonseed meals from glandless cotton varieties devoid of gossypol are a better source of protein, but they are not really available for animal feeding because of the low yield (Tanksley 1990). Heat processing can be used to inactivate gossypol to some extent.

Iron salts may also be effective in blocking the toxic effect of gossypol by forming a Fe-gossypol complex.

The use of glanded cottonseed meal as the only source of protein in grower-finisher diets is likely to reduce pig performance (Tanksley 1990) because of the low Lys content and availability. Lysine or Fe supplementation may partly alleviate the adverse effects of gossypol (Mello et al. 2012). In addition to Lys, supplementation with Thr and Trp or high-quality protein sources may resulted in a good pig performance (Tanksley 1990). Cottonseed meal may replace SBM as a primary protein source in the finisher diet without affecting the efficiency of growth or carcass traits (Qin et al. 2015). The digestibility of AA was greater for glandless than glanded cottonseed meal (Tanksley 1990), and glandless cottonseed meal supplemented with Lys could be used to replace at least 40% of the supplemental protein without any adverse effects on growing pigs (LaRue et al. 1985). It has been recommend earlier that a cottonseed meal diet should be supplemented with ferrous sulfate on a 1:1 with free gossypol (Haschek et al. 1989). Free gossypol can adversely affect the performance of reproducing pigs (Feedipedia 2012-2020), and Tanksley (1990) indicated that sow lactation diets should contain less than 50% cottonseed meal as the supplemental protein, and limiting it to 25% of the protein source is recommended.

Growth and reproductive performance of pigs can be reduced by $\geq 100$ ppm gossypol in their diets. Because of the high-fiber content and limited information, the use of cottonseed meal in weanling pig diets should be limited, even though replacing 40% of the supplemental protein with glandless cottonseed meal may have no adverse effects on growing pigs. Cottonseed meal can supply approximately 40–50 or 25% of the supplemental protein in growing and gestation or lactation diets, respectively, but all diets should be formulated based on Lys.

### Distillers Grains with Solubles, Dried

Distiller's grains can be defined as a cereal by-product of the distillation process, and the two primary sources are the brewery and fuel-ethanol plants (Liu 2011). Global production of bioethanol increased in recent decades, and the production of dried distiller's grains with solubles (DDGS) reached 38.9 mmt in 2019–2020 (ERS, USDA). Spiehs et al. (2002) reported that the average CP, lipid, crude fiber, and ash contents of 118 DDGS samples were 30.2, 10.9, 8.8, and 5.8%, respectively, on a dry matter (DM) basis. The CP contents of corn, sorghum, and wheat DDGS were 27.4, 31.5, and 40.7% (DM), respectively, and Lys contents of those DDGS were 0.76, 0.66, and 0.65%, respectively (Stein and Shurson 2009). The lipid content of "low-oil" DDGS has been reduced from 10.5 to 5.5% (Loar et al. 2012). The digestibility of GE in wheat DDGS in weanling pigs ranged from 82.6 to 86.4% (Wang et al. 2016), and the ileal and fecal digestible energy (DE) of wheat DDGS in growing pigs averaged 9.7 and 13.5 MJ/kg, respectively (Nyachoti et al. 2005). The estimated DE and ME values for corn DDGS were 15.6 and 14.2 MJ/kg, respectively (Corassa et al. 2017). The apparent ileal digestible Lys, Thr, and Ile in wheat DDGS were 43.8, 62.9, and 68.0%, respectively (Nyachoti et al. 2005). The SID CP and Lys in sorghum and corn DDGS seems to be similar in grower pigs, whereas the other SID AA values in sorghum DDGS were less than corn DDGS (Urriola et al. 2009). The DDGS is a good source of available P (Stein and Shurson 2009).

Although growth performance of weanling pigs reduced linearly as dietary DDGS increased from 0 to 20% (Wang et al. 2016), inclusion of up to 30% DDGS had no effect on growth performance or carcass traits of grower-finisher pigs when diets were formulated on the NE and ileal digestible AA basis (McDonnell et al. 2011). After comprehensive review, Stein and Shurson (2009) concluded that DDGS is an excellent source of energy and digestible P, and acceptable growth performance can be obtained by including up to 30% DDGS in diets for weanling to grower-finisher pigs. Many DDGS may contain relatively a high concentration of oxidized lipids and supplementation with 30% of peroxidized DDGS may negatively affect growth performance of weanling to finisher pigs (Song et al. 2014). Similarly, including up to 30% DDGS may result in softer pork, but such a problem can be alleviated by withdrawing DDGS at least three weeks before harvest (Xu et al. 2009) or using DDGS with less ether extract content and unsaturated FA (Sotak et al. 2015). Stein and Shurson (2009) indicated that up to 50% DDGS can be included in gestation diets and up to 30% DDGS can be included in lactation diets without adversely affecting sow and litter performance.

Growth performance of weanling pigs can be reduced by 20% DDGS, but acceptable growth performance can be obtained by feeding up to 30% DDGS in weanling to grower-finisher pigs by formulating diets appropriately. Acceptable carcass traits can be obtained by withdrawing DDGS for at least three weeks before harvest or feeding DDGS with less ether extract and unsaturated FA. Gestating and lactating sows can be fed diets containing up to 50 and 30% of DDGS, respectively, without adversely affecting their reproductive performance.

### Flaxseed (Linseed) Meal

Flax or linseed (*Linum usitatissimum*) is one of the oldest crops known to man (Aherne and Kennelly 1985; Bowland 1990), and now it is produced primarily for its oil (Church and Kellems 1998). The world production of flaxseed was 2.3 mmt in 1996 and 3.2 mmt in 2018, and Kazakhstan, Canada, Russia, and China are leading producers (FAO 2019b). The oil content of flaxseed ranges from 40 to 45% (Bowland 1990; Feedipedia 2012-2020), and flaxseed meal is a by-product of oil extraction from flaxseed. Flaxseed and flaxseed meal have been receiving a lot of attention in recent years because of their omega (T)-3 FA contents and their potential beneficial effects of those FA on human health (Huang et al. 2021). Flaxseed meal contains 1.4–5.6% ether extract, and the CP content averages 35–36% (Bowland 1990; NRC 2012), and its Lys content and SID Lys are 1.2% and 63–77%, respectively (NRC 2012). The variation in the crude fiber content (NRC 2012; Feedipedia 2012-2020) may affect its nutritional value. The major macro minerals and B-complex vitamins are comparable to other oilseed meals, and it is a good source of Se (Bowland 1990). Flaxseed meal contains a number of antinutritional factors such as linamarin and linatine (Bowland 1990). Linamarin, cyanogenic compound, is normally destroyed by the heat during the oil extraction, but linatine, which can act as an antagonist for vitamin $B_6$, may lead to its deficiency (Aherne and Kennelly 1985; Bowland 1990).

Although flaxseed meal is a potential protein supplement for pigs, its fiber content, presence of antinutritional factors, and its low Lys content may limit its use in pig diets (Feedipedia 2012-2020). On the other hand, flaxseed can be incorporated into pig diets to alter or improve the FA profile of pork (Huang et al. 2021). A low concentration of flaxseed meal (up to 3%) can be included in creep feed or weanling pig diets (Bowland 1990; Chiba 2001). Grower-finisher pig diets can contain 5–10% (Li et al. 2000b), 15% (Richter and Kohler 1997), or even 25% (Bowland 1990) flaxseed meal without any adverse effects, provided that the diet is balanced for Lys or digestible AA. Seerley (1991) suggested that flaxseed meal could replace up to 50% of the protein supplement. Flaxseed meal, however, may have an antithyroid effect, and its use in young pigs should be limited to 10% of the diet (Richter and Kohler 1997). On the other hand, the mucilage in flaxseed meal can absorb a large amount of water, and it may be beneficial in preventing constipation in sows

(Bowland 1990; Seerley 1991). The literature on the use of flaxseed meal for sows is rather limited, but, if properly balanced for Lys, at least, 10% flaxseed meal can be include in the sow diets (Bowland 1990).

Flaxseed meal should be limited to 10% or less in the diet for young pigs because of its fiber, antinutritional factors, and less Lys content. However, if appropriately balanced for AA, up to 15%, or even up to 25%, of flaxseed meal can be included in the grower-finisher diet without any adverse effects, and it can improve the FA profile of pork for human health. For gestating and lactating sows, at least 10% of flaxseed meal can be included in their diets.

### Palm Kernel Meal

The palm kernel is the edible seed of the oil palm fruit, and *Elaeis guineensis* is the one generally used for palm oil and palm kernel oil production (FAO 2002; Feedipedia 2012-2020). Palm oil, oil derived from the outer parts of the fruit, has numerous nonfood applications, but it is a major staple oil and an indispensable ingredient in some regions of the world, whereas palm kernel oil, derived from the kernel, is economically less important (Feedipedia 2012-2020). The world production of palm kernel was 16.0 mmt in 2015 and it increased to 20.1 mmt in 2019–2020 (FAS, USDA). Palm kernel meal is the main by-product of the palm kernel oil extraction process, and its world production has increased from 8.2 mmt in 2015–2016 to 10.0 mmt in 2019–2020 (Table 12.2). According to Feedipedia (2012-2020), it has large quantities of crude fiber (14–30%) with 10–18% lignin, and the neutron detergent and acid detergent fiber contents are 58–78 and 34–51%, respectively, whereas the ether extract content is 6–15%. The CP content is only 14–20%, and palm kernel meal may not be considered as a true protein supplement. The Lys content can be 2.0–3.8% of CP (Feedipedia 2012-2020), and the values reported by others (Sulabo et al. 2013b) fall within that range. Although palm kernel meal might be better suited for ruminant and some other species because of the nutrient contents (Feedipedia 2012-2020), its potential value in pig diets has been investigated over the years.

Although palm kernel meal is not very palatable to young pigs and it may reduce growth performance (Kim et al. 2001; Feedipedia 2012-2020), the inclusion of 5, 10, or 15% palm kernel meal in the diet has been shown to have no effect on overall growth performance of weanling pigs (Jaworski et al. 2014). Similarly, with β-glucanase supplementation, including 0, 4, 8, 12, or 16% palm kernel meal in the grower diet, increased feed intake linearly and had no effect on weight gain, blood urea N, and pH, proximate analysis, shear force, or water holding capacity of loin muscle (Lee et al. 2017), or no effect on pork quality with 12% palm kernel meal (An et al. 2017). The nutrient or total AA digestibility of palm kernel meal diets seemed to be lower in finisher pigs, but adding an enzyme complex counteracted the negative effect on growth performance of pigs (Kim et al. 2001). Stein et al. (2015) concluded that palm kernel meal can be included at 15 and 25% in diets for weanling pigs and grower-finisher pigs, respectively. Because of its low palatability and high dietary fiber content, palm kernel meal may not be appropriate for lactating sows; however, 30–40% of palm kernel meal can be fed to gestating sows (Feedipedia 2012-2020).

In some production systems, palm kernel meal can be economically important and viable feed ingredient for pigs. Although it is high in the fiber content and not palatable, perhaps, up to 15 and 25% palm kernel meal can be included in weanling and grower-finisher pig diets, respectively. Enzyme supplementation is likely to enhance the greater use of palm kernel meal in pig diets without deleterious effects on growth performance or meat quality. Gestating sows can utilize up to 30–40% of palm kernel meal in their diets.

### Peanut Meal and Whole Peanuts

Peanuts (*Arachis hypogaea* L.) have been grown extensively in tropical and subtropical regions (Aherne and Kennelly 1985), and the peanut production increased greatly as the demand for its oil increased. The world production reached 17.5 mmt in 1980–1981 (Aherne and Kennelly 1985), 27.1 mmt in 1997–1998, 41.1 mmt in 2015–2016, and 44.9 mmt in 2019–2020, and China and India are the largest producers (FAS, USDA). Peanut kernels are a rich source of lipids (40–50% oil) and also CP (26–34%; Feedipedia 2012-2020). The world production of its by-product, peanut meal, ranged from 6.6 to 7.7 mmt during 2015–2016 to 2019–2020 (Table 12.2). The CP content of peanut meal ranges from 41 to 50% (Seerley 1991; Feedipedia 2012-2020), and it is deficient in Lys and low in Met and Trp (Seerley 1991). The solvent-extracted meal may be better choice because mechanically extracted meals may contain 5–7% oil and tend to become rancid and produce soft fat in pigs (Seerley 1991; Feedipedia 2012-2020). The content of most minerals in peanut meal is lower than SBM, but it is a good source of Mg, S, and K (Aherne and Kennelly 1985). Peanut meal is a good source of niacin, pantothenic acid, and thiamin, but it can be deficient in others (Aherne and Kennelly 1985; Seerley 1991). Although antinutritional factors in peanuts are of limited concern (Church and Kellems 1998), their lipid content can cause some mold problems such as *Aspergillus flavus* and aflatoxin (Aherne and Kennelly 1985).

The optimum inclusion rate of roasted peanuts for weanling pigs was 5% (Haydon and Newton 1987), whereas 5% raw or roasted peanuts on an equal Lys basis (Newton and Haydon 1988) or 10, 15, or 20% of full-fat peanuts (Balogun and Koch 1979b) had no effect on growth performance or nutrient digestibility in growing pigs. Heat-processing of whole peanuts reduced trypsin inhibitor activity, but it had no effect on pig performance (Balogun and Koch 1979a). Feeding 20% whole peanuts reduced carcass quality (Balogun and Koch 1979b), but if balance for Lys, peanuts were as effective as added fat in improving feed efficiency. Haydon et al. (1990) reported that 12% raw and roasted peanuts can be substituted for 5% animal fat in the lactation diet. Many early studies indicated that peanut meal alone can reduce growth performance of pigs (Feedipedia 2012-2020). Supplementation of peanut meal diet with only Lys was partially effective, but addition of Met alleviated the growth depression (Aherne and Kennelly 1985; Feedipedia 2012-2020), even though growth performance seemed to be lower than that obtained with SBM (Chiba 2001). There was no difference in growth performance of pigs fed the SBM diet and those fed the diets containing 15–20% peanut meal and 3–4% blood (Ilori et al. 1984), indicating the importance of the dietary AA balance.

Weanling pigs can utilize approximately 5% of roasted peanuts. Grower-finisher pigs can utilize greater concentrations, but 20% whole peanuts may reduce carcass quality, even though they can improve feed efficiency. Whole peanuts are an excellent source of dietary lipids for sows. Because of its low Lys content, peanut meal alone is not a suitable protein supplement for weanling or grower-finisher pig diets. Using peanut meal in a combination with ingredients high in Lys might be the most effective way to incorporate peanut meal in pig diets.

### Safflower Meal

Safflower (*Carthamus tictorius*) is cultivated throughout tropical climates (Darroch 1990; Church and Kellems 1998). The world production of safflower seed was 0.84 mmt in 1996 (FAO 1997) and 0.63 mmt in 2018 (FAO 2019a), and Kazakhstan, United States, and Mexico are major producers (FAO 2019a). Safflower seeds contain approximately 40% hull and 36–40% oil (Seerley 1991).

A prepress solvent oil extraction process produces an undecorticated safflower meal with approximately 20–22% CP and 40% crude fiber. The quality of safflower meal is highly variable because of the amount of hulls and the extent of oil extraction. The decorticated meal has a greater CP (42–45%) and less crude fiber (15–16%) and better suited for animal feeding (Williams and Daniels 1973). Safflower meal is a poor source of Lys, Met, and Ile for pigs, and its available Lys is low (Darroch 1990). Safflower meal is relatively good source of Ca and P and a rich source of Fe, and it has a poor vitamin profile, except biotin, riboflavin, and niacin (Aherne and Kennelly 1985; Darroch 1990). Safflower meal contains two phenolic glucosides, matairesinol-β-glucoside, which gives a bitter flavor, and 2-hydroxyarctiin-β-glucoside, which has cathartic properties (Darroch 1990), but both of which can be removed by extraction with water or methanol, or by the addition of β-glucosidase.

A relative feeding value of the undecorticated safflower meal may be only 45–50% of SBM for pigs, even though the decortication improves its nutritional value (Williams and Daniels 1973; Darroch 1990). Considering their nutrient requirements and digestive capacity, safflower meal may not be a suitable protein source for weanling pigs (Chiba 2001). Feeding the diet containing 17% of decorticated safflower meal reduced growth performance and increased carcass fat in growing pigs (Feedipedia 2012-2020). Safflower meal should be supplemented with AA or protein sources high in Lys, and it should be restricted to pigs greater than 45 kg live weight (Williams and Daniels 1973; Williams and O'Rourke 1974). Darroch (1990) indicated that, if the Lys requirement is satisfied, up to 12% safflower meal can be included in the diet but it should not provide more than 5–10% or 12.5% of the supplemental protein in the growing pig diet (Chiba 2001). Although dehulled safflower meal can be included at up to 15% in the gestating sow diet, it should be limited to very low concentrations for the lactation diet (Darroch 1990).

Feeding safflower meal to weanling pigs is not recommended. Decorticated safflower meal may be used to supply approximately 5–10% of the supplemental protein in grower-finisher pig diets, or even slightly greater than 12% of safflower meal could be included in the grower-finisher diet, provided that the Lys requirement is met. Safflower meal can be included up to 15% in the diet of pregnant females, but it should be limited to very low concentrations for lactating sows.

### Sesame Meal

Sesame (*Sesamum indicum*) is one of the oldest vegetable oil crops cultivated by man (Feedipedia 2012-2020). The world production of sesame seed was 2.5 mmt in 1996 (FAO 1997) and 6.0 mmt in 2018 (FAO 2020). The major producers of sesame are Sudan, Myanmar, and India. Sesame is primarily grown for its edible seeds and oil with 65% of the seeds being used for oil production and 35% for food (Feedipedia 2012-2020). On average, sesame seeds contain 25% CP, 50% ether extract, 4% crude fiber, 5% ash and 11% N-free extract (Johnson et al. 1979). Dehulled and expeller-extracted sesame meal contains 42% CP and 6.5% fiber (Ravindran 1990; Yamauchi et al. 2006). The solvent extraction produces a meal that is greater in the CP (48.5%) and lower in oil (2.5%) than meals produced by the screw press or hydraulic methods (Seerley 1991; Feedipedia 2012-2020). Sesame meal is an excellent source of Met, Cys, and Trp, but it is low in Lys (Aherne and Kennelly 1985; Ravindran 1990). Sesame meal is a good source of Ca, P, Mg, and other minerals, and its vitamin contents are similar to other oilseed meals (Aherne and Kennelly 1985; Ravindran 1990). Oxalic and phytic acids in the hull may have adverse effects on availability of minerals and CP (Aherne and Kennelly 1985). Decortication of seeds removes most oxalates, but it has little effect on phytate (Ravindran 1990; Ravindran et al. 1994).

The inclusion of 1–3.5% ground sesame seeds may limit peroxidation and enhance feed intake and feed efficiency in pigs (Yamasaki et al. 2003). The information on the use of sesame meal in pig diets is rather limited. Because of the low Lys content and possible palatability problems associated with phytates and oxalates, sesame seeds should be limited to 5% of the diet for weanling pigs (Ravindran 1990). Sesame meal could be utilized by grower-finisher pigs, but the inclusion rate would depend on the type and quantity of other protein sources in the diet (Ravindran 1990). Earlier reports indicated that 15% sesame meal can be included in the grower pig diet (Feedipedia 2012-2020). Seerley (1991) pointed out that sesame meal should be blended with other protein sources high in Lys, and it can replace up to 10% of the SBM in grower-finisher diets. Li et al. (2000a) reported that the ileal digestibility of most AA tended to decrease and growth performance decreased linearly as dietary sesame meal increased, but up to 10% sesame meal can be included in the grower-finisher diet by considering the apparent ileal digestibility of CP and AA (Tartrakoon et al. 2001). By combining with some protein source high in Lys, sesame meal can replace up to 10% of SBM in sow diets (Seerley 1991).

Because of its low Lys content, palatability problems, and antinutritional factors, sesame meal should be limited to 5% of the weanling pig diet. By considering the its CP and ileal AA digestibility when formulating diets and combining it with other protein sources high in Lys, sesame meal can replace up to 10% of SBM in the grower-finisher and sow diets.

### Soybeans and Soybean Products

The existence of soybeans (*Glycine max*) dates back to at least 2838 B.C. (Aherne and Kennelly 1985). Soybean production has increased over the years, and its world production was 82.2 mmt in 1980–1981 (Aherne and Kennelly 1985), 117.3 mmt in 1992–1993, 156.1 mmt in 1997–1998, 316.6 mmt in 2015–2016, and 355.7 mmt in 2019–2020 (FAS, USDA). Brazil, United States, Argentina, and China are the main producers of soybeans (Table 12.3). The majority of soybeans are crushed into oil and SBM, and about 75% of SBM produced in the world is used for pig and poultry production (Stein et al. 2013). Soybean meal is the most widely used protein supplement in pig diets in many countries, and it is unsurpassed by other plant protein sources in terms of its production (Table 12.2) and feeding value. Consequently, it is the standard to which other protein sources are compared. China, United States, Argentina, and Brazil are the major SBM producing countries (Table 12.4). Soy protein concentrate is produced from dehulled and oil-extracted soybeans, whereas soy protein isolate is the most highly refined soy protein or most of the nonprotein components have been removed. A rapidl expanding area of interest in recent years has been bioengineering of soybeans, and new soybeans on the horizon include those with 80–100% increase in Met, 100–400% increase in Lys, reduced oligosaccharide contents, and altered FA contents (Haumann 1997). These developments on soybeans and also on cereal grains are likely to have considerable impacts on the way the nutritional needs of animals will be satisfied in the future.

The nutrient content of soybeans and soybean products has been summarized and presented over the years (e.g., Seerley 1991; Church and Kellems 1998; NRC 2012; Stein et al. 2013; Feedipedia 2012-2020). Whole soybeans contain 36–37% CP, whereas SBM contains 41–50% CP depending on the processing. Soybean meal is generally available in two forms, 44% CP meal and dehulled meal, which contains 48–50% CP. Soybean meal has an excellent AA balance, and it is high in Lys, Trp, and Thr, which are most often deficient in ccreal grains. Soy protein concentrate contains about 70% CP, whereas soy protein isolate contains about 90% CP on a DM basis. Whole soybeans contain 31–39% oil, which is usually removed by solvent oil extraction. Because of its low fiber content, the digestible and ME contents of SBM are greater than most other oilseed meals. Soybean meal is generally low in minerals and vitamins. Phytate P accounts for about two-thirds of the total P in SBM, and the availability of many mineral elements may be low because of the formation of phytate-protein-mineral complexes during the processing. Soybeans contain trypsin-chymotrypsin inhibitors and other antinutritional factors, but heat treatment, extrusion, and(or) ethanol extraction are effective in removing most of those growth inhibitors (Aherne and Kennelly 1985; Cromwell 1998; Stein et al. 2013). Heat treatment may reduce availability of Lys and other AA (Hulshof et al. 2017), but the range of heat treatments normally found during the processing has no effect on the nutrition value (Stein et al. 2013).

The ability of pigs to utilize raw soybeans may be related to their age and changes in their relative AA needs, and older pigs seem to tolerate raw soybeans better than younger pigs (Cromwell 1998). Young and grower-finisher pigs fed raw, uncooked soybeans have a reduced growth performance, but carcass traits were not affected when diets were supplemented with appropriate AA (Danielson and Crenshaw 1991). Although raw soybeans may have some detrimental effects during lactation, it can be fed to gestating sows as the only source of protein supplement without any adverse effects (Yen et al. 1991). Properly processed soybeans have high energy digestibility and a greater SID AA compared to raw soybeans (Feedipedia 2012-2020) and can be used to replace SBM in grower-finisher and reproducing pig diets (Danielson and Crenshaw 1991; Seerley 1991; Newkirk 2010). Because of the lipid, whole soybeans may improve growth performance of growing pigs and reproductive performance of lactating sows (Newkirk 2010). However, whole soybeans may have adverse effects on the carcass fat of growing pigs (Feedipedia 2012-2020). Although whole soybeans can be used in practical pig diets, the economics should be an important consideration in their use (Seerley 1991).

Soybean meal is simply the most widely used protein supplement in the United States and other countries, and a proper proportion of SBM and corn or other cereal grains make an excellent dietary AA pattern for pigs (Seerley 1991; Stein et al. 2013; Feedipedia 2012-2020). Commercially available SBM are quite consistent in terms of the nutrient content (Danielson and Crenshaw 1991; Seerley 1991) and can be fed to all classes of pigs. The inclusion rates are generally 30% in grower, finisher, and sow diets and slightly lower (20–25%) in weanling pig diets (Feedipedia 2012-2020). Moran et al. (2017) indicated that SBM in the diet for weanling pigs from a production unit, which were affected by porcine reproductive and respiratory syndrome (PRRS) virus, can be increased because the percentage of pigs removed for the medical treatment decreased linearly with increasing SBM. Fermented (Kim et al. 2010) or microbially enhanced or bioprocessed SBM (Sinn et al. 2017) can be a suitable alternative to fishmeal, dried skim milk, or plasma protein because of their beneficial effects on the immune system, weaning diarrhea, and others. In addition, fermented SBM promoted the expression of hepatic insulin-like growth fator-1 and enhanced growth performance of grower-finisher pigs (Fan et al. 2018). Feeding low-oligosaccharide SBM from weaning-to-finisher period reduced viscosity of the intestinal digesta, but it had no effect on growth performance, intestinal morphology, or carcass traits of pigs (Pangeni et al. 2017).

Growth performance of weanling pigs fed soy protein concentrate was similar to those fed the dried skim milk-based diet (Sohn et al. 1994a). Similarly, weanling pigs fed soy protein isolate (Sohn et al. 1994a) or moist extruded soy protein concentrate (Li et al. 1991) had similar performance to those fed the dried skim milk-based diet. A greater digestibility of N or AA over SBM is, perhaps, responsible for the response of pigs to those soybean products (Walker et al. 1986; Sohn et al. 1994b). However, Oliveira and Stein (2016) indicated that the SID of most AA and the standardized total tract digestibility of P are not different between soy protein concentrate and SBM, even though the concentration of DE or ME is greater in soy protein concentrate than SBM. Grinding soy protein concentrate to approximately 180 µm may maximize the SID of indispensable AA without affecting the DE or ME content, and diets based on SBM and soy protein concentrate as the main sources of protein, i.e., no fish meal or fish meal and spray-dried plasma protein, can be fed to pigs during the initial two week postweaning without affecting growth performance (Casas et al. 2017).

Gestating sows can perform rather well with raw soybeans as the only source of protein supplement. Properly processed whole soybeans can be used to replace soybean meal in pig diets. Soybean meal is the most widely used protein supplement, and a proper proportion of SBM and cereal grains make the most balanced AA source for all classes of pigs. Soy protein concentrate and isolate can be used successfully in weanling pig diets as a replacement for dried skim milk, fish meal, plasma protein, and others.

### Sunflower Seeds and Meal

Sunflower (*Helianthus anuus*) was developed as an oilseed in Southern Europe in the sixteenth century, and its oil is highly valued for its stability at high temperatures (Aherne and Kennelly 1985). The worldwide sunflower production was 13.2 mmt in 1980–1981 (Aherne and Kennelly 1985), and it has increased to 21.2 mmt in 1992–1993, 23.9 mmt in 1997–1998, 40.6 mmt in 2015–2016, and 51.0 mmt in 2019–2010 (FAS, USDA). Major sunflower producing countries are Ukraine, Russia, Argentina, and Turkey (FAS, USDA). Sunflower seeds contain approximately 38% oil, 17% CP, and 15% crude fiber (Dinusson 1990; Feedipedia 2012-2020). The world production of sunflower meal has increased from 16.5 mmt in 2015–2016 to 22.2 mmt in 2019–2020. The CP content of solvent-extracted meal can range from 36 to 44% (Aherne and Kennelly 1985). It is deficient in Lys, and González-Vega and Stein (2012) indicated that the SID of CP and Lys among sunflower seeds, sunflower meal, and dehulled sunflower meal were similar (González-Vega and Stein 2012). The SID CP and Lys in sunflower meals were 77.0 and 74.3%, respectively (Nørgaard et al. 2012b) or varied from 66.7 to 79.3% and 67.0 to 82.1%, respectively (Liu et al. 2015). Although Ca and P contents of sunflower meal are similar to other plant protein sources and low in trace elements, it tends to be high in the B vitamins and β-carotene (Aherne and Kennelly 1985). Unlike other major oilseeds and oilseed meals, sunflower seeds and meals seem to be relatively free of antinutritional factors (Dinusson 1990).

Although sunflower seeds can provide additional energy, the inclusion of sunflower seeds should be limited to 15% or lower in weanling pig diets because of the fiber content (Feedipedia 2012-2020). It has been mentioned earlier that sunflower seeds may increase growth performance of grower-finisher pigs. However, 25 and 50% sunflower seeds in the finisher diets reduced weight gain and carcass traits (Laudert and Allee 1975), and Marchello et al. (1984) indicated that grower-finisher diets should not contain more than 8–10 or 13% sunflower seeds. Although milk fat increased with sunflower seeds in gestation diets, pig survival rate or performance was not affected (Kepler et al. 1982). Because of its effect on feed intake, sunflower seeds should be limited to 25% of lactation diets (Kepler et al. 1982; Feedipedia 2012-2020). In addition to Lys, sunflower meal is deficient in Trp and Thr (Wahlstrom et al. 1985), and it may not be a suitable protein source for weanling pigs (Feedipedia 2012-2020). With Lys supplementation, up to 16% of sunflower meal can be included in the grower-finisher diet without adverse effects on growth performance or carcass traits (Shelton et al. 2001; de Araújo et al. 2014). Gestating sows and adult boars have lower Lys requirements, and sunflower meal can be used as the sole protein source in their diets, provided that diets contain sufficient digestible Lys (Dinusson 1990; Feedipedia 2012-2020). With Lys supplementation or in combination with other protein sources high in Lys, sunflower meal can be used in the lactation diet (Chiba 2001).

The use of sunflower seeds or meal is limited by its high level of fiber and its deficiency in Lys and several other AA. Sunflower seeds should be limited 10–15% of the weanling or grower-finisher diets. Feeding sunflower meal to young pigs should be avoided, but with supplementation with Lys, it can be included in the grower-finisher diet up to 16% without any adverse effects. Similarly, sunflower meal can be used as the protein supplement for reproducing pigs if their diets are balanced for Lys.

## Animal Protein Supplements

### *Animal Protein Sources in General*

Animal protein supplements are a good source of Lys and other AA, and the pattern of AA is often very similar to the dietary needs of the pig. Compared with plant sources, they are also a very good source of vitamins and minerals such as the B vitamins (especially, vitamin $B_{12}$) and Ca and P. However, animal protein supplements are more variable in the nutrient content, and they are subjected to high drying temperatures during processing for dehydration and sterilization. A proper heating is necessary to produce a quality product. Animal protein sources include meat meal and meat and bone meal. Meat meal is distinguished from meat and bone meal based on the P content. If the product contains more than 4% P, it is generally considered as meat and bone meal. For those meals, the Ca content should not be more than 2.2 times the actual P content. In addition, those meat by-product meals should not contain more than 14% pepsin indigestible residues and not more than 11% of the CP should be pepsin indigestible.

Because of many factors, considerable variations in the quality of animal by-product meals can be expected. Clearly distinguishing one animal by-product meal from other animal by-products may be difficult, and also there seem to be differences in the terminology used by various countries. Furthermore, animal by-product meals and other processed animal protein sources were considered as the vector of the bovine spongiform encephalopathy (BSE) epidemy in the Western Europe in the 1980–1990s. For that and other reasons, European Union and many other countries have some restrictions on the use of animal by-products. Readers are referred to various European Union regulations/legislations and Jędrejek et al. (2016) for the status of the use of animal by-products in the European feed industry. For those reasons, the description of meat meal and meat and bone meal and others or the discussion on the use of those products in pig diets in this chapter should be viewed with such uncertainties in mind.

### *Blood Meal*

Older methods of producing blood meal, such as oven drying and drum drying, may result in inconsistent products with varying degrees of contaminations, poor palatability, and low availability of Lys and other AA (Miller 1990; Campbell 1998; Feedipedia 2012-2020). Spray drying and flash drying procedures have improved both the palatability and Lys availability of blood meal (Miller 1990; Cromwell 1998). Blood meal contains mostly CP (about 90–95% of DM) and small amounts of lipids (less than 1% DM) and ash (less than 5% DM; Feedipedia 2012-2020). The composition of blood meal was reviewed by Miller (1990). Spray dried and flash dried blood meals contain 86 and 83% CP and 7.4 and 9.7% Lys, respectively. Similar values for blood meal have been reported by NRC (2012) and Feedipedia (2012-2020). The SID CP and Lys were estimated to be

76.4 and 78.9%, respectively, in grower pigs (Navarro et al. 2018). Blood meal is very high in Leu, which may increase the Ile requirement. Except highly available Fe, mineral concentrations are quite low in blood meal (Feedipedia 2012-2020), and it is a poor source of the vitamins. Like some other animal products, the utilization of blood meal is regulated in some countries for certain species (Feedipedia 2012-2020).

Blood meal is generally unpalatable, especially if overheated, and it should not be included in animal diets more than 5–6% (Feedipcdia 2012-2020). Wahlstrom and Libal (1977) reported that replacing SBM with 4% drum-dried blood meal on equivalent protein basis had adverse effects. However, replacing SBM with 6% of flash-dried blood had no effect on growth performance of grower-finisher pigs, and 3 or 6% flash-dried blood meal had greater N retention in weanling pigs than those fed the SBM diet (Parsons et al. 1985). Miller (1990) indicated that pigs fed diets containing 5% flash-dried blood meal during the starter, grower, and finisher phases performed as well as or better than those fed diets without blood meal (Miller 1990). It has been reported that, after a period of acclimation, spray-dried blood meal can actually stimulate feed intake in weanling pigs (Kats et al. 1994a,b). Blood meals of various origins (i.e., bovine, porcine, and avian) seem to be equally effective (Kats et al. 1994b), and DeRouchey et al. 2002) reported that spray-dried blood meal and blood cells and crystalline Lys had similar bioavailability in weanling pigs. More recently, Abonyi et al. (2016) reported that 75% of SBM in the weanling pig diets can be replaced by blood meal. Two and 5% blood meal can be used by weanling and older pigs, respectively (Miller 1990), but blood meal or blood cells in diets for growing pigs may increase the requirement for, again Ile (van Milgen et al. 2012).

It has been indicated earlier that blood meal may not be an appropriate source of protein for weanling pigs, but young pigs may be able to utilize 3 or 6% flash-dried or spray-dried blood meal. Depending on the status of Leu and Ile, up to 5 or 6% blood meal can be included in the diet for older pigs.

### Feather Meal

Although feathers have a high CP content, they are virtually indigestible in their natural state unless, e.g., disulfide bonds are broken (Papadopoulos 1985). The most widely used commercial feather product is hydrolyzed feather meal. Although feather meal is deficient in Met, Lys, His, and Trp, it is rich in many other AA (Han and Parsons 1991). The nutrient contents of feather meal ranged from 76.4 to 87.3% CP, 2.0 to 8.6% ether extract, 1.3 to 4.8% ash, 1.46 to 2.15% Lys, 4.07 to 5.30% Cys, and 0.45 to 0.61% Met (Han and Parsons 1991). Some hydrolyzed feather meals may also contain poultry blood. Four feather meals without blood contained 76.9–84.8 CP, 6.19–7.28% Lys, 5.78–11.83% ether extract, and 1.23–2.85% ash, whereas four different feather meals with blood contained 80.6–84.8 CP, 6.72–7.41% Lys, 5.7617–8.69% ether extract, and 1.11–3.02% ash (Sulabo et al. 2013a). The availability of nutrients may vary greatly, but with the possible exception of Lys, the availability of indispensable AA in feather meals seems to be similar to SBM for nonruminant species (Han and Parsons 1991). The SID of feather meals without blood ranged from 69.5 to 71.0 and 48.1 to 59.0% for CP and Lys, respectively, and the SID of those with blood ranged from 57.4 to 76.3 and 60.9 to 76.3% for CP and Lys, respectively (Sulabo et al. 2013a).

Hydrolyzed feather meal was effective as a source of nonspecific N to improve carcass quality of finisher pigs (Chiba et al. 1995). The rate and efficiency of weight gain did not differ in finisher pigs fed corn-SBM diets containing 0–12% feather meal, but carcass quality was reduced with 12% feather meal, indicating that up to 9% feather meal could be included in the finisher diet (Chiba et al. 1996). With Lys supplementation, feather meal may be used as the only source of protein supplement without decreasing carcass quality (Chiba et al. 1996). Apple et al. (2003) indicated that as much as 6% FM can be incorporated into isolysinic diets of grower-finisher pigs without adversely affecting growth performance or carcass and meat quality, and Seo et al. (2009) reported similar results. There has been some effort to replace SBM completely with feather meal. Finisher pigs fed the corn-feather meal diet supplemented with the necessary AA utilized feed and AA for body weight gain and lean gain as efficiently as those fed the corn-SBM diet (Divakala et al. 2009). Similarly, finisher pigs fed the diet formulated based on the determined SID AA content of feather meal with blood and supplemented with Lys and Trp utilized feed and Lys as efficiently as those fed the corn-SBM diets (Brotzge et al. 2014). Enzymatically treated feather meal had positive effect on growth performance and intestinal health (Pan et al. 2016), and it can be a promising source of protein for weanling and other pigs.

Because of the deficiency of important indispensable AA, feather meal should be incorporated into the pig diet based on AA, which may increase the CP content of diets and the excretion of urinary N into the environment. Supplementation with appropriate AA, therefore, would be the best way to use feather meal in pig diets. For optimum growth performance, feather meal should be limited to 6% of diets. However, with appropriate AA supplementation, it is possible to utilize feather meal as a sole source of protein supplement for older pigs.

### Fish Meal

World production of fish meal was 5.9 mmt in 1992–1993, 5.1 mmt in 1997–1998 (Chiba 2001), 4.5 mmt in 2015–2016, and 4.6 mmt in 2019–2020 (Table 12.2). The major producers are Peru, Viet Nam, Chile, China, and Thailand. Fish meal should not contain more than 10% moisture or 7% salt, and the amount of salt must be specified if it is greater than 3% (Chiba 2001). Although most of the oil is removed from fish meal (8–11%; Feedipedia 2012-2020), antioxidants are commonly included in the meal to prevent oxidation, overheating, and molding (Church and Kellems 1998). Fish meals are generally high in protein (50–75%) and indispensable AA that are deficient in many cereal grains (Seerley 1991; Church and Kellems 1998; NRC 2012), and the ash content is generally below 12% (Feedipedia 2012-2020). The average Lys, Thr, and Trp contents are 4.56, 2.58, and 0.63%, respectively (NRC 2012) and those range from 7.0 to 8.1, 3.1 to 4.6, and 0.8 to 1.2% of the CP, respectively (Feedipedia 2012-2020). Most mineral elements, especially Ca and P, and B vitamins are moderate to high compared with other protein sources (Seerley 1991; Church and Kellems 1998). The quality of fish meal may vary greatly depending on the quality of the raw materials, processing factors, and oxidation (Seerley 1991), and there may be some restrictions on its use in some countries (Feedipedia 2012-2020).

Fish meal is highly digestible and has a high biological value (Feedipedia 2012-2020). Addition of 5% fish meal promoted optimum pig performance during the first three week after weaning, whereas 10% fish meal was most effective during the last two week of the five-week weanling pig study (Gore et al. 1990). Similarly, weanling pigs fed the diets with 6% fish meal had greater feed intake and weight gain than those fed the control diet (Jones et al. 2018). Stoner et al. (1990) reported that the maximum weight gain in weanling pigs was obtained with 8% fish meal, but 12% fish meal was necessary to maximize their feed intake. Bergström et al. (1997) indicated that weanling pigs with the low health status seemed to respond more to fish meal than those with high-health status. Laksesvela (1961) reported earlier that most of the growth response to fish meal in grower-finisher pigs was obtained with the initial concentration of 6–8%, and carcass quality was not affected by

6–8% fish meal or less. The response of grower-finisher pigs to 5% fish meal was greater with the high-nutrient diet than the conventional diet (Pike et al. 1984). Cromwell (1998) suggested that the amount of fish meal in pig diets should not exceed 6–7% because of its potential of causing a fishy flavor in pork, and its use should be avoided in the diets of pigs that are approaching the slaughter weight.

The variation in the quality of fish meal is an important consideration in determining the optimum inclusion rate in pig diets. The results of studies seem to indicate that the optimum inclusion rate would be somewhere between 5 and 12% of the diet. Considering the organoleptic characteristic of pork, however, the use of fish meal should be limited to 6–7% of the finisher diet.

## Meat Meal and Meat and Bone Meal

Meat meal and meat and bone meal can be defined as the rendered product from mammalian tissues, exclusive of blood, hair, hoof, horn, hide trimmings, manure, and stomach and rumen contents (Chiba 2001). Meat meal consists mostly of waste meat trimmings and organs, whereas meat and bone meal may contain 21–61% bone. Bone protein is 83% collagen, which may account for 50–65% of the total CP, and contains largely indigestible keratin. If the product contains more than 4% P, again, it is considered as meat and bone meal. The average compositions of meat by-product meals are 56.4% CP, 3.20% Lys, 1.89% Thr, 0.40% Trp, 11.1% ether extract, 6.37% Ca, and 3.16% P for meat meal, and 50.1% CP, 2.59% Lys, 1.63% Thr, 0.30% Trp, 9.21% ether extract, 10.94% Ca, and 5.26% P for meat and bone meal (NRC 2012). The SID of CP, Lys, Thr, and Trp in finisher pigs were 63.5, 65.1, 65.9, and 78.5%, respectively, for meat meal (Kong et al. 2014), and 81.0, 77.0, 76.9, and 81.0%, respectively, for meat and bone meal (Navarro et al. 2018). The ME and N-corrected ME ranged from 1569 to 3308 kcal/kg DM and 1474 to 3361 kcal/kg DM, respectively (Adedokun and Adeola 2005). The apparent total tract digestibility of P (52.1–80.1%) and Ca (53.0–81.0%) in meat and bone meals differed greatly, and the standardized total tract digestibility of P ranged from 54.8 to 84.4% (Sulabo and Stein 2013). The use of meat by-product meals as a protein supplement in the animal feed for various species can be, again, regulated in many countries.

The N retention was improved by supplementation of meat by-product meal with either Lys and Met or Trp, but supplementation with all 3 AA was necessary to improve growth performance of weanling pigs (Leibholz 1982). Growth performance and DM digestibility in weanling pigs decreased linearly as replacement of SBM with meat meal increased from 0 to 100% (Evans and Leibholz 1979). Inclusion of 5–10% meat meal reduced the rate and efficiency of growth (Cromwell et al. 1991), but addition of L-Trp prevented the reduction of growth performance in grower-finisher pigs. Further supplementation with Ile or a combination of Ile and Thr had no effect. Lettner et al. (2001) indicated that up to 12% of meat meal can be included in the finisher diet, provided that the diet is balanced for the limiting ileal digestible AA. A low-ash meat and bone meal supported better rate and efficiency of growth than a high-ash meant and bone meal, but grower pigs fed the SBM diet had better growth performance and greater digestibility and N retention than those fed the meat and bone meal diets (Kennedy et al. 1974; Partanen et al. 1998). The efficiency of N utilization in finisher pigs decreased as dietary meat and bone meal increased from 10 to 20% (Partanen and Nasi 1994). Meat by-product meals can be used to replace one third of the SBM or protein supplement without any adverse effects on growing pigs (Seerley 1991) or sows (Cromwell 1998).

Meat meal and meal and bone meal should be included no greater than 10% of the pig diet because of the variation in the nutritional quality, palatability problems, and the Ca and P contents. More than 10% meat by-product meals can be included in grower-finisher pigs diets if diets are fortified with appropriate AA. It has been suggested that meat by-product meal can replace up to one-third of SBM or protein supplement grower-finisher and sow diets.

### Milk, Dried

Dried milk or milk powder is produced by evaporating milk to dryness, and dried milk products include dried whole milk and dried skim milk. The only difference between dried whole milk and dried skim milk is that most of the fat and fat-soluble vitamins are removed from the dried skim milk (Chiba 2001). Both milk products are very palatable and highly digestible protein supplements with an excellent balance of AA (Pond and Maner 1984). Dried skim milk contain: 36.8% CP, 2.42% Lys, 1.44% Thr, 0.44% Trp, 0.9% ether extract, 47.8% lactose, 1.27% Ca, and 1.06% P (NRC 2012). Both dried whole milk and dried skim milk are good sources of vitamins and minerals (Seerley 1991; Cromwell 1998). The only nutrients that tend to be deficient are the fat-soluble vitamins, Fe, and Cu, and depending on the species, Mg and Mn (Seerley 1991; Church and Kellems 1998). Under most instances, the value of milk products in human diets makes them too valuable to be used in animal diets.

Although they are nearly always too expensive for use as a feed ingredient (Mahan 2003), a certain amount of dried skim milk has been used in pig starter diets over the years. Research emphasis on this area has been mostly to replace dried skim milk with alternative protein supplements in weanling pig diets rather than evaluating the value of the dried milk per se. Dried whole milk and skim milk can be fed successfully to all classes of pigs. For instance, Yen et al. (2004) concluded that grower-finisher pigs fed diets containing 10% dried skim milk would have growth performance, carcass traits, and N digestibility similar to those fed typical corn–SBM diets. Their results indicate that dried milk can be used as a source of protein for older pigs when the price is favorable because of, e.g., its surplus (Yen et al. 2004).

Dried milk products are almost the perfect food for very young pigs and also for other classes of pigs. The use of these products in pig diets would be determined mostly based on the economic considerations.

### Plasma Protein

The plasma fraction of blood yields a fine, light tan powder containing 78% protein (spray-dried plasma protein), whereas the blood cell fraction yields a fine, dark red powder containing 92% protein (spray-dried blood cells; Campbell 1998; NRC 2012). Spray-dried plasma protein contains approximately 78% CP, and it is relatively high in Lys, Trp, and Thr (6.90, 1.41, and 4.47% respectively), but low in Met and Ile (0.79 and 2.29%, respectively; Campbell 1998; NRC 2012). Besides its AA content, globular proteins (including immunoglobulins) in dried plasma may stimulate feed intake and growth during the critical postweaning stage, and it has been shown to be an excellent source of protein for early-weaned pigs (Cromwell 1998). Spray-dried plasma seems to be highly digestible, and according to NRC (2012), the SID of CP, Lys, Trp, and Thr are 81, 87, 92, and 80%, respectively. The AA profile of plasma protein seems to match closely to the young pig's requirements. Plasma protein may have a positive effect on the immune system of the young pig.

Campbell (1998) summarized 25 studies that evaluated the nutritional value of spray-dried plasma protein. Essentially, all the experiments have shown that plasma protein can improve feed intake and weight gain of early-weaned pigs and reduce the postweaning lag. Similar results have been reported by others (e.g., Gottlob et al. 2007). Researchers reported earlier that 6% (Gatnau and Zimmerman 1992) or up to 10% plasma protein (Kats et al. 1994a) can be used during the early postweaning period. With more than 6% dietary plasma protein, however, Met may become the first limiting AA (Kats et al. 1994a). Pigs reared in an on-farm nursery setting responded more to plasma protein than those in a cleaner, off-site nursery (Bergström et al. 1997), indicating the effect of the nursery environment on its efficiency. Similarly, weanling pigs affected by PRRS virus seemed to have a greater response to plasma protein (Crenshaw et al. 2017). Several studies indicated that the high molecular weight fraction (globulin) may have a biological function, and it can affect the pig performance through the gut health (Campbell 1998; Balan et al. 2020). Plasma protein in the lactation diet had no beneficial effects on young sows, but it increased reproductive performance of older sows (Carter et al. 2018).

Spray dried plasma protein can increase growth performance of weanling pigs by, perhaps, enhancing feed intake and the immune function. The degree of response is dependent on the inclusion rate, age, and weight of pigs, health status, and environment. Spray-dried plasma protein may be included up to 6–10% of the diet for weanling pigs, but the diet may have to be supplemented with Met if it contains more than 6% plasma protein.

### *Poultry By-Product Meal*

Poultry by-product meal consists of the ground, rendered, or cleaned part of slaughtered poultry, such as heads, feet, undeveloped eggs, and intestines, exclusive of feathers (Church and Kellems 1998; Feedipedia 2012-2020). It cannot contain more than 16% ash and more than 4% acid-insoluble ash (Church and Kellems 1998). Poultry by-product meal contains 58% CP and 13% fat, and it has good AA balance and provides minerals and vitamins (Seerley 1991). Dong et al. (1993) reported that poultry by-product meals obtained from several manufacturers in North America differed in chemical composition (e.g., 55–74% CP, 10–19% ether extract, and 11–23% ash on a DM basis) and protein digestibility, indicating the range of product quality that can be expected in the market place. More recent data on the average composition of poultry by-product meal are 64.0% CP, 3.39% Lys, 2.35% Thr, 0.46% Trp, 12.0% ether extract, and 13.3% ash (NRC 2012). Rojas and Stein (2013) reported that the SID values for poultry by-product meal were 72.1% CP, 68.9% Lys, 67.1% Thr, and 72.7% Trp. The apparent total tract digestibility of GE was 89.2% (Rojas and Stein 2013).

Poultry by-product meal is generally palatable and can be included in the pig diet similar to meat meal (Feedipedia 2012-2020), and Seerley (1991) indicated that poultry by-product meal can replace some SBM, and it is an excellent source of protein for pigs. Rostagno et al. (2005) indicated that the amount of digestible Lys may be comparable to SBM, even though the value may depend on the processing method. Little research has been conducted on the use of poultry by-product meal in the pig diet (Chiba 2001; Feedipedia 2012-2020). Replacing plasma protein with 10% poultry by-product meal had no effect on overall growth performance of weanling pigs (Veum et al. 1999). Similarly, replacing plasma protein, blood meal, fish meal, and portion of SBM with 20% poultry by-product meal had essentially no effect on overall growth performance of weanling pigs (Zier et al. 2004). The low-ash poultry by-product meal seemed to be a better source of protein than the high-ash poultry by-product meal for weanling pigs (Keegan et al. 2004). Orozco-Hernandez et al. (2003) reported that increasing poultry by-product meal up to 7.5% in the weanling to finisher pig diets resulted in reduced growth performance. Similarly, finisher pigs fed diets containing

15% poultry meal as the sole protein source had reduced feed intake and weight gain compared with those fed the SBM diets (Shelton et al. 2001). It has been suggested that poultry by-product meal should be limited to 5–8% of pig diets because of the variability of the products on the market (Chiba 2001).

Poultry by-product meal can be an excellent source of protein for pigs. Some researchers demonstrated that up to 20% poultry by-product meal can be included in the pig diets without any adverse effects on growth performance. However, the composition or protein quality of poultry by-product meals on the market may vary considerably. For that reason, perhaps, it would be best to limit the use of poultry by-product meal to approximately 5–8% of pig diets.

### Whey, Dried

The global production of whey was estimated to be 200 mmt in 2011 (Arévalo et al. 2016). Generally, dried whey contains about 90, 20, 40, and 43% of the lactose, protein, Ca, and P that are originally present in milk. Dried whey contains at least 65% lactose, which has been shown to be the best sugar for baby pigs (Seerley 1991). Dried whey may not be considered as a protein supplement because of its CP content, but it normally contains about 13–17% of high-quality CP (Seerley 1991; Arévalo et al. 2016). The average composition of dried whey or whey powder to be 72.9% lactose, 11.6% CP, 0.88% Lys, 0.71% Thr, 0.20% Trp, 0.83% ether extract, and 8.0% ash (NRC 2012). Reis et al. (2007) reported that the ileal CP digestibility of sweet dried whey was 68.3%. Generally, AA availability of milk products is considered high, but their quality can be impaired by overheating. Dried whey is an excellent source of B vitamins, but it may be low in vitamins A and D, which are retained in the cheese (Leibbrandt and Benevenga 1991; Seerley 1991). Season, type of cheese produced, and geographical locations can have an effect on the mineral content of dried whey (Leibbrandt and Benevenga 1991). The type of cheese produced and the addition of salt-laden press drippings influence the salt content of whey (Leibbrandt and Benevenga 1991).

Although there have been some research effort to replace dried whey with some other ingredients, the research to evaluate its effect on weanling pig performance has been rather limited in recent years. However, the efficacy of high-quality dried whey in enhancing growth performance of weanling pigs has been well established in the past (e.g., Cera et al. 1988). Pancreatic enzyme activities in the digestive system seems to be greater in pigs fed the diets containing dried whey (Lindemann et al. 1986), and consequently it is highly digestible and can be utilized efficiently (Cera et al. 1988). By including either lactalbumin or lactose in the diet, Tokach et al. (1989) indicated that both the protein and carbohydrate fractions of dried whey are important. However, Mahan (1992) and Mahan et al. (1993) concluded that the lactose component of dried whey was primarily responsible for the beneficial effects of dried whey. It has been shown that edible-grade deproteinized whey and crystalline lactose can replace the lactose provided by high-quality dried whey without affecting pig performance (Nessmith et al. 1997). Tokach et al. (1995) reported that feeding the milk product (dried skim milk and edible dried whey) has beneficial effects on pigs not only during the starter phase but also during the grower-finisher phase. The inclusion rate of 10–30% is commonly used but 30–45% can be included without any adverse effects on weanling pig performance (Seerley 1991).

Dried whey is an excellent source of carbohydrate, i.e., lactose, and high-quality CP, and can be fed to all classes of pigs. However, it is primarily used for weanling pigs. Most diets offered during the early weaning phase contain some dried whey. Although 30–40% of dried whey can be included in the weanling pig diet, it is routinely included in weanling pig diets at 10–30%.

The optimum inclusion rate of dried whey in pig diets would be determined mostly by economic considerations.

## Other Protein Supplements or Feed Additives

Because of the projected increase in the global population and the expansion of animal production, potential protein sources, such as microalgae and insect meals, have been receiving a considerable interest in recent years. In addition, along with some yeast products, other proteinaceous substances, such as fish protein hydrolysates or peptides and antimicrobial peptides, have been receiving some attention. For instance, a wide variety of fish protein hydrolysates or peptides are considered as a bioactive ingredient and shown to have antioxidative, antihypertensive, immunomodulatory, and antimicrobial activities (Chalamaiah et al. 2012). Antimicrobial peptides have been reported to have beneficial effects on growth performance, nutrient digestibility, the intestinal microbiota, intestinal morphology, and immune functions in pigs (Xiao et al. 2015). Those hydrolysates or peptides are not included in the diet as a source of AA per se, and also they may not be considered as a protein supplement. Rather, they may be considered as a feed additive, even though, traditionally, feed additives have been defined as a nonnutritive substance added to feed to improve the efficiency of feed utilization, feed acceptance, or benefit the health or metabolism of the animal in some way. Regardless of the classification, a brief discussion on some of those proteinaceous feed ingredients in this chapter may be justified because they can be a source of dietary protein or AA.

### *Antimicrobial Peptides*

To prevent diseases and improve growth performance in pigs, subtherapeutic doses of antibiotics in feed have been used in the United States and Europe since 1950s. Public concerns on the antibiotic resistance have led to the ban of antibiotic as growth promoters in the European Union and others and the restricted use of dietary antibiotics in other countries. In recent years, considerable effort has been made to explore alternatives to antibiotics to ensure health and growth performance of pigs. Antimicrobial peptides can be promising novel antibiotics or alternatives to conventional antibiotics. Antimicrobial peptides are part of the innate defense peptides or peptides produced as a first line of defense by all living organisms (Hollmann et al. 2018). Those compounds with a variable AA composition (typically, 6–100 AA residues) can be mammalian (e.g., defensin), amphibian (e.g., magainin), insect (e.g., cecropin), plant (e.g., thionin), or microbial (e.g., gramicidin and nisin) origin and can be classified as antiviral, antibacterial, antifungal, or antiparasitic peptides (Xiao et al. 2015). For the complete list of antimicrobial peptides, please see, e.g., the online database (http://aps.unmc.edu/AP/).

Supplementation with 500 mg antimicrobial peptides/kg increased serum IgG, IgM, IgA, swine fever antibody, and total complement activity, indicating that dietary antimicrobial peptides enhanced the humoral immune response of weaning pigs (Yuan et al. 2015). Similarly, 400 mg antimicrobial peptides/kg reduced the secretion of IgA in the jejunum and reduced serum IgA, IgG, interleukin-1β, and interleukin-6 in newly weaned pigs challenged with *Escherichia coli* (Wu et al. 2012), and the authors concluded that the enhanced growth performance by antimicrobial peptide supplementation was due to increased immune status and reduced intestinal pathogens and increased N and energy retention. In addition, Ren et al. (2015) reported that antimicrobial peptides increased the population and function of T cells and reduced the apoptotic cells, indicating, again, the improved cellular immune function in weanling pigs. They recommended 1000 and 500 mg antimicrobial peptides/kg for the first two weeks and the entire four-week period, respectively. After reviewing a number of studies, Xiao et al. (2015) concluded that a single or composite antimicrobial peptides can improve the intestinal morphology, energy and nutrient digestibility, immune status, and growth performance, alleviate the intestinal injury and metabolic disturbances, and increased the survival rate of weanling pigs.

Because of their wide range of activities, antimicrobial peptides may have beneficial effects on the nutritional status, health, and growth performance of pigs. Although antimicrobial peptides may not be contributing protein or AA as such, it is clear that those peptides are effective in improving the health and growth performance of weanling pigs.

### *Fish Protein Hydrolysates and Peptides*

The fish industry is a pillar of their economy in many countries worldwide (Chalamaiah et al. 2012; Zamora-Sillero et al. 2018). Depending on the type of fish and processing method, a substantial portion of biomass is not used for human consumption (Gevaert et al. 2016; Zamora-Sillero et al. 2018). The inedible parts can be processed into various products, such as fish meal and fertilizer (Chalamaiah et al. 2012). In recent years, the production of protein hydrolysates has increased worldwide using both chemical and biological methods (Zamora-Sillero et al. 2018). Chalamaiah et al. (2012) summarized the proximate and AA composition of more than 50 fish protein hydrolysates, and their CP content ranged from 60 to 90%. Although the AA content varied depending on several factors, those fish protein hydrolysates are a good source of indispensable AA. Those hydrolyzed fish products have antioxidative, antimicrobial, antihypertensive, and other activities (Zamora-Sillero et al. 2018). Fish hydrolysates are now available worldwide as a supplement and a functional dietary ingredient (Gevaert et al. 2016). Various biological activities have been demonstrated in both animals and humans, but there seem to be no consistent regulatory guidelines on those hydrolysates (Gevaert et al. 2016).

Earlier, Stoner et al. (1985) reported that although feed intake or feed efficiency was not affected, pigs fed the diet supplemented with 3% fish protein hydrolysates had greater weight gain than those fed the corn-SBM diet. More recently, Thuy et al. (2016) reported that replacing fish meal with catfish protein hydrolysates was effective in improving growth performance of weanling pigs. Similarly, weanling pigs fed diets containing salmon hydrolysate had greater feed intake than those fed the diet containing soy protein concentrate, and they responded similarly to those fed the diet with fishmeal (Nørgaard et al. 2012a). Although salmon protein hydrolysates had no effect on growth performance of weanling pigs, their duodenal absorption area was improved by the hydrolysate supplementation (Opheim et al. 2016). Salmon protein hydrolysates differed in the raw materials resulted in no difference in growth performance, intestinal morphology, or intestinal microbiota in weanling pigs (Opheim et al 2016). The inclusion of 3% spray dried plasma protein and salmon protein hydrolysate had no effect on the growth performance of weanling pigs, and it had no subsequent carryover effects, and Tucker et al. (2011) concluded that salmon hydrolysate can be a viable alternative protein for weanling pigs. Fish peptides seemed to affect the fecal microbiome of weanling pigs, which could have long term or lasting beneficial effects on grower-finisher pigs (Poudel et al. 2020).2016).

The use of fish hydrolysates as a source of protein for young pigs is not really a new idea. Research over the years has clearly shown that, e.g., fish protein hydrolysates or peptides can effectively replace some special ingredients that are typically included in weanling pig diets to enhance their health and growth performance. Fish protein hydrolysates or peptides can be not only a viable source of protein for young pigs but also a functional dietary ingredient.

### Insect Meals

There are approximately 1 million known species of insects in the world and, perhaps, millions of others have not been identified (Sánchez-Muros et al. 2014). In recent years, insects have been receiving some attention as an important and sustainable alternative source of protein for animal feeds (Sogari et al. 2019). Sánchez-Muros et al. (2014) summarized the proximate composition of more than 60 species and the AA content of more than a dozen species. The CP content was > 40% for 28 species, > 60% for 20 species, and 70.1–77.1% CP for 4 species, and the CP digestibility ranged from 45 to 86%. Some insect species contain indigestible components such as chitin, and none of the species studied approached the Lys content of fish meal. However, some species contained greater amount of other indispensable AA compared with fish meal. It should be pointed out that only a handful of species were evaluated and also there could be millions of other species with different stages of development, which may differ in AA profiles considerably (Sánchez-Muros et al. 2014). Nevertheless, some insects can be a viable source of energy, CP, lipids, minerals, and vitamins (Rumpold and Schlüter 2013). Some insects can grow and reproduce easily, have great feed efficiency, and can be reared with bio-waste products. Although there are some challenges associated with the use of insects or insect meal as a source of protein for animal feeds, such as the mass production (Sánchez-Muros et al. 2014), the regulation (Sogari et al. 2019), and the consumer acceptance (Sogari et al. 2019), again, they can be an important source of protein and amino acid for food producing animals.

Although there have been some studies with fish and poultry and limited research with pigs (e.g., Newton et al. 1977), the use of insects or insect meals as a source of protein or AA for pig diets has not been explored until recently. It has been shown that black solder fly (*Hermetia illucens*) larvae meal or prepupae can be an effective replacement for SBM and grease in weanling pig diets (Spranghers et al. 2018). Biasato et al. (2019) concluded that a partially defatted black soldier fly meal can be used for weanling pigs without adverse effects on growth performance, nutrient digestibility, blood profile, gut morphology, or histological features. Although there was no difference in growth performance or fecal digestibility between the SBM and insect diets, the prepupae of the black soldier fly reduced group D Streptococcus slightly in weanling pigs (Spranghers et al. 2018). For finisher pig, Yu et al. (2019) indicated that black soldier fly may enhance mucosal immune system by altering bacterial composition and their metabolites. As dietary dried mealworm (*Tenebrio molitor*) increased, the N retention, DM and CP digestibility, and insulin-like growth factor increased linearly and decreased blood urea N in weanling pigs (Jin et al. 2016). The growth performance of weanling pigs improved up to 6% mealworm in their diet (Jin et al. 2016). In addition, weanling pigs fed the whole and peeled (legs removFeeding ed) crickets (*Teleogryllus testaceus*) had greater weight gain, feed efficiency, nutrient digestibility, and N retention than those fed the fishmeal diet (Miech et al. 2017).

The mass production, the regulatory systems, and the consumer acceptance are some of the challenges associated with the use of insects or insect meals for pig diets. Recent positive findings may provide a new perspective on insects or insect meals as a sustainable alternative source of protein or AA for pig production in the future.

### Microalgae

Algae are photosynthetic organisms that grow in a range of aquatic habitats, and there are two types of algae. Macroalgae are multicellular, plant-like form of algae and commonly referred to as seaweeds. Microalgae are unicellular photosynthetic microorganisms living in saline or freshwater environments that convert sunlight, water, and carbon dioxide to algal biomass. Microalgae can produce a wide range of compounds such as polysaccharides, lipids, pigments, proteins, vitamins, bioactive compounds, and antioxidants (Yaakob et al. 2014). There are distinct groups of microalgae such as eukaryotic diatoms (*Bacillariophyceae*), green algae (*Chlorophyceae*), and golden algae (*Chrysophyceae*), and also cyanobacteria (*Cyanophyceae* or blue-green algae; Madeira et al. 2017). There may be more than 100 000–800 000 species of microalgae, and only part of those species have been described in literature (Madeira et al. 2017). Culturing algae has been practiced since the late 1800s and early 1900s (Yaakob et al. 2014), and diverse approaches have been used to produce and characterize microalgae-derived protein products such as protein concentrates, hydrolysates, and bioactive peptides that can be used as animal feeds (Soto-Sierraa et al. 2018). Although the protein content of some species may be in the 20s, most species contain 40–70% CP, and many species of microalgae contain 2.5–7.0 g Lys/100 g CP and reasonable amount of other indispensable AA (Madeira et al. 2017).

Dietary microalgae supplementation improved nutrient digestibility and increased lymphocyte concentration but had no effect on growth performance or blood metabolites (Kibria and Kim 2019). Supplementation with 1 and 5% of partially defatted microalgae improved growth performance of weanling pigs (Urriola et al. 2018), but 10% defatted microalgae had no effect on growth performance, even though they reduced plasma urea N (Ekmay et al. 2014). Fish oil, which is high in T-3 FA, or microalgae reduced the stress response in weanling pigs following the immune stress challenge and both can differentially modulate the acute-phase response, indicating that microalgae are a viable source of ω-3 FA (Lee et al. 2019). Supplementation of the diet for young gilts with ω-3 FA from microalgae improved feed efficiency and reduced serum cholesterol, but it had no beneficial effects on the ovary (Otte et al. 2019). Newmann et al. (2018) indicated that spirulina (blue-green algae) can replace SBM in the growing pig diets completely, provide that the diet is balanced for AA. After reviewing several studies, Madeira et al. (2017) concluded that different microalgae had no effect on physical characteristics of pork such as color, pH, oxidative stability, cooking loss, or some others, and they can reduce ω-6 to ω-3 ratio or increase the eicosapentaenoic acid and docosahexaenoic acid contents of pork. Similarly, Huang et al. (2021) indicated that feeding microalgae to pigs can enhance the ω-3 FA content of pork.

Microalgae can be a good source of CP and AA, carbohydrates, lipids, important minerals and vitamins, carotenoids, and other nutrients, and they may improve the health status of young pigs. Microalgae can replace protein supplement in pig diets without adverse effects on growth performance or pork quality, and they can increase the ω-3 FA content of pork.

### Yeast Products

Yeast products consist of yeast biomass or dead yeast cells such as dried yeast, brewer's dried yeast, torula dried yeast, and whey yeast (Stone 2006; Shurson 2018). Dried yeast is grown on sugar substrates in aerobic bioreactors, and it is mostly used by the food industries (Stone 2006; Shurson 2018). Brewer's yeast is a by-product of the beer and ale industry (Stone 2006; Shurson 2018), whereas torula yeast is a special yeast species often grown on the wastewater from the paper industry

(NRC 1998, 2012; Stone 2006; Shurson 2018). Both brewer's yeast and torula yeast are used by the feed industry. Whey yeast, by-product from the cheese industry, has been also used by the feed industry, but its production and availability have diminished over the years (Stone 2006). Yeast products are a good source of CP or AA and contain approximately 40% CP, but approximately 20% of the CP in yeast is in the form of nucleic acids (Stone 2006). Although the total Lys, Met, and Thr contents of, e.g., brewer's yeast is better than those in SBM, their availability of indispensable AA is lower than SBM (Shurson 2018). Brewer's yeast is an excellent source of lipids, energy, P, and B vitamins (Stone 2006; Kim et al. 2014; Shurson 2018). The cell wall of the yeast cell consists of polysaccharides and specific components, such as β-D-glucans and α-D-mannans, may be responsible for the modulation of gut microbiota and immunity (Liu et al. 2018).

It has been shown that yeast extract or yeast-derived protein can partially replace blood plasma, and also yeast products can be a possible replacement for antimicrobial additives for weanling pig diets (Carlson et al. 2005; Pereira et al. 2012). LeMieux et al. (2010) indicated that dried brewer's yeast can be used in preweaning and weanling pig diets without any adverse effects on growth performance of young pigs. Yeast extracts contain nucleotides (Liu et al. 2018) and increasing yeast-based nucleotides in antibiotic-supplemented diets reduced energy and nutrient digestibility in weanling pigs (Waititu et al. 2016a). However, supplementation with nucleotide-rich yeast extract under unsanitary conditions can improve the immune response in weaning pigs (Waititu et al. 2016b; Waititu et al. 2017). For grower pigs, Zhang et al. (2019) indicated that brewer's yeast hydrolysate increased the digestibility of dry matter, energy, and N and improved growth performance. Supplementation of the gestation diet with yeast derivatives resulted in more favorable microbiota in sows and increased the colostrum availability and its energy content (Hasan et al. 2018). In addition, yeast derivatives in the gestation diet promoted beneficial microbiota in the neonate at one week of age (Hasan et al. 2018). Supplementation of gestation and lactation diets with yeast culture had beneficial effects on sow productivity by improving litter weight gain (Kim et al. 2008).

Yeast or yeast products can be used to replace special ingredients in weaning pig diets. In addition, because of their beneficial effects on the immune system, perhaps, yeast products can be a viable alternative to antimicrobial agents. Dietary yeast products may have beneficial effects on the microbiota of sows and their neonates and, possibly, reproductive performance.

## References

Abonyi, F. O., B. C. O. Omeke, and B. N. Marire. 2016. Evaluation and utilization of blood meal diets by weaner pigs reared under tropical environment. Niger. Vet. J. 37:72–81.

Adams, S., X. Kong, J. Hailong, G. Qin, F. L. Sossah, and D. Che. 2019. Prebiotic effects of alfalfa (Medicago sativa) fiber on cecal bacterial composition, short-chain fatty acids, and diarrhea incidence in weaning piglets. RSC Adv. 9:13586–13599.

Adedokun, S. A., and O. Adeola. 2005. Metabolizable energy value of meat and bone meal for pigs. J. Anim. Sci. 83:2519–2526.

Aherne, F. X., and J. M. Bell. 1990. Canola seed: Full-fat. In: P. A. Thacker, and R. N. Kirkwood, editors, Nontraditional feed sources for use in swine production. Butterworth-Heinemann, Boston, MA. p. 79–90.

Aherne, F. X., and J. J. Kennelly. 1985. Oilseed meals for livestock feeding. In: D. J. A. Cole, and W. Haresign, editors, Recent developments in pig nutrition. Butterworth-Heinemann, London, UK. p. 278–315.

An, J. Y., H. I. Yong, S. Y. Kim, H. B. Yoo, Y. Y. Kim, and C. Jo. 2017. Quality of frozen pork from pigs fed diets containing palm kernel meal as an alternative to corn meal. Kor. J. Food. Sci. Anim. Resour. 37:191–199.

Apple, J. K., C. B. Boger, D. C. Brown, C. V. Maxwell, K. G. Friesen, W. J. Roberts, and Z. B. Johnson. 2003. Effect of feather meal on live animal performance and carcass quality and composition of growing-finishing swine. J. Anim. Sci. 81:172–181.

de Araújo, W. A. G., L. F. T. Albino, H. S. Rostagno, M. I. Hannas, J. A. P. Luengas, F. C. de Oliveira Silva, T. A. Carvalho, and R. C. Maia. 2014. Sunflower meal and supplementation of enzyme complex in diets for growing and finishing pigs. Braz. J. Vet. Res. Anim. Sci. 51:49–59.

Arévalo, J. C. S., H. C. Vázquez, E. S. Amr, K. Mohammed, C. B. Cedeño, M. A. C. Vazquez, and A. C. Carlos. 2016. The use of sweet whey for weaning pigs. Global Vet. 16:52–56.

Balogun, T. F., and B. A. Koch. 1979a. Influence of trypsin inhibitor level and processing on the nutritional value of groundnuts for finishing pigs. Tropical Agric. 56:245–251.

Balogun, T. F., and B. A. Koch. 1979b. Raw or roasted groundnuts as a partial protein and energy source in rations for growing pigs. Tropical Agric. 56:135142.

Batterham, E. S., L. M. Anderson, D. R. Baigent, S. A. Beech, and R. Elliot. 1990. Utilization of ileal digestible amino acids by pigs. Br. J. Nutr. 64:679–690.

Bergström, J. R., J. L. Nelssen, M. D. Tokach, R. D. Goodband, S. S. Dritz, K. Q. Owen, and W. B. Nessmith, Jr. 1997. Evaluation of spray-dried animal plasma and select menhaden fish meal in transition diets of pigs weaned at 12 to 14 days of age and reared in different production systems. J. Anim. Sci. 75:3004–3009.

Biasato, I., M. Renna, F. Gai, S. Dabbou, M. Meneguz, G. Perona, S. Martinez, A. C. B. Lajusticia, S. Bergagna, L. Sardi, M. T. Capucchio, E. Bressan, A. Dama, A. Schiavone, and L. Gasco. 2019. Partially defatted black soldier fly larva meal inclusion in piglet diets: Effects on the growth performance, nutrient digestibility, blood profile, gut morphology and histological features. J. Anim. Sci. Biotechnol. 10:12.

Balan, P., M. Staincliffe, and P. J. Moughan. 2020. Effects of spray-dried animal plasma on the growth performance of weaned piglets - A review. J. Anim. Physiol. Anim. Nutr. doi: https://doi.org/10.1111/jpn.13435.

Bowland, J. P. 1990. Linseed meal. In: P. A. Thacker, and R. N. Kirkwood, editors, Nontraditional feed sources for use in swine production. Butterworth-Heinemann, Boston, MA. p. 213–236.

Brotzge, S. D., L. I. Chiba, C. K. Adhikari, H. H. Stein, S. P. Rodning, and E. G. Welles. 2014. Complete replacement of soybean meal in pig diets with hydrolyzed feather meal with blood by amino acid supplementation based on standardized ileal amino acid digestibility. Livest. Sci. 163:85–93.

Campbell, J. M. 1998. The use of plasma in swine feeds. Discoveries. A Quarterly Tech. Update. American Protein Corp., Ames, IA.

Canola Council of Canada. 2019. Canola meal feeding guide. 6th ed. Canola Council of Canada, Winnipeg, MB, Canada.

Carlson, M. S., T. L. Veum, and J. R. Turk. 2005. Effects of yeast extract versus animal plasma in weanling pig diets on growth performance and intestinal morphology. J. Swine Health Prod. 13:204–209.

Casas, G. A., C. Huang, and H. H. Stein. 2017. Nutritional value of soy protein concentrate ground to different particle sizes and fed to pigs. J. Anim. Sci. 95:827–836.

Cera, K. R., D. C. Mahan, and G. A. Reinhart. 1988. Effects of dietary dried whey and corn oil on weanling pig performance, fat digestibility and nitrogen utilization. J. Anim. Sci. 66:1438–1445.

Chalamaiah, M, B. Dinesh Kumar, R. Hemalatha, and T. Jyothirmayi. 2012. Fish protein hydrolysates: Proximate composition, amino acid composition, antioxidant activities and applications: A review. Food Chem. 135:3020–3038.

Chen, L., L.-X. Gao, L. Liu, Z.-M. Ding, and H.-F. Zhang. 2015. Effect of graded levels of fiber from alfalfa meal on apparent and standardized ileal digestibility of amino acids of growing pigs. J. Integr. Agric. 14:2598–2604.

Chen, L., L.-X. Gao, and H.-F. Zhang. 2014. Effect of graded levels of fiber from alfalfa meal on nutrient digestibility and flow of fattening pigs. J. Integr. Agric. 13:1746–1752.

Chiba, L. I. 2001. Protein supplements. In: A. J. Lewis, and L. L. Southern, editors, Swine nutrition. 2nd ed. CRC Press, Boca Raton, FL. p. 803–837.

Chiba, L. I. 2010a. By-product feeds: Animal origin. In: W. G. Pond, D. E. Ullrey, and C. Kirk-Baer, editors, Encyclopedia of animal science. Taylor & Francis, New York, NY. p. 169–174.

Chiba, L. I. 2010b. Feedstuffs: Protein sources. In: W. G. Pond, D. E. Ullrey, and C. Kirk-Baer, editors, Encyclopedia of animal science. Taylor & Francis, New York, NY. p. 416–421.

Chiba, L. I., H. W. Ivey, K. A. Cummins, and B. E. Gamble. 1995. Effects of hydrolyzed feather meal as a source of extra dietary nitrogen on growth performance and carcass traits of finisher pigs. Anim. Feed Sci. Technol. 53:1–16.

Chiba, L. I., H. W. Ivey, K. A. Cummins, and B. E. Gamble. 1996. Hydrolyzed feather meal as a source of amino acids for finisher pigs. Anim. Feed Sci. Technol. 57:15–24.

Church, D. C., and R. O. Kellems. 1998. Supplemental protein sources. In: R. O. Kellems, and D. C. Church, editors, Livestock feeds and feeding. Prentice Hall, Upper Saddle River, NJ. p. 135–163.

Corassa, A., I. P. A. da Silva Lautert, D. dos Santos Pina, C. Kiefer, A. P. S. Ton, C. M. Komiyama, A. B. Amorim, and A. de Oliveira Teixeira. 2017. Nutritional value of Brazilian distillers dried grains with solubles for pigs as determined by different methods. R. Bras. Zootec. 46:740–746.

Carter, S. D., M. D. Lindemann, L. I. Chiba, M. J. Estienne, and G. J. M. M. Lima. 2018. Effects of inclusion of spray-dried porcine plasma in lactation diets on sow and litter performance. Livest. Sci. 216:32–35.

Crenshaw, J. D., J. M. Campbell, J. Polo, and D. Bussières. 2017. Effects of a nursery feed regimen with spray-dried bovine plasma on performance and mortality of weaned pigs positive for porcine reproductive and respiratory syndrome virus. J. Swine Health Prod. 25:10–18.

Cromwell, G. L. 1998. Feeding swine. In: R. O. Kellems, and D. C. Church, editors, Livestock feeds and feeding. 4th ed. Prentice Hall, Upper Saddle River, NJ. p. 354–390.

Cromwell, G. L., T. S. Stahly, and H. J. Monegue. 1991. Amino acid supplementation of meat meal in lysine-fortified corn-based diets for growing-finishing pigs. J. Anim. Sci. 69:4898–4906.

Danielson, D. M., and J. D. Crenshaw. 1991. Raw and processed soybeans in swine diets. In: E. R. Miller, D. E. Ullrey, and A. J. Lewis, editors, Swine nutrition. Butterworth-Heinemann, Boston, MA. p. 573–584.

Darroch, C. S. 1990. Safflower meal. In: P.A. Thacker, and R. N. Kirkwood, editors, Nontraditional feed sources for use in swine production. Butterworths-Heinemann, Boston, MA. p. 373–382.

DeRouchey, J. M., M. D. Tokach, J. L. Nelssen, R. D. Goodband, S. S. Dritz, J. C. Woodworth, and B. W. James. 2002. Comparison of spray-dried blood meal and blood cells in diets for nursery pigs. J. Anim. Sci. 80:2879–2886.

Diarra, S. S. 2016. Effects of enzyme products in the diet on growth, dressing-out percent and organ weights of light pigs fed copra-meal-based diets. Anim. Prod. Sci. 57:683–689.

Dinusson, W. E. 1990. Sunflower meal. In: P. A. Thacker, and R. N. Kirkwood, editors, Nontraditional feed sources for use in swine production. Butterworth-Heinemann, Boston, MA. 465–472.

Divakala, K. C., L. I. Chiba, R. B. Kamalakar, S. P. Rodning, E. G. Welles, K. A. Cummins, J. Swann, F. Cespedes, and R. L. Payne. 2009. Amino acid supplementation of hydrolyzed feather meal diets for finisher pigs. J. Anim. Sci. 87:1270–1281.

Dong, F. M., R. W. Hardy, N. F. Haard, F. T. Barrows, B. A. Rasco, W. T. Fairgrieve, and I. P. Forster. 1993. Chemical composition and protein digestibility of poultry by-product meals for salmonid diets. Aquaculture 116:149–158.

Ekmay, R., S. Gatrell, K. Lum, J. Kim, and X. G. Lei. 2014. Nutritional and metabolic impacts of a defatted green marine microalgal (Desmodesmus sp.) biomass in diets for weanling pigs and broiler chickens. J. Agric. Food Chem. 62:9783–9791.

Evans, D. F., and J. Leibholz. 1979. Meat meal in the diet of the early-weaned pig. I. A comparison of meat meal and soyabean meal. Anim. Feed Sci. Technol. 4:33–42.

Fan, L., M. Dou, X. Wang, Q. Han, B. Zhao, J. Hu, G. Yang, X. Shi, and X. Li. 2018. Fermented corn-soybean meal elevated IGF1 levels in grower-finisher pigs. J. Anim. Sci. 96:5144–5151.

FAO. 1997. FAO yearbook. Production. Vol. 50. Food and Agriculture Organization of the United Nations, Rome, Italy.

FAO. 2002. Small-scale palm oil processing in Africa. FAO Agric. Serv. Bull. 148. FAO, United Nations, Rome, Italy.

FAO. 2019a. World production of safflower seeds in 2018. UN Food and Agriculture Organization, Statistics Division (FAOSTAT). United Nations, Rome, Italy.

FAO. 2019b. Flax (linseed) production in 2018. UN Food and Agriculture Organization, Statistics Division (FAOSTAT), United Nations, Rome, Italy.

FAO. 2020. Sesame seed production in 2018. UN Food and Agriculture Organization Corporate Statistical Database (FAOSTAT), United Nations, Rome, Italy.

Feedipedia. 2012-2020. Animal feed resources information system. INRAE, CIRAD, SAFZ, and FAO. https://www.feedipedia.org/ (Accessed 4 January 2020).

Gatnau, R., and D. R. Zimmerman. 1992. Determination of optimum levels of inclusion of spray-dried porcine plasma (SDPP) in diets for weanling pigs fed in practical conditions. J. Anim. Sci. 70(Suppl. 1):60 (Abstr.).

Gevaert, B., L. Veryser, F. Verbeke, E. Wynendaele, and B. De Spiegeleer. 2016. Fish hydrolysates: A regulatory perspective of bioactive peptides. Protein Pept. Lett. 23:1–9.

González-Vega, J. C., and H. H. Stein. 2012. Amino acid digestibility in canola, cottonseed, and sunflower products fed to finishing pigs. J. Anim. Sci. 90:4391–4400.

Gore, A. M., R. W. Seerley, and M. J. Azain. 1990. Menhaden fish meal and dried whey levels in swine starter diets. Special Publ. No. 67. Georgia Agric. Exp. Sta., Athens, GA.

Gottlob, R. O., J. M. DeRouchey, M. D. Tokach, J. L. Nelssen, R. D. Goodband, and S. S. Dritz. 2007. Comparison of whey protein concentrate and spray-dried plasma protein in diets for weanling pigs. Appl. Anim. Sci. 23:116–122.

Han, Y., and C. M. Parsons. 1991. Protein and amino acid quality of feather meals. Poult. Sci. 70:812–822.

Hasan, S., S. Junnikkala, O. Peltoniemi, L. Paulin, A. Lyyski, J. Vuorenmaa, and C. Oliviero. 2018. Dietary supplementation with yeast hydrolysate in pregnancy influences colostrum yield and gut microbiota of sows and piglets after birth. PLoS ONE 13:e0197586.

Haschek, W. M., V. R. Beasley, W. B. Buck, and J. H. Finnell. 1989. Cottonseed meal (gossypol) toxicosis in a swine herd. J. Am. Vet. Med. Assoc. 195:613–615.

Haumann, B. F. 1997. Bioengineered oilseed acreage escalating. Inform 8:804–811.

Haydon, K. D., and Newton, G. L. 1987. Effect of whole-roasted peanuts fed in either simple or complex diets on starter pig performance. Swine Rep., Univ. of Georgia, Athens, GA.

Haydon, K. D., G. L. Newton, C. R. Dove, and S. E. Hobbs. 1990. Effect of roasted or raw peanut kernels on lactation performance and milk composition of swine. J. Anim. Sci. 68:2591–2597.

Hill, G., D. Rozeboom, N. Trottier, D. Mahan, L. Adeola, T. Cline, D. Forsyth, and B. Richert. 1998. Tri-state swine nutrition guide. Bulletin 869. Ohio State Univ., Columbus, OH.

Hollmann, A., M. Martinez, P. Maturana, L. C. Semorile, and P. C. Maffia. 2018. Antimicrobial peptides: Interaction with model and biological membranes and synergism with chemical antibiotics. Front Chem. 6:204.

Huang, C., L. I. Chiba, and W. G. Bergen, 2021. Bioavailability and metabolism of omega-3 polyunsaturated fatty acids in pigs and omega-3 polyunsaturated fatty acid-enriched pork: A review. Livest. Sci. 243:104370.

Hulshof, T. G., A. F. B. van der Poel, W. H. Hendriks, and P. Bikker. 2017. Amino acid utilization and body composition of growing pigs fed processed soybean meal or rapeseed meal with or without amino acid supplementation. Animal 11:1125–1135.

Ilori, J. O., E. R. Miller, D. E. Ullrey, P. K. Ku, and M. G. Hogberg. 1984. Combinations of peanut meal and blood meal as substitutes for soybean meal in corn-based growing-finishing diets. J. Anim. Sci. 59:394–399.

Jang, J.-C., D. H. Kim, J. S. Hong, Y. D. Jang, and Y. Y. Kim. 2020. Effects of copra meal inclusion level in growing-finishing pig diets containing β-mannanase on growth performance, apparent total tract digestibility, blood rea nitrogen concentrations and pork quality. Animals 10:1840.

Jaworski, N. W., J. Shoulders, J. C. González-Vega, and H. H. Stein. 2014. Effects of using copra meal, palm kernel expellers, or palm kernel meal in diets for weanling pigs. Prof. Anim. Sci. 30:243–251.

Jędrejek, D., J. Levic, J. Wallace, and W. Oleszek. 2016. Animal by-products for feed: Characteristics, European regulatory framework, and potential impacts on human and animal health and the environment. J. Anim. Feed Sci. 25:189–202.

Jin, X. H., P. S. Heo, J. S. Hong, N. J. Kim, and Y. Y. Kim. 2016. Supplementation of dried mealworm (Tenebrio molitor larva) on growth performance, nutrient digestibility and blood profiles in weaning pigs. Asian-Australas. J. Anim. Sci. 29:979–986.

Johnson, L. A., T. M. Suleiman, and E. W. Lusas. 1979. Sesame protein: A review and prospectus. J. Am. Oil Chem. Soc. 56:463–468.

Jones, A. M., F. Wu, J. C. Woodworth, M. D. Tokach, R. D. Goodband, J. M. DeRouchey, and S. S. Dritz. 2018. Evaluating the effects of fish meal source and level on growth performance of nursery pigs. Transl. Anim. Sci. 23:144–155.

Jurgens, M. H., K. Bregendahl, J. Coverdale, and S. L. Hansen, 2012. Animal feeding and nutrition. 11th ed. Kendall Hunt Publishing Co., Dubuque, IA.

Kats, L. J., J. L. Nelssen, M. D. Tokach, R. D. Goodband, J. A. Hansen, and J. L. Laurin. 1994a. The effect of spray-dried porcine plasma on growth performance of the early-weaned pig. J. Anim. Sci. 72:2075–2081.

Kats, L. J., J. L. Nelssen, M. D. Tokach, R. D. Goodband, T. L. Weeden, S. S. Drits, J. A. Hansen, and K. G. Friesen. 1994b. The effects of spray-dried blood meal on growth performance of the early-weaned pig. J. Anim. Sci. 72:2860–2869.

Keegan, T. P, J. M. DeRouchey, J. L. Nelssen, M. D. Tokach, R. D. Goodband, and S. S. Dritz. 2004. The effects of poultry meal source and ash level on nursery pig performance. J. Anim. Sci. 82:2750–2756.

Kennedy, J. J., F. X. Aherne, D. L. Kelleher, and P. J. Caffrey. 1974. An evaluation of the nutritive value of meat-and-bone meal. 2. Effects of protein, ash and available lysine content on pig performance and nitrogen retention. Ir. J. Agric. Res. 13:11–16.

Kepler, M., G. W. Libal, and R. C. Wahlstrom. 1982. Sunflower seeds as a fat source in sow gestation and lactation diets. J. Anim. Sci. 55:1082–1086.

Kibria, S., and I. H. Kim. 2019. Impacts of dietary microalgae (Schizochytrium JB5) on growth performance, blood profiles, apparent total tract digestibility, and ileal nutrient digestibility in weaning pigs. J. Sci. Food Agric. 99:6084–6088.

Kim, S. W., M. Brandherm, M. Freeland, B. Newton, D. Cook, and I. Yoon. 2008. Effects of yeast culture supplementation to gestation and lactation diets on growth of nursing piglets. Asian-Australas. J. Anim. Sci. 21:1011–1014.

Kim, B. G., J. H. Lee, H. J. Jung, Y. K. Han, K. M. Park, and I. K. Han. 2001. Effect of replacement of soybean meal with palm kernel meal and copra meal on growth performance, nutrient digestibility and carcass characteristics of finishing pigs. Asian-Australas. J. Anim. Sci. 14:821–830.

Kim, B. G, Y. Liu, and H. H. Stein. 2014. Energy concentration and phosphorus digestibility in yeast products produced from the ethanol industry, and in brewers' yeast, fish meal, and soybean meal fed to growing pigs. J. Anim. Sci. 92:5476–5484.

Kim, S. W., E. van Heugten, F. Ji, C. H. Lee, and R. D. Mateo. 2010. Fermented soybean meal as a vegetable protein source for nursery pigs: I. Effects on growth performance of nursery pigs. J. Anim. Sci. 88:214–224.

Kong, C., H. G. Kang, B. G. Kim, and K. H. Kim. 2014. Ileal digestibility of amino acids in meat meal and soybean meal fed to growing pigs. Asian-Australas. J. Anim. Sci. 27:990–995.

Laksesvela, B. 1961. Graded levels of herring meal to bacon pigs, effect on growth rate, feed efficiency and bacon quality. J. Agric. Sci. (Camb.) 56:307–315.

LaRue, D. C., D. A. Knabe, and T. D. Tanksley, Jr. 1985. Commercially processed glandless cottonseed meal for starter, grower and finisher swine. J. Anim. Sci. 60:495–502.

Laudert, S. B., and G. L. Allee. 1975. Nutritive value of sunflower seed for swine. J. Anim. Sci. 41:318 (Abstr.).

Lee, J. H., H. B. Yoo, S. H. Do, J. S. Hong, and Y. Y. Kim. 2017. Effect of different supplementation levels of palm kernel meal with β-mannanase n growth performance, blood profiles, pork quality, and economic analysis in growing-finishing pigs. J. Anim. Sci. 95(Suppl. 2):97 (Abstr.).

Lee, A. V., L. You, S.-Y. Oh, Z. Li, A. Code, C. Zhu, R. E. Fisher-Heffernan, T. R. H. Regnault, C. F. M. De Lange, L.-A. Huber, and N. A. Karrow. 2019. Health benefits of supplementing nursery pig diets with microalgae or fish oil. Animals 9:80.

Leibbrandt, V. D., and N. J. Benevenga. 1991. Utilization of liquid whey in feeding swine. In: E. R. Miller, D. E. Ullrey, and A. J. Lewis, editors, Swine nutrition. Butterworth-Heinemann, Boston, MA. p. 559–571.

Leibholz, J. 1982. Meat meal in the diet of the early-weaned pig. IV. The supplementation of diets with tryptophan, lysine and methionine. Anim. Feed Sci. Technol. 7:27–34.

LeMieux, F. M., V. D. Naranjo, T. D. Bidner, and L. L. Southern. 2010. Effect of dried brewers yeast on growth performance of nursing and weanling pigs. Applied Anim. Sci. 26:70–75.

Lettner, F., W. Wetscherek, and S. Bickel. 2001. Meat meal for growing-finishing pigs. Die Bodenkultur. 52:247–254.

Li, D. F., J. L. Nelssen, P. G. Reddy, F. Blecha, R. D. Klemm, D. W. Giesting, J. D. Hancock, G. L. Allee, and R. D. Goodband. 1991. Measuring suitability of soybean products for early-weaned pigs with immunological criteria. J. Anim. Sci. 69:3299–3307.

Li, D., S. Y. Qiao, G. F. Yi, J. Y. Jiang, X. X. Xu, X. S. Piao, I. K. Han, and P. Thacker. 2000a. Performance of growing-finishing pigs fed sesame meal supplemented diets formulated using amino acid digestibilities determined by the regression technique. Asian-Australas. J. Anim. Sci. 13:213–219.

Li, D. F., G. F. Yi, S. Y. Qiao, C. T. Zheng, R. J. Wang, P. Thacker, X. S. Piao, and I. K. Han. 2000b. Nutritional evaluation of Chinese nonconventional protein feedstuffs for growing-finishing pigs - 1. Linseed meal. Asian-Australas. J. Anim. Sci. 13:39–45.

Lindemann, M. D., S. G. Cornelius, S. M. El Kandelgy, R. L. Moser, and J. E. Pettigrew. 1986. Effect of age, weaning and diet on digestive enzyme levels in the piglet. J. Anim. Sci. 62:1298–1307.

Little, K. L., B. M. Bohrer, T. Maison, Y. Liu, H. H. Stein, and D. D. Boler. 2015. Effects of feeding canola meal from high-protein or conventional varieties of canola seeds on growth performance, carcass characteristics, and cutability of pigs. J. Anim. Sci. 93:1284–1297.

Liu, K. 2011. Chemical composition of distillers grains, a review. J. Agric. Food Chem. 59:1508–1526.

Liu, Y., C. D. Espinosa, J. J. Abelilla, G. A. Casas, L. Vanessa Lagos, S. A. Lee, W. B. Kwon, J. K. Mathai, D. M. D.L. Navarro, N. W. Jaworski, and H. H. Stein. 2018. Non-antibiotic feed additives in diets for pigs: A review. Anim. Nutr. 4:113–125.

Liu, J. D., Q. Y. Li, Z. K. Zeng, P. Li, X. Xu, H. L. Wang, S. Zhang, and X. S. Piao. 2015. Determination and prediction of the amino acid digestibility of sunflower seed meals in growing pigs. Asian-Australas. J. Anim. Sci. 28:86–94.

Loar, R. E., II, K. Karges, and M. Giesemann. 2012. Corn oil extraction in the ethanol industry. Proc., 10th Mid-Atlantic Nutrition Conference, Baltimore, MD. March 28–29, 2012. pp. 106–112.

Ma, D. L, X. K. Ma, L. Liu, and S. Zhang. 2018. Chemical composition, energy, and amino acid digestibility in 7 cottonseed co-products fed to growing pigs. J. Anim. Sci. 96:1338–1349.

Mahan, D. C. 1992. Efficacy of dried whey and its lactalbumin and lactose components at two dietary lysine levels in postweaning pig performance and nitrogen balance. J. Anim. Sci. 70:2182–2187.

Mahan, D. C. 2003. Feeding milk products to swine. In: M. L. Eastridge, editor, Proc. 3rd Natl. Symp. Altern. Feeds for Livest. Poult. Ohio State Univ., Columbus, OH. p. 141–156.

Mahan, D. C., R. A. Easter, G. L. Cromwell, E. R. Miller, and T. L. Veum. 1993. Effect of dietary lysine levels formulated by altering the ratio of corn soybean meal with or without dried whey and lysine hydrochloric acid in diets for weanling pigs. J. Anim. Sci. 71:1848–1852.

Marchello, M. J., N. K. Cook, V. K. Johnson, W. D. Slanger, D. K. Cook, and W. E. Dinusson. 1984. Carcass quality, digestibility and feedlot performance of swine fed various levels of sunflower seed. J. Anim. Sci. 58:1205–1217.

McDonnell, P., C. J. O'Shea, J. J. Callan, and J. V. O'Doherty. 2011. The response of growth performance, nitrogen, and phosphorus excretion of growing-finishing pigs to diets containing incremental levels of maize dried distiller's grains with solubles. Anim. Feed Sci. Technol. 169:104–112.

Madeira, M. S., C. Cardoso, P. A. Lopes, D. Coelho, C. Afons, N. M. Bandarra, J. A. M. Prates. 2017. Microalgae as feed ingredients for livestock production and meat quality: A review. Livest. Sci. 205:111–121.

Mello, G., A. C. Laurentiz, R. S. Filardi, A. F. Bergamaschine, H. T. Okuda, M. M. Lima, and O. M. Junqueira. 2012. Cottonseed meal in diets for growing and finishing pigs. Arch. Zootecnia. 61:55–62.

Miech, P., J. Lindberg, Å. Berggren, T. Chhay, and A. Jansson. 2017. Apparent faecal digestibility and nitrogen retention in piglets fed whole and peeled Cambodian field cricket meal. J. Insect Food Feed 3:279–288.

Miller, E. R. 1990. Blood meal: Flash-dried. In: P. A. Thacker, and R. N. Kirkwood, editors, Nontraditional feed sources for use in swine production. Butterworth-Heinemann, Boston, MA. p. 53–60.

Moran, K., R. D. Boyd, C. Zier-Rush, P. Wilcock, N. Bajjalieh, and E. van Heugten. 2017. Effects of high inclusion of soybean meal and a phytase superdose on growth performance of weaned pigs housed under the rigors of commercial conditions. J. Anim. Sci. 95:5455–5465.

Mosenthin, R., W. C. Sauer, R. Blank, J. Huisman, and M. Z. Fan.2000. The concept of digestible amino acids in diet formulation for pigs. Livest. Prod. Sci. 64:265–280.

Navarro, D. M. D. L., J. K. Mathai, N. W. Jaworski, and H. H. Stein. 2018. Amino acid digestibility in six sources of meat and bone meal, blood meal, and soybean meal fed to growing pigs. Can. J. Anim. Sci. 98:860–867.

Nessmith, W. B., Jr., J. L. Nelssen, M. D. Tokach, R. D. Goodband, and J. R. Bergström. 1997. Effects of substituting deproteinized whey and(or) crystalline lactose for dried whey on weanling pig performance. J. Anim. Sci. 75:3222–3228.

Newkirk, R. 2010. Soybean: Feed industry guide. 1st ed. Canadian International Grains Institute, Winnipeg, MB, Canada.

Neumann, C., S. Velten, and F. Liebert. 2018. N balance studies emphasize the superior protein quality of pig diets at high inclusion level of algae meal (Spirulina platensis) or insect meal (Hermetia illucens) when adequate amino acid supplementation is ensured. Animals 8:172.

Newton, G. L., C. V. Booram, R. W. Barker, and O. M. Hale. 1977. Dried *Hermetia Illucens* larvae meal as a supplement for swine. J. Anim. Sci. 44:395–400.

Newton, G. L., and K. A. Haydon. 1988. Raw or roasted peanuts in growing-finishing diets. Swine Rep., Univ. of Georgia, Athens, GA.

Nørgaard, J. V., K. Blaabjerg, and H. D. Poulsen. 2012a. Salmon protein hydrolysate as a protein source in feed for young pigs. Anim. Feed Sci. Technol. 177:124–129.

Nørgaard, J. V., J. A. Fernández, and H. Jørgensen. 2012b. Ileal digestibility of sunflower meal, pea, rapeseed cake, and lupine in pigs. J. Anim. Sci. 90:203–205.

NRC. 1998. Nutrient requirements of swine. 10th rev. ed. Natl. Acad. Press, Washington, DC.

NRC. 2012. Nutrient requirements of swine. 11th rev. ed. Natl. Acad. Press, Washington, DC.

Nyachoti, C. M., J. D. House, B. A. Slominski, and I. R. Seddon. 2005. Energy and nutrient digestibilities in wheat dried distillers' grains with solubles fed to growing pigs. J. Sci. Food Agric. 85:2581–2586.

O'Doherty, J. V., and M. P. McKeon. 2000. The use of expeller copra meal in grower and finisher pig diets. Livest. Prod. Sci. 67:55–65.

Oliveira, M. S., and H. H. Stein. 2016. Digestibility of energy, amino acids, and phosphorus in a novel source of soy protein concentrate and in soybean meal fed to growing pigs. J. Anim. Sci. 94:3343–3352.

Opheim, M., M. L. Strube, H. Sterten, M. Øverland, and N. P. Kjos. 2016. Atlantic salmon (*Salmo salar*) protein hydrolysate in diets for weaning piglets - effect on growth performance, intestinal morphometry and microbiota composition. Arch. Anim. Nutr. 70:44–56.

Orozco-Hernandez, J. R., J. J. Uribe, S. G. Bravo, V. O. Fuentes-Hernandez, A. Aguilar, and O. H. Navarro 2003. Effect of poultry by-product meal on pig performance. J. Anim. Sci. 83(Suppl. 1):75. (Abstr.).

Otte, M. V., F. Moreira, I. Bianchi, J. Oliveira, Jr., R. E. Mendes, C. S. Haas, A. N. Anciuti, M. T. Rovani, B. G. Gasperin, T. Lucia, Jr. 2019. Effects of supplying omega-3 polyunsaturated fatty acids to gilts after weaning on metabolism and ovarian gene expression. J. Anim. Sci. 97:374–384.

Pan, L., X. K. Ma, H. L. Wang, X. Xu, Z. K. Zeng, Q. Y. Tian, P. F. Zhao, S. Zhang, Z. Y. Yang, and X. S. Piao. 2016. Enzymatic feather meal as an alternative animal protein source in diets for nursery pigs. Anim. Feed Sci. Technol. 212:112–121.

Pangeni, D., J. A. Jendza, L. Anil, X. Yang, and S. K. Baidoo. 2017. Effect of replacing conventional soybean meal with low-oligosaccharide soybean meal on growth performance and carcass characteristics of wean-to-finish pigs. J. Anim. Sci. 95:2605–2613.

Papadopoulos, M. C. 1985. Processed chicken feathers as feedstuff for poultry and swine. A review. Agric. Wastes 14:275–290.

Parsons, C. M. 1996. Digestible amino acids for poultry and swine. Anim. Feed Sci. Technol. 59:147–151.

Parsons, M. J., P. K. Ku, and E. R. Miller. 1985. Lysine availability in flash-dried blood meals for swine. J. Anim. Sci. 60:1447–1453.

Partanen, K., and M. Nasi. 1994. Nutritive value of meat and bone meal for growing pigs. Agric. Food Sci. Finland 3:449–455.

Partanen, K., H. Siljander-Rasi, T. Alaviuhkola, and N. Gilse van der Pals. 1998. Utilisation of reactive lysine from meat and bone meals of different ash content by growing-finishing pigs. Agric. Food Sci. Finland 7:1–11.

Pereira, C. M. C., J. L. Donzele, and F. C. O. Silva. 2012. Yeast extract with blood plasma in diets for piglets from 21 to 35 days of age. R. Bras. Zootec. 41:1676–1682.

Pike, I. H., M. K. Curran, M. Edge, and A. Harvey. 1984. Effect of nutrient density, presence of fish meal and method of feeding of unmedicated diets on early-weaned pigs. Anim. Prod. 39:291–297.

Pollmann, D. S., D. M. Danielson, M. A. Crenshaw, and E. R. Peo, Jr. 1981. Long-term effects of dietary additions of alfalfa and tallow on sow reproductive performance. J. Anim. Sci. 51:294–299.

Pond, W. G., and J. H. Maner. 1984. Swine production and nutrition. AVI Publ. Co. Inc., Westport, CT.

Poudel, P., C. L. Levesque, R. Samuel, and B. St-Pierre. 2020. Dietary inclusion of Peptiva, a peptidebased feed additive, can accelerate the maturation of the fecal bacterial microbiome in weaned pigs. BMC Vet. Res. 16:60.

Powley, J. S., P. R. Cheeke, D. C. England, T. P. Davidson, and W. H. Kennick. 1981. Performance of growing-finishing swine fed high levels of alfalfa meal: effects of alfalfa level, dietary additives and antibiotics. J. Anim. Sci. 53:308–316.

Qin, C., P. Huang, K. Qiu, W. Sun, L. Xu, X. Zhang, and J. Yin. 2015. Influences of dietary protein sources and crude protein levels on intracellular free amino acid profile in the longissimus dorsi muscle of finishing gilts. J. Anim. Sci. Biotechnol. 6:52.

Ravindran, V. 1990. Sesame meal. In: P. A. Thacker, and R. N. Kirkwood, editors, Nontraditional feed sources for use in swine production. Butterworth-Heinemann, Boston, MA. p. 419–427.

Ravindran, V., G. Ravindran, and S. Sivalogan. 1994. Total and phytate phosphorus contents of various foods and feedstuffs of plant origin. Food Chem. 50:133–136.

Reese, D. E., R. C. Thaler, M. C. Brumm, C. R. Hamilton, A. J. Lewis, G. W. Libal, and P. S. Miller. 1995. Nebraska and South Dakota Swine Nutrition Guide. Nebraska Cooperative Extension EC 95-273-C, Univ. of Nebraska, Lincoln, NE, US.

Reese, D. E., R. C. Thaler, M. C. Brumm, A. J. Lewis, P. S. Miller, and G. W. Libal. 2000. Swine nutrition guide. Faculty Papers and Publications in Animal Science. 694. http://digitalcommons.unl.edu/animalscifacpub/694 (Accessed 12 May 2019.)

Reese, D. E., E. van Heugten, H. H. Stein, J. DeRouchey, J. M. Benz, and J. F. Patience. 2010. Composition and usage rate of feed ingredients for swine diets. U.S. Pork Center of Excellence, Clive, IA.

Reis, S. T. C., L. G. Mariscal, B. A. Aguilera, and H. J. G. Cervantes. 2007. Protein and energy digestibility in diets for piglets, supplemental with three different kinds of dried whey. Vet. Mex. 38:141–151.

Ren, Z. H., W. Yuan, H. D. Deng, J. L. Deng, Q. X. Dan, H. T. Jin, C. L. Tian, X. Peng, Z. Liang, S. Gao, S. H. Xu, G. Li, and Y. Hu. 2015. Effects of antibacterial peptide on cellular immunity in weaned piglets. J. Anim. Sci. 93:127–134.

Richter, G., and H. Kohler. 1997. Suitability of extracted linseed meal for feeding to the suckling pig. Die Muhle Mischfuttertechnik 134:776–777.

Rojas, O. J., and H. H. Stein. 2013. Concentration of digestible and metabolizable energy and digestibility of amino acids in chicken meal, poultry byproduct meal, hydrolyzed porcine intestines, a spent hen–soybean meal mixture, and conventional soybean meal fed to weanling pigs. J. Anim. Sci. 91:3220–3230.

Rostagno, H. S., A. Teixeira, J. L. Donzele, P. C. Gomes, R. F. M. De Oliveira, D. C. Lopes, A. J. P. Ferreira, and S. L. Toledo Barreto. 2005. Brazilian tables for poultry and swine: Composition of feedstuffs and nutritional requirements. Universidade Federal de Viçosa, Viçosa, Brazil.

Rumpold, B. A., and O. K. Schlüter. 2013. Nutritional composition and safety aspects of edible insects. Mol. Nutr. Food Res. 57:802–823.

Sánchez-Muros, M.-J., F. G. Barroso, and F. Manzano-Agugliaro. 2014. Insect meal as renewable source of food for animal feeding: A review. J. Cleaner Prod. 65:16–27.

Schöne, F., B. Groppel, A. Hennig, G. Jahreis, and R. Lange. 1997a. Rapeseed meals, methimazole, thiocyanate and iodine affect growth and thyroid. Investigations into glucosinolate tolerance in the pig. J. Sci. Food Agric. 74:69–80.

Schöne, F., B. Rudolph, U. Kirchheim, and G. Knapp. 1997b. Counteracting the negative effects of rapeseed and rapeseed press cake in pig diets. Br. J. Nutr. 78:947–962.

Seerley, R. W. 1991. Major feedstuffs used in swine diets. In: E. R. Miller, D. E. Ullrey, and A. J. Lewis, editors, Swine nutrition. Butterworth-Heinemann, Boston, MA. p. 451–481.

Seo, S. H., B. Y. Jung, M. K. Lee, B. H. Lee, and I. K. Paik. 2009. The effect of dietary supplementation of feather meal on the performance and muscular taurine contents in growing-finishing pigs. Asian-Australas. J. Anim. Sci. 22:1407–1413.

Shelton, J. L., M. D. Hemann, R. M. Strode, G. L. Brashear, M. Ellis, F. K. McKeith, T. D. Bidner, and L. L. Southern. 2001. Effect of different protein sources on growth and carcass traits in growing-finishing pigs. J. Anim. Sci. 79:2428–2435.

Shurson, G. C. 2018. Yeast and yeast derivatives in feed additives and ingredients: Sources, characteristics, animal responses, and quantification methods. Anim. Feed Sci. Technol. 235:60–76.

Siebra, J. E. D. C., M. D. C. M. M. Ludke, J. V. Ludk, T. M. Bertol, and W. M. D. Junior. 2008. Bioeconomic performance of growing-finishing pigs fed diet with coconut meal. Rev. Bras. Zootec. 37:1996–2002.

Simpson, G. 2012. Comparative feed values for swine. Ontario Ministry of Agriculture, Food and Rural Affairs, Guelph, ON, Canada. http://www.omafra.gov.ca/english/livestock/swine/facts/03-003htm (Accessed 12 May 2019).

Sinn, S. M., W. R. Gibbons, M. L. Brown, J. M. DeRouchey, and C. L. Levesque. 2017. Evaluation of microbially enhanced soybean meal as an alternative to fishmeal in weaned pig diets. Animal 11:784–793.

Sogari, G., M. Amato, I. Biasato, S. Chiesa, and L. Gasco. 2019. The potential role of insects as feed: A multi-perspective review. Animals 9:119.

Sohn, K. S., C. V. Maxwell, D. S. Buchanan, and L. L. Southern. 1994a. Improved soybean sources for early-weaned pigs. I. Effects on performance and total tract amino acid digestibility. J. Anim. Sci. 72:622–630.

Sohn, K. S., C. V. Maxwell, L. L. Southern, and D. S. Buchanan. 1994b. Improved soybean sources for early-weaned pigs. II. Effects on ileal amino acid digestibility. J. Anim. Sci. 72:631–637.

Song, R., C. Chen, L. J. Johnston, B. J. Kerr, T. E. Weber, and G. C. Shurson. 2014. Effects of feeding diets containing highly peroxidized distillers dried grains with solubles and increasing vitamin E levels to wean-finish pigs on growth performance, carcass characteristics, and pork fat composition. J. Anim. Sci. 92:198–210.

Sotak, K. M., T. A. Houser, R. D. Goodband, M. D. Tokach, S. S. Dritz, J. M. DeRouchey, B. L. Goehring, G. R. Skaar, and J. L. Nelssen. 2015. The effects of feeding sorghum dried distillers grains with solubles on finishing pig growth performance, carcass characteristics, and fat quality. J. Anim. Sci. 93:2904–2915.

Soto-Sierraa, L., P. Stoykovab, and Z. L. Nikolov. 2018. Extraction and fractionation of microalgae-based protein products. Algal Res. 36:175–192.

Spiehs, M. J., M. H. Whitney, and G. C. Shurson. 2002. Nutrient database for distiller's dried grains with solubles produced from new ethanol plants in Minnesota and South Dakota. J. Anim. Sci. 80:2639–2645.

Spranghers, T., J. Michiels, J. Vrancx, A. Ovyn, M. Eeckhout, P. De Clercq, and S. De Smet. 2018. Gut antimicrobial effects and nutritional value of black soldier fly (*Hermetia illucens L.*) prepupae for weaned piglets. Anim. Feed Sci. Technol. 235:33–42.

Stein, H. H., G. A. Casas, J. J. Abelilla, Y. Liu, and R. C. Sulabo. 2015. Nutritional value of high fiber co-products from the copra, palm kernel, and rice industries in diets fed to pigs. J. Anim. Sci. Biotechnol. 6:56.

Stein, H. H., J. A. Roth, K. M. Sotak, and O. J. Rojas. 2013. Nutritional value of soy products fed to pigs. Swine Focus #004. Univ. of Illinois, Urbana-Champaign, IL.

Stein, H. H., and G. C. Shurson. 2009. Board-invited review: The use and application of distillers dried grains with solubles in swine diets. J. Anim. Sci. 87:1292–1303.

Stone, C. W. 2006. Yeast products in the feed industry: A practical guide for feed professionals. https://en.engormix.com/feed-machinery/articles/yeast-products-in-feed-industry-t33489.htm (Accessed 5 July 2019.)

Stoner, G. R., G. L. Allee, and J. L. Nelssen. 1985. Effects of fish protein hydrolysate and dried whey in starter pig diets. Swine Day. Kansas Agric. Exp. Stn., Kansas State Univ., Manhattan, KS. p. 67–73.

Stoner, G. R., G. L. Allee, J. L. Nelssen, M. E. Johnston, and R. D. Goodband. 1990. Effect of select menhaden fish meal in starter diets for pigs. J. Anim. Sci. 68:2729–2735.

Sulabo, R. C., L. I. Chiba, F. N. Almeida, S. D. Brotzge, R. L. Payne, and H. H. Stein. 2013a. Amino acid and phosphorus digestibility and concentration of digestible and metabolizable energy in hydrolyzed feather meal fed to growing pigs. J. Anim. Sci. 91:5829–5837.

Sulabo, R. C., W. S. Ju, and H. H. Stein. 2013b. Amino acid digestibility and concentration of digestible and metabolizable energy in copra meal, palm kernel expellers, and palm kernel meal fed to growing pigs. J. Anim. Sci. 91:1391–1399.

Sulabo, R. C., and H. H. Stein. 2013. Digestibility of phosphorus and calcium in meat and bone meal fed to growing pigs. J. Anim. Sci. 91:1285–1294.

Tanksley, T. D., Jr. 1990. Cottonseed meal. In: P. A. Thacker, and R. N. Kirkwood, editors, Nontraditional feed sources for use in swine production. Butterworth-Heinemann-Heinemann, Boston, MA. p. 139–151.

Tartrakoon, W., G. Thinggaard, T. Tartrakoon, N. Chalearmsan, T. Vearasilp, and U. Ter Meulen. 2001. Evaluation of feedstuffs quality for pigs in Thailand. 3. Use of apparent ileal crude protein and amino acid digestibilities of soybean, peanut and sesame meals in ration formulation for diets of growing finishing pigs. Thai J. Agric. Sci., 34:1–13.

Thacker, P. A. 1990a. Alfalfa meal. In: P. A. Thacker, and R. N. Kirkwood, editors, Nontraditional feed sources for use in swine production. Butterworth-Heinemann, Boston, MA. p. 1–11.

Thacker, P. A. 1990b. Canola meal. In: P. A. Thacker, and R. N. Kirkwood, editors, Nontraditional feed sources for use in swine production. Butterworth-Heinemann, Boston, MA. p. 69–78.

Thacker, P. A., and R. N. Kirkwood, Eds. 1990. Nontraditional feed sources for use in swine production. Butterworth-Heinemann, Boston, MA.

Thorne, P. J., D. J. A. Cole, and J. Wiseman. 1990. Copra meal. In: P. A. Thacker, and R. N. Kirkwood, editors, Nontraditional feed sources for use in swine production. Butterworth-Heinemann, Boston, MA. p. 123–129.

Thuy, N. T., M. Joseph, and N. C. Ha. 2016. Effects of replacing marine fishmeal with graded levels of Tra Catfish by-product protein hydrolysate on the performance and meat quality of pigs. S. Afr. J. Anim. Sci. 46:221–229.

Tokach, M. D., J. L. Nelssen, and G. L. Allee. 1989. Effect of protein and(or) carbohydrate fractions of dried whey on performance and nutrient digestibility of early weaned pigs. J. Anim. Sci. 67:1307–1312.

Tokach, M. D., J. E. Pettigrew, L. J. Johnston, M. Overland, J. W. Rust, and S. G. Cornelius. 1995. Effect of adding fat and(or) milk products to the weanling pig diet on performance in the nursery and subsequent grow-finish stages. J. Anim. Sci. 73:3358–3368.

Tucker, J. L., V. D. Naranjo, T. D. Bidner, and L. L. Southern. 2011. Effect of salmon protein hydrolysate and spray-dried plasma protein on growth performance of weanling pigs. J. Anim. Sci. 89:1466–1473.

Urriola, P. E., D. Hoehler, C. Pedersen, H. H. Stein, and G. C. Shurson. 2009. Amino acid digestibility of distillers dried grains with solubles, produced from sorghum, a sorghum-corn blend, and corn fed to growing pigs. J. Anim. Sci., 87:2574–2580.

Urriola, P. E., J. A. Mielke, Q. Mao, Y.-T. Hung, J. F. Kurtz, L. J. Johnston, G. C. Shurson, C. Chen, and M. Saqui-Salces. 2018. Evaluation of a partially de-oiled microalgae product in nursery pig diets. Transl. Anim. Sci. 2:169–183.

van Milgen, J., and J.-Y. Dourmad. 2015. Concept and application of ideal protein for pigs. J. Anim. Sci. Biotechnol. 6:15.

van Milgen, J., M. Gloaguen, N. Le Floc'h, L. Brossard, Y. Primot, and E. Corrent. 2012. Meta-analysis of the response of growing pigs to the isoleucine concentration in the diet. Animal 6:1601–1608.

Veum, T. L., A. K. M. Haque, and S. Bassi. 1999. Spray dried poultry by-product with or without spray dried animal plasma in weanling pig diets. Animal Sciences Departmental Report. Univ. of Missouri, Columbia, MO. p. 117–120.

Wahlstrom, R. C., and G. W. Libal. 1977. Dried blood meal as a protein source in diets for growing-finishing swine. J. Anim. Sci. 44:778–783.

Wahlstrom, R. C., G. W. Libal, and R. C. Thaler. 1985. Efficacy of supplemental tryptophan, threonine, isoleucine and methionine for weanling pigs fed a low-protein, lysine-supplemented, corn-sunflower meal diet. J. Anim. Sci. 60:720–724.

Waititu, S. M., J. M. Heo, R. Patterson, and C. M. Nyachoti. 2016a. Dietary yeast-based nucleotides as an alternative to in-feed antibiotics in promoting growth performance and nutrient utilization in weaned pigs. Can. J. Anim. Sci. 96:289–293.

Waititu, S. M., F. Yin, R. Patterson, J. C. Rodriguez-Lecompte, and C. M. Nyachoti. 2016b. Short-term effect of supplemental yeast extract without or with feed enzymes on growth performance, immune status and gut structure of weaned pigs challenged with Escherichia coli lipopolysaccharide. J. Anim. Sci. Biotechnol. 7:64.

Waititu, S. M., F. Yin, R. Patterson, A. Yitbarek, J. C. Rodriguez-Lecompte, and C. M. Nyachoti. 2017. Dietary supplementation with a nucleotide-rich yeast extract modulates gut immune response and microflora in weaned pigs in response to a sanitary challenge. Animal 11:2156–2164.

Walker, W. R., C. V. Maxwell, F. N. Owens, and D. S. Buchanan. 1986. Milk versus soybean protein sources for pigs: I. Effects of performance and digestibility. J. Anim. Sci. 63:505–510.

Wang, L.F., E. Beltranena, and R. T. Zijlstra. 2016. Diet nutrient digestibility and growth performance of weaned pigs fed wheat dried distillers grains with solubles (DDGS). Anim. Feed Sci. Technol. 218:26–32.

Wang, L.F., E. Beltranena and R.T. Zijlstra. 2017. Diet nutrient digestibility and growth performance of weaned pigs fed Brassica napus canola meal varying in nutritive quality. Anim. Feed Sci. Technol. 223:90–98.

Williams, K. C., and L. J. Daniels. 1973. Decorticated safflower meal as a protein supplement for sorghum and wheat based pig diets. Aust. J. Exp. Agric. Anim. Husb. 13:48–55.

Williams, K. C., and P. K. O'Rourke. 1974. Decorticated safflower meal as a protein supplement in diets fed either restrictively or ad libitum to barrow and gilt pigs over 45 kg live weight. Aust. J. Exp. Agric. Anim. Husb. 14:12–16.

Wu, S., F. Zhang, Z. Huang, H. Liu, C. Xie, J. Zhang, P. A. Thacker, and S. Qiao. 2012. Effects of the antimicrobial peptide cecropin AD on performance and intestinal health in weaned piglets challenged with Escherichia coli. Peptides 35:225–230.

Xiao, H., F. Shao, M. Wu, W. Ren, X. Xiong, B. Tan, and Y. Yin. 2015. The application of antimicrobial peptides as growth and health promoters for swine. J. Anim. Sci. Biotechnol. 6:19.

Xu, G., S. K. Baidoo, L. J. Johnston, D. Bibus, J. E. Cannon, and G. C. Shurson. 2009. The effects of feeding diets containing corn distillers dried grains with solubles (DDGS), and DDGS withdrawal period, on growth performance and pork quality in grower-finisher pigs. J. Anim. Sci. 88:1388–1397.

Yaakob, Z., E. Ali, A. Zainal, M. Mohamad, and M. Sobri Takriff. 2014. An overview: Biomolecules from microalgae for animal feed and aquaculture. J. Biol Res. (Thessalon). 21:6.

Yamasaki, S., L. H. Manh, R. Takada, L. T. Men, N. N. Xuan, D. V. A. K. Dung, and T. Taniguchi. 2003. Admixing synthetic antioxidants and sesame to rice bran for increasing pig performance in Mekong Delta, Vietnam. Jpn. Int. Res. Center Agric. Sci., Res. Highlights. pp. 38–39.

Yamauchi, K., M Samanya, K. Seki, N. Ijiri, and N. Thongwittaya. 2006. Influence of dietary sesame meal level on histological alterations of the intestinal mucosa and growth performance of chickens. J. Appl. Poult. Res., 15:266–273.

Yen, J. T., G. L. Cromwell, G. L. Allee, C. C. Calvert, T. D. Crenshaw, and E. R. Miller. 1991. Value of raw soybeans and soybean oil supplementation in sow gestation and lactation diets: A cooperative study. J. Anim. Sci. 69:656–663.

Yen, J. T., J. E. Wells, and D. N. Miller. 2004. Dried skim milk as a replacement for soybean meal in growing-finishing diets: Effects on growth performance, apparent total-tract nitrogen digestibility, urinary and fecal nitrogen excretion, and carcass traits in pigs. J. Anim. Sci. 82:3338–3345.

Yu, M., Z. Li, W. Chen, T. Rong, G. Wang, and X. Ma. 2019. Hermetia illucens larvae as a potential dietary protein source altered the microbiota and modulated mucosal immune status in the colon of finishing pigs. J. Anim. Sci. Biotechnol. 10:50.

Yuan, W., H. T. Jin, Z. H. Ren, J. L. Deng, Z. C. Zuo, Y. Wang, H. D. Deng, and Y. T. Deng. 2015. Effects of antibacterial peptide on humoral immunity in weaned piglets. Food Agric. Immunol. 26:682–689.

Zamora-Sillero, J., A. Gharsallaoui, and C. Prentice. 2018. Peptides from fish by-product protein hydrolysates and its functional properties: An overview. Marine Biotechnol. 20:118–130.

Zhang, J. Y., J. W. Park, and I. H. Kim. 2019. Effect of supplementation with brewer's yeast hydrolysate on growth performance, nutrients digestibility, blood profiles and meat quality in growing to finishing pigs. Asian-Australas. J. Anim. Sci. 32:1565–1572.

Zier, C. E., R. D. Jones, and M. J. Azain. 2004. Use of pet food-grade poultry by-product meal as an alternate protein source in weanling pig diets. J. Anim. Sci. 82:3049–3057.

# 13 Pulse Grains and Their Coproducts in Swine Diets

Li Fang Wang, Eduardo Beltranena, and Ruurd T. Zijlstra

## Introduction

Pulse grains are important starch and protein sources for humans, especially in developing countries. Approximately one-third of pulse grains are used in feed (Kelley et al. 2000), and use of pulse grains in livestock feeds may increase (van Barneveld et al. 2000).

Pulse crops are attractive for sustainable crop production in temperate climates (Miller et al. 2002). Growing pulse crops can reduce fertilizer demand for subsequent crops in rotations involving cereal grain and oilseeds, because Rhyzobia bacteria living in symbiotic association with pulse roots can fix atmospheric N. Diversification of cropping systems with pulses may prevent soil nutrient depletion, improve soil health by increasing microbial activity, and reduce disease and pest pressure (Krupinsky et al. 2002; Williams et al. 2014). World production of 11 pulse crops was 96 million MT in 2017 (Figure 13.1); dry bean, field pea, chickpea, and lentil accounted for 73% of pulse production (FAOSTAT 2019). India is the leading producer of pulse crops. Since 1980, the Canadian prairies have become a major producer of pulse crops (Carew et al. 2013). Nowadays, lead producer of field pea and lentil is Canada, of faba bean is China, and of lupin is Australia.

Pulses can be alternative feedstuffs to support sustainable swine production. Locally produced pulses can be cost effective to formulate swine diets (Bell and Wilson 1970), mitigate high feed cost (Zijlstra and Beltranena 2009; Woyengo et al. 2014a), and produce organic pork (Sundrum et al. 2000; Partanen et al. 2006). Pulses can replace animal protein sources such as meat and bone meal or fish meal, and plant protein sources such as soybean meal (SBM) or canola meal (Jezierny et al. 2010a; Stein et al. 2016). Cool season and tropical legume seeds are valuable sources of protein and energy in swine diets (Huisman and van der Poel 1994; Mekbungwan 2007; Jezierny et al. 2010a). Most lupin produced is used as feedstuff with less than 5% of global production consumed by humans (Písaříková and Zralý 2009; Watts 2011). The mix of pulses with rapeseed cake rich in sulfur amino acids (AA) can fully replace SBM in growing pig and sow diets and partially replace SBM in piglet diets (Hanczakowska and Swiatkiewicz 2013, 2014). Introducing novel pulses may pose a risk for consistent growth performance due to antinutritional factors (ANF) that should be managed using modern feed evaluation techniques.

## Nutritional Characteristics of Pulse Grains and Coproducts

To include pulse grains effectively in pig feed, their nutritional characteristics must be understood. Pulse grains contain two to three times more protein than cereals, but are low in fat and high in starch, except for lupin with little starch but more fat and protein (Abreu and Bruno-Soares 1998).

*Sustainable Swine Nutrition*, Second Edition. Edited by Lee I. Chiba.
© 2023 John Wiley & Sons Ltd. Published 2023 by John Wiley & Sons Ltd.

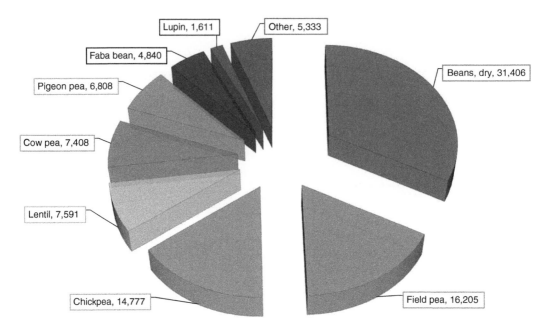

**Figure 13.1** Global pulse production in 2017 (1000 MT; Source: Food and Agriculture Organisation of the United Nations. Reproduced with permission.

Pulse grains have larger starch granules than cereal grains. Pulse grains also provide dietary fiber. The nonstarch polysaccharide (NSP) content was 17.7% in faba bean, 18.5% in field pea, and 32–40% in lupin (Gdala 1998).

For pulses, e.g., field pea, proteins are divided into two classes that account for 80% of seed storage proteins: salt-soluble globulins and water-soluble albumins. Globulins are divided into two groups based on their sedimentation coefficients: legumin of the 11S fraction and vicilin and convicilin of the 7S fraction (Tzitzikas et al. 2006). In comparison, soybean mainly contains glycinin of the 11S fraction and β-conglycinin of the 7S fraction (Hidayat et al. 2011). Yellow lupin seed contains five proteins: conglutins $\Upsilon$ and δ (unique to lupin), conglutin β1 and β2 (vicilin-like), and conglutin α (legumin-like). Conglutin $\Upsilon$, δ, and α possess interchain disulfide bonds (Esnault et al. 1991).

In pulse grains, arginine, leucine, lysine, aspartic acid, and glutamic acid account for approximately 50% of total AA (Sosulski and Holt 1980). Pulses are rich in lysine (7.3% of CP in field pea), leucine, and arginine but deficient in sulfur-containing AA and tryptophan (Gatel 1994; Iqbal et al. 2006); moreover, digestibility of methionine is poor. Thus, canola meal rich in sulfur-containing AA and threonine may complement pulse grains in swine diets (Partanen et al. 2001). Breeding for high sulfur-containing AA in pulse seed is promising (Molvig et al. 1997).

### Field Pea

In 59 field pea (*Pisum sativum*) samples, CP content ranged 13.7–30.7% of seed DM (average 22.3%). Total globulin content ranged 49.2–81.8% in pea protein. Vicilin was most abundant (26.3–52.0%), followed by legumin (5.9–24.5%) and convicilin (3.9–8.3%) (Tzitzikas et al. 2006). Albumin and globulin content in field pea is inversely related (Gueguen and Barbot 1988). Vicilin contains less sulfur than other proteins (Osborne and Harris 1907). In field pea, sulfur-containing AA are first-limiting and threonine is second-limiting (Holt and Sosulski 1979). Others reported that tryptophan was the first-limiting AA and sulfur-containing AA were second-limiting AA in field pea (Wang and Daun 2004). Arginine, aspartic acid, and glutamic acid were the most abundant AA in field pea (Holt and Sosulski 1979). Considering low concentration of methionine, threonine, and tryptophan, crystalline sources of these AA are required to fortify field pea diets (Stein and de Lange 2007).

The CP content in field pea varies greatly among and within cultivars (Gatel and Grosjean 1990). Protein and starch content in field pea were inversely correlated (Wang and Daun 2004). However, protein content does not accurately predict AA content in field pea. In field pea, essential AA content may decrease as CP content increases (Holt and Sosulski 1979; Igbasan et al. 1997; Partanen et al. 2001; Wang and Daun 2004).

Field pea contained 28–56% starch with approximately one-third as amylose (Gujska et al. 1994; Igbasan et al. 1997; Tzitzikas et al. 2006; Chung et al. 2008). However, green pea starch in Brazil was 61% as amylose and 39% as resistant starch (RS). Pea starch presented simple and composite granules with various shapes and sizes, B-type crystallinity pattern (Polesi et al. 2011). Field pea contained 17% dietary fiber, but low (2%) free lipid (Igbasan et al. 1997; Chung et al. 2008).

Variety and crop growing conditions may affect starch, fiber, and fat content of field pea (Wang and Daun 2004). Wrinkled, white-flowered field pea contained less starch (26.7 vs. 50%) than round field pea (Grosjean et al. 1999). Field pea samples grown in Australia over two seasons varied in proximate composition among genotypes but differed little between years (Black et al. 1998). Cultivars of field pea differ in seed coat color. But, seed coat color was not related to protein or starch content (Igbasan et al. 1997).

### Faba Bean

Faba bean (*Vicia faba*) typically contains more protein, but less starch than field pea. White-flowered faba bean contained 33% CP, whereas white-flowered field pea contained 26% CP (Partanen et al. 2001). Cultivars within flower color influenced starch, CP and NDF composition. White-flowered cultivars contained more vicine and convicine (Ivarsson and Neil 2018). Vicine and covicine content may be more relevant to pig performance than the low tannin content of modern faba cultivars (Cho et al. 2019).

Faba bean contains approximately 41% starch (Bach Knudsen 1997) and 0.3–0.5% P with 60% phytate-bound (Selle et al. 2003). Starch content of faba bean and field pea was greater when determined polarimetrically than enzymatically, e.g. 43.8 vs. 34.5% DM in faba beans and 50.9 vs. 39% DM in peas (Jezierny et al. 2017).

### Chickpea

Chickpea (*Cicer arietinum*) grain contains 20.5–30.5% CP, 32.8% starch, 8.8% crude fat, 12.5% total dietary fiber, and 1.29% chemically available lysine (Chavan et al. 1986; Wang et al. 2017). Globulin proteins in chickpea may suppress N metabolism by increasing N excretion, primarily as urinary N (Rubio et al. 1998). Vicilin, a major storage protein, contains much amides but little sulfur AA (Mandaokar et al. 1993). In chickpea, sulfur AA are first-limiting; valine, threonine, and tryptophan are also limiting (Khan et al. 1995). Tryptophan has been reported as the first-limiting AA (Chavan et al. 1986).

Chickpea has two types, Desi and Kabuli, with similar protein and essential AA content (Khan et al. 1995). Kabuli has larger seed (26 vs. 21 g/100 seeds). Desi contained more ADF (14.5 vs. 5%) and NDF (26.4 vs. 18.1%) and lower fat (3.7 vs. 5.1%) than Kabuli (Khan et al. 1995; Thacker et al. 2002). Desi contains up to 19.1% ADF (Visitpanich et al. 1985). Desi and Kabuli chickpea in Canada contained 42.9–46.3% total starch with 10.8–13.5% amylose, 6.5–7.1% free lipid and 0.9% bound lipid (Chung et al. 2008). Mexican chickpea cultivars varied little in chemical composition, averaged 19.5% CP and 5% fat (Sotelo et al. 1987). Chickpea starch in Brazil had 29% amylose and 32% RS, and starch granules had large oval shape and small spherical shape, with smooth surface and C-type crystallinity pattern (Polesi et al. 2011). Spring-planted chickpea has 8 g kg$^{-1}$ more CP than winter-planted chickpea (Singh et al. 1990).

### Lentil

Lentil (*Lens culinaris*) from Canada contained 28.7–31.5% CP and 46.0–47.1% total starch, but low amylose content (13.4–14.5%) and low free lipid content (2.0%) (Chung et al. 2008). Green lentil in Canada contained moderate amount of fiber (5% ADF and 15% NDF) (Landero et al. 2012).

### Lupin

Three major varieties of lupin exist: yellow lupin (*Lupinus luteus*) and white lupin (*Lupinus albus*) mainly in Europe and blue lupin (*Lupinus angustifolius*) in Australia (Sedlakova et al. 2016). Lupin seed contains (DM-based) 33–40% CP, 5–13% lipid, but slightly lower lysine (1.46%) and methionine (0.22%). Lysine and methionine are limiting AA in lupin seed (Aguilera et al. 1985; Sujak et al. 2006). Compared with French cultivars, Australian cultivars contained less methionine, tannin (0.3 vs. 0.6%), and fat (7 vs. 11%) (Eggum et al. 1993). Yellow lupin contains more CP (39–46.5%) than white lupin (34–38%) and blue lupin (33%), greater than faba bean (26%) and field pea (25%) (Brand et al. 2004; Sujak et al. 2006; Hanczakowska et al. 2017; Mierlita et al. 2018). Lupin seed contain 40% NSP, greater than other pulse grains, but negligible starch and more soluble dietary fiber (Eggum et al. 1993; Bach Knudsen 1997). Lupin seed is high in galactose, and its main storage carbohydrates are β-galactans (Perez-Maldonado et al. 1999). White lupin had high oil content (6.9–14.1%), with monounsaturated oleic acid being most abundant (48–56%), followed by polyunsaturated linoleic (19.9–20.4%) and linolenic fatty acids (8.0–10.9%) (Sujak et al. 2006; Mierlita et al. 2018; Rybinski et al. 2018).

### Antinutritional Factors in Pulse Grains

Pulse grains contain plant metabolites, e.g., protease inhibitors, saponins, lectins, glycosides, tannins, and alkaloids, which may affect nutrient digestion or feed intake in swine (Jezierny et al. 2011; Woyengo et al. 2017).

### Protease Inhibitors

Grain legumes, e.g., soybean, field pea, and faba bean contain protease inhibitors (Mikic et al. 2009). Trypsin inhibitor (TI) ranged 0.58–15.90 trypsin-inhibited units (TIU)/mg in field pea (Gatel 1994).

Most European spring pea cultivars contain less than 3 TIU/mg, but some ranged 6.1–8.4 TIU/mg (Masey O'Neill et al. 2012). Winter cultivars of field pea may contain as much as 11.2 TIU/mg. Growing conditions may influence pea TI content (Leterme et al. 1990). In Canada, one field pea cultivar ranged 2.22–7.66 TIU/mg. Cultivar of field pea accounted for more variability of TI than environment (Wang et al. 1998b). In Desi and Kabuli chickpea, trypsin inhibitor activity (TIA) was greater than chymotrypsin inhibitor activity. Desi cultivars have greater TIA than Kabuli chickpea (Singh and Jambunathan 1981). The TIA for one Canadian Kabuli chickpea sample was 5.2 mg g$^{-1}$ (Wang et al. 2017). Lupin contains negligible amounts of TI (Eggum et al. 1993; Petterson 2000; Písaříková and Zralý 2009).

Significant amounts of trypsin and chymotrypsin inhibitors from legumes can survive gastric and small intestine digestion in pigs (Rubio et al. 2006). Protease inhibitors of the Bowman-Birk family are proteins of low molecular weight (6–9 kDa) and inhibit serine proteases through competition with substrates for access to the active site of the enzyme. Bowman-Birk inhibitor proteins resist porcine gastrointestinal digestion. Survival rate of Bowman-Birk inhibitor proteins in chickpea-based diets at the ileum was 7.3% for TIA and 4.4% for chymotrypsin inhibitory activity (Clemente et al. 2008).

Inclusion of faba bean and field pea in diets for young pigs had minor effects on exocrine pancreatic secretions (Gabert et al. 1996b). Growing pigs can tolerate 4.5 mg g$^{-1}$ diet of trypsin and chymotrypsin inhibitors in chickpea and pigeon pea. These thresholds are unlikely exceeded in conventional swine diets containing pulse grains. Inclusion of pigeon pea meal may increase liver weight of growing pigs (Batterham et al. 1993). The TIA may reduce standardized ileal digestibility (SID) of AA in pigs (Woyengo et al. 2017; Messad et al. 2018). Pulses may exhibit pancreatic lipase inhibit activity (Lee et al. 2015). Chickpea had inhibitory activity against α-glucosidase, α-amylase and lipase that may impair the digestion of carbohydrates and lipids (Ercan and El 2016).

*Tannins*

Pulse grains contain condensed tannins that can precipitate protein by forming tannin-protein complex (Naczk et al. 2001). Field pea and faba bean contained more condensed tannin than chickpea (Perez-Maldonado et al. 1999), but condensed tannins are sometimes barely detectable in field pea. Genotype affected total phenolic content in field pea more than environment. Total phenolics in field pea ranged 162–325 mg catechin equivalent/kg DM among cultivars (Wang et al. 1998a). Brown-seeded cultivars of field pea contained appreciable quantities of tannins, whereas yellow- and green-seeded cultivars did not (Igbasan et al. 1997). Black chickpea contained 0.2% condensed tannin (Salgado et al. 2001). Color-flowered cultivars of field pea contained much more tannins (0.9%) than white-flowered ones (0.04%), and thus had declined nutrient digestion (Gdala and Buraczewska 1997; Grosjean et al. 1999; Smulikowska et al. 2001).

New cultivars of chickpea contained 0.01–0.05% tannin (Singhai and Shrivastava 2006). Recently, Canadian Kabuli chickpea contained 0.4% tannin (Wang et al. 2017).

High-tannin faba bean contains 1.6% tannins, most in the seed coat (5.8–7.6%) and little in the cotyledons (0.56–0.65%) (Cansfield et al. 1980; van der Poel et al. 1991). Color-flowered faba bean contained more condensed tannins (0.21–0.77% DM), whereas the white-flowered cultivar did not (Masey O'Neill et al. 2012; Jezierny et al. 2017; Ivarsson and Neil 2018). Color-flowered, high-tannin faba bean cultivars are more tolerant to frost and therefore have greater yield than zero-tannin cultivars (Henriquez et al. 2018).

Condensed tannin in faba bean, similar to tannins in sorghum grain (Marquardt et al. 1977), can reduce activity of trypsin but not chymotrypsin, and lower aminopeptidase activity in the small intestine (Jansman et al. 1994; Van Leeuwen et al. 1995). Condensed tannins in faba bean interact with dietary proteins rich in proline and histidine, and endogenous proteins in the digestive tract of pigs, thereby increasing excretion of endogenous proteins and reducing protein digestibility (Jansman et al. 1993a; Jansman et al. 1995). Dietary condensed tannin below 0.6% catechin equivalents may not cause systemic effects (Jansman et al. 1993b). Plant breeding can increase feeding value of faba bean by selecting low-tannin, white-flowered cultivar (van der Poel et al. 1992a).

### Alkaloids

Many lupin species contain alkaloids that are toxic or teratogenic to livestock (Panter et al. 2000). However, alkaloids in lupin may vary substantially. Alkaloid content ranged 0.01–0.99% in blue lupin and 0.03–0.15% in yellow lupin (Wasilewko et al. 1999; Hanczakowska et al. 2017). Quinolizidine alkaloid content ranged 9–634 mg kg$^{-1}$ in the three main lupin species in the Mediterranean basin (Musco et al. 2017). Pigs tolerate indole alkaloid gramine better than quinolizidine alkaloid that dominates in white lupin (Pastuszewska et al. 2001). Growing pigs can tolerate up to 0.5 g gramine/kg yellow lupin diets or 0.2 g alkaloids/kg blue lupin diets (Godfrey et al. 1985; Pastuszewska et al. 2001).

### Antigen and Immunologic Response

Young pigs fed diets containing field pea, faba bean, lupin, and chickpea can display systemic antibody responses to specific dietary proteins. Antibodies against β-conglutin of yellow lupin, vicilin of common vetch (*Vicia sativa*) and vicilin of red pea (*Lathyrus cicero*) were detected in sera of piglets after 28-day feeding (Seabra et al. 2001). Piglets fed legume-containing diets had greater plasma immunoglobulin (Ig) G titers to legume proteins than piglets fed casein (Salgado et al. 2002a). Immunoreactive polypeptides in ileal digesta of piglets fed pea, faba bean, and chickpea belonged mainly to proteins of the 7S family, and to other proteins including PA2 albumin and lectin in pea. Polypeptides from the 11S family were detected in piglets fed lupin or chickpea. Proteins of the 7S family were more immunogenic than proteins of the 11S family in weaned piglets (Salgado et al. 2003).

### Functional Properties of Pulse Grains

Pulse grains have nutritional and health benefits in humans (Rochfort and Panozzo 2007; Jukanti et al. 2012; Arnoldi et al. 2015; Gupta et al. 2017) that may also benefit pigs. Fermentable carbohydrates in pulse grains may modulate microbial population and composition, thus may prevent or provoke intestinal disorders in piglets (van der Meulen et al. 2010; Aumiller et al. 2015). Field pea fiber is highly fermentable, producing short-chain fatty acids (SCFA) and microbial protein, thus can modulate the gut environment and reduce N excretion (Jha et al. 2011; Jha and Leterme 2012). Pea fiber may improve intestinal barrier function in weaned piglets (Chen et al. 2013). Hulls of pea and faba bean may promote net fluid absorption in piglets during post weaning diarrhea (van der Meulen and Jansman 2010; Jansman et al. 2012). Hulls of field pea and faba bean may bind *E. coli* and enterotoxin, thereby reducing diarrhea prophylaxis and treatment (Becker et al. 2012). Lupin may increase bifidobacteria in caecum and reduce *Campylobacter* spp. excretion after one-week feeding (Jensen et al. 2013). However, diets containing lupin cannot protect pigs against developing swine dysentery or alleviating diarrhea (Sorensen et al. 2009; Hansen et al. 2010). Blue lupin may have functional properties that alter fat metabolism in pigs (Martins et al. 2005).

## Nutrient Digestion of Pulse Grains and Coproducts in Pigs

### *Energy*

#### *Field Pea*
Field pea is a good source of starch for pigs. For example, the DE and NE values of Canadian field pea averaged 3.72 and 2.58 Mcal/kg DM in 20-kg pigs and 3.92 and 2.72 Mcal NE/kg DM in 50-kg pigs, respectively. However, energy values varied among samples. In growing pigs, the DE value varied from 3.19 to 4.56 Mcal/kg DM and NE value from 2.22 to 3.08 Mcal/kg DM (Leterme et al. 2008). In gestating sows, the DE value of field pea in Western Canada ranged from 3.44 to 4.05 Mcal/kg DM and the NE from 2.40 to 2.84 Mcal/kg DM (Premkumar et al. 2008). Thus, using table values for energy to formulate diets is risky and rapid evaluation of DE of field pea is warranted (Wang and Zijlstra 2018). The DE value of field peas can be predicted by *in vitro* techniques (Montoya et al. 2010b).

#### *Chickpea*
Kabuli chickpea had 62.3% apparent ileal digestibility (AID) and 88.6% apparent total tract digestibility (ATTD) of GE, 3.78 Mcal DE (as fed) and 2.70 Mcal predicted NE/kg in 90-kg pigs (Wang et al. 2017). Fiber content affects energy value. High fiber chickpea had 5.5% units lower ATTD of GE (81.3 vs. 86.8%) and 0.21 Mcal/kg lower DE (3.56 vs. 3.87 Mcal/kg as fed) than low fiber (14.8 vs. 6.3% ADF) chickpea in 20-kg pigs (Visitpanich et al. 1985).

#### *Lentil*
Energy-yielding nutrients in lentil are highly digestible. For example, increasing inclusion of up to 30% lentil to replace 20% SBM and 10% wheat grain did not reduce ATTD of GE of diets in 9-kg pigs (Landero et al. 2012). Lentil had 7% units lower ATTD of GE and 0.54 Mcal lower DE/kg DM, but similar NE value (2.60 vs. 2.63 Mcal/kg DM) compared with SBM in 31-kg pigs (Woyengo et al. 2014b).

#### *Faba Bean*
Energy-yielding macronutrients in faba bean are also highly digestible. The ATTD of OM in faba bean ranged from 79.0 to 88.1% in 15-kg pigs (Jansman et al. 1993a). The ATTD of GE in zero-tannin faba bean was 88.5% and DE value was 3.47 Mcal/kg (as fed) in 60-kg pigs (Zijlstra et al. 2004b; Kiarie et al. 2013). Although their tannin content differed, dark-flowered faba bean and white-flowered faba bean had similar ATTD of DM (93.2 vs. 92.6%) in growing pigs (Mosenthin et al. 1993).

#### *Lupin*
Fiber content affects energy value of lupin. Blue lupin contains double the ADF than field pea; consequently, its ATTD of GE was 10% units lower (72 vs. 83%), and AID of GE was 20% units lower (57 vs. 77%) in growing pigs (Salgado et al. 2002c). Compared with wheat grain, more of the

DE of lupin was from hindgut fermentation (Taverner et al. 1983). In the pig gut, 28% ADF and 70% hemicellulose in lupin are degraded. Digestibility of fiber in lupin was higher in mature sows than growing pigs; thus, contribution of energy from hindgut fermentation of lupin is substantial in sows. Energy digestibility of blue lupin hulls, whole seed and dehulled seed was 40, 77 and 81% in growing pigs, but 78, 85 and 89% in sows, respectively (Noblet et al. 1998). Cultivar of lupin and growing region may cause variation in the DE value (Kim et al. 2009b).

## *Protein*

Pulse grains contain protease inhibitors that may affect protein digestion (Gatel 1994). Feeding pulse grains to pigs may increase ileal endogenous losses of serine-protease family trypsin or chymotrypsin (Salgado et al. 2002d). Digestibility of methionine was poor in pulse grains such as field pea, faba bean, and lupin (Partanen et al. 2001). The major storage protein, e.g. vicilin in field pea, can be broken down rapidly in the stomach and subsequently digested in the small intestine within three hours. However, lectin remained mostly intact throughout the stomach and small intestine. Pea albumins, rich in sulfur AA, were also not completely digested, contributing to lower digestibility of sulfur AA of field pea in pigs (Le Gall et al. 2007). *In vitro* digestion techniques are promising in predicting protein and AA digestibility of pulse grains in pigs (Swiech and Buraczewska 2001; Jezierny et al. 2010b).

### *Field Pea*
Field pea ANF may reduce nutrient utilization by increasing ileal endogenous N and AA losses. However, true ileal digestibility (TID) of CP in field pea was similar to SBM in 19- to 32-kg pigs. Due to lower TI content, field pea induced less specific ileal endogenous losses of CP (44 vs. 65 g kg$^{-1}$ DMI) and AA than SBM, and the SID of CP, methionine and tryptophan were lower for field pea than SBM in young pigs (Eklund et al. 2012; Petersen et al. 2014). The TID of indispensable AA, except for lysine, was similar between pea protein isolate and field pea, but pea protein isolate had lower ileal endogenous losses of AA, thus greater AID and SID of CP, isoleucine, leucine, lysine, and valine than field pea (Woyengo et al. 2015). Color- and white-flowered field pea had similar AID of CP, although they differed in tannin content (Kasprowicz and Frankiewicz 2004). The SID AA content may vary among field pea samples (Montoya et al. 2010a).

### *Chickpea*
Compared with defatted soybean, chickpea had 5%-unit lower SID of CP in 100-kg pigs (Rubio 2005). Kabuli chickpea had 66% SID of CP and 71.7% SID of lysine in 90-kg pigs (Wang et al. 2017), which was lower than that of field pea in growing pigs (Eklund et al. 2012).

### *Lentil*
Lentil had about 7–12%-units lower SID of AA than SBM. The SID of lysine was 81% compared with 93% for SBM (Woyengo et al. 2014b).

### *Faba Bean*
Faba bean had lower SID of CP than lupin and SBM. The SID of AA varied among cultivars, partly due to condensed tannin (Jezierny et al. 2011). Feeding faba bean or field pea to 8- or 18-kg pigs did not alter composition and flow of total, protein-bound and free AA in pancreatic juice (Gabert et al. 1996a). Condensed tannins can reduce activity of trypsin, but not chymotrypsin. Feeding 20%

faba bean hulls containing 3.5% condensed tannin reduced trypsin activity in ileal digesta, but not duodenal digesta of pigs. Reduced activity of proteases may minimally affect dietary protein digestion. Rather, binding of tannins to dietary proteins or increased excretion of endogenous proteins may reduce AID of CP (Jansman et al. 1994). The AID of CP of faba bean decreased from 74.3 to 68.8% in 20-kg pigs with increasing tannin content (Grala et al. 1993). White-flowered and dark-flowered faba bean, differing in tannin content (<0.1 vs. 0.58%), had similar TID of CP (89 vs. 88.7%) and AA in 35-kg pigs (Mosenthin et al. 1993). Faba bean containing 0.6–0.7% tannin had 10%-units lower AID of CP and of most AA than SBM in 25-kg pigs (Kasprowicz and Frankiewicz 2004). The AID of lysine, threonine, and methionine for zero-tannin faba bean was 85.9, 76.1, and 74.1%, respectively (Zijlstra et al. 2004b). Zero-tannin faba bean has SID of CP (76.7%) similar to, but greater SID of lysine (87.1%) than wheat grain or corn DDGS in 29-kg pigs (Kiarie et al. 2013). White-flowered faba bean had high SID of CP (83.7%) and lysine (90.5%), but low SID of methionine (58.2%) in 22-kg pigs (Presto et al. 2011).

*Lupin*

Compared with SBM, sweet lupin contains similar CP, but cannot completely substitute SBM because digestible tryptophan, isoleucine, and threonine are insufficient (Wunsche et al. 1990). Blue lupin had similar AID of isoleucine (86%) to SBM; however, isoleucine is limiting (Batterham and Andersen 1994). The TID of CP and lysine of blue lupin is 90.7 and 93.1%, respectively in 52-kg pigs (Hennig et al. 2008). White lupin had similar TID of CP as soybean protein (91 vs. 91.5%) in humans (Mariotti et al. 2002). In 9-kg pigs, blue lupin had similar AID of CP to SBM, field pea and faba bean (Salgado et al. 2002b). In 35-kg pigs, blue lupin had lower SID of CP than SBM and field pea (Norgaard et al. 2012). In 100-kg pigs, blue lupin had 5%-unit lower SID of CP and AA than defatted soybean, but similar SID as field pea (Rubio 2005). However, availability of lysine for blue lupin in growing pigs was only 37–65% using feed efficiency as criterion, compared with field pea at 93% and SBM at 89–98% (Batterham et al. 1984).

Fiber and alkaloid content in lupin may not affect protein digestibility. Blue lupin contained more ADF (22 vs. 15%) than white lupin, but also had greater SID of lysine (88 vs. 81%) (Mariscal-Landin et al. 2002). Alkaloids content differed within yellow lupin, but little variation exists in AID of CP (81.9–84.9%) and SID of CP (87.5–90.5%) in 13- to 34-kg pigs (Wasilewko et al. 1999).

*Starch*

Pulse grains contain less starch than cereal grains (41–45% vs. 57–68%). Larger pulse starch granules were embedded in the protein matrix; hence, pulse had lower AID of starch, e.g. 85% for faba bean, 85–90% for field pea and chickpea compared with 93–96% for cereal grains in growing and finishing pigs (Rubio et al. 2005; Tan et al. 2017). But, pulse starch can be almost completely digested in total tract of pigs (Tan et al. 2017). Others reported greater AID of starch (96.3–98.1%) for faba bean varying in tannin content (Jansman et al. 1993a). Starch determination method may yield different starch digestibility values. Using the polarimetric method resulted in greater digestibility values than enzymatic starch analysis (Jezierny et al. 2017). Flower color or tannin content of field pea had little correlation to AID of starch. The AID of starch of white- and colored-flowered field pea containing 0.04–0.6% tannin was 85–87%, and of faba bean containing 0.59–0.67% tannin was 82–86% in 34-kg pigs (Gdala and Buraczewska 1997). Undigested starch may have nearly equal energetic efficiency as digested starch considering adequate provision of SCFA and decreased energy loss due to decreased activity of pigs (Fouhse and Zijlstra 2018).

*Fiber*

Fiber in pulse grains is highly fermentable. Field pea hull is more fermentable than barley hull (Canibe and Bach Knudsen 2002). Field pea and chickpea had greater AID and ATTD of total dietary fiber (TDF) than corn in growing-finishing pigs (Tan et al. 2017). In 45-kg pigs, lupin hulls are more fermentable than field pea hulls, especially the degradation of cellulose (Stanogias and Pearce 1985). In 100-kg pigs, 65% NSP and 82% oligosaccharides of lupin were degraded at the distal ileum, greater than 56 and 43% for NSP and 51 and 69% for oligosaccharides in defatted soybean and chickpea, respectively (Rubio et al. 2005). Most oligosaccharides in field pea and faba bean are degraded in the small intestine. In 34-kg pigs, 83% α-galactosides in faba bean and up to 93% α-galactosides in white- or color-flowered field pea were fermented by the distal ileum (Gdala and Buraczewska 1997).

*Fat*

Chickpea and lupin had similar AID of stearic, oleic, linoleic, and linolenic fatty acids. However, the AID of unsaturated (C18:1, C18:2 and C18:3) fatty acids was greater than AID of saturated (C14:0, C16:0) fatty acids (71–92 vs. 48–71%) in 100-kg pigs (Rubio et al. 2005).

*Minerals*

The true total tract digestibility (TTTD) of P in field pea was 51–60% in 32-kg pigs (Stein et al. 2006a; Johnston et al. 2013). The STTD of P is greater for lupin kernel than lupin hull (72 vs. 52%) in growing-finishing pigs (Park et al. 2019).

**Feeding Pulse Grains and Their Coproducts to Pigs**

Pulse grains are feedstuffs that provide both energy and protein. Incorporating locally produced pulse grains into swine diets may provide cost and health benefits. Livestock production in Europe relies greatly on import of SBM. Because most soybean is genetically modified, concerns regarding its inclusion in feed may arise, especially in organic chains (Martelli et al. 2009).

*Field Pea*

Field pea is a starch and protein source in pig diets. In diets formulated to equal NE value and SID AA content, 30–40% field pea can entirely replace SBM without detrimental effects on growth but not G:F in weaned pigs (Smith et al. 2013; Landero et al. 2014; Hugman et al. 2018). Inclusion of 48–60% raw field pea in diets may reduce ADFI or ADG in weaned pigs (Stein et al. 2010). Inclusion of field pea should remain below 20% for early-stage weaned pigs. For growing-finishing pigs, inclusion up to 70% field pea in diets did not affect performance and carcass quality (Stein et al. 2006b; Stein and de Lange 2007). However, in diets formulated based on equal DE and AA, inclusion of 64% field pea replacing SBM and corn in diets reduced ADG and G:F in growing-finishing pigs (Brand et al. 2000).

Field pea can also serve as an energy source in pig diets. Inclusion of 18–20% field pea to replace corn in diets fed to 8-kg pigs did not affect ADFI, ADG and G:F (Stein et al. 2004), but may decrease

G:F in 12-kg pigs (Brooks et al. 2009). However, inclusion of 30% field pea to replace 22% corn and 8% SBM without balancing for NE and SID AA content for diets reduced ADFI and ADG in 6-kg weaned pigs (Friesen et al. 2006). Inclusion of 36% field pea to replace 25% corn and 12% SBM did not reduce growth performance in 22 to 110-kg pigs, but may increase loin depth (Stein et al. 2004), and may increase fat content in the longissimus muscle (Degola and Jonkus 2018).

### Faba Bean

Following plant breeding, zero-tannin faba bean (<0.05% condensed tannin) have a similar NE value and greater SID AA content than field pea, and thus can be considered for inclusion in diets for swine (van der Poel et al., 1992a; Zijlstra et al. 2008).

Faba bean is a viable protein source in diets fed to weaner pigs. Previously, 12.5% faba bean was included in diets as partial replacement for SBM and barley without influencing performance in 6-kg pigs (Onaghise and Bowland 1977). Recently, 40% zero-tannin faba bean fully replaced SBM in the starter diet without dietary adaptation to faba bean (Omogbenigun et al. 2006; Beltranena et al. 2009). Zero-tannin faba bean can fully replace field pea or SBM without reducing performance in growing-finishing pigs (Gunawardena et al. 2007b). Dietary inclusion of 30% faba bean to replace 16% SBM and 14% corn for 40-kg pigs, and 20% faba bean to replace 8% SBM and 12% corn for 120-kg pigs increased ADG (Prandini et al. 2011). However, without balancing for NE or SID AA content, inclusion of 30% faba bean replacing 15% SBM and barley or inclusion of 31.7% faba bean replacing 22% rapeseed meal and barley reduced ADG in growing but not finishing pigs (Castell 1976; Partanen et al. 2003).

Although color-flowered faba bean contained more condensed tannin and white-flowered faba bean contained more vicine and convicine, cultivar determines their nutritional value rather than flower color (Ivarsson and Neil 2018). Although condensed tannin can reduce nutrient digestibility, dietary inclusion of 30% high-tannin (0.2%) colored-flowered or low-tannin (0.003%) white-flowered faba bean had similar growth performance in 25-kg pigs (Flis et al. 1999). Inclusion of 30% colored-flowered faba bean replacing 14% SBM in diets with equal NE and SID AA content did not affect ADFI, ADG and G:F in 30- and 60-kg pigs (Smith et al. 2013), but lowered lean yield (White et al. 2015).

Feeding faba bean to pigs may affect meat quality. Dietary inclusion of 25% color-flowered faba bean to replace 15% SBM and 10% wheat increased fat content in the longissimus muscle, similar to feeding 28% field pea (Degola and Jonkus 2018). Feeding 10% faba bean with 0.5–0.7% tannin improved sensory characteristics of pork, e.g., juiciness, tenderness, and palatability (Milczarek and Osek 2016). Inclusion of 26% faba bean to replace 18% rapeseed meal and 8% barley in diets darkened meat color (Partanen et al. 2003). Slightly darker meat in pigs fed zero-tannin faba bean compared with SBM or field pea was also reported (Gunawardena et al. 2007a). Partial substitution of 9% SBM with 18% faba bean in diets fed to 56-kg pigs did not affect meat color, water-holding capacity, tenderness, and chemical composition (Gatta et al. 2013).

### Chickpea

Off-grade chickpea can be used as energy and protein sources in swine diets (Bampidis and Christodoulou 2011). Dietary inclusion of 15% Kabuli chickpea is viable for weaned pigs. However, inclusion of 30% Kabuli chickpea reduced ADG and G:F in 10-kg pigs despite formulating diets to equal NE and SID AA content (Wang et al. 2017). Without balancing for NE and AA, inclusion of 30% Kabuli or Desi chickpea to replace 18% SBM and wheat in diets reduced ADG in 20-kg pigs (Mustafa et al. 2000). Fiber content may affect DE value of chickpea, but not growth performance of pigs. Inclusion of 26% chickpeas with low (6.3% ADF) or high fiber (14.8% ADF) to replace 14% SBM did not affect ADG and G:F in 20-kg pigs (Visitpanich et al. 1985). For finishing pigs, inclusion of 30% Kabuli or Desi chickpeas to replace 12% SBM did not affect growth performance, dressing percentage, or lean yield (Mustafa et al. 2000). Over 350 chickpea cultivars have been released globally (Thudi et al. 2016), and their nutritional values require investigation.

## Lentil

Feeding value of lentil in pigs has not been well investigated. In diets balanced for equal NE and SID AA content, inclusion of 22.5% lentil to replace 15% SBM did not affect ADG and G:F in weaned pigs; however, inclusion of 30% lentil reduced ADG and G:F. Thus, lentil inclusion should not exceed 22.5% in diets for nursery pigs to maintain growth (Landero et al. 2012).

## Lupin

Sweet lupin has a lower bitter alkaloid content (Pearson and Carr 1976) and can be a cost-effective ingredient for pigs. However, species or cultivars differ considerably in their suitability as protein source for pigs. Pigs can utilize blue and yellow lupin well, but not white lupin (van Barneveld 1999; Písaříková and Zralý 2009). White lupin contains more alkaloids and may have lower N-retention than blue and yellow lupin (Gdala et al. 1996). High inclusion of white lupin can suppress feed intake and even halt growth. Reduction of alkaloid in white lupin may ameliorate reduced growth (Pearson and Carr 1977). Pigs may refuse to eat high alkaloid lupin (Hanczakowska et al. 2017). Lupin alkaloid may reduce feed palatability. Increasing inclusion up to 25% yellow lupin in diets decreased feed preference in 22.5-kg pigs. Nevertheless, inclusion of 25% yellow lupin to replace 21% SBM did not affect ADFI and ADG in 19-kg pigs (Bugnacka and Falkowski 2001). Feed preference for blue lupin may be similar to that of SBM, but better than rapeseed meal and sunflower meal in 18-kg pigs (Sola-Oriol et al. 2011). However, even with low alkaloid content, the value of lupin for pigs may be limited by other factors that suppress voluntary feed intake (Pearson and Carr 1977). Mean retention time of lupin diets was inversely related to feed intake. Diets containing white lupin seed had double mean retention time in 60-kg pigs and reduced feed intake, compared with diets of animal protein, field pea or blue lupin (Dunshea et al. 2001). Diets containing lupin may be limiting in methionine in growing pigs (Pearson and Carr 1979). Methionine supplementation in a diet containing 23.7% blue lupin increased ADG and G:F in 21-kg pigs (Leibholz 1984). White lupin increased ileum digesta viscosity, ammonia concentration in ileum and caecum digesta, and total bacteria and Enterobacteriaceae in ileal digesta in growing pigs (Kasprowicz-Potocka et al. 2017).

Considering high fiber and bitter alkaloid, high inclusion of lupin in diets fed to young pigs should be avoided. Inclusion of 25% white lupin replacing SBM increased diet ADF from 5 to 10%, consequently, decreased ADG and ADFI in 10-kg pigs (McNiven and Castell 1995). However, dietary inclusion of 31% white lupin to replace 20% SBM did not affect ADFI, but slightly reduced ADG in 18-kg pigs (Zraly et al. 2008). Inclusion of 26% blue lupin to replace whey or skim-milk did not affect ADFI and G:F in weaned pigs (Pearson and Carr 1976; Kim et al. 2012). However, inclusion of 31% blue lupin to replace 18% SBM reduced ADFI in 9-kg pigs (Kasprowicz-Potocka et al. 2013). Inclusion of 37% blue lupin to replace 27% SBM and 11% triticale decreased ADG and G:F in 20-kg pigs (Kasprowicz-Potocka et al. 2016b). Feeding value of blue lupin is superior to white lupin. Dietary inclusion of 37% blue lupin to replace 30% barley only slightly reduced ADG, but inclusion of 37% white lupin reduced ADFI 49% and ADG 80% in 15-kg pigs (Pearson and Carr 1979). For yellow lupin, 16–20% inclusion did not affect ADFI, ADG, and G:F in weaned pigs (Kim et al. 2008; Kasprowicz-Potocka et al. 2013); thus, 15% inclusion was recommended (Kim et al. 2008).

For growing pigs, including 15% white lupin did not reduce ADFI and ADG (King 1981; Flis et al. 1998; Kasprowicz-Potocka et al. 2017). Inclusion of 19% white lupin to replace 12% SBM did not affect ADFI, but decreased ADG and G:F in growing pigs (Donovan et al. 1993). Inclusion of 30% white lupin reduced ADFI and ADG in 35-kg pigs (Moore et al. 2016). Inclusion of 20% yellow lupin to replace 14% SBM and barley increased ADG and G:F in 30-kg pigs (Roth-Maier et al. 2004). Inclusion of 35% blue lupin to replace 24% SBM did not affect ADFI and ADG in 43-kg pigs (Kasprowicz-Potocka et al. 2016b). Inclusion of 25% blue lupin to replace 22% field pea and 3% barley decreased ADG in 30-kg pigs (Norgaard and Fernandez 2009).

For finishing pigs, 14% white lupin can be included in diets without affecting growth (Flis et al. 1998; Zraly et al. 2008), but may decrease G:F (Donovan et al. 1993). Inclusion of 20% blue lupin to replace 14% SBM did not affect ADFI and ADG in finishing pigs (Kasprowicz-Potocka et al. 2016b). Dietary inclusion of 15% blue lupin to replace 15% SBM in diets without balancing for NE and SID AA content did not affect ADFI, but reduced ADG and G:F in 60-kg pigs (Degola and Jonkus 2018). Inclusion of 30% Australian sweet lupin kernel in diets with equal NE decreased ADG and ADFI in 63-kg pigs (Kwak et al. 2000). Feeding blue or yellow lupin instead of SBM to sows may reduce litter weight of piglets (Hanczakowska et al. 2017).

Feeding lupin seed or kernel instead of animal protein may reduce dressing percentage because of increased gut fill and larger intestine weight (King et al. 2000). Inclusion of 30% white lupin reduced carcass weight and dressing percentage and increased C18:1 in backfat (Van Nevel et al. 2000). However, inclusion of 14–20% white lupin may not affect dressing percentage and lean yield (Donovan et al. 1993; Leikus et al. 2004). Inclusion of 20% white lupin to replace 10% blue lupin and 5% canola meal can be used as strategy to reduce feed intake, body fat, and backfat of pre-slaughter immune-castrated pigs (Moore et al. 2017). Inclusion of 20% white lupin may improve texture, juiciness, tenderness, and taste of meat (Zraly et al. 2006, 2007). However, inclusion of 35% blue lupin did not affect carcass and meat quality (Kim et al. 2011).

## Pulse Coproducts

Fractionation generates pulse coproducts (Zijlstra et al. 2004a; Zijlstra and Beltranena 2007). In weaned pigs, feeding faba bean or field pea protein concentrates had similar growth performance to feeding a blend of soy protein concentrate, corn gluten meal, and menhaden meal (Gunawardena et al. 2010b). Dietary inclusion of 9.5% yellow field pea protein concentrate to replace soy protein concentrate or 47% yellow field pea starch to replace cooked corn did not affect growth performance in weaned pigs (Parera et al. 2010). Dietary inclusion of 20% field pea starch to replace wheat increased sow ADFI and ADG during lactation and reduced weight loss during the first three weeks of lactation (Thingnes et al. 2013).

### Feed Formulation and Risk Management

Introducing pulse grains as feedstuff into swine diets is an opportunity to improve the economic sustainability of swine production, but also poses a risk that must be managed properly. This risk can be divided into a range of factors such as variation of chemical composition, ANF, and potential reductions of pork quality (de Lange 2000; Smits and Sijtsma 2007).

Apart from feed intake that is a major factor impacting growth and may be affected by ANF and fiber content (Seneviratne et al. 2010), feed quality evaluation for energy is the most important factor for successfully introducing new feedstuffs. Indisputably, the choice of energy evaluation system will alter the relative values placed upon feeds (Noblet et al. 1993). The feed industry in The Netherlands has been reliant on the NE system since 1970 (CVB 1993), partly to manage the risk of a wide ingredient matrix (Zijlstra and Payne 2007). Inclusion of high fiber or high protein feedstuffs into diets formulated to equal DE or ME reduces diet NE value. Subsequently in studies detecting reduced growth performance, the test feedstuff was blamed. Formulating diets to equal NE content may help maintaining growth performance (Zijlstra and Beltranena 2008).

Another risk associated with feedstuffs is nutrient variability due to cultivar, agronomic, weather, harvest, and storage conditions. With coproducts, processing introduces additional variability (Zijlstra et al. 2001). Predictions of nutritional value based on physical, chemical, and biological evaluations are warranted (Wang and Zijlstra 2018).

### Increase Feeding Values of Pulse Grains

#### Processing Methods to Reduce ANF

Pulse grains contain varying quantities of ANF. Pulse grains cultivars containing low ANF are nutritionally preferable (Singh 1988). Although breeding may increase the nutritive value of lentil, chickpea, field pea, and faba bean (Materne et al. 2011), ANF are still present in pulse grains. Processing of pulse grains can reduce or eliminate ANF. In general, thermal treatments (above 100 °C or greater) do reduce activity of enzyme inhibitors and lectin. Germination is most effective to reduce phytate in pulse grains, whereas dehulling can effectively reduce phenolics and tannins (Patterson et al. 2017).

Extrusion can reduce most ANF in pulse grains. Trypsin is heat labile; thus, extrusion can reduce TIA 85–91% in chickpea, 30–50% in faba bean, and 50–60% in field pea (Adamidou et al. 2011). Extrusion above 140 °C reduced condensed tannin, inhibitors of trypsin (3.8 down to 0.2 IU/mg), chymotrypsin (2.8 down to 1 IU/mg) and α-amylase, and lectin in field pea (Alonso et al. 1998), faba bean (Alonso et al. 2000; Leontowicz et al. 2001a; Leontowicz et al. 2001b), and lentil (Rathod and Annapure 2016). Soaking prior to extrusion may increase nutritive value of pulse grains (Abd El-Hady and Habiba 2003). Processing variables affect the nutritional value of pulse grains (Wiseman 2013). Extrusion of field pea reduced TIA from 2 to 0.18 TIU/mg at 129 °C, and not detectable at 135 or 142 °C (Frias et al. 2011). Extrusion at 105–111 °C with preconditioning slightly reduced phytate and tannin content in faba bean, chickpea, and field pea (Adamidou et al. 2011). Extrusion can reduce RS in pulse grains. Extrusion (135 °C, 22% moisture) reduced RS in field pea 89% (Zaworska et al. 2018). Extrusion (160 °C, 500 rpm) may decrease raffinose and stachyose in lentil, field pea, and chickpea (Berrios et al. 2010).

Dry heating chickpea at 140 °C longer than six hours inactivated TI, but reduced protein quality (Marquez et al. 1998). Heating faba bean starch at 80, 100, or 120 °C for 12 hours at 23% moisture increased crystallinity and gelatinization temperature, decreased slowly digestible starch, and increased RS at 80 and 100 °C, but decreased RS at 120 °C (Ambigaipalan et al. 2014).

Dehulling can reduce condensed tannin and polyphenols in faba bean (Alonso et al. 2000). Cooking is more effective in reducing ANF in chickpea than decortication. Cooking decreased oligosaccharide 30–34%, phytate 24–34.5%, polyphenols 58.7–62.2%, and TIA 53.6–59.9% (Attia et al. 1994). Boiling for three hours then drying for 20 hours at 70 °C destroyed 57% TI in chickpea (Sotelo et al. 1987). Boiling did not reduce tannin content in chickpea and lentil (Ummadi et al. 1995). Soaking/microwave cooking may reduce tannin and phytate content in Kabuli chickpea (Xu et al. 2016). Pressure cooking chickpea at 120 °C did not reduce protease inhibitors in the albumin fraction, but increased *in vitro* protein digestibility from 71.8 to 83.5% (Clemente et al. 1998). Exposing field pea to organic acids and oxides reduced TIA 80%, lectins 70%, and tannins 40% (Dvorak et al. 2005). Germination was less effective than autoclaving or microwave cooking in reducing TI, lectins, tannins, and saponins in chickpea, but more effective in reducing phytate, stachyose, and raffinose (El-Adawy 2002). Polyvinylpyrrolidone addition can reduce tannin content in faba bean hulls (Garrido et al. 1991).

## Processing to Increase Digestibility and Performance

### Dehulling
*Faba bean.* Dehulling decreases content of condensed tannin and fiber and increases the CP content of faba bean (Meijer et al. 1994). Dehulling increased the AID of DM and CP, but not starch, of faba bean by 3% units in 35-kg pigs (van der Poel et al. 1992b; Mariscal-Landin et al. 2002). Inclusion of 30% dehulled faba bean and SBM to replace 30% full fat soybean seed and wheat bran in diets did not affect ADG and ADFI in 10-kg pigs. However, when given a choice, pigs did prefer full fat soybean seed over dehulled faba bean (Emiola and Gous 2011).

*Lupin.* Dehulling may lower nutritional value of lupin for pigs, despite lower fiber, greater CP and lysine content and greater AID of AA and DE value in lupin kernel. Formulated to equal lysine/DE, 50% lupin kernel to replace 50% lupin reduced ADG in 20-kg pigs (Fernández and Batterham 1995). Lupin hull contains little alkaloids. Oligosaccharides in the lupin kernel may influence pigs more. Increasing inclusion up to 30% dehulled white or blue lupin linearly decreased ADG for both lupins and decreased ADFI for white lupin compared with lupin with hulls in 15- to 47-kg pigs. Pigs prefer SBM than dehulled lupin, but adding 5% lupin hulls to SBM diet did not affect feed preference and ADFI (Ferguson et al. 2003). Increasing inclusion of 24% dehulled blue lupin to replace up to 24% of milk products (whey and skim milk) in wheat-based diets with equal DE and SID AA fed to 6-kg weaned pigs reduced ADG mainly due to decreased ADFI. Thus, inclusion of dehulled lupin immediately after weaning should be limited to less than 18% to prevent reduced growth (Kim et al. 2012).

*Chickpea.* Decortication of Kabuli chickpea seed decreases dietary fiber, Ca, Zn, and polyphenols content and increases reducing sugars content and *in vitro* protein digestibility, but may not alter phytate content or TIA (Attia et al. 1994).

### Grinding
*Field pea.* Particle size reduction affects nutrient digestion. Increasing particle size linearly decreased ATTD of CP and GE, and DE of field pea in 28-kg pigs, also linearly decreased *in vitro* starch digestibility (Montoya and Leterme 2011). Mill settings can affect particle size and dispersion of AA among particle sizes. Loss of AA may occur if field pea is ground too fine (Tudor et al. 2009).

Mill type, e.g. hammer or disc-mill, with different settings and screen size altered particle size and its distribution. Hammer-milled field pea gave greater hydration and *in vitro* starch digestion rate than disc-milled field pea (Nguyen et al. 2015).

*Lupin.* Particle size influences nutrient and AA utilization and fermentation characteristics of lupin in the porcine gut. Decreasing particle size of blue lupin increased AID of CP, ATTD of GE, and decreased straight-chain fatty acids and increased branched-chain fatty acids in hindgut fermentation in 45-kg pigs (Kim et al. 2009a). Fine grinding (1-mm sieve) of blue lupin may increase AID of ether extract, but not ATTD of OM and NSP of diets in 20-kg pigs (Pieper et al. 2016).

### Soaking and Germination

Germination can modify chemical composition of seed, thereby influencing nutrient digestion. However, germination of lupin did not increase nutrient digestion in pigs. Germination of yellow or blue lupin increased CP and fiber content, but reduced ether extract, alkaloids, and oligosaccharide content. Germination of lupin did not affect SID of CP, but decreased SID of lysine and methionine compared with raw seed in 25-kg pigs (Chilomer et al. 2013).

Inclusion of 14.5% germinated seeds of yellow lupin to replace 14% SBM or inclusion of 15.5% germinated blue lupin to replace 13% SBM did not affect ADFI and ADG in 9-kg pigs. Inclusion of 26.5% germinated blue lupin to replace 19% SBM reduced ADFI in weaned pigs, indicating that greater level of germinated blue lupin seeds should be avoided in diets for weaned pigs (Kasprowicz-Potocka et al. 2013).

Soaking followed by pressure cooking (116 °C, 172 kPa, 20 minutes) can reduce RS in chickpea (Periago et al. 1997). Pressure cooking after soaking chickpea in tap water and alkaline solution of sodium bicarbonate reduced starch content, but increased *in vitro* starch digestibility (Rehman 2007).

### Autoclaving and Roasting

Autoclaving has little effect on nutritive value of faba bean in pigs, especially for low-tannin faba bean (Jansman et al. 1993a). Autoclaving faba bean 30 or 60 minutes did not increase SID of CP in 35-kg pigs (Ivan and Bowland 1976). Autoclaving (105 °C, 30 minutes) low-tannin faba bean did not increase AID of CP and starch in 15-kg pigs (Jansman et al. 1993a). Steam processing of faba bean cotyledons at 100 °C did not increase AID of CP in 35-kg pigs, indicating ANF such as TI and lectin present in cotyledons only play a minor role in depressing the digestion of protein in faba bean (van der Poel et al. 1992b).

Increasing temperature or prolonged time reduced lysine content in lupin and rendered lysine chemically unavailable. Inclusion of 30% oven-heated blue lupin at 105–150 °C for 15 minutes linearly decreased ADG, G:F and increased backfat thickness in growing pigs. Autoclaving lupin seed at 121 °C for 5–45 minutes linearly decreased ADG and G:F (Batterham et al. 1986a). Lupin has lower lysine availability than SBM. Autoclaving white lupin seed at 121 °C for five minutes did not affect growth of 20-kg pigs, indicating that low lysine availability might be not due to heat-labile ANF (Batterham et al. 1986b).

### Pelleting

Pelleting has various effects among pulse grains. Pelleting field pea at 75 °C did not affect AID of CP, starch, and GE but increased ATTD of GE in 69-kg pigs (Stein and Bohlke 2007). Steam-pelleting faba bean increased the SID of CP, isoleucine, leucine, methionine, valine, and starch and the predicted NE value in 54-kg pigs (Ruiz et al. 2017).

*Extrusion*

Extrusion is widely used in the food and pet food industry (Berrios 2012). With shear force, heat can reduce starch crystallinity, leading to better digestibility, particularly in young animals (Wiseman 2013). Extruding field pea and kidney bean starch at 20 or 24% moisture using a twin-screw extruder reduced crystallinity in starch. Extruded starch contained more rapidly digestible starch (RDS) and less RS. Field pea and kidney bean starch extruded with more moisture contained less RDS and more slowly digestible starch than that extruded with less moisture (Sharma et al. 2015). Extrusion (152–156 °C, 20% moisture) of faba bean increased *in vitro* protein and starch digestibility (Alonso et al. 2000). Extrusion temperature (135, 160, and 175 °C) had less effect on solubility and molecular weight of lentil proteins (Li and Lee 2000).

*Field Pea.* Extrusion may increase nutrient digestibility more than pelleting in field pea. Increasing extrusion temperature from 75 to 155 °C linearly increased AID of starch and GE and quadratically increased the SID of CP and ATTD of GE of field pea diets in 69-kg pigs. In comparison, pelleting at 75 °C increased ATTD of GE, but did not increase AID of starch and GE, or SID of CP. Field pea extruded at 75 °C had greater SID of CP and AA than that pelleted at 75 °C. However, extrusion or pelleting at 75 °C did not affect the ATTD of starch, NDF, and ADF (Stein and Bohlke 2007). Extrusion may especially increase ileal digestibility of tryptophan and cystine of field pea in growing pigs (Mariscal-Landin et al. 2002).

Extrusion can greatly reduce RS content in field pea, but not the content of raffinose, stachyose, and tannins (Hejdysz et al. 2016). Extrusion (135 °C, 22% moisture) of field pea reduced RS content and TIA, but did not increase AID of CP and ATTD of GE in 10-kg pigs. Extruding field pea may not increase growth performance in weaned pigs (Prandini et al. 2005; Tusnio et al. 2017). Weaned pigs fed raw field pea two-week postweaning can maintain growth but not G:F. The reduced G:F in weaned pigs could not be ameliorated by extrusion of the field pea (Hugman et al. 2018). However, inclusion of 25% extruded field pea increased ADFI, ADG, and G:F in 10-kg or 28- to 95-kg pigs compared with raw field pea (Zaworska et al. 2018).

*Faba Bean.* Extrusion may not increase the nutritive value of faba bean for pigs. Extruded zero-tannin faba bean starch diet has lower AID of starch than wheat diet (Wierenga et al. 2008). Extrusion of faba bean increased G:F in 8-kg pigs but not ADFI and ADG (Ruiz et al. 2018). Increasing inclusion up to 37.5% faba bean to replace 31% barley and 10% SBM in diets formulated to similar DE and lysine fed to 37-kg pigs did not affect ADFI, but linearly decreased ADG and G:F during the growing but not finishing phase. Extrusion at 120 °C for 30 seconds did not ameliorate reduced growth (O'Doherty and McKeon 2001).

*Chickpea.* Extrusion increased *in vitro* protein digestibility of chickpea flour (Milan-Carrillo et al. 2000). The optimal settings (150 °C, 26.5% moisture, 190 rpm) were suggested for extrusion of chickpea with a single-screw extruder (Milan-Carrillo et al. 2002). Increasing inclusion up to 30% extruded chickpea to replace 16% SBM and barley linearly increased ADG and G:F in 20-kg pigs (Christodoulou et al. 2006b) and did not influence the meat quality of pigs (Christodoulou et al. 2006a).

*Lentil.* Extrusion increased *in vitro* protein and starch digestibility of lentil with increasing temperature from 140 to 180 °C (Rathod and Annapure 2016).

*Lupin.* Extrusion of white lupin did not increase growth performance in 10-kg pigs (Prandini et al. 2005). Expansion (80 °C, 20 seconds) of blue lupin did not affect ATTD of GE, ether extract, Ca, and P but increased AID of lysine and methionine of lupin in 41-kg pigs. Expanding lupin seed did not affect growth performance in 54-kg pigs, but increased ADG in 84-kg pigs compared with raw lupin (Yang et al. 2007).

## *Fractionation*

Fractionation of field pea and faba bean has been conducted for four decades (Bramsnaes and Olsen 1979). Dehulling of field pea followed by fine grinding and air classification allows separation of fine [protein concentrate] and coarse (mainly starch) fractions that can be used in pig feeding (Wu and Nichols 2005). Protein concentrate from chickpea contains 67.8% CP and 17.3% lipid and has high nutritive value (Ulloa et al. 1988). Dry separation of cotyledon of field pea had lower protein recovery (29–42%), whereas wet separation allowed recovering 45–56% of seed proteins (Bergthaller et al. 2001).

Pulse protein concentrates are attractive nutritionally for substituting specialty protein sources in young pigs (Valencia et al. 2008; Gunawardena et al. 2010b). Field pea protein isolate is highly digestible, partly due to removal of ANF (Le Guen et al. 1995a, b). However, field pea protein isolate lacks the functional properties of plasma protein.

Field pea and faba bean starch concentrates are feedstuffs for young pigs (Gunawardena et al. 2010a, b). Extrusion may increase their nutritional value (Wierenga et al. 2008). Starch chemistry of isolates may influence glycemic responses in pigs (van Kempen et al. 2010).

## *Fermentation*

Fermentation can modify chemical composition of pulse grains and thereby nutrient digestibility. Fermentation of blue lupin seeds with bacteria and yeast under aerobic conditions increased content of CP, fiber, fat, and ash and AA and decreased content of oligosaccharides and phytate but increased the alkaloid content. Fermentation increased AID of CP of lupin in 25-kg pigs (Kasprowicz-Potocka et al. 2016a; Zaworska et al. 2017).

## *Enzyme Addition*

Vicine and convicine in faba bean cannot be eliminated by heat treatment but can be reduced by β-glucosidase addition. The hydrolysis products (divicine and isouramil) can decompose rapidly into inactive compounds (Arbid and Marquardt 1985). In yellow lupin, 80% α-galactosides can be digested in the small intestine of 18-kg pigs. Addition of 0.5% microbial α-galactosidase increased digestibility of α-galactoside to 97%. Supplementation of α-galactosidase increased digestibility of oligosaccharides in 11-kg pigs (Gdala et al. 1997). However, for cereal-white lupin-based diet, supplementation of α-galactosidase did not increase AID of CP and SID of AA in 8-kg pigs, but tended to increase AID of NDF (Pires et al. 2007).

The use of NSP-degrading enzymes may increase nutrient digestibility of fiber-rich ingredients (Zijlstra et al. 2010). Supplementation of microbial phytase in field pea diets can increase ATTD of P and Ca, thereby reducing P excretion (Stein et al. 2006a; Kahindi et al. 2015). Addition of fungal hemicellulose and cellulase to field pea diets did not alter performance in 7-kg pigs (Brooks et al. 2009). Similarly, supplementation of enzyme cocktail to blue lupin diets did not affect growth performance in 27-kg pigs, likely because the enzyme complex was not lupin NSP specific (Kim et al. 2011).

## Summary

For sustainable swine production, economics, environment, and societal acceptance are key components. As omnivores, pigs can effectively convert pulse grains in pork products (Zijlstra and Beltranena 2013). Locally produced pulse grains can be a component for sustainable organic farming. Feeding pulse grains to pigs may also pose challenges. First, nutrient composition variability exists within and among pulse grains. Thus, feed quality evaluation for energy and AA content and their digestibility is important, as is the system selected for evaluation (Williams et al. 1985; Font et al. 2006; Kim et al. 2009b; Montoya and Leterme 2012; Wang and Zijlstra 2018). Second, pulse grains commonly contain ANF that may reduce voluntary feed intake and nutrient digestion. Feed processing may increase nutritive value of pulse grains. Pulse grain with high fiber content may reduce dressing percentage. In conclusion, inclusion of pulse grains in pig diets may not only reduce feed cost per unit of pork produced but also provides challenges to achieve predictable growth performance and pork quality.

# References

Abd El-Hady, E. A., and R. A. Habiba. 2003. Efffect of soaking and extrusion conditions on antinutrients and protein digestibility of legume seeds. LWT-Food Sci. Technol. 36:285–293.

Abreu, J. F., and A. M. Bruno-Soares. 1998. Chemical composition, organic matter digestibility and gas production of nine legume grains. Anim. Feed Sci. Technol. 70:49–57.

Adamidou, S., I. Nengas, K. Grigorakis, D. Nikolopoulou, K. Jauncey. 2011. Chemical composition and antinutritional factors of field peas (*Pisum sativum*), chickpeas (*Cicer arietinum*), and faba beans (*Vicia faba*) as affected by extrusion preconditioning and drying temperatures. Cereal Chem. 88:80–86.

Aguilera, J.F., E. Molina, and C. Prieto. 1985. Digestibility and energy value of sweet lupin seed (*Lupinus albus* var Multolupa) in pigs. Anim. Feed Sci. Technol. 12:171–178.

Alonso, R., A. Aguirre, and F. Marzo. 2000. Effects of extrusion and traditional processing methods on antinutrients and *in vitro* digestibility of protein and starch in faba and kidney beans. Food Chem. 68:159–165.

Alonso, R., E. Orue, and F. Marzo. 1998. Effects of extrusion and conventional processing methods on protein and antinutritional factor contents in pea seeds. Food Chem. 63:505–512.

Ambigaipalan, P., R. Hoover, E. Donner, and G. Liu. 2014. Starch chain interactions within the amorphous and crystalline domains of pulse starches during heat-moisture treatment at different temperatures and their impact on physicochemical properties. Food Chem. 143:175–184.

Arbid, M. S. S., and R. R. Marquardt. 1985. Hydrolysis of the toxic constituents (vicine and convicine) in fababean (Vicia faba L.) food preparations following treatment with β-glucosidase. J. Sci. Food Agric. 36:839–846.

Arnoldi, A., G. Boschin, C. Zanoni, and C. Lammi. 2015. The health benefits of sweet lupin seed flours and isolated proteins. J. Funct. Foods 18:550–563.

Attia, R. S., A. M. E. Shehata, M. E. Aman, and M. A. Hamza. 1994. Effect of cooking and decortication on the physical properties, the chemical composition and the nutritive value of chickpea (*Cicer arietinum* L). Food Chem. 50:125–131.

Aumiller, T., R. Mosenthin, and E. Weiss. 2015. Potential of cereal grains and grain legumes in modulating pigs' intestinal microbiota - A review. Livest. Sci. 172:16–32.

Bach Knudsen, K. E. 1997. Carbohydrate and lignin contents of plant materials used in animal feeding. Anim. Feed Sci. Technol. 67:319–338.

Bampidis, V. A., and V. Christodoulou. 2011. Chickpeas (*Cicer arietinum* L.) in animal nutrition: A review. Anim. Feed Sci. Technol. 168:1–20.

Batterham, E. S., and L. M. Andersen. 1994. Utilization of ileal digestible amino acids by growing pigs - isoleucine. Br. J. Nutr. 71:531–541.

Batterham, E. S., L. M. Andersen, B. V. Burnham, and G. A. Taylor. 1986a. Effect of heat on the nutritional value of lupin (*Lupinus angustifolius*) seed meal for growing pigs. Br. J. Nutr. 55:169–177.

Batterham, E. S., L. M. Andersen, R. F. Lowe, and R. E. Darnell. 1986b. Nutritional-value of lupin (*Lupinus albus*) seed meal for growing pigs - availability of lysine, effect of autoclaving and net energy content. Br. J. Nutr. 56:645–659.

Batterham, E. S., R. D. Murison, ad L. M. Andersen. 1984. Availability of lysine in vegetable protein concentrates as determined by the slope-ratio assay with growing pigs and rats and by chemical techniques. Br. J. Nutr. 51:85–99.

Batterham, E. S., H. S. Saini, L. M. Andersen, and R. D. Baigent. 1993. Tolerance of growing pigs to trypsin and chymotrypsin inhibitors in chickpeas (*Cicer arietinum*) and pigeonpeas (*Cajanus cajan*). J. Sci. Food Agric. 61:211–216.

Becker, P. M., J. van der Meulen, A. J. M. Jansman, and P. G. van Wikselaar. 2012. *In vitro* inhibition of ETEC K88 adhesion by pea hulls and of LT enterotoxin binding by faba bean hulls. J. Anim. Physiol. Anim. Nutr. 96:1121–1126.

Bell, J. M., and A. G. Wilson. 1970. An evaluation of field peas as a protein and energy source for swine rations. Can. J. Anim. Sci. 50:15–23.

Beltranena, E., S. Hooda, and R. T. Zijlstra. 2009. Zero-tannin faba bean as a replacement for soybean meal in diets for starter pigs. Can. J. Anim. Sci. 89:489–492.

Bergthaller, W., B. H. Dijkink, H.C. Langelaan, and J. M. Vereijken. 2001. Protein from pea mutants as a co-product in starch separation - isolates from wet and dry separation: Yield, composition and solubility. Nahrung 45:390–392.

Berrios, J. D. 2012. Extrusion processing of main commercial legume pulses. In: M. Maskan, and A. Altan, editor, Advances in food extrusion technology. CRC Press-Taylor & Francis Group, Boca Raton, FL. p. 209–236.

Berrios, J. D., P. Morales, M. Camara, and M. C. Sanchez-Mata. 2010. Carbohydrate composition of raw and extruded pulse flours. Food Res. Int. 43:531–536.

Black, R. G., J. B. Brouwer, C. Meares, and L. Iyer. 1998. Variation in physico-chemical properties of field peas (*Pisum sativum*). Food Res. Int. 31:81–86.

Bramsnaes, F., and H. S. Olsen. 1979. Development of field pea and faba bean proteins. J. Am. Oil Chem. Soc. 56:450–454.

Brand, T. S., D. A. Brandt, and C. W. Cruywagen. 2004. Chemical composition, true metabolisable energy content and amino acid availability of grain legumes for poultry. S. Afr. J. Anim. Sci. 34:116–122.

Brand, T.S., D. A. Brandt, J. P. van der Merwe, and C. W. Cruywagen. 2000. Field peas (*Pisum sativum*) as protein source in diets of growing-finishing pigs. J. Appl. Anim. Res. 18:159–164.

Brooks, K. R., B. R. Wiegand, A. L. Meteer, G. I. Petersen, J. D. Spencer, J. R. Winter, J. A. Robb. 2009. Inclusion of yellow field peas and carbohydrase enzyme in nursery pig diets to improve growth performance. Prof. Anim. Sci. 25:17–25.

Bugnacka, D., and J. Falkowski. 2001. The effect of dietary levels of yellow lupin seeds (*Lupinus luteus* L.) on feed preferences and growth performance of young pigs. J. Anim. Feed Sci. 10:133–142.

Canibe, N., and K. E. Bach Knudsen. 2002. Degradation and physicochemical changes of barley and pea fibre along the gastrointestinal tract of pigs. J. Sci. Food Agric. 82:27–39.

Cansfield, P. E., R. R. Marquardt, and L. D. Campbell. 1980. Condensed proanthocyanidins of fababeans. J. Sci. Food Agric. 31:802–812.

Carew, R., W. J. Florkowski, and Y. Zhang. 2013. Review: Industry levy-funded pulse crop research in Canada: Evidence from the prairie provinces. Can. J. Plant Sci. 93:1017–1028.

Castell, A. G. 1976. Comparison of faba beans (*Vicia faba*) with soybean meal or field peas (*Pisum sativum*) as protein supplements in barley diets for growing-finishing pigs. Can. J. Anim. Sci. 56:425–432.

Chavan, J. K., S. S. Kadam, and D. K. Salunkhe. 1986. Biochemistry and technology of chickpea (*Cicer arietinum* L.) seeds. Crit. Rev. Food Sci. Nutr. 25:107–158.

Chen, H., X. Mao, J. He, B. Yu, Z. Huang, J. Yu, P. Zheng, and D. Chen. 2013. Dietary fibre affects intestinal mucosal barrier function and regulates intestinal bacteria in weaning piglets. Br. J. Nutr. 110:1837–1848.

Chilomer, K., M. Kasprowicz-Potocka, P. Gulewicz, and A. Frankiewicz. 2013. The influence of lupin seed germination on the chemical composition and standardized ileal digestibility of protein and amino acids in pigs. J. Anim. Physiol. Anim. Nutr. 97:639–646.

Cho, M., M. N. Smit, L. He, F. C. Kopmels, and E. Beltranena. 2019. Effect of feeding zero- or high-tannin faba bean cultivars and dehulling on growth performance, carcass traits and yield of saleable cuts of broiler chickens. J. Appl. Poultry Res. 28:1305–1323.

Christodoulou, V., J. Ambrosladis, E. Sossidou, V. Bampidis, J. Arkoudilos, B. Hucko, and C. Iliadis. 2006a. Effect of replacing soybean meal by extruded chickpeas in the diets of growing-finishing pigs on meat quality. Meat Sci. 73:529–535.

Christodoulou, V., V. A. Bampidis, E. Sossidou, J. Ambrosiadis, B. Hucko, C. Iliadis, and A. Kodes. 2006b. The use of extruded chlickpeas in diets for growing-finishing pigs. Czech J. Anim. Sci. 51:334–342.

Chung, H., Q. Liu, R. Hoover, T. D. Warkentin, and B. Vandenberg. 2008. *In vitro* starch digestibility, expected glycemic index, and thermal and pasting properties of flours from pea, lentil and chickpea cultivars. Food Chem. 111:316–321.

Clemente, A., E. Jimenez, M. C. Marin-Manzano, and L. A. Rubio. 2008. Active Bowman-Birk inhibitors survive gastrointestinal digestion at the terminal ileum of pigs fed chickpea-based diets. J. Sci. Food Agric. 88:513–521.

Clemente, A., R. Sanchez-Vioque, J. Vioque, J. Bautista, and F. Millan. 1998. Effect of cooking on protein quality of chickpea (*Cicer arietinum*) seeds. Food Chem. 62:1–6.

CVB. 1993. Net Energy of Feedstuffs for Swine. CVB Report No. 7. Centraal Veevoeder Bureau [Central Feedstuff Bureau], Lelystad, The Netherlands.

de Lange, C. F. M. 2000. Overview of determinants of the nutritional value of feed ingredients. In: Moughan, P.J., Verstegen, M.W.A., and Visser-Reyneveld, M.I., editors, Feed evaluation: Principles and practice. CABI, Wallingford, UK. p. 17–32.

Degola, L., and D. Jonkus. 2018. The influence of dietary inclusion of peas, faba bean and lupin as a replacement for soybean meal on pig performance and carcass traits. Agron. Res. 16:389–397.

Donovan, B. C., M. A. McNiven, T. A. van Lunen, D. M. Anderson, and J. A. Macleod. 1993. Replacement of soybean meal with dehydrated lupin seeds in pig diets. Anim. Feed Sci. Technol. 43:77–85.

Dunshea, F. R., N. J. Gannon, R. J. van Barneveld, B. P. Mullan, R. G. Campbell, and R. H. King. 2001. Dietary lupins (*Lupinus angustifolius* and *Lupinus albus*) can increase digesta retention in the gastrointestinal tract of pigs. Aust. J. Agric. Res. 52:593–602.

Dvorak, R., A. Pechova, L. Pavlata, J. Filipek, J. Dostalova, Z. Reblova, B. Klejdus, K. Kovarcik, and J. Poul. 2005. Reduction in the content of antinutritional substances in pea seeds (*Pisum sativum* L.) by different treatments. Czech J. Anim. Sci. 50:519–527.

Eggum, B. O., G. Tomes, R. M. Beames, and F. U. Datta. 1993. Protein and energy evaluation with rats of seed from 11 lupin cultivars. Anim. Feed Sci. Technol. 43:109–119.

Eklund, M., W. R. Caine, W. C. Sauer, G. S. Huang, G. Diebold, M. Schollenberger, and R. Mosenthin. 2012. True and standard-ized ileal digestibilities and specific ileal endogenous recoveries of crude protein and amino acid in soybean meal, rapeseed meal and peas fed to growing pigs. Livest. Sci. 145:174–182.

El-Adawy, T. A. 2002. Nutritional composition and antinutritional factors of chickpeas (*Cicer arietinum* L.) undergoing different cooking methods and germination. Plant. Food. Hum. Nutr. 57:83–97.

Emiola, I. A., and R. M. Gous. 2011. Nutritional evaluation of dehulled faba bean (*Vicia faba* cv. Fiord) in feeds for weaner pigs. S. Afr. J. Anim. Sci. 41, 79–86.

Ercan, P., and S. N. El. 2016. Inhibitory effects of chickpea and Tribulus terrestris on lipase, alpha-amylase and alpha-glucosidase. Food Chem. 205:163–169.

Esnault, M. A., A. Merceur, and J. Citharel. 1991. Characterization of globulins of yellow lupin seeds. Plant Physiol. Biochem. 29:573–583.

FAOSTAT. 2019. Food and agriculture data? FAO, Rome, Italy. http://www.fao.org/faostat/en/#data/QC. (Accessed 24 July 2019).

Ferguson, N. S., R. M. Gous, and P. A. Iji. 2003. Determining the source of anti-nutritional factor(s) found in two species of lupin (*L. albus* and *L. angustifolius*) fed to growing pigs. Livest. Prod. Sci. 84:83–91.

Fernández, J. A., and E. S. Batterham. 1995. The nutritive value of lupin seed and dehulled lupin seed meals as protein sources for growing pigs as evaluated by different techniques. Anim. Feed Sci. Technol. 53:279–296.

Flis, M., W. Sobotka, C. Purwin, and Z. Zdunczyk. 1999. Nutritional value of diets containing field bean (*Vicia faba* L.) seeds with high or low proanthocyanidin levels for pig. J. Anim. Feed Sci. 8:171–180.

Flis, M., W. Sobotka, and Z. Zdunczyk. 1998. Replacement of soyabean meal by white lupin cv. Bardo seeds and the effectiveness of β-glucanase and xylanase in growing-finishing pig diets. J. Anim. Feed Sci. 7:301–312.

Font, R., M. D. Rio-Celestino, and A. D. Haro-Bailon. 2006. The use of near-infrared spectroscopy (NIRS) in the study of seed quality components in plant breeding programs. Ind. Crop Prod. 24:307–313.

Fouhse, J. M., and R. T. Zijlstra. 2018. Impact of resistant vs. digested starch on starch energy value in the pig gut. Bioact. Carbohydr. Dietary Fibre 15:12–20.

Frias, J., S. Giacomino, E. Penas, N. Pellegrino, V. Ferreyra, N. Apro, M. O. Carrion, and C. Vidal-Valverde. 2011. Assessment of the nutritional quality of raw and extruded *Pisum sativum* L. var. Laguna seeds. LWT-Food Sci. Technol. 44:1303–1308.

Friesen, M. J., E. Kiarie, and C. M. Nyachoti. 2006. Response of nursery pigs to diets with increasing levels of raw peas. Can. J. Anim. Sci. 86:531–533.

Gabert, V. M., W. C. Sauer, S. Y. Li, and M. Z. Fan. 1996a. Exocrine pancreatic secretions in young pigs fed diets containing faba beans (*Vicia faba*) and peas (*Pisum sativum*): Concentrations and flows of total, protein-bound and free amino acids. J. Sci. Food Agric. 70:256–262.

Gabert, V. M., W. C. Sauer, S. Y. Li, M. Z. Fan, and M. Rademacher. 1996b. Exocrine pancreatic secretions in young pigs fed diets containing faba beans (*Vicia faba*) and peas (*Pisum sativum*): Nitrogen, protein and enzyme secretions. J. Sci. Food Agric. 70:247–255.

Garrido, A., A. Gomezcabrera, J. E. Guerrero, and J. M. van der Meer. 1991. Effects of treatment with polyvinylpyrrolidone and polyethylene-glycol on faba bean tannins. Anim. Feed Sci. Technol. 35:199–203.

Gatel, F. 1994. Protein quality of legume seeds for non-ruminant animals: A literature review. Anim. Feed Sci. Technol. 45:317–348.

Gatel, F., and F. Grosjean. 1990. Composition and nutritive value of peas for pigs: A review of European results. Livest. Prod. Sci. 26:155–175.

Gatta, D., C. Russo, L. Giuliotti, C. Mannari, P. Picciarelli, L. Lombardi, L. Giovannini, N. Ceccarelli, and L. Mariotti. 2013. Influence of partial replacement of soya bean meal by faba beans or peas in heavy pigs diet on meat quality, residual anti-nutritional factors and phytoestrogen content. Arch. Anim. Nutr. 67:235–247.

Gdala, J. 1998. Composition, properties, and nutritive value of dietary fibre of legume seeds. A review. J. Anim. Feed Sci. 7:131–150.

Gdala, J., and L. Buraczewska. 1997. Ileal digestibility of pea and faba bean carbohydrates in growing pigs. J. Anim. Feed Sci. 6:235–245.

Gdala, J., A. J. M. Jansman, L. Buraczewska, J. Huisman, and P. van Leeuwen. 1997. The influence of α-galactosidase supplementation on the ileal digestibility of lupin seed carbohydrates and dietary protein in young pigs. Anim. Feed Sci. Technol. 67:115–125.

Gdala, J., A. J. M. Jansman, P. van Leeuwen, J. Huisman, and M. W. A. Verstegen. 1996. Lupins (*L. luteus, L. albus, L. angustifolius*) as a protein source for young pigs. Anim. Feed Sci. Technol. 62:239–249.

Godfrey, N. W., A. R. Mercy, Y. Emms, and H. G. Payne. 1985. Tolerance of growing pigs to lupin alkaloids. Aust. J. Exp. Agric. 25:791–795.

Grala, W., A. J. M. Jansman, P. van Leeuwen, J. Huisman, G. J. M. van Kempen, and M. W. A. Verstegen. 1993. Nutritional value of faba beans (*Vicia faba* L.) fed to young pigs. J. Anim. Feed Sci. 2:169–179.

Grosjean, F., B. Barrier-Guillot, D. Bastianelli, F. Rudeaux, A. Bourdillon, and C. Peyronnet. 1999. Feeding value of three categories of pea (*Pisum sativum*, L.) for poultry. Anim. Sci. 69:591–599.

Gueguen, J., and J. Barbot. 1988. Quantitative and qualitative variability of pea (*Pisum sativum* L.) protein composition. J. Sci. Food Agric. 42:209–224.

Gujska, E., W. D. Reinhard, and K. Khan. 1994. Physicochemical properties of field pea, pinto and navy bean starches. J. Food Sci. 59:634–636.

Gunawardena, C. K., W. Robertson, M. Young, R. T. Zijlstra, and E. Beltranena. 2007a. Cut-out yield and pork quality of hogs fed zero-tannin fababean, field pea or soybean meal as protein sources. Banff Pork Seminar, Department of Agricultural, Food and Nutritional Science, University of Alberta, Banff, Alberta, Canada. p. A24.

Gunawardena, C. K., W. Robertson, M. Young, R. T. Zijlstra, and E. Beltranena. 2007b. Growth performance and carcass characteristics of growing-finishing hogs fed zero-tannin fababean, field pea or soybean meal as protein sources. Banff Pork Seminar, Department of Agricultural, Food and Nutritional Science, University of Alberta, Banff, Alberta, Canada. p. A23.

Gunawardena, C. K., R. T. Zijlstra, and E. Beltranena. 2010a. Characterization of the nutritional value of air-classified protein and starch fractions of field pea and zero-tannin faba bean in grower pigs. J. Anim. Sci. 88:660–670.

Gunawardena, C. K., R. T. Zijlstra, L. A. Goonewardene, and E. Beltranena. 2010b. Protein and starch concentrates of air-classified field pea and zero-tannin faba bean for weaned pigs. J. Anim. Sci. 88:2627–2636.

Gupta, R. K., G. Kriti, S. Akanksha, D. Mukul, I. A. Ansari, and P. D. Dwivedi. 2017. Health risks and benefits of chickpea (*Cicer arietinum*) consumption. J. Agric. Food Chem. 65:6–22.

Hanczakowska, E., J. Ksiezak, and M. Swiatkiewicz. 2017. Efficiency of lupine seed (*Lupinus angustifolium* and *Lupinus luteus*) in sow, piglet and fattener feeding. Agric. Food Sci. 26:1–15.

Hanczakowska, E., and M. Swiatkiewicz. 2013. Legume seeds and rapeseed press cake as substitutes for soybean meal in sow and piglet feed. Agric. Food Sci. 22:435–444.

Hanczakowska, E., and M. Swiatkiewicz. 2014. Legume seeds and rapeseed press cake as replacers of soybean meal in feed for fattening pigs. Ann. Anim. Sci. 14:921–934.

Hansen, C. F., N. D. Phillips, T. La, A. Hernandez, J. Mansfield, J. C. Kim, B. P. Mullan, D. J. Hampson, and J. R. Pluske. 2010. Diets containing inulin but not lupins help to prevent swine dysentery in experimentally challenged pigs. J. Anim. Sci. 88:3327–3336.

Hejdysz, M., S. A. Kaczmarek, and A. Rutkowski. 2016. Effect of extrusion on the nutritional value of peas for broiler chickens. Arch. Anim. Nutr. 70:364–377.

Hennig, U., W. Hackl, A. Priepke, A. Tuchscherer, W. B. Souffrant, and C. C. Metges. 2008. Comparison of ileal apparent, standardized and true digestibilities of amino acids in pigs fed wheat and lupine seeds. Livest. Sci. 118:61–71.

Henriquez, B., M. Olson, C. Hoy, M. Jackson, and T. Wouda. 2018. Frost tolerance of faba bean cultivars (*Vicia faba* L.) in central Alberta. Can. J. Plant Sci. 98:509–514.

Hidayat, M., M. Sujatno, N. Sutadipura, N. Setiawan, and A. Faried. 2011. Beta-conglycinin content obtained from two soybean varieties using different preparation and extraction methods. HAYATI J. Biosci. 18:37–42.

Holt, N. W., and F. W. Sosulski. 1979. Amino-acid composition and protein-quality of field peas. Can. J. Plant Sci. 59:653–660.

Hugman, J., E. Beltranena, J. K. Htoo, and R. T. Zijlstra. 2018. Growth performance of weaned pigs fed raw, cold-pelleted, steam-pelleted, or extruded field pea. J. Anim. Sci. 96:142–143.

Huisman, J., and A. F. B van der Poel. 1994. Aspects of the nutritional quality and use of cool season food legumes in animal feed, In: F. J. Muehlbauer, and W. J. Kaiser, editors, Expanding the production and use of cool season food legumes. Proceedings of the Second International Food Legume Research Conference on pea, lentil, faba bean, chickpea, and grasspea, Cairo, Egypt, 12–16 April 1992. Kluwer Academic Publishers Group, Dordrecht, The Netherlands. p. 53–76.

Igbasan, F. A., W. Guenter, and B. A. Slominski. 1997. Field peas: Chemical composition and energy and amino acid availabilities for poultry. Can. J. Anim. Sci. 77:293–300.

Iqbal, A., I. A. Khalil, N. Ateeq, and M. S. Khan. 2006. Nutritional quality of important food legumes. Food Chem. 97:331–335.

Ivan, M., and J. P. Bowland. 1976. Digestion of nutrients in small intestine of pigs fed diets containing raw and autoclaved faba beans. Can. J. Anim. Sci. 56:451–456.

Ivarsson, E., and M. Neil. 2018. Variations in nutritional and antinutritional contents among faba bean cultivars and effects on growth performance of weaner pigs. Livest. Sci. 212:14–21.

Jansman, A. J. M., H. Enting, M. W. A. Verstegen, and J. Huisman. 1994. Effect of condensed tannins in hulls of faba beans (*Vicia faba* L) on the activities of trypsin (EC-2.4.21.4) and chymotrypsin (EC-2.4.21.1) in digesta collected from the small intestine of pigs. Br. J. Nutr. 71:627–641.

Jansman, A. J. M., J. Huisman, and A. F. B. van der Poel. 1993a. Ileal and fecal digestibility in piglets of field beans (*Vicia faba* L.) varying in tannin content. Anim. Feed Sci. Technol. 42:83–96.

Jansman, A. J. M., J. van Baal, J. van der Meulen, and M. A. Smits. 2012. Effects of faba bean and faba bean hulls on expression of selected genes in the small intestine of piglets. J. Anim. Sci. 90:161–163.

Jansman, A. J. M., M. W. A. Verstegen, and J. Huisman. 1993b. Effects of dietary inclusion of hulls of faba beans (*Vicia faba* L) with a low and high content of condensed tannins on digestion and some physiological parameters in piglets. Anim. Feed Sci. Technol. 43:239–257.

Jansman, A. J. M., M. W. A. Verstegen, J. Huisman, and J. W. O. van den Berg. 1995. Effects of hulls of faba beans (*Vicia faba* L) with a low or high content of condensed tannins on the apparent ileal and fecal digestibility of nutrients and the excretion of endogenous protein in ileal-digesta and feces of pigs. J. Anim. Sci. 73:118–127.

Jensen, A. N., L. L. Hansen, D. L. Baggesen, and L. Molbak. 2013. Effects of feeding finisher pigs with chicory or lupine feed for one week or two weeks before slaughter with respect to levels of Bifidobacteria and Campylobacter. Animal 7:66–74.

Jezierny, D., R. Mosenthin, and E. Bauer. 2010a. The use of grain legumes as a protein source in pig nutrition: A review. Anim. Feed Sci. Technol. 157:111–128.

Jezierny, D., R. Mosenthin, N. Sauer, and M. Eklund. 2010b. *In vitro* prediction of standardised ileal crude protein and amino acid digestibilities in grain legumes for growing pigs. Animal 4:1987–1996.

Jezierny, D., R. Mosenthin, N. Sauer, S. Roth, H. P. Piepho, M. Rademacher, and M. Eklund. 2011. Chemical composition and standardised ileal digestibilities of crude protein and amino acids in grain legumes for growing pigs. Livest. Sci. 138:229–243.

Jezierny, D., R. Mosenthin, N. Sauer, K. Schwadorf, and P. Rosenfelder-Kuon. 2017. Methodological impact of starch determination on starch content and ileal digestibility of starch in grain legumes for growing pigs. J. Anim. Sci. Biotechnol. 8:1–8.

Jha, R., J. Bindelle, A. van Kessel, and P. Leterme. 2011. *In vitro* fibre fermentation of feed ingredients with varying fermentable carbohydrate and protein levels and protein synthesis by colonic bacteria isolated from pigs. Anim. Feed Sci. Technol. 165:191–200.

Jha, R., and P. Leterme. 2012. Feed ingredients differing in fermentable fibre and indigestible protein content affect fermentation metabolites and faecal nitrogen excretion in growing pigs. Animal 6:603–611.

Johnston, A. M., T. A. Woyengo, and C. M. Nyachoti. 2013. True digestive utilization of phosphorus in pea (*Pisum sativum*) fed to growing pigs. Anim. Feed Sci. Technol. 185:169–174.

Jukanti, A. K., P. M. Gaur, C. L. L. Gowda, and R. N. Chibbar. 2012. Nutritional quality and health benefits of chickpea (*Cicer arietinum* L.): A review. Br. J. Nutr. 108:S11–S26.

Kahindi, R. K., P. A. Thacker, and C. M. Nyachoti. 2015. Nutrient digestibility in diets containing low-phytate barley, low-phytate field pea and normal-phytate field pea, and the effects of microbial phytase on energy and nutrient digestibility in the low and normal-phytate field pea fed to pigs. Anim. Feed Sci. Technol. 203:79–87.

Kasprowicz-Potocka, M., P. Borowczyk, A. Zaworska, W. Nowak, A. Frankiewicz, P. Gulewicz. 2016a. The effect of dry yeast fermentation on chemical composition and protein characteristics of blue lupin seeds. Food Technol. Biotechnol. 54:360–366.

Kasprowicz-Potocka, M., K. Chilomer, A. Zaworska, W. Nowak, and A. Frankiewicz. 2013. The effect of feeding raw and germinated Lupinus luteus and Lupinus angustifolius seeds on the growth performance of young pigs. J. Anim. Feed Sci. 22:116–121.

Kasprowicz-Potocka, M., A. Zaworska, S. Kaczmarek, M. Hejdysz, R. Mikula, and A.Rutkowski. 2017. The effect of Lupinus albus seeds on digestibility, performance and gastrointestinal tract indices in pigs. J. Anim. Physiol. Anim. Nutr. 101:e216–e224.

Kasprowicz-Potocka, M., A. Zaworska, S. A. Kaczmarek, and A. Rutkowski. 2016b. The nutritional value of narrow-leafed lupine (*Lupinus angustifolius*) for fattening pigs. Arch. Anim. Nutr. 70:209–223.

Kasprowicz, M., and A. Frankiewicz. 2004. Apparent and standardized ileal digestibility of protein and amino acids of several field bean and pea varieties in growing pigs. J. Anim. Feed Sci. 13:463–473.

Kelley, T. G., P. Parthasarathy Rao, and H. Grisko-Kelley. 2000. The pulse economy in the mid-1990s: A review of global and regional developments. In: R. Knight, editor, Linking research and marketing opportunities for pulses in the 21st century: Proceedings of the Third International Food Legumes Research Conference. Springer, Dordrecht, The Netherlands. p. 1–29.

Khan, M. A., N. Akhtar, I. Ullah, and S. Jaffery. 1995. Nutritional evaluation of Desi and Kabuli chickpeas and their products commonly consumed in Pakistan. Int. J. Food Sci. Nutr. 46:215–223.

Kiarie, E., P. Lopez, C. Furedi, and C. M. Nyachoti. 2013. Amino acids and energy utilization in zero-tannin faba bean and co-fermented wheat and corn dried distillers grains with solubles fed to growing pigs. J. Anim. Sci. 91:1728–1735.

Kim, J. C., J. M. Heo, B. P. Mullan, and J. R. Pluske. 2012. Performance and intestinal responses to dehulling and inclusion level of Australian sweet lupins (*Lupinus angustifolius* L.) in diets for weaner pigs. Anim. Feed Sci. Technol. 172:201–209.

Kim, J. C., B. P. Mullan, J. M. Heo, C. F. Hansen, and J. R. Pluske. 2009a. Decreasing dietary particle size of lupins increases apparent ileal amino acid digestibility and alters fermentation characteristics in the gastrointestinal tract of pigs. Br. J. Nutr. 102:350–360.

Kim, J. C., B. P. Mullan, J. M. Heo, A. Hernandez, and J. R. Pluske. 2009b. Variation in digestible energy content of Australian sweet lupins (*Lupinus angustifolius* L.) and the development of prediction equations for its estimation. J. Anim. Sci. 87:2565–2573.

Kim, J.C., B. P. Mullan, R. R. Nicholls, and J. R. Pluske. 2011. Effect of Australian sweet lupin (*Lupinus angustifolius* L.) inclusion levels and enzyme supplementation on the performance, carcass composition and meat quality of grower/finisher pigs. Anim. Prod. Sci. 51:37–43.

Kim, J. C., J. R. Pluske, and B. P. Mullan. 2008. Nutritive value of yellow lupins (*Lupinus luteus* L.) for weaner pigs. Aust. J. Exp. Agric. 48:1225–1231.

King, R. H. 1981. Lupin seed meal (*Lupinus albus* cv Hamburg) as a source of protein for growing pigs. Anim. Feed Sci. Technol. 6:285–296.

King, R.H., F. R. Dunshea, L. Morrish, P. J. Eason, R. J. van Barneveld, B. P. Mullan, and R. G. Campbell. 2000. The energy value of Lupinus angustifolius and Lupinus albus for growing pigs. Anim. Feed Sci. Technol. 83:17–30.

Krupinsky, J. M., K. L. Bailey, M. P. McMullen, B. D. Gossen, and T. K. Turkington. 2002. Managing plant disease risk in diversified cropping systems. Agron. J. 94:198–209.

Kwak, B. O., H. J. Kim, and H. S. Park. 2000. The effect of different lupin kernel inclusion levels on the growth and carcass composition of growing and finishing pigs. Asian-Australas. J. Anim. Sci. 13:207–212.

Landero, J. L., E. Beltranena, and R. T. Zijlstra. 2012. The effect of feeding lentil on growth performance and diet nutrient digestibility in starter pigs. Anim. Feed Sci. Technol. 174:108–112.

Landero, J. L., L. F. Wang, E. Beltranena, and R. T. Zijlstra. 2014. Diet nutrient digestibility and growth performance of weaned pigs fed field pea. Anim. Feed Sci. Technol. 198:295–303.

Le Gall, M., L. Quillien, B. Seve, J. Gueguen, and J. P. Lalles. 2007. Weaned piglets display low gastrointestinal digestion of pea (*Pisum sativum* L.) lectin and pea albumin 2. J. Anim. Sci. 85:2972–2981.

Le Guen, M. P., J. Huisman, J. Guéguen, G. Beelen, and M. W. A. Verstegen. 1995a. Effects of a concentrate of pea antinutritional factors on pea protein digestibility in piglets. Livest. Prod. Sci. 44:157–167.

Le Guen, M. P., J. Huisman, and M. W. A. Verstegen. 1995b. Partition of the amino acids in ileal digesta from piglets fed pea protein diets. Livest. Prod. Sci. 44:169–178.

Lee, S. S., N. M. Esa, and S. P. Loh. 2015. *In vitro* inhibitory activity of selected legumes against pancreatic lipase. J. Food Biochem. 39:485–490.

Leibholz, J. 1984. A note on methionine supplementation of pig grower diets containing lupin seed meal. Anim. Prod. 38:515–517.

Leikus, R., K. Triukas, G. Svirmickas, and V. Juskiene. 2004. The influence of various leguminous seed diets on carcass and meat quality of fattening pigs. Czech J. Anim. Sci. 49:398–406.

Leontowicz, H., M. Leontowicz, H. Kostyra, G. Kulasek, M. A. Gralak, R. Krzeminski, and M. Podgurniak. 2001a. Effects of raw or extruded legume seeds on some functional and morphological gut parameters in rats. J. Anim. Feed Sci. 10:169–183.

Leontowicz, M., H. Leontowicz, M. Biernat, M. A. Gralak, H. Kostyra, and J. Czerwinski. 2001b. The influence of extrusion of faba bean seeds and supplementation of sulphur amino acids on performance, pancreatic trypsin activity, and morphological parameters of the jejunum in rats. J. Anim. Feed Sci. 10:323–329.

Leterme, P., Y. Beckers, and A. Thewis. 1990. Trypsin inhibitors in peas: varietal effect and influence on digestibility of crude protein by growing pigs. Anim. Feed Sci. Technol. 29:45–55.

Leterme, P., P. Kish, A. D. Beaulieu, and J. F. Patience. 2008. Variation in the digestible and net energy content of field peas in growing pigs, Banff Pork Seminar, Department of Agricultural, Food and Nutritional Science, University of Alberta, Banff, Alberta, Canada. p. A9.

Li, M., and A. C. Lee. 2000. Effect of extrusion temperature on the solubility and molecular weight of lentil bean flour proteins containing low cysteine residues. J. Agric. Food Chem. 48:880–884.

Mandaokar, A. D., K. R. Koundal, R. Kansal, and H. C. Bansal. 1993. Characterization of vicilin seed storage protein of chickpea (*Cicer arietinum* L). J. Plant Biochem. Biotechnol. 2:35–38.

Mariotti, F., M. E. Pueyo, D. Tome, and S. Mahe. 2002. The bioavailability and postprandial utilisation of sweet lupin (*Lupinus albus*)-flour protein is similar to that of purified soyabean protein in human subjects: a study using intrinsically N-15-labelled proteins. Br. J. Nutr. 87:315–323.

Mariscal-Landin, G., Y. Lebreton, and B. Seve. 2002. Apparent and standardised true ileal digestibility of protein and amino acids from faba bean, lupin and pea, provided as whole seeds, dehulled or extruded in pig diets. Anim. Feed Sci. Technol. 97:183–198.

Marquardt, R. R., A. T. Ward, L. D. Campbell, and P. E. Cansfield. 1977. Purification, identification and characterization of a growth inhibitor in faba beans (*Vicia faba* L. var Minor). J. Nutr. 107:1313–1324.

Marquez, M. C., V. Fernandez, and R. Alonso. 1998. Effect of dry heat on the *in vitro* digestibility and trypsin inhibitor activity of chickpea flour. Int. J. Food Sci. Technol. 33:527–532.

Martelli, G., L. Rizzi, R. Boccuzzi, R. Paganelli, and L. Sardi. 2009. Digestibility and nitrogen balance of nonconventional protein sources fed to pigs of two different genotypes. In: DiAlberto, P., and Costa, C., editors, New research on livestock science and dairy farming. Nova Science Publishers, Inc, Hauppauge. p. 23–39.

Martins, J. M., M. Riottot, M. C. de Abreu, A. M. Viegas-Crespo, M. J. Lanca, J. A. Almeida, J. B. Freire, and O. P. Bento. 2005. Cholesterol-lowering effects of dietary blue lupin (*Lupinus angustifolius* L.) in intact and ileorectal anastomosed pigs. J. Lipid Res. 46:1539–1547.

Masey O'Neill, H. V., M. Rademacher, I. Mueller-Harvey, E. Stringano, S. Kightley, and J. Wiseman. 2012. Standardised ileal digestibility of crude protein and amino acids of UK-grown peas and faba beans by broilers. Anim. Feed Sci. Technol. 175:158–167.

Materne, M., A. Leonforte, K. Hobson, J. Paull, and A. Gnanasambandam. 2011. Breeding for improvement of cool season food legumes. In: A. Pratap, and J. Kumar, editors, Biology and breeding of food legumes. CABI, Wallingford, UK. p. 49–62.

McNiven, M. A., and A. G. Castell. 1995. Replacement of soybean meal with lupin seed (*Lupinus albus*) in pig starter diets. Anim. Feed Sci. Technol. 52:333–338.

Meijer, M. M. T., J. J. M. Ogink, and W. M. J. van Gelder. 1994. Technological scale dehulling process to improve the nutritional value of faba beans. Anim. Feed Sci. Technol. 46:1–10.

Mekbungwan, A. 2007. Application of tropical legumes for pig feed. Anim. Sci. J. 78:342–350.

Messad, F., M. P. Letourneau-Montminy, E. Charbonneau, D. Sauvant, and F. Guay. 2018. Prediction of the amino acid digestibility of legume seeds in growing pigs: A meta-analysis approach. Animal 12:940–949.

Mierlita, D., D. Simeanu, I. M. Pop, F. Criste, C. Pop, C. Simeanu, and F. Lup. 2018. Chemical composition and nutritional evaluation of the lupine seeds (*Lupinus albus* L.) from low-alkaloid varieties. Rev. Chim. 69:453–458.

Mikic, A., V. Peric, V. Dordevic, M. Srebric, and V. Mihailovic. 2009. Anti-nutritional factors in some grain legumes. Biotechnol. Anim. Husb. 25:1181–1188.

Milan-Carrillo, J., C. Reyes-Moreno, E. Armienta-Rodelo, A. Carabez-Trejo, and R. Mora-Escobedo. 2000. Physicochemical and nutritional characteristics of extruded flours from fresh and hardened chickpeas (*Cicer arietinum* L). LWT-Food Sci. Technol. 33:117–123.

Milan-Carrillo, J., C. Reyes-Moreno, I. Camacho-Hernandez, and O. Rouzaud-Sandez. 2002. Optimisation of extrusion process to transform hardened chickpeas (*Cicer arietinum* L) into a useful product. J. Sci. Food Agric. 82:1718–1728.

Milczarek, A., and M. Osek. 2016. Partial replacement of soybean with low-tannin faba bean varieties (Albus or Amulet): Effects on growth traits, slaughtering parameters and meat quality of Pulawska pigs. Ann. Anim. Sci. 16:477–487.

Miller, P. R., B. G. McConkey, G. W. Clayton, S. A. Brandt, J. A. Staricka, A. M. Johnston, G. P. Lafond, B. G. Schatz, D. D. Baltensperger, and K. E. Neill. 2002. Pulse crop adaptation in the northern Great Plains. Agron. J. 94:261–272.

Molvig, L., L. M. Tabe, B. O. Eggum, A. E. Moore, S. Craig, S. D. Pencer, and T. J. V. Higgins. 1997. Enhanced methionine levels and increased nutritive value of seeds of transgenic lupins (*Lupinus angustifolius* L.) expressing a sunflower seed albumin gene. Proc. Natl. Acad. Sci. USA 94:8393–8398.

Montoya, C. A., A. D. Beaulieu, and P. Leterme. 2010a. Standardized ileal amino acid digestibility of field peas (Pisum sativum) in adult pigs. Banff Pork Seminar, Dept. of Agric. Food and Nutr. Sci., Univ. of Alberta, Banff, Alberta, Canada.

Montoya, C. A., A. D. Beaulieu, and P. Leterme. 2010b. Starch and energy digestibility of field peas (Pisium sativum). Banff Pork Seminar, Dept. of Agric. Food and Nutr. Sci., Univ. of Alberta, Banff, Alberta, Canada.

Montoya, C. A., and P. Leterme. 2011. Effect of particle size on the digestible energy content of field pea (*Pisum sativum* L.) in growing pigs. Anim. Feed Sci. Technol. 169:113–120.

Montoya, C. A., and P. Leterme. 2012. Validation of an *in vitro* technique for determining ileal starch digestion of field peas (*Pisum sativum*) in pigs. Anim. Feed Sci. Technol. 177:259–265.

Moore, K., B. Mullan, J. C. Kim, and F. Dunshea. 2016. The effect of *Lupinus albus* and calcium chloride on growth performance, body composition, plasma biochemistry and meat quality of male pigs immunized against gonadotrophin releasing factor. Animals 6:1–16.

Moore, K., B. Mullan, J. C. Kim, and F. Dunshea. 2017. The effect of *Lupinus albus* on growth performance, body composition and satiety hormones of male pigs immunized against gonadotrophin releasing factor. Animals 7:1–15.

Mosenthin, R., Sauer, W. C., Lien, K. A., De Lange, C. F. M. 1993. Apparent, true and real ileal protein and amino acid digestibilities in growing pigs fed 2 varieties of fababeans (*Vicia faba* L) different in tannin content. J. Anim. Physiol. Anim. Nutr. 70:253–265.

Musco, N., M. I. Cutrignelli, S. Calabro, R. Tudisco, F. Infascelli, R. Grazioli, V. Lo Presti, F. Gresta, and B. Chiofalo. 2017. Comparison of nutritional and antinutritional traits among different species (*Lupinus albus* L., *Lupinus luteus* L., *Lupinus angustifolius* L.) and varieties of lupin seeds. J. Anim. Physiol. Anim. Nutr. 101:1227–1241.

Mustafa, A. F., P. A.Thacker, J. J. McKinnon, D. A. Christensen, and V. J.Racz. 2000. Nutritional value of feed grade chickpeas for ruminants and pigs. J. Sci. Food Agric. 80:1581–1588.

Naczk, M., R. Amarowicz, R. Zadernowski, and F. Shahidi. 2001. Protein precipitating capacity of condensed tannins of beach pea, canola hulls, evening primrose and faba bean. Food Chem. 73:467–471.

Nguyen, G. T., M. J. Gidley, and P. A. Sopade. 2015. Dependence of *in vitro* starch and protein digestions on particle size of field peas (*Pisum sativum* L.). LWT-Food Sci. Technol. 63:541–549.

Noblet, J., H. Fortune, C. Dupire, and S. Dubois. 1993. Digestible, metabolizable and net energy values of 13 feedstuffs for growing pigs: effect of energy system. Anim. Feed Sci. Technol. 42:131–149.

Noblet, J., M. Mancuso, D. Bourdon, and R. van Barneveld. 1998. Energy value of lupin (*Lupinus angustifolius*) in growing pig and adult sow. Journees Recherche Porcine 30:239–243.

Norgaard, J. V., and J. A. Fernandez. 2009. The effect of reduced amino acid level and increasing levels of lupin on growth performance and meat content in organic reared pigs. J. Sci. Food Agric. 89:449–454.

Norgaard, J. V., J. A. Fernandez, and H. Jorgensen. 2012. Ileal digestibility of sunflower meal, pea, rapeseed cake, and lupine in pigs. J. Anim. Sci. 90:203–205.

O'Doherty, J. V., and M. P. McKeon. 2001. A note on the nutritive value of extruded and raw beans for growing and finishing pigs. Irish J. Agrric. Food. Res. 40:97–104.

Omogbenigun, F. O., R. T. Zijlstra, and E. Beltranena. 2006. Inclusion of zero-tannin fababean and substitution for soybean meal in nursery diets on weaned pig performance. Banff Pork Seminar, Department of Agricultural, Food and Nutritional Science, University of Alberta, Banff, Alberta, Canada. p. A12.

Onaghise, G. T. U., and J. P Bowland. 1977. Influence of dietary faba beans and cassava on performance, energy and nitrogen digestibility and thyroid activity of growing pigs. Can. J. Anim. Sci. 57:159–167.

Osborne, T. B., and I. F. Harris. 1907. The proteins of the pea (*Pisum Sativum*). J. Biol. Chem. 3:213–217.

Panter, K. E., D. R. Gardner, L. F. James, B. L. Stegelmeier, and R. J. Molyneux. 2000. Natural toxins from poisonous plants affecting reproductive function in livestock. In: A. T. Tu, and W. Gaffield, editor, Natural and selected synthetic toxins: Biological implications. Am. Chemical Soc., Washington, DC. p. 154–172.

Parera, N., R. P. Lázaro, M. P. Serrano, D. G. Valencia, and G. G. Mateos. 2010. Influence of the inclusion of cooked cereals and pea starch in diets based on soy or pea protein concentrate on nutrient digestibility and performance of young pigs. J. Anim. Sci. 88:671–679.

Park, S., E. Cho, H. Chung, K. Cho, S. Sa., B. Balasubramanian, T. Choi, and Y. Jeong. 2019. Digestibility of phosphorous in cereals and co-products for animal feed. Saudi J. Biol. Sci. 26:373–377.

Partanen, K., T. Alaviuhkola, H. Siljander-Rasi, and K. Suomi. 2003. Faba beans in diets for growing-finishing pigs. Agric. Food Sci. Finland 12, 35–47.

Partanen, K., H. Siljander-Rasi, and T. Alaviuhkola. 2006. Feeding weaned piglets and growing-finishing pigs with diets based on mainly home-grown organic feedstuffs. Agric. Food Sci. 15:89–105.

Partanen, K., J. Valaja, T. Jalava, and H. Siljander-Rasi. 2001. Composition, ileal amino acid digestibility and nutritive value of organically grown legume seeds and conventional rapeseed cakes for pigs. Agric. Food Sci. Finland 10:309–322.

Pastuszewska, B., S. Smulikowska, J. Wasilewko, L. Buraczewska, A. Ochtabinska, A. Mieczkowska, R. Lechowski, and W. Bielecki. 2001. Response of animals to dietary gramine. I. Performance and selected hematological, biochemical and histological parameters in growing chicken, rats and pigs. Arch. Anim. Nutr. 55:1–16.

Patterson, C. A., J. Curran, and T. Der. 2017. Effect of processing on antinutrient compounds in pulses. Cereal Chem. 94:2–10.

Pearson, G., and J. R. Carr. 1976. Lupin seed meal (*Lupinus angustifolius* cv. Uniwhite) as a protein supplement to barley-based diets for growing pigs. Anim. Feed Sci. Technol. 1:631–642.

Pearson, G., and J. R. Carr. 1977. A comparison between meals prepared from the seeds of different varieties of lupin as protein supplements to barley-based diets for growing pigs. Anim. Feed Sci. Technol. 2:49–58.

Pearson, G., and J. R. Carr. 1979. Methionine supplementation of weaner pig diets containing lupin seed meal. New Zeal. J. Exp. Agric. 7:99–101.

Perez-Maldonado, R. A., P. F. Mannion, and D. J. Farrell. 1999. Optimum inclusion of field peas, faba beans, chick peas and sweet lupins in poultry diets. I. Chemical composition and layer experiments. Br. Poult. Sci. 40:667–673.

Periago, M. J., G. Ros, and J. L. Casas. 1997. Non-starch polysaccharides and *in vitro* starch digestibility of raw and cooked chick peas. J. Food Sci. 62:93–96.

Petersen, G. I., Y. Liu, and H. H. Stein. 2014. Coefficient of standardized ileal digestibility of amino acids in corn, soybean meal, corn gluten meal, high-protein distillers dried grains, and field peas fed to weanling pigs. Anim. Feed Sci. Technol. 188:145–149.

Petterson, D. S. 2000. The use of lupins in feeding systems - Review. Asian-Australas. J. Anim. Sci. 13:861–882.

Pieper, R., M. Taciak, L. Pieper, E. Swiech, A. Tusnio, M. Barszcz, W. Vahjen, J. Skomial, and J. Zentek. 2016. Comparison of the nutritional value of diets containing differentially processed blue sweet lupin seeds or soybean meal for growing pigs. Anim. Feed Sci. Technol. 221:79–86.

Pires, V. M. R., C. M. A. Fontes, L. M. A. Ferreira, C. Guerreiro, L. F. Cunha, and J. P. B. Freire. 2007. The effect of enzyme supplementation on the true ileal digestibility of a lupin based diet for piglets. Livest. Sci. 109:57–59.

Písaříková, B., and Z.Zralý. 2009. Nutritional value of lupine in the diets for pigs (a review). Acta Vet. Brno 78:399–409.

Polesi, L. F., S. B. S. Sarmento, and C. B. P. dos Anjos. 2011. Composition and characterization of pea and chickpea starches. Braz. J. Food Technol. 14:74–81.

Prandini, A., M. Morlacchini, M. Moschini, G. Fusconi, F. Masoero, and G. Piva. 2005. Raw and extruded pea (*Pisum sativum*) and lupin (*Lupinus albus* var. Multitalia) seeds as protein sources in weaned piglets' diets: effect on growth rate and blood parameters. Ital. J. Anim. Sci. 4:385–394.

Prandini, A., S. Sigolo, M. Morlacchini, C. Cerioli, and F. Masoero. 2011. Pea (Pisum sativum) and faba bean (*Vicia faba* L.) seeds as protein sources in growing-finishing heavy pig diets: effect on growth performance, carcass characteristics and on fresh and seasoned Parma ham quality. Ital. J. Anim. Sci. 10:176–183.

Premkumar, R., A. Samaraweera, A. D. Beaulieu, and P. Leterme. 2008. Digestible and net energy content of field peas in gestating sows. Banff Pork Seminar, Department of Agricultural, Food and Nutritional Science, University of Alberta, Banff, Alberta, Canada. p. A10.

Presto, M. H., K. Lyberg, and J. E. Lindberg. 2011. Digestibility of amino acids in organically cultivated white-flowering faba bean and cake from cold-pressed rapeseed, linseed and hemp seed in growing pigs. Arch. Anim. Nutr. 65:21–33.

Rathod, R. P., and U. S. Annapure. 2016. Effect of extrusion process on antinutritional factors and protein and starch digestibility of lentil splits. LWT-Food Sci. Technol. 66:114–123.

Rehman, Z. U. 2007. Domestic processing effects on available carbohydrate content and starch digestibility of black grams (*Vigna mungo*) and chick peas (*Cicer arietium*). Food Chem. 100:764–767.

Rochfort, S., and J. Panozzo. 2007. Phytochemicals for health, the role of pulses. J. Agric. Food Chem. 55:7981–7994.

Roth-Maier, D. A., B. M. Bohmera, and F. X. Roth. 2004. Effects of feeding canola meal and sweet lupin (*L. luteus, L. angustifolius*) in amino acid balanced diets on growth performance and carcass characteristics of growing-finishing pigs. Anim. Res. 53:21–34.

Rubio, L. A. 2005. Ileal digestibility of defatted soybean, lupin and chickpea seed meals in cannulated Iberian pigs: I. Proteins. J. Sci. Food Agric. 85:1313–1321.

Rubio, L. A., G. Grant, P. Dewey, D. Brown, M. Annand, S. Bardocz, and A. Pusztai. 1998. Nutritional utilization by rats of chickpea (*Cicer arietinum*) meal and its isolated globulin proteins is poorer than that of defatted soybean or lactalbumin. J. Nutr. 128:1042–1047.

Rubio, L. A., M. M. Pedrosa, C. Cuadrado, E. Gelencser, A. Clemente, C. Burbano, and M. Muzquiz. 2006. Recovery at the terminal ileum of some legume non-nutritional factors in cannulated pigs. J. Sci. Food Agric. 86:979–987.

Rubio, L. A., M. M. Pedrosa, A. Perez, C. Cuadrado, C. Burbano, and M. Muzquiz. 2005. Ileal digestibility of defatted soybean, lupin and chickpea seed meals in cannulated Iberian pigs: II. Fatty acids and carbohydrates. J. Sci. Food Agric. 85:1322–1328.

Ruiz, U. S., G. C. Luna, L. F. Wang, E. Beltranena, and R. T. Zijlstra. 2017. Nutrient digestibility of mash, steam pelleted, and extruded barley and faba bean in growing pigs. J. Anim. Sci. 95:91.

Ruiz, U. S., G. C. Luna, L. F. Wang, E. Beltranena, and R. T. Zijlstra. 2018. Effects of feeding raw, steam-pelleted, or extruded faba bean on diet nutrient and energy digestibility and growth performance in weaned pigs. J. Anim. Sci. 96:139.

Rybinski, W., W. Swiecicki, J. Bocianowski, A. Borner, E. Starzycka-Korbas, and M. Starzycki. 2018. Variability of fat content and fatty acids profiles in seeds of a Polish white lupin (*Lupinus albus* L.) collection. Genet. Resour. Crop Evol. 65:417–431.

Salgado, P., J. B. Freire, R. B. Ferreira, M. Seabra, A. R. Teixeira, R. Toullec, and J. P. Lalles. 2002a. Legume proteins of the vicilin family are more immunogenic than those of the legumin family in weaned piglets. Food Agric. Immunol. 14:51–63.

Salgado, P., J. P. B. Freire, R. B. Ferreira, A. Teixeira, O. Bento, M. C.Abreu, R. Toullec, and J. P. Lalles. 2003. Immunodetection of legume proteins resistant to small intestinal digestion in weaned piglets. J. Sci. Food Agric. 83:1571–1580.

Salgado, P., J. P. B. Freire, M. Mourato, F. Cabral, R. Toullec, and J. P. Lalles. 2002b. Comparative effects of different legume protein sources in weaned piglets: Nutrient digestibility, intestinal morphology and digestive enzymes. Livest. Prod. Sci. 74:191–202.

Salgado, P., J. P. Lallès, R. Toullec, M. Mourato, F. Cabral, and J. P. B. Freire. 2001. Nutrient digestibility of chickpea (*Cicer arietinum* L.) seeds and effects on the small intestine of weaned piglets. Anim. Feed Sci. Technol. 91:197–212.

Salgado, P., J. M. Martins, F. Carvalho, M. Abreu, J. P. B. Freire, R. Toullec, J. P. Lallès, and O. Bento. 2002c. Component digestibility of lupin (*Lupinus angustifolius*) and pea (*Pisum sativum*) seeds and effects on the small intestine and body organs in anastomosed and intact growing pigs. Anim. Feed Sci. Technol. 98:187–201.

Salgado, P., L. Montagne, J. P. B. Freire, R.B. Ferreira, A. Teixeira, O. Bento, M. C. Abreu, R. Toullec, and J. P. Lalles. 2002d. Legume grains enhance ileal losses of specific endogenous serine-protease proteins in weaned pigs. J. Nutr. 132:1913–1920.

Seabra, M., S. Carvalho, J. Freire, R. Ferreira, M. Mourato, L. Cunha, F. Cabral, A. Teixeira, and A. Aumaitre. 2001. *Lupinus luteus, Vicia sativa* and *Lathyrus cicera* as protein sources for piglets: ileal and total tract apparent digestibility of amino acids and antigenic effects. Anim. Feed Sci. Technol. 89:1–16.

Sedlakova, K., E. Strakova, P. Suchy, J. Krejcarova, and I. Herzig. 2016. Lupin as a perspective protein plant for animal and human nutrition - a review. Acta Vet. Brno 85:165–175.

Selle, P. H., A. R. Walker, and W. L. Bryden. 2003. Total and phytate-phosphorus contents and phytase activity of Australian-sourced feed ingredients for pigs and poultry. Aust. J. Exp. Agric. 43:475–479.

Seneviratne, R. W., M. G. Young, E. Beltranena, L. A. Goonewardene, R. W. Newkirk, and R. T. Zijlstra. 2010. The nutritional value of expeller-pressed canola meal for grower-finisher pigs. J. Anim. Sci. 88:2073–2083.

Sharma, S., N. Singh, and B. Singh. 2015. Effect of extrusion on morphology, structural, functional properties and in vitro digestibility of corn, field pea and kidney bean starches. Starch-Starke 67:721–728.

Singh, K. B., P. C. Williams, and H. Nakkoul. 1990. Influence of growing season, location and planting time on some quality parameters of Kabuli chickpea. J. Sci. Food Agric. 53:429–441.

Singh, U. 1988. Antinutritional factors of chickpea and pigeonpea and their removal by processing. Plant. Food. Hum. Nutr. 38:251–261.

Singh, U., and R. Jambunathan. 1981. Studies on Desi and Kabuli chickpea (*Cicer arietinum* L.) cultivars - levels of protease inhibitors, levels of polyphenolic compounds and *in vitro* protein digestibility. J. Food Sci. 46:1364–1367.

Singhai, B., and S. K. Shrivastava. 2006. Nutritive value of new chickpea (*Cicer arietinum*) varieties. J. Food Agric. Environ. 4:48–53.

Smith, L. A., J. G. M. Houdijk, D. Homer, and I. Kyriazakis. 2013. Effects of dietary inclusion of pea and faba bean as a replacement for soybean meal on grower and finisher pig performance and carcass quality. J. Anim. Sci. 91:3733–3741.

Smits, C., and R. Sijtsma. 2007. A decision tree for co-product utilization. In: R. O. Ball, and R. T. Zijlstra, editors, Advances in pork production, Proc. Banff Pork Seminar. Vol. 18. University of Alberta, Edmonton, Alberta, Canada.

Smulikowska, S., B. Pastuszewska, E. Swiech, A. Ochtabinska, A. Mieczkowska, V. C. Nguyen, and L. Buraczewska. 2001. Tannin content affects negatively nutritive value of pea for monogastrics. J. Anim. Feed Sci. 10:511–523.

Sola-Oriol, D., E. Roura, and D. Torrallardona. 2011. Feed preference in pigs: effect of selected protein, fat, and fiber sources at different inclusion rates. J. Anim. Sci. 89:3219–3227.

Sorensen, M. T., E. M. Vestergaard, S. K. Jensen, C. Lauridsen, and S. Hojsgaard. 2009. Performance and diarrhoea in piglets following weaning at seven weeks of age: Challenge with E. coli O 149 and effect of dietary factors. Livest. Sci. 123:314–321.

Sosulski, F. W., and N. W. Holt. 1980. Amino acid composition and nitrogen-to-protein factors for grain legumes. Can. J. Plant Sci. 60:1327–1331.

Sotelo, A., F. Flores, and M. Hernandez. 1987. Chemical composition and nutritional value of Mexican varieties of chickpea (*Cicer arietinum* L). Plant. Food. Hum. Nutr. 37:299–306.

Stanogias, G., and G. R. Pearce. 1985. The digestion of fibre by pigs. 1. The effects of amount and type of fibre on apparent digestibility, nitrogen balance and rate of passage. Br. J. Nutr. 53:513–530.

Stein, H., and K. de Lange. 2007. Alternative feed ingredients for pigs. London Swine Conference, London, Canada.

Stein, H. H., G. Benzoni, R. A. Bohlke, and D. N. Peters. 2004. Assessment of the feeding value of South Dakota-grown field peas (*Pisum sativum* L.) for growing pigs. J. Anim. Sci. 82:2568–2578.

Stein, H. H., M. G. Boersma, and C. Pedersen. 2006a. Apparent and true total tract digestibility of phosphorus in field peas (*Pisum sativum* L.) by growing pigs. Can. J. Anim. Sci. 86:523–525.

Stein, H. H., and R. A. Bohlke. 2007. The effects of thermal treatment of field peas (*Pisum sativum* L.) on nutrient and energy digestibility by growing pigs. J. Anim. Sci. 85:1424–1431.

Stein, H. H., A. K. R. Everts, K. K. Sweeter, D. N. Peters, R. J. Maddock, D. M. Wulf, and C. Pedersen. 2006b. The influence of dietary field peas (*Pisum sativum* L.) on pig performance, carcass quality, and the palatability of pork. J. Anim. Sci. 84:3110–3117.

Stein, H. H., L. V. Lagos, and G. A. Casas. 2016. Nutritional value of feed ingredients of plant origin fed to pigs. Anim. Feed Sci. Technol. 218:33–69.

Stein, H. H., D. N. Peters, and B. G. Kim. 2010. Effects of including raw or extruded field peas (*Pisum sativum* L.) in diets fed to weanling pigs. J. Sci. Food Agric. 90:1429–1436.

Sujak, A., A. Kotlarz, and W. Strobel. 2006. Compositional and nutritional evaluation of several lupin seeds. Food Chem. 98:711–719.

Sundrum, A., L. Butfering, M. Henning, and K. H. Hoppenbrock. 2000. Effects of on-farm diets for organic pig production on performance and carcass quality. J. Anim. Sci. 78:1199–1205.

Swiech, E., and L. Buraczewska. 2001. *In vivo* and *in vitro* protein and amino acid digestibility of legume seeds in pig diets. J. Anim. Feed Sci. 10:159–162.

Tan, F. P., L. F. Wang, J. Gao, and E. Beltranena, T. Vasanthan, and R. T. Zijlstra. 2017. Comparative starch, fiber, and energy digestibility and characterization of undigested starch using confocal laser scanning of pulse and cereal grains in growing-finishing pigs. J. Anim. Sci. 95:91.

Taverner, M. R., D. M. Curic, and C. J. Rayner. 1983. A comparison of the extent and site of energy and protein digestion of wheat, lupin and meat and bone meal by pigs. J. Sci. Food Agric. 34:122–128.

Thacker, P. A., S. Y. Qiao, and V. J. Racz. 2002. A comparison of the nutrient digestibility of Desi and Kabuli chickpeas fed to swine. J. Sci. Food Agric. 82:1312–1318.

Thingnes, S. L., A. H. Gaustad, N. P. Kjos, H. Hetland, and T. Framstad. 2013. Pea starch meal as a substitute for cereal grain in diets for lactating sows: the effect on sow and litter performance. Livest. Sci. 157:210–217.

Thudi, M., A. Chitikineni, X. Liu, W. M. He, M. Roorkiwal, W. Yang, J. B. Jian, D. Doddamani, P. M. Gaur, A. Rathore, S. Samineni, R. K. Saxena, D. W. Xu, N. P. Singh, S. K. Chaturvedi, G. Y. Zhang, J. Wang, S. K. Datta, X. Xu, and R. K. Varshney. 2016. Recent breeding programs enhanced genetic diversity in both Desi and Kabuli varieties of chickpea (*Cicer arietinum* L.). Sci. Rep. 6:1–10.

Tudor, K. W., B. R. Wiegand, A. L. Meteer, H. L. Evans, and K. S. Roberts. 2009. Milling characteristics of corn and yellow, short-season field peas intended for swine diets. Prof. Anim. Sci. 25:8–16.

Tusnio, A., M. Taciak, M. Barszcz, E. Swiech, I. Bachanek, J. Skomial. 2017. Effect of replacing soybean meal by raw or extruded pea seeds on growth performance and selected physiological parameters of the ileum and distal colon of pigs. PLoS ONE 12:e0169467.

Tzitzikas, E. N., J. P. Vincken, J. De Groot, H. Gruppen, and R. G. F. Visser. 2006. Genetic variation in pea seed globulin composition. J. Agric. Food Chem. 54:425–433.

Ulloa, J. A., M. E. Valencia, and Z. H. Garcia. 1988. Protein concentrate from chickpea: Nutritive value of a protein concentrate from chickpea (*Cicer arietinum*) obtained by ultrafiltration and its potential use in an infant formula. J. Food Sci. 53:1396–1398.

Ummadi, P., W. L. Chenoweth, and M. A. Uebersax. 1995. The influence of extrusion processing on iron dialyzability, phytates and tannins in legumes. J. Food Process. Preserv. 19:119–131.

Valencia, D. G., M. P. Serrano, C. Centeno, R. Lázaro, and G. G. Mateos. 2008. Pea protein as a substitute of soya bean protein in diets for young pigs: Effects on productivity and digestive traits. Livest. Sci. 118:1–10.

van Barneveld, R. J. 1999. Understanding the nutritional chemistry of lupin (Lupinus spp.) seed to improve livestock production efficiency. Nutr. Res. Rev. 12:203–230.

van Barneveld, R. J., A. C. Edwards, and J. Huisman. 2000. Chemical and physical factors influencing the nutritional value and subsequent utilisation of food legumes by livestock. In: R. Knight, editor, Linking research and marketing opportunities for pulses in the 21st century. Springer, Dordrecht, The Netherlands. p. 661–670.

van der Meulen, J., and A. J. M. Jansman. 2010. Effect of pea and faba bean fractions on net fluid absorption in ETEC-infected small intestinal segments of weaned piglets. Livest. Sci. 133:207–209.

van der Meulen, J., H. Panneman, and A. J. M. Jansman. 2010. Effect of pea, pea hulls, faba beans and faba bean hulls on the ileal microbial composition in weaned piglets. Livest. Sci. 133:135–137.

van der Poel, A. F. B., L. M. W. Dellaert, A. Vannorel, and J. Helsper. 1992a. The digestibility in piglets of faba bean (*Vicia faba* L.) as affected by breeding towards the absence of condensed tannins. Br. J. Nutr. 68:793–800.

van der Poel, A. F. B., S. Gravendeel, and H. Boer. 1991. Effect of different processing methods on tannin content and *in vitro* protein digestibility of faba bean (*Vicia faba* L.). Anim. Feed Sci. Technol. 33:49–58.

van der Poel, A. F. B., S. Gravendeel, D. J. van Kleef, A. J. M. Jansman, and B. Kemp. 1992b. Tannin-containing faba beans (*Vicia faba* L.): effects of methods of processing on ileal digestibility of protein and starch for growing pigs. Anim. Feed Sci. Technol. 36:205–214.

van Kempen, T. A. T. G., P. R. Regmi, J. J. Matte, and R. T. Zijlstra. 2010. *In vitro* starch digestion kinetics, corrected for estimated gastric emptying, predict portal glucose appearance in pigs. J. Nutr. 140:1227–1233.

Van Leeuwen, P., A. J. M. Jansman, J. Wiebenga, J. F. J. G. Koninkx, and J. M. V. M. Mouwen. 1995. Dietary effects of faba bean (*Vicia faba* L) tannins on the morphology and function of the small intestinal mucosa of weaned pigs. Br. J. Nutr. 73:31–39.

Van Nevel, C., M. Seynaeve, G. Van De Voorde, S. De Smet, E. Van Driessche, and R. De Wilde. 2000. Effects of increasing amounts of *Lupinus albus* seeds without or with whole egg powder in the diet of growing pigs on performance. Anim. Feed Sci. Technol. 83:89–101.

Visitpanich, T., E. S. Batterham, and B. W. Norton. 1985. Nutritional value of chickpea (*Cicer arietinum*) and pigeonpea (*Cajanus cajan*) meals for growing pigs and rats. 1. Energy content and protein quality. Aust. J. Agric. Res. 36:327–335.

Wang, L. F., E. Beltranena, and R. T. Zijlstra. 2017. Nutrient digestibility of chickpea in ileal-cannulated finisher pigs and diet nutrient digestibility and growth performance in weaned pigs fed chickpea-based diets. Anim. Feed Sci. Technol. 234:205–216.

Wang, L. F., and R. T. Zijlstra. 2018. Prediction of bioavailable nutrients and energy. In: P.J. Moughan, and W. H. Hendriks, ediotrs, Feed evaluation science. Wageningen Academic Publishers, Wageningen, The Netherlands, p. 337–386.

Wang, N., and J. K. Daun. 2004. Effect of variety and crude protein content on nutrients and certain antinutrients in field peas (*Pisum sativum*). J. Sci. Food Agric. 84:1021–1029.

Wang, X., T. D. Warkentin, C. J. Briggs, B. D. Oomah, C. G. Campbell, and S. Woods. 1998a. Total phenolics and condensed tannins in field pea (*Pisum sativum* L.) and grass pea (*Lathyrus sativus* L.). Euphytica 101:97–102.

Wang, X., T. D. Warkentin, C. J. Briggs, B. D. Oomah, C. G. Campbell, and S. Woods. 1998b. Trypsin inhibitor activity in field pea (*Pisum sativum* L.) and grass pea (*Lathyrus sativus* L.). J. Agric. Food Chem. 46:2620–2623.

Wasilewko, J., J. Gdala, L. Buraczewska, and I. K. Han. 1999. Assessment of ileal digestibility of lupin amino acids and their use in formulating diets for pigs. J. Anim. Feed Sci. 8:13–26.

Watts, P. 2011. Global pulse industry: State of production, consumption and trade; marketing challenges and opportunities. In: Tiwari, B. K., Gowen, A., and McKenna, B., editors, Pulse foods. Academic Press, San Diego, p. 437–464.

White, G. A., L. A. Smith, J. G. M. Houdijk, D. Homer, I. Kyriazakis, and J. Wiseman. 2015. Replacement of soya bean meal with peas and faba beans in growing/finishing pig diets: Effect on performance, carcass composition and nutrient excretion. Anim. Feed Sci. Technol. 209:202–210.

Wierenga, K. T., E. Beltranena, J. L. Yanez, and R. T. Zijlstra. 2008. Starch and energy digestibility in weaned pigs fed extruded zero-tannin faba bean starch and wheat as an energy source. Can. J. Anim. Sci. 88:65–69.

Williams, C. M., J. R. King, S. M. Ross, M. A. Olson, C. F. Hoy, and K. J. Lopetinsky. 2014. Effects of three pulse crops on subsequent barley, canola, and wheat. Agron. J. 106:343–350.

Williams, P. C., S. L. Mackenzie, and P. M. Starkey. 1985. Determination of methionine in peas by near-infrared reflectance spectroscopy (NIRS). J. Agric. Food Chem. 33:811–815.

Wiseman, J. 2013. Influence of processing on the digestibility of amino acids and starch in cereals and legumes in non-ruminants. Anim. Prod. Sci. 53:1160–1166.

Woyengo, T.A., E. Beltranena, and R.T. Zijlstra. 2014a. Controlling feed cost by including alternative ingredients into pig diets: A review. J. Anim. Sci. 92, 1293–1305.

Woyengo, T. A., E. Beltranena, and R. T. Zijlstra. 2017. Effect of anti-nutritional factors of oilseed co-products on feed intake of pigs and poultry. Anim. Feed Sci. Technol. 233:76–86.

Woyengo, T. A., J. M. Heo, Y. L. Yin, and C. M. Nyachoti. 2015. Standardized and true ileal amino acid digestibilities in field pea and pea protein isolate fed to growing pigs. Anim. Feed Sci. Technol. 207:196–203.

Woyengo, T. A., R. Jha, E. Beltranena, A. Pharazyn, and R. T. Zijlstra. 2014b. Nutrient digestibility of lentil and regular-and low-oligosaccharide, micronized full-fat soybean fed to grower pigs. J. Anim. Sci. 92:229–237.

Wu, Y. V., and N. N. Nichols. 2005. Fine grinding and air classification of field pea. Cereal Chem. 82, 341–344.

Wunsche, J., U. Hennig, W. B. Souffrant, and F. Kreienbring. 1990. Studies of the prececal digestibility of crude protein and the absorption of the amino acids from lupines in pigs. Arch. Anim. Nutr. 40:831–839.

Xu, Y. X., A. Cartier, M. Obielodan, K. Jordan, T. Hairston, A. Shannon, and E. Sismour. 2016. Nutritional and anti-nutritional composition, and *in vitro* protein digestibility of Kabuli chickpea (*Cicer arietinum* L.) as affected by differential processing methods. J. Food Meas. Charact. 10:625–633.

Yang, Y. X., Y. G. Kim, S. Heo, S. J. Ohh, and B. J. Chae 2007. Effects of processing method on performance and nutrient digestibility in growing-finishing pigs fed lupine seeds. Asian-Australas. J. Anim. Sci. 20:1229–1235.

Zaworska, A., A. Frankiewicz, and M. Kasprowicz-Potocka. 2017. The influence of narrow-leafed lupin seed fermentation on their chemical composition and ileal digestibility and microbiota in growing pigs. Arch. Anim. Nutr. 71:285–296.

Zaworska, A., M. Kasprowicz-Potocka, A. Rutkowski, and D. Jamroz. 2018. The influence of dietary raw and extruded field peas (*Pisum sativum* L.) on nutrients digestibility and performance of weaned and fattening pigs. J. Anim. Feed Sci. 27:123–130.

Zijlstra, R. T., and E. Beltranena. 2007. New frontier in processing: Ingredient fractionation. In: Paterson, J. E., and Barker, J. A., editors, Manipulating pig production XI. Proceedings of the Eleventh Biennial Conference. Australasian Pig Science Association, Werribee, Vic, Australia. pp 216–222.

Zijlstra, R. T., and E. Beltranena. 2008. Variability of quality in biofuel co-products, Recent advances in animal nutrition 2008. 42nd University of Nottingham Feed Conference, Sutton Bonington Campus, Nottingham, UK, 2–4 September 2008. p. 313–326.

Zijlstra, R. T., and E. Beltranena. 2009. Regaining competitiveness: Alternative feedstuffs for swine. In: Ball, R.O., editor, Advances in pork production, Proc. Banff Pork Seminar. Vol. 20. University of Alberta, Edmonton, Alberta, Canada. p. 237–243.

Zijlstra, R. T., and E. Beltranena. 2013. Swine convert co-products from food and biofuel industries into animal protein for food. Anim. Frontiers 3:48–53.

Zijlstra, R. T., E. D. Ekpe, M. N. Casano, and J. F. Patience. 2001. Variation in nutritional value of western Canadian feed ingredients for pigs. Proc. 22nd Western Nutr. Conf., Saskatoon, SK, Canada. p. 12–24.

Zijlstra, R. T., A. G. V. Kessel, and M. D. Drew. 2004a. Ingredient fractionation: The value of value-added processing for animal nutrition. The worth of the sum of parts versus the whole. Proc. 25th Western Nutr. Conf., Saskatoon, SK, Canada. p. 41–53.

Zijlstra, R. T., K. Lopetinsky, and E. Beltranena. 2008. The nutritional value of zero-tannin faba bean for grower-finisher pigs. Can. J. Anim. Sci. 88:293–302.

Zijlstra, R. T., K. Lopetinsky, B. Dening, G. S. Bégin, and J. F. Patience. 2004b. The nutritional value of zero-tannin faba beans for grower-finisher pigs. Can. J. Anim. Sci. 84:792–793.

Zijlstra, R. T., A. Owusu-Asiedu, and P. H. Simmins. 2010. Future of NSP-degrading enzymes to improve nutrient utilization of co-products and gut health in pigs. Livest. Sci. 134:255–257.

Zijlstra, R. T., and R. L. Payne. 2007. Net energy system for pigs, In: J. E. Paterson, and J. A. Barker, Editors, Manipulating pig production XI. Proceedings of the Eleventh Biennial Conference of the Australasian Pig Science Association. Australasian Pig Science Association, Werribee, VIC, Australia. p. 80–90.

Zraly, Z., B. Pisarikova, M. Trckova, M. Dolezal, J. Thiemel, J. Simeonovova, and M. Juzl. 2008. Replacement of soya in pig diets with white lupine cv. Butan. Czech J. Anim. Sci. 53:418–430.

Zraly, Z., B. Pisarikova, M. Trckova, I. Herzig, M. Juzl, and J. Simeonovova. 2006. Effect of lupine and amaranth on growth efficiency, health, and carcass characteristics and meat quality of market pigs. Acta Vet. Brno 75:363–372.

Zraly, Z., B. Pisarikova, M. Trckova, I. Herzig, M. Juzl, and J. Simeonovova. 2007. The effect of white lupine on the performance, health, carcass characteristics and meat quality of market pigs. Vet. Med. 52:29–41.

# 14 Fiber in Swine Nutrition

J. Paola Lancheros, Charmaine D. Espinosa, Su A. Lee, Maryane S. Oliveira, and Hans H. Stein

## Introduction

Of all nutrients in diets for pigs, none have been more complex and difficult to characterize, analyze, and understand than fiber, and there are major gaps in our knowledge about dietary fiber and how fiber contributes to the overall provision of energy and nutrients for pigs. Because fiber cannot be digested, but only can contribute positively to the energy status of an animal after fermentation by microbes, a prerequisite for utilization of energy in fiber is that the animal has a large and diverse microbial population in the intestinal tract. Through evolution, some animal species have developed enhanced capacities for fiber fermentation, which resulted in differences among animal species in the development of their digestive systems. However, unlike pregastric and hindgut fermenters, pigs have a relatively small fermentation capacity, which by definition limits the ability of pigs to ferment fiber and subsequently utilize the energy from fiber. However, certain substances that are included in the fiber fraction of feed ingredients are easily fermentable, and the majority of the soluble fibers are fermented in the cecum of pigs (Urriola et al. 2010; Jaworski and Stein 2017), whereas the insoluble dietary fibers (IDFs) are harder to ferment. Accurate analysis of the different fiber fractions is, therefore, a prerequisite for being able to predict the utilization of fiber by pigs, but unfortunately, there is great confusion about fiber analysis and even when the same analysis is used, results may be different depending on the actual equipment used for the analysis. As a consequence, detailed analysis of fiber and its different components is important for understanding fiber and predicting the energy value to pigs of fiber in different feed ingredients. The objective of the current contribution is to summarize current knowledge about dietary fiber in terms of characterization, analysis, and fermentation of fiber. Postabsorptive metabolism of absorbed end products resulting from fiber fermentation in pigs is illustrated and the impact of dietary fiber on digestibility and absorption of other nutrients is discussed as well.

## Definition of Dietary Fiber

Dietary fiber is defined as the sum of "non-digestible soluble and insoluble carbohydrates with three or more monomeric units and lignin that are intrinsic and intact in plants, and isolated or synthetic non-digestible carbohydrates with three or more monomeric units determined to have physiological effects that are beneficial to human health" (FDA 2016). Total dietary fiber (TDF) include all non-digestible carbohydrates such as oligosaccharides, cellulose hemicelluloses, and resistant starch

(Bach Knudsen et al. 2013). Based on functional, chemical, and physical properties, TDF may be divided into soluble dietary fiber (SDF) and IDF.

The SDF may form a viscous gel due to its solubility in water and is composed of pectins, gums, inulin-type fructans, and other hemicelluloses like β-glucans (Lattimer and Haub 2010). SDF is partially or completely fermented in the cecum producing short-chain fatty acids (SCFA; Slavin 2013; Jaworski and Stein 2017). IDF does not form gels due to its water insolubility and is composed of cellulose, lignin, and insoluble hemicelluloses (Dreher 2001; Lattimer and Haub 2010). IDF may be fermented in the large intestine, where SCFAs are synthesized and after absorption contribute to the energy needs of the animal (Bach Knudsen 2011; Stipanuk and Caudill 2013). However, IDF is much less fermentable than SDF (Urriola et al. 2010).

### Structures of Dietary Fiber

Although the concept of defining dietary fiber is straight forward as indicated below, not all ingredients and diets contain all fiber components, and analytical difficulties exist in terms of quantifying each component. However, attempts to analyze all fiber components need to be made to account for all nutrients in feed ingredients (Navarro et al. 2018a).

#### Oligosaccharides

Naturally occurring oligosaccharides consist of galacto oligosaccharides, fructooligosaccharides, and mannan oligosaccharides (Figure 14.1). Synthetically produced transgalacto oligosaccharides are also available, but are primarily used in the human food sector and are usually not used in swine nutrition. Oligosaccharides typically contain between 3 and 60 monosaccharides and are separated from other fiber components based on their solubility in 80% ethanol (Englyst and Englyst 2005).

Galacto oligosaccharides, which are sometimes called α-galactosides, include raffinose, stachyose, and verbascose, and are primarily present in legumes with soybeans having the greatest concentration. Typical concentration of raffinose in soybean meal vary between 1 and 1.5%; stachyose varies between 4 and 8%, and verbascose is usually present at less than 0.5% (Choct et al. 2010; Ibáñez et al. 2020). Stachyose is present at lower concentrations in field peas and lupins, but the concentration of verbascose is between 1 and 3% in field peas (Wang and Daun 2004; Zaworska et al. 2018) and may be greater than 3% in lupins (Kaczmarek et al. 2016). Inclusion of galacto oligosaccharide is often limited in diets for young animals because they may cause diarrhea (Liying

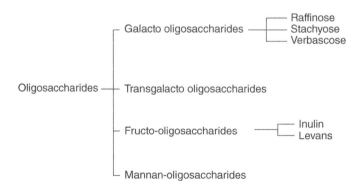

**Figure 14.1**  Oligosaccharides in feed ingredients.

et al. 2003), but fermentation of soybean meal effectively removes galacto oligosaccharide from the meal (Cervantes-Pahm and Stein 2010; Rojas and Stein 2013; Espinosa et al. 2020).

Fructooligosaccharides are sometimes referred to as fructans and consist of inulin and levans, which contain a molecule of sucrose and a varying number of fructose monosaccharides (Franck 2006). Inulin is naturally present in wheat, chicory, and other plants and may contain up to 60 monosaccharides. Inulin is sometimes produced commercially by extraction from chicory and may be included in diets for pigs as a prebiotic and may have immune protective effects in the small intestine of pigs (Hansen et al. 2010). Levans are longer chained and sometimes branched fructooligosaccharides that are produced by bacteria and fungi who secrete the levansucrase enzyme (Franck 2006).

Glucomannan is a water-soluble hemicellulose and is also known as mannan oligosaccharide. Alpha-(1-4)-linked mannose and glucose in a ratio of 1.6:1 are the main sugars in glucomannan (Katsuraya et al. 2003). Glucomannan is present in bacterial residues, konjac and mung bean plants, and yeast cell walls and exists with various branches or glycosidic linkages in the linear structure (Elbein 1969; Arvill and Bodin 1995; Chorvatovičová et al. 1999; Tokoh et al. 2002). For pigs, glucomannan derived from the cell wall of yeast has been used as an alternative to antibiotic growth promoters because this may eliminate pathogenic bacteria from the intestinal tract (Miguel et al. 2004). Average daily gain, average daily feed intake, and gain to feed of pigs fed diets containing supplemental glucomannan were increased compared with pigs fed a control diet (Miguel et al. 2004; Rosen 2006; Edwards et al. 2014).

*Nonstarch Polysaccharides*
Nonstarch polysaccharides (NSP) are composed of up to several hundred thousand monosaccharide units (Loy and Lundy 2019). In cereal grains, NSP provide rigidity to the outer layer and are key constituents needed for physiological processes during grain development (Hamaker et al. 2019). NSP do not contain α-(1-4)-linked glycosyl units that are characteristic for starch (Englyst et al. 2007), but instead contain 1. β-linkages that cannot be digested by animal enzymes. The NSP consist of cellulose and noncellulosic polysaccharides (Bach Knudsen et al. 2013), with the latter group sometimes called hemicellulose (Figure 14.2). Cellulose is present in all plant feed ingredients, and the majority of the noncellulosic polysaccharides in cereals and cereal grain coproducts are arabinoxylans and mixed linked 1. β-glucans although smaller amounts of xyloglucans may also be present (Bach Knudsen 2001). In contrast, the NSP in oilseed meals and other legumes primarily consist of cellulose, xyloglucan, and pectic polysaccharides (Choct et al. 2010; Navarro et al. 2019).

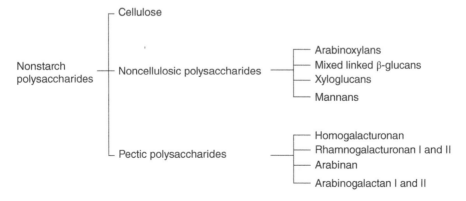

**Figure 14.2** Nonstarch polysaccharides in feed ingredients.

*Cellulose*

Cellulose is the most abundant organic compound in the biosphere, and it is the major polysaccharide of glucose in plants (Tymoczko et al. 2015; Held et al. 2015). Cellulose is composed of unbranched chains of β-(1-4)-linked D-glucose that bind between 7000 and 15 000 glucose units in a flat ribbon form. This particular conformation allows cellulose chains to maintain the conformation and form intramolecular hydrogen bonds between the hydroxyl groups at the C3 positions of neighboring glucose residues (Hamaker et al. 2019). This structure and the associated H bonds also allow the cellulose chains to stack together to form larger microfibrils, which facilitate the association with other carbohydrates and lignin moieties that allow the formation of a rigid strong cell wall structure (Tymoczko et al. 2015; Loy and Lundy 2019).

Cellulose contains both crystalline and amorphous domains, and the structure of the cellulose molecules influences the physical properties and chemical behavior. Reactants and enzymes may penetrate the amorphous region, and it is, therefore, primarily the amorphous regions of the cellulose that may be fermented. In contrast, the crystalline regions are largely resistant to fermentation (Ciolacu et al. 2011). In corn and corn distillers dried grains with solubles (**DDGS**), the crystallinity of cellulose is approximately 50% (Xu et al. 2009).

Hydrolysis of cellulose requires endoglucanases, which randomly hydrolyze β-(1-4) glycosidic bonds from the internal part of the amorphous region, forming new reducing or nonreducing ends that may be hydrolyzed by exoglucanases (Duan and Feng 2010; Zhang and Zhang 2013). Exoglucanases cleave from the reducing or nonreducing ends of cellulose (Zhang and Zhang 2013) to release oligosaccharides or disaccharides. To complete hydrolysis, two additional enzymes are needed: cellodextrinases and β-glucosidases. Cellodextrinases hydrolyze soluble cello-oligosaccharides, which generates cellobiose (Duan and Feng 2010), and β-glucosidases hydrolyze soluble cellodextrins and cellobiose to glucose (Zhang and Zhang 2013).

*Arabinoxylan*

Arabinoxylan is the main noncellulosic NSP in cereal grains and grain coproducts and often comprise 50–60% of all fiber in cereal grains and cereal grain coproducts (Jaworski et al. 2015). Arabinoxylan contains a xylose backbone and sidechains containing arabinose, D-glucuronic acid, and acetyl groups (Navarro et al. 2019). Xylose units may also be attached to arabinose monomers in the sidechains, and these xylose units may be further substituted with galactose. Strong heterogeneous intermolecular complexes are formed due to the arabinoxylan cross-linking. These complexes affect enzymatic degradation and potential encapsulation of nutrients within the cell wall (Lapierre et al. 2001; Pedersen et al. 2014). Several phenolic acids including ferulic acid, p-coumaric, dehydrodiferulic, and dehydrotriferulic acids may be covalently bound to cell wall arabinoxylans through ester linkages to the arabinose side chains. The phenolic acids are major barriers to fermentation of fiber because they link to lignin, which make the sidechains un-fermentable.

Arabinoxylans are present in the cell wall material from the endosperm and in the bran, but arabinoxylan from the bran has a lower arabinose/xylose ratio compared with arabinose from intact fiber, which includes bran and endosperm (Yadav et al. 2007). Therefore, the arabinoxylan that is located in the endosperm is more branched or have a less heterogeneous structure with more arabinose side chains than the arabinoxylan from the bran portion (Hamaker et al. 2019). The concentration of ferulic acid in the endosperm is also less than in the bran portion (Yadav et al. 2007). The phenolic acid content is high and in corn bran, it can be up to 5%, primarily ferulic acid (Saulnier and Thibault 1999).

Ferulic acid residues allow arabinoxylan chains to cross-link with each other through di and tri ferulic acid bridges, which together with linkages to lignin, and association with proteins limit the solubility of arabinoxylan (Navarro et al. 2019). Due to the complex structure of arabinoxylans nine different enzymes are needed for complete hydrolysis, and it is believed that at least three or four of these enzymes are not synthesized by microbes in the hindgut of pigs (Stein 2019). To increase fermentation of arabinoxylans, β-(1-4) endoxylanases may be added to swine diets to cleave the xylan backbone via hydrolysis of the β-(1-4) glycosidic bonds between xylose units (Dodd and Cann 2009). The debranching enzyme α-arabinofuranosidase also is needed to hydrolyze the α- (1,5) glycosidic bond between 2 L-arabinose units in the side chain, but this enzyme is not expressed by intestinal microbes, and therefore, needs to be added to the diets (Stein 2019). To de-lignify the arabinoxylans, ferulic and coumaric acid esterases are needed to hydrolyze the linkage between L-arabinose and ferulic acid or coumaric acid, respectively (de Vries 2003). However, these esterases are also not expressed by intestinal microbes and need to be added to the diets (Stein 2019).

*Beta-glucans*

Beta-glucans are homopolysaccharides of β-D-glucopyranosyl residues linked via a mixture of (1-3) and (1-4) glycosidic linkages, with blocks of (1-4) linked units (oligomeric cellulose-like segments) separated by (1-3) linkages (Izydorczyk 2017; Hamaker et al. 2019). Beta-glucans are present in quantities lower than arabinoxylans and cellulose, and polished rice, sorghum, and corn contain less than 1% beta glucans, whereas the concentration is approximately 2% in rye, and 2–5% in barley and oats (Jha and Berrocoso 2015; Navarro et al. 2019). In corn, β-glucan is composed of approximately 67.5% cellotrimer and 26.7% cellotetramer segments. The remaining part of the molecule consists of 0.3% cellobiose segments and 5.4% cellodextrin-like oligosaccharides containing more than four consecutive 4-O-linked glucose residues (Yoshida et al. 2014; Hamaker et al. 2019).

In the beta-glucan structure, the β-(1-3) linkages form coils that create an asymmetric conformation that prevents intermolecular alignment of chains into cellulose microfibrils. The asymmetric conformation allows the molecule to form water-soluble gel-like matrices that reinforce the cellulosic microfibrils in the wall (Burton and Fincher 2009).

*Pectic Polysaccharides*

Pectic polysaccharides are the major noncellulosic NSP in oilseeds and oilseed meals, which include homogalacturonan, rhamnogalacturonans, arabinans, arabinogalactans, and xylogalacturonans (Navarro et al. 2019). Homogalacturonan is a polymer of galacturonic acids that are substituted with O-methyl or O-acetyl groups (Caffall and Mohnen 2009). The more complex rhamnogalacturonan Type I has a backbone consisting of alternating units of galacturonic acid and rhamnose and sidechains containing galactose, arabinose, and O-acetyl units. Rhamnogalacturonan Type II is the most complex of the pectic polysaccharides with the galacturonan backbone having sidechains containing arabinose, galactose, fucose, glucuronic acid, rhamnose, and other monosaccharides. Due to the highly branched structure, rhamnogalacturonan Type II is largely resistant to fermentation by microbes in the hindgut of pigs (Albersheim et al. 2011). Arabinogalactans consist of a galactose backbone with arabinose sidechains (Type I) or both arabinose and galactose in the sidechains (Type II), whereas xylogalacturonans have a galacturonic acid back bone and sidechains of xylose (Caffall and Mohnen 2009).

*Xyloglucans*

Xyloglucans are mainly present in oilseed meals and pulse crops and consist of a glucose backbone with sidechains of xylose to which galactose, fucose, and arabinose may be attached (Bach Knudsen et al. 2013). After cellulose, xyloglucans are the quantitatively most important fiber component in oilseed meals.

*Lignin*

Lignin is not considered a polysaccharide, but is present in the cell wall and the sum of lignin and NSP equals the amount of TDF in a feed ingredient. Based on its structure and presence in the cell wall, lignin prevents biochemical degradation and physical damage of cell walls (Bach Knudsen 1997). Lignin is an organic polymer formed by phenyl propane units organized in a tridimensional structure. Coniferyl, sinapyl, and p-coumaryl alcohols, which can be transformed into guaiacyl (G unit), syringyl (S unit) and p-hydroxyphenyl (H unit), are the precursors needed to synthesize lignin (He et al. 2018), and these units are connected by aryl-ether and carbon–carbon linkages (Van Erven et al. 2019). The concentration of lignin in the corn grain is around 1.1% (weigh basis), which is lower than in other cereal grains (Bach Knudsen 1997; Hamaker et al. 2019).

Lignin is bound to coumaric acid or ferulic acid in arabinoxylans or pectic polysaccharides, but small amounts of lignin may be bound to other polysaccharides (Cuevas Montilla et al. 2011). Digestive or microbial enzymes cannot degrade lignin, and lignified parts of fiber are, therefore, unfermentable. As a consequence, fermentability of arabinoxylan from the bran of the corn grain is low due to the linkages between lignin and arabinoxylan in the cell wall.

### Analysis of Fiber in Feed Ingredients

Analysis of dietary fiber in feed ingredients may be divided into three main methods: (i) enzymatic–gravimetric methods; (ii) nonenzymatic–gravimetric methods; and (iii) enzymatic–chemical methods (Figure 14.3). Nonenzymatic–gravimetric methods include two procedures: The Weende method, which uses acid hydrolysis to extract sugars and starch, and alkaline hydrolysis to remove protein, some hemicelluloses, and lignin. The other procedure is the Van-Soest method, which quantifies neutral detergent fiber (NDF) and acid detergent fiber (ADF; Champ et al. 2003), also known as the detergent fiber analysis.

Enzymatic–gravimetric methods include the use of enzymes to remove protein and starch followed by precipitation and isolation of SDF and a correction for protein and ash in the residue (McCleary 2003). The most used enzymatic gravimetric methods to analyze feed ingredients are analyses that quantify IDF, SDF, TDF, and lignin.

In enzymatic–chemical methods, amylase is added to the sample to remove starch, followed by separation of soluble and insoluble fractions and determination of polysaccharides by chromatography (McCleary 2003). The enzymatic–chemical methods allow for quantification of the individual sugars included in NSP. Phenolic acids and lignin may also be analyzed.

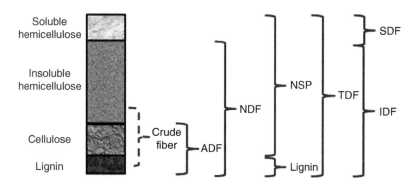

**Figure 14.3**   Fiber components included in each analysis for dietary fiber.

## Crude Fiber Analysis

The crude fiber analysis is a chemical-gravimetric method, which separates carbohydrates into crude fiber and nitrogen free extract. This method was developed at the Agricultural Experiment Station in Weende in Germany and is, therefore, included in the proximate analysis procedures known as the Weende Analyses (Fahey et al. 2019). Fiber is quantified by digesting a sample with 1.25% sulfuric acid and 1.25% sodium hydroxide and the residue left after digestion is quantified after drying (Bach Knudsen 2001). However, because of the alkali that is used to dissolve the protein, some lignin is also dissolved, and because of the sulfuric acid that is used, some of the hemicelluloses are also dissolved (Fahey et al. 2019). The crude fiber analysis, therefore, does not capture all fiber in a feed ingredient and the underestimation of hemicellulose and lignin may be more than 50% (Bach Knudsen 2001; Alyassin and Campbell 2019). As a consequence, the crude fiber analysis is not considered an accurate assessment of the fiber in a feed ingredient. Nevertheless, despite the inaccuracies associated with the crude fiber analysis, the amount of crude fiber needs to be declared for all swine diets that are commercially sold in the U.S. The official method for analyzing fiber in feed ingredients is AOAC method 962.09 (AOAC Int. 2007), but a fiber bag technique for analysis is also available (Method Ba6a.05; AOCS 2006).

## Detergent Fiber Analyses

A method to analyze ADF and lignin was developed by van Soest and coworkers in the 1960s, and the method has later been modified (van Soest et al. 1991; Mertens 2003). The main objective of the ADF analysis is to analyze cellulose and lignin, but the analysis does not include insoluble hemicelluloses (Jung 1997; Udén et al. 2005). In this method, protein is removed from the fibrous residue using cetrimonium bromide. Samples are then treated with $1 N H_2SO_4$ to remove nonfibrous compounds (Mertens 2003). At the end, samples are dried and the ADF fraction is calculated as the final weight of the sample divided by the initial sample weight.

Because the ADF method does not quantify total fiber due to the exclusion of hemicelluloses, the NDF method was developed (van Soest et al. 1991; Udén et al. 2005). NDF represents the insoluble fraction in fiber, which includes hemicellulose, cellulose, and lignin, but excludes soluble hemicelluloses, resistant starch, and low molecular weight sugars. The NDF analysis is performed using a chemical solubility-gravimetric method. Samples are treated with anionic detergent, which is mainly composed of sodium dodecyl sulfate, ethylenediaminetetraacetic disodium salt dehydrate, and sodium lauryl sulfate. These reagents extract the proteins, and fats are removed by hot water and acetone (Mertens 2003). During the extraction, thermostable α amylase is added to remove starch and this is followed by filtration and washing (Mongeau and Brooks 2003). Samples are then dried, and the NDF fraction is calculated as the final weight of the sample divided by the initial sample weight. Quantities of ADF and NDF in a feed ingredient may be analyzed using separate analyses – or they may be analyzed using a sequential procedure where the NDF is first quantified and the residue is then treated with the acid detergents.

The ADF residue may be further treated with $12M$ sulfuric acid for three hours, which results in the dissolution of cellulose and other nonlignin components. At the end of digestion, the residue is dried and lignin is calculated as the weight of the sample after the treatment divided by the weight of the sample before the treatment (Fukushima et al. 2015; Fahey et al. 2019). Lignin analyzed using this procedure is called acid detergent lignin (ADL). Another method to measure lignin includes the acid detergent permanganate lignin procedure, which typically yields values that are slightly greater, but correlated with, values for ADL (Fahey et al. 2019). Lignin may also be quantified using the Klason lignin procedure. In this analysis, samples are treated with amylase and amyloglucosidase to remove starch followed by a precipitation using 80% ethanol. Afterward, samples are treated with $12M$ sulfuric acid, and the nonhydrolyzed residue is collected and ashed and Klason lignin is determined as the difference in weight between the residue before ashing and after ashing (Hatfield et al. 1994; Bach Knudsen 1997).

Insoluble hemicelluloses may be calculated as the difference between ADF and NDF, and cellulose can be calculated as the difference between ADF and lignin (Jaworski and Stein 2017; Abelilla and Stein 2019). However, the detergent fiber analyses do not quantify any of the soluble hemicelluloses or the low molecular weight sugars and inaccuracies in results for NDF and ADF may be associated with extraction of tannins and chelation with Ca (Fahey et al. 2019).

### Total Dietary Fiber Analysis

There are several AOAC official methods for measuring TDF. The first AOAC official method for TDF (Method 985.29; AOAC Int. 2007) was an enzymatic–gravimetric method, which did not allow separation of dietary fiber into soluble and insoluble fractions. The basic principle of this method is that dried and defatted samples are treated with enzymes (α-amylase, amyloglucosidase, and protease) that mimic the digestive process in the small intestine. Digestible carbohydrates and protein are hydrolyzed and removed from the sample and the residue is weighed (Prosky et al. 1984; Mertens 2003). The residue also is analyzed for undigested proteins and ash, and the TDF is the weight of the filtered and dried residue less the weight of the protein and ash.

Subsequent modifications allowed for separating TDF into SDF and IDF (Method 991.43; AOAC Int. 2007). This method is an extension of Method 985.29 (AOAC Int. 2007), but includes an additional step. After the enzymatic digestion by heat-stable α-amylase (95 °C for 30 minutes), protease, and amyloglucosidase, IDF is filtered, and the residue is washed with warm distilled water. The combined solution of filtrate and water is precipitated with four volumes of 95% ethanol for SDF determination. The precipitate is then filtered and dried. Both SDF and IDF residues are corrected for protein and ash before final calculation of SDF and IDF. TDF is calculated as the sum of SDF and IDF.

The AOAC procedures for measuring TDF (e.g., AOAC Method 985.29 and AOAC Method 991.43) do not quantitatively measure undigestible low-molecular-weight sugars and only partially includes resistant starch although these fractions are also included in the total fiber fraction in a feed ingredient or diet. However, the method developed by McCleary (2007) allows for measuring low-molecular-weight sugars in addition to measuring TDF (including resistant starch) and this method was later adopted by the AOAC (Method 2009.01). In this procedure, samples are incubated with pancreatic α-amylase and amyloglucosidase (37 °C, pH 6) for 16 hours to hydrolyze nonresistant starch and protein is digested with protease. The pH is then adjusted to 4.3 and precipitated SDF and IDF are recovered in sintered glass crucibles. After washing with ethanol and drying, the weight of each fraction is recorded. One duplicate is analyzed for protein and ash. The TDF is the weight of the filtered and dried residue less the weight of the protein and ash. The ethanol wash solution can be concentrated, desalted, and analyzed by HPLC to determine low-molecular-weight sugars (McCleary 2007).

An extension of method 2009.01 was later suggested to quantify IDF, SDF, and TDF inclusive of the resistant starch as well as low-molecular-weight sugars, but this method includes resistant starch in the SDF fraction (Method 2011.25; McCleary et al. 2012; Nguyen et al. 2019a). Sample preparation

**Table 14.1** Current AOAC methods to analyze dietary fiber[a,b]

| Item | Enzymatic–gravimetric methods | | Enzymatic–gravimetric-liquid chromatographic methods | |
| --- | --- | --- | --- | --- |
| AOAC method: | 985.29 | 991.43 | 2009.01 | 2011.25 |
| Total dietary fiber | + | + | + | + |
| Insoluble dietary fiber | − | + | − | + |
| Soluble dietary fiber | − | + | − | + |
| Resistant starch | − | − | + | + |
| Low-molecular-weight sugars | − | − | + | + |

[a] + component measured in the method; − component not measured in the method.
[b] Adapted from AOAC Int. (2007); Fahey et al. (2019), and Nguyen et al. (2019a).

is similar to method 2009.01, and after incubation and digestion, samples are digested and filtered and the IDF is determined gravimetrically after correction for protein and ash in the residue. Ethanol is then added to the filtrate of the IDF, and SDF is captured by filtration and determined gravimetrically after correction for protein and ash in the precipitate. The TDF is calculated as the sum of IDF and SDF. The nonprecipitable part of the filtrate may also be recovered by concentrating the filtrate and analyzed by HPLC to measure low-molecular-weight sugars (McCleary et al. 2012).

AOAC Method 2011.25 is similar to AOAC Method 991.43, with the exception that Method 2011.25 simulates digestion in vivo and, therefore, TDF may be better quantified because resistant starch and low-molecular-weight sugars are included in the analysis (Nguyen et al. 2019a).

Thus, a total of 4 AOAC procedures may be used to analyze TDF, but each method will give different values because different parts of the fiber fraction is included (Table 14.1). However, it was recently demonstrated that method 991.43 and method 2011.25 do not give different estimates for TDF even when feed ingredients with relatively high concentrations of resistant starch are analyzed (Nguyen et al. 2019a). Whether this is a result of method 991.43 capturing most of the resistant starch or because method 2011.25 does not effectively quantify resistant starch is not known. It is, therefore, likely that further refinements in analyzing the TDF in feed ingredients will be developed in the future.

### Analysis for Nonstarch Polysaccharides

The TDF procedure includes both NSP and lignin, but NSP may be analyzed separately by quantifying all monosaccharides in the fiber using a chemical gravimetric method. The procedure is performed in three parallel extractions: (i) total NSP, (ii) noncellulosic polysaccharides, and (iii) insoluble NSP (Bach Knudsen 1997). This procedure starts with the removal of starch using an enzymatic procedure that includes (i) α-amylase to cleave α-(1-4) linkages to maltose, maltriose and maltodextrins and (ii) amyloglucosidase to cleave dextrins (Bach Knudsen 1997). Afterwards, solutes are removed using a sodium phosphate buffer at pH 7, and NSP is precipitated using ethanol. Sulfuric acid is then used to swell the cellulose and hydrolyze the NSP (Bach Knudsen 1997; Jaworski et al. 2015). Individual monosaccharide sugars are subsequently analyzed using gas chromatography. However, the monosaccharides first need to be derivatized (Chen et al. 2009), which is a chemical modification that allows for analysis of carbohydrates because of the instability of polysaccharides at high temperatures. Derivatization confer volatility to the carbohydrates without structural changes and improves chromatographic condition for detection (Izydorczyk 2017). There are different techniques for derivatization, but specifically for the analysis of monosaccharides in fiber, an acetylation of alcohols is used in the presence of 1-methyl imidazole as a catalyzer (Bach Knudsen 1997).

Uronic acids are analyzed separately using a colorimetric method (Scott 1979). Five standards with different concentrations of uronic acid are used to build a calibration curve. The standards and the samples react with 96% sulfuric acid in the presence of a solution composed of sodium chloride and boric acid as a catalyzer. Subsequently, 3,5 dimethyl phenol is added to form a chromogen because this reagent is selective for 5-formyl-2-furancarboxylic acid, which is formed from uronic acids (Scott 1979). After this reaction, the absorbance of the sample as well as the absorbance of the five standards is read on a spectrophotometer (Bach Knudsen 1997). Following quantification of monosaccharides and uronic acids, total NSP is calculated as the sum of the analyzed monosaccharides plus the uronic acids (Bach Knudsen 1997; Jaworski et al. 2015).

### Phenolic Acid Analyses

Phenolic acids are often associated with arabinoxylan and may link to lignin, which forms a major barrier for microbial fermentation in the hindgut (Wefers et al. 2017; Vangsøe et al. 2020). The major phenolic acids in cereal grains are coumaric acid and ferulic acid (Vangsøe et al. 2020). Analysis of ferulic acid and coumaric acid starts with treatment of the samples with $0.5M$ KOH to extract the free and bound ferulic and coumaric acids. Bound phenolic acids are then extracted by suspending the sample in sodium acetate buffer (pH 5.7). Quantification of the phenolic acids is performed using ultra-high performance liquid chromatography (Appeldoorn et al. 2010; Van Dongen et al. 2011).

### Fermentation of Dietary Fiber

Pigs lack endogenous enzymes to digest dietary fiber; therefore, microbes in the intestinal tract of pigs ferment dietary fiber to obtain energy (Anguita et al. 2006). Fermentation is a metabolic process where organic substrates are enzymatically hydrolyzed to yield energy in the absence of oxygen (Abdel-Rahman et al. 2013). The undigested portion of carbohydrates (i.e., nondigestible oligosaccharides, NSP, resistant starch) serves as a substrate for intestinal microbes, and this results in the production of SCFA, which can be utilized by pigs as a source of energy (Macfarlane and Macfarlane 2007). Dietary fiber is mostly fermented in the hindgut of the pig, particularly in the cecum and large intestine; however, the apparent ileal digestibility (**AID**) of NSP by pigs can range from zero to 45%, indicating that some fermentation can occur prior to the hindgut of the pig (Bach Knudsen et al. 2013; Abelilla and Stein 2019). Fermentation of nutrients in the gastrointestinal tract results in symbiotic advantages for the pig and the microbial population because microbes ferment dietary fiber to obtain ATP and synthesizes SCFA, which are absorbed and used as a source of energy by the pig. Fermentation, therefore, influences microbial ecology of pigs (Pieper et al. 2009) and may positively impact intestinal health. Production of SCFA in the hindgut is dependent on the substrate (i.e., composition and physicochemical characteristics, degree of lignification, and rate of passage) and microbial ecology in the gut (Jha and Berrocoso 2015), which indicates that the rate of hydrolysis of NSP may vary among feed ingredients. Substrates high in soluble fiber (e.g., beet pulp) have high fermentability, whereas substrates high in insoluble fiber (e.g., canola meal, corn fiber, oat bran, and wheat middlings) have lower fermentability (Jha and Berrocoso 2015; Navarro et al. 2018b, 2019). The composition of dietary fiber appears to affect fermentability more than the total amount of dietary fiber being consumed by pigs indicating that the pigs capacity for fermentation of dietary fiber is not easily overwhelmed (Navarro et al. 2018c, 2019).

### Synthesis of Short-Chain Fatty Acids

Short-chain fatty acids consist of carbon atoms and the SCFA produced from carbohydrate fermentation mainly comprises acetate, propionate, and butyrate (Miller and Wolin 1979). Intestinal microbes require an anaerobic environment, and microorganisms that produce SCFA in the absence of oxygen include *Acetobacter, Clostridium, Kluyveromyces, Propionibacterium,* and *Moorela* (Liang et al. 2012; Ehsanipour et al. 2016; Bhatia and Yang 2017). Fermentation is an energy conservation process where the substrate is less extensively metabolized than in aerobic oxidation. In anaerobic fermentation, only a small amount of energy is extracted for microbial growth and approximately 75% of energy in glucose is retained in SCFA (Bergman 1990). Microbes hydrolyze undigested carbohydrates into monosaccharides during fermentation, and the monosaccharides are subsequently absorbed into the microbial cell and channeled into the pathways of central metabolism (White 2000). Pentoses and hexoses are oxidized via the pentose phosphate pathway and the Emdem–Mayerhoff–Parnas/Entner–Doudoroff pathway, respectively (Jensen and Jørgensen 1994). The end product of both pathways is pyruvate, which is subsequently oxidized for synthesis of SCFA. Other end products of microbial fermentation of dietary fiber include hydrogen, methane, and carbon dioxide (Jensen and Jørgensen 1994).

Two pathways can be used to synthesize acetate from pyruvate (Table 14.2; Figure 14.4). In the first pathway, ATP is generated with no net $H^+$ used. In the second pathway, formate is also generated with a subsequent synthesis of ATP and methane, which allows microbes to get rid of 4 $H^+$.

For propionate synthesis, three different biochemical pathways may be used: (i) randomizing pathway, (ii) acrylate pathway, and (iii) propanediol pathway (Macy and Probst 1979; Hetzel et al. 2003; Scott et al. 2006). In the randomizing pathway, ATP is generated with 4 $H^+$ used per mol of propionate formed from succinate via succinyl-, methylmalonyl-, and propionyl-coA (Figure 14.5; Macy et al. 1978). In the acrylate pathway, no ATP is generated, but 4 $H^+$ is used;

**Table 14.2** Summary of pathways involved in short-chain fatty acid (SCFA) synthesis.

| Pathway | SCFA synthesized | ATP generation | Number of $H^+$ used | By-products |
|---|---|---|---|---|
| Acetate-I | Acetate | 1 | 0 | — |
| Acetate-II | Acetate | 1 | 4 | Methane |
| Randomizing | Propionate | 1 | 4 | — |
| Acrylate | Propionate | 0 | 4 | — |
| Propanediol | Propionate | 1 | 0 | — |
| Acetyl-CoA | Butyrate | 1 | 4 | — |
| Glutarate | Butyrate | 1 | 4 | — |
| 4-Aminobutyrate | Butyrate | 1 | 4 | — |
| Lysine | Butyrate | 1 | 8 | Ammonia |

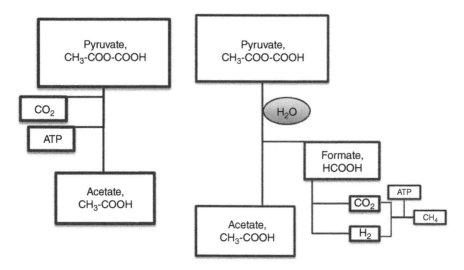

**Figure 14.4**  Two different pathways for synthesis of acetate. Source: Modified from Urriola et al. (2013).

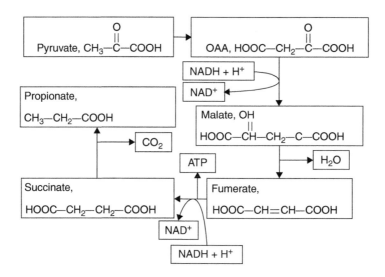

**Figure 14.5**  Synthesis of propionate via randomizing pathway. Source: Modified from Urriola et al. (2013).

therefore, this pathway is used to get rid of H⁺ to maintain redox balance (Figure 14.6). Lactate is converted to propionate with acrylyl CoA serving as an intermediate (Whanger and Matrone 1967). The metabolic reduction of lactate to propionate provides cells with a route to attain redox balance through acrylate and other metabolizable analogs (Prabhu et al. 2012). Bacteria related to *Roseburia* produce propionate from fucose and rhamnose via the propanediol pathway with propanediol as an intermediate (Figure 14.7; Scott et al. 2006). This pathway is present in the human colon, but it is not clear if this pathway is also used by pigs (Scott et al. 2006). In the propanediol pathway, one ATP is generated per mol of propionate formed (Walter et al. 1997).

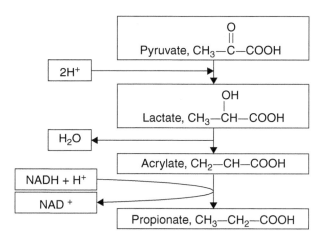

**Figure 14.6** Synthesis of propionate via acrylate pathway. Source: Modified from Urriola et al. (2013).

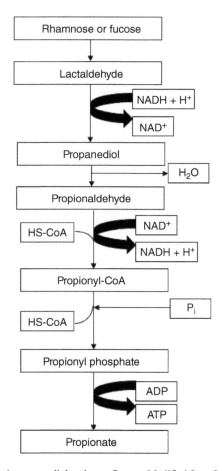

**Figure 14.7** Synthesis of propionate via propanediol pathway. Source: Modified from Scott et al. (2006).

For butyrate synthesis, four pathways can be used with crotonyl CoA as a common intermediate: (i) the acetyl-CoA, (ii) the glutarate, (iii) the 4-aminobutyrate, and (iv) the lysine pathway (Vital et al. 2014). The acetyl-CoA pathway is believed to be the most common pathway used for butyrate production using pyruvate/acetyl-CoA (derived from complex polysaccharides) as a substrate. The three remaining butyrate-producing pathways are less abundant, and these pathways are used by *Firmicutes, Spirochaetaceae,* and *Bacteroidetes* (Vital et al. 2014) which are present in the pig colon (Looft et al. 2014). In the acetyl-CoA pathway, glutarate pathway, and 4-aminobutyrate pathway, microbes generate one ATP and get rid of 4 $H^+$ per mole of butyrate formed to maintain redox balance (Figure 14.8–14.10; Vital et al. 2014). In the lysine pathway, one ATP is generated per mole of butyrate formed, and 2 moles of ammonia ($NH_3$) are excreted as well, which results in a loss of 8 $H^+$ (Figure 14.11; Perret et al. 2011).

The typical ratio of SCFA in feces is 60:20:20 acetate, propionate, and butyrate, respectively (Flint et al. 2012). However, different sources of dietary fiber may affect this ratio and acetate may range from 60 to 90%, propionate from 10 to 30%, and butyrate from 1 to 20% of the total SCFA synthesized (Titgemeyer et al. 1991; Marsono et al. 1993). As an example, the proportion of acetate was greater in feces from pigs fed diets containing DDGS compared with pigs fed diets without DDGS (Espinosa et al. 2019) although that is not always the case (Urriola and Stein 2010). However, SCFA recovered in feces represent the SCFA that was not absorbed by the pig, and it is uncertain if all SCFA are absorbed with the same efficiency, but if that is not the case, the ratio between synthesized acetate, propionate, and butyrate may be different from the ratio between the 3 SCFA that is present in the feces. Fecal bulk is heavily influenced by dietary fiber with increasing quantities of feces being

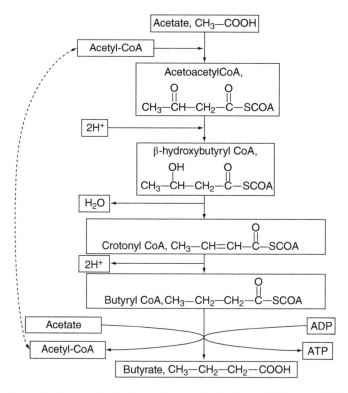

**Figure 14.8** Synthesis of butyrate via acetyl-CoA pathway. Source: Modified from Urriola et al. (2013).

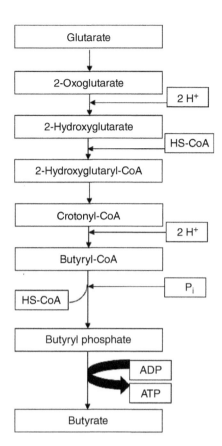

**Figure 14.9** Synthesis of butyrate via glutarate pathway. Source: Modified from Vital et al. (2014).

voided by pigs as dietary fiber concentration increases. It is, therefore, not unusual that the concentration of SCFA is less in feces from pigs fed a high-fiber diet than from pigs fed a low-fiber diet. However, if the concentration of each SCFA is multiplied by the total quantity of feces being voided, pigs fed high-fiber diets always excrete more SCFA than pigs fed low fiber diets (Jaworski et al. 2017).

### Absorption of Short-Chain Fatty Acids

The SCFA synthesized are released from the microbial cell into the intestinal lumen to be absorbed by pigs and mainly used as a source of energy. The majority of the production and absorption of SCFA in the gastrointestinal tract of pigs takes place in the large intestine (Barcroft et al. 1944; Elsden et al. 1946). Less than 1% of SCFA infused in the cecum are recovered in the feces of pigs (Jørgensen et al. 1997b) indicating that SCFA absorption in the large intestine is highly efficient. Short-chain fatty acids are absorbed through passive diffusion, anion exchange (carrier-mediated), or by transporter-mediated absorption (Kirat and Kato 2006; Wong et al. 2006). Passive diffusion requires the protonated form of each SCFA, but only 1% of total SCFA in the intestinal lumen is protonated (Cook and Sellin 1998). However, protons (from Na/H exchange, K$^+$H$^+$-ATPase, or bacterial metabolic activity) are exchanged at the apical epithelium where the pH is lower compared with the center of the lumen,

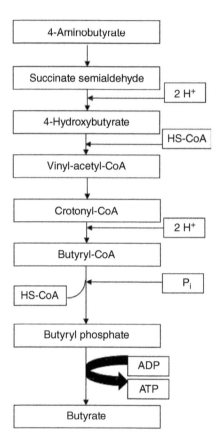

**Figure 14.10**   Synthesis of butyrate via 4-aminobutyrate pathway. Source: Modified from Vital et al. (2014).

and therefore, almost 60% of SCFA are protonated by the time they are present at the apical epithelium (Cook and Sellin 1998). The carrier-mediated mechanism exchanges bicarbonate for SCFA at the intestinal epithelium (Cook and Sellin 1998). Active SCFA transporters are located throughout the gastrointestinal tract, but abundance of transporters may be dependent on SCFA production (Hadjiagapiou et al. 2000; Gill et al. 2005). Active transporters of SCFA include monocarboxylate transporter (MCT-1) and sodium-coupled MCT-1 (Welter and Claus 2008; Thangaraju et al. 2008). The MCT-1 is a proton-coupled transporter of SCFA, lactate, and other monocarboxylates into colonocytes (Welter and Claus 2008). The sodium MCT-1, on the other hand, is a sodium-coupled electrogenic transporter with high affinity for butyrate (Thangaraju et al. 2008). The SLC5A8 form of the sodium-coupled MCT-1 was first identified as a tumor suppressor and is expressed on the apical epithelium, which thereby allows access to butyrate for transport (Ganapathy et al. 2008). The MCT-1 transporter is present in human and pig colonocytes (Ritzhaupt et al. 1998), but it is not clear if the SLC5A8 transporter is also present in pig colonocytes (Parada Venegas et al. 2019).

Absorption of other nutrients from the diet may also be facilitated during SCFA absorption (Yen 2001). Minerals bound to amino acids or SCFA can be absorbed through solvent drag, which involves movement of minerals through the tight junction pores (Goff 2018). Minerals suspended in the water can be absorbed when water passes through the pores within the protein meshwork forming the tight junction (Goff 2018).

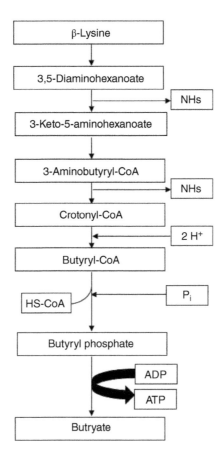

**Figure 14.11** Synthesis of butyrate via lysine pathway. Source: Modified from Vital et al. (2014).

## *Metabolism of Short-Chain Fatty Acids*

Uptake of individual SCFA by various tissues is regulated through the enzymatic activation of SCFA. The relative activities of key enzymes (i.e., acetyl-CoA synthetase, propionyl-CoA synthetase, and butyryl-CoA synthetase) in specific tissues ensure that butyrate is mainly metabolized in intestinal epithelium, whereas propionate and acetate are mainly metabolized in the liver (Ash and Baird 1973; Bergman 1990). Acetate, propionate, and butyrate are activated with coenzyme A, and this reaction is catalyzed by acetyl-CoA synthetase, propionyl-CoA synthetase, and butyryl-CoA synthetase, respectively (Bergman 1990). The activated SCFA are channeled into pathways of central metabolism to be metabolized and used as energy for maintenance and growth, as well as substrates for lipogenesis (Figure 14.12). Acetate accounts for at least 90% of the total SCFA in general circulation indicating selective metabolism of butyrate and propionate by colonocytes and hepatocytes, respectively (Robertson 2007).

The majority of the absorbed4 butyrate is oxidized to carbon dioxide and ketone bodies in the colonic mucosa, which indicates that colonocytes prefer butyrate as a source of energy instead of glucose and glutamine (Bergman 1990; Elia and Cummings 2007). The residual butyrate is mainly metabolized by the liver and used as a substrate for lipogenesis (Bergman 1990). Aside from

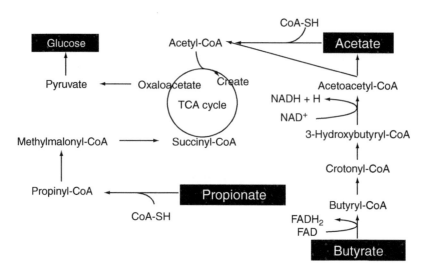

**Figure 14.12**    Oxidation of acetate, propionate, and butyrate. Source: Modified from Urriola et al. (2013).

providing energy, butyrate can regulate cell proliferation and differentiation, which subsequently may prevent colon cancer and adenoma development (Cook and Sellin 1998; Wong et al. 2006).

The activity of propionyl-CoA synthetase is high in the liver of rats and ruminants (Ash and Baird 1973), and as a result, the majority of the absorbed propionate is metabolized in the liver to be used as a substrate for gluconeogenesis (Wong et al. 2006). Biotin and vitamin $B_{12}$ act as coenzymes of propionyl-CoA carboxylase and methylmalonyl-CoA carboxylase, respectively (Marston et al. 1961); therefore, increased concentration of propionate can be detected in plasma if animals are deficient in vitamin $B_{12}$.

Acetate, which is the most abundant SCFA produced from carbohydrate fermentation, is absorbed and transported to the liver (Cook and Sellin 1998). Acetate may also be metabolized by the brain, as well as by skeletal and cardiac muscle (Elia and Cummings 2007). The activity of acetyl-CoA synthetase is high in adipose tissue and in the mammary gland, which favors acetate to be used as a substrate for fatty acid synthesis (Wong et al. 2006).

## Contribution of Energy from Fermentation

### *Fermentation and Digestibility of Dietary Fiber*

Because fiber is not digested by endogenous enzymes, digestion of dietary fiber refers to disappearance of fiber that are degraded by microbial fermentation in the intestinal tracts of pigs. Fermentation of dietary fiber varies among feed ingredients because of different types of fiber in the feed ingredients (Bindelle et al. 2009). The AID of dietary fiber in high fiber feed ingredients fed to pigs ranges from −10 to 62% indicating that some microbial activity may take place when pigs are fed certain diets (Bach Knudsen and Jørgensen 2001). However, the apparent total tract digestibility (**ATTD**) of cellulose varies between 23 and 65% in barley, 24 and 60% in wheat and wheat by-products, 10 and 84% in rye and rye fractions, and between 13 and 42% in bran and hulls of wheat, corn, and oats (Bach Knudsen and Jørgensen 2001). The ATTD of cellulose in a corn and soybean meal diet fed to growing pigs is 72.2%, but if high fiber ingredients including wheat middlings or soybean hulls are added to the diet, the ATTD is reduced to around 60% (Jaworski and Stein 2017). The ATTD of cellulose and hemicellulose in corn is less than in soybean meal or some high fiber ingredients including canola meal, copra expellers, and sugar beet pulp (Navarro et al. 2018b). The ATTD of NDF in corn, wheat, and sorghum is 63, 58, and 65%, respectively, and the ATTD of ADF is 56, 21, and 61%, respectively (Rodriguez et al. 2020). The ATTD of NDF in canola meal is only 39%, but the ATTD of NDF in corn DDGS, corn gluten meal, copra expellers, and sugar beet pulp ranged from 73 to 83% (Navarro et al. 2018b). The average ATTD of NDF in different sources of canola meal, 00-rapeseed meal, and 00-rapeseed expellers is 59.1, 59.5, and 61.4, respectively, and the ATTD of ADF was 41.4, 45.1, and 47.5%, respectively (Maison et al. 2015). The AID of TDF in corn DDGS varies from 11.4 to 30.8%; the ATTD of TDF varies from 29.3 to 57.0% (Urriola et al. 2010). The ATTD of SDF in most feed ingredients fed to pigs is relatively greater compared with IDF (Jaworski and Stein 2017; Navarro et al. 2018b).

Unlike endogenous losses of proteins, amino acids, fat, and minerals, endogenous fiber cannot be secreted into the intestinal tract of pigs because fiber by definition originate from plants. However, fiber may be analyzed in both the ileal and fecal outputs from pigs fed a fiber-free diet (Cervantes-Pahm et al. 2014a; Abelilla 2018). The reason for this observation is that certain nondietary fiber components in ileal digesta and feces are analyzed as fiber (Cervantes-Pahm et al. 2014a; Montoya et al. 2015, 2016). Indeed, mucin and microbial matter contribute to the ileal output analyzed as both SDF and IDF, but most of the mucin is fermented in the hindgut of pigs and the major endogenous contribution to fiber in the feces, therefore, is microbial matter, which is included in the IDF fraction (Montoya et al. 2015). Whereas this "endogenous" fiber is simply a result of the inability of current fiber analyses to distinguish between endogenous and microbial matter and unfermented plant material in ileal digesta and feces, the practical implication is that in fiber digestibility experiments, a correction for the "endogenous" fiber is needed. Values for fiber fermentation should, therefore, be expressed as standardized ileal digestibility (**SID**) or standardized total tract digestibility of fiber after correction for the endogenous components in ileal digesta and feces (Cervantes-Pahm et al. 2014a). A further consequence of the fermentation of mucin and other endogenous fractions in the hindgut is that some of the absorbed SCFA in the hindgut originate from nondietary substrates (Montoya et al. 2017). If the contribution of SCFA from fermentation of dietary fiber needs to be determined, it therefore, is necessary to correct for the SCFA produced via fermentation of nondietary substrates.

Soluble fiber is easily fermented by pigs, whereas insoluble fiber has a lower fermentability (Urriola et al. 2010; Acosta et al. 2020). Most of the SDF is fermented in the small intestine or the cecum, whereas IDF is primarily fermented in the hindgut (Jaworski and Stein 2017). Whereas the source of dietary fiber greatly influence fermentability of fiber (Cervantes-Pahm et al. 2014b; Navarro et al. 2018b; Zhao et al. 2020), the quantity of fiber from a specific source does not impact fiber fermentability (Navarro et al. 2019), and as a consequence, the greater inclusion rate of dietary fiber in a diet is, the greater is the contribution of SCFA to the total energy balance of the pig (Iyayi and Adeola 2015; Navarro et al. 2018a).

### *Amount of Short-Chain Fatty Acids Produced per Gram of Fermented Fiber*

Each gram of fermented fiber may yield different amounts of SCFA depending on the type of fiber that is fermented. Alpha galactosides such as raffinose and stachyose from soybeans yield more gases ($CH_4$ and $H_2$), cause flatulence, and produce less SCFA during fermentation in the large intestine than fermentation of cellulose and hemicellulose (Liener 1994). The relative synthesis of acetate, propionate, and butyrate vary slightly among different sources of fiber (Topping and

Clifton 2001; Zhao et al. 2020), but for practical purposes, the ratios between acetate, propionate, and butyrate may be assumed to be constant in pigs (de Lange 2008). Among the 3 SCFA, acetate is the most dominant, followed by propionate and butyrate. In vitro production rate of SCFA from colon digesta of pigs is greater if feed ingredients contain soluble fiber than if the substrate is only insoluble (den Besten et al. 2013). Fermentation of branched chained amino acids yields branched chain fatty acids (isobutyrate, isovalerate, and valerate) so the concentration of the branched chained fatty acids depends on the degree to which branched chained amino acids were fermented. In most circumstances, the production of the three branched chained fatty acids is less than 5% of the total SCFA production.

### *Moles of ATP Produced per Mole of Short-Chain Fatty Acids Absorbed and Metabolized*

Most of the SCFA that is synthesized in the intestinal tract is absorbed and used as an energy source or a regulator for energy metabolisms including fatty acid synthesis, lipolysis, cholesterol synthesis, gluconeogenesis, and glucose uptake (den Besten et al. 2013). When fully oxidized, acetate, propionate, and butyrate provide 10, 18, and 28 moles of ATP, respectively. However, because of the greater proportion of acetate produced compared with butyrate, acetate provides as much total energy as butyrate (Roediger 1982; Clausen and Mortensen 1995; Jørgensen et al. 1997a).

### Negative Effects of Fiber on Energy and Nutrient Digestibility

### *Possible Mechanisms*

The use of high fiber ingredients in pig diets has increased to reduce feed cost in diet formulation (Woyengo et al. 2014), but increased concentration of fiber in the diet may reduce digestibility of other nutrients. The possible negative effects of fiber on nutrient digestibility depends on the type of fiber and on inclusion rate. SDF is easily fermented, but because soluble fiber increases water-binding capacity in the intestine of pigs, the viscosity, and bulkiness of digesta may increase, which reduces the time of exposure to the digestive enzymes (Cervantes-Pahm et al. 2014a; Lindberg 2014). In contrast, insoluble fiber may increase the passage rate of the digesta, which may also reduce nutrient digestibility by decreasing exposure time to digestive enzymes (Lindberg 2014; Acosta et al. 2020). It is also possible that endogenous losses of energy and nutrients increase as dietary concentrations of fiber increases.

### *Effects on Energy Digestibility*

Increasing fiber concentration by increasing the inclusion of wheat bran or corn DDGS in the diet progressively decreased digestibility of dry matter, starch, and energy (Wilfart et al. 2007; Gutierrez et al. 2016; Acosta et al. 2020). The ATTD of gross energy was reduced by increasing the inclusion rate of wheat bran (Huang et al. 2015), palm kernel meal (Huang et al. 2018), and canola meal (Sanjayan et al. 2014). Inclusion of canola meal, corn germ meal, sugar beet pulp, and wheat middlings up to 30% also decreased ATTD of gross energy in diets fed to pigs (Navarro et al. 2018c). The reduction in dietary energy digestibility was associated with a reduction in dry matter and organic matter digestibility (Wilfart et al. 2007; Berrocoso et al. 2015). Adding a mixture of wheat bran, maize bran, soybean hulls, sugar beet pulp, canola meal, corn germ meal, and wheat middlings to increase dietary fiber concentrations also reduced total tract energy digestibility in pigs with a corresponding reduction in carbohydrate digestibility (Le Gall et al. 2009). Increasing dietary fiber by adding sugar beet pulp to diets decreased the ATTD of gross energy (Zhang et al. 2013; Zhang et al. 2018). The degree of energy reduction was calculated to be 1% for each 1% increase in NDF concentration (Le Gall et al. 2009). This negative effect of NDF on energy digestibility and energy concentrations in feed ingredients and diets fed to pigs has often been demonstrated (Noblet and Perez 1993; Sol et al. 2017; Zeng et al. 2019; Choi et al. 2020). The solubility of fiber influences energy digestibility due to differences in fermentability between soluble and insoluble fiber (Urriola et al. 2010; Jaworski and Stein 2017) and the presence of lignin in dietary fiber may reduce energy digestibility (Wenk 2001). The reduction in energy digestibility of diets is a consequence of (i) the substitution of digestible protein and carbohydrates such as starch with protein and carbohydrates bound to less digestible cell wall components of the fiber source, (ii) the influence of the physio-chemical characteristics of the fiber on the digestion and absorption processes of the dietary nutri-ents, and (iii) the physiological effects of fiber on the gastrointestinal tract (Le Gall et al. 2009).

### *Effects on Amino Acid Digestibility*

The effect of dietary fiber on amino acid digestibility depends on the type of fiber used in diets and the inclusion rate. Addition of 7.5% citrus pectin to a soybean meal-cornstarch based diet reduced the AID of protein and amino acids by 8.2–28.7 percentage units (Mosenthin et al. 1994). A reduc-tion in the SID of crude protein and amino acids was also observed when 4 or 8% apple pectin was added to a wheat-corn-soybean meal-based diet (Buraczewska et al. 2007). A linear decrease in ileal N digestibility was observed when purified NDF that was processed from wheat bran was added at increasing levels to a soy isolate-cornstarch based diet fed to pigs (Schulze et al. 1994). Adding 15% purified wheat NDF also reduced the AID of amino acids by 2–5.5 percentage units except for the AID of Cys, Ala, and Gly, which were reduced by 18, 16, and 12 percentage units, respectively (Lenis et al. 1996). Increasing the concentration of NDF from 2.72 to 4.16% by adding graded levels of soy hulls (3–9%) to soybean meal-cornstarch based diets also induced a linear or quadratic reduc-tion in AID and SID of most amino acids (Dilger et al. 2004). Likewise, the AID and SID of most amino acids were linearly reduced by increasing the concentration of TDF from 12.1 to 21.2% by adding alfalfa meal from 0 to 20% in corn-soybean meal-wheat bran diets (Chen et al. 2015). Addition of coarse rapeseed meal and rapeseed hulls also decreased the AID of amino acids in the diet containing wheat and barley fed to pigs (Pérez de Nanclares et al. 2017). However, when 10% cellulose and barley straw was added to a soybean meal-cornstarch based diet, the AID of amino acids except Leu and Gly was not reduced (Sauer et al. 1991). A reduction in AID of amino acids was also not observed when graded levels of purified cellulose (4.3–13.3 %) were added to a soy-bean meal-cornstarch based diet fed to young pigs (Li et al. 1994) and addition of cellulose (10%) to a semipurified diet reduced the AID of crude protein and amino acids (Cervantes-Pahm et al. 2014a; Liu et al. 2016). In contrast, when carboxymethylcellulose was added to the diets, SID of crude protein and amino acids increased (Larsen et al. 1994; Bartelt et al. 2002; Fledderus et al. 2007). Insoluble and poorly fermentable fiber such as cellulose impact protein digestibility through its water holding property whereas soluble fibers such as carboxymethylcellulose and pectin mediates its effects through its viscosity property. Summarized data indicated that each 1% increase in NDF in the diet resulted in a reduction of 0.03–0.08% of AID of crude protein (Dégen et al. 2007).

Dietary fiber can reduce the efficiency of amino acids utilization by impairing the digestion pro-cess, decreasing amino acid absorption or increasing endogenous amino acid losses (Mosenthin et al. 1994; Mathai et al. 2016). When a diet containing 20% purified wheat bran fiber was fed to pigs, an increase in ileal N flow was observed, but 59% of the N was of endogenous origin indicating increased synthesis of mucin as fiber increased (Schulze et al. 1995). Addition of graded levels of pea inner fibers to protein-free diets resulted in an exponential increase in ileal N flow, which was correlated with increased water-holding capacity of the diet (Leterme et al. 1998). The ileal flow of epithelial cells also increased exponentially with a corresponding linear increase in crude mucin and bacteria (Leterme et al. 1998). When a viscous and nonfermentable fiber (i.e., carboxymethylcellulose) was added, mucin secretion and endogenous N loss also increased, but there was no change in ileal bacterial population (Bartelt et al. 2002; Piel et al. 2005). However, an increase in some ileal populations of bacteria was observed by Owusu-Asiedu et al. (2006) when viscous and fermentable fibers such as guar gum was fed to pigs. In contrast, adding cellulose that is an insoluble and poorly fermentable fiber at 3.31–16.5% to the diet did not induce an increase in endogenous amino acid losses, which may be the reason for the absence of a reduction in the AID of amino acids of the diets when cellulose was added (Li et al. 1994). The level and the source of dietary fiber are important factors that influence endogenous amino acid losses (Sauer and Ozimek 1986), and inclusion of cellulose may reduce the AID of amino acids only if a certain threshold level is exceeded (Li et al. 1994).

### *Effects on Pancreatic Enzymes and Mucin Production*

Effects of fiber on pancreatic secretions and enzyme activity may be modulated by the physico-chemical properties of fiber, which results in differences in endogenous losses of nutrients. Barley or wheat-based diets increased bile and pancreatic juice secretions compared with cornstarch, casein, or cellulose-based diets without affecting enzyme output (Low 1989). Addition of flaxseed meal or oat hulls that are high in fiber to corn-soybean meal diets induced an increase in bile acid secretion by pigs (Ndou et al. 2019). However, when 40% wheat bran was added to isonitrogeous and isocaloric diets, chymotrypsin and trypsin secretions were greater than when pigs were fed diets without wheat bran (Langlois et al. 1986). In contrast, addition of pectin to soybean meal-based diets did not increase pancreatic secretions and did not affect secretions and enzyme activities of trypsin and chymotrypsin (Mosenthin et al. 1994). However, dietary carboxymethylcellulose may reduce pepsin activity in the stomach without affecting trypsin and chymotrypsin activities (Larsen et al. 1993).

The goblet cells of the gastrointestinal tract secrete mucin, which is a high molecular weight glycoprotein that lubricates the epithelial surface and protects the gut from physical abrasions, chemical aggressions, and microbial pathogenic attachments that may compromise gut health (Forstner and Forstner 1994; Tanabe et al. 2006). Mucin also plays an important role in digestion and absorption of nutrients and changes in mucin secretions may change the dynamics of absorption of dietary nutrients and endogenous molecules in the gut (Tanabe et al. 2006).

Dietary fiber may increase mucin secretion, and intestinal concentration of amino sugars in mucin (glucosamine and galactosamine) increased linearly as graded levels of wheat straw, corn cobs, and wood cellulose were added to protein-free diets fed to pigs (Mariscal-Landín et al. 1995). Supplementation of 5% citrus fiber to a purified diet also produced a significant increase in small intestinal mucin secretion (Satchithanandam et al. 1990). Because increasing dietary fiber increases the synthesis of mucin, which also contributes to the increment in endogenous loss of protein, the ideal Thr to Lys ratio for growing pigs may be different compared with pigs fed a diet containing less fiber (Mathai et al. 2016; Wellington et al. 2019).

In the stomach, the bulk forming property and the fermentability of fiber did not affect mucin secretion, but in the cecum, the fermentability of fructooligosaccharide and beet pulp increased mucin secretion (Tanabe et al. 2006). A similar observation was reported by Libao-Mercado et al. (2007) when the addition of pectin-stimulated mucin and mucosal protein synthesis in the colon, but not in the jejunum.

The mucin molecule is composed of a protein backbone with attached carbohydrate side chains. One of the two regions of the mucin molecule has a protein backbone composed of Pro, Ser, and Thr. This region is resistant to proteolytic digestion because 80% of the protein backbone is protected by oligosaccharides of which the carbohydrate components are fucose, galactose, N-acetyl galactosamine, N-acetyl glucosamine, and sialic acids (Montagne et al. 2004). Because of the proteolytically resistant region of mucin, it is poorly digested and, therefore, contributes significant amounts of endogenous crude protein and carbohydrates in the ileal digesta (Lee et al. 1988; Lien et al. 1997; Montoya et al. 2015). Therefore, values for the ileal digestibility of Thr, Pro, and Ser are relatively lower compared with the digestibility of other amino acids.

Endogenous protein and amino acids recovered from the ileal digesta are mostly from pancreatic enzymes, epithelial cells, bacterial cells, and mucin, whereas endogenous carbohydrates are mostly from mucin (Lien et al. 1997; Miner-Williams et al. 2009). Endogenous protein and amino acids that are not reabsorbed before the end of the ileum are utilized by microbes in the hind gut of pigs (Souffrant et al. 1993; Libao-Mercado et al. 2009). Very little mucin is recovered in the feces, which demonstrates that mucin in the hindgut is fully fermented (Lien et al. 2001; Montoya et al. 2015).

### Effects on Utilization of Other Nutrients

#### Carbohydrates
Addition of wheat bran to a barley-based diet did not affect starch digestibility (Högberg and Lindberg 2004) and adding 20% or 40% wheat bran to a cereal-based diet did not affect starch digestibility (Wilfart et al. 2007). Ninety nine percent of the starch was digested in the small intestine, and no starch was detected in the feces (Högberg and Lindberg 2004; Wilfart et al. 2007). In contrast, the addition of graded levels of wheat bran, maize bran, soybean hulls, or sugar beet pulp to increase dietary fiber reduced the ATTD of dietary fiber (Le Gall et al. 2009; Huang et al. 2018; Zhao et al. 2018). The ATTD of NDF, ADF, cellulose, and hemicellulose was reduced if lignocellulose was added to a barley-wheat-soybean meal-based diet (Barszcz et al. 2019). The ATTD of TDF was reduced by adding canola meal or wheat middlings from 0 to 30% to diets, but the ATTD of TDF was increased by adding sugar beet pulp, which is more easily fermented than other fiber sources (Navarro et al. 2019). The ATTD of sugars and dietary fiber was reduced by increasing the inclusion rate of rapeseed meal in diets fed to pigs (Pérez de Nanclares et al. 2019). The addition of guar gum also reduced glucose absorption from the jejunum by 50% (Rainbird et al. 1984). Similar observations were reported by Nunes and Malmlöf (1992) and Owusu-Asiedu et al. (2006) who observed that guar gum, but not cellulose, reduced plasma glucose concentration in pigs. The digesta viscosity induced by guar gum may have reduced the diffusion rate of glucose from the lumen to the epithelial cells causing a reduction in the absorption of glucose (Rainbird et al. 1984; Kritchevsky 1988).

#### Lipids
Dietary fiber may impede micelle formation and directly inhibit lipolytic activity, which may contribute to the reduction in digestibility of fat (Schneeman and Gallaher 2001). Addition of 20 or 40% wheat bran to a cereal-based diet reduced the ATTD of ether extract by 7–12% units compared with

the control diet (Wilfart et al. 2007). Addition of beet pulp or soybean hulls to a basal diet also reduced the AID and ATTD of fat, whereas addition of wheat bran did not affect digestibility (Graham et al. 1986; Lyu et al. 2018). In contrast, adding a combination of triticale, wheat, and wheat bran as a source of fiber to cereal-based diets improved the AID and ATTD of fat compared with the control diet (Högberg and Lindberg 2004). Increasing purified NDF from 3 to 11% in diets fed to pigs did not affect the ATTD and TTTD of fat (Kil et al. 2010). These observations suggest that the solubility of diets containing different sources of fiber influences fat digestibility because when a mixture of wheat bran, maize bran, soybean hulls, and sugar beet pulp was added at graded levels to a low-fiber diet, the ATTD of fat was not affected despite increasing levels of TDF in the diet (Le Gall et al. 2009). It is also possible that there is less stimulatory effect of purified NDF on fermentation of intestinal microbes, which produce the endogenous fat. The level of dietary fiber inclusion also influences lipid digestibility because decreasing fat digestibility was observed as coconut expeller, soybean hulls, or sugar beet pulp was added at graded levels to a low-fiber control diet (Canh et al. 1998).

*Minerals*

Dietary fiber is composed of polysaccharides that may bind minerals, but the results of studies on the effect of dietary fiber on mineral digestibility are not consistent. Addition of 6% cellulose depressed the apparent absorption of Ca, P, Mg, and K. Serum concentrations of Ca, P, Cu, and Zn per unit of mineral ingested were also lower in sows fed high fiber diets containing a combination of corn cobs and wheat bran or oats and oat hulls compared with corn-soybean meal diets (Girard et al. 1995). Likewise, insoluble fiber or phytate in diets may decrease the absorption of Ca or P due to less transit time in the gut (Nortey et al. 2007; Hill et al. 2008). In contrast, addition of oat hulls, soybean hulls, and alfalfa meal did not affect total tract Ca, P, Zn, or Mn digestibility (Moore et al. 1988). Likewise, the AID and ATTD of Ca, P, Mg, and Zn were not affected by addition of 6% inulin to diets fed to pigs (Vanhoof and De Schrijver 1996). The presence of hulls in rapeseed meal did not affect the ATTD of Ca and P (Bournazel et al. 2018). The AID of ash was also not affected by the addition of 20 or 40% wheat bran to a low fiber diet, but the ATTD of ash was reduced if high levels of wheat bran were added to the diet (Wilfart et al. 2007; Zhao et al. 2018).

### Effect of Dietary Fiber on Nitrogen Excretion and Manure Characteristics

A major impact of dietary fiber on nitrogen excretion in pigs is the shift of nitrogen excretion from the urine to the feces, which results in a reduction of the ratio between urine nitrogen excretion and fecal nitrogen excretion. With enhanced microbial fermentation in the hindgut because of the presence of fiber, the ammonia produced by the fermentation of dietary and endogenous protein is used for bacterial metabolism and growth (Zervas and Zijlstra 2002). Therefore, there is an overall reduction in the concentration of ammonia available for absorption in the blood to go to the liver for urea synthesis (Mroz et al. 2000; Zervas and Zijlstra 2002). As a consequence, urinary nitrogen excretion is reduced. The ratio between urine and fecal nitrogen excretion was reduced by increasing inclusion of corn cob, grass hay, lucerne hay, corn stover, and sunflower husk in diets fed to growing pigs (Mpendulo et al. 2018). However, the shift from urinary to fecal nitrogen excretion is dependent on different sources of fiber because increasing inclusion of sugar beet pulp in diets linearly decreased urinary to fecal nitrogen excretion ratio (Bindelle et al. 2009). The magnitude of response for the shift from urine to fecal nitrogen excretion also depends on the fiber sources because the urine to feces nitrogen excretion ratio in pigs fed diets containing barley is lower compared with diets containing corn or wheat; diets containing beet pulp has a lower urine to feces nitrogen excretion ratio than diets containing tapioca meal (Canh et al. 1997; Leek et al. 2007). Fermentable fiber such as pectin and potato starch has a stronger impact on shifting nitrogen excretion from urine to feces compared with poorly fermentable fiber such as cellulose (Pastuszewska et al. 2000). A gradual substitution of sugar beet pulp with oat hulls also increased the urinary to fecal nitrogen excretion ratio (Bindelle et al. 2009). This reduction in urinary nitrogen excretion is an advantage in the light of environmental concerns about ammonia emission from pig production systems (Aarnink and Verstegen 2007).

Increasing fiber in the diet linearly increases the amount of daily fecal matter excretion in pigs (Moeser and van Kempen 2002). However, there is a corresponding reduction in fecal dry matter with increasing fiber intake suggesting a significant contribution of water to the fecal bulk from pigs fed high-fiber diets (Canh et al. 1998). Manure pH is also reduced with the addition of dietary fiber to the diet and manure pH of pigs fed diets with soybean hulls and beet pulp was lower compared with pigs fed a control diet without soybean hulls and sugar beet pulp (Mroz et al. 2000). The manure from pigs fed diets containing 22% NDF from soybean hulls also had lower pH compared with the manure from pigs fed diets containing 6 or 12% NDF (Moeser and van Kempen 2002). The reduction in the pH of the manure was attributed to the presence of high concentrations of SCFA in the feces depending on the level and source of fiber (Canh et al. 1998; Nguyen et al. 2019b).

## References

Aarnink, A. J. A., and M. W. A. Verstegen. 2007. Nutrition, key factor to reduce environmental load from pig production. Livest. Sci. 109:194–203. doi:https://doi.org/10.1016/j.livsci.2007.01.112.

Abdel-Rahman, M. A., Y. Tashiro, and K. Sonomoto. 2013. Recent advances in lactic acid production by microbial fermentation processes. Biotechnol. Adv. 31:877–902. doi:https://doi.org/10.1016/j.biotechadv.2013.04.002.

Abelilla, J. J. 2018. Fermentation and energetic value of fiber in feed ingredients and diets fed to pigs. PhD. Diss. Univ. Illinois, Urbana-Champaign, IL, US. https://nutrition.ansci.illinois.edu/sites/default/files/DissertationAbelilla.pdf.

Abelilla, J. J., and H. H. Stein. 2019. Degradation of dietary fiber in the stomach, small intestine, and large intestine of growing pigs fed corn- or wheat-based diets without or with microbial xylanase. J. Anim. Sci. 97:338–352. doi:https://doi.org/10.1093/jas/sky403.

Acosta, J. A., H. H. Stein, and J. F. Patience. 2020. Impact of increasing the levels of insoluble fiber and on the method of diet formulation measures of energy and nutrient digestibility in growing pigs. J. Anim. Sci. 98. doi:https://doi.org/10.1093/jas/skaa130.

Albersheim, P., A. Darvill, K. Roberts, R. Sederoff, and A. Staehelin. 2011. Plant cell walls. Garland Science, Taylor & Francis Group, New York, NY.

Alyassin, M., and G. M. Campbell. 2019. Challenges and constraints in analysis of oligosaccharides and other fibre components. In: M. Alyassin, and G. M. Campbell, editors, The value of fibre engaging the second brain for animal nutrition. Wageningen Academic Publishers, Wageningen, The Netherlands. p. 257–272.

Anguita, M., N. Canibe, J. F. Pérez, and B. B. Jensen. 2006. Influence of the amount of dietary fiber on the available energy from hindgut fermentation in growing pigs: Use of cannulated pigs and in vitro fermentation. J. Anim. Sci. 84:2766–2778. doi:https://doi.org/10.2527/jas.2005-212.

AOAC Int. 2007. Official methods of analysis of AOAC Int. 18th ed. Rev. 2nd. Natl. Acad. Press, Gaithersburg, MD.

AOCS. 2006. Official methods and recommended practices of the AOCS. 5th. Am. Oil Chem. Soc., Champaign, IL.

Appeldoorn, M. M., M. A. Kabel, D. Van Eylen, H. Gruppen, and H. A. Schols. 2010. Characterization of oligomeric xylan structures from corn fiber resistant to pretreatment and simultaneous saccharification and fermentation. J. Agric. Food. Chem. 58:11294–11301. doi:https://doi.org/10.1021/jf102849x.

Arvill, A., and L. Bodin. 1995. Effect of short-term ingestion of konjac glucomannan on serum cholesterol in healthy men. Am. J. Clin. Nutr. 61:585–589. doi:https://doi.org/10.1093/ajcn/61.3.585.

Ash, R., and G. D. Baird. 1973. Activation of volatile fatty acids in bovine liver and rumen epithelium. Evidence for control by autoregulation. Biochem. J. 136:311–319. doi:https://doi.org/10.1042/bj1360311.

Bach Knudsen, K. E. 1997. Carbohydrate and lignin. contents of plant materials used in animal feeding. Anim. Feed Sci. Technol. 67:319–338. doi:https://doi.org/10.1016/S0377-8401(97)00009-6.

Bach Knudsen, K. E. 2001. The nutritional significance of "dietary fiber" analysis. Anim. Feed Sci. Technol. 90:3–20. doi:https://doi.org/10.1016/S0377-8401(01)00193-6.

Bach Knudsen, K. E. 2011. Triennial growth symposium: Effects of polymeric carbohydrates on growth and development in pigs. J. Anim. Sci. 89:1965–1980. doi:https://doi.org/10.2527/jas.2010-3602.

Bach Knudsen, K. E., and H. Jørgensen. 2001. Intestinal degradation of dietary carbohydrates – from birth to maturity. In: J. E. Lindberg, and B. Ogle, editors, Digestive physiology of pigs. Cabi Publishing, New York, NY. p. 109–120.

Bach Knudsen, K. E., H. N. Lærke, and H. Jørgensen. 2013. Carbohydrates and carbohydrate utilization in swine. In: L. I. Chiba, editor, Sustainable swine nutrition. John Wiley & Sons, Inc., Ames, IA. p. 109–135.

Barcroft, J., R. A. McAnally, and A. T. Phillipson. 1944. Absorption of volatile acids from the alimentary tract of the sheep and other animals. J. Exp. Biol. 20:120–129.

Barszcz, M., M. Taciak, A. Tuśnio, K. Čobanová, and L. U. Grešáková. 2019. The effect of organic and inorganic zinc source, used in combination with potato fiber, on growth, nutrient digestibility and biochemical blood profile in growing pigs. Livest. Sci. 227:37–43. doi:https://doi.org/10.1016/j.livsci.2019.06.017.

Bartelt, J., A. Jadamus, F. Wiese, E. Swiech, L. Buraczewska, and O. Simon. 2002. Apparent precaecal digestibility of nutrients and level of endogenous nitrogen in digesta of the small intestine of growing pigs as affected by various digesta viscosities. Arch. Tierernahr. 56:93–107. doi:https://doi.org/10.1080/00039420214182.

Bergman, E. N. 1990. Energy contributions of volatile fatty acids from the gastrointestinal tract in various species. Physiol. Rev. 70:567–590. doi:https://doi.org/10.1152/physrev.1990.70.2.567.

Berrocoso, J. D., D. Menoyo, P. Guzmán, B. Saldaña, L. Cámara, and G. G. Mateos. 2015. Effects of fiber inclusion on growth performance and nutrient digestibility of piglets reared under optimal or poor hygienic conditions. J. Anim. Sci. 93:3919–3931. doi:https://doi.org/10.2527/jas.2015-9137.

Bhatia, S. K., and Y. H. Yang. 2017. Microbial production of volatile fatty acids: Current status and future perspectives. Rev. Environ. Sci. Biotechnol. 16:327–345. doi:https://doi.org/10.1007/s11157-017-9431-4.

Bindelle, J., A. Buldgen, M. Delacollette, J. Wavreille, R. Agneessens, J. P. Destain, and P. Leterme. 2009. Influence of source and concentrations of dietary fiber on in vivo nitrogen excretion pathways in pigs as reflected by in vitro fermentation and nitrogen incorporation by fecal bacteria. J. Anim. Sci. 87:583–593. doi:https://doi.org/10.2527/jas.2007-0717.

Bournazel, M., M. Lessire, M. J. Duclos, M. Magnin, N. Même, C. Peyronnet, E. Recoules, A. Quinsac, E. Labussière, and A. Narcy. 2018. Effects of rapeseed meal fiber content on phosphorus and calcium digestibility in growing pigs fed diets without or with microbial phytase. Animal 12:34–42. doi:https://doi.org/10.1017/S1751731117001343.

Buraczewska, L., E. Święch, A. Tuśnio, M. Taciak, M. Ceregrzyn, and W. Korczyński. 2007. The effect of pectin on amino acid digestibility and digesta viscosity, motility and morphology of the small intestine, and on N-balance and performance of young pigs. Livest. Sci. 109:53–56. doi:https://doi.org/10.1016/j.livsci.2007.01.058.

Burton, R. A., and G. B. Fincher. 2009. (1,3;1,4)-β-D-Glucans in cell walls of the poaceae, lower plants, and fungi: a tale of two linkages. Mol. Plant. 2:873–882. doi:https://doi.org/10.1093/mp/ssp063.

Caffall, K. H., and D. Mohnen. 2009. The structure, function, and biosynthesis of plant cell wall pectic polysaccharides. Carbohydr. Res. 344:1879–1900. doi:https://doi.org/10.1016/j.carres.2009.05.021.

Canh, T. T., A. L. Sutton, A. J. A. Aarnink, M. W. A. Verstegen, J. W. Schrama, and G. C. M. Bakker. 1998. Dietary carbohydrates alter the fecal composition and pH and the ammonia emission from slurry of growing pigs. J. Anim. Sci. 76:1887–1895. doi:https://doi.org/10.2527/1998.7671887x.

Canh, T. T., M. W. A. Verstegen, A. J. A. Aarnink, and J. W. Schrama. 1997. Influence of dietary factors on nitrogen partitioning and composition of urine and feces of fattening pigs. J. Anim. Sci. 75:700–706. doi:https://doi.org/10.2527/1997.753700x.

Cervantes-Pahm, S. K., Y. Liu, A. Evans, and H. H. Stein. 2014a. Effect of novel fiber ingredients on ileal and total tract digestibility of energy and nutrients in semi-purified diets fed to growing pigs. J. Sci. Food Agric. 94:1284–1290. doi:https://doi.org/10.1002/jsfa.6405.

Cervantes-Pahm, S. K., Y. Liu, and H. H. Stein. 2014b. Comparative digestibility of energy and nutrients and fermentability of dietary fiber in eight cereal grains fed to pigs. J. Sci. Food Agric. 94:841–849. doi:https://doi.org/10.1002/jsfa.6316.

Cervantes-Pahm, S. K., and H. H. Stein. 2010. Ileal digestibility of amino acids in conventional, fermented, and enzyme-treated soybean meal and in soy protein isolate, fish meal, and casein fed to weanling pigs. J. Anim. Sci. 88:2674–2683.

Champ, M., A. M. Langkilde, F. Brouns, B. Kettlitz, and Y. L. B. Collet. 2003. Advances in dietary fibre characterisation. 1. Definition of dietary fibre, physiological relevance, health benefits and analytical aspects. Nutr. Res. Rev. 16:71–82. doi:https://doi.org/10.1079/NRR200254.

Chen, L., L. Gao, L. Liu, Z. Ding, and H. Zhang. 2015. Effect of graded levels of fiber from alfalfa meal on apparent and standardized ileal digestibility of amino acids of growing pigs. J. Integr. Agric. 14:2598–2604. doi:https://doi.org/10.1016/S2095-3119(14)60924-2.

Chen, Y., M. Y. Xie, Y. X. Wang, S. P. Nie, and C. Li. 2009. Analysis of the monosaccharide composition of purified polysaccharides in Ganoderma atrum by capillary gas chromatography. Phytochem. Anal. 20:503–510. doi:https://doi.org/10.1002/pca.1153.

Choct, M., Y. Dersjant-Li, J. McLeish, and M. Peisker. 2010. Soy oligosaccharides and soluble non-starch polysaccharides: A review of digestion, nutritive and anti-nutritive effects in pigs and poultry. Asian-Australas. J. Anim. Sci. 23:1386–1398. doi:https://doi.org/10.5713/ajas.2010.90222.

Choi, H., J. Y. Sung, and B. G. Kim. 2020. Neutral detergent fiber rather than other dietary fiber types as an independent variable increases the accuracy of prediction equation for digestible energy in feeds for growing pigs. Asian-Australas. J. Anim. Sci. 33:615–622. doi:https://doi.org/10.5713/ajas.19.0103.

Chorvatovičová, D., E. Machová, J. Šandula, and G. Kogan. 1999. Protective effect of the yeast glucomannan against cyclophosphamide-induced mutagenicity. Mutat. Res. Genet. Toxicol. Environ. Mutagen. 444:117–122. doi:https://doi.org/10.1016/S1383-5718(99)00102-3.

Ciolacu, D., F. Ciolacu, and V. I. Popa. 2011. Amorphous cellulose - structure and characterization. Cell. Chem. Tech. 45:13–21.

Clausen, M. R., and P. B. Mortensen. 1995. Kinetic studies on colonocyte metabolism of short chain fatty acids and glucose in ulcerative colitis. Gut 37:684–689. doi:https://doi.org/10.1136/gut.37.5.684.

Cook, S. I, and J. H. Sellin. 1998. Review article: Short chain fatty acids in health and disease. Aliment. Pharmacol. Ther. 12:499–507. doi:https://doi.org/10.1046/j.1365-2036.1998.00337.x.

Cuevas Montilla, E., S. Hillebrand, A. Antezana, and P. Winterhalter. 2011. Soluble and bound phenolic compounds in different bolivian purple corn (*Zea mays* L.) cultivars. J. Agric. Food. Chem. 59:7068–7074. doi:https://doi.org/10.1021/jf201061x.

Dégen, L., V. Halas, and L. Babinszky. 2007. Effect of dietary fibre on protein and fat digestibility and its consequences on diet formulation for growing and fattening pigs: A review. Acta Agric. Scand. Sect. A Anim. Sci. 57:1–9. doi:https://doi.org/10.1080/09064700701372038.

de Lange, C. F. M. 2008. Efficiency of utilization of energy from protein and fiber in the pig - a case for NE systems. Swine Nutrition Conference, Indianapolis, IN, USA. p. 58–72.

den Besten, G., K. van Eunen, A. K. Groen, K. Venema, D. J. Reijngoud, and B. M. Bakker. 2013. The role of short-chain fatty acids in the interplay between diet, gut microbiota, and host energy metabolism. J. Lipid Res. 54:2325–2340. doi:https://doi.org/10.1194/jlr.R036012.

de Vries, R. 2003. Regulation of *Aspergillus* genes encoding plant cell wall polysaccharide-degrading-enzymes; relevance for industrial production. Appl. Microbiol. Biotechnol. 61:10–20. doi:https://doi.org/10.1007/s00253-002-1171-9.

Dilger, R. N., J. S. Sands, D. Ragland, and O. Adeola. 2004. Digestibility of nitrogen and amino acids in soybean meal with added soyhulls. J. Anim. Sci. 82:715–724. doi:https://doi.org/10.2527/2004.823715x.

Dodd, D., and I. K. O. Cann. 2009. Enzymatic deconstruction of xylan for biofuel production. Glob. Change. Biol. Bioenergy. 1:2–17. doi:https://doi.org/10.1111/j.1757-1707.2009.01004.x.

Dreher, M. L. 2001. Dietary fiber overview. In: S. S. Cho, and M. L. Dreher, editors, Handbook of dietary fiber. Marcel Dekker Inc., New York, NY. p. 1–16.

Duan, C. J., and J. X. Feng. 2010. Mining metagenomes for novel cellulase genes. Biotechnol. Lett. 32:1765–1775. doi:https://doi.org/10.1007/s10529-010-0356-z.

Edwards, M. V., A. C. Edwards, P. Millard, and A. Kocher. 2014. Mannose rich fraction of *Saccharomyces cerevisiae* promotes growth and enhances carcass yield in commercially housed grower–finisher pigs. Anim. Feed Sci. Technol. 197:227–232. doi:https://doi.org/10.1016/j.anifeedsci.2014.08.004.

Ehsanipour, M., A. V. Suko, and R. Bura. 2016. Fermentation of lignocellulosic sugars to acetic acid by *Moorella thermoacetica*. J. Ind. Microbiol. Biotechnol. 43:807–816. doi:https://doi.org/10.1007/s10295-016-1756-4.

Elbein, A. D. 1969. Biosynthesis of a cell wall glucomannan in mung bean seedlings. J. Biol. Chem. 244:1608–1616.

Elia, M., and J. H. Cummings. 2007. Physiological aspects of energy metabolism and gastrointestinal effects of carbohydrates. Eur. J. Clin. Nutr. 61:S40 S74. doi:https://doi.org/10.1038/sj.ejcn.1602938.

Elsden, S. R., M. W. S. Hitchcock, R. A. Marshall, and A. T. Phillipson. 1946. Volatile acid in the digesta of ruminants and other animals. J. Exp. Biol. 22:191–202.

Englyst, K. N., and H. N. Englyst. 2005. Carbohydrate bioavailability. Br. J. Nutr. 94:1–11. doi:https://doi.org/10.1079/bjn20051457.

Englyst, K. N., S. Liu, and H. N. Englyst. 2007. Nutritional characterization and measurement of dietary carbohydrates. Eur. J. Clin. Nutr. 61:S19–S39. doi:https://doi.org/10.1038/sj.ejcn.1602937.

Espinosa, C. D., R. S. Fry, M. E. Kocher, and H. H. Stein. 2019. Effects of copper hydroxychloride and distillers dried grains with solubles on intestinal microbial concentration and apparent ileal and total tract digestibility of energy and nutrients by growing pigs. J. Anim. Sci. 97:4904–4911. doi:https://doi.org/10.1093/jas/skz340.

Espinosa, C. D., M. S. F. Oliveira, L. V. Lagos, T. L. Weeden, A. J. Mercado, and H. H. Stein. 2020. Nutritional value of a new source of fermented soybean meal fed to growing pigs. J. Anim. Sci. doi:https://doi.org/10.1093/jas/skaa357.

Fahey, G. C., L. Novotny, B. Layton, and D. R. Mertens. 2019. Critical factors in determining fiber content of feeds and foods and their ingredients. J. AOAC Int. 102:52–62. doi:https://doi.org/10.5740/jaoacint.18-0067.

FDA. 2016. CFR - Food labeling: Revision of the nutrition and supplement facts labels Code of Federal Regulations Title 21 No. Docket No. FDA–2012–N–1210. FDA, Washingtn, DC. p. 111.

Fledderus, J., P. Bikker, and J. W. Kluess. 2007. Increasing diet viscosity using carboxymethylcellulose in weaned piglets stimulates protein digestibility. Livest. Sci. 109:89–92. doi:https://doi.org/10.1016/j.livsci.2007.01.086.

Flint, H. J., K. P. Scott, P. Louis, and S. H. Duncan. 2012. The role of the gut microbiota in nutrition and health. Nat. Rev. Gastroenterol. Hepatol. 9:577–589. doi:https://doi.org/10.1038/nrgastro.2012.156.

Forstner, J. F., and G. G. Forstner. 1994. Gastrointestinal mucus. In: L. R. Johnson, editor, Physiology of the gastrointestinal tract. Raven Press, New York, NY. p. 1255.

Franck, A. 2006. Inulin. In: A. M. Stephen, G. O. Phillips, and P. A. Williams, editors, Food polysaccharides and their applications. CRC Press, Boca Raton, FL. p. 335–352.

Fukushima, R. S., M. S. Kerley, M. H. Ramos, J. H. Porter, and R. L. Kallenbach. 2015. Comparison of acetyl bromide lignin with acid detergent lignin and Klason lignin and correlation with *in vitro* forage degradability. Anim. Feed Sci. Technol. 201:25–37. doi:https://doi.org/10.1016/j.anifeedsci.2014.12.007.

Ganapathy, V., M. Thangaraju, E. Gopal, P. M. Martin, S. Itagaki, S. Miyauchi, and P. D. Prasad. 2008. Sodium-coupled monocarboxylate transporters in normal tissues and in cancer. AAPS J. 10:193–199. doi:https://doi.org/10.1208/s12248-008-9022-y.

Gill, R. K., S. Saksena, W. A. Alrefai, Z. Sarwar, J. L. Goldstein, R. E. Carroll, K. Ramaswamy, and P. K. Dudeja. 2005. Expression and membrane localization of MCT isoforms along the length of the human intestine. Am. J. Physiol. Cell Physiol. 289:C846–C852. doi:https://doi.org/10.1152/ajpcell.00112.2005.

Girard, C. L., S. Robert, J. J. Matte, C. Farmer, and G. P. Martineau. 1995. Influence of high fibre diets given to gestating sows on serum concentrations of micronutrients. Livest. Prod. Sci. 43:15–26. doi:https://doi.org/10.1016/0301-6226(95)00003-4.

Goff, J. P. 2018. *Invited review*: Mineral absorption mechanisms, mineral interactions that affect acid-base and antioxidant status, and diet considerations to improve mineral status. J. Dairy Sci. 101:2763–2813. doi:https://doi.org/10.3168/jds.2017-13112.

Graham, H., K. Hesselman, and P. Åman. 1986. The influence of wheat bran and sugar-beet pulp on the digestibility of dietary components in a cereal-based pig diet. J. Nutr. 116:242–251. doi:https://doi.org/10.1093/jn/116.2.242.

Gutierrez, N. A., N. V. L. Serão, and J. F. Patience. 2016. Effects of distillers' dried grains with solubles and soybean oil on dietary lipid, fiber, and amino acid digestibility in corn-based diets fed to growing pigs. J. Anim. Sci. 94:1508–1519. doi:https://doi.org/10.2527/jas.2015-9529.

Hadjiagapiou, C., L. Schmidt, P. K. Dudeja, T. J. Layden, and K. Ramaswamy. 2000. Mechanism(s) of butyrate transport in Caco-2 cells: Role of monocarboxylate transporter 1. Am. J. Physiol. Gastrointest. Liver. Physiol. 279:G775–G780. doi:https://doi.org/10.1152/ajpgi.2000.279.4.G775.

Hamaker, B. R., Y. E. Tuncil, and X. Shen. 2019. Carbohydrates of the kernel. In: S. O. Serna-Saldivar, editor, Corn:chemistry and technology. Elsevier Inc., Duxford, UK. p. 305–318.

Hansen, C. F., N. D. Phillips, T. La, A. Hernandez, J. Mansfield, J. C. Kim, B. P. Mullan, D. J. Hampson, and J. R. Pluske. 2010. Diets containing inulin but not lupins help to prevent swine dysentery in experimentally challenged pigs. J. Anim. Sci. 88:3327–3336. doi:https://doi.org/10.2527/jas.2009-2719.

Hatfield, R. D., H. J. G. Jung, J. Ralph, D. R. Buxton, and P. J. Weimer. 1994. A comparison of the insoluble residues produced by the Klason lignin and acid detergent lignin procedures. J. Sci. Food Agric. 65:51–58. doi:https://doi.org/10.1002/jsfa.2740650109.

He, Y., T. M. Mouthier, M. A. Kabel, J. Dijkstra, W. H. Hendriks, P. C. Struik, and J. W. Cone. 2018. Lignin composition is more important than content for maize stem cell wall degradation. J. Sci. Food Agric. 98:384–390. doi:https://doi.org/10.1002/jsfa.8630.

Held, M. A., N. Jiang, D. Basu, A. M. Showalter, and A. Faik. 2015. Plant cell wall polysaccharides: Structure and biosynthesis. In: K. G. Ramawat, and J.-M. Mérillon, editors, Polysaccharides bioactivity and biotechnology. Springer International Publishing AG Switzerland, Cham, Switzerland. p. 3–54.

Hetzel, M., M. Brock, T. Selmer, A. J. Pierik, B. T. Golding, and W. Buckel. 2003. Acryloyl-CoA reductase from *Clostridium propionicum*: An enzyme complex of propionyl-CoA dehydrogenase and electron-transferring flavoprotein. Eur. J. Biochem. 270:902–910. doi:https://doi.org/10.1046/j.1432-1033.2003.03450.x.

Hill, G. M., J. E. Link, M. J. Rincker, D. L. Kirkpatrick, M. L. Gibson, and K. Karges. 2008. Utilization of distillers dried grains with solubles and phytase in sow lactation diets to meet the phosphorus requirement of the sow and reduce fecal phosphorus concentration. J. Anim. Sci. 86:112–118. doi:https://doi.org/10.2527/jas.2006-381.

Högberg, A., and J. E. Lindberg. 2004. Influence of cereal non-starch polysaccharides on digestion site and gut environment in growing pigs. Livest. Prod. Sci 87:121–130. doi:https://doi.org/10.1016/j.livprodsci.2003.10.002.

Huang, Q., Y. B. Su, D. F. Li, L. Liu, C. F. Huang, Z. P. Zhu, and C. H. Lai. 2015. Effects of inclusion levels of wheat bran and body weight on ileal and fecal digestibility in growing pigs. Asian-Australas. J. Anim. Sci. 28:847–854. doi:https://doi.org/10.5713/ajas.14.0769.

Huang, C., S. Zhang, H. H. Stein, J. Zhao, D. Li, and C. Lai. 2018. Effect of inclusion level and adaptation duration on digestible energy and nutrient digestibility in palm kernel meal fed to growing-finishing pigs. Asian-Australas. J. Anim. Sci. 31:395–402. doi:https://doi.org/10.5713/ajas.17.0515.

Ibáñez, M. A., C. de Blas, L. Cámara, and G. G. Mateos. 2020. Chemical composition, protein quality and nutritive value of commercial soybean meals produced from beans from different countries: A meta-analytical study. Anim. Feed Sci. Technol. 267:114531. doi:https://doi.org/10.1016/j.anifeedsci.2020.114531.

Iyayi, E. A., and O. Adeola. 2015. Quantification of short-chain fatty acids and energy production from hindgut fermentation in cannulated pigs fed graded levels of wheat bran. J. Anim. Sci. 93:4781–4787. doi:https://doi.org/10.2527/jas.2015-9081.

Izydorczyk, M. S. 2017. Functional properties of cereal cell wall polysaccharides. In: A.-C. Eliasson, editor, Carbohydrates in food. Taylor & Francis Group, New York, NY. p. 193–246.

Jaworski, N. W., H. N. Lærke, K. E. Bach Knudsen, and H. H. Stein. 2015. Carbohydrate composition and in vitro digestibility of dry matter and nonstarch polysaccharides in corn, sorghum, and wheat and coproducts from these grains. J. Anim. Sci. 93:1103–1113. doi:https://doi.org/10.2527/jas.2014-8147.

Jaworski, N. W., A. Owusu-Asiedu, M. C. Walsh, J. C. McCann, J. J. Loor, and H. H. Stein. 2017. Effects of a 3 strain *Bacillus*-based direct-fed microbial and dietary fiber concentration on growth performance and expression of genes related to absorption and metabolism of volatile fatty acids in weanling pigs. J. Anim. Sci. 95:308–319. doi:https://doi.org/10.2527/jas.2016.0557.

Jaworski, N. W., and H. H. Stein. 2017. Disappearance of nutrients and energy in the stomach and small intestine, cecum, and colon of pigs fed corn-soybean meal diets containing distillers dried grains with solubles, wheat middlings, or soybean hulls. J. Anim. Sci. 95:727–739. doi:https://doi.org/10.2527/jas.2016.0752.

Jung, H. J. G. 1997. Analysis of forage fiber and cell walls in ruminant nutrition. J. Nutr. 127:810S–813S. doi:https://doi.org/10.1093/jn/127.5.810S.

Jensen, B. B., and H. Jørgensen. 1994. Effect of dietary fiber on microbial activity and microbial gas production in various regions of the gastrointestinal tract of pigs. Appl. Environ. Microbiol. 60:1897–1904.

Jha, R., and J. D. Berrocoso. 2015. Review: Dietary fiber utilization and its effects on physiological functions and gut health of swine. Animal 9:1441–1452. doi:https://doi.org/10.1017/S1751731115000919.

Jørgensen, J. R., M. R. Clausen, and P. B. Mortensen. 1997a. Oxidation of short and medium chain C2-C8 fatty acids in Sprague-Dawley rat colonocytes. Gut 40:400–405. doi:https://doi.org/10.1136/gut.40.3.400.

Jørgensen, H., T. Larsen, X. Q. Zhao, and B. O. Eggum. 1997b. The energy value of short-chain fatty acids infused into the caecum of pigs. Br. J. Nutr. 77:745–756. doi:https://doi.org/10.1079/BJN19970072.

Kaczmarek, S. A., M. Hejdysz, M. Kubis, M. Kasprowicz-Potocka, and A. Rutkowski. 2016. The nutritional value of yellow lupin (*Lupinus luteus* L.) for broilers. Anim. Feed Sci. Technol. 222:43–53. doi:https://doi.org/10.1016/j.anifeedsci.2016.10.001.

Katsuraya, K., K. Okuyama, K. Hatanaka, R. Oshima, T. Sato, and K. Matsuzaki. 2003. Constitution of konjac glucomannan: chemical analysis and $^{13}$C NMR spectroscopy. Carbohydr. Polym. 53:183–189. doi:https://doi.org/10.1016/S0144-8617(03)00039-0.

Kil, D. Y., T. E. Sauber, D. B. Jones, and H. H. Stein. 2010. Effect of the form of dietary fat and the concentration of dietary neutral detergent fiber on ileal and total tract endogenous losses and apparent and true digestibility of fat by growing pigs. J. Anim. Sci. 88:2959–2967. doi:https://doi.org/10.2527/jas.2009-2216.

Kirat, D., and S. Kato. 2006. Monocarboxylate transporter 1 (MCT1) mediates transport of short-chain fatty acids in bovine caecum. Exp. Physiol. 91:835–844. doi:https://doi.org/10.1113/expphysiol.2006.033837.

Kritchevsky, D. 1988. Dietary fiber. Ann. Rev. Nutr. 8:301–328. doi:https://doi.org/10.1146/annurev.nu.08.070188.001505.

Langlois, A., T. Corring, and J. A. Chayvialle. 1986. Effets de la consommation de son de blé sur la sécretion pancréatique exocrine et la teneur plasmatique de quelques peptides régulateurs chez le porc. Reprod. Nutr. 26:1178. doi:https://doi.org/10.1051/rnd:19860804.

Lapierre, C., B. Pollet, M.-C. Ralet, and L. Saulnier. 2001. The phenolic fraction of maize bran: evidence for lignin-heteroxylan association. Phytochemistry 57:765–772. doi:https://doi.org/10.1016/S0031-9422(01)00104-2.

Larsen, F. M., P. J. Moughan, and M. N. Wilson. 1993. Dietary fiber viscosity and endogenous protein excretion at the terminal ileum of growing rats. J. Nutr. 123:1898–1904. doi:https://doi.org/10.1093/jn/123.11.1898.

Larsen, F. M., M. N. Wilson, and P. J. Moughan. 1994. Dietary fiber viscosity and amino acid digestibility, proteolytic digestive enzyme activity and digestive organ weights in growing rats. J. Nutr. 124:833–841. doi:https://doi.org/10.1093/jn/124.6.833.

Lattimer, J. M., and M. D. Haub. 2010. Effects of dietary fiber and its components on metabolic health. Nutrients 2:1266–1289. doi:https://doi.org/10.3390/nu2121266.

Lee, S. P., J. F. Nicholls, A. M. Roberton, and H. Z. Park. 1988. Effect of pepsin on partially purified pig gastric mucus and purified mucin. Biochem. Cell Biol. 66:367–373. doi:https://doi.org/10.1139/o88-044.

Leek, A. B. G., J. J. Callan, P. Reilly, V. E. Beattie, and J. V. O'Doherty. 2007. Apparent component digestibility and manure ammonia emission in finishing pigs fed diets based on barley, maize or wheat prepared without or with exogenous non-starch polysaccharide enzymes. Anim. Feed Sci. Technol. 135: 86–99. doi:https://doi.org/10.1016/j.anifeedsci.2006.03.024.

Le Gall, M., M. Warpechowski, Y. Jaguelin-Peyraud, and J. Noblet. 2009. Influence of dietary fibre level and pelleting on the digestibility of energy and nutrients in growing pigs and adult sows. Animal 3:352–359. doi:https://doi.org/10.1017/S1751731108003728.

Lenis, N. P., P. Bikker, J. van der Meulen, J. Th. M. van Diepen, J. G. M. Bakker, and A. W. Jongbloed. 1996. Effect of dietary neutral detergent fiber on ileal digestibility and portal flux of nitrogen and amino acids and on nitrogen utilization in growing pigs. J. Anim. Sci. 74:2687–2699. doi:https://doi.org/10.2527/1996.74112687x.

Leterme, P., E. Froidmont, F. Rossi, and A. Théwis. 1998. The high water-holding capacity of pea inner fibers affects the ileal flow of endogenous amino acids in pigs. J. Agric. Food Chem. 46:1927–1934. doi:https://doi.org/10.1021/jf970955+.

Li, S., W. C. Sauer, and R. T. Hardin. 1994. Effect of dietary fiber level on amino acid digestibility in young pigs. Can. J. Anim. Sci. 74:1649–1656. doi:https://doi.org/10.4141/cjas94-044.

Liang, Z. X., L. Li, S. Li, Y. H. Cai, S. T. Yang, and J. F. Wang. 2012. Enhanced propionic acid production from Jerusalem artichoke hydrolysate by immobilized *Propionibacterium acidipropionici* in a fibrous-bed bioreactor. Bioprocess Biosyst. Eng. 35:915–921. doi:https://doi.org/10.1007/s00449-011-0676-y.

Libao-Mercado, A. J., C. L. Zhu, M. F. Fuller, M. Rademacher, B. Sève, and C. F. M. de Lange. 2007. Effect of feeding fermentable fiber on synthesis of total and mucosal protein in the intestine of the growing pig. Livest. Sci. 109:125–128. doi:https://doi.org/10.1016/j.livsci.2007.01.116.

Libao-Mercado, A. J. O., C. L. Zhu, J. P. Cant, H. Lapierre, J. N. Thibault, B. Sève, M. F. Fuller, and C. F. M. de Lange. 2009. Dietary and endogenous amino acids are the main contributors to microbial protein in the upper gut of normally nourished pigs. J. Nutr. 139:1088–1094. doi:https://doi.org/10.3945/jn.108.103267.

Lien, K. A., W. C. Sauer, and M. Fenton. 1997. Mucin output in ileal digesta of pigs fed a protein-free diet. Z. Ernährungswiss. 36:182–190. doi:https://doi.org/10.1007/BF01611398.

Lien, K. A., W. C. Sauer, and J. M. He. 2001. Dietary influences on the secretion into and degradation of mucin in the digestive tract of monogastric animals and humans. J. Anim. Feed Sci. 10:223–245. doi:https://doi.org/10.22358/jafs/67980/2001.

Liener, I. E. 1994. Implications of antinutritional components in soybean foods. Crit. Rev. Food Sci. Nutr. 34:31–67. doi:https://doi.org/10.1080/10408399409527649.

Lindberg, J. E. 2014. Fiber effects in nutrition and gut health in pigs. J. Anim. Sci. Biotechnol. 5:15. doi:https://doi.org/10.1186/2049-1891-5-15.

Liu, Z., S. Lv, S. Zhang, J. Liu, and H. Zhang. 2016. Effects of dietary cellulose levels on the estimation of endogenous amino acid losses and amino acid digestibility for growing pigs. Anim. Nutr. 2:74–78. doi:https://doi.org/10.1016/j.aninu.2016.04.001.

Liying, Z., D. Li, S. Qiao, E. W. Johnson, B. Li, P. A. Thacker, and I. K. Han. 2003. Effects of stachyose on performance, diarrhoea incidence and intestinal bacteria in weanling pigs. Arch. Anim. Nutr. 57:1–10. doi:https://doi.org/10.1080/0003942031000086662.

Looft, T., H. K. Allen, B. L. Cantarel, U. Y. Levine, D. O. Bayles, D. P. Alt, B. Henrissat, and T. B. Stanton. 2014. Bacteria, phages and pigs: The effects of in-feed antibiotics on the microbiome at different gut locations. ISME. J. 8:1566–1576. doi:https://doi.org/10.1038/ismej.2014.12.

Low, A. G. 1989. Secretory response of the pig gut to non-starch polysaccharides. Anim. Feed Sci. Technol. 23:55–65. doi:https://doi.org/10.1016/0377-8401(89)90089-8.

Loy, D. D., and E. L. Lundy. 2019. Nutritional properties and feeding value of corn and its coproducts. In: S. O. Serna-Saldivar, editor, Corn:chemistry and technology. Elsevier Inc., Duxford, UK.

Lyu, Z. Q., C. F. Huang, Y. K. Li, P. L. Li, H. Liu, Y. F. Chen, D. F. Li, and C. H. Lai. 2018. Adaptation duration for net energy determination of high fiber diets in growing pigs. Anim. Feed Sci. Technol. 241:15–26. doi:https://doi.org/10.1016/j.anifeedsci.2018.04.008.

Macfarlane, S., and G. T. Macfarlane. 2007. Regulation of short-chain fatty acid production. Proc. Nutr. Soc. 62:67–72. doi:https://doi.org/10.1079/PNS2002207.

Macy, J. M., L. G. Ljungdahl, and G. Gottschalk. 1978. Pathway of succinate and propionate formation in Bacteroides fragilis. J. Bacteriol. 134:84–91. doi:https://doi.org/10.1128/JB.134.1.84-91.1978.

Macy, J. M., and I. Probst. 1979. The biology of gastrointestinal bacteroides. Annu. Rev. Microbiol. 33:561–594. doi:https://doi.org/10.1146/annurev.mi.33.100179.003021.

Maison, T., Y. Liu, and H. H. Stein. 2015. Digestibility of energy and detergent fiber and digestible and metabolizable energy values in canola meal, 00-rapeseed meal, and 00-rapeseed expellers fed to growing pigs. J. Anim. Sci. 93:652–660. doi:https://doi.org/10.2527/jas.2014-7792.

Mariscal-Landín, G., B. Sève, Y. Colléaux, and Y. Lebreton. 1995. Endogenous amino nitrogen collected from pigs with end-to-end ileorectal anastomosis is affected by the method of estimation and altered by dietary fiber. J. Nutr. 125:136–146. doi:https://doi.org/10.1093/jn/125.1.136.

Marsono, Y., R. J. Illman, J. M. Clarke, R. P. Trimble, and D. L. Topping. 1993. Plasma lipids and large bowel volatile fatty acids in pigs fed on white rice, brown rice and rice bran. Br. J. Nutr. 70:503–513. doi:https://doi.org/10.1079/BJN19930144.

Marston, H. R., S. H. Allen, and R. M. Smith. 1961. Primary metabolic defect supervening on vitamin B12 deficiency in the sheep. Nature 190:1085–1091. doi:https://doi.org/10.1038/1901085a0.

Mathai, J. M., J. K. Htoo, J. E. Thomson, K. J. Touchette, and H. H. Stein. 2016. Effects of dietary fiber on the ideal standardized ileal digestible threonine:lysine ratio for twenty-five to fifty kilogram growing gilts. J. Anim. Sci. 94:4217–4230. doi:https://doi.org/10.2527/jas2016-0680.

McCleary, B. V. 2003. Dietary fibre analysis. Proc. Nutr. Soc. 62:3–9. doi:https://doi.org/10.1079/PNS2002204.

McCleary, B. V. 2007. An integrated procedure for the measurement of total dietary fibre (including resistant starch), non-digestible oligosaccharides and available carbohydrates. Anal. Bioanal. Chem. 389:291–308. doi:https://doi.org/10.1007/s00216-007-1389-6.

McCleary, B. V., J. W. DeVries, J. I. Rader, G. Cohen, L. Prosky, D. C. Mugford, M. Champ, and K. Okuma. 2012. Determination of insoluble, soluble, and total dietary fiber (CODEX definition) by enzymatic-gravimetric method and liquid chromatography: collaborative study. J. AOAC Int. 95:824–844. doi:https://doi.org/10.5740/jaoacint.cs2011_25.

Mertens, D. R. 2003. Challenges in measuring insoluble dietary fiber. J. Anim. Sci. 81:3233–3249. doi:https://doi.org/10.2527/2003.81123233x.

Miguel, J. C., S. L. Rodriguez-Zas, and J. E. Pettigrew. 2004. Efficacy of a mannan oligosaccharide (Bio-Mos®) for improving nursery pig performance. J. Swine Health Prod. 12:296–307.

Miller, T. L., and M. J. Wolin. 1979. Fermentations by saccharolytic intestinal bacteria. Am. J. Clin. Nutr. 32:164–172. doi:https://doi.org/10.1093/ajcn/32.1.164.

Miner-Williams, W., P. J. Moughan, and M. F. Fuller. 2009. Endogenous components of digesta protein from the terminal ileum of pigs fed a casein-based diet. J. Agric. Food Chem. 57:2072–2078. doi:https://doi.org/10.1021/jf8023886.

Moeser, A. J., and T. A. T. G. van Kempen. 2002. Dietary fibre level and enzyme inclusion affect nutrient digestibility and excreta characteristics in grower pigs. J. Sci. Food Agric. 82:1606–1613. doi:https://doi.org/10.1002/jsfa.1234.

Mongeau, R., and S. P. J. Brooks. 2003. Dietary fiber determination. In: B. Caballero, editor, Encyclopedia of food sciences and nutrition. Academic Press, Cambridge, MA. p. 1823–1833.

Montagne, L., C. Piel, and J. P. Lallès. 2004. Effect of diet on mucin kinetics and composition: Nutrition and health implications. Nutr. Rev. 62:105–114. doi:https://doi.org/10.1111/j.1753-4887.2004.tb00031.x.

Montoya, C. A., S. J. Henare, S. M. Rutherfurd, and P. J. Moughan. 2016. Potential misinterpretation of the nutritional value of dietary fiber: Correcting fiber digestibility values for nondietary gut-interfering material. Nutr. Rev. 74:517–533. doi:https://doi.org/10.1093/nutrit/nuw014.

Montoya, C. A., S. M. Rutherfurd, and P. J. Moughan. 2015. Nondietary gut materials interfere with the determination of dietary fiber digestibility in growing pigs when using the Prosky method. J. Nutr. 145:1966–1972. doi:https://doi.org/10.3945/jn.115.212639.

Montoya, C. A., S. M. Rutherfurd, and P. J. Moughan. 2017. Ileal digesta nondietary substrates from cannulated pigs are major contributors to in vitro human hindgut short-chain fatty acid production. J. Nutr. 147:264–271. doi:https://doi.org/10.3945/jn.116.240564.

Moore, R. J., E. T. Kornegay, R. L. Grayson, and M. D. Lindemann. 1988. Growth, nutrient utilization and intestinal morphology of pigs fed high-fiber diets. J. Anim. Sci. 66:1570–1579. doi:https://doi.org/10.2527/jas1988.6661570x.

Mosenthin, R., W. C. Sauer, and F. Ahrens. 1994. Dietary pectin's effect on ileal and fecal amino acid digestibility and exocrine pancreatic secretions in growing pigs. J. Nutr. 124:1222–1229. doi:https://doi.org/10.1093/jn/124.8.1222.

Mpendulo, C. T., M. Chimonyo, S. P. Ndou, and A. G. Bakare. 2018. Fiber source and inclusion level affects characteristics of excreta from growing pigs. Asian-Australas. J. Anim. Sci. 31:755–762. doi:https://doi.org/10.5713/ajas.14.0611.

Mroz, Z., A. J. Moeser, K. Vreman, J. T. van Diepen, T. van Kempen, T. T. Canh, and A. W. Jongbloed. 2000. Effects of dietary carbohydrates and buffering capacity on nutrient digestibility and manure characteristics in finishing pigs. J. Anim. Sci. 78:3096–3106. doi:https://doi.org/10.2527/2000.78123096x.

Navarro, D. M. D. L., E. M. A. M. Bruininx, L. de Jong, and H. H. Stein. 2018a. Analysis for low-molecular-weight carbohydrates is needed to account for all energy-contributing nutrients in some feed ingredients, but physical characteristics do not predict in vitro digestibility of dry matter. J. Anim. Sci. 96:532–544. doi:https://doi.org/10.1093/jas/sky010.

Navarro, D. M. D. L., E. M. A. M. Bruininx, L. de Jong, and H. H. Stein. 2018b. Effects of physicochemical characteristics of feed ingredients on the apparent total tract digestibility of energy, DM, and nutrients by growing pigs. J. Anim. Sci. 96:2265–2277. doi:https://doi.org/10.1093/jas/sky149.

Navarro, D. M. D. L., E. M. A. M. Bruininx, L. de Jong, and H. H. Stein. 2018c. The contribution of digestible and metabolizable energy from high-fiber dietary ingredients is not affected by inclusion rate in mixed diets fed to growing pigs. J. Anim. Sci. 96:1860–1868. doi:https://doi.org/10.1093/jas/sky090.

Navarro, D. M. D. L., E. M. A. M. Bruininx, L. de Jong, and H. H. Stein. 2019. Effects of inclusion rate of high fiber dietary ingredients on apparent ileal, hindgut, and total tract digestibility of dry matter and nutrients in ingredients fed to growing pigs. Anim. Feed Sci. Technol. 248:1–9. doi:https://doi.org/10.1016/j.anifeedsci.2018.12.001.

Ndou, S. P., E. Kiarie, N. Ames, and C. M. Nyachoti. 2019. Flaxseed meal and oat hulls supplementation: Impact on dietary fiber digestibility, and flows of fatty acids and bile acids in growing pigs. J. Anim. Sci. 97:291–301. doi:https://doi.org/10.1093/jas/sky398.

Nguyen, N., M. Jacobs, J. Li, C. Huang, D. F. Li, D. M. D. L. Navarro, H. H. Stein, and N. W. Jaworski. 2019a. Technical note: Concentrations of soluble, insoluble, and total dietary fiber in feed ingredients determined using Method AOAC 991.43 are not different from values determined using Method AOAC 2011.43 with the AnkomTDF Dietary Fiber Analyzer. J. Anim. Sci. 97:3972–3983. doi:https://doi.org/10.1093/jas/skz239.

Nguyen, Q. H., P. D. Le, C. Chim, N. D. Le, and V. Fievez. 2019b. Potential to mitigate ammonia emission from slurry by increasing dietary fermentable fiber through inclusion of tropical byproducts in practical diets for growing pigs. Asian-Australas. J. Anim. Sci. 32:574–584. doi:https://doi.org/10.5713/ajas.18.0481.

Noblet, J., and J. M. Perez. 1993. Prediction of digestibility of nutrients and energy values of pig diets from chemical analysis. J. Anim. Sci. 71:3389–3398. doi:https://doi.org/10.2527/1993.71123389x.

Nortey, T. N., J. F. Patience, P. H. Simmins, N. L. Trottier, and R. T. Zijlstra. 2007. Effects of individual or combined xylanase and phytase supplementation on energy, amino acid, and phosphorus digestibility and growth performance of grower pigs fed wheat-based diets containing wheat millrun. J. Anim. Sci. 85:1432–1443. doi:https://doi.org/10.2527/jas.2006-613.

Nunes, C. S., and K. Malmlöf. 1992. Effects of guar gum and cellulose on glucose absorption, hormonal release and hepatic metabolism in the pig. Br. J. Nutr. 68:693–700. doi:https://doi.org/10.1079/bjn19920126.

Owusu-Asiedu, A., J. F. Patience, B. Laarveld, A. G. Van Kessel, P. H. Simmins, and R. T. Zijlstra. 2006. Effects of guar gum and cellulose on digesta passage rate, ileal microbial populations, energy and protein digestibility, and performance of grower pigs. J. Anim. Sci. 84:843–852. doi:https://doi.org/10.2527/2006.844843x.

Parada Venegas, D., M. K. De la Fuente, G. Landskron, M. J. González, R. Quera, G. Dijkstra, H. J. M. Harmsen, K. N. Faber, and M. A. Hermoso. 2019. Short chain fatty acids (SCFAs)-mediated gut epithelial and immune regulation and its relevance for inflammatory bowel diseases. Front. Immunol. 10:277. doi:https://doi.org/10.3389/fimmu.2019.00277.

Pastuszewska, B., J. Kowalczyk, and A. Ochtabińska. 2000. Dietary carbohydrates affect caecal fermentation and modify nitrogen excretion patterns in rats. I. Studies with protein-free diets. Arch. Anim. Nutr. 53:207–225. doi:https://doi.org/10.1080/17450390009381948.

Pedersen, M. B., S. Dalsgaard, K. E. Bach Knudsen, S. Yu, and H. N. Lærke. 2014. Compositional profile and variation of distillers dried grains with solubles from various origins with focus on non-starch polysaccharides. Anim. Feed Sci. Technol. 197:130–141. doi:https://doi.org/10.1016/j.anifeedsci.2014.07.011.

Pérez de Nanclares, M., C. Marcussen, A. H. Tauson, J. Ø. Hansen, N. P. Kjos, L. T. Mydland, K. E. Bach Knudsen, and M. Øverland. 2019. Increasing levels of rapeseed expeller meal in diets for pigs: effects on protein and energy metabolism. Animal 13:273–282. doi:https://doi.org/10.1017/S1751731118000988.

Pérez de Nanclares, M., M. P. Trudeau, J. Ø. Hansen, L. T. Mydland, P. E. Urriola, G. C. Shurson, C. Piercey Åkesson, N. P. Kjos, M. Ø. Arntzen, and M. Øverland. 2017. High-fiber rapeseed co-product diet for Norwegian Landrace pigs: Effect on digestibility. Livest. Sci. 203:1–9. doi:https://doi.org/10.1016/j.livsci.2017.06.008.

Perret, A., C. Lechaplais, S. Tricot, N. Perchat, C. Vergne, C. Pellé, K. Bastard, A. Kreimeyer, D. Vallenet, A. Zaparucha, J. Weissenbach, and M. Salanoubat. 2011. A novel acyl-CoA beta-transaminase characterized from a metagenome. PLoS One 6:e22918. doi:https://doi.org/10.1371/journal.pone.0022918.

Piel, C., L. Montagne, B. Sève, and J. P. Lallès. 2005. Increasing digesta viscosity using carboxymethylcellulose in weaned piglets stimulates ileal goblet cell numbers and maturation. J. Nutr. 135:86–91. doi:https://doi.org/10.1093/jn/135.1.86.

Pieper, R., J. Bindelle, B. Rossnagel, A. Van Kessel, and P. Leterme. 2009. Effect of carbohydrate composition in barley and oat cultivars on microbial ecophysiology and proliferation of Salmonella enterica in an in vitro model of the porcine gastrointestinal tract. Appl. Environ. Microbiol. 75:7006–7016. doi:https://doi.org/10.1128/aem.01343-09.

Prabhu, R., E. Altman, and M. A. Eiteman. 2012. Lactate and acrylate metabolism by Megasphaera elsdenii under batch and steady-state conditions. Appl. Environ. Microbiol. 78:8564–8570. doi:https://doi.org/10.1128/aem.02443-12.

Prosky, L., N. G. Asp, I. Furda, J. W. Devries, T. F. Schweizer, and B. F. Harland. 1984. Determination of total dietary fiber in foods, food products, and total diets: Interlaboratory study. J. Assoc. Off. Anal. Chem. 67:1044–1052. doi:https://doi.org/10.1093/jaoac/67.6.1044.

Rainbird, A. L., A. G. Low, and T. Zebrowska. 1984. Effect of guar gum on glucose and water absorption from isolated loops of jejunum in conscious growing pigs. Br. J. Nutr. 52: 489–498. doi:https://doi.org/10.1079/BJN19840116.

Ritzhaupt, A., I. S. Wood, A. Ellis, K. B. Hosie, and S. P. Shirazi-Beechey. 1998. Identification and characterization of a monocarboxylate transporter (MCT1) in pig and human colon: Its potential to transport L-lactate as well as butyrate. J. Physiol. 513:719–732. doi:https://doi.org/10.1111/j.1469-7793.1998.719ba.x.

Robertson, M. D. 2007. Metabolic cross talk between the colon and the periphery: Implications for insulin sensitivity. Proc. Nutr. Soc. 66:351–361. doi:https://doi.org/10.1017/S0029665107005617.

Rodriguez, D. A., S. A. Lee, C. K. Jones, J. K. Htoo, and H. H. Stein. 2020. Digestibility of amino acids, fiber, and energy by growing pigs, and concentrations of digestible and metabolizable energy in yellow dent corn, hard red winter wheat, and sorghum may be influenced by extrusion. Anim. Feed Sci. Technol. 268:114602. doi:https://doi.org/10.1016/j.anifeedsci.2020.114602.

Roediger, W. E. W. 1982. Utilization of nutrients by isolated epithelial cells of the rat colon. Gastroenterology. 83:424–429. doi:https://doi.org/10.1016/S0016-5085(82)80339-9.

Rojas, O. J., and H. H. Stein. 2013. Concentration of digestible, metabolizable, and net energy and digestibility of energy and nutrients in fermented soybean meal, conventional soybean meal, and fish meal fed to weanling pigs. J. Anim. Sci. 91:4397–4405. doi:https://doi.org/10.2527/jas.2013-6409.

Rosen, G. D. 2006. Holo-analysis of the efficacy of Bio-Mos® in pig nutrition. Anim. Sci. 82:683–689. doi:https://doi.org/10.1079/ASC200684.

Sanjayan, N., J. M. Heo, and C. M. Nyachoti. 2014. Nutrient digestibility and growth performance of pigs fed diets with different levels of canola meal from Brassica napus black and Brassica juncea yellow. J. Anim. Sci. 92:3895–3905. doi:https://doi.org/10.2527/jas.2013-7215.

Satchithanandam, S., M. Vargofcak-Apker, R. J. Calvert, A. R. Leeds, and M. M. Cassidy. 1990. Alteration of gastrointestinal mucin by fiber feeding in rats. J. Nutr. 120:1179–1184. doi:https://doi.org/10.1093/jn/120.10.1179.

Sauer, W. C., R. Mosenthin, F. Ahrens, and L. A. den Hartog. 1991. The effect of source of fiber on ileal and fecal amino acid digestibility and bacterial nitrogen excretion in growing pigs. J. Anim. Sci. 69:4070–4077. doi:https://doi.org/10.2527/1991.69104070x.

Sauer, W. C., and L. Ozimek. 1986. Digestibility of amino acids in swine: Results and their practical applications. A review. Livest. Prod. Sci. 15: 367–388. doi:https://doi.org/10.1016/0301-6226(86)90076-X.

Saulnier, L., and J. F. Thibault. 1999. Ferulic acid and diferulic acids as components of sugar-beet pectins and maize bran heteroxylans. J. Sci. Food Agric. 79:396–402. doi:https://doi.org/10.1002/(SICI)1097-0010(19990301)79:3<396::AID-JSFA262>3.0.CO;2-B.

Schneeman, B. O., and D. D. Gallaher. 2001. Effects of dietary fiber on digestive enzymes. In: G. A. Spiller, editor, Dietary fiber in human nutrition. CRC Press, Washington, DC. p. 277–283.

Schulze, H., P. van Leeuwen, M. W. A. Verstegen, J. Huisman, W. B. Souffrant, and F. Ahrens. 1994. Effect of level of dietary neutral detergent fiber on ileal apparent digestibility and ileal nitrogen losses in pigs. J. Anim. Sci. 72:2362–2368. doi:https://doi.org/10.2527/1994.7292362x.

Schulze, H., P. van Leeuwen, M. W. A. Verstegen, and J. W. O. van den Berg. 1995. Dietary level and source of neutral detergent fiber and ileal endogenous nitrogen flow in pigs. J. Anim. Sci. 73:441–448. doi:https://doi.org/10.2527/1995.732441x.

Scott, R. W. 1979. Colorimetric determination of hexuronic acids in plant materials. Anal. Chem. 51:936–941. doi:https://doi.org/10.1021/ac50043a036.

Scott, K. P., J. C. Martin, G. Campbell, C. D. Mayer, and H. J. Flint. 2006. Whole-genome transcription profiling reveals genes up-regulated by growth on fucose in the human gut bacterium "Roseburia inulinivorans". J. Bacteriol. 188:4340–4349. doi:https://doi.org/10.1128/JB.00137-06.

Slavin, J. 2013. Fiber and prebiotics: Mechanisms and health benefits. Nutrients 5:1417–1435. doi:https://doi.org/10.3390/nu5041417.

Sol, C., L. Castillejos, S. López-Vergé, and J. Gasa. 2017. Prediction of the digestibility and energy contents of non-conventional by-products for pigs from their chemical composition and in vitro digestibility. Anim. Feed Sci. Technol. 234:237–243. doi:https://doi.org/10.1016/j.anifeedsci.2017.10.003.

Souffrant, W. B., A. Rérat, J. P. Laplace, B. Darcy-Vrillon, R. Köhler, T. Corring, and G. Gebhardt. 1993. Exogenous and endogenous contributions to nitrogen fluxes in the digestive tract of pigs fed a casein diet. III. Recycling of endogenous nitrogen. Reprod. Nutr. Dev. 33:373–382. doi:https://doi.org/10.1051/rnd:19930406.

Stein, H. H. 2019. Multi vs. single application of enzymes to degrade fibre in diets for pigs. In: G. González-Ortiz, M. R. Bedford, K. E. B. Knudsen, C. M. Courtin, and H. L. Classen, editors, The value of fibre – Engaging the second brain for animal nutrition. Wageningen Academic Publishers, Wageningen, the Netherlands. p. 117–124. doi:https://doi.org/10.3920/978-90-8686-893-3_6.

Stipanuk, M. H., and M. A. Caudill. 2013. Biochemical, physiological, and molecular aspects of human nutrition. 3rd ed. Elsevier Health Sciences, St. Louis, MO.

Tanabe, H., H. Ito, K. Sugiyama, S. Kiriyama, T. Morita. 2006. Dietary indigestible components exert different regional effects on luminal mucin secretion through their bulk-forming property and fermentability. Biosci. Biotechnol. Biochem. 70:1188–1194. doi:https://doi.org/10.1271/bbb.70.1188.

Thangaraju, M., G. Cresci, S. Itagaki, J. Mellinger, D. D. Browning, F. G. Berger, P. D. Prasad, and V. Ganapathy. 2008. Sodium-coupled transport of the short chain fatty acid butyrate by SLC5A8 and its relevance to colon cancer. J. Gastrointest. Surg. 12:1773–1782. doi:https://doi.org/10.1007/s11605-008-0573-0.

Titgemeyer, E. C., L. D. Bourquin, G. C. Fahey, Jr, and K. A. Garleb. 1991. Fermentability of various fiber sources by human fecal bacteria in vitro. Am. J. Clin. Nutr. 53:1418–1424. doi:https://doi.org/10.1093/ajcn/53.6.1418.

Tokoh, C., K. Takabe, J. Sugiyama, and M. Fujita. 2002. Cp/mas [13]c nmr and electron diffraction study of bacterial cellulose structure affected by cell wall polysaccharides. Cellulose 9:351–360. doi:https://doi.org/10.1023/A:1021150520953.

Topping, D. L., and P. M. Clifton. 2001. Short-chain fatty acids and human colonic function: Roles of resistant starch and nonstarch polysaccharides. Physiol. Rev. 81:1031–1064. doi:https://doi.org/10.1152/physrev.2001.81.3.1031.

Tymoczko, J. L., J. M. Berg, and L. Stryer. 2015. Biochemestry: A short course. 3rd ed. W. H. Freeman & Company, New York, NY.

Udén, P., P. H. Robinson, and J. Wiseman. 2005. Use of detergent system terminology and criteria for submission of manuscripts on new, or revised, analytical methods as well as descriptive information on feed analysis and/or variability. Anim. Feed Sci. Technol. 118:181–186. doi:https://doi.org/10.1016/j.anifeedsci.2004.11.011.

Urriola, P. E., S. K. Cervantes-Pahm, and H. H. Stein. 2013. Fiber in swine nutrition. In: L. I. Chiba, editor, Sustainable swine nutrition. John Wiley & Sons, Inc., Ames, IA. p. 255–276.

Urriola, P. E., G. C. Shurson, and H. H. Stein. 2010. Digestibility of dietary fiber in distillers coproducts fed to growing pigs. J. Anim. Sci. 88:2373–2381. doi:https://doi.org/10.2527/jas.2009-2227.

Urriola, P. E., and H. H. Stein. 2010. Effects of distillers dried grains with solubles on amino acid, energy, and fiber digestibility and on hindgut fermentation of dietary fiber in a corn-soybean meal diet fed to growing pigs. J. Anim. Sci. 88:1454–1462. doi:https://doi.org/10.2527/jas.2009-2162.

Van Dongen, F. E. M., D. Van Eylen, and M. A. Kabel. 2011. Characterization of substituents in xylans from corn cobs and stover. Carbohydr. Polym. 86:722–731. doi:https://doi.org/10.1016/j.carbpol.2011.05.007.

Van Erven, G., R. de Visser, P. de Waard, W. J. H. Van Berkel, and M. A. Kabel. 2019. Uniformly [13]C labeled lignin internal standards for quantitative pyrolysis–GC–MS analysis of grass and wood. ACS Sustain. Chem. Eng. 7:20070–20076. doi:https://doi.org/10.1021/acssuschemeng.9b05926.

Vanhoof, K., and R. De Schrijver. 1996. Availability of minerals in rats and pigs fed non-purified diets containing inulin. Nutr. Res. 16:1017–1022. doi:https://doi.org/10.1016/0271-5317(96)00101-7.

Vangsøe, C. T., N. P. Nørskov, M. F. Devaux, E. Bonnin, and K. E. Bach Knudsen. 2020. Carbohydrase complexes rich in xylanases and arabinofuranosidases affect the autofluorescence signal and liberate phenolic acids from the cell wall matrix in wheat, maize, and rice bran: An *in vitro* digestion study. J. Agric. Food Chem. 68:9878–9887. doi:https://doi.org/10.1021/acs.jafc.0c00703.

Van Soest, P. J., J. B. Robertson, and B. A. Lewis. 1991. Symposium: Carbohydrate methodology, metabolism, and nutritional implications in dairy cattle. J. Dairy Sci. 74:3583–3597. doi:https://doi.org/10.3168/jds.S0022-0302(91)78551-2.

Vital, M., A. C. Howe, and J. M. Tiedje. 2014. Revealing the bacterial butyrate synthesis pathways by analyzing (Meta)genomic data. MBio 5:e00889–00814. doi:https://doi.org/10.1128/mBio.00889-14.

Walter, D., M. Ailion, and J. Roth. 1997. Genetic characterization of the pdu operon: use of 1,2-propanediol in *Salmonella typhimurium*. J. Bacteriol. 179:1013–1022. doi:https://doi.org/10.1128/jb.179.4.1013-1022.1997.

Wang, N., and J. K. Daun. 2004. Effect of variety and crude protein content on nutrients and certain antinutrients in field peas (*Pisum sativum*). J. Sci. Food Agric. 84:1021–1029. doi:https://doi.org/10.1002/jsfa.1742.

Wefers, D., J. J. V. Cavalcante, R. R. Schendel, J. Deveryshetty, K. Wang, Z. Wawrzak, R. I. Mackie, N. M. Koropatkin, and I. Caan. 2017. Biochemical and structural analyses of two cryptic esterases in *Bacteroides intestinalis* and their synergistic activities with cognate xylanases. J. Mol. Biol. 429:2509–2527. doi:https://doi.org/10.1016/j.jmb.2017.06.017.

Wellington, M. O., A. K. Agyekum, K. Hamonic, J. K. Htoo, A. G. Van Kessel, and D. A. Columbus. 2019. Effect of supplemental threonine above requirement on growth performance of *Salmonella typhimurium* challenged pigs fed high-fiber diets. J. Anim. Sci. 97:3636–3647. doi:https://doi.org/10.1093/jas/skz225.

Welter, H., and R. Claus. 2008. Expression of the monocarboxylate transporter 1 (MCT1) in cells of the porcine intestine. Cell Biol. Int. 32:638–645. doi:https://doi.org/10.1016/j.cellbi.2008.01.008.

Wenk, C. 2001. The role of dietary fibre in the digestive physiology of the pig. Anim. Feed Sci. Technol. 90:21–33. doi:https://doi.org/10.1016/S0377-8401(01)00194-8.

Whanger, P. D., and G. Matrone. 1967. Metabolism of lactic, succinic and acrylic acids by rumen microorganisms from sheep fed sulfur-adequate and sulfur-deficient diets. Biochim. Biophys. Acta-Gen. Subj. 136:27–35. doi:https://doi.org/10.1016/0304-4165(67)90317-0.

White, D. 2000. The physiology and biochemistry of prokaryotes. 2nd ed. Oxford University Press, New York, NY.

Wilfart, A., L. Montagne, P. H. Simmins, J. van Milgen, and J. Noblet. 2007. Sites of nutrient digestion in growing pigs: Effects of dietary fiber. J. Anim. Sci. 85:976–983. doi:https://doi.org/10.2527/jas.2006-431.

Wong, J. M. W., R. de Souza, C. W. C. Kendall, A. Emam, and D. J. A. Jenkins. 2006. Colonic health: Fermentation and short chain fatty acids. J. Clin. Gastroenterol. 40:235–243. doi:https://doi.org/10.1097/00004836-200603000-00015.

Woyengo, T. A., E. Beltranena, and R. T. Zijlstra. 2014. Nonruminant nutrition symposium: Controlling feed cost by including alternative ingredients into pig diets: A review. J. Anim. Sci. 92:1293–1305. doi:https://doi.org/10.2527/jas.2013-7169.

Xu, W., N. Reddy, and Y. Yang. 2009. Extraction, characterization and potential applications of cellulose in corn kernels and distillers' dried grains with solubles (DDGS). Carbohydr. Polym. 76:521–527. doi:https://doi.org/10.1016/j.carbpol.2008.11.017.

Yadav, M. P., D. B. Johnston, A. T. Hotchkiss Jr., and K. B. Hicks. 2007. Corn fiber gum: a potential gum arabic replacer for beverage flavor emulsification. Food Hydrocoll. 21:1022–1030. doi:https://doi.org/10.1016/j.foodhyd.2006.07.009.

Yen, J. T. 2001. Anatomy of the digestive system and nutritional physiology. In: A. J. Lewis and L. L. Southern, editors, Swine nutrition. CRC Press, Washington, DC. p. 31–63.

Yoshida, T., Y. Honda, T. Tsujimoto, H. Uyama, and J. Azuma. 2014. Selective isolation of β-glucan from corn pericarp hemicelluloses by affinity chromatography on cellulose column. Carbohydr. Polym. 111:538–542. doi:https://doi.org/10.1016/j.carbpol.2014.04.050.

Zaworska, A., M. Kasprowicz-Potocka, A. Rutkowski, and D. Jamroz. 2018. The influence of dietary raw and extruded field peas (*Pisum sativum* L.) on nutrients digestibility and performance of weaned and fattening pigs. J. Anim. Feed Sci. 27:123–130. doi:https://doi.org/10.22358/jafs/91209/2018.

Zeng, Z., J. C. Jang, B. J. Kerr, G. C. Shurson, and P. E. Urriola. 2019. In vitro unfermented fiber is a good predictor of the digestible and metabolizable energy content of corn distillers dried grains with solubles in growing pigs. J. Anim. Sci. 97:3460–3471. doi:https://doi.org/10.1093/jas/skz221.

Zervas, S., and R. T. Zijlstra. 2002. Effects of dietary protein and fermentable fiber on nitrogen excretion patterns and plasma urea in grower pigs. J. Anim. Sci. 80: 3247–3256. doi:https://doi.org/10.2527/2002.80123247x.

Zhang, X., and Y. P. Zhang. 2013. Cellulases: Characteristics, sources, production, and applications. In: S. T. Yang, H. A. El-Enshasy and N. Thongchul, editors, Bioprocessing technologies in biorefinery for sustainable production of fuels, chemicals, and polymers. John Wiley & Sons, Inc, Hoboken, NJ. p. 131–146.

Zhang, W., D. Li, L. Liu, J. Zang, Q. Duan, W. Yang, and L. Zhang. 2013. The effects of dietary fiber level on nutrient digestibility in growing pigs. J. Anim. Sci. Biotechnol. 4:17. doi:https://doi.org/10.1186/2049-1891-4-17.

Zhang, Z. Y., S. Zhang, C. H. Lai, J. B. Zhao, J. J. Zang, and C. F. Huang. 2018. Effects of adaptation time and inclusion level of sugar beet pulp on nutrient digestibility and evaluation of ileal amino acid digestibility in pigs. Asian-Australas. J. Anim. Sci. 32:1414–1422. doi:https://doi.org/10.5713/ajas.18.0181.

Zhao, J., Y. Bai, G. Zhang, L. Liu, and C. Lai. 2020. Relationship between dietary fiber fermentation and volatile fatty acids' concentration in growing pigs. Animals 10:263. doi:https://doi.org/10.3390/ani10020263.

Zhao, J. B., P. Liu, C. F. Huang, L. Liu, E. K. Li, G. Zhang, and S. Zhang. 2018. Effect of wheat bran on apparent total tract digestibility, growth performance, fecal microbiota and their metabolites in growing pigs. Anim. Feed Sci. Technol. 239:14–26. doi:https://doi.org/10.1016/j.anifeedsci.2018.02.013.

# 15    Antinutritional Factors in Feedstuffs

Tofuko A. Woyengo

## Introduction

Feedstuffs of plant origin including cereal grains and oil seeds and their coproducts form the bulk of swine diets because they are relatively cheap. However, feedstuffs of plant origin contain antinutritional factors that can limit the utilization of dietary nutrients by pigs. Thus, definition of the antinutritional factors present in swine feedstuffs of plant origin and the negative effects of the antinutritional factors on dietary nutrient utilization and mechanisms by which the antinutritional factors reduce dietary nutrient utilization are critical to (i) optimize their utilization in diets for pigs and (ii) develop interventions.

The major antinutritional factors present in the feedstuffs of plant origin for pigs include (i) phytic acid, which is present in almost all feedstuffs of plant origin; (ii) trypsin inhibitors (TI), which are present in soybeans, pulses, and camelina and their coproducts; (iii) glucosinolates, which are present in feedstuffs of *Brassica* family including canola and camelina and their coproducts; (iv) cynogenic glucosides, which are present in flaxseed, sorghum, and cassava and their coproducts; (v) tannins, which are present in sorghum, canola, pulses, cottonseed, and barley and their coproducts; (vi) fiber, which is present in almost all plant feedstuffs for pigs; and (vii) mycotoxins, which can be present in almost all plant feedstuffs for pigs. This chapter will focus on negative effects of the fore-mentioned antinutritional factors on dietary nutrient utilization in pigs and the underlying mechanisms.

## Phytic Acid

Phytic acid, which consists of an inositol ring with six phosphate groups is the major storage form of P in plants. Naturally, phytic acid occurs as phytate (i.e., salt form) in spherical inclusions called globoids located within protein bodies (Prattley and Stanley 1982; Ockenden et al. 2004; Joyce et al. 2005), and it is concentrated in seeds, where it provides P required for seed germination and development (Centeno et al. 2001). Thus, phytic acid is present in almost all feedstuffs of plant origin that are used to formulate swine diets. The phytic acid content in these feedstuffs is variable and dependent on several factors, among them, type, variety, and growing conditions of crops from which the feedstuffs are derived (Steiner et al. 2007). Generally, however, it is highest in cereal milling coproducts (1.7–3%) followed by oilseed meals (1.6–2.4%) and then cereal grains (0.75–1.0%; Eeckhout and De Paepe 1994; Selle et al. 2003; Steiner et al. 2007). Its higher content in the cereal milling by-products than in the cereals themselves is attributable to its high content in the aleurone

*Sustainable Swine Nutrition*, Second Edition. Edited by Lee I. Chiba.
© 2023 John Wiley & Sons Ltd. Published 2023 by John Wiley & Sons Ltd.

and germ cells, which are part of the coproducts (Ravindran et al. 2006; Steiner et al. 2007). Phytic acid content in practical swine diets ranges from around 1–2% (Woyengo and Nyachoti 2011).

Monogastric animals such as pigs and poultry poorly degrade the phytic acid in the stomach and small intestine because they do not produce sufficient amount of phytase, which cleaves phosphate groups from phytic acid (Kornegay 2001). For example, Rutherfurd et al. (2004) and Cowieson et al. (2006) found its ileal digestibility in broilers to be as low as 10%. Phytic acid has negative charges, and hence it can bind positively charged molecules of dietary or endogenous origin in the gastrointestinal tract, leading to reduced nutrient utilization (Maenz 2001). However, phytic acid can be degraded by phytase that is produced by microorganisms in the hindgut. Hence, phytic acid can negatively affect nutrient utilization in the stomach and small intestine.

Phytic acid reduces the apparent mineral digestibility in pigs. The inclusion of phytic acid in cornstarch-casein-based diet for nursery pigs at 2% reduced apparent ileal digestibility of Ca, Mg, Na, and K by 41.3, 32.1, 67.4, and 6.5 percentage points, respectively (Woyengo et al. 2009). Also, an increase in the dietary level of phytic acid from 0.94 to 1.43% resulted in reduced ash digestibility in growing pigs by 2.2 percentage points (Kemme et al. 1999). Phytic acid reduces apparent mineral digestibility in pigs likely by reducing dietary mineral absorption and increasing endogenous secretion of the minerals in the gastrointestinal tract lumen. This is because (i) phytic acid reduced absorption of $^{59}Fe^{3+}$ (Kim et al. 1993) and $^{65}Zn$ (Rubio et al. 1994) in rats; (ii) decreased apparent ileal digestibility of sodium and magnesium in piglets to negative values (−18.2 and −3.0%, respectively; Woyengo et al. 2009), implying that phytic increased ileal endogenous losses of these minerals; and (iii) increased endogenous losses of calcium, iron, sodium (Cowieson et al. 2004) and of calcium, magnesium, manganese, and sodium (Cowieson et al. 2006) in broilers. Cations have positive charges at acidic, neutral and basic pH. Thus, phytic acid can (i) complex dietary cations, leading to their reduced digestibility and (ii) complex endogenous cations, leading to their increased secretion through negative feedback mechanisms, reduced re-absorption, and hence increased endogenous losses of the cations.

Amino acids have net positive charges at acidic pH and net negative charges at neutral pH, and hence phytic acid can directly bind amino acids of dietary and endogenous origin at acidic pH found in the stomach and bind amino acids of dietary and endogenous origin via cations at neutral pH found in the small intestine (Maenz 2001). Indeed, an increase in the level of phytic acid from 0 to 2% in diets for nursery pigs reduced gastric pepsin activity (Woyengo et al. 2010). Also, an increase in the level of phytic acid from 0.78 to 1.56% in diets for growing pigs reduced apparent ileal digestibility of amino acids (Liao et al. 2005). However, supplementation of a casein–cornstarch–casein-based (phytic acid-free) diet for nursery pigs with phytic acid at 2% did not affect ileal endogenous losses of amino acids (Woyengo et al. 2009). It should be noted that the amount of endogenous amino acids that appear at the terminal ileum is a function of their rate of production and absorption. Thus, phytic acid may increase the secretion of protein without effect on ileal flow of the amino acids if the amino acids are re-absorbed. However, energy is spent during the synthesis and reabsorption of endogenous protein. Hence, phytic acid may reduce dietary energy availability for production without affecting ileal flow of endogenous amino acids.

In addition to minerals and amino acids, phytic acid can decrease energy digestibility. For instance, the apparent ileal and total tract digestibilities of energy in nursery pigs were reduced by 7.5 and 6.1%, respectively, due to an increase in dietary phytic acid concentration from 0.78 to 1.56% (Liao et al. 2005). Phytic can reduce energy digestibility by reducing the digestibility carbohydrates, lipids, and protein, which are the energy-generating nutrients. Phytic acid reduced in vitro starch digestibility (Thompson et al. 1987), and activities of α-amylase, sucrose, and maltase in the duodenum of broilers (Liu et al. 2008), implying that phytic acid reduces the digestibility of carbohydrates partly by reducing the activity of digestive carbohydrases. Also, dietary phytic acid reduced apparent absorption of total lipid and cholesterol in mice (Lee et al. 2007) and increased fecal cholesterol and bile acid excretion in rats (Lee et al. 1997), indicating that phytic acid reduces apparent fat digestibility partly by reducing (re)absorption of lipids.

The inclusion of phytic acid in cornstarch-based diet (phytic acid free diet) at 2% reduced growth performance of nursery pigs (7.4–12.8 kg body weight) by 37% (Woyengo et al. 2012), which is attributed to reduced nutrient digestibility, increased endogenous nutrient losses, and increased requirement of dietary energy for maintenance. Thus, phytic acid negatively affects the nutrient utilization in pigs, and its negative effects should be alleviated.

## Trypsin Inhibitors

TI are bioactive proteins that inhibit the activity of trypsin or trypsin and chymotrypsin. TI occur naturally in many plants, and their function is to protect the plants from predators (Ryan 1990). Among the feedstuffs used to formulate swine diets, soybeans, pulses, and camelina and their coproducts contain appreciable amounts of TI. There are two major classes of TI: Kunitz and Bowman-Birk (Pusztai et al. 2004). Kunitz inhibits trypsin, whereas Bowman-Birk inhibits both trypsin and chymotrypsin (Pusztai et al. 2004). The major TI in soybeans and its coproducts is Kunitz, whereas the major TI in pulses is Bowman-Birk (Jezierny et al. 2010); TI present in camelina and its coproducts have not been characterized. Kunitz is more heat labile than Bowman-Birk (Jezierny et al. 2010).

Dietary TI reduces amino acid digestibility by binding to pancreatic trypsin and chymotrypsin in the small intestine to form inactive complexes that cannot digest the partially digested protein coming from the stomach (Jezierny et al. 2010). The partially undigested protein from the stomach stimulates I cells in the duodenum to secrete cholecystokinin (CCK), which in turn increases the secretion of these enzymes into the small intestine via a negative feedback mechanisms (Hara et al. 2000; Morisset 2008). The increased secretion of pancreatic enzymes results in increase in size of the pancreas (Pacheco et al. 2014), and hence increased expenditure of dietary energy in the pancreas. The increased secretion of pancreatic enzymes also results in increased requirement of dietary amino acids for synthesis of the pancreatic enzymes. Finally, the increased production of CCK can result in reduced voluntary feed intake because CCK inhibits feed intake (Ripken et al. 2015). Thus, TI can reduce performance of pigs by reducing amino acid digestibility, increasing dietary requirement of energy and amino acids for maintenance and by reducing the voluntary feed intake (Weller et al. 1990; Ripken et al. 2015).

Intestinal activities of trypsin and chymotrypsin of pigs were reduced due to an increase in the dietary level of TI by 8.27 TIU/mg through supplementation with purified TI (Yen et al. 1977). Digestibility of crude protein and lysine in growing pigs fed a diet containing defatted soybeans flour was increased from 38 to 77% and from 41 to 80%, respectively, due to reduction in dietary TI from 13.3 to 3.3 TIU/mg through autoclaving of the defatted soybean flour (Li et al. 1998). Similarly, apparent total tract digestibility of N was reduced by 45% due to an increase in the dietary level of TI from 0 to 23 TIU/mg through replacement of casein or heat-treated soy flour with raw soy flour (Struthers et al. 1983). However, pancreas weight of pigs was not affected by an increase in the dietary level of TI by 8.27 TIU/mg through supplementation with purified TI (Yen et al. 1977). Also, pancreas weight of pigs was unaffected by an increase in the dietary level of TI from 15 to 21.3 TIU/mg through replacement of soybean meal with raw soybeans (Myer et al. 1982). Additionally, pancreas weight of pigs was unaffected, whereas that for rats was increased by an increase in the dietary level of TI from 0 to 23 TIU/mg through replacement of casein or heat-treated soy flour with raw soy flour (Struthers et al. 1983). Results from studies with poultry have shown increased pancreas weight due to dietary TI (Chohan et al. 1993; Pacheco et al. 2014). Thus, dietary TI reduces protein digestibility in pigs. However, unlike in poultry and rats, dietary TI appears to have limited effect on the size of pancreas in pigs.

The effects of including TI-containing soybean products in diets for pigs on feed intake and growth performance have been extensively studied; results from these studies were recently reviewed by Woyengo et al. (2017). In summary, dietary TI can be increased up to 3.00 TIU/mg without compromising performance of pigs; dietary TI levels greater than 3.00 TIU/mg result in reduced growth performance of pigs. Soybean meal has relatively lower level of TI because most TI in it is inactivated during the desolventizing-toasting stage of oil extraction process. The TI in soybean meal ranged from 3 to 12 TIU/mg (Fan et al. 1995; Valencia et al. 2008; Baker et al. 2010; Woyengo et al. 2014). Thus, soybean meal can be included in diets for pigs (that do not contain any other TI-containing feedstuffs) at ≤25% without any negative effects of TI on growth performance of the pigs. However, the level of soybean meal in practical diets for pigs rarely exceeds 35%, implying that TI cannot limit the inclusion of soybean meal in practical swine diets if the TI level in the soybean meal is less than 9 TIU/mg.

Raw full-fat soybean products contain significant amounts of TI; the values ranged from 70 to 112 TIU/mg (Chohan et al. 1993; Leeson and Atteh 1996). Thus, the TI in full-fat soybean meal should be reduced by feed processing technologies such as extrusion, toasting, or micronizing before their inclusion in swine diets. The TI in extruded or micronized full-fat soybeans ranged from 3.2 to 8.2 TIU/mg (Fan et al. 1995; Baker et al. 2010; Woyengo et al. 2014). Also, Jezierny et al. (2010), after reviewing results from several studies, reported that the TI in pulses (field pea, faba bean and lupin) ranged from <0.4 to 8.55 TIU/mg. Thus, TI may not limit the inclusion of pulses or heat-processed full-fat soybeans in practical swine diets (that do not contain other TI-containing feedstuffs) if the TI levels in the pulses and heat-processed full-fat soybeans are within the fore-mentioned ranges. Also, inclusion in swine diets of full-fat soybean products from soybeans that has been bred to have very low level of TI (<9 TIU/mg) may not be limited by their TI. The TI in camelina meal ranged from 12 to 28 TIU/mg (Budin et al. 1995), and hence the level of inclusion of camelina meal in diets for pigs should be limited and should be partly be based on the TI level in the camelina meal.

## Glucosinolates

Glucosinolates are sulfur-containing compounds that are found in plants of *Brassica* family (Mithen 2001; Tripathi and Mishra 2007). The function of glucosinolates in plants of *Brassica* family is to protect the plants from pathogens and predators (Mithen 2001; Tripathi and Mishra 2007). Feedstuffs that are used to formulate swine diets that contain glucosinolates include canola, camelina, and carinata and their coproducts. Of these feedstuffs, canola coproducts are the major feedstuffs that are used to formulate swine diets. There are various species of canola, which include Brassica napus and Brassica Juncea. Glucosinolates are not toxic, but can be degraded to various toxic products such as goitrin, nitriles, and thiocynates by heat (Slominski and Campbell 1989; Newkirk and Classen 2002) or by enzyme known as myrosinase (Mithen 2001; Tripathi and Mishra 2007). Myrosinase is present in plants of *Brassica* family in close proximity to glucosinolates, but in compartments that are different from those containing glucosinolates (Mithen 2001). Myrosinase is also produced by microorganisms that reside in the gastrointestinal tract of animals (Bell 1993). Thus, glucosinolates can be degraded to toxic products by (i) heat during the production of oil and coproducts from oilseeds of Brassica family; (ii) heat when feed is subjected to heat processing such as extrusion and pelleting; (iii) feedstuff myrosinase in the mouth (during feed mastication) and gastrointestinal tract; and (iv) microbial myrosinase in the gastrointestinal tract.

Glucosinolates can be classified into two major groups depending on the amino acids from which they are derived; aliphatic and aromatic glucosinolates. Aliphatic glucosinolates are synthesized from aliphatic amino acids such as methionine, whereas aromatic glucosinolates are synthesized from aromatic amino acids such as tryptophan and phenylalanine (Tripathi and Mishra 2007). Aromatic glucosinolates are more heat labile than aliphatic glucosinolates (Jensen et al. 1995). The content and composition of glucosinolates in a feedstuff is dependent on species of the feedstuff and on how the feedstuff has been processed. The major aliphatic and aromatic glucosinolates in Napus canola products are progoitrin and glucobrassicin, respectively (Seneviratne et al. 2010; Woyengo et al. 2011; Lee and Woyengo 2018), whereas the major aliphatic glucosinolates in Juncea canola products are gluconapin (Dehghani 2013; Smit et al. 2014). The major aliphatic glucosinolates in camelina and carinata products are glucocamelinin and sinigrin, respectively (Lawrence and Anderson 2018). Composition of aromatic glucosinolates for Juncea canola coproducts has not been reported. Generally, canola and its coproducts have lower content of glucosinolates than camelina and carinata products because the former has been bred to contain less glucosinolates. For instance, the total glucosinolate content in Napus canola coproducts ranged from 4.1 to 14.2 µmol/g (Seneviratne et al. 2010; Landero et al. 2011; Woyengo et al. 2011; Velayudhan et al. 2017; Smit et al. 2014; Zhou et al. 2016), whereas that in camelina and carinata coproducts ranged from 20.6 to 36.3 µmol/g (Thacker and Widyaratne 2012; Kahindi et al. 2014; Rodriguez-Hernandez and Anderson 2018). Among the canola products, Napus canola meals have lower content of aliphatic glucosinolates and hence total glucosinolates than Juncea canola coproducts. The aliphatic glucosinolates content in solvent extracted Napus canola meal ranged from 3.0 to 4.3 µmol/g (Landero et al. 2011; Smit et al. 2014), whereas that in Juncea canola meal ranged from 10.4 to 12.5 µmol/g (Dehghani 2013; Landero et al. 2013; Smit et al. 2014). Cold-pressed coproducts have greater content of aromatic glucosinolates and hence total glcuosinolates than solvent extracted or expeller pressed coproducts because the former is subjected to less heat during oil extraction. For instance, total glucosinolate content in cold-pressed canola cake was 14.9 µmol/g (Lee and Woyengo 2018), which is greater than the values (3.8–5.9 µmol/g) that were reported for solvent extracted Napus canola meals (Landero et al. 2011; Smit et al. 2014).

The composition of glucosinolate degradation products is partly dependent on the type of glucosinolate and pH of the reaction medium. Results from in vitro studies show that progrotrin is degraded to goitrin at neutral pH (Leoni et al. 1993; Galletti et al. 2001) and to nitriles at acidic pH (Matusheski et al. 2006; Frandsen et al. 2019). Glucobrassicin is degraded to thiocynates at neutral pH and to nitriles at acidic pH (Chevolleau et al. 1997; Agerbirk et al. 1998). The effect of reaction medium pH (acidic vs. neutral pH) on composition of glucocamelinin and sinigrin degradation products has not been reported. Generally, glucosinolates reduce voluntary feed intake because they are bitter. Nitriles are toxic when they are consumed in large quantities, and hence they have to be detoxified by the liver and kidney, leading to increased metabolic activity in the liver and kidney that resulted in hyperplasia, hypertrophy, and necrosis of cells in these organs (Pearson et al. 1983; Roland et al. 1996). However, consumption of small amount of nitriles can be beneficial as they have antioxidant activity (Tanii et al. 2008). Goitrin and thiocyanates interfere with the synthesis of thyroid hormones, leading to the enlargement of thyroid gland and reduced amounts of thyroid hormones within the body (Felkcr et al. 2016). Also, goitrin and thiocyanates have to be detoxified by liver and kidneys, leading to hyperplasia, hypertrophy, and necrosis of cells in these organs. An increase in metabolic activity and hence size of visceral organs results in increased dietary energy expenditure in these organs at the expense of skeletal tissue deposition. Also, reduction in thyroid hormone production results in reduced metabolic processes that are required for normal growth, leading to reduced growth rate of animals because thyroid hormones are involved in the regulation of various metabolic processes in the body. Thus, the toxicity of glucosinolates can potentially vary depending on their content in the diet, composition of glucosinolates in the diet, and pH conditions in the gastrointestinal tract.

The effects of increasing the level of glucosinolates in diets for pigs through the dietary inclusion of feedstuffs of *Brassica* family on liver, thyroid gland weights, and growth performance have been extensively studied and reviewed (Bell 1993; Woyengo et al. 2017). In summary, liver and thyroid gland weights and growth performance of pigs are not affected by Napus canola-derived glucosinolates when their dietary concentration is less than 2.50 μmol/g, implying that pigs can tolerate up to 2.50 μmol/g of Napus canola-derived glucosinolates. However, growth performance of pigs was negatively affected by Juncea canola-derived glucosinolates when their dietary concentration was increased to 1.30 μmol/g likely because of relatively greater bitterness of gluconapin in Juncea canola coproducts (Landero et al. 2013). Thus, the dietary level of Juncea canola-derived glucosinolates that the pigs can tolerate is less than 2.50 μmol/g; this dietary needs is to be identified. The tolerable dietary levels of camelina- and carinata-derived glucosinolates have not been reported. Napus solvent extracted canola meal had total glucosinolate content that ranged from 3.3 to 8.7 μmol/g (Landero et al. 2011; Smit et al. 2014). Thus, Napus solvent extracted canola meal can be included in diets for pigs (that do not contain any other glucosinolate-containing feedstuffs) at ≤28% without any negative effects of glucosinolates on growth performance of the pigs. Napus expeller pressed canola meal and cold-pressed canola cake had total glucosinolate content that ranged from 8.0 to 15.0 μmol/g (Seneviratne et al. 2010; Woyengo et al. 2011; Velayudhan et al. 2017; Zhou et al. 2016; Lee and Woyengo 2018). Thus, Napus expeller pressed canola meal and cold-pressed canola cake can be included in diets for pigs (that do not contain any other glucosinolate-containing feedstuffs) at ≤16% without any negative effects of glucosinolates on growth performance of the pigs. Grow-finish pigs (≥25 kg body weight) have lower requirement of dietary protein, and thus, the inclusion of Napus solvent extracted canola meal in their diets may not be limited by glucosinolates. However, the inclusion of Napus solvent extracted canola meal in the diets for nursery pigs is limited by glucosinolates due to their higher dietary requirement of protein. Also, the inclusion of Napus expeller pressed canola meal and cold-pressed canola cake in the diets for both nursery and grow-finish pigs is limited by glucosinolates due to the relatively high content of glucosinolates in these canola coproducts. As previously mentioned, the composition of glucosinolates in Juncea canola, camelina, and carinata coproducts is different from that of Napus canola coproducts. Thus, the dietary tolerable levels of glucosinolates in the former may be different from that of the latter. However, the dietary tolerable levels of glucosinolates in Juncea canola, camelina, and carinata coproducts have not been reported. Because Juncea canola, camelina, and carinata coproducts have too high levels of glucosinolates, their inclusion in swine diets is definitely limited by the presence of glucosinolates in them.

### Cynogenic Glucosides

Cynogenic glucosides are amino acid-derived compounds that are found in various plants in which they serve as protective agents against herbivores (Poulton 1990; Gleadow and Møller 2014). Feedstuffs that are used to formulate swine diets that contain cynogenic glucosides include flaxseed products, sorghum, and cassava.

Like glucosinolates, cynogenic glucosides are not toxic, but can be degraded to toxic hydrogen cyanide by enzyme known as β-glycosidase (Poulton 1990). The β-glycosidase is present in plants tissues in close proximity to cynogenic glucosides, but in compartments that are different from those containing cynogenic glucosides (Gleadow and Møller 2014). Hydrogen cyanide is bitter and also produces a special type of odor that deters herbivores from consuming hydrogen cyanide-containing products (Gleadow and Møller 2014). Additionally, the ingestion of hydrogen cyanide can lead to hypoxia because it can bind and inactivate enzymes that are involved in oxygen metabolism (Petrikovics et al. 2015). Finally, high consumption of hydrogen cyanide can lead to enlargement and necrosis of liver and kidneys because it is toxic and hence it has to be detoxified by these organs (Rocha-e-Silva et al. 2010). Indeed, toxic vacuoles were formed in liver of quails due to the consumption of potassium cyanide at 3.0 mg/bird/day (Rocha-e-Silva et al. 2010). The tolerable level of cynogenic glucosides in diets for pigs has not be reported, and hence there is need to fill this gap in knowledge.

## Tannins

Tannins are polyphenolic compounds that naturally occur in many plants in which they serve as defense mechanism against predators and pathogens (Bennick 2002; Gilani et al. 2012). Swine feedstuffs that have appreciable amounts of tannins include brown-seeded sorghum, pulses, canola, and their coproducts. Tannins are darkish brown to brown in color. Tannins are concentrated in seed hulls (Smulikowska et al. 2001), and hence the reason why hulls of feedstuffs that are rich in tannins are brownish in color. There are two classes of tannins: hydrolysable and condensed tannins (Jansman et al. 1994a; Bennick 2002; Gilani et al. 2012). Hydrolysable tannins are more susceptible to acidic, alkaline, and enzymatic hydrolysis, whereas the condensed tannins are more resistant to the hydrolysis (Bennick 2002; Gilani et al. 2012). The major tannins found in feedstuffs that are used to formulate swine diets are condensed tannins (Gilani et al. 2012).

Tannins can bind to dietary nutrients, leading to their reduced availability for digestion and absorption. Energy and protein (amino acids) are the first and second most expensive components of swine diets. Starch is the major source of energy in practical swine diets. Tannins interacted with starch by forming hydrogen and hydrophobic bonds with amylose component of starch (Amoako and Awika 2016; Amoako and Awika 2019), and this resulted in reduced in vitro digestibility of starch (Amoako and Awika 2016). Tannins have particularly strong affinity for proteins; they interact with proteins mainly through hydrophobic and hydrogen bonding (Bennick 2002). For instance, canola tannins precipitated bovine serum albumin (Naczk et al. 1996), whereas sorghum tannins precipitated wheat proteins (Dunn et al. 2015).

In addition to binding dietary nutrients, tannins bind to endogenous proteins to form insoluble complexes. In the mouth, tannins interacted with salivary proteins to form insoluble complexes (Lu and Bennick 1998; Soares et al. 2018). In the stomach, tannins interact with pepsin to form insoluble complexes, leading to reduced activity of the pepsin. For instance, tannins precipitated pepsin in vitro (Sathe and Sze-tao 1997; Helal et al. 2014). In the small intestine, tannins can interact with digestive enzymes, leading to their reduced activity. For example, tannins inhibited the activity of chymotrypsin (Sathe and Sze-tao 1997) and amylase in vitro (Gonçalves et al. 2011; Links et al. 2015) and of trypsin in ileum of pigs (Jansman et al. 1994b). Also, tannins reduced aminopeptidase activity in small intestinal mucosa of weaned pigs (van Leeuwen et al. 1995). Additionally, tannins strongly interacted with porcine gastric mucins (Gombau et al. 2019), implying that tannins can increase secretion of mucin within the gastrointestinal tract and interfere with nutrient absorption.

Finally, tannins reduced cecal microbial activity in vitro (Biagia et al. 2010), implying that tannins can reduce organic matter fermentation in the gastrointestinal tract. Binding of tannic acid to pepsin in the stomach of rats increased pepsin and hydrochloric acid secretions (Mitjavila et al. 1973), implying the binding of tannins to endogenous protein and minerals result in an increase in their secretion via negative feedback mechanisms. Increased secretion of endogenous nutrients results in increased dietary requirement for the nutrients for maintenance and increased dietary requirement of energy for the synthesis of the nutrients, re-absorption, and post-absorptive metabolism of the nutrients (Nyachoti et al. 1997).

The effects of tannins in diets for pigs on digestibility and endogenous losses of protein have been determined in several studies. Porcine in vitro total tract digestibility of crude protein for low-tannin field pea (0.025% tannin content) was greater than that for high-tannin field pea (0.87% tannin content) by 9.74% (Smulikowska et al. 2001). Also, an increase in the level of tannins in diet for growing pigs from 0.04 to 0.52% through replacement of 50% low-tannin field pea with high-tannin field pea decreased true ileal digestibility of crude protein by 6.24% (Święch and Buraczewska 2005). For faba bean, an increase in the level of tannins in diet for growing pigs from <0.10 to 0.68% through replacement of 20% low-tannin hulls of faba bean with high-tannin hulls of faba bean decreased true ileal digestibility of crude protein by 15.7% (Jansman et al. 1994b). Similarly, replacement of 20% low-tannin hulls of faba bean with high-tannin hulls of faba bean increased ileal and fecal endogenous losses of crude protein by 40.9 and 28.1%, respectively (Jansman et al. 1995). With regard to sorghum, standardized ileal digestibility of CP and lysine was reduced by 10.6 and 25.7%, respectively, due to replacement of 89.9% low-tannin sorghum (0.003% tannin) with the high-tannin sorghum (0.64% tannin) in diets for growing pigs (de Souza et al. 2019). The reduced protein digestibility by tannins can be attributed to their binding to dietary protein, protein-digesting enzymes, and gastrointestinal mucosa.

The effects of tannins in diets for pigs on energy digestibility have also been investigated. An increase in the level of tannins in diet for growing pigs from <0.001 to 0.06% through replacement of 30% low-tannin field bean with high-tannin field bean resulted in decreased apparent total tract digestibility of gross energy in pigs by 2.84% (Flis et al. 1999). Apparent ileal digestibility of gross energy was reduced by 8.9% due to replacement of 89.9% low-tannin sorghum (0.003% tannin) with the high-tannin sorghum (0.64% tannin) in diets for growing pigs (de Souza et al. 2019). Also, Pan et al. (2016) included sorghum containing low-tannin (<0.16% tannin), medium-tannin (0.67–0.98% tannin), or high tannin (1.11–1.51% tannin) in diets for pigs at 96.9% (as a sole source of energy and protein in the diets) and observed negative correlation between tannin content in sorghum and apparent total tract digestibility of gross energy, digestible energy, and metabolizable energy values. In their study, the apparent total tract digestibility of gross energy, digestible energy, and metabolizable energy values for high-tannin sorghum was lower than those for medium-tannin sorghum, which in turn was lower than those for the low-tannin sorghum. The reduced dietary energy digestibility and value by tannins can be attributed to their binding to dietary carbohydrates and protein, carbohydrate- and protein-digesting enzymes, and gastrointestinal mucosa.

Despite the reduction in nutrient digestibility, dietary tannins can have beneficial effects in diets for weaned pigs. For example, the inclusion of tannins in diets for weaned pigs at 0.113, 0.225, and 0.450 resulted in linear increase in average daily gain and feed efficiency and tended to (i) increased viable count of lactobacillus in jejunum and (ii) decreased crypt depth in ileum (Biagia et al. 2010). Recently, the inclusion of tannins in a diet at 0.054% reduced diarrhea of weaned pigs that had been challenged with enterotoxigenic *Escherichia coli* (Girard et al. 2018).

From these studies, it is apparent that dietary tannins not only reduce the nutrient digestibility in pigs but also improves the gut health of weaned pigs. The intensity of interactions between tannins and nutrients or endogenous secretions is dependent on total content, size, and molecular structure of tannins (Hagerman 1992). However, information is lacking on the effects of total content and molecular structure of tannins on the growth performance and gut health of pigs. Thus, there is a need to fill this gap in knowledge in order to identify optimal dietary level of tannins for weaned pigs and dietary levels of tannins that other classes of pigs can tolerate.

### Fiber

Fiber is composed of nonstarch polysaccharides (**NSP**) and polyphenols such as lignin (Kumar et al. 2012). The NSP can either be soluble or insoluble. Fiber is indigestible by endogenous enzymes produced in the stomach and small intestine of monogastric animals such as pigs (Bedford 2000; Jha et al. 2011), but NSP can be fermented in the hindgut of these animals depending on the degree of lignification (Van Soest 1994). Fiber can physically limit the availability of other nutrients for small intestinal digestion (Schulze et al. 1994; Wilfart et al. 2007; Hooda et al. 2010). Also, fiber can bind nutrients in gastrointestinal tract to form complexes that are indigestible (Choct et al. 2010). Finally, some soluble NSP can increase digesta viscosity, and thereby reduce the flow rate of digesta in the gastrointestinal tract, and further limit nutrient digestion, whereas insoluble fiber can increase gastrointestinal digesta passage rate, leading to reduced time: (i) of interaction between enzymes and their substrate (undigested nutrients) and (ii) for absorption of nutrients (Bedford and Schulze 1998; Hooda et al. 2011). The reduction in nutrient digestibility by fiber can reduce growth performance of pigs. Also, because fiber is indigestible, its inclusion in diets for pigs dilutes dietary nutrient density that may reduce dietary nutrient intake. Additionally, the increased digesta viscosity due to dietary soluble fiber can promote satiety, leading to reduced voluntary feed intake (Scheenman 1985). Volatile fatty acids, which are the end products of fiber fermentation in the hindgut, stimulate the production of gastrointestinal peptide hormones (e.g., peptide tyrosine tyrosine and glucagon-like peptide-1) that inhibit food intake (Adam et al. 2014).

The effects of fiber on nutrient digestibility and growth performance of pigs have been investigated. Increasing dietary total NSP from 10 to 70 g/kg through the inclusion of guar gum (a soluble NSP) in a semi-purified diet of weaned pigs reduced daily gain by 40% (McDonald et al. 1999). Increasing total dietary fiber from 0 to 389 g/kg through the inclusion of sugar beet pulp (that contains insoluble and soluble NSP) in a purified diet also reduced feed intake and apparent total tract digestibility of N for growing pigs by 11 and 15%, respectively (Zhang et al. 2013). Similarly, the inclusion of 100 g insoluble fiber/kg from either cellulose or ground straw in diets for growing pigs reduced apparent total tract digestibility of N and amino acids, but not apparent ileal digestibility of N and amino acids (Sauer et al. 1991). However, increasing the dietary inclusion of total NSP from 103 to 138 g/kg using wheat bran (rich in insoluble NSP) in a complete diet for weaned pigs increased feed intake by 35%, whereas increasing the dietary inclusion of total NSP to 145 g/kg using sugar beet pulp in the diet did not affect feed intake of weaned pigs (Molist et al. 2009). Also, increasing dietary level of total NSP from 120 to 180 g/kg through the inclusion of cellulose (an insoluble NSP) in a complete diet for broiler chicks increased the feed intake by 9.3%, whereas increasing the dietary level of total NSP up to 190 g/kg through the dietary inclusion of wood shavings did not affect the feed intake of broiler chicks (Amerah et al. 2009). Thus, soluble fiber reduces feed intake and nutrient digestibility. However, the effect of insoluble fiber on feed intake is variable depending on the source and dietary inclusion level. Insoluble fiber may have limited effects on ileal nutrient digestibility. Pig feedstuffs that have relatively high content of soluble fiber are sugar beet pulp, oats, and rye, whereas pig feedstuffs that have high content of insoluble fiber are DDGS; corn, wheat, and rice brans; wheat millrun/middlings; and soybean hulls. Thus, the inclusion of these feedstuffs in swine diets can be limited by their high fiber content.

**Mycotoxins**

Mycotoxins are secondary metabolites that are produced by fungi (moulds) that grow on several feedstuffs or foodstuffs (Bryden 2012). Swine feedstuffs that fungi can grow on include cereal grains and oilseeds and their coproducts (Guerre 2016). Generally, however, cereal grains and their coproducts are more negatively affected by the fungal growth than oilseed and their coproducts (Guerre 2016). The DDGS has approximately three times higher levels of mycotoxins than parent cereal grains because during the production of ethanol and DDGS from the cereal grains, most starch is converted to ethanol and carbon dioxide to leave DDGS with higher content of all components (other than starch) that were originally in the parent cereal grains. Fungi can grow on the feedstuffs before harvest, during storage after harvest, and during storage after grain processing. The major genera of fungi that grow on the feedstuffs to produce mycotoxins include *Aspergillus* and *Fusarium* (Bryden 2012). There are several types the mycotoxins that are produced by these fungi when they grow of the feedstuffs. The major ones are aflatoxins, fumonisins, deoxynivalenol, and zearalenone (Binder et al. 2007; Bryden 2012; Escrivá et al. 2015). Aflatoxins are produced by various species of *Aspergillus* (Binder et al. 2007; Bryden 2012), whereas fumonisins, deoxynivalenol, and zearalenone are produced by various species of *Fusarium* (Bryden 2012; Escrivá et al. 2015). The content and composition of mycotoxins in feedstuffs varies widely depending on factors that affect the growth of fungi on the feedstuffs. The major factor that affects fungi growth on feedstuffs before harvest is climate. *Aspergillus* grow well in warm and wet weather conditions, whereas *Fusarium* grow well in cool and wet weather conditions. For instance, warm humid subtropical and tropical conditions favors the growth of *A. flavus* and *A. parasiticus* on corn ears leading to accumulation aflatoxins in corn grain, whereas temperate climate conditions favors the growth of *F. graminearum* on corn ears, resulting in the accumulation of deoxynivalenol in corn grain (Guerre 2016). Fumonisins frequently contaminate corn in most parts of the world except North America (Guerre 2016). The major factors that affect fungi growth on feedstuffs post-harvest are temperature and moisture. Stored feedstuffs contain microorganisms that they acquired pre-storage (Magan et al. 2003). Higher storage temperature and higher grain moisture content (>13%) favor the growth of these fungi on the feedstuffs (Magan et al. 2003). The effects of aflatoxins, fumonisins, deoxynivalenol, and zearalenone on performance and health of pigs are discussed later.

Consumption of aflatoxins results in reduced growth performance of pigs (Rustemeyer et al. 2010; Andretta et al. 2012). Upon ingestion, the aflatoxins are rapidly absorbed and metabolized/detoxified in liver. The increased metabolism in liver due to aflatoxin results in increased weight of the liver (Andretta et al. 2012), and hence reduced availability of nutrients for performance as previously discussed under phytic acid, glucosinolates, and tannins. Thus, the reduced performance of pigs due to consumption of aflatoxins is mainly due to hepatotoxicity. Consumption of deoxynivalenol results in increased intestinal permeability to toxins and pathogens (Pinton et al. 2010), compromised immune system (Gauthier et al. 2013), reduced nutrient digestibility (Jo et al. 2016), vomiting (Smith and MacDonald 1991), and hence reduced performance of pigs (Andretta et al. 2012). Fumonisins are structurally similar to sphingolipids, which are involved in cell growth, function, and apoptosis (Terciolo et al. 2019). The ingestion of fumonisins results in interference in sphingolipids synthesis, thereby impairing the functions liver (Andretta et al. 2012) and immune system (Devriendt et al. 2009), and hence reduced performance of pigs (Andretta et al. 2012). Zearalenone are structurally similar to oestrogens. Their ingestion results in their binding to oestrogen receptors, leading to increased uterus size (Andretta et al. 2012) and reduced reproductive performance of pigs (Bryden 2012).

It is thus apparent from this studies that dietary mycotoxins negatively affect pig performance. Because of the negative effects of mycotoxins on performance of pigs, U.S. Food and Drug Administration and European Union has come up with acceptable maximum dietary levels of various mycotoxins for various classes of pigs. Dietary mycotoxin levels below the acceptable maximum levels are assumed to have negligible effect on performance of pigs. However, results some recent studies have shown that pig performance can be negatively affected by dietary levels of mycotoxins that are below the acceptable maximum dietary levels. For example, Alizadeh et al. (2015) observed that contamination of diets for piglets with deoxynivalenol at 0.9 mg/kg negatively affected growth performance and various markers of gut health such as expression of different tight junction proteins, which is the European Union recommended upper limit of deoxynivalenol in diets for piglets. Also, Terciolo et al. (2019) observed histological alterations in the heart and the intestine due to contamination of diets for piglets with fumonisins at 3.7 mg/kg, which was less than the European Union recommended upper limit (5 mg/kg) of fumonisins in diets for piglets. Thus, there may be a need to re-identify the maximum tolerable levels of mycotoxins in diets for pigs.

## Summary

Plant feedstuffs for pigs contain various ANF including phytic acid, TI, glucosinolates, cynogenic glucosides, tannins, fiber, and mycotoxins that limit the dietary nutrient utilization in pigs. Phytic acid is an ANF that is present in almost all plant feedstuffs that are used to formulate practical swine diets. Phytic acid reduces dietary nutrient utilization through reduced digestibility and increased endogenous intestinal secretion of nutrients. Trypsin inhibitor is an ANF that is present in soybean coproducts, pulses, and camelina. The maximum tolerable TI in diets for pigs appears to be 3.00 TIU/mg. Pulses have relatively low content of TI. Also, solvent-extracted soybean meal has low content of TI because most TI in it is destroyed during the desolventizing-toasting stage of oil extraction. Hence, the inclusion of pulses and solvent extracted soybean meal in swine diets is generally not limited by TI. However, raw soybeans or expeller-pressed soybean coproducts and camelina coproducts have high content of TI, and hence they should be heat treated to inactivate most of the TI before their inclusion in swine diets or should be included in diets based on their actual TI content and maximum tolerable of TI in diets.

Glucosinolates are the major ANF in feedstuffs of *Brassica* family. Pigs can tolerate up to 2.50 µmol/g of Napus canola-derived glucosinolates in their diets. The conventional Napus solvent extracted canola meal has low concentration of glucosinolates (<9 µmol/g). Thus, up to 28% of Napus solvent extracted canola meal can be included in diets for pigs without major effects on nutrient utilization. However, Napus expeller-pressed canola meal and Napus cold-pressed canola cake can have high and variable concentration of glucosinolates (>9 µmol/g), and hence their level of inclusion in diets for pigs can be lower than that for Napus solvent extracted canola meal and should ideally be based on the actual glucosinolate concentration in the coproducts and maximum tolerable of glucosinolate in diets. The composition of glucosinolates in Juncea canola, camelina, and carinata coproducts is different from that of Napus canola coproducts, and hence the tolerable levels of glucosinolates in these coproducts in pigs should be established.

Cynogenic glucosides are the major ANF in flaxseed and cassava coproducts that limit nutrient utilization by pigs. Dietary fiber is an ANF that is present in almost all plant feedstuffs that are used to formulate practical swine diets. Soluble fiber present in rye, oats, and barley reduces dietary nutrient utilization by increasing digesta viscosity. Insoluble fiber reduces dietary nutrient utilization by increasing the passage rate of digesta in gastrointestinal tract. Tannins are ANF that are mainly present in sorghum and pulses and they reduces dietary nutrient utilization through reduced feed intake and digestibility and increased endogenous intestinal secretion of nutrients. Mycotoxins are ANF that can be present in almost all plant feedstuffs that are used to formulate practical swine diets depending on growing and storage conditions. Aflatoxins, fumonisins, and deoxynivalenol reduce the dietary nutrient utilization mainly by interfering with liver function. Deoxynivalenol additionally reduces dietary nutrient utilization by negatively affecting gastrointestinal integrity and causing the pigs to vomit.

# References

Adam, C. L., P. A. Williams, M. J. Dalby, K. Garden, L. M. Thomson, A. J. Richardson, S. W. Gratz, and A. W. Ross. 2014. Different types of soluble fermentable dietary fibre decrease food intake, body weight gain and adiposity in young adult male rats. Nutr. Metabol. 11:36. http://www.nutritionandmetabolism.com/content/11/1/36.

Agerbirk, N., C. E. Olsen, and H. Sørensen. 1998. Initial and final products, nitriles, and ascorbigens produced in myrosinase-catalyzed hydrolysis of indole glucosinolates. J. Agric. Food Chem. 46:1563–1571.

Alizadeh, A., S. Braber, P. Akbari, J. Garssen, and J. Fink-Gremmels. 2015. Deoxynivalenol impairs weight gain and affects markers of gut health after low-dose, short-term exposure of growing pigs. Toxins 7:2071–2095. doi:https://doi.org/10.3390/toxins7062071.

Amerah, A. M., V. Ravindran, and R. G. Lentle. 2009. Influence of insoluble fibre and whole wheat inclusion on the performance, digestive tract development and ileal microbiota profile of broiler chickens, Br. Poult. Sci. 50:366–375.

Amoako, D. B., and J. M. Awika. 2016. Polymeric tannins significantly alter properties and in vitro digestibility of partially gelatinized intact starch granule. Food Chem. 208:10–17.

Amoako, D. B., and J.M. Awika. 2019. Resistant starch formation through intrahelical V-complexes between polymeric proanthocyanidins and amylose. Food Chem. 285:326–333.

Andretta, I., M. Kipper, C. R. Lehnen, L. Hauschild, M. M. Vale, and P. A. Lovatto. 2012. Meta-analytical study of productive and nutritional interactions of mycotoxins in growing pigs. Animal 6:1476–1482.

Baker, K.M., B. G. Kim, and H. H. Stein. 2010. Amino acid digestibility in conventional, high-protein, or low-oligosaccharide varieties of full-fat soybeans and in soybean meal by weanling pigs. Anim. Feed Sci. Technol. 162: 66–73.

Bedford, M. R. 2000. Exogenous enzymes in monogastric nutrition - their current value and future benefits. Anim. Feed Sci. Technol. 86:1–13.

Bedford, M. R., and H. Schulze. 1998. Exogenous enzymes for pigs and poultry. Nutr. Res. Rev. 11:91–114.

Bell, J. M. 1993. Factors affecting the nutritional value of canola meal: A review. Can. J. Anim. Sci. 73:679–697.

Bennick, A. 2002. Interaction of plant polyphenols with salivary proteins. Crit. Rev. Oral Biol. Med. 13:184–196.

Biagia, G., I. Cipollini, B. R. Paulicks, and F. X. Roth. 2010. Effect of tannins on growth performance and intestinal ecosystem in weaned piglets. Arch. Anim. Nutr. 64:121–135.

Binder, E. M., L. M. Tan, L. J. Chin, J. Handl, and J. Richard. 2007. Worldwide occurrence of mycotoxins in commodities, feeds and feed ingredients. Anim. Feed Sci. Technol. 137:265–282.

Bryden, W. L. 2012. Mycotoxin contamination of the feed supply chain: Implications for animal productivity and feed security. Anim. Feed Sci. Technol. 173: 134–158.

Budin, J.T., W. M. Breene, and D. H. Putnam. 1995. Some compositional properties of camelina (Camelina sativa L. Crantz) seeds and oils. JAOCS. 72: 309–315.

Centeno, C., A. Viveros, A. Brenes, R. Canales, A. Lozano, and C. de la Cuadra. 2001. Effect of several germination conditions on total p, phytate p, phytase, and acid phosphatase activities and inositol phosphate esters in rye and barley. J. Agric. Food Chem. 49:3208–3215.

Chevolleau, S., N. Gasc, P. Rollin, and J. Tulliez. 1997. Enzymatic, chemical, and thermal breakdown of 3H-labeled glucobrassicin, the parent indole glucosinolate. J. Agric. Food Chem. 45:4290–4296.

Choct, M., Y. Dersjant-Li, J. McLeish, and M. Peisker. 2010. Soy oligosaccharides and soluble non-starch polysaccharides: A review of digestion, nutritive and anti-nutritive effects in pigs and poultry. Asian-Aust. J. Anim. Sci. 23:1386–1398.

Chohan, A. K., R. M. G. Hamilton, M. A. McNiven, and J. A. Macleod. 1993. High protein and low trypsin inhibitor varieties of full-fat soybeans in broiler chicken starter diets. Can. J. Anim. Sci. 73:401–409.

Cowieson A. J., T. Acamovic, and M. R. Bedford. 2004. The effects of phytase and phytic acid on the loss of endogenous amino acids and minerals from broiler chickens. Br. Poult. Sci. 45:101–108.

Cowieson A. J., T. Acamovic, and M. R. Bedford. 2006. Phytic acid and phytase: Implication for protein utilisation by poultry. Poult. Sci. 85:878–885.

de Souza, T. C. R., I. E. Á. Árres, E. R. Rodríguezc, and G. Mariscal-Landín. 2019. Effects of kafirins and tannins concentrations in sorghum on the ileal digestibility of amino acids and starch, and on the glucose and plasma urea nitrogen levels in growing pigs. Livest. Sci. 227:29–36.

Dehghani, Z. 2013. The effects of canola (Brassica napus) and juncea (Brassica juncea) meals in diets on broilers and turkeys. MS Thesis. Dalhousie Univ., Halifax, Nova Scotia, Canada.

Devriendt, B., M. Gallois, F. Verdonck, Y. Wache, D. Bimczok, I. P. Oswald, IB. M. Goddeeris, and E. Cox. 2009. The food contaminant fumonisin B(1) reduces the maturation of porcine CD11R1(+) intestinal antigen presenting cells and antigen-specific immune responses, leading to a prolonged intestinal ETEC infection. Vet. Res. 40:40. doi:https://doi.org/10.1051/vetres/2009023.

Dunn, K. L., L. Yang, A. Girard, S. Bean, and J. M. Awika. 2015. Interaction of sorghum tannins with wheat proteins and effect on in vitro starch and protein digestibility in a baked product matrix. J. Agric. Food Chem. 63:1234–1241.

Eeckhout, W., and M. De Paepe. 1994. Total phosphorus, phytate-phosphorus and phytase activity in plant feedstuffs. Anim. Feed Sci. Technol. 47:19–29.

Escrivá, L., G. Font, and L. Manyes. 2015. in vivo toxicity studies of fusarium mycotoxins in the last decade: A review. Food Chem. Toxicol. 78:185–206.

Fan, M.Z., W. C. Sauer, and C. F. M. de Lange. 1995. Amino acid digestibility in soybean meal, extruded soybean and full-fat canola for early-weaned pigs. Anim. Feed Sci. Technol. 52: 189–203.

Felker, P., R. Bunch, and A. M. Leung. 2016. Concentrations of thiocyanate and goitrin in human plasma, their precursor concentrations in brassica vegetables, and associated potential risk for hypothyroidism. Nutr. Rev. 74:248–258.

Flis, M., W. Sobotka, C. Purwin, and Z. Zdunczyk. 1999. Nutritional value of diets containing field bean (Vicia faba L.) seeds with high or low proanthocyanidin levels for pig. J. Anim. Feed Sci. 8:171–180.

Frandsen, H. B., J. C. Sørensen, S. K. Jensen, K. E. Markedal, M. S. Joehnke, H. Maribo, S. Sørensen, and H. Sørensen. 2019. Non-enzymatic transformations of dietary 2-hydroxyalkenyl and aromatic glucosinolates in the stomach of monogastrics. Food Chem. 291:77–86.

Galletti, S., R. Bernardi, O. Leoni, P. Rollin, and S. Palmieri. 2001. Preparation and biological activity of four epiprogoitrin myrosinase-derived products. J. Agric. Food Chem. 49:471–476.

Gauthier, T., Y. Waché, J. Laffitte, I. Taranu, N. Saeedikouzehkonani, Y. Mori, and I. P. Oswald. 2013. Deoxynivalenol impairs the immune functions of neutrophils. Mol. Nutr. Food Res. 57:1026–1036.

Gilani, G. S., C. W. Xiao, and K. A. Cockell. 2012. Impact of antinutritional factors in food proteins on the digestibility of protein and the bioavailability of amino acids and on protein quality. Br. J. Nutr. 108:S315–S332.

Girard, M., S. Thanner, N. Pradervand, D. Hu, C. Ollagnier, and G. Bee. 2018. Hydrolysable chestnut tannins for reduction of post-weaning diarrhea: Efficacy on an experimental ETEC F4 model. PLOS ONE. https://doi.org/10.1371/journal.pone.0197878.

Gleadow, R.M., and B. L. Møller. 2014. Cyanogenic glycosides: Synthesis, physiology, and phenotypic plasticity. Annu. Rev. Plant Biol. 65:155–185.

Gombau, J., P. Nadal, N. Canela, S. Gómez-Alonso, E. García-Romero, P. Smith, I. Hermosín-Gutiérrez, J. M. Canals, and F. Zamora. 2019. Measurement of the interaction between mucin and oenological tannins by Surface Plasmon Resonance (SPR); relationship with astringency. Food Chem. 275:397–406.

Gonçalves, R., N. Mateus, and V. de Freitas. 2011. Inhibition of α-amylase activity by condensed tannins. Food Chem.125:665–672.

Guerre, P. 2016. Worldwide mycotoxins exposure in pig and poultry feed formulations. Toxins 8:350. doi:https://doi.org/10.3390/toxins8120350.

Hagerman, A. E. 1992. Tannin protein interactions. ACS Symp. Ser. 506:236–247.

Hara, H., S. Ohyama, and T. Hira. 2000. Luminal dietary protein, not amino acids, induces pancreatic protease via CCK in pancreaticobiliary- diverted rats. Am. J. Physiol. Gastrointest. Liver Physiol. 278:G937–G945.

Helal. A., D. Tagliazucchi, E. Verzelloni, and A. Conte. 2014. Bioaccessibility of polyphenols and cinnamaldehyde in cinnamon beverages subjected to in vitro gastro-pancreatic digestion. J. Funct. Foods. 7:506–516.

Hooda, S., B. U. Metzler-Zebeli, T. Vasanthan, and R. T. Zijlstra. 2011. Effects of viscosity and fermentability of dietary fibre on nutrient digestibility and digesta characteristics in ileal-cannulated grower pigs. Br. J. Nutr. 106:664–674.

Hooda, S., J. J. Matte, T. Vasanthan, and R. T. Zijlstra. 2010. Dietary oat β-glucan reduces peak net glucose flux and insulin production, and modulates plasma incretin in portal-vein catheterized grower pigs. J. Nutr. 140:1564–1569.

Jansman, A. F. J., A. A. Frohlich, and R. R. Marquardt. 1994a. Production of proline-rich proteins by the parotid glands of rats is enhanced by feeding diets containing tannins from faba beans (Vicia faba L.). J. Nutr. 124:249–258.

Jansman, A. J., H. Enting, M. W. Verstegen, and J. Huisman. 1994b. Effect of condensed tannins in hulls of faba beans (Vicia faba L.) on the activities of trypsin (EC 2.4.21.4) and chymotrypsin (EC 2.4.21.1) in digesta collected from the small intestine of pigs. Br. J. Nutr. 71:627–641.

Jansman, A. J. M., M. W. A. Verstegen, J. Huisman, and J. W. O. Vandenberg. 1995. Effects of hulls of faba beans (viola-faba l) with a low or high content of condensed tannins on the apparent ileal and fecal digestibility of nutrients and the excretion of endogenous protein in ileal-digesta and feces of pigs. J. Anim. Sci. 73:118–127.

Jensen, S. K., Y. G. Liu, and B. O. Eggum. 1995. The effect of heat treatment on glucosinolates and nutritional value of rapeseed meal in rats. Anim. Feed Sci. Technol. 53:17–28.

Jezierny, D., R. Mosenthin, and E. Bauer. 2010. The use of grain legumes as a protein source in pig nutrition: A review. Anim. Feed Sci. Technol. 157:111–128.

Jha, R., J. Bindelle, B. Rossnagel, A. Van Kessel, and P. Leterme. 2011. in vitro evaluation of the fermentation characteristics of the carbohydrate fractions of hulless barley and other cereals in the gastrointestinal tract of pigs. Anim. Feed Sci. Technol. 163:185–193.

Jo, H., C. Kong, M. Song, and B. G. Kim. 2016. Effects of dietary deoxynivalenol and zearalenone on apparent ileal digestibility of amino acids in growing pigs. Anim. Feed Sci. Technol. 219:77–82.

Joyce, C., A. Deneau, K. Peterson, I. Ockenden, V. Raboy, and J. N. A. Lott. 2005. The concentrations and distributions of phytic acid phosphorus and other mineral nutrients in wild-type and low phytic acid Js-12-LPA wheat (Triticum aestivum) grain parts. Can. J. Bot. 83:1599–1607.

Kahindi, R. K., T. A. Woyengo, P. A. Thacker, and C. M. Nyachoti. 2014. Energy and amino acid digestibility of camelina cake fed to finishing pigs. Anim. Feed Sci. Technol. 193:93–101.

Kemme, P. A., A. W. Jongbloed, Z. Mroz, J. Kogut, and A. C. Beynen. 1999. Digestibility of nutrients in growing–finishing pigs is affected by Aspergillus niger phytase, phytate and lactic acid levels 2. Apparent total tract digestibility of phosphorus, calcium and magnesium and ileal degradation of phytic acid. Livest. Prod. Sci. 58:119–127.

Kim, Y., C. E. Carpenter, and A. W. Mahoney. 1993. Gastric acid production, iron status and dietary phytate alter enhancement by meat of iron absorption in rats. J. Nutr. 123:940–946.

Kornegay, E. T. 2001. Digestion of phosphorus and other nutrients: role of phytases and factors influencing their activity. In: M. R. Bedford, and G. G. Partridge, editors, Enzymes in farm animal nutrition. CABI Publishing, Wallingford, UK. p. 237–272.

Kumar, V., A. K. Sinha, H. P. S. Makkar, G. de Boeck, and K. Becker. 2012. Dietary roles of non-starch polysachharides in human nutrition: A review. Crit. Rev. Food Sci. Nutr. 52:899–935.

Landero, J. L., E. Beltranena, and R. T. Zijlstra. 2013. Diet nutrient digestibility and growth performance of weaned pigs fed solvent-extracted Brassica juncea canola meal. Anim. Feed Sci. Technol. 180:64–72.

Landero, J. L., E. Beltranena, M. Cervantes, A. Morales, and R. T. Zijlstra. 2011. The effect of feeding solvent-extracted canola meal on growth performance and diet nutrient digestibility in weaned pigs. Anim. Feed Sci. Technol. 170:136–140.

Lawrence, R. D, and J. L. Anderson. 2018. Ruminal degradation and intestinal digestibility of camelina meal and carinata meal compared with other protein sources. Prof. Anim. Sci. 34:10–18.

Lee, H. H., H. I. Rhee, S. H. Cha, S. Y. Lee, Y. S. Choi. 1997. Effects of dietary phytic acid on the lipid and mineral profiles in rats. Food Biotechnol. 6:261–264.

Lee, J. W., and T. A. Woyengo. 2018. Growth performance, organ weights, and blood parameters of nursery pigs fed diets containing increasing levels of cold-pressed canola cake1. J. Anim. Sci. 96:4704–4712.

Lee, S-H., H-J. Park, H-K. Chun, S-Y. Cho, H-J. Jung, S-M. Cho, D-Y. Kim, M-S. Kang, and H. S. Lillehoj. 2007. Dietary phytic acid improves serum and hepatic lipid levels in aged ICR mice fed a high-cholesterol diet. Nutr. Res. 27:505–510.

Leoni, O., F. Felluga, and S. Palmieri. 1993. The formation of 2-hydroxybut-3-enyl cyanide from (2S)-2-hydroxybut-3-enyl glucosinolate using immobilized myrosinase. Tetrahedron Lett. 34:7967–7970.

Leeson, S., and J. O. Atteh. 1996. Response of broiler chicks to dietary full-fat soybeans extruded at different temperatures prior to or after grinding. Anim. Feed Sci. Technol. 57: 239–245.

Li, S., W. C. Sauer, and W. R. Caine. 1998. Response of nutrient digestibilities to feeding diets with low and high levels of soybean trypsin inhibitors in growing pigs. J. Sci. Food Agric. 76:357–363.

Liao, S. F., A. K. Kies, W. C. Sauer, Y. C. Zhang, M. Cervantes, and J. M. He. 2005. Effect of phytase supplementation to a low- and a high-phytate diet for growing pigs on the digestibilities of crude protein, amino acids, and energy. J. Anim. Sci. 83:2130–2136.

Links, M. R., J. Taylor, M. C. Kruger, and J. R. N. Taylor. 2015. Sorghum condensed tannins encapsulated in kafirin microparticles as a nutraceutical for inhibition of amylases during digestion to attenuate hyperglycaemia. J. Funct. Foods. 12:55–63.

Liu, N., Y. J. Ru, F. D. Li, and A. J. Cowieson. 2008. Effect of diet containing phytate and phytase on the activity and messenger ribonucleic acid expression of carbohydrase and transporter in chickens. J. Anim. Sci. 86:3432–3439.

Lu, Y., and A. Bennick. 1998. Interaction of tannin with human salivary proline-rich proteins. Arch. Oral Biol. 43:717–728.

Maenz, D. D. 2001. Enzymatic and other characteristics of phytases as they relate to their use in animal feeds. In: M. R. Bedford, and G. G. Partridge, editors, Enzymes in farm animal nutrition. CABI Publishing, Wallingford, UK. p. 61–84.

Magan, N., R. Hope, V. Cairns, and D. Aldred. 2003. Post-harvest fungal ecology: Impact of fungal growth and mycotoxin accumulation in stored grain. Eur. J. Plant Pathol. 109:723–730.

Matusheski, N. V., R. Swarup, J. A. Juvik, R. Mithen, M. Bennett, and E. H. Jeffery. 2006. Epithiospecifier protein from broccoli (Brassica oleracea L. ssp. italica) inhibits formation of the anticancer agent sulforaphane. J. Agric. Food Chem. 54:2069–2076.

McDonald, D. E., D. W. Pethick, J. R. Pluske, and D. J. Hampson. 1999. Adverse effects of soluble non-starch polysaccharide (guar gum) on piglet growth and experimental colibacillosis immediately after weaning. Res. Vet. Sci. 67:245–250.

Mithen, R., 2001. Glucosinolates—biochemistry, genetics and biological activity. Plant GrowthRegul. 34:91–103.

Mitjavila, S., G. de Saint Blanquat, and R. Derache. 1973. Effet de l'acide tannique sur la sécrétion gastrique chez le rat. Ann. Nutr. Metab. 15:163–170.

Molist, F., A. Gomez de Segura, J. Gasa, R. G. Hermes, E. G. Manzanilla, M. Anguita, and J. F. Perez. 2009. Effects of the insoluble and soluble dietary fibre on the physicochemical properties of digesta and the microbial activity in early weaned piglets. Anim. Feed Sci. Technol. 149:346–353.

Morisset, J. 2008. Negative control of human pancreatic secretion: Physiological mechanisms and factors. Pancreas 37:1–12.

Myer, R. O., J. A. Froseth, and C. N. Coon. 1982. Protein utilization and toxic effects of raw beans (Phaseolus vulgaris) for young pigs. J. Anim. Sci. 55:1087–1098.

Naczk, M., D. Oickle, D. Pink, and F. Shahidi. 1996. Protein precipitating capacity of crude canola tannins: Effect of pH, tannin, and protein concentrations. J. Agric. Food Chem. 44: 2144–2148.

Newkirk, R. W., and H. L. Classen. 2002. The effects of toasting canola meal on body weight, feed conversion efficiency, and mortality in broiler chickens. Poult. Sci. 81:815–825.

Nyachoti, C. M., C. F. M. de Lange, B. W. McBride, and H. Schulze. 1997. Significance of endogenous gut nitrogen losses in the nutrition of growing pigs: A review. Can. J. Anim. Sci. 77:149–163.

Ockenden, I., J. A. Dorsch, M. M. Reid, L. Lin, L. K. Grant, V. Raboy, and J. N. A. Lott. 2004. Characterization of the storage of phosphorus, inositol phosphate and cations in grain tissues of four barley (Hordeum vulgare L.) low phytic acid genotypes. Plant Sci. 167:1131–1142.

Pacheco, W. J., C. R. Stark, P. R. Ferket, and J. Brake. 2014. Effects of trypsin inhibitor and particle size of expeller-extracted soybean meal on broiler live performance and weight of gizzard and pancreas. Poult. Sci. 93:2245–2252.

Pan, L., P. Li, X. K. Ma, Y. T. Xu, Q. Y. Tian, L. Liu, D. F. Li, and X. S. Piao. 2016. Tannin is a key factor in the determination and prediction of energy content in sorghum grains fed to growing pigs. J. Anim. Sci. 94:2879–2889.

Pearson, A. W., N. M. Greenwood, E. J. Butler, and G. R. Fenwick. 1983. Biochemical changes in layer and broiler chickens when fed on a high-glucosinolate rapeseed meal. Br. Poult. Sci. 24:417–427.

Petrikovics, I., M. Budai, K. Kovacs, and D. E. Thompson. 2015. Past, present and future of cyanide antagonism research: From the early remedies to the current therapies. World J. Methodol. 5: 88–100.

Pinton, P., C. Braicu, J-P. Nougayrede, J. Laffitte, I. Taranu, and I. P. Oswald. 2010. Deoxynivalenol impairs porcine intestinal barrier function and decreases the protein expression of claudin-4 through a mitogen-activated protein kinase-dependent mechanism. J. Nutr. 140:1956–1962.

Poulton, J.E. 1990. Cyanogenesis in plants: Review. Plant Physiol. 94: 401–405.

Prattley, C. A., and D. W. Stanley. 1982. Protein-phytate interactions in soybeans. I. Localisation of phytate in protein bodies and globoids. J. Food Biochem. 6:243–253.

Pusztai, A., S. Bardocz, and M. A. Martín-Cabrejas. 2004. The mode of action of ANFs on the gastrointestinal tract and its microflora. In: M. Muzquiz, G. D. Hill, C. Cuadrado, M. M. Pedrosa, and C. Burbano, editors, Recent advances of research in antinutritional factors in legume seeds and oilseeds. Wageningen Academic Publishers, Wageningen, The Netherlands, p. 87–100.

Ravindran, V., P. C. Morel, G. G. Partridge, M. Hruby, and J. S. Sands. 2006. Influence of an Escherichia coli-derived phytase on nutrient utilization in broiler starters fed diets containing varying concentrations of phytic acid. Poult. Sci. 85:82–89.

Ripken, D., N. van der Wielen, J. van der Meulen, T. Schuurman, R. F. Witkamp, H. F. J. Hendriks, and S. J. Koopmans. 2015. Cholecystokinin regulates satiation independently of the abdominal vagal nerve in a pig model of total subdiaphragmatic vagotomy. Physiol. Behav. 139:167–176.

Rocha-e-Silva, R.C., L. A. Cordeiro, and B. Soto-Blanco. 2010. Cyanide toxicity and interference with diet selection in quail (Coturnix coturnix). Comp. Biochem. Physiol. C. Toxicol. Pharmacol. 151: 294–297.

Rodriguez-Hernandez, K., and J. L. Anderson. 2018. Evaluation of carinata meal as a feedstuff for growing dairy heifers: Effects on growth performance, rumen fermentation, and total-tract digestibility of nutrients. J. Dairy Sci. 101:1206–1215.

Roland, N., S. Rabot, and L. Nugon-Baudon. 1996. Modulation of the biological effects of glucosinolates by inulin and oat fibre in gnotobiotic rats inoculated with a human whole faecal flora. Food Chem. Toxicol. 34:671–677.

Rubio, L. A., G. Grant, P. Dewey, I. Bremner, and A. Pusztai. 1994. The intestinal true absorption of 65Zn in rats is adversely affected by diets containing a faba bean (vicia faba l.) nonstarch polysaccharide fraction. J. Nutr. 124:2204–2211.

Rustemeyer, S. M., W. R. Lamberson, D. R. Ledoux, G. E. Rottinghaus, D. P. Shaw, R. R. Cockrum, K. L. Kessler, K. J. Austin, and K. M. Cammack. 2010. Effects of dietary aflatoxin on the health and performance of growing barrows. J. Anim. Sci. 88:3624–3630.

Rutherfurd, S. M., T. K. Chung, P. C. Morel, and P. J. Moughan. 2004. Effect of microbial phytase on ileal digestibility of phytate phosphorus, total phosphorus, and amino acids in a low-phosphorus diet for broilers. Poult. Sci. 83:61–68.

Ryan, C. A. 1990. Protease inhibitors in plants: genes for improving defenses against insects and pathogens. Annu. Rev. Pathol. 28:425–449.

Sathe, S. K., and K. W. C. Sze-tao. 1997. Effects of sodium chloride, phytate and tannin on in vitro proteolysis of phaseolin. Food Chem. 59:253–259.

Sauer, W. C., R. Mosenthin, F. Ahrens, and L. A. den Hartog. 1991. The effect of source of fiber on ileal and fecal amino acid digestibility and bacterial nitrogen excretion in growing pigs. J. Anim. Sci. 69:4070–4077.

Scheenman, B. O., 1985. Effects of nutrients and nonnutrients on food intake. Am. J. Clin. Nutr. 42:966–972.

Schulze, H., P. van Leeuwen, M. W. A. Verstegen, J. Huisman, W. B. Souffrant, and F. Ahrens. 1994. Effect of level of dietary neutral detergent fiber on ileal apparent digestibility and ileal nitrogen losses in pigs. J. Anim. Sci. 72:2362–2368.

Selle, P. H., A. R. Walker, and W. L. Bryden. 2003. Total and phytate-phosphorus contents and phytase activity of Australian-sourced feed ingredients for pigs and poultry. Aust. J. Exp. Agric. 43:475–475.

Seneviratne, R. W., M. G. Young, E. Beltranena, L. A. Goonewardene, R. W. Newkirk, and R. T. Zijlstra. 2010. The nutritional value of expeller-pressed canola meal for grower finisher pigs. J. Anim. Sci. 88:2073–2083.

Slominski, B. A., and L. D. Campbell. 1989. Indoleacetonitriles - thermal degradation products of indole glucosinolates in commercial rapeseed (Brassica napus) meal. J. Sci. Food Agric. 47:75–84.

Smit, M., R. W. Seneviratne, M. G. Young, G. Lanz, R. T. Zijlstra, and E. Beltranena. 2014. Feeding Brassica juncea or Brassica napus canola meal atincreasing dietary inclusions to growing-finishing gilts and barrows. Anim. Feed Sci. Technol. 198:176–185.

Smith, T. K., and E. J. MacDonald. 1991. Effect of fusaric acid on brain regional neurochemistry and vomiting behavior in swine. J. Anim. Sci. 69:2044–2049.

Smulikowska, S., B. Pastuszewska, E. Switch, A. Ochtabinska, A. Mieczkowska, V. C. Nguyen, and L. Buraczewska. 2001. Tannin content affects negatively nutritive value of pea for monogastrics. J. Anim. Feed Sci. 10:511–523.

Soares, S., I. García-Estévez, R. Ferrer-Galego, N. F. Brás, E. Brandão, M. Silva, N. Teixeira, F. Fonseca, S. F. Sousa, F. Ferreira-da-Silvad, N. Mateus, and Victor de Freitas. 2018. Study of human salivary proline-rich proteins interaction with food tannins. Food Chem. 243:175–185.

Steiner, T., R. Mosenthin, B. Zimmermann, R. Greiner, and S. Roth. 2007. Distribution of phytase activity, total phosphorus and phytate phosphorus in legume seeds, cereals and cereal by-products as influenced by harvest year and cultivar. Anim. Feed Sci. and Technol. 133:320–334.

Struthers, B. J., J. R. Macdonald, R. R. Dahlgren, and D. T. Hopkins. 1983. Effects on the monkey, pig and rat pancreas of soy products with varying levels of trypsin inhibitor and comparison with the administration of cholecystokinin. J. Nutr. 113:86–97.

Święch, E., and L. Buraczewska. 2005. True ileal digestibility of amino acids of pea seeds and soyabean products estimated in pigs, rats and in vitro. J. Anim. Feed Sci. 14:179–191.

Tanii, H., T. Higashi, F. Nishimura, Y. Higuchi, and K. Saijoh. 2008. Effects of cruciferous allyl nitrile on phase 2 antioxidant and detoxification enzymes. Med. Sci. Monit. 14:189–192.

Terciolo, C., A. P. Bracarense, P. C. M. C. Souto, A-M. Cossalter, L. Dopavogui, N. Loiseau, C. A. F. Oliveira, P. Pinton, and I. P. Oswald. 2019. Fumonisins at doses below EU regulatory limits induce histological alterations in piglets. Toxins 11:548. doi:https://doi.org/10.3390/toxins11090548.

Thacker, P. A., and G. Widyaratne. 2012. Effects of expeller pressed camelina meal and/or canola meal on digestibility, performance and fatty acid composition of broiler chickens fed wheat–soybean meal-based diets. Arch. Anim. Nutr. 66:402–415.

Thompson, L. U., C. L Button, and D. J. A. Jenkins. 1987. Phytic acid and calcium affect the in vitro rate of navy bean starch digestion and blood glucose response in humans. Am. J. Clin. Nutr. 46:467–473.

Tripathi, M. K., and A. S. Mishra. 2007. Glucosinolates in animal nutrition: A review. Anim. Feed Sci. Technol. 132:1–27.

Valencia, D.G., M. P. Serrano, R. Lázaro, R., M. A. Latorre, and G. G. Mateos. 2008. Influence of micronization (fine grinding) of soya bean meal and fullfat soya bean on productive performance and digestive traits in young pigs. Anim. Feed Sci. Technol. 147: 340–356.

van Leeuwen, P., A. J. Jansman, J. Wiebenga, J. F. Koninkx, and J. M. Mouwen. 1995. Dietary effects of faba-bean (Vicia faba L.) tannins on the morphology and function of the small-intestinal mucosa of weaned pigs. Br. J. Nutr. 73:31–39.

Van Soest, P. J. 1994. Nutritional ecology of the ruminant, 2nd ed. Cornell University Press, Ithaca, NY.

Velayudhan, D. E., K. Schuh, T. A. Woyengo, Sands J. S., and C. M. Nyachoti. 2017. Effect of expeller extracted canola meal on growth performance, organ weights and blood parameters of growing pigs. J. Anim. Sci. 95:302–307.

Weller, A., Smith, G.P., Gibbs, J., 1990. Endogenous cholecystokinin reduces feeding in young rats. Science 247:1589–1591.

Wilfart, A., L. Montagne, H. Simmins, J. Noblet, and J. van Milgen. 2007. Digesta transit in different segments of the gastrointestinal tract of pigs as affected by insoluble fibre supplied by wheat bran. Br. J. Nutr. 98:54–62.

Woyengo, T. A., A. J. Cowieson, O. Adeola, and C. M. Nyachoti. 2009. Ileal digestibility and endogenous flow of minerals and amino acids: responses to dietary phytic acid in piglets. Br. J. Nutr. 102:428–433.

Woyengo, T. A., and C. M. Nyachoti. 2011. Review: Supplementation of phytase and carbohydrases to diets for poultry. Can. J. Anim. Sci. 91: 177–192.

Woyengo, T. A., E. Beltranena, and R. T. Zijlstra. 2017. Effect of anti-nutritional factors in oilseed co-products on feed intake of pigs and poultry: Invited review. Anim. Feed Sci. Technol. 233:76–86.

Woyengo, T.A., R. Jha, E. Beltranena, A. Pharazyn, and R.T. Zijlstra. 2014. Nutrient digestibility of lentil and regular- and low-oligosaccharide, micronized full-fat soybean fed to grower pigs. J. Anim. Sci. 92: 229–237.

Woyengo, T. A., E. Kiarie, and C. M. Nyachoti. 2011. Growth performance, organ weights and blood parameters of broilers fed diets containing expeller extracted canola meal. Poult. Sci. 90:2520–2527.

Woyengo, T. A., O. Adeola, C. C. Udenigwe, and C. M. Nyachoti. 2010. Gastro-intestinal digesta pH, pepsin activity and soluble mineral concentration responses to supplemental phytic acid and phytase in piglets. Livest. Sci. 134:91–93.

Woyengo, T. A., D. Weihrauch, and C. M. Nyachoti. 2012. Effect of dietary phytic acid on performance and nutrient uptake in the small intestine of piglets. J. Anim. Sci. 90: 543–549.

Yen, J. T., A. H. Jensen, and J. Simon. 1977. Effect of dietary raw soybean and soybean trypsin inhibitor on trypsin and chymotrypsin activities in the pancreas and in small intestinal juice of growing swine. J. Nutr. 107:156–165.

Zhang, W., D. Li, L. Liu, J. Zang, Q. Duan, W. Yang, and L. Zhang. 2013. The effects of dietary fiber level on nutrient digestibility in growing pigs. J. Anim. Sci. Biotech. 4:17. http://www.jasbsci.com/content/4/1/17.

Zhou, X., E. Beltranena, and R. T. Zijlstra. 2016. Effects of feeding canola press-cake on diet nutrient digestibility and growth performance of weaned pigs. Anim. Feed Sci. Technol. 211:208–215.

# 16 Feed Processing Technology and Quality of Feed

Chad B. Paulk, Charles R. Stark, and Kara M. Dunmire

## Introduction

Swine feed cost represents 65–80% of total cost of production. Ingredients make up the greatest proportion of the cost, while feed manufacturing and delivery account for approximately 5% of the total diet cost. This has led producers to take advantage of feed manufacturing practices to maximize utilization of feedstuffs. In addition, variations in the quality of ingredients and finished feed can affect pig performance. Producers must routinely evaluate the quality of their ingredients and the manufacturing process in the feed mill to ensure a low cost to benefit ratio. A systematic evaluation of each process within the feed manufacturing facility will identify opportunities for improvement in manufacturing efficiency and reduced nutrient variation of finished feed, which ultimately results in a lower cost of swine production.

### Purchasing and Formulating Feed Ingredients

Alternative ingredients are frequently used in swine diet formulations to decrease diet cost. Using high-quality ingredients is of greatest importance when formulating which can be challenging with the use of variable alternative ingredients. Common coproducts that can have variable quality include soybean meal, dried distillers' grains with solubles (DDGS), bakery meal, and meat and bone meal (MBM). These ingredients undergo multistep processing before reaching the final product sold for feed usage. Differences within shipment variation, plant variety, growing conditions, and growing location can also play a role. Additional costs that come with alternative ingredients include added analytical costs to develop and maintain least-cost formulation matrix values, increased receiving time at the feed mill, additional feed mill storage, longer batching times from the addition of more ingredients to the scales, and reduced pellet mill throughput. Feed mill challenges can arise with changes in pellet quality, feed density, and palatability when alternative ingredients are included. Therefore cost shifts from ingredients to analytical verification and feed manufacturing which should be considered in purchasing decisions to determine actual system-wide benefit.

Soybeans are the most abundant oilseed worldwide that provide by-products including soybean meal and oil. Soybean meal used as a plant protein supplement is often formulated and sold by crude protein content, commonly 46% for swine diets. Although soybean meal has low variation in protein quality several processing steps are needed to achieve high-quality soybean meal. After dehulling and solvent extraction, soybean meal is further processed by heating to destroy antinutritional

*Sustainable Swine Nutrition*, Second Edition. Edited by Lee I. Chiba.
© 2023 John Wiley & Sons Ltd. Published 2023 by John Wiley & Sons Ltd.

factors. However, there can also be negative effects of over processing such as the Maillard browning reaction and amino acid (AA) digestibility loss. In the Maillard reaction a free AA (commonly Lys) will bind to a reducing sugar and toast or brown the soybean meal. Overprocessing or underprocessing of soybean meal can impact the nutritional value and ultimately performance. In a study analyzing 22 sources of soybean meal sourced from crush plants in the northern Midwest, central Midwest and eastern United States only minor differences were observed in CP and AA concentrations and digestibility, while digestible AA content remained constant regardless of geographical location of soybean crush plant (Sotak-Peper et al. 2017). The protein value of soybean meal in the United States was not influenced by location. Another study evaluated the variation of soybean meal from top producing countries of China, Argentina, Brazil, the United States, and India (Lagos and Stein 2017). Soybean meal sourced from Brazil and India had the highest concentrations of CP and AA with China having the lowest concentration. The United States and China had soybean meal with greater apparent ileal digestibility (AID) and standard ileal digestibility (SID) compared to Argentina, Brazil, and India. Soybean meal sourced from China had less SID CP and AA. The greatest concentrations of digestible AA soybean meal sourced from the U.S. and India followed by Brazil then Argentina. Soybean sourced from the United States, China, and Brazil had less variability among sources compared to Argentina and China. Therefore, sourcing soybean meal outside of a country draws more precaution than soybean meal within country, especially the United States.

Dried distiller's grains with solubles (DDGS) is a coproduct of ethanol production. The percent of fat is used to identify DDGS commonly ranging from 4.5 to 9%. Over time ethanol plants have extracted more oil during the process. These changes will influence DDGS inclusion. Stein et al. 2009 found that a 30% inclusion of DDGS had no influence in performance. When formulating diets using DDGS nutritionists need to consider the decreased metabolizable energy (ME) and net energy (NE) which is attributed to the increase in fiber content. There is approximately 12.0% acid detergent fiber (ADF) and 30.5% neutral detergent fiber (NDF) in DDGS. High dietary fiber content is also commonly associated with decreased nutrient utilization, low NE values, and reduced carcass yield. DDGS is primarily insoluble fiber that has an impact on digesta viscosity, rate of fermentation, and amount of VFA absorbed.

Energy values can easily be underestimated or overestimated influencing the caloric efficiency effecting growth performance. Maintenance energy requirement will be increase with the increase in fiber content. In a study evaluating medium-oil DDGS source and level on growth performance of finishing pigs, there were DDGS sources with 5.4% oil and 9.4% oil with each source included at 20 and 40% of the diet. For every 15% increase in DDGS energy intake decreased about 2%, while CP intake increases about 10% with approximately a 2% decrease and ADG and G:F (Graham 2014a). Negative effects of DDGS include decreases final body weight, carcass yield, hot carcass weight, backfat and loin eye depth, which were also observed in Graham 2014a. In a study of low-, medium-, and high-oil DDGS predictive energy equations indicated that on an as-fed basis, increasing the oil content of DDGS by 1% will result in a change of 60 kcal/kg and change in NE of 115 kcal/kg (Graham 2014b). Both inclusion level and duration of feeding should be considered when feeding DDGS.

Bakery meal is any combination of waste products that were rejected for human use including: candy, cookies, and crackers among others or it can single-sourced meal. Seasonal variation can add to the variability in oil, salt, sugar, and starch. A study of 46 sources of bakery meal from 5 regions in the United States; Iowa, Minnesota, the eastern U.S., eastern Midwest, and western Midwest U.S. (Liu et al. 2018). The high DM content observed in bakery meal compared to other feed ingredients is due to excessive drying of the waste products indicating a potential reason for the reduced Lys digestibility. There were only slight differences in chemical composition from the different regions.

Most notably, calcium and phosphorus were highest with bakery meal sourced from Minnesota compared to all other sources. Phosphorus, phytic acid, and phytate (P) were highest in the eastern and eastern Midwest bakery meal sources with the western Midwest source intermediate.

MBM used as an animal protein supplement is a by-product of carcass rendering and can also be a calcium and phosphorus source. MBM is a combination of fat trimmings, discarded meat and carcass, and bones. Further processing is done to grind, cook, and press this by-product into usable material. However, within the heat processing step crude protein, AA content, and mineral content can be destroyed. Collagen protein found in skin and connective tissue contributes to the high protein content. MBM is a regulated by-product with a minimum of 4% P with Ca not exceeding 2.2% that of P (AAFCO), less than 4% P is considered meat meal. In a study evaluating eight MBM sources CP values were between 45.7 and 57.2, acid hydrolyzed ether extract (AEE) values were between 11.6 and 15.2, and ash values were between 20.6 and 33.2 (Sulabo and Stein 2013). Variation was greatest for ash (13.7%) followed by AEE (10.5%) then CP (6.2%). The range of Ca and P was 5.1–11.0% and 2.6–5.3% with variations of 22 and 20%, respectively.

A diet formulation can only be as nutrient rich as the ingredients included. Challenges with variable alternative ingredients can be mitigated through approved suppliers to assure for manufacture of high-quality feed. Implementing ingredient specifications with a supplier can assist with receiving a consistent product. Communication with purchasing agents, suppliers, transporters, and receiving personnel is the foundation for producing high-quality finished feed, limiting product liability and variability, and lowering the cost of feed (Stark and Jones 2012). Product description, expected nutrient content, analytical methods, physical characteristics, and the basis for rejection should be included with a specification sheet. Considerations for alternative ingredient use in feed mills include mill design, mill capacity and storage, delivery time, and negative impacts on performance of unexpected formulation changes due to ingredients (Stark 2012a).

**Feed Ingredient Receiving**

Determining ingredient quality prior to diet formulation and feed processing is an important tool in precision feeding in swine production. Visual appearance and chemical analysis are commonly used to determine the quality of ingredients. Quality concerns include antinutritional factors, physical, chemical, and biological hazards. Collecting and retaining samples at receiving is important for testing prior to use to insure quality ingredients are used to produce high-quality finished feed.

Prior to sample collection, sampling equipment and sample storage procedures must be identified (AAFCO 2017). Proper labeling and clean sampling equipment are of utmost importance to prevent cross contamination. Equipment used to provide a representative sample include grain probes and sampling triers. Examples of probes and triers used for feed sampling are shown in Figure 16.1.

Grain probes and triers are available in many diameters and lengths from 0.5 to 1.375 in and 18 to 120 in, respectively. To determine the best sampling tool, length of the tool should reach the height or length of the container being sampled. Larger diameters should be used for whole grain ingredients and smaller diameters used for ground ingredients and complete feeds. When using a grain probe, the probe should be closed when inserted into the product and then twisted to reveal holes, allowing grain to flow in. With double tube grain probes, the tube should then be twisted to retain sample in the probe which can then be removed. Bags of dry feed, bags of free-flowing material, course textured feed, pellets, cube, wafers, powdered feeds, and bulk dry ingredients should be sampled using a trier or probe. Trailers, rail cars, hopper bottom trucks, and straight trucks should be collected with 10 full probes of material from 10 equally spaced sections of the container (Figure 16.2).

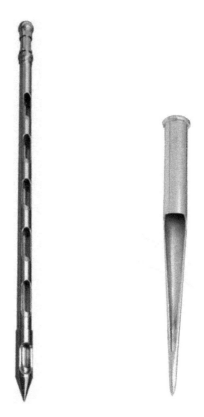

**Figure 16.1**    From left to right, and example of a double tube grain probe and bag trier, respectively (Gibson Company).

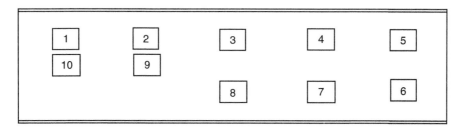

**Figure 16.2**    Diagram of truck and rail car sampling.

In vehicles with multiple compartments, the samples should be spaced evenly to collect 10 samples across all compartments. If samples cannot be collected using a probe, another option is cut stream sampling during unloading, depending on safety and accessibility. Equipment used for cut stream sampling includes pelican samplers or PVC pipe samplers. The sample should be taken in the middle of the load, and stream flow may need to be reduced while collecting so the sample does not overflow. Collection should be a side to side sweep of the sampler, collecting the entire stream. Bin sampling similar to trucks and rails with 10 full probes from 10 different areas spaced around the bin. Liquid ingredients should be sampled at the mixer's delivery line, or storage tanks. Before sample collection, liquid lines should be flushed to prime the line and discard the recovered material.

Liquid samples should be stored in plastic or glass containers (AAFCO 2017). Samples should be kept in the properly labeled bag and are best stored in a freezer or refrigerator.

Obtaining a representative sample for analysis is key to determine the quality of an ingredient. The samples odor and visual appearance can help decide quality and basis of rejection of some products. Whole grain ingredients such as corn can be visually inspected for excessive material from the field and indication of molds and should be tested to evaluate moisture and mycotoxins. Sensory inspection by personnel of by-product ingredients or processed ingredients such as DDGS, soybean meal, and fat is helpful in determining when running tests is critical. Soybean meal samples that appear dark and smell burnt can be an indication of Maillard browning reactions which can cause a decrease in amino acid availability. Energy content of DDGS and soybean meal can be affected by the amount fat extraction from DDGS and soybean meal. Fat source sensory inspection can indicate lipid oxidation with darker and rancid smelling the sample the greater chance of lipid oxidation. Testing for lipid oxidation can be analyzed using peroxide value by measuring the amount of peroxides, which are formed in early oxidation. The near inferred reflectance spectroscopy (NIRS) is a useful tool for on-site testing of ingredient quality. Sensory inspection becomes more challenging when considering minor and micro ingredients which is where supplier verification is important. The previously described ingredients should be rejected if the negative effects mentioned are displayed, however analysis beyond sensory inspection with NIRS should be performed to confirm.

In addition to sensory inspection, chemical analysis of a sample is needed to determine nutrient composition. Proximate analysis is often utilized because it provides dry matter, crude protein, crude fat, crude fiber, and ash.

Official methods used for crude protein include the LECO combustion method (AOAC 968.06) or Kjeldahl digestion method (AOAC 954.01). These analyses are crucial for protein meals. Additional analysis for protein content of ingredients include amino acid profile (AOAC 982.30 E(a,b,c) chp. 45.3.05,2006), available lysine (AOAC 975.44 chp 45.4.03,2006), protein solubility by potassium hydroxide (J Anim Sci, 69:2918-2924, 1991), trypsin inhibitor activity (soybean; AACC22-40, 2006), urease activity (AACC 22-90), and protein dispersibility index (AACC 46-24).

The Soxhlet extraction by ether or hexane is the official method for crude fat (ether extract) analysis (AOAC 920.39). Further analysis to determine fat and oil quality include peroxide value (AOAC 965.33), iodine value (AOCS Ja 14-91), thiobarbituric acid rancidity (AOCS Cd 19-90), color scale (AOCS Cc 13a-43), total carotenoids (AOAC 938.04), and fatty acids (AOAC 996.06).

Analysis to determine crude fiber content includes ceramic fiber filter method (AOAC 962.09) or crucible method (AOAC 978.10). ADF (AOAC 973.18), amylase treated NDF (AOAC 2002.04), and total detergent fiber (AOAC 985.29) are also official methods in determining fiber content of ingredients. Additional fiber analyses include cellulose (AOAC 973.18) and lignin (AOAC 973.18).

To determine the inorganic portion of feed ingredients, ash and minerals should be analyzed. Official ash analysis (AOAC 942.05) gives little information and should be further analyzed for mineral analysis which includes copper, iron, manganese, zinc, and calcium (AOAC 968.08).

Determining a cost-effective feed ingredient testing program will help meet nutrient needs while lowering production cost from a feed mill and pig performance standpoint. Determining the official analysis should be utilized will depend on the management question. However, when implementing a new supplier proximate analysis should be analyzed to assure ingredient quality with strict testing when it comes to by-product and alternative ingredients.

In addition to chemical analysis, rapid methods have been adapted to evaluate quality of ingredients. These methods help nutritionists and production systems save time and money. Determining the moisture content of an ingredient is important for making decisions pertaining to storage, shrink, and other feed milling processes. Options for rapid analysis of moisture include using a Dickey–John moisture analyzer or near infrared reflectance spectroscopy (NIRS). In addition, NIRS can be used to analyze other components such as protein, fat, fiber, starch, and amino acids. The accuracy of these results is dependent on many different factors. Literature states that there is a particle size by sample preparation method interaction of NIR ground samples and NIR unground samples compared to wet chemistry (Evans et al. 2019). However, if a sample is properly prepared, NIR results are comparable to wet chemistry methods. Crude protein results were influenced where wet chemistry and NIR ground samples were 20.3 and 19.6%, respectively, compared to the unground NIRS sample with 17.8% CP (Evans et al. 2019). Further, samples ground to 267 microns can provide more accurate proximate analysis results compared to particle sizes ranging from 693 to 3343 microns (Leiva et al. 2019). Ingredient quality can also be measured using NIR methods. Over processing of ingredients such as soybean meal can reduce reactive lysine content and negatively impact overall lysine digestibility by increasing antinutritional factors. Establishing accurate NIRS calibrations is imperative for predictability. Improved calibrations have shown that NIRS as a tool predicting reactive lysine compared to wet chemistry methods (Dunmire et al. 2019). Additionally, NIRS compared to official analytical methods were highly correlated for total Lys, available Lys, and Lys/CP. Advancements have been made with NIRS technology in feed mills to include in-line analyzers. In-line analyzers can provide results as ingredients pass along a sensor allowing for continuous monitoring and segregation based on nutrient value (Stark 2013).

Before receiving loads of grains, it is important to determine the risk of mycotoxins. In addition to analytical test, weekly reports are available to assess the risk of mycotoxin contamination in grains from a specific area. This can influence the how many loads need to be sampled. Mycotoxins are produced from molds and fungi that can develop during a wet harvest or when grain is stored. Known mycotoxins include aflatoxin, ochratoxin, deoxynivalenol (vomitoxin), T-2, zearalenone, and fumonisin. Presence of mycotoxins in complete feed can lead to growth performance and health challenges in swine. Testing for them at the feed mill can help mitigate these issues. Maximum acceptable levels have been established. High aflatoxin levels that can be found in corn, peanuts, and cottonseed can cause reduced gain and intake, liver damage, hemorrhaging, thymic atrophy, and reduced immunity cautionary levels are 0.02 ppm (Van Heutgen 2001). High ochratoxin levels that can be found in barley and oilseed crops can cause reduced gain and intake, kidney function, increased water consumption, polyuria, and reduced immunity cautionary levels are 0.2 ppm (Van Heutgen 2001). High deoxynivalenol (DON) vomitoxin levels that can be found in corn, wheat, barley, rye, and oats can cause reduced gain and intake, feed refusal, and vomiting cautionary levels are 1.0 ppm (Van Heutgen 2001). High T-2 levels that can be found in wheat and barley can cause reduced gain and intake, dermatitis, lymphoid necrosis, and reduced fertility cautionary levels are 0.5 ppm. High zearalenone levels that can be found in corn, wheat, barley, and rye can cause hyperestrogenism and infertility are 0.5 ppm (Van Heutgen 2001). High fumonisin levels that can be found in corn can cause reduced gain and intake, liver damage, and pulmonary edema cautionary levels are 5.0 ppm (Van Heutgen 2001). While the maximum allowances serve as a guideline, there may be effects at lower levels, there are many factors that determine the effects. Mycotoxin quick testing requires little training and sample preparation and is helpful for on-site rejection of ingredients. Mycotoxin testing in feed and ingredients can come from qualitative testing and quantitative testing and can range in skill intensity. Lateral flow devices or test strips can be used as a single mycotoxin test to define qualitatively or quantitively. Enzyme-linked immunosorbent assays (ELISA) can also be used for multimycotoxin testing but is more skill intensive. Another onsite option is the NIRS but cannot measure the mycotoxins themselves, can only establish low or high concentration and is only as sensitive as the calibration. Onsite laboratory testing provides quick results and alternative to the added time and cost associated with laboratory testing and results. Additional highly skill intensive methods include high-performance liquid chromatography (HPLC) and liquid chromatography couple with mass spectrometry (LC–MC) or tandem MS (LC–MC/MS). These methods provide a wide range of detailed results, require little sample, are time consuming and expensive.

## Particle Size Reduction in Feed

The importance of grinding cereal grains for use in swine feed became understood back in 1930s. As the primary ingredients in swine feed cereal grains were ground for improvements in nutrient digestibility. By breaking up the hard-protective outer layer nutrients become more available because of the increased surface area for digestive enzymes to act. Ingredients including soybean meal, rendered products, distillers dried grains, and wheat by-products are supplier ground ingredients that are ready for use in complete feed and do not require further feed mill processing. The importance of grinding was first researched using ground sorghum to demonstrate improvement in nutrient digestibility with reducing particle size. The idea was then expanded to pigs fed ground corn and sorghum with improvements found in feed efficiency (Aubel 1945, 1955). These preliminary experiments led to the application of the grinding process to all swine production systems.

There are two common mills that are used to reduce the particle size of ingredients, roller mills, and hammermills. Roller mills reduce particle size by crushing, in which a comprehensive force is applied to the ingredient. This process produces a small amount of fine material resulting in a fairly uniform particle size of the grain. Hammermills reduce the particle size of ingredients by impact grinding (Pfost 1976). This creates a more spherical particle shape and increases the amount of fine, pulverized particles, resulting in a less uniform particle size. Previous research has determined that the increased amount of nonuniform particles generated when grinding grains using a hammermill results in a greater angle of repose, indicating poorer flowability (Groesbeck et al. 2006). In addition to reducing handling characteristics, grinding grains increase costs associated with milling. These costs include (i) equipment cost; (ii) maintenance cost; (iii) the energy required to mill one ton of grain (kWh/ton); and (iv) the production rate per horsepower hour of operation. The initial equipment cost may seem expensive, but depending on the mill, the energy cost for one year can exceed the cost of a new piece of grinding equipment. The energy required to operate a roller mill or hammermill over its life expectancy will be 10–20 times more than the machine cost alone (Heimann 2014). Understanding the variety of factors that influence the efficiency and production rate of grinding can be very beneficial to feed manufacturers. Previous data have demonstrated that reducing the particle size of cereal grains leads to an increased energy requirement and decreased production rate (Gebhardt et al. 2018). However, the rate at which these factors change can be altered by grain, mill type, mill parameters, and mill maintenance. Roller mills and hammermills have different operating cost and capital investments. With the knowledge of the negative effects particle size reduction can have on milling efficiency, it is important to further investigate if these loses can be accounted for by improvements in animal performance.

Improvements in nutrient digestibility with decreasing particle size have been thoroughly researched. For example, reducing sorghum article size from 1262 to 471 µm improved apparent ileum and total tract digestibility of DM, starch, N, and GE in finishing pigs (Owsley et al. 1981). Acosta et al. 2015 did not observe any improvements in energy and nutrient digestibility when corn particle size was reduced from 700 to 300 µm with a hammer mill but did observe improvement when corn was ground using a roller mill. In contrast, Rojas and Stein (2015) observed a linear increase in the concentration of ME in corn as particle size was reduced from 865 to 339 µm using a hammermill. However, Bertol et al. (2017) developed a model that observed a break point between 523 and 524 μm; therefore, reducing the particle size below 523 μm did not further improve AMEn value of corn.

The standard method for determining the particle size is conducted using wire cloth sieves and sieve shakers (ANSI/ASAE S319.4 FEB 2009 R2012). It is important to note that there are several modifications used when determining particle size which will dramatically change the results. These modifications include the number of sieves used in a stack, length of agitation time, dispersion agents, and sieve agitators. Therefore, it is important to document how particle size is analyzed. It has been reported that there is no difference in particle size between 10- and 15-minutes agitation time with the use of dispersion agents and sieve agitators (Kalivoda 2016). Addition of a dispersion agent to hammer mill ground corn can reduce results by about 80 microns and the additions of sieve agitators reduced particle size by about 40 microns (Fahrenholz et al. 2010; Stark and Chewning 2012). Kalivoda et al. (2016) recommends when measuring particle size, the use of sieve agitators and sieving agent should be used to decrease mean particle size and increase the standard deviation. This will be a more accurate representation of particle size.

Nursery pig performance is highly variable with decreasing particle size. Diet texture and palatability have a significant influence on nursery pig intake when particle sizes are below 500 microns. Finely ground corn less than 325 microns has been observed to have a negative impact when fed to nursery pigs due to reduced feed intake with no impact on feed efficiency (De Jong et al. 2014a, b). Other research has demonstrated improved feed efficiency when corn particle size was reduced from 700 to 400 microns (Bokelman et al. 2014). Furthermore, Bokelman et al. 2014 and Gebhardt et al. 2018 found that nursery pigs preferred diets with corn ground at 525 or 700 μm consuming 67 or 80% of their daily intake but with diets containing corn ground to 267 or 400 microns pigs consumed 33 or 20% of their daily intake, respectively.

Based on multiple studies conducted prior to 2012, it was concluded that there was a 1.0–1.3% improvement in finishing pig feed efficiency for every 100 μm reduction in particle size when corn or sorghum was ground from approximately 1000–400 μm (Wondra 1995a; Paulk 2011; De Jong 2015). Bertol et al. (2017) observed a 6% improvement in feed efficiency when gilts were fed diets with corn ground from 904 to 483 μm. However, reducing particle size did not influence feed efficiency of barrows. In contrast to previous research, Gebhardt et al. (2018) did not observe improvements in finisher pig feed efficiency when corn was ground to reduced particle sizes (561–285 μm) using a roller mill. As previously mentioned, Bertol et al. (2017) and Rojas et al. (2015) determined increased ME of corn when the particle size was reduced. When formulating diets to be balanced for these increases in ME, there were no differences in finishing pig feed efficiency observed. It has been well documented that reducing the particle size of grains improves the feed efficiency of finishing pigs. However, previous research has demonstrated more variability in what may be considered as optimal particle size.

While reducing particle size improves animal performance, consideration must be given to the changes in stomach morphology. Ulceration of the esophageal region of the stomach has been extensively observed and researched with the reduction in particle size in finishing pigs (Maxwell et al. 1970; Wondra et al. 1995a; and Ayles et al. 1996) and lactating sows (Wondra et al. 1995b). The occurrence of ulcers with the decrease in particle size results from the increased mixing of stomach contents due to increased surface area. This mixing results in a decreased pH of stomach contents and allows continuous exposure of the esophageal region to digestive acids (Reimann et al. 1968; Maxwell et al. 1970). When gastric ulcers are advanced, it can cause death. Minimizing stomach ulcers without having negative effects on performance is challenge for swine producers.

Challenges of grinding are found both in the feed mill and on the farm. Decreasing particle size can influence the performance but must be balanced for increased cost and risk of ulceration.

## Feed Batching

Feed batching systems are known as the bottle neck of the feed mill. With the growth of the feed additive industry, this has increased the number of ingredients added to swine diets within the last decade. More ingredients mean more time needed to weigh ingredients accurately. Most commonly, there are several major ingredient scales as well as minor, totes, and micro scales. Under-addition of ingredients can lead to poor animal performance, while over-addition of ingredients can lead to deviation in inventory and added cost. Either way, the additional time needed to accurately batch feed far outweighs the cost of inaccuracy. Keeping scales within the upper and lower specification limits (1–2%) of the required quantities is key to getting precise diets in front of the pig (Stark and Jones 2015).

The speed and accuracy of the batching systems causes increased cost and headaches for feed mills. Smaller inclusion of ingredients such as concentrated enzymes, vitamins, and minerals requires greater scale resolution, finer control of equipment, and a higher degree of accuracy during weighing. This proposes a challenge when trying to weigh ingredients to the nearest 0.01 lbs (Stark 2016). Using loss-in-weight systems can help to improve weighing accuracy.

Increased consumer demands and government regulations have led to improvements for feed batching. These improvements include precision feed manufacturing, ingredient lot tracing, using bar code readers during hand additions, and summarizing process data (Stark 2014). Time, screw conveyor diameters, and the use of variable frequency drives (VFD) at multiple speeds determined the accuracy of scales (Stark and Jones 2015).

## Feed Mixing

The type of mixer used to mix ingredients greatly influences the time needed to create a uniform mix. Efficient mixer options include double shaft ribbon and paddle mixers. The order of scale discharge should be major scale, minor scale, and finally micro-scales. The location of discharge of micro-ingredients should also be considered to prevent discharge into dead zones or areas of the mixer where mixer paddled do not reach. The size of the batch of feed should never exceed the volume which the mixer is designed. However, the ingredient density must be considered since it can influence the uniformity of the mix. Commonly, high by-product diets will decrease density and therefore the batch size should decrease. As a general guideline, ribbons should always be visible (Stark 2016). Material buildup is a key indication that ribbons and paddles are not functioning properly.

For precision formulation to be successful, a uniform mix is required. To determine a uniform mix a coefficient of variation (CV) must be determined. A mixer uniformity test is often done by using a single source tracer (i.e. salt, trace minerals, or iron filings) as the indicator. Ten samples should be pulled in order from different spots in the mixer and the tracer analyzed to be tested for uniformity. The CV can be calculated by $CV\% = (\text{standard deviation/mean}) \times 100\%$. The feed industry standard is a CV of less than 10%. If results are between 10 and 15%, it is a good mix and mixing time should be increased by approximately 25%. With results of 15–20%, mixer time should be increased by 50%, and mixer wear and ingredient propriety should be addressed. Any results greater than 20% is considered poor and should be evaluated. Any changes in mixing should be validated by mixer uniformity.

To prevent drug carryover during mixing, the mixer should be subjected to effective cleanout procedures. The CGMP regulations require medicated feed manufacturers to use one or more of the approved cleanout procedure, such as cleaning, sequencing, and/or flushing to prevent unsafe contamination by drug carryover. The most effective cleanout procedure is considered the thorough cleaning of the feed manufacturing equipment. However, given its time-consuming nature and the down time needed to thoroughly clean the equipment, sequencing and flushing are the most commonly used in the feed industry. Sequencing is the predefined succession of feed manufacturing to prevent contamination of subsequent batches. This is usually done in high production facilities were they have enough production volume to predefine a weekly production schedule. When implementing flushing procedures, it is recommended to use flush size of 5–10% of the mixer's total capacity (Martinez et al. 2018).

A uniform mix can be assured through mixer testing. Factors that can influence ingredients distribution included ingredient characteristics and type of mixer. Additionally, mixer ribbons should be inspected weekly to minimize build up from addition of liquid ingredients. Mixer uniformity CVs should be done quarterly for validation. Mixer uniformity should be routinely monitored for its impact on animal performance and system-wide cost.

## Feed Pelleting

Pelleting swine diets has been well documented and is done for improvements in performance and feed handling. Benefits of pelleting include (Behnke 1994):

- Decreased feed wastage
- Reduced selective feeding
- Decreased ingredient segregation
- Less time and energy expended in prehension
- Destruction of pathogens
- Improved palatability

Pellets are produced via steam conditioning providing heat and moisture helping the material pass though the die. When heated, the starch and protein present create adhesive forces to establish the pellet. The temperature that the mash is heated to before pelleting is highly depended on ingredients present. For example, nursery pig diets with high inclusions of whey included would need to be pelleted at a lower temperature (about 145 °F) than a corn-soy finishing pig diet (about 185°F). Conditioner type, single pass or double pass, will determine the retention time with 30–45 seconds or 75–90 seconds, respectively. While longer retention times can improve pellet quality, this can also destroy some nutritional value and increase cost of pellet mill throughput. While higher retention times tend to improve pellet quality and pellet mill throughput.

Feeding pelleted diets to pigs has previously resulted in improvements in the rate of gain (Wondra et al. 1995a). However, there have also been a number of reports suggesting no significant benefits on growth rate due to pelleting (Skoch et al. 1983). Paulk (2009) summarized a series of 16 experiments and based on the average determined that pelleting diets for growing-finishing pigs resulted in approximately a 6 and 7% improvement in rate and efficiency of gain, respectively. There are a number of theories that try and explain the mechanism behind the improvement in growth performance do to pelleting. Skoch et al. (1983) proposed the idea that pelleting increased the bulk density of diets and reduced dustiness, therefore, making the diets more palatable. However, this is not supported by the inconsistencies observed with feed intake in pelleting trials. Although pelleting does not increase starch gelatinization to the extent of extrusion previous research has demonstrated an increase in gelatinized starch from approximately 6.1 to 11.7% in pelleted based diets (Lewis et al. 2015a). Rojas et al. (2015) observed an increase in apparent ileal digestible GE and starch and increased ME concentration in pigs fed pelleted diets compared to meal diets. However, the observed improvement in ME was less than the improvement in feed efficiency that has been previously reported. It is hypothesized that a majority of the improvement in feed efficiency results from a reduction in feed wastage (Hanrahan 1984). This is in agreement with previous research that demonstrated no difference in feed efficiency when pigs were fed pelleted diets that were ground into meal form vs meal diets (Lewis et al. 2015b).

When pelleting diets, feed mills must determine what their target pellet quality is. Ingredient composition, equipment design, and manufacturing parameters influence pellet quality as well as pellet mill production rate and energy consumption. Previous research observed an improvement in feed efficiency as percent fines decreased in pelleted finishing pig diets (Schell and van Heugten 1998; Nemechek et al. 2015; Stark 2014). However, Langdon (2015) demonstrated no difference in feed efficiency as pellet fines was decreased (60, 45, 30, 15, or 0%). These differences in response to pellet quality can potential be explained by feeder type and management. Nemechek et al. (2015) demonstrated that pigs fed pelleted diets with 50% fines had improved feed efficiency when feeders were adjusted to 12.7 mm opening vs a 25.4 mmm opening. Although feeder gap improved feed efficiency of poor-quality pellets, pigs fed good quality pellets had the best feed efficiency. Segregation of nutrients from fines to pellets is another area of concern with pellet quality.

Reducing the percent fines at the feeder is the objective of producing good-quality pellets. However, each production system is different, and variety of factors can influence the handling stress pellets undergo and the amount of fines generated. Although the amount of percent fines is the major concern feed mill require a methodology to estimate pellet quality for quality assurance purposes. Therefore, cooled pellet samples are collected at the feed mill to determine pellet durability index (PDI; ASAE 1997; S269.4) to estimate pellet quality. These methods will be discussed in the finished feed quality section below.

In addition to pellet quality, pellet cooling is important to drying pellets to decrease mold growth by reducing moisture. The goal is to decrease the moisture content adding during pelleting less than initial mash moisture content. Air flow, feed bed depth, ambient temperature, speed of the fan, and amount of fines blocking air flow will determine rate and uniformity of cooling (Stark 2012b).

## Post-Pellet Liquid Application

Certain scenarios during feed manufacturing make it beneficial to add liquid ingredients after diets have been pelleted. For example, mixer-added fat has been shown to decrease pellet quality (Gehring et al. 2011). In addition, certain feed additives are not heat stable and are denatured during the conditioning processes (De Jong et al. 2017). Post-pellet liquid application (PPLA) of fat or enzymes can occur at the pellet die while the pellets are hot or after the pellets have been cooled to alleviate these negative effects (Stark 2016). The liquid can be applied via spray nozzles or spinning discs. The goal of a PPLA system is to apply an accurate amount of liquid by weight to the pellets with a 360-degree uniform coating of material. The ideal PPLA system would include accurate ingredient metering, uniform ingredient distribution, liquid is absorbed into the pellet, and liquid and dry material contained within the contained within the coating system. When determining appropriate PPLA system equipment, it is important to consider the ingredient density, viscosity, pH, quantity, desired accuracy, and optimal temperature.

Method of application and control of metering are the two important factors to consider when determining a PPLA system. Liquid application typically occurs either after the pellet die or after pellets have been cooled. Applying liquid immediately after the die requires less expensive equipment that can be installed on existing equipment. However, the disadvantage to this system is the minimum amount of time the pellet has to absorb the liquid prior to cooling. When greater than 3% of liquid is applied prior to cooling, the liquid can be drawn off the surface of the pellet and end up in the cooler and the cooler air system. This can have a negative effect on the cooling system and the maintenance of cooling equipment will increase. When applying liquid to cooled pellets, a stand-alone system (spray in a screw, ribbon, or paddle conveyor; plenum or weir; rotating drum or reel; spinning disk; or batch mixer) is required. Both systems require liquid to be applied to a continuous flow, making it difficult to achieve target accuracy. The three principle methods used for measuring the flow rate of material in a continuous system are volumetric, mass flow, and loss-in-weight (Stark 2016). Regardless of which system is employed, regular cleaning, maintenance, and calibration of the system should be scheduled to ensure accurate and trouble-free operation. The use of PPLA systems requires additional monitoring but can help in accurate inclusion of liquid ingredients and improve the final feed quality.

**Feed Delivery**

Once feed manufacturing is complete, the final step is delivering feed. While this may seem simple, deliveries to the wrong bin happen frequently. Feed delivered to the wrong bin must either be fed to pigs at the wrong growth stage, blended with existing feed, or removed from bin all of which require a lot of time and money with potential losses in performance. Use of automation systems as a check system can help get the feed to the right place. Use of barcodes for loading and unloading and GPS tracking are a few tools that can be used to minimize incorrect feed delivery. Feed delivery equipment should be selected for uploading speed, durability, and minimize the degradation of pellet quality.

**Finished Feed Quality**

Nutrient value and pellet quality will determine the quality of finished feed. Using PDI and fines discussed previously are indication of pellet quality. There are several ways to determine PDI that involved different machines and modifications. The goal of all these methods is to determine the quality of the pellet after transportation and therefore the pellet that is present at the feeder for consumption. The traditional method is the KSU tumble box method in which a sample is rotated in a box for 10 minutes and can be modified with the addition of hex nuts to make the method more abrasive. The tumble box method is the most common method found in the feed mill because of the low equipment cost required. Another method involves Holman NHP 100 pellet tester in which forced air is used to agitate pellets and determine PDI. The Holmen methods are more aggressive and less time consuming compared to the tumble box and method with results that can be produced from 30 to 120 seconds. The use of an inline pellet tester, Holmen NHP 300, can produce results directly to a computer by collecting, cooling, and sieving samples directly from the pellet mill. Using a fully automated inline pellet tester can help with real time production parameter adjustments (Stark 2014). The use of PDI information can be used to create a model to determine pellet quality at the feeder and help nutritionist with information of effects of ingredients.

To determine nutrient value of finished feed, feed mills adapted use of table top near infrared (NIR) technology. Advantages if using NIR is quick, real time answers with minimal sample preparation, in-line options, precision and low cost after initial investment. In the past NIR has only been used for analysis of receiving ingredients; however, opportunities with finished feed creat an additional check for guaranteed analysis. Challenges with using an NIR machine include the importance of correct calibration equations used to predict nutrient content from wet chemistry analysis. Recently, companies have developed software to manage NIR machines and data remotely; however, there this requires time and additional cost for maintenance of calibrations and data management.

## Feed Mill Biosecurity and Animal Food Safety

The United States swine industry has made substantial gains in herd health by implementing farm biosecurity practices. Many of the primary routes of pathogen entry into the farm (i.e. other pigs, farm employees, visitors, air, etc.) have been minimized in high health systems, while less research has characterized the risk of feed as a transmission vector. However, there is increasing interest in risk associated with feed mills and opportunity for biosecurity protocols to be adapted. While not all pathogens are strong candidates for feed-based transmission, the knowledge about the role of feed and its transfer of pathogens is growing.

Feed mill biosecurity is important to the feed and animal-agriculture industries as a way to control the spread of feedborne disease and other hazards. Implementing a biosecurity plan to prevent or mitigate biological hazards in a feed mill is challenging because of differences in facility design, manufacturing operations, and significant risk factors among feed mills. However, biosecurity plans at the mill can minimize risk of introduction of biological hazards and limit potential economic losses from these hazards. The first step toward minimizing risk is to develop a feed mill biosecurity plan. Methods for developing a swine feed mill biosecurity plan are described by Cochrane et al. 2016. Other references, such as the AFIA guide for Developing Biosecurity Practices for Feed & Ingredient Manufacturing, and the K-State Swine Feed Mill Biosecurity Audit are helpful for facilities to determine opportunities for improving biosecurity. The followings are six key recommendations to consider when developing a biosecurity plan:

- Use receiving mats or funnels to limit pathogen entry into the receiving pit. The ingredient receiving pit is the single biggest entry point for contaminants into the feed manufacturing system. Magossi et al. (2019) reported that the pit was second to only employee shoes as the most unhygienic locations tested in 12 U.S. feed mills.
- Create lines of separation at all doors to minimize contamination from footwear. This involves employees and visitors changing shoes to keep exterior shoes on one side of the line and interior shoes on the other. Examples of how facilities may implement lines of separation are shown in Figures 16.1 and 16.2. In both examples, additional exits are available in case of emergency to satisfy OSHA requirements. If lines of separation cannot be developed, consider zoning to standardize traffic patterns, with foot baths or food-grade dry sanitizing powder placed in high traffic areas.
- Create cleaning and disinfection stations for delivery vehicles and feed trucks. Use wet-cleaning and sanitizers to remove debris from the tires, wheels, undercarriage, and exterior of ingredient trucks and feed delivery trucks prior to their entry into the mill. Similarly, create stations prior to entry of delivery trucks on and off of farms.

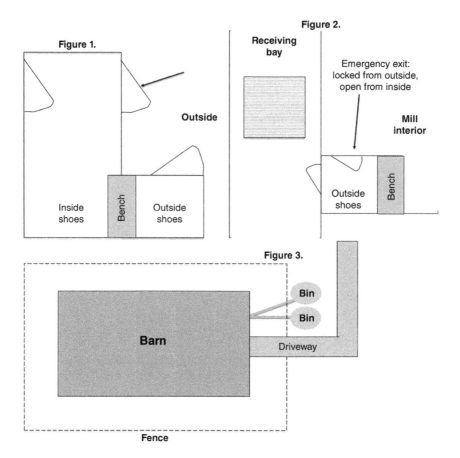

**Figure 16.3**   Delivering feed.

- Sanitize floors routinely. Sweep or vacuum all dirt and dust from floor, then mop with a 10% bleach solution or and EPA-approved FAD disinfectant on a weekly basis to limit the accumulation and spread of virus on nonfeed-contact surfaces
- Refrain from using dust, screenings, or similar materials as an ingredient or added back into feed production. These materials are frequently placed back in ground corn or an ingredient bin to minimize shrink. However, dust is consistently reported to carry high levels of pathogens, and should be composted or discarded, never fed to animals.
- When delivering feed, use cleaning and disinfection stations prior to entering and exiting farms. Alternatively, consider unloading feed across a line of segregation or fence into another feed truck or extend bin augers so bins can be filled on the exterior of the line of segregation, as shown in Figure 16.3.

## Conclusion and Future Trends

Feed processing is a rapidly changing field. As research and new technology influence the understanding of the effects of processing on animal performance, analytical methods for incoming ingredients, automation improvements, and further implementation of biosecurity, these areas will continue to evolve. Applying these changes to feed manufacturing and nutrition will continue to be a challenge for the future.

In the future, attention on feed safety will be a concern from both a food safety regulation standpoint and biosecurity. Therefore, complete traceability of feed and ingredients will require strict record-keeping. Diet formulation will be challenged with the push for medication free therefore introducing alternative ingredients having subsequent effects at the feed mill. Ingredient changes can affect pellet quality, all the way to down to mixing and batching depending on ingredient characteristic and inclusions. The variety of alternative ingredients will be the challenge. Narrowing the margin of variability of preprocessed ingredients (i.e. DDGS) at receiving will also be a focal point.

Relationships between mill managers, nutritionists, and decision makers are key for communicating the importance of quality, safe feed on the effects it can have on both pig performance and pocket performance. When implementing new technology, determining the cost–benefit should be decided with all parties to be affected to determine how the changes will affect all aspects of the process of getting feed to pigs. The ability for feed mills to adapt to these changes in technology, ingredients, and regulation will determine a successful system.

## References

ASAE. 1997. Cubes, pellets, and crumbles – definitions and methods for determining density, durability, and moisture content. ASAE Standard S269.4. Am. Soc. Agric. Biol. Eng., St. Joseph, MI.

ASAE. 2012. Method of determining and expressing fineness of feed materials by sieving. ANSI/ASAE Standard S319.4 R2012. Am. Soc. Agric. Biol. Eng., St. Joseph, MI.

Behnke, K. C. 1994. Factors affecting pellet quality. Proceedings of Maryland Nutrition Conference, 20–25 March, 1994. Dept. of Poult. Sci., College of Agric., Univ. of Maryland, College Park, MD, US.

Cochrane, R. A., S. S. Dritz, J. C. Woodworth, C. R. Stark, A. R. Huss, J. P. Cano, R. W. Thompson, A. C. Fahrenholz, and C. K. Jones. 2016. Feed mill biosecurity plans: A systematic approach to prevent biological pathogens in swine feed. J. Swine Health and Prod. 24:154–164.

De Jong, J. 2015. Feed processing challenges facing the swine industry. PhD. Diss. Kansas State Univ., Manhattan, KS, US.

De Jong, J. A., J. C. Woodworth, J. M. DeRouchey, R. D. Goodband, M. D. Tokach, S. S. Dritz, C. R. Stark, and C. K. Jones. 2017. Stability of four commercial phytase products under increasing thermal conditioning temperatures. Trans. Anim. Sci. 1:255–260.

Dunmire, K. M., J. Dhakal, K. Stringfellow, C. R. Stark, and C. B. Paulk. 2019. Evaluating soybean meal quality using near-infrared reflectance spectroscopy. Kansas Agric. Exp. Stn. Res. Rep. 5. https://doi.org/https://doi.org/10.4148/2378-5977.7864.

Evans, C., N. Frempong, C. Paulk, T. Nortey, C. Stark. 2019. Determining the influence of diet formulation, particle size, and analytical method on the crude protein predictability of near infrared reflectance spectroscopy. IPPE. February 11 to 13, 2019, Atlanta, GA, US.

Fahrenholz, A. C., L. J. McKinney, C. E. Wurth, and K. C. Behnke. 2010. The importance of defining the method in particle size analysis by sieving. Kansas State Univ. Agric. Exp. Stn. Rep. Prog. 1038. Swine Day 2010. Kansas Stae Univ., Manhattan, KS. US. p. 261.

Gehring, C. K., K. G. S. Lilly, L. K. Shires, K. R. Beaman, S. A. Loop, and J. S. Moritz. 2011. Increasing mixer-added fat reduces the electrical energy required for pelleting and improves exogenous enzyme efficacy for broilers. J. Appl. Poult. Res. 20:75–89.

Hanrahan, T. J. 1984. Effect of pellet size and pellet quality on pig performance. Anim. Feed Sci. Technol. 10:277.

Kalivoda, J. 2016. Effect of sieving methodology on determining particle size of ground corn, sorghum, and wheat by sieving. MS Thesis, Kansas State Univ., Manhattan, KS, US.

Lagos, L. V., and H. H. Stein. 2017. Chemical composition and amino acid digestibility of soybean meal produced in the United States, China, Argentina, Brazil, or India. J. Anim. Sci. 95:1626–1636.

Leiva, S., J. Sandoval, J. Flees, A. Calderon, G. Abascal Ponciano, K. Ordonez, D. Patino, L. Avila, M. Presume, W. Pacheco, and C. Starkey. 2019. Effect of particle size on near infrared reflectance spectroscopy (NIRS) nutrient analysis of ground corn. Poult. Sci. 98(E-Suppl. 1):3.

Lewis, L. L., C. R. Stark, A. C. Fahrenholz, J. R. Bergstrom, and C. K. Jones. 2015a. Evaluation of conditioning time and temperature on gelatinized starch and vitamin retention in a pelleted swine diet. J. Anim. Sci. 93:615–619.

Lewis, L. L., C. R. Stark, A. C. Fahrenholz, M. A. D. Goncalves, J. M. DeRouchey, and C. K. Jones. 2015b. Effects of pelleting conditioner retention time on nursery pig growth performance. J. Anim. Sci. 93:1098–1102.

Liu, Y., R. Jha, and H. H. Stein. 2018. Nutritional composition, gross energy concentration, and in vitro digestibility of dry matter in 46 sources of bakery meals. J. Anim. Sci. 96:4685–4692.

Magossi, G., N. Cernicchiaro, S. Dritz, T. Houser, J. Woodworth, C. Jones, and V. Trinetta. 2019. Evaluation of Salmonella presence in selected United States feed mills. Microbiology 8:e711.

Martinez, A., C. K. Jones, C. R. Stark, L. J. McKinney, K. C. Behnke, and C. B. Paulk. 2018. Evaluating flushing procedures to prevent nicarbazin carryover during medicated feed manufacturing. Anim. Feed Sci. Tech. 242:1–7.

Nemechek, J. E., M. D. Tokach, S. S. Dritz, E. D. Fruge, E. L. Hansen, R. D. Goodband, J. M. DeRouchey, and J. C. Woodworth. 2015. Effects of diet form and feeder adjustment on growth performance of nursery and finishing pigs. J. Anim. Sci. 93:4172–4180.

Paulk, C. B. 2009. Manipulation of processing technologies to enhance growth performance and(or) reduce production costs in pigs. MS Thesis. Kansas State Univ., Manhattan, KS, US.

Schell, T. C., and E. Van Heugten. 1998. The effect of pellet quality on growth performance of grower pigs. J. Anim. Sci. 76(Suppl. 1):185 (Abstr.).

Sotak-Peper, K. M., W. J. C. González-Vega, H. H. Stein. (2017). Amino acid digestibility in soybean meal sourced from different regions of the United States and fed to pigs, J. Anim. Sci. 95:771–778.

Stark, C. R. 2012a. Feed processing to maximize feed efficiency. In: J. F. Patience, editor, Feed efficiency in swine. Enfield Pub. Enfield, NH.

Stark, C. R. 2012b. Feed processing to improve poultry performance. Arkansas Nutr. Conf., Rogers, AR, US.

Stark, C. R. 2013. Feed processing to increase performance and profit. 34th Western Nutr. Conf. Saskatoon, SK, Canada.

Stark, C. R. 2014. Feed processing technology to improve feed efficiency in pigs and poultry. In: P. C. Garnsworthy and J. Wiseman, editors, Recent advances in animal nutrition 2014. Context Products Ltd., Leicestershire, UK.

Stark, C. R. 2016. Feed processing to improve swine and poultry performance. The 5th Global Feed and Food Congress, Antalya, Turkey.

Stark, C. R. and C. G. Chewning. 2012. The effect of sieve agitators and dispersing agent on the method of determining and expressing fineness of feed materials by sieving. Anim. Prod. Sci. 52:69–72.

Stark, C. R., and F. T. Jones. 2012. Quality assurance programs in feed manufacturing. Feedstuffs Ref. Issue 2013. Miller Publishing, Los Angeles, CA. p. 84–89.

Stark, C. R. and C. K. Jones. 2015. Feed processing to improve animal performance. Canada Eastern Anim. Nutr. Conf., Quebec, QC, Canada.

Sulabo, R. C., and H. H. Stein. 2013. Digestibility of phosphorus and calcium in meat and bone meal fed to growing pigs. J. Anim. Sci. 91:1285–1294.

Van Heutgen, E. 2001. Mycotoxins and other antinutritional factors in swine feeds. In: A. J. Lewis, and L. L. Souther, editors, Swine nutrition. 2nd ed. CRC Press, Boca Raton, FL, US.

# 17 Enzymes and Enzyme Supplementation of Swine Diets

Chan Sol Park and Olayiwola Adeola

## Introduction

Swine diets have been formulated based on feed ingredients originating from cereal grains and protein-rich by-products from the oil extraction of legume seeds. Although most nutrients in plant feed ingredients are digestible by pig endogenous enzymes, digestion and absorption of nutrients are often hindered by the presence of dietary antinutritional components in plant feed ingredients. Moreover, the use of alternative feed ingredients (e.g., by-products after milling processes of cereal grains or from the oil extraction of legume fruits or seeds other than soybeans) that contain high concentration of dietary antinutrients has received attention for reuse and production cost reduction.

Phytic acid is the major storage form of P in plant feed ingredients. Most plant feed ingredients contain considerable concentration of P; however, P in phytic acid is hardly digested and absorbed by pigs due to low endogenous phytase production and the low solubility of phytate, a salt form of phytic acid with cations, in the gastrointestinal tract (GIT). Therefore, majority of P in plant feed ingredients is undigested and excreted in feces, which may cause eutrophication. Moreover, because of the low digestibility of P in plant feed ingredients, inorganic P sources such as mono- or di-calcium phosphate have been used in swine diets which are finite sources from the earth and account for considerable portion of production cost. Reduction of nutrient utilization from the negative effects of dietary phytate and fiber results in oversupply of nutrients in swine diets, leading to the increased excretion of nutrients and total cost of production. To mitigate the negative effects of those antinutritional factors, exogenous enzymes have been used in swine diets.

The current global feed enzyme market is approximately $ 940 million US Dollars. The use of exogenous enzymes in animal feeds is expected to surpass 1 billion US Dollars by 2024, with phytase, carbohydrases, and proteases as the key animal feed enzymes for improving feed efficiency and nutrient utilization as well as reducing environmental waste. Phytase and carbohydrases are the two dominant exogenous enzymes in swine diets which degrade dietary phytin and carbohydrates, respectively. In addition, supradosing of phytase may improve the growth performance of pigs, which exceeds the expected response. Moreover, recent studies have suggested that oligosaccharides released from dietary fiber by exogenous carbohydrases may improve the gut health of pigs. In this chapter, various aspects of exogenous enzymes, mainly focusing on carbohydrases and phytase, in swine diets are discussed together with brief introduction of their substrates. Furthermore, future studies required to improve the understanding and implications of exogenous enzymes are suggested.

## Antinutritional Effects of Phytic Acid and Dietary Fiber

### *Phytic Acid*

Phytic acid in plant feed ingredients mostly presents as a structure consisting of an inositol ring and 6 phosphate esters (*myo*-inositol 1,2,3,4,5,6-hexakis dihydrogen phosphate; IP$_6$; Figure 17.1a). Five of 6 phosphate groups are equatorially oriented, but the phosphate group on carbon 2 is axially oriented, which makes it resistant to phytase activity (Adeola and Cowieson 2011). Completely dissociated phytic acid can carry 12 negative charges, all of which can bind to various cations, basic AA residues of proteins, or carbohydrates (Olukosi and Adeola 2012). Phytate refers to the salt of

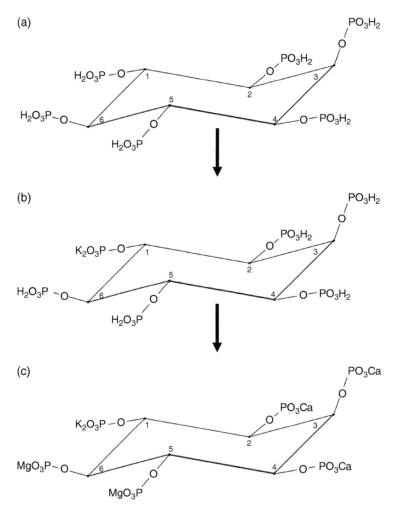

**Figure 17.1**　Hypothetical structure of phytic acid, phytate, and phytin in plants; phytic acid (a), *myo*-inositol 1,2,3,4,5,6-hexakis dihydrogen phosphate; potassium phytate (b), salt of phytic acid chelated with two K$^+$ replacing two protons at phosphate group on carbon 1; phytin (c), mixed salt of phytic acid chelated with K$^+$, Mg$^{2+}$, and Ca$^{2+}$ molecules replacing protons in phosphate groups.

phytic acid chelated with cations (Figure 17.1b). In plants, phytic acid generally presents as a form of mixed salt (i.e., phytin) with cations such as K, Mg, and Ca (Figure 17.1c).

Solubility of phytin is dependent, in part, on ratio among chelated cations. Due to the strong affinity of phytic acid for divalent cations (Rimbach et al. 2008), phytin with high ratio of $Ca^{2+}$ plus $Mg^{2+}$ to $K^+$ is less soluble than low ratio of $Ca^{2+}$ plus $Mg^{2+}$ to $K^+$. In general, legume seeds and their byproducts after oil extraction contain lower ratio of $Ca^{2+}$ plus $Mg^{2+}$ to $K^+$ than cereal grains. On the other hand, phytin is mainly located in aleurone layer of most cereal grains except for corn in which germ is the major site of phytin storage (O'Dell et al. 1972; Steiner et al. 2007), whereas legume seeds contain phytin throughout the cotyledon (Reddy et al. 1982). Therefore, dehulling process can remove a portion of phytin in several cereal grains and reduce antinutritional effects.

After ingestion of dietary phytin, cations are replaced with protons (hydrogen) in acidic conditions of stomach, leading to a liberation of partially protonated phytic acid (Schlemmer et al. 2001). Phosphate groups in liberated phytic acid can be partly degraded by intrinsic plant phytase (Rodehutscord et al. 2016). However, the activity of intrinsic plant phytase is limited due to the poor resistance of intrinsic plant phytase to proteolytic digestion and unfavorable pH in the stomach and small intestine (Rodehutscord and Rosenfelder 2016). In addition, intrinsic plant phytase is generally susceptible to heat damage, and therefore, heat processing of diets such as pelleting or extruding reduces the activity of intrinsic plant phytase (Schlemmer et al. 2001). Indeed, endogenous phytase is produced in the small intestine of pigs, mainly in the mucosal layer of jejunum (Hu et al. 1996). However, little P in phytic acid is digested and absorbed due to the poor solubility of phytate (Adeola and Cowieson 2011).

As pH of digesta gradually increases along the small intestine, phytic acid readily binds to di- and trivalent cations to form a stable complex, which reduces the solubility of phytic acid as well as cationic minerals including Ca, Mg, Fe, Zn, Cu, Mn, and Cd (Rimbach et al. 2008). Formation of mineral-phytate complex is resistant not only to intrinsic plant phytase and endogenous phytase but also to exogenous phytase. Maenz et al. (1999) reported that Zn-phytate complex was the most resistant to phytase activity at neutral pH, but Ca-phytate complex may be the most problematic due to the greatest concentration of Ca in diets compared to other dietary cations (Dersjant-Li et al. 2015). Dietary phytic acid may also increase the endogenous loss of Na (Woyengo et al. 2009). Cowieson et al. (2006) observed similar findings in broiler chickens and discussed that pH of digesta in the duodenum was decreased by phytic acid, leading to the increased secretion of $NaHCO_3$ from pancreas. Changes in endogenous Na secretion in the small intestine may be critical for absorption of other nutrients because it may impair the activity of Na-K-ATPase (Dersjant-Li et al. 2015). However, unlike poultry, reabsorption of Na may occur in the large intestine of pigs, and therefore, imbalance of electrolytes may only have a local effect in the small intestine, but not affect the whole physiological system of the body (Woyengo et al. 2009).

Phytic acid may bind to positively charged amino acid (AA) residues in proteins and form binary protein-phytate complex at acidic pH condition or bind to proteins together with Ca and form ternary protein-Ca-phytate complex at neutral pH condition (Selle et al. 2012). The formation of protein-phytate complex decreases the solubility of proteins (Kies et al. 2006a; Yu et al. 2012), which reduces AA digestibility. Moreover, phytic acid may reduce the activity of pepsin in the stomach by binding to pepsinogen (Woyengo et al. 2010). Endogenous losses of AA may be affected by phytic acid, which was consistently observed in broiler chickens (Cowieson et al. 2004; Cowieson and Ravindran 2007; Liu and Ru 2010). Possible mechanism suggested by Cowieson et al. (2009) includes gastric stimulation of pepsinogen and HCl secretion by protein-phytate complexes, leading to the increased secretion of mucin and pancreatic juice in the small intestine. The reabsorption of endogenous secretions may be compromised due to the increased secretion of $NaHCO_3$ stimulated by acidic digesta. However, Woyengo et al. (2009) reported that the endogenous losses of AA in pigs were not affected by the addition of phytic acid in purified diets. Woyengo and Nyachoti (2013) discussed that in pigs, although endogenous secretion of AA may be increased by dietary phytic acid, the reabsorption of those AA may not be affected.

Phytic acid may interact with α-amylase (Knuckles and Betschart 1987) or lipase (Knuckles 1988) and reduce the digestibility of gross energy (GE). Reduced activity of mucosal amylase, sucrase, and maltase in the duodenum or jejunum was observed in broiler chickens fed high phytic acid diets (Liu et al. 2008). However, negative effects of phytic acid on digestibility of GE or energy values (i.e., digestible energy or metabolizable energy) in pigs remain unclear due to variations in results of studies conducted with phytase.

### Dietary Fiber

Dietary fiber mainly resides in plant feed ingredients as a cell wall component. Major component of dietary fiber is nonstarch polysaccharides (NSP) in which monosaccharides (i.e., pyranose and furanose) are bound to each other at various positions and orientations, leading to the formation of complex structures (Bach Knudsen 2011). In general, the NSP in cereal grains mainly consist of arabinoxylan, β-glucan, and cellulose, whereas those in legume seeds mainly consist of cellulose, xylan, xyloglucan, pectin polysaccharides, and glycoproteins (Choct 1997; Bach Knudsen 2011).

Dietary fiber may physically prevent the digestion of nutrients or chemically interfere with digestive enzymes in the GIT (Adeola and Cowieson 2011). It has been reported that increasing concentration of dietary fiber decreased digestibility of nutrients (Le Goff and Noblet 2001; Dilger et al. 2004; Navarro et al. 2019). In addition, reduced digestibility of nutrients may be related to altered gastrointestinal motility by ingestion of dietary fiber. Gastrointestinal motility is mainly dependent on stimulation and response of enteric nervous system and viscosity of digesta. Wenk (2001) suggested that dietary fiber stimulates the peristaltic movement of the GIT and decreases the retention time of digesta. Wilfart et al. (2007) also reported that increasing dietary concentration of wheat bran, which contained insoluble NSP as a major component of dietary fiber, decreased the retention time of digesta in the small intestine. Reduced retention time of digesta in the small intestine may lead to the reduced digestibility of nutrients because nutrients in digesta may flow to the large intestine before being digested and absorbed. On the other hand, soluble NSP is known to increase the viscosity of digesta, leading to delayed gastric emptying (Johansen et al. 1996) and increased retention time of digesta (Owusu-Asiedu et al. 2006). Previous review studies suggested that the retention time of digesta may be increased by soluble NSP and reduced by insoluble NSP (Montagne et al. 2003; Zijlstra et al. 2010). However, results of previous studies conducted to verify the relationship among the solubility of NSP, viscosity of digesta, and retention time are inconsistent, especially when purified dietary fiber sources were used (Owusu-Asiedu et al. 2006; Hooda et al. 2011). Moreover, delayed gastric emptying of pigs by soluble NSP may lead to reductions of feed intake and growth performance (Zijlstra et al. 2010; Adeola and Cowieson 2011). Therefore, it can be concluded that the growth performance as well as digestibility of nutrients can be affected by complex interactions between insoluble and soluble NSP on gastrointestinal motility and viscosity of digesta.

Apart from the retention time of digesta, digestibility of nutrients may be affected by viscosity of digesta per se by interfering with absorption of nutrients through unstirred water layer of mucosa (Johnson and Gee 1981). In addition, increased viscosity of digesta may decrease villus height and ratio between villus height and crypt depth in the small intestine, leading to a reduction of nutrient absorption (McDonald et al. 2001). Although negative effects of viscosity of digesta were observed in poultry (Pekel et al. 2015; Konieczka and Smulikowska 2018), these effects are not consistent in pigs (Bartelt et al. 2002; Hooda et al. 2011; Lærke et al. 2015). Increased viscosity of digesta may also increase the endogenous losses of N and AA, which may result in an underestimation of digestibility values (Chen et al. 2017).

## Phytase

### Mode of Actions

Phytase (*myo*-inositol hexakis dihydrogen phosphate phosphohydrolases) hydrolyzes phosphate group from phytic acid in a stepwise manner. Most commercial phytases are classified as histidine acid phytases based on the catalytic mechanism (Greiner and Konietzny 2010). Depending on the location (i.e., carbon number) where phytases initiate the hydrolysis of phosphate ester, phytases can be divided into 3-phytase, 5-phytase, and 6-phytase with Enzyme Commission (EC) identifier 3.1.3.8, 3.1.3.72, and 3.1.3.26, respectively.

Commercial phytases are mostly either 3- or 6-phytase. Phytase produced by microbes has been considered as 3-phytase, whereas those produced by plants has been considered as 6-phytase; however, this is not a general rule (Greiner and Konietzny 2010). Commercial phytases are produced by fermentation of genetically modified microbes which enables the mass production (Greiner and Konietzny 2010). In general, fungal phytase produced by *Aspergillus niger*, which was the first phytase marketed in the animal feed industry, is considered as the first generation of phytase, and bacterial phytase produced by *Escherichia coli*, which was produced after fungal phytase, is considered as the new generation of phytase (Dersjant-Li et al. 2015).

Phytase hydrolyzes the phosphate group at the designated location first (i.e., carbon 3, 5, or 6), and then hydrolyzes the next phosphate groups by sequentially moving round phytic acid (Adeola and Cowieson 2011). Due to the axial orientation of phosphate group located on carbon 2, end products of catalytic reaction are 5 orthophosphates (Pi) and *myo*-inositol 2-phosphate (IP$_1$; Yu et al. 2012). However, depending on various conditions and characteristics of phytase, the reaction of phytase generally releases Pi and partially phosphorylated *myo*-inositol phosphates such as *myo*-inositol penta- (IP$_5$), tetra- (IP$_4$), or triphosphate (IP$_3$; Menezes-Blackburn et al. 2015). Even though phytase does not completely hydrolyze phosphate groups in IP$_6$, the solubility of phytic acid increases with decreasing number of phosphate groups in *myo*-inositol ring (Schlemmer et al. 2001), allowing phytic acid to be available to endogenous phytase and reducing the detrimental effects of phytic acid on other nutrients.

There is no standard unit to express the activity of phytase, but it is usually expressed as phytase unit (FTU), which defines the amount of phytase that liberates 1 μmol of Pi per minute from 0.0051 M sodium phytate at pH 5.5 and 37 °C (method 2000.12; AOAC 2000). The unit of phytase activity is often provided as FYT, but it is synonymous with FTU. Although FTU can provide the general information of phytase activity, it cannot represent the actual activity of phytase when fed to pigs because of the differences in in vivo and in vitro conditions.

Exogenous phytase mainly acts in the stomach and proximal small intestine where phytic acid presents as a soluble form. Rutherfurd et al. (2014) reported that 72–78% of phytate complexes was degraded before terminal jejunum when phytase was added in corn-soybean meal (SBM)-based diets at 1107 or 2215 FTU/kg, whereas 40% of phytate complexes was degraded in pigs fed the diet without phytase. Therefore, the activity of phytase in mild acidic condition and resistance to

**Table 17.1** Characteristics of seven commercial phytases[a,b].

| Item | Donor organism | Class | pH range[c] | Relative activity (%)[d] | | Residual activity (%)[d,e] | |
|------|----------------|-------|-------------|--------|--------|----------|----------|
| | | | | pH 3 | pH 7 | − pepsin | + pepsin |
| EC1 | *Escherichia coli* | 6-phytase | 4.0–5.0 | 92.5 | 0.8 | 95 | 93 |
| EC2 | *Escherichia coli* | 6-phytase | 3.5–5.0 | 101.3 | 2.2 | 98 | 98 |
| EC3 | *Escherichia coli* | 6-phytase | 3.0–5.0 | 82.8 | 1.7 | 92 | 92 |
| BSP | *Buttiauxella* spp. | 6-phytase | 3.0 | 235.1 | 0.5 | 87 | 85 |
| CB | *Citrobacter braakii* | 6-phytase | 3.0–4.5 | 145.7 | 0.6 | 93 | 92 |
| PL | *Peniophora lycii* | 6-phytase | 4.5–5.5 | 12.5 | 7.8 | 58 | 34 |
| AN | *Aspergillus niger* | 3-phytase | 4.5–5.5 | 64.2 | 7.0 | 81 | 47 |

[a] Adapted from Menezes-Blackburn et al. (2015).
[b] The in vitro assay was conducted with 350 μl of sodium phytate solution at 37 °C for 30 minutes.
[c] An 80% of the optimal activity.
[d] The activity of phytase at pH 5.5 was considered as 100%.
[e] The incubation was conducted at pH 3 for 45 minutes.

proteolytic digestion are critical for in vivo efficacy of phytase. Relevant characteristics of several commercial phytases were evaluated by Menezes-Blackburn et al. (2015), and a part of results is presented in Table 17.1. Most commercial phytases maintained 80% of their optimal activity in weak acidic condition ranging from pH 3.0 to 5.5; however, phytases produced from *Peniophora lycii* and *A. niger* (PL and AN in Table 17.1, respectively) had reduced activity at pH 3. All phytases had low activity at pH 7. Most phytases were resistant to pepsin except for those produced from *P. lycii* and *A. niger*. Due to those differences in optimal pH range and susceptibility to pepsin, the efficacy of phytase may vary among commercial products.

The efficacy of phytase also depends on the substrate specificity. Phytase that has broad substrate specificity can hydrolyze $IP_6$ into 5 Pi and $IP_1$, whereas phytase that has narrow substrate specificity can hydrolyze $IP_6$ into Pi and various partially phosphorylated *myo*-inositol phosphates (Greiner and Konietzny 2010). For example, if there are two phytases which have the same FTU but different substrate specificities, phytase with narrow substrate specificity would more rapidly remove $IP_6$ or $IP_5$ releasing Pi and produce various *myo*-inositol phosphates such as $IP_3$ or $IP_4$, whereas phytase with broad substrate specificity would mainly act on $IP_6$ and release the same amount of Pi by near complete hydrolysis. Therefore, even though the same amount of Pi is released by the two phytases, the phytase with narrow substrate specificity may produce more *myo*-inositol with fewer phosphate groups (i.e., $IP_3$ and $IP_4$) than the phytase with broad substrate specificity, which may be more beneficial to pigs because *myo*-inositol with fewer phosphate groups have greater solubility than $IP_5$ or $IP_6$ (Greiner and Konietzny 2010; Adeola and Cowieson 2011).

The composition, structure, and concentration of phytin vary among feed ingredients. Therefore, the efficacy of phytase relies on the available substrates in feed ingredients. Almeida et al. (2017) reported that the concentration of phytase to maximize the standardized total tract digestibility (STTD) of P ranged from 160 FTU/kg for SBM to 1219 FTU/kg for sunflower meal. In addition, Almeida and Stein (2012) found that the graded concentration of dietary phytase from 0 to 1500 FTU/kg quadratically increased the STTD of P in corn, corn high-protein distillers' dried grains, and corn germ, but not in corn distillers' dried grains with solubles (DDGS). The STTD of P in DDGS is relatively greater than other feed ingredients originating from plants because phytin in cereal grains is partly hydrolyzed during fermentation process to produce ethanol (Pedersen et al. 2007). Therefore, the efficacy of phytase in DDGS or diets containing high concentration of DDGS may be diminished; however, it should be noted that other previous studies observed the increased digestibility of P in DDGS (Xue and Adeola 2015; Almeida et al. 2017). In addition, heat processing of diets affects the activity of phytase. De Jong et al. (2017) revealed that the residual activity of four commercial phytases linearly decreased from 76.2 to 20.2% as the conditioning temperature increases from 65 to 95 °C.

Calcium can bind to phytic acid and form a Ca-phytate complex in the GIT, which is resistant to phytase activity. The formation of Ca-phytate complex is dependent on pH of digesta and molar ratio between Ca and phytic acid (Selle et al. 2009). Based on the extensive review of previous studies, Selle et al. (2009) reported that the formation of Ca-phytate complex generally occurs at pH 5. Therefore, the activity of phytase in the stomach, where Ca-phytate complex formation is unlikely, is critical for the efficacy of phytase. Even though phytase cannot completely hydrolyze the phosphate groups in phytic acid, the affinity of phytic acid to bind Ca ion decreases as the number of phosphate groups in *myo*-inositol ring decreases (Luttrell 1993), leading to the increased availability of phytic acid to endogenous and exogenous phytases. In addition, increased concentration of dietary Ca or Ca-to-P ratio (Ca:P) may reduce the efficacy of phytase. Brady et al. (2002) reported a greater apparent total tract digestibility (ATTD) of P with 362 FTU/kg in diets with Ca:P at 1.15:1 than in diets with Ca:P at 1.85:1. In addition, Beaulieu et al. (2007) observed the increased ATTD of P in diets prepared with Ca:P at 1.12:1 or 1.66:1 when phytase was added at 500 FTU/kg of diet, but not in diets prepared with Ca:P at 2.31:1. However, Adeola et al. (2006) showed that the beneficial effects of 1000 FTU/kg phytase on growth performance were not influenced by dietary Ca:P from 1.2 to 1.8:1, which linearly decreased the growth performance regardless of phytase addition. Poulsen et al. (2010) reported similar findings in which ATTD of P by phytase addition (750 FTU/ kg of diet) were not affected by increasing Ca:P ratio from 0.9 to 1.8:1 and discussed that negative effects of Ca on efficacy of phytase may be dependent on dietary concentration of Ca and P per se rather than their ratio. Considering the recommended Ca:P at 1 to 1.25:1 in NRC (2012), negative effects of Ca or Ca:P on phytase activity do not seem to be problematic if nutrient composition is properly controlled when formulating diets. Moreover, based on the results of aforementioned studies, the addition of phytase greater than 500 FTU/kg of diet may not interact with dietary Ca:P. However, care must be taken when standardized total tract digestible Ca and P are applied. It should be noted that phytase affects the STTD of both Ca and P in diets which needs to be considered in diet formulation (Stein et al. 2016).

### Studies on Nutrient Utilization

Previous studies have shown that the addition of phytase improved the STTD of P in various feed ingredients. Rojas et al. (2013) reported that the effect of phytase to improve STTD of P was greater in corn and hominy feed, all of which contained approximately 80% of P in phytate complexes, than in corn DDGS, which only contained 15% of P in phytate complexes. Specifically, the STTD of P in corn and hominy feed was improved by 51 and 61%, respectively, but that in corn DDGS was improved by 8% when phytase was added at 600 FTU/kg. On the other hand, an interaction between the effect of feed ingredient and phytase was not observed in Rodríguez et al. (2013) which showed the improved STTD of P by exogenous phytase at 500 FTU/kg in seven different feed ingredients including oilseeds or their meals after the oil extraction.

Previous studies conducted to determine the effects of phytase on digestibility of P and other nutrients in complete diets outnumber the studies with feed ingredients. In general, most studies conducted with graded concentration of phytase in diets containing low concentration of P reported linear or quadratic increases in ATTD of P regardless of phytase products used in studies. Some of previous studies reported that graded concentration of phytase in diets without inorganic P source quadratically increased the ATTD of P close to the values for positive control diets containing adequate concentration of P by inorganic P sources (Jendza et al. 2005; Zeng et al. 2011; Almeida et al. 2013; She et al. 2017; Arredondo et al. 2019). The concentration of phytase to maximize the ATTD of P in corn-SBM-based diets without inorganic P source was estimated at 1016 FTU/kg (Almeida et al. 2013) and 1107 FTU/kg (Arredondo et al. 2019) for weanling pigs and 801 FTU/kg (Almeida et al. 2013) for growing pigs, all of which were estimated by broken-line analysis. Increased ATTD of P by dietary phytase was also observed in both gestating and lactating sows (Jongbloed et al. 2004; Hanczakowska et al. 2009; Jang et al. 2014).

Although the effect of phytase on increasing the ATTD of P is evident in previous studies, the extent to increase the ATTD of P in diets varied among studies partly due to the different phytase characteristics used in experiments. The effects of four different phytases on ATTD of P in corn-SBM-based diets are presented in Figure 17.2. In the study reported by Kerr et al. (2010), the efficacy of four phytases was evaluated in separated four experiments, but ingredient composition of experimental diets was identical, and the mean initial body weight (BW) of pigs was also consistent among four experiments ranged from 82.4 to 89.3 kg. The ATTD of P quadratically increased when phytase originating from *A. niger* or *E. coli* was added at a range of 0 to 1000 FTU/kg, whereas the ATTD of P linearly increased when phytase originating from *E. coli* (different product from former phytase) or *P. lycii* was added.

The addition of phytase in diets may increase digestibility of other nutrients because of the antinutritional effects of phytic acid, which is referred to as extraphosphoric effect (Adeola and Cowieson 2011). Linear or quadratic increases in the ATTD of Ca were generally observed together with the ATTD of P. However, the effects of phytase on ATTD of Ca were not observed in other

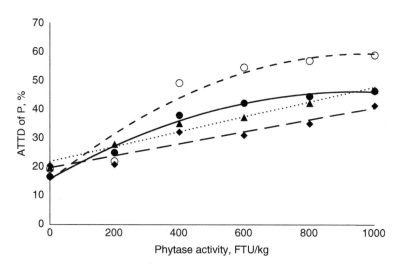

**Figure 17.2** Responses of apparent total tract digestibility (ATTD) of P to activity of phytase (phytase unit; FTU) in pigs fed corn-soybean meal-based diets; *Aspergillus niger* phytase (●, solid line), $y = 15.8 + 0.0639x + 0.0000335x^2$; *Escherichia coli* phytase 1 (○, short dash line), $y = 15.3 + 0.0900x + 0.0000461x^2$; *Escherichia coli* phytase 2 (▲, round dot line), $y = 21.9 + 0.0258x$; *Peniophora lycii* phytase (◆, long dash line), $y = 20.0 + 0.0206x$. Source: Adapted from Kerr et al. (2010).

studies (Zeng et al. 2011; She et al. 2017). These differences among results may be due to the differences in ingredient composition of diets or source of phytase used in experiments. Furthermore, results for the effects of phytase on ATTD of GE or AA varied among previous studies. Several studies reported that the ATTD of GE was increased by the addition of phytase, but the extent to increase digestibility values was less than responses in ATTD of P (Adedokun et al. 2015; Velayudhan et al. 2015; Arredondo et al. 2019). Liao et al. (2005) found that the apparent ileal digestibility (AID) of Arg, His, Phe, Thr, and Val was increased by the addition of phytase in wheat-SBM-canola meal-based diets but not in corn-SBM-, wheat-SBM-, or barley-pea-canola meal-based diets. Cowieson et al. (2017a) conducted meta-analysis based on data from 28 peer-reviewed papers to determine the effects of phytase on digestibility of AA in diets (Table 17.2). In the meta-analysis, majority of studies were conducted with corn-based diets with the addition of phytase ranging from 250 to 20 000 FTU/kg. The results showed that the AID of indispensable AA, except for Arg, Met, and Trp, was improved by the addition of phytase in diets.

Improved digestibility of minerals by the addition of phytase was also observed in previous studies; however, the results were variable. In general, the ATTD of K and Mg linearly or quadratically increased as dietary phytase concentration increases, but improvement in ATTD of other minerals was not consistent among studies (Adedokun et al. 2015; Velayudhan et al. 2015; Arredondo et al. 2019).

**Table 17.2**  Effect of phytase on apparent ileal digestibility (%) of amino acids in diets fed to pigs[a,b].

| Item | Control Mean | SE[c] | Phytase Mean | SE | P-value |
|---|---|---|---|---|---|
| Indispensable amino acid | | | | | |
| Arg | 84 | 1.3 | 86 | 1.2 | NS[d] |
| His | 77 | 1.3 | 78 | 1.2 | <0.01 |
| Ile | 78 | 1.3 | 80 | 1.2 | <0.01 |
| Leu | 79 | 1.3 | 81 | 1.2 | <0.001 |
| Lys | 78 | 1.3 | 80 | 1.2 | <0.01 |
| Met | 81 | 1.3 | 81 | 1.2 | NS |
| Phe | 77 | 1.3 | 81 | 1.2 | <0.001 |
| Thr | 69 | 1.2 | 72 | 1.2 | <0.001 |
| Trp | 74 | 1.5 | 76 | 1.4 | 0.09 |
| Val | 75 | 1.3 | 77 | 1.2 | <0.01 |
| Dispensable amino acid | | | | | |
| Ala | 73 | 1.3 | 75 | 1.2 | <0.05 |
| Asp | 75 | 1.3 | 77 | 1.2 | <0.001 |
| Cys | 70 | 1.4 | 71 | 1.3 | NS |
| Glu | 82 | 1.3 | 84 | 1.2 | <0.05 |
| Gly | 66 | 1.3 | 68 | 1.2 | <0.05 |
| Pro | 75 | 1.3 | 77 | 1.2 | <0.01 |
| Ser | 71 | 1.3 | 73 | 1.2 | <0.05 |
| Tyr | 77 | 1.3 | 79 | 1.2 | <0.001 |

[a] Adapted from Cowieson et al. (2017a).

[b] Data represents 925 observations originating from 28 peer-reviewed research articles published from 1994 to 2016. The acitivity of phytase in diets ranged from 250 to 20 000 phytase unit/kg. Body weight of pigs ranged from 6 to 71 kg with mean of 29.4 kg.

[c] SE = standard error.

[d] NS = no significance.

### Studies on Growth Performance

Improved growth performance by phytase is generally observed with increased digestibility of P. Therefore, the beneficial effects of phytase on growth performance of pigs can be mostly explained by increased digestibility of P in diets. In general, the average daily gain (ADG) and gain-to-feed ratio (G:F) of pigs linearly or quadratically increased as the concentration of phytase in diets increases, which is consistent with the responses in ATTD of P. However, the effects of phytase on average daily feed intake (ADFI) varied among previous studies, which were generally observed in experiments conducted for 42 days or more (O'Quinn et al. 1997; Adeola et al. 2004; Jendza et al. 2005; Torrallardona and Ader 2016). These observations may indicate that an improvement in ADG of pigs is not because of increased ADFI of pigs, but because of increased digestibility of P, and an improvement in ADFI is due to the accumulative effect of increased BW during experimental periods.

Improvements in bone parameters such as bone breaking strength, bone ash content, or the concentration of P in bone ash were observed together with increased growth performance in weaning pigs (Radcliffe and Kornegay 1998; Zeng et al. 2011) and growing-finishing pigs (O'Quinn et al. 1997; Jendza et al. 2005; Veum and Liu 2018). This can be also explained by increased digestibility of P in diets. Previous studies observed the linear or quadratic increases in the concentration of Ca and P in plasma when pigs were fed graded concentration of phytase (Adeola et al. 2004; Beaulieu et al. 2007). Therefore, increased absorption of Ca and P by phytase results in increased bone development. On the other hand, quantitative carcass traits do not seem to be affected by the addition of phytase in diets for finishing pigs (Dersjant-Li et al. 2018; Duffy et al. 2018). In sows, the addition of phytase to diets did not affect the performance of sows and litter (Nyachoti et al. 2006; Hill et al. 2008; Nasir et al. 2014).

### Supradosing of Phytase

Adeola and Cowieson (2011) defined "supradosing" of phytase as the addition of phytase greater than 2500 FTU/kg of diet, considering that the conventional concentration of phytase in diets ranges from 500 to 1000 FTU/kg. Previous studies have been reported that the ADG and G:F of pigs quadratically increased above the responses in pigs fed positive control diets when phytase was added up to 10 000 FTU/kg or greater (Braña et al. 2006; Kies et al. 2006b; Veum et al. 2006; Zeng et al. 2014). However, exact mechanisms of the supradosing effect are not completely understood. Adeola and Cowieson (2011) suggested three possible mechanisms: (i) increased liberation of Ca and P that establishes more favorable Ca:P for digestion and absorption; (ii) diminished antinutritional effects of phytic acid; and (iii) increased absorption of *myo*-inositol that potentially improves growth of pigs.

Phytase in diets mainly acts on $IP_6$ and $IP_5$, which readily forms an insoluble complex with Ca. Therefore, the conventional concentration of phytase (i.e., 500 to 1000 FTU/kg) may increase soluble Ca in the lumen more than the P released from phytic acid, leading to the increased Ca:P and decreased absorption of both Ca and P (Adeola and Cowieson 2011). However, supradosing of phytase may further release P from phytic acid and consequently establish adequate Ca:P in the lumen (Adeola and Cowieson 2011). Phytic acid has potential negative effects not only on Ca and P but also on other nutrients including proteins, carbohydrates, and other minerals. Therefore, supradosing of phytase may increase the digestibility of those nutrients, leading to an improvement in growth performance of pigs. In addition, increased uptake of *myo*-inositol may be responsible for increased growth performance of pigs. Cowieson et al. (2017b) reported that the plasma *myo*-inositol concentration in pigs fed the diet containing 1000 or 3000 FTU/kg phytase was not different from that in pigs fed the diet containing 0.2% *myo*-inositol at 360 minutes after ingestion. Beneficial effects of *myo*-inositol on growth performance were observed in broiler chickens (Cowieson et al. 2013; Żyła et al. 2013). Previous studies with mice found that *myo*-inositol induced the translocation of glucose transporter 4 (GLUT4) to the membrane surface of skeletal muscle cells (Dang et al. 2010; Yamashita et al. 2013), which may imply that *myo*-inositol exert insulin-like effects. Lu et al. (2019) also found that the addition of phytase at 2000 FTU/kg of diet increased the concentration of *myo*-inositol in plasma and GLUT4 in muscle plasma membrane. Therefore, increased absorption of *myo*-inositol in pigs may increase the protein synthesis due to the insulin mimetic effects and the stimulation of insulin signaling pathway (Bevan 2001).

## Carbohydrases

### Mode of Actions

Various carbohydrases have been developed and studied in swine nutrition. Carbohydrases refer to the enzymes that can hydrolyze the covalent bonds between two adjacent monosaccharides in their polymers. Most exogenously added feed carbohydrases are the enzymes that cannot be produced by pigs, but in some cases, specific carbohydrase is applied to enhance the enzymatic reaction together with the endogenous enzyme (e.g., exogenous amylase). Exogenous carbohydrases for swine diets are generally produced by fermentation of microbes from either classical or genetically modified strains (Paloheimo et al. 2010). Commercially available carbohydrases can be divided into three categories: (i) single-component carbohydrase; (ii) enzyme complex produced by blending two or more single-component carbohydrases; (iii) enzyme cocktail that contains multiple enzyme activities produced by a single fermentation (Masey O'Neill et al. 2014). Carbohydrases commonly used in swine diets include xylanase, β-glucanase, cellulase, β-mannanase, α-galactosidase, and pectinase.

Xylanase (endo-1,4-β-xylanase; EC 3.2.1.8) catalyzes the hydrolysis of (1→4)-β-xylosidic bond (Paloheimo et al. 2010). The substrate for xylanase is arabinoxylan that consist of (1→4)-β-xylan backbone with various branches mostly with arabinose at O atoms of carbon 2 or 3 in xylose monomers (Choct 1997). Other monomers found in fewer quantities in the branches include glucuronic acid, 4-*O*-methylglucuronic acid, acetic acid, ferulic acid, and *p*-coumaric acid (Collins et al. 2005). Branches of monomers not only exist directly on xylan backbone by either glycosidic linkage or esterification but also exist on O atom of carbon 5 in arabinose branch (Agger et al. 2010). In addition, dehydrodimer feruloyl residue on arabinose forms cross-linkage of arabinoxylan molecules, leading to a complex structure of arabinoxylan (Agger et al. 2010). Degree of branching and monosaccharide constituents in branches vary among feed ingredients and directly affect the physicochemical characteristics of arabinoxylan (Collins et al. 2005; Paloheimo et al. 2010). Because xylanase does not have the ability to hydrolyze branches (Collins et al. 2005), the efficacy of xylanase depends on components and structure of arabinoxylan, as well as access of xylanase to the xylan backbone. Therefore, the aid of other exogenous enzymes which hydrolyze branches of arabinoxylan may enhance the efficacy of xylanase. Previous in vitro studies showed that the addition of branch-degrading enzymes, such as arabinofuranosidase or feruloyl esterase, increased the hydrolyzed xylose when arabinoxylan was incubated with xylanase (Lei et al. 2016; Ravn et al. 2018). Hydrolysis of arabinoxylan reduces the molecular weight of polysaccharides and produces oligosaccharides, which may reduce the negative effects of dietary fiber and increase the proliferation of beneficial microflora in the large intestine.

A linear polysaccharide chain connected by (1→3) and (1→4) linkage between glucose monomers, β-glucan is the major component of soluble NSP in barley and oat (Bach Knudsen 1997; Choct 1997). β-Glucanase [endo-1,3(4)-β-glucanase; EC 3.2.1.6] acts on (1→3)- or (1→4)-β-glycosidic bonds of β-glucan. Due to the presence of (1→3)-β-glycosidic bonds, β-glucan in cereal grains is more soluble in water than the linear chain of (1→4)-β-glycosidic bonds (i.e., cellulose), thereby increasing the viscosity of digesta (Choct 1997). Åman and Graham (1987) reported that the solubility of β-glucan in barley is 54%, whereas that in oat is 80%. Because of the structural similarity between β-glucan and cellulose, β-glucanase is often mischaracterized as cellulase (Adeola and Cowieson 2011). However, cellulase refer to as a series of enzymes including endo-1,4-β-glucanase (EC 3.2.1.4), cellobiohydrolase (EC 3.2.1.91), and β-glucosidase (EC 3.2.1.21), all of which contribute to releasing glucose monomers from cellulose (Adeola and Cowieson 2011).

Beta-mannanase (endo-1,4-β-mannanase; EC 3.2.1.78) is applied to hydrolyze the polysaccharides mainly consisting of mannose including mannan, galactomannan, glucomannan, and galactoglucomannan (Jackson 2010). β-Mannanase catalyzes the hydrolysis of (1→4)-β-mannosidic bonds in mannan-based polysaccharides. The structure of mannan consists of only mannose as a linear chain, whereas that of glucomannan consists of glucose and mannose (Singh et al. 2018). Branches can be formed with galactose in both mannan and glucomannan by (1→6)-α-galactocyl bond (Singh et al. 2018). Mannan is mostly found in by-products of oil extraction particularly from palm and coconut (Bach Knudsen 1997). The concentration of mannan in copra expellers and palm kernel expellers used in the study reported by Kwon and Kim (2015) were 24.6 and 31.3%, respectively. Hsiao et al. (2006) reported that the average concentration of mannan in 22 sources of hulled SBM was 1.60% and that in 14 sources of dehulled SBM was 1.25%.

The hydrolysis of terminal α-galactose in raffinose-oligosaccharides and other polysaccharides containing α-galactose residues is catalyzed by α-galactosidase (EC 3.2.1.22). Raffinose-oligosaccharides including raffinose, stachyose, and verbascose are abundant in by-products of the oil extraction from legume seeds (Bach Knudsen 1997). Raffinose consists of sucrose (glucose and fructose) and one galactose residue in which one molecule of galactose is bound at carbon 6 of glucose. Stachyose contains two galactose molecules and sucrose in which galactose residues are bound at carbon 6 of the first galactose attached to glucose, whereas verbascose contains three galactose molecules attached in the same way (Martínez-Villaluenga et al. 2008). Grieshop et al. (2003) reported that the major raffinose-oligosaccharide in nine samples of SBM produced in the United States was stachyose ranged from 4.15 to 5.72% dry matter (DM) basis.

Pectinase generally refers to the class of enzymes that hydrolyzes pectin polysaccharides such as polygalacturonase, pectin transeliminase, or pectin methyl esterase (Adeola and Cowieson 2011). Pectin is the polysaccharide chain consisting of galacturonic acid (Jackson 2010). In general, by-products of the oil extraction from legume seeds contain greater concentration of pectin than cereal grains (Malathi and Devegowda 2001). Pectinase has been generally applied together with other carbohydrases as enzyme cocktails.

Among various carbohydrases, xylanase and β-glucanase have been predominant in livestock industry (Adeola and Cowieson 2011) because the majority of NSP in cereal grains including corn, wheat, and barley is arabinoxylan or β-glucan (Bach Knudsen 1997). β-mannanase has been occasionally applied especially when feed ingredients containing high concentration of mannan was used (Kim et al. 2017). Oligosaccharides released from NSP may be degraded further to monosaccharides which can be absorbed and used as energy sources for pigs. However, most carbohydrases cannot completely hydrolyze their substrates into monosaccharides possibly due to the substrate specificity or structural complexity. Nevertheless, exogenous carbohydrases have been applied mainly to reduce the antinutritional effects of NSP by reducing the molecular weight, thereby releasing the nutrients encapsulated by NSP or diminishing the interference with digestion and absorption of nutrients (Adeola and Cowieson 2011). Decreased molecular weight of NSP may also reduce the viscosity of digesta and subsequently increase digestibility of nutrients or GE (Bartelt et al. 2002; Kiarie et al. 2007; Passos et al. 2015; Zhang et al. 2017; Tiwari et al. 2018), although the effect of carbohydrases on viscosity of digesta is inconsistent among previous studies (Nørgaard et al. 2019).

### Studies on Nutrient Utilization

Although direct feeding of carbohydrases and measuring the responses of pigs can be considered as the most suitable and practical assessment, in vitro experiments have been used to determine the efficacy of carbohydrases due to its cost-effectiveness and methodological simplicity. In vitro DM or nutrient digestibility has been determined in previous studies to select feed ingredients suitable for a specific carbohydrase (Bindelle et al. 2011; Kong et al. 2015; Park et al. 2016a) or to evaluate the efficacy of various carbohydrases in selected feed ingredients (Zeng et al. 2018). Because experimental errors can be readily controlled in in vitro experiments, beneficial effects of carbohydrases on in vitro DM or nutrient digestibility were observed in previous studies. However, results of in vitro experiments have been generally inconsistent with the results of in vivo experiments (Park et al. 2016b; Torres-Pitarch et al. 2018; Woyengo et al. 2018).

Reduced antinutritional effects of dietary fiber by carbohydrases can result in improved digestibility of nutrients or GE in diets. Therefore, numerous studies have been conducted to evaluate the efficacy of carbohydrase using digestibility of nutrients or GE as responses. However, results were variable among studies. In general, the beneficial effects of carbohydrases on digestibility of nutrients or GE were observed in diets containing feed ingredients with high dietary fiber (Owusu-Asiedu et al. 2012; Kim et al. 2017; Koo et al. 2017; Tsai et al. 2017). However, other studies reported that there was no effect of carbohydrase on digestibility of nutrients or GE (Kwon and Kim 2015; Jang et al. 2016; Woyengo et al. 2018). This inconsistency among reports may be due to the differences in intrinsic characteristics of carbohydrases including microbial origins, substrate specificity, optimal pH condition, and activity or due to the differences in extrinsic factors such as ingredient composition of experimental diets, BW of pigs, or environmental conditions. Ndou et al. (2015) conducted an experiment with two diets based on either wheat or corn to evaluate the effects of five xylanases from different origins. A part of the results is shown in Figure 17.3. The addition of xylanase originating from *Bacillus substilis* (X3 in Figure 17.3) in wheat-based diets increased the AID of arabinoxylan, leading to an improvement in AID of GE, whereas there was no effect of this xylanase in corn-based diets. However, two different xylanases originating from *Trichoderma reesei* (X2 and X5 in Figure 17.3) increased the AID of arabinoxylan and GE in corn-based diets, which were not observed in wheat-based diets. These observations may indicate that each carbohydrase has its own suitable feed ingredients which contains specific structure and composition of dietary fiber favorable to its enzymatic reaction. Therefore, feed ingredients that contain adequate structure and amounts of substrates need to be considered when using carbohydrases to capture the beneficial effects.

The addition of carbohydrases in diets may increase the digestibility of AA by reducing the interference of dietary fiber with the proteolytic enzymes or by degrading dietary fiber surrounding proteins in feed ingredients. Ao eț al. (2010) and Owusu-Asiedu et al. (2012) reported that the AID

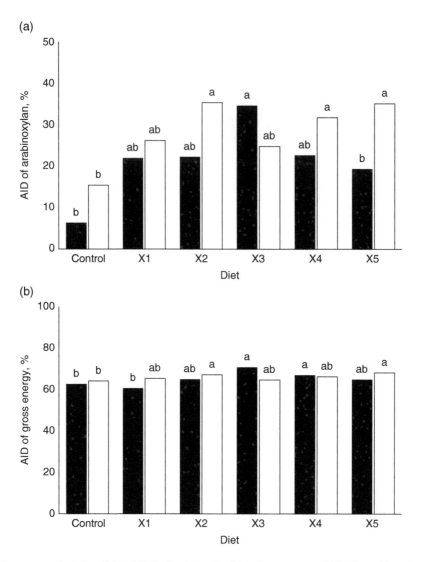

**Figure 17.3** The apparent ileal digestibility (AID) of arabinoxylan (a) and gross energy (b) in diets without (control) and with five xylanases: X1, *Fusarium verticilloide* xylanase; X2, *Aspergillus Clavatas* xylanase; X3, *Bacillus substilis* xylanase; X4, *Trichoderma reesei* xylanase; and X5, *Trichoderma reesei* xylanase (different product from X4). Xylanases were added into wheat-based (black bar) or corn-based diets (white bar). Interactions between diets and xylanase were observed in the AID of arabinoxylan (SEM = 3.5; P = 0.011) and gross energy (SEM = 1.9; P = 0.044). Within responses, means with different letter differ (P < 0.05). Source: Adapted from Ndou et al. (2015).

of several indispensable AA in diets increased when carbohydrase mixtures were added. Kiarie et al. (2010) also observed a linear increase in the standardized ileal digestibility of Met as dietary concentration of enzyme complex increases. However, Kong and Adeola (2012) did not observe the effect of β-glucanase on AID of AA in barley-SBM-based diets. Reduced molecular weight of dietary fiber may also prevent the excessive secretion of mucin, leading to a reduction of endogenous losses of N and AA. Cadogan and Choct (2015) observed that the secretion of mucin in the duodenum and ileum was reduced by xylanase when pigs were fed wheat-based diets. However, Yin et al. (2000) and Bartelt et al. (2002) found that endogenous loss of N was not affected by the addition of xylanase to diets.

Carbohydrases can be also used in diets for sows, although a few studies have been conducted to determine the effects of carbohydrases on digestibility of nutrients and GE in sows. In gestating sows, de Souza et al. (2007) reported that the addition of xylanase did not affect the AID and ATTD of DM and N in diets, but those values were improved in lactating sows. Similar findings were also observed in other studies, which showed beneficial effects of carbohydrases on digestibility of nutrients or GE (Upadhaya et al. 2016a; Kim et al. 2018; Zhou et al. 2018). Cozannet et al. (2018) conducted a meta-analysis of data collected from eight studies of lactating sows and reported that the ATTD of DM and GE was increased by the addition of enzyme complex containing xylanase and β-glucanase.

### *Studies on Growth Performance*

Similar to the results of digestibility studies, the effects of carbohydrases on growth performance are also inconsistent among studies. Improved digestibility of nutrients or GE by carbohydrases generally resulted in increased ADG or G:F in previous studies (Yoon et al. 2010; Hanczakowska et al. 2012; Kim et al. 2017; Koo et al. 2017). However, Passos et al. (2015) reported that the growth performance of pigs was not affected by the addition of xylanase despite of the improved AID of nutrients, which may indicate that the addition of carbohydrases is not effective if diets contain adequate concentration of nutrients or GE for pigs. The addition of carbohydrases in diets does not seem to affect the quantitative carcass traits of finishing pigs (Thacker et al. 2002; Yoon et al. 2010; Hanczakowska et al. 2012).

Previous studies with lactating sows that observed improvements in ATTD of nutrients or GE with added carbohydrases also reported improvements in performance of sows. In general, the addition of carbohydrases reduced the BW loss (Cozannet et al. 2018; Kim et al. 2018; Zhou et al. 2018) or back fat thickness loss (Upadhaya et al. 2016a) during lactation. Dietary carbohydrases do not seem to affect litter size (Upadhaya et al. 2016a; Kim et al. 2018; Zhou et al. 2018), but based on the results of meta-analysis, Cozannet et al. (2018) reported that the addition of enzyme complex containing xylanase and β-glucanase improved the litter size and BW of piglets at weaning.

### *Carbohydrases and Gut Health*

Majority of dietary fiber in diets is not digested in the small intestine and enters the large intestine. Moreover, as discussed earlier, increased intake of dietary fiber reduces the digestibility of nutrients, and consequently undigested nutrients also flow to the large intestine together with dietary fiber, all of which results in a provision of nutrients to microbiota. Therefore, microbial population in the large intestine is directly and indirectly driven by dietary fiber which may be either beneficial or detrimental to host animals.

Kiarie et al. (2013) suggested two possible roles of carbohydrase in microbial population: reduction of undigested nutrients and provision of short-chain oligosaccharides as prebiotics. Zhang et al. (2018a) evaluated the effects of five different xylanases in two diets (i.e., corn- or wheat-based diets) on microbial population in the ileum and cecum by high-throughput sequencing. The results of this experiment showed that each xylanase differently affected the microbial population depending on diets as well as sites of the GIT. For example, one xylanase increased the relative rRNA abundance of *Lactobacillus* spp. in the cecum of pigs fed corn-based diets, whereas the other two xylanases reduced that of *Lactobacillus* spp. in the cecum of pigs fed wheat-based diets. Zhang et al. (2018a) also reported that the diversity of microbial population increased in pigs fed diets containing xylanase compared with pigs fed diets without xylanase. *Lactobacillus* spp. is generally considered as beneficial bacteria for host animals due to the ability to decrease pH of lumen, which may suppress the proliferation of pathogenic microbes. Several studies observed that the addition of carbohydrases in diets increased the population of *Lactobacillus* spp. in the ileum (Chen et al. 2016) or feces (Kiarie et al. 2007; Clarke et al. 2019), whereas Reilly et al. (2010) and Li et al. (2019) found that the population of *Lactobacillus* spp. in the ileum was reduced by carbohydrases in diets.

Modulation of microbial population in the GIT may result in changes in intestinal barrier integrity and immune responses of host animals. Li et al. (2018, 2019) conducted an experiment to determine the effects of xylanase and enzyme complex containing xylanase, β-glucanase, and cellulase on the mRNA expression of tight junction proteins and cytokines in the ileum and colon of weanling pigs. Results of this experiment revealed that the addition of enzyme complex in diets increased the mRNA expression of claudin-3 in the ileum (Li et al. 2018), which was consistent with Tiwari et al. (2018) that reported increased mRNA expression of claudin and occludin in the jejunum by xylanase. Li et al. (2019) also reported that the addition of xylanase or enzyme complex in diets increased the mRNA expression of claudin-3 and occludin in the colon and decreased the concentration of interleukin-1β and tumor necrosis factor-α in plasma, which may imply a reduced inflammation in the GIT. Koo et al. (2017) found that the fecal score of weanling pigs were improved by feeding diets containing enzyme cocktail mainly consisting of xylanase, cellulase, and amylase, which may be due to the improved intestinal barrier integrity.

**Other Exogenous Enzymes**

Although exogenous enzyme market is dominated by carbohydrases and phytase, other exogenous enzymes have been occasionally used which include protease (EC 3.4.21.62), lipase (EC 3.1.1.3), and lysozyme (EC 3.2.1.17).

Protease and lipase have been generally applied to aid the enzymatic activity of their counterparts in the GIT. Previous studies conducted to evaluate the efficacy of protease showed increased ATTD of N in diets (Tactacan et al. 2016; Upadhaya et al. 2016b; Lei et al. 2017; Pan et al. 2017; Payling et al. 2017; Nørgaard et al. 2019). Yu et al. (2016) also reported that the standardized ileal digestibility of N and several indispensable AA were increased by the addition of protease in diets. On the other hand, the results of growth performance were not consistent among studies. Improvements in growth performance of pigs by dietary protease were observed in several studies (Tactacan et al. 2016; Upadhaya et al. 2016b; Payling et al. 2017), but not in other studies (Pan et al. 2017; Figueroa et al. 2019). Lei et al. (2017) showed that adding protease to diets reduced ADFI of pigs without affecting ADG, leading to an improvement in G:F. However, reduction of ADFI, ADG, and final BW of pigs was observed in other studies (Mc Alpine et al. 2012; O'Shea et al. 2014; Choe et al. 2017). Mc Alpine et al. (2012) discussed that the reason for reduced ADFI might be that increased digestion and absorption of AA stimulate the feedback mechanism of pigs.

A few studies have been conducted to test the effects of lipase on weanling pigs. Liu et al. (2018) showed that the ATTD of DM, N, fat, and GE in diets were improved by lipase at d 14 post weaning. Chen et al. (2014) reported that growth performance was improved by the addition of lipase to diets, but Liu et al. (2018) and Zhang et al. (2018b) did not observe any effects of lipase on growth performance.

Lysozyme (*N*-acetylmuramide glycanhydrolase) hydrolyzes the peptidoglycan of bacterial cell walls, and therefore considered as a potential alternative to antibiotics (Oliver and Wells 2015). Previous studies reported beneficial effects of dietary lysozyme on growth performance and intestinal morphology (Oliver and Wells 2013; Long et al. 2016).

## Future Studies for Exogenous Enzymes

The use of phytase in swine diets seems to be promising compared to carbohydrases. Due to the evident effects of phytase on P utilization, efforts have been made to estimate the matrix value (i.e., nutritional values given to phytase based on anticipated improvements in digestible P and other nutrients) to appropriately use phytase in swine diets (Adeola and Cowieson 2011). However, considering the variations in the structure of phytin among feed ingredients, fixed matrix value for complete diets may be compromised if various feed ingredients are used in diet formulations. Therefore, it may be more appropriate to estimate the matrix value for each feed ingredient for precision nutrition. Numerous studies have been conducted to determine the efficacy of phytase on digestibility of P and other nutrients in feed ingredients which provide a wide range of data to estimate the matrix value. In addition, the additivity of matrix values in feed ingredients needs to be further studied.

As discussed earlier, mode of actions for phytase supradosing is still unclear. Improvements in Ca utilization may be a trigger for beneficial effects of supradosing. Therefore, future research needs to focus on the effects of phytase on digestibility of Ca. Other possible mechanisms for supradosing of phytase also need to be investigated further with various responses including intestinal morphology, gut barrier integrity, humoral changes, or microbial population, all of which may contribute to better understanding of phytase supradosing.

It has been reported that the addition of exogenous carbohydrases increased the digestibility of substrate polysaccharides (Ndou et al. 2015; Kim et al. 2017). Nevertheless, the results of previous studies on digestibility of nutrients as well as growth performance were not consistent. This inconsistency may be due to the insufficient breakdown of substrates or insufficient substrates in diets. Considering the complexity of and variation in dietary fiber in feed ingredients, the structure of dietary fiber including the chemical bonds within and between polysaccharides as well as their 3-dimentional structure needs to be studied further for precise application of carbohydrases. Moreover, low molecular weight oligosaccharides released by carbohydrase may still exert the negative effects on digestion and absorption of nutrients. Therefore, further research is needed for better understanding of behaviors of oligosaccharides in the GIT.

Previous studies have shown the potential of carbohydrases as a gut health promotor (Zhang et al. 2018a; Li et al. 2018, 2019). However, exact mode of actions and responsible microorganisms have not been identified. Even though majority of low molecular weight oligosaccharides would be fermented by microbes, there may be direct effects on pigs such as stimulating the intestinal surface or exerting physiochemical changes in the lumen. In addition, relationship among carbohydrase, microorganisms, and host animals needs to be studied further for better application of carbohydrases (Kiarie et al. 2013).

## Summary

Exogenous enzymes have been routinely used in swine industry to attenuate the negative effects of antinutritional factors and consequently improve the nutritional values of diets and growth performance of pigs. Considerable amounts of phytic acid and dietary fiber reside in feed ingredients and exert antinutritional effects in swine diets. Phytic acid in diets can be hydrolyzed by exogenous phytase. The addition of phytase in diets increases the digestibility of P and possibly increases the digestibility of other nutrients. The addition of large doses of phytase further improves the digestibility of P and growth performance of pigs. The role of carbohydrases is to hydrolyze the dietary fiber that hinder the digestion and absorption of nutrients. Furthermore, oligosaccharides released by carbohydrase may improve the gut health of pigs by shifting microbial population beneficial to host animals. Further research is needed to precisely use exogenous enzymes for reducing the waste of nutrients and maximizing the benefits, all of which is prerequisite for sustainable swine nutrition.

# References

Adedokun, S. A., A. Owusu-Asiedu, D. Ragland, P. Plumstead, and O. Adeola. 2015. The efficacy of a new 6-phytase obtained from *Buttiauxella* spp. expressed in *Trichoderma reesei* on digestibility of amino acids, energy, and nutrients in pigs fed a diet based on corn, soybean meal, wheat middlings, and corn distillers' dried grains with solubles. J. Anim. Sci. 93:168–175.

Adeola, O., and A. J. Cowieson. 2011. Board-invited review: Opportunities and challenges in using exogenous enzymes to improve nonruminant animal production. J. Anim. Sci. 89:3189–3218.

Adeola, O., J. S. Sands, P. H. Simmins, and H. Schulze. 2004. The efficacy of an *Escherichia coli*-derived phytase preparation. J. Anim. Sci. 82:2657–2666.

Adeola, O., O. A. Olukosi, J. A. Jendza, R. N. Dilger, and M. R. Bedford. 2006. Response of growing pigs to *Peniophora lycii*- and *Escherichia coli*-derived phytases or varying ratios of calcium to total phosphorus. Anim. Sci. 82:637–644.

Agger, J., A. Viksø-Nielsen, and A. S. Meyer. 2010. Enzymatic xylose release from pretreated corn bran arabinoxylan: Differential effects of deacetylation and deferuloylation on insoluble and soluble substrate fractions. J. Agric. Food Chem. 58:6141–6148.

Almeida, F. N., and H. H. Stein. 2012. Effects of graded levels of microbial phytase on the standardized total tract digestibility of phosphorus in corn and corn coproducts fed to pigs. J. Anim. Sci. 90:1262–1269.

Almeida, F. N., R. C. Sulabo, and H. H. Stein. 2013. Effects of a novel bacterial phytase expressed in *Aspergillus Oryzae* on digestibility of calcium and phosphorus in diets fed to weanling or growing pigs. J. Anim. Sci. Biotechnol. 4:8.

Almeida, F. N., M. Vazquez-Añón, and J. Escobar. 2017. Dose-dependent effects of a microbial phytase on phosphorus digestibility of common feedstuffs in pigs. Asian-Australas. J. Anim. Sci. 30:985–993.

Åman, P., and H. Graham. 1987. Analysis of total and insoluble mixed-linked $(1 \rightarrow 3),(1 \rightarrow 4)$-β-D-glucans in barley and oats. J. Agric. Food. Chem. 35:704–709.

Ao, X., Q. W. Meng, L. Yan, H. J. Kim, S. M. Hong, J. H. Cho, and I. H. Kim. 2010. Effects of non-starch polysaccharide-degrading enzymes on nutrient digestibility, growth performance and blood profiles of growing pigs fed a diet based on corn and soybean meal. Asian-Australas. J. Anim. Sci. 23:1632–1638.

AOAC. 2000. Official methods of analysis. 17th ed. Assoc. Off. Anal. Chem., Arlington, VA.

Arredondo, M. A., G. A. Casas, and H. H. Stein. 2019. Increasing levels of microbial phytase increases the digestibility of energy and minerals in diets fed to pigs. Anim. Feed Sci. Technol. 248:27–36.

Bach Knudsen, K. E. 1997. Carbohydrate and lignin contents of plant materials used in animal feeding. Anim. Feed Sci. Technol. 67:319–338.

Bach Knudsen, K. E. 2011. Triennial growth symposium: Effects of polymeric carbohydrates on growth and development in pigs. J. Anim. Sci. 89:1965–1980.

Bartelt, J., A. Jadamus, F. Wiese, E. Swiech, L. Buraczewska, and O. Simon. 2002. Apparent precaecal digestibility of nutrients and level of endogenous nitrogen in digesta of the small intestine of growing pigs as affected by various digesta viscosities. Arch. Anim. Nutr. 56:93–107.

Beaulieu, A. D., M. R. Bedford, and J. F. Patience. 2007. Supplementing corn or corn-barley diets with an *E. coli* derived phytase decreases total and soluble P output by weanling and growing pigs. Can. J. Anim. Sci. 87:353–364.

Bevan, P. 2001. Insulin signalling. J. Cell Sci. 114:1429–1430.

Bindelle, J., R. Pieper, C. A. Montoya, A. G. Van Kessel, and P. Leterme. 2011. Nonstarch polysaccharide-degrading enzymes alter the microbial community and the fermentation patterns of barley cultivars and wheat products in an in vitro model of the porcine gastrointestinal tract. FEMS Microbiol. Ecol. 76:553–563.

Brady, S. M., J. J. Callan, D. Cowan, M. McGrane, and J. V. O'Doherty. 2002. Effect of phytase inclusion and calcium/ phosphorus ratio on the performance and nutrient retention of grower-finisher pigs fed barley/wheat/soya bean meal-based diets. J. Sci. Food Agric. 82:1780–1790.

Braña, D. V., M. Ellis, E. O. Castañeda, J. S. Sands, and D. H. Baker. 2006. Effect of a novel phytase on growth performance, bone ash, and mineral digestibility in nursery and grower-finisher pigs. J. Anim. Sci. 84:1839–1849.

Cadogan, D. J., and M. Choct. 2015. Pattern of non-starch polysaccharide digestion along the gut of the pig: Contribution to available energy. Anim. Nutr. 1:160–165.

Chen, L., L. X. Gao, Q. H. Huang, R. Q. Zhong, L. L. Zhang, X. F. Tang, and H. F. Zhang. 2017. Viscous and fermentable nonstarch polysaccharides affect intestinal nutrient and energy flow and hindgut fermentation in growing pigs. J. Anim. Sci. 95:5054–5063.

Chen, S. Y., Z. X. Liu, Y. D. He, C. Chu, and M. Q. Wang. 2014. Effects of coated lipase supplementation on growth, digestion and intestinal morphology in weaning piglets. J. Anim. Vet. Adv. 13:1093–1097.

Chen, Q., M. Li, and X. Wang. 2016. Enzymology properties of two different xylanases and their impacts on growth performance and intestinal microflora of weaned piglets. Anim. Nutr. 2:18–23.

Choct, M. 1997. Feed non-starch polysaccharides: Chemical structures and nutritional significance. Feed Milling Int. 13–26.

Choe, J., K. S. Kim, H. B. Kim, S. Park, J. Kim, S. Kim, B. Kim, S. H. Cho, J. Y. Cho, I. H. Park, J. H. Cho, and M. Song. 2017. Effects of protease on growth performance and carcass characteristics of growing finishing pigs. S. Afr. J. Anim. Sci. 47:697–703.

Clarke, L. C., T. Sweeney, S. K. Duffy, G. Rajauria, and J. V. O'Doherty. 2019. The variation in hectolitre weight of wheat grain fed with or without enzyme supplementation influences nutrient digestibility and subsequently affects performance in pigs. J. Anim. Physiol. Anim. Nutr. 103:583–592.

Collins, T., C. Gerday, and G. Feller. 2005. Xylanases, xylanase families and extremophilic xylanases. FEMS Microbiol. Rev. 29:3–23.

Cowieson, A. J., T. Acamovic, and M. R. Bedford. 2004. The effects of phytase and phytic acid on the loss of endogenous amino acids and minerals from broiler chickens. Br. Poult. Sci. 45:101–108.

Cowieson, A. J., T. Acamovic, and M. R. Bedford. 2006. Phytic acid and phytase: Implications for protein utilization by poultry. Poult. Sci. 85:878–885.

Cowieson, A. J., M. R. Bedford, P. H. Selle, and V. Ravindran. 2009. Phytate and microbial phytase: Implications for endogenous nitrogen losses and nutrient availability. World's Poult. Sci. J. 65:401–418.

Cowieson, A. J., A. Ptak, P. Maćkowiak, M. Sassek, E. Pruszyńska-Oszmałek, K. Żyła, S. Świątkiewicz, S. Kaczmarek, and D. Józefiak. 2013. The effect of microbial phytase and myo-inositol on performance and blood biochemistry of broiler chickens fed wheat/corn-based diets. Poult. Sci. 92:2124–2134.

Cowieson, A. J., and V. Ravindran. 2007. Effect of phytic acid and microbial phytase on the flow and amino acid composition of endogenous protein at the terminal ileum of growing broiler chickens. Br. J. Nutr. 98:745–752.

Cowieson, A. J., F. F. Roos, J. -P. Ruckebusch, J. W. Wilson, P. Guggenbuhl, H. Lu, K. M. Ajuwon, and O. Adeola. 2017b. Time-series responses of swine plasma metabolites to ingestion of diets containing myo-inositol or phytase. Br. J. Nutr. 118:897–905.

Cowieson, A. J., J. -P. Ruckebusch, J. O. B. Sorbara, J. W. Wilson, P. Guggenbuhl, L. Tanadini, and F. F. Roos. 2017a. A systematic view on the effect of microbial phytase on ileal amino acid digestibility in pigs. Anim. Feed Sci. Technol. 231:138–149.

Cozannet, P., P. G. Lawlor, P. Leterme, E. Devillard, P. -A. Geraert, F. Rouffineau, and A. Preynat. 2018. Reducing BW loss during lactation in sows: a meta-analysis on the use of a nonstarch polysaccharide-hydrolyzing enzyme supplement. J. Anim. Sci. 96:2777–2788.

Dang, N. T., R. Mukai, K. -I. Yoshida, and H. Ashida. 2010. D-pinitol and myo-inositol stimulate translocation of glucose transporter 4 in skeletal muscle of C57BL/6 mice. Biosci. Biotechnol. Biochem. 74:1062–1067.

De Jong, J. A., J. C. Woodworth, J. M. DeRouchey, R. D. Goodband, M. D. Tokach, S. S. Dritz, C. R. Stark, and C. K. Jones. 2017. Stability of four commercial phytase products under increasing thermal conditioning temperatures. Transl. Anim. Sci. 1:255–260.

de Souza, A. L. P., M. D. Lindemann, and G. L. Cromwell. 2007. Supplementation of dietary enzymes has varying effects on apparent protein and amino acid digestibility in reproducing sows. Livest. Sci. 109:122–124.

Dersjant-Li, Y., A. Awati, H. Schulze, and G. Partridge. 2015. Phytase in non-ruminant animal nutrition: A critical review on phytase activities in the gastrointestinal tract and influencing factors. J. Sci. Food Agric. 95:878–896.

Dersjant-Li, Y., P. Plumstead, A. Awati, and J. Remus. 2018. Productive performance of commercial growing and finishing pigs supplemented with a Buttiauxella phytase as a total replacement of inorganic phosphate. Anim. Nutr. 4:351–357.

Duffy, S. K., A. K. Kelly, G. Rajauria, L. C. Clarke, V. Gath, F. J. Monahan, and J. V. O'Dohearty. 2018. The effect of 25-hydroxyvitamin $D_3$ and phytase inclusion on pig performance, bone parameters and pork quality in finisher pigs. J. Anim. Physiol. Anim. Nutr. 102:1296–1305.

Dilger, R. N., J. S. Sands, D. Ragland, and O. Adeola. 2004. Digestibility of nitrogen and amino acids in soybean meal with added soyhulls. J. Anim. Sci. 82:715–724.

Figueroa, J. L., J. A. Martinez, M. T. Sanchez-Torres, J. L. Cordero, M. Martinez, V. M. Valdez, and A. Ruiz. 2019. Evaluation of reduced amino acids diets added with protected protease on productive performance in 25-100 kg barrows. Austral J. Vet. Sci. 51:53–60.

Greiner, R., and U. Konietzny. 2010. Phytases: Biochemistry, enzymology and characteristics relevant to animal feed use. In: M. R. Bedford, and G. G. Partridge, editors, Enzymes in farm animal nutrition. 2nd rev. ed. CAB Int, Wallingford, UK. p. 96–128.

Grieshop, C. M., C. T. Kadzere, G. M. Clapper, E. A. Flickinger, L. L. Bauer, R. L. Frazier, and G. C. Fahey, Jr. 2003. Chemical and nutritional characteristics of united states soybeans and soybean meals. J. Agric. Food Chem. 51:7684–7691.

Hanczakowska, E., M. Świątkiewicz, and I. Kühn. 2009. Effect of microbial phytase supplement to feed for sows on apparent digestibility of P, Ca and crude protein and reproductive parameters in two consecutive reproduction cycles. Medycyna Wet. 65:250–254.

Hanczakowska, E., M. Świątkiewicz, and I. Kühn. 2012. Efficiency and dose response of xylanase in diets for fattening pigs. Ann. Anim. Sci. 12:539–548.

Hill, G. M., J. E. Link, M. J. Rincker, D. L. Kirkpatrick, M. L. Gibson, and K. Karges. 2008. Utilization of distillers dried grains with solubles and phytase in sow lactation diets to meet the phosphorus requirement of the sow and reduce fecal phosphorus concentration. J. Anim. Sci. 86:112–118.

Hooda, S., B. U. Metzler-Zebeli, T. Vasanthan, and R. T. Zijlstra. 2011. Effects of viscosity and fermentability of dietary fibre on nutrient digestibility and digesta characteristics in ileal-cannulated grower pigs. Br. J. Nutr. 106:664–674.

Hsiao, H.-Y., D. M. Anderson, and N. M. Dale. 2006. Levels of β-mannan in soybean meal. Poult. Sci. 85:1430–1432.

Hu, H. L., A. Wise, and C. Henderson. 1996. Hydrolysis of phytate and inositol tri-, tetra-, and penta- phosphates by the intestinal mucosa of the pig. Nutr. Res. 16:781–787.

Jackson, M. E. 2010. Mannanase, alpha-galactosidase and pectinase. In: M. R. Bedford, and G. G. Partridge, editors, Enzymes in farm animal nutrition. 2nd rev. ed. CAB Int, Wallingford, UK. p. 54–84.

Jang, Y. D., M. D. Lindemann, E. van Heugten, R. D. Jones, B. G. Kim, C. V. Maxwell, and J. S. Radcliffe. 2014. Effects of phytase supplementation on reproductive performance, apparent total tract digestibility of Ca and P and bone characteristics in gestating and lactating sows. Rev. Colomb. Cienc. Pecu. 27:178–193.

Jang, Y. D., P. Wilcock, R. D. Boyd, and M. D. Lindemann. 2016. Effect of xylanase supplementation of a phytase-supplemented diet on apparent total tract digestibility in pigs. J. Anim. Sci. 94:260–263.

Jendza, J. A., R. N. Dilger, S. A. Adedokun, J. S. Sands, and O. Adeola. 2005. Escherichia coli phytase improves growth performance of starter, grower, and finisher pigs fed phosphorus-deficient diets. J. Anim. Sci. 83:1882–1889.

Johansen, H. N., K. E. Bach Knudsen, B. Sandström, and F. Skjøth. 1996. Effects of varying content of soluble dietary fibre from wheat flour and oat milling fractions on gastric emptying in pigs. Br. J. Nutr. 75:339–351.

Johnson, I. T., and J. M. Gee. 1981. Effects of gel-forming gums on the intestinal unstirred layer and sugar transport in vitro. Gut. 22:398–403.

Jongbloed, A. W., J. T. M. van Diepen, P. A. Kemme, and J. Broz. 2004. Efficacy of microbial phytase on mineral digestibility in diets for gestating and lactating sows. Livest. Prod. Sci. 91:143–155.

Kerr, B. J., T. E. Weber, P. S. Miller, and L. L. Southern. 2010. Effect of phytase on apparent total tract digestibility of phosphorus in corn-soybean meal diets fed to finishing pigs. J. Anim. Sci. 88:238–247.

Kiarie, E., C. M. Nyachoti, B. A. Slominski, and G. Blank. 2007. Growth performance, gastrointestinal microbial activity, and nutrient digestibility in early-weaned pigs fed diets containing flaxseed and carbohydrase enzyme. J. Anim. Sci. 85:2982–2993.

Kiarie, E., A. Owusu-Asiedu, P. H. Simmins, and C. M. Nyachoti. 2010. Influence of phytase and carbohydrase enzymes on apparent ileal nutrient and standardized ileal amino acid digestibility in growing pigs fed wheat and barley-based diets. Livest. Sci. 134:85–87.

Kiarie, E., L. F. Romero, and C. M. Nyachoti. 2013. The role of added feed enzymes in promoting gut health in swine and poultry. Nutr. Res. Rev. 26:71–88.

Kies, A. K., L. H. De Jonge, P. A. Kemme, and A. W. Jongbloed. 2006a. Interaction between protein, phytate, and microbial phytase. in vitro studies. J. Agric. Food Chem. 54:1753–1758.

Kies, A. K., P. A. Kemme, L. B. J. Šebek, J. Th. M. van Diepen, and A. W. Jongbloed. 2006b. Effect of graded doses and a high dose of microbial phytase on the digestibility of various minerals in weaner pigs. J. Anim. Sci. 84:1169–1175.

Kim, J. S., A. Hosseindoust, I. K. Ju, X. Yang, S. H. Lee, H. S. Noh, J. H. Lee, and B. J. Chae. 2018. Effects of dietary energy levels and β-mannanase supplementation in a high mannan-based diet during lactation on reproductive performance, apparent total tract digestibility and milk composition in multiparous sows. Ital. J. Anim. Sci. 17:128–134.

Kim, J. S., S. L. Ingale, A. R. Hosseindoust, S. H. Lee, J. H. Lee, and B. J. Chae. 2017. Effects of mannan level and β-mannanase supplementation on growth performance, apparent total tract digestibility and blood metabolites of growing pigs. Animal 11:202–208.

Knuckles, B. E. 1988. Effect of phytate and other myo-inositol phosphate esters on lipase activity. J. Food Sci. 53:250–252.

Knuckles, B. E., and A. A. Betschart. 1987. Effect of phytate and other myo-inositol phosphate esters on α-amylase digestion of starch. J. Food Sci. 52:719–721.

Kong, C., and O. Adeola. 2012. Supplementation of barley-based diets with β-glucanase for pigs: Energy and amino acid digestibility response. J. Anim. Sci. 90:74–76.

Kong, C., C. S. Park, and B. G. Kim. 2015. Effects of an enzyme complex on in vitro dry matter digestibility of feed ingredients for pigs. SpringerPlus 4:261.

Konieczka, P., and S. Smulikowska. 2018. Viscosity negatively affects the nutritional value of blue lupin seeds for broilers. Animal 12:1144–1153.

Koo, B., J. W. Kim, C. F. M. de Lange, M. M. Hossain, and C. M. Nyachoti. 2017. Effects of diet complexity and multicarbohydrase supplementation on growth performance, nutrient digestibility, blood profile, intestinal morphology, and fecal score in newly weaned pigs. J. Anim. Sci. 95:4060–4071.

Kwon, W. B., and B. G. Kim. 2015. Effects of supplemental beta-mannanase on digestible energy and metabolizable energy contents of copra expellers and palm kernel expellers fed to pigs. Asian-Australas. J. Anim. Sci. 28:1014–1019.

Lærke, H. N., S. Arent, S. Dalsgaard, and K. E. Bach Knudsen. 2015. Effect of xylanases on ileal viscosity, intestinal fiber modification, and apparent ileal fiber and nutrient digestibility of rye and wheat in growing pig. J. Anim. Sci. 93:4323–4335.

Le Goff, G., and J. Noblet. 2001. Comparative total tract digestibility of dietary energy and nutrients in growing pigs and adult sows. J. Anim. Sci. 79:2418–2427.

Lei, X. J., J. Y. Cheong, J. H. Park, and I. H. Kim. 2017. Supplementation of protease, alone and in combination with fructooligosaccharide to low protein diet for finishing pigs. Anim. Sci. J. 88:1987–1993.

Lei, Z., Y. Shao, X. Yin, D. Yin, Y. Guo, and J. Yuan. 2016. Combination of xylanase and debranching enzymes specific to wheat arabinoxylan improve the growth performance and gut health of broilers. J. Agric. Food Chem. 64:4932–4942.

Li, Q., N. K. Gabler, C. L. Loving, S. A. Gould, and J. F. Patience. 2018. A dietary carbohydrase blend improved intestinal barrier function and growth rate in weaned pigs fed higher fiber diets. J. Anim. Sci. 96:5233–5243.

Li, Q., S. Schmitz-Esser, C. L. Loving, N. K. Gabler, S. A. Gould, and J. F. Patience. 2019. Exogenous carbohydrases added to a starter diet reduced markers of systemic immune activation and decreased Lactobacillus in weaned pigs. J. Anim. Sci. 97:1242–1253.

Liao, S. F., W. C. Sauer, A. K. Kies, Y. C. Zhang, M. Cervantes, and J. M. He. 2005. Effect of phytase supplementation to diets for weanling pigs on the digestibilities of crude protein, amino acids, and energy. J. Anim. Sci. 83:625–633.

Liu, J. B., S. C. Cao, J. Liu, J. Pu, L. Chen, and H. F. Zhang. 2018. Effects of dietary energy and lipase levels on nutrient digestibility, digestive physiology and noxious gas emission in weaning pigs. Asian-Australas. J. Anim. Sci. 31:1963–1973.

Liu, N., and Y. J. Ru. 2010. Effect of phytate and phytase on the ileal flows of endogenous minerals and amino acids for growing broiler chickens fed purified diets. Anim. Feed Sci. Technol. 156:126–130.

Liu, N., Y. J. Ru, F. D. Li, and A. J. Cowieson. 2008. Effect of diet containing phytate and phytase on the activity and messenger ribonucleic acid expression of carbohydrase and transporter in chickens. J. Anim. Sci. 86:3432–3439.

Long, Y., S. Lin, J. Zhu, X. Pang, Z. Fang, Y. Lin, L. Che, S. Xu, J. Li, Y. Huang, X. Su, and D. Wu. 2016. Effects of dietary lysozyme levels on growth performance, intestinal morphology, non-specific immunity and mRNA expression in weanling piglets. Anim. Sci. J. 87:411–418.

Lu, H., I. Kühn, M. R. Bedford, H. Whitfield, C. Brearley, O. Adeola, and K. M. Ajuwon. 2019. Effect of phytase on intestinal phytate breakdown, plasma inositol concentrations, and glucose transporter type 4 abundance in muscle membranes of weanling pigs. J. Anim. Sci. 97:3907–3919.

Luttrell, B. M. 1993. The biological relevance of the binding of calcium ions by inositol phosphate. J. Biol. Chem. 268:1521–1524.

Maenz, D. D., C. M. Engele-Schaan, R. W. Newkirk, and H. L. Classen. 1999. The effect of minerals and mineral chelators on the formation of phytase-resistant and phytase-susceptible forms of phytic acid in solution and in a slurry of canola meal. Anim. Feed Sci. Technol. 81:177–192.

Malathi, V., and G. Devegowda. 2001. In vitro evaluation of nonstarch polysaccharide digestibility of feed ingredients by enzymes. Poult. Sci. 80:302–305.

Martínez-Villaluenga, C., J. Frias, and C. Vidal-Valverde. 2008. Alpha-galactosides: Antinutritional factors or functional ingredients? Crit. Rev. Food Sci. Nutr. 48:301–316.

Masey O'Neill, H. V., J. A. Smith, and M. R. Bedford. 2014. Multicarbohydrase enzymes for non-ruminants. Asian-Australas. J. Anim. Sci. 27:290–301.

Mc Alpine, P. O., C. J. O'Shea, P. F. Varley, and J. V. O'Doherty. 2012. The effect of protease and xylanase enzymes on growth performance and nutrient digestibility in finisher pigs. J. Anim. Sci. 90:375–377.

McDonald, D. E., D. W. Pethick, B. P. Mullan, and D. J. Hampson. 2001. Increasing viscosity of the intestinal contents alters small intestinal structure and intestinal growth, and stimulates proliferation of enterotoxigenic *Escherichia coli* in newly-weaned pigs. Br. J. Nutr. 86:487–498.

Menezes-Blackburn, D., S. Gabler, and R. Greiner. 2015. Performance of seven commercial phytases in an in vitro simulation of poultry digestive tract. J. Agric. Food Chem. 63:6142–6149.

Montagne, L., J. R. Pluske, and D. J. Hampson. 2003. A review of interactions between dietary fibre and the intestinal mucosa, and their consequences on digestive health in young non-ruminant animals. Anim. Feed Sci. Technol. 108:95–117.

Nasir, Z., J. Broz, and R. T. Zijlstra. 2014. Supplementation of a wheat-based diet low in phosphorus with microbial 6-phytase expressed in *Aspergillus oryzae* increases digestibility and plasma phosphorus but not performance in lactating sows. Anim. Feed Sci. Technol. 198:263–270.

Navarro, D. M. D. L., E. M. A. M. Bruininx, L. de Jong, and H. H. Stein. 2019. Effects of inclusion rate of high fiber dietary ingredients on apparent ileal, hindgut, and total tract digestibility of dry matter and nutrients in ingredients fed to growing pigs. Anim. Feed Sci. Technol. 248:1–9.

Ndou, S. P., E. Kiarie, A. K. Agyekum, J. M. Heo, L. F. Romero, S. Arent, R. Lorentsen, and C. M. Nyachoti. 2015. Comparative efficacy of xylanases on growth performance and digestibility in growing pigs fed wheat and wheat bran- or corn and corn DDGS-based diets supplemented with phytase. Anim. Feed Sci. Technol. 209:230–239.

Nørgaard, J. V., N. Malla, G. Dionisio, C. K. Madsen, D. Pettersson, H. N. Lærke, R. L. Hjortshøj, and H. Brinch-Pedersen. 2019. Exogenous xylanase or protease for pigs fed barley cultivars with high or low enzyme inhibitors. Anim. Feed Sci. Technol. 248:59–66.

NRC. 2012. Nutrient requirements of swine. 11th rev. ed. Natl. Acad. Press, Washington, DC.

Nyachoti, C. M., J. S. Sands, M. L. Connor, and O. Adeola. 2006. Effect of supplementing phytase to corn- or wheat-based gestation and lactation diets on nutrient digestibility and sow and litter performance. Can. J. Anim. Sci. 86:501–510.

O'Dell, B. L., A. R. de Boland, and S. R. Koirtyohann. 1972. Distribution of phytate and nutritionally important elements among the morphological components of cereal grains. J. Agric. Food Chem. 20:718–721.

Oliver, W. T., and J. E. Wells. 2013. Lysozyme as an alternative to antibiotics improves growth performance and small intestinal morphology in nursery pigs. J. Anim. Sci. 91:3129–3136.

Oliver, W. T., and J. E. Wells. 2015. Lysozyme as an alternative to growth promoting antibiotics in swine production. J. Anim. Sci. Biotechnol. 6:35.

Olukosi, O. A., and O. Adeola. 2012. Enzymes and enzyme supplementation of swine diets. In: L. I. Chiba, editor, Sustainable swine nutrition. John Wiley & Sons, Inc., Ames, IA. p. 277–294.

O'Shea, C. J., P. O. Mc Alpine, P. Solan, T. Curran, P. F. Varley, A. M. Walsh, and J. V. O'Doherty. 2014. The effect of protease and xylanase enzymes on growth performance, nutrient digestibility, and manure odour in grower–finisher pigs. Anim. Feed Sci. Technol. 189:88–97.

O'Quinn, P. R., D. A. Knabe, and E. J. Gregg. 1997. Efficacy of natuphos® in sorghum-based diets of finishing swine. J. Anim. Sci. 75:1299–1307.

Owusu-Asiedu, A., E. Kiarie, A. Péron, T. A. Woyengo, P. H. Simmins, and C. M. Nyachoti. 2012. Growth performance and nutrient digestibilities in nursery pigs receiving varying doses of xylanase and β-glucanase blend in pelleted wheat- and barley-based diets. J. Anim. Sci. 90:92–94.

Owusu-Asiedu, A., J. F. Patience, B. Laarveld, A. G. Van Kessel, P. H. Simmins, and R. T. Zijlstra. 2006. Effects of guar gum and cellulose on digesta passage rate, ileal microbial populations, energy and protein digestibility, and performance of grower pigs. J. Anim. Sci. 84:843–852.

Paloheimo, M., J. Piironen, and J. Vehmaanperä. 2010. Xylanases and cellulases as feed additives. In: M. R. Bedford, and G. G. Partridge, editors, Enzymes in farm animal nutrition. 2nd rev. ed. CAB Int, Wallingford, UK. p. 12–53.

Pan, L., Q. H. Shang, X. K. Ma, Y. Wu, S. F. Long, Q. Q. Wang, and X. S. Piao. 2017. Coated compound proteases improve nitrogen utilization by decreasing manure nitrogen output for growing pigs fed sorghum soybean meal based diets. Anim. Feed Sci. Technol. 230:136–142.

Park, C. S., I. Park, and B. G. Kim. 2016b. Effects of an enzyme cocktail on digestible and metabolizable energy concentrations in barley, corn, and wheat fed to growing pigs. Livest. Sci. 187:1–5.

Park, K. R., C. S. Park, and B. G. Kim. 2016a. An enzyme complex increases in vitro dry matter digestibility of corn and wheat in pigs. SpringerPlus 5:598.

Passos, A. A., I. Park, P. Ferket, E. von Heimendahl, and S. W. Kim. 2015. Effect of dietary supplementation of xylanase on apparent ileal digestibility of nutrients, viscosity of digesta, and intestinal morphology of growing pigs fed corn and soybean meal based diet. Anim. Nutr. 1:19–23.

Payling, L., I. H. Kim, M. C. Walsh, and E. Kiarie. 2017. Effects of a multi-strain *Bacillus* spp. direct-fed microbial and a protease enzyme on growth performance, nutrient digestibility, blood characteristics, fecal microbiota, and noxious gas emissions of grower pigs fed corn-soybean-meal-based diets-A meta-analysis. J. Anim. Sci. 95:4018–4029.

Pedersen, C., M. G. Boersma, and H. H. Stein. 2007. Digestibility of energy and phosphorus in ten samples of distillers dried grains with solubles fed to growing pigs. J. Anim. Sci. 85:1168–1176.

Pekel, A. Y., J. I. Kim, C. Chapple, and O. Adeola. 2015. Nutritional characteristics of camelina meal for 3-week-old broiler chickens. Poult. Sci. 94:371–378.

Poulsen, H. D., K. Blaabjerg, A. Strathe, P. Ader, and D. Feuerstein. 2010. Evaluation of different microbial phytases on phosphorus digestibility in pigs fed a wheat and barley based diet. Livest. Sci. 134:97–99.

Radcliffe, J. S., and E. T. Kornegay. 1998. Phosphorus equivalency value of microbial phytase in weanling pigs fed a maize-soyabean meal based diet. J. Anim. Feed Sci. 7:197–211.

Ravn, J. L., V. Glitsø, D. Pettersson, R. Ducatelle, F. Van Immerseel, and N. R. Pedersen. 2018. Combined endo-$\beta$-1,4-xylanase and $\alpha_{\text{-L}}$-arabinofuranosidase increases butyrate concentration during broiler cecal fermentation of maize glucurono-arabinoxylan. Anim. Feed Sci. Technol. 236:159–169.

Reddy, N. R., S. K. Sathe, and D. K. Salunkhe. 1982. Phytase in legumes and cereals. Adv. Food Res. 28:1–92.

Reilly, P., T. Sweeney, C. O'Shea, K. M. Pierce, S. Figat, A. G. Smith, D. A. Gahan, and J. V. O'Doherty. 2010. The effect of cereal-derived beta-glucans and exogenous enzyme supplementation on intestinal microflora, nutrient digestibility, mineral metabolism and volatile fatty acid concentrations in finisher pigs. Anim. Feed Sci. Technol. 158:165–176.

Rimbach, G., J. Pallauf, J. Moehring, K. Kraemer, and A. M. Minihane. 2008. Effect of dietary phytate and microbial phytase on mineral and trace element bioavailability - A literature review. Curr. Top. Nutraceutical Res. 6:131–144.

Rodehutscord, M., and P. Rosenfelder. 2016. Update on phytate degradation pattern in the gastrointestinal tract of pigs and broiler chickens. In: C. L. Walk, I. Kühn, H. H. Stein, M. T. Kidd, and M. Rodehutscord, editors, Phytate destruction - Consequences for precision animal nutrition. Wageningen Academic Publishers, Wageningen, Netherland. p. 15–32.

Rodehutscord, M., C. Rückert, H. P. Maurer, H. Schenkel, W. Schipprack, K. E. Bach Knudsen, M. Schollenberger, M. Laux, M. Eklund, W. Siegert, and R. Mosenthin. 2016. Variation in chemical composition and physical characteristics of cereal grains from different genotypes. Arch. Anim. Nutr. 70:87–107.

Rodríguez, D. A., R. C. Sulabo, J. C. González-Vega, and H. H. Stein. 2013. Energy concentration and phosphorus digestibility in canola, cottonseed, and sunflower products fed to growing pigs. Can. J. Anim. Sci. 93:493–503.

Rojas, O. J., Y. Liu, and H. H. Stein. 2013. Phosphorus digestibility and concentration of digestible and metabolizable energy in corn, corn coproducts, and bakery meal fed to growing pigs. J. Anim. Sci. 91:5326–5335.

Rutherfurd, S. M., T. K. Chung, and P. J. Moughan. 2014. Effect of microbial phytase on phytate P degradation and apparent digestibility of total P and Ca throughout the gastrointestinal tract of the growing pig. J. Anim. Sci. 92:189–197.

Schlemmer, U., K. -D. Jany, A. Berk, E. Schulz, and G. Rechkemmer. 2001. Degradation of phytate in the gut of pigs - pathway of gastrointestinal inositol phosphate hydrolysis and enzymes involved. Arch. Anim. Nutr. 55:255–280.

Selle, P. H., A. J. Cowieson, N. P. Cowieson, and V. Ravindran. 2012. Protein–phytate interactions in pig and poultry nutrition: A reappraisal. Nutr. Res. Rev. 25:1–17.

Selle, P. H., A. J. Cowieson, and V. Ravindran. 2009. Consequences of calcium interactions with phytate and phytase for poultry and pigs. Livest. Sci. 124:126–141.

She, Y., Y. Liu, J. C. González-Vega, and H. H. Stein. 2017. Effects of graded levels of an *Escherichia coli* phytase on growth performance, apparent total tract digestibility of phosphorus, and on bone parameters of weanling pigs fed phosphorus-deficient corn-soybean meal based diets. Anim. Feed Sci. Technol. 232:102–109.

Singh, S., G. Singh, and S. K. Arya. 2018. Mannans: An overview of properties and application in food products. Int. J. Biol. Macromol. 119:79–95.

Stein, H. H., L. A. Merriman, and J. C. González-Vega. 2016. Establishing a digestible calcium requirement for pigs. In: C. L. Walk, I. Kühn, H. H. Stein, M. T. Kidd, and M. Rodehutscord, editors, Phytate destruction - Consequences for precision animal nutrition. Wageningen Academic Publishers, Wageningen, Netherland. p. 207–216.

Steiner, T., R. Mosenthin, B. Zimmermann, R. Greiner, and S. Roth. 2007. Distribution of phytase activity, total phosphorus and phytate phosphorus in legume seeds, cereals and cereal by-products as influenced by harvest year and cultivar. Anim. Feed Sci. Technol. 133:320–334.

Tactacan, G. B., S. Cho, J. Cho, and I. H. Kim. 2016. Performance responses, nutrient digestibility, blood characteristics, and measures of gastrointestinal health in weanling pigs fed protease enzyme. Asian-Australas. J. Anim. Sci. 29:998–1003.

Thacker, P. A., J. G. Mcleod, and G. L. Campbell. 2002. Performance of growing-finishing pigs fed diets based on normal or low viscosity rye fed with and without enzyme supplementation. Arch. Anim. Nutr. 56:361–370.

Tiwari, U. P., H. Chen, S. W. Kim, and R. Jha. 2018. Supplemental effect of xylanase and mannanase on nutrient digestibility and gut health of nursery pigs studied using both in vivo and in vitro models. Anim. Feed Sci. Technol. 245:77–90.

Tsai, T., C. R. Dove, P. M. Cline, A. Owusu-Asiedu, M. C. Walsh, and M. Azain. 2017. The effect of adding xylanase or $\beta$-glucanase to diets with corn distillers dried grains with solubles (CDDGS) on growth performance and nutrient digestibility in nursery pigs. Livest. Sci. 197:46–52.

Torrallardona, D., and P. Ader. 2016. Effects of a novel 6-phytase (EC 3.1.3.26) on performance, phosphorus and calcium digestibility, and bone mineralization in weaned piglets. J. Anim. Sci. 94:194–197.

Torres-Pitarch, A., U. M. McCormack, V. E. Beattie, E. Magowan, G. E. Gardiner, A. M. Pérez-Vendrell, D. Torrallardona, J. V. O'Doherty, and P. G. Lawlor. 2018. Effect of phytase, carbohydrase, and protease addition to a wheat distillers dried grains with solubles and rapeseed based diet on in vitro ileal digestibility, growth, and bone mineral density of grower-finisher pigs. Livest. Sci. 216:94–99.

Upadhaya, S. D., S. I. Lee, and I. H. Kim. 2016a. Effects of cellulase supplementation to corn soybean meal-based diet on the performance of sows and their piglets. Anim. Sci. J. 87:904–910.

Upadhaya, S. D., H. M. Yun, and I. H. Kim. 2016b. Influence of low or high density corn and soybean meal-based diets and protease supplementation on growth performance, apparent digestibility, blood characteristics and noxious gas emission of finishing pigs. Anim. Feed Sci. Technol. 216:281–287.

Velayudhan, D. E., J. M. Heo, Y. Dersjant-Li, A. Owusu-Asiedu, and C. M. Nyachoti. 2015. Efficacy of novel 6-phytase from *Buttiauxella* sp. on ileal and total tract nutrient digestibility in growing pigs fed a corn-soy based diet. Anim. Feed Sci. Technol. 210:217–224.

Veum, T. L., D. W. Bollinger, C. E. Buff, and M. R. Bedford. 2006. A genetically engineered *Escherichia coli* phytase improves nutrient utilization, growth performance, and bone strength of young swine fed diets deficient in available phosphorus. J. Anim. Sci. 84:1147–1158.

Veum, T. L., and J. Liu. 2018. The effect of microbial phytase supplementation of sorghum-canola meal diets with no added inorganic phosphorus on growth performance, apparent total-tract phosphorus, calcium, nitrogen and energy utilization, bone measurements, and serum variables of growing and finishing swine. Livest. Sci. 214:180–188.

Wenk, C. 2001. The role of dietary fibre in the digestive physiology of the pig. Anim. Feed Sci. Technol. 90:21–33.

Wilfart, A., L. Montagne, H. Simmins, J. Noblet, and J. van Milgen. 2007. Digesta transit in different segments of the gastrointestinal tract of pigs as affected by insoluble fibre supplied by wheat bran. Br. J. Nutr. 98:54–62.

Woyengo, T. A., O. Adeola, C. C. Udenigwe, and C. M. Nyachoti. 2010. Gastro-intestinal digesta pH, pepsin activity and soluble mineral concentration responses to supplemental phytic acid and phytase in piglets. Livest. Sci. 134:91–93.

Woyengo, T. A., A. J. Cowieson, O. Adeola, and C. M. Nyachoti. 2009. Ileal digestibility and endogenous flow of minerals and amino acids: responses to dietary phytic acid in piglets. Br. J. Nutr. 102:428–433.

Woyengo, T. A., and C. M. Nyachoti. 2013. Review: Anti-nutritional effects of phytic acid in diets for pigs and poultry - current knowledge and directions for future research. Can. J. Anim. Sci. 93:9–21.

Woyengo, T. A., R. Patterson, and C. L. Levesque. 2018. Nutritive value of multienzyme supplemented cold-pressed camelina cake for pigs. J. Anim. Sci. 96:1119–1129.

Xue, P. C., and O. Adeola. 2015. Phosphorus digestibility response of growing pigs to phytase supplementation of triticale distillers' dried grains with solubles. J. Anim. Sci. 93:646–651.

Yamashita, Y., M. Yamaoka, T. Hasunuma, H. Ashida, and K. -I. Yoshida. 2013. Detection of orally administered inositol stereoisomers in mouse blood plasma and their effects on translocation of glucose transporter 4 in skeletal muscle cells. J. Agric. Food Chem. 61:4850–4854.

Yin, Y.-L., J. D. G. McEvoy, H. Schulze, U. Hennig, W. -B. Souffrant, and K. J. McCracken. 2000. Apparent digestibility (ileal and overall) of nutrients and endogenous nitrogen losses in growing pigs fed wheat (var. Soissons) or its by-products without or with xylanase supplementation. Livest. Prod. Sci. 62:119–132.

Yoon, S. Y., Y. X. Yang, P. L. Shinde, J. Y. Choi, J. S. Kim, Y. W. Kim, K. Yun, J. K. Jo, J. H. Lee, S. J. Ohh, I. K. Kwon, and B. J. Chae. 2010. Effects of mannanase and distillers dried grain with solubles on growth performance, nutrient digestibility, and carcass characteristics of grower-finisher pigs. J. Anim. Sci. 88:181–191.

Yu, G., D. Chen, B. Yu, J. He, P. Zheng, X. Mao, Z. Huang, J. Luo, Z. Zhang, and J. Yu. 2016. Coated protease increases ileal digestibility of protein and amino acids in weaned piglets. Anim. Feed Sci. Technol. 214:142–147.

Yu, S., A. Cowieson, C. Gilbert, P. Plumstead, and S. Dalsgaard. 2012. Interactions of phytate and myo-inositol phosphate esters ($IP_{1-5}$) including $IP_5$ isomers with dietary protein and iron and inhibition of pepsin. J. Anim. Sci. 90:1824–1832.

Zeng, Z. K., X. S. Piao, D. Wang, P. F. Li, L. F. Xue, L. Salmon, H. Y. Zhang, X. Han, and L. Liu. 2011. Effect of microbial phytase on performance, nutrient absorption and excretion in weaned pigs and apparent ileal nutrient digestibility in growing pigs. Asian-Australas. J. Anim. Sci. 24:1164–1172.

Zeng, Z. K., D. Wang, X. S. Piao, P. F. Li, H. Y. Zhang, C. X. Shi, and S. K. Yu. 2014. Effects of adding super dose phytase to the phosphorus-deficient diets of young pigs on growth performance, bone quality, minerals and amino acids digestibilies. Asian-Australas. J. Anim. Sci. 27:237–246.

Zeng, Z. K., J. L. Zhu, G. C. Shurson, C. Chen, and P. E. Urriola. 2018. Improvement of in vitro ileal dry matter digestibility by non-starch polysaccharide degrading enzymes and phytase is associated with decreased hindgut fermentation. Anim. Feed Sci. Technol. 246:52–61.

Zhang, S., J. Song, Z. Deng, L. Cheng, M. Tian, and W. Guan. 2017. Effects of combined α-galactosidase and xylanase supplementation on nutrient digestibility and growth performance in growing pigs. Arch. Anim. Nutr. 71:441–454.

Zhang, S., X. Zhang, H. Qiao, J. Chen, C. Fang, Z. Deng, and W. Guan. 2018b. Effect of timing of post-weaning supplementation of soybean oil and exogenous lipase on growth performance, blood biochemical profiles, intestinal morphology and caecal microbial composition in weaning pigs. Ital. J. Anim. Sci. 17:967–975.

Zhang, Z., H. M. Tun, R. Li, B. J. M. Gonzalez, H. C. Keenes, C. M. Nyachoti, E. Kiarie, and E. Khafipour. 2018a. Impact of xylanases on gut microbiota of growing pigs fed corn- or wheat-based diets. Anim. Nutr. 4:339–350.

Zhou, P., M. Nuntapaitoon, T. F. Pedersen, T. S. Bruun, B. Fisker, and P. K. Theil. 2018. Effects of mono-component xylanase supplementation on nutrient digestibility and performance of lactating sows fed a coarsely ground diet. J. Anim. Sci. 96:181–193.

Zijlstra, R. T., A. Owusu-Asiedu, and P. H. Simmins. 2010. Future of NSP-degrading enzymes to improve nutrient utilization of co-products and gut health in pigs. Livest. Sci. 134:255–257.

Żyła, K., R. Duliński, M. Pierzchalska, M. Grabacka, D. Józefiak, and S. Świątkiewicz. 2013. Phytases and *myo*-inositol modulate performance, bone mineralization and alter lipid fractions in the serum of broilers. J. Anim. Feed Sci. 22:56–62.

# 18 Feed Additives in Swine Diets

Cormac J. O'Shea

## Introduction

Feed additives may be defined as ingredients in a swine dietary formulation, which may not directly provide nutrients to the animal but may have functionality that brings about a positive change. The primary target of most feed additives is the gastrointestinal tract and could include enhancement of nutrient and energy utilization, modification of the immune status, and/or the enteric microbiota. The secondary target might be ancillary organs such as the liver or kidneys or even systemic change, but ultimately the intention is to improve the composition or output of an economically important trait such as weight gain, meat or milk composition. In other scenarios, the target might be to improve health outcomes for swine that may be compromised due to various environmental stressors such as management practices like weaning, disease, or unsuitable environmental conditions. Finally, feed additives may be included to modify the content or quantity of excreta with the purpose of mitigating nutrient excretion or gaseous emissions. In practice, where feed additives are employed, these target outcomes are intertwined. Feed additives may be included in dietary formulation to act directly on the animal or in the case of gestating and lactating animals, to effect a change indirectly in the progeny. Comparatively, far less research has been conducted on the latter. Hence, feed additives tend to be involved in a stage of production and challenge-dependent fashion. The vast majority of research into feed additives is focused on managing the postweaning period.

Experimental studies evaluating the role of certain feed additives have provided important evidence on the potential and limitations of their usage in feed formulations under a variety of conditions. Far less is known about the interactions, whether it is synergistic, additive, or antagonistic that may occur as a consequence of involving multiple additives. Furthermore, it is difficult to predict what these outcomes may be without, at least, evaluating the animal response.

In the eight years since the publication of the original edition of this chapter, there has been an explosion of research evaluating and reporting on feed additives that may serve as alternatives to prophylactic antibiotics. This reflects the continuation of the ban of in-feed antibiotics in the European Union and the increasing pressure in other countries to phase out their usage. The challenge continues to identify feed additives, which compare favorably with antibiotics in terms of efficacy to improve growth and health outcomes, but also technical considerations such as inclusion rates, functionality following feed processing, and under a variety of operating conditions. An important work is the redefining of already established feed additives investigation to solve additional problems related to swine production. Examples of this include the use of enzymes such as phytase to alleviate lower gastrointestinal tract infections in piglets (Brandão Melo et al. 2020; Perez-Palencia et al. 2021).

*Sustainable Swine Nutrition*, Second Edition. Edited by Lee I. Chiba.
© 2023 John Wiley & Sons Ltd. Published 2023 by John Wiley & Sons Ltd.

**Antimicrobial Agents**

Antimicrobials have held an important role in swine diets both in a prophylactic capacity to mitigate the onset of disease particularly in challenging production conditions, and as a therapeutic to resolve subclinical and clinical disease. In addition to these important roles, antimicrobials have been applied in swine diets as growth promoters. The earlier edition of this chapter has reviewed this feed additive in great detail and will not be substantially revisited in the present edition. The trend of reduced antimicrobial usage at prophylactic dosage rates in many parts of the world reflects the concerns that farm animals may serve as reservoir for antimicrobial resistance because of the shared role of many drugs in food production and animal health. As a consequence, their use have been heavily restricted in the European Union since 2006 [(Regulation (EC) No 1831/2003 of the European Parliament and of the council on additives for use in animal nutrition 2003] and subject to limited availability by veterinary prescription in the USA, Canada, and Australia. Despite these policies around the stewardship of antimicrobial use, the global outlook remains set to see antimicrobials as an important additive not just in the suckling and weaning phases but also in the grower-finisher phase (Lekagul et al. 2019). Antibiotics are considered to be an essential part of the toolkit for managing disease and poor growth performance particularly in weaned pig performance. Some of the reasons for this success and popularity is the excellent profile of in-feed antibiotics such as low dosage rate, predictable response rate, and functionality after diet preparation. Some authors have evaluated the consequences of not using in-feed antibiotics in the weaning and growing finishing phase. Diana et al. (2019; Table 18.1) assessed the growth performance, routine disease treatment requirements, and mortality of weaners and grower-finishers in a commercial setting and found minimal impact on lifetime growth performance when antibiotics were excluded although more veterinary treatments and slightly higher finisher mortality were observed in the untreated group. Regardless of these interesting findings, the decline of antibiotic use has stimulated much research into alternative feed additives to support swine production.

*Microbial Supplements*

Live or viable dietary microbial supplements, often termed probiotics, which exert a beneficial effect of some kind when consumed by the pig, have received increasing attention as alternatives to prophylactic in-feed antibiotics. Comprehensive reviews have been produced that cover definitions, mechanism of action, application in pig production (Kenny et al. 2010; Fouhse et al. 2016; Liao

**Table 18.1**  Effect of the inclusion of in-feed antibiotics [sulfadiazine-trimethoprim, 14.4 mg/kg body weight (BW) per day) on growth performance of pigs[a,b].

| Variables | First weaner stage | | | Second weaner stage | | | Finisher stage | | |
|---|---|---|---|---|---|---|---|---|---|
| | NOI | ABI | P-value | NOI | ABI | P-value | NOI | ABI | P-value |
| ADFI (g) | 585 | 647 | 0.048 | 1381 | 1440 | 0.589 | 1811 | 1819 | 0.984 |
| ADG (g) | 402 | 436 | 0.018 | 711 | 744 | 0.774 | 865 | 882 | 0.893 |
| FCR | 1.48 | 1.52 | 0.483 | 1.95 | 1.95 | 0.944 | 2.10 | 2.07 | 0.853 |
| BW (kg) | 21.9 | 23.0 | 0.032 | 41.4 | 43.3 | 0.218 | 99.4 | 101 | 0.483 |

[a] NOI = no in-feed antibiotics, ABI = in-feed antibiotics, ADG = average daily gain, ADFI = average daily feed intake, and FCR = feed conversion ratio.

[b] Source: Diana et al. (2019) Springer Nature / CC BY 4.0.

and Nyachoti 2017; Barba-Vidal et al. 2019), and technical, safety, and regulatory considerations (Salminen et al. 1998; A.V. Wright 2005). Microbial supplements commonly reported in the literature include bacteria and yeasts, while bacteriophages are less commonly, but increasingly studied (Kim et al. 2014; Kim et al. 2017) and are included in dietary formulations with the purpose of colonizing and/or influencing the contents of the gastrointestinal tract with the consequent benefits that might bring for the pig. Attempts to deliver live microbials via waterlines are understudied at the present time, and the limited reports available show little efficacy in mitigating infectious challenges in weaned pigs (Walsh et al. 2012a,b). It follows that a microbial supplement must therefore possess the technical properties to remain viable during sections of the feed production process, during storage and then post ingestion, survive the harsh environment of the upper gastrointestinal tract to reach the lower gastrointestinal tract. Once there, the microbial supplement must influence the composition of the resident microbiota and/or the metabolic profile or the gastrointestinal tract contents. In order for a microbial supplement to colonize the lower gastrointestinal tract, it must coexist with or displace a portion of the resident microbiota. The enteric microbiota of the pig may be generally considered to have a stage of life-specific profile and be either dynamic or stable at key stages of life, with the greatest instability seen around weaning. Thereafter, the microbiota transitions to a relatively stable period characterized by greater diversity throughout adulthood (Chen et al. 2017; Wang et al. 2019). Environmental exposure to microorganisms has a key role in establishing and accelerating to a mature enteric microbiota (Vo et al. 2017). Hence, the most effective junction to introduce microbial supplements is before the establishment of a stable microbiota, and the literature mostly focuses on the benefits of such in the post-weaned period. Here there is persuasive evidence for an improvement in health and growth variables. The role that probiotics may place in the swine industry was placed in the context by Kenny et al. (2010) who concluded that a beneficial response to probiotics was very much dependent on the environment. Farms with suboptimal conditions, in terms of health and hygiene, would likely elicit the greatest improvements, whereas farms with a record of high productivity, health, and welfare could expect negligible benefit. Kenny et al. (2010) also discuss an indirect role for promoting lactic acid bacteria through modification of the maternal diet, with the sow thus providing a potentially greater source of the beneficial microbes when compared with the problematic direct feeding of newly weaned pigs with poor appetites.

A sensible starting point to identify candidate probiotics is profiling the microbial community of sow's milk on the general assumption that these microbes will have a beneficial or benign role in the microbiota of the piglet. Martín et al. (2009) reported on the presence of various microbes in sow's milk and evaluated their probiotic potential using a range of criteria including capacity to survive the gastrointestinal processes and confer a benefit for the pig.

More recently, studies have attempted to associate the enteric microbiota using metagenomic methodology with important production traits in pigs (Si et al. 2020; Vigors et al. 2020). Those authors report various taxonomic ranks down to species level that are associated with feed efficiency measurements. The potential of dietary microbial supplements composed of single species or combinations to mitigate experimental infectious disease challenges has been reported for various pathogenic microorganisms, including *Escherichia coli* K88 (Pan et al. 2017), *Salmonella enterica* serovar Typhimurium (Casey et al. 2007; Gebru et al. 2010). A review by Liao and Nyachoti (2017; see Table 18.2) summarized the outcomes in the literature of feeding microbial supplements at each production phase, and it can be concluded that there are benefits for the key performance variables, particularly at weaning stage.

Bacteriophages, viral particles which predate upon and multiply with bacteria, are comparatively understudied relative to bacteria and yeasts as microbial supplements but in recent times are receiving more attention. A review on the evidence for bacteriophages to mitigate *Salmonella* and

**Table 18.2** Response of growth performance variables to dietary inclusion of microbial supplements at different pig production phases[a,b].

| Microorganisms | ADFI | ADG | FCR | Age group |
|---|---|---|---|---|
| B. *subtilis* | NS | S(+) | S(−) | Growing-finishing pigs |
| C. *butyricum* | | | | |
| L. *acidophilus* | S(−) | S(+) | NS | Weaned piglets |
| S. *cerevisiae* | | | | |
| L. *acidophilus* | NS | S(+) | NS | Growing pigs |
| S. *cerevisiae* | | | | |
| B. *subtilis* | | | | |
| L. *plantarum* ATCC 4336 | NS | S(+) | NS | Weaned piglets |
| L. *fermentum* DSM 20016 | | | | |
| E. *faecium* ATCC 19434 | | | | |
| E. *faecium* EK13 | — | NS | — | Newborn piglets |
| Bi. *longum* AH1206 | NS | NS | NS | Neonatal piglets |
| B. *licheniformis* | — | S(+) | S(−) | Weaned piglets |
| B. *subtilis* | NS | S(+) | S(−) | Growing pigs |
| B. *licheniformis* | | | | |
| B. *subtilis* | NS | S(+) | S(−) | Grower-finisher pigs |
| B. *licheniformis* | | | | |
| B. *subtilis* MA139 | NS | S(+) | S(−) | Weaned piglets |
| B. *toyonensis* | S(+) | S(+) | S(−) | Weaning piglets |
| B. *licheniformis* | NS | NS | S(−) | Growing–finishing pigs |
| B. *subtilis* | | | | |
| S. *cerevisiae* subsp. *boulardii* CNCM I-1079 | — | — | S(−) | Weaned piglets |

[a] ADG = average daily gain, FCR = feed conversion ratio, ADFI = average daily feed intake, S(+) = significantly increased, S(−) = significantly decreased, NS = non-significant, and — = not studied.

[b] Adapted from Liao and Nyachoti (2017).

*Escherichia coli* infections has been provided by Zhang et al. (2015). Kim et al. (2017) reported an improvement in growth performance and a reduction in ileal coliform bacteria of weaned pigs offered a bacteriophage cocktail. The use of bacteriophages in weaning diets has shown improvement in diarrheal score, reductions in intestinal coliforms and *Clostridium* spp., and improvements in growth comparable with more commonly used growth promoters (Hosseindoust et al. 2017).

### Fermentable Carbohydrates

Fermentable carbohydrates encompass a broad group of complex carbohydrates that are not degraded, or only partially degraded by endogenous enzymes produced by the pig during passage through the gastrointestinal tract. Consequently, fermentable carbohydrates are available for utilization by the enteric microbiota. This utilization primarily occurs in the large intestine, but there may be some microbial fermentation initiated earlier in the tract, largely dependent on the solubility and degradability of the carbohydrate. As a consequence of this, the phylogeny or metabolic activity of the microbiota can be altered, which may have consequences for the pig. A primary goal of using fermentable carbohydrates in diets is to promote microbes that will preferentially utilize such substrates as energy sources (Gibson et al. 1995; Zimmermann et al. 2001). Various studies profiling the influence of fermentable carbohydrates have shown that the abundance of lactic acid-producing bacteria such as *Lactobacillus* spp. and *Bifidobacteria* spp., is enhanced and are associated with improvements in markers of gastrointestinal health and growth performance outcomes especially in young pigs. A consequence of promoting lactic acid bacteria may be a decline in other less desirable groups, such as Clostridia spp. and the *Enterobacteriaceae*. Dotsenko et al. (2018) reported how arabinoxylooligosaccharides and xylooligosaccharides, fermentable carbohydrates associated mostly with wheat directly increase lactic acid bacteria and suppressed the growth of *Clostridium perfringens* in an *in vitro* study.

An additional objective of including a dietary fermentable carbohydrate is to increase the ratio of carbohydrate to protein that is available for fermentation in the large intestine and hence reduce the formation of protein degradation products (Smith and Macfarlane 1998; Awati et al. 2006). Protein fermentation is associated with diarrhea in newly weaned piglets (Heo et al. 2008), possibly through the generation of toxic metabolites such as ammonia (Blachier et al. 2007). While protein fermentation occurs across the gut microbiome, it has been associated with potentially pathogenic microbes such as enterobacteria and clostridia (Vince and Burridge 1980). More recently, advanced methodologies have revealed the breadth of taxa which utilize undigested protein in the large intestine (Amaretti et al. 2019). Short-chain fatty acids (SCFA) are the main by-product that arise as a consequence of microbial fermentation of carbohydrates in the lumen and have several important roles both locally, including serving as an energy source for epithelial colonocytes (Evans et al. 1992). The SCFA produced in the greatest concentrations are acetic acid, propionic acid, and butyric acid. The branched-chain fatty acids (BCFA), isovaleric acid, isobutyric acid, and valeric acid are present in comparatively smaller concentrations (~collectively around 5% of the SCFA pool). While protein fermentation contributes generally to the SCFA pool, it is the main contributor to the formation of BCFA (Marounek et al. 2002) and therefore a relative decrease in those SCFA is considered a marker of an enhanced carbohydrate:protein substrate ratio in the large intestine.

In addition to the BCAA, ammonia is a by-product that arises from protein fermentation. Ammonia is undesirable in the lumen where it is toxic to epithelial cells (Blachier et al. 2007). Pié et al. (2007) reported that BCFA and ammonia were associated with several proinflammatory molecules in weaned pigs and demonstrated how a combination of dietary fermentable carbohydrates sources led to a decrease in ammonia and BCFA in lumen contents and an increase in lactic, acetic, and propionic acids. Hence, the role of integrating fermentable carbohydrates into the diet can be viewed as multipurpose, modifying the composition of the microbiota but also the profile of microbial metabolic end products.

All carbohydrates that escape degradation in the upper gastrointestinal tract may be considered fermentable, but the extent of microbial degradation and hence function varies significantly. Therefore, cellulose will be comparatively undegraded and have less influence on the quantity and composition of fermentation end products than, for example, an oligofructose such as galacto- (Difilippo et al. 2015) and fructooligosaccharides (Ayuso et al. 2020). However, there are additional functional consequences such as modifying the rate of digesta transit through the gut, influencing the rate of digesta disappearance from the gut and as a direct stimulant for the lumen, and influencing the mucosa and stimulating immunity. In this regard, the degradability of the fermentable carbohydrate is relevant in addition to other physico-chemical properties. Thorough reviews on these subjects are available (Jha and Berrocoso 2015; Williams et al. 2019).

In recent years, fermentable carbohydrates that may be labeled as prebiotic but are included at low levels in the diet have received greater attention. The lower inclusion rates are relevant because one presumed mode of action is not to meaningfully influence digestion dynamics or the composition of fermentation end- products, but rather to influence the enteric microbiota and/or the immune system on the pig that will have downstream consequences for important production variables.

Some examples of these include the algal polysaccharides, such as beta-glucans and sulfated poly-saccharides, which have demonstrated impressive improves in pig performance, particularly when directly provided to weaned pigs but also for progeny after feeding to gestating and lactating sows (O'Doherty et al. 2017). A compelling application of such prebiotics, which have immune-modulating properties, is in feeding sows during gestation and lactation to influence the physiology and performance of the piglets possibly permanently. Sows fed green algae extracts rich in sulfated polysaccharides had altered IgA antibodies in milk at day 7 and 21 of lactation (Bussy et al. 2019). Heim et al. (2015) found a significantly greater bodyweight in pigs at slaughters from sows fed laminarin (1 g/day) during the final stages of gestation and for the duration of lactation.

### Fermentable Carbohydrates and the Environment

Fermentable carbohydrates and, to a lesser extent, probiotics have occupied a role outside the weaner phase as a dietary mechanism to improve emissions from pigs. Due to the previously discussed impact of fermentable carbohydrates in modifying digestive processes and altering the profile of microbial end products, these feed additives can influence the concentrations and proportions of compounds that are associated with environmental pollution in intensive swine production. Many by-products of protein fermentation are associated with the noxious smell of pig manure including cadaverine, putrescine, the BCAA, and ammonia (Le et al. 2005). While the degradation products of fermentable carbohydrates will also contribute to odor formation, generally the provision of such in the diet has been reported to have a beneficial impact on odor emissions. There are of course excep-tions to this; Lynch et al. (2008) found an increase in odor emissions in finishing pigs fed 20% sugar beet pulp, attributed to the greatly increased SCFA production stimulated by the high content of fer-mentable pectin in the beet pulp. Several excellent reviews are available on the topic of fermentable carbohydrates in pig production with a focus on gaseous emissions (Sutton et al. 1999; Nahm 2003).

Microbial supplements and fermentable carbohydrates are often considered collectively because of a shared purpose in modifying in some way the microbiota of the pig. Furthermore, these addi-tives are often integrated into a diet in tandem, with the purpose of providing the exogenous live microbes with an energy supply in the gut. The concept of combining pro- and pre-biotics, termed "synbiotics," is reviewed by Zimmermann et al. (2001).

## Minerals

Trace minerals traditionally included to satisfy physiological requirements for growth and mainte-nance can also have functional benefits beyond minimum requirements. Of these, zinc (Zn) and cop-per (Cu) are well established, and for Zn at least, are now under scrutiny because of contribution to environmental pollution. Other minerals such as selenium may have condition-specific application at supranutritional inclusion rates to manage various challenges in the field and have attracted attention in recent years to evaluate a role for supplementing beyond the recommended inclusion rate (Liu et al. 2018). Selenium, Zn, and Cu all have important roles to play in mitigating oxidative stress, a state where physiological concentrations of oxides may be heightened (Klotz et al. 2003). Like with other farm production systems, changes in the health status, nutritional sufficiency, or ambient conditions may vary scenarios, which can induce physiological imbalance in redox status.

### Zinc Oxide

Zinc (Zn) is an essential mineral serving a range of physiological functions, chiefly as a compo-nent of various metalloenzymes involved in DNA regulation and enzymatic digestive processes.

The nutritional requirement for Zn by weaned pigs has been reported to be between 80 and 100 ppm (van Heugten et al. 2003; Pettigrew 2006). Zinc has been integrated to weaned pig diets at concentrations beyond nutritional requirements in an extra-nutritional role due to the beneficial response observed in mitigating the negative consequences of the post-wean period. In the aspiration of reducing or replacing the use of in-feed antibiotics, Zn emerged as a product that consistently demonstrates efficacy in the post-wean setting (Pettigrew 2006). Pharmacological doses of Zn improve fecal consistency and have been demonstrated to increase appetite, feed efficiency, and growth rate in piglets (Poulsen 1995; Pérez et al. 2011). When fed at levels in excess of dietary requirements, zinc has multiple modes of action both directly on the lumen epithelium (Medani et al. 2012) and through its effect on the microbiota. Zinc has antimicrobial and microbiostatic effects on the enteric microbiota. These effects may be genus and species specific, with some reports that gram-positive bacteria may be more susceptible than gram negative (Højberg et al. 2005). Starke et al. (2014) have shown that the microbiota is sensitive in a similarly selective manner. Those authors profiled the microbial community of weaned pigs offered high dietary ZnO and reported a decrease in cell numbers of *Lactobacillus* spp. numbers and *Streptococcus* spp., while *Enterobacteriaceae* and *Bifidobacterium* spp. were unaffected. The presence of high levels of dietary ZnO has also been shown to suppress concentrations of SCFA and other microbial metabolites such as ammonia (Starke et al. 2014; O'Shea et al. 2014a) demonstrating an impact on microbial metabolic activity. Hence, the presence of ZnO throughout the gastrointestinal tract has profound, suppressive impacts on the gut microbiota and its activities at least in the immediate post-wean period. A reduction in microbial fermentation due to the suppression of the microbiota may lead to greater opportunities for host capture of nutrients (Højberg et al. 2005).

High dietary ZnO has been associated with an improvement in fecal score, both where diarrhea is reported (Trckova et al. 2015) and in nonclinical scenarios (O'Shea et al. 2014a). This desiccating effect on feces likely reflects both the suppression of the gut microbiota and also direct effects on the colonic epithelium. The secretion of fluid from colonocytes is an innate defense mechanism that can contribute to dehydration in newly weaned pigs. High dietary zinc has been shown to have a profound effect on the morphology and gene expression of newly weaned pig colonocytes, increasing mucin-producing cells and modulation of cytokine expression associated with inflammation (Liu et al. 2014a).

The use of pharmacological doses of Zn has consequences for the composition of the enteric microbiota, as microorganisms particularly sensitive are selected against, whilst resistant strains may benefit from a competitive advantage. Evidence for resistance to Zn has been reported, and this problem may be compounded by greater prevalence or emergence of multiresistant microorganisms such as *E. coli* (Yazdankhah et al. 2014). While it is commonplace for Zn dosage rates of 2500–3000 ppm to be seen in weaning pig diets, particularly where inorganic compounds such as zinc oxide (ZnO) is used, there is a lot of interest and recent research to indicate efficacy at lower inclusion rates when the mineral is provided in different forms. The range of such "next-generation" Zn compounds now available is expanding and are generally characterized as either being chelated to an organic molecule and hence more readily available for absorption from the gastrointestinal tract or are presented in a different format, such as encapsulation or in the nanoparticle range (Pei et al. 2019). It follows that this enhanced bioavailability facilitates the inclusion of organic Zn at lower inclusion rates. Morales et al. (2012) reported improved growth performance in newly weaned pigs offered a proprietary Zn compound at 110 ppm in contrast with conventional Zn oxide included at 3000 ppm. Improvements in growth performance at similar low concentrations have been observed for other various chelates such as zinc glycine (Wang et al. 2010) and zinc chitosan compounds (Xie et al. 2010). Hence, the future of Zn use as a feed additive will likely see increasing prevalence of such alternative compounds at lower concentration rates.

### Copper

Copper is required for hemoglobin synthesis and involvement in the redox balance at approximately 5–6 ppm. At greater levels, between 100 and 200 ppm, a growth-promoting effect is observed (Cromwell et al. 1998; NRC 2012). Copper has antimicrobial effects in the gastrointestinal tract, which has been associated with improvements in growth performance and management of diarrhea in the weaning stage. Højberg et al. (2005) reported decreased counts of coliforms in cecal and colon contents of weaned piglets at a dosage rate of 175 ppm. The compound form of copper may be an important consideration. Van Kuijk et al. (2019) reported that piglets preferred copper hydroxychloride when compared with sulfated forms. Espinosa et al. (2017) reported an improvement in the growth performance and diarrhea score of weaned piglets offered diets containing copper hydroxychloride, which were not associated with changes to digestibility. Therefore, the improvements observed with supranutritional doses of copper may be primarily related to reducing the microbial load in newly weaned piglets.

### Legislation and Future Usage of Heavy Metals

The usage of zinc and copper may become limited in many regions. Excessive accumulation of heavy metals disrupts the homeostasis of the soil microbiome and plant development. While this contamination tends to occur in focal locations and may be related to other industries aside from agriculture (Tóth et al. 2016; Wyszkowska et al. 2013), concerns of environmental pollution herald changes in policy. For example, in the European Union, concerns about the accumulation of heavy metals in manure-applied soil have led to plans to phase out usage of such minerals beyond what is required for nutritional needs. Circumvention of these restrictions if they occur elsewhere will lie in optimizing the use of heavy metals at lower concentrations in more effective forms such as organic chelates or encapsulation.

### Selenium

Selenium is a trace mineral in pig diets with a range of physiological functions. Dietary selenium is required in the synthesis of selenoproteins, which have a role in regulating redox balance. It has a narrow band between deficiency and toxicity when provided as an inorganic form such as selenium selenite. In recent years, with the availability of organic forms reported to have greater safety profiles, there has been interest in a role for selenium as a functional feed additive. Application of organic selenium in a functional capacity may have the greatest potential where the environmental challenges are pronounced, such as variations in temperature. A key role for organic selenium that has been investigated is in the preservation of meat quality variables, particularly during challenging growth conditions, which may affect meat composition. Typically, selenium is included in the diet in the range of 0.2–0.3 ppm. Liu et al. (2018) reported that a dietary Se yeast at 1 ppm improved some markers of oxidative stress in heat stress pigs. In addition to improving meat composition, Se has also reported to mitigate the effect of a high temperature challenge on intestinal epithelial integrity (Liu et al. 2016), improving compromised epithelial barrier function associated with a heat stress.

## Acidifiers

The commencement of the weaning period is characterized by the abrupt, premature separation of the piglet from the sow and consequently the cessation of lactose-rich sow's milk. This rapid change from one form of nutrition to the other has profound consequences for gastrointestinal physiology and biochemistry. Appropriate acidity at distinct regions of the gastrointestinal tract is considered an important variable that influences the digestibility of the post wean diet, inhibits the passage of potential pathogens through the stomach, and is in part an indicator of lactic acid bacteria activity (Cranwell et al. 1976). Barrow et al. (1977) reported a higher stomach pH in weaned pigs when compared with contemporaries remaining with the sow and a lag of eight days until gastric acidity was comparable between suckling and weaned groups. A strong association between lactic acid concentrations and pH reported by that same study suggested an important role for lactose in influencing gastric acidity. In a similar, more recent study, Montagne et al. (2007) did not observe the same trend in stomach pH but did report on a stabilization of a lower cecal and colonic pH in the latter stages of weaning. Snoeck et al. (2004) showed a similar decrease in pH along several regions of the gastrointestinal tract as weaning progressed, with some pigs having a stomach pH as low as 1.6–1.7 at one to two weeks postweaning. In that study, *E. coli* F4 fimbriae were rapidly degraded at that pH range but persisted for longer periods at > pH 3. The diet may have an important role to play in the acidification of the gastrointestinal tract postweaning. Various ingredients, such as inorganic calcium and phosphorus sources, have very high acid-binding capacities, indicating their presence in the diet may impede acidification in the gut (Lawlor et al. 2005). Consequently, there has been a lot of interest in using various acids as feed additives to redress the assumed impact of weaning on gastrointestinal tract pH. The mode of action may be to enhance acidification of various regions along the gastrointestinal tract, but this is unlikely to be the only mode of action responsible for the growth response observed (Ravindran and Kornegay 1993)

*Organic Acids*
Organic acids such as the SCFA and their salts are added in the diet either individually or combined to bring about a beneficial change in gastrointestinal function through reducing pH (Partanen and Mroz 1999; Mroz 2005). Organic acids may also directly influence the enteric microbiota through disrupting the cell membranes of sensitive microbes and, thus, favoring those that prefer a more acidic microenvironment (Mroz 2005), which can lead to the improved growth performance of weaned pigs (Table 18.3). The SCFA are also produced along the gastrointestinal tract as a by-product of microbial degradation of dietary components and have pleiotropic roles in the gut and systemically. Locally, the SCFA provide a direct source of energy for epithelial cells, influence the composition of the microbiota (Zhang et al. 2018), and directly or indirectly modify the immune status of intestinal tissue (Grilli et al. 2016). In more recent years, thanks to detailed investigations, more complex roles for SCFA have been described. Butyrate, for example, has been reported to have multiple sites of action, directly inhibiting bacteria virulence, but also acting as a substrate for

**Table 18.3** Effect of various dietary organic acids on the growth performance of weaned pigs[a,b].

| Variable | Formic acid | | Fumaric acid | | Citric acid | | Potassium diformate | |
|---|---|---|---|---|---|---|---|---|
| | − | + | − | + | − | + | − | + |
| ADFI (g) | 667 | 710 | 613 | 614 | 534 | 528 | 764 | 823 |
| ADG (g) | 387 | 428 | 358 | 374 | 382 | 396 | 479 | 536 |
| FCR | 1.64 | 1.60 | 1.59 | 1.55 | 1.67 | 1.60 | 1.60 | 1.54 |

[a] ADFI: average daily feed intake, ADG: average daily gain, and FCR: feed conversion ratio.
[b] Adapted from Mroz (2005).

epithelial cells in the lumen, accelerating cellular repair, and thought to have a role on gut epithelia through participation in a hormone-neuro-immuno pathway. Increased butyric acid may be achieved directly through dietary provision of a butyrate salt, such as sodium butyrate or through introducing a fermentable carbohydrate, that will enhance butyric acid production endogenously in the lower gastrointestinal tract (Guilloteau et al. 2010). Dietary butyrate has been shown to enhance growth performance in weaned pigs (Manzanilla et al. (2006).

Other examples of organic acids associated with pig production include formic, benzoic, citric, and lactic acid. These may be added singly or in cocktails to achieve similar purposes of acidification. Hosseindoust et al. (2017) reported an improvement in growth performance accompanied by reductions in intestinal coliform and *Clostridia* spp. using a cocktail of citric, formic, and lactic acid in weaned pigs. Feeding weaning pigs a combination of formic and lactic acid brought about a decrease in stomach pH and a decrease in Enterobacteria in stomach contents (Hansen et al. 2007).

Benzoic acid has multiple purposes, depending on the stage of production. In weaned piglets, dietary benzoic acid is provided as a growth promoter, bringing about beneficial changes in gut morphology and markers of protein fermentation (Diao et al. 2014) and improving growth performance. Dietary benzoic acid at 0.5% of the diet improved growth performance, markedly improving weight gain and feed efficiency in the first two weeks post wean (Torrallardona et al. 2007). Diao et al. (2014) reported how piglets fed benzoic acid had decreased intestinal pH at 14 days and 42 days postweaning and a decrease in *Escherichia coli* counts at both time periods also. Those authors also reported an increase in propionic acid and a decrease in ammonia at different locations along the lower gastrointestinal tract. Provision of benzoic acid may also increase lactic acid producing bacteria. Diao et al. (2014) reported an increase in ileal *Bifidobacteria* spp and caecal *Bacillus* spp in 14-day-old weaned piglets. In finisher pigs, benzoic acid has been reported to improve ammonia emissions in a dose-dependent manner due in part to the strong acidifying effect of benzoic acid on urine pH (Murphy et al. 2011). Inclusion of benzoic acid at 0.3 and 0.5% brought about improvements in bodyweight gain in weanlings and grower-finishers due to an improvement in feed conversion efficiency (Zhai et al. 2017).

*Enzymes*

The use of exogenous enzymes to modify important economic outcomes in intensive pig production has a long history relative to many feed additives, particularly in the case of phytase (Jongbloed and Kemme 1990; Simons et al. 1990) and are now routinely found included in dietary formulations for various purposes across different phases of production. It is now clear that dietary enzymes work best in scenarios where ingredients and/or the nutrients and energy supplied are suboptimal (Gagne et al. 2002). Exogenous enzymes therefore offer the greatest potential to enhance the use of by- and coproducts that otherwise would underperform relative to traditional mainstream ingredients. In current usage, the role sought for enzymes is to reduce feed costs while maintaining animal growth performance. In feed formulation, a nutrient value is now commonly assigned to the enzyme, particularly for phytase and increasingly for other enzymes, which then allows expensive energy and amino acid sources to be proportionately reduced (Cowieson and Roos 2016).

The response to enzymes is subject to many factors, including the technical efficacy of the product that must remain active following feed processing, storage and transit through the acidic and enzymatic conditions of the gastrointestinal tract (Dersjant-Li et al. 2015). The magnitude of response for parameters such as growth performance, destruction of the target substrate or bone mineralisation in the case of phytase is influenced by the composition of the diet, available nutrients and energy, the age of the pig and various environmental conditions.

*Phytase*

Plant materials stores P as phytic acid (*myo*-inositol 1,2,3,4,5,6-hexakis [dihydrogen phosphate] [InsP$_6$]) and its salts, a form that is unavailable to pigs due to low levels of endogenous enzymatic activity. Phytase describes a family of phosphatase enzymes that catalyze the hydrolysis of phytic acid and its salts in a stepwise manner in the upper gastrointestinal tract to progressively smaller phosphated inositols and ultimately increase P availability to the pig. A review, that provides more detail on this topic, is provided by Dersjant-Li et al. (2015). The optimization of dietary P is an important objective in finisher pig diets to mitigate P excretion and the subsequent risk of leaching (Nelson et al. 2005; Liu et al. 2012; Bai et al. 2014). Furthermore, the future supply of economically extractable P is finite, and although there is uncertainty of when this source will run out, the appropriate usage of dietary P will continue to be a priority for nutritionists (Cordell and White 2011). Phytase has been implemented successfully in pig diets across the growth stages with important improvements seen in bone composition, bone strength and growth performance. Phytase supplementation in suboptimal P diets allows improvements in P excretion while maintaining growth performance (Varley et al. 2010; Vigors et al. 2014). New directions for phytase include the promising evidence for super-dosing phytase to achieve higher levels of phytate destruction and boost performance further (Lu et al. 2019; Moran et al. 2017) and as a tool to mitigate post-wean health challenges. Weaner pigs consume and excrete comparatively far less P, but there are important reasons to optimise P nutrition in this phase also as high levels can contribute to scouring (Varley et al. 2011a). Varley et al. (2011b) reported an improvement in feed conversion ratio and Ca and P utilisation during the weaner phase when phytase was added to diets containing suboptimal P, while Moran et al. (2017) reported an improvement in fecal score with the addition of phytase.

*Carbohydrases and Proteases*

While enhancing the availability of dietary protein has been a long-term preoccupation for nutritionists, commercially proteases have really only gained a foothold as a single enzymatic additive in pig diets in the 15 years (Cowieson and Roos 2016). Carbohydrases and proteases signify important advances in trying to optimize the carbohydrates and proteins of the diet more recalcitrant to digestion, which may therefore be fermented in the large intestine and can be problematic for the reasons discussed earlier. Duarte et al. (2019) recently demonstrated a beneficial response to xylanase and protease both singly and when combined in newly weaned pigs.

## Interaction Between Enzymes

The potential to improve the availability of a range of nutrients simultaneously is attractive to maximize the use of diet ingredients, particularly where co and by-products are used. There is interest on the potential additivity or interactions between several enzymes that target different substrates. Evidence for reduced phytase activity in the presence of endogenous proteases (Kumar et al. 2003; Morales et al. 2011) has spurred research into the interactions between phytase and commercial proteases (Dersjant-Li et al. 2015). The potential for a commercial protease and a carbohydrase blend of xylanases and $\beta$-glucanases to interact with phytase was investigated in finisher pigs offered diets based on wheat distillers grains and rapeseed meal (Torres-Pitarch et al. 2018). Those authors reported that phytase maintained growth performance, carcass composition, and bone mineral density in a low P diet regardless of the addition of proteases or carbohydrases. In contrast, O'Shea et al. (2014b) reported a negative impact on appetite and bodyweight gain in finisher pigs following supplementation of both protease and xylanase to a similar by-product-based diet. While there is an abundance of literature evaluating the response and mode of action of single enzymes, there is much to be done in the future to understand the complex interactions that may be induced when enzymes are combined.

### *Flavors*

Stimulating an appropriate appetite is an important consideration across all phases of production but is a particularly important priority for the lactating sow and the newly weaned piglet due to inappetence that can occur at these times. Pigs have highly developed olfactory and taste senses (Hellekant and Danilova 1999; Brunjes et al. 2016), and the organoleptic properties of the feed are a critical factor. Preference for, or aversion to a feed, is a complex process that can lead to feed refusal or diminished appetite if dietary ingredients with strong odors or flavors are present (Mawson et al. 1993; Gaultier et al. 2011; Michiels et al. 2012). The purpose of adding flavors to diets is to enhance palatability or mask off-flavors or odors in feed. While the control of appetite and hence bodyweight gain is an important objective for gestating sows, a common challenge reported in newly farrowed sows is inappetence, which has origins in modern genotypes and environmental factors (Eissen et al. 2000). Stimulating appetite through the addition of flavors such as anise, but not butyrate (Wang et al. 2014) and raspberry and vanilla (Silva et al. 2018) has been reported as increased feed consumption in sows with subsequent benefits for milk production and the growth of piglets. There is also a role to play in using supplementation of the sow's diet with a flavor to stimulate the feeding behavior of the progeny through sensory conditioning (Figueroa et al. 2013; Blavi et al. 2016). The abrupt, premature transition from liquid milk to solid feed leads to a well-described reduction in appetite in weaned piglets (Dong and Pluske 2007). This can be alleviated through the addition of flavors that may simulate some of the characteristics of sow milk (Araújo et al. 2010) or simply a flavor that enhances the palatability and appeal of the weaning diet (Sterk et al. 2008). Millet et al. (2008) found no benefit of providing a flavored creep feed to suckling piglets on subsequent weaning performance and similarly Sulabo et al. (2010) found no benefit in growth response to supplementing either creep diet or weaner diet with a flavor.

Assuming enhanced intake is the solitary mode of action is probably not accurate for all the feed additives, which may be considered as flavors, due to other various properties such as antioxidant, anti-inflammatory, or antimicrobial capacity. For example, Mellencamp et al. (2009) showed an improvement in piglet performance following the supplementation of sow diets with oregano oil, which was attributed in part to enhanced feed intake during lactation. However, Tan et al. (2015) also reported improvements in piglet performance following supplementation of sows with oregano oil but only modest improvements in sow intake in the third week of lactation. Those authors attributed the improvements in part to changes in redox status. Amrik and Bilkei (2004) reported an improvement in sow mortality and increased farrowing rate following supplementation of pre-farrowing and lactation diets with oregano. There is encouraging evidence for flavors to be a beneficial additive to boost appetite, but studies have been variable reflecting the complex nature of sensory response, the wide array of flavors available, and the uncertainty around mode of action.

## Phytogenic Compounds

The term "phytogenic," meaning derived from plants, has appeared with greater frequency in literature concerned with pig nutrition in the past decade. The increased interest in phytogenic feed additives relates to the possible application of these compounds in place of in-feed antibiotics. The term phytogenic as it relates to pig nutrition refers to a broad panel of compounds, with multiple modes of action. Furthermore, it is relatively common to experimentally test blends of phytogenic compounds rather than single extracts, or to pair phytogenic compounds with other feed additives such as organic acids (Omonijo et al. 2018). This makes it difficult to provide an overview of efficacy in any meaningful way. Nonetheless, the subject has been reviewed in greater detail elsewhere and further reading is available (Windisch et al. 2008; Karásková et al. 2015; Yang et al. 2015). An excellent review on the shortcomings of current studies and the future direction of phytogenic compound research have been provided by Blanco-Penedo et al. (2017) and Omonijo et al. (2018). Phytogenic compounds related to pig nutrition, which have been reported in the literature, include essential oils and various botanical and herbal extracts. A wide array of factors such as harvesting conditions and extraction processes affect the concentration and properties of these extracts and may partially explain some of the variability in response seen when fed to pigs. Furthermore, researchers frequently combine phytogenic compounds or pair them with other additives that can mask the mechanistic link (Yang et al. 2019). Nonetheless, there are some persuasive data available, and it is clear from the volume of publications emerging in recent years that while there remains more research to be undertaken, phytogenic compounds are increasingly playing a role in diet formulations. The target phase for phytogenic compounds has predominately focused on the post-weaning period, but the impact on later growth phases has also been investigated with a view to improve other challenges related to pig production including gaseous emissions and meat quality. The mode of action, which has been reported from pig and in vitro experimentation, is varied and includes impacts on gut motility, peristalsis and relaxation (Magalhães et al. 1998; Zhai et al. 2018), immunomodulatory (Liu et al. 2014b), and antimicrobial effects (Si et al. 2006; de Nova et al. 2019; Yang et al. 2019).

The literature on phytogenic compounds and the response of growth performance in weaned pigs is persuasive. Turmeric, garlic, and capsicum offered individually at 10 ppm to weaned pigs challenged with pathogenic *E. coli* had improved diarrhea score, increased ileal villi height, and modified a range of immune variables (Liu et al. 2013, 2014b). In another study, Li et al. (2012) reported how pigs fed essential oils, thymol, and cinnamaldehyde had improved growth performance and diarrhea score during the weaner-grower phase along with changes to serum immunity variables. Those authors also reported a decrease in fecal *E coli* counts. Castillo et al. (2006) reported an increase in the abundance of *Lactobacillus* spp. relative to *Enterobacteria* spp. in weaned pigs offered an experimental diet containing carvacrol, cinnamaldehyde, and capsicum oleoresin extracts. A recent study showed that offering a phytogenic supplement containing essential oils from *Thymus vulgaris*, *Origanum vulgare*, and *Coriandrum* sp. and a plant extract of *Castanea sativa* to pigs challenged with *Lawsonia intracellularis,* an important intracellular pathogen reduced the fecal shedding of the microbe (Draskovic et al. 2018). In contrast to the improvements seen in these infectious challenge studies, Hagmüller et al. (2006) found no improvement of feeding pigs a thymol-rich phytogenic compound on measurements of hemolytic *E. coli* shedding following a challenge. A study reported by Ahmed et al. (2013) incorporating antibiotics as a positive control showed a beneficial response of resveratrol, but not an essential oil blend, on the growth performance of *Salmonella enterica* serovar Typhimurium-challenged piglets. While both phytogenic treatments modified the immune status and fecal microbiota, the antibiotic-treated group had the best growth performance.

The impact of phytogenic compounds on the composition of the enteric microbiota and implications for fermentation have stimulated research into the potential impact on growth performance and noxious gaseous emissions of grower-finisher pigs. Piglets offered a phytogenic supplement containing fenugreek, clove, and cinnamon had improved growth performance and a reduction in fecal ammonia nitrogen and hydrogen sulfide (Cho et al. 2006). Grower-finisher pigs offered diets containing an essential oil blend (primarily caraway and lemon), dried herbs, and spices (rosemary and thyme), and quillaja saponins had improved bodyweight gain and decreased ammonia, but not nitrous oxide or methane emissions (Bartoš et al. 2016).

The potential for essential oils to beneficially impact the sensory qualities and to modify the antioxidant capacity of meat has been summarized in a review by Zhai et al. (2018). Some improvements have been reported, but the results are generally variable; therefore, a consensus for phytogenic compounds as this pertains to essential oils is not evident at this stage.

*Polyphenols*

Polyphenols are secondary plant metabolites, and as the name implies contain multiple phenol structures and includes compound groups such as flavonoids, stilbenes, lignans, and phenolic acids. Dietary polyphenols induce various biological responses that are of interest to pig nutritionists including immunomodulatory and antioxidant properties. However, polyphenols are also characterized by low availability, rapid metabolism, and sometimes poor water solubility, and depending on the target tissue or system, may limit their application (Biasutto et al. 2014). Excellent general reviews on the definitions, structures, bioavailability, and biological response are available (Manach et al. 2004, 2005; El Gharras 2009). The utility and limitations of polyphenols for pigs and poultry have been reviewed by Mahfuz et al. (2021). The general consensus of those authors were for a role for polyphenols in pig production as feed additives with multiple properties relevant to pig health and metabolism.

*Mycotoxin Binders*

Fungi are abundant at varying concentrations in plant material used for animal feed and replicate rapidly in the field and during storage under appropriate conditions (Bryden 2012). Mycotoxins are metabolites generated as a consequence of fungal activity. Various studies show widespread levels of fungi and toxin contamination in excess of levels that are considered safe for animal health (González Pereyra et al. 2008; Pleadin et al. 2012). The aflatoxins, fumonisins, zearalenone, trichothecenes, and ochratoxin A are the most relevant found in feedstuffs (Di Gregorio et al. 2014). Mycotoxin contamination of primary ingredients and feed is a significant global problem because following consumption can induce various deleterious effects in pigs with particular phases of production such as dry sows and weaned piglets being more vulnerable (Bryden 2012). At a general level, fungal growth results in musty odors and increased dusting in feed, reducing palatability and appetite. While undoubtedly there is a critical role to play in the prevention of fungal growth and the concentration of mycotoxins in the feed production supply chain, there are feed additives available to mitigate the physiological impact on animals. There are a wide range of dietary mineral additives that have been studied for adsorbent capacity, including aluminosilicates, bentonites, zeolites, sepiolite, and diatomite, which have been reviewed in detail for their role in pig diets by Di Gregorio et al. (2014).

### Interactions Between Feed Additives

Feed additives tend to be developed in a singular manner, with the intention of being integrated into a complete but simplistic feed formulation with the purpose of inducing an added benefit for the target animal. In commercial practice, however, the reality is that several feed additives may be added for various reasons. It is difficult to predict the outcome of these multiplications. Several examples of these have been cited in this chapter on the possible additivity, synergy, or otherwise in response to delivering multiple additives in the diet.

### Summary

In summary, the past decade has seen a surge in the availability and subsequently research of feed additives that may have general or very specific functions in pig diets in an age and environmental specific manner. Under certain environmental conditions, impressive responses in growth rate or other important variables have been reported, which compare favorably with the improvements associated with in-feed antibiotics. Important research priorities for the future include understanding the potential additivity or interaction between various feed additives and characterizing the other purposes, for which both novel and well-defined feed additives may be utilized.

### References

Ahmed, S. T., M. E. Hossain, G. M. Kim, J. A. Hwang, H. Ji, and C. J. Yang. 2013. Effects of resveratrol and essential oils on growth performance, immunity, digestibility and fecal microbial shedding in challenged piglets. Asian-Australas. J. Anim. Sci. 26:683–690.

Amaretti, A., C. Gozzoli, M. Simone, S. Raimondi, L. Righini, V. Pérez-Brocal, R. García-López, A. Moya, and M. Rossi. 2019. Profiling of protein degraders in cultures of human gut microbiota. Front. Microbiol.10:2614.

Amrik, B., and G. Bilkei. 2004. Influence of farm application of oregano on performances of sows. Can. Vet. J. 45:674–677.

Araújo, W. A. G., A. S. Ferreira, D. Renaudeau, P. C. Brustolini, and B. A. N. Silva. 2010. Effects of diet protein source on the behavior of piglets after weaning. Livest. Sci. 132:35–40.

Awati, A., B. A. Williams, M. W. Bosch, W. J. J. Gerrits, and M. W. A. Verstegen. 2006. Effect of inclusion of fermentable carbohydrates in the diet on fermentation end-product profile in feces of weanling piglets. J. Anim. Sci. 84:2133–2140.

Ayuso, M., J. Michiels, S. Wuyts, H. Yan, J. Degroote, S. Lebeer, C. Le Bourgot, E. Apper, M. Majdeddin, N. Van Noten, C. Vanden Hole, S. Van Cruchten, M. Van Poucke, L. Peelman, and C. Van Ginneken. 2020. Short-chain fructo-oligosaccharides supplementation to suckling piglets: Assessment of pre- and post-weaning performance and gut health. PLOS One 15:e0233910.

Bai, Z. H., L. Ma, W. Qin, Q. Chen, O. Oenema, and F. S. Zhang. 2014. Changes in pig production in China and their effects on nitrogen and phosphorus use and losses. Environ. Sci. Technol. 48:12742–12749.

Barba-Vidal, E., S. M. Martín-Orúe, and L. Castillejos. 2019. Practical aspects of the use of probiotics in pig production: A review. Livest. Sci. 223:84–96.

Barrow, P. A., R. Fuller, and M. J. Newport. 1977. Changes in the microflora and physiology of the anterior intestinal tract of pigs weaned at 2 days, with special reference to the pathogenesis of diarrhea. Infect. Immun. 18:586.

Bartoš, P., A. Dolan, L. Smutný, M. Šístková, I. Celjak, M. Šoch, and Z. Havelka. 2016. Effects of phytogenic feed additives on growth performance and on ammonia and greenhouse gases emissions in growing-finishing pigs. Anim. Feed Sci. Technol. 212:143–148.

Biasutto, L., A. Mattarei, N. Sassi, M. Azzolini, M. Romio, C. Paradisi, and M. Zoratti. 2014. Improving the efficacy of plant polyphenols. Anticancer Agents Med. Chem. 14:1332–1342.

Blachier, F., F. Mariotti, J. F. Huneau, and D. Tome. 2007. Effects of amino acid-derived luminal metabolites on the colonic epithelium and physiopathological consequences. Amino Acids 33:547–652.

Blanco-Penedo, I., C. Fernández González, L. M. Tamminen, A. Sundrum, and U. Emanuelson. 2017. Priorities and future actions for an effective use of phytotherapy in livestock-outputs from an expert workshop. Front. Vet. Sci. 4:248.

Blavi, L., D. Solà-Oriol, J. J. Mallo, and J. F. Pérez. 2016. Anethol, cinnamaldehyde, and eugenol inclusion in feed affects postweaning performance and feeding behavior of piglets. J. Anim. Sci. 94:5262–5271.

Brandão Melo, A. D., A. C. D. F. de Oliveira, P. da Silva, J. B. O. Santos, R. de Morais, G. R. de Oliveira, B. Wernick, P. L. D. O. Carvalho, S. M. B. Artoni, and L. B. Costa. 2020. 6-phytase and/or endo-β-xylanase and -glucanase reduce weaner piglet´s diarrhea and improve bone parameters. Livest. Sci. 238:104034.

Brunjes, P. C., S. Feldman, and S. K. Osterberg. 2016. The pig olfactory brain: A primer. Chem. Senses 41:415–425.

Bryden, W. L. 2012. Mycotoxin contamination of the feed supply chain: Implications for animal productivity and feed security. Anim. Feed Sci. Technol. 173:134–158.

Bussy, F., L. G. Matthieu, H. Salmon, J. Delaval, M. Berri, and N. C. Pi. 2019. Immunomodulating effect of a seaweed extract from Ulva armoricana in pig: Specific IgG and total IgA in colostrum, milk, and blood. Vet. Anim. Sci. 7:100051.

Casey, P. G., G. E. Gardiner, G. Casey, B. Bradshaw, P. G. Lawlor, P. B. Lynch, F. C. Leonard, C. Stanton, R. P. Ross, G. F. Fitzgerald, and C. Hill. 2007. A five-strain probiotic combination reduces pathogen shedding and alleviates disease signs in pigs challenged with *Salmonella enterica* serovar Typhimurium. Appl. Environ. Microbiol. 73:1858–1863.

Castillo, M., S. M. Martin-Orue, M. Roca, E. G. Manzanilla, I. Badiola, J. F. Perez, and J. Gasa. 2006. The response of gastrointestinal microbiota to avilamycin, butyrate, and plant extracts in early-weaned pigs. J. Anim. Sci. 84:2725–2734.

Chen, L., Y. Xu, X. Chen, C. Fang, L. Zhao, and F. Chen. 2017. The maturing development of gut microbiota in commercial piglets during the weaning transition. Front. Microbiol. 8:1688.

Cho, J. H., Y. Chen, H. J. Kim, O. S. Kwon, K. S. Shon, I.-S. Kim, S. J. Kim, and A. Asamer. 2006. Effects of essential oils supplementation on growth performance, IgG concentration and fecal noxious gas concentration of weaned pigs. Asian-Australas. J. Anim. Sci. 19:80–85.

Cordell, D. and S. White. 2011. Peak phosphorus: Clarifying the key issues of a vigorous debate about long-term phosphorus security. Sustainability 3:2027–2049.

Cowieson, A. J., and F. F. Roos. 2016. Toward optimal value creation through the application of exogenous mono-component protease in the diets of non-ruminants. Anim. Feed Sci. Technol. 221:331–340.

Cranwell, P. D., D. E. Noakes, and K. J. Hill. 1976. Gastric secretion and fermentation in the suckling pig. Br. J. Nutr. 36:71–86.

Cromwell, G. L., M. D. Lindemann, H. J. Monegue, D. D. Hall, and D. E. Orr, Jr. 1998. Tribasic copper chloride and copper sulfate as copper sources for weanling pigs. J. Anim. Sci. 76:118–123.

Dersjant-Li, Y., A. Awati, H. Schulze, and G. Partridge. 2015. Phytase in non-ruminant animal nutrition: a critical review on phytase activities in the gastrointestinal tract and influencing factors. J. Sci. Food Agric. 95:878–896.

Diana, A., L. Boyle, F. Leonard, C. Carroll, E. Sheehan, D. Murphy, and E. Garcia Manzanilla. 2019. Removing prophylactic antibiotics from pig feed: How does it affect their performance and health? BMC Vet. Res. 15:67.

Diao, H., P. Zheng, B. Yu, J. He, X. B. Mao, J. Yu, and D. W. Chen. 2014. Effects of dietary supplementation with benzoic acid on intestinal morphological structure and microflora in weaned piglets. Livest. Sci. 167:249–256.

Difilippo, E., M. Bettonvil, R. Willems, S. Braber, J. Fink-Gremmels, P. V. Jeurink, M. H. C. Schoterman, H. Gruppen, and H. A. Schols. 2015. Oligosaccharides in urine, blood, and feces of piglets fed milk replacer containing galacto-oligosaccharides. j. Agric. Food Chem. 63:10862–72.

Di Gregorio, M. C., D. V. D. Neeff, A. V. Jager, C. H. Corassin, Á. C. D. P. Carão, R. D. Albuquerque, A. C. D. Azevedo, and C. A. F. Oliveira. 2014. Mineral adsorbents for prevention of mycotoxins in animal feeds. Toxin Rev. 33:125–135.

Dong, G. Z. and J. R. Pluske. 2007. The low feed intake in newly-weaned pigs: Problems and possible solutions. Asian-Australas. J. Anim. Sci. 20:440–452.

Dotsenko, G., A. S. Meyer, N. Canibe, A. Thygesen, M. K. Nielsen, and L. Lange. 2018. Enzymatic production of wheat and ryegrass derived xylooligosaccharides and evaluation of their in vitro effect on pig gut microbiota. Biomass Conv. Biorefinery 8:497–507.

Duarte, M. E., F. X. Zhou, W. M. Dutra, and S. W. Kim. 2019. Dietary supplementation of xylanase and protease on growth performance, digesta viscosity, nutrient digestibility, immune and oxidative stress status, and gut health of newly weaned pigs. Anim. Nutr. 5:351–358.

Draskovic, V., J. Bosnjak-Neumuller, M. Vasiljevic, B. Petrujkic, N. Aleksic, V. Kukolj, and Z. Stanimirovic. 2018. Influence of phytogenic feed additive on Lawsonia intracellularis infection in pigs. Prev. Vet. Med. 151:46–51.

EC. 2003. Regulation (EC) No 1831/2003 of the European Parliament and of the councel on additives for use in animal nutrition. Off. J. Eur. Union L268:29-43.151:46–51.

Eissen, J. J., E. Kanis, and B. Kemp. 2000. Sow factors affecting voluntary feed intake during lactation. Livest. Prod. Sci. 64:147–165.

El Gharras, H. 2009. Polyphenols: Food sources, properties and applications - a review. Int. J. Food Sci. Technol. 44:2512–2518.

Espinosa, C. D., R. S. Fry, J. L. Usry, and H. H. Stein. 2017. Copper hydroxychloride improves growth performance and reduces diarrhea frequency of weanling pigs fed a corn-soybean meal diet but does not change apparent total tract digestibility of energy and acid hydrolyzed ether extract. J. Anim. Sci. 95:5447–5454.

Evans, G.S., N. Flint, A. S. Somers, B. Eyden, and C.S. Potten. 1992. The development of a method for the preparation of rat intestinal epithelial cell primary cultures. J. Cell. Sci. 101:219–231.

Figueroa, J., D. Solà-Oriol, L. Vinokurovas, X. Manteca, and J. F. Pérez. 2013. Prenatal flavour exposure through maternal diets influences flavour preference in piglets before and after weaning. Anim. Feed Sci. Technol. 183:160–167.

Fouhse, J. M., R. T. Zijlstra, and B. P. Willing. 2016. The role of gut microbiota in the health and disease of pigs. Anim. Front. 6:30–36.

Gagne, F., J. Matte, G. Barnett, and C. Pomar. 2002. The effect of microbial phytase and feed restriction on protein, fat and ash deposition in growing-finishing pigs. Can. J. Anim. Sci. 82:551–558.

Gaultier, A., M. C. Meunier-Salaün, C. H. Malbert, and D. Val-Laillet. 2011. Flavour exposures after conditioned aversion or preference trigger different brain processes in anaesthetised pigs. Eur. J. Neurosci. 34:1500–1511.

Gebru, E., J. S. Lee, J. C. Son, S. Y. Yang, S. A. Shin, B. Kim, M. K. Kim, and S. C. Park. 2010. Effect of probiotic-, bacteriophage-, or organic acid-supplemented feeds or fermented soybean meal on the growth performance, acute-phase response, and bacterial shedding of grower pigs challenged with Salmonella enterica serotype Typhimurium1. J. Anim. Sci. 88:3880–3886.

Gibson, G. R., E. R. Beatty, X. Wang, and J. H. Cummings. 1995. Selective stimulation of bifidobacteria in the human colon by oligofructose and inulin. Gastroenterol. 108:975–982.

González Pereyra, M. L., C. M. Pereyra, M. L. Ramírez, C. A. R. Rosa, A. M. Dalcero, and L. R. Cavaglieri. 2008. Determination of mycobiota and mycotoxins in pig feed in central Argentina. Lett. Appl. Microbiol. 46:555–561.

Grilli, E., B. Tugnoli, C. J. Foerster, and A. Piva. 2016. Butyrate modulates inflammatory cytokines and tight junctions components along the gut of weaned pigs. J. Anim. Sci. 94(suppl_3):433–436.

Guilloteau, P., Martin, L., Eeckhaut, V., Ducatelle, R., Zabielski, R., and van Immerseel, F. 2010. From the gut to the peripheral tissues: the multiple effects of butyrate. Nutr. Res. 23:366–384.

Hagmuller W., M. Jugl-Chizzola, K. Zitterl-Eglseer, C. Gabler, J. Spergser, R. Chizzola, and F. Chlodwig. 2006. The use of Thymi Herba as feed additive (0.1%, 0.5%, 1.0%) in weanling piglets with assessment of the shedding of haemolysing E. coli and the detection of thymol in the blood plasma. Berl. Munch. Tierarztl. Wochenschr. 119:50–54.

Hansen, C. F., G. Sorensen, and M. Lyngbye. 2007. Reduced diet crude protein level, benzoic acid and inulin reduced ammonia, but failed to influence odour emission from finishing pigs. Livest. Sci. 109:228–231.

Heim, G., T. Sweeney, C. O'shea, D. Doyle, and J. O'doherty. 2015. Effect of maternal dietary supplementation of laminarin and fucoidan, independently or in combination, on pig growth performance and aspects of intestinal health. Anim. Feed Sci. Technol. 204:28–41.

Hellekant, G. and V. Danilova. 1999. Taste in domestic pig, Sus scrofa. J. Anim. Physiol. Anim. Nutr. 82:8–24.

Heo, J.-M., J.-C. Kim, C. F. Hansen, B. P. Mullan, D. J. Hampson, and J. R. Pluske. 2008. Effects of feeding low protein diets to piglets on plasma urea nitrogen, faecal ammonia nitrogen, the incidence of diarrhoea and performance after weaning. Arch. Anim. Nutr. 62:343–358.

Højberg, O., N. Canibe, H. D. Poulsen, M. S. Hedemann, and B. B. Jensen. 2005. Influence of dietary zinc oxide and copper sulfate on the gastrointestinal ecosystem in newly weaned piglets. Appl. Environ. Microbiol. 71:2267–2277.

Hosseindoust, A., S.-H. Lee, J. S. Kim, Y. H. Choi, H. S. Noh, J.-B. Lee, P. K. Jha, I. K. Kwon, and B. J. Chae. 2017. Dietary bacteriophages as an alternative for zinc oxide or organic acids to control diarrhoea and improve the performance of weanling piglets. Vet. Med. (Praha). 62:53–61.

Jha, R. and J. D. Berrocoso. 2015. Review: Dietary fiber utilization and its effects on physiological functions and gut health of swine. Animal 9:1441–1452.

Jongbloed, A. W. and P. A. Kemme. 1990. Effect of pelleting mixed feeds on phytase activity and the apparent absorbability of phosphorus and calcium in pigs. Anim. Feed Sci. Technol. 28:233–242.

Karásková, K., P. Suchý, and E. Straková. 2015. Current use of phytogenic feed additives in animal nutrition: A review. Czech J. Anim. Sci. 60:521–530.

Kenny, M., H. Smidt, E. Mengheri, and B. Miller. 2010. Probiotics – do they have a role in the pig industry? Animal 5:462–470.

Kim, K. H., S. L. Ingale, J. S. Kim, S. H. Lee, J. H. Lee, I. K. Kwon, and B. J. Chae. 2014. Bacteriophage and probiotics both enhance the performance of growing pigs but bacteriophage are more effective. Anim. Feed Sci. Technol. 196:88–95.

Kim, J. S., A. Hosseindoust, S. H. Lee, Y. H. Choi, M. J. Kim, J. H. Lee, I. K. Kwon, and B. J. Chae. 2017. Bacteriophage cocktail and multi-strain probiotics in the feed for weanling pigs: effects on intestine morphology and targeted intestinal coliforms and Clostridium. Animal 11:45–53.

Klotz, L.-O., K.-D. Kröncke, D. P. Buchczyk, and H. Sies. 2003. Role of copper, zinc, selenium and tellurium in the cellular defense against oxidative and nitrosative stress. J. Nutr. 133:1448S–1451S.

Kumar, V., A. Miasnikov, J. S. Sands, and P. H. Simmins. 2003. In vitro activities of three phytases under different pH and protease challenges. Manipulating Pig Production. Aust. Pig Sci. Assoc. Vitoria, Australia:164.

Lawlor, P. G., P. B. Lynch, P. J. Caffrey, J. J. O'Reilly, and M. K. O'Connell. 2005. Measurements of the acid-binding capacity of ingredients used in pig diets. Ir. Vet. J. 58:447–452.

Le, P. D., A. J. A. Aarnink, N. W. M. Ogink, P. M. Becker, and M. W. A. Verstegen. 2005. Odour from animal production facilities: its relationship to diet. Nutr. Res. Rev. 18:3–30.

Lekagul, A., V. Tangcharoensathien, and S. Yeung. 2019. Patterns of antibiotic use in global pig production: a systematic review. Vet. Anim. Sci. 7:100058.

Li, S. Y., Y. J. Ru, M. Liu, B. Xu, A. Péron, and X. G. Shi. 2012. The effect of essential oils on performance, immunity and gut microbial population in weaner pigs. Livest. Sci. 145:119–123.

Liao, S. F. and M. Nyachoti. 2017. Using probiotics to improve swine gut health and nutrient utilization. Anim. Nutr. 3:331–343.

Liu, F., P. Celi, J. J. Cottrell, S. S. Chauhan, B. J. Leury, and F. R. Dunshea. 2018. Effects of a short-term supranutritional selenium supplementation on redox balance, physiology and insulin-related metabolism in heat-stressed pigs. J. Anim. Physiol. Anim. Nutr. 102:276–285.

Liu, J., H. Aronsson, L. Bergström, and A. Sharpley. 2012. Phosphorus leaching from loamy sand and clay loam topsoils after application of pig slurry. SpringerPlus 1:53.

Liu, P., R. Pieper, J. Rieger, W. Vahjen, R. Davin, J. Plendl, W. Meyer, and J. Zentek. 2014a. Effect of dietary zinc oxide on morphological characteristics, mucin composition and gene expression in the colon of weaned piglets. PLOS One 9:e91091–e91.

Liu, Y., M. Song, T. M. Che, J. A. Almeida, J. J. Lee, D. Bravo, C. W. Maddox, and J. E. Pettigrew. 2013. Dietary plant extracts alleviate diarrhea and alter immune responses of weaned pigs experimentally infected with a pathogenic Escherichia coli. J. Anim. Sci. 91:5294–5306.

Liu, Y., M. Song, T. M. Che, D. Bravo, C. W. Maddox, and J. E. Pettigrew. 2014b. Effects of capsicum oleoresin, garlic botanical, and turmeric oleoresin on gene expression profile of ileal mucosa in weaned pigs. J. Anim. Sci. 92:3426–3440.

Liu, F., J. J. Cottrell, J. B. Furness, L. R. Rivera, F. W. Kelly, U. Wijesiriwardana, R. V. Pustovit, L. J. Fothergill, D. M. Bravo, P. Celi, B. J. Leury, N. K. Gabler, and F. R. Dunshea. 2016. Selenium and vitamin E together improve intestinal epithelial barrier function and alleviate oxidative stress in heat-stressed pigs. Exp. Physiol. 101:801–810.

Lu, H., A. J. Cowieson, J. W. Wilson, K. M. Ajuwon, and O. Adeola. 2019. Extra-phosphoric effects of super dosing phytase on growth performance of pigs is not solely due to release of myo-inositol. J. Anim. Sci. 97:3898–3906.

Lynch, M. B., C. J. O'Shea, T. Sweeney, J. J. Callan, and J. V. O'Doherty. 2008. Effect of crude protein concentration and sugar-beet pulp on nutrient digestibility, nitrogen excretion, intestinal fermentation and manure ammonia and odour emissions from finisher pigs. Animal 2:425–434.

Magalhães, P. J. C., D. N. Criddle, R. A. Tavares, E. M. Melo, T. L. Mota, and J. H. Leal-Cardoso. 1998. Intestinal myorelaxant and antispasmodic effects of the essential oil of Croton nepetaefolius and its constituents cineole, methyl-eugenol and terpineol. Phytother. Res. 12:172–177.

Mahfuz, S., Q. Shang, and X. Piao. 2021. Phenolic compounds as natural feed additives in poultry and swine diets: a review. J. Anim. Sci. Biotechnol. 12:48.

Manach, C., A. Scalbert, C. Morand, C. Rémésy, and L. Jiménez. 2004. Polyphenols: food sources and bioavailability. Am. J. Clin. Nutr. 79:727–747.

Manach, C., G. Williamson, C. Morand, A. Scalbert, and C. Rémésy. 2005. Bioavailability and bioefficacy of polyphenols in humans. I. Review of 97 bioavailability studies. Am. J. Clin. Nutr. 81:230S–242S.

Manzanilla, E. G., M. Nofrarías, M. Anguita, M. Castillo, J. F. Perez, S. M. Martín-Orúe, C. Kamel, and J. Gasa. 2006. Effects of butyrate, avilamycin, and a plant extract combination on the intestinal equilibrium of early-weaned pigs1. J. Anim. Sci. 84:2743–2751.

Marounek, M., T. Adamec, V. Skivanova, and N. I. Latsik. 2002. Nitrogen and in vitro fermentation of nitrogenous substrates in caecal contents of the pig. Acta Vet. Brno 71:429–433.

Martín, R., S. Delgado, A. Maldonado, E. Jiménez, M. Olivares, L. Fernández, O. J. Sobrino, and J. M. Rodríguez. 2009. Isolation of lactobacilli from sow milk and evaluation of their probiotic potential. J. Dairy Res. 76:418–425.

Mawson, R., R. K. Heaney, Z. Zdunczyk, and H. Kozlowska. 1993. Rapeseed meal-glucosinolates and their antinutritional effects. Part II. Flavour and palatability. Food Nahrung 37:336–344.

Medani, M., V. A. Bzik, A. Rogers, D. Collins, R. Kennelly, D. C. Winter, D. J. Brayden, and A. W. Baird. 2012. Zinc sulphate attenuates chloride secretion in human colonic mucosae in vitro. Eur. J. Pharmacol. 696:166–171.

Mellencamp, M., M. Evelsizer, R. Dvorak, J. Hedges, M. Motram, and D. Cadogan. 2009. Oregano essential oil in gestation and lactation diets improves sow and piglet performance. Proc. Allen D. Leman Swine Conference. p. 192.

Michiels, J., J. Missotten, A. Ovyn, N. Dierick, D. Fremaut, and S. De Smet. 2012. Effect of dose of thymol and supplemental flavours or camphor on palatability in a choice feeding study with piglets. Czech J. Anim. Sci. 57:65–74.

Millet, S., M. Aluwé, D. L. De Brabander, and M. J. van Oeckel. 2008. Effect of seven hours intermittent suckling and flavour recognition on piglet performance. Arch. Anim. Nutr. 62:1–9.

Montagne, L.,G. Boudry, C. Favier, I. L. Huërou-Luron, I. J.-P. Lallès, and B. Sève. 2007. Main intestinal markers associated with the changes in gut architecture and function in piglets after weaning. Br. J. Nutr. 97:45–57.

Morales, G., F. Moyano, and L. Márquez. 2011. in vitro assessment of the effects of phytate and phytase on nitrogen and phosphorus bioaccessibility within fish digestive tract. Anim. Feed Sci. Technol. 170:209–221.

Morales, J., G. Cordero, C. Piñeiro, and S. Durosoy. 2012. Zinc oxide at low supplementation level improves productive performance and health status of piglets. J. Anim. Sci. 90:436–438.

Moran, K., R. D. Boyd, C. Zier-Rush, P. Wilcock, N. Bajjalieh, and E. van Heugten. 2017. Effects of high inclusion of soybean meal and a phytase superdose on growth performance of weaned pigs housed under the rigors of commercial conditions. J. Anim. Sci. 95:5455–5465.

Mroz, Z. 2005. Organic acids as potential alternative to antibiotic growth promoters for pigs. Adv Pork Prod 16:169–182.

Murphy, D. P., J. V. O'Doherty, T. M. Boland, C. J. O'Shea, J. J. Callan, K. M. Pierce, and M. B. Lynch. 2011. The effect of benzoic acid concentration on nitrogen metabolism, manure ammonia and odour emissions in finishing pigs. Anim. Feed Sci. Technol. 163:194–199.

de Nova, P. J. G., A. Carvajal, M. Prieto, and P. Rubio. 2019. In vitro susceptibility and evaluation of techniques for understanding the mode of action of a promising non-antibiotic citrus fruit extract against several pathogens. Front. Microbiol. 10:884.

Nahm, K. H. 2003. Influences of fermentable carbohydrates on shifting nitrogen excretion and reducing ammonia emission of pigs. Crit. Rev. Environ. Sci. Technol. 33:165–186.

Nelson, N. O., J. E. Parsons, and R. L. Mikkelsen. 2005. Field-scale evaluation of phosphorus leaching in acid sandy soils receiving swine waste. J. Environ. Qual. 34:2024–2035.

NRC. 2012. Nutrient requirements of swine. 11th rev. ed. Natl. Acad. Press, Washington, DC.

Omonijo, F. A., L. Ni, J. Gong, Q. Wang, L. Lahaye, and C. Yang. 2018. Essential oils as alternatives to antibiotics in swine production. Anim. Nutr. 4:126–136.

O'Doherty, J., M. Bouwhuis, and T. Sweeney. 2017. Novel marine polysaccharides and maternal nutrition to stimulate gut health and performance in post-weaned pigs. Anim. Prod. Sci. 57:2376.

O'Shea, C. J., P. McAlpine, T. Sweeney, P. F. Varley, and J. V. O'Doherty. 2014a. Effect of the interaction of seaweed extracts containing laminarin and fucoidan with zinc oxide on the growth performance, digestibility and faecal characteristics of growing piglets. Br. J. Nutr. 111:798–807.

O'Shea, C. J., P. O. Mc Alpine, P. Solan, T. Curran, P. F. Varley, A. M. Walsh, and J. V. O. Doherty. 2014b. The effect of protease and xylanase enzymes on growth performance, nutrient digestibility, and manure odour in grower–finisher pigs. Anim. Feed Sci. Technol. 1890:88–97.

Pan, L., P. F. Zhao, X. K. Ma, Q. H. Shang, Y. T. Xu, S. F. Long, Y. Wu, F. M. Yuan, and X. S. Piao. 2017. Probiotic supplementation protects weaned pigs against enterotoxigenic Escherichia coli K88 challenge and improves performance similar to antibiotics. J. Anim. Sci. 95:2627–2639.

Partanen, K. H., and Z. Mroz. 1999. Organic acids for performance enhancement in pig diets. Nutr. Res. Rev. 12:117–145.

Pei, X., Z. Xiao, L. Liu, G. Wang, W. Tao, M. Wang, J. Zou, and D. Leng. 2019. Effects of dietary zinc oxide nanoparticles supplementation on growth performance, zinc status, intestinal morphology, microflora population, and immune response in weaned pigs. J. Sci. Food Agric. 99:1366–1374.

Pérez, V. G., A. M. Waguespack, T. D. Bidner, L. L. Southern, T. M. Fakler, T. L. Ward, M. Steidinger, and J. E. Pettigrew. 2011. Additivity of effects from dietary copper and zinc on growth performance and fecal microbiota of pigs after weaning. J. Anim. Sci. 89:4140425.

Perez-Palencia, J. Y., R. S. Samuel, and C. L. Levesque. 2021. Supplementation of protease to low amino acid diets containing superdose level of phytase for wean-to-finish pigs: effects on performance, postweaning intestinal health and carcass characteristics. Transl. Anim. Sci. 5(2):txab088.

Pettigrew, J. E. 2006. Reduced use of antibiotic growth promoters in diets fed to weanling pigs: Dietary tools, part 1. Anim. Biotechnol. 17:207–215.

Pié, S., A. Awati, S. Vida, I. Falluel, B. A. Williams, and I. P. Oswald.2007. Effects of added fermentable carbohydrates in the diet on intestinal proinflammatory cytokine-specific mrna content in weaning piglets. J. Anim. Sci. 85:673–83.

Pleadin, J., M. Zadravec, N. Perši, A. Vulić, V. Jaki, and M. Mitak. 2012. Mould and mycotoxin contamination of pig feed in northwest Croatia. Mycotoxin Res. 28:157–162.

Poulsen, H. D. 1995. Zinc oxide for weanling piglets. Acta Agric. Scand. Sec. A Anim. Sci. 45:159–167.

Ravindran, V., and E. T. Kornegay. 1993. Acidification of weaner pig diets: A review. J. Sci. Food Agric. 62:313–322.

Salminen, S., A. von Wright, L. Morelli, P. Marteau, D. Brassart, W. M. de Vos, R. Fondén, M. Saxelin, K. Collins, G. Mogensen, S.-E. Birkeland, and T. Mattila-Sandholm. 1998. Demonstration of safety of probiotics — a review. Int. J. Food Microbiol. 44:93–106.

Si, J., L. Feng, J. Gao, Y. Huang, G. Zhang, J. Mo, S. Zhu, W. Qi, J. Liang, and G. Lan. 2020. Evaluating the association between feed efficiency and the fecal microbiota of early-life Duroc pigs using 16S rRNA sequencing. AMB Express 10:115–115.

Si, W., J. Gong, C. Chanas, S. Cui, H. Yu, C. Caballero, and R. M. Friendship. 2006. In vitro assessment of antimicrobial activity of carvacrol, thymol and cinnamaldehyde towards Salmonella serotype Typhimurium DT104: effects of pig diets and emulsification in hydrocolloids. J. Appl. Microbiol. 101:1282–1291.

Silva, B. A. N., R. L. S. Tolentino, S. Eskinazi, D. V. Jacob, F. S. S. Raidan, T. V. Albuquerque, N. C. Oliveira, G. G. A. Araujo, K. F. Silva, and P. F. Alcici. 2018. Evaluation of feed flavor supplementation on the performance of lactating high-prolific sows in a tropical humid climate. Anim. Feed Sci. Technol. 236:141–148.

Simons, P. C. M., H. A. J. Versteegh, A. W. Jongbloed, P. A. Kemme, P. Slump, K. D. Bos, M. G. E. Wolters, R. F. Beudeker, and G. J. Verschoor. 1990. Improvement of phosphorus availability by microbial phytase in broilers and pigs. Br. J. Nutr. 64:525–540.

Smith, E. A. and G. T. Macfarlane. 1998. Enumeration of amino acid fermenting bacteria in the human large intestine: effects of pH and starch on peptide metabolism and dissimilation of amino acids. FEMS Microbiol. Ecol. 25:355–368.

Snoeck, V., E. Cox, F. Verdonck, J. J. Joensuu, and B. M. Goddeeris. 2004. Influence of porcine intestinal pH and gastric digestion on antigenicity of F4 fimbriae for oral immunisation. Vet. Microbiol. 98(1):45–53.

Starke, I. C., R. Pieper, K. Neumann, J. Zentek, and W. Vahjen. (2014). The impact of high dietary zinc oxide on the development of the intestinal microbiota in weaned piglets. FEMS Microbiol. Ecol. 87:416–27.

Sterk, A., P. Schlegel, A. J. Mul, M. Ubbink-Blanksma, and E. M. A. M. Bruininx. 2008. Effects of sweeteners on individual feed intake characteristics and performance in group-housed weanling pigs1. J. Anim. Sci. 86:2990–2997.

Sulabo, R., M. Tokach, J. M. Derouchey, S. Dritz, R. Goodband, and J. Nelssen. 2010. Influence of feed flavors and nursery diet complexity on preweaning and nursery pig performance. J. Anim. Sci. 88:3918–3926.

Sutton, A. L., K. B. Kephart, M. W. Verstegen, T. T. Canh, and P. J. Hobbs. 1999. Potential for reduction of odorous compounds in swine manure through diet modification. J. Anim. Sci. 77:430–439.

Tan, C., H.-K. Wei, H. Sun, J. Ao, G. Long, S. Jiang, and J. Peng. 2015. Effects of dietary supplementation of oregano essential oil to sows on oxidative stress status, lactation feed intake of sows, and piglet performance. BioMed Res. Int. 2015:1–9.

Torrallardona, D., I. Badiola, and J. Broz. 2007. Effects of benzoic acid on performance and ecology of gastrointestinal microbiota in weanling piglets. Livest. Sci. 108:210–213.

Torres-Pitarch, A., U. M. McCormack, V. E. Beattie, E. Magowan, G. E. Gardiner, A. M. Pérez-Vendrell, D. Torrallardona, J. V. O'Doherty, and P. G. Lawlor. 2018. Effect of phytase, carbohydrase, and protease addition to a wheat distillers dried grains with solubles and rapeseed based diet on in vitro ileal digestibility, growth, and bone mineral density of grower-finisher pigs. Livest. Sci. 216:94–99.

Tóth, G., T. Hermann, M. R. Da Silva, and L. Montanarella. (2016). Heavy metals in agricultural soils of the european union with implications for food safety. Enviorn. Int. 88:299–309.

Trckova, M., A. Lorencova, K. Hazova, and Z. Sramkova Zajacova. 2015. Prophylaxis of post-weaning diarrhoea in piglets by zinc oxide and sodium humate. Vet. Med. 60:351–360.

van Heugten, E., J. W. Spears, E. B. Kegley, J. D. Ward, and M. A. Qureshiet. 2003. Effects of organic forms of zinc on growth performance, tissue zinc distribution, and immune response of weanling pigs. J. Anim. Sci. 82:2063–2071.

Van Kuijk, S. J. A., M. A. Fleuren, A. P. J. Balemans, and Y. Han. 2019. Weaned piglets prefer feed with hydroxychloride trace minerals to feed with sulfate minerals. Trans. Anima. Sci. 3:709–716.

Varley, P. F., J. J. Callan, and J. O'Doherty. 2010. Effect of phosphorus level and phytase inclusion on the performance, bone mineral concentration, apparent nutrient digestibility, and on mineral and nitrogen utilisation in finisher pigs. Ir. J. Agric. Food Res. 49:141–152.

Varley, P. F., B. Flynn, J. J. Callan, and J. V. O'Doherty. 2011a. Effect of phytase level in a low phosphorus diet on performance and bone development in weaner pigs and the subsequent effect on finisher pig bone development. Livest. Sci. 138:152–158.

Varley, P. F., J. J. Callan, and J. V. O'Doherty. 2011b. Effect of dietary phosphorus and calcium level and phytase addition on performance, bone parameters, apparent nutrient digestibility, mineral and nitrogen utilization of weaner pigs and the subsequent effect on finisher pig bone parameters. Anim. Feed Sci. Technol. 165:201–209.

Vigors, S., T. Sweeney, C. J. O'Shea, J. A. Browne, and J. V. O'Doherty. 2014. Improvements in growth performance, bone mineral status and nutrient digestibility in pigs following the dietary inclusion of phytase are accompanied by modifications in intestinal nutrient transporter gene expression. Br. J. Nutr. 112:688–697.

Vigors, S., J. O'Doherty, and T. Sweeney. 2020. Colonic microbiome profiles for improved feed efficiency can be identified despite major effects of farm of origin and contemporary group in pigs. Animal 14:1–9.

Vince, A. J., and S. M. Burridge. 1980. Ammonia production by intestinal bacteria: The effects of lactose, lactulose and glucose. J. Med. Microbiol. 13:177–191.

Vo, N., T. C. Tsai, C. Maxwell, and F. Carbonero. 2017. Early exposure to agricultural soil accelerates the maturation of the early-life pig gut microbiota. Anaerobe 45:31–39.

Walsh, M. C., M. H. Rostagno, G. E. Gardiner, A. L. Sutton, B. T. Richert, and J. S. Radcliffe. 2012a. Controlling Salmonella infection in weanling pigs through water delivery of direct-fed microbials or organic acids. Part I: Effects on growth performance, microbial populations, and immune status. J. Anim. Sci. 90:261–271.

Walsh, M. C., M. H. Rostagno, G. E. Gardiner, A. L. Sutton, B. T. Richert, and J. S. Radcliffe. 2012b. Controlling Salmonella infection in weanling pigs through water delivery of direct-fed microbials or organic acids: Part II. Effects on intestinal histology and active nutrient transport1. J. Anim. Sci. 90:2599–2608.

Wang, Y., J. W. Tang, W. Q. Ma, J. Feng, and J. Feng. 2010. Dietary zinc glycine chelate on growth performance, tissue mineral concentrations, and serum enzyme activity in weanling piglets. Biol. Trace Elem. Res. 133:325–334.

Wang, J., M. Yang, S. Xu, Y. Lin, L. Che, Z. Fang, and D. Wu. 2014. Comparative effects of sodium butyrate and flavors on feed intake of lactating sows and growth performance of piglets. Anim. Sci. J. 85:683–689.

Wang, X., T. Tsai, F. Deng, X. Wei, J. Chai, J. Knapp, J. Apple, C. V. Maxwell, J. A. Lee, Y. Li, and J. Zhao. 2019. Longitudinal investigation of the swine gut microbiome from birth to market reveals stage and growth performance associated bacteria. Microbiome 7:109.

Williams, B. A., D. Mikkelsen, B. M. Flanagan, and M. J. Gidley. 2019. "Dietary fibre": moving beyond the "soluble/insoluble" classification for monogastric nutrition, with an emphasis on humans and pigs. J. Anim. Sci. Biotechnol. 10:45.

Windisch, W., K. Schedle, C. Plitzner, and A. Kroismayr. 2008. Use of phytogenic products as feed additives for swine and poultry1. J. Anim. Sci. 86(Suppl. 14):E140–E148.

Wright, A. V. 2005. Regulating the safety of probiotics - the European approach. Curr. Pharm. Des. 11:17–23.

Wyszkowska, J., A. Borowik, M. Kucharski, and J. Kucharski. 2013. Effect of cadmium, copper and zinc on plants, soil microorganisms and soil enzymes. J. Elementol. 18:769–796.

Xie, Z. J., Y. M. Zhu, D. MeiDan, and X. Y. Han. 2010. Effects of chitosan-zinc on growth performance, serum hormone and biochemical indices of weanling piglets. Chinese J. Anim. Nutr. 22:1355–1360.

Yang, C., M. A. K. Chowdhury, Y. Huo, and J. Gong. 2015. Phytogenic compounds as alternatives to in-feed antibiotics: Potentials and challenges in application. Pathogens 4:137–156.

Yang, C., L. Zhang, G. Cao, J. Feng, M. Yue, Y. Xu, B. Dai, Q. Han, and X. Guo. 2019. Effects of dietary supplementation with essential oils and organic acids on the growth performance, immune system, fecal volatile fatty acids, and microflora community in weaned piglets. J. Anim. Sci. 97:133–143.

Yazdankhah, S., K. Rudi, and A. Bernhoft. 2014. Zinc and copper in animal feed – development of resistance and co-resistance to antimicrobial agents in bacteria of animal origin. Microb. Ecol. Health Dis. 25:25862.

Zimmermann, B., E. Bauer, and R. Mosenthin. 2001. Pro- and prebiotics in pig nutrition: Potential modulators of gut health? J. Anim. Feed Sci. 10:47–56.

Zhai, H., W. Ren, S. Wang, J. Wu, P. Guggenbuhl, and A.-M. Kluenter. 2017. Growth performance of nursery and grower-finisher pigs fed diets supplemented with benzoic acid. Anim. Nutr. 3:232–235.

Zhai, H., H. Liu, S. Wang, J. Wu, and A.-M. Kluenter. 2018. Potential of essential oils for poultry and pigs. Anim. Nutr. 4:179–196.

Zhang, J., Z. Li, Z. Cao, L. Wang, X. Li, S. Li, and Y. Xu. 2015. Bacteriophages as antimicrobial agents against major pathogens in swine: A review. J. Anim. Sci. Biotechnol. 6:39.

Zhang, Y., K. Yu, H. Chen, Y. Su, and W. Zhu. 2018. Caecal infusion of the short-chain fatty acid propionate affects the microbiota and expression of inflammatory cytokines in the colon in a fistula pig model. Microbiol. Biotechnol. 11:859–868.

# 19 Digestibility and Availability of Nutrients in Feed Ingredients

Su A. Lee and Hans H. Stein

## Introduction

To formulate cost-effective diets, nutritional evaluation of feed ingredients is an essential part. Most nutrients used in feed formulation and for requirement estimates need to be expressed as digestible rather than total or available contents (NRC 2012). Therefore, determination of nutritional value in feed ingredients and nutrient requirements based on digestible nutrients are required for an accurate formulation of swine diets. Because nutrients have different digestibility values in different ingredients, it is necessary that the digestibility of each nutrient in each ingredient is determined. However, to formulate diets accurately, digestibility values for nutrients in different feed ingredients must be expressed in such a way that these values are additive in mixed diets. Because digestibility can be expressed in different ways, and it is, therefore, necessary that only values that are expressed in such a way that they are additive in a mixed diet are used in diet formulation (Stein et al. 2005; Xue et al. 2014; She et al. 2018b).

The objective of the present contribution is to provide clarity for determination of values for digestibility by discussing different methods, terminologies, and specific considerations. Digestibility and availability of amino acids (AA), carbohydrates, lipids, minerals, and vitamins will be reviewed.

## Bioavailability of Nutrients

Bioavailability is defined as "the degree to which an ingested nutrient in a particular source is absorbed in a form that can be utilized in metabolism by the animal" (Ammerman et al. 1995). It is difficult to directly measure values for the bioavailability of nutrients, but there are a number of methods that allow for indirect estimates for bioavailability.

Historically, bioavailability has been estimated by calculating the efficiency of utilization of nutrients (Batterham 1974), the biological value (Mitchell 1924), the relative bioavailability using the slope-ratio method (Finney 1952; Littell et al. 1997), or the digestibility of the nutrient (Dietrich and Grindley 1914).

The efficiency of utilization is calculated in a way that is similar to calculating feed efficiency. As an example, the utilization of crystalline Lys was calculated by dividing differences in gain between pigs fed diets with no crystalline Lys or with crystalline Lys by differences in feed intake of pigs (Batterham 1974).

The term "biological value" of nutrients is defined as a percentage retention of the absorbed nutrients (absorbed nutrient – nutrient urine excretion). The biological value has been considered important to estimate utilization of nutrients including AA, Ca, and P after absorption (Hart et al. 1909; Mitchell 1924).

*Sustainable Swine Nutrition*, Second Edition. Edited by Lee I. Chiba.
© 2023 John Wiley & Sons Ltd. Published 2023 by John Wiley & Sons Ltd.

The relative bioavailability is calculated using the slope-ratio method as the ratio of the slope of the response to the test ingredient to the slope of the response to a standard ingredient (Littell et al. 1997). Response criteria may include whole-body protein deposition (Batterham 1992), AA oxidation (Moehn et al. 2005), growth performance (Chung and Baker 1992b), bone characteristics (Ross et al. 1984), or urine excretion or retention (Cho et al. 1980; Opapeju et al. 2012). There are three assumptions for the slope-ratio assay that are essential for obtaining meaningful estimates (Littell et al. 1997): (i) the response to nutrients is linear (i.e., nutrients need to be provided below the requirement), (ii) the y-intercepts for regression lines are the same ("common intercept"), and (iii) the response to the basal diet is equal to the common intercept.

For practical feed formulation, however, bioavailability values for macronutrients that are estimated by the efficiency, biological value, or relative bioavailability procedures are not recommended because these values are not additive in a mixed diet and the procedures used to estimate these values are time consuming and tedious. Instead, the digestibility of macronutrients has been measured and used as an indication of the quantities of nutrients that are available to the animal. However, for many macronutrients, specifically vitamins, digestibility cannot be determined and relative bioavailability is often used.

### Use of Digestibility

#### *Direct Method vs. Difference Method*

The direct method is the simplest and easiest method to use if digestibility of nutrients is determined in feed ingredients. Test feed ingredients are included as the sole source of a nutrient in a test diet. Therefore, the digestibility of the nutrient in the test diet containing the test feed ingredient is considered the digestibility of the nutrient in the test feed ingredient. Values for the digestibility of nutrients are calculated from the following equation:

$$\text{Digestibility}(\%) = \left[\left(\text{Intake} - \text{Output}\right) / \text{Intake}\right] \times 100.$$

The difference method is used to determine digestibility of nutrients or energy in feed ingredients that may induce problems if fed as the only source of the nutrient. This may be because of low palatability of the ingredient, low concentration of the nutrient in the ingredient, or a high concentration of antinutritional factors. In this case, a basal diet and a test diet containing both a basal and a test ingredient are formulated (Fan and Sauer 1995a,b). A basal diet can contain one or more ingredients that provide nutrients of interest in the diet, but if more than one ingredient in the basal diet contains the test nutrient, the ratio between these ingredients needs to be constant in the basal diet and in all test diets. Values for the digestibility of nutrients are then calculated by difference using the following equation (adopted from Kong and Adeola 2014):

$$D_{\text{Ing}}(\%) = \left[D_{\text{Test}} - \left(D_{\text{Basal}} \times P_{\text{Basal}}\right)\right] / P_{\text{Ing}},$$

where $D_{\text{Ing}}$ is the digestibility of the test nutrient in the test ingredient, $D_{\text{Test}}$ is the digestibility of the test nutrient in the test diet, and $D_{\text{Basal}}$ is the digestibility of the test nutrient in the basal diet. $P_{\text{Basal}}$ is the proportional contribution of the test nutrient from the basal diet to the test diet and $P_{\text{Ing}}$ is the proportional contribution of the test nutrient from the test ingredient to the test diet. $P_{\text{Basal}}$ and $P_{\text{Ing}}$ are expressed as coefficients and the sum of the proportional contributions must be 1.00.

The proportional contributions can be calculated from the following equation (adopted from Kong and Adeola 2014):

$$\text{Proportional contribution} = [(\text{Nutrient in test ingredient} \times \text{Inclusion rate}) / 100] / \text{Nutrient in test diet},$$

where nutrient and inclusion rates are expressed on a percentage basis.

A prerequisite for getting accurate results with the difference procedure is that analyzed nutrients in the basal diet and in the test ingredient add up to the analyzed concentrations in the test diet. If that is not the case, accurate results cannot be calculated with this procedure. Additivity of nutrients in diets is the most important assumption for using the difference procedure. However, errors from diet formulation, subsampling of feed, analysis inaccuracy, and variation in nutrients among different sources of feed ingredient do not always allow for additivity of analyzed nutrients in complete diets. Specifically, this has been a problem for Ca in complete diets fed to pigs and broilers (Walk 2016; Wu et al. 2018). Therefore, calculation of digestibility values based on calculated nutrients rather than analyzed nutrients in test diets is sometimes used to get more accurate results (Lee et al. 2019b). Another consideration when using the difference procedure is that the basal diet should not have lower digestibility of the test nutrient than the test ingredient, and in general, the greater the contribution of the test ingredient to the nutrients in the test diet is, the more accurate results will be calculated. Dry matter (DM) should sometimes be used in calculations if the test ingredient and the basal diet contain different amounts of moisture (Kong and Adeola 2014). The greater standard error of the mean associated with the use of the indirect procedure compared with the direct procedure will require more replications to maintain a certain power of an experiment (Oliveira et al. 2020b).

The regression procedure is also based on the difference procedure, but with several diets containing the test ingredient. In this procedure, test ingredients are included in diets by replacing ingredients in the basal diet with multiple inclusion rates (Park and Adeola 2020; Wang et al. 2020). Differences in digestibility among diets are subsequently extrapolated to 100% replacement (Adeola 2001).

Because more than 1 diet is needed to determine digestibility values in a test ingredient if the difference procedure is used, the direct procedure is preferred because it is easier to use only one diet and because fewer replications are needed to get a certain power of the experiment. However, feeding test diets that contain nutrients below the requirement to pigs for a long period of time have been thought to impact digestibility values, but recent data demonstrated that is not the case (Bolarinwa and Adeola 2016; Jaworski et al. 2016; Zhao et al. 2017; Oliveira et al. 2020b). There may, however, be feed ingredients with very high concentration of fiber or with very low palatability that requires use of the difference procedure (Sulabo et al. 2013; Almeida et al. 2014).

### Total Collection vs. Partial Collection (Index Method)

Fecal or digesta samples can be quantitatively collected (Adeola 2001). Ingestible markers that have specific colors (i.e., chromic oxide, indigo carmine, and ferric oxide) are used to distinguish initiation and termination of collections.

Fecal or digesta samples can also be partially collected using an index or an indicator (Adeola 2001). The partial collection is used when complete quantitative collection of ileal digesta or fecal samples is not possible. Requisites for the index include easy to analyze, not absorbed from the intestinal tract, nonessential, not toxic to the animals, and uniform in diets, digesta, and feces (Adeola 2001). Acid insoluble ash (provided from diatomaceous earth), chromium (provided as $Cr_2O_3$), and titanium (provided as $TiO_2$) are the three indexes that are most frequently used. Values for the digestibility of nutrients are then calculated from the following equation (Stein et al. 2007):

$$\text{Digestibility}(\%) = [1 - (\text{Nutrient}_{\text{Output}} / \text{Nutrient}_{\text{Input}}) \times (\text{Index}_{\text{Input}} / \text{Index}_{\text{Output}})] \times 100,$$

where concentrations of nutrients and index are in %.

The concentration of the indicators needs to be stabilized in the digesta or fecal samples before collection is initiated, and an adaptation period of three to seven days is needed for the indexes to be stabilized in fecal output (Clawson et al. 1955; Adeola 2001; Jang et al. 2014). However, in ileal digestibility experiments, an adaptation period of three to four days is adequate (Kim et al. 2020). In experiments to determine total tract digestibility, 5–10 days of adaptation is usually used, but the required time for adaptation appears to depend on the fiber concentrations of the diet (Choi and Kim 2019). One advantage of using the index method is that feed intake is not required in the calculation of digestibility and pigs can, therefore, be fed on an ad libitum basis.

**Amino Acid Digestibility**

The term "AA digestibility" refers to the digestion and absorption of AA into the body. Dietary protein or peptides are degraded into free AA or small peptides throughout the intestinal tract by enzymatic hydrolysis and microbial fermentation. However, free AA and small peptides can only be absorbed before the end of the small intestine of pigs, which means that there is no absorption of AA in the hindgut of pigs, whereas the unabsorbed AA and peptides are fermented by bacteria in the hindgut. Therefore, total tract digestibility of AA does not reflect actual AA absorption. For this reason, ileal digestibility is the most accurate estimation of AA digestibility to avoid the interference of microbial modification in the hindgut (Sauer and Ozimek 1986).

Ileal fluids from pigs can be collected at the end of the small intestine using techniques that have been reviewed and discussed (Gabert et al. 2001; Moughan 2003). However, surgical installation of a T-shaped cannula in the distal ileum of pigs (i.e., 10–15 cm prior to the ileocecal valve) is the procedure of choice in most countries in the world. This procedure has proven to minimize trial-to-trial variation and thus provides accurate output. The standard procedure for installation of the cannula and ileal digesta collection has been explained (Stein et al. 1998). The T-cannula only allows for partial collection of ileal digesta and thus requires the use of an index. Therefore, the ileal digestibility of AA is calculated with concentrations of AA and index in diet and ileal digesta (Stein et al. 1998).

*Endogenous Losses of AA*

The ileal digesta contain undigested dietary AA along with endogenous AA that are secreted into the intestinal tract of pigs. The endogenous proteins mainly originate from mucoproteins, epithelial cells, digestive enzymes including gastric secretions, pancreatic juice, and bile acids, and serum albumin in forms of peptides, free AA, amines, and urea (Low and Zebrowska 1989; Moughan and Schuttert 1991; Tamminga et al. 1995). Sixty percent of total endogenous losses are from the intestinal secretions that consist of sloughed cells, mucin, and glycoconjugates secreted by the enterocytes and bile acids (Low and Zebrowska 1989; Lien et al. 1997). Only 8–10% of the total endogenous losses are from saliva and gastric, pancreatic, and bile secretions. However, 70–80% of the endogenous proteins that are secreted into the intestinal tract are hydrolyzed and reabsorbed into the body before the end of the distal ileum (Souffrant et al. 1993; Krawielitzki et al. 1994; Fan and Sauer 2002b). The remaining endogenous losses of AA consist mainly of deconjugated bile salts and mucin glycoprotein that are highly resistant to proteolysis and thus not reabsorbed (Taverner et al. 1981; Moughan and Schuttert 1991; Lien et al. 1997). The fact that mucin glycoprotein is rich in Pro, Glu, Asp, Ser, and Thr and because 90% of the AA in bile acids is Gly result in a greater content of these AA in endogenous losses and lower digestibility compared with other AA. The lower digestibility of these AA can also be explained by the slow absorption into the body (Taverner et al. 1981). Greater Pro synthesis in the intestines by pyrroline-5-carboxylate reductase than degradation by Pro oxidase is also a possible reason for the high concentration of Pro in endogenous protein (Mariscal-Landin et al. 1995). The AA composition of endogenous protein has been reported (Wünsche et al. 1987; Boisen and Moughan 1996; Stein et al. 1999; Kong et al. 2014; Park and Adeola 2020).

The endogenous AA that are secreted into the intestinal tract may be categorized into non-diet specific (i.e., basal endogenous) and diet specific (Jansman et al. 2002; Stein et al. 2007). Basal endogenous losses of AA are inevitable losses of AA that are independent of the diet or feed ingredient, but are increased with increasing dry matter intake of pigs (Stein et al. 2007). Therefore, the basal endogenous losses of AA are usually expressed as g per kg dry matter intake (DMI). However, basal endogenous losses of AA that are measured in g/kg DMI decrease by increased DMI of pigs (Stein et al. 1999; Moter and Stein 2004). This decrease in the basal endogenous losses may be because the relative contribution of the fasting endogenous losses decreases as DMI increases (Moter and Stein 2004). Therefore, it is suggested that the basal endogenous losses of AA be measured in growing pigs and lactating sows fed close to ad libitum intake and in gestating sows restricted in feed intake to reflect commercial conditions (Moter and Stein 2004).

Basal endogenous losses of AA can be determined by feeding a protein-free diet to pigs and subsequently measuring the ileal outflow of AA (Stein et al. 2007). Feeding the protein-free diet to pigs may create an imbalance of AA and overestimate values for the endogenous losses of Gly and Pro. Alternative procedures to estimate basal endogenous losses of AA have been discussed (Stein et al. 2007), but the N-free diet is by far the most commonly used procedure because it is simple and practical (Stein et al. 2007; Adeola et al. 2016). Values for basal endogenous losses of AA from 101 experiments that were conducted at the University of Illinois from 2010 to 2020 are summarized (Table 19.1). Because there are variations in the endogenous losses of AA among experiments, it is recommended to determine the basal endogenous losses of AA every time the SID of AA is measured (Stein et al. 2007; Park et al. 2013).

The basal endogenous losses of AA are calculated from the following equation:

$$\text{IAA}_{\text{End}} = \left[ \text{AA}_{\text{Output}} \times \left(\text{Ileal digesta outflow / Feed intake}\right)\right],$$

where $\text{IAA}_{\text{End}}$ is the basal endogenous losses of AA determined at the distal ileum that is expressed as g/kg DMI. $\text{AA}_{\text{Output}}$ expressed in % DM is the concentration of the AA in the ileal digesta and ileal digesta outflow and feed intake are expressed as kg of DM. The index is used because quantitative collection of ileal digesta is impossible if pigs are fitted with a T-cannula. By assuming that the amount of index fed to pigs is equal to the amount of index excreted in the digesta the following calculation can be completed:

**Table 19.1**  Summary of values for the basal endogenous losses of AA from growing pigs fed N-free diets[a,b].

| Item | N | Average | SD | CV, % | Min | Max |
|---|---|---|---|---|---|---|
| Initial BW, kg | 100 | 35.27 | 23.16 | 65.66 | 7.21 | 106.60 |
| Basal endogenous losses of AA, g/kg DM intake | | | | | | |
| CP | 95 | 20.19 | 5.04 | 24.96 | 10.10 | 36.16 |
| Indispensable AA | | | | | | |
| Arg | 101 | 0.72 | 0.23 | 31.78 | 0.32 | 1.53 |
| His | 101 | 0.21 | 0.07 | 31.73 | 0.11 | 0.55 |
| Ile | 101 | 0.36 | 0.10 | 28.83 | 0.17 | 0.64 |
| Leu | 101 | 0.59 | 0.17 | 28.40 | 0.32 | 1.07 |
| Lys | 101 | 0.48 | 0.17 | 34.94 | 0.20 | 1.13 |
| Met | 101 | 0.10 | 0.04 | 43.31 | 0.05 | 0.36 |
| Phe | 101 | 0.37 | 0.11 | 30.78 | 0.10 | 0.71 |
| Thr | 101 | 0.63 | 0.17 | 26.41 | 0.34 | 1.21 |
| Trp | 101 | 0.28 | 1.48 | 526.88 | 0.03 | 15.00 |
| Val | 101 | 0.53 | 0.15 | 27.90 | 0.10 | 0.93 |
| Dispensable AA | | | | | | |
| Ala | 101 | 0.70 | 0.16 | 23.13 | 0.38 | 1.22 |
| Asp | 100 | 0.88 | 0.23 | 26.45 | 0.47 | 1.84 |
| Cys | 100 | 0.24 | 0.09 | 39.48 | 0.12 | 0.70 |
| Glu | 100 | 1.07 | 0.30 | 27.76 | 0.20 | 2.32 |
| Gly | 97 | 1.93 | 0.50 | 25.87 | 0.96 | 3.90 |
| Pro | 69 | 6.40 | 2.70 | 42.15 | 1.53 | 14.75 |
| Ser | 100 | 0.62 | 0.48 | 77.44 | 0.32 | 5.15 |
| Tyr | 74 | 0.29 | 0.08 | 26.66 | 0.16 | 0.50 |

[a] Data were from 101 experiments conducted at University of Illinois from 2010 to 2020.
[b] AA = amino acid, SD = standard deviation, CV = coefficient of variation; Min = minimum, Max = maximum, BW = body weight, DM = dry matter, and CP = crude protein.

$$\text{Feed intake} \times \text{Index}_{\text{Input}} = \text{Ileal digesta outflow} \times \text{Index}_{\text{Output}},$$

$$\text{and thus, Ileal digesta outflow} / \text{Feed intake} = \text{Index}_{\text{Input}} / \text{Index}_{\text{Output}},$$

where feed intake and ileal digesta outflow are expressed as kg of DM and concentrations of index are expressed in % DM.

By modifying the previous equations, the basal endogenous losses of AA can be calculated with concentrations of AA and index in the diet and the ileal digesta without knowing the amount of feed intake and ileal digesta outflow (Stein et al. 2007):

$$\text{IAA}_{\text{End}} = \left[ \text{AA}_{\text{Output}} \times \left( \text{Index}_{\text{Input}} / \text{Index}_{\text{Output}} \right) \right],$$

where $\text{IAA}_{\text{End}}$ is in g/kg DMI and concentrations of AA and index are in % DM.

Fiber and antinutritional factors in feed ingredients may induce diet-specific endogenous losses of AA (Sève et al. 1994; Boisen and Moughan 1996; Jansman et al. 2002). The diet-specific losses may vary depending on the type of feed ingredients used. For example, purified ingredients, including casein, have no fiber or antinutritional factors, and therefore are not likely to induce secretions of specific endogenous AA. Therefore, total endogenous losses of AA from pigs fed a casein diet are close to the basal endogenous losses of AA. However, some feed ingredients that have high concentrations of fiber and antinutritional factors may increase the specific endogenous losses of AA by pigs as reviewed previously (Tamminga et al. 1995; Boisen and Moughan 1996; Nyachoti et al. 1997; Jansman et al. 2002). Variation in total quantities of endogenous proteins that have been reported in the literature also has been published (Jansman et al. 2002; Park et al. 2013).

Diet-specific endogenous losses of AA cannot be directly determined. However, the homoarginine technique and the N15 isotope dilution technique may be used to estimate total endogenous losses of some AA. The total endogenous losses of all other AA are then calculated using a ratio among AA in the endogenous protein loss that is assumed to be always constant. The AA composition in the endogenous protein loss, however, is not always constant (Stein et al. 1999), and these techniques may, therefore, not yield accurate estimates for endogenous losses of AA. The techniques are also tedious and expensive to use, and total endogenous losses are, therefore, not commonly determined (Stein et al. 2007).

### *Apparent Ileal Digestibility of AA*

The apparent ileal digestibility (AID) of AA is calculated by subtracting ileal AA outflow from AA intake (Stein et al. 2007):

$$AID(\%) = \left[ (AA\,intake - Ileal\,AA\,outflow) / AA\,intake \right] \times 100,$$

where AA intake and ileal AA outflow are expressed in g. The AA intake (a) and ileal AA outflow (b) are calculated from the following equations:

$$(a)\ AA\,intake = 1,000 \times (AA_{Input} \times Feed\,intake) / 100$$

$$(b)\ Ileal\,AA\,outflow = 1,000 \times (AA_{Output} \times Ileal\,digesta\,outflow) / 100,$$

where AA intake and ileal AA outflow are expressed in g, AA concentrations are expressed as % of DM, and feed intake and ileal digesta outflow are expressed in kg DM. By plotting (a), (b), and "Ileal digesta outflow/Feed intake = $Index_{Input}/Index_{Output}$" into the first equation, the AID of AA can be calculated from the following equation (Stein et al. 2007):

$$AID(\%) = \left[ 1 - (AA_{Output} / AA_{Input} \times Index_{Input} / Index_{Output}) \right] \times 100,$$

where concentrations of AA and index are expressed on a percentage basis.

Values for AID do not always give accurate information about AA absorption because both unabsorbed AA and the endogenous AA are included in the total ileal AA output. The endogenous AA contribute a relatively greater portion to the total ileal AA output from pigs if concentrations of AA in ileal digesta are below the requirement than if they are at or above the requirement. Therefore, values for AID of AA are affected by the level of dietary AA and are underestimated in pigs that are fed low-protein diets (Donkoh and Moughan 1994; Fan and Sauer 2002a; Stein et al. 2005). This underestimation results in values for AID that are measured in individual feed ingredients not being additive in mixed diets that contain multiple feed ingredients (Stein et al. 2005; Xue et al. 2014). The lack of additivity is only observed if a low-protein feed ingredient (i.e., cereal grain) is included in the mixed diet (Xue et al. 2014). The problem with nonadditivity is a function of the endogenous losses being included in values for AID and to solve this problem, some of the endogenous losses need to be removed from the equation.

By correcting values for the AID of AA for total endogenous losses, values for true ileal digestibility (TID) of AA are calculated. However, from a practical diet formulation point of view, it is important that digestibility values represent the ingredient-specific endogenous losses because these losses should be debited against the ingredient. Values for TID of AA are sometimes used in academic exercises to estimate AA absorption, but should not be used in practical feed formulation.

### *Standardized Ileal Digestibility of AA*

Because the basal endogenous losses of AA are independent of ingredients and diet composition, these losses may be disregarded in the calculation of digestibility. By doing that values for standardized ileal digestibility (**SID**) of AA are calculated using the following equation (Stein et al. 2007):

$$SID(\%) = \left[ AID + \left( IAA_{End} / AA_{Input} \right) \right],$$

where AID is in %, $IAA_{End}$ is in g/kg DMI, and $AA_{Input}$ is in g/kg DM.

Because values for SID are calculated by excluding the basal endogenous losses of AA, these values are not affected by dietary AA concentration and values for SID of AA are, therefore, additive in mixed diets (Stein et al. 2005; Xue et al. 2014). Therefore, it is recommended to use SID values rather than AID values in practical formulation (Mosenthin et al. 2000; Jansman et al. 2002; Stein et al. 2007; NRC 2012). Use of SID values for AA in diet formulation rather than total AA or AID of AA also may reduce urine N excretion because diets containing multiple feed ingredients may result in less balanced AA if formulated based on total AA or the AID of AA (Lee et al. 2017). The use of values for SID of AA in diet formulation for pigs has been adopted rapidly throughout the world (Stein 2017). Therefore, SID values in most feed ingredients fed to pigs have been published and requirements for AA are also expressed on an SID basis (NRC 2012; Rostagno et al. 2017).

### *Factors Affecting Digestibility of AA*

#### *Use of Crystalline AA*
If a test diet does not contain sufficient AA, crystalline AA may be included in diets during the adaptation period before collecting ileal digesta from pigs (Pedersen et al. 2007; Strang et al. 2016). Crystalline AA mixtures are used because it is assumed that AA in the crystalline forms are rapidly absorbed and thus are 100% digestible (Chung and Baker 1992a). Addition of a crystalline AA mixture to diets containing AA below requirements during both the adaptation and collection periods, however, does not influence the AID or SID of CP and AA (Oliveira et al. 2020a). The observation that AID and SID values are not influences by dietary crystalline AA also confirms that crystalline AA are completely absorbed before the end of the ileum (Oliveira et al. 2020a). The implication is that crystalline AA can be added to test diets used in digestibility experiments if the crystalline AA are disregarded in the calculations of AID or SID values.

#### *Fiber and Anti-Nutritional Factors*
Values for AID of AA are affected by dietary protein because of the interference of endogenous protein in the ileal digesta. By definition, values for the SID of AA are supposed to be not influenced by dietary protein, but the SID of AA decreased by increasing protein levels in diets containing corn

and soybean meal (Zhai and Adeola 2011), which may be a result of increases in diet-specific endogenous losses of AA.

Antinutritional factors include dietary fiber, trypsin inhibitors, lectins, and tannins in feed ingredients that may reduce ileal digestibility of AA by pigs (de Lange et al. 1989; Jansman et al. 1994; Le Guen et al. 1995; Schulze et al. 1995; Yu et al. 1996). Dietary fiber may increase diet-specific ileal endogenous losses of AA and thus decrease the digestibility of AA (Mosenthin et al. 1994; Schulze et al. 1994; Lenis et al. 1996). However, the negative effect of fiber depends on the type of fiber that is used because different fibers have different physicochemical properties (Chen et al. 2017).

*Effects of Heat Treatment*

Heat treatment has been used to improve the nutritional value of feed ingredients that contain antinutritional factors. However, overheating the feed ingredients may result in decreases in both concentrations of AA and the digestibility of AA in corn coproducts and soy products (Fontaine et al. 2007; González-Vega et al. 2011; Kim et al. 2012a; Oliveira et al. 2020c). This decrease is a result of Maillard reactions between reducing sugars and AA that may take place during heating of ingredients in the presence of moisture (Maillard 1912). Lysine is the most susceptible AA to the Maillard reaction because Lys has an ε-amino group that is highly reactive, and, therefore, cross-linkages between the ε-amino group and reducing sugars are easily formed (Brestenský et al. 2014).

*Dietary Fat*

Inclusion of fat in diets reduces gastric emptying and the passage rate of digesta, which may result in increased time for dietary proteins to be hydrolyzed by digestive enzymes, and as a consequence, digestibility of AA is increased (Imbeah and Sauer 1991; Li and Sauer 1994; Cervantes-Pahm and Stein 2008; Kil and Stein 2011). However, dietary fat does not affect endogenous losses of AA from pigs (de Lange et al. 1989).

*Effect of Age and Physiological State of Animals*

Amino acids in sow milk are highly digestible by young pigs (Mavromichalis et al. 2001), but AA in soybeans are not well digested by piglets (Caine et al. 1997). The reduction in digestibility of AA is because young animals have low activities of some digestive enzymes in early life (Hartman et al. 1961; Moughan 1991). Digestibility of AA increases with increased age of pigs (Wilson and Leibholz 1981; Li and Sauer 1994). However, it is not clear if there are differences in the digestive capacity of pigs between growing-finishing pigs and sows. Values for the SID of AA in various feed ingredients are not different between growing pigs and lactating sows if they are provided at the same level of feed intake (Stein et al. 2001).

*Effect of Feed Intake*

The AID of protein and AA increases as feed intake of pigs increases and reaches a plateau when pigs are fed diets with an amount that is close to twice the maintenance energy requirement (Sauer et al. 1982; Haydon et al. 1984; Albin et al. 2001; Moter and Stein 2004; Goerke et al. 2012). The reason for the increase in the AID of AA as AA intake increases is that a proportion of basal endogenous losses of AA in the ileal digesta is greater if feed intake is low than if feed intake is greater (Butts et al. 1993; Stein et al. 1999; Moter and Stein 2004). However, SID of AA decreases as feed intake of pigs increases (Moter and Stein 2004), because SID values are calculated by correcting AID for the basal endogenous losses. The implication of this observation is that the SID of AA in feed ingredients should be measured in pigs fed at the same level of feed intake as pigs fed under commercial conditions (Stein et al. 2001; Moter and Stein 2004).

Pigs that are used in digestibility experiments are often fed diets on a restricted basis, which may impact the digestibility of AA. However, feeding once or twice per day does not influence the AID of AA if pigs are fed close to ad libitum intake and AID values obtained in pigs fed three times the maintenance requirement for metabolizable energy are not different from values obtained in pigs fed on an ad libitum basis (Chastanet et al. 2007).

## Lipid Digestibility

The amount of energy available to pigs in lipids is greater compared with that available in carbohydrates and proteins. Therefore, the primary role of dietary lipids has been to increase energy concentration in swine diets. Dietary lipids in feed ingredients that are commonly used in swine diets are predominantly present in the form of triglycerides (Stahly 1984). However, all feed ingredients contain different amounts of lipids and different fatty acid compositions (Sauvant et al. 2004; NRC 2012).

Lipids are not soluble in the aqueous environment of the gastrointestinal tract of pigs, and therefore, digestion and absorption of dietary lipids are different compared with other nutrients. Because lipids are hydrophobic, they must be emulsified by bile acids to be hydrolyzed by digestive enzymes. After digestion, the monoacylglycerols, diacylglycerols, free fatty acids, and glycerols are packed with bile salt to form micelles to increase movement in the lumen and absorption into the enterocytes.

Most absorbed lipids in growing animals are directly deposited in adipose tissue and, to a lesser extent, oxidized to yield energy in the form of ATP. Because it is assumed that all absorbed lipids are bioavailable in the body, bioavailability is most often determined based on digestibility of dietary lipids.

Lipid digestibility may be measured both at the end of the ileum (i.e., ileal digestibility) and over the entire intestinal tract (i.e., total tract digestibility). However, the ileal digestibility of lipids is more accurate than total tract digestibility because dietary lipids are mostly digested and absorbed before the end of the ileum (Nørgaard and Mortensen 1995) and most endogenous lipids are excreted into the large intestine (Jørgensen et al. 2000). In the large intestine of pigs, there is also significant microbial hydrogenation of unsaturated fatty acids, which results in changes in the fatty acids profiles (Jørgensen et al. 1993; Duran-Montgé et al. 2007), and there is a net synthesis of endogenous lipids in the large intestine (Shi and Noblet 1993; Reis de Souza et al. 1995). Therefore, the AID of lipids is often greater than the apparent total tract digestibility (ATTD) of lipids (Kil et al. 2010; Kim et al. 2013; Espinosa et al. 2019). However, the difference between AID and ATTD of lipids may be negligible in pigs fed diets with low concentrations of fiber (Jørgensen et al. 2000; Kil et al. 2010) because low dietary fiber results in low microbial fermentation activity in the hindgut and thus a low synthesis of microbial lipids.

## *Endogenous Losses of Lipids*

The ileal digesta and fecal samples contain both undigested dietary lipids and lipids of endogenous origin, which consist of bile acids, sloughed cells, and microbial fat (Sambrook 1979). Therefore, as is the case for the AID of AA, the apparent digestibility of lipids increases as the concentrations of dietary lipids increase (Jørgensen et al. 1993; Jørgensen and Fernández 2000; Kil et al. 2010). The apparent digestibility of lipids is underestimated if the concentration of dietary lipids is low because of large contributions of endogenous lipids to the total output of lipids. Values for the standardized or true digestibility of lipids that are corrected for the endogenous losses, however, are not affected by increasing dietary lipid intake (Kil et al. 2010), and values for true or standardized digestibility of lipids are, therefore, additive in mixed diets.

### Factors Affecting Digestibility of Lipid

Lipid digestibility in diets fed to pigs may be affected by several factors including lipid analysis, chemical and physical characteristics of fatty acids, dietary factors, and animal factors.

### Lipid Analysis

Lipids in diets, digesta, and feces may be analyzed using different extraction methods, which result in differences according to the lipid analysis method used (Boisen and Verstegen 2000). Crude fat concentration may be increased if an acid hydrolysis step is used prior to ether extraction (Stoldt 1952). Acid hydrolysis is more critical for lipid analysis in fecal samples than in diet samples because it is more difficult for ether to extract the mineral fatty acid soaps that are formed in the intestinal tract if they are not acid hydrolyzed prior to analysis. However, the mineral fatty acid soaps can be extracted by solvents if hydrolyzed (Just 1982), and acid hydrolysis, therefore, changes values for lipid digestibility (Ji et al. 2008). The ATTD of lipids in soybean oil was 7.3% greater if acid hydrolysis was not used before ether extraction than if diet and fecal samples were acid hydrolyzed (Agunbiade et al. 1992). Therefore, it is important to specify the analysis procedures for crude fat analysis (Kim et al. 2013), and it is recommended to use acid hydrolysis prior to ether extraction to prevent the overestimation of lipid digestibility (Just 1982).

### Chemical Characteristics of Lipids

Chemical characteristics of dietary lipids are mostly related to characteristics of fatty acids including degree of saturation, chain length, and existing form. These characteristics may largely influence the micelle formation, lipid solubility in the intestinal tract and thus digestibility of lipids in pigs.

Unsaturated fatty acids are more digestible compared with lipids with fewer double bonds (Jørgensen et al. 2000; Li et al. 2017; Wang et al. 2020). The reason for this is that unsaturated fatty acids may better support formation of micelles compared with saturated fatty acids (Freeman et al. 1968; Stahly 1984). Unsaturated fatty acids may also help absorption of saturated fatty acids by increasing the formation of micelles (Powles et al. 1993), which indicates that the ratio of unsaturated to saturated fatty acids in feed ingredients is an important factor for lipid and energy digestibility (Stahly 1984; Powles et al. 1995; Kellner and Patience 2017).

Lipid peroxidation is the oxidative degradation of lipids (Kerr et al. 2015). Polyunsaturated fatty acids are highly reactive to oxidation because of multiple double bonds. Peroxidation of oil may reduce lipid digestibility (Lindblom et al. 2018) because lipid peroxidation makes fatty acids more saturated by hydrogenation (DeRouchey et al. 2004; Liu et al. 2014a) and lipid peroxidation metabolites (e.g., aldehydes) that interfere with lipid digestibility are produced (Lindblom et al. 2018).

Lipid digestibility is also affected by chain length of fatty acids. Short-chain fatty acids are more soluble in the intestinal tract than longer-chain fatty acids and therefore may be directly absorbed without micelle formation (Ramírez et al. 2001). Short-chain fatty acids may also be more easily assembled into micelles than longer-chain fatty acids (Bach and Babayan 1982; Stahly 1984). As a consequence, fatty acids that contain less than 14 carbons in the chain are more digestible in weanling pigs than fatty acids that contain more carbons in the chain (Cera et al. 1989; Straarup et al. 2006). However, it is not always possible to demonstrate an impact of chain length of fatty acids on digestibility of lipids in growing pigs (Jørgensen et al. 2000).

Pancreatic lipase hydrolyzes triglycerides at the sn-1 and sn-3 positions, and the remaining monoglyceride has a greater potential to be incorporated into micelles than free fatty acids, which results in greater digestibility of the fatty acid attached at the sn-2 position. Free fatty acids also form insoluble mineral soaps in the intestinal tract (Ramírez et al. 2001), which reduces digestibility and they may disturb micelle formation (Dierick and Decuypere 2004). Therefore, increasing free fatty acids (i.e., non-esterified fatty acids) in dietary lipids decreases digestibility of lipids in weanling (Swiss and Bayley 1976; Powles et al. 1994) and growing-finishing pigs (Powles et al. 1993; Jørgensen and Fernández 2000). However, there was no effect of increasing free fatty acids on lipid digestibility in diets fed to weanling pigs if the free fatty acids were provided from choice white grease or soybean oil (DeRouchey et al. 2004; Kerr and Shurson 2017).

### Physical Properties of Lipids

Dietary lipids may be provided in intact forms that are present in cereal grains, pulse crops, and coproducts or free forms in which fats were extracted from plant or animal sources. However, extracted oil is more digestible by pigs than intact oil because there are no cell walls that interfere with the digestion of lipids (Adams and Jensen 1984; Li et al. 1990; Duran-Montgé et al. 2007; Kil et al. 2010; Kim et al. 2013). The reason for this may be that extracted oil is more accessible to bile acids and digestive enzymes in the intestinal tract. Therefore, digestibility of lipids in intact oils may be increased by rupturing lipid cell walls. Pelleting diets, in which most dietary fats were in intact forms, increased the digestibility of lipids in weanling and growing pigs (Noblet and Champion 2003; Xing et al. 2004). The true digestibility of lipids in corn distillers dried grains with solubles was greater than in corn germ, which indicates the production process including steeping and fermentation contributed to the increased digestibility (Kim et al. 2013).

### Dietary Fiber

Because dietary fiber increases microbial fermentation in the hindgut and, therefore, synthesis of microbial fat, which will be excreted in feces, the calculated ATTD of dietary fat is reduced as dietary fiber increases. However, if values for AID or TID of fat, rather than ATTD of fat, are calculated, the negative impact of fiber on calculated digestibility values is avoided. Dietary fiber may negatively affect lipid digestibility because increasing dietary fiber results in increased passage rate of digesta and decreased solubility of lipids in the intestinal tracts of pigs (Stahly 1984). Another reason for the reduction in lipid digestibility is increased losses of epithelial cells, bile acids, and microorganisms, which increases the endogenous losses of lipids and decreases digestibility of lipids (Bach Knudsen et al. 1991; de Lange 2000).

Different physicochemical characteristics of fiber may have different impact on lipid digestibility. Lipid digestibility was decreased in poultry and pigs by supplementing dietary soluble dietary fiber (Smits and Annison 1996; Ndou et al. 2019) because of increased digesta viscosity, which interferes with enzymatic hydrolysis, micelle formation and thus absorption of lipids. In contrast, increasing purified cellulose, which is an insoluble fiber, did not change lipid digestibility in pigs (Kil et al. 2010).

### Interactions Between Dietary Lipid and Minerals

Lipid digestibility may also be reduced by dietary minerals. Especially, Ca and Mg tend to form insoluble soaps with long-chain fatty acids in the intestinal tract and, therefore, may negatively influence lipid digestion (Stahly 1984). High concentrations of Ca in the diets decreased lipid digestibility in weanling and growing pigs (Jørgensen et al. 1992; Han and Thacker 2006), but that was not the case for the ileal digestibility (Jørgensen et al. 1992). The reason for this difference may be that insoluble soap is mainly formed in the large intestine. In contrast, addition of fat to diets fed to pigs did not affect the ATTD of Mg, Zn, Mn, Na, and K and instead increased the ATTD of Ca, P, and S if supplemental oils were from tallow, palm oil, corn oil, or soybean oil (Merriman et al. 2016).

### Effects of Feed Additives on Lipid Digestibility

Pigs drink water at a rate of at least 2.5 times as much as the amount of feed DM consumed (NRC 2012), and the intestinal tract of pigs is, therefore, an aqueous environment. Because dietary lipids need to be emulsified for digestion and absorption and because young animals produce a limited amount of bile salts, lipid digestibility is depressed in the early stage of life of pigs.

Use of emulsifiers, either lecithin or lysolecithin, increased lipid digestibility in diets for weanling pigs containing supplemental oils (Jones et al. 1992; Reis de Souza et al. 1995; Jin et al. 1998). However, the use of emulsifiers did not always increase lipid digestibility (Soares and Lopez-Bote 2002; Dierick and Decuypere 2004; Xing et al. 2004; Kerr and Shurson 2017). The inconsistent results may be a result of variations among fat sources, which may have different chemical characteristics and levels of dietary lipids (Jones et al. 1992; Dierick and Decuypere 2004) and variations among sources of emulsifiers (Wieland et al. 1993; Dierick and Decuypere 2004). Exogenous lipase in diets may improve lipid digestibility, but this hypothesis was not yet confirmed in diets fed to growing pigs (Dierick and Decuypere 2004; Liu et al. 2018).

An antimicrobial agent (i.e., carbadox), which depresses microbial activity, improved lipid digestibility in diets fed to pigs (Partanen et al. 2001; Wang et al. 2005). Likewise, the supplementation of copper hydroxychloride to diets for pigs increased the ATTD of lipids (Espinosa et al. 2019). This increase in lipid digestibility may be due to a reduction in endogenous loss of fat due to reduced synthesis of microbial fat if copper hydroxylchloride or other antimicrobials are used. Increased microbial activity increases the endogenous losses of lipids by promoting irreversible bile catabolism (Smits and Annison 1996), and microbial hydrogenation may reduce lipid digestibility by converting unsaturated fatty acids to saturated fatty acids in the small intestine (Yen 2001).

### Effect of Age of Animals

Because newly weaned pigs have consumed liquid diets until weaning, introduction of solid feeds to weanling pigs may decrease lipid digestibility because of limited pancreatic lipase activity (Lindemann et al. 1986; Kerr et al. 2015). The low digestibility of fat in weanling pigs is more pronounced if pigs are fed diets containing animal origin oil instead of plant oils (Cera et al. 1989; Jones et al. 1992; Jin et al. 1998). This indicates that weanling pigs have a lower digestibility of saturated fatty acids than of unsaturated fatty acids. However, the difference in digestibility between animal origin and plant origin fat disappears as pigs become older (Agunbiade et al. 1992; Jørgensen and Fernández 2000). Lipid digestibility gradually increases during the post-weaning period (Cera et al. 1989; Soares and Lopez-Bote 2002; Straarup et al. 2006) and pigs may reach a maximum capacity for digestion of lipids at approximately 40 kg (Wiseman and Cole 1987; Agunbiade et al. 1992).

## Digestible Carbohydrates

### Digestibility of Monosaccharides and Disaccharides

Glucose and galactose are easily absorbed via energy-dependent transporters before the end of the small intestine and fructose is absorbed via facilitated diffusion (Englyst and Hudson 2000). However, arabinose, xylose, and mannose are absorbed in the small intestine via a passive transport process, but concentrations of these monosaccharides in food and feed are quantitatively small (Englyst and Hudson 2000; IOM 2001). Most monosaccharides are believed to be 100% digestible (Englyst and Hudson 2000; Abelilla and Stein 2019b).

Sucrose (i.e., glucose + fructose) and maltose (i.e., glucose + glucose) are present in small amounts in most plant feed ingredients fed to pigs with the exception that soybean meal contains 8–9% sucrose (Cervantes-Pahm and Stein 2010; Navarro et al. 2018). Lactose (i.e., glucose + galactose) is a disaccharide that is present only in milk products. These disaccharides are digested by the brush-border enzymes, and the monosaccharides that are hydrolyzed from the disaccharides are absorbed into the bloodstream. It is assumed that dietary sucrose, maltose, and lactose are 100% digested and absorbed in the small intestine (van Beers et al. 1995), but digestibility of disaccharides is rarely measured.

### Starch Digestibility

Plant feed ingredients are rich sources of starch (Wiseman 2006; Bach Knudsen et al. 2006; NRC 2012; Stein et al. 2016a). Because most swine diets are formulated based on plant feed ingredients, commercially fed pigs usually consume diets that are high in starch. Starch exists in the form of granules that contain amylose and amylopectin polymers (Navarro et al. 2019). Amylose consists of glucose units that are mostly linked in a linear way with $\alpha(1\text{-}4)$ glycosidic bonds. Amylopectin is also composed of glucose units, but the glucose molecules are linked with both $\alpha(1\text{-}4)$ and $\alpha(1\text{-}6)$ glycosidic linkages, which makes amylopectin a large and highly branched polymer. Starch in most cereal grains contains apparently 25% amylose and 75% amylopectin (BeMiller 2007).

Dietary starch is digested in the mouth by salivary amylase and in the small intestine by $\alpha$-amylase, isomaltase, and maltase and is absorbed into the bloodstream as glucose molecules. There are several factors that affect the rate and extent of starch digestion: (i) the crystallinity of the starch granules, (ii) the source of starch, (iii) the amylose-to-amylopectin ratio, and (iv) the type and extent of processing of feed ingredients containing starch (Cummings et al. 1997; Englyst and Hudson 2000; Svihus et al. 2005). Amylose is more resistant to enzymatic digestion than amylopectin (Svihus et al. 2005). There is limited information on the digestibility of starch in feed ingredients fed to pigs, but it is believed that the AID of starch in most cereal grains is greater than 90% (Bach Knudsen et al. 2006; Sun et al. 2006; Wiseman 2006). However, depending on the feed ingredients, up to 20% of dietary starch may escape enzymatic digestion in the small intestine (Sun et al. 2006; Stein and Bohlke 2007; Cervantes-Pahm et al. 2014). Undigested starch (i.e., resistant starch) will enter the hindgut where it will be rapidly fermented by microorganisms, and the ATTD of starch, therefore, is close to 100% (Canibe and Bach Knudsen 1997; Sun et al. 2006; Stein and Bohlke 2007; Gunawardena et al. 2010). Short-chain fatty acids that are by-products of fermentation will be absorbed into the body as energy sources, but this will result in a reduced efficiency of energy utilization compared with absorption of glucose in the small intestine. Because of the microbial fermentation of resistant starch in the hindgut, values for total tract digestibility of starch are not representative of the digestibility of starch. Therefore, it is more accurate to measure the ileal digestibility of starch than total tract digestibility. There is no endogenous secretion of starch into the small intestine and, therefore, the AID of starch will closely represent digestion and absorption of dietary starch.

Extruding feed ingredients or mixed diets increases the AID of starch in pigs because of gelatinization of starch (Sun et al. 2006; Stein and Bohlke 2007; Rojas et al. 2016; Rodriguez et al. 2020). The gelatinized starch is highly digestible because during the process of gelatinization the starch granules are hydrolyzed, and there are more space between molecules that allows better access by digestive enzymes (Ai 2013). Starch digestibility may be increased by pelleting (Rojas et al. 2016), but that is not always the case (Svihus et al. 2005; Stein and Bohlke 2007).

In pigs fed diets containing high fiber feed ingredients, there was disappearance of some of the dietary fiber prior to the distal ileum, which indicates that some microbial fermentation takes place in the small intestine (Bach Knudsen and Jørgensen 2001; Urriola and Stein 2010; Jaworski and Stein 2017). It is, therefore, possible that some of the dietary starch that disappears in the small intestine may be fermented rather than digested and absorbed as glucose. However, more research is needed to demonstrate to which extend fermentation of starch in the small intestine contributes to small intestinal starch disappearance in pigs fed diets containing different levels of dietary fiber.

## Nondigestible Carbohydrates

### Digestibility of Oligosaccharides

Raffinose, stachyose, and verbascose are the three main oligosaccharides that are naturally present in feed ingredients, such as beans, legumes, cotton seeds, and molasses (Navarro et al. 2019). Raffinose consists of one unit of glucose, one unit of fructose, and one unit of galactose. Stachyose and verbascose have a structure that is similar to raffinose with the exception that they contain two or three units of galactose, respectively. The glucose, fructose, and galactose units are linked with combinations of $\alpha(1\text{-}2)$ and $\alpha(1\text{-}6)$ glycosidic linkages and raffinose, stachyose, and verbascose are commonly called $\alpha$-galactosides. The enzyme $\alpha$-galactosidase is needed to hydrolyze the glycosidic linkages between the sugar molecules in the oligosaccharides, but pigs do not secrete this enzyme in the intestinal tract, and $\alpha$-galactosides are, therefore, not digested in the small intestine (Caine et al. 1997). However, 50–80% of $\alpha$-galactosides disappeared prior to the distal ileum (Bengala-Freire et al. 1991; Caine et al. 1997; Smiricky et al. 2002), which may be a result of the fermentation taking place in the small intestine. The total tract digestibility of $\alpha$-galactosides is 100% because $\alpha$-galactosides that are not fermented in the small intestine are rapidly fermented in the large intestine (Smiricky-Tjardes et al. 2003).

Depending on feed ingredients, swine diets may contain fructans, levans, and mannan-oligosaccharides, but there is limited information about the digestibility of these oligosaccharides.

Synthetic oligosaccharides including fructooligosaccharides and transgalactooliogosaccharides may also be used in swine diets because these may have prebiotic effects in the intestinal tract by increasing microbial fermentation activity. These oligosaccharides are partly fermented in the small intestine (Smiricky-Tjardes et al. 2003), but there is no information on total tract digestibility.

## Digestibility of Dietary Fiber

Dietary fiber is defined as carbohydrates that are not digested by body enzymes, but are completely or partially fermented (De Vries 2004). Non-starch polysaccharides, resistant starch, non-digestible oligosaccharides, and sugar alcohols are classified as dietary fiber (Englyst and Englyst 2005; Englyst et al. 2007). Because these carbohydrates are mostly building blocks for plant cell walls, dietary fiber is always present in plant-based feed ingredients in various forms and amounts (Cummings et al. 1997; Navarro et al. 2019). However, some carbohydrates that are not present in plant cell walls are also classified as dietary carbohydrates because they have similar physiological characteristics as plant cell-wall carbohydrates (De Vries 2004). Therefore, dietary fiber includes non-starch polysaccharides that contain both plant cell-wall components and non-cell-wall components (Bach Knudsen et al. 2013).

Cellulose and hemicelluloses are the most common non-starch polysaccharides in cell walls, but lignin is also considered part of the fiber in feed ingredients. Cellulose consists of glucose molecules that are linearly linked by $\beta$(1-4) linkages, which cannot be hydrolyzed by digestive enzymes in pigs. The most common hemicelluloses in cereal grains are arabinoxylan (BeMiller 2007; Jaworski et al. 2015; Navarro et al. 2019). Arabinoxylans consist of a xylose backbone and side chains that are composed of arabinose, mannose, galactose, and glucose (Cummings and Stephen 2007). Hemicelluloses in oilseed meal may be linked with uronic acids that are derivatives from galactose (galacturonic acid), and this hemicellulose-uronic acid complex is likely to form salts with metal ions such as calcium and zinc (Southgate and Spiller 2001; Cummings and Stephen 2007). Because some of the hemicelluoses are soluble, they are more fermentable in the large intestine of pigs than cellulose.

Other noncellulosic polysaccharides that are considered dietary fiber include pectins, gums, non-digestible oligosaccharides, and resistant starches. The main components of pectins are linear polymers of galacturonic acids that are derived from galactose and these are linked by $\alpha$(1-4) linkages. Pectins may also contain side chains of rhamnose, galactose, and arabinose (Cummings and Stephen 2007). Gums are natural plant polysaccharides, but may also be produced by fermentation. Plant gums are naturally occurring when plants or shrubs that are physically damaged, or as a part of the seed endosperm (BeMiller 2007). An example of an exudate gum is gum arabic and an example of a gum from seed endosperm is guar gum. Xanthan gum and pullulan are examples of gums produced from fermentation.

Dietary fiber may also be classified into soluble dietary fiber and insoluble dietary according to solubility (Lee et al. 1992). Total dietary fiber is defined as carbohydrates and lignin resistant to enzymatic digestion by endogenous enzymes (Trowell 1976). Insoluble dietary fiber includes lignin, cellulose, and some of the hemicelluloses that are components of the cell wall, and soluble dietary fiber are the non-cell wall components along with the remaining hemicelluloses.

The ileal digestibility of dietary fiber is generally low, but disappearance of dietary fiber in the small intestine has been reported, which indicates that microbial fermentation in the small intestine is somewhat efficient (Bach Knudsen and Jørgensen 2001; Urriola and Stein 2010; Gutierrez et al. 2013; Jaworski and Stein 2017; Abelilla and Stein 2019a). However, the only end products from microbial fermentation of dietary fiber are the short-chain fatty acids, which are easily absorbed and from a quantitative point of view, it is, therefore, not important if fiber is fermented in the small intestine or the large intestine. Therefore, the total tract digestibility of dietary fiber can be used to estimate disappearance of dietary fiber in the intestinal tract of pigs.

The total tract digestibility of total dietary fiber is less than 70% depending on the fiber components in feed ingredients (Bach Knudsen and Jørgensen 2001; Urriola et al. 2010; Jaworski and Stein 2017). The total tract digestibility of soluble dietary fiber is greater than of insoluble dietary fiber because soluble dietary fiber is highly fermentable in the hindgut (Urriola et al. 2010; Urriola and Stein 2010; Jaworski and Stein 2017). Therefore, the greater the concentration of soluble dietary fiber is, the greater is digestibility of total dietary fiber.

There is limited secretion of carbohydrates into the intestinal tract, but endogenous secretions and microbial matter may be analyzed as fiber, and therefore contribute to misinterpretation of data for digestibility of fiber (Cervantes-Pahm et al. 2014; Montoya et al. 2015; Abelilla 2018). Both soluble and insoluble contributions of endogenous secretions to ileal or total tract output of dietary fiber have been described (Montoya et al. 2015). The origin of these secretions are primarily mucin and microbial matter including chitin (Montoya et al. 2015). Mucin mostly consists of substances that are characterized as soluble dietary fiber, which is fermented in the hindgut (Montoya et al. 2015). In contrast, microbial matter is primarily analyzed as insoluble dietary fiber and is not fermented in the hindgut and most endogenous secretions that are analyzed in the feces are, therefore, insoluble fiber (Montoya et al. 2015). As a consequence of the endogenous secretion of compounds analyzed as fiber, a correction for these secretions is needed to calculate the digestibility of fiber. Endogenous secretions analyzed as fiber may be quantified by feeding a fiber-free diet (Cervantes-Pahm et al. 2014). By correcting values for the AID or ATTD for the endogenous secretions analyzed as fiber, values for the SID or standardized total tract digestibility (**STTD**) of dietary fiber may be calculated and these values are more representative of the disappearance of fiber in the intestinal tract than values fore AID or ATTD of dietary fiber (Cervantes-Pahm et al. 2014; Montoya et al. 2015, 2016).

### Bioavailability and Digestibility of Minerals

Excellent reviews on comparative mineral bioavailability are available (Nelson and Walker 1964; Ammerman and Miller 1972; Peeler 1972; Cantor et al. 1975a,b; Cromwell 1992; Ammerman et al. 1995; Richards et al. 2010; Vitti and Kebreab 2010). Most of these reviews have dealt with the relative bioavailability of minerals in mineral supplements and less information has been published about the availability of minerals in plant feed ingredients. Recently, a review of mineral absorption and interactions among minerals was published (Goff 2018). However, with recent evidence demonstrating that endogenous losses of minerals contribute to the total mineral output from animals (Petersen and Stein 2006; González-Vega et al. 2013), a greater emphasis has been placed on determining digestibility of minerals rather than relative bioavailability.

### *Effects of Phytate and Exogenous Phytase on Digestibility of Mineral in Pigs*

Corn is unique among the cereal grains because 90% of the phytate is located in the germ portion of the kernel (Reddy et al. 1982). Wheat and rice germ also have considerable concentration of phytate, but the hull portion is also rich in phytate. Thus, wheat bran and rice bran have high concentrations of phytate (Halpin and Baker 1987; Casas and Stein 2015). Soybeans are different from most other oilseeds in that phytic acid is contained in protein bodies distributed throughout the seed. Therefore, soy protein isolates contain more phytate than soybean meal (Erdman 1979).

Phytic acid in peanuts, cottonseed, and sunflower seeds is concentrated only in crystalloid and globoid substructures.

Phytic acid is charged negatively under neutal pH conditions in the aqueous environment of the intestinal tract of pigs and di- and trivalent cations (e.g., Ca, Mg, Fe, Zn, Cu, and Mn) tend to chelate to phytic acid to maintain a stable complex. Therefore, it is likely that the bioavailability of minerals in plant feed ingredients is lower compared with animal feed ingredients. Supplementation of microbial or feed ingredient-derived phytase significantly improves the utilization of phytate-P by pigs (Almeida and Stein 2010; Broomhead et al. 2018). On the other hand, the use of exogenous phytase in diets fed to pigs may also release the minerals that are chelated to phytate, which results in an increase in digestibility. The digestibility of Ca, K, Na, Mg, Cu, Mn, and Zn is often increased by use of microbial phytase at levels ranging from 250 to 15 000 FTU/kg diet (Adeola et al. 1995; Kies et al. 2006; Zeng et al. 2014; Arredondo et al. 2019). However, phytase differentially affects the digestibility of minerals, and it is, therefore, necessary that the effects of phytase on each mineral is assessed.

*Phosphorus*

Excretion of P that is caused by excess dietary P or low digestibility of P may increase environmental pollution (Knowlton et al. 2004). Therefore, there has been an attempt to formulate diets for pigs based on digestible P because that results in reduced excretion of P (Almeida and Stein 2010). Historically, the relative bioavailability of P in feed ingredients with monosodium phosphate or monocalcium phosphate as the standard was used to estimate P availability (NRC 1998). However, values for the relative bioavailability of P are not additive in mixed diets, and these values vary depending on the standard used and the bioavailability values for the standards are usually greater than P digestibility (Petersen et al. 2011). It is also less costly to obtain digestibility values than bioavailability values. Therefore, it is recognized that values for digestible P need to be used in diet formulation (NRC 2012).

Total tract digestibility is used to determine the digestibility of P because although P is absorbed in the small intestine there is no net absorption or secretion of P into the large intestine (Fan et al. 2001; Shen et al. 2002; Ajakaiye et al. 2003; Bohlke et al. 2005; Dilger and Adeola 2006). It is also easier and less expensive to measure the total tract digestibility of P than to measure the ileal digestibility of P.

Values for the ATTD of P are usually influenced by the dietary level of P as is the case for the AID values for AA (Fan et al. 2001; Stein et al. 2008; Kim et al. 2012b). The reason for this is that both undigested dietary P and endogenous P are excreted in the fecal output (Fernández 1995; Fan et al. 2001; Akinmusire and Adeola 2009; Zhai and Adeola 2012). As a consequence, ATTD of P usually underestimates P digestibility if dietary P is low. Therefore, values for the ATTD of P need to be corrected for endogenous loss of P and STTD of P is calculated. Values for the STTD of P are calculated from the following equation (Almeida and Stein 2010):

$$\text{STTD}(\%) = [\text{P intake} - (\text{P output} - \text{daily basal endogenous loss of P})] / \text{P intake},$$

where intake, output, and daily basal endogenous loss of P are in g/d.

Values for the STTD of P from individual feed ingredients are additive in mixed diets (Fan and Sauer 2002a; Fang et al. 2007; Kwon 2016; She et al. 2018b). The STTD of P has been determined in most feed ingredients that are commonly used for pigs and are also included in feed ingredient tables and requirement estimates (NRC 2012; Stein et al. 2016a; Rostagno et al. 2017).

*Endogenous Loss of P*

The main sources of endogenous P are epithelial intestinal cells that contain phospholipids, digestive fluids and enzymes, and microbes in the intestinal tract (Fan et al. 2001). The endogenous loss of P may be measured using the regression method, which results in estimation of total endogenous loss of P (Akinmusire and Adeola 2009). Using this procedure, the digestible P (g/kg DM intake) is regressed against P intake from diets with different concentration of P that is expressed as g/kg DMI. The slope and the y-intercept of the regression are considered the values for true total tract digestibility of P and endogenous loss of P, respectively.

Basal endogenous loss of P has been estimated using a P-free diet (Petersen and Stein 2006). Using this procedure, P in the feces from pigs fed the P-free diet is used to correct values for the ATTD of P, and thus the STTD of P is calculated. The P-free diets used in most experiments that have measured the STTD of P are based on gelatin as the protein source and cornstarch that do not contain P. The basal endogenous loss of P is calculated using the following equation (Almeida and Stein 2010):

$$\text{Basal endogenous loss}(\text{mg}/\text{kg DM intake}) = (1{,}000 \times \text{P output})/$$
$$\text{Feed intake},$$

where P output that is calculated using the concentration of P and fecal output is in g/d and feed intake is in kg DM/d.

Values for the basal endogenous loss of P by growing pigs have been constant among a large number of experiments. The average basal endogenous loss of P that was summarized by NRC was 190 mg/kg DM intake (NRC 2012), and the average of currently published experiments is 186 mg/kg DM intake (Table 19.2). Increasing cellulose in P-free diets does not influence the basal endogenous loss of P, but the basal endogenous loss of P may differ among different fiber sources (Son and Kim, 2015; Son, 2016; Bikker et al., 2016). Because of the constant contribution of endogenous P to total P output, it is not necessary to measure the basal endogenous loss of P every time the STTD of P in feed ingredients is determined (NRC 2012).

**Table 19.2** Summary of values for the basal endogenous losses of P and Ca from growing pigs fed P-free and Ca-free diets, respectively[a].

| Item | N | Average | SD | CV, % | Min | Max |
|---|---|---|---|---|---|---|
| Phosphorus[b] | | | | | | |
|   Initial BW, kg | 33 | 34.8 | 16.3 | 46.7 | 11.0 | 78.7 |
|   Basal endogenous losses of P, mg/kg DM intake | 33 | 186 | 40.2 | 21.6 | 111 | 325 |
| Calcium[c] | | | | | | |
|   Cornstarch-based diet | | | | | | |
|     Initial BW, kg | 2 | 18.5 | 1.1 | 5.7 | 17.7 | 19.2 |
|     Basal endogenous losses of Ca, mg/kg DM intake | 2 | 172 | 68.6 | 40.0 | 123 | 220 |
|   Corn-based diet | | | | | | |
|     Initial BW, kg | 7 | 16.3 | 3.4 | 20.9 | 10.2 | 19.8 |
|     Basal endogenous losses of Ca, mg/kg DM intake | 7 | 433 | 108.5 | 23.3 | 329 | 659 |

[a] SD = standard deviation, CV = coefficient of variation, Min = minimum, Max = maximum, BW = body weight, and DM = dry matter.
[b] Data for P were from 33 experiments that were conducted from 2006 to 2018.
[c] Data for Ca were from 9 experiments that were conducted from 2015 to 2019.

*Phytate, Inorganic P, and Exogenous Phytase*

Phytic acid or phytate consists of an inositol ring and 6 phosphates that are attached by ester bonds. Phytate is a primary form of P stored in grains and oilseeds, which results in a high concentration of phytate bound P relative to total P (NRC 2012). Swine diets mostly contain plant feed ingredients, and thus dietary P is largely unavailable to pigs because pigs do not produce a sufficient amount of phytase to hydrolyze the ester bonds. As a consequence, most dietary P is supplied by mineral supplements. The two most used mineral supplements in swine diets are dicalcium phosphate (**DCP**) and monocalcium phosphate (**MCP**) and DCP and MCP produced in North America contain approximately 18.5 and 21.0% P, respectively (NRC 2012).

Phosphorus can also be provided by plants when exogenous phytase that releases P from the phytate is used. Exogenous phytase increases P digestibility in plant feed ingredients used in a number of studies (Simons et al. 1990; Selle and Ravindran 2008; Almeida and Stein 2012). Details on the use of phytase are discussed in Chapter 17.

*Interaction Between P and Ca*

Dietary P is not affecting Ca digestibility, but increasing dietary Ca reduces the ATTD of P in growing pigs and gestating sows (Stein et al. 2011; González-Vega et al. 2016; Lee et al. 2020). Reduction in P digestibility caused by Ca is likely a result of precipitation of a nondigestible Ca-P complex, but more research is needed to demonstrate the exact mechanism behind this hypothesis.

*Effect of Age and Physiological State of Animals*     The STTD of P in most feed ingredients commonly used in swine diets has been determined in recent years (NRC 2012). However, most experiments used growing pigs and subsequently these values were applied to all categories of pigs, including sows. Results of several studies have demonstrated that values for the digestibility of P and Ca in growing pigs are greater than in lactating and gestating sows when they were fed the same diet (Kemme et al. 1997; Lee et al. 2021). The reason for the reduced digestibility of Ca and P by gestating sows is that gestating sows have a very low requirement for Ca and P. The ATTD of P and Ca is greater in lactating sows compared with gestating sows because of the greater requirement for P and Ca by lactating sows to produce milk. However, digestibility of P and Ca in growing pigs is not influenced by the dietary concentration, whereas Ca and P homeostasis is controlled by urine excretion (González-Vega et al. 2016).

### Calcium

The relative bioavailability of Ca in some Ca sources is over 90% if limestone is used as the standard, but there are few data for the bioavailability of Ca in feed ingredients (Ross et al. 1984; Kuznetsov et al. 1987; NRC 1998). The lack of data for the bioavailability of Ca is likely a result of Ca being less expensive than other feed ingredients. Therefore, Ca in feed ingredients has been considered less important than other nutrients. However, limestone is often used as a flow agent in feed ingredients or a carrier in nutritional supplements, which may lead to an oversupply of Ca in diets for pigs (Walk 2016; Wu et al. 2018), which has negative effects on digestibility of P (Stein et al. 2011; Lee et al. 2020) and on growth performance of pigs (González-Vega et al. 2016; Lagos et al. 2019; Vier et al. 2019). To prevent excess Ca in diets, therefore, values for the digestibility of Ca by pigs in feed ingredients have been determined (González-Vega et al. 2015b; Stein et al. 2016b; Zhang and Adeola 2017; Lee et al. 2019b).

Total tract digestibility of Ca is measured because there was no net absorption of Ca from the hindgut of pigs (González-Vega et al. 2014). The ATTD of Ca is, however, influenced by the dietary level of Ca because of endogenous Ca secretion (González-Vega et al. 2013). Therefore, the endogenous loss of Ca has been determined by using the regression method (González-Vega et al. 2013; Zhang and Adeola 2017) or by using a Ca-free diet (González-Vega et al. 2015b; Lee et al. 2019b). Calculation of the ATTD of Ca, endogenous loss of Ca, and the STTD of Ca is similar to calculations for ATTD and STTD of P.

*Endogenous Loss of Ca*
The main sources of endogenous loss of Ca are body fluids, epithelial intestinal cells, digestive fluids and enzymes, and microbes in the intestinal tract (González-Vega and Stein 2014). Total endogenous loss of Ca that is estimated by the regression method ranged from 160 to 316 mg/kg DM intake depending on the feed ingredients used (González-Vega et al. 2013; Zhang and Adeola 2017). The endogenous loss of Ca may also be determined using a cornstarch-based Ca-free diet, which results in a loss of Ca ranging from 123 to 220 mg/kg DM intake (González-Vega et al. 2015a, b; Table 19.2). Because corn does not contain Ca, a corn-based Ca-free diet has also been used and the basal endogenous loss of Ca using this diet ranges from 329 to 659 mg/kg DM intake (González-Vega et al. 2015a; Merriman 2016; Blavi et al. 2017; Lee et al. 2019b; Sung et al. 2020). The main reason for the difference in values for the basal endogenous loss of Ca that are obtained using the two types of Ca-free diets is the presence of phytate and fiber in corn. Endogenous origin Ca can be chelated to dietary phytate from corn, and, therefore, use of microbial phytase has reduced the basal endogenous loss of Ca because the endogenous Ca released from the Ca-phytate complex by microbial phytase was reabsorbed (Lee et al. 2019a, b).

*Phytate and Exogenous Phytase*
Calcium in swine diets is mostly provided by limestone, DCP, and MCP because most grains and oilseed meals have low Ca concentration (NRC 2012). Because swine diets are formulated based mostly on plant feed ingredients, these feed ingredients also provide phytate. However, Ca from limestone and the phytate from plant ingredients can form a non-digestible Ca-phytate complex in the intestinal tract of pigs (González-Vega et al. 2015a, b; Lee et al. 2019b). Therefore, addition of phytase to the diet releases Ca from the complex, which results in an increase in the digestibility of Ca in limestone (González-Vega et al. 2015b; Blavi et al. 2017; Lee et al. 2019b), but phytase does not increase the digestibility of Ca in MCP or DCP (González-Vega et al. 2015b; Lee et al. 2019b).

## Sodium and Chloride

Sodium and Cl are minerals involved in acid–base balance in the body and in urinary excretion of cations and anions (Remer 2000). Therefore, suitable bioassays for bioavailability are difficult to develop. Using information from requirement studies, Na in NaCl, $Na_2SO_4$, $NaHCO_3$, Na acetate, and Na citrate is thought to be equally bioavailable and Cl in NaCl, KCl, $NH_4Cl$, and $CaCl_2 \cdot 2H_2O$ is also equally bioavailable for pigs. Defluorinated phosphates are generally rich in Na, and the Na in defluorinated phosphate was 83% available for pigs relative to Na in NaCl (Miller 1980).

Most Na and Cl are provided by salt in swine diets, and true absorption of Na and Cl is believed to be close to 100% (Groff et al. 1995). The ATTD of Na in diets containing salt, cereal grains, oilseed meals, animal products, and mineral premix fed to weanling and growing pigs ranges from 77.9 to 94.0 (Kies et al. 2006; Sauer et al. 2009; Jolliff and Mahan 2012; Arredondo et al. 2019), and the digestibility of Na is, therefore, greater than that of other minerals. Although the ATTD of Na is generally high, it may be increased by the addition of microbial phytase to the diet (She et al. 2018a; Arredondo et al. 2019). There is limited information about the ATTD of Cl in diets fed to pigs, but it is likely that the ATTD of Cl is greater than 70% (Sauer et al. 2009).

### Magnesium

Magnesium is one of the bone minerals and is used as a cofactor for a number of enzymes in the body. It is believed that the amount of Mg provided by corn and soybean meal in diets is adequate to meet the requirement of growing pigs (Svajgr et al. 1969; Krider et al. 1975). However, impurity of limestone and calcium phosphates also provides Mg as magnesium oxide or magnesium phosphate (Spiropoulos 1985; Baker 1989). Therefore, most commercial mineral premixed do not contain Mg, but if synthetic diets are used, Mg needs to be added.

Bioavailability of Mg for pigs in feed-grade sources of MgO, $MgSO_4$, and $MgCO_3$ is estimated to be 100% relative to the Mg in reagent-grade MgO, and Mg in grains and concentrates was 50–60% available compared with MgO (Miller 1980). The true absorption efficiency by chickens of Mg from soybean meal is 60%, which is about the same as that estimated for $MgSO_4 \cdot 7H_2O$ (Guenter and Sell 1974). The ATTD of Mg in diets containing cereal grains, oilseed meals, or milk products fed to weanling or growing pigs ranges from 15 to 50% (Kemme et al. 1999; Kies et al. 2006; Sauer et al. 2009; Jolliff and Mahan 2012; Merriman et al. 2016). However, because Mg is a divalent cation, dietary Mg may be chelated to phytate, and addition of microbial phytase to diets, therefore, increases the ATTD of Mg (She et al. 2018a; Arredondo et al. 2019).

### Potassium

Potassium is the third most abundant mineral in the body. Most K is present in muscle tissues (Stant et al. 1969), and K is involved in maintaining the electrolyte balance and neuromuscular function. Potassium is also used in the Na-K pump in cell membranes for the active transport process.

Corn-soybean meal diets are rich in K, and K supplements are, therefore, not added to commercial diets for pigs. Definitive information on K bioavailability from various K sources is lacking, but K in $K_2CO_3$, $KHCO_3$, $K_2HPO_4$, K acetate, and K citrate is assumed to be 100% available relative to the K in KCl (Peeler 1972). Potassium in $K_2CO_3$, $KHCO_3$, corn, and soybean meal is 100% bioavailable relative to K acetate (Combs and Miller 1985; Combs et al. 1985). Among response criteria that have been used to estimate the bioavailability of K are blood K, urinary K, and K retention, but only K retention (balance trials) seems to respond linearly to K intake (Combs and Miller 1985).

The ATTD of K in diets containing plant and animal ingredients fed to weanling and growing pigs is greater than 73% (Kies et al. 2006; Jolliff and Mahan 2012; Merriman et al. 2016). If microbial phytase is added to the diets, the ATTD of K may increase to above 90% (She et al. 2018a; Arredondo et al. 2019).

### Copper

Copper is needed in the synthesis of hemoglobin and several enzymes needed to produce energy. Dietary Cu is provided by natural feed ingredients and mineral supplements including Cu sulfate.

Copper bioavailability, like that of Zn, is difficult to quantify accurately because Cu accumulation in tissues (primarily liver) increases only slightly (and curvilinearly) between deficient levels and a dietary level of up to 250 mg Cu/kg diet. Beyond this level, Cu accumulates rapidly and generally in a linear fashion. Copper in CuO has a lower bioavailability in chickens and pigs than Cu from Cu sulfate (Cromwell et al. 1989; Baker et al. 1991; Aoyagi and Baker 1993a; Baker and Ammerman 1995a). The relative bioavailability of Cu in a number of Cu sources has been estimated in chickens with analytical-grade $CuSO_4 \cdot 5H_2O$ used as the standard (Aoyagi and Baker 1993b, 1994, 1995). Copper in feces is utilized no better than 30% relative to that provided as $CuSO_4 \cdot 5H_2O$ (Izquierdo and Baker 1986). The ATTD of Cu in swine diets is less than the ATTD of most other minerals and sometimes negative (Kies et al. 2006; Burkett et al. 2009), although values between 15 and 40% are most often observed (Adeola and Orban 1995; Burkett et al. 2009; Liu et al. 2014b). The ATTD of Cu in chelated Cu is greater than in nonchelated sources (Burkett et al. 2009; Liu et al. 2014b). Absorption of Cu is reduced if $Na_2S$ (Barber et al. 1961; Cromwell et al. 1978), roxarsone (Czarnecki and Baker 1985; Edmonds and Baker 1986), or reducing agents such as cysteine or ascorbic acid (Baker and Czarnecki-Maulden 1987; Aoyagi and Baker 1994) are included in the diet. Dietary Cu may be chelated to phytate, and microbial phytase, therefore, increases the ATTD of Cu (She et al. 2018a; Arredondo et al. 2019).

## *Iodine*

Iodine is needed for the synthesis of thyroid hormones that control the metabolism in the body. Little definitive work on bioavailability and digestibility of I has been conducted. However, in research with rats, it was estimated that I in KI, Ca $(IO_3)_2 \cdot 2H_2O$, $KIO_3$, CuI, and ethylenediaminedihydriodide $(C_2H_8N_2 \cdot 2HI)$ is close to 100% bioavailable relative to NaI (Miller 1980; Miller and Ammerman 1995). It is, therefore, assumed that the I present in iodized salt is 100% available relative to NaI.

## *Iron*

Iron is primarily used for synthesis of red blood cells and for the transfer of oxygen from the lungs to body tissues. Baby pigs reared in confinement are uniquely susceptible to Fe deficiency anemia because of their rapid growth rate and lack of placental or mammary Fe transfer from dam to offspring. Thus, it is standard practice to give newborn pigs Fe injections during the first few days of life. Over 90% of the Fe from Fe-dextran (100- or 200-mg injections) is incorporated into hemoglobin (Braude et al. 1962; Miller 1980). Fe-dextran doses administered orally during the first 12 hours after birth (prior to gut closure) also promote efficient Fe incorporation into hemoglobin (Harmon et al. 1974; Thorén-Tolling 1975; Cornelius and Harmon 1976). Injection of Fe is usually the preferred mode of administration, but data from 20 commercial swine farms in Ontario, Canada, indicated that 34% of weaned pigs and 61% of pigs that had been weaned for three weeks are deficient in Fe (Perri et al. 2016).

There is no effective means of increasing placental or mammary transfer of Fe from the sow to her offspring. Oral administration of Fe to sows or administration via injection does not increase pig Fe stores at birth or Fe concentration in milk sufficiently to prevent anemia in the offspring (Pond et al. 1961). Feeding of Fe to gestating sows or lactating sows at high levels may elicit hemoglobin responses in the nursing pigs (Peters and Mahan 2008; Li et al. 2018), but this is primarily due to consumption of Fe from the feces of the sow rather than being a result of absorption of Fe from the milk.

Pig studies on Fe bioavailability have revealed essentially 100% relative bioavailability of Fe in $FeSO_4 \cdot 2H_2O$, $FeSO_4 \cdot 7H_2O$, Fe citrate, and Fe choline citrate relative to Fe in $FeSO_4 \cdot H_2O$. Iron in oxides of Fe is almost unavailable, whereas Fe in carbonates is variable in bioavailability, depending on where the carbonates are mined (Harmon et al. 1969; Henry and Miller 1995). Deposition of Fe in liver of pigs was greater if Fe was provided as $FeSO_4 \cdot H_2O$ compared with Fe that was bound to protein, indicating that chelated Fe may not always be more bioavailable than non-chelated Fe (Thomaz et al. 2014; Ma et al. 2018).

Commercial dicalcium phosphate and defluorinated phosphates are rich in Fe, containing 2.5–3.0% $FePO_4 \cdot 2H_2O$ because of the impurity of the sources (Baker 1989). Bioavailability of Fe in these products generally ranges from 35 to 85% (Ammerman and Miller 1972; Kornegay 1972; Deming and Czarnecki-Maulden 1989).

The Fe in grains and oilseed meals is complexed with phytate, fiber, or protein. As such, its bioavailability is expected to be lower than in $FeSO_4 \cdot 7H_2O$, but data for availability of Fe in these sources are limited. Without definitive data, one should assume that cereal grain and oilseed Fe sources are no more than 50% available, relative to $FeSO_4 \cdot H_2O$.

The true digestibility of Fe by humans consuming a mixed animal- and plant-based diet is assumed to be around 15% (Groff et al. 1995), but many factors can affect the digestibility of Fe. Dietary ascorbic acid, Cys, and organic acids including citrate or lactate may double the efficiency of Fe absorption. On the other hand, dietary phytate, oxalates, and excess dietary Zn (in the presence of phytate) decrease the absorption of Fe to less than half that occurring without these antagonizing factors. Digestibility of Fe is much greater when consumed by Fe-deficient animals than by Fe-adequate animals. The ATTD of Fe in Fe-adequate diets containing plant feed ingredients and Fe sulfate is less than 40% (Burkett et al. 2009; Liu et al. 2014b; Arredondo et al. 2019).

*Manganese*

Manganese is a trace mineral that is involved in metabolism of nutrients, bone formation, blood clotting, and maintenance of immune systems. Manganese deficiency is a greater problem in poultry than in swine. The bioavailability of manganese has, therefore, been extensively studied using avian models (Southern and Baker 1983a, b; Halpin and Baker 1987; Tufarelli and Laudadio 2017). Tissues (primarily bone) accumulate Mn linearly, and this is used as a basis for assessing Mn bioavailability. Relative to $MnSO_4 \cdot H_2O$, Mn bioavailability is approximately 100% in $MnCl_2$, 75% in MnO, 55% in $MnCO_3$, and 30% in $MnO_2$ (Henry 1995). Availability of Mn in a protein-Mn or a Met-Mn complex is similar to that in $MnSO_4 \cdot H_2O$ (Baker et al. 1986; Fly et al. 1989). The Mn in corn, soybean meal, wheat bran, and fish meal is considered totally unavailable for both poultry and swine (Baker et al. 1986), and the Mn in rice bran is only minimally bioavailable.

Values for the ATTD of Mn in plant feed ingredients fed to pigs ranges from 14.7 to 27.3% (Adeola and Orban 1995; Liu et al. 2014b; Merriman et al. 2016). In practical diets, excesses of either Fe or Co in corn-soybean meal diets have minimal influence on Mn utilization, but excesses of Mn reduce Fe absorption from the gut. Excess dietary P reduces Mn utilization substantially in poultry (Baker and Wedekind 1988; Wedekind and Baker 1990a, b; Wedekind et al. 1991a, b; Baker and Oduho 1994). Increasing dietary F also decreases retention of Mg in body tissues (Tao et al. 2005).

Even with a highly available source of Mn such as $MnSO_4 \cdot H_2O$, gut absorption of Mn by chicks is only about 2–3% of the ingested dose (Wedekind et al. 1991a, b), but elimination of all fiber and phytate from the diet nearly doubles the absorption efficiency of Mn (Halpin et al. 1986). However, in pigs fed a corn-soybean meal diet supplemented with $MnSO_4$, the ATTD of Mn was 35% and this value was increased to 50% if microbial phytase was added to the diet (Arredondo et al. 2019).

### Selenium

Selenium is a mineral that is an essential component of antioxidant enzymes. Because most feed ingredients that are used in swine diets have a low concentration of Se (NRC 2012), supplemental Se is included in commercial mineral premixes.

Selenium in $Na_2SeO_3$ and $Na_2SeO_4$ is well utilized, and based on Se accumulation in tissues, it is estimated that Se in selenomethionine is more than 100% bioavailable in poultry and pigs relative to Se in $Na_2SeO_3$ (Mathias et al. 1967; Cantor et al. 1975a, b; Mahan and Moxon 1978). Selenium from cereal grains has high digestibility (60–80%), and even greater digestibility of Se from alfalfa meal has been reported, whereas digestibility of Se is less than 40% in fish and poultry by-product meals. Poultry data provided Se bioavailability estimates in soybean meal of only 18% for restoration of glutathione peroxidase activity, but 60% for prevention of exudative diathesis (Cantor et al. 1975b). The bioavailability of Se in animal-derived feed ingredients fed to broiler chicken averages 28% relative to $Na_2SeO_3$, whereas plant feed ingredients have Se bioavailability of 47% (Wedekind et al. 1998). Selenium-enriched yeast has a bioavailability of 159% relative to $Na_2SeO_3$. Mahan and Parrett (1996) also compared Se-enriched yeast to $Na_2SeO_3$ as Se sources for grower-finisher pigs. Based on Se retention in the body, the Se in Se-enriched yeast was more bioavailable than the Se in $Na2SeO_3$, but the opposite was true when serum GSH was used as the criterion for determining bioavailability.

Selenium absorption from the gut is relatively efficient and 63% of an administered dose was absorbed by pigs (Wright and Bell 1966). There is little definitive information on the ATTD of Se in feed ingredients fed to pigs, but it seems that the true digestibility of Se is approximately 41 and 73%, respectively, for sodium selenite and Se-enriched yeast fed to growing pigs (Kim and Mahan 2001). A variety of arsenic compounds as well as Cys, Met, Cu, tungsten (W), Hg, Cd, and Ag decrease the efficiency of Se absorption from the gut (Baker and Czarnecki-Maulden 1987; Lowry and Baker 1989).

### Zinc

Zinc is needed for the immune system, structure and function of the skin, the senses of smell and taste, and synthesis of insulin. Dietary Zn in swine diets is mostly provided by inorganic sources (i.e., Zn oxide or Zn sulfate) in mineral premixes. Zinc, like many other trace elements, is poorly utilized by nonruminant animals fed conventional corn-soybean meal diets. Indeed, the dietary requirement for Zn is three to four times greater in animals fed these diets than in those fed phytate-free (e.g., egg white) diets (Norii and Suzuki 2002; Gibson et al. 2018). In the presence of phytate and fiber, excess Ca decreases Zn absorption and retention (Morgan et al. 1969). However, use of pharmacological levels of Zn (i.e., 3000 mg/kg diet) may interfere with Ca and P digestibility by competing for the same transports for absorption or by forming an indigestible Ca-phytate-Zn complex (Blavi et al. 2017). High concentrations of dietary Zn oxide also reduce the ability of microbial phytase to increase the ATTD of P in diets fed to pigs (Blavi et al. 2017). Whereas excess Zn can exacerbate Cu and Fe deficiency, excesses of either Cu or Fe have minimal effects on Zn utilization (Southern and Baker 1983c; Bafundo et al. 1984).

Little swine data exist for the relative bioavailability of Zn in Zn-containing supplements. Miller et al. (1981) reported that the bioavailability of Zn in Zn dust (99.3% Zn) was high for pigs relative to analytical-grade ZnO. Feed-grade ZnO for pigs, however, has a bioavailability of only 56–68% (Hahn and Baker 1993; Wedekind et al. 1994), relative to that in a feed-grade $ZnSO_4 \cdot H_2O$ standard. The ATTD of Zn in inorganic Zn and feed ingredients ranges from 7 to 40% (Burkett et al. 2009; Liu et al. 2014b; Merriman et al. 2016), but the ATTD of Zn is increased if dietary phytate is removed or if a chelated source is included in diets fed to pigs (Liu et al. 2014b). However, addition of microbial phytase to a corn-soybean meal-based diet did not increase the ATTD of Zn (Arredondo et al. 2019), indicating that there is limiting binding of Zn to phytate in a corn-soybean meal diet.

Many of the factors that affect the efficiency of Fe utilization also apply to Zn. Thus, low intakes of Zn and dietary-reducing agents, such as ascorbic acid and Cys, increase Zn absorption. Stress or trauma or both (e.g., surgery and burns) decrease the efficiency of Zn absorption in humans (Groff et al. 1995).

Because high levels of feed-grade ZnO are often used in the United States for growth promotion in weanling pigs (Hahn and Baker 1993; Hill et al. 2000), the issue of bioavailability of Zn in ZnO products has become more important. Based on the review by Baker and Ammerman (1995b), it may be concluded that the Zn in $ZnSO_4 \cdot H_2O$, $ZnCO_3$, $ZnCl_2$, analytical-grade ZnO, Zn-Met, and Zn acetate is highly bioavailable relative to analytical-grade $ZnSO_4 \cdot 7H_2O$, and all of these sources may, therefore, supply Zn in diets fed to swine. Weight gain and bone Zn accumulation of animals fed Zn-deficient diets are the best measures of bioavailability of Zn, but soft-tissue Zn, plasma Zn, and plasma alkaline phosphatase activity generally give poor fits when regressed against supplemental Zn intake (Wedekind et al. 1992).

The Zn in soy products is poorly utilized (Edwards and Baker 2000). Soybean meal, soy protein concentrate, and soy protein isolate have Zn bioavailabilities (relative to $ZnSO_4 \cdot 7H_2O$) of 34, 18, and 25%, respectively. Utilization of Zn is greater in animal ingredients than in plant-source ingredients, but some animal products may contain factors that antagonize Zn utilization (Baker and Halpin 1988).

*Chromium*

Chromium is needed for nutrient and nucleic acid metabolism. Chromium helps insulin bind to cells in the body for controlling blood concentration of glucose. However, there is limited information about concentrations of Cr in feed ingredients used in swine diets (NRC 2012), and the digestibility or bioavailability of Cr has rarely been reported. Chromium became of interest in swine nutrition when Page et al. (1993) reported that Cr tripicolinate supplementation increased carcass merits in finishing pigs. Subsequently, it was observed that Cr addition to diets for gestation sows may increase litter size (Lindemann et al. 1995, 2004). However, use of Cr tripicolinate did not increase the performance of pigs under thermal and immune stresses (Kim et al. 2009, 2010).

Relative to Cr tripicolinate, the bioavailability of Cr in Cr propionate, Cr-Met, and Cr yeast is 13.1, 50.5, and 22.8%, respectively, in the growing pig (Lindemann et al. 2008). There is no definitive information on the ATTD of Cr in supplemental Cr or complete diets fed to pigs.

**Bioavailability and Digestibility of Vitamins**

Vitamins are essential nutrients that are required for normal metabolic function of the body. Vitamins may be provided by feed ingredients either in the functional form or as the precursor to the functional form. Microbes in the hindgut may also synthesize some of the water-soluble vitamins. However, commercial diets for breeding and growing pigs are supplemented with crystalline sources of vitamins above the requirements of pigs (Flohr et al. 2016). There are two primary concerns regarding vitamin bioavailability: (i) stability in vitamin and vitamin-mineral premixes and in diets and supplements and (ii) utilization efficiency from plant and animal feed ingredients. Reviews by Wornick (1968), Zhuge and Klopfenstein (1986), Baker (1995, 2001), and Veum (2004) provide details of factors affecting the stability of crystalline vitamins in diets and premixes. Cautions to minimize loss of efficacy may include avoiding heat, light, oxygen, and moisture.

There are many pitfalls when bioavailability of vitamin is estimated. Body stores often preclude developing a distinct deficiency during the course of a conventional growth trial. Even if a deficiency can be produced, one must deal with the vexing question of whether the responding criterion (usually weight gain) increased because of the vitamin being supplied or perhaps because of increased diet intake that results from adding the unknown ingredient to an often less-than-voraciously palatable purified diet. Because water-soluble B vitamins respond better insofar as growth is concerned, they are, in many respects, easier to evaluate than fat-soluble vitamins.

Certain conclusions about the proper bioassay methodology for maximum efficacy and extrapolative values of results for assessing vitamin bioavailability are evident (i) a pretest period to obtain desired deficiency states is generally necessary, (ii) the activity of a key enzyme of which the vitamin is a component or cofactor is generally a less-desirable dependent variable than weight gain, (iii) precursor materials (e.g., Met for choline and Trp for niacin) must be carefully considered, and (iv) use of specific vitamin inhibitors may assist in establishing the veracity of assessed bioavailability values.

### Vitamin A

Vitamin nomenclature policy (Anonymous 1979) dictates that the term vitamin A be used for all β-ionone derivatives, other than pro-vitamin A carotenoids that exhibit the biological activity of all-*trans* retinol (i.e., vitamin A alcohol or vitamin A$_1$). Esters of all-*trans* retinol should be referred to as retinyl esters. Vitamin A is fat-soluble and exists in many different forms – retimoids, carotenoids, and retinol (i.e., retinyl esters).

Vitamin A is important for vision because it is needed in the synthesis of pigments in the retina of the eyes. Retinoic acid, another form of Vitamin A, is also needed for the growth, development, and maintenance of the immune system by differentiating and proliferating T cells.

Vitamin A is present in animal tissues, whereas most plant materials contain only pro-vitamin A carotenoids, which must be split in the intestinal tract to form vitamin A. Digestive enzymes including pepsin, proteolytic enzymes, lipases, and esterase can release vitamin A and carotenoids from foods (Sauberlich 1985). Because the retinyl esters and carotenoids are hydrophobic, absorption depends on the micelle formation in the small intestine. Absorption efficiency of the retinols and retinyl esters is around 75%, but it is variable for carotenoids (Combs 2012a). Absorption efficiency of vitamin A is relatively constant regardless of dietary levels, but higher doses of carotenoids are absorbed much less efficiently than lower doses (Erdman et al. 1988). Crystalline β-carotene has better absorption efficiency compared with the intact form of β-carotene that exists in protein- or fiber complexes (Rao and Rao 1970; Rodriguez and Irwin 1972). Dietary pectin reduces absorption of β-carotene in broiler chickens (Erdman et al. 1986). After absorption, vitamin A is transported as retinol in blood, but it is stored as retinyl palmitate mostly in the liver.

Vitamin A esters are more stable in feeds and premixes than retinol because retinol has the hydroxyl group and the four double bonds on the side chain, which is susceptible to oxidative losses. Vitamin A can be esterified, but this does not completely protect the vitamin from oxidative loses. Current commercial sources of vitamin A are generally "coated" esters (e.g., acetate or palmitate) that contain an added antioxidant such as ethoxyquin or butylated hyroxytoluene.

The bioavailability of vitamin A precursor materials is lower in pigs compared with rats because pigs are less efficient in converting carotenoid precursors to active vitamin A (Ullrey 1972). Relative to all-trans retinyl palmitate, bioefficacies (wt/wt) were from 7 to 14% in corn carotenes fed to pigs. Thus, at best, carotenoid precursors in corn and corn coproducts have no more than 261 international unit (**IU**)/mg vitamin A activity when consumed by swine. Quantification of vitamin A bioavailability is difficult, but accumulation of vitamin A in the liver may be the most acceptable method (Erdman et al. 1988; Chung et al. 1990).

## *Vitamin D*

Vitamin D is referred as "the sunshine vitamin" because exposure to ultraviolet light will produce the active form of Vitamin D. Vitamin D is mainly involved in maintaining blood levels of Ca. The term *vitamin D* is appropriate for all steroids having cholecalciferol biological activity (Ross et al. 2011). Vitamin D indicates both ergocalciferol (vitamin $D_2$) and cholecalciferol (vitamin $D_3$) that exist in nature.

Feed ingredients used in swine diets may contain vitamin D, but commercially, vitamin $D_3$ is available as a spray-dried product (frequently in combination with vitamin A) or as gelatin-coated beadlets. One IU is equal to $0.025\,\mu g$ of cholecalciferol (Anonymous 1979). These products are quite stable if stored as the vitamin itself at room temperature. However, in complete feeds and mineral-vitamin premixes, up to 20% of vitamin D activity can be lost after four to six months of storage at room temperature (Baker 2001).

Vitamin D precursors are present in plant and animal feed ingredients as ergosterol 7-dehydrocholesterol, but the precursor needs to be convented into active forms by ultraviolet irradiation. Bioavailability of vitamin $D_2$ and $D_3$ is believed to be equal for pigs, but $D_3$ may be more bioactive than $D_2$ (Horst et al. 1982). Hydroxylated forms of cholecalciferol [$25(OH)D_3$, $1\alpha(OH)D_3$, $1,25(OH)_2D_3$], particularly $1\alpha$-hydroxylated products, are assumed to be more bioavailable than vitamin $D_3$ because the steps of converting into active forms are eliminated in the body, but research to demonstrate the bioavailability of these metabolites is needed.

## *Vitamin E*

Vitamin E has antioxidant functions in cells. Vitamin E can only be synthesized by photosynthetic organisms including plants. Vitamin E is the generic term for all tocol and tocotrienol derivatives having $\alpha$-tocopherol biological activity. There are eight naturally occurring forms of vitamin E: $\alpha$-, $\beta$-, $\gamma$-, and $\delta$-tocopherols and $\alpha$-, $\beta$-, $\gamma$-, and $\delta$-tocotrienols. Among these, D-$\alpha$-tocopherol possesses the greatest biological activity (Bieri and McKenna 1981). An IU of vitamin E is the activity of 1 mg of DL-$\alpha$-tocopheryl acetate. The DL-form of tocopheryl acetate (i.e., tocopherol) has less bioavailability for swine compared with the D-form (Anderson et al. 1995; Shelton et al. 2014). Compared with $\alpha$-tocopherol, $\beta$-tocopherol and $\gamma$-tocopherol have only 40 and 10% activity, respectively (Bieri and McKenna 1981). Alpha-tocotrienol is the only other natural form to possess activity, which on the rating scale used above was listed by Bieri and McKenna (1981) as containing a biopotency of 25%.

Plant ingredients are richer in vitamin E bioactivity than animal-source feed ingredients. Plant oils are particularly rich in bioactive vitamin E, although corn and corn oil contain about six times more γ-tocol than α-tocol (Ullrey 1981). Fat-extracted soybean meal has very little vitamin E activity.

Because vitamin E is fat-soluble, absorption takes place in the small intestine with fat droplets. Depending on the intake of vitamin E, the absorption efficiency ranges from 20 to 70%, but γ-tocopherol is only 85% efficient relative to α-tocopherol (Combs 2012b).

Vitamin E can be destroyed by oxidation, and this process is accelerated by heat, moisture, unsaturated fat, and trace minerals. Losses of 50–70% may occur in alfalfa hay stored at 32 °C for 12 weeks and losses of up to 30% may occur during dehydration of alfalfa (Livingston et al. 1968). Treatment of high-moisture grains with organic acids also greatly enhances vitamin E destruction (Young et al. 1975; Young et al. 1977, 1978). However, even mildly alkaline conditions of vitamin E storage are also detrimental to vitamin E stability. For example, finely ground limestone or MgO coming in direct contact with vitamin E can markedly reduce bioavailability of vitamin E, and vitamin E may also be destroyed by fermentation of grains.

### Vitamin K

Vitamin K is needed for blood coagulation. It is also involved in the chelation of Ca in bones. There are three forms of vitamin K: phylloquinones ($K_1$) in plants, menaquinones ($K_2$) formed by microbial fermentation, and menadiones ($K_3$), which are synthetic. All three forms of vitamin K are biologically active, but because vitamin K has a rapid turnover rate, the metabolites are quantitatively excreted in the urine.

Because animals cannot synthesize vitamin K, supplementation of vitamin K by natural sources or vitamin premixes is necessary for pigs, even though vitamin K may be synthesized via microbial fermentation. Only water-soluble forms of menadione (vitamin $K_3$) are used to supplement swine diets. Because menadione is water soluble, it can be absorbed well even in low-fat diets. The commercially available forms of $K_3$ supplements are menadione sodium bisulfite (MSB), menadione sodium bisulfate complex (MSBC), and menadione dimethyl pyrimidinol bisulfite (MPB). These forms contain 52, 33, and 45.5% menadione, respectively. Stability of these $K_3$ supplements in premixes and diets is impaired by moisture, choline chloride, trace elements, and alkaline conditions. The MSBC and MPB forms of vitamin K can lose almost 80% of bioactivity if stored for three months in a vitamin-trace mineral premix containing choline, but losses are less if stored in a similar premix containing no choline (Baker 2001).

Coated $K_3$ supplements are generally more stable than uncoated supplements. Bioactivity of MPB is greater than either MSB or MSBC for broiler chickens (Griminger 1965; Charles and Huston 1972), and MPB also have high bioactivity in diets fed to pigs (Seerley et al. 1976). Menadione nicotinamide bisulfite (MNB; 45.7% menadione, 32% nicotinamide) can be used as a source of vitamin K for young broiler chickens and pigs (Oduho et al. 1993; Marchetti et al. 2000).

### Thiamine (Vitamin B$_1$)

Thiamine is needed for carbohydrate metabolism, branched chain amino acid metabolism, and neural functions (Sauberlich 1985; Koh et al. 2015). Thiamine occurs predominantly as free forms in plants, but thiamine in animal tissues is in phosphorylated forms. Thiamine is available to the food and feed industries as thiamine·HCl (89% thiamine) or thiamine·NO$_3$ (92% thiamine). These compounds are stable up to 100 °C and are readily soluble in water (NRC 1987). An IU of thiamine activity is equivalent to 3 µg of crystalline thiamine·HCl.

Because thiamine contains a free amino group, heat processing can reduce thiamine bioactivity via the Maillard reaction if moisture is present as well. Similarly, alkaline treatment may reduce thiamine activity. The thiamine contained in swine feed ingredients is present largely in phosphorylated forms, either as protein-phosphate complexes or as thiamine mono-, di-, or triphosphates. Phosphatase and pyrophosphatase are used to release thiamin from the phosphate complex, and thiamin is absorbed in the small intestine via both active transport and passive diffusion.

Some raw ingredients (e.g., fish) contain thiaminase, which can destroy thiamine in diets to which it may be added. While thiaminase is of particular concern in the nutrition of cats and fur-bearing animals, it is of little consequence in modern swine feeding. Thiamine in fishmeal is lost to the fish solubles fraction when fishmeal is produced. Thus, fishmeal contains essentially no bioavailable thiamine. Similarly, as a result of the high-temperature processing, meat meals and pelleted feeds contain very little bioavailable thiamine.

Grains and soybean meal contain thiamine (Chen et al. 2019). The AID of thiamine was greater than 60% in wheat, whole-grain bread, and boiled potatoes and was greater than 90% if roast pork provides most thiamin in diets fed to pigs (Roth-Maier et al. 1998). Therefore, even with losses of bioactivity due to heat or lengthy storage, practical diets for swine will rarely be deficient in thiamine. However, in most cases, commercial diets for pigs are fortified with synthetic thiamine (Flohr et al. 2016).

### Riboflavin (Vitamin B$_2$)

Riboflavin is essential for metabolism of carbohydrates, AA, and lipids and for cellular antioxidant protection. This vitamin is stable to heat, but relatively labile to light, alkali, and oxygen, which results in reduced bioavailability. In feedstuffs, riboflavin exists primarily as a nucleotide coenzyme, in which form the bioavailability is probably less than 100%. Proteolytic enzymes and brush border phosphatases are needed to liberate a free form of riboflavin to be absorbed via active transport. Bioavailability of riboflavin in a corn-soybean meal diet is estimated at 60% relative to crystalline riboflavin in chickens (Chung and Baker 1990). The AID of riboflavin in roast pork and roast beef fed to pigs was greater than 50% (Roth-Maier et al. 1998).

Crystalline riboflavin may lose activity in vitamin-mineral premixes over time and high-temperature storage enhances the activity loss (Zhuge and Klopfenstein 1986). Excess dietary levels of tetracycline, Fe, Zn, Cu, ascorbate, and caffeine may reduce the bioavailability of riboflavin (Sauberlich 1985). However, growth performance data failed to prove negative effects of any nutrients on the bioavailability of crystalline riboflavin in broiler chickens (Patel and Baker 1996).

### Niacin (Vitamin B$_3$)

Niacin is essential to biosynthesize the pyridine nucleotides [i.e., NAD(H) and NADP(H)] that are involved in the metabolism of energy and macronutrients. Supplementation of crystalline niacin to niacin-deficient diets increased feed intake, N retention, and weight gain of pigs, but energy retention was not affected by niacin supplementation (Ivers and Veum 2012). Yeast and some animal feed ingredients are great sources of niacin. The term *niacin* is the generic descriptive term for pyridine 3-carboxylic acid and derivatives delivering nicotinamide activity. Thus, pyridine 3-carboxylic acid per se is properly referred to as nicotinic acid (Anonymous 1979).

Niacin is stable to heat, oxygen, moisture, or light when added to feed or premixes. By-products of cereal grains and oilseeds contain more niacin compared with grains or oilseeds (Chen et al. 2019). Much of the niacin in plant feed ingredients, mostly nicotinamide nucleotides, is unavailable because they are bound to protein (Yen et al. 1977). Approximately 85–90% of the niacin in cereal grains and 40% of niacin in oilseeds are in a bound form and thus unavailable (Ghosh et al. 1963). The AID of niacin in wheat by growing pigs was approximately 55% (Wauer et al. 1999). Hydrolyzing with alkaline may release protein- or carbohydrate-bound niacin in these ingredients and niacin in corn is, therefore, more available in pigs after soaking corn in an alkali solution (Kodicek et al. 1959). Meat and milk products, on the other hand, do not contain bound niacin, but instead contain free nicotinic acid and nicotinamide.

Because nicotinic acid is synthesized from Trp and because all common feed ingredients contain both Trp and nicotinic acid, it is very difficult to estimate the bioavailability of niacin in feed ingredients per se. It is estimated that 50 mg of Trp yields 1 mg of nicotinic acid (Baker et al. 1973; Czarnecki et al. 1983). It is possible that excess Leu in corn-soybean meal swine diets antagonizes Trp or nicotinic acid or impairs the metabolic conversion of Trp to nicotinic acid (Anonymous 1986). However, data from studies with broiler chickens indicate that excess Leu has no effect on either Trp conversion to niacin or niacin bioavailability (Lowry and Baker 1998). Excess dietary Leu may decrease plasma and hypothalamic concentrations of serotonin that is synthesized from Trp, indicating that Trp uptake into the brain can be reduced by excess Leu (Wessels et al. 2016; Kwon et al. 2019). However, there is scarce information about the antagonistic effects of excess dietary Leu on niacin synthesis in pigs.

Iron, on the other hand, is required in two metabolic reactions in the pathway of Trp to nicotinate mononucleotide. The conversion efficiency of Trp to niacin is reduced in broiler chickens that are deficient in Fe (Oduho et al. 1994).

Niacin in nicotinamide is approximately 120% as bioavailable as niacin in nicotinic acid as a precursor for nicotinamide adenine dinucleotide synthesis (Baker et al. 1976; Oduho and Baker 1993). It has, however, also been suggested that niacin and nicotinamide are equally bioavailable in chickens (Bao-Ji and Combs 1986; Ruiz and Harms 1988).

### Pantothenic Acid (Vitamin B₅)

*Pantothenic Acid (Vitamin $B_5$)*

Pantothenic acid (**PA**) is used in the synthesis of coenzyme A. The two forms in which PA is generally sold are D- and DL-Ca pantothenate, but the DL-form needs to converted to the D-isomer to be utilized by the body (Staten et al. 1980). Whereas 92% of D-Ca pantothenate can generate PA, only 46% of DL-Ca pantothenate can generate PA activity. Crystalline PA is relatively stable to heat, oxygen, and light, but it can lose activity rapidly when exposed to moisture.

Most foods and feed ingredients contain PA in the form of coenzyme A or acyl-carrier protein. Thus, absorption may depend on the digestibility of protein complexes. Pantothenic acid in corn and soybean meal diets is 100% bioavailable, but it is only 60% for PA in barley, wheat, and sorghum in broiler chickens (Southern and Baker 1981). Values for the bioavailability of PA in a typical American diet for adults range from 40 to 60% (Yu and Kies 1993), but processing including freezing, canning, or refining may decrease bioavailability of PA in foods (Sauberlich 1985). The AID of pantothenic acids in wheat by growing pigs was approximately 60% (Wauer et al. 1999).

## Vitamin B₆

*Vitamin B*$_6$

The three naturally occurring forms of vitamin B$_6$ are pyridoxine, pyridoxamine, and pyridoxal. This vitamin is important for many reactions in the metabolism of AA and synthesis of red blood cells and neurotransmitters. Corn and soybean meal contain significant quantities of vitamin B (Chen et al. 2019), and supplemental vitamin B$_6$ is not generally necessary in practical swine diets. Because corn and soybean meal-based diets are not deficient in vitamin B$_6$, it is difficult to estimate bioavailability of supplemental vitamin B$_6$ if added to these diets, and synthetic diets are, therefore, needed to determine vitamin B$_6$ bioavailability. The AID of vitamin B$_6$ in wheat fed to pigs was approximately 68%, but the AID of pyridoxine and pyridoxal in wheat was approximately 88 and 61%, respectively (Wauer et al. 1999). The AID of pyridoxamine in wheat was close to 0% (Wauer et al. 1999).

Compared with crystalline vitamin B$_6$, values for the bioavailability of vitamin B$_6$ in corn and soybean meal is 40 and 60%, respectively (Yen et al. 1976). Treating corn with moderate heat (i.e., 80–120 °C) may increase vitamin B$_6$ bioavailability, but greater heat treatment (i.e., 160 °C) may decrease availability (Yen et al. 1976). Corn contains vitamin B$_6$ as forms of pyridoxal and pyridoxamine, but vitamin B$_6$ in corn is more susceptible to heat compared with pyridoxine (Schroeder 1971). Plant feed ingredients may contain B$_6$ in forms of either pyridoxine glucoside or pyridoxallysine, but both compounds have minimal B$_6$ bioactivity (Gregory and Kirk 1981; Trumbo et al. 1988).

Minerals in the form of carbonates or oxides in premixes can reduce the bioactivity of vitamin B$_6$ (Verbeeck 1975). High temperatures can also enhance the loss of activity. Retention of B$_6$ activity after three months of storage at room temperature is 76%, but activity decreased to 45% after three months of storage at 37 °C (Baker 2001). Vitamin B$_6$ activity in pelleted complete diets that are stored at room temperature for three months can be reduced by 20% (Baker 2001).

## Biotin (Vitamin B₇)

*Biotin (Vitamin B*$_7$*)*

Commercial D-biotin has no specific unit of activity. Thus, 1 g of D-biotin equals 1 g of activity. Much of the biotin in feed ingredients exists in a bound form, ε-N-biotinyl-L-Lys (biocytin). Crystalline biotin is absorbed from the small intestine, but the bioavailability of biotin in biocytin varies and is dependent on the digestibility of the proteins in which it is present (Baker 1995). Pelleting or heat treatment has little effect on biotin activity in feeds, but oxidative rancidity severely reduces biotin bioavailability.

Avidin, a glycoprotein in egg albumen, binds biotin and makes biotin less available for the body (Hamilton and Veum 2012). Proper heat treatment of egg white will denature avidin and prevent it from binding biotin. Bioavailability of biotin in corn fed to broiler chickens is twice as high as in wheat, barley, and sorghum (Anderson and Warnick 1970; Frigg 1976; Anderson et al. 1978). Bioavailable biotin concentrations are 0.11, 0.08, 0.09, and 0.04 mg/kg for corn, barley, sorghum, and wheat, respectively (Anderson et al. 1978). In laying hens, biotin in soybean meal and meat and bone meal is 100 and 86% available, respectively (Buenrostro and Kratzer 1984).

Because of the high concentrations of bioavailable biotin in corn and soybean meal, a response to supplemental biotin in pigs fed a corn-soybean meal-based diets is not expected. However, supplemental biotin to sows may increase conception rate, decrease weaning-to-estrus interval, and improve both foot health and hair coat, particularly in advanced parities (Bryant et al. 1985). Addition of 0.33 mg biotin/kg to a corn-soybean meal diet throughout gestation and lactation increases the number of pigs weaned, but does not improve foot health (Lewis et al. 1991). However, supplementation of corn-soybean meal diets for gilts or sows with 0.44–0.70 mg biotin/kg does not always increase N and energy digestibility or animal performance (Watkins et al. 1991; Hamilton and Veum 2012).

### Folacin (Vitamin $B_9$)

The term *folacin* is the accepted generic term for folic acid and related compounds exhibiting "folacin" activity. Over 150 forms of folacin exist in foods. Chemically, folic acid consists of a pteridine ring, *para*-aminobenzoic acid (PABA), and glutamic acid. Animal cells cannot synthesize PABA, nor can they attach glutamic acid to pteroic acid (i.e., pteridine attached to PABA). Therefore, folic acid must be supplied in the diet of non-ruminant animals. The folacin in feeds and foods exists largely as poly-glutamates (Pietrzik et al. 2010). In plants, folacin exists as a poly-glutamate conjugate containing a γ-linked polypeptide chain of (primarily) seven glutamic acid residues. Intestinal proteases do not cleave the glutamate residues from this compound. Instead, a group of intestinal enzymes known as conjugases (folyl poly-glutamate hydrolases) remove all, but the last glutamate residue. The mono-glutamyl form than poly-glutamates is thought to be absorbed into the enterocyte (Pietrzik et al. 2010). Most of the folic acid taken up by the brush border is reduced to tetrahydrofolate ($FH_4$) and then methylated to $N^5$-methyl-$FH_4$, which is the predominant form of folate in blood plasma. The majority of the $N^5$-methyl-$FH_4$ in plasma is bound to protein.

Like thiamin, folic acid has a free amino group (on the pteridine ring), which makes it sensitive to heat treatment with moisture and reducing sugars. Intestinal conjugase inhibitors may be present in certain beans and pulses, which may reduce folacin absorption (Krumdieck et al. 1973; Bailey 1988). Storage of feeds and premixes can also reduce folacin activity (Verbeeck 1975).

It is difficult to observe a response to folacin supplementation from growing pigs fed conventional corn-soybean meal diets, but vitamin premixes used in swine diets usually contain folic acid (Flohr et al. 2016). Reproductive performance may be improved by supplementation of folacin to diets fed to gestating-lactating sows (Lindemann and Kornegay 1989; Matte et al. 1992), although that is not always the case (Pharazyn and Aherne 1987; Easter et al. 1983; Harper et al. 1994).

### Cyanocobalamin (Vitamin $B_{12}$)

Vitamin $B_{12}$ is present in the tissues of animals and is involved in energy and AA metabolisms. This vitamin can be synthesized by bacteria, and plant feed ingredients are usually devoid in vitamin $B_{12}$. Both animal and fermented feed ingredients contain vitamin $B_{12}$ as methylcobalamin or adenosyl-cobalamin that are bound to proteins. As in humans, but unlike in sheep and horses, an "intrinsic" factor is required for gut absorption of $B_{12}$ by pigs. The bioavailability of vitamin $B_{12}$ ranges from 12 to 33% in dairy products fed to pigs compared with synthetic vitamin $B_{12}$ (Dalto et al. 2018). Bioavailability of vitamin $B_{12}$ by pigs decreases if dietary vitamin $B_{12}$ increases, which indicates that there is a negative correlation between intake and bioavailability (Matte et al. 2010).

Cyanocobalamin, or vitamin $B_{12}$, is available in crystalline form, where 1 U.S. Pharmacopoeia unit is considered equivalent to 1 μg of the vitamin. Crystalline vitamin $B_{12}$ is believed to be stable in feeds and premixes because crystalline $B_{12}$ is stable in long-term storage (Macek and Feller 1952).

## Choline

In animal nutrition, choline is categorized as a B-vitamin, even though the quantity required far exceeds the "trace organic nutrient" definition of a vitamin. Choline is involved in a number of metabolisms in the body including (i) phospholipid synthesis, (ii) acetyl choline formation, and (iii) transmethylation of homocysteine to Met. Broiler chickens fed a choline-free diet prioritized phospholipid and acetylcholine synthesis over transmethylation. If one half to two-thirds of the choline needed for maximal growth is supplied as choline, as in practical diets, synthetic choline and betaine are equally efficacious (Lowry et al. 1987; Dilger et al. 2007).

In mammalian, but not in avian species, excess Met can replace the dietary need for choline. In crystalline form, choline chloride (74.6% choline) is hygroscopic. Therefore, a separate premix for choline is needed to prevent other vitamins in the general premix from the damage derived from high humidity. Crude plant oils (e.g., corn and soybean oil) contain choline as phospholipid-bound phosphatidyl choline. The bioavailability of choline in this form is at least 100% relative to choline chloride (Emmert et al. 1996). Alkaline treatment and "bleaching" are applied when refined plant oils are produced, which removes most phospholipids, including phospholipid-bound choline, and refined plant oils, therefore, do not contain choline.

Values for the bioavailability of choline in soybean meal, peanut meal, and canola meal fed to broiler chickens are estimated at 83, 76, and 24%, respectively, relative to crystalline choline chloride (Molitoris and Baker 1976a; Emmert and Baker 1997). Excess protein in diets fed to broiler chickens may increase the dietary requirement for choline (Molitoris and Baker 1976b). Because choline is involved in the metabolism of lipids, more choline is needed to minimize liver fat content compared with the amount of choline needed to maximize growth (Anderson et al. 1979).

An assessment of bioavailability of choline is difficult because all common feed ingredients used in swine diets supply both choline and Met, which is likely to confound results. However, use of a transmethylation inhibitor or an inhibitor of Met methylation of aminoethanol in choline biosynthesis (i.e., 2-amino-2-methyl-1-propanol) may be useful (Molitoris and Baker 1976a; Anderson et al. 1979; Lowry et al. 1987).

Pigs fed a corn-soybean meal diet often do not respond to choline supplementation, which is likely because soybean meal is so rich in choline (NCR-42 Committee on Swine Nutrition 1980). However, gestating sows benefit from choline addition to corn-soybean meal diets (Kornegay and Meacham 1973; Stockland and Blaylock 1974; NCR-42 Committee on Swine Nutrition 1976).

## Ascorbic Acid (Vitamin C)

Ascorbic acid is a water-soluble vitamin that is needed in the antioxidant protection of cells. Vitamin C exists mostly as ascorbic acid, but also as dehydroascorbic acid in plant and animals foods and feeds. There is little concern about the bioavailability of ascorbic acid because pigs can synthesize vitamin C. Nonetheless, vitamin C is often included in vitamin premixes for use in purified swine diets because of its antioxidant and putative antistress properties. Plasma cortisol, an indicator of stress, is likely reduced by supplementation of vitamin C in diets fed to weanling pigs that are changed with a lipopolysaccharide (Eicher et al. 2006).

Vitamin C activity may be lost during storage, but coating ascorbate with ethylcellulose may reduce the loss of potency. Both pelleting and extrusion may reduce bioactivity of supplemental ascorbate (Baker 2001). Vitamin C is susceptible to oxidation that transforms ascorbic acid (reduced form) to dehydroascorbic acid, which in turn can be further irreversibly oxidized to diketogulonic acid (Parsons et al. 2011). Both reduced and oxidized forms of ascorbate retain scurvy-preventing ascorbate activity, but diketogulonic acid has no activity because the reaction resulting in generation of this compound is not reversible. Both ascorbate and dehydroascorbate are heat labile, particularly when heat is applied in the presence of trace minerals such as Cu, Fe, or Zn.

## References

Abelilla, J. J. 2018. Fermentation and energetic value of fiber in feed ingredients and diets fed to pigs. PhD. Diss. Univ. Illinois, Urbana-Champaign, IL, US. http://hdl.handle.net/2142/101027.

Abelilla, J. J., and H. H. Stein. 2019a. Degradation of dietary fiber in the stomach, small intestine, and large intestine of growing pigs fed corn- or wheat-based diets without or with microbial xylanase. J. Anim. Sci. 97:338–352. doi: https://doi.org/10.1093/jas/sky403.

Abelilla, J. J., and H. H. Stein. 2019b. Fate of pentoses in the small intestine and hindgut of growing pigs. J. Anim. Sci. 97(Suppl. 2):95. (Abstr.). doi: https://doi.org/10.1093/jas/skz122.171.

Adams, K. L., and A. H. Jensen. 1984. Comparative utilization of in-seed fats and the respective extracted fats by the young pig. J. Anim. Sci. 59:1557–1566. doi: https://doi.org/10.2527/jas1984.5961557x.

Adeola, O. 2001. Digestion and balance techniques in pigs. In: A. J. Lewis and L. L. Southern, editors, Swine nutrition. CRC Press, Washington, DC. p. 903–916.

Adeola, O., B. V. Lawrence, A. L. Sutton, and T. R. Cline. 1995. Phytase-induced changes in mineral utilization in zinc-supplemented diets for pigs. J. Anim. Sci. 73:3384–3391. doi: https://doi.org/10.2527/1995.73113384x.

Adeola, O., and J. I. Orban. 1995. Chemical composition and nutrient digestibility of pearl millet (Pennisetum glaucum) fed to growing pigs. J. Cereal Sci. 22:177–184. doi: https://doi.org/10.1016/0733-5210(95)90048-9.

Adeola, O., P. C. Xue, A. J. Cowieson, and K. M. Ajuwon. 2016. Basal endogenous losses of amino acids in protein nutrition research for swine and poultry. Anim. Feed Sci. Technol. 221:274–283. doi: https://doi.org/10.1016/j.anifeedsci.2016.06.004.

Agunbiade, J. A., J. Wiseman, and D. J. A. Cole. 1992. Utilization of dietary energy and nutrients from soya bean products by growing pigs. Anim. Feed Sci. Technol. 36:303–318. doi: https://doi.org/10.1016/0377-8401(92)90064-D.

Ai, Y. F. 2013. Structures, properties, and digestibility of resistant starch. PhD. Diss. Iowa State Univ. Ames, IA. https://lib.dr.iastate.edu/etd/13558/.

Ajakaiye, A., M. Z. Fan, T. Archbold, R. R. Hacker, C. W. Forsberg, and J. P. Phillips. 2003. Determination of true digestive utilization of phosphorus and the endogenous phosphorus outputs associated with soybean meal for growing pigs. J. Anim. Sci. 81:2766–2775. doi: https://doi.org/10.2527/2003.81112766x.

Akinmusire, A. S., and O. Adeola. 2009. True digestibility of phosphorus in canola and soybean meals for growing pigs: Influence of microbial phytase. J. Anim. Sci. 87:977–983. doi: https://doi.org/10.2527/jas.2007-0778.

Albin, D. M., M. R. Smiricky, J. E. Wubben, and V. M. Gabert. 2001. The effect of dietary level of soybean oil and palm oil on apparent ileal amino acid digestibility and postprandial flow patterns of chromic oxide and amino acids in pigs. Can. J. Anim. Sci. 81:495–503. doi: https://doi.org/10.4141/A00-104.

Almeida, F. N., J. K. Htoo, J. Thomson, and H. H. Stein. 2014. Digestibility by growing pigs of amino acids in heat-damaged sunflower meal and cottonseed meal. J. Anim. Sci. 92:585–593. doi: https://doi.org/10.2527/jas.2013-6769.

Almeida, F. N., and H. H. Stein. 2010. Performance and phosphorus balance of pigs fed diets formulated on the basis of values for standardized total tract digestibility of phosphorus. J. Anim. Sci. 88:2968–2977. doi: https://doi.org/10.2527/jas.2009-2285.

Almeida, F. N., and H. H. Stein. 2012. Effects of graded levels of microbial phytase on the standardized total tract digestibility of phosphorus in corn and corn coproducts fed to pigs. J. Anim. Sci. 90:1262–1269. doi: https://doi.org/10.2527/jas.2011-4144.

Ammerman, C. B., D. H. Baker, and A. J. Lewis. 1995. Bioavailability of nutrients for animals: Amino acids, minerals, and vitamins. Academic Press, San Diego, CA.

Ammerman, C. B., and S. M. Miller. 1972. Biological availability of minor mineral ions: A review. J. Anim. Sci. 35:681–694. doi: https://doi.org/10.2527/jas1972.353681x.

Anderson, J. O., and R. E. Warnick. 1970. Studies of the need for supplemental biotin in chick rations. Poult. Sci. 49:569–578. doi: https://doi.org/10.3382/ps.0490569.

Anderson, L. E., Sr., R. O. Myer, J. H. Brendemuhl, and L. R. McDowell. 1995. Bioavailability of various vitamin E compounds for finishing swine. J. Anim. Sci. 73:490–495. doi: https://doi.org/10.2527/1995.732490x.

Anderson, P. A., D. H. Baker, and S. P. Mistry. 1978. Bioassay determination of the biotin content of corn, barley, sorghum and wheat. J. Anim. Sci. 47:654–659. doi: https://doi.org/10.2527/jas1978.473654x.

Anderson, P. A., D. H. Baker, P. A. Sherry, and J. E. Corbin. 1979. Choline-methionine interrelationship in feline nutrition. J. Anim. Sci. 49:522–527. doi: https://doi.org/10.2527/jas1979.492522x.

Anonymous. 1979. Nomenclature policy: generic descriptors and trivial names for vitamins and related compounds. J. Nutr. 109:8–15. doi: https://doi.org/10.1093/jn/120.1.12.

Anonymous. 1986. Pellagragenic effect of excess leucine. Nutr. Rev. 44:26–27. doi: https://doi.org/10.1111/j.1753-4887.1986.tb07552.x.

Aoyagi, S., and D. H. Baker. 1993a. Bioavailability of copper in analytical-grade and feed-grade inorganic copper sources when fed to provide copper at levels below the chick's requirement. Poult. Sci. 72:1075–1083. doi: https://doi.org/10.3382/ps.0721075.

Aoyagi, S., and D. H. Baker. 1993b. Nutritional evaluation of copper-lysine and zinc-lysine complexes for chicks. Poult. Sci. 72:165–171. doi: https://doi.org/10.3382/ps.0720165.

Aoyagi, S., and D. H. Baker. 1994. Copper-amino acid complexes are partially protected against inhibitory effects of L-cysteine and L-ascorbic acid on copper absorption in chicks. J. Nutr. 124:388–395. doi: https://doi.org/10.1093/jn/124.3.388.

Aoyagi, S., and D. H. Baker. 1995. Effect of microbial phytase and 1,25-dihydroxycholecalciferol on dietary copper utilization in chicks. Poult. Sci. 74:121–126. doi: https://doi.org/10.3382/ps.0740121.

Arredondo, M. A., G. A. Casas, and H. H. Stein. 2019. Increasing levels of microbial phytase increases the digestibility of energy and minerals in diets fed to pigs. Anim. Feed Sci. Technol. 248:27–36. doi: https://doi.org/10.1016/j.anifeedsci.2019.01.001.

Bach, A. C., and V. K. Babayan. 1982. Medium-chain triglycerides: An update. Am. J. Clin. Nutr. 36:950–962. doi: https://doi.org/10.1093/ajcn/36.5.950.

Bach Knudsen, K. E., B. Borg Jensen, J. O. Andersen, and I. Hansen. 1991. Gastrointestinal implications in pigs of wheat and oat fractions: 2. Microbial activity in the gastrointestinal tract. Br. J. Nutr. 65:233–248. doi: https://doi.org/10.1079/BJN19910083.

Bach Knudsen, K. E., and H. Jørgensen. 2001. Intestinal degradation of dietary carbohydrates-from birth to maturity. In: J. E. Lindberg and B. Ogle, editors, Digestive physiology of pigs. Cabi Publishing, New York, NY. p. 109–120.

Bach Knudsen, K. E., H. N. Lærke, and H. Jørgensen. 2013. Carbohydrates and carbohydrate utilization in swine. In: L. I. Chiba, editor, Sustainable swine nutrition. John Wiley & Sons, Inc., Ames, IA. p. 109–135.

Bach Knudsen, K. E., H. N. Lærke, S. Steenfeldt, M. S. Hedemann, and H. Jørgensen. 2006. In vivo methods to study the digestion of starch in pigs and poultry. Anim. Feed Sci. Technol. 130:114–135. doi: https://doi.org/10.1016/j.anifeedsci.2006.01.020.

Bafundo, K. W., D. H. Baker, and P. R. Fitzgerald. 1984. The iron-zinc interrelationship in the chick as influenced by Eimeria Acervulina infection. J. Nutr. 114:1306–1312. doi: https://doi.org/10.1093/jn/114.7.1306.

Bailey, L. B. 1988. Factors affecting folate bioavailability. Food Technol. 42:206–210.

Baker, D. H. 1989. Phosphorus sources for poultry. Multi-State Poult. Newsl. 1:5–6.

Baker, D. H. 1995. Vitamin bioavailability. In: C. B. Ammerman, D. H. Baker, and A. J. Lewis, editors, Bioavailability of nutrients for animals: Amino acids, minerals, and vitamins. Academic Press, San Diego, CA. p. 399–431.

Baker, D. H. 2001. Bioavailability of minerals and vitamins. In: A. J. Lewis and L. L. Southern, editors, Swine nutrition. CRC Press, Washington, DC. p. 357–379.

Baker, D. H., N. K. Allen, and A. J. Kleiss. 1973. Efficiency of tryptophan as a niacin precursor in the young chick. J. Anim. Sci. 36:299–302. doi: https://doi.org/10.2527/jas1973.362299x.

Baker, D. H., and C. B. Ammerman. 1995a. Copper bioavailability. In: C. B. Ammerman, D. H. Baker, and A. J. Lewis, editors, Bioavailability of nutrients for animals: Amino acids, minerals, and vitamins. Academic Press, San Diego, CA. p. 127–157.

Baker, D. H., and C. B. Ammerman. 1995b. Zinc bioavailability. In: C. B. Ammerman, D. H. Baker, and A. J. Lewis, editors, Bioavailability of nutrients for animals: Amino acids, minerals, and vitamins. Academic Press, San Diego, CA. p. 367–399.

Baker, D. H., and G. L. Czarnecki-Maulden. 1987. Pharmacologic role of cysteine in ameliorating or exacerbating mineral toxicities. J. Nutr. 117:1003–1010. doi: https://doi.org/10.1093/jn/117.6.1003.

Baker, D. H., and K. M. Halpin. 1988. Zinc antagonizing effects of fish meal, wheat bran and a corn-soybean meal mixture when added to a phytate- and fiber-free casein-dextrose diet. Nutr. Res. 8:213–218. doi: https://doi.org/10.1016/S0271-5317(88)80025-3.

Baker, D. H., K. M. Halpin, D. E. Laurin, and L. L. Southern. 1986. Manganese for poultry – a review. Proc. of the Arkansas Nutr. Conf., Little Rock, Arkansas (18–19 September 1986), p. 1–6.

Baker, D. H., J. Odle, M. A. Funk, and T. M. Wieland. 1991. Research note: Bioavailability of copper in cupric oxide, cuprous oxide, and in a copper-lysine complex. Poult. Sci. 70:177–179. doi: https://doi.org/10.3382/ps.0700177.

Baker, D. H., and G. W. Oduho. 1994. Manganese utilization in the chick: Effects of excess phosphorus on chicks fed manganese-deficient diets. Poult. Sci. 73:1162–1165. doi: https://doi.org/10.3382/ps.0731162.

Baker, D. H., and K. J. Wedekind. 1988. Manganese utilization in chicks as affected by excess calcium and phosphorus ingestion. Proc. of the Maryland Nutr. Conf., College Park, MD, p. 29–34.

Baker, D. H., J. T. Yen, A. H. Jensen, R. G. Teeter, E. N. Michel, and J. H. Burns. 1976. Niacin activity in niacinamide and coffee. Nutr. Rep. Int. 14:115–120.

Bao-Ji, C., and G. F. Combs. 1986. Evaluation of biopotencies of nicotinamide and nicotinic acid for broiler chickens. Poult. Sci. 65(Suppl. 1):24. (Abstr).

Barber, R. S., J. P. Bowland, R. Braude, K. G. Mithcell, and J. W. G. Porter. 1961. Copper sulphate and copper sulphide (CuS) as supplements for growing pigs. Br. J. Nutr. 15:189–197. doi: https://doi.org/10.1079/BJN19610024.

Batterham, E. S. 1974. The effect of frequency of feeding on the utilization of free lysine by growing pigs. Br. J. Nutr. 31:237–242. doi: https://doi.org/10.1079/BJN19740029.

Batterham, E. S. 1992. Availability and utilization of amino acids for growing pigs. Nutr. Res. Rev. 5:1–18. doi: https://doi.org/10.1079/NRR19920004.

BeMiller, J. 2007. Carbohydrate chemistry for food scientist. 2. AACC International, Inc., St. Paul, MN.

Bengala-Freire, J., A. Aumaitre, and J. Peiniau. 1991. Effects of feeding raw and extruded peas on ileal digestibility, pancreatic enzymes and plasma glucose and insulin in early weaned pigs. J. Anim. Physiol. Anim. Nutr. 65:154–164. doi: https://doi.org/10.1111/j.1439-0396.1991.tb00253.x.

Bieri, J. G., and M. C. McKenna. 1981. Expressing dietary values for fat-soluble vitamins: Changes in concepts and terminology. Am. J. Clin. Nutr. 34:289–295. doi: https://doi.org/10.1093/ajcn/34.2.289.

Bikker, P., C. M. C. van der Peet-Schwering, W. J. J. Gerrits, V. Sips, C. Walvoort, and H. van Laar. 2017. Endogenous phosphorus losses in growing-finishing pigs and gestating sows. J. Anim. Sci. 95:1637–1643. doi:10.2527/jas.2016.1041.

Blavi, L., D. Sola-Oriol, J. F. Perez, and H. H. Stein. 2017. Effects of zinc oxide and microbial phytase on digestibility of calcium and phosphorus in maize-based diets fed to growing pigs. J. Anim. Sci. 95:847–854. doi: https://doi.org/10.2527/jas.2016.1149.

Bohlke, R. A., R. C. Thaler, and H. H. Stein. 2005. Calcium, phosphorus, and amino acid digestibility in low-phytate corn, normal corn, and soybean meal by growing pigs. J. Anim. Sci. 83:2396–2403. doi: https://doi.org/10.2527/2005.83102396x.

Boisen, S., and P. J. Moughan. 1996. Dietary influences on endogenous ileal protein and amino acid loss in the pig - A review. Acta Agric. Scand. Sect. A Anim. Sci. 46:154–164. doi: https://doi.org/10.1080/09064709609415866.

Boisen, S., and M. W. A. Verstegen. 2000. Developments in the measurement of the energy content of feeds and energy utilisation in animals. In: P. J. Moughan, M. W. A. Verstegen, and M. I. Visser-Reyneveld, editors, Feed evaluation: Principles and practice. Wageningen Press, Wageningen, the Netherlands. p. 57–76.

Bolarinwa, O. A., and O. Adeola. 2016. Regression and direct methods do not give different estimates of digestible and metabolizable energy values of barley, sorghum, and wheat for pigs. J. Anim. Sci. 94:610–618. doi: https://doi.org/10.2527/jas.2015-9766.

Braude, R., A. G. Chamberlain, M. Kotarbińska, and K. G. Mitchell. 1962. The metabolism of iron in piglets given labelled iron either orally or by injection. Br. J. Nutr. 16:427–449. doi: https://doi.org/10.1079/BJN19620043.

Brestenský, M., S. Nitrayová, J. Heger, P. Patráš, J. Rafay, and A. Sirotkin. 2014. Methods for determination reactive lysine in heat-treated foods and feeds. J. Microbiol. Biotechnol. Food Sci. 4:13–15. doi: https://doi.org/10.15414/jmbfs.2014.4.1.13-15.

Broomhead, J. N., P. A. Lessard, R. M. Raab, and M. B. Lanahan. 2018. Effects of feeding corn-expressed phytase on the live performance, bone characteristics, and phosphorus digestibility of nursery pigs. J. Anim. Sci. 97:1254–1261. doi: https://doi.org/10.1093/jas/sky479.

Bryant, K. L., E. T. Kornegay, J. W. Knight, H. P. Veit, and D. R. Notter. 1985. Supplemental biotin for swine. III. Influence of supplementation to corn- and wheat-based diets on the incidence and severity of toe lesions, hair and skin characteristics and structural soundness of sows housed in confinement during four parities. J. Anim. Sci. 60:154–162. doi: https://doi.org/10.2527/jas1985.601154x.

Buenrostro, J. L., and F. H. Kratzer. 1984. Use of plasma and egg yolk biotin of white leghorn hens to assess biotin availability from feedstuffs. Poult. Sci. 63:1563–1570. doi: https://doi.org/10.3382/ps.0631563.

Burkett, J. L., K. J. Stalder, W. J. Powers, K. Bregendahl, J. L. Pierce, T. J. Baas, T. Bailey, and B. L. Shafer. 2009. Effect of inorganic and organic trace mineral supplementation on the performance, carcass characteristics, and fecal mineral excretion of phase-fed, grow-finish swine. Asian-Australas. J. Anim. Sci. 22:1279–1287. doi: https://doi.org/10.5713/ajas.2009.70091.

Butts, C. A., P. J. Moughan, W. C. Smith, G. W. Reynolds, and D. J. Garrick. 1993. The effect of food dry matter intake on endogenous ileal amino acid excretion determined under peptide alimentation in the 50 kg liveweight pig. J. Sci. Food Agric. 62:235–243. doi: https://doi.org/10.1002/jsfa.2740620306.

Caine, W. R., W. C. Sauer, S. Tamminga, M. W. A. Verstegen, and H. Schulze. 1997. Apparent ileal digestibilities of amino acids in newly weaned pigs fed diets with protease-treated soybean meal. J. Anim. Sci. 75:2962–2969. doi: https://doi.org/10.2527/1997.75112962x.

Canibe, N., and K. E. Bach Knudsen. 1997. Digestibility of dried and toasted peas in pigs. 1. Ileal and total tract digestibilities of carbohydrates. Anim. Feed Sci. Technol. 64:293–310. doi: https://doi.org/10.1016/S0377-8401(96)01032-2.

Cantor, A. H., M. L. Langevin, T. Noguchi, and M. L. Scott. 1975a. Efficacy of selenium in selenium compounds and feedstuffs for prevention of pancreatic fibrosis in chicks. J. Nutr. 105:106–111. doi: https://doi.org/10.1093/jn/105.1.106.

Cantor, A. H., M. L. Scott, and T. Noguchi. 1975b. Biological availability of selenium in feedstuffs and selenium compounds for prevention of exudative diathesis in chicks. J. Nutr. 105:96–105. doi: https://doi.org/10.1093/jn/105.1.96.

Casas, G. A., and H. H. Stein. 2015. Effects of microbial phytase on the apparent and standardized total tract digestibility of phosphorus in rice coproducts fed to growing pigs. J. Anim. Sci. 93:3441–3448. doi: https://doi.org/10.2527/jas.2015-8877.

Cera, K. R., D. C. Mahan, and G. A. Reinhart. 1989. Apparent fat digestibilities and performance responses of postweaning swine fed diets supplemented with coconut oil, corn oil or tallow. J. Anim. Sci. 67:2040–2047. doi: https://doi.org/10.2527/jas1989.6782040x.

Cervantes-Pahm, S. K., Y. Liu, A. Evans, and H. H. Stein. 2014. Effect of novel fiber ingredients on ileal and total tract digestibility of energy and nutrients in semi-purified diets fed to growing pigs. J. Sci. Food Agric. 94:1284–1290. doi: https://doi.org/10.1002/jsfa.6405.

Cervantes-Pahm, S. K., and H. H. Stein. 2008. Effect of dietary soybean oil and soybean protein concentration on the concentration of digestible amino acids in soybean products fed to growing pigs. J. Anim. Sci. 86:1841–1849. doi: https://doi.org/10.2527/jas.2007-0721.

Cervantes-Pahm, S. K., and H. H. Stein. 2010. Ileal digestibility of amino acids in conventional, fermented, and enzyme-treated soybean meal and in soy protein isolate, fish meal, and casein fed to weanling pigs. J. Anim. Sci. 88:2674–2683. doi: https://doi.org/10.2527/jas.2009-2677.

Charles, O. W., and T. M. Huston. 1972. The biological activity of vitamin K materials following storage and pelleting. Poult. Sci. 51:1421–1427. doi: https://doi.org/10.3382/ps.0511421.

Chastanet, F., A. A. Pahm, C. Pedersen, and H. H. Stein. 2007. Effect of feeding schedule on apparent energy and amino acid digestibility by growing pigs. Anim. Feed Sci. Technol. 132:94–102. doi: https://doi.org/10.1016/j.anifeedsci.2006.03.012.

Chen, L., L. X. Gao, Q. H. Huang, R. Q. Zhong, L. L. Zhang, X. F. Tang, and H. F. Zhang. 2017. Viscous and fermentable non-starch polysaccharides affect intestinal nutrient and energy flow and hindgut fermentation in growing pigs. J. Anim. Sci. 95:5054–5063. doi: https://doi.org/10.2527/jas2017.1662.

Chen, Y. F., C. F. Huang, L. Liu, C. H. Lai, and F. L. Wang. 2019. Concentration of vitamins in the 13 feed ingredients commonly used in pig diets. Anim. Feed Sci. Technol. 247:1–8. doi: https://doi.org/10.1016/j.anifeedsci.2018.10.011.

Cho, E. S., D. W. Andersen, L. J. Filer, Jr., and L. D. Stegink. 1980. D-methionine utilization in young miniature pigs, adult rabbits, and adult dogs. J. Parenter. Enteral Nutr. 4:544–547. doi: https://doi.org/10.1177/0148607180004006544.

Choi, H., and B. G. Kim. 2019. A low-fiber diet requires a longer adaptation period before collecting feces of pigs compared with a high-fiber diet in digestibility experiments using the inert marker method. Anim. Feed Sci. Technol. 256:114254. doi: https://doi.org/10.1016/j.anifeedsci.2019.114254.

Chung, T. K., and D. H. Baker. 1990. Riboflavin requirement of chicks fed purified amino acid and conventional corn-soybean meal diets. Poult. Sci. 69:1357–1363. doi: https://doi.org/10.3382/ps.0691357.

Chung, T. K., and D. H. Baker. 1992a. Apparent and true amino acid digestibility of a crystalline amino acid mixture and of casein: comparison of values obtained with ileal-cannulated pigs and cecectomized cockerels. J. Anim. Sci. 70:3781–3790. doi: https://doi.org/10.2527/1992.70123781x.

Chung, T. K., and D. H. Baker. 1992b. Utilization of methionine isomers and analogs by the pig. Can. J. Anim. Sci. 72:185–188. doi: https://doi.org/10.4141/cjas92-024.

Chung, T. K., J. W. Ekdman, and D. H. Baker. 1990. Hydrated sodium calcium aluminosilicate: Effects on zinc, manganese, vitamin A, and riboflavin utilization. Poult. Sci. 69:1364–1370. doi: https://doi.org/10.3382/ps.0691364.

Clawson, A. J., J. T. Reid, B. E. Sheffy, and J. P. Willman. 1955. Use of chromium oxide in digestion studies with swine. J. Anim. Sci. 14:700–709. doi: https://doi.org/10.1093/ansci/14.3.700.

Combs, G. F. 2012a. Chapter 5 - Vitamin A. In: G. F. Combs, editor, The vitamins. Academic Press, San Diego, CA. p. 93–138. doi: https://doi.org/10.1016/B978-0-12-381980-2.00005-0.

Combs, G. F. 2012b. Chapter 7 - Vitamin E. In: G. F. Combs, editor, The vitamins. Academic Press, San Diego, CA. p. 181–211. doi: https://doi.org/10.1016/B978-0-12-381980-2.00007-4.

Combs, N. R., and E. R. Miller. 1985. Determination of potassium availability in $K_2CO_3$, $KHCO_3$, corn and soybean meal for the young pig. J. Anim. Sci. 60:715–719. doi: https://doi.org/10.2527/jas1985.603715x.

Combs, N. R., E. R. Miller, and P. K. Ku. 1985. Development of an assay to determine the bioavailability of potassium in feedstuffs for the young pig. J. Anim. Sci. 60:709–714. doi: https://doi.org/10.2527/jas1985.603709x.

Cornelius, S., and B. Harmon. 1976. Sources of oral iron for neonatal piglets. J. Anim. Sci. 42:1351–1351. (Abstr.)

Cromwell, G. L. 1992. The biological availability of phosphorus in feedstuffs for pigs. Pig News Info. 13:75N–78N. (Abstr.)

Cromwell, G. L., V. W. Hays, and T. L. Clark. 1978. Effect of copper sulfate, copper sulfide and sodium sulfide on performance and copper stores of pigs. J. Anim. Sci. 46:692–698. doi: https://doi.org/10.2527/jas1978.463692x.

Cromwell, G. L., T. S. Stahly, and H. J. Monegue. 1989. Effects of source and level of copper on performance and liver copper stores in weanling pigs. J. Anim. Sci. 67:2996–3002. doi: https://doi.org/10.2527/jas1989.67112996x.

Cummings, J. H., M. B. Roberfroid, H. Andersson, C. Barth, A. Ferro-Luzzi, Y. Ghoos, M. Gibney, K. Hermonsen, W. P. T. James, O. Korver, D. Lairon, G. Pascal, and A. G. S. Voragen. 1997. A new look at dietary carbohydrate: Chemistry, physiology and health. Eur. J. Clin. Nutr. 51:417–423. doi: https://doi.org/10.1038/sj.ejcn.1600427.

Cummings, J. H., and A. M. Stephen. 2007. Carbohydrate terminology and classification. Eur. J. Clin. Nutr. 61:S5–S18. doi: https://doi.org/10.1038/sj.ejcn.1602936.

Czarnecki, G. L., and D. H. Baker. 1985. Reduction of liver copper concentration by the organic arsenical, 3-nitro-4-hydroxyphenylarsonic acid. J. Anim. Sci. 60:440–450. doi: https://doi.org/10.2527/jas1985.602440x.

Czarnecki, G. L., K. M. Halpin, and D. H. Baker. 1983. Precursor (amino acid): product (vitamin) interrelationship for growing chicks as illustrated by tryptophan-niacin and methionine-choline. Poult. Sci. 62:371–374. doi: https://doi.org/10.3382/ps.0620371.

Dalto, D. B., I. Audet, C. L. Girard, and J.-J. Matte. 2018. Bioavailability of vitamin B$_{12}$ from dairy products using a pig model. Nutrients. 10:1134. doi: https://doi.org/10.3390/nu10091134.

de Lange, C. F. M. 2000. Characterisation of the non-starch polysaccharides. In: P. J. Moughan, M. W. A. Verstegen, and M. I. Visser-Reyneveld, editors, Feed evaluation: Principles and practice. Wageningen Press, Wageningen, the Netherlands. p. 77–92.

de Lange, C. F. M., W. C. Sauer, R. Mosenthin, and W. B. Souffrant. 1989. The effect of feeding different protein-free diets on the recovery and amino acid composition of endogenous protein collected from the distal ileum and feces in pigs. J. Anim. Sci. 67:746–754. doi: https://doi.org/10.2527/jas1989.673746x.

De Vries, J. W. 2004. Dietary fiber: The influence of definition on analysis and regulation. J. AOAC Int. 87:682–706. doi: https://doi.org/10.1093/jaoac/87.3.682.

Deming, J. G., and G. L. Czarnecki-Maulden. 1989. Iron bioavailability in calcium and phosphorus sources. J. Anim. Sci. 67(Suppl. 1):253. (Abstr).

DeRouchey, J. M., J. D. Hancock, R. H. Hines, C. A. Maloney, D. J. Lee, H. Cao, D. W. Dean, and J. S. Park. 2004. Effects of rancidity and free fatty acids in choice white grease on growth performance and nutrient digestibility in weanling pigs. J. Anim. Sci. 82:2937–2944. doi: https://doi.org/10.2527/2004.82102937x.

Dierick, N. A., and J. A. Decuypere. 2004. Influence of lipase and/or emulsifier addition on the ileal and faecal nutrient digestibility in growing pigs fed diets containing 4% animal fat. J. Sci. Food Agric. 84:1443–1450. doi: https://doi.org/10.1002/jsfa.1794.

Dietrich, W., and H. S. Grindley. 1914. Coefficients of digestibility of some common rations for swine. University of Illinois Agricultural Experiment Station, Urbana, IL.

Dilger, R. N., and O. Adeola. 2006. Estimation of true phosphorus digestibility and endogenous phosphorus loss in growing pigs fed conventional and low-phytate soybean meals. J. Anim. Sci. 84:627–634. doi: https://doi.org/10.2527/2006.843627x.

Dilger, R. N., T. A. Garrow, and D. H. Baker. 2007. Betaine can partially spare choline in chicks but only when added to diets containing a minimal level of choline. J. Nutr. 137:2224–2228. doi: https://doi.org/10.1093/jn/137.10.2224.

Donkoh, A., and P. J. Moughan. 1994. The effect of dietary crude protein content on apparent and true ileal nitrogen and amino acid digestibilities. Br. J. Nutr. 72:59–68. doi: https://doi.org/10.1079/BJN19940009.

Duran-Montgé, P., R. Lizardo, D. Torrallardona, and E. Esteve-Garcia. 2007. Fat and fatty acid digestibility of different fat sources in growing pigs. Livest. Sci. 109:66–69. doi: https://doi.org/10.1016/j.livsci.2007.01.067.

Easter, R. A., P. A. Anderson, E. J. Michel, and J. R. Corley. 1983. Response of gestating gilts and starter, grower and finisher swine to biotin, pyridoxine, folacin and thiamine additions to corn-soybean meal diets. Nutr. Rep. Int. 28:945–954.

Edmonds, M. S., and D. H. Baker. 1986. Toxic effects of supplemental copper and roxarsone when fed alone or in combination to young pigs. J. Anim. Sci. 63:533–537. doi: https://doi.org/10.2527/jas1986.632533x.

Edwards, H. M., III, and D. H. Baker. 2000. Zinc bioavailability in soybean meal. J. Anim. Sci. 78:1017–1021. doi: https://doi.org/10.2527/2000.7841017x.

Eicher, S. D., C. A. McKee, J. A. Carroll, and E. A. Pajor. 2006. Supplemental vitamin C and yeast cell wall β-glucan as growth enhancers in newborn pigs and as immunomodulators after an endotoxin challenge after weaning. J. Anim. Sci. 84:2352–2360. doi: https://doi.org/10.2527/jas.2005-770.

Emmert, J. L., and D. H. Baker. 1997. A chick bioassay approach for determining the bioavailable choline concentration in normal and overheated soybean meal, canola meal and peanut meal. J. Nutr. 127:745–752. doi: https://doi.org/10.1093/jn/127.5.745.

Emmert, J. L., J. L. Garrow, and D. H. Baker. 1996. Development of an experimental diet for determining bioavailable choline concentration and its application in studies with soybean lecithin. J. Anim. Sci. 74:2738–2744. doi: https://doi.org/10.2527/1996.74112738x.

Englyst, K. N., and H. N. Englyst. 2005. Carbohydrate bioavailability. Br. J. Nutr. 94:1–11. doi: https://doi.org/10.1079/BJN20051457.

Englyst, K. N., and G. J. Hudson. 2000. Carbohydrates. In: J. S. Garrow, W. P. T. James, and A. Ralph, editors, Human nutrition and dietetics. Churchill Livingston, Edinburgh, UK. p. 61–76.

Englyst, K. N., S. Liu, and H. N. Englyst. 2007. Nutritional characterization and measurement of dietary carbohydrates. Eur. J. Clin. Nutr. 61:S19–S39. doi: https://doi.org/10.1038/sj.ejcn.1602937.

Erdman, J. W., C. L. Poor, Jr., and J. M. Dietz. 1988. Processing and dietary effects on the bioavailability of vitamin A, carotenoids and vitamin E. Food Technol. 42:214–219.

Erdman, J. W., Jr. 1979. Oilseed phytates: Nutritional implications. J. Am. Oil Chem. Soc. 56:736–741. doi: https://doi.org/10.1007/BF02663052.

Erdman, J. W., Jr., G. C. Fahey, Jr., and C. B. White. 1986. Effects of purified dietary fiber sources on β-carotene utilization by the chick. J. Nutr. 116:2415–2423. doi: https://doi.org/10.1093/jn/116.12.2415.

Espinosa, C. D., R. S. Fry, M. E. Kocher, and H. H. Stein. 2019. Effects of copper hydroxychloride and distillers dried grains with solubles on intestinal microbial concentration and apparent ileal and total tract digestibility of energy and nutrients by growing pigs. J. Anim. Sci. 97:4904–4911. doi: https://doi.org/10.1093/jas/skz340.

Fan, M. Z., T. Archbold, W. C. Sauer, D. Lackeyram, T. Rideout, Y. X. Gao, C. F. M. de Lange, and R. R. Hacker. 2001. Novel methodology allows simultaneous measurement of true phosphorus digestibility and the gastrointestinal endogenous phosphorus outputs in studies with pigs. J. Nutr. 131:2388–2396. doi: https://doi.org/10.1093/jn/131.9.2388.

Fan, M. Z., and W. C. Sauer. 1995a. Determination of apparent ileal amino acid digestibility in barley and canola meal for pigs with the direct, difference, and regression methods. J. Anim. Sci. 73:2364–2374. doi: https://doi.org/10.2527/1995.7382364x.

Fan, M. Z., and W. C. Sauer. 1995b. Determination of apparent ileal amino acid digestibility in peas for pigs with the direct, difference and regression methods. Livest. Prod. Sci. 44:61–72. doi: https://doi.org/10.1016/0301-6226(95)00057-R.

Fan, M. Z., and W. C. Sauer. 2002a. Additivity of apparent ileal and fecal phosphorus digestibility values measured in single feed ingredients for growing-finishing pigs. Can. J. Anim. Sci. 82:183–191. doi: https://doi.org/10.4141/A01-072.

Fan, M. Z., and W. C. Sauer. 2002b. Determination of true ileal amino acid digestibility and the endogenous amino acid outputs associated with barley samples for growing-finishing pigs by the regression analysis technique. J. Anim. Sci. 80:1593–1605. doi: https://doi.org/10.2527/2002.8061593x.

Fang, R. J., T. J. Li, F. G. Yin, Y. L. Yin, X. F. Kong, K. N. Wang, Z. Yuan, G. Y. Wu, J. H. He, Z. Y. Deng, and M. Z. Fan. 2007. The additivity of true or apparent phosphorus digestibility values in some feed ingredients for growing pigs. Asian-Australas. J. Anim. Sci. 20:1092–1099. doi: https://doi.org/10.5713/ajas.2007.1092.

Fernández, J. A. 1995. Calcium and phosphorus metabolism in growing pigs. I. Absorption and balance studies. Livest. Prod. Sci. 41:233–241. doi: https://doi.org/10.1016/0301-6226(94)00063-D.

Finney, D. J. 1952. Statistical method in biological assay. Charles Griffin and Co. Ltd., London, UK.

Flohr, J. R., J. M. DeRouchey, and J. C. Woodworth. 2016. A survey of current feeding regimens for vitamins and trace minerals in the US swine industry. J. Swine Health Prod. 24:290–303.

Fly, A. D., O. A. Izquierdo, K. R. Lowry, and D. H. Baker. 1989. Manganese bioavailability in a MN-methionine chelate. Nutr. Res. 9:901–910. doi: https://doi.org/10.1016/S0271-5317(89)80035-1.

Fontaine, J., U. Zimmer, P. J. Moughan, and S. M. Rutherfurd. 2007. Effect of heat damage in an autoclave on the reactive lysine contents of soy products and corn distillers dried grains with solubles. Use of the results to check on lysine damage in common qualities of these ingredients. J. Agric. Food Chem. 55:10737–10743. doi: https://doi.org/10.1021/jf071747c.

Freeman, C. P., D. W. Holme, and E. F. Annison. 1968. The determination of the true digestibilities of interesterified fats in young pigs. Br. J. Nutr. 22:651–660. doi: https://doi.org/10.1079/BJN19680076.

Frigg, M. 1976. Bioavailability of biotin in cereals. Poult. Sci. 55:2310–2318.

Gabert, V. M., H. Jørgensen, and C. M. Nyachoti. 2001. Bioavailability of amino acids in feedstuffs for swine. In: A. J. Lewis and L. L. Southern, editors, Swine nutrition. CRC Press, Washington, DC. p. 151–186.

Ghosh, H. P., P. K. Sarkar, and B. C. Guha. 1963. Distribution of the bound form of nicotinic acid in natural materials. J. Nutr. 79:451–453. doi: https://doi.org/10.1093/jn/79.4.451.

Gibson, R. S., V. Raboy, and J. C. King. 2018. Implications of phytate in plant-based foods for iron and zinc bioavailability, setting dietary requirements, and formulating programs and policies. Nutr. Rev. 76:793–804. doi: https://doi.org/10.1093/nutrit/nuy028.

Goerke, M., M. Eklund, N. Sauer, M. Rademacher, H.-P. Piepho, C. Börner, and R. Mosenthin. 2012. Effect of feed intake level on ileal digestibilities of crude protein and amino acids in diets for piglets. J. Sci. Food Agric. 92:1261–1266. doi: https://doi.org/10.1002/jsfa.4692.

Goff, J. P. 2018. Invited review: Mineral absorption mechanisms, mineral interactions that affect acid–base and antioxidant status, and diet considerations to improve mineral status. J. Dairy Sci. 101:2763–2813. doi: https://doi.org/10.3168/jds.2017-13112.

González-Vega, J. C., B. G. Kim, J. K. Htoo, A. Lemme, and H. H. Stein. 2011. Amino acid digestibility in heated soybean meal fed to growing pigs. J. Anim. Sci. 89:3617–3625. doi: https://doi.org/10.2527/jas.2010-3465.

González-Vega, J. C., Y. Liu, J. C. McCann, C. L. Walk, J. J. Loor, and H. H. Stein. 2016. Requirement for digestible calcium by eleven- to twenty-five-kilogram pigs as determined by growth performance, bone ash concentration, calcium and phosphorus balances, and expression of genes involved in transport of calcium in intestinal and kidney cells. J. Anim. Sci. 94:3321–3334. doi: https://doi.org/10.2527/jas.2016-0444.

González-Vega, J. C., and H. H. Stein. 2014. Invited review: Calcium digestibility and metabolism in pigs. Asian-Australas. J. Anim. Sci. 27:1–9. doi: https://doi.org/10.5713/ajas.2014.r.01.

González-Vega, J. C., C. L. Walk, Y. Liu, and H. H. Stein. 2013. Endogenous intestinal losses of calcium and true total tract digestibility of calcium in canola meal fed to growing pigs. J. Anim. Sci. 91:4807–4816. doi: https://doi.org/10.2527/jas.2013-6410.

González-Vega, J. C., C. L. Walk, Y. Liu, and H. H. Stein. 2014. The site of net absorption of Ca from the intestinal tract of growing pigs and effect of phytic acid, Ca level and Ca source on Ca digestibility. Arch. Anim. Nutr. 68:126–142. doi: https://doi.org/10.1080/1745039X.2014.892249.

González-Vega, J. C., C. L. Walk, and H. H. Stein. 2015a. Effect of phytate, microbial phytase, fiber, and soybean oil on calculated values for apparent and standardized total tract digestibility of calcium and apparent total tract digestibility of phosphorus in fish meal fed to growing pigs. J. Anim. Sci. 93:4808–4818. doi: https://doi.org/10.2527/jas.2015-8992.

González-Vega, J. C., C. L. Walk, and H. H. Stein. 2015b. Effects of microbial phytase on apparent and standardized total tract digestibility of calcium in calcium supplements fed to growing pigs. J. Anim. Sci. 93:2255–2264. doi: https://doi.org/10.2527/jas.2014-8215.

Gregory, J. F., III, and J. R. Kirk. 1981. The bioavailability of vitamin B$_6$ in foods. Nutr. Rev. 39:1–8. doi: https://doi.org/10.1111/j.1753-4887.1981.tb06700.x.

Griminger, P. 1965. Relative vitamin K potency of two water-soluble menadione analogues. Poult. Sci. 44:210–213. doi: https://doi.org/10.3382/ps.0440210.

Groff, J. L., S. S. Gropper, and S. M. Hunt. 1995. Advanced nutrition and human metabolism. West Publishing, St. Paul, MN.

Guenter, W., and J. L. Sell. 1974. A method for determining "true" availability of magnesium from foodstuffs using chickens. J. Nutr. 104:1446–1457. doi: https://doi.org/10.1093/jn/104.11.1446.

Gunawardena, C. K., R. T. Zijlstra, and E. Beltranena. 2010. Characterization of the nutritional value of air-classified protein and starch fractions of field pea and zero-tannin faba bean in grower pigs. J. Anim. Sci. 88:660–670. doi: https://doi.org/10.2527/jas.2009-1980.

Gutierrez, N. A., B. J. Kerr, and J. F. Patience. 2013. Effect of insoluble-low fermentable fiber from corn-ethanol distillation origin on energy, fiber, and amino acid digestibility, hindgut degradability of fiber, and growth performance of pigs. J. Anim. Sci. 91:5314–5325. doi: https://doi.org/10.2527/jas.2013-6328.

Hahn, J. D., and D. H. Baker. 1993. Growth and plasma zinc responses of young pigs fed pharmacologic levels of zinc. J. Anim. Sci. 71:3020–3024. doi: https://doi.org/10.2527/1993.71113020x.

Halpin, K. M., and D. H. Baker. 1987. Mechanism of the tissue manganese-lowering effect of corn, soybean meal, fish meal, wheat bran, and rice bran. Poult. Sci. 66:332–340. doi: https://doi.org/10.3382/ps.0660332.

Halpin, K. M., D. G. Chausow, and D. H. Baker. 1986. Efficiency of manganese absorption in chicks fed corn-soy and casein diets. J. Nutr. 116:1747–1751. doi: https://doi.org/10.1093/jn/116.9.1747.

Hamilton, C. R., and T. L. Veum. 2012. Effects of cecal oxytetracycline infusion, and dietary avidin and biotin supplementation on the biotin status of nongravid gilts. J. Anim. Sci. 90:3821–3832. doi: https://doi.org/10.2527/jas.2011-4831.

Han, Y. K., and P. A. Thacker. 2006. Effects of the calcium and phosphorus ratio in high zinc diets on performance and nutrient digestibility in weanling pigs. J. Anim. Vet. Adv. 5:5–9.

Harmon, B. G., S. G. Cornelius, J. Totsch, D. H. Baker, and A. H. Jensen. 1974. Oral iron dextran and iron from steel slats as hematinics for swine. J. Anim. Sci. 39:699–702. doi: https://doi.org/10.2527/jas1974.394699x.

Harmon, B. G., D. E. Hoge, A. H. Jensen, and D. H. Baker. 1969. Efficacy of ferrous carbonate as a hematinic for young swine. J. Anim. Sci. 29:706–710. doi: https://doi.org/10.2527/jas1969.295706x.

Harper, A. F., M. D. Lindemann, L. I. Chiba, G. E. Combs, D. L. Handlin, E. T. Kornegay, and L. L. Southern. 1994. An assessment of dietary folic acid levels during gestation and lactation on reproductive and lactational performance of sows: a cooperative study. J. Anim. Sci. 72:2338–2344. doi: https://doi.org/10.2527/1994.7292338x.

Hart, E. B., E. V. McCollum, and J. G. Fuller. 1909. The role of inorganic phosphorus in the nutrition of animals. The University of Wisconsin Agricultural Experiment Station, Madison, WI.

Hartman, P. A., V. W. Hays, R. O. Baker, L. H. Neagle, and D. V. Catron. 1961. Digestive enzyme development in the young pig. J. Anim. Sci. 20:114–123. doi: https://doi.org/10.2527/jas1961.201114x.

Haydon, K. D., D. A. Knabe, and T. D. Tanksley, Jr. 1984. Effects of level of feed intake on nitrogen, amino acid and energy digestibilities measured at the end of the small intestine and over the total digestive tract of growing pigs. J. Anim. Sci. 59:717–724. doi: https://doi.org/10.2527/jas1984.593717x.

Henry, P. R. 1995. Manganese bioavailability. In: C. B. Ammerman, D. H. Baker, and A. J. Lewis, editors, Bioavailability of nutrients for animals: Amino acids, minerals, and vitamins. Academic Press, San Diego, CA. p. 239–256.

Henry, P. R., and E. R. Miller. 1995. Iron bioavailability. In: C. B. Ammerman, D. H. Baker, and A. J. Lewis, editors, Bioavailability of nutrients for animals: Amino acids, minerals, and vitamins. Academic Press, San Diego, CA. p. 169–199.

Hill, G. M., G. L. Cromwell, T. D. Crenshaw, C. R. Dove, R. C. Ewan, D. A. Knabe, A. J. Lewis, G. W. Libal, D. C. Mahan, G. C. Shurson, L. L. Southern, and T. L. Veum. 2000. Growth promotion effects and plasma changes from feeding high dietary concentrations of zinc and copper to weanling pigs (regional study). J. Anim. Sci. 78:1010–1016. doi: https://doi.org/10.2527/2000.7841010x.

Horst, R. L., J. L. Napoli, and E. T. Littledike. 1982. Discrimination in the metabolism of orally dosed ergocalciferol and cholecalciferol by the pig, rat and chick. Biochem. J. 204:185–189. doi: https://doi.org/10.1042/bj2040185.

Imbeah, M., and W. C. Sauer. 1991. The effect of dietary level of fat on amino acid digestibilities in soybean meal and canola meal and on rate of passage in growing pigs. Livest. Prod. Sci. 29:227–239. doi: https://doi.org/10.1016/0301-6226(91)90068-2.

IOM. 2001. Dietary reference intakes: Proposed definition of dietary fiber. National Academies Press, Washington, DC.

Ivers, D. J., and T. L. Veum. 2012. Effect of graded levels of niacin supplementation of a semipurified diet on energy and nitrogen balance, growth performance, diarrhea occurrence, and niacin metabolite excretion by growing swine. J. Anim. Sci. 90:282–288. doi: https://doi.org/10.2527/jas.2011-4035.

Izquierdo, O. A., and D. H. Baker. 1986. Bioavailability of copper in pig feces. Can. J. Anim. Sci. 66:1145–1148. doi: https://doi.org/10.4141/cjas86-127.

Jang, Y. D., M. D. Lindemann, J. H. Agudelo-Trujillo, C. S. Escobar, B. J. Kerr, N. Inocencio, and G. L. Cromwell. 2014. Comparison of direct and indirect estimates of apparent total tract digestibility in swine with effort to reduce variation by pooling of multiple day fecal samples. J. Anim. Sci. 92:4566–4576. doi: https://doi.org/10.2527/jas2013-6570.

Jansman, A. J. M., A. A. Frohlich, and R. R. Marquardt. 1994. Production of proline-rich proteins by the parotid glands of rats is enhanced by feeding diets containing tannins from faba beans (*Vicia faba* L.). J. Nutr. 124:249–258. doi: https://doi.org/10.1093/jn/124.2.249.

Jansman, A. J. M., W. Smink, P. van Leeuwen, and M. Rademacher. 2002. Evaluation through literature data of the amount and amino acid composition of basal endogenous crude protein at the terminal ileum of pigs. Anim. Feed Sci. Technol. 98:49–60. doi: https://doi.org/10.1016/S0377-8401(02)00015-9.

Jaworski, N. W., H. N. Laerke, K. E. B. Knudsen, and H. H. Stein. 2015. Carbohydrate composition and in vitro digestibility of dry matter and nonstarch polysaccharides in corn, sorghum, and wheat and coproducts from these grains. J. Anim. Sci. 93:1103–1113. doi: https://doi.org/10.2527/jas.2014-8147.

Jaworski, N. W., D. W. Liu, D. F. Li, and H. H. Stein. 2016. Wheat bran reduces concentrations of digestible, metabolizable, and net energy in diets fed to pigs, but energy values in wheat bran determined by the difference procedure are not different from values estimated from a linear regression procedure. J. Anim. Sci. 94:3012–3021. doi: https://doi.org/10.2527/jas.2016-0352.

Jaworski, N. W., and H. H. Stein. 2017. Disappearance of nutrients and energy in the stomach and small intestine, cecum, and colon of pigs fed corn-soybean meal diets containing distillers dried grains with solubles, wheat middlings, or soybean hulls. J. Anim. Sci. 95:727–739. doi: https://doi.org/10.2527/jas.2016.0752.

Ji, F., D. P. Casper, P. K. Brown, D. A. Spangler, K. D. Haydon, and J. E. Pettigrew. 2008. Effects of dietary supplementation of an enzyme blend on the ileal and fecal digestibility of nutrients in growing pigs. J. Anim. Sci. 86:1533–1543. doi: https://doi.org/10.2527/jas.2007-0262.

Jin, C. F., J. H. Kim, I. K. Han, H. J. Jung, and C. H. Kwon. 1998. Effects of various fat sources and lecithin on the growth performances and nutrient utilization in pigs weaned at 21 days of age. Asian-Australas. J. Anim. Sci. 11:176–184. doi: https://doi.org/10.5713/ajas.1998.176.

Jolliff, J. S., and D. C. Mahan. 2012. Effect of dietary inulin and phytase on mineral digestibility and tissue retention in weanling and growing swine. J. Anim. Sci. 90:3012–3022. doi: https://doi.org/10.2527/jas.2011-4424.

Jones, D. B., J. D. Hancock, D. L. Harmon, and C. E. Walker. 1992. Effects of exogenous emulsifiers and fat sources on nutrient digestibility, serum lipids, and growth performance in weanling pigs. J. Anim. Sci. 70:3473–3482. doi: https://doi.org/10.2527/1992.70113473x.

Jørgensen, H., and J. A. Fernández. 2000. Chemical composition and energy value of different fat sources for growing pigs. Acta Agric. Scand. Sect. A Anim. Sci. 50:129–136. doi: https://doi.org/10.1080/090647000750014250.

Jørgensen, H., V. M. Gabert, M. S. Hedemann, and S. K. Jensen. 2000. Digestion of fat does not differ in growing pigs fed diets containing fish oil, rapeseed oil or coconut oil. J. Nutr. 130:852–857. doi: https://doi.org/10.1093/jn/130.4.852.

Jørgensen, H., K. Jakobsen, and B. O. Eggum. 1992. The influence of different protein, fat and mineral levels on the digestibility of fat and fatty acids measured at the terminal ileum and in faeces of growing pigs. Acta Agric. Scand. Sect. A Anim. Sci. 42:177–184. doi: https://doi.org/10.1080/09064709209410125.

Jørgensen, H., K. Jakobsen, and B. O. Eggum. 1993. Determination of endogenous fat and fatty acids at the terminal ileum and on faeces in growing pigs. Acta Agric. Scand. Sect. A Anim. Sci. 43:101–106. doi: https://doi.org/10.1080/09064709309410151.

Just, A. 1982. The net energy value of crude fat for growth in pigs. Livest. Prod. Sci. 9:501–509. doi: https://doi.org/10.1016/0301-6226(82)90054-9.

Kellner, T. A., and J. F. Patience. 2017. The digestible energy, metabolizable energy, and net energy content of dietary fat sources in thirteen- and fifty-kilogram pigs. J. Anim. Sci. 95:3984–3995. doi: https://doi.org/10.2527/jas.2017.1824.

Kemme, P. A., A. W. Jongbloed, Z. Mroz, and A. C. Beynen. 1997. The efficacy of *Aspergillus niger* phytase in rendering phytate phosphorus available for absorption in pigs is influenced by pig physiological status. J. Anim. Sci. 75:2129–2138. doi: https://doi.org/10.2527/1997.7582129x.

Kemme, P. A., A. W. Jongbloed, Z. Mroz, J. Kogut, and A. C. Beynen. 1999. Digestibility of nutrients in growing-finishing pig is affected by *Aspergillus niger* phytase, phytate and lactic acid levels: 2. Apparent total tract digestibility of phosphorus, calcium and magnesium and ileal degradation of phytic acid. Livest. Prod. Sci. 58:119–127. doi: https://doi.org/10.1016/S0301-6226(98)00202-4.

Kerr, B. J., T. A. Kellner, and G. C. Shurson. 2015. Characteristics of lipids and their feeding value in swine diets. J. Anim. Sci. Biotechnol. 6:30. doi: https://doi.org/10.1186/s40104-015-0028-x.

Kerr, B. J., and G. C. Shurson. 2017. Determination of ether extract digestibility and energy content of specialty lipids with different fatty acid and free fatty acid content, and the effect of lecithin, for nursery pigs. Prof. Anim. Sci. 33:127–134. doi: https://doi.org/10.15232/pas.2016-01561.

Kies, A. K., P. A. Kemme, L. B. J. Šebek, J. T. M. van Diepen, and A. W. Jongbloed. 2006. Effect of graded doses and a high dose of microbial phytase on the digestibility of various minerals in weaner pigs. J. Anim. Sci. 84:1169–1175. doi: https://doi.org/10.2527/2006.8451169x.

Kil, D. Y., T. E. Sauber, D. B. Jones, and H. H. Stein. 2010. Effect of the form of dietary fat and the concentration of dietary neutral detergent fiber on ileal and total tract endogenous losses and apparent and true digestibility of fat by growing pigs. J. Anim. Sci. 88:2959–2967. doi: https://doi.org/10.2527/jas.2009-2216.

Kil, D. Y., and H. H. Stein. 2011. Dietary soybean oil and choice white grease improve apparent ileal digestibility of amino acids in swine diets containing corn, soybean meal, and distillers dried grains with solubles. Rev. Colomb. Cienc. Pecu. 24:248–253.

Kim, B. G., D. Y. Kil, and H. H. Stein. 2013. In growing pigs, the true ileal and total tract digestibility of acid hydrolyzed ether extract in extracted corn oil is greater than in intact sources of corn oil or soybean oil. J. Anim. Sci. 91:755–763. doi: https://doi.org/10.2527/jas.2011-4777.

Kim, B. G., D. Y. Kil, Y. Zhang, and H. H. Stein. 2012a. Concentrations of analyzed or reactive lysine, but not crude protein, may predict the concentration of digestible lysine in distillers dried grains with solubles fed to pigs. J. Anim. Sci. 90:3798–3808. doi: https://doi.org/10.2527/jas.2011-4692.

Kim, B. G., J. W. Lee, and H. H. Stein. 2012b. Energy concentration and phosphorus digestibility in whey powder, whey permeate, and low-ash whey permeate fed to weanling pigs. J. Anim. Sci. 90:289–295. doi: https://doi.org/10.2527/jas.2011-4145.

Kim, B. G., S. A. Lee, K. R. Park, and H. H. Stein. 2020. At least 3 days of adaptation are required before indigestible markers (chromium, titanium, and acid insoluble ash) are stabilized in the ileal digesta of 60-kg pigs, but values for amino acid digestibility are affected by the marker. J. Anim. Sci. 98:skaa027. doi: https://doi.org/10.1093/jas/skaa027.

Kim, B. G., M. D. Lindemann, and G. L. Cromwell. 2009. The effects of dietary chromium(III) picolinate on growth performance, blood measurements, and respiratory rate in pigs kept in high and low ambient temperature. J. Anim. Sci. 87:1695–1704. doi: https://doi.org/10.2527/jas.2008-1218.

Kim, B. G., M. D. Lindemann, and G. L. Cromwell. 2010. The effects of dietary chromium(III) picolinate on growth performance, vital signs, and blood measurements of pigs during immune stress. Biol. Trace. Elem. Res. 135:200–210. doi: https://doi.org/10.1007/s12011-009-8503-x.

Kim, Y. Y., and D. C. Mahan. 2001. Effects of high dietary levels of selenium-enriched yeast and sodium selenite on macro and micro mineral metabolism in grower-finisher swine. Asian-Australas. J. Anim. Sci. 14:243–249. doi: https://doi.org/10.5713/ajas.2001.243.

Knowlton, K. F., J. S. Radcliffe, C. L. Novak, and D. A. Emmerson. 2004. Animal management to reduce phosphorus losses to the environment. J. Anim. Sci. 82(E. Suppl):E173–E195. doi: https://doi.org/10.2527/2004.8213_supplE173x.

Kodicek, E., R. Braude, S. K. Kon, and K. G. Mitchell. 1959. The availability to pigs of nicotinic acid in tortilla baked from maize treated with lime-water. Br. J. Nutr. 13:363–384. doi: https://doi.org/10.1079/BJN19590047.

Koh, F., K. Charlton, K. Walton, and A.-T. McMahon. 2015. Role of dietary protein and thiamine intakes on cognitive function in healthy older people: A systematic review. Nutrients 7:2415–2439. doi: https://doi.org/10.3390/nu7042415.

Kong, C., and O. Adeola. 2014. Evaluation of amino acid and energy utilization in feedstuff for swine and poultry diets. Asian-Australas. J. Anim. Sci. 27:917–925. doi: https://doi.org/10.5713/ajas.2014.r.02.

Kong, C., D. Ragland, and O. Adeola. 2014. Ileal endogenous amino acid flow response to nitrogen-free diets with differing ratios of corn starch to dextrose in pigs. Asian-Australas. J. Anim. Sci. 27:1124–1130. doi: https://doi.org/10.5713/ajas.2014.14232.

Kornegay, E. T. 1972. Availability of iron contained in defluorinated phosphate. J. Anim. Sci. 34:569–572. doi: https://doi.org/10.2527/jas1972.344569x.

Kornegay, E. T., and T. N. Meacham. 1973. Evaluation of supplemental choline for reproducing sows housed in total confinement on concrete or in dirt lots. J. Anim. Sci. 37:506–509. doi: https://doi.org/10.2527/jas1973.372506x.

Krawielitzki, K., F. Kreienbring, T. Zebrowska, R. Schadereit, and J. Kowalczyk. 1994. Estimation of N absorption, secretion and reabsorption in different intestinal sections of growing pigs using the N-15 isotope dilution method. In: Proc. of Digestive Physiology in Pigs, (4–6 October 1994) vol. I., Bad Doberan, Germany. p. 79–82.

Krider, J. L., J. L. Albright, M. P. Plumlee, J. H. Conrad, C. L. Sinclair, L. Underwood, R. G. Jones, and R. B. Harrington. 1975. Magnesium supplementation, space and docking effects on swine performance and behavior. J. Anim. Sci. 40:1027–1033. doi: https://doi.org/10.2527/jas1975.4061027x.

Krumdieck, C. L., A. J. Newman, and C. E. Butterworth, Jr. 1973. A naturally occurring inhibitor of folic acid conjugase (petroylopolyglutamyl hydrolase) in beans and other pulses. Am. J. Clin. Nutr. 24:460. (Abstr).

Kuznetsov, S. G., B. D. Kal'nitskii, and A. P. Bataeva. 1987. Biological availability of calcium from chemical compounds for young pigs. Soviet Agric. Sci. 3:32–34.

Kwon, W. B. 2016. Nutritional evaluation of inorganic phosphate sources and phosphorus digestibility in pigs. M. S. thesis. Konkuk University, Seoul, Republic of Korea. http://www.dcollection.net/handler/konkuk/000002060175.

Kwon, W. B., K. J. Touchette, A. Simongiovanni, K. Syriopoulos, A. Wessels, and H. H. Stein. 2019. Excess dietary leucine in diets for growing pigs reduces growth performance, biological value of protein, protein retention, and serotonin synthesis. J. Anim. Sci. 97:4282–4292. doi: https://doi.org/10.1093/jas/skz259.

Lagos, L. V., S. A. Lee, G. Fondevila, C. L. Walk, M. R. Murphy, J. J. Loor, and H. H. Stein. 2019. Influence of the concentration of dietary digestible calcium on growth performance, bone mineralization, plasma calcium, and abundance of genes involved in intestinal absorption of calcium in pigs from 11 to 22 kg fed diets with different concentrations of digestible phosphorus. J. Anim. Sci. Biotechnol. 10:47. doi: https://doi.org/10.1186/s40104-019-0349-2.

Le Guen, M. P., J. Huisman, J. Guéguen, G. Beelen, and M. W. A. Verstegen. 1995. Effects of a concentrate of pea antinutritional factors on pea protein digestibility in piglets. Livest. Prod. Sci. 44:157–167. doi: https://doi.org/10.1016/0301-6226(95)00053-4.

Lee, S. A., H. Jo, C. Kong, and B. G. Kim. 2017. Use of digestible rather than total amino acid in diet formulation increases nitrogen retention and reduces nitrogen excretion from pigs. Livest. Sci. 197:8–11. doi: https://doi.org/10.1016/j.livsci.2016.12.013.

Lee, S. A., L. V. Lagos, M. R. Bedford, and H. H. Stein. 2020. Increasing calcium from deficient to adequate concentration in diets for gestating sows decreases digestibility of phosphorus and reduces serum concentration of a bone resorption biomarker. J. Anim. Sci. 98:skaa076. doi: https://doi.org/10.1093/jas/skaa076.

Lee, S. A., L. V. Lagos, C. L. Walk, and H. H. Stein. 2019a. Basal endogenous loss, standardized total tract digestibility of calcium in calcium carbonate, and retention of calcium in gestating sows change during gestation, but microbial phytase reduces basal endogenous loss of calcium. J. Anim. Sci. 97:1712–1721. doi: https://doi.org/10.1093/jas/skz048.

Lee, S. A., L. V. Lagos, C. L. Walk, and H. H. Stein. 2019b. Standardized total tract digestibility of calcium varies among sources of calcium carbonate, but not among sources of dicalcium phosphate, but microbial phytase increases calcium digestibility in calcium carbonate. J. Anim. Sci. 97:3440–3450. doi: https://doi.org/10.1093/jas/skz176.

Lee, S. A., C. L. Walk, and H. H. Stein. 2021. Comparative digestibility and retention of calcium and phosphorus in normal- and high-phytate diets fed to gestating sows and growing pigs. Anim. Feed Sci. Technol. 280:115084. doi: https://doi.org/10.1016/j.anifeedsci.2021.115084.

Lee, S. C., L. Prosky, and J. W. D. Vries. 1992. Determination of total, soluble, and insoluble dietary fiber in foods—enzymatic-gravimetric method, MES-TRIS buffer: Collaborative study. J. AOAC Int. 75:395–416. doi: https://doi.org/10.1093/jaoac/75.3.395.

Lenis, N. P., P. Bikker, J. van der Meulen, J. T. van Diepen, J. G. Bakker, and A. W. Jongbloed. 1996. Effect of dietary neutral detergent fiber on ileal digestibility and portal flux of nitrogen and amino acids and on nitrogen utilization in growing pigs. J. Anim. Sci. 74:2687–2699. doi: https://doi.org/10.2527/1996.74112687x.

Lewis, A. J., G. L. Cromwell, and J. E. Pettigrew. 1991. Effects of supplemental biotin during gestation and lactation on reproductive performance of sows: a cooperative study. J. Anim. Sci. 69:207–214. doi: https://doi.org/10.2527/1991.691207x.

Li, D. F., R. C. Thaler, J. L. Nelssen, D. L. Harmon, G. L. Allee, and T. L. Weeden. 1990. Effect of fat sources and combinations on starter pig performance, nutrient digestibility and intestinal morphology. J. Anim. Sci. 68:3694–3704. doi: https://doi.org/10.2527/1990.68113694x.

Li, S., and W. C. Sauer. 1994. The effect of dietary fat content on amino acid digestibility in young pigs. J. Anim. Sci. 72:1737–1743. doi: https://doi.org/10.2527/1994.7271737x.

Li, Y., W. Yang, D. Dong, S. Jiang, Z. Yang, and Y. Wang. 2018. Effect of different sources and levels of iron in the diet of sows on iron status in neonatal pigs. Anim. Nutr. 4:197–202. doi: https://doi.org/10.1016/j.aninu.2018.01.002.

Li, Z. C., Y. B. Su, X. H. Bi, Q. Y. Wang, J. Wang, J. B. Zhao, L. Liu, F. L. Wang, D. F. Li, and C. H. Lai. 2017. Effects of lipid form and source on digestibility of fat and fatty acids in growing pigs. J. Anim. Sci. 95:3103–3109. doi: https://doi.org/10.2527/jas.2016.1268.

Lien, K. A., W. C. Sauer, and M. Fenton. 1997. Mucin output in ileal digesta of pigs fed a protein-free diet. Z. Ernährungswiss. 36:182–190. doi: https://doi.org/10.1007/BF01611398.

Lindblom, S. C., N. K. Gabler, and B. J. Kerr. 2018. Influence of feeding thermally peroxidized soybean oil on growth performance, digestibility, and gut integrity in growing pigs. J. Anim. Sci. 96:558–569. doi: https://doi.org/10.1093/jas/sky004.

Lindemann, M. D., S. D. Carter, L. I. Chiba, C. R. Dove, F. M. LeMieux, and L. L. Southern. 2004. A regional evaluation of chromium tripicolinate supplementation of diets fed to reproducing sows. J. Anim. Sci. 82:2972–2977. doi: https://doi.org/10.2527/2004.82102972x.

Lindemann, M. D., S. G. Cornelius, S. M. El Kandelgy, R. L. Moser, and J. E. Pettigrew. 1986. Effect of age, weaning and diet on digestive enzyme levels in the piglet. J. Anim. Sci. 62:1298–1307. doi: https://doi.org/10.2527/jas1986.6251298x.

Lindemann, M. D., G. L. Cromwell, H. J. Monegue, and K. W. Purser. 2008. Effect of chromium source on tissue concentration of chromium in pigs. J. Anim. Sci. 86:2971–2978. doi: https://doi.org/10.2527/jas.2008-0888.

Lindemann, M. D., and E. T. Kornegay. 1989. Folic acid supplementation to diets of gestating-lactating swine over multiple parities. J. Anim. Sci. 67:459–464. doi: https://doi.org/10.2527/jas1989.672459x.

Lindemann, M. D., C. M. Wood, A. F. Harper, E. T. Kornegay, and R. A. Anderson. 1995. Dietary chromium picolinate additions improve gain:feed and carcass characteristics in growing-finishing pigs and increase litter size in reproducing sows. J. Anim. Sci. 73:457–465. doi: https://doi.org/10.2527/1995.732457x.

Littell, R. C., P. R. Henry, A. J. Lewis, and C. B. Ammerman. 1997. Estimation of relative bioavailability of nutrients using SAS procedures. J. Anim. Sci. 75:2672–2683. doi: https://doi.org/10.2527/1997.75102672x.

Liu, J. B., S. C. Cao, J. Liu, J. Pu, L. Chen, and H. F. Zhang. 2018. Effects of dietary energy and lipase levels on nutrient digestibility, digestive physiology and noxious gas emission in weaning pigs. Asian-Australas. J. Anim. Sci. 31:1963–1973. doi: https://doi.org/10.5713/ajas.18.0087.

Liu, P., B. J. Kerr, C. Chen, T. E. Weber, L. J. Johnston, and G. C. Shurson. 2014a. Methods to create thermally oxidized lipids and comparison of analytical procedures to characterize peroxidation. J. Anim. Sci. 92:2950–2959. doi: https://doi.org/10.2527/jas.2012-5708.

Liu, Y., Y. L. Ma, J. M. Zhao, M. Vazquez-Añón, and H. H. Stein. 2014b. Digestibility and retention of zinc, copper, manganese, iron, calcium, and phosphorus in pigs fed diets containing inorganic or organic minerals. J. Anim. Sci. 92:3407–3415. doi: https://doi.org/10.2527/jas.2013-7080.

Livingston, A. L., J. W. Nelson, and G. O. Kohler. 1968. Stability of α-tocopherol during alfalfa dehydration and storage. J. Agric. Food Chem. 16:492–495. doi: https://doi.org/10.1021/jf60157a016.

Low, A. G., and T. Zebrowska. 1989. Digestion in pigs. In: H. D. Bock, B. O. Eggum, A. G. Low, O. Simon and T. Zebrowska, editors, Protein metabolism in farm animals. Oxford Univ. Press, Oxford, UK. p. 53–121.

Lowry, K. R., and D. H. Baker. 1989. Amelioration of selenium toxicity by arsenicals and cysteine. J. Anim. Sci. 67:959–965. doi: https://doi.org/10.2527/jas1989.674959x.

Lowry, K. R., and D. H. Baker. 1998. Effect of excess leucine on niacin provided by either tryptophan or niacin. FASEB J. 3:A666. (Abstr).

Lowry, K. R., O. A. Izquierdo, and D. H. Baker. 1987. Efficacy of betaine relative to choline as a dietary methyl donor. Poult. Sci. 66(Suppl. 1):135. (Abstr).

Ma, Y. L., M. D. Lindemann, S. F. Webb, and G. Rentfrow. 2018. Evaluation of trace mineral source and preharvest deletion of trace minerals from finishing diets on tissue mineral status in pigs. Asian-Australas. J. Anim. Sci. 31:252–262. doi: https://doi.org/10.5713/ajas.17.0189.

Macek, T. J., and B. A. Feller. 1952. Crystalline vitamin $B_{12}$ in pharmaceutical preparations. J. Am. Pharm. Assoc. 41:285–288. doi: https://doi.org/10.1002/jps.3030410602.

Mahan, D. C., and A. L. Moxon. 1978. Effects of adding inorganic or organic selenium sources to the diets of young swine. J. Anim. Sci. 47:456–466. doi: https://doi.org/10.2527/jas1978.472456x.

Mahan, D. C., and N. A. Parrett. 1996. Evaluating the efficacy of selenium-enriched yeast and sodium selenite on tissue selenium retention and serum glutathione peroxidase activity in grower and finisher swine. J. Anim. Sci. 74:2967–2974. doi: https://doi.org/10.2527/1996.74122967x.

Maillard, L. C. 1912. Action of amino acids on sugars. Formation of melanoidins in a methodical way. Comptes Rendus Acad. Sci. 154:66–68.

Marchetti, M., M. Tassinari, and S. Marchetti. 2000. Menadione nicotinamide bisulphite as a source of vitamin K and niacin activities for the growing pig. Anim. Sci. 71:111–117. doi: https://doi.org/10.1017/S135772980005493X.

Mariscal-Landin, G., B. Sève, Y. Colléaux, and Y. Lebreton. 1995. Endogenous amino nitrogen collected from pigs with end-to-end ileorectal anastomosis is affected by the method of estimation and altered by dietary fiber. J. Nutr. 125:136–146. doi: https://doi.org/10.1093/jn/125.1.136.

Mathias, M. M., D. E. Hogue, and J. K. Loosli. 1967. The biological value of selenium in bovine milk for the rat and chick. J. Nutr. 93:14–20. doi: https://doi.org/10.1093/jn/93.1.14.

Matte, J. J., C. L. Girard, and G. J. Brisson. 1992. The role of folic acid in the nutrition of gestating and lactating primiparous sows. Livest. Prod. Sci. 32:131–148. doi: https://doi.org/10.1016/S0301-6226(12)80032-7.

Matte, J. J., F. Guay, N. Le Floc'h, and C. L. Girard. 2010. Bioavailability of dietary cyanocobalamin (vitamin B$_{12}$) in growing pigs. J. Anim. Sci. 88:3936–3944. doi: https://doi.org/10.2527/jas.2010-2979.

Mavromichalis, I., T. M. Parr, V. M. Gabert, and D. H. Baker. 2001. True ileal digestibility of amino acids in sow's milk for 17-day-old pigs. J. Anim. Sci. 79:707–713. doi: https://doi.org/10.2527/2001.793707x.

Merriman, L. A. 2016. Factors affecting the digestibility of calcium in feed ingredients and requirements for digestible calcium by pigs. PhD. Diss. Univ. Illinois, Urbana-Champaign, IL, US. http://hdl.handle.net/2142/90614.

Merriman, L. A., C. L. Walk, C. M. Parsons, and H. H. Stein. 2016. Effects of tallow, choice white grease, palm oil, corn oil, or soybean oil on apparent total tract digestibility of minerals in diets fed to growing pigs. J. Anim. Sci. 94:4231–4238. doi: https://doi.org/10.2527/jas.2016-0682.

Miller, E. R. 1980. Bioavailability of minerals. Proc. Minnesota Nutr. Conf., St. Paul, MN, US. p. 144–154.

Miller, E. R., and C. B. Ammerman. 1995. Iodine bioavailability. In: C. B. Ammerman, D. H. Baker, and A. J. Lewis, editors, Bioavailability of nutrients for animals: Amino acids, minerals, and vitamins. Academic Press, San Diego, CA. p. 157–167.

Miller, E. R., P. K. Ku, J. P. Hitchcock, and W. T. Magee. 1981. Availability of zinc from metallic zinc dust for young swine. J. Anim. Sci. 52:312–315. doi: https://doi.org/10.2527/jas1981.522312x.

Mitchell, H. 1924. The nutritive value of proteins. Physiol. Rev. 4:424–478.

Moehn, S., R. F. P. Bertolo, P. B. Pencharz, and R. O. Ball. 2005. Development of the indicator amino acid oxidation technique to determine the availability of amino acids from dietary protein in pigs. J. Nutr. 135:2866–2870. doi: https://doi.org/10.1093/jn/135.12.2866.

Molitoris, B. A., and D. H. Baker. 1976a. Assessment of the quantity of biologically available choline in soybean meal. J. Anim. Sci. 42:481–489. doi: https://doi.org/10.2527/jas1976.422481x.

Molitoris, B. A., and D. H. Baker. 1976b. Choline utilization in the chick as influenced by levels of dietary protein and methionine. J. Nutr. 106:412–418. doi: https://doi.org/10.1093/jn/106.3.412.

Montoya, C. A., S. J. Henare, S. M. Rutherfurd, and P. J. Moughan. 2016. Potential misinterpretation of the nutritional value of dietary fiber: Correcting fiber digestibility values for nondietary gut-interfering material. Nutr. Rev. 74:517–533. doi: https://doi.org/10.1093/nutrit/nuw014.

Montoya, C. A., S. M. Rutherfurd, and P. J. Moughan. 2015. Nondietary gut materials interfere with the determination of dietary fiber digestibility in growing pigs when using the Prosky method. J. Nutr. 145:1966–1972. doi: https://doi.org/10.3945/jn.115.212639.

Morgan, D. P., E. P. Young, I. P. Earle, R. J. Davey, and J. W. Stevenson. 1969. Effects of dietary calcium and zinc on calcium, phosphorus and zinc retention in swine. J. Anim. Sci. 29:900–905. doi: https://doi.org/10.2527/jas1969.296900x.

Mosenthin, R., W. C. Sauer, and F. Ahrens. 1994. Dietary pectin's effect on ileal and fecal amino acid digestibility and exocrine pancreatic secretions in growing pigs. J. Nutr. 124:1222–1229. doi: https://doi.org/10.1093/jn/124.8.1222.

Mosenthin, R., W. C. Sauer, R. Blank, J. Huisman, and M. Z. Fan. 2000. The concept of digestible amino acids in diet formulation for pigs. Livest. Prod. Sci. 64:265–280. doi: https://doi.org/10.1016/S0301-6226(99)00139-6.

Moter, V., and H. H. Stein. 2004. Effect of feed intake on endogenous losses and amino acid and energy digestibility by growing pigs. J. Anim. Sci. 82:3518–3525. doi: https://doi.org/10.2527/2004.82123518x.

Moughan, P. J. 1991. Towards an improved utilization of dietary amino acids by the growing pig. In: W. Haresign and D. J. A. Cole, editors, Recent advances in animal nutrition. Butterworth-Heinemann, Oxford, UK. p. 45–64.

Moughan, P. J. 2003. Amino acid availability: aspects of chemical analysis and bioassay methodology. Nutr. Res. Rev. 16:127–141. doi: https://doi.org/10.1079/Nrr200365.

Moughan, P. J., and G. Schuttert. 1991. Composition of nitrogen-containing fractions in digesta from the distal ileum of pigs fed a protein-free diet. J. Nutr. 121:1570–1574. doi: https://doi.org/10.1093/jn/121.10.1570.

Navarro, D. M. D. L., J. J. Abelilla, and H. H. Stein. 2019. Structures and characteristics of carbohydrates in diets fed to pigs: A review. J. Anim. Sci. Biotechnol. 10:39. doi: https://doi.org/10.1186/s40104-019-0345-6.

Navarro, D. M. D. L., E. M. A. M. Bruininx, L. de Jong, and H. H. Stein. 2018. Analysis for low-molecular-weight carbohydrates is needed to account for all energy-contributing nutrients in some feed ingredients, but physical characteristics do not predict in vitro digestibility of dry matter. J. Anim. Sci. 96:532–544. doi: https://doi.org/10.1093/jas/sky010.

NCR-42 Committee on Swine Nutrition. 1976. Effect of supplemental choline on reproductive performance of sows: A cooperative regional study. J. Anim. Sci. 42:1211–1216. doi: https://doi.org/10.2527/jas1976.4251211x.

NCR-42 Committee on Swine Nutrition. 1980. Effect of supplemental choline on performance of starting, growing and finishing pigs: A cooperative regional study. J. Anim. Sci. 50:99–102. doi: https://doi.org/10.2527/jas1980.50199x.

Ndou, S. P., E. Kiarie, M. C. Walsh, N. Ames, C. F. M. de Lange, and C. M. Nyachoti. 2019. Interactive effects of dietary fibre and lipid types modulate gastrointestinal flows and apparent digestibility of fatty acids in growing pigs. Br. J. Nutr. 121:469–480. doi: https://doi.org/10.1017/S0007114518003434.

Nelson, T. S., and A. C. Walker. 1964. The biological evaluation of phosphorus compounds: A summary. Poult. Sci. 43:94–98. doi: https://doi.org/10.3382/ps.0430094.

Noblet, J., and M. Champion. 2003. Effect of pelleting and body weight on digestibility of energy and fat of two corns in pigs. J. Anim. Sci. 81(Suppl):140. (Abstr.).

Nørgaard, I., and P. B. Mortensen. 1995. Digestive processes in the human colon. Nutrition 11:37–45.

Norii, T., and H. Suzuki. 2002. Influences of dietary protein levels and phytate contents on zinc requirement in rats. Int. J. Food Sci. Nutr. 53:317–323. doi: https://doi.org/10.1080/09637480220138089.

NRC. 1987. Vitamin tolerance of animals. Natl. Acad. Press, Washington, DC.

NRC. 1998. Nutrient requirements of swine. 10th rev. ed. Natl. Acad. Press, Washington, DC.

NRC. 2012. Nutrient requirements of swine. 11th rev. ed. Natl. Acad. Press, Washington, DC.

Nyachoti, C. M., C. F. M. de Lange, B. W. McBride, and H. Schulze. 1997. Significance of endogenous gut nitrogen losses in the nutrition of growing pigs: A review. Can. J. Anim. Sci. 77:149–163. doi: https://doi.org/10.4141/A96-044.

Oduho, G. W., and D. H. Baker. 1993. Quantitative efficacy of niacin sources for chicks: Nicotinic acid, nicotinamide, NAD and tryptophan. J. Nutr. 123:2201–2206. doi: https://doi.org/10.1093/jn/123.12.2201.

Oduho, G. W., T. K. Chung, and D. H. Baker. 1993. Menadione nicotinamide bisulfite is a bioactive source of vitamin K and niacin activity for chicks. J. Nutr. 123:737–743. doi: https://doi.org/10.1093/jn/123.4.737.

Oduho, G. W., Y. Han, and D. H. Baker. 1994. Iron deficiency reduces the efficacy of tryptophan as a niacin precursor. J. Nutr. 124:444–450. doi: https://doi.org/10.1093/jn/124.3.444.

Oliveira, M. S. F., J. J. Abelilla, N. W. Jaworski, J. K. Htoo, and H. H. Stein. 2020a. Crystalline amino acids do not influence calculated values for standardized ileal digestibility of amino acids in feed ingredients included in diets for pigs. J. Anim. Sci. 98:skaa333. doi: https://doi.org/10.1093/jas/skaa333.

Oliveira, M. S. F., J. K. Htoo, and H. H. Stein. 2020b. The direct and difference procedures result in similar estimates for amino acid digestibility in feed ingredients fed to growing pigs. J. Anim. Sci. 98. doi: https://doi.org/10.1093/jas/skaa225.

Oliveira, M. S. F., M. K. Wiltafsky, S. A. Lee, W. B. Kwon, and H. H. Stein. 2020c. Concentrations of digestible and metabolizable energy and amino acid digestibility by growing pigs may be reduced by autoclaving soybean meal. Anim. Feed Sci. Technol. 269:114621. doi: https://doi.org/10.1016/j.anifeedsci.2020.114621.

Opapeju, F. O., J. K. Htoo, C. Dapoza, and C. M. Nyachoti. 2012. Bioavailability of methionine hydroxy analog-calcium salt relative to DL-methionine to support nitrogen retention and growth in starter pigs. Animal. 6:1750–1756. doi: https://doi.org/10.1017/S1751731112000869.

Page, T. G., L. L. Southern, T. L. Ward, and D. L. Thompson, Jr. 1993. Effect of chromium picolinate on growth and serum and carcass traits of growing-finishing pigs. J. Anim. Sci. 71:656–662. doi: https://doi.org/10.2527/1993.713656x.

Park, C. S., and O. Adeola. 2020. Basal ileal endogenous losses of amino acids in pigs determined by feeding nitrogen-free diet or low-casein diet or by regression analysis. Anim. Feed Sci. Technol. 267:114550. doi: https://doi.org/10.1016/j.anifeedsci.2020.114550.

Park, C. S., S. I. Oh, and B. G. Kim. 2013. Prediction of basal endogenous losses of amino acids based on body weight and feed intake in pigs fed nitrogen-free diets. Rev. Colomb. Cienc. Pecu. 26:186–192.

Parsons, H. T., T. Yasmin, and S. C. Fry. 2011. Alternative pathways of dehydroascorbic acid degradation in vitro and in plant cell cultures: Novel insights into vitamin C catabolism. Biochem. J. 440:375–385. doi: https://doi.org/10.1042/bj20110939.

Partanen, K., T. Jalava, J. Valaja, S. Perttilä, H. Siljander-Rasi, and H. Lindeberg. 2001. Effect of dietary carbadox or formic acid and fibre level on ileal and faecal nutrient digestibility and microbial metabolite concentrations in ileal digesta of the pig. Anim. Feed Sci. Technol. 93:137–155. doi: https://doi.org/10.1016/S0377-8401(01)00288-7.

Patel, K. P., and D. H. Baker. 1996. Supplemental iron, copper, zinc, ascorbate, caffeine and chlortetracycline do not affect riboflavin utilization in the chick. Nutr. Res. 16:1943–1952. doi: https://doi.org/10.1016/S0271-5317(96)00217-5.

Pedersen, C., M. G. Boersma, and H. H. Stein. 2007. Energy and nutrient digestibility in NutriDense corn and other cereal grains fed to growing pigs. J. Anim. Sci. 85:2473–2483. doi: https://doi.org/10.2527/jas.2006-620.

Peeler, H. T. 1972. Biological availability of nutrients in feeds: Availability of major mineral ions. J. Anim. Sci. 35:695–712. doi: https://doi.org/10.2527/jas1972.353695x.

Perri, A. M., R. M. Friendship, J. C. S. Harding, and T. L. O'Sullivan. 2016. An investigation of iron deficiency and anemia in piglets and the effect of iron status at weaning on post-weaning performance. J. Swine Health Prod. 24:10–20.

Peters, J. C., and D. C. Mahan. 2008. Effects of neonatal iron status, iron injections at birth, and weaning in young pigs from sows fed either organic or inorganic trace minerals. J. Anim. Sci. 86:2261–2269. doi: https://doi.org/10.2527/jas.2007-0577.

Petersen, G. I., C. Pedersen, M. D. Lindemann, and H. H. Stein. 2011. Relative bioavailability of phosphorus in inorganic phosphorus sources fed to growing pigs. J. Anim. Sci. 89:460–466. doi: https://doi.org/10.2527/jas.2009-2161.

Petersen, G. I., and H. H. Stein. 2006. Novel procedure for estimating endogenous losses and measurement of apparent and true digestibility of phosphorus by growing pigs. J. Anim. Sci. 84:2126–2132. doi: https://doi.org/10.2527/jas.2005-479.

Pharazyn, A., and F. X. Aherne. 1987. Folacin requirement of the lactating sow. In: 66th Annual Feeders Day Report. University of Alberta. p. 16–18.

Pietrzik, K., L. Bailey, and B. Shane. 2010. Folic acid and L-5-methyltetrahydrofolate. Clin. Pharmacokinet. 49:535–548. doi: https://doi.org/10.2165/11532990-000000000-00000.

Pond, W. G., R. S. Lowrey, J. H. Maner, and J. K. Loosli. 1961. Parenteral iron administration to sows during gestation or lactation. J. Anim. Sci. 20:747–750. doi: https://doi.org/10.2527/jas1961.204747x.

Powles, J., J. Wiseman, D. J. A. Cole, and B. Hardy. 1993. Effect of chemical structure of fats upon their apparent digestible energy value when given to growing/finishing pigs. Anim. Sci. 57:137–146. doi: https://doi.org/10.1017/S000335610000670X.

Powles, J., J. Wiseman, D. J. A. Cole, and B. Hardy. 1994. Effect of chemical structure of fats upon their apparent digestible energy value when given to young pigs. Anim. Sci. 58:411–417. doi: https://doi.org/10.1017/S0003356100007364.

Powles, J., J. Wiseman, D. J. A. Cole, and S. Jagger. 1995. Prediction of the apparent digestible energy value of fats given to pigs. Anim. Sci. 61:149–154. doi: https://doi.org/10.1017/S1357729800013631.

Ramírez, M., L. Amate, and A. Gil. 2001. Absorption and distribution of dietary fatty acids from different sources. Early Hum. Dev. 65(Supple):S95–S101. doi: https://doi.org/10.1016/S0378-3782(01)00211-0.

Rao, C. N., and B. S. N. Rao. 1970. Absorption of dietary carotenes in human subjects. Am. J. Clin. Nutr. 23:105–109. doi: https://doi.org/10.1093/ajcn/23.1.105.

Reddy, N. R., S. K. Sathe, and D. K. Salunkhe. 1982. Phytates in legumes and cereals. Adv. Food Res. 28:1–92.

Reis de Souza, T., J. Peiniau, A. Mounier, and A. Aumaitre. 1995. Effect of addition of tallow and lecithin in the diet of weanling piglets on the apparent total tract and ileal digestibility of fat and fatty acids. Anim. Feed Sci. Technol. 52:77–91. doi: https://doi.org/10.1016/0377-8401(94)00705-E.

Remer, T. 2000. Acid-base in renal failure: Influence of diet on acid-base balance. Semin. Dial. 13:221–226. doi: https://doi.org/10.1046/j.1525-139x.2000.00062.x.

Richards, J. D., J. Zhao, R. J. Harrell, C. A. Atwell, and J. J. Dibner. 2010. Trace mineral nutrition in poultry and swine. Asian-Australas. J. Anim. Sci. 23:1527–1534. doi: https://doi.org/10.5713/ajas.2010.r.07.

Rodriguez, D. A., S. A. Lee, C. K. Jones, J. K. Htoo, and H. H. Stein. 2020. Digestibility of amino acids, fiber, and energy by growing pigs, and concentrations of digestible and metabolizable energy in yellow dent corn, hard red winter wheat, and sorghum may be influenced by extrusion. Anim. Feed Sci. Technol. 268:114602. doi: https://doi.org/10.1016/j.anifeedsci.2020.114602.

Rodriguez, M. S., and M. I. Irwin. 1972. A conspectus of research on vitamin A requirements of man. J. Nutr. 102:909–968. doi: https://doi.org/10.1093/jn/102.7.909.

Rojas, O. J., E. Vinyeta, and H. H. Stein. 2016. Effects of pelleting, extrusion, or extrusion and pelleting on energy and nutrient digestibility in diets containing different levels of fiber and fed to growing pigs. J. Anim. Sci. 94:1951–1960. doi: https://doi.org/10.2527/jas2015-0137.

Ross, A. C., C. L. Taylor, A. L. Yaktine, and H. B. Del Valle. 2011. Dietary reference intakes for calcium and vitamin D. National Academies Press, Washington, DC. doi: https://doi.org/10.17226/13050.

Ross, R. D., G. L. Cromwell, and T. S. Stahly. 1984. Effects of source and particle size on the biological availability of calcium in calcium supplements for growing-pigs. J. Anim. Sci. 59:125–134. doi: https://doi.org/10.2527/jas1984.591125x.

Rostagno, H. S., L. F. T. Albino, M. I. Hannas, J. L. Donzele, N. K. Sakomura, F. G. Perazzo, A. Saraiva, M. L. T. D. Abreu, P. B. Rodrigues, R. F. D. Oliveira, S. L. D. T. Barreto, and C. O. Brito. 2017. Brazilian tables for poultry and swine: Composition of feedstuffs and nutritional requirements. 4th ed.. Fed. Univ. Viçosa, Viçosa, MG, Brazil.

Roth-Maier, D. A., M. Kirchgessner, W. Erhardt, J. Henke, and U. Hennig. 1998. Comparative studies for the determination of precaecal digestibility as a measure for the availability of B-vitamins. J. Anim. Physiol. Anim. Nutr. 79:198–209. doi: https://doi.org/10.1111/j.1439-0396.1998.tb00643.x.

Ruiz, N., and R. H. Harms. 1988. Comparison of the biopotencies of nicotinic acid and nicotinamide for broiler chicks. Br. Poult. Sci. 29:491–498. doi: https://doi.org/10.1080/00071668808417075.

Sambrook, I. E. 1979. Studies on digestion and absorption in the intestines of growing pigs: 8. Measurements of the flow of total lipid, acid-detergent fibre and volatile fatty acids. Br. J. Nutr. 42:279–287. doi: https://doi.org/10.1079/BJN19790114.

Sauberlich, H. E. 1985. Bioavailability of vitamins. Prog. Food Nutr. Sci. 9:1–33.

Sauer, W., M. Cervantes, J. Yanez, B. Araiza, G. Murdoch, A. Morales, and R. T. Zijlstra. 2009. Effect of dietary inclusion of benzoic acid on mineral balance in growing pigs. Livest. Sci. 122:162–168. doi: https://doi.org/10.1016/j.livsci.2008.08.008.

Sauer, W. C., A. Just, and H. Jørgensen. 1982. The influence of daily feed intake on the apparent digestibility of crude protein, amino acids, calcium and phosphorus at the terminal ileum and overall in pigs. J. Anim. Physiol. Anim. Nutr. 48:177–182. doi: https://doi.org/10.1111/j.1439-0396.1982.tb01389.x.

Sauer, W. C., and L. Ozimek. 1986. Digestibility of amino acids in swine: Results and their practical applications. A review. Livest. Prod. Sci. 15:367–388. doi: https://doi.org/10.1016/0301-6226(86)90076-X.

Sauvant, D., J. M. Perez, and G. Tran. 2004. Tables of composition and nutritional value of feed materials: Pigs, poultry, cattle, sheep, goats, rabbits, horses and fish. 2nd ed. Wageningen Academic Publishers, Wageningen, The Netherlands.

Schroeder, H. A. 1971. Losses of vitamins and trace minerals resulting from processing and preservation of foods. Am. J. Clin. Nutr. 24:562–573. doi: https://doi.org/10.1093/ajcn/24.5.562.

Schulze, H., H. S. Saini, J. Huisman, M. Hessing, W. van den Berg, and M. W. A. Verstegen. 1995. Increased nitrogen secretion by inclusion of soya lectin in the diets of pigs. J. Sci. Food Agric. 69:501–510. doi: https://doi.org/10.1002/jsfa.2740690415.

Schulze, H., P. van Leeuwen, M. W. A. Verstegen, J. Huisman, W. B. Souffrant, and F. Ahrens. 1994. Effect of level of dietary neutral detergent fiber on ileal apparent digestibility and ileal nitrogen losses in pigs. J. Anim. Sci. 72:2362–2368. doi: https://doi.org/10.2527/1994.7292362x.

Seerley, R. W., O. W. Charles, H. C. McCampbell, and S. P. Bertsch. 1976. Efficacy of menadione dimethylpyrimidinol bisulfite as a source of vitamin K in swine diets. J. Anim. Sci. 42:599–607. doi: https://doi.org/10.2527/jas1976.423599x.

Selle, P. H., and V. Ravindran. 2008. Phytate-degrading enzymes in pig nutrition. Livest. Sci. 113:99–122. doi: https://doi.org/10.1016/j.livsci.2007.05.014.

Sève, B., G. Mariscal-Landin, Y. Colleaux, and Y. Lebreton. 1994. Ileal endogenous amino acid and amino sugar flows in pigs fed graded levels of protein or fibre. In: Proc. of Digestive Physiology in Pigs (4–6 October 1994), vol. I. Bad Doberan, Germany. p. 35–38.

She, Y., J. C. Sparks, and H. H. Stein. 2018a. Effects of increasing concentrations of an Escherichia coli phytase on the apparent ileal digestibility of amino acids and the apparent total tract digestibility of energy and nutrients in corn-soybean meal diets fed to growing pigs. J. Anim. Sci. 96:2804–2816. doi: https://doi.org/10.1093/jas/sky152.

She, Y., Q. Y. Wang, H. H. Stein, L. Liu, D. F. Li, and S. Zhang. 2018b. Additivity of values for phosphorus digestibility in corn, soybean meal, and canola meal in diets fed to growing pigs. Asian-Australas. J. Anim. Sci. 31:1301–1307. doi: https://doi.org/10.5713/ajas.17.0547.

Shelton, N. W., S. S. Dritz, J. L. Nelssen, M. D. Tokach, R. D. Goodband, J. M. DeRouchey, H. Yang, D. A. Hill, D. Holzgraefe, D. H. Hall, and D. C. Mahan. 2014. Effects of dietary vitamin E concentration and source on sow, milk, and pig concentrations of α-tocopherol. J. Anim. Sci. 92:4547–4556. doi: https://doi.org/10.2527/jas.2014-7311.

Shen, Y., M. Z. Fan, A. Ajakaiye, and T. Archbold. 2002. Use of the regression analysis technique to determine the true phosphorus digestibility and the endogenous phosphorus output associated with corn in growing pigs. J. Nutr. 132:1199–1206. doi: https://doi.org/10.1093/jn/132.6.1199.

Shi, X. S., and J. Noblet. 1993. Contribution of the hindgut to digestion of diets in growing pigs and adult sows: Effect of diet composition. Livest. Prod. Sci. 34:237–252. doi: https://doi.org/10.1016/0301-6226(93)90110-4.

Simons, P. C. M., H. A. J. Versteegh, A. W. Jongbloed, P. A. Kemme, P. Slump, K. D. Bos, M. G. E. Wolters, R. F. Beudeker, and G. J. Verschoor. 1990. Improvement of phosphorus availability by microbial phytase in broilers and pigs. Br. J. Nutr. 64:525–540. doi: https://doi.org/10.1079/BJN19900052.

Smiricky-Tjardes, M. R., C. M. Grieshop, E. A. Flickinger, L. L. Bauer, and G. C. Fahey, Jr. 2003. Dietary galactooligosaccharides affect ileal and total-tract nutrient digestibility, ileal and fecal bacterial concentrations, and ileal fermentative characteristics of growing pigs. J. Anim. Sci. 81:2535–2545. doi: https://doi.org/10.2527/2003.81102535x.

Smiricky, M. R., C. M. Grieshop, D. M. Albin, J. E. Wubben, V. M. Gabert, and G. C. Fahey, Jr. 2002. The influence of soy oligosaccharides on apparent and true ileal amino acid digestibilities and fecal consistency in growing pigs. J. Anim. Sci. 80:2433–2441. doi: https://doi.org/10.2527/2002.8092433x.

Smits, C. H. M., and G. Annison. 1996. Non-starch plant polysaccharides in broiler nutrition - Towards a physiologically valid approach to their determination. Worlds Poult. Sci. J. 52:203–221. doi: https://doi.org/10.1079/Wps19960016.

Soares, M., and C. J. Lopez-Bote. 2002. Effects of dietary lecithin and fat unsaturation on nutrient utilisation in weaned piglets. Anim. Feed Sci. Technol. 95:169–177. doi: https://doi.org/10.1016/S0377-8401(01)00324-8.

Son, A. R. 2016. Effects of dietary fiber on digestibility of nutrients and endogenous losses of phosphorus in pigs. PhD. Diss. Konkuk University, Seoul, Republic of Korea. http://www.dcollection.net/handler/konkuk/000002217016.

Son, A. R., and B. G. Kim. 2015. Effects of dietary cellulose on the basal endogenous loss of phosphorus in growing pigs. Asian-Australas. J. Anim. Sci. 28:369–373. doi: https://doi.org/10.5713/ajas.14.0539.

Souffrant, W. B., A. Rérat, J. P. Laplace, B. Darcy-Vrillon, R. Köhler, T. Corring, and G. Gebhardt. 1993. Exogenous and endogenous contributions to nitrogen fluxes in the digestive tract of pigs fed a casein diet. III. Recycling of endogenous nitrogen. Reprod. Nutr. Dev. 33:373–382. doi: https://doi.org/10.1051/rnd:19930406.

Southern, L. L., and D. H. Baker. 1981. Bioavailable pantothenic acid in cereal grains and soybean meal. J. Anim. Sci. 53:403–408. doi: https://doi.org/10.2527/jas1981.532403x.

Southern, L. L., and D. H. Baker. 1983a. Eimeria acervulina infection in chicks fed deficient or excess levels of manganese. J. Nutr. 113:172–177. doi: https://doi.org/10.1093/jn/113.1.172.

Southern, L. L., and D. H. Baker. 1983b. Excess manganese ingestion in the chick. Poult. Sci. 62:642–646. doi: https://doi.org/10.3382/ps.0620642.

Southern, L. L., and D. H. Baker. 1983c. Zinc toxicity, zinc deficiency and zinc-copper interrelationship in Eimeria Acervulina-infected chicks. J. Nutr. 113:688–696. doi: https://doi.org/10.1093/jn/113.3.688.

Southgate, D. A. T., and G. A. Spiller. 2001. Polysaccharide food additives that contribute to dietary fiber. In: G. A. Spiller, editor, CRC handbook of dietary fiber in human nutrition. CRC Press, Boca Raton, FL. p. 27–31.

Spiropoulos, J. 1985. Small scale production of lime for building. F. Vieweg, Braunschweig, Germany.

Stahly, T. S. 1984. Use of fats in diets for growing pigs. In: J. Wiseman, editor, Fats in animal nutrition. Butterworths, London, UK.

Stant, E. G., Jr., T. G. Martin, and W. V. Kessler. 1969. Potassium content of the porcine body and carcass at 23, 46, 68 and 91 kilograms live weight. J. Anim. Sci. 29:547–556. doi: https://doi.org/10.2527/jas1969.294547x.

Staten, F. E., P. A. Anderson, D. H. Baker, and P. C. Harrison. 1980. The efficacy of D,L-pantothenic acid relative to D-pantothenic acid in chicks. Poult. Sci. 59:1664. (Abstr.)

Stein, H. H. 2017. Procedures for determining digestibility of amino acids, lipids, starch, fibre, phosphorus, and calcium in feed ingredients fed to pigs. Anim. Prod. Sci. 57:2317–2324. doi: https://doi.org/10.1071/AN17343.

Stein, H. H., O. Adeola, G. L. Cromwell, S. W. Kim, D. C. Mahan, and P. S. Miller. 2011. Concentration of dietary calcium supplied by calcium carbonate does not affect the apparent total tract digestibility of calcium, but decreases digestibility of phosphorus by growing pigs. J. Anim. Sci. 89:2139–2144. doi: https://doi.org/10.2527/jas.2010-3522.

Stein, H. H., and R. A. Bohlke. 2007. The effects of thermal treatment of field peas (Pisum sativum L.) on nutrient and energy digestibility by growing pigs. J. Anim. Sci. 85:1424–1431. doi: https://doi.org/10.2527/jas.2006-712.

Stein, H. H., C. T. Kadzere, S. W. Kim, and P. S. Miller. 2008. Influence of dietary phosphorus concentration on the digestibility of phosphorus in monocalcium phosphate by growing pigs. J. Anim. Sci. 86:1861–1867. doi: https://doi.org/10.2527/jas.2008-0867.

Stein, H. H., S. W. Kim, T. T. Nielsen, and R. A. Easter. 2001. Standardized ileal protein and amino acid digestibility by growing pigs and sows. J. Anim. Sci. 79:2113–2122. doi: https://doi.org/10.2527/2001.7982113x.

Stein, H. H., L. V. Lagos, and G. A. Casas. 2016a. Nutritional value of feed ingredients of plant origin fed to pigs. Anim. Feed Sci. Technol. 218:33–69. doi: https://doi.org/10.1016/j.anifeedsci.2016.05.003.

Stein, H. H., L. A. Merriman, and J. C. González-Vega. 2016b. Establishing a digestible calcium requirement for pigs. In: C. L. Walk, I. Kühn, H. H. Stein, M. T. Kidd and M. Rodehutscord, editors, Phytate destruction - consequences for precision animal nutrition. Wageningen Academic Publishers, Wageningen, the Netherlands. p. 207–216. doi: https://doi.org/10.3920/978-90-8686-836-0_13.

Stein, H. H., C. Pedersen, A. R. Wirt, and R. A. Bohlke. 2005. Additivity of values for apparent and standardized ileal digestibility of amino acids in mixed diets fed to growing pigs. J. Anim. Sci. 83:2387–2395. doi: https://doi.org/10.2527/2005.83102387x.

Stein, H. H., B. Sève, M. F. Fuller, P. J. Moughan, and C. F. M. de Lange. 2007. Invited review: Amino acid bioavailability and digestibility in pig feed ingredients: Terminology and application. J. Anim. Sci. 85:172–180. doi: https://doi.org/10.2527/jas.2005-742.

Stein, H. H., C. F. Shipley, and R. A. Easter. 1998. Technical note: A technique for inserting a T-cannula into the distal ileum of pregnant sows. J. Anim. Sci. 76:1433–1436. doi: https://doi.org/10.2527/1998.7651433x.

Stein, H. H., N. L. Trottier, C. Bellaver, and R. A. Easter. 1999. The effect of feeding level and physiological status on total flow and amino acid composition of endogenous protein at the distal ileum in swine. J. Anim. Sci. 77:1180–1187. doi: https://doi.org/10.2527/1999.7751180x.

Stockland, W. L., and L. G. Blaylock. 1974. Choline requirement of pregnant sows and gilts under restricted feeding conditions. J. Anim. Sci. 39:1113–1116. doi: https://doi.org/10.2527/jas1974.3961113x.

Stoldt, W. 1952. Vorschlag zur vereinheitlichung der fettbestimmung in lebensmitteln. Fette Seifen Anstrich. 54:206–207. doi: https://doi.org/10.1002/lipi.19520540406.

Straarup, E. M., V. Danielsen, C.-E. Høy, and K. Jakobsen. 2006. Dietary structured lipids for post-weaning piglets: fat digestibility, nitrogen retention and fatty acid profiles of tissues. J. Anim. Physiol. Anim. Nutr. 90:124–135. doi: https://doi.org/10.1111/j.1439-0396.2005.00573.x.

Strang, E. J. P., M. Eklund, P. Rosenfelder, N. Sauer, J. K. Htoo, and R. Mosenthin. 2016. Chemical composition and standardized ileal amino acid digestibility of eight genotypes of rye fed to growing pigs. J. Anim. Sci. 94:3805–3816. doi: https://doi.org/10.2527/jas2016-0599.

Sulabo, R. C., W. S. Ju, and H. H. Stein. 2013. Amino acid digestibility and concentration of digestible and metabolizable energy in copra meal, palm kernel expellers, and palm kernel meal fed to growing pigs. J. Anim. Sci. 91:1391–1399. doi: https://doi.org/10.2527/jas.2012-5281.

Sun, T., H. N. Lærke, H. Jørgensen, and K. E. B. Knudsen. 2006. The effect of extrusion cooking of different starch sources on the in vitro and in vivo digestibility in growing pigs. Anim. Feed Sci. Technol. 131:67–86. doi: https://doi.org/10.1016/j.anifeedsci.2006.02.009.

Sung, J. Y., S. Y. Ji, and B. G. Kim. 2020. Amino acid and calcium digestibility in hatchery byproducts fed to nursery pigs. Anim. Feed Sci. Technol. 270:114703. doi: https://doi.org/10.1016/j.anifeedsci.2020.114703.

Svajgr, A. J., E. R. Peo, Jr., and P. E. Vipperman, Jr. 1969. Effects of dietary levels of manganese and magnesium on performance of growing-finishing swine raised in confinement and on pasture. J. Anim. Sci. 29:439–443. doi: https://doi.org/10.2527/jas1969.293439x.

Svihus, B., A. K. Uhlen, and O. M. Harstad. 2005. Effect of starch granule structure, associated components and processing on nutritive value of cereal starch: A review. Anim. Feed Sci. Technol. 122:303–320. doi: https://doi.org/10.1016/j.anifeedsci.2005.02.025.

Swiss, L. D., and H. S. Bayley. 1976. Influence of the degree of hydrolysis of beef tallow on its absorption in the young pig. Can. J. Physiol. Pharmacol. 54:719–727. doi: https://doi.org/10.1139/y76-100.

Tamminga, S., H. Schulze, J. van Bruchem, and J. Huisman. 1995. The nutritional significance of endogenous N-losses along the gastro-intestinal tract of farm animals. Arch. Anim. Nutr. 48:9–22. doi: https://doi.org/10.1080/17450399509381824.

Tao, X., Z. R. Xu, and Y. Z. Wang. 2005. Effect of excessive dietary fluoride on nutrient digestibility and retention of iron, copper, zinc, and manganese in growing pigs. Biol. Trace. Elem. Res. 107:141–151. doi: https://doi.org/10.1385/BTER:107:2:141.

Taverner, M. R., I. D. Hume, and D. J. Farrell. 1981. Availability to pigs of amino acids in cereal grains. 1. Endogenous levels of amino acids in ileal digesta and feces of pigs given cereal diets. Br. J. Nutr. 46:149–158. doi: https://doi.org/10.1079/BJN19810017.

Thomaz, M. C., P. H. Watanabe, L. A. F. Pascoal, M. M. Assis, U. S. Ruiz, A. B. Amorim, S. Z. Silva, V. V. Almeida, G. M. P. Melo, and R. A. Robles-Huaynate. 2014. Inorganic and organic trace mineral supplementation in weanling pig diets. Rev. Colomb. Cienc. Pecu. 87:1071–1081. doi: https://doi.org/10.1590/0001-3765201520140154.

Thorén-Tolling, K. 1975. Studies on the absorption of iron after oral administration in piglets. Acta Vet. Scand. Suppl. 54:1–121.

Trowell, H. 1976. Definition of dietary fiber and hypotheses that it is a protective factor in certain diseases. Am. J. Clin. Nutr. 29:417–427. doi: https://doi.org/10.1093/ajcn/29.4.417.

Trumbo, P. R., J. F. Gregory, III, and D. B. Sartain. 1988. Incomplete utilization of pyridoxine-β-glucoside as vitamin $B_6$ in the rat. J. Nutr. 118:170–175. doi: https://doi.org/10.1093/jn/118.2.170.

Tufarelli, V., and V. Laudadio. 2017. Manganese and its role in poultry nutrition: An overview. J. Exp. Biol. Agric. Sci. 5:749–754. doi: https://doi.org/10.18006/2017.5(6).749.754.

Ullrey, D. E. 1972. Biological availability of fat-soluble vitamins: Vitamin A and carotene. J. Anim. Sci. 35:648–657. doi: https://doi.org/10.2527/jas1972.353648x.

Ullrey, D. E. 1981. Vitamin E for swine. J. Anim. Sci. 53:1039–1056. doi: https://doi.org/10.2527/jas1981.5341039x.

Urriola, P. E., G. C. Shurson, and H. H. Stein. 2010. Digestibility of dietary fiber in distillers coproducts fed to growing pigs. J. Anim. Sci. 88:2373–2381. doi: https://doi.org/10.2527/jas.2009-2227.

Urriola, P. E., and H. H. Stein. 2010. Effects of distillers dried grains with solubles on amino acid, energy, and fiber digestibility and on hindgut fermentation of dietary fiber in a corn-soybean meal diet fed to growing pigs. J. Anim. Sci. 88:1454–1462. doi: https://doi.org/10.2527/jas.2009-2162.

van Beers, E. H., H. A. Büller, R. J. Grand, A. W. C. Einerhand, and J. Dekker. 1995. Intestinal brush border glycohydrolases: Structure, function, and development. Crit. Rev. Biochem. Mol. Biol. 30:197–262. doi: https://doi.org/10.3109/10409239509085143.

Verbeeck, J. 1975. Vitamin behavior in premixes. Feedstuffs. 47:45–48.

Veum, T. L. 2004. Feed supplements: Crystalline vitamins. In: W. G. Pond and A. W. Bell, editors, Encyclopedia of animal science. Marcel Dekker, New York, NY. p. 372–378.

Vier, C. M., S. S. Dritz, M. D. Tokach, J. M. DeRouchey, R. D. Goodband, M. A. D. Gonçalves, U. A. D. Orlando, J. R. Bergstrom, and J. C. Woodworth. 2019. Calcium to phosphorus ratio requirement of 26- to 127-kg pigs fed diets with or without phytase. J. Anim. Sci. 97:4041–4052. doi: https://doi.org/10.1093/jas/skz257.

Vitti, D. M., and E. Kebreab. 2010. Phosphorus and calcium utilization and requirements in farm animals. CAB International, London, UK. doi: https://doi.org/10.1079/9781845936266.0001.

Walk, C. L. 2016. The influence of calcium on phytase efficacy in non-ruminant animals. Anim. Prod. Sci. 56:1345–1349. doi: https://doi.org/10.1071/AN15341.

Wang, L., L. Wang, J. Zhou, T. Gao, X. Liang, Q. Hu, B. Huang, Z. Lyu, L. J. Johnston, and C. Lai. 2020. Comparison of regression and fat-free diet methods for estimating ileal and total tract endogenous losses and digestibility of fat and fatty acids in growing pigs. J. Anim. Sci. doi: https://doi.org/10.1093/jas/skaa376.

Wang, Y., Z. Yuan, H. Zhu, M. Ding, and S. Fan. 2005. Effect of cyadox on growth and nutrient digestibility in weanling pigs. South Afr. J. Anim. Sci. 35:117–125.

Watkins, K. L., L. L. Southern, and J. E. Miller. 1991. Effect of dietary biotin supplementation on sow reproductive performance and soundness and pig growth and mortality. J. Anim. Sci. 69:201–206. doi: https://doi.org/10.2527/1991.691201x.

Wauer, A., G. I. Stangl, M. Kirchgessner, W. Erhardt, J. Henke, U. Hennig, and D. A. Roth-Maier. 1999. A comparative evaluation of ileo-rectal anastomosis techniques for the measurement of apparent precaecal digestibilities of folate, niacin and pantothenic acid. J. Anim. Physiol. Anim. Nutr. 82:80–87. doi: https://doi.org/10.1111/j.1439-0396.1999.00227.x.

Wedekind, K. J., and D. H. Baker. 1990a. Effect of varying calcium and phosphorus level on manganese utilization. Poult. Sci. 69:1156–1164. doi: https://doi.org/10.3382/ps.0691156.

Wedekind, K. J., and D. H. Baker. 1990b. Manganese utilization in chicks as affected by excess calcium and phosphorus ingestion. Poult. Sci. 69:977–984. doi: https://doi.org/10.3382/ps.0690977.

Wedekind, K. J., R. S. Bever, and G. F. Combs. 1998. Is selenium addition necessary in pet foods? FASEB J. 12:A823. (Abstr.)

Wedekind, K. J., A. E. Hortin, and D. H. Baker. 1992. Methodology for assessing zinc bioavailability: efficacy estimates for zinc-methionine, zinc sulfate, and zinc oxide. J. Anim. Sci. 70:178–187. doi: https://doi.org/10.2527/1992.701178x.

Wedekind, K. J., A. J. Lewis, M. A. Giesemann, and P. S. Miller. 1994. Bioavailability of zinc from inorganic and organic sources for pigs fed corn-soybean meal diets. J. Anim. Sci. 72:2681–2689. doi: https://doi.org/10.2527/1994.72102681x.

Wedekind, K. J., M. R. Murphy, and D. H. Baker. 1991a. Manganese turnover in chicks as affected by excess phosphorus consumption. J. Nutr. 121:1035–1041. doi: https://doi.org/10.1093/jn/121.7.1035.

Wedekind, K. J., E. C. Titgemeyer, A. R. Twardock, and D. H. Baker. 1991b. Phosphorus, but not calcium, affects manganese absorption and turnover in chicks. J. Nutr. 121:1776–1786. doi: https://doi.org/10.1093/jn/121.11.1776.

Wessels, A. G., H. Kluge, F. Hirche, A. Kiowski, J. Bartelt, E. Corrent, and G. I. Stangl. 2016. High leucine intake reduces the concentration of hypothalamic serotonin in piglets. J. Anim. Sci. 94:26–29. doi: https://doi.org/10.2527/jas.2015-9728.

Wieland, T. M., X. Lin, and J. Odle. 1993. Utilization of medium-chain triglycerides by neonatal pigs: Effects of emulsification and dose delivered. J. Anim. Sci. 71:1863–1868. doi: https://doi.org/10.2527/1993.7171863x.

Wilson, R. H., and J. Leibholz. 1981. Digestion in the pig between 7 and 35 d of age. 3. The digestion of nitrogen in pigs given milk and soya-bean proteins. Br. J. Nutr. 45:337–346. doi: https://doi.org/10.1079/Bjn19810109.

Wiseman, J. 2006. Variations in starch digestibility in non-ruminants. Anim. Feed Sci. Technol. 130:66–77. doi: https://doi.org/10.1016/j.anifeedsci.2006.01.018.

Wiseman, J., and D. J. A. Cole. 1987. The digestible and metabolizable energy of two fat blends for growing pigs as influenced by level of inclusion. Anim. Sci. 45:117–122. doi: https://doi.org/10.1017/S0003356100036709.

Wornick, R. C. 1968. The stability of micro-ingredients in animal feed products. Feedstuffs. 40:25–28.

Wright, P. L., and M. C. Bell. 1966. Comparative metabolism of selenium and tellurium in sheep and swine. Am. J. Physiol. 211:6–10. doi: https://doi.org/10.1152/ajplegacy.1966.211.1.6.

Wu, F. Z., M. D. Tokach, S. S. Dritz, J. C. Woodworth, J. M. DeRouchey, R. D. Goodband, M. A. D. Goncalves, and J. R. Bergstrom. 2018. Effects of dietary calcium to phosphorus ratio and addition of phytase on growth performance of nursery pigs. J. Anim. Sci. 96:1825–1837. doi: https://doi.org/10.1093/jas/sky101.

Wünsche, J., U. Herrmann, M. Meinl, U. Hennig, F. Kreienbring, and P. Zwierz. 1987. Einfluß exogener Faktoren auf die präzäkale Nährstoff- und Aminosäurenresorption, ermittelt an Schweinen mit Ileo-Rektal-Anastomosen. Arch. Anim. Nutr. 37:745–764. doi: https://doi.org/10.1080/17450398709428245.

Xing, J. J., E. van Heugten, D. F. Li, K. J. Touchette, J. A. Coalson, R. L. Odgaard, and J. Odle. 2004. Effects of emulsification, fat encapsulation, and pelleting on weanling pig performance and nutrient digestibility. J. Anim. Sci. 82:2601–2609. doi: https://doi.org/10.2527/2004.8292601x.

Xue, P. C., D. Ragland, and O. Adeola. 2014. Determination of additivity of apparent and standardized ileal digestibility of amino acids in diets containing multiple protein sources fed to growing pigs. J. Anim. Sci. 92:3937–3944. doi: https://doi.org/10.2527/jas2014-7815.

Yen, J. T. 2001. Anatomy of the digestive system and nutritional physiology. In: A. J. Lewis and L. L. Southern, editors, Swine nutrition. CRC Press, Washington, DC. p. 31–63.

Yen, J. T., A. H. Jensen, and D. H. Baker. 1976. Assessment of the concentration of biologically available vitamin B-6 in corn and soybean meal. J. Anim. Sci. 42:866–870. doi: https://doi.org/10.2527/jas1976.424866x.

Yen, J. T., A. H. Jensen, and D. H. Baker. 1977. Assessment of the availability of niacin in corn, soybeans and soybean meal. J. Anim. Sci. 45:269–278. doi: https://doi.org/10.2527/jas1977.452269x.

Young, L. G., A. Lun, J. Pos, R. P. Forshaw, and D. E. Edmeades. 1975. Vitamin E stability in corn and mixed feed. J. Anim. Sci. 40:495–499. doi: https://doi.org/10.2527/jas1975.403495x.

Young, L. G., R. B. Miller, D. E. Edmeades, A. Lun, G. C. Smith, and G. J. King. 1977. Selenium and vitamin E supplementation of high moisture corn diets for swine reproduction. J. Anim. Sci. 45:1051–1060. doi: https://doi.org/10.2527/jas1977.4551051x.

Young, L. G., R. B. Miller, D. E. Edmeades, A. Lun, G. C. Smith, and G. J. King. 1978. Influence of method of corn storage and vitamin E and selenium supplementation on pig survival and reproduction. J. Anim. Sci. 47:639–647. doi: https://doi.org/10.2527/jas1978.473639x.

Yu, B. H., and C. Kies. 1993. Niacin, thiamin, and pantothenic acid bioavailability to humans from maize bran as affected by milling and particle size. Plant Food Hum. Nutr. 43:87–95. doi: https://doi.org/10.1007/BF01088100.

Yu, F., P. J. Moughan, and T. N. Barry. 1996. The effect of cottonseed condensed tannins on the ileal digestibility of amino acids in casein and cottonseed kernel. Br. J. Nutr. 75:683–698. doi: https://doi.org/10.1079/BJN19960173.

Zeng, Z. K., D. Wang, X. S. Piao, P. F. Li, H. Y. Zhang, C. X. Shi, and S. K. Yu. 2014. Effects of adding super dose phytase to the phosphorus-deficient diets of young pigs on growth performance, bone quality, minerals and amino acids digestibilities. Asian-Australas. J. Anim. Sci. 27:237–246. doi: https://doi.org/10.5713/ajas.2013.13370.

Zhai, H., and O. Adeola. 2011. Apparent and standardized ileal digestibilities of amino acids for pigs fed corn- and soybean meal-based diets at varying crude protein levels. J. Anim. Sci. 89:3626–3633. doi: https://doi.org/10.2527/jas.2010-3732.

Zhai, H., and O. Adeola. 2012. True total-tract digestibility of phosphorus in monocalcium phosphate for 15-kg pigs. J. Anim. Sci. 90:98–100. doi: https://doi.org/10.2527/jas.51536.

Zhang, F., and O. Adeola. 2017. True is more additive than apparent total tract digestibility of calcium in limestone and dicalcium phosphate for twenty-kilogram pigs fed semipurified diets. J. Anim. Sci. 95:5466–5473. doi: https://doi.org/10.2527/jas2017.1849.

Zhao, J., Z. Li, M. Lyu, L. Liu, X. Piao, and D. Li. 2017. Evaluation of available energy and total tract digestibility of acid-hydrolyzed ether extract of cottonseed oil for growing pigs by the difference and regression methods. Asian-Australas. J. Anim. Sci. 30:712–719. doi: https://doi.org/10.5713/ajas.16.0546.

Zhuge, Q., and C. F. Klopfenstein. 1986. Factors affecting storage stability of vitamin A, riboflavin, and niacin in a broiler diet premix. Poult. Sci. 65:987–994. doi: https://doi.org/10.3382/ps.0650987.

# 20  Swine Nutrition and Environment

Ming Z. Fan, Brian J. Kerr, Steven L. Trabue, Xindi Yin, Zeyu Yang, and Weijun Wang

## Introduction

Global pork production has steadily increased over the past several decades with pork representing about 35% of the overall meat production, thereby contributing significantly to the world food security, trade, and economy (VanderWaal and Deen 2018). Evidently, pathogens and swine herd health are regarded as the top risk factors that can potentially disrupt normal pork production (VanderWaal and Deen 2018). Pigs are also recognized intermediate hosts for zoonotic diseases such as the generation of pandemic influenza virus, and a recently emerged Influenza A virus subtype G4 EA H1N1 virus in swine has been shown in increasing its human infectivity and raised concerns for the possible generation of pandemic viruses (Sun et al. 2020). Porcine reproductive and respiratory syndrome (PRRS), the porcine epidemic diarrhea virus (PED), and African swine fever (ASF) are some of the most challenging porcine viral pathogens currently facing the global pork industry. Development of efficacious vaccines has been associated with limited success for control of PED (Park and Shin 2018) and PRRS (Montaner-Tarbes et al. 2019; Wei et al. 2019). So far, there are no reports of commercially available vaccines for ASF, and this is likely due to the fact that ASF viruses are not only large genome DNA based and have a very large particle size but also have some of the most complex viral structure (Wang et al. 2019a). While gene editing for control of these porcine viral diseases has been demonstrated (Whitworth et al. 2016; Hübner et al. 2018; Whitworth et al. 2019), regulatory framework and consumer acceptance would be conceived as the main hurdles for practical applications (Proudfoot et al. 2019). It should be emphasized that stringent biosecurity measures, including high-quality swine barn physical infrastructure and facilities, are some of the prerequisite and practical mitigation strategies to sustain swine herd health, viable swine nutrition, and production operations.

Since the last edition of the Sustainable Swine Nutrition edited by Chiba (2013), unprecedented global COVID-19 pandemic challenges and disruptive cell-based new meat production technologies have emerged, which need to be touched on in the context. Meanwhile, considerable progress has been made in many fronts of multidisciplinary areas of applied and basic pig nutrition, digestive and metabolic physiology, metabolism, gut microbiome, and biotechnology, which warrants critically reviewing and effectively implementing these new developments into swine nutrition and feeding and production practices.

The main concerned negative impacts of intensive swine production on the footprint of the environment, including limited efficiency of utilization and excessive manure nutrient excretions, emission of greenhouse gases (GHG) and odorants, and the persistence of antimicrobial resistance (AMR), are still to be mitigated (Figure 20.1). These environmental concerns may be now even more challenging for the Canadian and US pork industries because in-feed antimicrobial uses have been banned in both countries since December 2018.

Being consistent with our previous contribution to this book by Fan (2013), herein we have focused more on reviewing recent conceptual understanding about various dietary and physiological factors affecting gut microbiota-microbiome in mediating efficiency of digestive and postabsorptive utilization of dietary monomer and polymer nutrient elements. These include carbon (C), nitrogen (N), sulfur (S), calcium (Ca), phosphorus (P), magnesium (Mg), the electrolytes of sodium (Na) and potassium (K), as well as the swine dietary required main microminerals of copper (Cu), iron (Fe), manganese (Mn), selenium (Se), and zinc (Zn) in weanling and feeder pigs. Comprehensive nutritional and nondietary strategies are summarized and recommended for adoption and further commercialization development to maximize pork production profit margins and tackling the environmental sustainability concerns including AMR and GHG facing the global pork production industry.

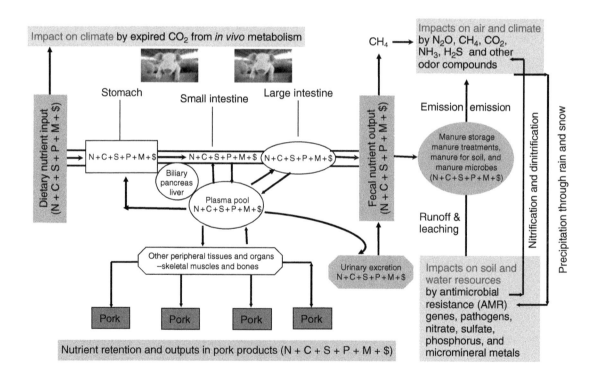

N: nitrogen compounds; C: carbon-containing compounds; S: sulfur-containing compounds; P: phosphorus compounds; M: other concerned micromineral metals; $N_2O$: nitrous oxide; $CH_4$: methane; $CO_2$: carbon dioxide; $NH_3$: ammonia; $H_2S$: hydrogen sulfide and other volatile sulfides; and $: economic impacts

**Figure 20.1** Schematic illustration of the major sustainable issues associated with enhancing mass-energy conversion of intensive pork production. Source: Adapted from Fan et al. (2006).

**Emerging Contemporary Issues Impacting Pork Industry**

Concurrently, the unprecedented new acute respiratory syndrome (SARS) pneumonia, termed coronavirus disease-2019 (COVID-19), is caused by the new coronavirus SARS-CoV-2 with its onset in Wuhan of China in December 2019 (Wu et al. 2020; Zhou et al. 2020b). COVID-19 has been causing enormous damages in losses of a large number of human lives and disruption of global social and economic activities (Bar-On et al. 2020), being regarded as the most severe pandemic occurred within the past century ever since the 1918 Influenza Pandemic. This SARS-CoV-2 might have been originated from wild bats (Zhou et al. 2020a,b), and pangolins were evidenced as its possible intermediate hosts before it jumped species to infect humans (Lam et al. 2020). The global scientific community still seeks to resolve the path of the SARS-CoV-2 insurgency.

Similar to SARS-CoV, SARS-CoV-2 virus particle recognizes and binds to the human host epithelial apical receptor, i.e., angiotensin-converting enzyme 2 (ACE2) via their spike trimeric glycoprotein (S) subunit-1 (S1) receptor-binding domain (RBD) (Lan et al. 2020; Shang et al. 2020; Zhou et al. 2020a). Biological roles of ACE2 in animals and humans are well established to be, such as a membrane surface hydrolase, and the chaperone, with a molecular weight at about 120 kDa, for control of trafficking of other epithelial transmembrane proteins (e.g., $Na^+$-dependent neutral amino acid transporter $B^oAT1$ in gut) for the apical membrane anchoring (Yan et al. 2020). Expression of ACE2 is widely reported in oral, respiratory, as well as intestinal and renal polarized epithelia with anchoring on the apical membrane surface (Ren et al. 2006; Yang et al. 2016; Xu et al. 2020). Each full-length ACE2 consists of an N-terminal peptidase domain and a C-terminal transmembrane helix with an extended intracellular segment (Yan et al. 2020). The cryo-EM structure analyses have further unraveled that a multiunit complex consisting of the SARS-CoV-2 spike glycoprotein two subunit S1 RBD in binding to the corresponding peptidase domains of a dimerized ACE2 unit and then in overtopping further binding to another dimerized transmembrane protein unit (e.g., $B^oAT1$) in polarized human epithelia through noncovalent and nonionized hydrogen bonds and van der Waals forces (Yan et al. 2020). Clearly, this SARS-CoV-2 binding to the human ACE2 is in a much strong affinity in support of its much high infectivity in comparison with the SARS-CoV (Shang et al. 2020; Yan et al. 2020). Respiratory and gastrointestinal manifestations of CoV-19 infection symptoms and SARS-CoV-2 virus load and shed via patients' respiratory and fecal routes were demonstrated (Cheung et al. 2020; Zhou et al. 2020b). Understanding the basic cellular and molecular mechanisms of how SARS-CoV-2 infects humans can help educate general public and train swine producers and pork value chain workers to clearly know its high contagiousness and its primary routes of community transmission through proximity physical contact. Thus, before effective vaccination is developed, stringent social distancing and thorough strict personal hygiene are essential to contain the spreading of COVID-19.

Liu et al. (2020) have further demonstrated that SARS-CoV-2 was effectively transmitted via human respiratory droplets in aerosols. Their results indicate that good room ventilation, open space, effective sanitization of protective apparel, and proper use and disinfection of toilet areas can all effectively decrease the level of SARS-CoV-2 in aerosols for transmission (Liu et al. 2020). Morawska and Milton (2020) revealed and recognized the significant potential for airborne spread of COVID-19 through inhalation exposure to viruses in microscopic respiratory droplets, i.e., microdroplets, at short to medium distances for up to several meters, or room scale. And thus they are advocating for the use of preventive measures to mitigate this route of airborne transmission (Morawska and Milton 2020). Under this context, while it has been well adopted in most Asian countries, simply wearing a nonmedical grade effective facial mask or covering needs to be further implemented by the broader global community, including pork producers and pork packing plant workers, to contain COVID-19 via the community transmission route of human respiratory droplets and microscopic in aerosols.

There has been reporting of a high variability in clinical manifestations of the COVID-19, ranging broadly from asymptomatic infection to acute respiratory failure and death, which is now further shown to be associated with the CD4[+] and CD8[+]T cells as well as B cells based adaptive immunity specifically to neutralize SARS-CoV-2's ability in docking to the ACE2 (Braun et al. 2020; Cao et al. 2020; Grifoni et al. 2020). Thus, T cell and B based adaptive immunity responses specific to block SARS-CoV-2's binding to host ACE2 are likely the underlying mechanisms for the individual human susceptibility and the between species vulnerability to infection by SARS-CoV-2.

Furthermore, Shi et al. (2020) observed that SARS-CoV-2 replicated poorly in chickens and ducks. We have here further compared the coding AA sequence alignments of the full-length ACE2 genes in some common domesticated animals of pigs, dogs, cats, chicken, ferrets, and bovine species in comparisons with humans, respectively (Figure 20.2). Chicken, representing poultry and other avian species, generally have a low ACE2 homology (66%) to humans. We have then compared the AA sequence alignments and homologies of the N-terminus based peptidase (PD) receptor domain of the ACE2 corresponding to the RBD of SARS-CoV-2 virus spike glycoprotein two subunit S1, in pigs, dogs, cats, chicken, ferrets, and bovine species in comparisons with humans, respectively (Table 20.1). Chicken again have a low ACE2 receptor domain PD homology (50%) to humans (Table 20.1). These ACE2 genomics comparisons, consistent with the above-cited literature report (Abdel-Moneim and Abdelwhab 2020), suggest that poultry and other avian species are likely not susceptible to infection and thus not to play a role in community transmission of SARS-CoV-2 as in humans.

Although cats and dogs have a relatively higher (about 84–85%) ACE2 full-length homology to the humans, all of the other compared mammalians, including pigs, dogs, cats, chicken, ferrets, and bovine species, are in similar ranges (~80%) of the ACE2 homologies to humans (Figure 20.2). Moreover, bovine species have the highest ACE2 receptor domain PD homology (90%) to humans. And this is followed by the porcine at 80%, whereas the other three mammalian species of dogs, cats, and ferrets have shared the identical ACE2 receptor domain PD sequence and its homology (70%) to humans (Table 20.1).

These genomics comparisons of the SARS-CoV-2 receptor ACE2-binding domain or motif across species are consistent with the report by Shi et al. (2020) and the review by Abdel-Moneim and Abdelwhab (2020) that ferrets, cats, monkeys, hamsters, tree shrews, transgenic mice, and fruit bats were permissive to infection by SARS-CoV-2, and cats were susceptible to airborne infection by SARS-CoV-2. The comparison analyses by Wan et al. (2020) shown that while conventional rodents (mice and rats) are not possible intermediate SARS-CoV-2 hosts, several other mammalian species including cats, civet, ferrets, pigs, and nonhuman primates are possible intermediate hosts and/or transmitters to SARS-CoV-2. Whereas in the same report by Shi et al. (2020), SARS-CoV-2 was shown replicated poorly in dogs and pigs. However, Sit et al. (2020) reported that dogs could be infected with SARS-CoV-2. A zoo tiger infection and mink to human transmission of COVID-19 have been reported (Abdel-Moneim and Abdelwhab 2020; Medicalxpress 2020; USDA 2020). Tigers and cats are in the same family, while ferrets and minks are closely related and ferrets are used as a relevant animal model for studying viral pathogenesis in humans (Enkirch and von Messling 2015) and COVID-19 vaccine development. Meanwhile, there is a lack of reports of porcine and bovine-based clinical manifestations of COVID-19. On the other hand, differential immunity responses among the species would ultimately account for their clinical responses. Clinical studies have further shown that clinical manifestations of COVID-19 in humans are significantly associated with adaptive immunity responses including circulating blood levels of intermediate total lymphocyte, CD3[+], CD4[+], and

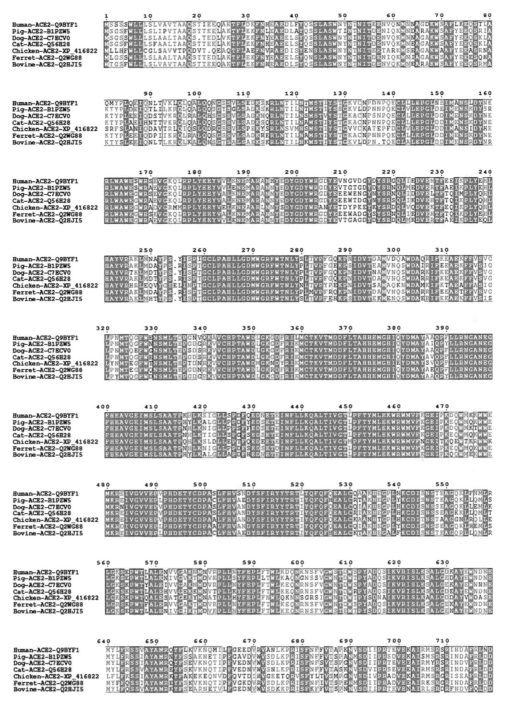

**Figure 20.2** The coded amino acid (AA) sequence alignments of the angiotensin-converting enzyme 2 (ACE2) of the host polarized epithelial apical membrane surface receptor recognized in binding to the SARS-CoV-2 virus particle in humans, pigs, dogs, cats, chicken, ferrets, and bovine species. Fully conserved residues are shaded in red. The residues with the global similarity scores over 0.7 calculated using Reiser are boxed in blue. The sequence alignment was performed using Clustal Omega (https://www.ebi.ac.uk/Tools/msa/clustalo/), and the figure was prepared using ESPript (http://espript.ibcp.fr/ESPript/ESPript/). Each sequence is labeled by the source, enzyme name (ACE2), and uniport number/accession number. Overall, the human sequence had 81.37, 83.46, 85.22, 65.92, 82.61, and 80.97% homologies to the known sequences of pigs, dogs, cats, chicken, ferrets, and bovine species, respectively.

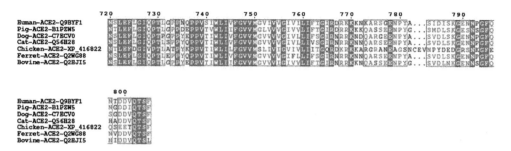

**Figure 20.2** (Continued).

**Table 20.1** Amino acid (AA) sequence alignments and homologies of the N-terminus based peptidase (PD) domain of angiotensin-converting enzyme 2 (ACE2) of host polarized epithelia, recognized as the apical membrane surface ACE2 receptor domain, in binding to the receptor binding domain (RBD) of SARS-CoV-2 virus spike glycoprotein two subunit S1, in pigs, dogs, cats, chicken, ferrets, and bovine species in comparisons with humans, respectively.

| Species[a] | AA numbering from N-terminus in ACE2 | | | | | | | | | | Homology (%)[b] |
|---|---|---|---|---|---|---|---|---|---|---|---|
| | 24 | 30 | 35 | 37 | 38 | 41 | 42 | 83 | 353 | 393 | |
| Pig[c] | L | E | E | E | D | Y | Q | Y | K | R | 80 |
| Dog[c] | L | E | E | E | E | Y | Q | Y | K | R | 70 |
| Cat[c] | L | E | E | E | E | Y | Q | Y | K | R | 70 |
| Chicken[c] | E | A | R | E | D | Y | E | F | K | R | 50 |
| Ferret[c] | L | E | E | E | E | Y | Q | Y | K | R | 70 |
| Bovine[c] | Q | E | E | E | D | Y | Q | Y | K | R | 90 |
| Human | Q | D | E | E | D | Y | Q | Y | K | R | 100 |

[a] The partial sequences labeled by the species' source, the enzyme name (ACE2), and uniport number/accession number are pig-ACE2-B1PZW5; dog-ACE2-C7ECV0; cat-ACE2-Q56H28; chicken-ACE2-XP_416822; ferret-ACE2-Q2WG88; and human-ACE2-Q9BYF1, respectively.

[b] The concerned species' ACE2 receptor PD domain AA sequence homology (%) to the human's, respectively.

[c] The concerned species' ACE2 receptor PD domain AA residues in boldface that are different from the humans.

CD8[+]T cell counts as well as pro-inflammatory cytokines of IL-6 and IL-8 (Zhang et al. 2020). Thus, our ACE2 species genomics comparisons along with literature reports suggest that cats, dogs, ferrets, pigs, and bovine species may be all biologically susceptible to infection by SARS-CoV-2 or serve as possible intermediate SARS-CoV-2 hosts as humans, whereas clinical manifestations of COVID-19 in porcine and bovine species and potential roles of two domestic food animal species in transmission of SARS-CoV-2 still need to be further documented.

In summary, the on-going global COVID-19 pandemics have been causing widespread damages and challenges until successful vaccination strategies are developed. Although pigs are not yet clinically manifested by COVID-19, swine researchers, pork producers, and particularly pork packing plant workers need to be very vigilant in personal protection, environment hygiene, and appropriate social distancing, including properly handling our pets of cats and dogs as well as other domesticated animals of ferrets, minks, and bovine species, in protecting against transmission of this very contagious and deadly viral disease and its disruptive negative effects on pork production value chains.

Two other societal factors have also emerged and may have long-term opposite impacts on global pork production. One factor being the recent series of high-profile human epidemiological studies have questioned the long time misconception of causal associations between some of the fatal chronical human diseases, including cancers and cardiovascular failures, and consumptions of red meats including pork (Vernooij et al. 2019; Zeraatkar et al. 2020). Carroll and Doherty (2019) have concluded that the public have long been misinformed of consumption of red and processed meat being bad for human health. Thus, new efforts should be continuously made to promote health aspects of pork production, such as providing high-quality protein and bioavailable trace minerals for human nutrition and public health.

The other societal factor is the calling for balancing human diets with much less animal protein especially red meats to be associated with reduced food animal production for ethical concerns about animal welfare (Maes et al. 2016; Carroll and Doherty 2019) and concerns of effects of high level of meat consumption on the environment for developing much more sustainable global food systems (Krishna Bahadur et al. 2018; Poore and Nemecek 2018; Willett et al. 2019). Thus, swine production needs to become much more environmentally sustainable to be continuously acceptable as an increasing component of the global food systems.

In a different direction, plant protein-based diets for vegetarian and vegan are gradually gaining acceptance (Medawar et al. 2019). Cultured meat concept has been conceived as newly emerging disruptive meat production technologies (Datar 2010; Sharma et al. 2015). These emerging societal changes in consumer behavior and food production systems will inevitably influence the global pork production trend and scale for the next coming decades. Therefore, pork industry needs to further enhance its animal welfare standards for public acceptance, efficiency of energy and mass conversion in pig growth and production for profit margins, as well as environmental footprint to stay competitive and viable as an industry for the long run.

## Carbon Nutrient Utilization and Environment Impacts

Carbon is fundamental in metabolic energy-contributing ingredients, namely starch, proteins, oils and fats, and nonstarch polysaccharides (NPS), and is an essential component ($CO_2$) measured during indirect calorimetry experimentation (Campos et al. 2014; Huntley et al. 2018). Efficiency of C is not, however, evaluated in most swine nutrient balance trials, which generally focus on energy, N, or mineral utilization. However, given the basis of C as the key element in energy metabolism with $CO_2$ as an end product, and it is also a volatile organic carbon gas in GHG (i.e., methane emissions), its input–output relationship is important when considering an overall environmental impact (Figure 20.1). Thus, any dietary or animal factor that influences the digestion and retention of dietary C will have a direct environmental impact.

Unlike excess intake of dietary N, dietary intake of energy, and thus C, above that needed for maintenance and lean tissue growth can be stored in adipose tissue. As summarized in the Table 20.3, the body is much less efficient (20%) in capturing and retaining dietary C (a typical diet contains about 40% carbon, Kerr et al. 2006) where, on a live body basis, the body of a pig contains approximately 24% C (B. Kerr, unpublished data), with body protein and lipid contributing to approximately 53 and 76% of the retained C, respectively (Kleiber 1961). Up to 25% of total dietary C intake may be lost as the GHG $CH_4$ and $CO_2$ in manure during swine feeding and via post-feeding fermentation under liquid manure slurry storage (Table 20.3). It is less conventional to directly measure C than to determine DM and GE in swine nutrition research. While there are limited data on C digestibility and/or retention in pigs, the use of dry matter (DM), organic matter (OM), or gross energy (GE) digestibility can effectively be used as a surrogate for C digestibility (Kerr et al. 2013, 2015, 2017). Ji et al. (2008) and Rho et al. (2018) contrastingly observed that the total tract DM and OM digestibility could overestimate or underestimate the corresponding GE digestibility in grower pig diets by 1–2% units, respectively, likely due to differences in their research conditions and proximate nutrient composition of their study diets. Similar responses in dietary DM and GE digestibility values were reported in weaning pigs (Omogbenigun et al. 2004; Zuo et al. 2015; Acosta et al. 2020). Because of this relationship, an improvement in DM, OM, or GE digestibility results directly in an improvement in C digestibility, and because little energy is lost in urine independent of N-containing compounds (Calloway and Margen 1971; Putnam 1971; van Milgen et al. 2001), any dietary manipulation that improves DM or GE digestibility or improves the efficiency of growth in terms of feed to gain ratio will be advantageous for reducing the amount of C released into the environment.

Up to 15% of total dietary C intake may be lost as $CO_2$ and $CH_4$ during swine feeding and liquid swine manure fermentation (Table 20.2). As a potent GHG, $CH_4$ mainly originates from anaerobic microbial fermentation during the storage of liquid manure slurry from swine production facilities (Figure 20.1) (Mackie et al. 1998; Fan et al. 2006). Emission of GHG from intensive swine production operations is believed to account for a significant portion of the total GHG generated within the animal production sectors (Mackie et al. 1998). Furthermore, while the overall animal production sector is documented to contribute to about 14.5% of the total human-induced GHG emissions, swine, poultry, dairy, and beef, respectively, contribute to 9, 8, 20, and 41% to the sector's total emissions (Gerber et al., 2013). Consequently, improvements in ingredient or dietary DM digestibility are key to improving C balance in the pig and to mitigate GHG emission from swine production.

Extensive research and summarization of the DE content of ingredients play a key role in this topic, with ingredients having a greater DE:GE ratio (i.e., energy digestibility) being advantageous to C balance compared to ingredients with a low DE:GE ratio (NRC 2012). In addition, feed processing techniques (e.g., grinding) can increase the surface area of the diet allowing for increased efficacy of digestive enzymes, which on an energy basis may increase the DE by 30–45 kcal/kg diet (or approximately 0.90% units of energy or DM digestibility) for each 100 µm reduction in particle size (Hancock and Behnke 2001; Lundbald 2009; Richert and DeRouchey 2010; NRC 2012), an improvement which would effectively reduce fecal C loss by the animal by a similar percentage unit.

**Table 20.2** Chemical names and composition of major environmentally concerned gas emissions from swine production facilities.

| Chemical name | Chemical formula | Chemical name | Chemical formula |
|---|---|---|---|
| Acetic acid | $C_2H_4O_2$ | Hydrogen sulfide | $H_2S$ |
| Propionic acid | $C_3H_6O_2$ | Methane thiol | $CH_4S$ |
| Butyric acid | $C_4H_6O_2$ | Carbonyl sulfide | COS |
| Isobutyric acid | $C_4H_6O_2$ | Dimethyl sulfide | $C_2H_6S$ |
| Isovaleric acid | $C_5H_{10}O_2$ | Dimethyl disulfide | $C_2H_6S_2$ |
| Valeric acid | $C_5H_{10}O_2$ | Ammonia | $NH_3$ |
| Isocaproic acid | $C_6H_{12}O_2$ | Indole | $C_8H_7N$ |
| Caproic acid | $C_6H_{12}O_2$ | 3-Methyl indole | $C_9H_9N$ |
| Heptanoic acid | $C_7H_{14}O_2$ | Trimethyl amine | $C_3H_9N$ |
| Phenol | $C_6H_6O$ | Nitrous oxide | $N_2O$ |
| Cresol | $C_7H_8O$ | Methane | $CH_4$ |
| Ethyl phenol | $C_8H_{10}O$ | Carbon dioxide | $CO_2$ |

*Utilization of Lipids*

Cholesterol and its ester products of both animal and plant origins are very low in quantity; these lipid compounds do not yield metabolic energy ATP and are largely involved in forming membrane structure and serve as precursors for the biosynthesis of regulatory steroid hormones and metabolites such as bile acids and vitamin D. Fats and oils as supplements in commercial swine diets are used to enhance net energy density and to reduce diet dustiness. The apparent ileal and fecal digestibility has been reported to be approximately 90% for highly saturated fats, higher at 94–97% for oils such as rapeseed oil and coconut oil, and near complete (97–98%) in oils high in polyunsaturated fatty acids such as fish oil in grower pigs fed semi-purified diets with an estimated low dietary neutral-detergent fiber (NDF) (Jørgensen et al. 2000). Thus, total tract crude fat digestion in swine is almost complete at about 100% digestibility when fed semi-purified diets with very low dietary fiber content.

Studies by Ji et al. (2008) reported the apparent fecal crude fat digestibility ranged from 78 to 81% in grower pigs fed typical corn and soybean meal-based diets, containing approximately 9% dietary NDF, with dietary supplementation of beta-glucanase-protease in enhancing the apparent ileal and fecal digestibility values of DM, CP, and minerals but not improving the crude fat digestibility. However, observed high crude fat digestibility of 91–95% in the grower pigs fed semi-purified diets with dietary crude fat primarily originating from supplementing soy oil and dietary NDF at 24–25%. Kerr et al. (2013) reported that inclusion of fiber-rich corn-derived distillers' dried grains with soluble (DDGS) at 30% in replacing corn and increasing dietary NDF to 12.4% resulted in much lower and a demonstrated pattern of postnatal developmental decreases in the apparent fecal crude fat digestibility values in the weanling pigs at 71%, grower pigs at 59–62%; and in the finisher pigs at 45%, respectively. Dietary supplementation of various exogenous enzymes, including phytase, cellulases and hemi-cellulases, did not improve the crude fat digestibility in the starter and finisher swine (Kerr et al. 2013). Being consistently to the study by Kerr et al. (2013), Rho et al. (2018) reported low crude fat digestibility of 64–68% in the liquid feeding grower pigs with the dietary inclusion of DDGS at 30% and dietary NDF content at about 16% without significant effects of steeping and exogenous enzyme cocktail supplementation. Clearly, these variable dietary NDF contents (i.e., 9 vs. 12–16% vs. 24–25%) did not consistently affect the crude fat digestibility changes among the above studies. Thus, pigs fed diets with a significant level of inclusion of high-fiber and high-fat coproducts such as the corn-derived DDGS will likely be associated with much reduced digestive utilization of crude fat.

From a classic view, high levels of dietary fiber, as antinutritive factors, might have contributed to disruption of lipid digestion process involving the formation of lipid droplets and lipase hydrolysis in the gut lumen and potentially caused a much higher hepatic-biliary endogenous lipid secretion loss. Other factors might have further contributed to the low fat digestibility in the starter and finisher pigs fed 30% DDGS-based diets reported by Kerr et al. (2013) and Rho et al. (2018). Plant lignin has been well established to negatively affect nutrient digestibility as a group of antinutritive factors particularly in monogastric animals (Crampton and Maynard 1938; Jung et al. 1997; Huang et al. 2017). Sewalt et al. (1997) and Vermaas et al. (2015) further reported that lignin, freed up or isolated from various sources such as biomass and/or plant feed pretreatment, could effectively bind to and precipitate enzymes, thereby negatively affecting enzyme activities. Pedersen et al. (2014) reported that Klason lignin content was enriched and averaged at 2.5% in their surveyed corn DDGS samples, and this lignin would be in a much more freed up and reactive form and at the level, when included via DDGS at 30%, much higher than an anticipated lignin content in typical corn and soybean swine diets. Furthermore, while feed antibiotics are typically not used in commercial grower-finisher swine diets, it should be pointed out that the antibiotic Tylan was included in the starter and finisher pigs in the study by Kerr et al. (2013), and the feed antibiotics could have inhibited gut symbiotic microbial growth the potential contributions of gut microbial lipases to lipid digestion, leading to the relatively low fecal lipid digestibility values of these pigs in their studies.

Thus, it can be conceived that several strategies may be further developed, such as the use of reduced level of DDGS inclusion, e.g., at 10–15% with much reduced level of lignin, dietary supplementation of exogenous lipases and counteracting treatment of potential negative effects of lignin would improve crude fat digestibility, dietary net energy density, and growth performances of pigs fed the DDGS-based diets and mitigate environmental footprint from intensive swine operation.

On other hand, because lipids are susceptible to peroxidation where it has been shown that swine diets with significant lipid peroxidation will negatively influence swine growth performances (Kerr et al. 2015; Shurson et al. 2015). However, it is not known if dietary lipid peroxidation will negatively affect lipid digestion (Shurson et al. 2015). Therefore, compound swine diet storage and valid shelf life span may be also an important consideration in organizing sustainable swine production.

### *Utilization of Starch*

As major C-polymer nutrient group, dietary starch primarily from cereal grains is well digested in weanling pigs at 87% at the ileal level via the host starch digestive enzymes and reaches 94% at the fecal level by additional microbial fermentation (Omogbenigun et al. 2004). Dietary supplementation of a complete exogenous feed enzyme mixture, including various hemi-cellulases, invertase, phytase, protease, alpha-amylase, and cellulase, increased the ileal starch digestibility from 87 to 95% by improving 8% units and the fecal starch digestibility from 94 to 99% with an improvement of 5% units, respectively, in weanling pigs (Omogbenigun et al. 2004). Wierenga et al. (2008) reported that, albeit of small magnitudes, ileal starch digestibility was higher (95.8 vs. 93.0%), whereas total tract starch digestibility was lower (98.2 vs. 98.4%) in extruded wheat grain than in extruded zero-tannin faba bean in weaning pigs. Clearly, their results also showed that extrusion resulted relatively high starch digestibility at the ieal and fecal levels in weaning pigs (Wierenga et al. 2008). In a semi-purified diet-based study, Fouhse et al. (2015) shown that different types of starch, i.e., different types of resistant starch (RS), not only affected their ileal starch digestibility but also impacted growth performance endpoints and various gut physiology indices such as luminal pH, fermentation, and microbiota and mucosal morphology however did not affect total tract starch digestibility in weaning pigs. Results of this weanling pig study are consistent with the well-established concept that starch digestion capacity is not fully established in weanling pigs (Fan 2013). Chen et al. (2020) further shown that multitherapeutic antimicrobials could enhance growth performances of weaning pigs through improving the ieal terminal starch hydrolytic enzyme activity. Thus, efficiency and dynamics, such as speed and site, of dietary starch digestion in weanling pigs can be further maximized through effective mitigation of various dietary strategies to unleash their positive impacts on growth performances and footprint of environment.

On the other hand, the study by Ji et al. (2008) reported the dietary starch digestibility was 98% at the distal ileal level and 99% at the fecal level in grower pigs without additional significant effects by dietary supplementation of exogenous beta-glucanase and protease. In the study by Rideout et al. (2008), ileal digestibility of normal corn starch was 99.4%; and ileal digestibility of three other tested RS ranged from 57.2 to 72.9%, whereas the total tract digestibility of the normal corn starch and the three tested RS was at 100% through microbial fermentation in the hindgut of grower pigs. Furthermore, differential chemical and physical properties were responsible for differential ileal starch digestibility values and their effects on CP and macromineral utilization and intestinal microbial fermentation patterns in terms of some of the key metabolites such as an odor-causing compound indole and the cell proliferating substrate butyrate in cecal digesta in grower pigs (Ridcout et al. 2008). Rosenfelder-Kuon et al. (2017) determined the ileal starch digestibility from eight different genotypes of barley, rye, triticale, and wheat each, on average, the ileal starch digestibility was at 92.7, 95.0, 97.3, and 92.2% for barley, rye, triticale, and wheat, respectively. These *in vivo* ileal starch digestibility data further confirm the established notion that the host enzyme starch digestion capacity in the small intestine is fully established in the grower-finisher pig (Fan 2013). Hence, while starch digestion is complete at the fecal level in grower-finisher pigs, hindgut fermentation of a small quantity of residual starch and/or dietary supplemental RS in the feeder pigs may exert additional benefits to improve metabolic physiology and mitigate odor impact of intensive swine production.

### *Utilization of Nonstarch Polysaccharides*

Utilization efficiency of swine feeds, particularly containing high-fiber agricultural coproducts such as canola meal, wheat bran and shorts, and the biofuel industry coproducts, such as dried distiller's grains with solubles (DDGS), is largely limited by poor fiber digestibility variable at about 23 – 60% (Fan 2013; Kerr and Shurson 2013; Woyengo et al. 2014; Jha and Berrocoso 2015), which limits pork profit margin and contributes to negative impacts of swine production on environment. Trying to improve fiber digestion by the addition of NSP-degrading enzymes into swine diets is not a new concept (Chesson 1987; Bedford 2000; Kerr and Shurson 2013). While the efficacy of these fiber degradation exogenous enzymes to improve NSP and energy digestibility has been inconsistent and variable depending upon trial conditions (Zijlstra et al. 2010; Kerr and Shurson 2013). Variable experimental conditions such as growth stages of pigs, basal diet composition, and nature of exogenous fiber enzymes may be responsible for these inconsistencies.

In the study by Omogbenigun et al. (2004) with weaning pigs fed corn, barley, wheat-wheat screenings, canola meal, blood plasma, pea and soybean meal-based diets, dietary supplementation of multienzymes of cellulase, galactanase, mannanase, and pectinase increased fecal digestibility of dietary NSP from 49 to 67% and DM from 76 to 80%, respectively, and improved average daily gain and feed conversion efficiency without affecting average daily feed intake. Duarte et al. (2019) with weaning pigs (initial BW at 7.2 kg) fed corn, DDDS, blood plasma, poultry meal and soybean meal-based diets, dietary supplementation of xylanase alone or with protease improved average daily gain and jejunal morphology (i.e., increasing villus height and decreasing crypt depth), and reduced ileal digesta viscosity, however, did not affect feed conversion efficiency and the apparent ieal DM, GE, and CP digestibility. It should be pointed out that while plasma protein (3 or 3.3%) was used in both weaning pig studies, ZnO at the 250 ppm pharmacological level was used in the study by Duarte et al. (2019), whereas no dietary antimicrobials were used in the study by Omogbenigun et al. (2004). Yin et al. (2018) reported that dietary supplementation of therapeutic aureomycin improved growth performances and gut permeability but decreased the total tract DM digestibility in weaning pigs fed corn- and SBM-based diets. These studies are generally consistent with a literature review and a meta-analysis based study in concluding that supplementation with multienzyme complexes, including mannanase and xylanase alone or in combination with β-glucanase and/or protease, had most consistent effects on piglet growth and nutrient digestibility with a mean difference improvement on the apparent total tract digestibility of GE being 1.6% in weaned pigs (Torres-Pitarch et al. 2017). Weaning pig gut microbiota are undergoing development, thus dietary supplementations of suitable exogenous fiber enzymes as well as experimental conditions such as use of antimicrobials and other gut modifiers would effectively improve digestive utilization of dietary fiber and DM in weanling pigs. Experimental conditions such as initial BW, study duration, exogenous fiber enzyme stability, and use of antimicrobials may also affect experimental endpoint responses and should have been designed and controlled to be relevant to industrial practices and further considered for comparisons between reported studies. Fan et al. (2001) shown that weanling pigs fed on SBM-based and semi-purified low-fiber diets with dietary NDF ranging from 1 to 5%, fecal DM digestibility could reach 96–92%, respectively, thus representing the highest possible fecal DM digestibility in weanling pigs when dietary fiber is very low or well digested. As indicated earlier, any improvement in GE and DM digestibility due to fiber in responses to exogenous fiber enzyme supplementation should have a direct impact on C balance as well to mitigate the negative environment impacts associated with GHS.

In the study by Ji et al. (2008) with grower pigs fed corn and soybean meal based diets, dietary supplementation of beta-glucanase increased fecal digestibility of total dietary fiber (TDF) from 61 to 66% and DM from 87.4 to 88.5%, respectively. Passos et al. (2015) observed that dietary supplementation of xylanase increased ileal digestibility of dietary DM, OM, GE, total mineral, and NDF without affecting growth performance endpoints and jejunal morphology in grower pigs fed corn- and SBM-based diets. This is likely due to the fact the basal diets were not designed to be inadequate in net energy density and limiting essential AA contents in this study (Passos et al. 2015). However, Liu et al. (2012) did not report significant improvement in dietary fiber digestibility in the wheat bran (NDF, 67–69%; and ADF, 55%) and the soybean hull (NDF, 65–71%; and ADF, 60–62%) based diets, when the diets were supplemented with cellulase and xylanase in grower pigs. Ajakaiye et al. (2003) shown grower pigs fed on SBM-based and semi-purified low-fiber diets with dietary NDF ranging from 1 to 5%, fecal DM digestibility could reach 96–92%, respectively, hence representing the highest possible fecal DM digestibility in grower pigs when dietary fiber is very low or well digested. Therefore, inconsistent responses in efficacy of fiber digestibility and growth performances to various dietary supplementation of cellulase and hemicellulases in grower pigs are likely affected by the composition of basal diets and sources of dietary fiber.

As a major biofuel coproducts, DDGS have high and variable total dietary fiber contents (Stein and Shurson 2009; Pedersen et al. 2014). At their established upper dietary inclusion of 30%, Urriola et al. (2010) observed that the fecal digestibility values of TDF and NDF in DDGS for grower pigs were generally low and variable, ranging from 29 to 57% for TDF and ranging from 52 to 66% for NDF, respectively. Soluble fiber in the DDGS and other pig feeds was low and was near compete digestion at the fecal levels (Urriola et al. 2010; Ji et al. 2008; Liu et al. 2012). Studies have been conducted to examine effects of exogenous fiber enzyme cocktails on improving efficiency of dietary fiber utilization in feeder pigs. Rho et al. (2018) did not observe significant improvements in the fecal digestibility of NDF (50–54%), DM (77–79%), and GE (79–81%) in the liquid feeding grower pigs fed corn and SBM-based diets with DDGS supplemented at 30% under steeping and exogenous fiber cocktail supplementation. Whereas Jakobsen et al. (2015) shown that fermentation and dietary supplementation of mixtures of xylanase and β-glucanase or cellulase and xylanase independently improved fecal NSP digestibility in DDGS in feeder pigs. Furthermore, Kerr et al. (2013) reported much low fecal digestibility values of dietary NDF (27–39%) and ADF (31–40%) in the weanling and grower pigs and also relatively low fecal digestibility values of dietary NDF (35–47%) and ADF (44–57%) in the late finisher pigs fed corn and SBM-based meal form of diets with DDGS supplementation at 30%. Dietary supplementation of various exogenous cellulases and hemi-cellulases from a number of vendors did not improve the fecal NDF and ADF digestibility values in the feeder pigs and the late finisher pigs (Kerr et al. 2013). Pedersen et al. (2014) reported that Klason lignin content was averaged at 2.5% in DDGS samples and this lignin would be in a much more freed up and reactive form; and at the level, when included via DDGS at 30%, much higher than an anticipated lignin content in typical corn and soybean swine diets. Sewalt et al. (1997) and Vermaas et al. (2015) documented that lignin could effectively bind to and precipitate enzymes, thereby negatively affecting cellulase and hemi-cellulase activities. Furthermore, antibiotic Tylan was included in the starter and finisher pigs in the study by Kerr et al. (2013), and the feed antibiotics could have inhibited gut symbiotic microbial growth the potential contributions of gut microbial cellulase and hemi-cellulase to fiber digestion, leading to the relatively low fecal fiber digestibility values of these pigs in their studies. Thus, relatively high dietary level of lignin or polyphenolic tannin associated with DDGS inclusion at 30% is likely, in part, to result in poor efficiency of digestive utilization of dietary fiber, DM, GE, and carbon in feeder pig feeding.

On other hand, diet type and level of dietary fiber digestive utilization also impact manure output, where it has been shown that an increase in dietary fiber, which if not fully digested, result in increased manure DM output (Graham et al. 1986; Canh et al. 1998b; Kreuzer et al. 1998; Burkhalter et al. 2001), which therefore has a direct impact on manure C output (Kerr et al. 2006). This has been most studied in diets containing DDGS (Trabue and Kerr 2014; van Weelden et al. 2016a,b; Saqui-Salces et al. 2017; Kerr et al. 2017, 2018) where it has been shown that there is a consistent increase in manure C content with DDGS inclusion in the diet. The impact of diet type on C-based emissions (VOC and GHG) has also been most studied in diets containing DDGS, which while some individual gas emissions have been shown to be affected by DDGS inclusion, there appears to be little to no apparent effect of DDGS addition on total C released or GHG equivalence (Trabue and Kerr 2014; Trabue et al. 2016; Kerr et al. 2018). Given that lignin may affect microbiota, the potential impact of Klason lignin level in association with DDGS inclusion in diets on biogenesis and emission of GHG *in vivo* and during post-feeding manure slurry storage in feeder pig feeding need to be further quantified in future studies.

Limited and inconsistent efficacy of fiber8 enzymes, including cellulases and hemi-cellulases, in swine diet supplementation can be generally reflected from above review, and this is further reflected by low market share of fiber enzymes on the global feed enzyme market (Adeola and Cowieson 2011). While exogenous fiber enzymes are expected to be functional in ruminal acidic pH environment in ruminant feeding, fiber enzyme stability has been recognized as a major limiting factor responsible for their limited *in vivo* efficacy in ruminant nutrition (Beauchemin et al. 2003; Krause et al. 2003), and primary hydrolytic sites of exogenous fiber enzymes would be in the small and large intestines in swine (Kidder and Manners 1978). Optimal exogenous fiber enzyme activity pH would range from slightly acidic to slightly alkaline pH in the small and large intestines in pigs (Kidder and Manners 1978). Resistance to irreversible inactivation by very low gastric acidic pH such as pH = 2–4, as well as intestinal luminal proteases, including pepsin, trypsin, and chymotrypsin, would be desirable properties for exogenous fiber enzymes (Li et al. 2013; Wang et al. 2019). The importance and efficacy of exogenous fiber enzyme stability has been further demonstrated reported by Zhang et al. (2018) in the transgenic pigs expressing phytase alone with selected β-glucanase and xylanase that are stable in passaging through the gastric environment. Apart from enhanced Ca and P digestibility, these transgenic pigs had improved fecal DM digestibility and growth performance endpoints including average daily gain, average daily feed intake, and feed conversion efficiency and shortened days to market by 19–21% when fed corn and soybean meal-based diets further supplemented with other agriculture by-products of wheat bran, rice bran meal, cotton seed meal, and rapeseed meal (Zhang et al. 2018). Hence, development of exogenous fiber enzymes that are highly stable in the gastric-gut lumen environment is the future direction to further enhance their application efficacy in swine nutrition to mitigate C utilization efficiency.

### Gut Microbiota and Utilization of Carbon Nutrients

Gut microbiota contributes to digestion and degradation of dietary carbon nutrients in pigs (Xiao et al. 2016). Wang et al. (2019b) carried out metagenomic analyses of fecal microbial diversity and restructured 360 high-quality bacterial and archaea genomes in piglets fed cereal grain-based commercial type of weaning diets going through the first to the 3rd wk of weaning transition by using the Illumina MiSeq and the Hiseq 2500 PE125 platforms. Fecal alpha diversity was low at the starting of the 21-days weaning, increased during the first wk of weaning, and reached and maintained its plateau at end of the two to three week of weaning (Wang et al. 2019b). And this observation is consistent with the report by Frese et al. (2015) that piglet fecal alpha-diversity was increased by weaning vs. suckling and reached its plateau at end of the two week of post-weaning by using the Illumina MiSeq platform. During weaning transition, while young pigs are well known to lose their host gut endogenous lactase hydrolytic capacity, several genera of bacteria only within the *Firmicutes* phylum, including members of *Lactobacillus*, *Subdoligranulum*, and *Ruminococcus,* have now been identified to catalog extracellular β-galactosidase (GH2 – BbgIII) and intracellular β-galactosidase (GH2 – BbgI, LacM, and BbgIV) or β-galactosidase (GH42 – LacA) genes (Wang et al. 2019b). Owing to the intrinsic limitations of (i) partial sequencing 16S rRNA gene R3/R4 regions with the MiSeq platform and (ii) high error rates in short-read shot gun sequencing and assembly of regions with repetitive sequences including rRNA operons with the Hiseq 2500 PE125 platform, taxa of bacteria responsible for lactose degradation were primarily identified at the genus level within the *Firmicutes* phylum, and only *Lactobacillus delbrueckii* was identified at the level of species with declining abundances during the weaning transition (Wang et al. 2019b). Xue et al. (2020) shown in adult rats fed excessive level of dietary lactose from 30 to 50% for three week induced colonic dysbiosis, abnormal microbial fermentation, and disorder of ion transport in colon, leading to lactose intolerance and diarrhea. Bacterial degradation of lactose under the anaerobic environment of lower gut *in vivo* and swine manure slurry storage *in vitro* in lagoon would lead to further fermentation of glucose and galactose, contributing to biogenesis of volatile short-chain fatty acids (Table 20.2) and the typical swine odor impact, as discussed by Mackie et al. (1998). Thus *Firmicutes* bacteria contribute to intestinal microbial lactose degradation in the weanling pigs fed dry whey powder-supplemented diets.

Furthermore, *Firmicutes* phylum of bacteria have now been characterized in weaning pig gut to have abundant extracellular pullulanases (GH13-Amy12) as endo-acting α-(1→6)-amylases, then much less abundant extracellular neopullulanase2 as endo-acting α-(1→4)-amylase as well as extracellular α-(1→4)-glucan branching enzyme (GH13-GlgB) genes; and clearly, this α-(1→4)-glucan branching enzyme (GH13-GlgB) services as a glucanotransferase complementing to catalyze glucan chain conversion from the α-(1→4)-glucosidic linear linkage to the α-(1→6)-glucosidic branched linkage, thus effectively degrading polymer starch into gluco-oligosaccharides within the *Firmicutes* bacteria (Wang et al. 2019b). Whereas weaning pig gut *Bacteroidetes* phylum of bacteria have also harbored a distinct and more complete set of pathway genes for starch degradation, *including* extracellular endo-acting α-(1→4)-amylases (GH13-Amy1 and 4) and extracellular neopullulanase (GH13-*SusG*) as endo-acting α-(1→4)/(1→6)-amylases as well as the periplasmic exo-acting neopullulanase (GH13-*SusA*) and the wide spectrum specificity periplasmic α-glucosidase

(GH97-*SusB*), thus digesting starch into free D-glucose for further catabolism by these bacteria (Wang et al. 2019b). The periplasmic exo-acting α-glucosidases of the *Bacteroidetes* phylum of bacteria would further cooperatively degrade gluco-oligosaccharides that are supplied through the hydrolysis by *Firmicutes* phylum of bacteria extracellular α-amylase and α-(1→4)-glucanotransferase, into free D-glucose (Wang et al. 2019b). Thus, *Bacteroidetes* phylum of bacteria could be benefited from *Firmicutes* phylum of bacteria in harvesting metabolic energy from starch in the intestinal environment. While young pigs are also well known to still further develop their host pancreatic α-amylase and the gut terminal starch digestion enzyme complex maltase-glucoamylase (MGA) and sucrase-isomaltase (SI) hydrolytic capacity, intestinal *Firmicutes* and *Bacteroidetes* phyla of bacterial starch degradation pathway enzymes, as revealed by Wang et al. (2019b), would contribute to additional starch digestion in weaning pigs. Bacterial fermentation of glucose derived from the starch degradation, under anaerobic condition of lower gut *in vivo* and swine manure slurry storage *in vitro* in lagoon, contributing to biogenesis of volatile short-chain fatty acids (Table 20.2) and the typical swine odor impact (Mackie et al. 1998). These gene-centric gut metagenomic analyses provide further genetic evidence for the additional prebiotic effects in weaning pigs benefited from feeding cereal grain-based weaning diets. Future studies need to be conducted with a sequencing platform for higher resolution to identify specific bacterial species and/or strains engaged in starch degradation within the *Firmicutes* and *Bacteroidetes* phyla.

Gut microbiota play important roles in the degradation of dietary fiber in pigs (Varel and Yen 1997; Xiao et al. 2016). By deep metagenome sequencing of faecal DNA from 287 pigs with the Illumina GAIIx and the Hiseq 2000 platforms, Xiao et al. (2016) identified 7.7 million nonredundant genes representing 719 metagenomic species; and with functional metabolic pathways' genes, including fiber degradation, previously found and reported in the human gut metagenomic catalogue by Qin et al. (2010), 96% were also present in the porcine catalogue, further supporting the notion to the use of porcine gut microbiome for human biomedical research. Weaning pig Gram-positive gut commensal *Firmicutes* phylum of bacteria would degrade fructan by possessing intracellular β-fructofuranosidase genes (ScrA/ScrB, GH32) and extracellular fructansucrase genes (Inu, GH68, and FruA, GH32) (Wang et al. 2019b). Within the *Firmicutes* phylum, e.g., several genera of bacteria including *Subdoligranulum variabile, Faecalibacterium prausnitzii, Eubacterium,* and unclassified *Clostridiales* possessed the intracellular β-fructofuranosidase (ScrA), while FruA, ScrB, and Inu were presented only in *Lactobacilli* to process ingested fructan in the weaning pig (Wang et al. 2019b). *Bacteroidetes* phylum of bacteria were also identified to degrade fructan in the gut by expressing β-(2→6) endo-fructanases (GH32) to act in extracellular, periplasmic, and intracellular spaces in the weaning pig (Wang et al. 2019b). Liu et al. (2012) reported that dietary supplementation of chicory root powder rich in fructose polymer inulin increased colonic *Bacillus* and *Prevotella* bacteria by quantitative PCR analyses. These gene-centric gut metagenomic analyses provide genetic evidence for the well-documented prebiotic effects of gut microbiota in weaning pigs utilizing fructan supplemented diets. Furthermore, bacterial fermentation of fructose derived from the fructan degradation, under anaerobic condition of lower gut *in vivo* and swine manure slurry storage *in vitro* in lagoon, contributing to biogenesis of volatile short-chain fatty acids (Table 20.2) and the typical swine odor impact (Mackie et al. 1998). Specific porcine gut symbiotic bacteria engaged in degrading frucan and exerting frucan-based prebiotic effects still need to be further established at the levels of species and/or strains through using metagenomic platforms at a much higher resolution.

Arumugam et al. (2011) classified human gut microbiota into three main bacterial enterocyte phenotypes of *Bacteroides* as enterotype-1, *Prevotella* as enterotype-2, and *Ruminococcus* as enterotype-3. Several studies have further identified *Prevotella* genus of bacteria are more specific to dietary fiber digestion in pigs. Ivarsson et al. (2014) observed ileal microbiota in grower pigs fed diets supplemented with different sources of pectin and arabinoxylan were significantly increased with *Prevotella* bacteria when analyzed by quantitative PCR. Burrough et al. (2015) reported that colonic microbiota α-diversity, i.e., bacterial richness, was not affected, whereas there was a decreased *Firmicutes:Bacteriodetes* ratio with significantly lower abundance of *Lactobacillus* and higher abundance of *Prevotella, leading to increased abundances of bacterial genes* involved in fiber degradation with postweaned pigs were fed corn and soybean meal-based diets with DDGS replacing corn at 30% for five week and sampled at nine week of age through sequencing the V3-V4 region of the 16S rRNA genes using the Illumina MiSeq platforms. Mach et al. (2015) showed significant correlations between body weight and luminal secretory IgA and fecal *Prevotella* abundance in piglets between the ages of 14 days of suckling and 70 days of postweaning growth. In the study by Ramayo-Caldas et al. (2016), performances of 518 healthy large white piglets, weaned on cereal-based cereal grain based typical commercial diets at 28 d of age and none of their sows and piglets received antibiotic therapy, were measured and their fecal microbiota were characterized at the phylum level by pyrosequencing the V3-V4 region of the 16S rRNA gene in a Roche 454 GS FLX at the end of postweaning growth at ages of 60–70 days. Phylogenetic network analysis was applied to identify linkages between pig fecal microbiota groups at the phylum level of growth traits such as final body weight (BW) and average daily gain (ADG); there was a strong coexclusion ($r = -0.67$) between fecal abundances of *Prevotella* and *Ruminococcus* genera; and there was a strong co-occurrence ($r = 0.67$) between fecal abundances of *Prevotella* and *Mitsuokella* genera (PEB) (Ramayo-Caldas et al. 2016). Network and statistical analyses further revealed that rather than different α-diversity (i.e., richness of microbiota), the specific PEB genera of *Prevotella* and *Mitsuokella* bacteria were significantly linked to both final BW ($P = 0.005$) and ADG ($P = 0.027$) during the postweaning period (Ramayo-Caldas et al. 2016). Unfortunately, feed conversion, DM, and dietary fiber digestibility and their association with measured fecal microbiota abundances at phylum levels were not determined in this study. *Prevotella* bacteria possess abundant plant cell wall fiber degradation pathway enzymes, as discussed by Burrough et al. (2015) and reported by Crespo-Piazuelo et al. (2019), thus weanling pigs with gut microbiota rich in *Prevotella* bacteria would have higher efficiency of dietary fiber and energy digestibility, leading to higher final BW and ADG during the postweaning growth.

Wang et al. (2012) established a metagenomic plasmid screening library was constructed, resulting in a number of screened and assembled insert sequences of fiber degradation genes likely derived from *Prevotella* bacteria in the cecal microbiota of grower pigs fed crystalline cellulose-enriched diets for four week. One of the identified insert novel porcine gut bacterial cellulase gene, referred to as p4818Cel5_2A, was overexpressed and characterized as a monomodular GH5-endoglucanase to exhibit relatively high activities toward a wide variety of plant cell wall polysaccharides, including cellulosic substrates of avicel and solka-Floc® and the hemicelluloses of β-(1→4)/(1→3)-glucans, xyloglucan, glucomannan, and galactomannan; and hydrolysis product analyses further revealed that this enzyme was a processive endo-β-(1→4)-glucanase capable of hydrolyzing cellulose into cellobiose and cellotriose as the primary end products (Wang et al. 2019c). Given that this newly characterized monomodular processive GH5-endoglucanase p4818Cel5_2A might have been derived from *Prevotella* and/or Clostridia species (Wang et al. 2012, 2019c), this cellulase gene may be further developed to be a porcine gut microbiome biomarker to guide dietary and genetic strategies for enhancing dietary DM and fiber digestibility, feed conversion efficiency, and growth in feeder pigs. As pointed out by Wang et al. (2019c), this enzyme may also be further tailored as an efficient biocatalyst candidate for potential industrial application including as an exogenous multiproperty cellulase supplement in swine nutrition.

On other hand, when Burrough et al. (2015) predicted significantly enhanced abundances in fecal microbial fiber degradation pathway genes, including cellulases and hemi-cellulases, in feeding postweaned pigs with diets supplemented with 30% DDGS. Furthermore, bacterial fermentation of glucose and hemi-cellulose sugars such as arabinose, xylose, and mannose derived from the degradation of cellulose and hemi-celluloses, under anaerobic condition of lower gut *in vivo* and swine manure slurry storage *in vitro* in lagoon, contributing to biogenesis of volatile short-chain fatty acids (Table 20.2) and the typical swine odor impact (Mackie et al. 1998). Polyphenolic lignin is high in fiber feeds such as DDGS and is central to crosslinking with hemicelluloses and pectin in the formation porosity and macrofibril structure, limiting enzyme access and causing recalcitrance of plant cell wall materials to biodegradation (Zhang and Lynd 2004). Huang et al. (2017) reviewed lignin also has antimicrobial property thus inhibiting microbial enzymes in digesting fiber in the gut. While contribution of anaerobic bacterial fermentation to polyphenolic lignin degradation has been recognized (Bugg et al. 2011; Tian et al. 2014; Xie et al. 2016), there is scarcity of lignin degradation pathway enzyme genes reported in metagenomic sequencing and cataloguing studies in pigs (Xiao et al. 2016; Wang et al. 2019), humans (Qin et al. 2010), cows Hess et al. (2011), and pandas (Zhu et al. 2011). By using a metagenomic screening library, Fang et al. (2012) identified and characterized lignin-degrading enzyme genes encoding laccase and multicopper oxidase gene from panda fecal samples, suggesting the presence of lignin degrading enzymes in the gut environment. Kadosh et al. (2020) shown that a single phenolic metabolite gallic acid, formed from 3-dehydroshikimate derived from polyphenols by the action of the bacterial enzyme shikimate dehydrogenase (SDH)20 specifically from the two bacterial strains of *Lactobacillus plantarum* and *Bacillus subtilis*, abolishes the somatic mutant p53 protein tumor suppression activity via changes of binding to chromatin, affecting jejunal transcriptome pattern and causing ileum and colon of the study mice becoming hyperproliferative dysplasia. However, when treated with antibiotics, this gallic acid effect could be effectively counteracted (Kadosh et al. 2020). These results suggest polyphenols and their subsequent metabolism by microbiota could not only enhance bowel tumor formation risks in humans but also reduce terminal nutrient digestion and absorptive functions in food animal including pigs. Thus, porcine gut microbiome and microbiota can be further modulated, at specific bacterial species and gene levels, to further enhance swine fiber utilization, performances, and negative C impacts on environment.

The pivotal roles of gut microbiota to hosts are further defined by the concepts of gut eubiosis vs. dysbiosis, and these concepts have been widely reported in studies for human nutrition and health management. Eubiosis refers to the eubiotic status under which gut microbiota are balanced with beneficial bacterial species in maintaining healthy host, whereas dysbiosis is a status under which healthy or "good" bacteria are longer in control of colonization and proliferation of pathogenic or "bad" bacteria, thus leading to negative health development in host (Iebba et al. 2016). With pyrosequencing and Illumina-based MiSeq partial sequencing of 16S rRNA genes, changes in bacterial taxonomy, as affected by gut dysbiosis, are characterized at the phylum level and the identified ones are mainly *Firmicutes, Bacteroidetes,* and *Proteobacteria*. Gut dysbiosis, as approximately indicated by increasing *Firmicutes/Bacteroidetes* ratio, has been associated with human obesity (Ley et al. 2006; Costantini et al. 2017). Out of the *Firmicutes* phylum, *Clostridium difficile* has been recognized as the opportunistic pathogenic bacterial species that can cause symptoms ranging from diarrhea to life-threatening inflammation of the colon in humans (Mylonakis et al. 2001; Iebba et al. 2016). As a major genus in the *Bacteroidetes* phylum, there is a large number of *Prevotella* bacteria spp., and they play important roles in fiber degradation, whereas some *Prevotella* bacteria spp. are intestinal pathobionts (Precup and Vodnar 2019). For example, *Prevotella copri* is recognized for the development of chronic inflammatory diseases such as rheumatoid arthritis

(Larsen 2017; Maeda and Takeda 2019). Under this context, as analyzed by fecal shotgun metagenomics, Zhao et al. (2018) conducted a randomized human clinical study and identified 15 most relevant positive bacterial responders in promoting short-chain fatty acid production and attenuating type-2 diabetes mellitus (T2DM) marked by improvement in hemoglobin A1c levels partly via increasing glucagon-like peptide-1 production. Their identified 15 most relevant positive bacterial responders for managing T2DM include two *Bifidobacterium* spp. under the *Actinobacteria* phylum occupying the two top spots, 12 species all within the *Firmicutes* phylum and only one *Sutterella* sp. within the *Proteobacteria* phylum (Zhao et al. 2018). They further shown that enhancing these T2DM-positive bacterial responders decreased bacterial producers of the metabolically and environmentally (i.e., odor impact) detrimental compounds such as indole and hydrogen sulfide (Zhao et al. 2018). Thus, bacterial contributions to the modulation of gut dysbiosis for eubiosis and improving host physiological functions and health status need to be clarified and established at level of species and/or strains.

Gut dysbiosis, as induced by high-fat feeding in the mice model, increased the proportion of an LPS-containing microbiota in the gut and caused mild increases (1.5–3 folds) in plasma LPS concentration two to three times, resulting in low-grade metabolic endotoxemia (Cani et al. 2007) and inflammation via the glucagon-like polypeptide-2 (GLP-2) (Cani et al. 2008, 2009) and initiating metabolic syndromes such as obesity and insulin resistance (Cani et al. 2007, 2012). In contrast, Boutagy et al. (2016) reviewed that under sepsis caused by pathogenic bacterial infections, circulating blood plasma LPS would reach to 10–50 times higher than the plasma LPS activity level of the metabolic endotoxemia defined by Cani et al. (2007). Gram-negative rather than Gram-positive bacteria have been recognized as a source of endotoxin (Mani et al. 2012; Fan and Archbold 2015; Boutagy et al. 2016). The determination of endotoxin activity in blood circulation and biological samples is quantitatively conducted by the limulus amebocyte lysate (LAL) based chromogenic method (Munford 2016). Endotoxin activity (EU) is an arbitrarily defined unit and indicates endotoxicity in analyzed biological samples (EU/mL or g). The FDA (2015) has defined 1 EU of endotoxin activity is equal to one International Units (IU) and represents the activity contained in 0.2 ng of the pure endotoxin from the U.S. Reference Standard Endotoxin Lot EC-2 (USP standard reference material). Munford (2016) reviewed that most vendors of endotoxin assay kits include their own preparation of an *E. coli* LPS as the assay standard, practically, at 1 EU = 0.067–0.100 ng LPS (~0.25 pmol, assuming average LPS molecular weight at = ~4000). For example, Kaliannan et al. (2013) adopted the GenScript assay kit in their study, in which 1 EU = 0.10 ng of *E. coli* LPS or equivalent to the amount of LPS present in approximately 105 *E. coli* bacteria (GenScript 2019). A careful review of endotoxemia research literature has shown that most authors would indicate their vendors of the LAL-based chromogenic endotoxin assay kits but would not define their endotoxin activity unit. Furthermore, Kaliannan et al. (2013) demonstrated in rodents that intestinal alkaline phosphatase (AP) could effectively detoxify endotoxin and prevent metabolic syndrome associated with metabolic endotoxemia, and this was also achieved through modulating gut microflora (Malo et al. 2010; Lallès 2014). Kaliannan et al. (2015) again in rodents shown that omega-3 polyunsaturated fatty acids and multiwide spectrum antibiotics could also attenuate metabolic endotoxemia and systemic inflammation for human health management through modulation of gut AP functionality. Economopoulos et al. (2016) in a mice model study shown that early childhood exposure to antibiotics causing metabolic syndrome later on in adulthood was associated alternations in gut microbiota, and this occurrence could be effectively prevented through orally coadministrating exogenous AP, suggesting that gut endogenous AP is not sufficient to prevent gut dysbiosis caused by antibiotics treatment. Fuke et al. (2019) reviewed various dietary factors in the regulation of systemic endotoxemia. Gut dysbiosis-induced endotoxemia and systemic inflammation would have

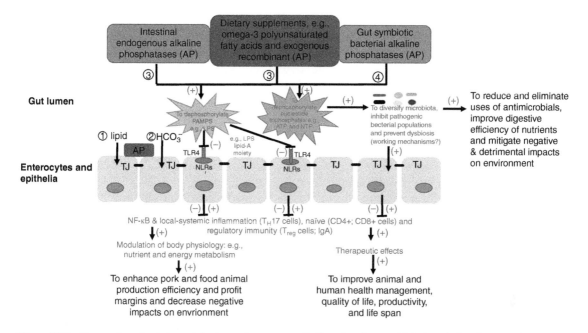

**Figure 20.3** Proposed mechanistic contributions of exogenous alkaline phosphatases (AP), intestinal endogenous AP, and gut symbiotic bacterial origins of AP as essential gut modifiers primarily to (1) the digestive dephosphorylation of proinflammatory pathogen-associated molecular patterns (PAMPs) compounds, including endotoxins of lipopolysaccharides (LPS) and deactivation of their binding to cell membrane surface toll-like receptors (e.g., TLR4) and signaling intracellular nucleotide-binding oligomerization domain-like (NOD) receptors (NLRs) and the digestive dephosphorylation of nucleotide triphosphates (e.g., ATP, GTP) in diversifying gut luminal microbiota, preventing dysbiosis, altering bile acid metabolism and regulatory roles, and exerting prebiotic effects and secondarily to (2) modulation of gut epithelial tight junction (TJ) paracellular permeability, nuclear factor kappa-light-chain-enhancer of activated B cells (NF-κB) based gut local and systemic inflammation, naïve (e.g., via cytotoxic CD4 and CD8 T cells), and regulatory (e.g., via affecting IgA and IL-17a expressing T helper, i.e., TH17; and regulatory T, Treg, cells) immunity in affecting digestive and postabsorptive utilization and metabolism of energy and nutrients in pigs with implications for management of pig and human health. Source: Concepts adapted from Lallès (2014), Fan and Archbold (2015), Honda and Littman (2016), Hang et al. (2019), and Helmink et al. (2019).

implications in affecting growth and efficiency of nutrient utilization in pigs (Gabler and Spurlock 2008; Mani et al. 2012). Roles of gut endogenous and exogenous AP and dietary factors in modulation of host gut dysbiosis and whole-body physiology are further illustrated in Figure 20.3. More research is needed to quantify potential impacts of gut dysbiosis-associated metabolic endotoxemia on C utilization efficiency in pork production.

Several lines of evidence suggest that pigs, particularly young pigs during weaning transition, are susceptible to develop and experience gut dysbiosis, metabolic endotoxemia, systemic inflammation, and various adverse physiological responses. First, when analyzed at the fecal level, majority of the bacteria in the pig intestinal microbiota are from the two Phyla of *Firmicutes* and *Bacteroidetes* (Isaacson and Kim 2012; Frese et al. 2015; Slifierz et al. 2015; Zhao et al. 2015; Xiao et al. 2016). And there are clear and distinct postnatal differences and changes in composition of microbiota along the longitudinal axis from the proximal to the end of the intestine with ileum and cecum having a relatively high percentage of bacteria in the phylum *Proteobacterium* such as *Enterobacteriaceae*, a large family of pathogenic bacteria, including well-known pathogens such as *Escherichia coli* and *Salmonella* existing in young (suckling and weanling) pigs vs. postweaned feeder (grower-finisher) pigs (Isaacson and Kim 2012; Fouhse et al. 2015; Slifierz et al. 2015; Zhao et al. 2015; Crespo-Piazuelo et al. 2018). Thus, young pigs especially weaning pigs are vulnerable to gut dysbiosis in colonizing pathogenic bacteria and developing diarrhea and growth check (Lackeyram et al. 2010). Second, Laugerette et al. (2015) optimally measured and compared endotoxin activity in a large number of plasma samples in healthy fasting and postprandial pigs (8.5 ± 1.3 EU/mL, $n = 186$); rodents (0.9 ± 0.2 EU/mL, $n = 295$); and humans (0.73 ± 0.05 EU/mL, $n = 903$). Clearly, the porcine plasma endotoxin activity is several folds of the values measured in the rodents and humans under the same condition (Laugerette et al. 2015). Plasma endotoxin activity was reported at 4–5 EU/mL and at about 8 EU/mL under the rodent metabolic endotoxemia status induced by high-fat feeding and/or LPS injection (Cani et al. 2007; Kaliannan et al. 2013). Metabolic endotoxemia is defined to be two to three times of higher endotoxin activity than a normal level (Kaliannan et al. 2013). Comparatively, the study by Laugerette et al. (2015) supports the notion that pigs, as a species, experience spontaneous low-grade metabolic endotoxin in comparisons with rodents and humans, and the porcine results of this study are consistent with the porcine studies by Mani et al. (2013a,b) with plasma endotoxin activity analyzed by using the rFC-based fluorescence assay. Third, studies by Burello et al. (2017, 2019) and Kiffer-Moreira et al. (2014) collectively point to the fact that porcine gut AP enzyme affinity is much lower (i.e., having a larger $K_m$ value) in comparison with humans in hydrolyzing natural substrates such as LPS and triphosphate nucleotides (e.g., ATP), thus leading to spontaneous metabolic endotoxin. Finally, weaning transition reduced gut AP enzyme affinity and capacity, contributing to gut dysbiosis and development of diarrhea and growth check in young pigs (Lackeyram et al. 2010). Through chronic challenges with extra LPS injections to mimic pathogenic bacterial infectious inflammation, Campos et al. (2014) and Huntley et al. (2018) demonstrated that weaned and grower pigs did not show changes in DM and energy digestibility, however, experienced decreases in feed intake and energy deposition with increases in ME requirements for maintenance under immune activation and inflammation. Furthermore, Mani et al. (2013b) shown that pigs in the genetic line divergently selected for low residual feed intake and high feed conversion efficiency were associated with significantly low plasma endotoxin activity as well ileal and hepatic AP activity in contrasting to pigs in the high residual feed intake genetic line analyzed by using the recombinant Factor C (rFC)-based fluorescence assay kit, while endotoxin activity unit EU was not defined in terms of equivalent to what amount of free LPS in the study. This study further shown that pigs of the high residual feed intake genetic line were associated with gut-dysbiosis derived metabolic endotoxemia marked by several folds of higher plasma endotoxin activity, suggesting porcine pan-genomic controlling components (Mani et al. 2013a). This rFC-based assay is relatively new and has been approved by FDA for pharm industry application since 2018, and thus is believed to be environmentally more sustainable than the classic LAL-based assay (Piehler et al. 2020). Piehler et al. (2020) compared both the LAL- and the rFC-based endotoxin activity assays and concluded that the rFC-based assay was reliable and suitable for routine bacterial endotoxin testing. Therefore, dietary, genomic, and metagenomic strategies can be collectively developed to mitigate porcine gut dysbiosis, metabolic endotoxemia, efficiency of pork production, and C footprint impacts on environment.

Pathogenic bacteria mediate their effects through the pathogen-associated molecular pattern (PAMPs, e.g., LPS, ATP) compounds' direct interactions, as typical physiological ligands, with cellular membrane surface toll-like receptors (e.g., TLR4) and the intracellular NOD receptors (NLRs) (Shaw et al. 2008), exerting proinflammatory and immune responses and causing infectious diseases (Poltorak et al. 1998; Lallès 2010; Fan and Archbold 2015). Intestinal endogenous AP is found on the apical membrane of the small and the large intestines and plays essential roles in protecting gut health through catalysis of hydrolytic dephosphorylation of gut luminal endotoxin LPS (Poelstra et al. 1997), ultimately detoxifying endotoxins, preventing gut dysbiosis and disorders, and maintaining whole-body normal metabolism (Kaliannan et al. 2013; Estaki et al. 2014). Deficiency in intestinal endogenous AP activity leads to chronic bowel inflammatory disorders (Bilski et al. 2017). Exogenous recombinant bovine and human gut AP supplements in their coated forms have been developed and used in rodent studies and human clinical research for various human health management therapeutic purposes (Eriksson et al. 2003; Estaki et al. 2014; Fawley and Gourlay 2016; Rader 2017). Several lines of evidence from rodent studies using knock-out mice in combination with exogenous recombinant AP supplementation demonstrated that endogenous and/or exogenous AP could effectively prevent gut dysbiosis by enhancing commensal bacterial population, particularly the gram-positive *Firmicutes* phyla of bacteria, and reducing proliferation of specific pathogenic bacteria (Malo et al. 2010, 2014). As shown in Figure 20.3, gut symbiotic bacteria also contribute to the overall AP activity in the lumen (Malo 2015); however, this bacterial AP activity is mainly from the gram-negative but not the gram-positive bacteria (Cheng and Costerton 1977). It was further demonstrated that endogenous and exogenous AP did not directly kill gram-negative bacteria (Malo et al. 2010). Rather, AP enzymes indirectly exert their impact on diversifying gut commensal bacterial populations while reducing pathogenic bacterial population through enzymatically decreasing luminal concentrations of ATP and other nucleotide triphosphates (Lallès 2014; Malo et al. 2014). However, exact working biological mechanisms of AP in restoring and maintaining gut eubiosis remain to be further elucidated. Meanwhile, there is a scarcity of reports of using metagenomic analyses such as shotgun sequencing and full-length 16S rRNA gene sequencing for quantifying gut microbiota responses at the bacterial species and strain levels in animals and humans receiving exogenous AP supplementation orally.

However, exogenous recombinant bovine and human AP supplements currently available on the market will not be feasible for potential commercial application in swine and other livestock applications for several reasons. First, these mammalian recombinant AP enzymes are not thermo-stable, and majority of these enzyme activities would be lost during the feed pelleting temperature and duration (e.g., 80–90 °C for 7–10 minutes) (Fan and Archbold 2015). Second, these mammalian recombinant AP enzymes are gastric acidic pH and gastric pepsin labile and would lose their activities during the passage through the gastric phase (Eriksson et al. 2003; Millán 2006b; Fan and Archbold 2015; Archbold et al. 2017). Third, although recombinant bovine AP are very high in their activities, their enzyme affinity is relatively very low and are effectively inhibited by luminal free L-Phe (Millán 2006a; Millán 2006b). Thus, these facts are rendering these bovine and human exogenous AP enzymes from being effective in detoxification of physiological levels of proinflammatory molecules including LPS, flagellin, and nucleotides triphosphate (e.g., ATP) as well as in restoring and maintaining gut eubiosis in the porcine gut lumen.

Exogenous recombinant AP supplements have been conceptually recognized to complement the porcine gut endogenous AP functional deficiency, including AP contributions from the gut symbiotic bacteria as shown in Figure 20.3, particularly in weanling pigs (Fan and Archbold 2015; Melo et al. 2016). However, there is a scarcity of reported research on the application of exogenous feed AP in weanling swine nutrition. Other commercially available feed phosphatases such as microbial phytases are specific to the degradation of phytate but not biochemically effective for the dephosphorylation-based detoxification of endotoxin for attenuating metabolic endotoxemia, as analyzed by using the chromogenic substrate $p$-nitrophenyl phosphate for assessing phytase for AP functional kinetics (Forsberg et al. 2014). Moreover, microbial phytases are typically only functional under the gastric acidic environment and will lose most of their activity once moved into the small and the large intestinal lumen, thus not expected to play significant physiological roles as AP in the gut. Therefore, it is imperative to develop functional exogenous recombinant AP as potentially novel feed enzyme gut modifiers to enhance gut eubiosis, physiology, as well as feed conversion efficiency and performances in pigs especially weanling pigs.

Alternatively, dietary long-chain omega-3 polyunsaturated fatty acids (ω-3-PUFA) have been reported to positively modulate gut AP functionality=, dysbiosis, and enhance physiology with implications for both human health management and pig nutrition and pork production (Costantini et al. 2017; Holscher 2020). Dietary inclusion of ω-3-PUFA has been documented to regulate whole-body physiology through conversion into bioactive autacoids, including the resolvins E series, the resolvins D series, and protectins in various types of cells (Calder 2012; Fan 2013). Resolvins and protectins are very potent cellular regulators and are anti-inflammatory and can resolve inflammation via cell membrane surface and intracellular receptors partly through regulating gut AP functionality and microbiota (Calder 2012; Fan 2013). In rodent models, Campbell et al. (2010) showed that resolvin-E1 derived from ω-3-PUFA could effectively attenuated bowel inflammatory responses through enhancing gut AP functionality, while Kaliannan et al. (2015) demonstrated that increased dietary ω-3-PUFA contents and their ratios to ω-6-PUFA were effective to manage metabolic endotoxemia and low-grade systemic inflammation associated with gut dysbiosis by upregulating gut AP expression, which could also be manifested by the use of a wide spectrum of antibiotic treatment. However, in the human clinical study by Watson et al. (2018), daily ingestion of 4 g of ω-3-PUFA for 8 wk in healthy middle-aged volunteers did not affect gut bacterial diversity and microbiota composition. In contrast, Kaliannan et al. (2015) in the transgenic mice study shown that ω-3-PUFA could attenuate metabolic syndrome endpoints associated with metabolic endotoxemia and systemic inflammation for human health management through effectively enhancing gut AP functionality and suppressing gut LPS-producing bacteria identified at the phylum level. Call et al. (2020) in a preterm porcine parenteral nutrition model observed that the inclusion of ω-3-PUFA lipids could re-shape hepatic-gut bile acid pools and prevent cholestasis; however, rather than analyzing microbiota change responses in the bile duct exit in the duodenal region, they conducted the MiSeq-based colonic digesta microbiota profiling and reported decreases in relative abundance of several *Firmicutes* Phylum Gram-positive anaerobes, including *Clostridrium* XIVa, and higher abundance of Gram-negative *Enterobacteriaceae* bacteria in the phylum *Proteobacterium*. Furthermore, gut microbiota and microbiome have been targeted for developing immune-based therapies for some of the critical human diseases including auto-immune diseases and cancers (Cryan et al. 2019; Honda and Littman 2016; Helmink et al. 2019; Skelly et al. 2019); however, limited research has been reported to investigate roles of ω-3-PUFA in the treatment of these critical illness beyond the metabolic diseases. More human clinical studies in this topic area will help further elucidate optimal dietary levels and ω-3/ω-3-PUFA ratios in both human health management swine nutrition and pork production applications.

Several studies have been conducted to investigate effects of dietary inclusion of ω-3-PUFA on pig nutrition, performances, and physiological responses. Bazinet et al. (2003) reported that dietary inclusion of ω-3-PUFA reduced whole-body C-18 fatty acid oxidation but did not affect performance endpoints in segregated early-weaned pig responses. Huber et al. (2018) reported that dietary inclusion of ω-3-PUFA variable from fish oil (1.25–5%) did not improve growth performance, plasma acute-phase protein, and jejunal morphology endpoints in weanling pigs. Lee et al. (2019) reported that dietary inclusion of ω-3-PUFA variable from fish oil and oil (at 1.25) from algae supplement did not improve growth performances in weanling pigs. In contrast, Mani et al. (2013a) shown that dietary inclusion of ω-3-PUFA could reduce postprandial plasma endotoxemia by 50% partly through blocking transepithelial LPS transport but they did not report performance changes. McAfee et al. (2019) in a sow to weaning pig study shown dietary inclusion of gradient levels of ω-3-PUFA from a well-protected commercial fish oil (0–1%) supplemental product Gromega and at 1% dietary inclusion, d-3 weaning pig BW and some plasma physiological endpoints were improved, and ethoxyquin as the antioxidant was also included in their study diets. Gut AP expression and microbiota responses were not examined in the above reviewed pig studies. Several discrepancies in experimental conditions may be responsible for the inconsistent responses to the dietary inclusion of ω-3-PUFA in the pig studies. Anti-oxidation protection of their testing oil sources was not described and it appeared to be that oils were mixed with their study diets daily in the study by Mani et al. (2013a). Dietary supplementation of vitamin E was used to preserve ω-3-PUFA in the weanling pig studies (Huber et al. 2018; Lee et al. 2019). Dietary inclusion levels and ratios of ω-6/ω-3-PUFA were variable and seemed yet to be optimized for enhancing swine gut and whole body physiology and nutrition. Roszkos et al. (2020) reviewed that ω-6/ω-3-PUFA share the same biochemical pathway for their elongation, justifying the notion to establish an optimal dietary ratios of ω-6/ω-3-PUFA. While transgenic pigs capable of synthesizing ω-3-PUFA have been developed Lai et al. (2006), their commercial application for pork production may be limited by the regulatory approval process. While flaxseed and algae are the practical sources of ω-3-PUFA in swine nutrition application and effective antioxidation protection are warranted in diet formulation (Lai et al. 2006), Gromega is the well protected and commercially available feed grade ω-3-PUFA supplement (McAfee et al. 2019). Thus, further research is needed to establish the potential role and efficacy of ω-3-PUFA in the modulation of porcine gut dysbiosis, physiology, nutrition, and performances also for counteracting footprint of intensive pork production on environment.

Other feed additives (e.g., acidifiers, minerals, pre- and probiotics, direct-fed microbials, and plant extracts) may also have an effect on gut microbiota, intestinal fermentation, and total tract GE and nutrient (e.g., DM) digestibility (Fan 2013; Liu et al. 2018), but these effects have not been well quantified. A unique seaweed, classified as red macroalgae *Asparagopsis taxiformis*, has been demonstrated effective in modulating rumen microbiota and reducing methane ($CH_4$) production and emission as a feed additive in cattle feeding (Roque et al. 2019). Future research should be carried out to examine the potential of this red macroalgae *Asparagopsis taxiformis*-based seaweed as a functional supplement in swine nutrition and feedings and its efficacy on mitigating biogenesis and emission of methane from swine manure slurry resulted from feeding this supplement in swine manure storage facilities.

### Nitrogen Nutrient Utilization and Environment Impacts

Efficiency of N utilization, including N intake, excretion, and retention, is one of the most studied nutrients in swine nutrition. Retention of dietary N in the body is far from 100% efficient, ranging from 30 to 60% of N intake (Table 20.3), with pigs consuming diets formulated to be highly digestible and with amino acid levels closely matching their respective requirements resulting in the highest level of N efficiency, while pigs consuming practical diets are reflective of lower levels of N retention (Chung and Baker 1992; Kirchgessner et al. 1994; Otto et al. 2003a; van Kempen et al. 2003; van Milgen and Dourmad 2015). Of the N that excreted, approximately 30% is excreted in the feces, including the gastrointestinal endogenous N, and 70% N excreted in the urine. Factors that can influence the digestion and retention of dietary N include both dietary (diet composition, exogenous enzymes, crystalline amino acids, feed processing, and exogenous feed additives) and animal (pig age or BW, gender, genetic background, stage of production, health, and environmental conditions) effects. Several detrimental environmental aspects associated N nutrients are illustrated in Figure 20.1, including $NH_3$ and $N_xO$ emission during swine manure slurry storage as well as excessive nitrate leaching resulted from swine manure soil application (Mackie et al. 1998; Fan et al. 2006). Microbial catabolism of tryptophan contributes to the biogenesis and manure emission

**Table 20.3**  Estimation of percentages of nutrient flows in swine production (% of intake)[a].

| Nutrient elements | Partitioning (%) of nutrient flow | | | |
|---|---|---|---|---|
| | Body | Respiration | Manure | Gas |
| *Energy nutrient elements* | | | | |
| Carbon[b,c] | 20 | 55 | 10 | 15 |
| Nitrogen[d,e] | 35 | Neg | 25 | 40 |
| Sulfur[f–i] | 40 | Neg | 55 | 5 |
| *Macromineral elements* | | | | |
| Calcium[e,j,k] | 55–67 | Neg | 33–45 | Neg |
| Phosphorus[e,j,k] | 39–67 | Neg | 33–61 | Neg |
| Magnesium[j] | 29–38 | Neg | 62–71 | Neg |
| Sodium[k] | 15 | Neg | 85 | Neg |
| Potassium[k] | 10 | Neg | 90 | Neg |
| *Micromineral elements* | | | | |
| Copper[e,j,l] | 0–48 | Neg | 52–100 | Neg |
| Iron[k,l] | 16–46 | Neg | 54–84 | Neg |
| Manganese[j–l] | 23–42 | Neg | 58–77 | Neg |
| Zinc[j–l] | 4–50 | Neg | 50–96 | Neg |

[a] Neg = negligible.
[b] Kerr et al. (2006).
[c] Kerr et al. (2018).
[d] Kirchgessner et al. (1991).
[e] Li et al. (2011).
[f] Trabue et al. (2016).
[g] Trabue et al. (2014).
[h] Trabue et al. (2019a).
[i] Trabue et al. (2019b).
[j] Adeola (1995).
[k] Forsberg et al. (2013).
[l] Liu et al. (2014).

of the N-containing key-odor volatile compounds of indole and 3-methyl indole, as summarized in Table 20.3, while microbial catabolism of phenylalanine and tyrosine leads to other odor-causing volatile compounds cresol phenols (Mackie et al. 1998; Rideout et al. 2004). In addition, host and microbial catabolism of nitrogenous S-amino acids cysteine and methionine as well as other organic nitrogenous S-compounds leads to volatile and toxic sulfides as compiled in Table 20.3 and reviewed by Mackie et al. (1998) and Carbonero et al. (2012). Thus, inefficient utilization of nitrogenous nutrients bears important negative impacts on environment.

### Digestive Utilization of Nitrogen Nutrients

A wide variety of ingredients can be used in feed formulation relative to their price in relation to their digestible AA content in addition to how different ingredients and their nutritional profile complement each other to meet the nutritional needs of an animal (NRC 2012). As expected, it is possible to increase the digestibility of the diet, and thus reduce fecal N loss, through the use of ingredients that are highly digestible (e.g., skim milk, soy isolates, and plasma proteins) in contrast to ingredients that are less digestible (e.g., heat-damaged proteins). This is limited, however, because of their cost compared to other feedstuffs and potential limitations to use in the diet (e.g., high whey in finishing pigs, high fiber in nursery pigs). Fiber associated with plant feed ingredients limits true ileal CP and AA digestibility (Fan and Sauer 2002; Fan et al. 2006). Surprisingly, Adeola and Ragland (2016) and Rho et al. (2017) reported much higher standardized ileal CP and AA digestibility values in conventional DDGS and high-CP DDGS than the values reported in DDGC and other coproducts in the study by Soares et al. (2012), as well as higher and/or comparable true ileal CP and AA digestibility values in SBM in the study by Fan et al. (1995) for growing pigs. The determination of standardized ileal CP and AA digestibility values obtained through measuring the basal ileal CP and AA endogenous outputs by using a N-free low-fiber diet would, in principle, have led to underestimated rather than overestimated ileal CP and AA digestibility values in the studies by Soares et al. (2012), Adeola and Ragland (2016), and Rho et al. (2017). Crosbie et al. (2020) reported standardized ileal CP and AA digestibility values in insect larvae meals and slightly lower than the standardized and/or true ileal CP and AA digestibility values in SBM (Fan et al. 1995; NRC 1998, 2012). Thus, intrinsic factors and differences in research conditions may be causing discrepancies in the determination of ileal CP and AA digestibility values among feed ingredients.

Prescribed therapeutic antimicrobials are used in swine industry. Fan et al. (2020) attempted to improve understanding of the modes of action of in-feed therapeutic antibiotics in improving gut health and growth performance of weaning pigs by examining effects of dietary supplementation of therapeutic aureomycin on the ileal CP and AA digestibility, the intestinal terminal protein gut terminal protein hydrolase aminopeptidase N (APN) activity kinetics and the APN gene in weaning pigs fed crystalline Lys-supplemented corn and SBM- based low-CP diets. Their results suggest that dietary improved growth performances of weaning pigs by dietary therapeutic aureomycin were not attributed to improved efficiency of the digestive and postabsorptive utilization of dietary CP, and increased apparent ileal Lys digestibility was noted independent of the gut terminal protein hydrolase APN kinetics as well as the APN and the cationic AA transporter-1 (CAT-1) protein expression at protein and mRNA abundances (Fan et al. 2020). Both porcine epidemic diarrhea virus (PEDV) and gastroenteritis virus (TGEV) are transmissible viruses of high morbidity and mortality in neonatal and weanling pigs, and APN has been recognized as the gut enterocytic apical membrane receptor for both PEDV and TGEV Whitworth et al. 2019). Thus, the lack of effects of dietary supplementation of therapeutic aureomycin on the small intestinal APN kinetics and gene expression, indicating that therapeutic aureomycin would have no efficacy against PEDV and TGEV infection in young pigs.

One method to improve the digestibility of the diet is through the utilization of exogenous enzymes. While most commercially available enzymes are targeted toward improving the digestibility of fiber-based complexes (Bedford and Schulze 1998; Kerr and Shurson 2013), specifically targeting proteins with proteases has shown promise. Omogbenigun et al. (2004) shown that dietary supplementation of proteases with multicarbohydrate enzymes in weanling pigs fed cereal grain, canola meal, and soybean meal-based diets with wheat-by-products increased the apparent ileal CP digestibility and improved growth performance end points. Zuo et al. (2015) reported dietary exogenous protease supplementation increased growth performances, gut development, total tract protein digestibility, and gut health status of weaned piglets when fed low-digestible plant protein supplements. As reviewed by Lee et al. (2018), the addition of a protease in diets not containing other enzymes was shown to improve the apparent ileal AA digestibility for most AA and resulted in an improvement in feed efficiency. When other enzymes were added to the diet, however, the benefit of the added protease is negligible (Lee et al. 2018). Hence, if a diet is already formulated with good-quality ingredients high in AA digestibility in feeder pigs, the benefit of the added protease may be limited.

Jeon et al. (2019) shown that dietary supplementation of beta-mannanase improved standardized ileal digestibility of some essential AA and growth performances in grower pigs. In contrast, Ji et al. (2008) did not observe improvement in the standardized ileal CP and AA digestibility in growing pigs supplemented with combined protease and beta-glucanase. In the multienzyme transgenic pig study by Zhang et al. (2019), the coexpression of NSP enzymes of beta-glucanase and xylanase with phytase not only improved total tract mineral digestibility but also improved total tract CP digestibility, leading to significant improvement in growth performances and days required to reaching market BW in feeder pigs fed agriculture by-product-based high-fiber diets. Tessier et al. (2020) in the rodent study using $^{15}$N and $^{2}$H intrinsic labeling demonstrated that sunflower protein isolate after removal of fiber had real CP and AA digestibility (at about 97%) similar to goat whey protein. Thus, fiber intrinsically limits ileal CP and AA digestibility in plant feeds. Ingredient processing for fiber removal and use of efficacious exogenous NSP enzymes would further improve ileal CP and AA digestibility in plant feeds.

### Postdigestive Utilization of Nitrogen Nutrients

While excess AA have minimal effects on pig performance, they can have a large influence on N loss because excess AA are not stored but are deaminated in the body and their N component excreted as urinary urea (Kerr and Easter 1995; NRC 2012). Thus, a key to minimizing N loss is to supply N (i.e., AA) as closely to their requirement as possible for productive purposes. While this appears to be a seemingly easy task, on a practical basis, this can be a challenge due to differences in AA requirements at specific points in time (e.g., lean tissue gain, maintenance, milk, conceptus, immune functions, etc.) over the life of an animal and due to the biological variation between animals and groups of animals for each parameter of interest (NRC 2012). Given these challenges, the most common approach to determining a specific AA is a modeling approach. To this end, an enormous amount of research time and effort has been taken to estimate digestible AA requirements for swine based on maintenance, lean deposition rates, growth patterns, dietary fiber levels, and health status as summarized by the Committee on Nutritional Requirement of Swine in 2012 (NRC 2012). Once specific AA requirements are known, it is then through the feed that these needs must be met, which for AA, standardized ileal AA digestibility has a distinct advantage over total or fecal AA digestibility (Stein et al. 2007; NRC 2012). Consequently, extensive research and a summary of that research has reported the digestibility of AA in many feed ingredients (NRC 2012). In addition, there has been a tremendous increase in the availability of crystalline AA (isoleucine, lysine, methionine, threonine, tryptophan, and valine) which are 100% available for metabolic purposes (Leibholz et al. 1986; Izquierdo et al. 1988), and depending upon prices relative to other feedstuffs, can be used in diet formulation to closely match diet composition to animal needs. Thus, the linkage of known requirements with an understanding of the digestible AA concentration of feed ingredients can improve the ability to meet needs for all AA, while potentially minimizing excess AA intake which will ultimately minimize N excretion.

Fates of postdigestive utilization of absorbed CP compounds, including AA and non-AA nitrogenous compounds such as nucleotides, are the gastrointestinal endogenous fecal excretion, nonreversible catabolism in providing ATP, and excretion as $NH_3$ and urea as well as various biosynthesis (Fan et al. 2006). Visceral organs use disproportionally large amounts of absorbed CP compounds to support their active metabolism, referred to as first-pass and first-priority utilization and metabolism (Fan et al. 2006). Environmental, nutritional, and physiological stressors can effectively affect visceral organ metabolism, thus remodeling and shifting whole body requirements for digestible CP and AA. As compared in Table 20.4, fractional synthesis rates (FSR) of hepatic and other visceral organs were dramatically augmented in 24-day-old weaning pigs fed a corn and soybean meal-based diet experiencing an experimental recovery jugular catheterization surgical trauma (Bregendahl et al. 2004, 2008), indicating this physiological stressor would dramatically enhance CP and AA required for maintenance in pigs. For example, when shifting grower pigs from SBM-based diet to high-fiber hulled barley-based diets, their distal ileal endogenous Thr and Met losses were increased from 5 – 12% to 31% of the NRC (1998) recommended true ileal digestible Thr and Met requirements, whereas their distal ileal endogenous CP AND Lys losses were only augmented from 5–8% to 17–18% of the NRC (1998) recommended true ileal digestible CP and Lys requirements (Fan and Sauer 2002; Fan et al. 2006). These comparisons are consistent with the portal tracer balance study by Stoll et al. (1998) in showing more than 50% of the enteral Met and Thr intake was first-pass utilized by the viscera in pigs. Cemin et al. (2018) shown that His requirement in terms of its ratio to dietary Lys was 31%, lower in weanling pigs fed crystalline AA supplemented and low-SBM diets than the 34% compiled by NRC (2012) for weanling pigs fed corn-SBM-based diets. These comparisons suggest that dietary digestible essential AA requirements may need to be much higher in pigs fed high-fiber by-products based diets or altered if compositions of basal diets are different, compared with the AA requirements for pigs fed corn-SBM-based diets that NRC (1998, 2012) are based on.

In this context, Yap et al. (2020) demonstrated that limitation of dietary essential AA supply, particularly Thr and Trp, would sensitively affect whole body metabolic status likely due to their essential regulatory roles. Since December 2018, Canada and US pork industry has banned the

**Table 20.4** Comparisons of fractional synthesis rates (FSR) of hepatic and other visceral organs in 24-day-old weaning pigs fed a corn and soybean meal-based diet receiving vs. not receiving the experimental recovery jugular catheterization surgical trauma.

| Organ | FSR (%/d)[a] | | |
| --- | --- | --- | --- |
| | Surgical trauma[b] | No surgical trauma[c] | Change of FSR (%)[d] |
| Pancreas | 141.7 ± 18.0 | 93.8 ± 13.9 | 51 |
| Liver[e] | 87.0 ± 5.5 | 48.9 ± 5.9 | 78 |
| Spleen[e] | 45.9 ± 2.1 | 35.0 ± 3.0 | 31 |
| Lungs[e] | 37.6 ± 2.4 | 25.8 ± 2.9 | 46 |
| Kidney[e] | 50.1 ± 3.8 | 33.50 ± 2.9 | 49 |
| Stomach | 52.9 ± 4.4 | 40.6 ± 3.4 | 30 |
| Proximal jejunum | 89.0 ± 14.1 | 79.3 ± 5.8 | 12 |
| Distal jejunum | 76.7 ± 8.1 | 68.4 ± 5.4 | 12 |
| Cecum | 55.4 ± 4.2 | 47.0 ± 4.1 | 18 |
| Colon | 47.4 ± 3.7 | 42.7 ± 3.8 | 11 |

[a] Values are means ± SE, $n = 5$, as measured by intraperitoneal injection of a flooding dose of phenylalanine (Bregendahl et al. (2004, 2008).

[b] Data from the study by Bregendahl et al. (2008) in which the weaning pigs received an experimental recovery jugular catheterization surgical trauma including the surgery on the d 5 postweaning and followed by 3-d post-surgical recovery prior to sacrificing for the terminal sampling.

[c] Data from the study by Bregendahl et al. (2004) in which the weaning pigs did not receive an experimental recovery jugular catheterization surgical trauma and were subjected to the terminal sacrificing for sampling on the d 8 postweaning.

[d] Percentage (%) changes in FSR values in the surgical trauma group of piglets reported by Bregendahl et al. (2008) relative to nonsurgical trauma reported by Bregendahl et al. (2004).

[e] Indicating organs with significant relative FSR changes ($P < 0.05$) between the surgical trauma group and the nonsurgical trauma group of weaning piglets, as compared by the pooled t-tests.

over-counter use of feed antimicrobials as growth promotants. Jayaraman et al. (2015) and Kahindi et al. (2017) shown that under immune activation triggered by unsanitary conditions, dietary standardized ileal digestible sulfur-AA and Thr requirements and their ratios to standardized ileal digestible need to be higher in weanling pigs. Mathai et al. (2016) and Wellington et al. (2018) reported that high dietary fiber or immune activation by LPS injection increased dietary standardized ileal digestible Thr requirements. And reduction in voluntary feed intake under the immune activation and high dietary fiber intake might have also contributed to the reported changes in these essential AA intake in the above studies. Furthermore, standardized ileal Thr digestibility under the high-fiber diet feeding might have underestimated true ileal Thr digestibility simply because the basal ileal endogenous Thr loss typically measured at low dietary fiber N-free basal diet condition would be much lower than the ileal endogenous Thr loss value estimated under high dietary fiber conditions (Fan and Sauer 2002). Thus, dietary digestible essential AA requirements in pigs need to be further defined for different basal diet composition and feeding environment to precisely guide swine diet formulation for optimal growth performances and efficiency of N utilization.

### *Diet Formulation to Minimize Nitrogen Losses*

Formulating swine diets using only natural feedstuffs in order to meet the first limiting AA as defined in the NRC (2012) results in large excesses of dispensable and indispensable AA. However, depending on ingredient and crystalline AA cost and availability, total dietary CP levels can be drastically reduced thereby limiting excess N (or AA) intake. In an extensive review of the literature, Kerr et al. (2003) summarized that for each one percentage unit reduction in CP, N excretion was reduced by approximately 8%, with no impact on pig performance. An example of how this type of diet formulation can impact total N loss is that to meet the needs of a 40-kg pig by feeding a diet containing no crystalline AA (only corn, SBM, and DDGS), the diet would be approximately 23% CP, while a diet containing 6 crystalline AA (a diet with no SBM in the diet) would be approximately 12% CP, which based on the review by Kerr et al. (2003) would result in a 88% drop in total manure N excreted. Thus, swine manure N can be dramatically reduced by low-CP feeding, but this potential reduction is modulated by feed ingredient availability and cost relationships.

Not only does feeding low-CP, AA-supplemented diets affect N excretion into the environment, but combined with its potential to reduce manure pH (Sutton et al. 1999; Shriver et al. 2003; Kerr et al. 2006), feeding low-CP diets also results in reduced aerial ammonia emissions as well (Latimier et al. 1993, Pfeiffer 1993; Otto et al. 2003b; Kerr et al. 2006; Panetta et al. 2006; Leek et al. 2007; Le et al. 2009), in a direct relationship to the reduction in total manure N excreted. Besides an improvement in reducing N emissions into the environment, this reduction may also affect pig performance as well where it has been shown that pigs kept in an ammonia-contaminated environment, at 50 ppm, have been shown to have greater pulmonic weight and percentage of pulmonic weight, have lungs that contain 51% more bacteria, and grow slower than pigs kept in a room with filtered air without added ammonia (Drummond et al. 1978, 1980). Hence, low-CP feeding can effectively reduce ammonia emission out of swine production facilities into environment and improve pig health and welfare.

The impact of feeding reduced CP, and AA-supplemented diets on the emission of volatile organic compounds are inconclusive as it has been shown that lowering dietary CP decreases (Hobbs et al. 1996; Hayes et al. 2004; Le et al. 2007; Leek et al. 2007), increases (Cromwell et al. 1999; Otto et al. 2003b), or does not affect (Obrock-Hegel et al. 1997; Sutton et al. 1999; Clark et al. 2005; Le et al. 2008, 2009) volatile organic compounds or odor emissions. The effect of feeding reduced CP, and AA-supplemented diets on greenhouse gas emissions are likewise inconclusive (Clark et al. 2005; Velthof et al. 2005; Kerr et al. 2006; Le et al. 2009; van Weelden et al. 2016b). These reviewed responses are likely due to the fact that changes in emission of ammonia and other volatile organic compounds because of low-CP feeding have caused very limited reduction of other key swine odor compounds such as indole, skatole, and sulfides. Although biogenesis and emission of greenhouse gas $CH_4$ is large as reviewed by Mackie et al. (1998), but emission in the form of $N_xO$ may be limited during swine manure slurry storage. Future research should be conducted to investigate impacts of low-CP feeding on biogenesis and emission of $N_xO$ from crop field application of resulting swine manure slurry.

Research has also evaluated the ability to shift urinary N excretion into fecal N loss (i.e., incorporation into microbial mass). In general, increasing or supplementing the dietary content of resistant starch, nondigestible oligosaccharides, or nonstarch polysaccharides (NSP), or inulin has been shown to increase hind-gut bacterial proliferation thereby increasing the proportion of N excreted in the feces compared to the urine (Canh et al. 1997, 1998a,b; Younes et al. 1997; Bakker and Dekker 1998; Zervas and Zijlstra 2002; Rideout et al. 2004). While this has no impact on the net loss of N from the animal, the shifting of N loss from urine into fecal microbial mass has environmental impact where it has been shown to reduce ammonia emissions from manure storage facilities (van de Peet-Schwering et al. 1999; Aarnink and Verstegen 2007; Jha and Berrocoso 2016). While the addition of certain fiber results in a reduction in ammonia losses, these same fibers generally increase manure volatile fatty acid concentrations (Canh et al. 1987, 1998a,b; Sutton et al. 1999; DeCamp et al. 2001; Shriver et al. 2003). In contrast, the addition of dietary fiber has been shown to have little impact on odor and greenhouse gas emissions (Gralapp et al. 2002; Galassi et al. 2004; Velthof et al. 2005; Trabue and Kerr 2014; Trabue et al. 2016; Kerr et al. 2018). The above-reviewed responses are also likely due to the fact that changes in the emission of ammonia, volatile organic compounds, and GHG because of dietary inclusion of fermentable NSP have limited reduction impacts on biogenesis and emission of other key swine odor compounds such as indole, skatole, and sulfides as well as the major GHG, $CH_4$ and $N_xO$.

### *Dysbiosis, Endotoxemia, and Utilization of Nitrogen Nutrients*

Postnatal pig growth is marked by very high efficiency of protein and energy/carbon (C) deposition, at 89–91%, and fractional growth rate at birth and colostral suckling (Le Dividich et al. 1994). Efficiency of CP deposition in neonatal pigs is maintained at about 80–82% during the rest neonatal suckling and/or feeding on liquid milk replacers (Fan et al. 2006), which is uniquely supported by high small intestinal glucose and AA absorption rate through expressing glucose and AA exchangers and transporters along the entire gut crypt-villus axis (Fan et al. 2004; Yang et al. 2011, 2016, 2017) and high visceral and skeletal muscle protein synthetic capacity and efficiency (Yang et al. 2008; Columbus et al. 2015; Rudar et al. 2019). Then this is followed by a dramatic decline in postweaned efficiency of whole-body N utilization and fractional growth rate growth being responsible for low profit margin and some of the main environmental sustainability issues facing the global pork production (Fan et al. 2006; Fan et al. 2008). While the gastrointestinal endogenous CP loss and protein turnover are two the recognized negative factors contributing to the postweaned poor efficiency of whole-body N utilization in pigs (Bergen 2008; Fan et al. 2008), decreases in the fractional synthesis rate (FSR, %/d) in the skeletal muscle, i.e., the largest body protein pool, are mainly responsible for the low efficiency of whole-body N utilization and growth in postweaned pigs (Davis et al. 2008; Rudar et al. 2019). The reduced skeletal muscle FSR in postweaned pigs is largely due to declined muscle cell regulatory responses and sensitivity to various nutrient (e.g., leucine) and hormonal (e.g., insulin) stimuli, which is then further causally associated with postweaning porcine gut dysbiosis and spontaneous metabolic endotoxemia due to the intrinsic gut AP functional deficiency in pigs, which has been well deliberated in an earlier section of this chapter.

In a number of acute and chronic immune activation pig studies through the injection of purified LPS, infectious endotoxemia and inflammation-induced whole-body protein catabolic state (Webel et al. 1997), decreased feed intake and dietary protein and energy intake and deposition but increased energy requirement for maintenance (Rakhshandeh and de Lange 2012; Campos et al. 2014; Huntley et al. 2018), altered muscle fiber type (McGilvray et al. 2019a) and enhanced sulfur-AA and Thr requirements without affecting ileal AA digestibility and partial efficiency of retention of these AA (Rakhshandeh et al. 2014; Wellington et al. 2018; McGilvray et al. 2019b), however, decreased partial efficiency of Trp retention (de Ridder et al. 2012). These differential responses in requirements and partial efficiency for body deposition among Met/Cys, Thr, and Trp, as activated by infectious endotoxemia and inflammation through acute and chronic LPS injection, reflect their different roles in contributing to whole body maintenance and body protein deposition. These results are consistent with what Fan et al. (2006) reviewed that the first-pass and first-priority utilization by the viscera could effectively remodel, shift, and repartition whole-body CP and AA metabolism and utilization.

In a postweaned pig study by Rudar et al. (2016), effects of feeding excess level of dietary CP and excess amount of the well-recognized signaling essential AA Leu on whole-body protein retention and metabolic status were examined in the feeder pigs with and without LPS injection. In our view, the normal pigs in their study would represent spontaneous porcine metabolic endotoxemia associated gut dysbiosis and the LPS-injected pigs would be associated with induced infectious endotoxemia and inflammation (Rudar et al. 2016). Importantly, their data suggest the notion that excess dietary CP and Leu were not effective to ameliorate the negative effects of endotoxemia on whole-body protein anabolism and to restore protein deposition rate. To the best of our knowledge, upregulation of intestinal AP functionality would be potentially the correct direction of developing dietary strategies to attenuate negative effects of endotoxemia on whole-body protein metabolism and utilization (Figure 20.3), as demonstrated through the rodent studies by Kaliannan et al. (2013, 2015) in which enhancing intestinal AP functionality directly via exogenous AP supplementation or indirectly via dietary ω-3-PUFA supplementation in mitigating endotoxemia-mediated metabolic syndrome and in restoring muscle sensitivity to insulin.

Apart from playing the direct and essential role in intestinal detoxifying PAMPs compounds, biological mechanisms of AP in modulation of whole-body physiology appear to be related to remodeling microbiota and microbiome from dysbiosis to eubiosis and improving host immunity and metabolism, as illustrated in Figure 20.3, which need to be further elucidated. In their classic knock-out mice study, Malo et al. (2010) clearly shown intestinal commensal gut microbiota, particularly the gram-positive *Firmicutes* phylum of bacteria, could be effectively recovered while proliferation of pathogenic bacteria such as *Salmonella typhimurium* were inhibited by AP after antibiotic treatment. Malo et al. (2014) further proposed a working mechanism that intestinal AP dephosphorylated luminal nucleotide triphosphates, thus leading to the "gut lumen pH-lowering effects" in affecting gut microbiota; however, this theory was not supported by their data then rejected by them. Nevertheless, the Malo et al. (2014) study did directly link intestinal AP dephosphorylation of luminal nucleotide triphosphates to microbiota recovery.

Intestinal digesta has a significant amount of DNA/RNA and nucleotides as nonprotein nitrogenous (NPN) compounds that are originated from dietary ingredient DNA, dietary supplemental nucleotides (Liu et al. 2018), sloughed-off host gut epithelial DNA/RNA, and luminal dead bacterial genomic DNA/RNA. Several digestive enzymes are involved in the degradation and recycling of DNA/RNA and nucleotides for the salvage pathway of utilizing of recycled nucleotides in luminal bacterial and fast-turn-over host cellular (e.g., intestinal epithelia and immune cells) *de novo* DNA synthesis proliferation, including host pancreatic ribonuclease and deoxyribonuclease to convert DNA/RNA into oligonucleotides, host intestinal mucosal and symbiotic bacterial phosphodiesterase to convert oligonucleotides into free nucleotide phosphates (e.g., ATP, ADP and AMP) and host intestinal and/or symbiotic bacterial AP and nucleotidase for the final dephosphorylation of luminal nucleotide phosphates into free nucleosides (e.g., inosine, adenosine, guanosine, cytidine, 3-methyluridine, uridine, and thymidine) for transmembrane absorption by bacteria and host intestinal epithelia (Whitt and Savage 1988; Fan et al. 1999). It has been well documented that biosynthesis of nucleosides is a multistep biochemical pathway intensively requiring trophic AA (i.e., Gln, Asp, and Gly) and the metabolic energy ATP for cell cycles (Lane and Fan 2015); thus, bioavailability of nucleosides could become a limiting factor for maintaining gut symbiotic microbial eubiosis and host physiology. Although host gut epithelia express mammalian AP and nucleotidase and gram-negative bacteria widely express AP (Cheng and Costerton 1977; Whitt and Savage 1988), a number of gut commensal bacteria, particularly the gram-positive bacteria such as *Firmicutes* phylum, do not express bacterial AP for the terminal dephosphorylation of luminal nucleotide phosphates into free nucleosides (Cheng and Costerton 1977). More importantly, free nucleosides rather than nucleotide phosphates are then effectively transported by membrane nucleoside transporters for the intracellular salvage pathway of further utilization by gut bacteria and host (Martinussen et al. 2010; Pastor-Anglada and Pérez-Torras 2018). Thus, nucleotidase and/or AP are needed to further enhance bioavailability of nucleosides for maintaining gut symbiotic microbial eubiosis and host physiology. Wang et al. (2020) shown that inosine could be an effective alternative substrate for enhancing human T-cell-based immunity and immune therapy, while Mager et al. (2020) identified specific gut bacterial species in releasing inosine into systemic circulation for modulation of immune therapy for cancers. Therefore, development of efficacious exogenous supplemental AP could improve the efficiency of DNA/RNA and nucleotide-based NPN utilization, whole-body physiological function, and immunity in swine nutrition and production.

Development and commercialization of effective exogenous supplemental AP as gut modifier in swine nutrition will have two additional important implications. First, dietary supplementation of exogenous AP, in principle, will not lead to increases in AMR, simply because this potentially emerged new enzyme gut modifier can potentially restore gut eubiosis but will not kill any bacteria. Second, decreased intestinal dysbiosis will not only prevent or minimize subclinical mild and septic inflammation but may also reduce microbial biogenesis of some of the key odor-causing compounds including ammonia, volatile sulfides, phenols such as phenol and 4-methylphenol (or *p*-cresol), indole and 3-methyl indole (or skatole), as well as several VFAs (butanoic, 3-methylbutanoic, and pentanoic acid) (Trabue et al. 2011). Albeit of their relatively low concentrations at the ppm level in fresh swine manure, indole, skatole, and sulfides are recognized as some of the key volatile compounds causing the featured swine manure odor, and skatole is also responsible for off-flavor of pork products (Jensen et al. 1995; Mackie et al. 1998; Rideout et al. 2004; Trabue et al. 2011; Fan 2013). Skatole was among the three compounds, including skatole as well as 4-methylphenol (or *p*-cresol) and 4-ethylphenol, accounting for up to 93 percent of the summed odor activity value following land application of swine manure slurry (Parker et al. 2013).

Some bacterial Trp catabolites, particularly indole and tryptamin, play important roles in regulation of gut and whole body physiology via affecting enteroendocrine secretion such as GLP-1 and innate immunity (Chimerel et al. 2014; Roager and Licht 2018), whereas out of the volatile sulfides, hydrogen sulfide ($H_2S$) also affects gut and whole body physiology via GLP-1 and peptide YY (PYY) secretion (Bala et al. 2014). Biogenesis of indole and skatole is carried out by specific genera and species of bacteria and some of which may be pathogenic (Yokoyama and Carlson 1979; Jensen et al. 1995; Mackie et al. 1998; Vhile et al. 2012; Roager and Licht 2018; Zhao et al. 2018), while some of the bacterial spp. identified for catabolizing sulfur-AA into $H_2S$ have also been reviewed (Carbonero et al. 2012). Interestingly, some bacterial species identified for catabolizing Trp into indole and skatole as well as degrading sulfur-AA into $H_2S$ are also causally associated with developing inflammatory bowel diseases in humans (Carbonero et al. 2012; Roager and Licht 2018).

Dietary supplementation of chicory root inulin could effectively reduce skatole content but did not affect the contents of indole, *p*-cresol, total volatile sulfides including $H_2S$ and total ammonia/ ammonium in fresh feces of grower pigs fed corn- and SBM-based diets (Rideout et al. 2004). Through shotgun metagenomic sequencing, Zhao et al. (2018) shown reduction in abundances of the key bacterial genes encoding for tryptophanase and sulfite reductase, respectively, responsible for prebiotic-fiber feeding in association with decreased indole and $H_2S$ contents in presence of their identified responder species and/or strain of bacteria in the human gut; however, skatole response was not reported in their study. These reviewed research reports of gut bacterial catabolism of Trp and sulfur-AA are consistent with the research findings that pigs, under acute and chronic infectious endotoxemia and immune activation by LPS injection, are associated with increased requirements and/or reduced partial efficiency of retention for Trp and sulfur-AA (de Ridder et al. 2012; Rakhshandeh et al. 2014). Both feed grade crystalline Trp and sulfur-AA are widely used in global commercial swine diet formulation. It appears to be that dietary supplementation of suitable types and levels of prebiotic fibers such as inulin is likely to attenuate skatole biogenesis, however, may be limited to transform gut microbiota & microbiome to remodel and control biogenesis of indole, phenols, and sulfides. Thus, development and application of exogenous supplemental AP as a gut modifier would potentially modulate gut microbiota/microbiome and mitigate biogenesis and emission of swine manure odor impact associated with indole, skatole, phenols, and sulfides, address concerns of off-flavor pork by concumers, and enhance efficiency of Trp, aromatic AA (i.e., Phe/Tyr), and sulfur-AA utilization in swine nutrition, improving profit margins for swine producers.

**Sulfur Compound Utilization and Environment Impacts**

Sulfur (S) in swine diets includes organic S such as sulfur AA and peptides such as glutathione as well as inorganic sulfate compounds largely included in swine diet formulation as sulfate-based trace mineral supplements. Total dietary S content in swine diets and its requirements by swine are not a concern thus are not defined (NRC 2012). For example, Ji et al. (2008) reported that dietary total S content was at 0.22–0.23% (on DM basis) in crystalline Lys-Thr-Met supplemented low-CP corn- and SBM-based diets for grower pigs. The Canadian Food Inspection Agency has set the total dietary S content in commercial swine diets to be at the 0.40% (on air-dry basis with 88% dry matter) ceiling level after consultation with various Canadian swine industry stake holders (Table 20.5). And this is largely due to safety concerns for pigs in barn facilities, environment, swine producers, swine manure slurry handling workers, and general public. Simply, a significant proportion of S in swine manure slurry is known converted into volatile sulfides that are not only causing strong odor impact but also are very toxic to pigs and humans (Mackie et al. 1998; Rideout et al. 2004). Thus, efforts need to be made to minimize total swine manure S excretion and the negative impacts of volatile sulfide emission from intensive swine production on environment and swine production facility nearby community.

**Table 20.5** Summary of the National Research Council (NRC, 2012) recommended levels and the maximum levels of macro-minerals and micro-minerals in swine diets regulated by the Canadian Food Inspection Agency (CFIA 2016).

| Mineral elements | Requirement levels by NRC (2012)[a] | Maximum levels by CFIA[b] |
|---|---|---|
| Sulfur (%) | — | 0.4 |
| Calcium (%) | 0.80 | 2.0 |
| Phosphorus (%) | 0.50 | 1.0 |
| Magnesium (%) | 0.04 | 0.3 |
| Sodium (%) | 0.20 | 1.1 |
| Chloride (%) | 0.20 | — |
| Potassium (%) | 0.28 | 3 |
| Copper (ppm)[c] | 5.0 | 125 |
| Cobalt (ppm)[c] | — | 1 |
| Iodine (ppm)[c] | 0.14 | 4 |
| Iron (ppm)[c] | 100 | 750 |
| Manganese (ppm)[d] | 3.0 | 125 or 1000 |
| Selenium (ppm)[c] | 0.25 | 1 |
| Zinc (ppm)[e] | 80 | 150 |

[a] Levels (i.e., total chemical content except otherwise specified) of minerals in diets (at 90% dry matter) for the phase-II weaning pig (5–12 kg body weight) in representing some of the highest mineral requirement levels recommended by the NRC (1998, 2012).

[b] Maximum levels (i.e., total chemical content except otherwise specified) of minerals in commercial swine diets (at 88% dry matter) regulated by the CFIA (2016).

[c] The NRC (1998, 2012) recommended value represents a supplemental level of the concerned available micro-mineral in swine diets but does not include the concerned micro-mineral contributed from other dietary components, whereas the CFIA (2016) regulated value represents maximum total chemical content of the concerned micro-mineral allowed in swine diets.

[d] The NRC (1998, 2012) recommended value represents a supplemental level of the available Mn in swine diets but does not include Mn contributed from other dietary components, whereas the CFIA (2016) regulated value represents maximum total chemical content of Mn (125 ppm for starter pigs of 3–25 kg; and 1000 ppm for grower-finisher swine of 25–135 kg body weight, respectively) allowed in swine diets.

[e] The NRC (1998, 2012) recommended value represents a supplemental level of the available Zn in swine diets but does not include Zn contributed from other dietary components, whereas the CFIA (2016) regulated value represents maximum total chemical content of Zn allowed, including in the form of insoluble ZnO that potentially has an antimicrobial property, in swine diets.

Sulfur AA metabolism has been reviewed in detail (du Vigneaud 1952; Baker 2006) with meeting the S-AA requirements of being a critical component in optimizing lean and efficient growth (NRC 2012). Sulfur can also be taken up by the animals in drinking water, where high water sulfate levels have been shown to increase the incidence of diarrhea and reduce pig performance (CAST 1974; Anderson and Stothers 1978; Paterson et al. 1979; Veenhuizen et al. 1992; Anderson et al. 1994; NRC 1998, 2012). In contrast to S-AA requirements, inorganic S requirements have received little attention, other than being required under unique circumstances (Lovett et al. 1986). Thus, total S requirements are not a concern in practical swine nutrition and diet formulation.

As is the case with most nutrients, excess S in the form of AA (Met and Cys) or sulfate forms of minerals (iron sulfate, manganese sulfate, and zinc sulfate) that are subsequently excreted via urine can be further converted by manure microbes to form volatile sulfur compounds (VSC), with being $H_2S$ most commonly reported during pig feeding and during the swine manure slurry storage (Banwart and Bremner 1975; Mackie et al. 1998), as illustrated in Figure 20.1. This is especially important given that the retention of total sulfur intake has been suggested to reach only up to 65% (Shurson et al. 1998). Increased dietary S has been shown to increase S-containing odorants in feces or manure (Sutton et al. 1999; van Heugten and van Kempen 2002; Eriksen et al. 2010), although this may be modulated by dietary CP levels (Liu et al. 2017). While certain ingredients contain concentrated levels of total S (e.g., blood meal, canola meal, fish meal, dried whey, etc., NRC 2012), the total S level in DDGS has been most studied because its concentration is abnormally high due to the high contents of AA during the removal of starch during fermentation but also due to residual sulfate left from the use of $H_2SO_4$ in the steeping processing of corn for ethanol production (Kerr et al. 2006; Linde et al. 2008; NRC 2012). In light of this, there has been extensive research on the addition of DDGS to swine diets where it has been shown that supplemental dietary DDGS will either not affect (Spiehs et al. 2012) or increase manure S excretion (Trabue and Kerr 2014; Kerr et al. 2018; Trabue et al. 2019b). Ji et al. (2008) shown that dietary supplementation of β-glucanase-protease enzyme blend product did not consistently improve the apparent ileal total S (68–71%) and standardized ileal sulfur-AA (Cys, 78–81% vs. Met, 85–88%) digestibility, however, improved the apparent total tract S digestibility (82 vs. 85%) in grower pigs fed crystalline Lys-Thr-Met supplemented low-CP corn- and SBM based diets with total dietary S content at 0.22–0.23% (on DM basis). These results indicate large intestinal microbial fermentation does contribute significantly to S digestion and absorption and total tract S digestibility is reasonably high in pigs. Thus, formulation of low-S diets through using low-S ingredients and nonsulfate trace mineral supplements will lead to increased whole-body S retention efficiency and manure S excretion in swine.

In contrast to the well-described relationship between N excretion and ammonia emission (Latimier et al. 1993; Pfeiffer 1993; Panetta et al. 2006; Leek et al. 2007; Le et al. 2009), the relationship between total S excretion and volatile sulfides' emissions is inconclusive. Li et al. (2011) reported that DDGS supplemented to the diet resulted in manure with increased $H_2S$ emissions which is supported by Kerr et al. (2018) who reported increased total volatile S compounds due to dietary supplementation of DDGS. In contrast, Spiehs et al. (2012) and Trabue et al. (2019b) reported no effect of DDGS supplementation on total reduced S emissions, and Trabue and Kerr (2014) reported a reduction in $H_2S$ emissions (largely attributed to crust formation). There is more agreement on GHG emissions in responses to dietary DDGS supplementation, where Li et al. (2011), Trabue and Kerr (2014), Kerr et al. (2018), and Trabue et al. (2019b) have all reported no impact of DDGS on GHG emissions. In an experiment evaluating impact of inorganic S intake on total manure S excretion and sulfides' emissions, Trabue et al. (2019a) indicated that for each 0.10% increase in dietary total S content, total manure S increased by 10% and $H_2S$ emission increased by 8%. This is largely attributed to S being retained in the body by 40%, with 55% ending up in the manure and only 5% being emitted as volatile S-containing sulfides compounds. Ji et al. (2008) reported the apparent total tract S digestibility was 82% in grower pigs fed a CP corn- and SBM-based diet with total dietary S content measured at 0.22%, on DM basis; and this would correspond to 2.2 g total S intake/kg DM diet vs. 0.40 g fecal S output/kg DM diet, respectively. Whereas in the study by Rideout et al. (2004), fresh fecal output of total volatile sulfides as the hydrogen sulfide or S was estimated to be 83.2 mg $H_2S$ or 78.3 mg S/kg DM diet, respectively. Collectively, results of the studies by Rideout et al. (2004) and Ji et al. (2008) would suggest about 3.6% of total dietary S and about 20% of total fecal S were, respectively, used for the biogenesis of total volatile sulfides in the gastrointestinal tract and ended in the fresh feces of growing pigs fed a typical CP corn and SBM based diet. These estimates of total volatile sulfides in fresh feces of the growing pigs are consistent with the 5% emission of the total swine manure S as volatile S-containing sulfides' compounds reported by Trabue et al. (2019a). Thus, only a small fraction of swine fecal and manure total S is converted into volatile sulfides via microbial activities, potentially causing odor impacts.

Dietary supplementation of prebiotic chicory root inulin could effectively reduce skatole but not total volatile sulfide content in fresh feces of grower pigs fed the corn and SBM-based diets (Rideout et al. 2004). Through shotgun metagenomic sequencing, Zhao et al. (2018) associated the key bacterial gene encoding sulfite reductase with prebiotic-fiber feeding induced reduction of biogenesis of hydrogen sulfide in presence of their identified responder species of bacteria in the gut of human testing subjects demonstrating improvements in type-II diabetic symptoms. Some of the bacterial spp. identified for converting inorganic sulfates into $H_2S$ and/or other volatile sulfides have been reviewed by Carbonero et al. (2012). Nevertheless, volatile sulfides are recognized as a group of the key volatile odor-causing compounds responsible for the featured swine manure odor (Rideout et al. 2004; Trabue et al. 2011). Thus, manipulation of microbiota/microbiome in porcine gut and in swine manure slurry during storage may be more effective than merely reducing total dietary S inputs and manure S excretion in the control of biogenesis and emission of volatile sulfides and their environmental impacts.

## Macromineral Utilization and Environment Impacts

The requirements for macrominerals including calcium (Ca), phosphorus (P) and digestible P, magnesium (Mg), sodium (Na), chlorine (Cl), and potassium (K), in g/kg diet (on 90% DM air-dry basis), are well established for swine under various production and physiological phenotypes of conditions (NRC 1998, 2012), reflecting developmental changes in contents of these macrominerals in nonfat body weight (Mahan and Shields 1998). These macromineral requirements for the phase-II weanling pigs corn- and SBM-based diets, as recommended by NRC (2012), represent some of the highest dietary contents of these nutrient recommended by the NRC Swine Committee. Comparatively, these dietary contents of macrominerals recommended by the NRC (2012) for the phase-II weanling pigs are further listed with the CFIA (2016) regulated upper limit of dietary levels of these minerals in Canadian commercial swine diets (Table 20.5). Noticeably, the CFIA allowed upper total swine dietary levels of Ca at 2% vs. P at 1% are about twofolds higher than the NRC suggested total swine dietary Ca and P contents.

Both Ca and P are major components of the skeletal system, with approximately 98 and 75% of whole body Ca and P, respectively, stored in bone mass. Only 20 to 50% of the Ca or P consumed, however, is retained in the body (Kornegay and Harper 1997), resulting in a large portion of these minerals end up in the manure with excessive manure P being responsible for surface water eutrophication as a major recognized environmental concern (Table 20.3). A key to minimizing excretion of Ca and P is reducing the amount of Ca and P fed by having accurate Ca and P requirements, either through the summarization of empirical data or through a modeling approach. As discussed in detail by the Committee on Nutrient Requirements of Swine (NRC 2012), there was limited empirical data meeting defined standards from which to determine Ca and P requirements, whereupon the Committee elected to use a modeling approach to estimating Ca and P requirements. This modeling approach estimated P retention based on the close relationship between whole-body P mass and whole-body N mass. After which, application of basal endogenous fecal and urinary P losses was determined and through estimating the efficiency of using standardized total tract digestible (STTD) P intake for whole-body P retention, a STTD P requirement was determined, with the Ca requirement calculated using fixed ratios between STTD P and total Ca (NRC 2012). Once this modeling process was completed, the redefined estimates of P requirements resulted in a 15 to 20% drop in the STTD P needs in the diet alone compared to the previous swine NRC ((NRC, 1998)). Corn and SBM-specific total tract endogenous endogenous Ca and P losses by Fan and Archbold (2012) were likely much higher the basal endogenous fecal Ca (0.91 ± 0.20) and P (1.31 ± 0.15) losses, suggesting true digestible Ca and P requirements should be further established.

An another key in minimizing excess Ca and P intake, and therefore minimizing Ca and P losses, is knowing the digestibility of these minerals within a feedstuff, similar as that for other nutrients the animals consumes. It is well known that Ca and P digestibility can be affected by mineral and feedstuff sources, Ca:P ratios, other mineral levels, and body weight (NRC 2012). Furthermore, dietary Ca and P are closely interrelated to each other in their digestive and postabsorptive utilization and metabolism. Mahan and Shields (1998) reported that pig whole-body Ca:P ratio decreased from birth to 20 kg then remained relatively constant to 90 kg BW, as reflected by the declining dietary Ca and P requirement levels and Ca/P ratios from weanling to later finishing pigs (NRC 2012). Current optimal dietary Ca/P ratios are largely defined to be between 1:1 and 1.25:1 for total Ca-to-total P as well as 2:1 and 3:1 for total Ca-to-available P, respectively, through empirical performance trials in pigs fed cereal grains and SBM-based diets (NRC 1998, 2012). Stein et al. (2011) also observed decreases in the apparent P but not in the apparent Ca digestibility by increasing dietary Ca intake levels with feed grade limestone in growing pigs. Fan and Archbold (2012) attempted to establish true digestible Ca to true digestible P ratio in growing pigs through measuring true Ca and P digestibility and retention; however, true Ca and P digestibility and retention responses were not consistently affected by dietary Ca intake levels with dietary inclusion of increasing feed grade limestone. As discussed by Gonzalez-Vega and Stein (2014), variable dietary Ca/P ratios may interfere with Ca and P solubility, transport, and their gastrointestinal endogenous recycling and losses. Lee et al. (2019, 2020) shown P digestibility and basal endogenous P loss were affected in gestation sows compared with growing pigs, and increasing dietary Ca/P ratios enhanced P retention and improved bone resorption. Gestation sows are fed with restricted diets especially during their late gestation to control their weight gain and fat deposition. However, essential dietary contents, including Ca and P, in the gestation sow diets have been not listed for considering this dramatic reduced feed intake in gestation sows (NRC 1998, 2020). For example, dietary total contents of Ca (0.75%) and P (0.60%), on 90%-DM air-dry basis, are recommended for gestation sows with an estimated 1.85 kg/day feed intake (NRC 1998, 2020). It is conceivable that the CFIA (2016) allowed upper total swine dietary levels of Ca at 2% and P at 1% would genuinely reflect much higher actual Ca and P contents that are likely formulated for late gestation sow diets by Canadian swine industry. Thus, optimal digestible Ca to digestible P ratios still need to be defined in pigs. Dietary Ca and P levels and the Ca/P ratios for gestation sows need to be correctly established by considering their variable restricted feed intake levels to meet the sows' Ca and P requirements and to minimize bone mineral resorption, thus reducing excessive swine manure P excretion.

In cereal grains, grain by-products, and oilseed meals, approximately 60 to 75% of the P is organically bound in the form of phytate or phytin–P which is poorly available to the pig. As a result, the biological availability of P in cereal grains is variable, ranging from less than 15% in corn to approximately 50% in wheat. In oilseed meals, P availability is also relatively low, being 32% for canola meal and 48% for soybean meal (NRC 2012). For animal protein by-products meals, their P will have a higher digestibility, with the P in meat and bone meal being approximately 70%, although this can be variable. Lastly, the P in inorganic P supplements tends to be highly available, with the P in monocalcium phosphate being 88% digestible (NRC 2012).

Utilization of exogenous phytase to release phytin-P has been extensively shown to improve P digestibility, being affected by the amount and type of phytase added, feedstuff composition of the diet which the phytase acts on, the Ca and P levels (and ratios) of the diet, and body weight of the animal (Gonzalez-Vega and Stein 2014). Microbial phytases hydrolyze dephosphorylation of phytate (IP, myo-inositol hexakisphosphate), and depending upon which of the 6-C positions on the inositol ring phytase initiate its first dephosphorylation hydrolysis, microbial phytases are grouped into 3-, 5-, and 6-phytases (Ariza et al. 2013). Most of earlier commercially available microbial phytases are 3- and 5-phytases. Shen et al. (2005) used the original Natuphos phytase (not containing 6-phytase) in an *in vitro* study in showing that 85–89% of the total phytate P could be hydrolyzed within two hours under the gastric pH and the rest P, presumably IP6, was further completely hydrolyzed with extended incubations for 4–10 hours, which would not occur *in vivo* in pigs, because gastric digesta retention time in stomach is at about two hours. Dersjant-Li et al. (2020) shown that dietary supplementation of 6-phytase effectively improved growth performances of weanling pigs without inorganic phosphate supplementation, and this was also due to the improvements of energy and AA digestibility.

Less is known about the digestibility of Ca in commonly used feedstuffs, but because of their phytic acid content in plant-based feedstuffs, the Ca bioavailability tends to be low. However, given the total content of Ca in most feedstuffs is low, this has little effect on the overall dietary Ca digestibility. Calcium digestibility in animal protein by-products meals is moderately high, and the digestibility of Ca in most inorganic mineral sources is also relatively high (NRC 2012). In a study with the transgenic enviropig™ expressing a conventional microbial 3- and 5-phytase, Meidinger et al. (2013) shown that the complete removal of inorganic P and partial removal of inorganic Ca in formulating low-Ca and low-P diets dramatically improved efficiency of P retention largely through reducing fecal P excretion in weanling, grower, and finisher pigs fed corn and SBM-based diets. Zhang et al. (2018) in the transgenic pigs expressing a conventional microbial 3- and 5-phytase alone with β-glucanase and xylanase shown enhanced Ca and P digestibility, improved fecal DM digestibility, growth performances, and shortened days to market by 19–21% when fed corn and SBM-based diets further supplemented with other agriculture by-products of wheat bran, rice bran meal, cotton seed meal, and rapeseed meal. Because of regulatory approval limitation, commercial applications of these transgenic phytase pig lines seem not realistic. Nevertheless, taken together, the use of exogenous phytases can reduce dietary levels of Ca and P, thereby lowering P excretion by 30 to 60%.

Adequate dietary level of magnesium (Mg) is essential in swine and commercial swine dietary ingredients typically provide total Mg content to meet the NRC level at 0.04%, on 90%-DM airdry basis (NRC 1998, 2012). However, when formulating purified or semi-purified diets, supplemental Mg needs to be added to the diet formulation. Ji et al. (2008) reported total dietary Mg at 0.16% in their corn-SBM-based diets but only at Mg 0.03% (DM basis) in their casein-based diet with inclusion of supplemental Mg for grower pigs. Adeola (1995) determined the apparent total tract Mg digestibility at 39–42% in inorganic Mg supplemented corn-SBM-based diets and dietary phytase supplementation did not improve the Mg digestibility and retention. Ji et al. (2008) measured the apparent ileal (12–13%) and total tract Mg digestibility (33–35%) values in grower pigs fed in their corn-SBM-based diets that were much lower than the values (50–60%) reviewed by NRC (1998, 2012). Wu et al. (2019) reported the much lower true fecal Mg digestibility at about 11% in corn fed grower pigs. Kovácsné Gaál et al. (2004) reviewed and concluded that dietary Mg supplementation improved Mg digestibility, physiological status, and performances in several food animal species including sows, cows, broilers, and laying hens. Meanwhile, the CFIA (2016) has regulated upper total swine dietary level of Mg at 0.30%, and this may again suggest that much higher supplemental Mg is being used in commercial swine diets by Canadian swine industry. Thus, research is needed to further establish true total tract digestible Mg requirements in swine to potentially improve and optimize swine nutrition and performances.

Sodium ($Na^+$), chloride ($Cl^-$), and potassium ($K^+$) are the major electrolytes for maintaining whole-body physiological functions such as acid–base balance, osmolarity, membrane potential, as well as absorption, resorption, and transport of nutrients including nutrient solutes, ions, and water (NRC 1998, 2012). Whole-body homeostasis and control of the major electrolytes are largely achieved by the $Na^+/K^+$-ATPase and the second $Na^+$-ATPase (e.g., Adeola et al. 1989; Fan et al. 2001) at the consumption of a significant amount of metabolic energy ATP. The $Na^+/K^+$-ATPase and the second $Na^+$-ATPase are expressed on the basolateral membrane in the intestinal epithelia (Fan et al. 2001). Dietary Na, Cl, and K requirements in swine were largely established via performance trials (NRC 1998, 2012). NRC (2012) recommended dietary NaCl supplementation is from 0.18 to 0.50% (air-dry basis) much higher in weanling pigs and lower in grower-finisher swine. The CFIA (2016) has regulated the upper dietary Na level at 1.1% while no limit is given to Cl, and this may again allow some flexibility for Canadian commercial swine and feed industry. Plant-based swine feeds are rich in K but low in Na and Cl contents; thus, typical commercial swine diets are not supplemented for K but added with NaCl. Ji et al. (2008) and Meidinger et al. (2013) reported total dietary K at 0.89–0.92% and 0.62–0.63 % (DM basis) in corn-SBM-based grower and finisher diets, respectively, that are much higher than the level of 0.30% (air-dry basis) compiled by NRC (2012) for pigs. Meanwhile, the CFIA (2016) has regulated the upper total swine dietary level of K at 3.0%. And K is the recognized mineral fertilizer for crops.

Furthermore, Ji et al. (2008) observed the apparent ileal Na digestibility at about −500% and Cl digestibility at about −60%, suggesting no net absorption of Na and Cl in the small intestine and considerable levels of endogenous Na and Cl recycling within the stomach-small intestine to support gut functions, whereas the fecal digestibility reached 88% for Na and 95% for Cl in the grower pig. Meidinger et al. (2013) observed the whole-body Na and K deposition was low at about 10–15% in conventional finisher swine, whereas the phytase-finisher pigs improved K deposition by about 16% at the cost of Na, and this is consistent with the summary of Table 20.3. Poor efficiency of whole-body Na, Cl, and K retention and excessive manure excretion of these major electrolytes may have the following implications, as discussed by Meidinger et al. (2013), including (i) alkaline manure pH and ammonia emission; (ii) excessive manure slurry associated with high level of water consumption; and (iii) a significant amount of metabolic energy consumption and $CO_2$-C production associated with processing these excessive electrolytes. Thus, there is a huge potential to further improve postabsorptive utilization of whole-body Na, Cl, and K electrolytes to enhance swine production efficiency and associated footprints to environment.

## Micromineral Utilization and Environment Impacts

The supplemental levels of microminerals, including copper (Cu), cobalt (Co), iodine (I), iron (Fe), manganese (Mn), selenium (Se), and zinc (Zn) ppm in diet (on 90% DM air-dry basis), are well established for swine under various production and physiological phenotypes of conditions (NRC 1998, 2012), also reflecting developmental changes in contents of these microminerals in nonfat body weight (Mahan and Shields 1998). These supplemental levels of microminerals for the phase-II weanling pigs corn and SBM-based diets, as recommended by NRC (2012), represent some of the highest supplemental levels of microminerals recommended by the NRC Swine Committee. Whereas since 2016, the CFIA has regulated upper limits of total dietary levels for all of these trace minerals in Canadian commercial swine diets (Table 20.5).

Pharmacological levels of dietary Cu ion (at about 250 ppm) and insoluble ZnO (at 2500–3000 ppm) had been widely used for as growth-promoting additives in weaning pig diets (NRC 1998, 2012; Shannon and Hill 2019). However, these pharmacological levels of dietary Cu ion and ZnO applications are associated with concerns of AMR and their potential impacts on inhibiting surface soil microbial activity (e.g., Fan 2013). European Union countries regulated dietary Cu and Zn levels in compound diets for swine (e.g., Zn at 120–150 ppm) (Brugger and Windisch 2017). The CFIA (2016) also regulated total dietary contents of Cu to be at 125 ppm and 150 ppm upper limits; however, there is no specifications of the Zn compound forms to address these public and environmental safety concerns. Swain et al. (2016) reviewed recent progress of developing low-dose Nano-zinc for swine nutrition applications. On other hand, $B_{12}$ is widely included in vitamin premix in swine diet formulations, thus Co deficiency and supplementation is no longer an issue. Iodine is typically supplemented in iodinated salt in iodine-deficient regions. While there are no major public health and environmental safety concerns about supplementing high dietary levels of Se in swine diets, Se has been increasingly recognized as an effective antioxidant in swine nutrition and human food and health. Mahan et al. (1999) shown that dietary supplementation of higher levels of Se especially organic Se could effectively increase Se contents in pork products. Thus, the CFIA (2016) regulated dietary upper Se level at 1 pm is much higher than the supplemental Se levels of (0.15–0.30 ppm) compiled by NRC (2012), to give the Canadian swine and feed industry the needed flexibility in developing high-Se swine diets.

Out of these seven essentially required microminerals, contents of Cu, Fe, Mn, and Zn in common swine feed ingredients of plant origins are substantial as compiled by NRC (1998, 2012). However, there is less consistent and reliable data about true bioavailability of these trace minerals in common swine feed ingredients, largely because of technical difficulties to obtain these estimates and little economical incentives to measure true trace mineral bioavailability. Adeola (1995), Ji et al. (2008), and Liu et al. (2014) reported very variable apparent total tract digestibility of Cu, Fe, Mn, and Zn in weanling and grower pigs fed corn and SBM-based diets. Cargo-Froom et al. (2019) estimated and compared true total tract digestibility of trace minerals in plant and animal protein-based diets for dogs and shown trace minerals in plant proteins were much more highly digestible than previously thought. Yang (2019) and Yang et al. (2019) shown that the apparent total tract digestibility of Cu, Fe, Mn, Se, and Zn was dramatically affected by their assay dietary contents due to the impacts of the gastrointestinal endogenous fecal losses of these trace minerals in postweaned pigs fed corn and SBM-based diets. True total tract digestibility of Cu, Fe, Mn, Se, and Zn associated with corn- and SBM-based diets was determined by the regression analysis and was much higher than the apparent total tract digestibility of these trace minerals in weanling and grower pigs fed the corn and SBM based diets previously reported in the literature (Adeola 1995; Ji et al. 2008; Liu et al. 2014). They further shown that the gastrointestinal endogenous fecal losses of these trace minerals dominate whole-body requirements of these trace minerals in pigs (Yang 2019; Yang et al. 2019). Thus, the gastrointestinal endogenous fecal losses and true total tract digestibility of trace minerals associated with diets should be measured to help define true digestible trace mineral requirements and to guide swine diet formulation.

It has been well documented that hepatic-biliary secretion is the major route of the gastrointestinal endogenous trace mineral excretion. Dietary and physiological factors that influence the visceral organ turn-over rates, such as the experimental surgical trauma as compiled in Table 20.4, would inevitably trigger changes in trace mineral requirements under field conditions. Therefore, the gastrointestinal endogenous fecal losses and true total tract digestibility of trace minerals associated with commercial swine diets under various challenging conditions should be further investigated to help define true digestible trace mineral requirements and to more accurately guide swine diet formulation and production operation.

## Summary

Global pork production provides about 35% of the overall meat consumed in the world in support of human nutrition and public health. Stringent biosecurity measures, including high quality swine barn physical infrastructure and facilities as well as effective vaccination for disease prevention, are some of the pivotal pre-requisite and practical mitigation strategies to sustain swine herd health, swine nutrition and viable pig production operations.

The occurring and on-going global Covid-19 pandemics have been causing widespread damages and challenges to pork industry and society until successful social and vaccination strategies are fully developed and implemented on the global scale. Pork industry needs to further enhance its animal welfare standards for public acceptance and efficiency of energy and nutrient mass conversion in pig growth and production to improve profit margins and to be socially and economically sustainable.

Comprehensive nutritional and non-dietary strategies are reviewed, summarized and recommended for adoption and further commercialization to tackle the environmental sustainability concerns, including excessive manure nutrient excretion, AMR threat to public health and GHG emissions facing the global pork production industry. Pigs are biologically susceptible to incur gut dysbiosis, thus developing metabolic and septic endotoxemia and inflammation. Modulation and optimization of porcine gut microbiota and microbiome will further enhance nutrition, physiology and sustainable pork production; and with pigs as relevant large animal models, these efforts will also inevitably contribute greatly to development of new strategies for human health management as well as disease prevention and treatment.

## References

Aarnink, A. J. A., and M. W. A. Verstegen. 2007. Nutrition, key factor to reduce environmental load from pig production. Livest. Sci. 109:194–203.

Abdel-Moneim, A. S., and E. M. Abdelwhab. 2020. Evidence for sars-cov-2 infection of animal hosts. Pathogens 9:529. doi: https://doi.org/10.3390/pathogens9070529.

Acosta, J. A., N. K. Gabler, and J. F. Patience. 2020. The effect of lactose and a prototype Lactobacillus acidophilus fermentation product on digestibility, nitrogen balance, and intestinal function of weaned pig. Transl. Anim. Sci. 4:1–14. doi: https://doi.org/10.1093/tas/txaa045.

Adeola, O. 1995. Digestive utilization of minerals by weanling pigs fed copper- and phytase-supplemented diets. Can. J. Anim. Sci. 75:603–610.

Adeola, O., and A. J. Cowieson. 2011. BOARD-INVITED REVIEW: opportunities and challenges in using exogenous enzymes to improve nonruminant animal production. J. Anim. Sci. 89:3189–3218.

Adeola, O., and D. Ragland. 2016. Comparative ileal amino acid digestibility of distillers' grains for growing pigs. Anim. Nutr. 2:262–266. doi: https://doi.org/10.1016/j.aninu.2016.07.008.

Adeola, O., L. G. Young, B. W. McBride, and R. O. Ball. 1989. *In vitro* Na⁺,K⁺-ATPase (EC 3.6.1.3)-dependent respiration and protein synthesis in skeletal muscle of pigs fed at three dietary protein levels. Br. J. Nutr. 61:453–465. doi: https://doi.org/10.1079/bjn19890135.

Ajakaiye, A., M. Z. Fan, T. Archbold, R. R. Hacker, C. W. Forsberg, and J. P. Phillips. 2003. Determination of true phosphorus digestibility and the endogenous loss associated with soybean meal for growing-finishing pigs. J. Anim. Sci. 81:2766–2775.

Anderson, D., and S. Stothers. 1978. Effects of saline water high in sulfates, chlorides and nitrates on the performance of young weanling pigs. J. Anim. Sci. 47:900–907.

Anderson, J., D. Anderson, and J. Murphy. 1994. The effect of water quality on nutrient availability for grower/finisher pigs. Can. J. Anim. Sci. 74:141–148.

Archbold, T., N. A. Burello, W. Wang, D. P. Bureau, and M. Z. Fan. 2017. Characterization of stability of the young porcine intestinal alkaline phosphatase as a candidate exogenous biocatalyst. FASEB J. 31(No. 1 Supplement):644.21.

Ariza, A., O. V. Moroz, E. V. Blagova, J. P. Turkenburg, J. Waterman, S. M. Roberts, J. Vind, C. Sjøholm, S. F. Lassen, L. De Maria, V. Glitsoe, L. K. Skov, and K. S. Wilson. 2013. Degradation of phytate by the 6-phytase from Hafnia alvei: A combined structural and solution study. PLoS One 8:e65062. doi: https://doi.org/10.1371/journal.pone.0065062.

Arumugam, M., J. Raes, E. Pelletier, D. Le Paslier, T. Yamada, D. R. Mende, G. R. Fernandes, J. Tap, T. Bruls, J.-M. Batto, M. Bertalan, N. Borruel, F. Casellas, L. Fernandez, L. Gautier, T. Hansen, M. Hattori, T. Hayashi, M. Kleerebezem, K. Kurokawa, M. Leclerc, F. Levenez, C. Manichanh, H. B. Nielsen, T. Nielsen, N. Pons, J. Poulain, J. Qin, T. Sicheritz-Ponten, S. Tims, D. Torrents, E. Ugarte, E. G. Zoetendal, J. Wang, F. Guarner, O. Pedersen, W. M. de Vos, S. Brunak, J. Doré, MetaHIT Consortium, J. Weissenbach, S. D. Ehrlich, and P. Bork. 2011. Enterotypes of the human gut microbiome. Nature 473:174–180.

Baker, D. H. 2006. Comparative species utilization and toxicity of sulfur amino acids. J. Nutr. 136:1670S–1675S.

Bakker, G. C. M., and R. A. Dekker. 1998. Effect of source and amount of non-starch polysaccharides on the site of excretion of nitrogen in pigs. J. Anim. Sci. 76 (Suppl. 1):665. (Abstr.).

Bala, V., S. Rajagopal, D. P. Kumar, A. D. Nalli, S. Mahavadi, A. J. Sanyal, J. R. Grider, and K. S. Murthy. 2014. Release of GLP-1 and PYY in response to the activation of G protein-coupled bile acid receptor TGR5 is mediated by Epac/PLC-ε pathway and modulated by endogenous $H_2S$. Front. Physiol. 5:420. doi: https://doi.org/10.3389/fphys.2014.00420.

Banwart, W. L., and J. M. Bremner. 1975. Identification of sulfur gases evolved from animal manures. J. Environ. Qual. 4:363–366.

Bar-On YM, A. Flamholz, R. Phillips, and R. Milo. 2020. SARS-CoV-2 (COVID-19) by the numbers. elife 9:e57309. doi: https://doi.org/10.7554/eLife.57309.

Bazinet, R. P., E. G. McMillan, R. Seebaransingh, A. M. Hayes, and S. C. Cunnane. 2003. Whole body ß-oxidation of 18:2ω6 and 18:3ω3 in the pig varies markedly with weaning strategy and dietary 18:3ω3. J. Lipid. Res. 44:314–319.

Beauchemin, K. A., D. Colombatto, D. P. Morgavi, and W. Z. Yang. 2003. Use of exogenous fibrolytic enzymes to improve feed utilization by ruminants. J. Anim. Sci. 81:E37–E47.

Bedford, M. R., and H. Schulze. 1998. Exogenous enzymes for pigs and poultry. Nutr. Res. Rev. 11:91–114. doi:10.1079/NRR19980007.

Bedford, M. R. 2000. Exogenous enzymes in monogastric nutrition-their current value and future benefits. Anim. Feed Sci. Tech. 86:1–13.

Bergen, W. G. 2008. Measuring in vivo intracellular protein degradation rates in animal systems. J. Anim. Sci. 86(E. Suppl):E3–E12.

Bregendahl K., L. Liu, J. P. Cant, H. S. Bayley, B. W. McBride, L. P. Milligan, J. T. Yen, and M. Z. Fan. 2004. Fractional protein synthesis rates measured by an intraperitoneal injection of a flooding dose of l-[ring-$^2H_5$]phenylalanine in pigs. J. Nutr. 134:2722–2728.

Bregendahl, K., X. Yang, L. Liu, J. T. Yen, T. C. Rideout, Y. Shen, G. Werchola, and M. Z. Fan. (2008). Fractional protein synthesis rates are similar when measured by intraperitoneal or intravenous flooding doses of l-[ring-$^2H_5$]phenylalanine in combination with a rapid regimen of sampling in piglets. J. Nutr. 138:1976–1981.

Braun, J., L. Loyal, M. Frentsch, D. Wendisch, P. Georg, F. Kurth, S. Hippenstiel, M. Dingeldey, B. Kruse, F. Fauchere, E. Baysal, M. Mangold, L. Henze, R. Lauster, M. Mall, K. Beyer, J. Roehmel, J. Schmitz, S. Miltenyi, M. A. Mueller, M. Witzenrath, N. Suttorp, F. Kern, U. Reimer, H. Wenschuh, C. Drosten, V. M. Corman, C. Giesecke-Thiel, L. Sander, and A. Thiel. 2020. Presence of SARS-CoV-2-reactive T cells in COVID-19 patients and healthy donors. https://doi.org/10.1101/2020.04.17.20061440.

Bilski, J., A. Mazur-Bialy, D. Wojcik, J. Zahradnik-Bilska, B. Brzozowski, M. Magierowski, T. Mach, K. Magierowska, and T. Brzozowsk. 2017. The role of intestinal alkaline phosphatase in inflammatory disorders of gastrointestinal tract. Mediators Inflamm. doi: https://doi.org/10.1155/2017/9074601.

Boutagy, N. E., R. P. McMillan, M. I. Frisard, and M. W. Hulvera. 2016. Metabolic endotoxemia with obesity: is it real and is it relevant? Biochimie 124:11–20. doi: https://doi.org/10.1016/j.biochi.2015.06.020.

Brugger, D., and W. M. Windisch. 2017. Strategies and challenges to increase the precision in feeding zinc to monogastric livestock. Anim. Nutr. 3:103–108.

Bugg, T. D. H., M. Ahmad, E. M. Hardiman, and R. Singh. 2011. The emerging role for bacteria in lignin degradation and bioproduct formation. Curr. Opin. Biotechnol. 22:394–400.

Burello, N. A., T. Archbold, W. Wang, A. K. Shoveller, and M. Z. Fan. 2017. Piglets are vulnerable to intestinal inflammation and infection because of expression of low-affinity jejunal alkaline phosphatase activities. FASEB J. 31(Suppl. 1):315.7.

Burello, N. A., H. Zhang, W. Wang, T. Archbold, R. Tsao, and M. Z. Fan. 2019. Comparative characterization of intestinal alkaline phosphatase Kinetics in young piglets and human Caco-2 cells. J. Anim. Sci. 97(Suppl. 3):282–283.

Burkhalter, T. M., N. R. Merchen, L. L. Bauer, S. M. Murray, A. R. Patil, J. L. Brent, Jr., and G. C. Fahey, Jr. 2001. The ratio of insoluble to soluble fiber components in soybean hulls affects ileal and total-tract nutrient digestibilities and fecal characteristics of dogs. J. Nutr. 131:1978–1985.

Burrough, E. R., B. L. Arruda, J. F. Patience, and P. J. Plummer. 2015. Alterations in the colonic microbiota of pigs associated with feeding distillers dried grains with solubles. PLoS ONE 10: e0141337. doi:10.1371/journal.pone.0141337.

Calder, P. C. 2012. Long-chain fatty acids and inflammation. Proc. Nutr. Soc. 28:1–6.

Call, L., T. Molina, B. Stoll, G. Guthrie, S. Chacko, J. Plat, J. Robinson, S. Lin, C. Vonderohe, M. Mohammad, D. Kunichoff, S. Cruz, P. Lau, M. Premkumar, J. Nielsen, Z. Fang, O. Olutoye, T. Thymann, R. Britton, P. Sangild, and D. Burrin. 2020. Parenteral lipids shape gut bile acid pools and microbiota profiles in the prevention of cholestasis in preterm pigs. J. Lipid Res. 61:1038–1051. doi: https://doi.org/10.1194/jlr.RA120000652.

Calloway, D. H., and S. Margen. 1971. Variation in endogenous nitrogen excretion and dietary nitrogen utilization as determinants of human protein requirement. J. Nutr. 101:205–216.

Campbell, E. L., C. F. MacManus, D. J. Kominsky, S. Keely, L. E. Glover, B. E. Bowers, M. Scully, W. J. Bruyninckx, and S. P. Colgan. 2010. Resolvin E1-induced intestinal alkaline phosphatase promotes resolution of inflammation through LPS detoxification. Proc. Natl. Acad. Sci. USA. 107:14298–14303.

Campos, P. H. R. F. Labussière, J., Hernández-García, S. Dubois, D. R., and J. Noblet. 2014. Effects of ambient temperature on energy and nitrogen utilization in lipopolysaccharide-challenged growing pigs. J. Anim. Sci. 92:4909–4920.

Cani, P. D., J. Amar, M. A. Iglesias, M. Poggi, C. Knauf, D. Bastelica, A. M. Neyrinck, F. Fava, K. M. Tuohy, C. Chabo, A. Waget, E. Delmee, B. Cousin, T. Sulpice, B. Chamontin, J. Ferrieres, J.-F. Tanti, G. R. Gibson, L. Casteilla, N. M. Delzenne, M. C. Alessi, and R. Burcelin. 2007. Metabolic endotoxemia initiates obesity and insulin resistance. Diabetes 56:1761–1772.

Cani, P. D., R. Bibiloni, C. Knauf, A. Waget, A. M. Neyrinck, N. M. Delzenne, and R. Burcelin. 2008. Changes in gut microbiota control metabolic endotoxemia-induced inflammation in high-fat diet–induced obesity and diabetes in mice. Diabetes 57:1470–1481. doi: https://doi.org/10.2337/db07–1403.

Cani, P. D., S. Possemiers, T. Van de Wiele, Y. Guiot, A. Everard, O. Rottier, L. Geurts, D. Naslain, A. Neyrinck, D. M. Lambert, G. G. Muccioli, and N. M. Delzenne. 2009. Changes in gut microbiota control inflammation in obese mice through a mechanism involving GLP-2-driven improvement of gut permeability. Gut 58:1091–1103. doi: https://doi.org/10.1136/gut.2008.165886.

Cani, P. D., M. Osto, L. Geurts, and A. Everard. 2012. Involvement of gut microbiota in the development of low-grade inflammation and type 2 diabetes associated with obesity. Gut Microbes 3:279–288. doi: https://doi.org/10.4161/gmic.19625.

Canh, T. T., A. J. A. Aarnink, M. W. A. Verstegen, and J. W. Schrama. 1998a. Influence of dietary factors on the pH and ammonia emission of slurry from growing-finishing pigs. J. Anim. Sci. 76:1123–1130.

Canh, T. T., A. L. Sutton, A. J. A. Aarnink, M. W. A. Verstegen, J. W. Schrama, and G. C. M. Bakker. 1998b. Dietary carbohydrates alter the fecal composition and pH and the ammonia emission from slurry of growing pigs. J. Anim. Sci. 76:1887–1895.

Cao, Y., Su B., Guo X., Sun, W., Y. Deng, L. Bao, Q. Zhu, X. Zhang, Y. Zheng, C. Geng, X. Chai, R. He, X. Li, Q. Lv, H. Zhu, W. Deng, Y. Xu, Y. Wang, L. Qiao, Y. Tan, L. Song, G. Wang, X. Du, N. Gao, J. Liu, J. Xiao, X. D. Su, Z. Du, Y. Feng, C. N. Qin, C. Qin, R. Jin, and X. S. Xie. 2020. Potent neutralizing antibodies against SARS-CoV-2 identified by high-throughput single-cell sequencing of convalescent patients' B cells. Cell. doi: https://doi.org/10.1016/j.cell.2020.05.025.

Carbonero, F., A. C. Benefiel, A. H. Alizadeh-Ghamsari, and H. R. Gaskins. 2012. Microbial pathways in colonic sulfur metabolism and links with health and disease. Front. Physiol. 3:448. doi: https://doi.org/10.3389/fphys.2012.00448.

Cargo-Froom, C., G. Pfeuti, M. Z. Fan, and A. K. Shoveller. (2019). Apparent and true digestibility of macro and micro nutrients in animal and vegetable ingredient based adult maintenance dog food. J. Anim. Sci. 97:1010–1019.

Carroll, A. E., and T. S. Doherty. 2019. Meat consumption and health: food for thought. Ann. Intern. Med. 171:767–768. doi: https://doi.org/10.7326/M19-2620.

CAST (Council for Agricultural Science and Technology). 1974. Quality of Water for Livestock. Vol. I. Report No. 26.

Cemin, H. S., C. M. Vier, M. D. Tokach, S. S. Dritz, K. J. Touchette, J. C. Woodworth, J. M. DeRouchey, and R. D. Goodband 2018. Effects of standardized ileal digestible histidine to lysine ratio on growth performance of 7- to 11-kg nursery pigs. J. Anim. Sci. 96:4713–4722. doi: https://doi.org/10.1093/jas/sky319.

CFIA. 2016. Consultation summary on maximum nutrient values in swine feeds. Canadian Food Inspection Agency, Ottawa, ON, Canada.

Chimerel, C., E. Emery, D. K. Summers, U. Keyser, F. M. Gribble, and F. Reimann. 2014. Bacterial metabolite indole modulates incretin secretion from intestinal enteroendocrine L cells. Cell Rep. 9:1202–1208. doi: https://doi.org/10.1016/j.celrep.2014.10.032.

Chen, J., M. Wang, T. L. Archbold, W. Wang, W. Fan, X. Yin, L. Cheng, and M. Z. Fan. 2020. Ileal terminal starch hydrolytic activity is increased in association with improved growth and feed efficiency in weaning pigs fed a therapeutic multi-antimicrobial-supplemented diet. J. Anim. Sci. 98(Suppl. 4):178. (Abstr.).

Cheng, K. J. and J. W. Costerton. 1977. Alkaline phosphatase activity of rumen bacteria. Appl. Environ. Micro. 34:586–590.

Chesson, A. 1987. Supplementary enzymes to improve the utilization of pig and poultry diets. In: W. Haresign, and D. J. A. Cole, editors, Recent advances in animal nutrition. Butterworths, London, UK. p. 71–89.

Cheung, K. S., I. F. N. Hung, P. P. Y. Chan, K. C. Lung, E. Tso, R. Liu, Y. Y. Ng, M. Y. Chu, T. W. H. Chung, A. R. Tam, C. C. Y. Yip, K.-H. Leung, A. Y.-F. Fung, R. R. Zhang, Y. Lin, H. M. Cheng, A. J. X. Zhang, K. K. W. To, K.-H. Chan, K.-Y. Yuen, and W. K. Leung. 2020. Gastrointestinal manifestations of sars-cov-2 infection and virus load in fecal samples from a Hong Kong cohort: systematic review and meta-analysis. Gastroenterology 159:81–95.

Chiba, I. L. 2013. Sustainable swine nutrition. Wiley-Blackwell, John Wiley & Sons, Inc., Hoboken, NJ.

Chung, T. K., and D. H. Baker. 1992. Ideal amino acid pattern for 10-kilogram pigs. J. Anim. Sci. 70:3102–3111.

Clark, O. G., S. Moehn, I. Edeogu, J. Price, and J. Leonard. 2005. Manipulation of dietary protein and nonstarch polysaccharide to control swine manure emissions. J. Environ. Qual. 34:1461–1466.

Columbus, D. A., J. Steinhoff-Wagner, A. Suryawan, H. V. Nguyen, A. H. Hernandez-Garcia, M. L. Fiorotto, and T. A. Davis. 2015. Impact of prolonged leucine supplementation on protein synthesis and lean growth in neonatal pigs. Am. J. Physiol. 309:E601–E610. doi: https://doi.org/10.1152/ajpendo.00089.2015.

Costantini, L., R. Molinari, B. Farinon, and N. Merendino. 2017. Impact of omega-3 fatty acids on the gut microbiota. Int. J. Mol. Sci. 18:2645. doi: https://doi.org/10.3390/ijms18122645.

Crampton, E. W., and L. A. Maynard. 1938. The relation of cellulose and lignin content to the nutritive value of animal feeds. J. Nutr. 15:383–395. doi: https://doi.org/10.1093/jn/15.4.383.

Crespo-Piazuelo, D., J. Estellé, M. Revilla, L. Criado-Mesas, Y. Ramayo-Caldas, C. Óvilo, A. I. Fernández, M. Ballester, and J. M. Folch. 2018. Characterization of bacterial microbiota compositions along the intestinal tract in pigs and their interactions and functions. Sci. Rep. 8: 12727. doi.org/10.1038/s41598-018-30932-6.

Crespo-Piazuelo, D. L. J. Migura-Garcia L. Estellé M. Criado-Mesas A. Revilla M. Castelló J. M. Muñoz A. I. García-Casco M. B. Fernández, and J. M. Folch. 2019. Association between the pig genome and its gut microbiota composition. Sci. Rep. 9:8791. doi: https://doi.org/10.1038/s41598-019-45066-6.

Cromwell, G. L., L. W. Turner, R. S. Gates, J. L. Taraba, M. D. Lindemann, S. L. Traylor, W. A. Dozier III, and H. J. Monegue. 1999. Manipulation of swine diets to reduce gaseous emissions from manure that contribute to odor. J. Anim. Sci. 77 (Suppl. 1):69. (Abstr.).

Crosbie, M., C. Zhu, A. K. Shoveller, and L.-A. Huber. 2020. Standardized ileal digestible amino acids and net energy contents in full fat and defatted black soldier fly larvae meals (Hermetia illucens) fed to growing pigs. Transl. Anim. Sci. 4:1–10. doi: https://doi.org/10.1093/tas/txaa104.

Cryan, J. F., K. J. O'Riordan, C. S. M. Cowan, K. V. Sandhu, T. F. S. Bastiaanssen, M. Boehme, M. G. Codagnone, S. Cussotto, C. Fulling, A. V. Golubeva, K. E. Guzzetta, M. Jaggar, C. M. Long-Smith, J. M. Lyte, J. A. Martin, A. Molinero-Perez, G. Moloney, E. Morelli, E. Morillas, R. O'Connor, J. S. Cruz-Pereira, V. L. Peterson, K. Rea, N. L. Ritz, E. Sherwin, S. Spichak, E. M. Teichman, M. van de Wouw, A. P. Ventura-Silva, S. E. Wallace-Fitzsimons, N. Hyland, G. Clarke, and T. G. Dinan. 2019. The Microbiota-Gut-Brain Axis. Physiol. Rev. 99:1877–2013. doi: 10.1152/physrev.00018.2018.

Datar, I. 2010. Possibilities for an in vitro meat production system. Innov. Food Sci. Emerg. Technol. 11:13–22. doi: https://doi.org/10.1016/j.ifset.2009.10.007.

Davis, T. A., A. Suryawan, R. A. Orellana, H. V. Nguyen, and M. L. Fiorotto. 2008. Postnatal ontogeny of skeletal muscle protein synthesis in pigs. J. Anim. Sci. 86(E. Suppl):E13–E18.

de Ridder, K., C. L. Levesque, J. K. Htoo, and C. F. M. de Lange. 2012. Immune system stimulation reduces the efficiency of tryptophan utilization for body protein deposition in growing pigs. J. Anim. Sci. 90:3485–3491. doi: https://doi.org/10.2527/jas.2011-4830.

DeCamp, S. A., B. E. Hill, S. L. Hankins, D. C. Kendall, B. T. Richert, A. L. Sutton, D. T. Kelly, M. L. Cobb, D. W. Bundy, and W. J. Powers. 2001. Effects of soybean hulls in a commercial diet on pig performance, manure composition, and selected air quality parameters in swine facilities. J. Anim. Sci. 79 (Suppl. 1):252. (Abstr.).

Dersjant-Li, Y., B. Villca, V. Sewalt, A. de Kreij, L. Marchal, D. E. Velayudhan, R. A. Sorg, T. Christensen, R. Mejldal, I. Nikolaev, S. Pricelius, H. S. Kim, S. Haaning, J. F. Sørensen, and R. Lizardo 2020. Functionality of a next generation biosynthetic bacterial 6-phytase in enhancing phosphorus availability to weaned piglets fed a corn-soybean meal-based diet without added inorganic phosphate. Anim. Nutr. 6:24–30. doi: https://doi.org/10.1016/j.aninu.2019.11.003.

Drummond, J. G., S. E. Curtis, and J. Simon. 1978. Effects of atmospheric ammonia on pulmonary bacterial clearance in the young pig. Am. J. Vet. Res. 39:211–212.

Drummond, J. G., S. E. Curtis, J. Simon, and H. W. Norton. 1980. Effects of aerial ammonia on growth and health of young pigs, J. Anim. Sci. 50:1085–1091.

Duarte, M. E., F. X. Zhou, M. W. Dutra Jr., and S. W. Kim. 2019. Dietary supplementation of xylanase and protease on growth performance, digesta viscosity, nutrient digestibility, immune and oxidative stress status, and gut health of newly weaned pigs. Anim. Nutr. 5:351–358. doi: https://doi.org/10.1016/j.aninu.2019.04.005.

du Vigneaud, V. 1952. A trial of research in sulfur chemistry and metabolism. Cornel Univ. Press, Ithaca, NY.

Economopoulos, K. P., N. L. Ward, C. D. Phillips, A. Teshager, P. Patel, M. M. Mohamed, S. Hakimian, S. B. Cox, R. Ahmed, O. Moaven, K. Kaliannan, S. N. Alam, J. F. Haller, A. M. Goldstein, A. K. Bhan, M. S. Malo, and R. A. Hodin. 2016. Prevention of antibiotic-associated metabolic syndrome in mice by intestinal alkaline phosphatase. Diabetes Obes. Metab. 18:519–527. doi: https://doi.org/10.1111/dom.12645.

Enkirch, T., and V. von Messling. 2015. Ferret models of viral pathogenesis. Virology 479–480:259–70.

Eriksson, H. J., W. R. Verweij, K. Poelstra, W. L. Hinrichs, G. J. de Jong, G. W. Somsen, and H. W. Frijlink. 2003. Investigations into the stabilisation of drugs by sugar glasses: II. Delivery of an inulin-stabilised alkaline phosphatase in the intestinal lumen via the oral route. Int. J. Pharm. 257:273–281. doi: https://doi.org/10.1016/s0378-5173(03)00152-2.

Eriksen, J., A. Adamsen, J. Norgaard, H. Poulsen, B. Jensen, and S. Petersen, S. 2010. Emissions of sulfur-containing odorants, ammonia, and methane from pig slurry: Effects of dietary methionine and benzoic acid. J. Environ. Qual. 39:1097–1107.

Estaki, M., D. DeCoffe, and D. L. Gibson. 2014. Interplay between intestinal alkaline phosphatase, diet, gut microbes and immunity. World J. Gastroenterol. 20:15650–15656.

Fan, M. Z. 2013. Swine nutrition and environment. In: L. I. Chiba, editor, Sustainable swine nutrition, Wiley-Blackwell, John Wiley & Sons, Inc., Hoboken, NJ. p. 365–411.

Fan, M. Z., O. Adeola, and E. K. Asem. 1999. Characterization of brush border membrane-bound alkaline phosphatase activity in different segments of the porcine small intestine. J. Nutr. Biochem. 10:299–305.

Fan, M. Z., S. W. Kim, T. J. Applegate, and M. Cervantes. 2008. Nonruminant nutrition symposium: understanding protein synthesis and degradation and their pathway regulations. J. Anim. Sci. 86 (E. Suppl.): E1–E2.

Fan, M. Z., and T. Archbold. 2012. Effects of dietary true digestible calcium to phosphorus ratio on growth performance and efficiency of calcium and phosphorus utilization in growing pigs fed corn and soybean meal-based diets. J. Anim. Sci. 90:254–256.

Fan, M. Z., and T. Archbold. 2015. Novel and disruptive biological strategies for resolving gut health challenges in monogastric food animal production. Anim. Nutr. 1:138–143.

Fan, M. Z., T. Archbold, W. C. Sauer, D. Lackeyram, T. Rideout, Y. Gao, C. F. M. de Lange, and R. R. Hacker. 2001a. Novel methodology allows measurement of true phosphorus digestibility and the gastrointestinal endogenous phosphorus outputs in studies with pigs. J. Nutr. 131:2388–2396.

Fan, M. Z., L. I. Chiba, P. D. Matzat, X. Yang, Y. L. Yin, Y. Mine, and H. Stein. 2006. Measuring synthesis rates of nitrogen-containing polymers by using stable isotope tracers. J. Anim. Sci. 84(E. Suppl):E79–E93.

Fan, M. Z., J. Matthews, N. M. P. Etienne, B. Stoll, D. Lackeyram, and D. G. Burrin. 2004. Expression of brush border L-glutamate transporters in epithelial cells along the crypt-villus axis in the neonatal pig. Am. J. Physiol. 287:G385–G398.

Fan, M. Z., and W. C. Sauer. 2002. Determination of true ileal amino acid digestibility and the endogenous amino acid outputs associated with barley samples for growing-finishing pigs by the regression analysis technique. J. Anim. Sci. 80:1593–1605.

Fan, M. Z., W. C. Sauer, and M. I. McBurney. 1995. Estimation by regression analysis of endogenous amino acid levels in digesta collected from the distal ileum of pigs. J. Anim. Sci. 73:2319–2328.

Fan, W., X. Yin, W. Wang, J. Zhang, L.-A. Huber, N. Karrow, Y. Mine, and M. Z. Fan. 2020. Dietary therapeutic aureomycin improves ileal lysine digestibility independent of gut aminopeptidase n and cationic aa transporter-1 expression in weaning pigs fed crystalline lysine supplemented low-crude protein diets. Global Animal Nutrition Summit (GANS), August 11–15, 2020, Guelph, ON, Canada. p. 53.

Fang, W., Z. Fang, P. Zhou, F. Chang, Y. Hong, X. Zhang, H. Peng, and Y. Xiao. 2012. Evidence for lignin oxidation by the giant panda fecal microbiome. PLOS ONE 7:e50312. doi: https://doi.org/10.1371/journal.pone.0050312.

Fawley, J. and D. Gourlay. 2016. Intestinal alkaline phosphatase: a summary of its role in clinical disease. J. Surg. Res. 202:225–234.

FDA. 2015. Endotoxin testing recommendations for single-use intraocular ophthalmic devices. Food and Drug Administration, Washington, DC.

Frese, S. A., K. Parker, C. C. Calvert, and D. A. Mills. 2015. Diet shapes the gut microbiome of pigs during nursing and weaning. Microbiome 3:28. doi: https://doi.org/10.1186/s40168-015-0091-8.

Forsberg, C. W., R. G. Meidinger, D. Murray, N. D. Keirstead, M. A. Hayes, M. Z. Fan, J. Ganeshapillai, M. A. Monteiro, S. P. Golovan, and J. P. Phillips. 2014. Phytase properties and locations in tissues of transgenic pigs secreting phytase in the saliva. J. Anim. Sci. 92:3375–3387.

Fouhse, J. M., M. G. Gänzle, P. R. Regmi, T. A. van Kempen, and R. T. Zijlstra. 2015. High amylose starch with low in vitro digestibility stimulates hindgut fermentation and has a bifidogenic effect in weaned pigs. J. Nutr. 145:2464–2470. doi: https://doi.org/10.3945/jn.115.214353.

Fuke, N., N. Nagata, H. Suganuma, and T. Ota. 2019. Regulation of gut microbiota and metabolic endotoxemia with dietary factors. Nutrients 11:2277. doi: https://doi.org/10.3390/nu11102277.

Gabler, N. K. and M. E. Spurlock. 2008. Integrating the immune system with the regulation of growth and efficiency. J. Anim. Sci. 86:E64–E74. doi: https://doi.org/10.2527/jas.2007-0466.

Galassi, G., G. M. Crovetto, L. Rapetti, and A. Tamburini. 2004. Energy and nitrogen balance in heavy pigs fed different fibre sources. Livest. Prod. Sci. 85:253–262.

GenScript. 2019. ToxinSensor™ chromogenic LAL endotoxin assay kit. Genscript Biotech Corporation, Piscataway, NJ.

Gerber, P.J., Steinfeld, H., Henderson, B., Mottet, A., Opio, C., Dijkman, J., Falcucci, A. and Tempio, G. 2013. Tackling climate change through livestock – A global assessment of emissions and mitigation opportunities. Food and Agriculture Organization of the United Nations (FAO), Rome.

González-Vega, J. C., and H. H. Stein. 2014. Invited review – calcium digestibility and metabolism in pigs. Asian-Australas J. Anim. Sci. 27: 1–9. doi:10.5713/ajas.2014.r.01.

Graham, H., K. Hesselman, and J. Lindberg. 1986. The influence of wheat bran and sugar-beet pulp on the digestibility of dietary components in a cereal-based pig diet. J. Nutr. 116:242–251.

Gralapp, A. K., W. J. Powers, M. A. Faust, and D. S. Bundy. 2002. Effects of dietary ingredients on manure characteristics and odorous emissions from swine. J. Anim. Sci. 80:1512–1519.

Grifoni, A., D. Weiskopf, S. I. Ramirez, J. Mateus, J. M. Dan, C. R. Moderbacher, S. A. Rawlings, A. Sutherland, L. Premkumar, R. S. Jadi, D. Marrama, A. M. de Silva, A. Frazier, A. F. Carlin, J. A. Greenbaum, B. Peters, F. Krammer, D. M. Smith, S. Crotty, and A. Sette. 2020. Targets of T cell responses to SARS-CoV-2 coronavirus in humans with COVID-19 disease and unexposed individuals. Cell. doi: https://doi.org/10.1016/j.cell.2020.05.015.

Hancock, J. D., and K. C. Behnke. 2001. Use of ingredient and diet processing technologies (grinding, mixing, pelleting, and extruding) to produce quality feeds for pigs. In: A. J. Lewis, and L. L. Southern, editors, Swine nutrition. 2nd ed. CRC Press, Boca Raton, FL. p. 469–497.

Hayes, E. T., A. B. G. Leek, T. P. Curran, V. A. Dodd, O. T. Carton, V. E. Beattie, and J. V. O'Doherty. 2004. The influence of diet crude protein level on odour and ammonia emissions from finishing pig houses. Bioresource Tech. 91:309–315.

Helmink, A., M. A. Beth, W. Khan, A. Hermann, V. Gopalakrishnan, and J. A. Wargo. 2019. The microbiome, cancer, and cancer therapy. Nat. Med. 25:377–388. doi: https://doi.org/10.1038/s41591-019-0377-7.

Hess, M., A. Sczyrba, R. Egan, T.-W. Kim, H. Chokhawala, G. Schroth, S. Luo, D. S. Clark, F. Chen, T. Zhang, R I. Mackie, L. A. Pennacchio, S. G. Tringe, A. Vise, T. Woyke, Z. Wang, and E. M. Rubin. 2011. Metagenomic discovery of biomass-degrading genes and genomes from cow rumen. Science 331:463–467.

Hobbs, P. J., B. F. Pain, R. M. Kay, and P. A. Lee. 1996. Reduction of odorous compounds in fresh pig slurry by dietary control of crude protein. J. Sci. Food Agric. 71:508–514.

Holscher, H. D. 2020. Gut microbes: Nuts about fatty acids. J. Nutr. 150:652–653. doi: https://doi.org/10.1093/jn/nxaa045.

Honda, K., and D. R. Littman. 2016. The microbiota in adaptive immune homeostasis and disease. Nature 535:75–84. doi: https://doi.org/10.1038/nature18848.

Huber, L. A., S. Hooda, R. E. Fisher-Heffernan, N. A. Karrow, and C. F. M. de Lange. 2018. Effect of reducing the ratio of omega-6-to-omega-3 fatty acids in diets of low protein quality on nursery pig growth performance and immune response. J. Anim. Sci. 96:4348–4359. doi: https://doi.org/10.1093/jas/sky296.

Hübner, A., B. Petersen, G. M. Keil, H. Niemann, T. C. Mettenleiter, and W. Fuchs. 2018. Efficient inhibition of African swine fever virus replication by CRISPR/Cas9 targeting of the viral p30 gene (CP204L). Sci. Rep. 8:1449.

Huang, Q., X. Liu, G. Zhao, T. Hu, and Y. Wang. 2017. Potential and challenges of tannins as an alternative to in-feed antibiotics for farm animal production. Anim. Nutr. 4:137–150. doi: https://doi.org/10.1016/j.aninu.2017.09.004.

Huntley, N. F., C. M. Nyachoti, and J. F. Patience. 2018. Lipopolysaccharide immune stimulation but not β-mannanase supplementation affects maintenance energy requirements in young weaned pigs. J. Anim. Sci. Biotechnol. 9:47. doi: https://doi.org/10.1186/s40104-018-0264-y.

Izquierdo, O. A., C. M. Parsons, and D. H. Baker, 1988. Bioavailability of lysine in L-lysine-HCL. J. Anim. Sci. 66:2590–2597.

Jakobsen, G. V., B. B. Jensen, K. E. Bach Knudsen, and N. Canibe. 2015. Impact of fermentation and addition of non-starch polysaccharide-degrading enzymes on microbial population and on digestibility of dried distillers grains with solubles in pigs. Livest. Sci. 178: 216–22. doi.org/10.1016/j.livsci.2015.05.028.

Jayaraman, B., J. Htoo, and C. M. Nyachoti. 2015. Effects of dietary threonine_lysine ratioes and sanitary conditions on performance, plasma urea nitrogen, plasma-free threonine and lysine of weaned pigs. Anim. Nutr. 1:283–288. doi: https://doi.org/10.1016/j.aninu.2015.09.003.

Jensen, M. T., R. P. Cox, and B. B. Jensen. 1995. Microbial production of skatole in the hindgut of pigs given different diets and its relation to skatole deposition in backfat. Anim. Sci. 61:293–304.

Jeon, S. M., A. Hosseindoust, Y. H. Choi, M. J. Kim, K. Y. Kim, J. H. Lee, D. Yong, K. Beo, G. Kim, and B. J. Chaea. 2019. Comparative standardized ileal amino acid digestibility and metabolizable energy contents of main feed ingredients for growing pigs when adding dietary beta-mannanase. Anim. Nutr. 5:359–365. doi: https://doi.org/10.1016/j.aninu.2019.07.001.

Jha, R. and J. D. Berrocoso. 2015. Review: Dietary fiber utilization and its effects on physiological functions and gut health of swine. Animal 9:1441–1452.

Jha, R., and J. F. D. Berrocoso. 2016. Dietary fiber and protein fermentation in the intestine of swine and their interactive effects on gut health and on the environment: A review. Anim. Feed Sci. Tech. 212:18–26.

Ji, F., D. P. Casper, P. K. Brown, D. A. Spangler, K. D. Haydon, and J. E. Pettigrew. 2008. Effects of dietary supplementation of an enzyme blend on the ileal and fecal digestibility of nutrients in growing pigs. J. Anim. Sci. 86:1533–1543.

Jørgensen, H., V. M. Gabert, M. S. Hedemann, and S. K. Jensen. 2000. Digestion of fat does not differ in growing pigs fed diets containing fish oil, rapeseed oil or coconut oil1. J. Nutr. 130:852–857.

Jung, H. G., D. R. Mertens, and A. J. Payne. 1997. Correlation of acid detergent lignin and Klason lignin with digestibility of forage dry matter and neutral detergent fiber. J. Dairy Sci. 80:1622–1628. doi: https://doi.org/10.3168/jds.S0022-0302(97)76093-4.

Kadosh, E., I. Snir-Alkalay, A. Venkatachalam, S. May, A. Lasry, E. Elyada, A. Zinger, M. Shaham, G. Vaalani, M. Mernberger, T. Stiewe, E. Pikarsky, M. Oren, and Y. Ben-Neriah. 2020. The gut microbiome switches mutant p53 from tumour-suppressive to oncogenic. Nature 586:133–138. doi.org/10.1038/s41586-020-2541-0.

Kahindi, R. A., Regassaa J., Htoob, and M. Nyachoti. 2017. Optimal sulfur amino acid to lysine ratio for post weaning piglets reared under clean or unclean sanitary conditions. Anim. Nutr. 3:380–385. doi: https://doi.org/10.1016/j.aninu.2017.08.004.

Kaliannan, K., S. R. Hamarneh, K. P. Economopoulos, N. S. Alam, O. Moaven, P. Patel, N. S. Maloa, M. Raya, S. M. Abtahia, N. Muhammada, A. Raychowdhurya, A. Teshagera, M. M. R. Mohameda, A. K. Mossa, R. Ahmeda, S. Hakimiana, S. Narisawab, J. L. Millánb, E. Hohmannc, H. S. Warrenc, A. K. Bhand, M. S. Maloa, and R. A. Hodina. 2013. Intestinal alkaline phosphatase prevents metabolic syndrome in mice. Proc. Natl. Acad. Sci. USA 110:7003–7008.

Kaliannan, K., B. Wang, X. Y. Li, K. J. Kim, and J. X. Kang. 2015. A host-microbiome interaction mediates the opposing effects of omega-6 and omega-3 fatty acids on metabolic endotoxemia. Sci. Rep. 5:1–17. doi: https://doi.org/10.1038/srep11276/.

Krishna Bahadur, K. C., G. M. Dias, A. Veeramani, C. J Swanton, D. Fraser, D. Steinke, E. Lee, H. Wittman, J. M. Farber, K. Dunfield, K. McCann, M. Anand, M. Campbell, N. Rooney, N. E. Raine, R. Van Acker, R. Hanner, S. Pascoal, S. Sharif, T. G. Benton, and E. D. G. Fraser. 2018. When too much isn't enough: Does current food production meet global nutritional needs? PLOS One 13:e0205683.

Kerr, B. J., and R. A. Easter. 1995. Effect of feeding reduced protein, amino acid-supplemented diets on nitrogen and energy balance in grower pigs. J. Anim. Sci. 73:3000–3008.

Kerr, B. J., L. L. Southern, T. D. Bidner, K. G. Friesen, and R. A. Easter. 2003. Influence of dietary protein level, amino acid supplementation, and dietary energy levels on growing-finishing pig performance and carcass composition. J. Anim. Sci. 81:3075–3087. doi:10.2527/2003.81123075x.

Kerr, B. J., T. A. Kellner, and G. C. Shurson. 2015. Characteristics of lipids and their feeding value in swine diets. J. Anim. Sci. Biotechnol. 6:30. doi: https://doi.org/10.1186/s40104-015-0028-x.

Kerr, B. J., and G. C. Shurson. 2013. Strategies to improve fiber utilization in swine. J. Anim. Sci. *Biotechnol.* 4:11

Kerr, B. J., S. L. Trabue, and D. S. Andersen. 2017. Narasin effects on energy, nutrient, and fiber digestibility in corn-soybean meal or corn-soybean meal-dried distillers' grains with solubles diets fed to 16-, 92-, and 141-kg pigs. J. Anim. Sci. 95:4030–4036.

Kerr, B. J., S. L. Trabue, M. B. van Weelden, D. S. Andersen, and L. M. Pepple. 2018. Impact of narasin on manure composition, microbial ecology, and gas emissions from finishing pigs fed either a corn-soybean meal or a corn-soybean meal-dried distillers grains with solubles diets. J. Anim. Sci. 95:4030–4036.

Kerr, B. J., T. E. Weber, and G. C. Shurson. 2013. Evaluation of commercially available enzymes, probiotics, or yeast on apparent nutrient digestion and growth in nursery and finishing pigs fed diets containing corn-dried distillers grains with solubles. Prof. Anim. Sci. 29:508–517.

Kerr, B. J., C. J. Ziemer, S. L. Trabue, J. D. Crouse, and T. B. Parkin. 2006. Manure composition of swine as affected by dietary protein and cellulose concentrations. J. Anim. Sci. 84:1584–1592.

Kleiber, M. 1961. The fire of life, an introduction to animal energetic. R. E. Krieger Publishing Co., Malabar, FL.

Kornegay, E. T., and A. F. Harper. 1997. Environmental nutrition: Nutrient management strategies to reduce nutrient excretion of swine. Prof. Anim. Sci. 13:99–111.

Krause, D. O., S. E. Denman, R. I. Mackie, A. L Rae, G. T. Attwood, and C. S. McSweeney. 2003. Opportunities to improve fiber degradation in the rumen: microbiology, ecology, and genomics. FEMS Microbiol. Rev. 27:663–693.

Kreuzer, M., A. Machmuller, M. M. Gerdemann, H. Hanneken, and M. Wittmann. 1998. Reduction of gaseous nitrogen loss from pig manure using feeds rich in easily-fermentable non-starch polysaccharides. Anim. Feed Sci. Technol. 73:1–19.

Kidder, D. E., and M. J. Manners 1978. Digestion in the pig. Scientechnica Bristol, Kingston Press, Bath, UK.

Kiffer-Moreira, T., C. R. Sheen, K. C. Gasque, M. Bolean, P. Ciancaglini, A. van Elsas, M. F. Hoylaerts, and J. L. Millan. 2014. Catalytic signature of a heat-stable, chimeric human alkaline phosphatase with therapeutic potential. PLOS One 9:e89374.

Kirchgessner, M., M. Kreuzer, H. Muller, and W. Windisch. 1991. Release of methane and carbon dioxide by the pig. Agribiol. Res. 44:103–113.

Kirchgessner, M., W. Windisch, and F. X. Roth. 1994. The efficiency of nitrogen conversion in animal nutrition. Nova Acta Leopoldina 70:393–412.

Kovácsné Gaál, K., O. Sáfár, L. Gulyás, and P. Stadler. 2004. Magnesium in animal nutrition. J. Am. Coll. Nutr. 23:754S–757S. doi: https://doi.org/10.1080/07315724.2004.10719423.

Lackeyram, D., C. Yang, T. Archbold, K. C. Swanson, and M. Z. Fan 2010. Early weaning reduces small intestinal alkaline phosphatase expression in pigs. J. Nutr. 140:461–8.

Lallès, J. P. 2010. Intestinal alkaline phosphatase: multiple biological roles in maintenance of intestinal homeostasis and modulation by diet. Nutr. Rev. 68:323–332.

Lallès, J. P. 2014. Luminal ATP: The missing link between intestinal alkaline phosphatase, the gut microbiota, and inflammation? Am. J. Physiol. 306:G824–G825.

Lai, L., J. X. Kang, R. Li, J. Wang, W. T. Witt, H. Y. Yong, Y. Hao, D. M. Wax, C. N. Murphy, A. Rieke, M. Samuel, M. L. Linville, S. W. Korte, R. W. Evans, T. E. Starz, R. S. Prather, and Y. Dai. 2006. Generation of cloned transgenic pigs rich in omega-3 fatty acids. Nat Biotechnol. 24:435–436. doi: https://doi.org/10.1038/nbt1198.

Lam, T. T., M. H. Shum, H. C. Zhu, Y. G. Tong, X. B. Ni, Y. S. Liao, W. Wei, W. Y. Cheung, W. J. Li, L. F. Li, G. M. Leung, E. C. Holmes, Y. L. Hu, and Y. Guan. 2020. Identifying SARS-CoV-2 related coronaviruses in Malayan pangolins. Nature 583:282–285. doi: https://doi.org/10.1038/s41586-020-2169-0.

Lan, J., J. Ge, J. Yu, S. Shan, H. Zhou, S. Fan, Q. Zhang, X. Shi, Q. Wang, L. Zhang, and Z. Wang. 2020. Structure of the SARS-CoV-2 spike receptor-binding domain bound to the ACE2 receptor. Nature 581:215–220. doi: https://doi.org/10.1038/s41586-020-2180-5.

Lane, A. N., and T. W.-M. Fan. 2015. Regulation of mammalian nucleotide metabolism and biosynthesis. Nucleic Acids Res. 43:2466–2485. doi: https://doi.org/10.1093/nar/gkv047.

Larsen, J. M. 2017. The immune response to *Prevotella* bacteria in chronic inflammatory disease. Immunology 151:363–374.

Latimier, P., J. Y. Dourmad, A. Corlouer, J. Chauvel, J. le Pan, M. Gautier, and D. Lesaicherre. 1993. Effect of three protein feeding strategies, for growing-finishing pigs, on growth performance and nitrogen output in the slurry. J. Rech. Porcine en France 25:295–300.

Laugerette, F., G. Pineau, C. Vors, and M. C. Michalski. 2015. Endotoxemia analysis by the limulus amoebocyte lysate assay in different mammal species used in metabolic studies. J. Anal. Bioanal. Tech. 6:4.

Le, P. D., A. J. A. Aarnink, and A. W. Jongbloed. 2009. Odour and ammonia emission from pig manure as affected by dietary crude protein level. Livest. Sci. 121:267–274.

Le, P. D., A. J. A. Aarnink, A. W. Jongbloed, C. M. C. van der Peet-Schwering, N. W. M. Ogink, and M. W. A. Verstegen. 2008. Interactive effects of dietary crude protein and fermentable carbohydrate levels on odour from pig manure. Livest. Sci. 114:48–61.

Le, P. D., A. J. A. Aarnink, A. W. Jongbloed, C. M. C. van der Peet-schwering, N. W. M. Ogink, and M. W. A. Verstegen. 2007. Effects of dietary crude protein level on odour from pig manure. Animal 1:734–744.

Lee, S. A., M. R. Bedford, and C. L. Walk. 2018. Meta-analysis: explicit value of mono-component proteases in monogastric diets. Poul. Sci. 97: 2078–2085. doi.org/10.3382/ps/pey042.

Lee, S. A., L. V. Lagos, M. R. Bedford, and H. H. Stein. 2020. Increasing calcium from deficient to adequate concentration in diets for gestating sows decreases digestibility of phosphorus and reduces serum concentration of a bone resorption biomarker. J. Anim. Sci. 98:skaa076. doi: https://doi.org/10.1093/jas/skaa076.

Lee, S. A., L. V. Lagos, C. L. Walk, and H. H. Stein. 2019. Basal endogenous loss, standardized total tract digestibility of calcium in calcium carbonate, and retention of calcium in gestating sows change during gestation, but microbial phytase reduces basal endogenous loss of calcium1. J. Anim. Sci. 97:1712–1721. doi: https://doi.org/10.1093/jas/skz048.

Leek, A. B. G., E. T. Hayes, T. P. Curran, J. J. Callan, V. E. Beattie, V. A. Dodd, and J. V. O'Doherty. 2007. The influence of manure composition on emissions of odour and ammonia from finishing pigs fed different concentrations of dietary crude protein. Bioresource Tech. 98:3431–3439.

Le Dividich, J., P. Herpin, and M. Rosario-Ludovino. 1994. Utilization of colostral energy by the newborn pig. J. Anim. Sci. 72:2082–2089.

Leibholz, J., R. J. Love, Y. Mollah, and R. R. Carter, 1986. The absorption of dietary L-lysine and extruded L-lysine in pigs. Anim. Feed Sci. Technol. 15: 141–148.

Ley, R., P. Turnbaugh, S. J. Klein, and I. Gordon. 2006. Human gut microbes associated with obesity. Nature 444:1022–1023. doi: https://doi.org/10.1038/4441022a.

Li, Y., F. Hua, X. Wang, H. Cao, D. Liua, and D. Yao. 2013. A rational design for trypsin-resistant improvement of *Armillariella tabescens*-mannanase MAN47 based on molecular structure evaluation. J. Biotechnol. 163:401–407.

Li, W, W. Powers, and G. M. Hill. 2011. Feeding distillers dried grains with solubles and organic trace mineral sources to swine and the resulting effect on gaseous emissions. J. Anim. Sci. 89.3286–3299.

Linde, M., M. Galbe, and G. Zacchi. 2008. Bioethanol production from non-starch carbohydrate reside in process streams a dry-mill ethanol plant. Bioresour. Technol. 34:6505–6511.

Liu, S., J.-Q. Ni, J. Radcliffe, and C. Vonderohe. 2017. Hydrogen sulfide emissions from a swine building affected by dietary crude protein. J. Environ. Mange. 204:136–143.

Liu, H., E. Ivarsson, J. Dicksved, T. Lundh, and J. E. Lindberg. 2012. Inclusion of chicory (Cichorium intybus L.) in pigs' diets affects the intestinal microenvironment and the gut microbiota. Appl. Environ. Microbiol. 78:4102–4109. doi: https://doi.org/10.1128/AEM.07702-11.

Liu, Y., Y. L. Ma, J. M. Zhao, M. Vazquez-Añón, and H. H. Stein. 2014. Digestibility and retention of zinc, copper, manganese, iron, calcium, and phosphorus in pigs fed diets containing inorganic or organic minerals. J. Anim. Sci. 92:3407–3415 doi: https://doi.org/10.2527/jas2013-7080.

Liu, Q., W. M. Zhang, Z. J. Zhang, Y. J. Zhang, Y. W. Zhang, L. Chen, and S. Zhuang. 2016. Effect of fiber source and enzyme addition on the apparent digestibility of nutrients and physicochemical properties of digesta in cannulated growing pigs Anim. Feed Sci. Technol. 216: 262–272. doi.org/10.1016/j.anifeedsci.2016.04.002.

Liu, S., J.-Q. Ni, J. Radcliffe, and C. Vonderohe. 2017. Hydrogen sulfide emissions from a swine building affected by dietary crude protein. J. Environ. Mange. 204:136–143.

Liu, Y., C. D. Espinosa, J. J. Abelilla, G. A. Casas, L. V. Lagos, S. A. Lee, W. B. Kwon, J. K. Mathai, D. M. D. L. Navarro, N. W. Jaworski, and H. H. Stein. 2018. Non-antibiotic feed additives in diets for pigs: A review. Anim. Nutr. 4:113–125. doi: https://doi.org/10.1016/j.aninu.2018.01.007.

Liu, Y., Z. Ning, Y. Chen, M. Guo, Y. Liu, N. K. Gali, Li Sun, Y. Duan, J. Cai, D. Westerdahl, X. Liu, K. Xu, K.F. Ho, H. Kan, Q. Fu, and K. Lan. 2020. Aerodynamic analysis of SARS-CoV-2 in two Wuhan hospitals. Nature. doi: https://doi.org/10.1038/s41586-020-2271-3.

Lovett, T. D., M. T. Coffey, R. D. Miles, and G. E. Combs. 1986. Methionine, choline and sulfate interrelationships in the diet of weanling swine. J. Anim. Sci. 63:467–471.

Lundbald, K. K. 2009. Effect of diet conditioning on physical and nutritional quality of feed for pigs and chickens. PhD Diss., Norwegian Univ. of Life Sci., Aas, Norway.

Iebba, V., V. Totino, A. Gagliardi, F. Santangelo, F. Cacciotti, M. Trancassini, C. Mancini, C. Cicerone, E. Corazziari, F. Pantanella, and S. Schippa. 2016. Eubiosis and dysbiosis: the two sides of the microbiota. New Microbiol. 39:1–12.

Isaacson, R., and H. B. Kim. 2012. The intestinal microbiome of the pig. Anim. Health Res. Rev. 13:100–109. doi: https://doi.org/10.1017/S1466252312000084.

Ivarsson, E., S. Roos, H. Y. Liu, and J. E. Lindberg. 2014. Fermentable non-starch polysaccharides increases the abundance of *Bacteroides–Prevotella–Porphyromonas* in ileal microbial community of growing pigs. Animal 8:1777–1787. doi: https://doi.org/10.1017/S1751731114001827.

Mach, N., M. Berri, J. Estellé, F. Levenez, G. Lemonnier, C. Denis, J. Leplat, C. Chevaleyre, Y. Billon, J. Doré, C. Rogel-Gaillard, and P. Lepage. 2015. Early-life establishment of the swine gut microbiome and impact on host phenotypes. Environ. Microbiol. Rep. 7:554–69. doi: https://doi.org/10.1111/1758-2229.12285.

Mackie, R. I., P. G. Stroot, and V. H. Varel. 1998. Biochemical identification and biological origin of key odor components in livestock waste. J. Anim. Sci. 76:1331–1342.

Maeda, Y., and K. Takeda. 2019. Host–microbiota interactions in rheumatoid arthritis. Exp. Mol. Med. 51:150. doi: https://doi.org/10.1038/s12276-019-0283-6.

Maes, D., L. Pluym, and O. Peltoniemi. 2016. Impact of group housing of pregnant sows on health. Porcine Health Manag. 2:17. doi: https://doi.org/10.1186/s40813-016-0032-3.

Mager, L. F., R. Burkhard, N. Pett, N. Cooke, K. Brown, H. Ramay, S. Paik, J. Stagg, R. A. Groves, M. Gallo, I. A. Lewis, M. B. Geuking, and K. D. McCoy. 2020. Microbiome-derived inosine modulates response to checkpoint inhibitor immunotherapy. Science 369:1481–1489. doi: https://doi.org/10.1126/science.abc3421.

Mahan, D. C., and R. G. Shields Jr. 1998. Macro- and micromineral composition of pigs from birth to 145 kilograms of body weight. J. Anim. Sci. 76:506–512. doi: https://doi.org/10.2527/1998.762506x.

Mahan, D. C., T. R. Cline, and B. Richert. 1999. Effects of dietary levels of selenium-enriched yeast and sodium selenite as selenium sources fed to growing-finishing pigs on performance, tissue selenium, serum glutathione peroxidase activity, carcass characteristics, and loin quality. J. Anim. Sci. 77:2172–2179.

Malo, M. S. 2015. A high level of intestinal alkaline phosphatase is protective against type 2 diabetes mellitus irrespective of obesity. EBioMedicine 2:2016–2023.

Malo, M. S., S. N. Alam, G. Mostafa, S. J. Zeller, P. V. Johnson, N. Mohammad, K. T. Chen, A. K. Moss, S. Ramasamy, A. Faruqui, S. Hodin, P. S. Malo, F. Ebrahimi, B. Biswas, S. Narisawa, J. L. Millan, H. S. Warren, J. B. Kaplan, C. L. Kitts, E. L. Hohmann, and R. A. Hodin. 2010. Intestinal alkaline phosphatase preserves the normal homeostasis of gut microbiota. Gut 59:1476–1484.

Malo, M. S., O. Moaven, N. Muhammad, B. Biswas, S. N. Alam, K. P. Economopoulos, S. S. Gul, S. R. Hamarneh, N. S. Malo, A. Teshager, M. M. R. Mohamed, Q. Tao, S. Narisawa, J. L. Millán, E. L. Hohmann, H. S. Warren, S. C. Robson, and R. A. Hodin. 2014. Intestinal alkaline phosphatase promotes gut bacterial growth by reducing the concentration of luminal nucleotide triphosphates. Am. J. Physiol. 306:G826–G838. doi: https://doi.org/10.1152/ajpgi.00357.2013.

Mani, V., A. J. Harris, A. F. Keating, T. E. Weber, J. C. M. Dekkers, and N. K. Gabler. 2013a. Intestinal integrity, endotoxin transport and detoxification in pigs divergently selected for residual feed intake. J. Anim. Sci. 91:2141–2150. doi: https://doi.org/10.2527/jas2012-6053.

Mani, V., J. H. Hollis, and N. K. Gabler. 2013b. Dietary oil composition differentially modulates intestinal endotoxin transport and postprandial endotoxemia. Nutr. Metab. 10:6. doi: nutritionandmetabolism.com/content/10/1/6.

Mani, V., T. E. Weber, L. H. Baumgard, and N. K. Gabler. □ 2012. Growth and development symposium: endotoxin, inflammation, and intestinal function in livestock. J. Anim. Sci. 90:1452–1465. https://doi.org/10.2527/jas.2011-4627.

Martinussen, J., C. Sørensen, C. B. Jendresen, and M. Kilstrup. 2010. Two nucleoside transporters in *Lactococcus lactis* with different substrate specificities. Microbiology 156:3148–3157. doi: https://doi.org/10.1099/mic.0.039818-0.

Mathai, J. K., J. K. Htoo, J. E. Thomson, K. J. Touchette, and H. H. Stein. 2016. Effects of dietary fiber on the ideal standardized ileal digestible threonine:lysine ratio for twenty-five to fifty kilogram growing gilts. J. Anim. Sci. 94:4217–4230. doi: https://doi.org/10.2527/jas.2016-0680.

McAfee, J. M., H. G. Kattesh, M. D. Lindemann, B. H. Voy, C. J. Kojima, N. C. Burdick Sanchez, J. A. Carroll, B. E. Gillespie, and A. M. Saxton. 2019. Effect of omega-3 polyunsaturated fatty acid (n-3 PUFA) supplementation to lactating sows on growth and indicators of stress in the postweaned pig. J. Anim. Sci. 97:4453–4463. doi: https://doi.org/10.1093/jas/skz300.

McGilvray, W. D., B. Johnson, H. Wooten, A. R. Rakhshandeh, and A. Rakhshandeh. 2019a. Immune system stimulation reduces the efficiency of whole-body protein deposition and alters muscle fiber characteristics in growing pigs. Animals 9:323. doi: https://doi.org/10.3390/ani9060323.

McGilvray, W. D., H. Wootenm, A. R. Rakhshandeh, A. Petry, and A. Rakhshandeh. 2019b. Immune system stimulation increases dietary threonine requirements for protein deposition in growing pigs1. J. Anim. Sci. 97:735–744. doi: https://doi.org/10.1093/jas/sky468.

Medawar, E., S. Huhn, A. Villringer, and A. V. Witte. 2019. The effects of plant-based diets on the body and the brain: A systematic review. Transl. Psychiatry 9:226. doi: https://doi.org/10.1038/s41398-019-0552-0.

Medicalxpress. 2020. Dutch farm worker likely got COVID-19 from mink: Minister. https://medicalxpress.com/news/2020-05-dutch-farm-worker-covid-mink.html.

Melo, A. D. B., H. Silveira, F. B. Luciano, C. Andrade, L. B. Costa, and M. H. Rostagno. 2016. Intestinal alkaline phosphatase: Potential roles in promoting gut health in weanling piglets and its modulation by feed additives — a review. Asian-Aus. J. Anim. Sci. 29:16–22. doi: https://doi.org/10.5713/ajas.15.0120.

Meidinger, R. G., A. Ajakaiye, M. Z. Fan, J. Zhang, J. P. Phillips, and C. W. Forsberg. 2013. Digestive utilization of phosphorus from plant based diets in the Cassie line of transgenic Yorkshire pigs that secrete phytase in the saliva. J. Anim. Sci. 91:1307–1320.

Millán, J. L. 2006a. Alkaline phosphatases: Structure, substrate specificity and functional relatedness to other members of a large superfamily of enzymes. Purinergic Signal 2:335–341.

Millán, J. L. 2006b. Mammalian alkaline phosphatases: from biology to applications in medicine and biotechnology. Wiley-Vch, Weinheim, Germany.

Montaner-Tarbes, S., H. A. Del Portillo, M. Montoya, and L. Fraile. 2019. Key gaps in the knowledge of the porcine respiratory reproductive syndrome virus (PRRSV). Front Vet. Sci. 6:38. doi: https://doi.org/10.3389/fvets.2019.00038.

Morawska, L., and D. K. Milton. 2020. It is time to address airborne transmission of COVID-19. Clin. Infect. Dis. doi: https://doi.org/10.1093/cid/ciaa939.pdf.

Munford, R. S. 2016. Endotoxemia - menace, marker, or mistake? J. Leukoc. Biol. 100: 687–698. doi: https://doi.org/10.1189/jlb.3RU0316-151R:10.1189/jlb.3RU0316-151R.

Mylonakis, E., E. T. Ryan, and S. B. Calder. 2001. *Clostridium difficile*–associated diarrhea. Arch. Intern. Med. 161:525–533.

NRC. 1998. Nutrient requirements of swine. 10th rev. ed. Natl. Acad. Press, Washington, DC.

NRC. 2012. Nutrient eequirements of swine. 11th rev. ed. Natl. Acad. Press, Washington, DC.

Obrock-Hegel, C. E. 1997. The effects of reducing dietary crude protein concentration on odor in swine facilities. M.S. Thesis. University of Nebraska, Lincoln.

Omogbenigun, F. O, C. M. Nyachoti, and B. A. Slominski. 2004. Dietary supplementation with multienzyme preparations improved nutrient utilization and growth performance in weaned pigs. J. Anim. Sci. 82:1053–1061.

Otto, E. R., M. Yokoyama, P. K. Ku, N. K. Ames, and N. L. Trottier. 2003a. Nitrogen balance and ileal amino acid digestibility in growing pigs fed diets reduced in protein concentration. J. Anim. Sci. 81:1743–1753.

Otto, E. R., M. Yokoyama, R. D. von Bermuth, T. van Kempen, and N. L. Trottier. 2003b. Ammonia, volatile fatty acids, phenolics and odor offensiveness in manure from growing pigs fed diets reduced in protein concentration. J. Anim. Sci. 81:1754–1763.

Panetta, D. M., W. J. Powers, H. Xin, B. J. Kerr, and K. J. Stalder. 2006. Nitrogen excretion and ammonia emissions from pigs fed modified diets. J. Environ. Qual. 35:1297–1308.

Park, J. E., and H. J. Shin. (2018). Porcine epidemic diarrhea vaccine efficacy evaluation by vaccination timing and frequencies. Vaccine 36:2760–2763. doi: https://doi.org/10.1016/j.vaccine.2018.03.041.

Parker, D. B. J. Gilley, B. W., K.-H. Kim, G. Galvind, S. L. Bartelt-Hunte, X. Lie, and D. D. Snowe. 2013. Odorous VOC emission following land application of swine manure slurry. Atmos. Environ. 66:91–100. doi: https://doi.org/10.1016/j.atmosenv.2012.01.001.

Passos, A. A., I. Park, P. Ferket, E. von Heimendahl, and S. W. Kim. 2015. Effect of dietary supplementation of xylanase on apparent ileal digestibility of nutrients, viscosity of digesta, and intestinal morphology of growing pigs fed corn and soybean meal based diet. Anim. Nutr. 1:19–23. doi: https://doi.org/10.1016/j.aninu.2015.02.006.

Pastor-Anglada, M., and S. Pérez-Torras. 2018. Emerging roles of nucleoside transporters. Front. Pharmacol. 9:606. doi: https://doi.org/10.3389/fphar.2018.00606.

Paterson, D. W., R. C. Wahlstrom, G. W. Libal, and O. E. Olson. 1979. Effects of sulfate in water on swine reproduction and young pig performance. J. Anim. Sci. 49:664–667.

Pedersen, M. B., S. Dalsgaard, K. E. Bach Knudsen, S. Yu, and H. N. Lærke. 2014. Compositional profile and variation of distillers dried grains with solubles from various origins with focus on non-starch polysaccharides. Anim. Feed Sci. Technol. 197: 130–141. DOI: 10.1016/j.anifeedsci.2014.07.011.

Pfeiffer, A. von. 1993. Protein reduced feeding concepts, a contribution to reduced ammoniac emissions in pig fattening. Zuchtungskunde 65:431–443.

Piehler, M., R. Roeder, S. Blessing, and J. Reich. 2020. Comparison of LAL and rFC assays—participation in a proficiency test program between 2014 and 2019. Microorganisms 8:418. doi: https://doi.org/10.3390/microorganisms8030418.

Poelstra, K., W. W. Bakker, P. A. Klok, J. A. Kamps, M. J. Hardonk, and D. K. Meijer. 1997. A physiologic function for alkaline phosphatase: endotoxin detoxification. Lab. Invest. 76:319–327.

Poltorak, A, X. He, I. Smirnova, M. Y. Liu, C. Van Huffe, X. Du, D. Birdwell, E. Alejos, M. Silva, C. Galanos, M. Freudenberg, P. Ricciardi-Castagnoli, B. Layton, and B. Beutler. 1998. Defective LPS signaling in C3H/HeJ and C57BL/10ScCr mice: mutations in Tlr4 gene. Science 282:2085–2088.

Poore, J., and T. Nemecek. 2018. Reducing food's environmental impacts through producers and consumers. Science 360:987–992. doi: https://doi.org/10.1126/science.aaq0216.

Precup, G., and D.-C. Vodnar. 2019. Gut *Prevotella* as a possible biomarker of diet and its eubiotic versus dysbiotic roles: A comprehensive literature review. Br. J. Nutr. 122:131–140. doi: https://doi.org/10.1017/S0007114519000680.

Proudfoot, C., S. Lillico, and C. Tait-Burkard. 2019. Genome editing for disease resistance in pigs and chickens. Anim. Front. 9:6–12. doi: https://doi.org/10.1093/af/vfz013.

Putnam, D. F. 1971. Composition and concentrative properties of human urine. NASA, Washington, DC.

Qin, J., R. Li, J. Raes, J. Raes, M. Arumugam, K. S. Burgdorf, C. Manichanh, T. Nielsen, N. Pons, F. Levenez, T. Yamada, D. R. Mende, J. Li, J. Xu, S. Li, D. Li, J. Cao, B. Wang, H. Liang, H. Zheng, Y. Xie, J. Tap, P. Lepage, M. Bertalan, J. M. Batto, T. Hansen, D. Le Paslier, A. Linneberg, H. B. Nielsen, E. Pelletier, P. Renault, T. Sicheritz-Ponten, K. Turner, H. Zhu, C. Yu, S. Li, M. Jian, Y. Zhou, Y. Li, X. Zhang, S. Li, N. Qin, H. Yang, J. Wang, S. Brunak, J. Doré, F. Guarner, K. Kristiansen, O. Pedersen, J. Parkhill, J. Weissenbach; MetaHIT Consortium, P. Bork, S. D. Ehrlich, and J. Wang. 2010. A human gut microbial gene catalogue established by metagenomic sequencing. Nature 464:59–65. doi: https://doi.org/10.1038/nature08821.

Rader, B. A. 2017. Alkaline phosphatase, an unconventional immune protein. Front. Immunol. 8:897. doi: https://doi.org/10.3389/fimmu.2017.00897.

Rakhshandeh, A., and C. F. M. de Lange. 2012. Evaluation of chronic immune system stimulation in growing pigs. 6:305–310. doi: https://doi.org/10.1017/S1751731111001522.

Rakhshandeh, A., J. Htoo, N. Karrow, S. Miller, and C. De Lange, 2014. Impact of immune system stimulation on the ileal nutrient digestibility and utilization of methionine plus cysteine intake for whole-body protein deposition in growing pigs. Br. J. Nutr. 111:101–110. doi: https://doi.org/10.1017/S0007114513001955.

Ramayo-Caldas, Y., N. Mach, P. Lepage, F. Levenez, C. Denis, G. Lemonnier, J. J. Leplat, Y. Billon, M. Berri, J. Doré, Rogel- C. Gaillard, and J. Estellé. 2016. Phylogenetic network analysis applied to pig gut microbiota identifies an ecosystem structure linked with growth traits. ISME J. 10:2973–2977. doi: 10.1038/ismej.2016.77.

Ren, X., J. Glende, M. Al-Falah, V. de Vries, C. Schwegmann-Wessels, X. Qu, L. Tan, T. Tschernig, H. Deng, H. Y. Naim, and G. Herrler. 2006. Analysis of ACE2 in polarized epithelial cells: surface expression and function as receptor for severe acute respiratory syndrome-associated coronavirus. J. Gen. Virol. 87:1691–1695. doi: https://doi.org/10.1099/vir.0.81749-0.

Rho, Y., D. Wey, C. Zhu, E. Kiarie, K. Moran, E. van Heugten, and C. F. M. de Lange. 2018. Growth performance, gastrointestinal and digestibility responses in growing pigs when fed corn-soybean meal-based diets with corn DDGS treated with fiber degrading enzymes with or without liquid fermentation. J. Anim. Sci. 96:5188–5197. doi: https://doi.org/10.1093/jas/sky369.

Rho, Y., C. Zhu, E. Kiarie, and C. F. M. de Lange. 2017. Standardized ileal digestible amino acids and digestible energy contents in high-protein distiller's dried grains with solubles fed to growing pigs. J. Anim. Sci. 95:3591–3597. doi: https://doi.org/10.2527/jas.2017.1553.

Richert, B. T., and J. M. DeRouchey. 2010. Swine feed processing and manufacturing. In: D. J. Meisinger, editor, National swine nutrition guide. U.S. Pork Center of Excellence, Ames, IA. p. 245–250.

Rideout, T. C., M. Z. Fan, J. P. Cant, C. Wagner-Riddle, and P. Stonehouse. 2004. Excretion of major odor-causing and acidifying compounds in response to dietary supplementation of chicory inulin in growing-finishing pigs. J. Anim. Sci. 82:1678–1684.

Rideout, T. C., Q. Liu, P. Wood, and M. Z Fan. 2008. Nutrient utilization and intestinal fermentation are differentially affected by the consumption of resistant starch varieties and conventional fibers in pigs. Br. J. Nutr. 99:984–992.

Roager, H. M., and T. R. Licht, 2018. Microbial tryptophan catabolites in health and disease. Nat. Commun. 9:3294. doi: https://doi.org/10.1038/s41467-018-05470-4.

Roque, B. M., C. G. Brooke, J. Ladau, T. Polley, L. J. Marsh, N. Najafi, P. Pandey, L. Singh, R. Kinley, J. K. Salwen, E. Eloe-Fadrosh, E. Kebreab, and M. Hess. 2019. Effect of the macroalgae *Asparagopsis taxiformis* on methane production and rumen microbiome assemblage. Anim. Microbiome 1:3. doi: https://doi.org/10.1186/s42523-019-0004-4.

Rosenfelder-Kuon, P., E. J. P. Strang, H. K. Spindler, M. Eklund, and R. Mosenthin. 2017. Ileal starch digestibility of different cereal grains fed to growing pigs. J. Anim. Sci. 95:2711–2717. doi: https://doi.org/10.2527/jas.2017.1450.

Roszkos, R., T. Tóth, and M. Mézes. 2020. Practical use of n-3 fatty acids to improve reproduction parameters in the context of modern sow nutrition. Animals 10:1141. doi: https://doi.org/10.3390/ani10071141.

Rudar, M., M. L. Fiorotto, and T. A. Davis. 2019. Regulation of muscle growth in early postnatal life in a swine model. Ann. Rev. Anim. Biosci. 7:309–335. doi: https://doi.org/10.1146/annurev-animal-020518-115130.

Rudar, M., C. L. Zhu, and C. F. M. de Lange. 2016. Effect of supplemental dietary leucine and immune system stimulation on whole-body nitrogen utilization in starter pigs. J. Anim. Sci. 94:2366–2377. doi: https://doi.org/10.2527/jas2015-0120.

Saqui-Salces, M., Z. Luo, P. E. Urriola, B. J. Kerr, and G. C. Shurson. 2017. Effect of dietary fiber and diet particle size on nutrient digestibility and gastrointestinal secretory function in growing pigs, J. Anim. Sci. 95:2640–2648. doi.org/10.2527/jas.2016.1249.

Sewalt, V. J. H., W. G. Glasser, and K. A. Beauchemin. 1997. Lignin impact on fiber degradation. 3. Reversal of inhibition of enzymatic hydrolysis by chemical modification of lignin and by additives. J. Agric. Food Chem. 45:1823–1828. doi: https://doi.org/10.1021/jf9608074.

Shang, J., G. Ye, K. Shi, Y. Wan, C. Luo, H. Aihara, Q. Geng, A. Auerbach, and F. Li 2020. Structural basis of receptor recognition by SARS-CoV-2. Nature 581:221–224. doi: https://doi.org/10.1038/s41586-020-2179-y.

Shannon, M. C., and G. M. Hill. 2019. Trace mineral supplementation for the intestinal health of young monogastric animals. Front. Vet. Sci. 6:73. doi: https://doi.org/10.3389/fvets.2019.00073.

Sharma, S., S. S. Thind, and A. Kaur. 2015. in vitro meat production system: why and how? J. Food Sci. Technol. 52:7599–7607. doi: https://doi.org/10.1007/s13197-015-1972-3.

Shaw, M. H., T. Reimer, Y.-G. Kim, and G. Nuñez. 2008. Nod-like receptors (nlrs): Bona fide intracellular microbial sensors. Curr. Opin. Immunol. 20:377–382. doi: https://doi.org/10.1016/j.coi.2008.06.001.

Shen, Y., Y. L. Yin, E. R. Chavez, and M. Z. Fan. 2005. Methodological aspects of measuring phytase activity and phytate phosphorus content in selected cereal grains and digesta and feces of pigs. J. Agric. Food Chem. 53:853–859.

Shi, J., Z. Wen, G. Zhong, H. Yang, C. Wang, B. Huang, R. Liu, X. He, L. Shuai, Z. Sun, Y. Zhao, P. Liu, L. Liang, P. Cui, J. Wang, X. Zhang, Y. Guan, W. Tan, G. Wu, H. Chen, and Z. Bu. 2020. Susceptibility of ferrets, cats, dogs, and other domesticated animals to SARS–coronavirus 2. Science 368:1016–1020. doi: https://doi.org/10.1126/science.abb7015.

Shriver, J. A., S. D. Carter, A. L. Sutton, B. T. Richert, B. W. Senne, and L. A. Pettey. 2003. Effects of adding fiber sources to reduced crude protein, amino acid-supplemented diets on nitrogen excretion, growth performance, and carcass traits of finishing pigs. J. Anim. Sci. 81:492–502.

Shurson, J., M. Whitney, and R. Nicolai. 1998. Nutritional manipulation of swine diets to reduce hydrogen sulfide emissions. In: Proc. 59th Minnesota Nutr. Conf. IPC Tech. Symp., Bloomington, MN. pp. 219–240.

Shurson, G. C., B. J. Kerr, and A. R. Hanson. 2015. Evaluating the quality of feed fats and oils and their effects on pig growth performance. J. Anim. Sci. Biotechnol. 6:10. doi: https://doi.org/10.1186/s40104-015-0005-4.

Sit, T. H. C., C. J. Brackman, S. M. Ip, K. W. S. Tam, P. Y. T. Law, E. M. W. To, V. Y. T. Yu, L. D. Sims, D. N. C. Tsang, D. K. W. Chu, R. A. P. M. Perera, L. L. M. Poon, and M. Peiris. 2020. Infection of dogs with SARS-CoV-2. Nature 586:776–778. doi: 10.1038/s41586-020-2334-5.

Skelly, A. N., Y. Sato, S. Kearney, and K. Honda. 2019. Mining the microbiota for microbial and metabolite-based immunotherapies. Nat. Rev. Immunol. 19:305–323. doi: https://doi.org/10.1038/s41577-019-0144-5.

Slifierz, M. J., R. M. Friendship, and J. S. Weese. 2015. Longitudinal study of the early-life fecal and nasal microbiotas of the domestic pig. BMC Micro. 15:184. doi: https://doi.org/10.1186/s12866-015-0512-7.

Soares, J. A., H. H. Stein, V. Singh, G. S. Shurson, and J. E. Pettigrew. 2012. Amino acid digestibility of corn distillers dried grains with solubles, liquid condensed solubles, pulse dried thin stillage, and syrup balls fed to growing pigs. J. Anim. Sci. 90:1255–1261. doi: https://doi.org/10.2527/jas.2010-3691.

Spiehs, M. J., M. H. Whitney, G. C. Shurson, R. E. Nicolai, J. A. Renteria-Flores, and D. B. Parker. 2012. Odor and gas emissions and nutrient excretion from pigs fed diets containing dried distillers' grains with solubles. Applied Eng. Ag. 28:431–437.

Stein, H. H., B. Sève, M. F. Fuller, P. J. Moughan, and C. F. M. de Lange. 2007. Amino acid bioavailability and digestibility in pig feed ingredients: terminology and application. J. Anim. Sci. 85:172–180.

Stein, H. H., and G. C. Shurson. 2009. Board-invited review: the use and application of distillers dried grains with solubles in swine diets. J. Anim. Sci. 87:1292–1303.

Stein, H. H., O. Adeola, G. L. Cromwell, S. W. Kim, D. C. Mahan, and P. S. Miller. 2011. North Central Coordinating Committee on Swine Nutrition (NCCC-42). Concentration of dietary calcium supplied by calcium carbonate does not affect the apparent total tract digestibility of calcium, but decreases digestibility of phosphorus by growing pigs. J. Anim. Sci. 89:2139–44. doi: 10.2527/jas.2010-3522.

Stoll, B., J. Henry, P. J. Reeds, H. Yu, F. Jahoor, and D. G. Burrin. 1998. Catabolism dominates the first-pass intestinal metabolism of dietary essential amino acids in milk protein-fed piglets. J. Nutr. 128:606–614.

Sun, H., Y. Xiao, J. Liu, D. Wang, F. Li, C. Wang, C. Li, J. Zhu, J. Song, H. Sun, Z. Jiang, L. Liu, X. Zhang, K. Wei, D. Hou, J. Pu, Y. Sun, Q. Tong, Y. Bi, K. C. Chang, S. Liu, G. F. Gao, and J. Liu. 2020. Prevalent Eurasian avian-like H1N1 swine influenza virus with 2009 pandemic viral genes facilitating human infection. Proc. Natl. Acad. Sci. USA. 29:201921186. doi: https://doi.org/10.1073/pnas.1921186117.

Sutton, A. L., K. B. Kephart, M. W. A. Verstegen, T. T. Canh, and P. J. Hobbs. 1999. Potential for reduction of odorous compounds in swine manure through diet modification. J. Anim. Sci. 77:430–439.

Swain, P. S., S. B. N. Rao, D. Rajendran, G. Dominic, and S. Selvaraju. 2016. Nano zinc, an alternative to conventional zinc as animal feed supplement: A review. Anim. Nutr. 2:134–141. doi: https://doi.org/10.1016/j.aninu.2016.06.003.

Tessier, R., N. Khodorova, J. Calvez, R. Kapel, A. Quinsac, J. Piedcoq, D. Tomé, and C. Gaudichon. 2020. $^{15}$N- and $^2$H-Intrinsic labeling demonstrate that real digestibility in rats of proteins and amino acids from sunflower protein isolate is almost as high as that of goat whey. J. Nutr. 150:450–457. doi: https://doi.org/10.1093/jn/nxz279.

Tian, J.-H., A.-M. Pourcher, T. Bouchez, E. Gelhaye, and P. Peu. 2014. Occurrence of lignin degradation genotypes and phenotypes among prokaryotes. Appl. Microbiol. Biotechnol. 98:9527–9544. doi: https://doi.org/10.1007/s00253-014-6142-4.

Torres-Pitarch, A., D. Hermans, E. G. Manzanilla, J. Bindelle, N. Everaert, Y. Beckers, D. Torrallardona, G. Bruggeman, G. E. Gardiner, and P. G. Lawlor. 2017. Effect of feed enzymes on digestibility and growth in weaned pigs: A systematic review and meta-analysis. Anim. Feed Sci. Technol. 233:145–159.

Trabue, S., and B. J. Kerr. 2014. Emissions of greenhouse gases, ammonia, and hydrogen sulfide from pigs fed standard diets and diets supplemented with dried distillers grains with solubles. J. Environ. Qual. 43:1176–1186.

Trabue, S., B. Kerr, B. Bearson, and C. Ziemer. 2011. Swine odor analyzed by odor panels and chemical techniques. J. Environ. Qual. 40:1510–1520. doi: https://doi.org/10.2134/jeq2010.0522.

Trabue, S., B. Kerr, and K. Scoggin. 2016. Odor and odorous compound emissions from manure of swine fed standard and dried distillers grains with soluble supplemented diets. J. Environ. Qual. 45:915–923.

Trabue, S. L., B. J. Kerr, and K. D. Scoggin. 2019a. Swine diets impact manure characteristics and gas emissions: Part I sulfur level. Sci. Tot. Envt. 687:800–807.

Trabue, S. L., B. J. Kerr, and K. D. Scoggin. 2019b. Swine diets impact manure characteristics and gas emissions: Part II sulfur source. Sci. Tot. Envt. 689:1115–1124.

Urriola, P. E., H. H. Stein, and G. C. Shurson. 2010. Digestibility of dietary fiber in distillers coproducts fed to growing pigs. J. Anim. Sci. 88:2373–2381.

USDA. 2020. USDA statement on the confirmation of covid-19 in a tiger in New York. US Dept. of Agric., Washington, DC.

van Heugten, E., and T. A. van Kempen. 2002. Growth performance, carcass characteristics, nutrient digestibility and fecal odorous compounds in growing-finishing pigs fed diets containing hydrolyzed feather meal. J. Anim. Sci. 2002 Jan;80(1):171–178.

van Kempen, T. A. T. G., D. H. Baker, and E. van Heugten. 2003. Nitrogen losses in metabolism trials. J. Anim. Sci. 81:2649–2650.

van Milgen, J., J. Noblet, and S. Dubois. 2001. Energetic efficiency of starch, protein and lipid utilization in growing pigs. J. Nutr. 131:1309–1318.

van Milgen, J., and J. Y. Dourmad. 2015. Concept and application of ideal protein for pigs. J. Anim. Sci. Biotech 6:15.

van de Peet-Schwering, C., A. Aarnink, H. B. Rom, and J. Dourmad. 1999. Ammonia emissions from pig houses in the Netherlands, Denmark and France. Lives. Prod. Sci. 58:265–269.

van Weelden, M. B., D. S. Andersen, B. J. Kerr, S. Trabue, and L. Pepple. 2016a. Impact of fiber source and feed particle size on swine manure properties related to spontaneous foam formation during anaerobic decomposition. Bioresourc. Technol. 202:84–92.

van Weelden, M. B., D. S. Andersen, B. J. Kerr, S. L. Trabue, K. A. Rosentrater, L. M. Pepple, and T.M.B. dos Santos. 2016b. Impact of dietary carbohydrate and protein source and content on swine manure foaming properties. Trans. ASABE. 59:923–932.

VanderWaal, K., and J. Deen. 2018. Global trends in infectious diseases of swine. Proc. Natl. Acad. Sci. 115:11495–11500. doi: https://doi.org/10.1073/pnas.1806068115.

Varel, V. H., and J. T. Yen. 1997. Microbial perspective on fiber utilization by swine. J. Anim. Sci. 75:2715–2722.

Veenhuizen, M. F., G. C. Shurson, and E. M. Kohler. 1992. Effect of concentration and source of sulfate on nursery pig performance and health. JAVMA 201:1203–1208.

Velthof, G. L., J. A. Nelemans, O. Oenema, and P. J. Kuikman. 2005. Gaseous nitrogen and carbon losses from pig manure derived from different diets. J. Environ. Qual. 34:698–706.

Vermaas, J. V., L. Petridis, X. Qi, R. Schulz, B. Lindner and J. C. Smith. 2015. Mechanism of lignin inhibition of enzymatic biomass deconstruction. Biotechnol. Biofuels 8:217. doi: https://doi.org/10.1186/s13068-015-0379-8.

Vernooij, R. W. M., D. Zeraatkar, M. A. Han, R. E. Dib, M. Zworth, K. Milio, D. Sit, Y. Lee, H. Gomaa, C. Valli, M. J Swierz, Y. Chang, S. E. Hanna, P. M. Brauer, J. Sievenpiper, R. de Souza, P. Alonso-Coello, M. M. Bala, G. H. Guyatt, and B. C. Johnston. 2019.

Patterns of red and processed meat consumption and risk for cardiometabolic and cancer outcomes: A systematic review and meta-analysis of cohort studies. Annu. Intern. Med. 171:732–741. doi: https://doi.org/10.7326/M19-1583.

Vhile, S. G., N. P. Kjos, H. Sørum, and M. Overland. 2012. Feeding Jerusalem artichoke reduced skatole level and changed intestinal microbiota in the gut of entire male pigs. Animal 6:807–814.

Wan, Y., J. Shang, R. Graham, R. S. Baric, and F. Li 2020. Receptor recognition by the novel coronavirus from Wuhan: an analysis based on decade-long structural studies of SARS coronavirus. J. Virol. 94:e00127–20. doi: https://doi.org/10.1128/JVI.00127-20.

Wang, N., D. Zhao, J. Wang, Y. Zhang, M. Wang, Y. Gao, F. Li, J. Wang, Z. Bu, Z. Rao, and X. Wang. 2019a. Architecture of African swine fever virus and implications for viral assembly. Science 366:640–644 doi: https://doi.org/10.1126/science.aaz1439.

Wang, T., J. N. R. Gnanaprakasam, X. Chen, S. Kang, X. Xu, H. Sun, L. Liu, H. Rodgers, E. Miller, T. A. Cassel, Q. Sun, S. Vicente-Muñoz, M. O. Warmoes, P. Lin, Z. L. Piedra-Quintero, M. Guerau-de-Arellano, K. A. Cassady, S. G. Zheng, J. Yang, A. N. Lane, X. Song, T. W.-M. Fan, and R. Wang. 2020. Inosine is an alternative carbon source for CD8+-T-cell function under glucose restriction. Nat. Metab. 2:635–647. doi: https://doi.org/10.1038/s42255-020-0219-4.

Wang, W., H. Hu, R. T. Zijlstra, J. Zheng, and M. G. Gänzle. 2019b. Metagenomic reconstructions of gut microbial metabolism in weanling pigs. Microbiome 7:48. doi: https://doi.org/10.1186/s40168-019-0662-1.

Wang, W., T. Archbold, M. S. Kimber, J. Li, J. S. Lam, and M. Z. Fan. 2012. The porcine gut microbial metagenomic library for mining novel cellulases established from growing pigs fed cellulose-supplemented high-fat diets. J. Anim. Sci. 90:400–402.

Wang, W., T. Archbold, J. S. Lam, M. S. Kimber, and M. Z. Fan. (2019c). Processive endoglucanase with multi-substrate specificity is characterized from porcine gut microbiota. Sci. Rep. 9:13630. doi: https://doi.org/10.1038/s41598-019-50050-1.

Watson, H., S. Mitra, F. C. Croden, M. Taylor, H. M. Wood, S. L. Perry, J. A. Spencer, P. Quirke, G. J. Toogood, C. L. Lawton, L. Dye, P. M. Loadman, and M. A. Hull. 2018. A randomised trial of the effect of omega-3 polyunsaturated fatty acid supplements on the human intestinal microbiota. Gut 67:1974–1983. doi: https://doi.org/10.1136/gutjnl-2017-314968.

Webel, D. M., B. N. Finck, D. H. Baker, and R. W. Johnson. 1997. Time course of increased plasma cytokines, cortisol, and urea nitrogen in pigs following intraperitoneal injection of lipopolysaccharide. J. Anim. Sci. 75:1514–1520.

Wellington, M. O., J. K. Htoo, A. G. Van Kessel, and D. A. Columbus. 2018. Impact of dietary fiber and immune system stimulation on threonine requirement for protein deposition in growing pigs. J. Anim. Sci. 96:5222–5232. doi: https://doi.org/10.1093/jas/sky381.

Wei, C., A. Dai, J. Fan, Y. Li, A. Chen, X. Zhou, M. Luo, X. Yang, and J. Liu. 2019. Efficacy of type 2 PRRSV vaccine against challenge with the chinese lineage 1 (NADC30-like) PRRSVs in pigs. Sci. Rep. 9:10781. doi: https://doi.org/10.1038/s41598-019-47239-9.

Whitworth, K. M., R. R. Rowland, C. L. Ewen, B. R. Trible, M. A. Kerrigan, A. G. Cino-Ozuna, M. S. Samuel, J. E. Lightner, D. G. McLaren, A. J. Mileham, K. D. Wells, and R. S. Prather. 2016. Gene-edited pigs are protected from porcine reproductive and respiratory syndrome virus. Nat. Biotechnol. 34:20–22. doi: https://doi.org/10.1038/nbt.3434

Whitworth, K. M., R. R. R. Rowland, V. Petrovan, M. Sheahan, A. G. Cino-Ozuna, Y. Fang, R. Hesse, A. Mileham, M. S. Samuel, K. D. Wells, and R. S. Prather. 2019. Resistance to coronavirus infection in amino peptidase N-deficientpigs. Transgenic Res. 28:21–32. doi: https://doi.org/10.1007/s11248-018-0100-3.

Wierenga, K. T., E. Beltranena, J. L. Yáñez, and R. T. Zijlstra. 2008. Starch and energy digestibility in weaned pigs fed extruded zero-tannin faba bean starch and wheat as an energy source. Can. J. Anim. Sci. 88:65–69.

Willett, W., J. Rockström, B. Loken, M. Springmann, T. Lang, S. Vermeulen, T. Garnett, D. Tilman, F. DeClerck, A. Wood, M. Jonell, M. Clark, L. J. Gordon, J. Fanzo, C. Hawkes, R. Zurayk, J. A. Rivera, W. De Vries, L. Majele Sibanda, A. Afshin, A. Chaudhary, M. Herrero, R. Agustina, F. Branca, A. Lartey, S. Fan, B. Crona, E. Fox, V. Bignet, M. Troell, T. Lindahl, S. Singh, S. E.Cornell, K. Srinath Reddy, S. Narain, S. Nishtar, and C. J. L. Murray. 2019. Food in the anthropocene: the EAT–Lancet Commission on healthy diets from sustainable food systems. Lancet 393:447–492. doi: https://doi.org/10.1016/S0140-6736(18)31788-4.

Whitt, D. D., and D. C. Savage. 1988. Influence of indigenous microbiota on activities of alkaline phosphatase, phosphodiesterase I, and thymidine kinase in mouse enterocytes. Appl. Environ. Microbiol. 54:2405–2410. doi: https://doi.org/10.1128/AEM.54.10.2405-2410.1988.

Woyengo, T. A., E. Beltranena, and R. T. Zijlstra. 2014. Controlling feed cost by including alternative ingredients into pig diets: A review. J. Anim. Sci. 92:1293–1305.

Wu, F, S. Zhao, B. Yu, Y. M. Chen, W. Wang, Z. G. Song, Y. Hu, Z. W. Tao, J.H. Tian, Y. Pei, M. L. Yuan, Y. L. Zhang, F. H. Dai, Y. Liu, Q. M. Wang, J. J. Zheng, L. Xu, E. C. Holmes, and Y. Z. Zhang. 2020. A new coronavirus associated with human respiratory disease in China. Nature 579:265–269. doi: https://doi.org/10.1038/s41586-020-2008-3.

Wu, H., X. Yin, Z. Yang, T. Archbold, W. Fan, W. Wang, L. Xu, and M. Z. Fan. 2019. Determination of true digestibility and the endogenous outputs of magnesium in corn for growing pigs by using the regression analysis technique. J. Anim. Sci. 97(Suppl. 3): 285.

Xiao, L., J. Estellé, P. Kiilerich, Y. Ramayo-Caldas, Z. Xia, Q. Feng, S. Liang, A. Ø. Pedersen, N. J. Kjeldsen, C. Liu, E. Maguin, J. Doré, N. Pons, E. Le Chatelier, E. Prifti, J. Li, H. Jia, X. Liu, X. Xu, S. D. Ehrlich, L. Madsen, K. Kristiansen, C. Rogel-Gaillard, and J. Wang. 2016. A reference gene catalogue of the pig gut microbiome. Nat. Microbiol. 1:16161. doi: https://doi.org/10.1038/nmicrobiol.2016.161.

Xie, S., A. J. Ragauskas, and J. S. Yuan. 2016. Lignin conversion: opportunities and challenges for the integrated biorefinery. Ind. Biotechnol. 12:161–167. doi: https://doi.org/10.1089/ind.2016.0007.

Xue, H., M. Zhang, J. Ma, T. Chen, F. Wang, and X. Tang. 2020. Lactose-induced chronic diarrhea results from abnormal luminal microbial fermentation and disorder of ion transport in the colon. Front. Physiol. 11:877. doi: https://doi.org/10.3389/fphys.2020.00877.

Xu, H., L. Zhong, J. Deng, J. Peng, H. Dan, X. Zeng, T. Li, and Q. Chen. 2020. High expression of ACE2 receptor of 2019-nCoV on the epithelial cells of oral mucosa. Int. J. Oral. Sci. 12:8. doi: https://doi.org/10.1038/s41368-020-0074-x.

Yan, R., Y. Zhang, Y. Li, L. Xia, Y. Guo, and Q. Zhou 2020. Structural basis for the recognition of the SARS-CoV-2 by full-length human ACE2. Science 367:1444–1448. doi: https://doi.org/10.1126/science.abb2762.

Yang, C., D. M. Albin, Z. Wang, B. Stoll, D. Lackeyram, K. C. Swanson, Y. L. Yin, K. A. Tappenden, Y. Mine, R. Y. Yada, D. G. Burrin, and M. Z. Fan. 2011. Apical Na+-D-glucose co-transporter 1 (SGLT1) activity and protein abundance are expressed along the jejunal crypt-villus axis in the neonatal pig. Am. J. Physiol. 300:G60–G70.

Yang, C., X. Yang, and M. Z. Fan. 2017. Apical Na+-dependent neutral amino acid exchanger ASCT2 (ATB0) and mTOR-signaling components are expressed along the entire jejunal crypt-villus axis in young pigs fed a liquid milk replacer. Can. J. Anim. Sci. 97:19–29.

Yang, C., X. Yang, D. Lackeyram, T. C. Rideout, Z. Wang, B. Stoll, Y. Yin, D. G. Burrin, and M. Z. Fan. 2016. Expression of apical Na+-L-glutamine co-transport activity, B0-system neutral amino acid co-transporter (B0AT1) and angiotensin converting enzyme 2 along the jejunal crypt-villus axis in young pigs fed a liquid formula. Amino Acids 48:1491–1508.

Yang, X., C. Yang, A. Farberman, T. C. Rideout, C. F. M. de Lange, J. France, and M. Z. Fan. 2008. The Mammalian target of rapamycin-signaling pathway in regulating metabolism and growth. J. Anim. Sci. 86(E. Suppl):E36–E50.

Yang, Z., H. Wu, T. Archbold, X. Yin, W. Fan, and M. Z. Fan. 2019. Intestinal responses and the determination of true total tract trace mineral digestibility and the endogenous losses in weanling pigs by the regression analysis technique. J. Anim. Sci. 97(Suppl. 3):287–288. (Abstr.).

Yang, Z. 2019. Determination of True Micro-Mineral Digestibility and Their Endogenous Losses in Weanling Pigs by the Regression Analysis. M.S. Thesis, Univ. of Guelph, Guelph, ON, Canada.

Yap, Y. W., P. M. Rusu, A. Y. Chan, B. C. Fam, A. Jungmann, S. M. Solon-Biet, C. K. Barlow, D. J. Creek, C. Huang, R. B Schittenhelm, B. Morgan, D. Schmoll, Bente Kiens, M. D. W. Piper, M. Heikenwälder, S. J. Simpson, S. Bröer, S. Andrikopoulos, O. J., Müller, and A. J. Rose. 2020. Restriction of essential amino acids dictates the systemic metabolic response to dietary protein dilution. Nat. Commun. 11:2894. doi: https://doi.org/10.1038/s41467-020-16568-z.

Yin, X., T. Archbold, W. Fan, N. Burello, Z. Yang, H. Wu, M. Scolaro, K. Zhou, W. Wang, and M. Z. Fan. 2018. Responses in gut permeability, total tract dry matter digestibility and growth performances in weanling pigs fed the diet supplemented with the antibiotic aureomycin. The 14th Intl. Symp. Digestive Physiology of Pigs. Aug. 21–24, 2018. Brisbane, Australia. p. 48.

Yokoyama, M. T., and J. R. Carlson. 1979. Microbial metabolites of tryptophan in the intestinal tract with special reference to skatole. Am. J. Clin. Nutr. 32:173–8. doi: https://doi.org/10.1093/ajcn/32.1.173.

Younes, H., C. Remesy, S. Behr, and C. Demigne. 1997. Fermentable carbohydrate exerts a urea-lowering effect in normal and nephrectomized rats. Am. J. Physiol. 272:G515–521.

Zeraatkar, D., G. H. Guyatt, P. Alonso-Coello, M. M. Bala, M. Rabassa, M. A. Han, R. W. M. Vernooij, C. Valli, R. E. Dib, and B. C. Johnston. 2020. Red and processed meat consumption and risk for all-cause mortality and cardiometabolic outcomes. Annu. Intern. Med. 172:511–512. doi: https://doi.org/10.7326/L20-0070.

Zervas, S., and R. T. Zijlstra. 2002. Effects of dietary protein and fermentable fiber on nitrogen excretion patterns and plasma urea in grower pigs. J. Anim. Sci. 80:3247–3256.

Zhang, Y.-H. P., and L. R. Lynd. 2004. Toward an aggregated understanding of enzymatic hydrolysis of cellulose: Noncomplexed cellulase systems. Biotechnol. Bioeng. 88:797–824.

Zhang, X., Z. Li, H. Yang, D. Liu, G. Cai, G. Li, J. Mo, D. Wang, C. Zhong, H. Wang, Y. Sun, J. Shi, E. Zheng, F. Meng, M. Zhang, X. He, R. Zhou, J. Zhang, M. Huang, R. Zhang, N. Li, M. Z. Fan, Y. Yang, and Z. Wu. 2018. Novel transgenic pigs with enhanced growth and reduced environmental impact. eLIFE:7. doi: https://doi.org/10.7554/eLife.34286.

Zhang, X., Y. Tan, Y. Ling, G. Lu, F. Liu, Z. Yi, X. Jia, M. Shi, S. Xu, J. Chen, W. Wang, B. Chen, L. Jiang, S. Yu, J. Lu, J. Wang, M. Xu, Z. Yuan, Q. Zhang, X. Zhang, G. Zhao, S. Wang, S. Chen, and H. Lu. 2020. Viral and host factors related to the clinical outcome of COVID-19. Nature doi: https://doi.org/10.1038/s41586-020-2355-0.

Zhao, W., Y. Wang, S. Liu, J. Huang, Z. Zhai, C. He, J. Ding, J. Wang, H. Wang, W. Fan, J. Zhao, and H. Meng. 2015. The dynamic distribution of porcine microbiota across different ages and gastrointestinal tract segments. PLOS One 10:e0117441. doi: https://doi.org/10.1371/journal.pone.0117441.

Zijlstra, R. T., A. Owusu-Asiedu, and P. H. Simmins. 2010. Future of NSP-degrading enzymes to improve nutrient utilization of co-products and gut health in pigs. Livest. Sci. 134:255–257.

Zhao, L, F. Zhang, X. Ding, G. Wu, Y. Y. Lam, X. Wang, H. Fu, X. Xue, C. Lu, J. Ma, L. Yu, C. Xu, Z. Ren, Y. Xu, S. Xu, H. Shen, X. Zhu, Y. Shi, Q. Shen, W. Dong, R. Liu, Y. Ling, Y. Zeng, X. Wang, Q. Zhang, J. Wang, L. Wang, Y. Wu, B. Zeng, H. Wei, M. Zhang, Y. Peng, and C. Zhang. 2018. Gut bacteria selectively promoted by dietary fibers alleviate type 2 diabetes. Sci. 359:1151–1156. doi: 10.1126/science.aao5774.

Zhou, H., X. Chen, T. Hu, H. Song, Y. Liu, P. Wang, D. Liu, J. Yang, E. C. Holmes, A. C. Hughes, Y. Bi, and W. Shi. 2020a. A novel bat coronavirus closely related to SARS-CoV-2 contains natural insertions at the s1/s2 cleavage site of the spike protein. Curr. Biol. 30:2196–2203.e3. doi: https://doi.org/10.1016/j.cub.2020.05.023.

Zhou, P., X. L. Yang, X. G. Wang, B. Hu, L. Zhang, W. Zhang, H. R. Si, Y. Zhu, B. Li, C. L. Huang, H. D. Chen, J. Chen, Y. Luo, H. Guo, R. D. Jiang, M. Q. Liu, Y. Chen, X. R. Shen, X. Wang, X. S. Zheng, K. Zhao, Q. J. Chen, F. Deng, L. L. Liu, B. Yan, F. X. Zhan, Y. Y. Wang, G. F. Xiao, and Z. L. Shi. 2020b. A pneumonia outbreak associated with a new coronavirus of probable bat origin. Nature 579:270–273. doi: https://doi.org/10.1038/s41586-020-2012-7.

Zhu, L., Q. Wu, J. Dai, S. Zhang, and F. Wei. 2011. Evidence of cellulose metabolism by the giant panda gut microbiome. PNAS 108:17714–17719. doi: https://doi.org/10.1073/pnas.1017956108.

Zuo, J., L. Baoming, L. Long, T. Li, L. Lahaye, C. Yang, and D. Feng. 2015. Effect of dietary supplementation with protease on growth performance, nutrient digestibility, intestinal morphology, digestive enzymes and gene expression of weaned piglets. Anim. Nutr. 1:276–282.

# 21     Swine Housing Systems, Behavior, and Welfare

Peter J. Lammers, Mark S. Honeyman, Rachel M. Park, and Monique D. Pairis-Garcia

## Introduction

Swine were domesticated and adapted by early agricultural societies as they grew quickly and gave birth to large, resilient litters (Kittawornrat and Zimmerman 2010). Over the last century, scientific advancements have prioritized improving the efficiency and productivity of domesticated swine-herds by changing management practices, housing systems, nutrition, and genetics. Domestication coupled with tremendous fluctuations to the pig's natural environment has resulted in changes to the behavioral needs, as well as the development of abnormal and unwanted behaviors as an indirect effect on the species. This chapter begins with a summary of pig behavior and follows with a focused discussion on how vbehavior and performance are impacted by housing systems throughout all production phases.

## Pig Behavior

Before one can understand the etiology of abnormal behaviors, natural behavior must first be defined. Natural behaviors consist of a set or series of behaviors inherent and genetically encoded to the animal (Rollin 1993). Animals will prioritize those behaviors in which they are highly motivated to perform. Although domestication alters the behavioral repertoire of any species, comparing motivational behaviors between wild and domesticated breeds helps scientist to quantify how management factors influence behavior and behavioral development.

Wild pigs' most common habitats include forests and scrub brush areas but are capable of living in a variety of habitats including mountains, swamps, and farmland (Frädrich 1974). An ideal environment for pigs is one that has a high-energy food source present, cover from predators, and protection from inclement weather (D'Eath and Turner 2009). Depending on the environmental conditions, home ranges for wild pigs can range anywhere from 100 to 6000 hectares (Kurz and Marchinton 1972; Pine and Gerdes 1973; Martin 1975; Barrett 1978; Mauget 1979; Janeau and Spitz 1984). Floor space provided in commercial swine systems is drastically different, with minimum floor area recommendations ranging from 0.55 to 3.25 m$^2$ per animal based on body weight and production stage (Salak-Johnson et al. 2012).

Social groupings vary dependent on sex and stage of development. Sounders are groups of female pigs composed of two to four sows that are genetically related along with their young from the most recent litter (Macdonald 1984; Graves 1984; Gabor et al. 1999). A sounder will operate as a unit

---

*Sustainable Swine Nutrition*, Second Edition. Edited by Lee I. Chiba.
© 2023 John Wiley & Sons Ltd. Published 2023 by John Wiley & Sons Ltd.

sharing their home range and synchronizing their activities from foraging to parturition (Stolba and Wood-Gush 1989).

When piglets are weaned from the sounder, they will break off and form juvenile groups that will share the same home range as the sounder but will engage in activity and rest separately (Cousse et al. 1994; Stolba and Wood-Gush 1989). Boars will also separate from these juvenile groups as they reach sexual maturity and form small bachelor groups (~3 boars/bachelor group). As boars age, they will become more solitary (Frädrich 1974) and will interact with adult females only during mating (Gonyou 2001).

Aggression is uncommon in groups of familiar pigs (Mendl 1995), as a linear dominance hierarchy is established (Jensen and Wood-Gush 1984; Gonyou 2001) where subordinate pigs avoid conflict by physically removing themselves from the area. In modern pig production, pigs are not maintained in family social groupings but rather housed with same age and production stage conspecifics (D'Eath and Turner 2009). Compared to wild herds, pigs in modern production environments have less space allocated per animal and will reside in larger group sizes. This results in increased aggression as pigs have less ability to engage in mutual avoidance, and therefore, fights can be more intense and last for a longer duration in these environments (D'Eath and Turner 2009).

During active hours, wild pigs spend the majority of their time foraging for food (approximately 75%; Frädrich 1974). Foraging includes rooting, grazing, and exploring with the snout and is performed to obtain suitable and diverse foodstuffs (Stolba and Wood-Gush 1989). Diets vary based on habitat and may include agricultural crops, wild plant seeds, roots, nuts, invertebrates, carrion, tree bark, grasses, forbs, and berries (Dardaillon 1987; Focardi et al. 2000; Genov 1981, Graves 1984; Herrero et al. 2006; Leaper et al. 1999). Although vegetation does compose ~90% of pig's overall diet, these animals cannot thrive on grass alone (Graves 1984).

Pigs in commercial environments are provided a concentrated feed composed primarily of finely ground corn and soybean meal. Feed is provided one to two times a day for breeding stock or ad libitum for nursery and finisher pigs and substrate is not typically provided. Because pigs can consume their required nutrients in less time when fed concentrated diets and do not have access to substrate, foraging behavior is redirected and can result in abnormal behaviors including sham chewing, bar biting, ear chewing, and tail biting (Fraser 1975; Rushen 1985; Wood-Gush and Vestergaard 1989; Lawrence and Terlouw 1993; Schrøder-Petersen and Simonsen 2001).

In addition to foraging, pigs are highly motivated to perform thermoregulatory behaviors. The body weight of a pig strongly influences the ambient temperature at which pigs are comfortable (Wathes and Whittemore 2006). The impact of thermal environment on nutrient utilization by pigs at different production stages is beyond the scope of this chapter but is well summarized by Noblet et al. (2001). During hot seasons, pigs will adapt a stable home range with several water holes (Eisenberg and Lockhart 1972) to regulate temperature. In colder weather, pigs will seek warmth through huddling and insulating day beds with additional vegetation (Signoret et al. 1975; Ruhe 1997). In commercial settings, the thermal environment is maintained at the room level primarily through the control of airflow, humidity, and supplemental heat. Sprinklers or drip lines combined with fans are commonly used to cool pigs during hot weather with evaporative cooling cells becoming increasingly common. Heat lamps, heating pads, and creep areas are routinely used within farrowing crates to create a warmer microclimate for newborn pigs. During cold weather, supplemental heat is often added to pig barns through various heaters and control of ventilation, but bedding substrate is not normally provided.

Lastly, the pregnant gilt or sow will perform nest building behavior prior to parturition, and this is considered one of the most highly motivated behaviors performed. Nesting behavior is primarily performed by rooting, pawing, and manipulating materials with the snout (D'Eath and Turner 2009).

Nesting material may include twigs, hay, grass, and leaves. Sows may travel up to 6.5 km to find substrates (Jensen 1986). Nesting behaviors occur within 24 hours of farrowing and peak 6–12 hours prior to parturition (Jensen 1986; Cronin et al. 1994; Algers and Uvnas-Moberg 2007; Wischner et al. 2009). Nest building behaviors are a complex set of behaviors driven by hormonal regulation (Jensen 1993; Algers and Uvnas-Moberg 2007; Johnson and Marchant-Forde 2011) and environmental feedback (Burne et al. 2000). In order to successfully perform this behavior, the sow requires both materials to manipulate (e.g. straw) and physical space within the housing environment. Traditional stalls lack both of these resources, which may contribute to longer farrowing durations and higher stillborn rates (Thodberg et al. 2002; Condous et al. 2016).

One to two weeks postfarrowing, sows will rejoin the sounder (Petersen et al. 1989), and it is common for the group to synchronize lactation behavior. In current commercial farrowing facilities, sows are typically provided 3.6 m² of space and do not have nest building materials provided. Sows that are not provided nest building material may develop abnormal oral behaviors during parturition such as bar biting (biting/chewing stall bars or chains), sham chewing, or manipulation of their drinker (Lawrence and Terlouw 1993).

## Evaluating Welfare

Metrics used to evaluate welfare are divided into three broad categories: animal, environment, and resource-based measures (Mench 2003; Rushen et al. 2011). Animal-based measures are gathered through direct assessment of the animal on farm, while environment-based measures evaluate features of the housing facilities. Resource-based measures are derived from evaluation of records, documents, and interviews with farm personnel. Table 21.1 provides specific examples of measures for each category. By combining findings from metrics within each category, thorough conclusions about the animal welfare experienced on farm can be determined. Verification programs have adapted this type of evaluation framework to assess animal welfare on farm and at processing facilities.

## Gestation Systems

After a gilt weans her first litter, she will typically spend about 85% of her reproductive lifetime within gestation housing. Gestation housing includes boar housing, breeding facilities, and housing for pregnant gilts and sows. On many farms, replacement gilt development also occurs under the gestation housing umbrella. However, this section will focus on housing for pregnant sows and gilts because this is the largest segment and primary driver of gestation housing system design decisions.

### *Individual Gestation Stalls*

In the United States, 97% of the swine breeding herd is housed in facilities without outside access, and over 75% of all gestating sows and gilts are housed in individual stalls (USDA 2015). Gestation stalls are usually tubular metal frames with a feeder and drinker positioned at one end. The flooring of the stall is commonly concrete, with 76% of gestation stalls having either full or partial slats (USDA 2015). Although there are variations in dimensions on commercial farms, a typical gestation stall is 2.2 m long, 0.6 m wide, and 1.0 m high. These dimensions restrict the ability of most sows to

**Table 21.1**  Example metrics for evaluating swine welfare on farm[a].

| Animal based[b] | Environment based[c] | Resource based[d] |
| --- | --- | --- |
| Eye condition | Ammonia concentrations measured in barn during site visit | Ammonia concentrations recorded weekly by producer |
| Thermal comfort (i.e., location of pigs in relation to one another) | Ambient air temperature of barn during site visit | Regular monitoring, maintenance, and timely repair of heating, cooling, and ventilation equipment documented |
| Body condition score | Space allowance | Feed delivery records |
| | Waters and feeders in good state of repair | Employee training documented |
| Lesion score | Identify pen protrusions that could injure an animal | Zero tolerance of abuse policy documented |
| | | Written emergency procedures available |
| Lameness/gait score | Euthanasia equipment readily available and operational | Mortality and culling records with reasons outlined |

[a] Based on Mench (2003), Rushen et al. (2011), and NPB (2019).
[b] Animal-based measures are measures gathered from direct assessment of the animal on farm.
[c] Environment-based measures evaluate features of the housing facilities.
[d] Resource-based measures are derived from evaluation of records, documents and interviews with farm personnel.

turn around (McGlone et al. 2004a; Li and Gonyou 2006). Gestation stalls provide the opportunity for individual feeding regimes as well as allowing a small number of workers to effectively manage a large number of sows. Gestation stalls prevent some physical altercations between sows (McGlone et al. 2004b; Rhodes et al. 2005; McGlone 2013), but impacts of aggressive interactions between neighbors may be exasperated by the inability for the less dominant animal to flee (Marchant-Forde 2010). Gestation stalls prevent sows from performing exploratory behavior and regulating their thermal environment by moving to preferred micro-climates within a barn or pen (Rhodes et al. 2005; McGlone 2013). Stereotypies—a behavior pattern that is repetitive, invariant, and has no obvious goal or function (Mason 1991)—occur more frequently in individually housed sows (McGlone et al. 2004a, 2004b; Rhodes et al. 2005; McGlone 2013), but it is not possible to determine from available studies whether stereotypies are caused by management factors associated with gestation management (i.e. limit feeding and lack of physical substrate to manipulate) or by gestation stalls directly.

Group housing is a broad alternative strategy to gestation stalls with numerous variants. The main differences in group housing systems include the number of sows in a group, space allocated per sow, and feed delivery method. In the literature, sow number per group can range from three animals (Barnett 1997) to as many as 200 (Barbari 2000) with space per pig usually ranging between $1\,m^2$/pig and $5\,m^2$/pig (Schau et al. 2013; Verdon et al. 2015). Three general types of feeding systems for dry feed are used in group housed sows. They include floor feeding, individual feeding stalls, and electronic sow feeders.

### *Group Housing with Floor Feeding*

Delivering feed manually or automatically onto the floor pen is one of the simplest methods for feeding group housed sows. Floor fed sows (Figure 21.1) are either kept in small groups or feed droppers are spread throughout the pen to accommodate approximately six to eight sows per feed delivery point (Marchant-Forde 2009). Variation in feed intake and body condition among group housed sows is

**Figure 21.1**    Group housed sows with floor feeding. Source: Photograph: Provided by Monique D. Pairis-Garcia.

most pronounced in floor-feeding systems as aggressive sows will outcompete subordinate pen mates (Verdon et al. 2015). Salak-Johnson et al. (2007) reported that increasing floor space from 1.4 to 2.3 m² improved body condition score and reduced lesions on the head, neck, shoulder, rear, and hind legs in floor fed sows housed in pens of five. Similarly, Hemsworth et al. (2016) demonstrated that providing more floor space immediately after mixing group housed sows reduced aggressive behavior but sows adapted to reduced floor space over time. Aggression was highest during feeding, which in that study occurred four times per day via dropping feed onto the floor of the pen (Hemsworth et al. 2016).

### Group Housing with Feeding Stalls

Aggression around feed can be reduced in group housed sows by providing each sow with an individual feeding stall, particularly those which have a gated full-body length stall (Barnett et al. 1992; Barnett 1997; Andersen et al. 1999). Self-closing free access stalls allow sows the opportunity to choose whether to socialize with pen cohorts or separate themselves from the group (Harmon and Levis 2003). Usually, these feeding/loafing stalls can lock an animal in to facilitate caretaker duties including vaccinating, estrus detection, inspection, and other activities. While generally less competitive than floor feeding systems, gaining access to preferred feeding stalls can also lead to aggression between group housed sows (Verdon et al. 2015).

### Group Housing with Electronic Sow Feeders

Electronic sow feeders (Figure 21.2) allow individualized feeding of sows in a protected environment. Typically, 40–50 sows can be fed from one feeding station sequentially. After a sow enters the feeding area, a protective gate closes behind her to prevent other sows from entering. Individual sows are identified via transponders in the ear tag and feed is dispensed from the computer controlled feeder. Frequency and amount of feed dispensed will vary across designs but generally small

**Figure 21.2** Group housed sows with bedding and electronic sow feeder. Source; Photograph: Provided by Monique D. Pairis-Garcia.

amounts of feed are delivered until the sow either exits the feeder – thereby ending the dispensing of feed until a later visit to the feeder – or the sow has received her entire feed allotment. The system typically cycles every 24 hours and if maintained and managed well can provide valuable information to animal caretakers regarding sow activity and feed consumption. Gilts and sows unfamiliar with electronic sow feeders require training, but this is generally easily accomplished with well-designed training pens and protocols.

Large groups of sows may be able to more efficiently use limited floor space as communal areas can be shared. Work by Li et al. (2018) indicates that providing 1.5 m² per sow in an electronic sow feeder system is adequate for maintaining reproductive performance and welfare of gestating sows. Increasing floor space allowance from 1.5 to 2.1 m² did not affect performance or skin lesion scores of sows (Li et al. 2018). Because electronic sow feeders require sows to eat sequentially, aggressive behavior may still occur particularly around the entrance to the feeder. Vulva injuries may be more frequent when electronic sow feeders are used, especially if sufficient space is not provided between the sow that is eating and the sows waiting to enter the feeder (McGlone et al. 2004b; Rhodes et al. 2005; McGlone 2013; Verdon et al. 2015).

### Static vs. Dynamic Groups

Pigs establish a social hierarchy within a group and frequent mixing of groups disrupt that order and usually results in more aggression and injury as the social hierarchy must be re-established with the introduction or removal of any pig. Under dynamic group management, the number of sows per pen is kept fairly constant by adding pregnant sows that are early in their gestation cycle to replace sows removed prior to parturition or due to culling. Alternatively, static group management occurs when individual sows are grouped together early in gestation with no additions to the pen occurring as sows are culled or removed for farrowing. Dynamic systems allow for maximal use of space and are akin to a continuous flow strategy for growing pigs. Static systems require batching groups of sows based on expected farrowing date and are analogous to an all-in-all-out management system.

Housing sows in dynamic gestation groups increased total injury score and reduced nonagonistic social interactions compared to gestation sows kept in static groups (Anil et al. 2006). In that study, group size was kept at 50 sows per pen, sows were provided 1.72 m² of floor space, and feed was delivered via an electronic sow feeder. Although farrowing performance and longevity were not impacted between groups, welfare was reduced in the dynamic system (Anil et al. 2006). In another study, it was demonstrated that maintaining static groups of females for 2 gestation cycles reduced aggression between the first gestation period and the second (Campler et al. 2019). In that study, groups of 19 or 20 sows were housed in 6.8×5.5 m pens and fed using an electronic sow feeder (Campler et al. 2019). In the second gestation cycle, there was more space per sow as nine sows did not rebreed in synch with the group or were otherwise removed (Campler et al. 2019).

### Provision of Bedding and Other Substrates

Providing straw or other bedding could potentially increase sow comfort and improve overall welfare in all gestation systems as it is an important stimulus and outlet for exploration, foraging, rooting, and chewing behavior (Tuyttens 2005). In practice, group pens may provide the most realistic opportunity for incorporating bedding. Reproductive performance was similar between systems using individual gestation stalls and group pens with feeding stalls in bedded hoop barns (Lammers et al. 2007). Because of lower construction costs for a hoop barn as compared to a mechanically ventilated gestation facility, weaned pig costs was found to be reduced for the group housed sows (Lammers et al. 2008).

The wide range of factors associated with gestation sow housing systems makes it difficult to objectively compare overall welfare across qualitatively different systems. While group housing systems allow more freedom of movement and social interaction, they can also be associated with more aggression around feeding and uneven body conditioning. Individual gestation stalls may afford a very high level of protection against aggression during feeding, but gestation stalls in existing barns are often simply too narrow to support good welfare for late parity and otherwise large sows. Existing reviews support the conclusion that sow productivity can be similar across well-managed systems and that welfare of sows can also be improved in most cases ((McGlone et al. 2004b; Rhodes et al. 2005; McGlone 2013).

## Farrowing Systems

Farrowing systems utilized in swine production can vary in design based on the needs of the sow, piglet, and caretaker. In general, farrowing systems include traditional stalls, modified stalls, pens, groups, and outdoor systems. Regardless of the system used, the producer must balance the behavioral needs of the sow and piglet while supporting the management and productivity of the system. This section will outline the advantages and disadvantages of the four most commonly used systems to date, addressing the needs of the sow, piglet and producer.

### Traditional Stalls

Traditional farrowing stalls (Figure 21.3) were widely adapted in the 1960s in an effort to minimize piglet mortality (Roberston et al. 1966), as piglet mortality specific to crushing may be reduced when using a stall that confines and controls sow movement (Kilbride et al. 2012). Table 21.2

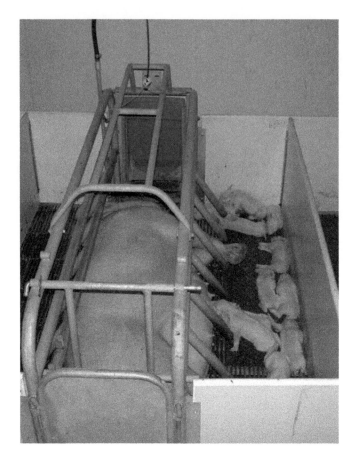

**Figure 21.3** Traditional farrowing stall (2.0×0.61 m) with metal bars running the length of the stall. Source: Photograph: Courtesy of Steven Moeller, The Ohio State University, Columbus, OH, US.

summarizes reported mortality rates for different farrowing systems. Although this this may be beneficial for the piglet, the sow's movement is restricted and prevents her from performing nest building behaviors.

### Modified Stalls

In an effort to improve physical space, modified stalls have been implemented. Modified stalls consist of a traditional farrowing stall that can be adjusted and opened to allow the increased movement of the sow. Modified stall designs include ellipsoid, turn-around and 360° Farrower™ stalls. Although these systems provide benefit for the sow from a behavioral standpoint, sows are still typically confined during the active stage of nest building and are not provided with nest building materials (Baxter et al. 2012). Therefore, although sows may have access to more space during lactation, modified stalls do not address the behavioral needs of the sow during parturition. From the piglet and producer perspective, modified stalls may not dramatically decrease piglet mortality (Table 21.2), but this system does allow flexibility for the producer, as they can confine the sow when needed (Weber 2000).

**Table 21.2** Reported mortality rates of farrowing systems[a].

| | Mortality, % | | |
|---|---|---|---|
| System | Total[b] | Liveborn[c] | Crushing[d] |
| Traditional stalls | 18.1–18.3 | 6.1–11.7 | 4.6–5.5 |
| Modified stalls | 16.3–17.4 | 11.4–11.7 | NR[e] |
| Simple pens | 16.4–20.7 | 10.2–14.2 | NR |
| Designed pens | 15.0–16.6 | 11.8 | NR |
| Group | 11.8–28.0 | 10.9–22.6 | 5.6–6.0 |
| Outdoor | 17.1–20.4 | 12.8–16.8 | 8.9 |

[a] Vosough Ahmadi et al. (2011), Baxter et al. (2012), Chidgey et al. (2015), Kilbride et al. (2012), Baxter et al. (2009), Weber et al. (2009), Li et al. (2010), and Mazzoni et al. (2018).
[b] Still born and live born deaths combined.
[c] Excluding still born deaths.
[d] Live born deaths from crushing, excluding deaths from unknown causes, %.
[e] NR = not reported.

### Individual Pens

To address the behavioral needs of the sow, pens or free farrowing systems have been utilized to facilitate sow movement and provide suitable nesting material prior to and during parturition. Pen systems consist of an open space designed in a manner to encourage the sow to lie down in a specific location within the pen. Elaborate pen systems are designed to include designated areas for farrowing, nesting (creep area for piglets), and elimination (Cronin et al. 2000; Figure 21.4). More simplified pen examples include hillside, sloped, and mushroom pens (E.g. Werribee, Schmid and pigSAFE pens; Vosough Ahmadi et al. 2011; Guy et al. 2012). Pen systems allow sows to express a greater range of behaviors through increased physical space. In addition, if the pen is designed in such a way to provide substrate (i.e. solid flooring), sows can fulfill their motivation to nest build (Cronin et al. 1994).

**Figure 21.4** Pen system design with designated areas for farrowing and elimination. A creep area for the piglets is located in the corner of the pen. Substrate is provided for nest building. Source: Photograph: Provided by Monique D. Pairis-Garcia.

Although pen systems can provide additional space to the litter and sow during lactation, piglet mortality within the first three days of life can increase. Both simple and elaborate pen designs have protective structures (i.e. railings, protrusions) to protect piglets from being crushed when the sow lies down (McGlone and Blecha 1987; Blackshaw et al. 1994). Although these designs help minimize crushing, total mortality rates can be as high as 20.7% (15–20.7%; Vosough Ahmadi et al. 2011; Baxter et al. 2012). This high mortality rate may be associated with challenges faced by gilts in pen systems. For example, work conducted by Cronin et al. (2000) observed that gilts were more likely to farrow in unintended areas (e.g. elimination areas), resulting in increased piglet death due to hypothermia. Therefore, pen design and animal experience can both be critical factors influencing piglet mortality.

### Group Farrowing

Group farrowing systems offer increased space and nesting material, as well as a diversified social environment for the sow. Group farrowing systems can be in either an indoor or outdoor setting and consist of a common shared space and individual farrowing areas (Figure 21.3). Individual farrowing areas may include paddocks, A-frames, runs, or boxes and should support the sow's desire to isolate herself from the group prior to farrowing (Figures 21.4, 21.5, and 21.6). As with pen systems, group housing in farrowing may result in higher piglet mortality; however, this is dependent upon house design, sow experience, and management.

From the producer perspective, farrowing systems need to be profitable, as well as sustainable in an industry with fluid profit and loss cycles (Johnson and Marchant-Forde 2009). In addition, labor costs and worker safety must also be considered from both a practical and financial standpoint. Traditional and modified stalls allow caretakers to easily access both the sow and piglets for routine care (processing, treatment). In these systems, sows can be individually evaluated from a nutritional

(a)                                            (b)

**Figure 21.5** Indoor group (a) and outdoor group (b) systems with communal shared space and individual areas for farrowing. Source: Photograph: Provided by Monique D. Pairis-Garcia.

**Figure 21.6**  Outdoor huts designed to allow sows to isolate themselves from the group to farrow. Source: Photograph: Provided by Monique D. Pairis-Garcia.

and health standpoint and feed can be adjusted based on body condition and lactation stage (Roberston et al. 1966).

Labor costs may be increased within pen farrowing systems compared to traditional stalls, as sows in loose housing may be more difficult to manage when intervention is required. Sow temperament is an additional concern within the pen environment as there are minimal opportunities for restraint from a management standpoint. Previous work suggests that caretakers may be less likely to assist sows within pen farrowing compared to traditional stalls, which could result in welfare concerns for both the sow and the piglet (Cronin et al. 2000).

Table 21.3 summarizes farrowing systems relative to floor space per sow. Indoor group farrowing systems have the highest labor costs and greatest economic footprint for a farrowing system (Kerr et al. 1988). This high economic impact is due to capital investment, labor input, and high piglet mortality. An example of an indoor group system is a deep-bedded system developed in Sweden described by Algers(1991) and Honeyman (1995). In an attempt to replicate this system in the United States, the farrowing rate and group lactation were successful, but the free-choice farrowing boxes had high piglet mortality with most mortalities occurring in the first week after birth and due to crushing (29% prewean mortality over 9 farrowings) (Honeyman and Kent 2001). Lammers et al. (2004) describe several indoor, group farrowing systems used in the Midwest U.S. Ultimately, for group farrowing to be a successful system, genetic selection valuing maternal behavior and highly trained labor are critical.

Outdoor farrowing systems are able to meet or exceed similar space and social requirements of indoor group housed systems while remaining economically efficient (Baxter et al. 2012). This farrowing system does require more labor from caretakers to manage sows across large distances, although

**Table 21.3**  Comparison of space allowance for across systems[a].

| System | Floor space, m²/sow and litter | Stocking density, sows/100 m² area |
|---|---|---|
| Traditional stalls | 3.5–3.6 | 27–28 |
| Modified stalls | 4.3–5.3 | 18–23 |
| Simple pens | 3.4–10.5 | 9–29 |
| Designed pens | 7.1–8.9 | 11–14 |
| Group | Variable | |
| Outdoor | Variable | |

[a] Based on Vosough Ahmadi et al. (2011), Baxter et al. (2012), and Pairis-Garcia (2015).

some operations keep labor low by using layouts that streamline operations. The use of specialized equipment (e.g. all-terrain vehicles, trailers, temporary holding pens, huts, and electric fence) can aid in allowing caretakers to handle and manage pigs safely and effectively (Pairis-Garcia 2015; Honeyman et al. 2001). The advent of low cost GPS systems may also prove useful in outdoor systems.

In outdoor systems, the farrowing hut is the major modifier of the environment for the sow and piglet. These simple structures can be made from wood, steel, or plastic. Various floorless hut types (five) were compared in Iowa (Honeyman and Roush 2002) over four years. Prewean mortality was impacted by hut design width. Superior huts had a defined and larger sow area with designated space that provided pliglets space protection from the sow. Clean, absorptive, plentiful bedding is critical in outdoor farrowing systems. Bedding allows the sow to perform nest building behavior and creates a microenvironment for the piglets that is dry and draft free. Outdoor farrowing requires lower capital investment with minimal maintenance costs for producers suggesting this could be a sustainable alternative.

A key to managing group farrowing systems in either indoor or outdoor systems is to group sows so that all sows within the group farrow within a narrow time frame. This prevents older, larger piglets from stealing milk from younger litters as well as facilitates batching of labor activities. Flexible floor layouts and temporary fencing support altering the size of group-farrowing pens as needed. A mix of indoor and outdoor systems may be a model for sustainable, alternative production of natural pork (Honeyman 1996; Honeyman 2005).

Any farrowing system must address the needs of the sow, piglet, and producer in an effort to optimize pig welfare while ensuring the economic productivity of the system (Johnson and Marchant-Forde 2011). For any farrowing system design, creating an environment that supports the sow's natural behavior while minimizing piglet mortality should be a priority. The system should also be designed in a way to allow the producer to provide the care needed for the pigs in an environment that is safe and effective.

## Growing Pig Housing

Pigs grow rapidly and convert feed efficiently to weight gain from birth to market. The rapid growth is maximized when environmental and nutritional needs of the pig are met. Because of this rapid growth, the environmental and nutritional needs or requirements of the growing pig quickly change over time. Housing systems and the accompanying management approaches are designed to meet and adjust to these dynamic needs.

Unlike other domestic mammalian species, pigs are covered with minimal hair. Pigs historically relied on subcutaneous fat layers for insulation and protection. This subcutaneous fat has been effectively removed through decades of animal breeding and selection for leanness. Thus, the modern pig is even more dependent on the housing system for protection and environmental buffering.

Housing systems for the weaned, growing, and finishing pig have evolved rapidly with "modern" confinement barns beginning in the 1960s and becoming widespread. After about 1990, multisite production systems that emphasized biosecurity, pig flow, and all-in/all-out management became prevalent. The multisite systems were adopted to enhance pig health and improve pig performance. Multisite systems feature (Site 1) sow farms that produce piglets often weaned at or prior to 21 days. Site 2 locations are nursery barns where the weaned pigs are reared. After five to seven weeks or around 23 kg live weight, pigs are moved to finishing sites for feeding to market weight (Site 3; Harris 2000). Traditionally, flooring in nursery facilities is slotted plastic or metal with solid pen dividers to minimize the spread of infectious diseases between pens. Finishing barns almost always have concrete slats for flooring and metal gating to separate pens.

Wean-to-finish barns were introduced and widely adopted after 2000. These facilities placed the weaned pig directly into finishing barns and reduced or eliminated some labor costs as well as additional transport stress for the pig. Today, wean-to-finish barns are not quite as popular because pig companies are recognizing that managing newly weaned pigs requires a different set of management skills than managing a finishing pig. There is continued interest in placing finishing barns near crop fields so that the manure can be used as fertilizer for growing crops.

### *Group Size and Stocking Density*

Growing pigs are able to perform well in a wide range of group sizes. Typical group size is approximately 20–30 grow-finish pigs per pen; however, research confirms that pig performance is sustained in larger group sizes if stocking density remains constant. Wolter et al. (2001) compared performance of groups of 25, 50, and 100 wean-to-finish pigs per pen uniformly allocated 0.68 m$^2$ floor space and 4.3 cm feeder space per pig. Pigs were raised from 5.9 kg to a mean pen weight of 116 kg. No growth performance or carcass traits were impacted by group size and mortality rates were similar across treatments (Wolter et al. 2001). Morbidity was higher for pens of 25 pigs as compared to pens of 50 pigs but was not statistically different from morbidity in pens of 100 pigs (Wolter et al. 2001). Similarly, Schmolke et al. (2003) reported no difference in growth performance or incidence of tail biting in pens of 10, 20, 40, or 80 pigs (initial body weight 23 kg, final body weight 95 kg) when all pigs were allotted 0.76 m$^2$ floor space and 3.2 cm feeder space per pig.

Because of the small size of weaned pigs compared to finishing pigs, when weaned pigs are placed in a finishing barn, there is extra space in the pen. This may be addressed by overstocking or double stocking the pens until the pigs grow larger to utilize the available space more efficiently. Wolter et al. (2002a) examined the performance of pigs in double-stocked pens for 70 days following weaning at 17 days of age in a pair of experiments. In the first experiment, floor and feed space was kept uniform across pens, but pens were either stocked with 52 or 104 weaned pigs. This resulted in a floor space and feeder allowance of 0.65 m$^2$ and 4 cm for the pigs in single-stocked pens and 0.325 m$^2$ and 2 cm for pigs kept in the double-stocked pens. In the second experiment, smaller pens and group sizes were used—27 pigs/pen, 0.64 m$^2$/pig, and 5.6 cm feeder space/pig vs 54 pigs/pen, 0.32 m$^2$/pig, and 2.8 cm feeder space/pig. Double stocking reduced growth rate and increased culling in pigs in both experiments (Wolter et al. 2002a).

In a subsequent study, pens of 108 weaned pigs (0.313 m²/pig) were provided either 2 or 4 cm of feeder space/pig for eight weeks (Wolter et al. 2002b). Growth performance was not impacted by feeder space allowance for the first six weeks following weaning. However, for weeks six to eight and for the entire eight-week study period, growth rate was lower, and morbidity was higher for pigs allocated 2 cm of feeder space (Wolter et al. 2002b).

Laskoski et al. (2019) reported that group size and feeder space allowance impacts growth rate and incidence of lesions to the ear and tail of nursery pigs. Newly weaned pigs (20.5 days of age, 5.6 kg BW) were assigned to pens of 15, 20, 25, or 30 pigs for a 42-day long feeding trial. Total feeder space provided was uniform for each pen resulting in feeder space per pig of 4.3, 3.2, 2.6, and 2.1 cm/pig, respectively. Space allowance within the pen was kept constant (0.23 m²/pig) across treatments. It is impossible to determine if the reduced growth rate and higher incidence of ear and tail lesions is best attributed to the number of pigs per pen or feeder space allowance per pig as these factors are confounded (Laskoski et al. 2019).

Space allowance (0.52 m²/pig vs 0.78 m²/pig) but not group size (18 vs 108 pigs/pen) has been shown to negatively impact the growth performance of 37 kg pigs fed for seven to eight weeks (Street and Gonyou 2008). Pigs with more space grew faster and more efficiently than crowded pigs. Pigs housed in large groups had poorer scores for lameness and leg lesion scores, but morbidity levels were not different across treatments. Crowding pigs increased lameness in large groups but decreased lameness in small groups. The negative impacts of reduced floor space were most dramatic as pigs grew larger (Street and Gonyou 2008).

Space allowance also has an impact on the growth performance in pigs raised from 27 to 138 kg (Johnston et al. 2017). In that multistate study, pigs were allotted 0.71, 0.80, 0.89, 0.98, or 1.07 m²/pig. Feeder space across locations ranged from 4.8 to 9.7 cm/pig but was held constant across floor space treatments within each location. Increasing floor space allowance up to 0.89 m²/pig increased final bodyweight and increased ADG and ADFI, but salivary cortisol concentrations and lesion score were not impacted by stocking density (Johnston et al. 2017). The negative consequences of overstocking were mostly observed as pigs approached market weight (Johnston et al. 2017).

Floor space allowance is one of the most significant factors affecting growth performance, economic return, and animal welfare for growing pigs. Providing inadequate space clearly reduces growth rate and welfare (Wolter et al. 2002a; Street and Gonyou 2008; Johnston et al. 2017), but it is recognized that space requirement changes as an animal grows. Expressing space requirement in terms of space per animal (m²/pig) or in terms of weight density (kg/m²) is straight forward, but is not precise over a wide range of body weights. Alternatively, an allometric approach using the following equation:

$$A = k \times BW^{0.667}$$

where	A: area in m²
k = space allowance coeffienct
BW = body weight in kg

was evaluated in a review paper by Gonyou et al. (2006). Based on growth performance from published work, Gonyou et al. (2006) suggest that the critical value of k is between 0.0317 and 0.0348. Table 21.4 summarizes space allowance (m²/pig) for various body weights using the allometric approach.

**Table 21.4** Recommended space allowance for growing pigs[a].

| Pig body weight, kg | Space allowance, m²/head[b] |
|---|---|
| 15 | 0.19–0.21 |
| 45 | 0.40–0.44 |
| 75 | 0.56–0.62 |
| 105 | 0.71–0.78 |
| 135 | 0.84–0.92 |

[a] Based on the following equation $A = k \times BW^{0.667}$ (Gonyou et al. 2006).
[b] Lower values calculated with k = 0.0317; upper values calculated with k = 0.0348

### Feeders and Drinkers or Waterers

Perhaps one of the more important but often overlooked features of swine barns are the feeders and waterers (drinkers). These devices are responsible for delivering high-quality nutrients to the pigs with minimal waste. This is no minor feat in swine barns as pigs are remarkably good at wasting feed and water!

Feed waste occurs primarily due to feeder design and feeder adjustment. Feed waste in growing pigs ranges from 2 to 12% (van Heutgen 2010). Monitoring and adjusting feeders to reduce waste is a critical component of growing pig management. Reducing feed flow so that 25–50% of the feed pan is covered with feed is generally recommended (DeRouchey and Richert 2010; van Heutgen 2010).

Reducing water waste is also a worthwhile effort. Waste water dilutes manure, fills manure storage, and increases water costs. In a summary of water waste in pig barns by waterer type with conventional nipple waterers as the base for comparison, swinging nipples reduced water waste by 11%, managing height and flow by 16–26%, bite-style nipples by 8–22%, and bowls or cups by 9–31% (Muhlbauer et al. 2010). Wet–dry feeders provide both feed and water in the same pan and may reduce water wastage while maintaining or improving growth performance (van Heutgen 2010; Muhlbauer et al. 2010; Meyers et al. 2013; Li et al. 2017). Total time spent eating was significantly reduced in finishing pigs fed using a wet–dry feeder as compared to a dry feeder (Li et al. 2017). Because each pig required less time to consume adequate feed, competition for access to a given space at a wet–dry feeder should be less than at a dry feeder (Li et al. 2017). This has obvious implications for improving animal welfare as feeding systems with high levels of competition may be detrimental to individual pig performance and welfare even if pen average outcomes are similar (Botermans and Svendsen 2000; Georgsson and Svendsen 2002).

### Mixing

Growing pigs incurs stress when moved or mixed, which may result in aggression or fighting. These events should be minimized and closely monitored; however, some changes to group dynamics are unavoidable, such as weaning. Younger pigs seem to adjust to changes in groups with less fighting than older pigs. Mixing pigs over about 34 to 45 kg should be avoided when possible, as because fighting will be intense. If mixing is unavoidable, some tips that may help include allow extra floor, feeder and waterer space, move all pigs to a new space rather than add pigs to an existing group, add bedding if possible, and check pigs frequently. One advantage of the wean-to-finish system is that pig moves and mixing can be minimized.

## Bedded Systems

Partially in response to the major shift to confinement swine housing, a counter/alternative production system movement has flourished since the mid-1990s that is driven by consumer demand for certain pork quality and credence attributes (Honeyman et al. 2006). These "pork niche market" production systems typically feature bedding, minimal environmental controls, natural ventilation, more space allocated per pig, no antibiotics, certain swine breeds, and specified feed ingredients. Many use bedded hoop barns or other alternative structures (Honeyman and Harmon 2003; Honeyman 2005) and are smaller operations linked in marketing networks. This trend is analogous to other livestock production trends of cage-free eggs, free-range chickens, grass-fed beef, or organic dairy. Although this is a small part of the US pork market, the niche pork production is consumer-responsive, seems vigorous, growing, and is often a forecast of upcoming trends.

The use of bedding such as small-grain straws or cornstalks is beneficial to absorb fecal and urinary wetness and provide the pig with the ability to modify their environment to reduce draft and heat loss in cold environments. The bedded pack actually composts in place, aerated by the rooting pigs, and can generate heat, which is beneficial in cold environments. During hot periods, the pigs will frequent wetter parts of the pack that are not composting. Bedded areas can also be sprinkled to help pigs stay cool. Bedding adds cost while providing the benefits described. In addition, bedding provides a stimulating continually changing environment for the pigs and lessens aggression and stereotypical behaviors. Clean dry bedding is critical for success in these systems.

## Summary

Acknowledging the natural environment, diet, social groupings, and behavior of wild pigs allows for critical evaluation of currently used production systems and their impact on swine welfare. Behavioral and biological motivations must be considered when choosing housing environments and management strategies to create an environment suitable for these animals. Doing so may mitigate the performance of abnormal and unwanted behaviors and ultimately supports good animal welfare within a sustainable swine production system.

## References

Algers, B. 1991. Group housing of farrowing sows, health aspects of a new system. Proc. 7th Int. Congress on Animal Hygiene. Liepzig, Germany. p. 851–857.

Algers, B., and K. Uvnas-Moberg. 2007. Maternal behavior in pigs. Horm. Behav. 52:78–85. doi: https://doi.org/10.1016/j.yhbeh.2007.03.022.

Andersen, I. L., E. Bøe, and A. L. Kristiansen. 1999. The influence of different feeding arrangements and food type on competition at feeding in pregnant sows. Appl. Anim. Behav. Sci. 65:91–104. doi: https://doi.org/10.1016/S0168-159(99)00058-1.

Anil, L., S. S. Anil, J. Deen, S. K. Baidoo, and R. D. Walker. 2006. Effect of group size and structure on welfare and performance of pregnant sows in pens with electronic sow feeders. Can. J. Vet. Res. 70:128–136.

Barbari M. 2000. Analysis of reproductive performance of sows in relation to housing systems. In: Proc. 1st Int. Conf. on Swine Housing. Des Moines, IA, US. p. 188–196.

Barnett, J. L., P. H. Hemsworth, G. M. Cronin, E. A. Newman, T. H. McCallum, and D. Chilton. 1992. Effects of pen size, partial stalls and method of feeding on welfare-related behavioural and physiological responses of group-housed pigs. Appl. Anima. Behav. Sci. 91:207–220. doi: https://doi.org/10.1016/S0168-159(05)80116-9.

Barnett, J. L. 1997. Modifying the design of pens with individual feeding places affects the welfare of pigs. In: Proc. 5th Int. Livest. Environ. Symp. p. 613–618

Barrett, R. 1978. The feral hog at Dye Creek ranch, California. 46:283–355. doi: https://doi.org/10.3733/hilg.v46n09p283.

Baxter, E. M., S. Jarvis, L. Sherwood, S. K. Robson, E. Ormandy, M. Farish, K. M. Smurthwaite, R. Roehe, A. B. Lawrence, and S. A. Edwards. 2009. Indicators of piglet survival in outdoor farrowing systems. Livest. Sci. 124:266–276.

Baxter, E. M., A. B. Lawrence, and S. A. Edwards. 2012. Alternative farrowing accommodation: welfare and economic aspects of existing farrowing and lactation systems for pigs. Animal 6:96–117. doi: https://doi.org/10.1017/S1751731111001224.

Blackshaw, J. K., A. W. Blackshaw, F. J. Thomas, and F. W. Newman. 1994. Comparison of behaviour patterns of sows and litters in a farrowing crate and a farrowing pen. Appl. Anim. Behav. Sci. 39:281–295. doi: https://doi.org/10.1016/0168-1591(94)90163-5.

Botermans, J. A. M. and Svendsen, J. 2000. Effect of feeding environment on performance, injuries and behaviour in growing-finishing pigs: group-based studies. Acta Agric. Scand. Sect. A Anim. Sci. 50: 237–249. doi: https://doi.org/10.1080/090647000750069430.

Burne, T. H. J., P. J. E. Murfitt, and C. L. Gilbert. 2000. Deprivation of straw bedding alters PGF2 induced nesting behavior in female pigs. Appl. Anim. Behav. Sci. 69:215–225. doi: https://doi.org/10.1016/S0168-1591(00)00135-0.

Campler, M., M. Pairis-Garcia J. Kieffer, and S. Moeller. 2019. Sow behavior and productivity in a small stable group-housing system. J. Swine Health Prod. 27:76–86.

Chidgey, K. L., P. C.H. Morel, K. J. Stafford, and I. W. Barugh. 2015. Sow and piglet productivity and sow reproductive performance in farrowing pens with temporary crating or farrowing crates on a commercial New Zealand pig farm. Livest. Sci. 173:87–94.

Condous, P. C., K. J. Plush, A. J. Tilbrook, and W. H. E. van Wettere. 2016. Reducing sow confinement during farrowing and in early lactation increases piglet mortality. J. Anim. Sci. 94:3022–3029. doi: https://doi.org/10.2527/jas.2015-0145.

Cousse, S., F. Spitz, M. Hewison, and G. Janeau. 1994. Use of space by juveniles in relation to their postnatal range, mother, and siblings: An example in the wild boar, Sus scrofa. Can. J. Zool. 72:1691–1694. doi: https://doi.org/10.1139/z94-227.

Cronin, G. M., J. A. Smith, F. M. Hodge, and P. H. Hemsworth. 1994. The behavior of primiparous sows around farrowing in response to restraint and straw bedding. Appl. Anim. Behav. Sci. 39:269–280. doi: https://doi.org/10.1016/0168-1591(94)90162-7.

Cronin, G. M., J. L. Barnett, F. M. Hodge, J. A. Smith, and T. H. Mccallum. 2000. A comparison of piglet production and survival in the Werribee farrowing pen and conventional farrowing crates at a commercial farm. Aus. J. Exp. Agric. 40:17–23. doi: https://doi.org/10.1071/EA99124.

Dardaillon, M. 1987. Seasonal feeding habits of the wild boar in a Mediterranean Wetland, the Camargue (Southern France) Acta Theriol. 32:389–401. doi: https://doi.org/10.4098/AT.arch.87-28.

D'Eath, R. B., and S. P. Turner. 2009. The natural behaviour of the pig. In: J. N. Marchant-Forde. The welfare of pigs. Springer Publishing, New York, NY. p. 13–45. doi: https://doi.org/10.1007/978-1-4020-8909-1.

DeRouchey, J. M., and B. T. Richert. 2010. Feeding systems for swine. PIG 07-01-09. Natl. Swine Nutr. Guide. http://porkgateway.org/resource/feeding-systems-for-swine/.

Eisenberg, J. F., and M. Lockhart. 1972. An ecological reconnaissance of Wilpattu National Park, Ceylon. Smithsonian Inst. Press, Washington, DC.

Focardi, S., D. Capizzi, and D. Monetti. 2000. Competition for acorns among wild boar (Sus scrofa) and small mammals in a Mediterranean woodland. J. Zool. 250:329–334. doi: https://doi.org/10.1111/j.1469-7998.2000.tb00777.x.

Frädrich, H. 1974. A comparison of behaviour in the *Suidae*. In: V. Geist, and F. Walther, editors, The behaviour of ungulates and its relation to management. International Union for Conservation of Nature and Natural Resources Publication No. 24. IUCN Publishers, Morges, Switzerland. p. 133–143.

Fraser, D. 1975. The nursing and suckling behavior of pigs. IV. The effect of interrupting the suckling stimulus. Br. Vet. J. 131:549. doi: https://doi.org/10.1016/S0007-1935(17)35187-4.

Gabor, T. M., E. C. Hellgren, R. A. Van Den Bussche, and N. J. Silvy. 1999. Demography, sociospatial behaviour and genetics of feral pigs (Sus scrofa) in a semi-arid environment. J. Zool. 247:311–322.

Genov, P. 1981. Food composition of wild boar in North-Eastern and Western Poland. Acta Theriol. 26:185–205.

Georgsson, L., and J. Svendsen. 2002. Degree of competition at feeding differentially affects behavior and performance of group-housed growing-finishing pigs of different relative weights. J. Anim. Sci. 80:376–383. doi: https://doi.org/10.2527/2002.802376x.

Gonyou, H. W. 2001. The social behaviour of pigs. In: L. J. Keeling, and H. W. Gonyou, editors, Social behaviour in farm animals. CABI, Wallingford, Oxon, UK. p. 47–176.

Gonyou, H. W., M. C. Brumm, E. Bush, J. Deen, S. A. Edwards, T. Fangman, J. J. McGlone, M. Meunier-Salaun, R. B. Morrison, H. Spoolder, P. L. Sundberg, and A. K. Johnson. 2006. Application of broken-line analysis to assess floor space requirements of nursery and grower-finisher pigs expressed on an allometric basis. J. Anim. Sci. 84:229–235 doi: https://doi.org/10.2527/2006.841229x.

Graves, H. B. 1984. Behavior and ecology of wild and feral swine (Sus scrofa). J. Anim. Sci. 58:481–492. doi: https://doi.org/10.2527/jas1984.582482x.

Guy, J. H., P. J. Cain, Y. M. Seddon, E. M. Baxter, and S. A. Edwards. 2012. Economic evaluation of high 460 welfare indoor farrowing systems for pigs. Anim. Welf. 21:19–24. doi: https://doi.org/10.7120/096272812X13345905673520.

Harmon, J. D. and D. G. Levis. 2003. Sow housing options for gestation. Pork Information Gateway factsheet 01-01-03. Natl. Pork Board, Des Moines, IA, US.

Harris, D. L. 2000. Multi-site pig production. Iowa State University Press, Ames, IA. US.

Hemsworth P. H., R. S. Morrison, A. J. Tilbrook, K. L. Butler, M. Rice, and S. J. Moeller. 2016. Effects of varying floor space on aggressive behavior and cortisol concentrations in group-housed sows. J. Anim. Sci. 94:4809–4818. doi: https://doi.org/10.2527/jas2016-0583.

Herrero, J., A. Garcia-Serrano, S. Couto, V. M. Ortuno, and R. Garcia-Gonzalez. 2006. Diet of wild boar Sus scrofa L. and crop damage in an intensive agroecosystem. Eur. J. Wildlife Res. 52:245–250.

Honeyman, M. S. 1996. Sustainability issues of U.S. swine production. J. Anim. Sci. 74:1410–1417.

Honeyman, M. S. 1995. Västgötmodellen: Sweden's sustainable alternative for swine production. Am. J. Alt. Agric. 10:129–132.

Honeyman, M. S., and D. Kent. 2001. Performance of a Swedish deep-bedded feeder pig production system in Iowa. Am. J. Alt. Agric. 16:50–56.

Honeyman, M. S., J. J. McGlone, J. B. Kliebenstein, and B. E. Larson. 2001. Outdoor pig production. PIH-145. Pork Industry Handbook. Purdue Univ., W. Lafayette, IN, US. p. 9.

Honeyman, M. S., and W. B. Roush. 2002. The effects of outdoor farrowing hut type on prewean piglet mortality in Iowa. Am. J. Alt. Agri. 17:92–95.

Honeyman, M. S. and J. D. Harmon. 2003. Performance of finishing pigs in hoop structures and confinement during winter and summer. J. Anim. Sci. 81:1663–1670. doi: https://doi.org/10.2527/2003.8171663x.

Honeyman, M. S. 2005. Extensive bedded indoor and outdoor pig production systems in USA: Current trends and effects on animal care and product quality. Livest. Prod. Sci. 94:15–24.

Honeyman, M. S., R. S. Pirog, G. H. Huber, P. J. Lammers, and J. R. Hermann. 2006. The United States pork niche market phenomenon. J. Anim. Sci. 84:2269–2275.

Janeau, G., and F. Spitz. 1984. L'espace chez le sanglier (Sus scrofa L.) Occupation et mode d'utilisation. Gibier Faune Sauvage. 1:76–89.

Jensen, P., and D. G. M. Wood-Gush. 1984. Social interactions in a group of free-ranging sows. Appl. Anim. Behav. Sci. 12:327–337. doi: https://doi.org/10.1016/0168-1591(84)90125-4.

Jensen, P. 1986. Observations on the maternal behaviour of free-ranging domestic pigs. Appl. Anim. Behav. 16:131–142. doi: https://doi.org/10.1016/0168-1591(86)90105-X.

Jensen, P. 1993. Nest building in domestic sows: the role of external stimuli. Anim. Behav. 45:351–358. doi: https://doi.org/10.1006/anbe.1993.1040

Johnson, A. K., and J. N. Marchant-Forde. 2009. Welfare of pigs in the farrowing environment. Anim. Sci. Pub. 103:141–188. doi: https://doi.org/10.1007/978-1-4020-8909-1_5.

Johnson, A. K., and J. N. Marchant-Forde. 2011. Farrowing systems for the sow and her piglets. Anim. Sci. White Papers, Tech. Rep. Fact Sheets. 17. https://lib.dr.iastate.edu/ans_whitepapers/17.

Johnston, L. J., D. W. Rozeboom, R. D. Goodband, S. J. Moeller, M. C. Shannon, and S. J. Scheick. 2017. Effect of floor space allowances on growth performance of finishing pigs marketed at 138 kilograms. J. Anim. Sci. 95:4917–4925. doi: https://doi.org/10.2527/jas2017.1870.

Kerr, S. G. C., D. G. M. Wood-Gush, H. Moser, and C. T. Whittemore. 1988. Enrichment of the production environment and the enhancement of welfare through the use of the Edinburgh Family Pen System of pig production. Res. Dev. Ag. 5:171–186.

Kilbride, A. L., M. Mendl, P. Statham, S. Held, M. Harris, S. Cooper, and L. E. Green. 2012. A cohort study of preweaning piglet mortality and farrowing accommodation on 112 commercial pig farms in England. Prev. Vet. Med. 104:281–291. doi: https://doi.org/10.1016/j.prevetmed.2011.11.011.

Kittawornrat, A., and J. J. Zimmerman. 2010. Toward a better understanding of pig behavior and pig welfare. Anim. Health Res. Rev. 12:25–32. doi: https://doi.org/10.1017/S1466252310000174.

Kurz, J. C., and R. L. Marchinton. 1972. Radiotelemetry studies of feral hogs in South Carolina. J. of Wildl. Manag. 36:1240–1248. doi: https://doi.org/10.2307/3799254.

Laskoski, F., J. E. G. Faccin, C. M. Vier, M. A. D. Gonçalves, U. A. D. Orlando, R. Kummer, A. P. G. Mellagi, M. L. Bernardi, I. Wentz, and F. P. Bortolozzo. 2019. Effects of pigs per feeder hole and group size on feed intake onset, growth performance, and ear and tail leisons in nursery pigs with consistent space allowance. J. Swine Health Prod. 27:12–18.

Lammers, P.J., M. S. Honeyman, and J. D. Harmon. 2004. Alternative systems for farrowing in cold weather. Agricultural engineers digest AED47. MidWest Plan Service, Ames IA, US.

Lammers, P. J., M. S. Honeyman, J. W. Mabry, and J. D. Harmon. 2007. Performance of gestating sows in bedded hoop barns and confinement stalls. J. Anim. Sci. 85:1311–1317. doi: https://doi.org/10.2527/jas.2006-437.

Lammers, P. J., M. S. Honeyman, J. B. Kliebenstein, and J. D. Harmon. 2008. Impact of gestation housing system on weaned pig production cost. Appl. Eng. Agric. 24:245–249.

Lawrence, A. B., and E. M. Terlouw. 1993. A review of behavioral factors involved in the development and continued performance of stereotypic behaviors in pigs. J. Anim. Sci. 71(10):2815–2825. doi: https://doi.org/10.2527/1993.71102815x.

Leaper, R., G. Massei, M. L. Gorman, and R. Aspinall. 1999. The feasibility of reintroducing Wild Boar (Sus scrofa) to Scotland. Mamm. Rev. 29:239–259. doi: https://doi.org/10.1046/j.1365-2907.1999.2940239.x

Li, Y. Z., and H. W. Gonyou. 2006. Effects of stall width and sow size on behavior of gestating sows. Can. J. Anim. Sci. 87:129–138.

Li, Y.Z., S. Q. Cui, X. J. Yang, L. J. Johnston, and S. K. Baidoo. 2018. Minimal floor space allowance for gestating sows kept in pens with electronic sow feeders on fully slatted floors. J. Anim. Sci. 96:4195–4208. doi: https://doi.org/10.1093/jas/sky282.

Li, Y., L. Johnston, and A. Hillbrands. 2010. Pre-weaning mortality of piglets in a bedded group-farrowing system. J. Swine. Health Prod. 18:75–80.

Li, Y. Z., K. A. McDonald, and H. W. Gonyou. 2017. Determining feeder space allowance across feed forms and water availability in the feeder for growing-finishing pigs. J. Swine Health Prod. 25:174–182.

Macdonald, D. W. 1984. The encyclopaedia of mammals. 3rd ed. Oxford University Press, London, UK. doi: https://doi.org/10.1093/acref/9780199206087.001.0001.

Mason, G. J. 1991. Stereotypies: A critical review. Animal Behav. 41:1015–1037.

Mauget, R. 1979. Mise en évidence, par captures-recaptures et radiotracking, du domaine vital chez le sanglier (Sus scrofa L.) en forêt de Chizé. Biol. Behav. 4:25–41.

Marchant-Forde, J. N. 2010. Social behavior in swine and its impact on welfare. In Proc. 21st IPVS Congress. Vancouver, Canada. p. 36–39.

Marchant-Forde, J. N. 2009. Welfare of dry sows. In: J. N. Marchant-Forde, editor, The welfare of pigs. Springer, New York, NY. p. 95–139.

Martin, J. T. 1975. Movement of feral pigs in North Canterbury, New Zealand. J. Mammol. 56:914–915.

Mazzoni, C., A. Scollo, F. Righi, E. Bigliardi, M. Di Ianni, M. Bertocchi, E. Parmingiani, and C. Brescianti. 2018. Effects of three different designed farrowing crates on neonatal piglets crushing: Preliminary study. Ital. J. Anim. Sci. 17: 505–510. doi: https://doi.org/10.1080/1828051X.2017.1385428.

McGlone, J. J., and F. Blecha. 1987. An examination of behavioral, immunological and productive traits in four management systems for sows and piglets. Appl. Anim. Behav. Sci. 18:269–286. doi: https://doi.org/10.1016/0168-1591(87)90222-X.

McGlone, J. J., B. Vines, A. C. Rudine, and P. DuBois. 2004a. The physical size of gestating sows. J. Anim. Sci. 82:2421–2427. doi: https://doi.org/10.2527/2004.8282421x.

McGlone, J. J., E. H. von Borell, J. Deen, A. K. Johnson, D. G. Levis, M. Meunier-Salaün, J. Morrow, D. Reeves, J. L. Salak-Johnson, and P. L. Sundberg. (2004b). Review: Compilation of the scientific literature comparing housing systems for gestating sows and gilts using measures of physiology, behavior, performance, and health. Prof. Anim. Sci. 20:105–117.

McGlone, J. J. 2013. Review: Updated scientific evidence on the welfare of gestating sows kept in different housing systems. Prof. Anim. Sci. 29:189–198

Mench, J.A. 2003. Assessing animal welfare at the farm level: a United States perspective. Anim. Welfare 12:493–503.

Mendl, M. 1995. The social behaviour of non-lactating sows and its implications for managing sow aggression. Pig Vet. J. 34:9–20.

Meyers. A. J., R. D. Goodband, M. D. Tokach, S. S. Dritz, J. M. DeRouchey, and J. L. Nelssen. 2013. The effects of diet form and feeder design on the growth and performance of finishing pigs. J. Anim. Sci. 91:3420–3428. doi:https://doi.org/10.2527/jas2012-5612.

Muhlbauer, R.V., L. B. Moody, R. R. Burns, J. Harmon, and K. Stalder. 2010. Water consumption and conservation techniques currently available for swine production. National Pork Board, Des Moines, IA.

NPB. 2019. Common Swine Industry Audit. National Pork Board. Des Moines, IA.

Noblet, J., J. Le Dividich, and J. Van Milgen. 2001. Thermal environment and swine nutrition. In: A. J. Lewis, and L. L. Southern, editors, Swine nutrition. 2nd ed. CRC Press, Boca Raton, FL. p. 519–544.

Pairis-Garcia, M. 2015. Alternative farrowing options in the swine industry. PIG 17-01-01. Pork Information Gateway. U.S. Pork Center of Excellence, Clive, IA, US.

Petersen, H. V., K. Vestergaard, and P. Jensen. 1989. Integration of piglets into social groups of free-ranging domestic pigs. Appl. Anim. Behav. Sci. 23:223–236. doi: https://doi.org/10.1016/0168-1591(89)90113-5.

Pine, D. and G. L. Gerdes. 1973. Wild pigs in Monterey County, California. California Fish and Game 59:126–137.

Rhodes, R. T, M. C. Appleby, K. Chinn, L. Douglas, L. D. Firkins, K. A. Houpt, C. Irwin, J. J. McGlone, P. Sundberg, L. Tokach, and R. W. Wills. 2005. A comprehensive review of housing for pregnant sows. J. Am. Vet. Med. Assoc. 227:1580–1590.

Roberston, J. B., R. Laird, K. S. Hall, R. J. Forsyth, J. M. Thompson, and J. Walker Love. 1966. A comparison of two indoor farrowing housing systems for sows. Anim. Prod. 8:171–178. doi: https://doi.org/10.1017/S0003356100034553.

Rollin, B. E. 1993. Animal welfare, science, and value. J. Agric. Envir. Ethics. 6(Suppl. 2):44–50.

Ruhe, F. 1997. On the construction of resting places among wild boar (Sus scrofa L) during a period of severe frost. Z. Jagdwiss. 43:116–119. doi: https://doi.org/10.1007/BF02241420.

Rushen, J. P. 1985. Stereotypies, aggression and the feeding schedules of tethered sows. App. Anim. Behav. Sci. 14:137–147. doi: https://doi.org/10.1016/0168-1591(85)90025-5.

Rushen, J., A. Butterworth, and J. C. Swanson. 2011. Animal behavior and well-being symposium: Farm animal welfare assurance: Science and application. J. Anim. Sci. 89:1219–1228. doi: https://doi.org/10.2527/jas.2010-3589.

Salak-Johnson, J. L., S. R. Niekamp, S. L. Rodriguez-Zas, and S. E. Curtis. 2007. Space allowance for dry, pregnant sows in pens: Body condition, skin lesions, and performance. J. Anim. Sci. 85:1758–1769. doi: https://doi.org/10.2527/jas.2006-510.

Salak-Johnson, J., J. Cassady, M. B. Wheeler, and A. Johnson. 2012. Swine. In: S. E. Vaughn, editor. FASS guide For the care and use of agricultural animals in agricultural research and teaching. Federation of Animal Sciences Societies, Champaign, IL. p. 132–152.

Schau, D. J., J. D. Brue, and K. A. Rosentrater 2013. Review of housing options for gestating sows. In: Agricultural and biosystems engineering conference proceedings and presentations. Iowa State Univ., Ames, IA. p. 1542–1567.

Schröder-Petersen, D. L., and H. B. Simonsen. 2001. Tail biting in pigs. Vet. J. 162:196–210. doi: https://doi.org/10.1053/tvjl.2001.0605.

Schmolke, S. A., Y. Z. Li, and H. W. Gonyou. 2003. Effect of group size on performance of growing-finishing pigs. J. Anim. Sci. 81:874–878. doi: https://doi.org/10.2527/2003.814874x.

Signoret, J. P., B. A. Baldwin, S. Fraser, and E. S. E. Hafez. 1975. In: E. S. E. Hafez, editor, The behaviour of domestic animals. Tindall and Cox, Bailliere, London, UK. p. 295–329.

Street, B. R., and H. W. Gonyou. 2008. Effects of finishing pigs in two group sizes and at two floor space allocations on production, health, behavior, and physiological variables. J. Anim. Sci. 86:982–991. doi: https://doi.org/10.2527/jas.2007-0449.

Stolba, A., and D. G. M. Wood-Gush. 1989. The behaviour of pigs in a semi-natural environment. Anim. Prod. 48:419–425. doi: https://doi.org/10.1017/S0003356100040411.

Thodberg, K., K. H. Jensen, and M. S. Herskin. 2002. Nest building and farrowing in sows: relation to the reaction pattern during stress, farrowing environment and experience. Appl. Anim. Behav. Sci. 77:21–42. doi: https://doi.org/10.1016/S0168-1591(02)00026-6.

Tuyttens, F. A. M. 2005. The importance of straw for pig and cattle welfare: A review. Appl. Anim. Behav. Sci. 92:261–282.

USDA. 2015. Swine 2012 Part 1 Baseline reference of swine health and management in the United States, 2012. Natl. Anim. Health Monitoring System, USDA, Washngton, DC.

van Heutgen, E., 2010. Growing-Finishing Swine Nutrient Recommendations and Feeding Management. PIG 07-01-09. National Swine Nutrition Guide. http://porkgateway.org/resource/growing-finishing-swine-nutrient-recommendations-and-feeding-management/.

Verdon, M., C. F. Hansen, J.-L. Rault, E. Jongman, L. U. Hansen, K. Plush, and P. H. Hemsworth. 2015. Effects of group housing on sow welfare: A review. J. Anim. Sci. 93:1999–2017. doi: https://doi.org/10.2527/jas2014-8742.

Vosough Ahmadi, B., A. W. Stott, E. M. Baxter, A. B. Lawrence, and S. A. Edwards. 2011. Animal welfare and economic optimisation of farrowing systems. Anim. Welf. 20:57–67.

Wathes, C., and C. Whittemore. 2006. Environmental management of pigs. In: I. Kyriazakis and C. Whittemore editors, Whittemore's science and practice of pig production. 3rd ed. Blackwell Publishing, Ames, IA. p. 533–592.

Weber, R. 2000. Alternative housing systems for farrowing and lactating sows. In: H. J., Blokhuis, E. D. Ekkel, and B. Wechsler, editors, Improving health and welfare in animal production. EEAP. Wageningen Pers, The Hague, The Netherlands. p. 109–115.

Weber, R., N. M. Keil, M. Fehr, and R. Horat. 2009. Factors affecting piglet mortality in loose farrowing systems on commercial farms. Livest. Sci. 124:216–222. doi: https://doi.org/10.1016/j.livsci.2009.02.002.

Wischner, D., N. Kemper, and J. Krieter. 2009. Nest-building behaviour in sows and consequences for pig husbandry. Livest. Sci. 124:1–8. doi: https://doi.org/10.1016/j.livsci.2009.01.015.

Wood-Gush, D.G.M., and K. Vestergaard. 1989. Exploratory behavior and the welfare of intensively kept animals. J. Agric. Ethics. 2:161–169. doi: https://doi.org/10.1007/BF01826929.

Wolter, B. F., M. Ellis, S. E. Curtis, N. R. Augspurger, D. N. Hamilton, E. N. Parr, and D. M. Webel. 2001. Effect of group size on pig performance in a wean-to-finish production system. J. Anim. Sci. 79:1067–1073. doi: https://doi.org/10.2527/2001.7951067x.

Wolter, B. F., M. Ellis, J. M. DeDecker, S. E. Curtis, G. R. Hollis, R. D. Shanks, E. N. Parr, and D. M. Webel. 2002a. Effects of double stocking and weighing frequency on pig performance in wean-to-finish production systems. J. Anim. Sci. 80:1442–1450. doi: https://doi.org/10.2527/2002.8061442x.

Wolter, B. F., M. Ellis, S. E. Curtis, E. N. Parr, and D. M. Webel. 2002b. Effects of feeder-trough space and variation in body weight within a pen of pigs on performance in a wean-to-finish production system. J. Anim. Sci. 80:2241–2246. doi: https://doi.org/10.2527/2002.8092241x.

# 22    Feeding Reproducing Swine and Neonatal Pigs

Lee J. Johnston

## Introduction

The central objective of this chapter is to provide requirements for nutrient composition of diets fed to reproducing swine. Nutrient requirements are derived from the biological needs for body maintenance, productive functions (e.g., muscle growth, bone growth, milk production, and semen production), and activity of the pig. As the pig becomes larger physically, nutrient demands for body maintenance will increase. Likewise pigs with the genetic ability to develop more muscle faster or produce larger quantities of milk will require more nutrients to support these functions and thus have higher nutrient requirements than pigs with lower genetic potential. So, nutrient requirements are strictly a function of the pig's biological needs for optimal growth and productivity. The National Research Council's Subcommittee on Swine Nutrition periodically updates the nutrient requirements of swine (NRC 2012).

The nutrient requirements listed in this chapter are based on the National Research Council's publication "Nutrient Requirements of Swine" (NRC 2012) and the National Swine Nutrition Guide (NSNG 2010). The requirements listed represent minimum levels required for optimal production under ideal conditions. Typically, practicing nutritionists increase requirements by 10 to 15% because, in the real world of pig production, conditions on commercial farms do vary. Feed mixing errors, nutrient variability of natural feedstuffs, antinutritional factors present in feed ingredients, undetected environmental or disease challenges could change nutrient requirements of pigs. This safety margin is not included in the nutrient requirements presented in this chapter. So, the nutrient requirements presented in this chapter may need to be adjusted to support the desired pig performance on any particular farm.

## Breeding Swine

### Replacement Gilts

Proper nutrition and management of gilts during their development will set the stage for a long, productive life once that female enters the breeding herd. Most research efforts in gilt development have focused on feeding during the growing-finishing period with the primary emphasis to encourage a high proportion of gilts to express pubertal estrus at an early age. A successful gilt development program must generate females with a high farrowing rate and large litters at first farrowing.

*Sustainable Swine Nutrition*, Second Edition. Edited by Lee I. Chiba.
© 2023 John Wiley & Sons Ltd. Published 2023 by John Wiley & Sons Ltd.

Additionally, proper gilt nutrition increases the probability of the breeding female staying productive in the herd for a long period of time.

*Establishing Gilt Development Targets*
The primary challenge in a gilt development program is establishing the proper growth targets and nutritional requirements to ensure maximal long-term productivity of sows after they enter the breeding herd. Rozeboom (2015) published an extensive review of literature regarding the conditioning of replacement gilts for optimal reproductive performance. He clearly highlighted the challenges of identifying specific targets for age, growth rate, and body composition of gilts at first mating that result in optimal reproductive performance of gilts because these traits are highly inter-related. For example, selection of a specific age at first mating comes with a specific body composition. Changing the age also likely changes the body composition. Thus, traits are confounded. Slowed growth (510 g/d) caused by restricted feeding during the growing-finishing period delays puberty (Beltranena et al. 1991), but increasing growth rate above 600 g/d in ad libitum fed gilts did not hasten pubertal estrus. Kummer et al. (2009) found that increasing growth rates from 577 to 724 g/d hastened expression of puberty and decreased the incidence of anestrous gilts. In contrast, Patterson et al. (2010) observed no effect of growth rates ranging from about 440 to 800 g/d on age at puberty. Slow growth during the development period of gilts is related negatively to size of the first litter (Tummaruk et al. 2001). But, very fast growth may be detrimental to proper skeletal development of gilts (Williams et al. 2004; Orth 2007) and lifetime productivity of sows (Jorgensen and Sorensen 1998; Johnston et al. 2007; Knauer et al. 2011). It appears that both slow growth and very rapid growth of gilts are detrimental to reproductive performance. Growth rate of gilts from birth to selection for breeding should be maintained between 600 and 800 g/d to optimize lifetime productivity.

Body composition at puberty and the time of first mating could influence first litter and lifetime performance of females. Retrospective analysis of gilt development data and the resulting reproductive performance of those females indicate a positive association between body fat at selection and productive life as a breeding sow (Lopez-Serrano et al. 2000; Stalder et al. 2005; Johnston et al. 2007). These studies seem to suggest that fatter gilts at selection will have increased lifetime productivity after entering the breeding herd. In contrast, Rozeboom et al. (1996) manipulated body composition of gilts at first mating in a controlled experiment and found no influence on the proportion of gilts that were retained through four parities. Similarly, Edwards (1998) concluded after reviewing several studies that modification of the gilt's fat reserves as she enters the breeding herd through protein restriction or increased feeding levels rarely demonstrates an improvement over several parities.

Gilts should weigh at least 135 kg at first mating but not exceed 155 kg to enhance lifetime productivity. Williams et al. (2005) found that the number of pigs born over three parities was depressed when females did not weigh 135 kg at first mating. This target range in body weight at first mating provides a reasonable balance between enough body mass and maturity for adequate skeletal development and tissue reserves without allowing the female to get too big. Our retrospective data (Johnston et al. 2007) suggest that allowing gilts to get too heavy can compromise lifetime productivity. Similarly, Dourmad et al. (1994) suggested that excessive body weight reduces sow longevity. In addition, excessive body weight and size can create sow welfare concerns if sows become too large to fit comfortably in existing farrowing and gestating stalls.

*Nutrient Requirements for Gilt Development*
The targets for growth rate (600 to 800 g/d) and body weight at first mating (135 to 155 kg) were used to establish nutrient requirements for developing gilts listed in Tables 22.1 and 22.2. Estimates of voluntary feed intake of gilts were adapted from the NSNG (2010). The expected feed intakes and

**Table 22.1** Energy and amino acid requirements of replacement gilts (as-fed basis)[a,b].

| Item | Body weight, kg | | | |
|---|---|---|---|---|
| | 20–50 | 50–80 | 80–110 | 110–140 |
| Estimated feed intake, kg/d[c] | 1.5 | 2.1 | 2.5 | 2.7 |
| Diet ME, kcal/kg | 3265 | 3265 | 3265 | 3265 |
| ME intake, kcal/d | 4.9 | 6.8 | 8.2 | 8.8 |
| Expected weight gain, g/d | 774 | 872 | 882 | 806 |
| SID amino acid, % | | | | |
| Arginine | 0.45 | 0.39 | 0.33 | 0.28 |
| Histidine | 0.34 | 0.29 | 0.25 | 0.21 |
| Isoleucine | 0.52 | 0.44 | 0.39 | 0.33 |
| Leucine | 1.00 | 0.85 | 0.74 | 0.62 |
| Lysine | 0.99 | 0.84 | 0.73 | 0.61 |
| Methionine | 0.29 | 0.24 | 0.21 | 0.18 |
| Methionine + Cystine | 0.56 | 0.48 | 0.42 | 0.36 |
| Phenyalanine | 0.59 | 0.51 | 0.44 | 0.37 |
| Phenyalanine + Tyrosine | 0.93 | 0.80 | 0.69 | 0.59 |
| Threonine | 0.60 | 0.52 | 0.46 | 0.41 |
| Tryptophan | 0.17 | 0.15 | 0.13 | 0.11 |
| Valine | 0.64 | 0.55 | 0.48 | 0.41 |

[a] Based on NRC (2012) growth model with no safety margin included. Assumed Fat-free Lean (FFL) gain of 325 g/d.

[b] SID = standardized ileal digestible.

[c] Adapted from National Swine Nutrition Guide (2010).

the target growth rates listed above were used in the growing pig model of NRC (2012) to establish requirements. Feeding the developing gilt is very similar with a few exceptions to feeding slaughter progeny through the early and middle portions of the growth period except that a more moderate growth rate is desired. Most nutritionists recommend higher levels of dietary calcium and phosphorus (usually 0.1% higher) throughout the growing phase to optimize bone mineralization based on research of Nimmo et al. (1981). This recommendation is broadly observed throughout the swine industry. However, Rozeboom (2015) suggested that findings of other research groups do not support this recommendation and called for further research to establish mineral requirements for optimal skeletal development in replacement gilts. Levels of vitamins and trace minerals are increased dramatically in the last phase of growth compared to that of slaughter progeny. This increase reflects the expectation that these females will soon be performing more like gestating sows than market hogs so nutrient levels are increased in anticipation of this new level of performance. This also reflects the common practice in commercial pork production to switch developing gilts to diets containing the breeder vitamin-trace mineral premix when they are selected to enter the gilt pool. Vitamin and trace mineral requirements listed in Table 22.2 reflect this change for gilts from 110 to 140 kg body weight.

## Gestating Sows

Feeding sows during gestation focuses on minimizing embryo and fetal losses and preparing the sow for farrowing and lactation. Proper preparation of the sow for farrowing and lactation dictates that caloric intake and the associated body weight gain is controlled to prevent sows from getting too fat during gestation. In the very early stages of gestation immediately after conception, the first

**Table 22.2** Mineral and vitamin requirements of replacement gilts (as-fed basis)[a,b].

| Item | Body weight, kg | | | |
|---|---|---|---|---|
| | 20–50 | 50–80 | 80–110 | 110–140 |
| Estimated feed intake, kg/d[c] | 1.5 | 2.1 | 2.5 | 2.7 |
| Diet ME, kcal/kg | 3,265 | 3,265 | 3,265 | 3,265 |
| ME intake, kcal/d | 4.9 | 6.8 | 8.2 | 8.8 |
| Expected weight gain, g/d | 774 | 872 | 882 | 806 |
| Minerals, % or Amt/kg[d] | | | | |
| Calcium, % | 0.67 | 0.59 | 0.53 | 0.68 |
| Phosphorus, STTD, % | 0.31 | 0.27 | 0.25 | 0.30 |
| Sodium, % | 0.10 | 0.10 | 0.10 | 0.15 |
| Chlorine, % | 0.08 | 0.08 | 0.08 | 0.12 |
| Magnesium, % | 0.04 | 0.04 | 0.04 | 0.06 |
| Potassium, % | 0.22 | 0.19 | 0.18 | 0.20 |
| Copper, mg | 3.9 | 3.4 | 3.2 | 10.0 |
| Iodine, mg | 0.13 | 0.14 | 0.14 | 0.14 |
| Iron, mg | 60.4 | 48.8 | 42.7 | 80.0 |
| Manganese, mg | 1.9 | 2.0 | 2.1 | 25.0 |
| Selenium, mg | 0.19 | 0.16 | 0.15 | 0.15 |
| Zinc, mg | 56 | 53.1 | 51 | 100.0 |
| Vitamins, Amt/kg[d] | | | | |
| Vitamin A, IU | 1248 | 1330 | 1338 | 4000 |
| Vitamin D, IU | 144 | 153 | 154 | 800 |
| Vitamin E, IU | 11.0 | 11.0 | 11.0 | 44.0 |
| Vitamin K (menadione), mg | 0.50 | 0.50 | 0.50 | 0.50 |
| Biotin, mg | 0.05 | 0.05 | 0.05 | 0.20 |
| Choline, g | 0.29 | 0.31 | 0.31 | 1.25 |
| Folacin, mg | 0.29 | 0.31 | 0.31 | 1.30 |
| Niacin, avail., mg | 28.8 | 30.7 | 30.9 | 10.0 |
| Pantothenic acid, mg | 7.6 | 7.4 | 7.2 | 12.0 |
| Riboflavin, mg | 2.3 | 2.2 | 2.0 | 3.8 |
| Thiamin, mg | 1.0 | 1.0 | 1.0 | 1.0 |
| Vitamin $B_6$, mg | 1.0 | 1.0 | 1.0 | 1.0 |
| Vitamin $B_{12}$, ug | 9.0 | 6.2 | 5.1 | 15.0 |

[a] Based on NRC (2012) growth model with no safety margin included.

[b] ME = metabolizable energy; Amt = amount; STTD = standardized total tract digestible; avail. = available.

[c] Adapted from National Swine Nutrition Guide (2010).

[d] Trace mineral and vitamin requirement estimates are increased to that of gestating sows in the final phase of the gilt development period.

objective is to provide conditions that will ensure maximal survival of embryos and favor a large litter size at the subsequent farrowing. Growth of the developing fetuses in conjunction with increasing nutrient stores in the sow's body through continued growth of young sows or replenishment of nutrient stores lost during the previous lactation for older sows are the main objectives during mid-gestation (day 30–75). In late gestation, fetal growth continues at a very rapid rate and mammary development occurs in preparation for the upcoming lactation. Proper feeding programs will satisfy these nutritional needs and ensure continued reproductive performance of sows at a reasonable cost.

*Establishing Nutrient Needs*
Variations in body size and condition, productivity, stage of gestation, health status, and environmental circumstances dictate different daily amounts of nutrients to be fed to satisfy the sow's

requirements. Nutrient requirements for gestating sows can be broken down into three basic components: maintenance, fetal growth, and maternal weight gain. Each of these components can be estimated individually and the components then summed to establish the sow's total daily nutrient need.

Nutrient requirements for maintenance are influenced primarily by body weight of the sow and the environment in which the female is housed. Older, heavier sows have increased nutrient needs and will require more feed to maintain their body than younger, lighter sows. Maintenance energy requirements account for 75–85% of the sow's total energy requirement. In general, for every 23 kg increase in sow body weight, daily metabolizable energy needs increase about 470 kcal that requires about 0.15 kg of additional feed (assumes a corn-soybean meal-based diet). In addition to body weight, maintenance energy requirements are influenced by the effective ambient temperature experienced by the sow. The effective ambient temperature is not necessarily the thermometer reading but is the temperature that the sow experiences. Use of bedding provides insulation so reduced temperatures do not feel as cold to the sow. Conversely, wet conditions make the sow feel colder than the thermometer reading due to evaporative heat loss. Under commercial conditions, we are most concerned about temperatures that fall below the sow's thermoneutral, or comfort zone. These cooler temperatures require increased energy and hence feed intake to maintain the sow's core body temperature without the need for the sow to mobilize her own body tissues. Generally speaking, for every 5.5 °C drop below 18 °C, individually housed sows should receive an additional 360 g of a corn-soybean meal diet to satisfy their maintenance energy requirement. Group housing and use of bedding material can help the sow conserve body heat so increased feeding levels need not be implemented until temperatures fall below 10–13 °C.

Growth of the products of conception and the associated nutrient needs for that growth are fairly resistant to nutritional manipulations at feed intakes typical of production settings. Under conditions of adequate energy intakes ranging from 6 to 10 Mcal of ME daily, changes in weight of fetuses are relatively small (Noblet et al. 1990). Similarly, feeding level has little influence on body composition of fetuses. However, feeding sows supplemental fat in late gestation can improve the survival rate of piglets after birth.

Conceptually, maternal weight gain is supported by "extra" nutrients available after needs for maintenance and fetal growth are satisfied. Maternal weight gain accounts for about 15 to 25% of the sow's total energy needs. The composition of this weight gain is determined primarily by parity of the sow and diet composition (Pettigrew and Yang 1997). Similarly, the amount of maternal weight gain desired will depend on age of the sow (Tables 22.3 and 22.4). Nulliparous (Parity 0) and primiparous (Parity 1) sows are still growing so more weight gain should be allowed compared with older sows that have reached their mature body weight. Gains in maternal body weight provide a reservoir of nutrients for the upcoming lactation should the demands of milk production exceed nutrient intake from feed. However, excessive weight gains can predispose the sow to poor performance during lactation due to depressed voluntary feed intake of sows (Weldon et al. 1994; Sinclair et al. 2001) and reduced sow longevity (Dourmad et al. 1994).

*Fetal Imprinting*
Supply of nutrients to the fetuses during pregnancy can influence size at birth and ultimately pig performance from birth to market weight. This phenomenon, called fetal imprinting, is the 'physiological 'setting' by an early stimulus or insult at a 'sensitive' period, resulting in long-term consequences for function" (Davies and Norman 2002). Extreme reductions in nutrient intake by pregnant sows or reduced nutrient supply to individual fetuses causes intrauterine growth retardation (IUGR) and seems to be responsible for development of low birth weight pigs (runt pigs) at farrowing (Foxcroft

**Table 22.3**  Energy and amino acid requirements for gestating sows (as-fed basis)[a,b].

| Item | Parity[c] | | | |
|---|---|---|---|---|
| | 0 | 1 | 2 | > 2 |
| Litter size, total born | 12.5 | 14.5 | 13 | 15 |
| Body weight at breeding, kg | 155 | 155 | 195 | 195 |
| Assumed feed intake, kg/day[d] | 2.3 | 2.3 | 2.1 | 2.1 |
| Assumed total weight gain, kg | 64 | 68 | 40 | 45 |
| Dietary ME, Mcal/kg | 3.3 | 3.3 | 3.3 | 3.3 |
| SID amino acid, %: | | | | |
| Lysine | 0.54 | 0.55 | 0.39 | 0.40 |
| Threonine | 0.38 | 0.39 | 0.31 | 0.32 |
| Methionine | 0.15 | 0.16 | 0.11 | 0.11 |
| Methionine + cysteine | 0.35 | 0.36 | 0.27 | 0.28 |
| Tryptophan | 0.10 | 0.10 | 0.08 | 0.08 |
| Isoleucine | 0.30 | 0.31 | 0.22 | 0.23 |
| Valine | 0.38 | 0.39 | 0.29 | 0.30 |
| Arginine | 0.28 | 0.29 | 0.20 | 0.21 |
| Histidine | 0.18 | 0.19 | 0.13 | 0.13 |
| Leucine | 0.49 | 0.51 | 0.37 | 0.38 |
| Phenylalanine | 0.30 | 0.30 | 0.22 | 0.23 |
| Phenylalanine + tyrosine | 0.52 | 0.53 | 0.39 | 0.40 |

[a] Based on NRC (2012) gestating sow model with no safety margin included. Assumes one diet is limit-fed throughout gestation under thermoneutral conditions.

[b] ME = metabolizable energy; SID = standardized ileal digestible.

[c] Parity 0 means female in her first pregnancy.

[d] Feed allotment assumes 5% feed wastage and should be adjusted to achieve a desired body condition or weight gain.

et al. 2006). These low birth weight pigs display lower survival, slower postnatal growth rates, lighter carcass weights at a given harvest age with more carcass fat and less lean compared to pigs with average or heavy birth weights (Milligan et al. 2002; Gondret et al. 2006). There seems to be a consensus that a critical threshold in birth weight exists related to piglet survival to weaning. Various research groups have determined independently that preweaning mortality increases dramatically when birth weight of piglets is below 0.95 kg (Calderon-Diaz et al. 2017), 1.00 kg (Zeng et al. 2019) or 1.11 kg (Feldpausch et al. 2016). It seems that piglets need to weigh at least 1 kg at birth to enhance their odds of survival to weaning. A few studies (Cromwell et al. 1989; Dwyer et al. 1994; Wu et al. 2006) have investigated the effects of elevated nutrient intake during gestation on birth weight and postnatal performance of progeny working under the basic premise that increased nutrient intake would minimize occurrence of low birth weight pigs. In general, provision of nutrients above recommended levels did not elicit important, lasting effects in the offspring. So, a more targeted approach is necessary to reduce the incidence of low weight pigs at birth and(or) improve their postnatal survival.

Recently, attention has been focused on trace mineral supplementation of sow diets, specifically zinc supplementation. Vallet et al. (2014) supplemented corn-soybean meal diets fed to pregnant gilts from day 80 of gestation until farrowing with 0.07% zinc sulfate. Zinc sulfate supplementation was in addition to 150 ppm added zinc in the basal diet. Additional zinc had no effect on piglet birth weight or weaning weight of pigs. However, the stillbirth rate declined as birth interval of pigs increased and more importantly, preweaning mortality of pigs weighing less than 1 kg at birth was reduced substantially by supplemental zinc sulfate. Subsequently, Holen et al. (2020) evaluated the effects of 365 ppm or 595 ppm dietary zinc in diets that contained 125 ppm Zn. Additional zinc was

**Table 22.4** Mineral and vitamin requirements for gestating sows (as-fed basis)[a,b].

| Item | % or Amt/kg |
|---|---|
| Minerals: | |
|   Calcium, % | 0.68 |
|   Phosphorus, STTD, % | 0.30 |
|   Sodium, % | 0.15 |
|   Chlorine, % | 0.12 |
|   Magnesium, % | 0.06 |
|   Potassium, % | 0.20 |
|   Copper, mg | 10 |
|   Iodine, mg | 0.14 |
|   Iron, mg | 80 |
|   Manganese, mg | 25 |
|   Selenium, mg | 0.15 |
|   Zinc, mg | 100 |
| Vitamins: | |
|   Vitamin A, IU | 4000 |
|   Vitamin D, IU | 800 |
|   Vitamin E, IU | 44 |
|   Vitamin K (menadione), mg | 0.50 |
|   Biotin, mg | 0.20 |
|   Choline, g | 1.25 |
|   Folacin, mg | 1.30 |
|   Niacin, avail., mg | 10 |
|   Pantothenic acid, mg | 12 |
|   Riboflavin, mg | 3.75 |
|   Thiamin, mg | 1.00 |
|   Vitamin $B_6$, mg | 1.00 |
|   Vitamin $B_{12}$, µg | 15.00 |

[a] Based on NRC (2012) with no safety margin included.
[b] Amt = amount; STTD = standardized total tract digestible; avail. = available.

supplied by zinc sulfate in the last 40 days of gestation to mixed parity sows on a commercial farm. Zinc supplementation decreased overall preweaning mortality of piglets from 15% to 13.2 and 12.2%, respectively. Much of the decline in mortality was realized in piglets that weighed less than 1 kg at birth. In this birth weight category, mortality declined from 38.3 to 36.4 and 28.1% with increasing zinc supplementation. The mechanism(s) responsible for improved survival of light birth weight pigs is not clear at present. However, this research suggests that renewed attention on trace mineral nutrition of reproducing sows warrants attention. Since the placenta is the link between maternal nutrient supply and the developing fetus, researchers have focused recently on improving placental development and function. Proper placental development is central to successful fetal development and reducing incidence of low-birth weight piglets (Wu et al. 2010). High concentrations of dietary arginine and its catabolic products, nitric oxide and polyamines, seem to improve placental growth and function resulting in improved litter size, decreased within-litter variation in piglet birth weight, and decreased proportion of runt pigs compared to feeding a standard corn-soybean meal-based diet to gilts from days 30 to 114 of gestation (Wu et al. 2010). Arginine may be considered a "functional amino acid." "Functional amino acids" may be essential or nonessential amino acids that play a specialized role in bodily functions above and beyond the traditional requirements for body maintenance, growth, and reproduction (Wu and Kim 2007).

Amino acids other than arginine have received some attention as possible modulators of piglet development *in utero* and subsequent piglet performance. The amino acid, leucine, and its metabolite, β-hydroxy-β-methylbutryate (HMB), have been studied to determine possible effects on piglet birth weight and postnatal performance. Supplementation of sow diets with HMB in late gestation increases birth weight of pigs, improves postweaning growth performance (Tatara et al. 2007, 2012), and increases carcass lean percentage of market pigs (Tatara et al. 2007). Improved postnatal performance of pigs may be due to increase circulating concentrations of growth hormone in pigs produced by HMB-treated sows (Tatara et al. 2007, 2012; Wan et al. 2017). Rehfeldt et al. (2001) showed that exogenous administration of growth hormone preferentially increased birth weight of small pigs. So, HMB may be useful to preferentially increase weight of small pigs at birth that would decrease within-litter birth weight variation and improve survival of low birth weight pigs. Clarke et al. (2019) used pregnant mice as a model for sows to test this hypothesis and reported no beneficial effects of HMB on birth weight of pups. Lack of positive results may mean that HMB is not effective in reducing within-litter variation in birth weights or that the mouse model is not an appropriate model for sows. Supplementation of early gestation diets with L-proline may increase litter size and improve piglet birth weight presumably through improvements in functionality of the placenta; however, this positive response was specific to second parity females (Gonzalez-Anover and Gonzalez-Bulnes 2017). The concept of functional amino acids is relatively new and may become an integral part of establishing dietary requirements for pregnant sows in the future.

*Precision Feeding*

Amino acid needs of the pregnant sow are not constant throughout gestation. Kim et al. (2009) suggested that both the quantity of amino acids required and the ratio among individual amino acids varies as pregnancy progresses, which dictates a different amino acid profile in early and late gestation diets. Formulating diets for gestating sows according to this approach improved sow weight gain throughout gestation and eliminated the backfat loss observed in control sows. The practical implementation of this nutritional approach will require alteration of feed storage bins and feed delivery lines since sows in all stages of gestation currently are fed from the same feed bin on most commercial farms.

*Feeding to Improve Sow Welfare*

In most pork production systems, gains in sow body weight need to be controlled during gestation to prevent the sow from getting too fat. In geographical areas where high energy dense diets are economically priced, feed intake of sows usually is restricted to limit caloric intake. Consequently, sows experience hunger much of each day that encourages stereotypic behaviors to develop (Lawrence and Terlouw 1993; Brouns et al. 1994). Odberg (1978) defined stereotypic behaviors as those motions that are repeated regularly, serve no obvious function, and are apparently useless to the animal.

Even though increasing feed intake is effective in reducing stereotypic behaviors (Bergeron et al. 2000), it is not practical because of the increased body weight gain that ensues. An alternative approach is to decrease the caloric density of the diet by including high levels of fibrous ingredients and allowing greater feed intake. Several researchers (Brouns et al. 1994; Bergeron et al. 2000; Danielsen and Vestergaard 2001) have demonstrated that diets containing between 50 and 89% fibrous ingredients fed at levels to satisfy the sow's nutrient requirements of maintenance, fetal growth, and moderate sow body weight gain significantly reduce stereotypic behaviors. Diets for gestating sows should contain at least 30% NDF (Meunier-Salaun et al. 2001), contain high levels of fermentable fiber, and be fed at levels to ensure nutrient requirements for optimal reproduction are satisfied if a goal is to minimize expression of stereotypic behaviors. There may be one level of nutrient or feed intake required to support acceptable biological performance and a higher level required for improved sow welfare (Johnston and Holt 2006). Currently, diets for gestating sows in North America typically are not formulated with an eye toward improving sow behavior.

*Late Gestation Feeding*

Traditionally, gestating sows have received a set amount of feed designed to meet their nutritional requirements and control gain in body weight. Throughout most of gestation, the feeding level remained constant. However, rapid growth of fetuses and the products of conception in the last third of gestation occurs (Ullrey et al. 1965; Kim 2010). This rapid increase in fetal growth causes sows to fall into negative nutrient balance if feeding levels remain constant (Shields and Mahan 1983; Trottier 1991). To avoid negative nutrient balance, "bump feeding" may be implemented with the hope that nutrient balance will be restored and piglet birth weights might be improved. Bump feeding is simply a 50 to 100% increase in feed allotment from about day 75 of gestation until farrowing. While this practice will correct the negative nutrient balance, excessive increases in body fat may result if feed intake is increased above that needed for the developing fetuses. Increased body fatness of sows at farrowing consistently reduces voluntary feed intake of sows during the subsequent lactation (Weldon et al. 1994; Sinclair et al. 2001; Kim et al. 2015). Goncalves et al. (2016) recently separated the effects of increased energy intake and increased amino acid intake during late gestation using 1,100 sows under commercial conditions. They reported increased energy intake increased body weight gain of sows with a greater response when amino acid intake was increased as well. Increasing net energy intake by 50% did not affect birth weight of the total litter but did increase birth weight of live pigs by only 30 g and significantly increased the stillbirth rate in sows. Increasing amino acid intake by 86% reduced stillbirth rate and improved preweaning survival of pigs. Taken together, there was not strong evidence that increased energy and(or) amino acid intake in late gestation resulted in more pigs produced at weaning. In general, the practice of bump feeding during the last 40 days of gestation has limited value in improving the efficiency of pork production.

*Transition Sow Feeding*

Until recently, swine producers and nutritionists have not had sound scientific information to understand the special needs of sows as they transition from gestation to lactation. Space to fully discuss this topic is limited here so readers are referred to an excellent recent review by Theil (2015). The transition period is defined as the 10 days before and 10 days after farrowing. During this time, a multitude of changes are underway in the sow that include fetal growth, mammary development, colostrum production, the farrowing process, milk production, neonate nutrition, and metabolic changes, to name a few. All these changes suggest special attention should be paid to the nutrition and feeding management of sows in this period.

During the last 7–10 days of gestation, feeding management of sows must focus on preparation for farrowing. Farrowing is an energy-demanding process that can be lengthy, especially when litters are large. Sows must have a ready supply of metabolic energy to drive muscular contractions responsible for timely expulsion of fetuses at birth. If sows are energy-deprived, the farrowing process will slow which may increase the birth interval. Increased birth intervals are associated with an increase in stillbirth rates (Oliviero et al. 2010; Vallet et al. 2010). Increasing the feeding level in the last week of gestation will provide the sow with needed energy for parturition. This practice is different than bump feeding because it is of a very short duration. Feyera et al. (2018) demonstrated that the time between the last meal and onset of farrowing was correlated positively with farrowing duration, birth interval, and stillbirth rate. When this interval was greater than three hours, farrowing duration increased and when six or more hours elapsed between the last meal and farrowing probability of stillbirths and farrowing assistance increased significantly. This observation suggests that sows be fed several meals throughout the day in an attempt to reduce the time from their last meal until the onset of farrowing. Alternatively, sows could be allowed ad libitum access to feed during this period so that they could consume feed at will and hopeful within three hours of the onset of farrowing. Offering feed ad libitum during the week before farrowing does not have detrimental effects on sows that are not overconditioned (less than 22 mm backfat; Cools et al. 2014) and encourages higher feed intake than the traditional practice of restrictive feeding during gestation. Elevated feed intake during this period might enhance colostrum yield and colostrum intake of newborn pigs as demonstrated by Decaluwe et al. (2014).

Changing diet formulation to include elevated concentrations of fiber during the transition period may benefit the sow. In a large study conducted on a commercial farm, Feyera et al. (2017) observed a statistically significant reduction in the stillbirth rate from 8.8 to 6.6% when sows received a high-fiber supplement based on sugar beet pulp and soybean hulls from day 102 of pregnancy to farrowing. Other researchers have noted similar trends in stillbirth rate due to dietary fiber additions, but the differences were not statistically significant (Krogh et al. 2015). Dietary fiber increases water holding capacity of digesta that softens fecal matter reducing the incidence and severity of constipation (Oliviero et al. 2009; Tan et al. 2015). Feyera et al. (2017) speculated that softer stools would reduce blockages in the pelvic region of sows allowing decreased farrowing duration that would reduce stillbirth rate, but they did not measure this directly. In support of this idea, Loisel et al. (2013) found that a high fiber diet fed to sows from day 106 of pregnancy to farrowing decreased the birth interval for the first 5 pigs born but did not shorten total farrowing duration.

Dietary fiber alters metabolism of sows in ways that might benefit the farrowing process. Inclusion of fiber in sow diets decreases reliance on rapid starch digestion and absorption of glucose in favor of slower fermentation of fiber in the hindgut and absorption of short-chain fatty acids as energy sources (de Leeuw et al. 2005; Serena et al. 2009). This slower digestion of feed ingredients may provide a prolonged source of energy for sows during the farrowing process that would be beneficial because sows typically consume very little if any feed once farrowing commences. Use of high fiber diets combined with frequent feeding or ad libitum feeding during the transition period may help maintain a sufficient energy supply to sows during the farrowing process as suggested by Feyera et al. (2018).

Dietary fiber can influence colostrum production and composition. Loisel et al. (2013) compared a high fiber diet (23% dietary fiber) based on wheat, barley, and soybean meal and included soy hulls, wheat bran, sugar beet pulp, and sunflower meal to a similar wheat-barley-soybean meal diet that contained 13% dietary fiber for sows in late gestation. Yield of colostrum was not affected by diet nor was weight gain of piglets through day 21 postpartum. However, colostrum intake of low birth weight pigs (< 900 g) was greater, and preweaning mortality of all piglets was lower in litters suckling sows fed the high fiber compared with the low fiber diet. Similarly, Krogh et al. (2015) reported no effect of high fiber diets (alfalfa meal or sugar beet pulp) on colostrum yield and piglet weight gain when fed to sows in late gestation. But, they did not observe a significant reduction in piglet mortality due to high fiber diets. Increasing dietary fiber in late gestation sow diets may be beneficial to sow and piglet performance, but details of fiber type, level of dietary inclusion, and period of supplementation need to be established for routine use in commercial production.

*Water Needs*

A complete discussion of water for pigs is presented in Chapter 2. Water is essential for gestating sows, and sows should be offered as much water as they choose to consume. Because gestating sows are limit-fed, they often exhibit a behavioral trait called polydipsia, which means they consume "excess" water to satiate their hunger (Patience 2013). Caretakers may be motivated to limit water intake to reduce this behavior, but restricted water intake can predispose sows to urinary tract infections (Almond and Stevens 1995), which compromises sow health and production.

### Lactating Sows

Feeding sows during lactation focuses on maximizing litter performance through high milk production with minimal mobilization of maternal body tissues. Conserving maternal body tissues during lactation improves the likelihood that sows will come into estrus and conceive a large litter quickly after weaning. The basic approach to achieving these objectives is to maximize voluntary intake of feed that contains a sufficient concentration of nutrients to satisfy the sow's daily nutrient requirements.

*Establishing Nutrient Needs*

The energy and nutrient requirements of the lactating sow depend on the sow's body weight, milk yield and composition, and to some degree the environment in which she is housed. Just like gestating sows, lactating sows require energy and nutrients to maintain body weight and body functions. Older, heavier sows have increased nutrient needs for maintenance compared with younger, lighter sows. Because sows are typically allowed ad libitum access to feed during lactation in an effort to maximize nutrient intake, maintenance requirements for energy and other nutrients are satisfied easily.

In contrast to maintenance requirements, energy and nutrient requirements to support milk production can be more difficult to satisfy. The goal is to satisfy requirements for milk by nutrients supplied in the diet. However, if feed intake of sows is suppressed by the environment, health status, or genetics, or nutrient concentration of the diet is too low, sows will fall into negative nutrient balance and mobilize nutrients in body tissues to meet nutrient needs for milk production. Energy and nutrient needs for milk production are directly related to the quantity and composition of the milk secreted by the sow. Nutrient composition of milk can be altered slightly by genetics or diet, but these subtle differences are not considered in practical feeding programs. So, quantity of milk produced is the primary factor considered in determining nutrient needs for milk production. Under practical conditions, milk production of sows is not measured directly but indirectly by weight gain of the suckling litter. Regression equations have been developed to relate litter weight gain to milk production (NRC 2012). Using these equations and an assumed standard composition of milk, energy and nutrient requirements for milk production can be predicted (NSNG 2010; NRC 2012). The nutrient requirements for milk production and body maintenance are summed to establish the total energy and nutrient requirements for lactating sows. The goal is to satisfy these requirements by formulating diets that include sufficient concentrations of nutrients and encourage maximal feed intake by sows. If feed intake or nutrient concentration in the diet are limiting, sows will mobilize body tissues to meet the deficient nutrient intake and(or) reduce milk production.

Energy and nutrient requirements for lactating sows in four different situations are presented in Table 22.5. Selected situations are presented because it is beyond the scope of this chapter to provide nutrient requirements for all possible situations. The requirements presented in Table 22.5 were generated by the requirement prediction model for lactation presented by NRC (2012). Mineral and

**Table 22.5**  Energy and amino acid requirements of lactating sows (as-fed basis)[a,b].

| Item | Parity | | | |
|---|---|---|---|---|
| | 1 | 1 | 2+ | 2+ |
| Post-farrowing sow wt., kg | 180 | 180 | 220 | 220 |
| Assumed sow wt. change, kg | −18 | −3.5 | −19.2 | −4.0 |
| Assumed feed intake, kg/day[c] | 5.9 | 5.9 | 6.6 | 6.6 |
| Assumed piglet wt. gain, g/d | 270 | 230 | 270 | 230 |
| Litter size weaned | 11 | 10 | 12 | 11 |
| Litter wean wt., kg[d] | 77.0 | 62.0 | 85.0 | 68.0 |
| Dietary ME, Mcal/kg | 3.3 | 3.3 | 3.3 | 3.3 |
| ME intake, Mcal/d | 18.6 | 18.6 | 20.7 | 20.7 |
| SID amino acid, % | | | | |
| Lysine | 0.87 | 0.77 | 0.85 | 0.76 |
| Threonine | 0.55 | 0.49 | 0.54 | 0.48 |
| Methionine | 0.23 | 0.20 | 0.23 | 0.20 |
| Methionine + cysteine | 0.47 | 0.41 | 0.46 | 0.40 |
| Tryptophan | 0.17 | 0.14 | 0.17 | 0.14 |
| Isoleucine | 0.49 | 0.43 | 0.48 | 0.42 |
| Valine | 0.74 | 0.65 | 0.73 | 0.64 |
| Arginine | 0.46 | 0.43 | 0.45 | 0.42 |
| Histidine | 0.34 | 0.30 | 0.34 | 0.30 |
| Leucine | 1.00 | 0.86 | 0.98 | 0.85 |
| Phenylalanine | 0.48 | 0.42 | 0.47 | 0.41 |
| Phenylalanine + tyrosine | 0.99 | 0.86 | 0.98 | 0.85 |

[a] Based on NRC (2012) lactating sow model with no safety margin included. Sows are offered ad libitum access to diets under thermoneutral conditions. wt. = weight.

[b] wt. = weight; ME = metabolizable energy; SID = standardized ileal digestible.

[c] Assumed intake includes 5% feed wastage.

[d] 21-day lactation.

vitamin requirements (Table 22.6) are similarly based on NRC (2012) predictions. There has been little recent research reported to establish mineral and vitamin requirements of modern sows.

### Feeding Management around Farrowing

Feeding management right before and after farrowing is focused on transitioning the sow from limited feed intake of late gestation to ad libitum consumption during lactation. This transition period was discussed more thoroughly in the Gestation section of this chapter. Immediately after farrowing, sows should be allowed progressively increasing amounts of feed so that they are provided ad libitum intake by day 4 postpartum or sooner. The goal during lactation is to implement feed formulations, feeding management practices, and environmental controls to ensure maximal voluntary feed intake of sows (Johnston 1993). Restricting feed intake of lactating sows at any time during lactation increases sow weight loss, reduces litter performance, and can compromise subsequent reproductive performance (Koketsu et al. 1996).

### Nutrition and Subsequent Reproduction

Nutrition during lactation clearly influences subsequent reproductive performance. Feed restriction in early or late lactation increases sow body weight and backfat loss, decreases gonaodotrophic support for ovarian function, increases weaning-to-estrus interval, and decreases ovulation rate after weaning (Zak et al. 1997a). If feed restriction occurs near the end of lactation, survival of embryos after mating may be compromised which could reduce subsequent litter size (Zak et al. 1997a).

**Table 22.6**  Mineral and vitamin requirements of lactating sows (as-fed basis)[a,b].

| Item | % or Amt/kg |
|------|-------------|
| Minerals | |
| Calcium, % | 0.80 |
| Phosphorus, STTD, % | 0.40 |
| Sodium, % | 0.20 |
| Chlorine, % | 0.16 |
| Magnesium, % | 0.06 |
| Potassium, % | 0.20 |
| Copper, mg | 20.00 |
| Iodine, mg | 0.14 |
| Iron, mg | 80 |
| Manganese, mg | 25 |
| Selenium, mg | 0.15 |
| Zinc, mg | 100 |
| Vitamins | |
| Vitamin A, IU | 2,000 |
| Vitamin D, IU | 800 |
| Vitamin E, IU | 44 |
| Vitamin K (menadione), mg | 0.50 |
| Biotin, mg | 0.20 |
| Choline, g | 1.00 |
| Folacin, mg | 1.30 |
| Niacin, avail., mg | 10.00 |
| Pantothenic acid, mg | 12.00 |
| Riboflavin, mg | 3.75 |
| Thiamin, mg | 1.00 |
| Vitamin $B_6$, mg | 1.00 |
| Vitamin $B_{12}$, μg | 15.00 |

[a] Based on NRC (2012) with no safety margin included.
[b] STTD = standardized total tract digestible; avail. = available.

Feed restriction seems to inhibit normal follicular development and compromises oocyte maturation (Zak et al. 1997b).

Feed restriction causes reduced intake of energy, protein, vitamins and minerals. However, it seems that reduced protein (amino acid) intake and the associated loss of body protein is the most detrimental to subsequent reproductive functions (King 1987; Clowes et al. 2003). Low lysine intake as a result of feeding a low protein diet increases muscle degradation, decreases frequency of luteinizing hormone pulses in late lactation, and decreases follicular support for proper oocyte development (Yang et al. 2000a, 2000c). To avoid this problem, one may theorize that very high protein diets seem warranted. However, Yang et al. (2000b) demonstrated that very high protein diets can reduce subsequent litter size but this response is not consistent (Tritton et al. 1996).

*Feedstuffs*
Compared with other phases of the reproductive cycle, lactation places the greatest energy and nutrient demands on the sow. Consequently, lactation diets must be composed of high energy, nutrient dense ingredients. Corn and other cereal grains are commonly used in high proportions to supply energy to the diet and protein concentrates such as soybean meal, canola meal, or other oilseed meals are used as natural sources of the required amino acids. Crystalline amino acids are also appropriate to supply specific needs.

Supplemental fat such as choice white grease, tallow, or soybean oil, can be added to lactation diets to increase energy density of the diet and ultimately energy intake of the sow. Fat additions to lactation diets can increase fat content of sow's milk (Pettigrew 1981) and daily fat output in milk (Lauridsen and Danielsen 2004) and can improve growth rate of suckling pigs (Pettigrew 1981; Lauridsen and Danielsen 2004). However, there are practical limits to fat inclusion level due to increased potential for bridging of feed in bulk bins and feed hoppers and increased incidence of fat randicity as fat inclusion rate increases. (See Chapter 3)

Fibrous feed ingredients such as sugar beet pulp, wheat straw, soy hulls, or wheat midds are generally not appropriate, with one exception, for lactation diets due to their relatively low energy density. Inclusion of fibrous feed ingredients reduces energy intake of sows which is contrary to the nutritional goals for lactation. Fibrous ingredients may be included in diets before and a few days after farrowing to reduce incidence of constipation. While this practice may help the sow feel better, it is difficult to document improvements in sow performance.

*Water Needs*

Intake of water is a critical component of a successful lactation. Milk is about 80% water (Hurley 2015) and is primarily responsible for the high water requirement of lactating sows. To meet these water requirements, lactating sows must have ad libitum access to water and be encouraged to consume as much water as possible. Fraser and Phillips (1989) showed a positive relationship between water intake of sows and weight gain of piglets in the first three days postpartum. Sow water intake accounted for 30 to 40% of the variation in piglet weight gain in the first 3 days postpartum. Similarly, most piglet deaths occurred in litters suckling sows with low water intake (≤ 6 liters/day) compared to sows with high water intake. Caretakers need to manage sows and their environment to encourage the highest water intake possible for optimal lactation performance. Sows with greater activity after farrowing will consume more water than lethargic sows (Fraser and Phillips 1989). Thus, stockpeople need to ensure sows stand after farrowing which will encourage water intake. Water flow from drinkers must be at least 1 liter per minute. Leibbrandt et al. (2001) demonstrated that restricted water flow (70 vs. 700 ml/min) reduced sow feed intake and increased body weight loss of sows during lactation. Jeon et al. (2006) reported that sows consumed more feed and water when temperature of the drinking water during lactation was 15°C compared with 22°C under heat stress conditions. Method of supplying water to lactating sows can be important. Peng et al. (2007) observed higher feed intake and heavier pigs at weaning with lower water wastage when a wet-dry sow feeder was compared to a dry feeder with separate water nipple. Stockpeople need to consider the important positive effects water intake has on sow performance and seek methods to enhance water intake of sows.

### Neonatal Pigs

The primary source of nutrition for neonatal pigs is milk provided by the sow. However, a supplemental source of nutrition can be offered in the form of solid feed commonly called "creep feed." Creep feed is so named because it is presented to suckling pigs in the creep area (area that suckling pigs can access but sows cannot) of the farrowing/lactation stall or pen. In addition to creep feed, a supplemental source of water is also commonly provided for suckling pigs. Creep feed is particularly important in times of low milk production by the sow or in late lactation when nutrient needs of piglets exceed supply of nutrients from milk.

*Creep Feeding*

Numerous factors influence the success of a creep feeding program under commercial conditions. Lactation length can influence the utility of creep feeding. van der Meulen et al. (2010) found pigs weaned at seven weeks of age responded to creep feeding with greater feed intake and daily weight gain than pigs weaned at four weeks of age. Presumably as lactation lengths get shorter (less than 28 days), the response to creep feeding declines, but this issue has not been studied intensely. Milk production peaks about day 18 of lactation then declines as lactation progresses (Hansen et al. 2012). This decline in milk production occurs coincident with increasing nutrient demands of suckling pigs. So, milk production becomes limiting to piglet growth and creep feed can help erase the nutrient deficit.

Intuitively, creep feeding makes sense and should improve the performance of piglets and sows because additional resources (nutrients in feed) are being supplied to lactation. Creep feeding can improve preweaning performance of pigs (Heo et al. 2018; Lee and Kim 2018), introduce suckling pigs to solid feed that aids transition to exclusively solid feed at weaning (Collins et al. 2013; Pluske 2016; Muns and Magowan 2018), and advance gut development ahead of the stressful weaning event (Pluske 2016). One might speculate that creep feeding could preserve sow body condition that may enhance postweaning reproductive performance of sows, but this effect has not been documented. However, responses to creep feeding practices are notoriously inconsistent (Pluske 2016). There are several reports that show no improvement in preweaning growth (Sulabo et al. 2010a; Collins et al. 2013; Huting et al. 2017), sow body condition (Sulabo et al. 2010a), or postweaning pig performance (Sulabo et al. 2010a) as a result of supplying creep feed to suckling pigs. One reason for this inconsistent response is that creep feed intake is quite variable. Not all pigs in a litter consume creep feed when offered. And, some pigs consume large amounts of creep feed while some littermates consume very little or none. So, researchers have categorized piglets into those that consume creep feed (eaters) compared with those that do not (noneaters). Several groups reported that growth performance of eaters is improved postweaning (Bruininx et al. 2002; Sulabo et al. 2010a,c; Collings et al. 2013) compared with noneaters. This suggests practitioners need to focus on getting pigs to eat the creep feed because there appears to be a benefit if piglets do consume creep feed. Low intake of creep feed can actually be detrimental to postweaning pig performance if allergenic ingredients such as soybean meal are included in both the creep feed and the postweaning diet (Miller et al. 1984; Li et al. 1991).

The consistently inconsistent intake of creep feed has prompted many attempts to improve creep feed intake by piglets. Obviously composition of the creep feed could influence palatability of the feed and how attractive it is to young pigs. Several researchers have compared simple and complex diet formulations to determine the most efficacious approach. Collins et al. (2013) found that pigs consumed more of a simple creep feed based primarily on cereals than a complex diet based on animal by-products from 9 days of age until weaning. In contrast, others have reported a positive relationship between diet complexity and creep feed intake by suckling pigs (Fraser et al. 1994; Pajor et al. 2002). The use of flavoring agents has been explored as a way to attract pigs to the creep feed and encourage intake with positive effects (Adeleye et al. 2014) or no effects (Sulabo et al. 2010b). Flavoring agents have been included in the sow's diet during pregnancy in hopes that fetal pigs *in utero* will be imprinted and will recognize those flavors as familiar when the same flavor is included in the creep feed. Figueroa et al. (2013) tested amniotic fluid as an olfactory agent and found that piglets were more attracted to amniotic fluid from their own dam than from an alien dam. This demonstrated that newborn piglets are influenced by odors associated with their mothers. In a follow-up study, these researchers fed sows in the last two weeks of pregnancy flavoring agents such as anise, milky-cheese, or garlic flavors and found that neonates were attracted to the flavor that their dams received prepartum, but this preference did not translate into increased consumption of flavored creep feeds. Similarly, Sulabo et al. (2010b) found no carryover effects on postweaning pig performance of including the same flavor compound in creep feed and postweaning pig diets.

The method of presenting creep feed to pigs can affect intake. Many producers provide creep feed to pigs on a flat floor surface or in a flat pan to encourage early intake of creep feed. This approach takes advantage of pigs' exploratory behaviors by allowing them to root in and manipulate the feed and coincidently consume some feed. Initial interactions of piglets with feed are driven by exploratory behaviors, but later consumption is driven by nutrient needs (Pluske 2016). Once pigs start to consume some feed, intake typically increases in a curvilinear fashion (Sulabo et al. 2010a). The primary challenge with this presentation method is that pigs will urinate and defecate in the feed pan soiling the feed. This dictates that stockpeople offer limited quantities of feed often and keep the feed clean and fresh to encourage sustained intake. Creep feeding in practice can add substantial labor requirements so the benefits of creep feeding need to be documented to justify the additional labor expense. Creep feed can be offered in liquid, gruel, meal, or pelleted forms. Most commonly, it is offered as a dry meal or pellet. Clark et al. (2016) determined that suckling pigs had greater creep feed intake when offered large (12.7 mm diameter) compared with small (3.2 mm diameter) pellets. Pigs consuming the larger creep pellet also grew faster and were more efficient in the first week postweaning than pigs that consumed the smaller pellet. Feed form seems to influence pigs' response to creep feed. Likely, there is less feed wastage with pelleted compared with meal forms of creep feed.

Many factors influence the success of a creep feeding program. Scientific reports indicate quite variable responses ranging from no effect to some small positive influences on piglet performance pre- and postweaning. Each farm needs to evaluate the utility of creep feeding in their situation. If creep feed is used, the feed should be offered to piglets in small quantities frequently so feed stays fresh and clean. Soiled, spoiling creep feed surely will not encourage meaningful intake. A low flat pan or surface that allows pigs to root in the feed would be preferred for initial presentation of feed to piglets.

*Supplemental Water for Neonates*
One might question the need to supply supplemental water for suckling pigs that rely almost exclusively on milk for nutrition. Fraser et al. (1988) found that average water intake (measured as water disappearance) of suckling pigs was 31 ml/pig/day one day after birth and increased to 53 ml/pig/day on day 4 postpartum. In their study, water intake of piglets in the first 24 hours of life ranged from 3 to 157 ml/pig. This suggests that suckling piglets do consume water early in life even when milk is readily available. These same researchers observed that newborn piglets used a bowl drinker or push-lever drinker that had water dripping about 55 hours sooner than a nipple drinker or push-lever drinker that was not dripping water (Phillips and Fraser 1991). Evidently, piglets need to see or smell water and have it readily accessible to encourage early intake. Provision of water to neonates that is easy to find and readily accessible may help combat dehydration of piglets and reduce preweaning death loss.

### Boars

Boars have a significant influence on the swine breeding program, but relatively little attention is typically paid to them, likely because they represent a relatively small proportion of the total pig population. Not only do boars provide a source of genetic improvement but they also influence farrowing rate and litter size (Whitney and Baidoo 2010). Nutrition serves as an important factor in determining reproductive performance of the breeding herd and overall animal well-being. Nutritional status affects libido, structural soundness and longevity, sperm production, and semen quality. Factors that may affect nutrient requirements include type of mating system used, age and stage of maturity, body condition, environmental conditions, and frequency of ejaculation.

If boars are to be used in natural mating systems, a primary goal is to minimize mature body weight so that boars can mate smaller sows and gilts. Overfeeding boars can impair libido and could also lead to reproductive problems and decreased length of service in the herd; therefore, limit feeding is required. Nutrition for stud boars used for artificial insemination should be focused on optimizing sperm production and quality of semen, while ensuring overall well-being of the boar. Welfare issues such as lameness are important because this affects the boar's desire and ability to mount dummy sows for collection. Reduced emphasis relative to natural mating is placed on minimizing mature body size of boars in artificial insemination centers. However, safety of workers handling excessively large boars must be considered.

Young breeding boars (one to two years old) are still growing. Feeding programs must allow for moderate weight gain of about 180 to 250 g/day (Whitney and Baidoo 2010). The goal is to restrict energy intake to achieve a gradual growth rate while maintaining high amino acid, vitamin, and mineral intakes to preserve fertility and libido. As boars become heavier and older, their growth rates decrease and proportion of nutrients provided for body maintenance increases. Mature boars should be fed to meet body maintenance needs while still allowing for optimal reproductive performance. Body weight and body condition are critical determinants of a boar's maintenance requirements. Furthermore, boars that are subjected to cold temperatures or have extremely high libido should be provided an increased feed allowance. Kemp et al. (1989a) estimated extra thermoregulatory heat production of 3.8 kcal/kg$^{0.75}$/°C/day is needed when temperatures are below 20°C.

*Influence of Nutrition on Reproductive Performance*

Nutrition affects boar libido, sperm output, and semen quality. Severe and prolonged restrictions in feeding levels, resulting in significant losses in body weight, cause boars to refuse service (Stevermer et al. 1961). Similarly, feeding low protein diets, especially when energy intake is restricted, reduces boar's interest in mating due to decreased blood levels of estradiol-17β (Louis et al. 1994a,b). However, short-term restrictions in feeding level or nutrient intake have minimal effects on libido (Ju et al. 1985; Kemp et al. 1989b). In some artificial insemination centers, boars receive diets with elevated protein concentration (5 to 10% above requirements) to enhance libido. However, excessive weight gain makes pigs more lethargic and may also decrease a boar's physical ability to mount a sow by reducing locomotive soundness and balance (Westendorf and Richter 1977).

Providing a lower plane of nutrition (50 to 70% of requirements) reduces semen volume and total sperm production (Beeson et al. 1953; Kemp et al. 1989b), while increasing the nutrient levels back to required levels returns semen volume and sperm production to normal. Historical data, however, indicate that boars can tolerate widely varying levels of nutrition for short periods of time without detrimental effects on the quality of the sperm produced (Stevermer et al. 1961).

*Nutrient Requirements of Boars*

Nutrient requirements of breeding boars are based on NRC (2012) and modified slightly according to the National Swine Nutrition Guide (Whitney and Baidoo 2010; Table 22.7). The NRC (2012) nutrient concentrations were used to compute modern requirement levels for young and mature boars based on production data (body weight range, caloric density, and feed intake) from the National Swine Nutrition Guide (Whitney and Baidoo 2010).

**Table 22.7**  Nutrient requirements of breeding boars (as-fed basis)[a,b].

| Item | % or Amt/kg | Amt/d | Amt/d |
|---|---|---|---|
| Body weight, kg | — | 135–185 | 185–300 |
| Diet ME, kcal/kg | 3300 | 3300 | 3300 |
| ME intake, kcal/d | — | 8085 | 9075 |
| Expected feed intake, kg/d[c] | — | 2.45 | 2.75 |
| SID amino acid | | | |
| Arginine | 0.20% | 4.9 g | 5.5 g |
| Histidine | 0.17% | 4.2 g | 4.7 g |
| Isoleucine | 0.31% | 7.7 g | 8.6 g |
| Leucine | 0.33% | 8.1 g | 9.1 g |
| Lysine | 0.51% | 12.5 g | 14.0 g |
| Methionine | 0.08% | 2.0 g | 2.2 g |
| Methionine + cystine | 0.25% | 6.1 g | 6.9 g |
| Phenylalanine | 0.36% | 8.8 g | 9.9 g |
| Phenylalanine + tyrosine | 0.58% | 14.2 g | 16.0 g |
| Threonine | 0.22% | 5.4 g | 6.0 g |
| Tryptophan | 0.20% | 4.9 g | 5.5 g |
| Valine | 0.27% | 6.6 g | 7.4 g |
| Minerals | | | |
| Calcium | 0.75% | 18.4 g | 20.6 g |
| Phosphorus, STTD | 0.33% | 8.1 g | 9.1 g |
| Sodium | 0.15% | 3.7 g | 4.1 g |
| Chlorine | 0.12% | 2.9 g | 3.3 g |
| Magnesium | 0.04% | 0.9 g | 1.0 g |
| Potassium | 0.20% | 4.9 g | 5.5 g |
| Copper | 5.0 mg | 12.2 mg | 13.8 mg |
| Iodine | 0.14 mg | 0.34 mg | 0.4 mg |
| Iron | 80 mg | 196.0 mg | 220.0 mg |
| Manganese | 20.0 mg | 49.0 mg | 55.0 mg |
| Selenium | 0.30 mg | 0.7 mg | 0.8 mg |
| Zinc, mg | 50.0 mg | 122.5 mg | 137.5 mg |
| Vitamins | | | |
| Vitamin A | 4,000 IU | 9,800 IU | 11,000 IU |
| Vitamin D | 200 IU | 490 IU | 550 IU |
| Vitamin E | 44 IU | 108 IU | 121 IU |
| Vitamin K (menadione) | 0.50 mg | 1.2 mg | 1.4 mg |
| Biotin | 0.20 mg | 0.5 mg | 0.55 mg |
| Choline | 1.25 g | 3.06 g | 3.40 g |
| Folacin | 1.30 mg | 3.2 mg | 3.6 mg |
| Niacin, avail. | 10.0 mg | 24.5 mg | 27.0 mg |
| Pantothenic acid | 12.0 mg | 29.4 mg | 33.0 mg |
| Riboflavin | 3.75 mg | 9.2 mg | 10.3 mg |
| Thiamin | 1.0 mg | 2.4 mg | 2.7 mg |
| Vitamin B6 | 1.0 mg | 2.4 mg | 2.7 mg |
| Vitamin B12 | 15 ug | 36.8 ug | 41.2 ug |

[a] Based on NRC (2012) with no safety margin included.

[b] Amt = amount; ME = metabolizable energy; SID = standardized ileal digestible; STTD = standardized total tract digestible; avail. = available.

[c] Based on National Swine Nutrition Guide (2010) and assumes 5% feed wastage.

*Water Needs*

As with other class of breeding swine, boars should have ad libitum access to water for optimal performance. Readers are referred to Chapter 2 of this book for a complete discussion of water quantity and quality for all classes of swine.

## Summary

Dietary nutrient requirements are presented for breeding swine in this chapter. For any given class of pig, one estimate is presented for each nutrient. Users must realize that there is inherent variation in the true requirement for any individual pig due to variation in pig genetics, health status of the pig, environmental conditions the pig experiences, and feedstuffs consumed by the pigs. Requirement estimates presented here are minimal estimates under ideal production conditions. These estimates need to be adjusted based on characteristics of any specific production system to achieve optimal pig performance.

## References

Adeleye, O. O., J. H. Guy, and S. A. Edwards. 2014. Exploratory behaviour and performance of piglets fed novel flavoured creep in two housing systems. Anim. Feed Sci. Technol. 191:91–97.

Almond, G. W., and J. B. Stevens. 1995. Urinalysis techniques for swine practitioners. Comp. Cont. Ed. Prac. Vet. 17:121–129.

Beeson, W.M., E.W. Crampton, T.J. Cunha, N.R. Ellis, and R.W. Leucke. 1953. Nutrient requirements of swine. National Research Council, Washington, DC.

Beltranena, E., F. X. Aherne, G. R. Foxcroft, and R. N. Kirkwood. 1991. Effects of pre- and postpubertal feeding on production traits at first and second estrus in gilts. J. Anim. Sci. 69:886–893.

Bergeron, R., J. Bolduc, Y. Ramonet, M. C. Meunier-Salaun, and S. Robert. 2000. Feeding motivation and stereotypies in pregnant sows fed increasing levels of fiber and/or food. Appl. Anim. Behav. Sci. 70:27–40.

Brouns, F., S. A. Edwards, and P. R. English. 1994. Effect of dietary fiber and feeding system on activity and oral behavior of group housed gilts. Appl. Anim. Behav. Sci. 39:215–223.

Bruininx, E., G. Binnendijk, C. Van der Peet-Schwering, J. Schrama, L. Den Hartog, H. Everts, and A. Beynen. 2002. Effect of creep feed consumption on individual feed intake characteristics and performance of group-housed weanling pigs. J. Anim. Sci. 80:1413–1418.

Calderon-Diaz, J. A., L. A. Boyle, A. Diana, F. C. Leonard, J. P. Moriarty, M. C. McElroy, S. McGettrick, D. Kelliher, and E. G. Manzanilla. 2017. Early life indicators predict mortality, illness, reduced welfare and carcass characteristics in finisher pigs. Prev. Vet. Med. 146:94–102.

Clark, A. B., J. A. DeJong, J. M. DeRouchey, M. D. Tokach, S. S. Dritz, R. D. Goodband, and J. C. Woodworth. 2016. Effects of creep feed pellet diameter on suckling and nursery pig performance. J. Anim. Sci. 94(Suppl. 2):100–101 (Abstr.).

Clarke, A. S., N. Horn, G. C. Shurson, C. Faulk, and L. J. Johnston. 2019. Beta-hydroxy-beta-methylbutyrate (HMB) supplementation to mouse dams in gestation does not affect birth weight variation or growth of offspring. J. Anim. Sci. 97(Suppl. 3):110.

Clowes, E. J., F. X. Aherne, G. R. Foxcroft, and V. E. Baracos. 2003. Selective protein loss in lactating sows is associated with reduced litter growth and ovarian function. J. Anim. Sci. 81:753–764.

Collings, R., L. J. Harvey, L. Hooper, R. Hurst, T. J. Brown, J. Ansett, M. King, and S. J. Fairweather-Tait, 2013. The absorption of iron from whole diets: a systematic review. Am. J. Clin. Nutr. 98:65–81.

Collins, C. L., R. S. Morrison, R. J. Smits, D. J. Henman, F. R. Dunshea, and J. R. Pluske. 2013. Interactions between piglet weaning age and dietary creep feed composition on lifetime growth performance. Anim. Prod. Sci. 53:1025–1032.

Cools, A., D. Maes, R. Decaluwe, J. Buyse, T. A. T. G. van Kempen, A. Liesegang, and G. P. J. Janssens. 2014. Ad libitum feeding during the peripartal period affects body condition, reproduction results and metabolism of sows. Anim. Reprod. Sci. 145:130–140.

Cromwell, G. L., D. D. Hall, A. J. Clawson, G. E. Combs, D. A. Knabe, C. V. Maxwell, P. R. Noland, D. E. Orr, Jr., and T. J. Prince. 1989. Effects of additional feed during late gestation on reproductive performance of sows: A cooperative study. J. Anim. Sci. 67:3–14.

Danielsen, V., and E. M. Vestergaard. 2001. Dietary fiber for pregnant sows: Effect on performance and behavior. Anim. Feed Sci. Technol. 90:71–80.

Davies, M. J., and R. J. Norman. 2002. Programming and reproductive functioning. Trends Endocrinol. Metab. 13:386–392.

Decaluwe, R., D. Maes, A. Cools, B. Wuyts, S. De Smet, B. Marescau, P. P. De Deyn, and G. P. J. Janssens. 2014. Effect of peripartal feeding strategy on colstrum yield and composition in sows. J. Anim. Sci. 92:3557–3567.

de Leeuw, J. A., A. W. Jongbloed, H. A. M. Spoolder, and M. W. A. Verstegen. 2005. Effects of hindgut fermentation of non-starch polysaccharides on the stability of blood glucose and insulin levels and physical activity in empty sows. Livest. Prod. Sci. 96:165–174.

Dwyer, C. M., N. C. Strickland, and J. M. Fletcher. 1994. The influence of maternal nutrition on muscle fiber number development in the porcine fetus and on subsequent postnatal growth. J. Anim. Sci. 72:911–917.

Dourmad, J. Y., M. Etienne, A. Prunier, and J. Noblet. 1994. The effect of energy and protein intake of sows on their longevity: A review. Livest. Prod. Sci. 40:87–97.

Edwards, S. A. 1998. Nutrition of the rearing gilt and sow. In: J. Wiseman, M. A. Varley, and J. P. Chadwick, editors, Progress in pig science. Nottingham University Press, Nottingham, UK. p. 361–382.

Feldpausch, J. A., J. Jourquin, J. R. Bergstrom, C. D. Bokenkroger, J. L. Nelssen, M. J. Ritter, D. L. Davis, and J. M. Gonzalez. 2016. Birth weight threshold for identifying piglets at-risk for preweaning mortality. J. Anim. Sci. 94(Suppl. 2):34.

Feyera, T., C. K. Hojgaard, J. Vinther, T. S. Bruun, and P. K. Theil. 2017. Dietary supplement rich in fiber fed to late gestating sows during transition reduces rate of stillborn piglets. J. Anim. Sci. 95:5430–5438.

Feyera, T., T. F. Pedersen, U. Krogh, L. Foldager, and P. K. Theil. 2018. Impact of sow energy status during farrowing on farrowing kinetics, frequency of stillborn piglets, and farrowing assistance. J. Anim. Sci. 96:2320–2331.

Figueroa, J., D. Sola-Oriol, L. Vinokurovas, X. Manteca, and J. F. Perez. 2013. Prenatal flavor exposure through maternal diets influences flavor preference in piglets before and after weaning. Anim. Feed Sci. Technol. 183:160–167.

Foxcroft, G. R., W. T. Dixon, S. Novak, C. T. Putman, S. C. Town, and M. D. A. Vinsky. 2006. The biological basis for prenatal programming of postnatal performance in pigs. J. Anim. Sci. 84(E. Suppl):E105–E112.

Fraser, D., J. J. R. Feddes, and E. A. Pajor. 1994. The relationship between creep feeding behavior of piglets and adaptation to weaning: Effect of diet quality. Can. J. Anim. Sci. 74:1–6.

Fraser, D. and P. A. Phillips. 1989. Lethargy and low water intake by sows during early lactation: a cause of low piglet weight gains and survival? Appl. Anim. Behav. Sci. 24:13–22.

Fraser, D., P. A. Phillips, B. K. Thompson, and W. B. Peeters Weem. 1988. Use of water by piglets in the first days after birth. Can. J. Anim. Sci. 68:603–610.

Goncalves, M. A. D., K. M. Gourley, S. S. Dritz, M. D. Tokach, N. M. Bello, J. M. DeRouchey, J. C. Woodworth, and R. D. Goodband. 2016. Effects of amino acids and energy intake during late gestation of high-performing gilts and sows on litter and reproductive performance under commercial conditions. J. Anim. Sci. 94:1993–2003.

Gondret, F., L. Lefaucheur, H. Juin, I. Louveau, and B. Lebret. 2006. Low birth weight is associated with enlarged muscle fiber area and impaired meat tenderness of the longissimus muscle in pigs. J. Anim. Sci. 84:93–103.

Gonzalez-Anover, P., and A. Gonzalez-Bulnes. 2017. Maternal age modulates the effects of early-pregnancy L-proline supplementation on the birth-weight of piglets. Anim. Reprod. Sci. 181:63–68.

Hansen, A. V., A. B. Strathe, E. Kebreab, J. France, and P. K. Theil. 2012. Predicting milk yield and composition in lactating sows: A Bayesian approach. J. Anim. Sci. 90:2285–2298.

Heo, P. S., D. H. Kim, J. C. Jang, J. S. Hong, and Y. Y. Kim. 2018. Effects of different creep feed types on pre-weaning and post-weaning performance and gut development. Asian-Australas J. Anim. Sci. 12:1956–1962.

Holen, J. P., P. E. Urriola, M. Schwartz, J. C. Jang, G. C. Shurson, and L. J. Johnston. 2020. Effects of supplementing late-gestation sow diets with zinc on preweaning mortality of pigs under commercial rearing conditions. Transl. Anim. Sci. 4:515–530.

Hurley, W. L. 2015. Composition of sow colostrum and milk. In: The gestating and lactating sow, C. Farmer, editor, Wageningen Acad. Publishers, Wageningen, The Netherlands. p. 193–229.

Huting, A. M. S., K. Almond, I. Wellock, and I. Kyriazakis. 2017. What is good for small piglets might not be good for big piglets: The consequences of cross-fostering and creep feed provision on performance to slaughter. J. Anim. Sci. 95:4926–4944.

Jeon, J. H., S. C. Yeon, Y. H. Choi, W. Min, S. Kim, P. J. Kim, and H. H. Chang. 2006. Effects of chilled drinking water on the performance of lactating sows and their litters during high ambient temperatures under farm conditions. Livest. Sci. 105:86–93.

Johnston, L. J. 1993. Maximizing feed intake of lactating sows. Compendium Cont. Ed. Pract. Vet. 15:133–141.

Johnston, L. J., C. Bennett, R. J. Smits, and K. Shaw. 2007. Identifying the relationship of gilt rearing characteristics to lifetime sow productivity. In: J. E. Paterson, and J. A. Barker, editors, Manipulating pig prod. XI. Aust. Pig Sci. Assoc., South Perth, WA, Australia. p. 39.

Johnston, L., and J. Holt. 2006. Improving pig welfare - The role of dietary fiber. Proc. 67th Minnesota Nutr. Conf., Univ. of Minnesota, Minneapolis, MN, US. p. 171–182.

Jorgensen, B. and M. T. Sorensen. 1998. Different rearing intensities of gilts: II. Effects on subsequent leg weakness and longevity. Livest. Prod. Sci. 54:167–171.

Ju, J. C., S. P. Cheng, and H. T. Yen. 1985. Effects of amino acid additions in diets on semen characteristics of boars. J. Chinese Soc. Anim. Prod. 14:27–35.

Kemp, B., M. W. A. Verstegen, L. A. den Hartog, and H. J. G. Grooten. 1989a. The effect of environmental temperature on metabolic rate, and partitioning of energy intake in breeding boars. Livest. Prod. Sci. 23:329.

Kemp, B., L. A. den Hartog, and H. J. G. Grooten. 1989b. The effect of feeding level on semen quantity and quality of breeding boars. Anim. Reprod. Sci. 20:245.

Kim, S. W. 2010. Recent advances in sow nutrition. R. Bras. Zootec. 39:303–310

Kim, S. W., W. L. Hurley, G. Wu, and F. Ji. 2009. Ideal amino acid balance for sows during gestation and lactation. J. Anim. Sci. 87(Suppl.):E123–E132.

Kim, J. S., X. Yang, D. Pangeni, and S. K. Baidoo. 2015. Relationship between backfat thickness of sows during late gestation and reproductive efficiency at different parities. Acta Agric. Scand., Anim. Sci. 65:1–8.

King, R. H. 1987. Nutritional anestrus in young sows. Pig News and Inform. 8:15–22

Knauer, M. T., J. P. Cassady, D. W. Newcom, M. T. See. 2011. Phenotypic and genetic correlations between gilt estrus, puberty, growth, composition, and structural conformation traits with first-litter reproductive measures. J. Anim. Sci. 89:935–942.

Koketsu, Y., G. D. Dial, J. E. Pettigrew, and V. L. King. 1996. Feed intake pattern during lactation and subsequent reproductive performance of sows. J. Anim. Sci. 74:2875–2884.

Krogh, U., T. S. Bruun, C. Amdi, C. Flummer, J. Poulsen, and P. K. Theil. 2015. Colostrum production in sows fed different sources of fiber and fat during late gestation. Can. J. Anim. Sci. 95:211–223.

Kummer, R., M. L. Bernardi, A. C. Schenkel, W. S. Amaral Filha, I. Wentz, and F. P. Bortolozzo. 2009. Reproductive performance of gilts with similar age but with different growth rate at the onset of puberty stimulation. Reprod. Dom. Anim. 44:255–259.

Lauridsen, C., and V. Danielsen. 2004. Lactational dietary fat levels and sources influence milk composition and performance of sows and their progeny. Livest. Prod. Sci. 91:95–105.

Lawrence, A. B., and E. M. C. Terlouw. 1993. A review of behavioral factors involved in the development and continued performance of stereotypic behaviors in pigs. J. Anim. Sci. 71:2815–2825.

Lee, S. I. and I. H. Kim. 2018. Creep feeding improves growth performance of suckling piglets. R. Bras. Zootec. 47:e20170081.

Leibbrandt, V. D., L. J. Johnston, G. C. Shurson, J. D. Crenshaw, G. W. Libal, and R. D. Arthur. 2001. Effect of nipple drinker water flow rate and season on performance of lactating sows. J. Anim. Sci. 79:2770–2775.

Li, D. F., J. L. Nelssen, P. G. Reddy, F. Blecha, R. Klemm, and R. D. Goodband. 1991. Interrelationship between hypersensitivity to soybean proteins and growth performance in early-weaned pigs. J. Anim. Sci. 69:4062–4069.

Loisel, F., C. Farmer, P. Ramaekers, and H. Quesnel. 2013. Effects of high fiber intake during late pregnancy on sow physiology, colostrum production, and piglet performance. J. Anim. Sci. 91:5269–5279.

Lopez-Serrano, M., N. Reinsch, H. Looft, and E. Kalm. 2000. Genetic correlations of growth, backfat thickness and exterior with stayability in Large White and Landrace sows. Livest. Prod. Sci. 64:121–131.

Louis, G. F., A. J. Lewis, W. L. Weldon, P. S. Miller, R. J. Kittok, and W.W. Stroup. 1994a. Calcium levels for boars and gilts. J. Anim. Sci. 72:2038–2050.

Louis, G. F., A. J. Lewis, W. L. Weldon, P. S. Miller, R. J. Kittok, and W. W. Stroup. 1994b. The effect of protein intake on boar libido, semen characteristics and plasma hormone concentrations. J. Anim. Sci. 72:2051–2060.

Meunier-Salaun, M. C., S. A. Edwards, and S. Robert. 2001. Effect of dietary fiber on the behavior and health of the restricted fed sow. Anim. Feed Sci. Technol. 90:53–69.

Miller, B. G., T. J. Newby, C. R. Stokes, and F. J. Bourne. 1984. Influence of diet on postweaning malabsorption and diarrhea in the pig. Res. Vet. Sci. 36:187–193.

Milligan, B. N., D. Fraser, and D. L. Kramer. 2002. Within-litter birth weight variation in the domestic pig and its relation to preweaning survival, weight gain, and variation in weaning weights. Livest. Prod. Sci. 76:181–191.

Muns, R. and E. Magowan. 2018. The effect of creep feed intake and starter diet allowance on piglets' gut structure and growth performance after weaning. J. Anim. Sci. 96:3815–3823.

Nimmo, R. D., E. R. Peo, Jr., B. D. Moser, and A. J. Lewis. 1981. Effect of level of dietary calcium-phosphorus during growth and gestation on performance, blood and bone parameters of swine. J. Anim. Sci. 52:1330–1342.

Noblet, J., J. Y. Dourmad, and M. Etienne. 1990. Energy utilization in pregnant and lactating sows: modeling of energy requirements. J. Anim. Sci. 68:562–572.

NRC. 2012. Nutrient Requirements of Swine. 11th rev. ed. Natl. Acad. Press, Washington, DC.

NSNG. 2010. National swine nutrition guide. U.S. Pork Center of Excellence. Ames, IA.

Odberg, F. O. 1978. Abnormal behaviors: Stereotypies. In: 1st World Congress on Ethology Applied to Zootechnics, Madrid, Spain, p. 475–480.

Oliviero, C., M. Heinonen, A. Valros, and O. Peltoniemi. 2010. Environmental and sow-related factors affecting the duration of farrowing. Anim. Reprod. Sci. 119:85–91.

Oliviero, C., T. Kokkonen, M. Heinonen, S. Sankari, and O. Peltoniemi. 2009. Feeding sows with high fibre diet around farrowing and early lactation: Impact on intestinal activity, energy balance related parameters and litter performance. Res. Vet. Sci. 86:314–319.

Orth, M. 2007. Optimizing skeletal health: The impact of nutrition. Proc. Minnesota Nutr. Conf., Univ. of Minnesota, Minneapolis, MN, US. p. 139–143.

Pajor, E. A., D. M. Weary, C. Caceres, D. Fraser, and D. L. Kramer. 2002. Alternative housing for sows and litters Part 3. Effects of piglet diet quality and sow-controlled housing on performance and behaviour. Appl. Anim. Behav. Sci. 76:267–277.

Patience, J. F. 2013. Water in swine nutrition. In: L. I. Chiba, editor, Sustainable swine nutrition. Wile-Blackwell, John Wiley & Sons, Hoboken, NJ, US. p. 3–22.

Patterson, J. L., E. Beltranena, and G. R. Foxcroft. 2010. The effect of gilt age at first estrus and breeding on third estrus on sow body weight changes and long-term reproductive performance. J. Anim. Sci. 88:2500–2513.

Peng, J. J., S. A. Somes, and D. W. Rozeboom. 2007. Effect of system of feeding and watering on performance of lactating sows. J. Anim. Sci. 85:853–860.

Pettigrew, J. E. 1981. Supplemental dietary fat for peripartal sows: A review. J. Anim. Sci. 53:107–117.

Pettigrew, J. E., and H. Yang. 1997. Protein nutrition of gestating sows. J. Anim. Sci. 75:2723–2730.

Pluske, J. R. 2016. Invited review: Aspects of gastrointestinal tract growth and maturation in the pre-and postweaning period of pigs. J. Anim. Sci. 94:399–411.

Rehfeldt, C., G. Kuhn, G. Nurnberg, E. Kanitz, F. Schneider, M. Beyer, K. Nurnberg, and K. Ender. 2001. Effects of exogenous somatotropin during early gestation on maternal performance, fetal growth, and compositional traits in pigs. J. Anim. Sci. 79:1789–1799.

Phillips, P. A. and D. Fraser. 1991. Discovery of selected water dispensers by newborn pigs. Can. J. Anim. Sci. 71:233–236.

Rozeboom, D. W. 2015. Conditioning of the gilt for optimal reproductive performance. In: The gestating and lactating sow, C. Farmer ed., Wageningen Acad. Publishers, Wageningen, The Netherlands. p. 13–26.

Rozeboom, D. W., J. E. Pettigrew, R. L. Moser, S. G. Cornelius, and S. M. El Kandelgy. 1996. Influence of gilt age and body composition at first breeding on sow reproductive performance and longevity. J. Anim. Sci. 74:138–150.

Serena, A., H. Jorgensen, and K. E. Bach Knudsen. 2009. Absorption of carbohydrate-derived nutrients in sows as influenced by types and contents of dietary fiber. J. Anim. Sci. 87:136–147.

Shields, R. G. and D. C. Mahan. 1983. Effects of pregnancy and lactation on the body composition of first-litter female swine. J. Anim. Sci. 57:594–603.

Sinclair, A. G., V. C. Bland, and S. A. Edwards. 2001. The influence of gestation feeding strategy on body composition of gilts at farrowing and response to dietary protein in a modified lactation. J. Anim. Sci. 79:2397–2405.

Stalder, K. J., A. M. Saxton, G. E. Conatser, and T. V. Serenius. 2005. Effect of growth and compositional traits on first parity and lifetime reproductive performance in U.S. Landrace sows. Livest. Prod. Sci. 97:151–159.

Stevermer, E. J., M. F. Kovacs, R. C. Hoekstra, and H. L. Self. 1961. Effect of feed intake on semen characteristics and reproductive performance of mature boars. J. Anim. Sci. 20:858–865.

Sulabo, R. C., J. Y. Jacela, M. D. Tokach, S. S. Dritz, R. D. Goodband, J. M. DeRouchey, and J. L. Nelssen. 2010a. Effects of lactation feed intake and creep feeding on sow and piglet performance. J. Anim. Sci. 88:3145–3153.

Sulabo, R. C., M. D. Tokach, J. M. DeRouchey, S. S. Dritz, R. D. Goodband, and J. L. Nelssen. 2010b. Influence of feed flavors and nursery diet complexity on preweaning and nursery pig performance. J. Anim. Sci. 88:3918–3926.

Sulabo, R. C., M. D. Tokach, S. S. Dritz, R. D. Goodband, J. M. DeRouchey, and J. L. Nelssen. 2010c. Effects of varying creep feeding duration on the proportion of pigs consuming creep feed and neonatal pig performance. J. Anim. Sci. 88:3154–3162.

Tan, C. Q., H. K. Wei, H. Q. Sun, G. Long, J. T. Ao, S. W. Jiang, and J. Peng. 2015. Effects of supplementing sow diets during two gesttions with konjac flour and *Saccharomyces boulardii* on constipation in peripartal period, lactation feed intake and piglet performance. Anim. Feed Sci. Technol. 210:254–262.

Tatara, M. R., E. Sliwa, and W. Krupski. 2007. Prenatal programming of skeletal development in the offspring: Effects of maternal treatment with β-hydroxy-β-methylbutyrate (HMB) on femur properties in pigs at slaughter age. Bone 40:1615–1622.

Tatara, M. R., W. Krupski, B. Tymczyna and T. Studziński. 2012. Effects of maternal administration with alpha-ketoglutarate (AKG) and β-hydroxy-β-methylbutyrate (HMB) on prenatal programming of skeletal properties in the offspring. Nutr. Metab. 9:39–51.

Theil, P. K. 2015. Transition feeding of sows. In: C. Farmer editor. The Gestating and Lactating Sow. Wageningen Acad. Publishers, Wageningen, The Netherlands. p. 147–172.

Tritton, S. M., R. H. King, R. G. Campbell, A. C. Edwards, and P. E. Hughes. 1996. The effects of dietary protein and energy levels of diets offered during lactation on lactational and subsequent reproductive performance of first-litter sows. Anim. Sci. 62:573–579.

Trottier, N. L. 1991. Relationship between energy intake, backfat thickness and reproductive performance of sows. M. S. Thesis, McGill Univ., Montreal, QC, Canada.

Tummaruk, P., N. Lundeheim, S. Einarsson, and A. M. Dalin. 2001. Effect of birth litter size, birth parity number, growth rate, backfat thickness and age at first mating of gilts on their reproductive performance as sows. Anim. Reprod. Sci. 66:225–237

Ullrey, D. E., J. I. Sprague, D. E. Becker, and E. R. Miller. 1965. Growth of the swine fetus. J. Anim. Sci. 24:711–717.

Vallet, J. L., J. R. Miles, T. M. Brown-Brandl, and J. A. Nienaber. 2010. Proportion of the litter farrowed, litter size, and progesterone and estradiol effects on piglet birth intervals and stillbirths. Anim. Reprod. Sci. 119:68–75.

Vallet, J. L., L. A. Rempel, J. R. Miles, and S. K. Webel. 2014. Effect of essential fatty acid and zinc supplementation during pregnancy on birth intervals, neonatal piglet brain myelination, stillbirth, and preweaning mortality. J. Anim. Sci. 92:2422–2432.

van der Meulen, J., S. J. Koopmans, R. A. Dekker, and A. Hoogendoorn. 2010. Increasing weaning age of piglets from 4 to 7 weeks reduces stress, increases post-weaning feed intake but does not improve intestinal functionality. Animal 4:1653–1661.

Wan, H., J. Zhu, C. Wu, P. Zhou, Y. Shen, Y. Lin, S. Xu, L. Che, B. Feng, J. Li, Z. Fang, and D. Wu. 2017. Transfer of β-hydroxy-β-methylbutyrate from sows to their offspring and its impact on muscle fiber type transformation and performance in pigs. J. Anim. Sci. Biotechnol. 8:438–449.

Weldon, W. C., A. J. Lewis, G. F. Louis, J. L. Kovar, M. A. Giesemann, and P. S. Miller. 1994. Postpartum hypophagia in primiparous sows: I. Effects of gestation feeding level on feed intake, feeding behavior, and plasma metabolite concentration during lactation. J. Anim. Sci. 72:387–394.

Westendorf, P., and L. Richter. 1977. Nutrition of the boar. Ubersicht fur Tierernahrung. 5:161–184.

Whitney, M. H., and S. K. Baidoo. 2010. Breeding boar nutrient recommendations and feeding management. PIG 07-01-13. In: National Swine Nutrition Guide, U.S. Pork Center of Excellence, Ames, IA, US.

Williams, N. H., J. Patterson, and G. Foxcroft. 2005. Non-negotiables in gilt development. Adv. Pork Prod. 16:281–289.

Williams, B., D. Waddington, D. H. Murray, and C. Farquharson. 2004. Bone strength during growth: Influence of growth rate on cortical porosity and mineralization. Calcif. Tissue Int. 74:236–245.

Wu, G., F. W. Bazer, R. C. Burghardt, G. A. Johnson, S. W. Kim, X. L. Li, M. C. Satterfield, and T. E. Spencer. 2010. Impacts of amino acid nutrition on pregnancy outcome in pigs: Mechanisms and implications for swine production. J. Anim. Sci. 88:E195–E204.

Wu, G., F. W. Bazer, J. M. Wallace, and T. E. Spencer. 2006. Board-Invited Review: Intrauterine growth retardation: Implications for the animal sciences. J. Anim. Sci. 84:2316–2337.

Wu, G., and S. W. Kim. 2007. Functional amino acids in animal production. In: W. G. Pond, and A. W. Bell, editors, Encyclopedia Anim. Sci. doi:https://doi.org/10.1081/E-EAS-120043422.

Yang, H., G. R. Foxcroft, J. E. Pettigrew, L. J. Johnston, G. C. Shurson, A. N. Costas, and L. J. Zak. 2000a. Impact of dietary lysine intake during lactation on follicular development and oocyte maturation after weaning in primiparous sows. J. Anim. Sci. 78:993–1000.

Yang, H., J. E. Pettigrew, L. J. Johnston, G. C. Shurson, and R. D. Walker. 2000b. Lactational and subsequent reproductive responses of lactating sows to dietary lysine (protein) concentrations. J. Anim. Sci. 78:348–357.

Yang, H., J. E. Pettigrew, L. J. Johnston, G. C. Shurson, J. E. Wheaton, M. E. White, Y. Koketsu, A. F. Sower, and J. A. Rathmacher. 2000c. Effects of dietary lysine intake during lactation on blood metabolites, hormones, and reproductive performance in primiparous sows. J. Anim. Sci. 78:1001–1009.

Zak, L. J., J. R. Cosgrove, F. X. Aherne, and G. R. Foxcroft. 1997a. Pattern of feed intake and associated metabolic and endocrine changes differentially affect postweaning fertility in primiparous lactating sows. J. Anim. Sci. 75:208–216.

Zak, L. J., X. Xu, R. T. Hardin, and G. R. Foxcroft. 1997b. Impact of different patterns of feed intake during lactation in the primiparous sow on follicular development and oocyte maturation. J. Reprod. Fertil. 110:99–106.

Zeng, Z. K., P. E. Urriola, J. R. Dunkelberger, J. M. Eggert, R. Vogelzang, G. C. Shurson, and L. J. Johnston. 2019. Implications of early life indicators for survival rate, subsequent growth performance, and carcass characteristics of commercial pigs. J. Anim. Sci. 97:3313–3325.

# 23    Feeding Weanling, Grower, and Finisher Swine

Robert D. Goodband, Mariana B. Menegat, and Hayden E. Williams

## Introduction

Efficient and profitable swine production from weaning to market depends upon an understanding of the concepts of genetics, health, management, environment, and nutrition. Nutrition represents 60–75% of the total cost of pork production. Thus, a thorough knowledge of the principles of swine nutrition is essential to maintain productivity and profitability in a swine enterprise. Typically in commercial production, diets are categorized into two groupings: weaner (weaning to approximately 23 kg) and growing-finishing diets (23 kg to market). While many of the same basic concepts and rationale are similar between the two, the formulations will vary. Typically, weaner pig diets are generally broken into three dietary phases and often contain several specialty protein sources included to stimulate feed intake. Growing-finishing pig diets are fed generally to 5 different phases and based on grain and oilseed meals to economically meet nutrient requirements. However, regardless of the classifications and formulations, there are numerous ways in which weaner and growing-finishing pigs can profitably fed. Because there is no single way to best feed pigs, this indicates that the pig is a relatively resilient animal and can thrive in multiple environments.

## Weaner Pigs

The biology of weanling pigs must be considered for a successful nutritional program in the nursery. Young pigs have high protein deposition, low feed intake, high lactase activity, and low amylase, maltase, sucrase, and lipase activities. This means that newly weaned pigs are able to easily digest lactose and specialty protein sources, but have limited ability to digest plant protein sources, sugars, and to utilize fat. In general, it is important to provide adequate amino acid levels from highly digestible protein sources because weanling pigs have a high capacity for protein deposition in relation to feed intake.

Feed intake is a key determinant of performance and health status of weanling pigs. While the majority of pigs begin consuming feed within the first 24 hours after weaning, approximately 30% take between 24 and 60 hours to start on feed (Bruinix et al. 2001). Weanling pigs are in a highly energy-dependent stage of growth, which means that any increase in feed intake results in improvements in growth rate and lean deposition. Moreover, feed intake is important to sustain an adequate gut structure for nutrient absorption (Pluske et al. 1996) and to reduce the occurrence of diarrhea in weanling pigs (Madec et al. 1998).

*Sustainable Swine Nutrition*, Second Edition. Edited by Lee I. Chiba.
© 2023 John Wiley & Sons Ltd. Published 2023 by John Wiley & Sons Ltd.

The most important aspect to maximize feed intake is to have feed available and offered ad libitum as soon as pigs are weaned. There are several strategies to encourage feed intake of weanling pigs. One strategy is to use feeding boards or mats to supply adequate feeding space. Another strategy is to offer a gruel with a mixture of feed with water. With both of these strategies, feed should also be available in the feeders and the strategies should be used temporarily during the first few days after weaning as to not discourage consumption of feed in the feeders. The boards, mats, and gruels should be appropriately managed to prevent feed spoilage and disease transmission.

Feeding behavior after weaning is also stimulated by providing creep feed, while pigs are nursing to ease the transition from milk to solid diets (Bruinix et al. 2002). The creep diet does not have to be offered for a long period before weaning, but it must be a highly palatable and digestible diet (Sulabo et al. 2009, 2010). A viable strategy is to offer a creep diet for 3 to 5 days before weaning to increase the proportion of pigs consuming creep feed and improve feed intake after weaning. Also, using similar ingredients in initial nursery diets can stimulate weanling feed intake in the early postweaning.

One of the goals of the nutritional program in the nursery is to prepare pigs for grow-finish diets. Diets in the grow-finish period are relatively simple (grain-oil seed-based) and less expensive compared to nursery diets. Although the use of specialty ingredients results in excellent performance in the nursery, benefits generally do not result in further improvement in grow-finish performance. Thus, specialty ingredients should be paid for in the nursery without projections of improved finishing performance. The goal is to gradually remove specialty high-cost ingredients from nursery diets and replace them with typical lower cost ingredients, such as grains and oil seed meals, as quickly as possible.

### Phase Feeding

The phase feeding program is often matched to the weight of piglets at weaning using a feed budget. Generally, as pigs become heavier at weaning, the amount of the initial nursery diets is reduced. As an example, a 3-phase feeding program can be used in the nursery, consisting of phase 1 (5–7 kg), phase 2 (7–12 kg), and phase 3 (12–23 kg) diets. An intensive care diet can be used in the nursery for low weaning weight pigs (3–5 kg), health-challenged pigs, as well as be offered as a creep feed during the nursing phase.

Intensive care and phase 1 diets are more complex to improve feed intake and provide high-quality feed ingredients to weanling pigs, such as specialty protein and lactose sources. Intensive care and phase 1 diets are commonly provided as pellets or crumbles because of the impact of specialty ingredients on feed flow ability. These diets are more expensive, but their use is typically justified considering the small amount of feed used. Diet complexity rapidly reduces in phase 2 and 3 diets as feed intake becomes established. Phase 2 and 3 diets can be provided either as meal, pellets, or crumbles. These diets represent most of the total feed use in the nursery and have a significant impact on total feed cost up to 23 kg. Adhering to the feed budget guidelines in Table 23.1 helps to optimize performance in the nursery and minimize overfeeding of expensive initial nursery diets.

### Diet Complexity

Diet complexity refers to the use of highly digestible specialty feed ingredients in nursery diets. Complex diets are typically fed to weanling pigs to provide high-quality feed ingredients and

**Table 23.1**  Feed budget recommendations for nursery diets according to weaning weight[a].

| Item | Weaning weight, kg | | | | | | |
|---|---|---|---|---|---|---|---|
| | 4.5 | 5.0 | 5.5 | 6.0 | 6.5 | 7.0 | 7.5 |
| Diet, kg per pig | | | | | | | |
| Intensive care | 1.0 | 0.5 | 0.5 | — | — | — | — |
| Phase 1 | 2.70 | 2.25 | 1.80 | 1.40 | 1.0 | 0.50 | 0.50 |
| Phase 2 | | | | 5–7 | | | |
| Phase 3 | | | | 20–23 | | | |

[a] Adapted from Menegat et al. (2019).

improve intake in the early postweaning period. As complex diets are more expensive, diet complexity should be rapidly reduced during the course of the nursery.

Nutritional strategies in the nursery have been of great interest because it is generally assumed that pigs that grow faster in the nursery also grow faster in the finisher. However, not all dietary efforts to improve performance in the nursery are rewarded with improvements in growth rate in the finisher. An important distinction to make is whether the dietary effort is able to induce a fundamental or a transitory change in the nursery pig.

Weaning age is able to induce a fundamental change in the pig. The enhancement of nursery performance by increasing weaning age is typically maintained into the finisher as a consequence of increasing weaning weight (Main et al. 2004), but most importantly as a consequence of a physiological change in the pig, like improvements in digestive and immune functions (Moeser et al. 2007; Smith et al. 2010; McLamb et al. 2013).

Most nutritional strategies induce a transitory change in the pig, with improvements in performance while being fed in the nursery, but not necessarily in the subsequent finisher period. Diet complexity typically generates this type of response, with significant improvements in feed intake and growth rate while the complex diet is fed, but no performance or additive advantages thereafter (Whang et al. 2000; Wolter et al. 2003; Skinner et al. 2014; Collins et al. 2017). This is also the case of amino acid concentration (Main et al. 2008), fat (Tokach et al. 1995), antibiotics (Skinner et al. 2014), or milk replacers (Wolter and Ellis, 2001) in nursery diets. In contrast, lactose is able to improve nursery performance with further improvements in finisher performance (Tokach et al. 1995). Therefore, the value of diet complexity should consider the benefit gained during the feeding period but not projected additional benefit in the subsequent nursery or finisher periods.

*Intensive Care Diet*
The intensive care diet is typically fed to pigs from 3 to 5 kg body weight. The purpose is to provide nutritional support for piglets that require intensive care, which typically are early-weaned, low-weight, or health-challenged pigs. This population of pigs typically represent approximately 10 to 12% of all nursery pigs. The intensive care diet can also be offered as a creep feed during the nursing phase.

The intensive care diet must provide high-quality feed ingredients to stimulate feed intake and match the digestive capabilities of weanling pigs. Typically, the intensive care diet contains high amounts of specialty protein sources such as fermented soybean meal, enzyme-treated soybean meal, soy protein concentrate, spray-dried plasma protein, or fish meal, among others. The level of lactose is also high, with at least 18% and up to 30% lactose.

*Phase 1 Diet*

The phase 1 diet is typically fed to pigs from weaning at approximately 5 until approximately 7 kg body weight. During this phase, it is important to provide high-quality feed ingredients to stimulate feed intake and match the digestive capabilities of weanling pigs.

Newly weaned pigs are able to easily digest lactose and specialty proteins but have limited ability to digest plant proteins and utilize fat. Pigs also have a hypersensitivity reaction to soybean meal induced by allergenic proteins and indigestible carbohydrates of soybeans. Pigs experience a transitory period of poor nutrient absorption and low growth performance following the first exposure to a diet with high amounts of soybean meal (Li et al. 1990). The effects are transitory and pigs develop tolerance after 7–10 days (Engle, 1994). The best approach to alleviate this problem is to gradually expose weanling pigs to increasing levels of soybean meal in subsequent nursery diets to allow pigs to gradually overcome the hypersensitivity reaction. The early exposure to soybean meal reduces the potential for delayed-type hypersensitivity reaction and allows for greater levels in subsequent nursery diets without an impact on growth performance.

Soybean meal is commonly included up to 16–18% in the phase 1 diet. Other specialty protein sources often used in combination are fermented soybean meal, enzyme-treated soybean meal, soy protein concentrate, or fish meal, among others. Typically, the sources are used in combination to achieve the adequate amino acid profile in the diet and because most specialty protein sources cannot be the sole protein source in the diet without affecting palatability or performance. Feed grade amino acids can also represent a large proportion of the pig's amino acid supply, thus reducing the need for high amounts of the specialty protein sources.

Lactose is the carbohydrate component derived from milk and provides an easily digestible source of energy for pigs. The phase 1 diet typically contains around 15–18% lactose to improve growth rate of weanling pigs (Tokach et al. 1995; Mahan et al. 2004). Common sources of lactose are crystalline lactose, whey permeate, and dried whey, with whey products also providing a highly digestible source of amino acids. However, the addition of lactose products in the diet influences feed processing. The use of high amounts of lactose in pelleted diets can increase friction in the diet during the pelleting process, and in meal, diets can increase bridging and reduce flow ability in bins and feeders.

Fat is not easily utilized by weanling pigs for growth performance. In the early postweaning period, weanling pigs seem to require a more digestible fat source rich in unsaturated and short-chain fatty acids for an efficient energy utilization (Gu and Li, 2003). Vegetable oils such as soybean oil and coconut oil are high-quality sources of energy for weanling pigs (Weng, 2016), but cost often limits the use in nursery diets. Animal fat sources of good quality like choice white grease or beef tallow are usually more cost effective to use in nursery diets. The addition of 3–4% fat is mainly used to improve the pelleting process of phase 1 diets with high levels of lactose.

*Phase 2 Diet*

The phase 2 diet is typically fed to pigs from 7 to 12 kg body weight. During this phase, feeding behavior is already established and, thus, diet complexity is reduced. The phase 2 diet is typically based on grain and soybean meal with low levels of specialty protein sources and lactose.

Soybean meal is often included at up to 20 to 24% of the diet. Like the phase 1 diet, other specialty protein sources are often used in combination and a greater portion of the amino acids in the diet can be used from feed-grade amino acids. The level of lactose is reduced to around 5–7%. Added fat begins to be utilized by the pig to improve growth performance and can be included at 1 to 3% of the diet (Tokach et al. 1995).

*Phase 3 Diet*

The phase 3 diet is typically fed to pigs from 12 to 23 kg body weight. During this phase, feed consumption is the greatest accounting for more than 50% of the total feed cost in the nursery. The phase 3 diet is typically based on grain and soybean meal with no specialty protein or lactose sources. Fat is well utilized by the pig to improve growth performance and can be included at 1 to 3% of the diet.

### Amino Acid Requirements

The amino acid requirements of nursery pigs have changed significantly over time. As pigs have been growing faster with lower feed intake, the amino acid requirements as percentage of the diet have changed.

Current statistical modelling techniques have been applied to determine the dose–response to individual amino acids. Recently, the requirements of lysine (Nemechek et al. 2012; Nichols et al. 2018) and ratios for methionine and cysteine, threonine, tryptophan, isoleucine, valine, and histidine relative to lysine have been estimated for nursery pigs (Table 23.2; Gonçalves et al. 2015; Jayaraman et al. 2015; Clark et al. 2017a,b; Kahindi et al. 2017; Cemin et al. 2018). By doing so, it is possible to determine the requirement to maximize growth performance and also predict the change in growth performance at a particular amino acid level.

In the past, phase 1 nursery diets have been formulated to very high amounts of standardized ileal digestible lysine. A more practical approach is to feed more moderate levels of lysine in initial nursery diets (1.35 to 1.40% standardized ileal digestible lysine from weaning to 12 kg) that are typically composed of more expensive specialty protein sources. The moderate lysine levels allow the reduction in crude protein and savings in diet cost. This approach leads to an excellent overall growth performance in the nursery as long as the lysine levels in late nursery diets are adequate (1.25–1.30% SID Lys from 12 to 23 kg). Thus, feeding lower lysine in early nursery and adequate levels in late nursery allows the reduction in crude protein and savings in diet cost while maintaining growth performance throughout the nursery period (Nemechek et al. 2018).

**Table 23.2** Standardized ileal digestible lysine and other amino acid estimates for weanling pigs[a].

| Item | Weight, kg | | |
| --- | --- | --- | --- |
| | Weaning to 7 | 7–12 | 12–23 |
| Lysine, % | 1.40 | 1.35 | 1.30 |
| Amino acid to lysine ratio, %[b] | | | |
| Methionine | 28 | 28 | 28 |
| Methionine + cysteine | 56 | 56 | 56 |
| Threonine | 63 | 63 | 63 |
| Tryptophan | 19+ | 19+ | 19+ |
| Isoleucine | 55 | 55 | 55 |
| Valine | 70 | 70 | 70 |
| Histidine | 32 | 32 | 32 |

[a] Adapted from Nemecheck et al. (2018).
[b] Suggested minimum standardized ileal digestible amino acid ratios relative to lysine.

## Crude Protein

The lysine level influences the crude protein content of the diet. However, the crude protein level in nursery diets can pose an additional burden on weanling pigs. High-protein diets increase the quantity of undigested proteins and their microbial fermentation in the hindgut, which predisposes the occurrence of postweaning diarrhea (Heo et al. 2013). A practical approach is to feed low-protein, amino acid-fortified diets to decrease the burden imposed by high-protein diets to the gut of weanling pigs. Typically, initial nursery diets may contain up to 22% crude protein, but using feed-grade amino acids the crude protein level can be reduced to approximately 18%. The reduction in crude protein allows to minimize the inclusion of soybean meal that cause a hypersensitivity reaction in pigs, as well as the inclusion of specialty protein sources that increase diet cost. The low-protein diets should be supplemented with feed-grade amino acids to meet the amino acid requirements and support growth performance.

The maximum addition of feed-grade amino acids is often dictated by the ratio of lysine to crude protein and by the most limiting amino acid in the diet. Lysine to crude protein ratio is used to ensure a sufficient supply of nitrogen for synthesis of nonessential amino acids in low-protein, amino acid-fortified diets. For nursery pigs, up to a ratio of standardized ileal digestible lysine to crude protein of 6.35 or total lysine to crude protein of approximately 7.20 is recommended (Millet et al. 2018a,b). Also, the inclusion of feed-grade amino acids is determined by the next limiting amino acid in the diet, which can be isoleucine or histidine when all five first-limiting amino acids are available for feed-grade inclusion. However, low-protein diets fortified with the five first-limiting amino acids are typically able to meet the histidine requirements for nursery pigs (Cemin et al. 2018).

## Calcium and Phosphorus

Calcium and phosphorus are essential for growth performance of nursery pigs. These minerals are involved in skeletal structure development and maintenance, lean tissue deposition, muscle contraction, and many other physiological functions. Phosphorus levels in nursery diets typically have a low safety margin above its requirement because of environmental and economic concerns. Calcium levels, on the other hand, are typically high in nursery diets due to the unaccounted-for contribution of calcium from premix carriers, release from phytase, variability of calcium estimation in feed ingredients, and no environmental and economic concerns regarding calcium.

### Phosphorus Requirements

Phosphorus requirement estimates are typically expressed as digestible phosphorus (Table 23.3). Recently, the phosphorus requirements for nursery pigs have been determined by dose–response models, allowing for a more precise estimation of phosphorus levels to maximize growth performance and optimize economics while minimizing phosphorus excretion (Vier et al. 2017).

The phosphorus requirements of nursery pigs appear to be greater than the NRC (2012) recommendations for digestible phosphorus as a percentage of the diet (Vier et al. 2017; Wu et al. 2018b). The variation on phosphorus requirements depends on the goal, but typically the phosphorus requirements to optimize phosphorus retention (bone ash) are greater than to maximize growth (Vier et al. 2017; Wu et al. 2018b). A practical approach consists of maintaining phosphorus levels at approximately 140% and 130% of the NRC (2012) recommendations of digestible phosphorus for nursery pigs between 7 to 12 kg and 12 to 23 kg, respectively.

**Table 23.3** Mineral estimates for nursery pigs[a].

| Item | Days after weaning | | |
|---|---|---|---|
| | 0–7 | 7–12 | 12–23 |
| Total calcium, %[a] | 0.85 | 0.80 | 0.70 |
| STTD calcium, %[b,c] | 0.60 | 0.58 | 0.48 |
| STTD phosphorus, %[d] | 0.63 | 0.56 | 0.43 |
| Sodium, %[e] | 0.40 | 0.35 | 0.28 |
| Chloride, %[e] | 0.50 | 0.45 | 0.32 |

[a] From NRC (2012).
[b] STTD = standardized total tract digestible.
[c] From González-Vega et al. (2016).
[d] From Vier et al. (2017) and Wu et al. (2018b).
[e] From NRC (2012) and Shawk et al. (2018a,b).

*Calcium Requirements*
Calcium requirement estimates are typically expressed as total calcium (Table 23.3). Total calcium accounts for the analyzed calcium content of ingredients. Recently, values for calcium digestibility in feed ingredients have been determined (González-Vega et al. 2015a,b; Merriman et al. 2016), allowing the requirements for digestible calcium to be estimated (González-Vega et al. 2016).

Nursery diets with excessive calcium levels have a negative impact on growth performance (González-Vega et al. 2016). The negative impact of excessive dietary calcium on growth performance is even more evident in diets with marginal or deficient phosphorus levels (González-Vega et al. 2016; Wu et al. 2018a).

*Calcium:Phosphorus Ratio*
Nursery diets with wide calcium:phosphorus ratios or excessive calcium and marginal or deficient phosphorus concentrations interfere with phosphorus absorption (Reinhardt and Mahan, 1986). Consequently, growth performance of nursery pigs is negatively affected by wide calcium:phosphorus ratios (González-Vega et al. 2016; Wu et al. 2018a). Diets with adequate phosphorus levels allow the calcium:phosphorus ratio to be on the upper range, with a decrease in growth performance around 1.9:1 to 2:1. However, diets with marginal phosphorus levels require a narrow calcium:phosphorus ratio (Reinhardt and Mahan, 1986; Qian et al. 1996; Wu et al. 2018a).

*Phytase*
Phytase is an enzyme that catalyzes the release of phosphorus from phytate. The addition of exogenous microbial phytase to nursery diets is a common practice to efficiently and economically enhance phosphorus utilization. Moreover, the use of phytase above conventional levels (500 to 1,000 FTU/kg) seems to have the potential to improve growth performance of nursery pigs beyond what is expected with adequate phosphorus levels (Zeng et al. 2014). The use of high levels of phytase, known as "super-dosing" (approximately 2,000 to 3,000 FTU), is also becoming more common in nursery diets (Gourley et al. 2018; Laird et al. 2018 Wu, et al. 2019).

**Sodium and Chloride**

Sodium and chloride are involved in nutrient absorption, electrolyte balance, and regulation of pH. Salt is the most common source of sodium and chloride, but it is often not included in sufficient quantities to meet the requirements of sodium and chloride of nursery pigs. Sodium and chloride concentrations are often overlooked in nursery diets because some commonly used ingredients contain high levels of sodium, particularly dried whey (approximately 1% sodium) and spray-dried plasma protein (approximately 3% sodium).

The requirements of sodium and chloride are greater for nursery pigs and abruptly decrease for grow-finish pigs. Recently, the requirements of sodium and chloride for nursery pigs have been determined (Shawk et al. 2018a,b) and indicated that the NRC (2012) requirement estimates are accurate (Table 23.3). Diets need as much as 0.5 to 0.6% added salt to meet the requirements of nursery pigs. Nursery diets deficient in salt result in decreased growth performance due to reduced feed intake and poor feed efficiency (Shawk et al. 2018a,b).

## Zinc and Copper

Zinc and copper are trace minerals required at concentrations of 80 to 100 mg/kg and 5 to 6 mg/kg, respectively, to meet the requirements of nursery pigs (Table 23.4). However, the addition of zinc and copper at quantities greater than the requirement exerts a beneficial effect on growth performance of nursery pigs (Liu et al. 2018). Greater quantities of zinc and copper are often referred to as growth promoting or pharmacological levels.

Pharmacological level of dietary zinc between 2,000 and 3,000 mg/kg is a common recommendation to initial nursery (up to 7 kg body weight) diets to reduce postweaning diarrhea and improve growth performance (Hill et al. 2000; Shelton et al. 2011). These effects have been consistently demonstrated with dietary zinc provided as zinc oxide (ZnO) (Hill et al. 2001; Hollis et al. 2005; Walk et al. 2015), while zinc sulfate (ZnSO4) has greater potential to induce toxicity (Hahn and Baker, 1993). Organic sources of zinc with greater bioavailability have not consistently demonstrated the same benefits as zinc oxide when organic zinc is added at lower levels (Hahn and Baker, 1993; Carlson et al. 2004; Hollis et al. 2005). However, pharmacological levels of zinc appear to interfere with calcium and phosphorus absorption, prompting the use of phytase or greater levels of calcium and phosphorus in nursery diets to ameliorate this effect (Blavi et al. 2017).

**Table 23.4** Zinc and copper requirement estimates and recommended pharmacological levels for nursery pigs[a].

| | Days after weaning | | |
|---|---|---|---|
| Item | 0–7 | 7–12 | 12–23 |
| Requirement estimates[1] | | | |
| Zinc, mg/kg | 100 | 100 | 80 |
| Copper, mg/kg | 6 | 6 | 5 |
| Pharmacological levels[2] | | | |
| Zinc, mg/kg[3] | 3,000 | 2,000 | — |
| Copper, mg/kg | — | — | up to 250 |

[a] From NRC (2012).
[b] From Hill et al. (2001) and Shelton et al. (2011).
[c] Pharmacological levels of zinc should only be used for short time (up to 12 kg). Maximum tolerance level of 1,000 ppm for long-term use.
— Not recommended to use.

Pharmacological levels of dietary copper between 125 and 250 mg/kg are commonly used in the diet to enhance fecal consistency and improve growth performance of nursery pigs (Bikker et al. 2016). The most commonly used source of dietary copper is copper sulfate (CuSO4; Cromwell et al. 1998), but tribasic copper chloride (TBCC) is as effective as copper sulfate in promoting growth performance (Cromwell et al. 1998; Coble et al. 2017). Organic sources of copper with greater bioavailability, such as Cu-amino acid chelate, also seem to have the potential to influence growth performance (Pérez et al. 2011; Carpenter et al. 2018).

A typical recommendation is to use pharmacological levels of zinc in initial nursery diets fed to pigs up to 12 kg and then replace zinc by pharmacological levels of copper for the remaining nursery period (12–23 kg; Table 23.4). Additive effects of using pharmacological levels of zinc and copper are not common (Hill et al. 2000), but might occur to some degree (Pérez et al. 2011). In diets with in-feed antimicrobials, the use of pharmacological levels of zinc or copper seems to have an additive effect in growth performance (Stahly et al. 1980; Hill et al. 2001).

The use of pharmacological levels of zinc and copper poses an environmental concern because of the greater excretion of minerals in swine waste and ultimately in the soil fertilized with swine manure (Jondreville et al. 2003). In addition, the implication of pharmacological levels of zinc and copper as a cause of increasing antimicrobial resistance is a rising concern (Yazdankhah et al. 2014). Therefore, regulations have been implemented in some countries restricting or prohibiting the use of zinc or copper as growth promoters. Thus, there is an appeal for prudent use of pharmacological levels of zinc and copper in swine production.

## Grower and Finisher Pigs

In the United States, with current market weights of 130 kg and greater, the period from 23 to 130 kg represents 90% of the feed a pig will consume from weaning to market. Therefore, establishing the nutrient requirements of grow-finish pigs is important from a growth performance and economic standpoint.

### *Economics of a Nutritional Program*

Before a growing-finishing pig nutrition program can be developed, the production system's goals and objectives must be clearly defined. Different economic measurements can be used to determine economic success, and this can differ between production systems. The four different economic measurements include evaluating total dietary cost, feed cost per unit of gain, income over feed cost, and income over feed and facility cost. Total dietary costs and feed cost per unit of gain can be used as economic evaluators when the focus is to reduce variable costs. Whereas income over feed cost and income over feed and facility cost are accurate methods to determine total profitability as they consider total revenue, dietary costs, and facilities costs.

*Feed Cost*
Feed cost only takes into consideration the cost of the diets for comparison between one nutrition program and another. This method is the simplest and has its greatest and best application when there is no expected change in pig performance associated with the nutritional program. However, because changes in ingredients or nutrient levels often change pig performance, it should rarely be used as the main evaluation of economic competitiveness of a feeding program.

*Feed Cost per Unit of Gain*

Feed cost per kilogram of gain is calculated by multiplying feed efficiency by the feed cost per kilogram. The best application of this method is for comparison between nutritional programs when there is an expected change in feed efficiency without a change in growth rate.

Feed cost per kg of gain, $/kg gain = Feed efficiency × Feed cost, $/kg

*Income Over Feed Cost*

Income over feed cost is a margin of profit calculated by subtracting feed cost from the revenue, usually on a per pig basis. Revenue per pig is often estimated by multiplying hot carcass weight by hot carcass weight price, or by multiplying total weight gain by live weight price. Feed cost per pig is estimated by multiplying total feed intake by feed cost. Facility cost can be also added to feed cost to estimate the income over feed and facility cost. Typically, facility cost in the United States is around $0.10 to 0.12 per pig per day. Income over feed costs and income over feed and facility costs are accurate methods to determine the economic value of a nutritional program. The best application is for systems that run on a fixed-time basis and for comparison between nutritional programs when there is an expected or possible change in both feed efficiency and growth rate.

Income over feed cost, $/pig = Revenue, $/pig – Feed cost, $/pig
Income over feed and facility cost, $/pig = Revenue, $/pig – (Feed cost + Facility cost, $/pig)

**Use of Different Economic Assessments to Determine Feeding Regimens: Example on Fats or Oils**

Energy is the most expensive component of the diet. The use of fat in the diet increases dietary energy and has direct impact on growth rate, feed efficiency, and carcass criteria. The use of fat should be based on an economic analysis to determine the most economical dietary energy level considering the value of incremental changes in energy on production indicators and the market price.

The example in Table 23.5 illustrates the use of feed cost, feed cost per kilogram of gain, and income over feed cost, to determine the economics of added dietary fat. The example is a hypothetical comparison between two nutritional programs with or without 5% added fat in grow-finish diets. The assumption is that diets with added fat are approximately 20% more expensive but result in 10% improvement in feed efficiency and 5% increase in average daily gain in a system running on a fixed-time basis.

In this example, considering feed cost or feed cost per kilogram of gain, the nutritional program without added fat would be more economical. However, taking into account the extra weight gain and improvement in feed efficiency with added fat, there is a $0.76 per pig advantage in income over feed cost with a nutritional program with added fat to grow-finish diets. This would be the interpretation in a system that runs on a fixed-time basis. However, if the system runs on a fixed-weight basis (has unlimited time in the finishing phase) and could take longer to achieve a heavier carcass weight, then feed cost per pig would also be an adequate indicator of the economic value of the nutritional program.

**Key Concepts of Diet Formulation**

Once a production system's economic goals have been set, the first and most important step in diet formulation is to set the energy concentration. Energy is the most expensive component of the diet,

**Table 23.5** An example on how to determine the economic value of added dietary fat in finisher diets[a,b].

| Assumptions | |
|---|---|
| Diets without added fat: | Diets with 5% added fat: |
| Feed cost: $176/ton or $0.176/kg | Feed cost: $209/ton or $0.209/kg |
| Initial weight: 23 kg | Initial BW: 23 kg |
| F/G: 2.8 | F/G: 2.5 |
| ADG: 820 g/day | ADG: 860 g/day |
| 130 days in the grow-finisher | 130 days in the grow-finisher |
| Final weight: 23 kg + (820 g/day × 130 days) = 130 kg | Final weight: 23 kg + (860 g/day × 130 days) = 135 kg |

| Calculations | |
|---|---|
| • Feed cost: $176/ton or $0.176/kg | • Feed cost: $209/ton or $0.209/kg |
| • Feed cost per kg of gain = 2.8 × $0.176 = $0.493/kg gain | • Feed cost per kg of gain = 2.5 × $0.209 = $0.523/kg gain |
| • Revenue = 130 kg × $1.54/kg = $200.70 | • Revenue = 135 kg × $1.54/kg = $207.90 |
| • Feed cost = 130 - 23 kg = 107 kg × $0.493 = $52.75 | • Feed cost = 135 - 23 kg = 112 kg × $0.523 = $58.58 |
| • Income over feed cost = $200.70 – $52.75 = $147.95/pig | • Income over feed cost = $207.90 – $58.58 = $149.93/pig |
| | • Added value = $149.93 – $147.95 = $1.98/pig |

[a] Live weight = $1.54 US$/kg.
[b] Income over feed cost = revenue – feed cost.

and the concentrations of other nutrients are set relative to the energy concentration of the diet. Incremental changes in dietary energy have direct impact on important production indicators, such as diet cost, growth performance, and carcass criteria. The energy level used in formulation should be based on an in-depth analysis to determine the most economical level considering the value of incremental changes in energy on production indicators (Figure 23.1) and market price.

After the dietary energy level is determined, the lysine:calorie ratio should be set. The lysine requirement should be expressed as a ratio to energy instead of a dietary lysine percentage because changes in dietary energy density affect feed intake. Thus, the lysine to calorie ratio is used to ensure the right amount of lysine is provided in diets that vary in energy density. The lysine requirements

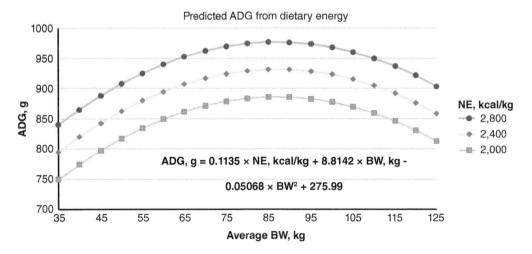

**Figure 23.1** Equation to predict growth rate of grow-finish pigs fed varying energy levels (Nitikanchana et al. 2015) / with permission of Oxford University Press. (ADG = average daily gain; NE = net energy; and BW = body weight.)

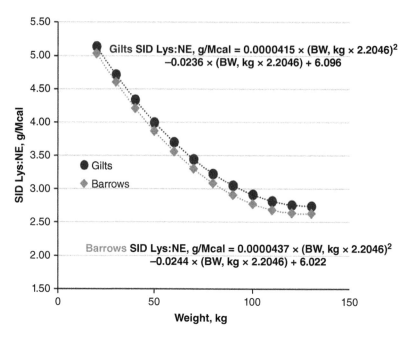

**Figure 23.2**    Suggested standardized ileal digestible lysine to net energy rations for gilts and barrows (Menegat et al. 2019) / with permission of Kansas State University. (SID = standardized ileal digestible; NE = net energy; and BW = body weight.}

can be estimated from research data conducted at universities (Figure 23.2), genetic suppliers, feed companies. or from research conducted within the production system.

Once the lysine:calorie ratio is determined, the levels of other amino acids are set based on a ratio to lysine. Amino acid ratio is a means of expressing the requirements for amino acids relative to the requirement for lysine. Lysine is used as a reference because it is typically the first-limiting amino acid in most swine diets, and the proper concentration of lysine and other amino acids is essential for protein synthesis. Amino acid ratios are often used because the requirements for amino acids remain relatively constant relative to lysine for a given stage of growth. In diet formulation, the use of amino acid to lysine ratios makes establishing the levels for other amino acids relatively easy.

Next, a digestible phosphorous concentration should be set as a ratio to energy content of the diet. Phosphorus is essential for growth and development, but excess supplementation can lead to unnecessary expenses and increased phosphorus excretion in swine waste. Like amino acids, when formulating for phosphorus levels, dietary energy concentration should be considered and a digestible phosphorus:calorie ratio used (Figure 23.3). Thus, when using a standard ratio, the dietary level of phosphorus can be adjusted to account for changes in feed intake. Also, as feed efficiency and lean growth improve, the dietary requirement concentration will increase (Vier et al. 2017). Furthermore, when using phytase, phosphorous release values should be to ensure accurate phosphorus levels are being formulated.

The dietary calcium levels are set relative to phosphorus in a calcium:phosphorus ratio. The ratio of calcium to phosphorus is important because of the close association of both minerals (Table 23.6). The dietary concentration of calcium affects the absorption and retention of phosphorus, particularly in diets with excess calcium and marginal phosphorus. Diets with excess calcium are not

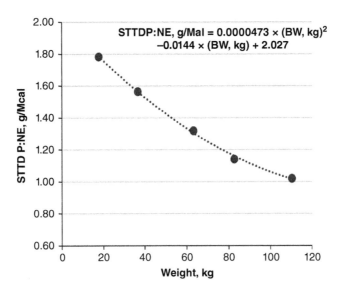

**Figure 23.3** Suggested standardized total tract digestible phosphorus (STTD) to net energy rations for finishing pigs adapted from (Vier et al. 2017). (SID = standardized ileal digestible; NE = net energy; and BW = body weight.}

**Table 23.6** Calcium and phosphorus estimates for grower-finisher pigs[a,b].

| Item | Grower-finisher pig weight, kg | | | |
| --- | --- | --- | --- | --- |
| | 23–60 | 60–80 | 80–100 | 100–130 |
| Total calcium, %[c] | 0.65 | 0.60 | 0.55 | 0.50 |
| STTD calcium, % | 0.49 | 0.43 | 0.38 | 0.33 |
| STTD phosphorus, %[d] | 0.38 | 0.35 | 0.30 | 0.26 |
| Available phosphorus, % | 0.33 | 0.29 | 0.25 | 0.21 |
| STTD Ca:STTD P | 1.30 | 1.30 | 1.30 | 1.27 |

[a] Adapted from Vier et al. (2017) and Menegat et al. (2019).
[b] STTD = standardized total tract digestible.
[c] Indication of maximum calcium levels for each phase.
[d] Indication of phosphorous levels in grower-finisher diets to optimize growth performance (Vier et al. 2017).

uncommon in swine diets because, unlike phosphorus, there are no economic and environmental concerns regarding dietary calcium.

Finally, the levels of vitamins, trace minerals, and any other ingredients are set in the diet formulation. These trace nutrients are generally added in fixed amounts to provide the necessary vitamins and trace minerals.

## Factors Affecting Nutrient Requirements

Several factors affect the estimation of nutrient requirements in pigs. In fact, any factor that influences performance and feed intake is likely to affect nutrient requirements estimates. Generally, improvements in growth performance or productivity and decreases in feed intake are associated with a demand for nutrient fortification in the diet to meet the requirements. Some of the most important factors are as follows:

- Breed, sex, and genetics
- Energy concentration of the diet
- Environmental temperature
- Health status of the herd
- Stocking density
- Feed processing
- Feeding strategy and degree of competition for feed
- Variability of nutrient content and availability in ingredients
- Presence of molds, toxins, or antinutritional factors in the diet
- Inclusion of growth promoters or feed additives in the diet

### Determining Lysine Requirements

The most common approaches to estimate lysine and other nutrient requirements are the empirical and factorial methods (Hauschild et al. 2010).

Empirical data from studies conducted in production systems or at universities is one option to determine amino acid requirements. This is typically performed as dose–response studies where a nutrient concentration is tested and the level associated with the best or most economical performance is determined through statistical analysis. The response curve adapts to the shape of the data and indicates the nutrient level that maximizes or minimizes a performance criterion. In the United States, because of their size and scope, many production systems conduct research in their own facilities and under production conditions. The challenge is that with private production systems conducting their own internal research, the results may not be available to the public.

Factorial methods to determine lysine estimates can be modeled based on growth rate and feed intake data or protein accretion curves (Schinckel and De Lange 1996). This method is considered an option to determine farm-specific lysine recommendations because it reflects the conditions found in a commercial production system. The weighing and ultrasound scanning start earlier in the grower period and are collected past the normal market weights to ensure the lysine requirements at the beginning and end of the period are accurately estimated. Although this method is conducted under commercial facilities within the production system and does not require investment in research facilities, personnel expertise is required to conduct the measurements and perform the modeling with precision.

### Determining other Amino Acid Requirements

The amino acid requirements are usually estimated by dose–response studies, and the most common approach is to express the requirements as an amino acid ratio relative to lysine. Amino acid ratio is a means of expressing the requirements for amino acids relative to the requirement for lysine. Lysine is used as a reference because it is typically the first-limiting amino acid in most swine diets, and the proper concentration of lysine and other amino acids is essential for protein synthesis (Table 23.7).

In dose–response experiments, the first limiting amino acid in the diet must be the amino acid for which the requirement is wanted to be estimated and the second limiting amino acid must be lysine.

**Table 23.7** Standardized ileal digestible lysine and other amino acid estimates for growing-finishing pigs[a].

| Amino acid[b] | Weight range, kg | | | |
|---|---|---|---|---|
| | 23–60 | 60–80 | 80–100 | 100–130 |
| Lysine, %[c] | 1.12 | 0.88 | 0.78 | 0.70 |
| Ratio relative to lysine, % | | | | |
| Methionine | 28 | 29 | 29 | 30 |
| Methionine + Cysteine | 56 | 58 | 58 | 59 |
| Threonine | 61 | 61 | 63 | 66 |
| Tryptophan | 18+ | 18+ | 18+ | 18+ |
| Isoleucine | 55 | 55 | 55 | 55 |
| Valine | 70 | 70 | 70 | 70 |
| Histidine | 32 | 32 | 32 | 32 |

[a] Adapted from Menegat et al. (2019).
[b] Standardized ileal digestible amino acids.
[c] Assumes a diet with 2525 kcal net energy/kg.

The supply of other amino acids and nutrients should meet or slightly exceed the requirements to avoid being a limiting factor. Using this approach, the requirement is determined at the point which both the tested amino acid and lysine are equally limiting and can, therefore, be expressed relative to lysine (Simongiovanni et al. 2012).

The requirements of amino acids relative to lysine are often depicted as a diminishing returns model. This model can be used to determine which ratio provides 97 to 99% of the maximum performance and indicates the most economical amino acid ratio. Thus, the optimum amino acid ratio should be set by balancing the value accrued in performance to the incremental cost to increase the ratio.

Finally, like in weaner pig diets, the amounts of calcium and phosphorus are determined, and levels of vitamins, minerals, and other ingredients are set in the diet formulation.

## References

Bikker, P., A. W. Jongbloed, and J. van Baal. 2016. Dose-dependent effects of copper supplementation of nursery diets on growth performance and fecal consistency in weaned pigs. J. Anim. Sci. 94(Suppl. 3):181–186. doi: https://doi.org/10.2527/jas.2015-9874.

Blavi, L., D. Sola-Oriol, J. F. Perez, and H. H. Stein. 2017. Effects of zinc oxide and microbial phytase on digestibility of calcium and phosphorus in maize-based diets fed to growing pigs. J. Anim. Sci. 95:847–854. doi: https://doi.org/10.2527/jas.2016.1149.

Bruininx, E. M. A. M, C. M. C. Van Der Peet-Schwering, J. W. Schrama, P. F. G. Vereijken, P. C. Vesseur, H. Everts, L. A. den Hartog, and A. C. Beynen. 2001. Individually measured feed intake characteristics and growth performance of group-housed weanling pigs: Effects of sex, initial body weight, and body weight distribution within groups. J. Anim. Sci. 79:301–308. doi: https://doi.org/10.2527/2001.792301x.

Bruininx, E. M. A. M., G. P. Binnendijk, C. M. C. Van Der Peet-Schwering, J. W. Schrama, L. A. Den Hartog, H. Everts, and A. C. Beynen. 2002. Effect of creep feed consumption on individual feed intake characteristics and performance of group-housed weanling pigs. J. Anim. Sci. 80:1413–1418. doi: https://doi.org/10.2527/2002.8061413x.

Carlson, M. S., C. A. Boren, C. Wu, C. E. Huntington, D. W. Bollinger, and T. L. Veum. 2004. Evaluation of various inclusion rates of organic zinc either as polysaccharide or proteinate complex on the growth performance, plasma, and excretion of nursery pigs. J. Anim. Sci. 82:1359–1366. doi: https://doi.org/10.2527/2004.8251359x.

Carpenter, C. B., J. C. Woodworth, J. M. DeRouchey, M. D. Tokach, R. D. Goodband, S. S. Dritz, F. Wu, and J. L. Usry. 2018. Effects of increasing copper from tri-basic copper chloride or a copper-methionine chelate on growth performance of nursery pigs. Trans. Anim. Sci. txy091. doi: https://doi.org/10.1093/tas/txy091.

Cemin, H. S., C. M. Vier, M. D. Tokach, S. S. Dritz, K. J. Touchette, J. C. Woodworth, J. M. DeRouchey, and R. D. Goodband. 2018. Effects of standardized ileal digestible histidine to lysine ratio on growth performance of 7- to 11-kg nursery pigs. J. Anim. Sci. 96:4713–4722. doi: https://doi.org/10.1093/jas/sky319.

Clark, A. B., M. D. Tokach, J. M. DeRouchey, S. S. Dritz, R. D. Goodband, J. C. Woodworth, K. J. Touchette, and N. M. Bello. 2017a. Modeling the effects of standardized ileal digestible isoleucine to lysine ratio on growth performance of nursery pigs. Transl. Anim. Sci. 1:437–447. doi: https://doi.org/10.2527/tas2017.0048.

Clark, A. B., M. D. Tokach, J. M. DeRouchey, S. S. Dritz, R. D. Goodband, J. C. Woodworth, K. J. Touchette, and N. M. Bello. 2017b. Modeling the effects of standardized ileal digestible valine to lysine ratio on growth performance of nursery pigs. Transl. Anim. Sci. 1:448–457. doi: https://doi.org/10.2527/tas2017.0049.

Coble, K. F., J. M. DeRouchey, M. D. Tokach, S. S. Dritz, R. D. Goodband, J. C. Woodworth, and J. L. Usry. 2017. The effects of copper source and concentration on growth performance, carcass characteristics, and pen cleanliness in finishing pigs. J. Anim. Sci. 95:4052–4059. doi: https://doi.org/10.2527/jas2017.1624.

Collins, C. L., J. R. Pluske, R. S. Morrison, T. N. McDonald, R. J. Smits, D. J. Henman, I. Stensland, and F. R. Dunshea. 2017. Post-weaning and whole-of-life performance of pigs is determined by live weight at weaning and the complexity of the diet fed after weaning. Anim. Nutr. 3:372–379. doi: https://doi.org/10.1016/j.aninu.2017.01.001.

Cromwell, G. L., M. D. Lindemann, H. J. Monegue, D. D. Hall, and D. E. Orr Jr. 1998. Tribasic copper chloride and copper sulfate as copper sources for weanling pigs. J. Anim. Sci. 76:118–123. doi: https://doi.org/10.2527/1998.761118x.

Engle, M. J. 1994. The role of soybean meal hypersensitivity in postweaning lag and diarrhea in piglets. J. Swine Health Prod. 2:7–10.

Gonçalves, M. A. D., S. Nitikanchana, M. D. Tokach, S. S. Dritz, N. M. Bello, R. D. Goodband, K. J. Touchette, J. L. Usry, J. M. DeRouchey, and J. C. Woodworth. 2015. Effects of standardized ileal digestible tryptophan:lysine ratio on growth performance of nursery pigs. J. Anim. Sci. 93:3909–3918. doi: https://doi.org/10.2527/jas.2015-9083.

González-Vega, J. C., C. L. Walk, and H. H. Stein. 2015a. Effects of microbial phytase on apparent and standardized total tract digestibility of calcium in calcium supplements fed to growing pigs. J. Anim. Sci. 93:2255–2264. doi: https://doi.org/10.2527/jas.2014-8215.

González-Vega, J. C., C. L. Walk, and H. H. Stein. 2015b. Effect of phytate, microbial phytase, fiber, and soybean oil on calculated values for apparent and standardized total tract digestibility of calcium and apparent total tract digestibility of phosphorus in fish meal fed to growing pigs. J. Anim. Sci. 93:4808–4818. doi: https://doi.org/10.2527/jas.2015-8992.

González-Vega, J. C., Y. Liu, J. C. McCann, C. L. Walk, J. J. Loor, and H. H. Stein. 2016. Requirement for digestible calcium by eleven- to twenty-five-kilogram pigs as determined by growth performance, bone ash concentration, calcium and phosphorus balances, and expression of genes involved in transport of calcium in intestinal and kidney cells. J. Anim. Sci. 94:3321–3334. doi: https://doi.org/10.2527/jas.2016-0444.

Goodband, B., M. Tokach, S. Dritz, J. DeRouchey, and J Woodworth. 2014. Practical starter pig amino acid requirements in relation to immunity, gut health and growth performance. J. Anim. Sci. & Biotech. 5:12–23. doi: https://doi.org/10.1186/2049-1891-5-12.

Gourley, K. M., J. C. Woodworth, J. M. DeRouchey, S. S. Dritz, M. D. Tokach, and R. D. Goodband. 2018. Effect of high doses of Natuphos E 5,000 G phytase on growth performance of nursery pigs. J. Anim. Sci. 96:570–578. doi: https://doi.org/10.1093/jas/sky001.

Gu, X. and D. Li. 2003. Fat nutrition and metabolism in piglets: A review. Anim. Feed Sci. Technol. 109:151–170. doi: https://doi.org/10.1016/S0377-8401(03)00171-8.

Hahn, J. D. and D. H. Baker. 1993. Growth and plasma zinc responses of young pigs fed pharmacologic levels of zinc. J. Anim. Sci. 71:3020–3024. doi: https://doi.org/10.2527/1993.71113020x.

Hauschild, L., C. Pomar, and P. A. Lovatto. 2010. Systematic comparison of the empirical and factorial methods used to estimate the nutrient requirements of growing pigs. Animal. 4:714–723. doi: https://doi.org/10.1017/S1751731109991546.

Heo, J. M., F. O. Opapeju, J. R. Pluske, J. C. Kim, D. J. Hampson, and C. M. Nyachoti. 2013. Gastrointestinal health and function in weaned pigs: A review of feeding strategies to control post-weaning diarrhoea without using in-feed antimicrobial compounds. J. Anim. Phys. and Anim. Nutr. 97:207–37. doi: https://doi.org/10.1111/j.1439-0396.2012.01284.x.

Hill, G. M., D. C. Mahan, S. D. Carter, G. L. Cromwell, R. C. Ewan, R. L. Harrold, A. J. Lewis, P. S. Miller, G. C. Shurson, and T. L. Veum. 2001. Effect of pharmacological concentrations of zinc oxide with or without the inclusion of an antibacterial agent on nursery pig performance. J. Anim. Sci. 79:934–941. doi: https://doi.org/10.2527/2001.794934x.

Hill, G. M., G. L. Cromwell, T. D. Crenshaw, C. R. Dove, R. C. Ewan, D. A. Knabe, A. J. Lewis, G. W. Libal, D. C. Mahan, G. C. Shurson, L. L. Southern, and T. L. Veum. 2000. Growth promotion effects and plasma changes from feeding high dietary concentrations of zinc and copper to weanling pigs (regional study). J. Anim. Sci. 78:1010–1016. doi: https://doi.org/10.2527/2000.7841010x.

Hollis, G. R., S. D. Carter, T. R. Cline, T. D. Crenshaw, G. L. Cromwell, G. M. Hill, S. W. Kim, A. J. Lewis, D. C. Mahan, P. S. Miller, H. H. Stein, and T. L. Veum. 2005. Effects of replacing pharmacological levels of dietary zinc oxide with lower dietary levels of various organic zinc sources for weanling pigs. J. Anim. Sci. 83:2123–2129. doi: https://doi.org/10.2527/2005.8392123x.

Jayaraman, B., J. Htoo, and C. M. Nyachoti. 2015. Effects of dietary threonine:lysine ratioes and sanitary conditions on performance, plasma urea nitrogen, plasma-free threonine and lysine of weaned pigs. Animal Nutrition. 1:283–288. doi: https://doi.org/10.1016/j.aninu.2015.09.003.

Jondreville, C., P. S. Revy, and J. Y. Dourmad. 2003. Dietary means to better control the environmental impact of copper and zinc by pigs from weaning to slaughter. Lvst. Prod. Sci. 84:147–156. doi: https://doi.org/10.1016/j.livprodsci.2003.09.011.

Kahindi, R., A. Regassa, J. Htoo, and M. Nyachoti. 2017. Optimal sulfur amino acid to lysine ratio for post weaning piglets reared under clean or unclean sanitary conditions. Animal Nutrition. 3:380–385. doi: https://doi.org/10.1016/j.aninu.2017.08.004.

Laird, S., I. Kühn, and H. M. Miller. 2018. Super-dosing phytase improves the growth performance of weaner pigs fed a low iron diet. Anim. Feed Sci. Technol. 242:150–160. doi: https://doi.org/10.1016/j.anifeedsci.2018.06.004.

Li, D. F., J. L. Nelssen, P. G. Reddy, F. Blecha, J. D. Hancock, G. L. Allee, R. D. Goodband, and R. D. Klemm. 1990. Transient hypersensitivity to soybean meal in the early-weaned pig. J. Anim. Sci. 68:1790–1799. doi: https://doi.org/10.2527/1990.6861790x.

Liu, Y., C. D. Espinosa, J. J. Abelilla, G. A. Casas, L. V. Lagos, S. A. Lee, W. B. Kwon, J. K. Mathai, D. M. D. L. Navarro, N. W. Jaworski, and H. H. Stein. 2018. Non-antibiotic feed additives in diets for pigs: A review. Animal Nutrition. 4:113–125. doi: https://doi.org/10.1016/j.aninu.2018.01.007.

Madec, F., N. Bridoux, S. Bounaix, and A. Jestin. 1998. Measurement of digestive disorders in the piglet at weaning and related risk factors. Preventive Veterinary Medicine. 35:53–72. doi: https://doi.org/10.1016/S0167-5877(97)00057-3.

Mahan, D. C., N. D. Fastinger, and J. C. Peters. 2004. Effects of diet complexity and dietary lactose levels during three starter phases on postweaning pig performance. J. Anim. Sci. 82:2790–2797. doi: https://doi.org/10.2527/2004.8292790x.

Main, R. G., S. S. Dritz, M. D. Tokach, R. D. Goodband, and J. L. Nelssen. 2008. Determining an optimum lysine:calorie ratio for barrows and gilts in a commercial finishing facility. J. Anim. Sci. 86:2190–2207. doi: https://doi.org/10.2527/jas.2007-0408.

Main, R. G., S. S. Dritz, M. D. Tokach, R. D. Goodband, and J. L. Nelssen. 2004. Increasing weaning age improves pig performance in a multi-site production system. J. Anim. Sci. 82:1499–1507. doi: https://doi.org/10.2527/2004.8251499x.

McLamb, B. L., A. J. Gibson, E. L. Overman, C. Stahl, and A. J. Moeser. 2013. Early weaning stress in pigs impairs innate mucosal immune responses to enterotoxigenic E. coli challenge and exacerbates intestinal injury and clinical disease. PLoS ONE. 8:e59838. doi: https://doi.org/10.1371/journal.pone.0059838.

Menegat, M. B., R. D. Goodband, J. M. DeRouchey, M. D. Tokach, J. C. Woodworth, and S. S. Dritz. 2019. Kansas State University Swine Nutrition Guide: General Nutrition Principles. Kansas State Univ., Manhattan, KS.

Merriman, L. A., C. L. Walk, and H. H. Stein. 2016. The effect of microbial phytase on the apparent and standardized total tract digestibility of calcium in feed ingredients of animal origin. J. Anim. Sci. 94(Suppl. 2):110. doi: https://doi.org/10.2527/msasas2016-240.

Millet, S., M. Aluwé, A. Van den Broeke, F. Leen, J. De Boever, and S. De Campeneere. 2018a. Review: Pork production with maximal nitrogen efficiency. Animal. 12:1060–1067. doi: https://doi.org/10.1017/S1751731117002610.

Millet, S., M. Aluwé, A. Van den Broeke, J. De Boever, B. de Witte, L. Douidah, A. van den Broeke, F. Leen, C. de Cuyper, B. Ampe, and S. De Campeneere. 2018b. The effect of crude protein reduction on performance and nitrogen metabolism in piglets (four to nine weeks of age) fed two dietary lysine levels. J. Anim. Sci. 96:3824–3836. doi: https://doi.org/10.1093/jas/sky254.

Moeser, A. J., K. A. Ryan, P. K. Nighot, and A. T. Blikslager. 2007. Gastrointestinal dysfunction induced by early weaning is attenuated by delayed weaning and mast cell blockade in pigs. American Journal of Physiology and Gastrointestinal Liver Physiology. 293:413–421. doi: https://doi.org/10.1152/ajpgi.00304.2006.

National Research Council. 2012. Nutrient requirements of swine. 11th Revised Edition. The National Academies Press, Washington, DC. doi: https://doi.org/10.17226/13298.

Nemechek, J. E., A. M. Gaines, M. D. Tokach, G. L. Allee, R. D. Goodband, J. M. DeRouchey, J. L. Nelssen, J. L. Usry, G. Gourley, and S. S. Dritz. 2012. Evaluation of standardized ileal digestible lysine requirement of nursery pigs from seven to fourteen kilograms. J. Anim. Sci. 90: 4380–4390.

Nemechek, J. E., F. Wu, M. D. Tokach, S. S. Dritz, R. D. Goodband, J. M. DeRouchey, and J. C. Woodworth. 2018. Effect of standardized ileal digestible lysine on growth and subsequent performance of weanling pigs. Transl. Anim. Sci. 2:156–161. doi: https://doi.org/10.1093/tas/txy011.

Nichols, G. E., C. M. Vier, A. B. Lerner, M. B. Menegat, H. S. Cemin, C. K. Jones, J. M. DeRouchey, M. D. Tokach, B. D. Goodband, J. C. Woodworth, and S. S. Dritz. 2018. Effects of standardized ileal digestible lysine on 7-15 kg nursery pigs growth performance. J. Anim. Sci. 96(Suppl. 2):258–259. doi: https://doi.org/10.1093/jas/sky073.480.

Nitikanchana, S., S. S. Dritz, M. D. Tokach, J. M. DeRouchey, R. D. Goodband, and B. J. White. 2015. Regression analysis to predict growth performance from dietary net energy in growing-finishing pigs. J. Anim. Sci. 93:2826–2839. doi: https://doi.org/10.2527/jas.2015-9005.

Pérez, V. G., A. M. Waguespack, T. D. Bidner, L. L. Southern, T. M. Fakler, T. L. Ward, M. Steidinger, and J. E. Pettigrew. 2011. Additivity of effects from dietary copper and zinc on growth performance and fecal microbiota of pigs after weaning. J. Anim. Sci. 89:414–425. doi: https://doi.org/10.2527/jas.2010-2839.

Pluske, J. R., I. H. Williams, and F. X. Aherne. 1996. Maintenance of villous height and crypt depth in piglets by providing continuous nutrition after weaning. Animal Science. 62(1):131–144. doi: https://doi.org/10.1017/S1357729800014417.

Qian, H., E. T. Kornegay, and D. E. Conner, Jr. 1996. Adverse effects of wide calcium:phosphorus ratios on supplemental phytase efficacy for weanling pigs fed two dietary phosphorus levels. J. Anim. Sci. 74:1288–1297. doi: https://doi.org/10.2527/1996.7461288x.

Reinhart, G. A. and D. C. Mahan. 1986. Effect of various calcium: Phosphorus ratios at low and high dietary phosphorus for starter, grower and finishing swine. J. Anim. Sci. 63:457–466. doi: https://doi.org/10.2527/jas1986.632457x.

Schinckel, A. P. and C. F. M. de Lange. 1996. Characterization of growth parameters needed as inputs for pig growth models J.Anim. Sci. 74:2021–2036. doi: https://doi.org/10.2527/1996.7482021x.

Shawk, D. J., M. D. Tokach, R. D. Goodband, S. S. Dritz, J. C. Woodworth, J. M. DeRouchey, A. B. Lerner, F. Wu, C. M. Vier, M. M. Moniz, and K. N. Nemechek. 2018a. Effects of sodium and chloride source and concentration on nursery pig growth performance. J. Anim. Sci. sky429. doi: https://doi.org/10.1093/jas/sky429.

Shawk, D. J., R. D. Goodband, M. D. Tokach, S. S. Dritz, J. M. DeRouchey, J. C. Woodworth, A. B. Lerner, and H. E. Williams. 2018b. Effects of added dietary salt on pig growth performance. Transl. Anim. Sci. 2:396–406. doi: https://doi.org/10.1093/tas/txy085.

Shelton, N. W., M. D. Tokach, J. L. Nelssen, R. D. Goodband, S. S. Dritz, J. M. DeRouchey, and G. M. Hill. 2011. Effects of copper sulfate, tri-basic copper chloride, and zinc oxide on weanling pig performance. J. Anim. Sci. 89:2440–2451. doi: https://doi.org/10.2527/jas.2010-3432.

Simongiovanni A., E. Le Gall, Y. Primot, and E. Corrent. 2012. Estimating amino acid requirements through dose-response experiments in pigs and poultry. Ajinomoto Eurolysine Technical Note. Available at: http://ajinomoto-eurolysine.com/estimating-amino-acid-requirements.html.

Skinner, L. D., C. L. Levesque, D. Wey, M. Rudar, J. Zhu, S. Hooda, and C. F. M. De Lange. 2014. Impact of nursery feeding program on subsequent growth performance, carcass quality, meat quality, and physical and chemical body composition of growing-finishing pigs. J. Anim. Sci. 92:1044–1054. doi: https://doi.org/10.2527/jas.2013-6743.

Smith, F., J. E. Clark, B. L. Overman, C. C. Tozel, J. H. Huang, J. E. F. Rivier, A. T. Blisklager, and A. J. Moeser. 2010. Early weaning stress impairs development of mucosal barrier function in the porcine intestine. Am. J.Physiol. Gastrointest. Liver Physiol. 298:352–363. doi: https://doi.org/10.1152/ajpgi.00081.2009.

Stahly, T. S., G. L. Cromwell, and H. J. Monegue. 1980. Effect of single additions and combinations of copper and antibiotics on the performance of weanling pigs. J. Anim. Sci. 51:1347–1351. doi: https://doi.org/10.2527/jas1981.5161347x.

Sulabo, R. C., J. R. Bergstrom, M. D. Tokach, J. M. DeRouchey, R. D. Goodband, J. L. Nelssen, and S. S. Dritz. 2009. Effects of creep diet complexity on individual consumption characteristics and growth performance of neonatal and weanling pigs. Kansas Agric. Exp. Stn. Res. Rep. 1020:51–64.

Sulabo, R. C., M. D. Tokach, S. S. Dritz, R. D. Goodband, J. M. DeRouchey, and J. L. Nelssen. 2010. Effects of varying creep feeding duration on the proportion of pigs consuming creep feed and neonatal pig performance. J. Anim. Sci. 88:3154–3162. doi: https://doi.org/10.2527/jas.2009-2134.

Tokach, M. D., J. E. Pettigrew, L. J. Johnston, M. Øverland, J. W. Rust, and S. G. Cornelius. 1995. Effect of adding fat and(or) milk products to the weanling pig diet on performance in the nursery and subsequent grow-finish stages. J. Anim. Sci. 73:3358–3368. doi: https://doi.org/10.2527/1995.73113358x.

Vier, C. M., F. Wu, M. B. Menegat, H. Cemin, S. S. Dritz, M. D. Tokach, M. A. D. Goncalves, U.A.D. Orlando, J.C. Woodworth, R.D. Goodband, and J.M. DeRouchey. 2017. Effects of standardized total tract digestible phosphorus on performance, carcass characteristics, and economics of 24 to 130 kg pigs. Anim. Prod. Sci. 57:2424–2424. doi: https://doi.org/10.1071/ANv57n12Ab071.

Walk, C. L., P. Wilcock, and E. Magowan. 2015. Evaluation of the effects of pharmacological zinc oxide and phosphorus source on weaned piglet growth performance, plasma minerals and mineral digestibility. Animal. 9:1145–1152. doi: https://doi.org/10.1017/S175173111500035X.

Weng, R. C. 2016. Dietary fat preference and effects on performance of piglets at weaning. Asian-Australas. J. Anim. Sci. 30:834–842. doi: https://doi.org/10.5713/ajas.16.0499.

Whang, K. Y., F. K. Mckeith, S. W. Kim, and R. A. Easter. 2000. Effect of starter feeding program on growth performance and gains of body components from weaning to market weight in swine. J. Anim. Sci. 78:2885–2895. doi: https://doi.org/10.2527/2000.78112885x.

Wolter, B. F. and M. Ellis. 2001. The effects of weaning weight and rate of growth immediately after weaning on subsequent pig growth performance and carcass characteristics. Can. J. Anim. Sci. 81:363–369. doi: https://doi.org/10.4141/A00-100.

Wolter, B. F., M. Ellis, B. P. Corrigan, J. M. Dedecker, S. E. Curtis, E. N. Parr, and W. M. Webel. 2003. Impact of early postweaning growth rate as affected by diet complexity and space allocation on subsequent growth performance of pigs in a wean- to-finish production system. J. Anim. Sci. 81:353–359. doi: https://doi.org/10.2527/2003.812353x.

Wu, F., J. Liao, M. D. Tokach, S. S. Dritz, J. C. Woodworth, R. D. Goodband, J. M. DeRouchey, C. I. Vahl, H. I. Calderón-Cartagena and D. L. Van De Stroet. 2019. A retrospective analysis of seasonal growth patterns of nursery and finishing pigs in commercial production. J. Swine Health Prod. 27:19–33.

Wu, F., J. C. Woodworth, M. D. Tokach, J. M. DeRouchey, S. S. Dritz, and R. D. Goodband. 2018b. Standardized total tract digestible phosphorus requirement of 13- to 28-lb pigs fed diets with or without phytase. Kansas Agric. Exp. Stn. Res. Rep. 4:9. doi: https://doi.org/10.4148/2378-5977.7665.

Wu, F., M. D. Tokach, S. S. Dritz, J. C. Woodworth, J. M. DeRouchey, R. D. Goodband, M. A. D. Gonçalves, and J. R. Bergstrom. 2018a. Effects of dietary calcium to phosphorus ratio and addition of phytase on growth performance of nursery pigs. J. Anim. Sci. 96:1825–1837. doi: https://doi.org/10.1093/jas/sky101.

Wu, F., J. Woodworth, M. Tokach, S. Dritz, J. DeRouchey, R. Goodband, and J. Bergstrom. 2019. Standardized total tract digestible phosphorus requirement of 6 to 13 kg pigs fed diets without or with phytase. Animal 13:2473–2482. doi: https://doi.org/10.1017/S1751731119000922.

Yazdankhah, S., K. Rudi, and A. Bernhoft. 2014. Zinc and copper in animal feed – development of resistance and co-resistance to antimicrobial agents in bacteria of animal origin. Microb. Ecol. Health Disease 25:25862–25869. doi: https://doi.org/10.3402/mehd.v25.25862.

Zeng, Z. K., D. Wang, X. S. Piao, P. F. Li, H. Y. Zhang, C. X. Shi, and S. K. Yu. 2014. Effects of adding super dose phytase to the phosphorus-deficient diets of young pigs on growth performance, bone quality, minerals and amino acids digestibilities. Asian-Australas. J. Anim. Sci. 27:237–246. doi: https://doi.org/10.5713/ajas.2013.13370.

# 24 Organic Swine Production and Nutrition

Sandra A. Edwards

## Introduction

### The Principles of Organic Production

Organic agriculture is a farming system that sets out to create its products in an environmentally and socially responsible manner. Its underlying belief system goes back many years to the early twentieth century, but was formalized worldwide with the establishment of the International Federation of Organic Agriculture Movements (IFOAM) in 1972. The IFOAM, which now has membership from more than 120 countries, has agreed a formal definition of organic agriculture:

Organic Agriculture is a production system that sustains the health of soils, ecosystems and people. It relies on ecological processes, biodiversity and cycles adapted to local conditions, rather than the use of inputs with adverse effects. Organic Agriculture combines tradition, innovation and science to benefit the shared environment and promote fair relationships and a good quality of life for all involved.

This definition reflects the four guiding Principles of Organic Agriculture: the Principles of Health, Ecology, Fairness and Care, which have been laid out in detail by IFOAM (2005). While organic pig production is still a niche activity in comparison with the large-scale conventional production that exists in most countries, often representing less than 1% of the pig population (Früh et al. 2014; Edwards and Leeb 2018), its popularity is increasing in many parts of the world as growing consumer awareness of healthy eating, environmental protection and animal welfare generate greater market demand.

### The Regulatory Framework

In order to achieve organic certification for marketing purposes, many countries now have formal standards for production. In the European Union, these were first harmonized under legislation in 1991 and are currently governed by European Council Regulation 2018/848 (European Council 2018) and European Commission Regulation 889/2008 (European Commission 2008). In individual countries, the standards laid out in these Regulations may be implemented and inspected

---

by a governmental body, or by one or more approved private certification bodies (Früh et al. 2014). While the EU Regulations set the minimum requirements for such certification, individual certification bodies may impose additional requirements. For example, while EU Regulations require that all pigs have outdoor access, which may be a fenced concrete run, the UK Soil Association requires pigs to be kept with permanent access to pasture or vegetated range (Soil Association 2020).

In North America, organic production in the United States is regulated under the Organic Food Production Act of 1990 (USDA 1990) and the National Organic Program 2010 (NOP 2010). In comparison with the EU Regulations, more emphasis is placed on feeding and medication practices, although close confinement is prohibited and pigs should have outdoor access. A summary of the regulatory framework in other countries can be found in Blair (2018).

### Standards Relating to Pig Production

Organic certification is a whole farm process, and every enterprise must be compliant. Where the farm includes a pig enterprise, there are a number of key requirements that differentiate this from conventional production practice. The following overview is based on the IFOAM norms (IFOAM 2014), and details may differ between different countries and Certification Bodies.

Organic pigs must be born and raised on a certified organic holding, and only a limited number of replacement breeding stock may be brought in from conventional farms. Breeds must be adapted to local conditions and show good resistance to disease. Some schemes recommend the use of traditional native breeds. Pigs can only receive feed that is certified as having been grown according to organic standards. Processing methods involving chemical agents, such as the use of solvent extraction, are banned. Lists of approved ingredients exist in some schemes, and there are restrictions on feed additives that prevent the use of synthetic or antimicrobial growth promoters, synthetic amino acids, and many enzymes. Vitamins and trace elements should come from natural sources where possible. Piglets should receive mother's milk and not be weaned before six weeks of age.

To allow expression of natural behavior, animals should have daily access to pasture or to a soil-based open air exercise area whenever conditions permit. Stocking densities should not lead to soil degradation or water pollution. All animals should live in group housing with sufficient space for free movement. Mutilations are not permitted, with the exception of castration of male pigs and the nose-ringing of sows in some countries. Prophylactic veterinary treatments are not permitted and, to treat sick animals, preference should be given to natural medicines and treatments such as homeopathy. If synthetic allopathic veterinary drugs or antibiotics are repeatedly used, the animal will lose its organic status. However, it is emphasized that such medicines should not be withheld in case of need. All of these standards have implications for the nutritional management of the pigs, and these will be reviewed in the following sections.

Readers are referred to, among others, several websites for further information on the organic swine production and nutrition. The first website contains useful information on the background and principles of organic agriculture (https://www.ifoam.bio/en). The second website gives access to a large range of papers and reports on organic farming, including many on pigs (https://orgprints. org/). Another website describes the outcomes from a recent EU project on "Improved Contribution of Local Feed to Support 100% Organic Feed Supply to Pigs and Poultry" (http://www. organicresearchcentre.com/icopp/).

## Factors Affecting the Nutritional Requirements of Organic Pigs

### Genetics

The basic biology of growth and reproduction in the pig is the same for animals reared in organic or conventional systems. Therefore, the various models that can be used to calculate nutrient requirements of conventional pigs (e.g. NRC 2012; van Milgen et al. 2007) will still be applicable in organic circumstances. However, such models require information on either the genetic characteristics or the expected production performance of the animals and this may well be different in organic systems. Where conventional breeds drawn from intensive production are used in organic farms, their characteristics are well documented, although environmental and dietary factors that are discussed subsequently may predispose lower performance in practice. However, many organic farms use traditional indigenous breeds (Kelly et al. 2001; Früh et al. 2014), and information on their genetic characteristics is often lacking. These traditional breeds are often slower growing, earlier maturing and more obese than modern genetic lines (Kelly et al. 2007) and may not show the same degree of prolificacy. Where this is the case, a lower dietary protein: energy requirement will pertain.

### Outdoor Living

The requirement that organic pigs should have access to pasture or to a soil-based open air exercise area will also have major influences on the nutrient requirements. This will result from greater exercise, thermal challenge, and foraging opportunities (Buckner et al. 1994; Close and Poornan 1993; Edwards 2003). The greater space and environmental complexity afforded by outdoor living might be expected to increase the level of activity and hence the energy requirement of organic pigs relative to those in intensive housing. Surprisingly, the proportion of daily time spent active does not seem to differ greatly in either sows or growing pigs, although it will depend on climatic conditions and the animals' level of hunger and motivation to forage. Season and reproductive state have been shown to influence activity level in outdoor sows (Buckner et al. 1998). When animals are active, the distance traveled during locomotion is likely to be higher under outdoor conditions, and the additional energy requirement for such locomotion will depend on body weight, walking speed, and distance traveled. Precise data on this subject are lacking and estimates vary widely, but it has been calculated that a 200 kg sow walking 1 km/day at 4 km/h would require an extra 1.5 MJ/day to compensate for this activity (Edwards 2003).

The effect of climate on the nutritional requirement of outdoor pigs will depend on the season and geographic region in which they are located. Hot climatic conditions will reduce activity and feed intake, although these effects can be mitigated by provision of shades and wallows. High environmental temperatures represent a particular challenge for the lactating sow because of her high metabolic activity for milk production. Cold climatic conditions can also be mitigated by the provision of well-designed shelter and deep bedding. However, in temperate regions, outdoor pigs may often be exposed to climatic conditions below their lower critical temperature, especially in the case of young pigs or those on restricted feeding. The additional energy required to compensate for low temperature has been calculated as 15–18 kJ/kg of metabolic body weight per day for each degree below lower critical temperature (Close and Poornan 1993).

A further influence on nutritional requirements from outdoor living is the ability to forage and obtain nutrients from herbage and soil (Figure 24.1). Any inputs obtained in this way will modify the quantity and nutrient balance of the supplementary diet that must be supplied for efficient

**Figure 24.1**   Sows grazing in a pasture system.

feeding. The nutrients obtained from foraging will depend on the voluntary intake of the pigs, the quantity and type of plant species available, and the seasonal changes in their composition (Edwards 2003). The quantity of herbage available will depend on the place of the pigs in the crop rotation of the farm, the season, and whether pigs are fitted with nose rings to prevent rooting and rapid sward destruction. In organic systems, it is common to cultivate legume or mixed grass/legume swards to benefit from their nitrogen-fixing properties. Legume herbage will have a higher protein content than grasses and is also reported to be more palatable (Aubé et al. 2019a). For all plant species, the nutrient digestibility will depend on the growth stage and level of fiber, since fiber can only be processed by less efficient microbial fermentation in the hindgut, as opposed to enzymic digestion in the foregut. For example, the apparent organic matter digestibility of a grazed grass-clover sward by pregnant sows varied from 0.8 to 0.5 at different times of year (Rivera-Ferre et al. 2001). Data on the intake of grazed herbage by pigs are relatively scarce. Reported values range from 1 to 4 kg DM per day for pregnant sows, 0.5–1.0 kg DM/day for lactating sows and 0.1–0.4 kg DM/day in growing/finishing pigs (Edwards 2003; Table 24.1).

Intake may be influenced by the level of supplementary feed on offer, particularly in growing pigs. The contribution that unrestricted grazed herbage can make to the daily nutrient requirements has been estimated as up to 50% of the maintenance energy requirement and more than 80% of amino acid requirements for pregnant sows, but only 5–10% of energy requirement and a negligible contribution to amino acid requirements in growing pigs (Edwards 2003). Lactating sows have a high daily nutrient requirement due to the demands of milk production, and intake limitations will mean that grazed herbage can therefore only make up a lesser proportion of requirement. Eskildsen et al. (2019) tested the hypothesis that the combination of increased energy requirement for exercise and thermoregulation, and substantial protein intake from legume herbage, meant that a lower protein to energy ratio in the concentrate diet would be appropriate for organic lactating sows. They found that milk yield and daily litter gain were unaffected by a 13% reduction in protein level.

Grazed herbage also has the potential to make a significant contribution to the vitamin, mineral, and trace element requirements of pigs. The extent of this will depend on plant species, with legumes generally having a higher mineral and trace element content, but also on stage of maturity, soil characteristics, and fertilizing regimen. Furthermore, even if herbage is not present, pigs that are able to root may obtain substantial levels of minerals from ingestion of soil and even energy and protein from roots and soil fauna (Edwards 2003).

**Table 24.1** Reported values for the intake of grazed herbages by pigs[a].

| Herbage type | Season | Pig type | Herbage intake/day | Reference |
|---|---|---|---|---|
| Lucerne (*Medicago sativa*) | Summer | Pregnant gilt | 3.2–4.2 kg DM | Honeyman and Roush (1999) |
| Perennial ryegrass (*Lolium perenne*) + white clover (*Trifolium repens*) | Early summer<br>Late summer | Pregnant sow | 2.4 kg DM<br>3.7 kg DM | Sehested et al. (1999) |
| Perennial ryegrass (*Lolium perenne*) + white clover (*Trifolium repens*) | Spring<br>Summer | Pregnant sow | 1.1 kg OM<br>1.5 kg OM | Rivera Ferre et al. (2001) |
| Perennial ryegrass (*Lolium perenne*) + white clover (*Trifolium repens*) | Spring/summer | Pregnant sow | 2.4 kg DM | Sehested et al. (2004) |
| Perennial ryegrass (*Lolium perenne*) + white clover (*Trifolium repens*) + red clover (*Trifolium pratense*) + timothy (*Phleum pratense*) | | | 3.0 kg DM | |
| Red clover (*Trifolium pretense*) + grasses (*Phleum pratense, Lotus corniculatus, Bromus riparius, Poa pratensis*) | Summer | Pregnant sows | 1.3 kg DM | Aubé et al. (2021) |
| Perennial + Italian ryegrass (*Lolium perenne + Lolium multiflorum*) | Spring | Pregnant sows<br>Lactating sows | 2.6 kg DM<br>1.0 kg DM | Gannon (1996) |
| Grass (undefined) | Summer<br>Autumn | Lactating sows | 0.4 kg DM<br>1.0 kg DM | Jurjanz and Roinsard (2014) |
| Perennial ryegrass (*Lolium perenne*) + white clover (*Trifolium repens*) | Spring | Growing pigs (50–60 kg) | 0.1 kg OM | Mowat et al. (2001) |
| Grass + clover (unspecified) | Summer/autumn | Growing pigs (30–112 kg) | 0.4 kg DM | Gustafson and Stern (2003) |
| Perennial ryegrass (*Lolium perenne*) + white clover (*Trifolium repens*) | Spring to autumn | Growing pigs (24–240 kg) | 0.4 kg OM | Hodgkinson et al. (2017) |
| Lucerne (*Medicago sativa*) + grasses (*Festuce arundinacea + Cebadilla criolla*) | Spring<br>Summer<br>Spring<br>Summer | Growing pigs (30–70 kg)<br><br>Growing pigs (70–100 kg) | 0.15 kg DM<br>0.04 kg DM<br>0.5 kg DM<br>0.3 kg DM | Riart (2002) |
| Lucerne (*Medicago sativa*) | Autumn | Growing pigs (58–90 kg) | 0.4 kg DM | Jakobsen et al. (2015) |
| Kikuyu grass (*Pennissetum clandestinum*) | Early winter | Growing pigs (27 kg) | 0.13 kg OM | Kanga et al. (2012) |

[a] DM = dry matter, and OM = organic matter.

## Management Practices

Two other notable aspects in the management of organic pigs will affect nutrient requirements. The first is the extended lactation period dictated by the requirement not to wean piglets at less than 40 days of age. While this will impose additional demands on the sow for milk production, the greater weaning age and weight of the piglets mean that less sophisticated postweaning diets can be used. However, the prohibition on the use of additives that will control proliferation of pathogenic microbes in the gut means that diet formulation for the weaned piglet must give greater consideration to avoiding characteristics predisposing digestive upset. The prohibition on growth promoting agents at any stage may also result in lower performance levels in comparison with pigs in conventional production which commonly receive such compounds, and adjustments to the nutrient specification of diets for growing pigs may need to take account of this.

### Factors Affecting Ingredient Choice in Organic Diets

The composition of diets fed to organic pigs may, in some cases, differ substantially from those used in conventional production. These differences arise because of the need to use organically grown ingredients, preferably produced on-farm or in the local region, the prohibition on use of some raw materials commonly used in conventional diets and, most significantly, the requirement in many organic standards to include a certain level of roughage in the diet for all pigs.

The raw materials used in diet formulation should ideally be produced on the same farm as the pig enterprise, facilitating closed nutrient cycles for sustainability of the overall system. However, many farms find it difficult to produce all of the feedstuffs required and may purchase a proportion of their diets or raw materials from external suppliers of organic feed. While organic pigs must be fed on ingredients that are certified as being produced according to organic standards, some regions still experience difficulty in sourcing sufficient quantities of such materials. In consequence, some certification schemes still include a derogation to use a small proportion of feed from nonorganic sources, although GMO products are prohibited in all cases. For example, in the EU, young piglets up to 35 kg may be fed up to 5% of nonorganic protein feed under an exemption in place until 31 December 2025 (European Council 2018). In addition to the requirement for organic provenance, some schemes produce a more specific list of permitted and excluded raw materials (Blair 2018).

### *Ingredients for Energy*

The primary source of energy in most organic pig diets comes from cereals and their processing by-products. The possibilities for home-grown cereals depend on the growing conditions in the geographic location, meaning that maize or wheat is widely used in North America and Southern Europe, whereas barley or rye is often better suited in more Northern European countries. The nature of the cereal available will affect the level of dietary fiber and hence energy density in the diet. Cereal by-products from human food processing industries are also important raw materials. These include wheat brans and maize hominy and gluten products, as well as spent grains from the alcohol distillation process. Since these by-products are typically the residue after starch removal, they tend to be higher in fiber and protein content and can be very variable in composition depending on the exact processing procedure. A more detailed discussion of different cereal ingredients and by-products can be found in Blair (2018).

Another very important source of energy for organic pigs is root crops and vegetables, which are often part of an organic rotation. Many of these root crops have the potential for very high energy yields per hectare, and they may be harvested and stored for controlled feeding or foraged in situ by the pigs. Widely used root crops include potatoes, grown worldwide (Edwards and Livingstone 1990), fodder beet and Jerusalem artichokes in Europe (Chambers et al. 1986; Kelly et al. 2007; Kongsted et al. 2013), and sweet potato and cassava in the tropics (Machin and Nyvold 1992). A further source of ingredients comes from the growing and processing of vegetables for human consumption. Vegetables, such as carrots, swedes, and parsnips, grown for human consumption but surplus or unfit for purpose can be utilized, as can by-products such as peelings. Potato residues from starch extraction processes and sugar beet pulps after sugar extraction also provide useful feeds.

Root crops and their by-products are generally low in dry matter, and this can give rise to very variable voluntary intake by pigs, depending on the stage of production. Because of their bulk, they are most widely used for dry sows and finishing pigs. Information on nutrient composition and likely intakes can be found in various sources (Thacker and Kirkwood 1990; Edwards 2002; Blair 2018).

### Ingredients for Protein

Obtaining adequate sources of good-quality protein still poses the greatest feeding challenge for many organic pig farms (Smith et al. 2014). Protein sources of animal origin, which have better digestibility and amino acid composition than vegetable sources, are generally banned from organic pig diets. Exceptions are organically produced milk products and products from sustainable fisheries that are permitted in many schemes. The major protein sources are oilseeds and legumes. Oilseeds may be used as the complete seed but are more commonly used as the meals remaining after oil extraction for human consumption or industrial processes. Only meals derived from mechanical pressing processes are permitted, since solvent extraction processes are banned. This means that oil and energy content is generally higher, and protein content lower, for organic meals than those widely used in conventional pig production. The most commonly used oilseeds are soya, in regions with suitable growing conditions, and rapeseed in more northerly locations. However, cottonseed, linseed, sunflower seed, and other oilseeds may sometimes be available. An important characteristic of many of these seeds is the presence of a range of different antinutritive factors, such as protease inhibitors, lectins, tannins, and glucosinolates, which can impair digestion and give rise to toxic effects at high levels. Some of these factors can be inactivated by processing but other persist and limit inclusion levels in the diet. For a more detailed discussion of these see Blair (2018).

Legumes make up an important part of organic crop rotations because of their nitrogen-fixing characteristic, which is important for the maintenance of soil fertility in the absence of synthetic fertilizers. In consequence, they are a widely available source of home-grown protein ingredients for pig diets. Beans, peas, and lupins are the most widely grown legume seeds. Their protein content is generally lower than that of oilseed meals, and they too contain a range of similar antinutritional factors. Leguminous forages can also make an important contribution to protein supply and are discussed in the next section.

All of the vegetable protein sources suffer from the drawback of suboptimal amino acid composition for pig feeding. Compared to animal proteins, they are deficient in the key essential acids of lysine and methionine. In conventional pig diets, synthetically produced amino acids are commonly used to rectify this deficiency, but these are generally not permitted in organic diets. As a result, it is often necessary to feed an excess level of crude protein in the diet in order to achieve the required levels of these first-limiting amino acids. This can give rise to digestive problems in weaned piglets and to increased nitrogen excretion which poses a pollution risk, as discussed in later sections.

### Roughages

Most organic standards require that pigs receive forage or roughage in their diet, and this is one of the biggest nutritional differences from conventional production. It allows the animals to express more natural feeding behaviors and improves satiety in animals, such as pregnant sows, which are given restricted concentrate feed. The additional fiber can also improve gut health by preventing gastric ulcers and stimulating beneficial microflora. Furthermore, high-fiber diets during gestation

**Figure 24.2** Finishing pigs offered silage in a rack in a housing-with-run system.

reduce constipation and enhance farrowing progress, as well as increasing sow feed intake capacity during the lactation period, which can increase milk production and piglet survival and growth (Meunier-Salaun et al. 2001). Some pigs obtain roughage by grazing at pasture but, in seasons when herbage is not growing or when pigs are kept in housing with outdoor runs, additional roughage needs to be supplied (Figure 24.2). This can be achieved by incorporating dried forage in their compound diet, by allowing them *ad libitum* or restricted daily access to forages and providing separate supplementary concentrate, or by feeding a complete mixed diet of forage and concentrate in long troughs. In the latter case, significant problems with differential selection of diet components and high feed wastage can occur if the complete diet is fed close to appetite.

Roughages can include fresh grass or whole-crop cereals, or these can be processed for use all year round by drying in the form of hay, or preserving moist as silage or haylage. Green vegetable wastes from other farm enterprises, such as brassicas grown for human consumption, may also be available. The conservation method can affect pig preference for forages. For example, Aubé et al. (2019b) demonstrated that, when given a choice, sows preferred haylage over fresh forage, with dried hay least preferred. When offered separately, the voluntary intakes also reflected these preferences, possibly because of the higher fiber content and lower soluble carbohydrate content of hay. Intake can also be affected by the botanical composition of the forage, with legume forages generally preferred over grasses (Aubé et al. 2019a), and by other processing characteristics such as chop length and quality of fermentation of silages (Edwards et al. 1994). Only preservatives permitted in organic standards can be used when preparing silage.

The extent to which roughages make a significant contribution to the nutrient requirements of the pig depends on their nutritive value and voluntary intake. Forages are high in fiber, which cannot be utilized through enzymic digestion in the pig but is dependent on fermentation by the hind gut microbiota, a process which is energetically less efficient. Inclusion of roughage in the diet of can also reduce the ileal digestibility of nutrients (Andersson and Lindberg 1997a, 1997b). The nutritive value will obviously depend on the species and growth stage of the forage. Thus, maize silage or whole crop cereal silage, where the starch-containing grain is included, can make a much higher energy contribution than forages based only on vegetation (Carlson et al. 1999; Galassi et al. 2017). Legume forages, such as lucerne and clover, will have a higher protein content, while herbage taken at an earlier growth stage will have a lower fiber content and higher concentration of sugars which enhance palatability and energy level. Typical nutrient values for different forages can be found in various publications (Kephart et al. 1990;

Edwards 2002; Kyntäjä et al. 2014; Blair 2018). Attention also needs to be paid to possible antinutritive factors, such as tanins and saponins in lucerne, or isoflavones with estrogenic properties in red clover.

Voluntary intake of forages is primarily influenced by liveweight and stage of production but can show very large individual variation within a stage. Pregnant sows can consume as much as 20–50% of their dry matter intake from forage (Bikker et al. 2014a), but in lactation, when metabolic demands on the sow are high, expecting the sow to obtain a significant part of her diet from low value forages will result in excessive condition loss, poor milk yield, and subsequent rebreeding problems (Kongsted et al. 2000). In this stage, forages are better used as a supplement rather than replacement of concentrate diets. The same is true for young piglets, whose digestive tract has low capacity and poorly developed fermentation ability. However, as pigs get older and their intake capacity increases, the amount of forage in the diet can be increased (Bikker et al. 2014b). When forage replaces concentrate in the diet, the daily nutrient intake of the pig is decreased and growth will be slower. While this may be disadvantageous in the early stages of rearing, in the later stages, it can help to improve carcass quality by preventing over fatness in less improved breeds. When growing pigs are offered both concentrate and forage *ad libitum*, the intakes of forage are low, typically only 2–5% of daily dry matter intake (Kelly et al. 2007). If concentrate intake is restricted by 20–25%, intakes of forage can be increased to 5–10% of dry matter, but growth rate will be reduced by 10–15% and a leaner carcass will be produced (Wallenbeck et al. 2014; Presto Åkerfeldt et al. 2019).

### Vitamins, Minerals, and Additives

Organic principles indicate that minerals, trace elements, and vitamins should be sourced from natural foraging activities and feedstuffs whenever possible. Significant amounts may be ingested from soil and fresh herbage (Edwards 2003), but this may not always be adequate to prevent deficiencies. In particular, pigs kept in indoor systems with runs are likely to benefit from supplements. Supplements containing approved minerals sources and vitamins derived from raw materials occurring naturally in feedstuffs, or synthetic vitamins identical to natural vitamins, are therefore permitted by most schemes (e.g. European Council 2018). Any additives intended to stimulate growth or production are prohibited in organic diets. However, certain enzymes and probiotic organisms, if added for the purposes of improving health and nutrient utilization rather than growth, may be permitted. These must be extracted from natural sources and cannot be derived from genetic engineering technology.

### Diet Formulations

The sophistication of diet formulation will depend on the scale of the organic pig enterprise and the technical knowledge of the farmer or his advisors. Diet formulation relies on good knowledge of the requirements of the pig, its feed intake, and the composition of the available raw materials. Larger and nutritionally knowledgeable organic enterprises will use standard least cost formulation software to derive the best diet from the ingredients available, typically formulating four to six different diets to match the different production stages. On small units with more traditional breeds, feeding may be simplified by using the same diet for dry sows and finishing pigs, and for lactating sows and growing pigs, although this will involve compromises in formulation and some loss in efficiency of feed use. Where large amounts of separate forages are fed, the compound diets need to be modified to complement the contribution from these forages on a case-by-case basis.

While nutrient composition data are readily available for the commonly used conventional feedstuffs, they are still lacking for the wider range of raw materials likely to be used in organic diets and particularly for forages. Because of the fertilizing regimes used in organic agronomy, organically grown cereals often contain a lower protein content and lower energy level because of a higher fiber content. It is therefore advantageous to use feed composition databases based on organically produced materials where possible, such as the feed table with data on chemical composition and nutritive value of organic feedstuffs produced in the EU ICOPP project, which is available online (Kyntäjä et al. 2014). Information on a range of nontraditional ingredients can also be found in various publications (Thacker and Kirkwood 1990; Edwards 2002, Blair 2018). Where least cost formulation facilities are not available, or nutritional expertise is lacking on smaller farms, a range of example diets for different production stages can be found in published material (Edwards 2002; Blair 2018).

## Organic Production Systems and Nutritional Influences on Their Objectives

Given the differences between organic and conventional production systems that have been discussed in the preceding sections, it is valid to ask to what extent these are successful in practice in achieving the stated objectives of organic production, i.e. in reducing environmental impact, improving animal health, and producing a more nutritious product for the human food chain (Edwards and Leeb 2018). Any assessment of this is hampered by the great diversity that exists in organic pig enterprises, both between and even within geographic regions (Früh et al. 2014). At the extremes, some farms maintain all pigs in outdoor paddocks throughout the year (Figure 24.1), sometimes in conjunction with natural woodland or agroforestry, while others house all animals in permanent buildings with an outdoor concrete run and imported supply of roughage (Figure 24.2). Many farms have a combination of these two approaches, either seasonally, with winter housing, or by production stage, with outdoor sows and housed growing pigs. Shelter in paddock systems may be provided by moveable huts or by tents, while in North America hoop barns with deep straw bedding are also common. As a result of such diversity, the variation in any measured outcome assessment may be greater within organic production systems than in a comparison of organic with conventional production (Rudolph et al. 2018; Leeb et al. 2019).

### Environmental Impact

An important objective of organic farming is long-term sustainability based on closed nutrient cycles. This implies that the majority of the food used for the pigs should be grown on-farm or in local nutrient sharing arrangements. Where the pig enterprise is integrated into a farm rotation plan involving cereals and legumes, both energy and protein feedstuffs are generated. A typical European rotation might be a root crop, spring wheat, winter cereal, spring peas, and spring cereal undersown with grass/clover to produce a ley for subsequent pig grazing (Martins et al. 2002). In a detailed evaluation of nutrient flows in a Danish organic farming system with pigs as the only livestock, Kristensen and Kristensen (1997) demonstrated that the farm could not be totally self-sufficient, but needed to import 25% of its pig feed and also some additional manure (45 kg N/ha). The problems in self-sufficiency arise because of a conflict between the area of cropping necessary to supply the nutrient demands of the pig and the area in grass/clover ley to fix adequate nitrogen in the soil to provide fertility for subsequent cereal crops. Furthermore, the limited ability of the pigs to utilize dietary nutrients in pasture contributed to the net shortfall in feed supply. Such challenges could be partly addressed by including ruminant livestock in the overall model to give mixed species livestock production.

When considering organic feed supply at the regional level rather than for an individual farm, significant challenges still exist. An EU project (ICOPP) recently explored the possibility for using entirely organically produced local feed to supply poultry and pigs with the required level of nutrients in different phases of production and support high animal health and welfare. It was concluded that, for the countries involved in this project, there was a self-sufficiency rate for organic concentrate feed for livestock of only 69%. The self-sufficiency rate for crude protein was 56%, with the supply gap for certain essential amino acids being just above 50% for lysine and about 40% for methionine.

Another challenge identified by the modeling of Kristensen and Kristensen (1997) was the inefficient use of manure from the pigs for fertilizing crop growth if pigs are not kept in housed systems where manure can be captured, stored, and applied with optimal distribution and at optimal times to match crop requirement. Pigs kept on pasture show uneven spatial distribution of excreta, creating hot-spots with high soil concentrations of nitrogen and phosphorus (Watson et al. 2003). Depending on temperature and rainfall, these are prone to production of polluting gaseous emissions of ammonia and nitrous oxide and the leaching of nutrients into waterways where they cause eutrophication. The destruction of vegetation resulting from pig rooting activity exacerbates this problem, as it reduces the capture and stabilization of nutrients by herbage.

The overall environmental impact of a production system is generally evaluated using the standardized framework of life cycle assessment (LCA). This calculates the total energy use and deleterious outcomes that result from the production of a unit of output; in the case of a pig enterprise, this is typically 1 kg of pig liveweight or pigmeat produced. While a variety of different outcome parameters can be calculated, those most commonly used in the comparison of different pig production systems are the greenhouse gas emissions (GHGE), acidification potential (AP), and eutrophication potential (EP). Nitrogen excretion from animals, and the associated ammonia emission, is a major contributor to AP, while EP is influenced by both nitrogen and phosphorus excretion. GHGEs contributing to global warming include carbon dioxide, methane, and nitrous oxide emanating from animal respiration and from the decomposition of manures. However, it is actually the environmental burdens arising from the production of the feedstuffs required for the animals that are the biggest contributors to total impact. Initial comparisons of pigmeat produced in organic and conventional systems relied on the modeling of typical scenarios for each production type (Basset-Mens and Van Der Werf 2005; Williams et al. 2006; Halberg et al. 2010). These studies, with the exception of Williams et al. (2006), indicated higher environmental burdens per unit of pigmeat produced by organic systems, although Halberg et al. (2010) noted that, due to the integration of grass clover in their organic crop rotations, organic systems had an estimated net soil carbon sequestration which, when included in the LCA, gave lower GHGE compared with conventional pig production. Differences between studies in the environmental impact indicators were mainly attributable to differing assumed levels of animal performance, with the lower reproductive efficiency and poorer feed conversion ratio in organic systems outweighing other possible agronomic benefits. Subsequently, two large EU studies have published LCA calculations based on actual data collected from a range of individual farms. Dourmad et al. (2014) evaluated 15 contrasting pig production systems from five European countries, with data collected on 5–10 farms per system. Animal housing, feed production, manure storage, and manure spreading all resulted in higher absolute impact values in organic systems than in conventional ones. Compared with conventional systems, impacts per kg liveweight produced for climate change (GHGE 2.4 kg $CO_2$-eq; +4%), acidification (AP 57 g $SO_2$-eq; +29%), eutrophication (EP 16 g $PO_4$-eq; −16%), energy use (18.1 MJ; +11%), and land occupation (9.1 m²; +121%) were nearly all higher for organic systems. The variation within organic systems was also greater than for conventional systems, possibly reflecting their greater diversity of production methods. Halberg et al. (2010) compared Danish scenarios in which the typical Danish system of sows kept in huts on grassland and finishing pigs raised in indoor-with-run housing, with the alternatives of also rearing the fattening pigs on grassland all year round, or in a bedded tent with restricted access to a grazing area. Higher burdens were calculated for all indicators in the free-range finishing system. Using data collected on 64 individual organic pig farms across Europe, Rudolph et al. (2018) found no difference between systems in GHGE, but higher AP in indoor-with-run systems than in systems with mixed indoor and paddock housing and lower EP in these mixed systems than in all-paddock systems. However, variation was mostly higher within than between systems. Overall, the values obtained in this more extensive organic survey for the different impact categories were comparable to those of Dourmad et al. (2014), apart from slightly higher EPs.

Nutritional solutions to reduce environmental impact in pigs kept on pasture have been identified (Edwards 2008), and many are also applicable to housed pigs. However, not all of these can be exploited in organic farms. While significant reductions in excreted nutrients can be made in conventional production by raw material selection for higher digestibility, optimization of amino acid balance toward "ideal protein" by use of synthetic amino acids, and the use of synthetic enzymes such as phytase to increase nutrient availability, these approaches are not available to organic farmers. Reduction in nutrient oversupply in feeds can be achieved if a more detailed "precision feeding" approach is utilized. This involves the use of a greater number of diets to better match the changing requirements of liveweight, production stage, and seasonal climatic differences. Such a strategy requires detailed knowledge of the metabolic requirements of the animals and is often limited by the practicalities of handling many different diets in a smaller organic farming situation where infrastructure for feed storage and distribution is limited. However, in organic systems, as in conventional outdoor systems, a significant improvement in the efficiency of nutrient utilization can be made by careful management to reduce feed wastage. For example, providing feed in well-designed troughs that reduce spillage and wildlife access will significantly reduce feed requirement in comparison with scattering feed on the ground.

### Animal Welfare

While the organic standards contain many requirements that might be expected to promote good welfare (Edwards et al. 2014; Edwards and Leeb 2018), there have been relatively few objective comparisons of welfare outcome measures in organic and conventional production systems. Literature review has highlighted potentially greater problems in organic systems of ectoparasites and endoparasites in sows, neonatal mortality in piglets, postweaning diarrhea in weaned piglets, and endoparasites in finishing pigs.

Looking specifically at the welfare of pigs on organic farms using selected animal-based outcome measures, Dippel et al. (2014) reported that the prevalence of health and welfare problems of organic pigs from 101 farms across six EU countries was generally similar to or lower than values previously reported in similar assessments on conventional farms, with the exception of poor body condition in sows. A follow-up study of 74 organic farms in eight different European countries gathered data in relation to different organic management systems (Leeb et al. 2019). This concluded that European organic pigs kept in indoor-with-run, paddock, and mixed husbandry system showed a low prevalence of health and welfare problems, but identified respiratory health and diarrhea in weaners and fatteners kept indoors and neonatal piglet mortality in all systems as areas needing attention.

Several of the health and welfare challenges identified for organic pigs have links to nutrition. Better attention to the nutritional requirements of the sow, taking into account the additional demands of outdoor living, climatic challenges, and extended lactations, could reduce the prevalence of thin sows and might beneficially influence piglet survival through increased piglet birth weight and milk production. Postweaning diarrhea can be influenced by a number of known dietary risk factors and strategies to promote gut health (Lallès et al. 2007). Despite the greater weaning age of organic pigs, the prohibition on the use of prophylactic veterinary treatments, antimicrobial feed additives, and synthetic enzymes which can favorably modify the gut microflora increases the risk of dysbiosis in the gut once the protective influence of maternal milk is removed. The limitations on the use of animal-derived raw materials with high protein quality, and the prohibition on the use of synthetic amino acids, result in higher inclusions of less digestible plant proteins in weaner diets and increase the risk of undigested protein passing down the gastrointestinal tract, where it can exacerbate digestive problems. Risk of diarrhea can therefore be reduced by the use of lower protein diets, dietary ingredients containing prebiotic fermentable carbohydrates, and permitted acidifying agents. Furthermore, endoparasite infestation in both sows and growing pigs is also subject to nutritional influences. A high intake of insoluble dietary carbohydrates, i.e. dietary fibers as found in many roughages that are resistant to digestion and fermentation in the gut, may increase the severity of infection with gastrointestinal nematodes, such as Oesophagostomum. In contrast, inclusion of readily fermentable carbohydrates, such as inulin, may have the opposite effect (Petkevicius et al. 2001).

### Product Quality

Many consumers purchase organic products because they believe them to have superior properties in relation to human nutrition and health. There are very few objective assessments of this in the case of pigmeat. Srednicka-Tober et al. (2016) reported that the concentrations of saturated fatty acids, deemed detrimental to health, were lower in organic pork, whereas concentrations of polyunsaturated fatty acids, perceived as beneficial to health, were found to be higher. Such differences may again be related to nutritional influences, in particular the ingestion of forage by organic pigs and the higher vegetable oil content in diets with cold-pressed rather than chemically extracted soya. Whilst the composition of many feed ingredients may be favorably modified as a result of organic agronomic practices, notably with an increased content of antioxidants (Barański et al. 2014), whether consumption of these results in any beneficial changes in pigmeat composition remains to be demonstrated.

### Summary

Organic pig production, while currently very small in scale compared to conventional production, is growing in popularity as it is perceived by consumers to better meet their wish for healthy and ethically produced meat. Constraints imposed by organic production standards on feeding, housing, and management practices necessitate different nutritional approaches and reduce biological performance. The greatest nutritional challenges come from the requirements for outdoor living and forage feeding. Great diversity in production methods and sustainability outcomes exists both between and within countries at the present time, indicating a need for future research on ways to optimize organic pig production.

# References

Andersson, C., and J. E. Lindberg. 1997a. Forages in diets for growing pigs 1. Nutrient apparent digestibilities and partition of nutrient digestion in barley-based diets including lucerne and white-clover meal. Anim. Sci. 65:483–491.

Andersson, C., and J. E. Lindberg. 1997b. Forages in diets for growing pigs 2. Nutrient apparent digestibilities and partition of nutrient digestion in barley-based diets including red-clover and perennial ryegrass meal. Anim. Sci. 65:493–500.

Aubé, L., F. Guay, R.Bergeron, G. Bélanger, G. F. Tremblay, and N. Devillers. 2019a. Sows' preferences for different forage mixtures offered as fresh or dry forage in relation to botanical and chemical composition. Animal 13:2885–2895.

Aubé, L., Guay, F., Bergeron, R., Bélanger, G., Tremblay, G.F. and Devillers, N., 2019b. Sows prefer forages conserved as haylage over hay and fresh forage. *Livestock Science* 228: 93–96.

Aubé, L., F. Guay, R. Bergeron, G. Bélanger, G. F. Tremblay, S. A. Edwards, J. H. Guy, and N. Devillers. 2021. Feed restriction and type of forage influence performance and behaviour of outdoor gestating sows. Animal 15:100346.

Barański, M., D. Srednicka-Tober, N. Volakakis, C. Seal, R. Sanderson, G. B. Stewart, C. Benbrook, B. Biavati, E. Markellou, C. Giotis, J. Gromadzka-Ostrowska, E. Rembiałkowska, K. Skwarło-Sońta, R. Tahvonen, D. Janovská, U. Niggli, P. Nicot, and C. Leifert. 2014. Higher antioxidant and lower cadmium concentrations and lower incidence of pesticide residues in organically grown crops: a systematic literature review and meta-analyses. Br. J. Nutr. 112:794–811.

Basset-Mens, C., and H. M. G. Van Der Werf. 2005. Scenario-based environmental assessment of farming systems: The case of pig production in France. Agric. Ecosystems Environ. 105:127–144.

Bikker, P., G. P. Binnendijk, H. M. Vermeer, and C. M. C. van der Peet-Schwering. 2014a. Silage in diets for organic sows in gestation. In: G. Rahmann, and U. Assoy, editors, Building Organic Bridges. Proc. 4th ISOFAR Sci. Conf. Johann Heinrich von Thünen-Institut, Braunschweig, Germany. p. 819–822.

Bikker, P., G. Binnendijk, H. Vermeer, and C. Van Der Peet-Schwering. 2014b. Grass silage in diets for organic growing-finishing pigs. In: G. Rahmann, and U. Assoy, editors, Building organic bridges. Proc. 4th ISOFAR Sci. Conf. Johann Heinrich von Thünen-Institut, Braunschweig, Germany. p. 815–818.

Blair, R. B. 2018. Nutrition and feeding of organic pigs. 2nd ed. CABI, Wallingford, UK.

Buckner, L. J., J. M. Bruce, and S. A. Edwards. 1994. Modelling the energy system of outdoor sows. Pig News Info. 15:125N–128N.

Buckner, L. J., S. A. Edwards, and J. M. Bruce. 1998. Behaviour and shelter use by outdoor sows. Appl. Anim. Behav. Sci. 57:69–80.

Carlson, D., H. N. Lærke, H. D. Poulsen, and H. Jorgensen. 1999. Roughages for growing pigs, with emphasis on chemical composition, ingestion and faecal digestibility. Acta Agric. Scand. Sec. A – Anim. Sci. 49:129–136.

Chambers, J., B. Hardy, and O. Pugh. 1986. The use of an electronically controlled sow feeder to supply a compound feed as a balancer to sows grazing fodder beet. In: Proc. Br. Soc. Anim. Sci. p. 121.

Close, W. H., and P. K. Poornan. 1993. Outdoor pigs – their nutrient requirements, appetite and environmental responses. In: W. Haresign, and D. J. A Cole, editors, Recent advances in animal nutrition. Nottingham University Press, Nottingham, UK. p. 175–196.

Dippel, S., C. Leeb, D. Bochicchio, M. Bonde, K. Dietze, S. Gunnarsson, K. Lindgren, A. Sundrum, S. Wiberg, C. Winckler, and A. Prunier. 2014. Health and welfare of organic pigs in Europe assessed with animal-based parameters. Org. Agric. 4:149–161.

Dourmad, J. Y., J. Ryschawy, T. Trousson, M. Bonneau, J. Gonzalez, H. W. J. Houwers, M. Hviid, C. Zimmer, T. L. T. Nguyen, and L. Morgensen. 2014. Evaluating environmental impacts of contrasting pig farming systems with life cycle assessment. Animal 8:2027–2037.

Edwards, S. A., 2002. Feeding organic pigs. A handbook of raw materials and recommendations for feeding practice. University of Newcastle, Callaghan, NSW, Australia. p. 59.

Edwards, S. A. 2003. Intake of nutrients from pasture by pigs. Proc. Nutr. Soc. 62:257–265.

Edwards, S. A., 2008. Nutritional approaches to reducing the environmental impact in outdoor pigs. In: P. C. Garnsworthy, and J. Wiseman, editors, Recent advances in animal nutrition 2007. Nottingham University Press, Nottingham, UK. p. 295–309.

Edwards, S. A., and R. M. Livingstone. 1990. The use of potatoes and potato products in swine rations. In: P. A. Thacker, and R. N. Kirkwood, editors, Non-traditional feed sources for use in swine production. Butterworths, Oxford, UK. p. 305–314.

Edwards, S. A., A. Prunier, M. Bonde, and E. A. Stockdale. 2014. Organic pig production in Europe – animal health, welfare and production challenges. Org. Agric. 4:79–81.

Edwards, S. A., and C. Leeb. 2018. Organic pig production. In: A. Matthew, editor. Achieving sustainable production of pig meat Volume 1: Safety, quality and sustainability. Burleigh Dodds Science Publishing, Cambridge, UK.

Edwards, S. A., J. Weddell, C. Fordyce, A. Cadenhead, and J. Rooke. 1994. Intake and digestibility of silage by pregnant sows and effects of treatment with Maxgrass. Proc. Br. Soc. Anim. Prod. p. 154.

Eskildsen, M., D. V. Sampego, U. Krogh, T. Larsen, and P. Theil. 2019. Effect of dietary protein level on milk yield, milk composition and blood metabolites in organic sows on pasture summer and winter. In: M. L. Chizzotti, editor, Energy and protein metabolism and nutrition. EAAP Sci. Ser. Vol. 138. Wageningen Academic Publishers, Wageningen, The Netherlands. p. 331–333.

European Commission. 2008. Commission Regulation (EC) No 889/2008 of 5 September 2008 laying down detailed rules for the implementation of Council Regulation (EC) No 834/2007 on organic production and labelling of organic products with regard to organic production, labelling and control. Off. J. Eur. Union L250:1–84.

European Council. 2018. Regulation (EU) 2018/848 of the European Parliament and of the Council of 30 May 2018 on organic production and labelling of organic products and repealing Council Regulation (EC) No 834/2007. Off. J. Eur. Union L150:1–92.

Früh, B., D. Bochicchio, S. Edwards, L. Hegelund, C. Leeb, A. Sundrum, S. Werne, S. Wilberg, and A. Prunier. 2014. Description of organic pig production in Europe. Org. Agric. 4:83–92.

Galassi, G., L. Malagutti, L. Rapetti, G. M. Crovetto, C. Zanfi, D. Capraro, and M. Spanghero. 2017. Digestibility, metabolic utilisation and effects on growth and slaughter traits of diets containing whole plant maize silage in heavy pigs. Ital. J. Anim. Sci. 16:122–131.

Gannon, M. A. 1996. The energy balance of pigs outdoors. PhD Diss. University of Nottingham, Nottingham, UK.

Gustafson, G. M., and S. Stern. 2003. Two strategies for meeting energy demands of growing pigs at pasture. Livest. Prod. Sci. 80:167–174.

Halberg, N., J. E. Hermansen, I. S. Kristensen, J. Eriksen, N. Tvedegaard, and B. M. Petersen. 2010. Impact of organic pig production systems on $CO_2$ emission, C sequestration and nitrate pollution. Agron. Sustainable Dev. 30:721–731.

Hodgkinson, S. M., C. Polanco, L. Aceiton, and I. F. Lopez. 2017. Pasture intake and grazing behaviour of growing European wild boar (Sus scrofa L.) and domestic pigs (Sus scrofa domesticus, Landrace×Large White) in a semi-extensive production system. J. Agric. Sci. 155:1659–1668.

Honeyman M. S., and W. B. Roush. 1999. Supplementation of mid gestation swine grazing alfalfa. Am. J. Alt. Agric. 14:103–108.

IFOAM. 2005. Principles of organic agriculture. Int. Fed. Org. Agric. Movements, Bonn, Germany.

IFOAM. 2014. The IFOAM NORMS for organic production and processing version 2014. Int. Fed. Org. Agric. Movements, Bonn, Germany.

Jakobsen, M., A. G. Kongsted, and J. E. Hermansen. 2015. Foraging behaviour, nutrient intake from pasture and performance of free-range growing pigs in relation to feed CP level in two organic cropping systems. Animal 9:2006–2016.

Jurjanz, S., and A. Roinsard. 2014. Ingestion d'herbe et de sol par des truies au pâturage. Alt. Agric. 125:27–28.

Kanga, J. S., A. T.Kanengoni, O. G. Makgothi, and J. J. Baloyi. 2012. Estimating pasture intake and nutrient digestibility of growing pigs fed a concentrate-forage diet by n-alkane and acid-insoluble markers. Trop. Anim. Health Prod. 47:1797–1802.

Kelly, H. R. C., H. M. Browning, A. P. Martins, G. P. Pearce, C. Stopes, and S. A. Edwards. 2001. Breeding and feeding pigs for organic production. In: M. Hovi, and T. Baars, editors, Proc. NAHWOA Workshop, Wageningen 24–27 March. University of Reading, Reading, UK. p. 86–93.

Kelly, H. R. C., H. M. Browning, J. E. L. Day, A. P. Martins, G. P. Pearce, C. Stopes, and S. A. Edwards. 2007. The effect of breed type, housing and feeding system on performance of growing pigs managed under organic conditions. J. Sci. Food Agric. 87:2794–2800.

Kephart, K. B., G. R. Hollis, and D. M. Danielson. 1990. Forages for swine. PIH-126. Pennsylvania State University, University Park, PA, US.

Kongsted, A. G., K. Horsted, and J. E. Hermansen. 2013. Free-range pigs foraging on Jerusalem artichokes (Helianthus tuberosus L.) – Effect of feeding strategy on growth, feed conversion and animal behaviour. Acta Agric. Scand. Sect. A Anim. Sci. 63:76–83.

Kongsted, A. G., J. Larcher, and V. A. Larsen. 2000. Silage for outdoor lactating sows. In: J. E. Hermansen, V. Lund, and E. Thuen, editors, Ecological animal husbandry in the Nordic countries. Proceedings from NJF-seminar No. 303. Danish Research Centre for Organic Agriculture, Tjele, Denmark. p. 125–130.

Kristensen, I. S., and T. Kristensen. 1997. Animal production and nutrient balances on organic farming systems. Prototypes. In: Proc. 3rd ENOF Workshop: Resource use in organic farming, Ancona, Italy. p. 189–202.

Kyntäjä, S., K.Partanen, H. Siljander-Rasi, and T. Jalava. 2014. Tables of composition and nutritional values of organically produced feed materials for pigs and poultry. MTT Report 164. MTT Agrifood Res. Finland, Jokioinen, Finland.

Lallès, J.-P., P. Bosi, H. Smidt, and C. R. Stokes. 2007. Nutritional management of gut health in pigs around weaning. Proc. Nutr. Soc. 66:260–268.

Leeb, C., G. Rudolph, D. Bochicchio, S. A. Edwards, B. Früh, M. Holinger, D. Holmes, G. Illmann, D. Knop, A. Prunier, C. Winckler, and S. Dippel. 2019. Effects of three husbandry systems on health, welfare and productivity of organic pigs. Animal 13:2025–2033.

Machin, D., and S. Nyvold. 1992. Roots, tubers, plantains and bananas in animal feeding. FAO Animal Production and Health Paper 95. FAO, Rome, Italy.

Martins, A., H. Kelly, J. Day, C. Stopes, H. Browning, and S. Edwards. 2002. Optimising organic pig production – a guide to good practice. Agric. Dev. Adv. Serv., Kings, Lynn, UK.

Meunier-Salaun, M. C., S. A. Edwards, and S. Robert. 2001. Effect of dietary fibre on the behaviour and health of the restricted-fed sow. Anim. Feed Sci. and Technol. 90:53–69.

Mowat, D., C. A. Watson, R. W. Mayes, H. Kelly, H. Browning, and S. A. Edwards. 2001. Herbage intake of growing pigs in an outdoor organic production system. Proc. Br. Soc. Anim. Sci. p. 169.

NOP. 2010. Agricultural marketing service 7, code of federal regulations, Part 205. National Organic Program, USDA, Washington, DC.

NRC. 2012. Nutrient requirements of swine. 11th rev. ed. Natl. Acad. Press, Washington, DC.

Petkevicius, S., K. E. Bach Knudsen, P. Nansen, and D. Murrell. 2001. The effect of dietary carbohydrates with different digestibility on the populations of Oesophagostomum dentatum in the intestinal tract of pigs. Parasitology 123:315–324.

Presto Åkerfeldt, M., J. Nihlstrand, M. Neil, N. Lundeheim, H. K. Andersson, and A. Wallenbeck. 2019. Chicory and red clover silage in diets to finishing pigs—influence on performance, time budgets and social interactions. Org. Agric. 9:127–138.

Riart, G. R. 2002. Some aspects of outdoor pig production in Argentina. PhD. Diss. Univ. of Aberdeen, Aberdeen, Scotland, UK.

Rivera Ferre, M. G., S. A. Edwards, R. W. Mayes, I. Riddoch, and B.F de Hovell. 2001. The effects of season and level of concentrate on the voluntary intake and digestibility of herbage by outdoor sows. Anim. Sci. 72:501–510.

Rudolph, G., S. Hörtenhuber, D. Bochicchio, G. Butler, R. Brandhofer, S. Dippel, J.-Y. Dourmad, S. Edwards, B. Früh, M. Meier, A. Prunier, C. Winckler, W. Zollitsch, and C. Leeb. 2018. Effect of three husbandry systems on environmental impact of organic pigs. Sustainability 10:3796.

Sehested, J., K. K. Breinhild, K. Søegaard, L. Vognsen, H. H. Hansen, J. A. Fernandez, V. Danielsen, and V. F. Kristensen, 1999. Use of n-alkanes to estimate grass intake and digestibility in sows. In: H. Dove, and S. W. Coleman, editor, Nutritional ecology of herbivores. Satellite Symposium: Emerging Techniques for Studying the Nutrition of Free Ranging Herbivores. Cambridge University Press, Cambridge, UK.

Sehested, J., K. Søegaard, V. Danielsen, A. Roepstorff, and J. Monrad. 2004. Grazing with heifers and sows alone or mixed: Herbage quality, sward structure and animal weight gain. Livest. Prod. Sci. 88:223–238.

Smith, J., C. Gerrard, and J. Hermansen. 2014. Improved contribution of local feed to support 100% organic feed supply to pigs and poultry: Synthesis report. ICOPP Consortium. http://orgprints.org/28078.

Soil Association. 2020. Soil association standards farming and growing. Ver. 18.3. Soil Association, Bristol, UK.

Srednicka-Tober, D., M. Baranski, C. Seal, R. Sanderson, C. Benbrook, H. Steinshamn, J. Gromadzka-Ostrowska, E. Rembialkowska, K. Skwarlo-Sonta, M. Eyre, G. Cozzi, M. K. Larsen, T. Jordon, U. Niggli, T. Sakowski, P. C. Calder, G. C. Burdge, S. Sotiraki, A. Stefanakis, H. Yolcu, S. Stergiadis, E. Chatzidimitriou, G. Butler, G. Stewart, and C. Leifert. 2016. Composition differences between organic and conventional meat: a systematic literature review and meta-analysis. Br. J. Nutr. 115:994–1011.

Thacker, P. A., and R. N. Kirkwood, editors. 1990. Non-traditional feed sources for use in swine production. Butterworths, Stoneham, MA, US.

USDA. 1990. Organic foods production act of 1990. Title 21 of Food, Agriculture, Conservation, and Trade Act of 1990. USDA, Washington, DC.

van Milgen, J., J. Noblet, M. Étienne, A. Valancogne, S. Dubois, and J. Dourmad. 2007. InraPorc: A model and decision support tool for the nutrition of growing pigs and sows. J. Anim. Sci. 85(Suppl. 1):440.

Wallenbeck, A., M. Rundgren, and M. Presto. 2014. Inclusion of grass/clover silage in diets to growing/finishing pigs–influence on performance and carcass quality. Acta Agric. Scand. Sect. A Anim. Sci. 64:145–153.

Watson, C. A., T. Atkins, S. Bento, A. C. Edwards, and S. A. Edwards. 2003. Appropriateness of nutrient budgets for environmental risk assessment: A case study of outdoor pig production. Eur J. Agron. 20:117–126.

Williams, A. G., E. Audsley, and D. L. Sandars. 2006. Determining the environmental burdens and resource use in the production of agricultural and horticultural commodities. Defra Research Project IS0205. Cranfield University, Bedford, UK.

# 25    Swine Nutrition and Pork Quality

Jason K. Apple

## Introduction

A number of factors affect pork quality, with swine genetics, preslaughter handling, and pork carcass chilling having the greatest impacts. However, there is a growing body of literature suggesting that dietary modifications may offset the negative effects of genetic predisposition and/or pig handling on pork quality, and, in some cases, actually enhance pork quality traits of well-handled pigs of good quality genotypes.

Pork quality typically refers to the measurement of muscle pH, color, firmness, marbling or intramuscular fat (IMF) content, shelf-life, and cooked pork palatability. Yet, domestic and international consumers may include environmental, ethical, and animal welfare aspects of pork production to the definition of pork quality, whereas today's pork processors may opt to include fat color, firmness, and composition, as well as nutrient composition (i.e., quantity and quality of protein, vitamins, and minerals) and microbiological safety (i.e., absence of pathogenic bacteria and chemical residues) to their definition of pork quality. The focus of this chapter will be on how dietary modifications affect technological (i.e., pH, color, firmness, water-holding capacity, etc.) and eating qualities of pork, as well as emphasis on the dietary manipulation of pork fat and fresh belly quality, and the ramifications of "natural" vs. "conventional" pig production systems on pork quality.

### Manipulating Postmortem Metabolism and Pork Quality

The cessation of blood circulation at exsanguinations restricts oxygen availability to muscles, thereby shifting muscle metabolism from aerobic metabolism of lipids to anaerobic metabolism of muscle glycogen reserves. The end-product of anaerobic postmortem muscle metabolism is lactic acid, and, accumulation of lactic acid in muscle causes postmortem muscle pH to decline from approximately 7.1 to 7.3 to an ultimate pH value of 5.4 to 5.7. There are three basic pork quality defects associated with abnormal postmortem pH decline: 1) pale, soft, and exudative (PSE) pork; 2) dark, firm, and dry (DFD) pork; and 3) red, soft, and exudative pork (RSE).

When muscle pH declines rapidly (falls below 5.8 to 6.0 within the first hour postmortem) due to excessive lactic acid accumulation, the high intramuscular acidity coupled with elevated muscle temperature results in muscle protein denaturation, and the ultimate development of PSE pork. Rapid postmortem glycogenolysis can be attributed to a number of factors, including genetic predisposition, preslaughter stress and activation of the sympathetic-adrenal medullar axis, or both acting in concert. Conversely, when muscle glycogen reserves are low, the accumulation of lactic

*Sustainable Swine Nutrition*, Second Edition. Edited by Lee I. Chiba.
© 2023 John Wiley & Sons Ltd. Published 2023 by John Wiley & Sons Ltd.

acid is greatly curtailed, leading to ultimate pH values in excess of 6.0 being established early in the postmortem period and the development of DFD pork. The low antemortem muscle glycogen reserves are typically the synergistic effect of a combination of stressors increasing energy demands prior to slaughter.

Muscle from pigs with a mutation in the protein kinase adenosine monophosphate-activated γ-2-subunit gene (Milan et al. 2000), often referred to as the Rendement Napole (RN-) gene (Monin and Sellier, 1985), have abnormally high concentrations of muscle glycogen; therefore, excessive intra-muscular lactic acid accumulation can cause ultimate muscle pH values to decline to isoelectric point (pI) of 5.0 to 5.3 (pH value where the net charge of the contractile proteins is zero; Hamm, 1986). Even though the color of RSE pork is virtually normal at the pI, the equal number of positive and negative charges of the contractile proteins at the pI causes proteins to be attracted to each other, thereby reducing the amount of water attracted to myofibrillar proteins, which leads to excessive moisture losses (i.e., decreased water-holding capacity) and reduced protein functionality of fresh pork.

### Preslaughter Feed Withdrawal

Obviously, manipulating preslaughter muscle glycogen reserves could lead to improvements in fresh pork color and water-holding capacity. Withholding feed from pigs for 16 to 24 h before slaughter effectively reduces *longissimus* muscle (LM) glycogen concentrations and elevates initial (45-min) and ultimate (24-h) muscle pH values, which equates to improvements in the water-holding capacity (WHC) of fresh pork (Partanen et al. 2007; Sterten et al. 2009). On the other hand, feed withdrawal periods of less than 16 h before slaughter have no appreciable effects on muscle glycogen reserves, postmortem pH decline, or WHC (Faucitano et al. 2006). Interestingly, Bidner et al. (2004) discovered that withholding feed from RN- pigs for 30 to 60 h prior to slaughter did not reduce LM glycogen levels, indicating that preslaughter fasting, alone, was not an effective method of manipulating the abnormally high muscle glycogen reserves in pigs with this genetic mutation.

Preslaughter fasting periods of 16 to 48 h have been shown to produce darker (lower L* values), more desirable colored pork (Sterten et al. 2009); however, traditionally feed withdrawal periods of 16 to 36 h do not affect the redness (a* values) or yellowness (b* values) of fresh pork (Faucitano et al. 2006; Partanen et al. 2007). In addition to the beneficial effects on pork quality, preslaughter feed withdrawal also decreases pig mortality during transportation and lairage, reduces carcass contamination with pathogenic bacteria in response to puncture of the gastrointestinal tract, and produces less waste to be rendered and/or disposed (Murray et al. 2001).

### Glycogen-Reducing Diets

In Europe, research has shown that feeding high-fat (17 to 19%), high-protein (19 to 25% CP) diets formulated with very low levels (< 5%) of digestible carbohydrates will effectively reduce total glycogen – including proglycogen and macroglycogen – concentrations in pork LM at slaughter (Rosenvold et al. 2003; Bee et al. 2006). More importantly, 45-min, but not 24-h, postmortem muscle pH was elevated, and drip loss percentages were decreased in response to feeding these glycogen-reducing diets (Rosenvold et al. 2001, 2002). The effects of glycogen-reducing diets on pork color are inconsistent (Rosenvold et al. 2001, 2002; Bee et al. 2006), and there is little information to indicate cooked pork palatability is altered by feeding pigs high-protein/high-fat/low-carbohydrate diets prior to slaughter (Rosenvold et al. 2001; Bee et al. 2006). It should be noted that Leheska et al. (2002) fed pigs an ultra-high protein (33.7% CP), high-fat (19.6% crude fat) diet for 2 to 14 d before slaughter and failed to observe an effect of this glycogen-reducing diet on muscle glycogen reserves, pH decline, or any fresh pork quality attribute; however, the dietary fiber content wasn't nearly as great in the lone US study as it was in the European studies, and may explain the discrepancy in results of Leheska et al. (2002) to the European research.

### Other Dietary Modifications to Alter Postmortem Metabolism

#### Magnesium Supplementation
Both long-term (Otten et al. 1992) and short-term (D'Souza et al. 1998) magnesium (Mg) supplementation has also been shown to effectively reduce the stress response of pigs prior to slaughter and reduce the incidence of PSE pork (D'Souza et al. 1998, 2000). More importantly, it has been regularly demonstrated that supplementing swine diets with Mg for as little as a week before slaughter will improve the WHC of fresh pork, regardless of Mg source (Figure 25.1; Apple, 2007). Furthermore, research has shown that fresh pork color can be improved by either long- (Apple et al. 2000) or short-duration (D'Souza et al. 2000) Mg supplementation.

#### Creatine Monohydrate Supplementation
Creatine is produced naturally in the liver, kidneys and pancreas from glycine, arginine, and methionine, and it increases the bioavailability of phosphocreatine for cellular ATP production. Based on human research, Berg and Allee (2001) hypothesized that supplementing swine diets with creatine monohydrate might reduce the incidence of PSE pork by increasing muscle phosphocreatine levels

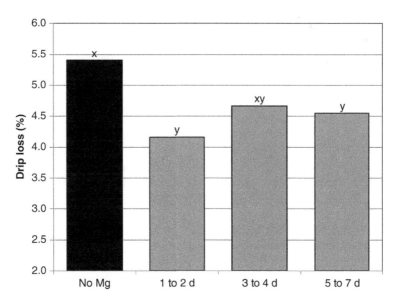

**Figure 25.1** Meta-analysis of the effect of duration of preslaughter magnesium (Mg) supplementation on longissimus drip loss percentage (adapted from Apple, 2007). [x,y]Bars lacking a common letter differ ($P < 0.05$).

and sparing muscle glycogen. However, the effects of supplemental creatine monohydrate on pork quality are erratic and unreliable. For instance, Young et al. (2005) reported that initial LM pH values were increased by 5 d of creatine monohydrate supplementation, but others have reported no effect of preslaughter creatine-supplementation on initial or ultimate LM pH or fresh pork color (James et al. 2002; Rosenvold et al. 2007). In fact, two studies have shown a couple of studies have shown that pork became lighter (higher L* values), less red (lower a* values) and more yellow (higher b* values) in color in response to supplemental creatine monohydrate (Stahl et al. 2001; Young et al. 2005). On the other hand, there is evidence that feeding finishing pigs supplemental creatine will reduce LM drip loss percentages (James et al. 2002; Young et al. 2005).

## Dietary Modifications to Increase Intramuscular Fat Content

The IMF content of pork plays an important role in consumers' perceptions of cooked pork tenderness, flavor, and juiciness (Lonergan et al. 2007), and it is suggested that an IMF content between 2.5 and 3.0% is necessary for consumer acceptability of cooked pork (DeVol et al. 1988). Moreover, the majority of the United States' import partners prefer pork with IMF contents of at least 4% (NPPC marbling score of 4). The adaption of leaner swine genotypes by US pig producers over the past two decades has reduced IMF contents to as low as 1.0% (Gil et al. 2008); thus the onus of increasing the marbling/IMF content in today's pork is squarely on the shoulders of swine nutrition.

### *Dietary Protein and Amino Acid Effects on Pork IMF*

Increasing the crude protein (**CP**) and/or lysine levels in swine diets has been repeatedly shown to increase pork carcass lean yields (Grandhi & Cliplef 1997) and the moisture content of pork (Friesen et al. 1994; Goerl et al. 1995); however, IMF content is reduced by increasing the proportions of CP and/or lysine (Goodband et al. 1990, 1993; Grandhi & Cliplef 1997). In fact, Goerl et al. (1995) found that the IMF content of the LM decreased 71.3% as the dietary CP level increased from 10 to 22%, and linear reductions in LM marbling scores were observed as dietary lysine content increased from 0.54 to 1.04% (Friesen et al. 1994) and 0.8 to 1.4% (Johnston et al. 1993).

On the other hand, one strategy shown to effectively increase the IMF content of pork is reducing the CP and/or lysine content of swine diets (Figure 25.2). When dietary CP levels were reduced in grower and finisher diets, IMF was increased 13.7 to 64.7% (Wood et al. 2004; Teye et al. 2006a), whereas reducing the dietary lysine content in diets of growing-finishing pigs elevated IMF content 66.7 to 136.8% (Blanchard et al. 1999; Cameron et al. 1999). However, long-term exposure of pigs to CP- or lysine-deficient diets will have severe detrimental effects on gain and feed conversion efficiency.

Interestingly, two studies from the University of Illinois (Cisneros et al. 1996; Bidner et al. 2004) demonstrated that feeding lysine-reduced diets over the last 5 to 6 weeks of the finishing period had virtually no detrimental impact on performance and increased IMF concentrations but not to the same extent as other studies. When pigs were fed reduced CP diets supplemented with synthetic amino acids (**AA**) to meet SID requirements for lysine, threonine, methionine, tryptophan, isoleucine, and valine during the finishing phase, Apple et al. (2017) reported that LM marbling scores and IMF content increased linearly with decreasing dietary CP with minimal effects on live pig performance. In a subsequent experiment, feeding AA-supplemented reduced-CP diets during the

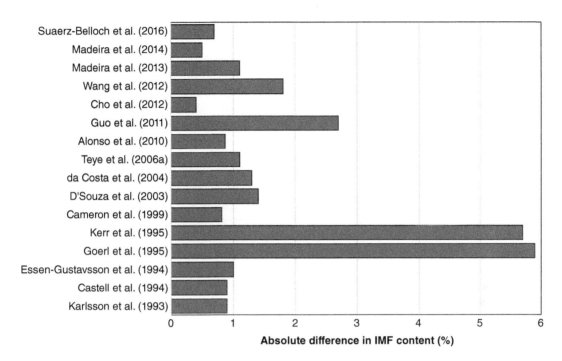

**Figure 25.2**   Effect of reduced crude protein (CP), with or without amino acid supplementation, on longissimus muscle (LM) intramuscular fat (IMF) content. Values are the absolute difference in reported LM IMF between means for the lowest dietary CP level and the "control" dietary CP level.

entirety of the growing-finishing period again resulted in linear increases in IMF content with decreasing dietary CP in the LM of barrows, but not gilts, and the only reductions in performance were noted in the treatment with the greatest CP reductions during the last finishing phase when ractopamine hydrochloride (RAC) was included the diets (Young et al. 2013).

### Dietary Energy Content and Sources on Pork IMF

#### Feed and Energy Intake

Even though restricting feed intake of finishing pigs does not affect muscle pH (Cameron et al. 1999; Lebret et al. 2001) or fresh pork (Cameron et al. 1999; Sterten et al. 2009), feed restrictions of 75 to 80% *ad libitum* have been repeatedly shown to reduce IMF content between 8 and 27%, (Lebret et al. 2001; Daza et al. 2007). Interestingly, reducing the energy density of swine finishing diets does not alter IMF or any other fresh pork quality attribute (Lee et al. 2002), and there is no evidence that the grain source incorporated in the diet affects marbling scores (Carr et al. 2005b; Sullivan et al. 2007).

#### Fats and Oils

Fats and oils have been used for decades to increase the caloric density of swine diets and aid in feed pelleting; however, the effects of dietary fat levels and/or sources on pork IMF content are inconsistent. Miller et al. (1990) reported that LM marbling scores were reduced by feeding pigs diets formulated with 10% sunflower or canola oil, and Myer et al. (1992) noted a linear reduction in marbling with increasing dietary canola oil. Conversely, Apple et al. (2008b) observed that LM IMF content increased with increasing dietary corn oil, whereas LM IMF content was increased by approximately 25% by feeding swine diets formulated with 5% beef tallow (Eggert et al. 2007). For the most part, however, few studies have demonstrated an effect of dietary fats/oil on marbling scores (Engel et al. 2001; Apple et al. 2008a) or IMF content (Morel et al. 2006).

Conjugated linoleic acid (CLA) refers to a mixture of positional (c8, c10, c9, c11, c10, c12, and c13) and genometric (*cis/cis, cis/trans, trans/cis,* and *trans/trans*) conjugated isomers of linoleic acid. Most synthetic CLA sources contain approximately 65% CLA isomers, primarily comprised of the *cis*-9/*trans*-11 and *trans*-10/*cis*-12 isomers, and, since July, 2009, CLA is being marketed under the trade-name of Lutalin (BASF SE, Ludwigschafen, Germany) for inclusion into swine and broiler diets. More importantly, supplementation of swine diets with CLA appears to increase LM marbling scores and/or IMF content. Dugan et al. (1999) reported that marbling scores were increased 11.3% by CLA supplementation. Furthermore, Joo et al. (2002), Sun et al. (2004), and Martin et al. (2008b) reported increases in IMF content ranging from as little as 12% to as much as 44%, whereas Wiegand et al. (2001) demonstrated that incorporating 0.75% CLA in diets increased the IMF content in the LM of halothane-negative, halothane-carriers, and halothane-positive pigs by 17.8, 19.2, and 16.6%, respectively.

### *Vitamin A Supplementation*

A derivative of vitamin A, retinoic acid, is involved in the regulation of adipose cell differentiation and proliferation (Pairault et al. 1988); thus, retinoic acid deficiencies may directly increase intramuscular adipocyte proliferation and IMF content. In fact, feeding cattle vitamin A-deficient diets increased LM marbling scores and/or IMF content without affecting performance or carcass composition (Oka et al. 1998; Gorocica-Buenfil et al. 2007). D'Souza et al. (2003) also demonstrated that feeding vitamin A-restricted diets during the grower and finisher phases increased IMF content by almost 54%, whereas Olivares et al. (2009) noted that feeding diets supplemented with 100,000 IU of vitamin A actually increased IMF in pigs with the genetic propensity for IMF, but not in high-lean genotypes. The evidence indicates that both vitamin A deficiencies and supranutritional dietary inclusion of vitamin A can increase pork IMF/marbling is promising; however, dietary inclusion levels, feeding durations, and interactive effects with other feedstuffs and feed additives are largely unknown, especially when fed to growing-finishing pigs.

### *Ractopamine Hydrochloride*

Ractopamine (RAC) hydrochloride (Paylean; Elanco Animal Health, Greenfield, IN) is a feed supplement that redirects nutrients to improved pig growth performance and carcass lean muscle yields (Apple et al. 2007b). More importantly, the beneficial effects of RAC are not typically accompanied by changes in muscle pH, firmness, or WHC of fresh pork (Apple et al. 2007a). Even though there is great deal of anecdotal misinformation concerning the effect of RAC on LM marbling scores and/ or IMF content, the fact is that most research has not detected differences in marbling scores between RAC- and control-fed pigs (Carr et al. 2005b; Patience et al. 2009; Rincker et al. 2009), and, in many cases, marbling scores were actually increased by including 5 (Watkins et al. 1990), 10 (Apple et al. 2008a), or 20 mg/kg (Carr et al. 2009) of RAC in swine finishing diets.

## Dietary Modifications on Pork Fat Quality

The fatty acids in pork muscle and fat may be obtained from *de novo* lipogenesis from nonlipid substrates and the absorption of exogenous fatty acids from the pig's diet. Glucose from the digestion of corn and barley, for example, will increase the proportion of saturated fatty acids (SFA) at the expense of polyunsaturated fatty acids (PUFA) derived from the oil fraction of the grain sources (Lampe et al. 2006). However, as indicated previously, fat is routinely incorporated in swine diets to increase the energy density of the diet and reduce the proportion of dietary cereal grains, especially corn.

### Dietary Fat Source and Pork Fat Quality

The quality of the dietary fat source included in swine diets is dependent on a number of factors, including iodine value (**IV**; a measure of the chemical unsaturation of the fat), titre (temperature at which a fat is completely solid), and melting point (temperature at which a fat is completely lique-fied). Highly saturated fat sources, like tallow and lard, will have IV of 30 to 70 g of I/100 g of fat, titres of 32 to 47°C, and melting points of 45 to 50°C. Conversely, unsaturated oils from soybeans, canola, corn, sunflower, and safflower seeds will typically have IV greater than 100 g of I/100 g of fat, titre of less than 30°C, and melting points of 20°C, or less. In addition, fat digestibility by the pig apparently increases as the SFA content of the fat source decreases (Averette Gatlin et al. 2005). Therefore, the fatty acid composition of pork fat depots will typically reflect the quality (i.e., fatty acid composition) of the fat and/or oil formulated in the diets. For example, the fatty acid composition of fat from pigs fed tallow will tend to have lower proportions of PUFA and lower IV, whereas feeding oils will elevate the proportions of PUFA at the expense of SFA and monounsaturated fatty acids (MUFA; Table 25.1).

Although there are apparent health benefits associated with the consumption of PUFA, increasing the polyunsaturation of pork fat depots leads to the development of soft fat. According to Whittington et al. (1986), pork fat with a $18:2_{n-6}$ content greater than 15% is classified as soft; thus, it is not surprising that feeding pigs polyunsaturated fat sources high in $18:2_{n-6}$ content will also cause soft fat (Miller et al. 1990; Myer et al. 1992) and pork bellies (Apple et al. 2007a, 2008b). Soft fat and pork bellies cause carcass handling and fabrication difficulties, reduced bacon yields, unattractive products, reduced shelf-life, and, more importantly, discrimination by domestic consumers and export partners. Research has shown that belly thickness and firmness increased as the IV of the dietary fat source decreased from 80 to 20 (Averette Gatlin et al. 2003); so, feeding animal fats does not appear to depress fat and belly firmness/hardness (Engel et al. 2001) as severely as feeding plant oils. Interestingly, Shackelford et al. (1990) reported that bacon from pigs fed sunflower, safflower, or canola oil received much lower sensory scores for crispiness, chewiness, saltiness, flavor, and overall palatability than bacon from pigs fed diets devoid of added fat and diets formulated with tallow. Moreover, Teye et al. (2006b) observed that pigs fed soybean oil-formulated diets product soft bacon and a greater number of low-quality, soft bacon slices.

There is growing evidence that between 50 and 60% of the change in the fatty acid composition of pork fat caused by manipulating the dietary fat source, inclusion level, or both, occurs during the first 14 to 35 d on the particular dietary fat source and diminishes with a longer time on feed (Wiseman & Agunbiade 1998). Apple et al. (2009a, b) reported that the fatty acid profile of the LM, subcutaneous fat, and carcass composite samples was altered substantially within the first 17.4 kg of BW gain, with IV of pork fat increasing almost 12 points during the first feeding phase in pigs

**Table 25.1**    Percentage change between dietary fat sources and no added fat controls in fatty acid composition of subcutaneous fat and longissimus muscle.

| Item | Beef tallow[a] | Poultry fat[a] | Soybean oil[a] | Corn oil[b] | Canola oil[c] | Yellow grease[d] | Choice white grease[e] |
|---|---|---|---|---|---|---|---|
| **Subcutaneous fat** | | | | | | | |
| Total SFA | −3.43 | −8.33 | −12.93 | −14.40 | −22.25 | −0.60 | −5.86 |
| Palmitic acid (16:0) | −4.60 | −6.42 | −11.70 | −12.80 | −20.62 | −1.67 | −5.68 |
| Stearic acid (18:0) | −3.38 | −11.79 | −15.45 | −17.91 | −28.36 | +1.99 | −8.79 |
| Total MUFA | +4.80 | +1.69 | −11.17 | −12.47 | +2.93 | −2.44 | +2.42 |
| Oleic acid (18:1 *cis-9*) | +3.86 | −0.76 | −10.81 | −10.41 | +4.85 | −4.12 | +2.51 |
| Total PUFA | −8.94 | +10.46 | +50.26 | +91.50 | +86.73 | +8.38 | +10.73 |
| Linoleic acid (18:2 *n-6*) | −9.96 | +10.43 | +46.53 | +97.28 | +44.44 | +6.62 | +10.63 |
| Linolenic acid (18:3 *n-3*) | −1.37 | +13.70 | +180.82 | +30.19 | +162.50 | +13.56 | +4.76 |
| Iodine value | −1.35 | +5.84 | +18.43 | +23.72 | — | — | — |
| **Longissimus muscle** | | | | | | | |
| Total SFA | −0.98 | −2.12 | −2.83 | −0.78 | — | +3.34 | −9.71 |
| Palmitic acid (16:0) | −2.09 | −2.26 | −3.41 | −0.65 | — | +2.28 | −2.41 |
| Stearic acid (18:0) | +1.05 | −1.40 | −2.27 | −1.26 | — | +6.36 | −2.49 |
| Total MUFA | −1.45 | −2.55 | −6.58 | −5.70 | — | −4.87 | −0.33 |
| Oleic acid (18:1 *cis-9*) | −0.52 | −2.23 | −5.35 | −2.81 | — | −6.85 | −0.16 |
| Total PUFA | +13.95 | +22.64 | +40.93 | +31.82 | — | +6.86 | +16.37 |
| Linoleic acid (18:2 *n-6*) | +13.67 | +24.58 | +45.13 | +41.60 | — | +6.20 | +14.65 |
| Linolenic acid (18:3 *n-3*) | +17.14 | +20.00 | +148.57 | +18.18 | — | +12.31 | +5.56 |
| Iodine value | +3.35 | +5.70 | +10.83 | +5.71 | — | — | — |

[a] Fat sources included at 5% as-fed basis (Apple et al. 2009a,b).
[b] Fat source included at 4% as-fed basis (Apple et al. 2008c).
[c] Fat source included at 5% as-fed basis (Myer et al. 1992).
[d] Fat source included at 4% as-fed basis (Averette Gatlin et al. 2002).
[e] Fat source included at 4% as-fed basis (Engel et al. 2001).

fed 5% soybean oil. In addition, Anderson et al. (1972) found that the half-life of linolenic acid (**18:3** *n-3*) in pork subcutaneous fat was almost 300 d; thus, the economic savings associated with increased efficiency during the grower phases when high levels of fat are traditionally fed may also cause irreparably damage to the fat quality of pigs at slaughter. Moreover, it is doubtful that removing all fat from the late-finishing diet or replacing an unsaturated fat source with tallow or a hydrogenated fat source will have dramatic effects on pork fat quality (Apple et al. 2009b).

### *By-products of Biofuel Production*

In an attempt to reduce reliance on fossil fuels, considerable efforts have been made to generate biofuels from renewable resources. Ethanol production from corn, as well as sorghum and wheat, has increased substantially over the past 10 yr, leading to substantial supplies of dried distillers' grains with solubles (**DDGS**), which can be incorporated in swine diets. The crude fat content of DDGS ranges between 10 and 15% (Rausch & Belyea 2006), and the fat from DDGS has a high proportion of unsaturated fatty acids; thus, it is not surprising that feeding pigs high levels of DDGS increases the PUFA content and IV of pork subcutaneous fat (Xu et al. 2010a,b; White et al. 2009). More recently, Graham et al. (2014) also reported increased IV of jowl, belly, and backfat as dietary inclusion of DDGS increased from 0 to 40%; yet, pork fat IV of pigs fed DDGS containing 9.4

or 12.1% was substantially greater than that from pigs fed DDGS containing 5.4% crude fat, regardless of dietary inclusion level. Benz et al. (2010) reported that IV of backfat, jowl subcutaneous fat, and belly fat increased 1.3, 1.6, and 2.2 g I/100 g fat, respectively, for every 10% increase in dietary DDGS.

The degree of polyunsaturation of fat in fresh pork bellies increases linearly with the amount of DDGS included in swine diets (Whitney et al. 2006; Xu et al. 2010a, b; White et al. 2009), which leads to soft, pliable, undesirable fresh pork bellies (Whitney et al. 2006; Weimer et al. 2008; Widmer et al. 2008). Moreover, Weimer et al. (2008) reported greater fat-lean separation with increased dietary DDGS, and Xu et al. (2010b) noted linear reductions in bacon fattiness and tenderness with increased dietary inclusion rates of DDGS, even though DDGS did not affect the crispiness, flavor, or overall acceptability of cooked bacon (Widmer et al. 2008; Xu et al. 2008b). It should be noted that replacing DDGS during the finishing phase(s) with saturated fat sources (e.g. like tallow, butter, and palm kernel oil) does not alleviate the deleterious effects of DDGS on fatty acid composition, fresh belly quality characteristics, or cooked bacon palatability attributes (Browne et al. 2013a, b; Lee et al. 2013; Shircliff et al. 2019).

Any new or recycled animal fat or vegetable oil can be reacted with an alcohol, in the presence of a catalyst, to produce methyl esters more commonly referred to as biodiesel. Crude glycerol/glycerin is a by-product of biodiesel production, and, like DDGS, has received a great deal of interest as an energy source in swine diets. Mourot et al. (1994) and Della Casa et al. (2009) found that including 5 to 10% crude glycerol in swine diets increased the proportion of 18:1 cis-9 and all MUFA in pork backfat, whereas Mourot et al. (1994) and Lammers et al. (2008) observed reductions in 18:2 n-6 in subcutaneous fat and muscle. More importantly, the reduction in fat polyunsaturation associated with feeding glycerol resulted in firmer pork bellies (Schieck et al. 2010).

### Conjugated Linoleic Acid

Supplementing swine diets with CLA routinely increases the proportions of SFA, especially palmitic (**16:0**) and stearic acid (**18:0**), in both pork fat (Dugan et al. 2003; Sun et al. 2004) and muscle (Eggert et al. 2001; Sun et al. 2004; Martin et al. 2008b). Moreover, several studies have shown that dietary CLA reduced the 18:1 cis-9 and total MUFA composition of pork fat (Dugan et al. 2003; Sun et al. 2004) and muscle (Sun et al. 2004; Martin et al. 2008b); yet, others have reported that proportions of 18:1 cis-9 and MUFA in pork fat and muscle were not changed (Eggert et al. 2001) or even elevated (Joo et al. 2002) by CLA-supplementation. There are conflicting results, however, on the impact of CLA on PUFA composition. With the exception of Thiel-Cooper et al. (2001), Averette Gatlin et al. (2002) and Martin et al. (2009), who reported that CLA-supplementation increased fresh LM 18:2 n-6 content, most research has demonstrated that supplementing swine diets with CLA either reduces (Joo et al. 2002; Sun et al. 2004) or has no effect (Eggert et al. 2001) on the PUFA composition of pork lean and fat. More importantly, the increases in SFA and concomitant decreases in PUFA lead to reductions in IV (Eggert et al. 2001; Averette Gatlin et al. 2002; Larsen et al. 2009) and firmer pork fat (Dugan et al. 2003) and fresh bellies (Eggert et al. 2001; Larsen et al. 2009).

### Other Dietary Modifications on Pork Fat Quality

Feeding diets with depressed CP/lysine levels increases the proportions of SFA and MUFA at the expense of PUFA (Wood et al. 2004; Teye et al. 2006a) and reduces IV of fresh pork LM (Grandhi

& Cliplef 1997). More recently, feeding growing-finishing pigs reduced-CP diets supplemented with food-grade AA to meet the SID requirements for lysine, threonine, tryptophan, methionine, isoleucine, valine, and histidine reduced the proportions of total PUFA, particularly linoleic acid ($18:2\,n\text{-}6$), and increased the proportions of total MUFA, especially oleic acid ($18:1\,cis\text{-}9$), in jowl subcutaneous fat and LM (Young et al. 2014, Cook et al. 2015a), as well as in the subcutaneous and intermuscular fat layers of fresh pork bellies (Cook et al. 2015b). Interestingly, the isocaloric diets fed in the Young et al. (2014) met ME requirements for growing-finishing swine, whereas diets in the Cook et al. (2015a, b) experiment were formulated to meet NE requirements; thus, feeding AA-supplemented reduced-CP diets enhanced de novo lipid synthesis regardless of energy formulation basis.

Furthermore, formulating swine diets with high-oil corn will increase the proportions of $18:2\,n\text{-}6$ and all PUFA in pork fat (Rentfrow et al. 2003), whereas feeding pigs high-linoleic acid corn (Della Casa et al. 2010) or high-oleic acid high-oil corn (Rentfrow et al. 2003) obviously increase the concentrations of $18:2\,n\text{-}6$ and $18:1\,cis\text{-}9$, respectively, in fresh pork. On the other hand, Skelley et al. (1975) reported that backfat from corn-fed pigs was firmer than barley-fed pigs, and Sather et al. (1999) observed that fat from wheat-fed pigs was harder than fat from corn-fed pigs; however, belly firmness does not appear to be affected by the dietary grain source (Skelley et al. 1975; Carr et al. 2005b).

Fat firmness and fresh pork belly firmness are reduced substantially as feed intake is reduced to 70 to 85% of ad libitum (Haydon et al. 1989). The reductions in fat/belly firmness are likely in response to increased proportions of $18:1\,cis\text{-}9$, all MUFA, and $18:2\,n\text{-}6$ in the pork fat and muscle caused by restricting feed intake (Wood et al. 1996; Daza et al. 2007). Moreover, Daza et al. (2007) demonstrated that the activity of lipogenic enzymes was depressed when feed was restricted during the grower phase and the activity of these enzymes declines in the finisher phase, even in pigs with ad libitum access to feed.

Fat from pigs fed RAC has an elevated proportion of PUFA at the expense of SFA and the observed increases in polyunsaturation can be attributed to greater deposition of $18:2\,n\text{-}6$ and $18:3\,n\text{-}3$ (Carr et al. 2005b; Apple et al. 2008a). Moreover, Mills et al. (1990) demonstrated that RAC depressed de novo lipogenesis in pork fat; thus, the fatty acid content of pork subcutaneous fat would be a direct reflection of the fatty acid composition of the late-finishing diet, especially if the diet was formulated with any added fat/oil. Interestingly, RAC does not appear to alter pork belly firmness (Carr et al. 2005a; Apple et al. 2007a; Scramlin et al. 2008), bacon quality (Scramlin et al. 2008), or bacon palatability (Jeremiah et al. 1994).

Carnitine is a vitamin-like compound involved in the transportation of long-chain fatty acids across the inner mitochondrial membrane for $\beta$-oxidation within the mitochondria; thus, it was not surprising that supplementing swine diets with L-carnitine could improve growth efficiency and carcass leanness (Owen et al. 2001; Chen et al. 2008), without affecting fresh pork quality (Apple et al. 2008b). However, Apple et al. (2008c) reported that supplementing swine diets with L-carnitine reduced the proportions of PUFA in backfat samples and increased the proportion of MUFA in the LM – but not the IV of either pork fat or muscle. This led the authors to hypothesize that L-carnitine may stimulate the desaturation of $18:2\,n\text{-}6$ into $18:1\,cis\text{-}9$ via $\Delta^9$ desaturase, leading to the observed reductions in PUFA with the concomitant increases in MUFA.

### Dietary Modifications on Lipid and Color Stability

It would be expected that any dietary modification that increases the PUFA content of pork would also increase the susceptibility of pork to lipid oxidation. In fact, feeding swine diets containing canola oil (Leskanich et al. 1997), fish oil (Leskanich et al. 1997), soybean oil (Morel et al. 2006), linseed oil (Morel et al. 2006), or high-oil corn (Guo et al. 2006) increased thiobarbituric acid reactive substances

(TBARS) value during refrigerated storage. Thus, a great deal of research has focused on either the feeding of antioxidants, especially vitamin E, as well as stimulating endogenous antioxidative enzymes via mineral supplementation.

### Vitamin E

Vitamin E ($\alpha$-tocopherol) is a radical-chain-breaking antioxidant that protects cell membrane integrity (Morrissey et al. 1993) and retards lipid and myogoblin oxidation (Faustman et al. 1989), especially during refrigerated storage and/or retail display. So, it is not surprising that incorporating supranutritional levels of vitamin E in swine growing-finishing diets may be the most widely recognized nutritional modification to improve pork quality.

Research has repeatedly shown that feeding pigs an additional 100 to 200 mg/kg of dl-$\alpha$-tocopherol acetate effectively delays the onset of lipid oxidation in fresh whole-muscle pork cuts (Monahan et al. 1994; Boler et al. 2009) and ground pork (Phillips et al. 2001; Boler et al. 2009), as well as precooked (Guo et al. 2006) and cured pork products (Coronado et al. 2002). Furthermore, because lipid oxidation is positively correlated to pigment oxidation, vitamin E supplementation of cattle finishing diets not only slows the rate of discoloration but actually improves the color stability of fresh beef (Faustman et al. 1989), and early studies indicated that supplementing swine diets with dl-$\alpha$-tocopherol acetate also improved fresh pork color stability (Monahan et al. 1994). However, the vast majority of research has failed to detect any benefits of elevating the levels of vitamin E in swine diets with either dl-$\alpha$-tocopherol acetate (Phillips et al. 2001; Guo et al. 2006) or the natural-occurring stereoisomer, d-$\alpha$-tocopheryl acetate (Boler et al. 2009), on fresh pork color or color stability during refrigerated storage.

### Vitamin C

Vitamin C has antioxidant properties, and pigs typically produce adequate amounts of this water-soluble vitamin from D-glucose in the liver; yet, subcutaneous injections of vitamin C immediately before slaughter have been shown to reduce the incidence of PSE carcasses (Cabadaj et al. 1983), whereas feeding ascorbic acid within 4 h of slaughter produced darker, redder pork (Peeters et al. 2006). However, neither short-term (Ohene-Adjei et al. 2001; Pion et al. 2004) nor long-term vitamin C supplementation (Eichenberger et al. 2004; Gebert et al. 2006) affected pork color or WHC. Furthermore, there is no evidence to suggest that supplementing swine diets with vitamin C improves the oxidative stability of LM lipids during storage or retail display (Gebert et al. 2006), and, in fact, both Ohene-Adjei et al. (2001) and Eichenberger et al. (2004) reported that feeding pigs diets formulated with elevated levels of vitamin C actually increased TBARS values of LM chops during refrigerated storage. Circulating ascorbic acid levels return quickly to basal levels soon after dietary supplementation ends (Pion et al. 2004); therefore, timing of vitamin C supplementation appears to be critical to eliciting any beneficial effects of vitamin C on pork quality.

### Mineral Supplementation

#### Selenium
Selenium (Se) is a component of the endogenous antioxidant enzyme glutathione peroxidase, and a number of studies have shown that serum glutathione peroxidase activity is increased by

supplementing swine diets with either sodium selenite or a selenium-yeast compound (Mahan et al. 1999; Zhan et al. 2007). Yet, the increased glutathione activity associated with supplemental Se does not equate into changes in fresh color and WHC (Mahan et al. 1999; Wolter et al. 1999) or, more importantly, lipid stability during storage of fresh pork (Wolter et al. 1999; Han & Thacker 2006).

*Manganese*

Manganese (Mn) and Mg are both divalent, transition metal cations that may be interchangeable in several biological functions; however, Mn is a required for the activation of superoxide dismutase, which is involved in the breakdown of superoxide free radicals; so, it was not surprising that TBARS values of fresh LM chops were reduced by dietary Mn supplementation (Apple et al. 2005), and the LM from pigs fed diets supplemented with 350 mg/kg of Mn were less discolored after 2 and 4 d of simulated retail display than the LM from nonsupplemented pigs (Sawyer et al. 2007). Additional benefits of supplementing swine diets with Mn include increased LM pH and visual color scores and reduced L* values of fresh pork LM (Apple et al. 2005, 2007c). Although the Mn results appear promising, more research is warranted because very little is known about the effects of Mn supplementation on pork palatability, and all of the research to date has been conducted by a single research group.

*Vitamin-Trace Mineral Removal*

In a survey of swine nutritionists that represented over 2 million sows, Flohr et al. (2016) reported that fat-soluble (A, D, E, and K) and water-soluble vitamins (biotin, choline, folic acid, niacin, pantothenic acid, pyridoxine, riboflavin, thiamin, and vitamins $B_{12}$ and C) were included in finishing diets at rates of 180 to 670% and 70 to 380% of NRC (2012) requirements, respectively. In addition, finishing diets inclusion rates for copper and manganese were 22.0 to 28.1 and 9.3 to 12.6 times NRC (2012) requirements, respectively, whereas the other trace minerals (iron, iodine, selenium, and zinc) incorporated in finishing diets ranged between 1.4 and 3.0 times NRC (2012) recommendations. There is growing sentiment, however, that reducing vitamins and minerals, especially during the late-finishing period, will reduce not only production costs but also excretion of phosphorus and other mineral elements into the environment (McGlone, 2000). Even though Ma et al. (2012) reported that LM a* values and drip losses decreased linearly as the duration of mineral depletion increased from 0 to 6 weeks before slaughter, there is little evidence to suggest that removing all vitamins and trace minerals during the late-finishing phase will affect fresh pork color, marbling, or firmness, as well as WBSF values (Mavromichalis et al. 1999; Choi et al. 2001; Shelton et al. 2004). The lone disadvantage of vitamin and trace mineral removal may be that TBARS values were elevated during refrigerated storage by vitamin/trace mineral removal (Choi et al. 2001), whereas fortifying finishing diets with 150, 200, and 250% of the NRC (1998) vitamin and trace minerals during the last few weeks before slaughter substantially reduced TBARS values during as much as 3 weeks of refrigerated storage (Choi et al. 2001; Hamman et al. 2001).

**Dietary Modifications on Cooked Pork Palatability**

Even though fresh pork color is the single most important factor in the purchasing decision of a consumer, their perception of cooked pork palatability will impact whether or not they purchase pork again. Therefore, it is vitally important that palatability is either not affected by, or improved by, any dietary modification.

*Crude Protein/Lysine*

Warner–Bratzler shear force (WBSF) values of cooked LM chops increased almost 23% as CP content increased from 10 to 22% in the finishing diet (Goerl et al. 1995). Furthermore, Goodband et al. (1990, 1993) reported linear increases in WBSF values in cooked LM and SM chops as dietary lysine levels were elevated from 0.6 to 1.4%, whereas Apple et al. (2004) observed a linear increase in WBSF values as the lysine-to-energy ratio of the late-finishing diet increased from 1.7 (0.56 to 0.59% lysine) to 3.1 g/Mcal (1.02 to 1.08% lysine). Goodband et al. (1990) also noted decreased sensory panel myofibrillar and overall tenderness scores, whereas Castell et al. (1994) reported decreased pork flavor scores, with increasing dietary lysine levels; yet, for the most part, elevating dietary lysine levels in swine diets does not affect juiciness, flavor intensity, or tenderness scores of cooked pork (Goodband et al. 1993; Castell et al. 1994; Grandhi & Cliplef 1997).

*Energy Content and Sources*

Reducing the energy density in diets of growing-finishing swine does not affect the palatability of pork (Lee et al. 2002); however, LM chops from pigs fed ad libitum received greater tenderness scores and had lower WBSF values than pork from pigs fed at 75 (Cameron et al. 1999), 80 (Blanchard et al. 1999), or 82 ad libitum (Ellis et al. 1996), even though total and soluble muscle collagen contents (Wood et al. 1996; Lebret et al. 2001) and myofibrillar fragmentation index (an indicator of postmortem proteolysis; Cameron et al. 1999) were not affected by dietary intake. Furthermore, a number of studies have shown that pork from pigs with ad libitum access to grower-finisher diets was rated higher for pork flavor (Blanchard et al. 1999; Cameron et al. 1999), flavor-liking, juiciness, and overall acceptability (Ellis et al. 1996; Cameron et al. 1999) by trained sensory panelists.

The cereal grain source included in swine diets can create palatability differences. For example, cooked chops from wheat-fed pigs received higher flavor scores than chops from sorghum-fed pigs (McConnell et al. 1975), whereas LM chops from pigs fed a 33:67% or 67:33% mixture of yellow and white corns received higher juiciness and flavor scores than chops from pigs fed yellow corn or white corn and barley, respectively (Lampe et al. 2006). Furthermore, McConnell et al. (1975) reported that the LM from wheat-fed pigs had lower WBSF values and higher tenderness scores than the LM from sorghum-fed pigs, and Robertson et al. (1999) noted that sensory panelists rated LM chops from barley-fed pigs more tender than chops from pigs fed corn or barley with triticale. Conversely, WBSF values were similar among pigs fed yellow corn, white corn, wheat, barley, or triticale (Skelley et al. 1975; Carr et al. 2005b; Lampe et al. 2006; Sullivan et al. 2007), and neither trained sensory (Carr et al. 2005b; Sullivan et al. 2007) nor consumer panels (Jeremiah et al. 1999) detected tenderness, juiciness, flavor, or overall acceptability differences in response to varying the cereal grain source included in swine growing and/or finishing diets.

Feeding canola oil and/or fish oil has been shown to impart more abnormal odors and off-flavors, thereby reducing the overall acceptability of cooked pork (Miller et al. 1990; Tikk et al. 2007). However, there is no effect of dietary fat source on WBSF values (Miller et al. 1990; Engel et al. 2001; Apple et al. 2008a, b) or sensory panelists' evaluations of tenderness, juiciness, or flavor intensity (Miller et al. 1990; Engel et al. 2001; Tikk et al. 2007). Neither WBSF values nor palatability ratings of cooked LM chops have been affected by feeding pigs DDGS- (Whitney et al. 2006; Widmer et al. 2008; Xu et al. 2010b) or glycerol-formulated diets (Lammers et al. 2008; Della Casa et al. 2009). In addition, it does not appear that supplementing swine diets with CLA affects WBSF values (Dugan et al. 1999), palatability scores (Dugan et al. 1999, 2003; Wiegand et al. 2001; Larsen et al. 2009), flavor attributes (Averette Gatlin et al. 2006), or flavor volatile profiles (Martin et al. 2008a) of cooked pork LM chops or bacon.

### Compensatory Gain

Compensatory growth is the accelerated growth rate that occurs in pigs having ad libitum access to feed after a period of restricted feeding. The increase in protein degradation during the period of restricted feed intake does not appear to decrease during the realimentation period, which led Kristensen et al. (2002) to hypothesize that high antemortem proteolytic activity would lead to a more rapid post-mortem muscle tenderization. Interestingly, both Kristensen et al. (2002) and Therkildsen et al. (2002) found that the activities of both $\mu$- and m-calpain, but not calpastatin, were increased in the LM from pigs afforded ad libitum access to feed following a period of restricted feed intake, and Therkildsen et al. (2002) noted that the longer the period of ad libitum feed intake prior to slaughter, the greater the $\mu$-calpain activity. Total collagen content does not appear to be affected by compensatory growth, but there is evidence that the proportion of soluble collagen in the LM is actually increased by feed restriction followed by ad libitum feed intake (Kristensen et al. 2002, 2004; Therkildsen et al. 2002). However, WBSF values and sensory panel tenderness scores were only improved in pork from pigs with confirmed compensatory growth (Kristensen et al. 2002); in other words, in studies where the length or severity of the feed restriction was insufficient to cause a significant reduction in the growth rate, the period of ad libitum intake had little to no effect on cooked pork palatability, especially tenderness (Therkildsen et al. 2002; Kristensen et al. 2004; Heyer & Lebret 2007).

### Vitamin D$_3$

Because of the well-established association between calcium and meat tenderness, it is generally accepted that increasing muscle calcium concentrations will increase postmortem calpain degradation of the cytoskeletal proteins and improve cooked meat tenderness. Vitamin D is involved in intercellular calcium mobilization and regulation, and feeding supranutritional levels of vitamin D$_3$ to feedlot cattle was shown to elevate blood and muscle calcium levels and, more importantly, improve cooked beef tenderness (Swanek et al. 1999). Even though plasma and muscle calcium concentrations were increased over 125% by supplementing swine finishing diets with vitamin D$_3$ (Wiegand et al. 2002; Lahucky et al. 2007), neither pork WBSF values (Wiegand et al. 2002; Swigert et al. 2004; Wilborn et al. 2004), sensory panel tenderness ratings (Swigert et al. 2004; Wilborn et al. 2004), nor any other palatability attribute (Swigert et al. 2004; Wilborn et al. 2004) have been altered by supplemental vitamin D$_3$. Interestingly, there is evidence that indicate that supplementing swine diets with supranutritional levels of vitamin D$_3$ can cause improvements in fresh pork quality, including increased initial and ultimate muscle pH values, subjective color scores, and LM a* values, along with reductions in L* values and drip loss percentages (Wilborn et al. 2004; Swigert et al. 2004; Lahucky et al. 2007).

### Ractopamine Hydrochloride

Even though RAC does not affect cooked pork juiciness or flavor (Carr et al. 2005a, b; Patience et al. 2009; Rincker et al. 2009), there are a number of studies demonstrating that feeding as little as

5 mg/kg of RAC will increase WBSF values of cooked pork (Patience et al. 2009; Rincker et al. 2009) and reduce sensory panel tenderness scores (Carr et al. 2005a, b; Patience et al. 2009). Now it might seem odd that the section on the effects of RAC on cooked pork palatability is as mentioned previously, but it is generally accepted that to optimize the effect of RAC on live pig performance, the lysine content of the late-finishing diet must be increased to as high 1.0% (Webster et al. 2007). Thus, the increases in WBSF are a result of the increased dietary lysine content of RAC-supplemented diets and not dietary RAC per se. Moreover, Xiong et al. (2006) found that WBSF values of LM chops from RAC-fed pigs were greater than those of control-fed pigs after 2, 4, and 7 d of postmortem aging, but WBSF did not differ when chops were aged 10, 14, or 21 d postmortem, suggesting that incorporating RAC in swine finishing diets delays the postmortem tenderization process.

**Organic Pork Production**

Over the past decade, "natural" pork niche markets have seen considerable growth in response to consumers' perception that pork is more nutritious and wholesale if it came from free-ranging pigs in environmentally friendly facilities. More importantly, these consumers are willing to pay premiums for "natural" pork products, even though in their opinion it may be less palatable, if the "small," family-farmer shares similar viewpoints. According to Honeyman et al. (2006), as many as 750,000 pigs are slaughtered annually for the US. "natural" pork market, but demand still exceeds supplies at this time. There are a number of issues confronting the "natural" pork markets, with the cost of production and year-round supply of pork as the two most pressing concerns. Stender et al. (2009) estimated that the cost of production ranged between $66.50 and $99.00/cwt of pork produced, whereas Honeyman et al. (2006) observed that the production costs associated with producing "organic" pork were 400 to 500% greater than those for "natural" pork production. In addition, farrowing outdoors during the Winter is a necessity for year-round pork production, but colder temperatures lead to greater death losses prior to weaning (Stender et al. 2009), lower growth rates, and greater feed consumption, which equates to reductions in production efficiency (Bee et al. 2004; Gentry et al. 2004).

Honeyman et al. (2006) indicated that pork quality was extremely important to the sustainability of the niche "natural" pork market. The initial and ultimate pH values of pork are similar between indoor- and outdoor-raised pigs; however, meta-analysis of 33 published studies indicated that outdoor pig production reduces the WHC of fresh pork (Figure 25.3). In fact, several studies have demonstrated that rearing pigs outdoors causes increases in drip loss percentages (Gentry et al. 2002a, b; Galián et al. 2008) and expressible moisture percentages (Kim et al. 2009) when compared to indoor-reared pigs. Meta-analysis also indicated that pork from pigs reared outdoors is darker (lower L* values) and tended to be more yellow (greater b* values) than pork from pigs reared indoors (Figure 25.3). Interesting, several researchers reported increases of 20%, or more, in the redness (a*) values of pork from pigs raised outdoors (Gentry et al. 2002b, 2004; Hoffman et al. 2003), but meta-analysis indicated that a* values did not differ between pork from indoor- and outdoor-reared pigs.

Most research, as well as the meta-analysis, has demonstrated that pork from pigs conventionally reared indoors has more marbling (Gentry et al. 2002a, 2004) and IMF (Högberg et al. 2002; Hoffman et al. 2003; Bee et al. 2004) than pork from their contemporaries reared outdoors. Moreover, neither the meta-analysis (Figure 25.4) nor the results of a number of studies detected an effect of indoor vs. outdoor pig production on WBSF values of cooked pork (Hoffman et al. 2003; Oksbjerg et al. 2005; Galián et al. 2008). It is not surprising that pork from outdoor-reared pigs

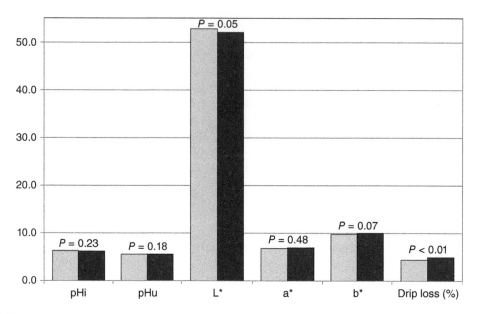

**Figure 25.3** Meta-analysis of studies (n = 33) comparing the fresh pork quality attributes of pigs raised indoors (light-colored bars) and outdoors (dark-colored bars). The quality attributes include initial muscle pH (pHi), ultimate muscle pH (pHu), instrumental color coordinates (L*, a*, and b*), and drip loss percentage.

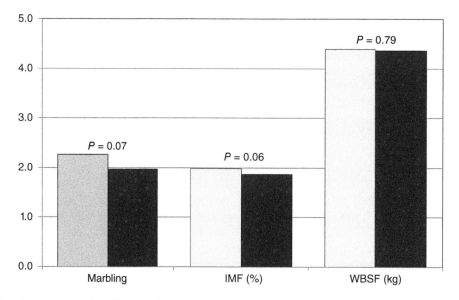

**Figure 25.4** Meta-analysis of studies (n = 33) comparing the marbling scores, intramuscular fat percentage (IMF), and Warner–Bratzler shear force (WBSF) values of pork from pigs raised indoors (light-colored bars) and outdoors (dark-colored bars).

received lower juiciness ratings than pork from pigs reared indoors (Gentry et al. 2002a; Jonsäll et al. 2002), but sensory panelists have deemed pork from pigs reared indoors and outdoors similar for flavor, tenderness, and overall acceptability (Jonsäll et al. 2002; Gentry et al. 2002a, b; Kim et al. 2009).

The subcutaneous fat from outdoor-reared pigs typically has lower proportions of 16:0, 18:0, and/ or all SFA and greater proportions of $18:2_{n-6}$, $18:3_{n-3}$, and/or all PUFA than fat from indoor-reared pigs (Bee et al. 2004; Hansen et al. 2006). Bee et al. (2004), Oksbjerg et al. (2005), and González and Tejeda (2007) reported that the proportions of 16:0 were lower in the LM from outdoor-reared pigs, and it is apparent that rearing pigs outdoors reduces the proportions of 18:0 and all SFA in the LM (**Table 25.2**). Kim et al. (2009) observed an 11.7% reduction in $18:1^{cis9}$ and a 12.2% reduction in all MUFA in the LM from free-range pigs, but outdoor rearing typically only reduces the MUFA composition of the LM approximately 5% (refer to Table 25.2). It was interesting that both Högberg et al. (2002) and Galián et al. (2008) observed robust decreases in the PUFA composition of the LM from outdoor-reared pigs because raising pigs outdoors appears to increase the proportions of $18:2_{n-6}$, $18:3_{n-3}$, and the sum of all SFA (Oksbjerg et al. 2005; González & Tejeda 2008; Kim et al. 2009). The elevation in unsaturated fatty acids in the LM from outdoor-reared pigs is representative of the PUFA content of the polar (phospholipid) lipid fraction associated with the cell membrane because, as previously pointed out, the IMF content (neutral lipid fraction) of the LM is usually substantially lower in muscle of outdoor-reared pigs.

The greatest majority of the research on "natural" pork production has been conducted in Europe, and there is limited information concerning free-range pork production being generated in the United States. In most of the mentioned studies, pigs were reared on alfalfa pastures, so there are no studies detailing the effects of other forages on pork quality, especially cooked pork palatability. Gentry et al. (2002b) noted that pork from pigs born and reared outdoors had greater discoloration during retail display; otherwise, there is almost no information concerning lipid stability during

**Table 25.2** Comparison of the fatty acid composition of the longissimus muscle intramuscular fat from natural, outdoor-reared pigs to indoor, conventionally reared pigs (results are presented as a percentage change from the indoor pigs).

| Reference | [a]16:0 | [a]18:0 | [a]ΣSFA | [a]18:1[cis9] | [a]ΣMUFA | [a]18:2$_{n-6}$ | [a]18:3$_{n-3}$ | [a]ΣPUFA |
|---|---|---|---|---|---|---|---|---|
| Högberg et al. (2001) | −1.9 | −9.8* | −5.3* | +3.1* | +2.1 | +6.9 | +19.1* | +1.5 |
| Högberg et al. (2002) | +3.4 | −0.4 | +2.5 | +5.4* | +7.3* | −30.9* | −47.6* | −30.1* |
| Hoffman et al. (2003) | +0.5 | −21.7* | −8.3 | −5.4 | −6.9 | +44.8* | +9.9 | +33.7* |
| Bee et al. (2004) | −1.9* | −0.6 | −2.2 | −3.3* | −3.6* | +29.1* | +45.2* | +31.3* |
|  | −1.0* | +0.2 | −1.2* | −6.7* | −6.8* | +27.2* | +55.6* | +33.1* |
| Högberg et al. (2004) | +3.0 | +2.2 | +2.8 | −3.2 | −1.9 | −7.2 | 0.0 | −6.9 |
|  | +0.4 | +4.9 | +1.8 | 0.0 | −1.4 | +4.0 | +8.0 | +4.0 |
|  | −0.8 | +2.4 | +0.3 | −0.7 | −0.9 | +8.4 | +27.6* | +10.1 |
| Oksbjerg et al. (2005) | −1.3 | −1.6 | −0.6 | −2.9* | −3.5* | +12.5 | +13.3 | +12.4 |
|  | −3.8* | −24.0* | −6.9* | −7.5* | −8.4* | +39.7 | −2.2 | +38.2* |
| González & Tejeda (2007) | −5.8* | −10.6 | −7.2* | +3.1 | +2.8* | +31.0* | +94.1* | +28.3 |
| Galián et al. (2008) | +17.4* | −10.6* | +10.7* | −4.2* | −4.1* | −14.7 | −3.7 | −21.4 |
| Kim et al. (2009) | −7.8* | −0.6 | −6.9* | −11.7* | −12.2* | +54.1* | +31.9 | +58.3* |

[a] 16:0 = palmitic acid; 18:0 = stearic acid; ΣSFA = sum of all saturated fatty acids; 18:1cis9 = oleic acid; ΣMUFA = sum of all monounsaturated fatty acids; 18:2n−6 = linoleic acid; 18:3n−3 = linolenic acid; and ΣPUFA = sum of all polyunsaturated fatty acids.
* An asterisk indicates that the change differed from the indoor, conventionally reared pigs.

storage, especially considering the increases in pork polyunsaturation typically noted in free-range pigs. It is likely that animal welfare legislation will necessitate both swine nutritionists and meat scientist to step back 50 yr to test nutritional modifications for improving the efficiency and quality of "natural" pork in coming years.

# References

Alonso, V., M. del Mar Campo, L. Provincial, P. Roncalés, and J. A. Beltrán. 2010. Effect of protein level in commercial diets on pork meat quality. Meat Sci. 85:7–14.

Anderson, D. B., R. G. Kauffman, and N. J. Benevenga. 1972. Estimate of fatty acid turnover in porcine adipose tissue. Lipids 7:488–489.

Apple, J. K. 2007. Effects of nutritional modifications on the water-holding capacity of fresh pork: a review. J. Anim. Breeding Gen. 124(Suppl. 1):43–58.

Apple, J. K., C. V. Maxwell, B. E. Bass, J. W. S. Yancey, R. L. Payne, and J. Thomson. 2017. Effects of reducing dietary CP levels and replacement with crystalline AA on growth performance, carcass composition, and fresh pork quality of finishing pigs fed ractopamine hydrochloride. J. Anim. Sci. 95:4971–4985.

Apple, J. K., C. V. Maxwell, D. C. Brown, K. G. Friesen, R. E. Musser, Z. B. Johnson, and T. A. Armstrong. 2004. Effects of dietary lysine and energy density on performance and carcass characteristics of finishing pigs fed ractopamine. J. Anim. Sci. 82:3277–3287.

Apple, J. K., C. V. Maxwell, B. deRodas, H. B. Watson, and Z. B. Johnson. 2000. Effect of magnesium mica on performance and carcass quality of growing-finishing swine. J. Anim. Sci. 78:2135–2143.

Apple, J. K., C. V. Maxwell, D. L. Galloway, S. Hutchison, and C. R. Hamilton. 2009a. Interactive effects of dietary fat source and slaughter weight in growing-finishing swine: I. Growth performance and longissimus muscle fatty acid composition. J. Anim. Sci. 87:1407–1422.

Apple, J. K., C. V. Maxwell, D. L. Galloway, C. R. Hamilton, and J. W. S. Yancey. 2009b. Interactive effects of dietary fat source and slaughter weight in growing-finishing swine: II. Fatty acid composition of subcutaneous fat. J. Anim. Sci. 87:1423–1440.

Apple, J. K., C. V. Maxwell, J. T. Sawyer, B. R. Kutz, L. K. Rakes, M. E. Davis, Z. B. Johnson, and T. A. Armstrong. 2007a. Interactive effect of ractopamine and dietary fat source on quality characteristics of fresh pork bellies. J. Anim. Sci. 85:2682–2690.

Apple, J. K., C. V. Maxwell, B. R. Kutz, L. K. Rakes, J. T. Sawyer, Z. B. Johnson, T. A. Armstrong, S. N. Carr, and P. D. Matzat. 2008a. Interactive effect of ractopamine and dietary fat source on pork quality characteristics of fresh pork chops during simulated retail display. J. Anim. Sci. 86:2711–2722.

Apple, J. K., P. J. Rincker, F. K. McKeith, T. A. Armstrong, S. N. Carr, and P. D. Matzat. 2007b. Review: meta-analysis of the ractopamine response in finishing swine. Prof. Anim. Sci. 23:179–196.

Apple, J. K., W. J. Roberts, C. V. Maxwell, Jr., L. K. Rakes, K. G. Friesen, and T. M. Fakler. 2007c. Influence of dietary inclusion level of manganese on pork quality during retail display. Meat Sci. 75:640–647.

Apple, J. K., W. J. Roberts, C. V. Maxwell, Jr., C. B. Boger, K. G. Friesen, L. K. Rakes, and T. M. Fakler. 2005. Influence of dietary manganese source and supplementation level on pork quality during retail display. J. Muscle Foods 16:207–222.

Apple, J. K., J. T. Sawyer, C. V. Maxwell, J. C. Woodworth, J. W. S. Yancey, and R. E. Musser. 2008b. Effect of L-carnitine supplementation on the performance and pork quality traits of growing-finishing swine fed three levels of corn oil. J. Anim. Sci. 86(E-Suppl. 2):37.

Apple, J. K., J. T. Sawyer, C. V. Maxwell, J. C. Woodworth, J. W. S. Yancey, and R. E. Musser. 2008c. Effect of L-carnitine supplementation on the fatty acid composition of subcutaneous fat and LM from swine fed three levels of corn oil. J. Anim. Sci. 86(E-Suppl. 2):38.

Averette Gatlin, L., M. T. See, J. A. Hansen, and J. Odle. 2003. Hydrogenated dietary fat improves pork quality of pigs from two lean genotypes. J. Anim. Sci. 81:1989–1997.

Averette Gatlin, L., M. T. See, D. K. Larick, X. Lin, and J. Odle. 2002. Conjugated linoleic acid in combination with supplemental dietary fat alters pork quality. J. Nutr. 132:3105–3112.

Averette Gatlin, L., M. T. See, K. K. Larick, and J. Odle. 2006. Descriptive flavor analysis of bacon and pork loin from lean-genotype gilts fed conjugated linoleic acid and supplemental fat. J. Anim. Sci. 84:3381–3386.

Averette Gatlin, L., M. T. See, and J. Odle. 2005. Effects of chemical hydrogenation of supplemental fat on relative apparent lipid digestibility in finishing swine. J. Anim. Sci. 83:1890–1898.

Bee, G., C. Biolley, G. Guex, W. Herzog, S. M. Lonergan, and E. Huff-Lonergan. 2006. Effects of available dietary carbohydrate and preslaughter treatment on glycolytic potential, protein degradation, and quality traits of pig muscles. J. Anim. Sci. 84:191–203.

Bee, G., G. Guex, and W. Herzog. 2004. Free-range rearing of pigs during the winter: adaptations in muscle fiber characteristics and effects on adipose tissue composition and meat quality traits. J. Anim. Sci. 82:1206–1218.

Benz, J. M., S. K. Linneen, M. D. Tokach, S. S. Dritz, J. L. Nelssen, J. M. DeRouchey, R. D. Goodband, R. C. Sulabo, and K. J. Prusa. 2010. Effects of dried distillers grains with solubles on carcass fat quality of finishing pigs. J. Anim. Sci. 88:3666–3682.

Berg, E. P., and G. L. Allee. 2001. Creatine monohydrate supplemented in swine finishing diets and fresh pork quality: I. A controlled laboratory experiment. J. Anim. Sci. 79:3075–3080.

Bidner, B. S., M. Ellis, D. P. Witte, S. N. Carr, and F. K. McKeith. 2004. Influence of dietary lysine level, pre-slaughter fasting, and rendement napole genotype of fresh pork quality. Meat Sci. 68:53–60.

Blanchard, P. J., M. Ellis, C. C. Warkup, B. Hardy, J. P. Chadwick, and G. A. Deans. 1999. The influence of rate of lean and fat tissue development on pork eating quality. Anim. Sci. 68:477–485.

Boler, D. D., S. R. Gabriel, H. Yan, R. Balsbaugh, D. C. Mahan, M. S. Brewer, F. K. McKeith, and J. Killefer. 2009. Effect of different dietary levels of natural-source vitamin E in grow-finish pigs on pork quality and shelf life. Meat Sci. 83:723–730.

Browne, N. A., J. K. Apple, B. E. Bass, C. V. Maxwell, J. W. S. Yancey, T. M. Johnson, and D. L. Galloway. 2013a. Phase-feeding dietary fat to growing-finishing pigs fed dried distillers' grains with soluble: I. Growth performance, pork carcass characteristics, and fatty acid composition of subcutaneous fat depots. J. Anim. Sci. 91:1493–1508.

Browne, N. A., J. K. Apple, C. V. Maxwell, J. W. S. Yancey, T. M. Johnson, D. L. Galloway, and B. E. Bass. 2013b. Phase-feeding dietary fat to growing-finishing pigs fed dried distillers' grains with soluble: II. Fresh belly and bacon quality characteristics. J. Anim. Sci. 91:1509–1521.

Cabadaj, R., J. Pleva, and P. Mala. 1983. Vitamin C in the prevention of PSE and DFD meat. Folia Veterinaria 27:81–87.

Cameron, N. D., J. C. Penman, A. C. Fisken, G. R. Nute, A. M. Perry, and J. D. Wood. 1999. Genotype with nutrition interactions for carcass composition and meat quality in pig genotypes selected for components of efficient lean growth rate. Anim. Sci. 69:69–80.

Carr, S. N., D. N. Hamilton, K. D. Miller, A. L. Schroeder, D. Fernández-Dueñas, J. Killefer, M. Ellis, and F. K. McKeith. 2009. The effect of ractopamine hydrochloride (Paylean®) on lean crcass yields and pork quality characteristics of heavy pigs fed normal and amino acid fortified diets. Meat Sci. 81:533–539.

Carr, S. N., D. J. Ivers, D. B. Anderson, D. J. Jones, D. H. Mowrey, M. B. England, J. Killefer, P. J. Rincker, and F. K. McKeith. 2005a. The effects of ractopamine hydrochloride on lean carcass yields and pork quality characteristics. J. Anim. Sci. 83: 2886–2893.

Carr, S. N., P. J. Rincker, J. Killefer, D. H. Baker, M. Ellis, and F. K. McKeith. 2005b. Effects of different cereal grains and ractopamine hydrochloride on performance, carcass characteristics, and fat quality in late-finishing pigs. J. Anim. Sci. 83:223–230.

Castell, A. G., R. L. Cliplef, L. M. Poste-Flynn, and G. Butler. 1994. Performance, carcass and pork characteristics of castrates and gilts self-fed diets differing in protein content and lysine:energy ratio. Can. J. Anim. Sci. 74:519–528.

Chen, Y. J., I. H. Kim, J. H. Cho, J. S. Yoo, Q. Wang, Y. Wang, and Y. Huang. 2008. Evaluation of dietary l-carnitine or garlic powder on growth performance, dry matter and nitrogen digestibilities, blood profiles and meat quality in finishing pigs. Anim. Feed Sci. Technol. 141:141–152.

Cho, S. B., I. K. Han, Y. Y. Kim, S. K. Park, O. H. Hwang, C. W. Choi, S. H. Yang, K. H. Park, D. Y. Choi, and Y. H. Yoo. 2012. Effect of lysine to digestible energy ratio on growth performance and carcass characteristics in finishing pigs. Asian-Aust. J. Anim. Sci. 25:1582–1587.

Choi, S. C., B. J. Chae, and I. K. Han. 2001. Impacts of dietary vitamins and trace minerals on growth and pork quality in finishing pigs. Asian-Aust. J. Anim. Sci. 14:1444–1449.

Cisneros, F., M. Ellis, D. H. Baker, R. A. Easter, and F. K. McKeith. 1996. The influence of short-term feeding of amino acid-deficient diets and high dietary leucine levels on the intramuscular fat content of pig muscle. Anim. Sci. 63:517–522.

Cook, D. G., J. K. Apple, C. V. Maxwell, A. N. Young, D. L. Galloway, H. J. Kim, and T. C. Tsai. 2015a. Effects of amino acid supplementation of reduced crude protein (RCP) diets formulated on a NE basis on belly fatty acid deposition in swine. J. Anim. Sci. 93(Suppl. 2):166.

Cook, D. G., J. K. Apple, C. V. Maxwell, A. N. Young, D. L. Galloway, H. J. Kim, and T. C. Tsai. 2015b. Effects of amino acid supplementation of reduce crude protein (RCP) diets formulated on a NE basis on the fatty acid composition of the LM and jowl subcutaneous fat. J. Anim. Sci. 93(Suppl. 2):188–189.

Coronado, S. A., G. R. Trout, F. R. Dunshea, and N. P. Shah. 2002. Effect of dietary vitamin E, fishmeal and wood and liquid smoke on the oxidative stability of bacon during 16 weeks' frozen storage. Meat Sci. 62:51–60.

da Costa, N, C. McGillivray, Q. Bai, J. D. Wood, G. Evans, and K. C. Chang. 2004. Restriction of dietary energy and protein induces molecular changes in young porcine skeletal muscles. J. Nutr. 134:2191–2199.

Daza, A., A. I. Rey, D. Menoyo, J. M. Bautista, A. Olivares, and C. J. López-Bote. 2007. Effect of level of feed restriction during growth and/or fattening on fatty acid composition and lipogenic enzyme activity in heavy pigs. Anim. Feed Sci. Technol. 138:61–74.

Della Casa, G., D. Bochicchio, V. Faeti, G. Marchetto, E. Poletti, A. Garavaldi, A. Panciroli, and N. Brogna. 2009. Use of pure glycerol in fattening heavy pigs. Meat Sci. 81:238–244.

Della Casa, G., D. Bochicchio, V. Faeti, G. Marchetto, E. Poletti, A. Rossi, A. Panciroli, A. L. Mordenti, and N. Brogna. 2010. Performance and fat quality of heavy pigs fed maize differing in linoleic acid content. Meat Sci. 84:152–158.

DeVol, D. L., F. K. McKeith, P. J. Bechtel, J. Navakofski, R. D. Shanks, and T. R. Carr. 1988. Variation in composition and palatability traints and relationships between muscle characteristics and palatability in a random sample of pork carcasses. J. Anim. Sci. 66:385–395.

D'Souza, D. N., D. W. Pethick, F. R. Dunshea, J. R. Pluske, and B. P. Mullan. 2003. Nutritional manipulation increases intramuscular fat levels in the *Longissimus* muscle of female finisher pigs. Aust. J. Agric. Res. 54:745–749.

D'Souza, D. N., R. D. Warner, B. J. Leury, and F. R. Dunshea. 1998. The effect of dietary magnesium aspartate supplementation on pork quality. J. Anim. Sci. 76:104–109.

D'Souza, D. N., R. D. Warner, B. J. Leury, and F. R. Dunshea. 2000. The influence of dietary magnesium supplement type, and supplementation dose and duration, on pork quality and the incidence of PSE pork. Aust. J. Agric. Res. 51:51, 185–189.

Dugan, M. E. R., J. L. Aalhus, L. E. Jeremiah, J. K. G. Kramer, and A. L. Schaefer. 1999. The effects of feeding conjugated linoleic acid on subsequent pork quality. Can. J. Anim. Sci. 79:45–51.

Dugan, M. E. R., J. L. Aalhus, D. C. Rolland, and L. E. Jeremiah. 2003. Effects of feeding different levels of conjugated linoleic acid and total oil to pigs on subsequent pork quality and palatability. Can. J. Anim. Sci. 83:713–720.

Eggert, J. M., M. A. Belury, A. Kempa-Steczko, S. E. Mills, and A. P. Schinckel. 2001. Effects of conjugated linoleic acid on the belly firmness and fatty acid composition of genetically lean pigs. J. Anim. Sci. 79:2866–2872.

Eggert, J. M., A. L. Grant, and A. P. Schinckel. 2007. Factors affecting fat distribution in pork carcasses. Prof. Anim. Sci. 23:42–53.

Eichenberger, B., H. P. Pfirter, C. Wenk, and S. Gebert. 2004. Influence of dietary vitamin E and C supplementation on vitamin E and C content and thiobarbituric acid reactive substances (TBARS) in different tissues of growing pigs. Arch. Anim. Nutr. 58:195–208.

Ellis, M., A. J. Webb, P. J. Avery, and P. J. Brown. 1996. The influence of terminal sire genotype, sex, slaughter weight, feeding regime and slaughter-house on growth performance and carcass and meat quality in pigs and on the organoleptic properties of fresh pork. Anim. Sci. 62:521–530.

Engel, J. J., J. W. Smith, II, J. A. Unruh, R. D. Goodband, P. R. O'Quinn, M. D. Tokach, and J. L. Nelssen. 2001. Effects of choice white grease or poultry fat on growth performance, carcass leanness, and meat quality characteristics of growing-finishing pigs. J. Anim. Sci. 79:1491–1501.

Essén-Gustavsson, A. K. L. Karlson, and A. C. Enfält. 1994. Intramuscular fat and muscle fibre lipid contents in halothane-gene-free pigs fed high or low protein diets and its relation to meat quality. Meat Sci. 38:269–277.

Faucitano, L., L. Saucier, J. A. Correa, S. Methot, A. Giguere, A. Foury, P. Mormede, and R. Bergeron. 2006. Effect of feed texture, meal frequency and pre-slaughter fasting on carcass and meat quality, and urinary cortisol in pigs. Meat Sci. 74:697–703.

Faustman, C., R. G. Cassens, D. M. Schaefer, D. R. Buege, S. N. Williams, and K. K. Scheller. 1989. Improvement of pigment and lipid stability in Holstein steer beef by dietary supplementation with vitamin E. J. Food Sci. 54:858–862.

Flohr, J. R., J. M. DeRouchey, J. C. Woodworth, M. D. Tokach, R. D. Goodband, and S. S. Dritz. 2016. A survey of current feeding regimens for vitamins and trace minerals in the US swine industry. J. Swine Health Prod. 24:290–303.

Friesen, K. G., J. L. Nelssen, R. D. Goodband, M. D. Tokach, J. A. Unruh, D. H. Kropf, and B. J. Kerr. 1994. Influence of dietary lysine on growth and carcass composition of high-lean-growth gilts fed from 34 to 72 kilograms. J. Anim. Sci. 72:1761–1770.

Galián, M., A. Poto, M. Santaella, and B. Pelnado. 2008. Effects of the rearing system on the quality traits of the carcass, meat and fat of the Chato Murciano pig. Anim. Sci. J. 79:487–497.

Gebert, S., B. Eichenberger, H. P. Pfirter, and C. Wenk. 2006. Influence of different vitamin C levels on vitamin E and C content and oxidative stability in various tissues and stored *m. longissimus dorsi* of growing pigs. Meat Sci. 73:362–367.

Gentry, J. G., J. J. McGlone, J. R. Blanton, Jr., and M. F. Miller. 2002a. Alternate housing systems for pigs: influences on growth, composition, and pork quality. J. Anim. Sci. 80:1781–1790.

Gentry, J. G., J. J. McGlone, M. F. Miller, and J. R. Blanton, Jr. 2002b. Diverse birth and rearing environment effects on pig growth and meat quality. J. Anim. Sci. 80:1707–1715.

Gentry, J. G., J. J. McGlone, M. F. Miller, and J. R. Blanton, Jr. 2004. Environmental effects on pig performance, meat quality, and muscle characteristics. J. Anim. Sci. 82:209–217.

Gil, M., M. I. Delday, M. Gispert, M. Font i Furnols, C. M. Maltin, G. S. Plastow, and R. Klont. 2008. Relationships between biochemical characteristics and meat quality of *Longissimus thoracis* and *Semimembranosus* muscles in five porcine lines. Meat Sci. 80:927–933.

Goerl, K. F., S. J. Eilert, R. W. Mandigo, H. Y. Chen, and P. S. Miller. 1995. Pork characteristics as affected by two populations of swine and six crude protein levels. J. Anim. Sci. 73:3621–3626.

González, E., and J. F. Tejeda. 2007. Effects of dietary incorporation of different antioxidant extracts and free-range rearing on fatty acid composition and lipid oxidation of Iberian pig meat. Animal 1:1060–1067.

Goodband, R. D., J. L. Nelssen, R. H. Hines, D. H. Kropf, G. R. Stoner, R. C. Thaler, A. J. Lewis, and B. R. Schricker. 1993. Interrelationships between porcine somatotropin and dietary lysine on growth performance and carcass characteristics of finishing swine. J. Anim. Sci. 71:663–672.

Goodband, R. D., J. L. Nelssen, R. H. Hines, D. H. Kropf, R. C. Thaler, B. R. Schricker, G. E. Fitzner, and A. J. Lewis. 1990. The effects of porcine somatotropin and dietary lysine on growth performance and carcass characteristics of finishing swine. J. Anim. Sci. 68:3261–3276.

Gorocica-Buenfil, M. A., F. L. Fluharty, T. Bohn, S. J. Schwartz, and S. C. Loerch. 2007. Effect of low vitamin A diets with high-moisture or dry corn on marbling and adipose tissue fatty acid composition of beef steers. J. Anim. Sci. 85:3355–3366.

Graham, A. B., R. D. Goodband, M. D. Tokach, S. S. Dritz, J. M. DeRouchey, S. Nitikanchana, and J. J. Updike. 2014. The effects of low-, medium-, and high-oil distillers dried grains with solubles on growth performance, nutrient digestibility, and fat quality in finishing pigs. J. Anim. Sci. 92:3610–3623.

Grandhi, R. R., and R. L. Cliplef. 1997. Effects of selection for lower backfat, and increased levels of dietary amino acids to digestible energy on growth performance, carcass merit and meat quality in boars, gilts, and barrows. Can. J. Anim. Sci. 77:487–496.

Guo, Q., B. T. Richert, J. R. Burgess, D. M. Webel, D. E. Orr, M. Blair, A. L. Grant, and D. E. Gerrard. 2006. Effects of dietary vitamin E and fat supplementation on pork quality. J. Anim. Sci. 84:3089–3099.

Guo, Z., R. Tang, W. Wang, D. Liu, and K. Wang. 2011. Effects of dietary protein/carbohydrate ratio on fat deposition and gene expression of peroxisome proliferator activated receptor γ and heart fatty acid-binding protein of pigs. Livst. Sci. 140:111–116.

Hamm, R. 1986. Functional properties of the myofibrillar system and their measurements. In: P. J. Bechtel, editor, Muscle as Food. Academic Press, Inc., Orlando, FL. p. 135–199.

Hamman, L. L., J. G. Gentry, C. B. Ramsey, J. J. McGlone, and M. F. Miller. 2001. The effect of vitamin-mineral nutritional modulation on pork quality of halothane carriers. J. Muscle Food 12:37–51.

Han, Y. K., and P. A. Thacker. 2006. Effect of l-carnitine, selenium-enriched yeast, jujube fruit and hwangto (red clay) supplementation on performance and carcass measurements of finishing pigs. Asian-Austlas. J. Anim. Sci. 19:217–223.

Hansen, L. L., C. Claudi-Magnussen, S. K. Jensen, and H. J. Anderson. 2006. Effect of organic pig production systems on performance and meat quality. Meat Sci. 74:605–615.

Haydon, K. D., T. D. Tanksley, Jr., and D. A. Knabe. 1989. Performance and carcass composition of limit-fed growing-finishing swine. J. Anim. Sci. 67:1916–1925.

Heyer, A., and B. Lebret. 2007. Compensatory growth response in pigs: effects on growth performance, composition of weight gain at carcass and muscle levels, and meat quality. J. Anim. Sci. 85:769–778.

Hoffman, L. C., E. Styger, M. Muller, and T. S. Brand. 2003. The growth and carcass and meat characteristics of pigs raised in a free-range or conventional housing system. S. Afr. J. Anim. Sci. 33:166–175.

Högberg, A., J. Pickova, J. Babol, K. Andersson, and P. C. Dutta. 2002. Muscle lipids, vitamins E and A, and lipid oxidation as affected by diet and RN genotype in female and castrated male Hampshire crossbred pigs. Meat Sci. 60:411–420.

Högberg, A., J. Pickova, P. C. Dutta, J. Babol, and A. C. Bylund. 2001. Effect of rearing system on muscle lipids of gilts and castrated male pigs. Meat Sci. 58:223–229.

Högberg, A., J. Pickova, S. Stern, K. Lundström, and A. C. Bylund. 2004. Fatty acid composition and tocopherol concentrations in muscle of entire male, castrated male and female pigs, reared in an indoor or outdoor housing system. Meat Sci. 68:659–665.

Honeyman, M. S., R. S. Pirog, G. H. Huber, P. J. Lammers, and J. R. Hermann. 2006. The United States pork niche market phenomenon. J. Anim. Sci. 84:2269–2275.

James, B. W., R. D. Goodband, J. A. Unruh, M. D. Tokach, J. L. Nelssen, S. S. Dritz, P. R. O'Quinn, and B. S. Andrews. 2002. Effect of creatine monohydrate on finishing pig growth performance, carcass characteristics and meat quality. Anim. Feed Sci. Technol. 96:135–145.

Jeremiah, L. E., R. O. Ball, J. K. Merrill, P. Dick, L. Stobbs, L. L. Gibson, and B. Uttaro. 1994. Effects of feed treatment and gender on the flavour and texture profiles of cured and uncured pork cuts. I. Ractopamine treatment and dietary protein level. Meat Sci. 31:1–20.

Jeremiah, L. E., A. P. Sather, and E. J. Squires. 1999. Gender and diet influences on pork palatability and consumer acceptance. I. Flavor and texture profiles and consumer acceptance. J. Muscle Food 10:10, 305–316.

Johnston, M. E., J. L. Nelssen, R. D. Goodband, D. H. Kropf, R. H. Hines, and B. R. Schricker. 1993. The effects of porcine somatotropin and dietary lysine on growth performance and carcass characteristics of finishing swine fed to 105 or 127 kilograms. J. Anim. Sci. 71:2986–2995.

Jonsäll, A., L. Johansson, K. Lundström, K. H. Andersson, A. N. Nilsen, and E. Risvik. 2002. Effects of genotype and rearing system on sensory characteristics and preference for pork (*M. longissimus dorsi*). Food Qual. Pref. 13:73–80.

Joo, S. T., J. I. Lee, Y. L. Ha, and G. P. Park. 2002. Effects of dietary conjugated linoleic acid on fatty acid composition, lipid oxidation, color, and water-holding capacity of pork loin. J. Anim. Sci. 80:108–112.

Karlsson, A., A. –C. Enfält, B. Essén-Gustavsson, K. Lundström, L. Rydhmer, and S. Stern. 1993. Muscle histochemical and biochemical properties in relation to meat quality during selection for increased lean tissue growth rate in pigs. J. Anim. Sci. 71:930–938.

Kerr, B. J., F. K. McKeith, and R. A. Easter. 1995. Effect on performance and carcass characteristics of nursery to finisher pigs fed reduced crude protein, amino acid-supplemented diets. J. Anim. Sci. 73:433–440.

Kim, D. H., P. N. Seong, S. H. Cho, J. H. Kim, J. M. Lee, C. Jo, and D. G. Lim. 2009. Fatty acid composition and meat quality of organically reared Korean native black pigs. Livest. Sci. 120:96–102.

Kristensen, L., M. Therkildsen, M. D. Aaslyng, N. Oksbjerg, P. P. Purslow, and P. Ertbjerg. 2004. Compensatory growth improves meat tenderness in gilts but not in barrows. J. Anim. Sci. 82:3617–3624.

Kristensen, L., M. Therkildsen, B. Riis, M. T. Sørensen, N. Oksbjerg, P. P. Purslow, and P. Ertbjerg. 2002. Dietary-inducted changes of muscle growth rate in pigs: effects on in vivo and postmortem muscle proteolysis and meat quality. J. Anim. Sci. 80:2862–2871.

Lahucky, R., I. Bahelka, U. Kuechenmeister, K. Vasickova, K. Nuernberg, K. Ender, and G. Nuernberg. 2007. Effects of dietary supplementation of vitamins $D_3$ and E on quality characteristics of pigs and *longissimus* muscle antioxidative capacity. Meat Sci. 77:264–268.

Lammers, P. J., B. J. Kerr, T. E. Weber, K. Bregendahl, S. M. Lonergan, K. J. Prusa, D. U. Ahn, W. C. Stoffregen, W. A. Dozier, III, and M. S. Honeyman. 2008. Growth performance, carcass characteristics, meat quality, and tissue histology of growing pigs fed crude glycerin-supplemented diets. J. Anim. Sci. 86:2962–2970.

Lampe, J. F., J. W. Mabry, and T. Baas. 2006. Comparison of grain sources for swine diets and their effect on meat and fat quality traits. J. Anim. Sci. 84:1022–1029.

Larsen, S. T., B. R. Wiegand, F. C. Parrish, Jr., J. E. Swan, and J. C. Sparks. 2009. Dietary conjugated linoleic acid changes belly and bacon quality from pigs fed varied lipid sources. J. Anim. Sci. 87:285–295.

Lebret, B., H. Juin, J. Noblet, and M. Bonneau. 2001. The effects of two methods of increasing age at slaughter on carcass and muscle traits and meat sensory quality in pigs. Anim. Sci. 72:87–94.

Lee, C. Y., H. P. Lee, J. H. Jeong, K. H. Baik, S. K. Jin, J. H. Lee, and S. H. Sohnt. 2002. Effects of restricted feeding, low-energy diet, and implantation of trenbolone acetate plus estradiol on growth, carcass traits, and circulating concentrations of insulin-like growth factor (IGF)-I and IGF-binding protein-3 in finishing barrows. J. Anim. Sci. 80:84–93.

Lee, J. W., D. Y. Kil, B. D. Keever, J. Killefer, F. K. McKeith, R. C. Sulabo, and H. H. Stein. 2013. Carcass fat quality of pigs is not improved by adding corn germ, beef tallow, palm kernel oil, or glycerol to finishing diets containing distillers dried grains with solubles. J. Anim. Sci. 91:2426–2437.

Leheska, J. M., D. M. Wulf, J. A. Clapper, R. C. Thaler, and R. J. Maddock. 2002. Effects of high-protein/low-carbohydrate swine diets during the final finishing phase on pork muscle quality. J. Anim. Sci. 80:137–142.

Leskanich, C. O., K. R. Matthews, C. C. Warkup, R. C. Noble, and M. Hazzledine. 1997. The effect of dietary oil containing (*n*-3) fatty acids on the fatty acid, physiochemical, and organoleptic characteristics of pig meat and fat. J. Anim. Sci. 75:673–683.

Lonergan S.M., K. J. Stalder, E. Huff-Lonergan, T. J. Knight, R. N. Goodwin, K. J. Prusa, and D. C. Beitz. 2007. Influence of lipid content on pork sensory quality within pH classification. J. Anim. Sci. 85:1074–1079.

Ma, Y. L., M. D. Lindemann, G. L. Cromwell, R. B. Cox, G. Rentfrow, and J. L. Pierce. 2012. Evaluation of trace mineral source and preharvest deletion of trace minerals from finishing diets for pigs on growth performance, carcass characteristics, and pork quality. J. Anim. Sci. 90:3833–3841.

Madeira, M. S., C. M. Alfaia, P. Costa, P. A. Lopes, J. P. C. Lemos, R. J. B. Bessa, and J. A. M. Prates. 2014. The combination of arginine and leucine supplementation of reduced crude protein diets for boars increases eating quality of pork. J. Anim. Sci. 92:2030–2040.

Madeira, M. S., P. Costa, C. M. Alfaia, P. A. Lopes, R. J. B. Bessa, J. P. C. Lemos, and J. A. M. Prates. 2013. The increase in intramuscular fat promoted by dietary lysine restriction in lean but not in fatty pig genotypes improves pork sensory attributes. J. Anim. Sci. 91:3177–3187.

Mahan, D. C., T. R. Cline, and B. Richert. 1999. Effects of dietary levels of selenium-enriched yeast and sodium selenite sources fed to growing-finishing pigs on performance, tissue selenium, serum glutathione peroxidase activity, carcass characteristics, and loin quality. J. Anim. Sci. 77:2172–2179.

Martin, D., T. Antequera, E. Muriel, A. I. Andrex, and J. Ruiz. 2008a. Oxidative changes of fresh pork loin from pig, caused by dietary conjugated linoleic acid and monounsaturated fatty acids, during refrigerated storage. Food Chem. 111:730–737.

Martin, D., T. Antequera, E. Muriel, A. I. Andrex, and J. Ruiz. 2009 Quantitative changes in the fatty acid profile of lipid fractions of fresh loin from pigs as affected by dietary conjugated linoleic acid and monounsaturated fatty acids during refrigerated storage. J. Food Compos. Anal. 22:102–111.

Martin, D., E. Muriel, E. Gonzalez, J. Viguera, and J. Ruiz. 2008b. Effect of dietary conjugated linoleic acid and monounsaturated fatty acids on productive, carcass and meat quality traits of pigs. Livest. Sci. 117:155–164.

Mavromichalis, I., J. D. Hancock, I. H. Kim, B. W. Seene, D. H. Kropf, G. A. Kennedy, R. H. Hines, and K. C. Behnke. 1999. Effects of omitting vitamin and trace mineral premixes and(or) reducing inorganic phosphorus additions on growth performance, carcass characteristics, and muscle quality in finishing pigs. J. Anim. Sci. 77:2700–2708.

McConnell, J. C., G. C. Skelley, D. L. Handlin, and W. E. Johnston. 1975. Corn, wheat, milo and barley with soybean meal or roasted soybeans on feedlot performance, carcass traits and pork acceptability. J. Anim. Sci. 41:1021–1030.

McGlone, J. J. 2000. Deletion of supplemental minerals and vitamins during the late finishing period does not affect pig weight and feed intake. J. Anim. Sci. 78:2797–2800.

Milan, D., J. T. Jeon, C. Looft, V. Amarger, A. Robic, M. Thelander, C. Rogel-Gaillard, S. Paul, N. Iannuccelli, L. Rask, H. Ronne, K. Lundström, N. Reinsch, J. Gellin, E. Kalm, P. L. Roy, P. Chardon, and L. Andersson. 2000. A mutation in PRAKG3 associated with excess glycogen content in pig skeletal muscle. Science 288:1248–1251.

Miller, M. F., S. D. Schackelford, K. D. Hayden, and J. O. Reagan. 1990. Determination of the alteration in fatty acid profiles, sensory characteristics and carcass traits of swine fed elevated levels of monounsaturated fats in the diet. J. Anim. Sci. 68:1624–1631.

Mills, S. E., C. Y. Liu, Y. Gu, and A. P. Schinckel. 1990. Effects of ractopamine on adipose tissue metabolism and insulin binding in finishing hogs. Interaction with genotype and slaughter weight. Domest. Anim. Endocrinol. 7:215–264.

Monahan, F. J., A. Asghar, J. I. Gray, D. J. Buckley, and P. A. Morrissey. 1994. Effect of oxidized dietary lipid and vitamin E on the colour stability of pork chops. Meat Sci. 37:205–215.

Monin, G., and P. Sellier. 1985 Pork of low technological quality with a normal rate of muscle pH fall in the immediate post-mortem period – the case of the Hampshire breed. Meat Sci. 13:49–63.

Morel, P. C. H., J. C. McIntosh, and J. A. M. Janz. 2006. Alteration of the fatty acid profile of pork by dietary manipulation. Asian-Aust. J. Anim. Sci. 19:431–437.

Morrissey, P. A., P. J. Sheehy, and P. Gaynor. 1993. Vitamin E. Int. J. Vit. Nutr. Res. 63:260–264.

Mourot, J., A. Aumaitre, A. Mounier, P. Peiniaua, and A. C. Françoisb. 1994. Nutritional and physiological effects of dietary glycerol in the growing pigs. Consequences on fatty tissues and post mortem muscular parameters. Livest. Prod. Sci. 38:237–244.

Murray, A., W. Robertson, F. Nattress, and A. Fortin. 2001. Effect of pre-slaughter overnight feed withdrawal on pig carcass and muscle quality. Can. J. Anim. Sci. 81:89–97.

Myer, R. O., J. W. Lamkey, W. R. Walker, J. H. Brendemuhl, and G. E. Combs. 1992. Performance and carcass characteristics of swine when fed diets containing canola oil and copper to alter the unsaturated:saturated ratio of pork fat. J. Anim. Sci. 70:1417–1423.

NRC. 1998. *Nutrient requirements of swine*. 10th rev. ed. Natl. Acad. Press, Washington, DC, US.

NRC. 2012. *Nutrient requirements of swine*. 11th rev. ed. Natl. Acad. Press, Washington, DC, US.

Ohene-Adjei, S., T. Bertol, Y. Hyun, M. Ellis, S. Brewer, and F. K. McKeith. 2001. The effect of dietary supplemental vitamin E and C on odors and color changes in irradiated pork. J. Anim. Sci. 79(Suppl. 1):443.

Oka, A., Y. Maruo, T. Miki, T. Yamasaki, and T. Saito. 1998. Influence of vitamin A on the quality of beef from the Tajima strain of Japanese Black cattle. Meat Sci. 48:159–167.

Oksbjerg, N., K. Strudsholm, G. Lindahl, and J. E. Heremansen. 2005. Meat quality of fully or partly outdoor reared pigs in organic production. Acta Agric. Scand. Section A 55:106–112.

Olivares, A., A. Daza, A. I. Rey, and C. J. Lopez-Bote. 2009. Interactions between genotype, dietary fat saturation and vitamin A concentration on intramuscular fat content and fatty acid composition in pigs. Meat Sci. 82:6–12.

Otten, W., A. Berrer, S. Hartmann, T. Bergerhoff, and H. M. Eichinger. 1992. Effects of magnesium fumarate supplementation on meat quality in pigs. Pages 117–120 in *Proc. 38th International Congress of Meat Science and Technology*, Clermont-Ferrand, France.

Owen, K. Q., H. Ji, C. V. Maxwell, J. L. Nelssen, R. D. Goodband, M. D. Tokach, G. C. Tremblay, and S. I. Koo. 2001. Dietary l-carnitine suppresses mitrochondrial branched-chain keto acid dehydrogenase activity and enhances protein accretion and carcass characteristics of swine. J. Anim. Sci. 79:3104–3112.

Pairault, J., A. Quignard-Boulange, I. Dugail, and F. Lasnier. 1988. Differential effects of retinoic acid upon early and late events in adipose conversion of 3T3 preadipocytes. Exp. Cell Res. 177:27–36.

Partanen, K., H. Siljander-Rasi, M. Honkavaara, and M. Ruusunen. 2007. Effects of finishing diet and pre-slaughter fasting time on meat quality in crossbred pigs.Agric. Food Sci. 16:245–258.

Patience, J. F., P. Shand, Z. Pietraski, J. Merrill, G. Vessie, K. A. Ross, and A. D. Beaulieu. 2009. The effect of ractopamine supplementation at 5 ppm of swine finishing diets on growth performance, carcass composition and ultimate pork quality. Can. J. Anim. Sci. 89:53–66.

Peeters, E., B. Driessen, and R. Geers. 2006. Influence of supplemental magnesium, tryptophan, vitamin C, vitamin E, and herbs on stress response and pork quality. J. Anim. Sci. 84:1827–1838.

Phillips, A. L., C. Faustman, M. P. Lynch, K. E. Govoni, T. A. Hoagland, and S. A. Zinn. 2001. Effect of dietary $\alpha$-tocopherol supplementation on color and lipid stability in pork. Meat Sci. 58:389–393.

Pion, S. J., E. van Heugten, M. T. See, D. K. Larick, and S. Pardue. 2004. Effects of vitamin C supplementation on plasma ascorbic acid and oxalate concentrations and mat quality in swine. J. Anim. Sci. 82:2004–2012.

Rausch, K. D., and R. L. Belyea. 2006. The future of coproducts from corn processing. Appl. Biochem. Biotechnol. 128:47–86.

Rentfrow, G., T. E. Sauber, G. L. Allee, and E. P. Berg. 2003. The influence of diets containing either conventional corn, conventional corn with choice white grease, high oil corn, or high oil high oleic corn on belly/bacon quality. Meat Sci. 64:459–466.

Rincker, P. J., J. Killefer, P. D. Matzat, S. N. Carr, and F. K. McKeith. 2009 The effect of ractopamine and intramuscular fat content on sensory attributes of pork from pigs of similar genetics. J. Muscle Foods 20:79–88.

Rosenvold, K., H. C. Bertram, and J. F. Young. 2007. Dietary creatine monohydrate has no effect on pork quality of Danish crossbred pigs. Meat Sci. 76:160–164.

Rosenvold, K., B. Essén-Gustavsson, and H. J. Andersen. 2003. Dietary manipulation of pro- and macroglycogen in porcine skeletal muscle. J. Anim. Sci. 81:130–134.

Rosenvold, K., H. N. Lærke, S. K. Jensen, A. H. Karlsson, K. Lundström, and H. J. Andersen. 2001. Strategic finishing feeding as a tool in the control of pork quality. Meat Sci. 59:397–406.

Rosenvold, K., H. N. Lærke, S. K. Jensen, A. H. Karlsson, K. Lundström, and H. J. Andersen. 2002. Manipulation of critical quality indicators and attributes in pork through vitamin E supplementation, muscle glycogen reducing finishing feeding and pre-slaughter stress. Meat Sci. 62:485–496.

Sather, A. P., L. E. Jeremiah, and E. J. Squires. 1999. Effects of castration on live performance, carcass yields, and meat quality of male pigs fed wheat or corn based diets. J. Muscle Foods 10:245–259.

Sawyer, J. T., A. W. Tittor, J. K. Apple, J. B. Morgan, C. V. Maxwell, L. K. Rakes, and T. M. Fakler. 2007. Effects of supplemental manganese on performance of growing-finishing pigs and pork quality during retail display. J. Anim. Sci. 85:1046–1053.

Schieck, S. J., G. C. Shurson, B. J. Kerr, and L. J. Johnston. 2010. Evaluation of glycerol, a biodiesel coproduct, in grow-finish pig diets to support growth and pork quality. J. Anim. Sci. 88:3927–3935.

Scramlin, S. M., S. N. Carr, C. W. Parks, D. M. Fernandez-Duenas, C. M. Leick, F. K. McKeith, and J. Killefer. 2008. Effect of ractopamine level, gender, and duration of ractopamine on belly and bacon quality traits. Meat Sci. 80:1218–1221.

Shackelford, S. D., M. F. Miller, K. D. Haydon, N. V. Lovegren, C. E. Lyon, and J. O. Reagan. 1990. Acceptability of bacon as influenced by the feeding of elevated levels of monounsaturated fats to growing-finishing swine. J. Food Sci. 55:621–624.

Shelton, J. L., L. L. Southern, F. M. LeMieux, T. D. Bidner, and T. G. Page. 2004. Effects of microbial phytase, low calcium and phosphorus, and removing the dietary trace mineral premix on carcass traits, pork quality, plasma metabolites, and tissue mineral content in growing-finishing pigs. J. Anim. Sci. 82:2630–2639.

Shircliff, K. E., G. L. Allee, and B. R. Wiegand. 2019. Added fat fed with 30% dried distillers grains with solubles to pigs alters fatty acid composition in 4 fat depots but does not change carcass composition or quality. Appl. Anim. Sci. 35:550–562.

Skelley, G. C., R. F. Borgman, D. L. Handlin, J. C. Acton, J. C. McConnell, F. B. Wardlaw, and E. J. Evans. 1975. Influence of diet on quality, fatty acids and acceptability of pork. J. Anim. Sci. 41:1298–1304.

Stahl, C. A., G. L. Allee, and E. P. Berg. 2001. Creatine monohydrate supplemented in swine finishing diets and fresh pork quality: II. Commercial applications. J. Anim. Sci. 79:3081–3086.

Stender, D., J. Kliebenstein, R. Ness, J. Mabry, and G. Huber. 2009 Costs, returns, production and financial efficiency of niche pork production in 2008. Accessed Jan. 2020. https://www.leopold.iastate.edu/files/pubs-and-papers/2009-09-costs-returns-production-and-financial-efficiency-niche-pork-production-2008.pdf.

Sterten, H., T. Frøystein, N. Oksbjerg, A. C. Rehnberg, A. S. Ekker, and N. P. Kjos. 2009. Effect of fasting prior to slaughter on technological and sensory properties of the loin muscle (*M. longissimus dorsi*) of pigs. Meat Sci. 83:351–357.

Suárez-Belloch, J., M. A. Latorre, and J. A. Guada. 2016. The effect of protein restriction during the growing period on carcass, meat and fat quality of heavy barrows and gilts. Meat Sci. 112:16–23.

Sullivan, Z. M., M. S. Honeyman, L. R. Gibson, and K. J. Prusa. 2007. Effect of triticale-based diets on pig performance and pork quality in deep-bedded hoop barns. Meat Sci. 76:428–437.

Sun, D., Z. Zhu, S. Qiao, S. Fan, and D. Li. 2004. Effects of conjugated linoleic acid levels and feeding intervals on performance, carcass traits and fatty acid composition of finishing barrows. Arch. Anim. Nutr. 58:277–286.

Swanek, S. S., J. B. Morgan, F. N. Owens, D. R. Gill, C. A. Stratia, H. G. Dolezal, and F. K. Ray. 1999. Vitamin $D_3$ supplementation of beef steers increases longissimus tenderness. J. Anim. Sci. 77:874–881.

Swigert, K. S., F. K. McKeith, T. C. Carr, M. S. Brewer, and M. Culbertson. 2004. Effects of dietary $D_3$, vitamin E, and magnesium supplementation on pork quality. Meat Sci. 67:81–86.

Teye, G. A., P. R. Sheard, F. M. Whittington, G. R. Nute, A. Stewart, and J. D. Wood. 2006a. Influence of dietary oils and protein level on pork quality. 1. Effects on muscle fatty acid composition, carcass, meat and eating quality. Meat Sci. 73:157–165.

Teye, G. A., J. D. Wood, F. M. Whittington, A. Stewart, and P. R. Sheard. 2006b. Influence of dietary oils and protein level on pork quality. 2. Effects on properties of fat and processing characteristics of bacon and frankfurter-style sausages. Meat Sci. 73:166–177.

Therkildsen, M., B. Riis, A. Karlsson, L. Kristensen, P. Ertbjerg, P. P. Purslow, M. D. Aaslyng, and N. Oksbjerg. 2002. Compensatory growth response in pigs, muscle protein turn-over and meat texture: effects of restriction/realimentation period. Animal Sci. 75:367–377.

Thiel-Cooper, R. L., F. C. Parrish, Jr., J. C. Sparks, B. R. Wiegand, and R. C. Ewan. 2001. Conjugated linoleic acid changes swine performance and carcass composition. J. Anim. Sci. 79:1821–1828.

Tikk, K., M. Tikk, M. D. Aaslyng, A. H. Karlsson, G. Lindahl, and A. H. Andersen. 2007. Significance of fat supplemented diets on pork quality – connections between specific fatty acids and sensory attributes of pork. Meat Sci. 77:275–286.

Wang, J., S. M. Zhao, X. L. Song, H. B. Pan, W. Z. Li, Y. Y. Zhang, S. Z. Gao, and D. W. Chen. 2012. Low protein diet up-regulate intramuscular lipogenic gene expression and down-regulate lipolytic gene expression in growth-finishing pigs. Livest. Sci. 148:1119–128.

Watkins, L. E., D. J. Jones, D. H. Mowrey, D. B. Anderson, and E. L. Veenhuizen. 1990. The effect of various levels of ractopamine hydrochloride on the performance and carcass characteristics of finishing swine. J. Anim. Sci. 68:3588–3595.

Webster, M. J., R. D. Goodband, M. D. Tokach, J. L. Nelssen, S. S. Dritz, J. A. Unruh, K. R. Brown, D. E. Real, J. M. DeRouchey, J. C. Woodworth, C. N. Groesbeck, and T. A. Marsteller. 2007. Interactive effects between ractopamine hydrochloride and dietary lysine on finishing pig growth performance, carcass characteristics, pork quality, and tissue accretion. Prof. Anim. Sci. 23:597–611.

Weimer, D., J. Stevens, A. Schinckel, M. Latour, and B. Richert. 2008. Effects of feeding increasing levels of distillers dried grains with solubles to grow-finish pigs on growth performance and carcass quality. J. Anim. Sci. 86(E-Suppl. 3):85.

White, H. M., B. T. Richert, J. S. Radcliffe, A. P. Schinckel, J. R. Burgess, S. L. Koser, S. S. Donkin, and M. A. Latour. 2009. Feeding conjugated linoleic acid partially recovers carcass quality in pigs fed dried corn distillers grains with solubles. J. Anim. Sci. 87:157–166.

Whitney, M. H., G. C. Shurson, L. J. Johnston, D. M. Wulf, and B. C. Shanks. 2006. Growth performance and carcass characteristics of grower-finisher pigs fed high-quality corn distillers dried grain with solubles originating from modern Midwestern ethanol plant. J. Anim. Sci. 84:3356–3363.

Whittington, F. M., N. J. Prescott, J. D. Wood, and M. Enser. 1986. The effect of dietary linoleic acid on the firmness of backfat in pigs of 85 kg live weight. J. Sci. Food Agric. 37:753–761.

Widmer, M. R., L. M. McGinnis, D. M. Wulf, and H. H. Stein. 2008 Effects of feeding distillers dried grains with solubles, high-protein distillers grains, and corn germ to growing-finishing pigs on pig performance, carcass quality, and the palatability of pork. J. Anim. Sci. 86:1819–1831.

Wiegand, B. R., F. C. Parrish, Jr., J. E. Swan, S. T. Larsen, and T. J. Baas. 2001. Conjugated linoleic acid improves feed efficiency, decreases subcutaneous fat, and improves certain aspects of meat quality in Stress-Genotype pigs. J. Anim. Sci. 79:2187–2195.

Wiegand, B. R., J. C. Sparks, D. C. Beitz, F. C. Parrish, R. L. Horst, A. H. Trenkle, and R. C. Ewan. 2002. Short-term feeding of vitamin $D_3$ improves color but does not change tenderness of pork-loin chops. J. Anim. Sci. 80:2116–2121.

Wilborn, B. S., C. R. Kerth, W. F. Owsley, W. R. Jones, and L. T. Frobish. 2004. Improving pork quality by feeding supranutritional concentrations of vitamin $D_3$. J. Anim. Sci. 82:218–224.

Wiseman, J., and J. A. Agunbiade. 1998. The influence of changes in dietary fat and oils on fatty acid profiles of carcass fat in finishing pigs. Livest. Prod. Sci. 54:217–227.

Wolter, B., M. Ellis, F. K. McKeith, and K. D. Miller. 1999. Influence of dietary selenium source on growth performance, and carcass and meat quality characteristics in pigs. Can. J. Anim. Sci. 79:119–121.

Wood, J. D., S. N. Brown, G. R. Nute, F. M. Whittington, A. M. Perry, S. P. Johnson, and M. Enser. 1996. Effects of breed, feed level and conditioning time on the tenderness of pork. Meat Sci. 44:105–112.

Wood, J. D., G. R. Nute, R. I. Richardson, O. Southwood, G. Plastow, R. Mansbridge, N. da Costa, and K. C. Chang. 2004. Effects of breed, diet and muscle on fat deposition and eating quality in pigs. Meat Sci. 67:651–667.

Xiong, Y. L., M. J. Grower, C. Li, C. A. Elmore, G. L. Cromwell, and M. D. Lindemann. 2006. Effect of dietary ractopamine on tenderness and postmortem protein degradation of pork muscle. Meat Sci. 73:600–604.

Xu, G., S. K. Baidoo, L. J. Johnston, D. Bibus, J. E. Cannon, and G. C. Shurson. 2010a. The effects of feeding diets containing corn distillers dried grains with solubles, and withdrawal period of distillers dried grains with solubles, on growth performance and pork quality in grower-finisher pigs. J. Anim. Sci. 88:1388–1397.

Xu, G., S. K. Baidoo, L. J. Johnston, D. Bibus, J. E. Cannon, and G. C. Shurson. 2010b. Effects of feeding diets containing increasing content of corn distillers dried grains with solubles to grower-finisher pigs on growth performance, carcass composition, and pork fat quality. J. Anim. Sci. 88:1398–1410.

Young, A. N., J. K. Apple, J. W. S. Yancey, J. J. Hollenbeck, T. M. Johnson, B. E. Bass, T. C. Tsai, C. V. Maxwell, M. D. Hanigan, J. S. Radcliffe, B. T. Richert, J. S. Popp, R. Ulrich, and G. Thoma. 2013. Effects of amino acid supplementation of reduced crude protein (RCP) diets on LM quality of growing-finishing swine. J. Anim. Sci. 91(Suppl. 2):97–98.

Young, A. N., J. K. Apple, J. W. Yancey, T. M. Johnson, T. C. Tsai, and C. V. Maxwell. 2014. Effects of amino acid supplementation of reduced crude protein (RCP) diets on fatty acid composition of subcutaneous fat and muscle. J. Anim. Sci. 92(E-Suppl. 2):619.

Young, J. F., H. C. Bertram, K. Rosenvold, G. Lindahl, and N. Oksbjerg. 2005. Dietary creatine monohydrate affects quality attributes of Duroc but not Landrace pork. Meat Sci. 70:717–725.

Zhan, X. A., M. Wang, R. Q. Zhao, W. F. Li, and Z. R. Xu. 2007. Effects of different selenium source on selenium distribution, loin quality and antioxidant status in finishing pigs. Anim. Feed Sci. Technol. 132:202–211.

# Index

---